Pinus is a remarkable genus comprising at least 111 tree species with a very large distribution range in the northern hemisphere. Where they occur, pines usually form the dominant vegetation cover and are extremely important components of ecosystems. They also provide a wide range of products for human use. In many cases exploitation and other human pressures are threatening the survival of natural pine forests, although pines are now also widely grown in commercial plantations, both within and outside their natural ranges. This book presents a definitive review of pine ecology and biogeography written by forty of the world's leading authorities on this important genus. In the face of increasing human pressure and global climate change, it provides an essential source of reference for all those concerned with the management of natural and planted pine forests.

Ecology and Biogeography of *Pinus*

Ecology and Biogeography of *Pinus*

Edited by **David M. Richardson**

Institute for Plant Conservation, University of Cape Town, South Africa

CAMBRIDGE
UNIVERSITY PRESS

PUBLISHED BY THE PRESS SYNDICATE OF THE UNIVERSITY OF CAMBRIDGE
The Pitt Building, Trumpington Street, Cambridge, United Kingdom

CAMBRIDGE UNIVERSITY PRESS
The Edinburgh Building, Cambridge CB2 2RU, UK   http://www.cup.cam.ac.uk
40 West 20th Street, New York, NY 10011-4211, USA   http://www.cup.org
10 Stamford Road, Oakleigh, Melbourne 3166, Australia
Ruiz de Alarcón 13, 28014 Madrid, Spain

First published 1998
First paperback edition 2000

Typeface 8.5/11.5pt Swift Regular   *System* QuarkXPress® [SE]

A catalogue record for this book is available from the British Library

Library of Congress Cataloguing in Publication data

Ecology and biogeography of *Pinus*/edited by David M. Richardson.
    p. cm.
  Includes bibliographical references and index.
  ISBN 0 521 55176 5 (hb)
  1. Pine–Ecology.   2. Pine–Geographical distribution.
  I. Richardson, D. M. (David M.), 1958-   .
  QK494.5.P66E27   1998
  585'.217–dc21   97-11056   CIP

ISBN 0 521 55176 5 hardback
ISBN 0 521 78910 9 paperback

Transferred to digital printing 2004

This volume is dedicated to the memory of William Burke Critchfield (1923–1989) and Nicholas Tiho Mirov (1893–1980), whose dedication to thorough studies on various aspects of pines greatly improved our knowledge of this remarkable genus. Their influence lives on in this book.

Contents

x / Contents

Contributors

Agee, J.K. Division of Ecosystem Science and Conservation, College of Forest Resources, Box 352100, University of Washington, Seattle, WA 98195, USA

Barbéro, M. Institut Mediterranéen d'Ecologie et de Paléoécologie, Faculté des Sciences et Techniques de St Jerôme, Case 461, Avenue Escadrille-Niémen, 13397 Marseille Cédex 20, France

Bennett, K.D.[1] Department of Plant Sciences, University of Cambridge, Downing Street, Cambridge CB2 3EΛ, UK

Birks, H.J.B. Botanical Institute, University of Bergen, Allegaten 41, N-5007 Bergen, Norway; and Environmental Change Research Centre, University College London, 26 Bedford Way, London WC1 OAP, UK

Cwynar, L.C. Department of Biology, University of New Brunswick, Fredericton, NB, E3B 6E1, Canada

de Groot, P. Natural Resources Canada, Canadian Forest Service, Great Lakes Forestry Centre, PO Box 490, Sault Ste. Marie, Ontario P6A 5M7, Canada

Enquist, B.J. Department of Biology, University of New Mexico, Albuquerque, NM 87131, USA

Graham, A. Department of Biological Sciences, Kent State University, PO Box 5190, Kent, OH 44242-0001, USA

Harrington, T.C. Deptartment of Plant Pathology, Iowa State University, 351 Bessy Hall, Ames, IA 50011, USA

Higgins, S.I.[2] Institute for Plant Conservation, Botany Department, University of Cape Town, Rondebosch 7700, South Africa

Keeley, J.E.[3] Division of Environmental Biology, National Science Foundation, Arlington, VA 22230, USA

Kremenetski, C.V. Laboratory of Evolutionary Geography, Institute of Geography, Russian Academy of Sciences, Staomonetny pereulok, 29 Moscow 109017, Russia

Lanner, R.M.[4] Department of Forest Resources, Utah State University, Logan, UT 84322-5215, USA

Lavery, P.B. Plantation Audit Services, Box 1182, Box Hill 3128, Australia

Ledig, F.T. Institute of Forest Genetics, Pacific Southwest Research Station, USDA Forest Service, 2480 Carson Road, Placerville, CA 95667, USA

Le Maitre, D.C. CSIR Division of Water, Environment and Forestry Technology, PO Box 320, Stellenbosch 7599, South Africa

Liston, A. Department of Botany and Plant Pathology, 2082 Cordley Hall, Oregon State University, Corvallis, OR 97331-2902, USA

Liu, Kam-biu Department of Geography and Anthropology, Louisiana State University, Baton Rouge, LA 70803-4105, USA

Loisel, R. Institut Mediterranéen d'Ecologie et de Paléoécologie, Faculté des Sciences et Techniques de St Jerôme, Case 461, Avenue Escadrille-Niémen, 13397 Marseille Cédex 20, France

MacDonald, G.M. Department of Geography, University of California at Los Angeles, 405 Higard Ave., Los Angeles, CA 90095-1524, USA

Mead, D.J. Department of Plant Science, Lincoln University, Lincoln, Canterbury, New Zealand

Millar, C.I. Institute of Forest Genetics, USDA Forest Service, Pacific Southwest Research Station, PO Box 245, Berkeley, CA 94701, USA

Nowicki, T.E.[5] Department of Geological Sciences, University of Cape Town, Rondebosch 7700, South Africa

Perry, J.P., Jr 306 N Front Street, Hertford, NC 27944, USA

Price, R.A. Department of Botany, University of Georgia, Athens, GA 30602, USA

Quézel, P. Institut Mediterranéen d'Ecologie et de Paléoécologie, Faculté des Sciences et Techniques de St Jerôme, Case 461, Avenue Escadrille-Niémen, 13397 Marseille Cédex 20, France

Read, D.J. Department of Animal and Plant Sciences, University of Sheffield, Sheffield S10 2TN, UK

Richardson, D.M. Institute for Plant Conservation, Botany Department, University of Cape Town, Rondebosch 7701, South Africa

Romane, F. Centre Nationale de la Recherche Scientifique, Centre d'Ecologie Fonctionelle et Evolutive, 1919 Route de Mende, 34293 Montpellier Cédex 5, France

Rundel, P.W. Department of Biology, University of California at Los Angeles, 900 Veteran Ave, Los Angeles, CA 90024-1786, USA

Scholes, M.C. Department of Botany, University of the Witwatersrand, PO Wits 2050, South Africa

Stevens, G.C. [5]Department of Biology, University of New Mexico, Albuquerque, NM 87131, USA

Strauss, S.H. Department of Forest Science, FSL 020, Oregon State University, Corvallis, OR 97331-7501, USA

Turgeon, J.J. Natural Resources Canada, Canadian Forest Service, Great Lakes Forestry Centre, PO Box 490, Sault Ste. Marie, Ontario P6A 5M7, Canada

Van Devender, T.R. Arizona-Sonora Desert Museum, 2021 N Kinney Road, Tucson, AZ 85743-8918, USA

Whitlock, C. Department of Geography, University of Oregon, Eugene, OR 97403, USA

Willis, K.J.[6] Department of Plant Sciences, University of Cambridge, Downing Street, Cambridge CB2 3EA, UK

Wingfield, M.J.[7] Tree Pathology Cooperative Programme, Department of Microbiology and Biochemistry, University of the Orange Free State, PO Box 339, Bloemfontein 9300, South Africa

Yoder, B.J. Department of Forest Science, Peavy Hall 154, Oregon State University, Corvallis, OR 97331-7501, USA

Zedler, P.H.[8] Biology Department, San Diego State University, San Diego, CA 92182-0057, USA

Current addresses of contributors are:

[1] Uppsala University Geocentrum, Villavagen 16, S-752 36 Uppsala, Sweden
[2] National Botanical Institute, Kirstenbosch, Private Bag X7, Claremont 7735, South Africa
[3] US Geological Survey, Sequoia-Kings Canyon Field Station, 47050 Generals Highway, Three Rivers, CA 93271-9651, USA
[4] 2651 Bedford Ave., Placerville, CA 95667, USA
[5] Cape Cod Museum of Natural History, PO Box 1710, Brewster, MA 02631, USA
[6] School of Geography, University of Oxford, Mansfield Road, Oxford, OX1 3TB, UK
[7] Forestry and Agricultural Biotechnology Institute, Faculty of Biological Sciences and Agriculture, University of Pretoria, Pretoria 0002, South Africa
[8] Institute for Environmental Studies and Arboretum, University of Wisconsin – Madison 70 Science Hall, 550 N. Park St., Madison, WI 53706, USA

Preface and acknowledgements

The main aim of this book is to provide, within the confines of one manageable volume, an informative overview of the current understanding of the ecology and biogeography of the genus *Pinus*. This is clearly a very ambitious objective.

The idea of compiling this book first came to me about a decade ago but for several years I was unable to devote time to the project. Also, I felt that it would be rather presumptuous for a South African ecologist with very little first-hand experience of pines in their natural habitats to initiate a synthesis of the kind that I felt was needed. At that time, my knowledge of pines was that of a forestry graduate with training in the silviculture of pines as exotics, and of a researcher with an interest in the ecology and management of pines as invasive alien species in South African fynbos. My experience of pines in their natural range was confined to brief encounters with a handful of species in the Mediterranean Basin. My infatuation with pines came about rather differently to that of any other 'pinophile' that I know. Since this has influenced how I approached the compilation of this volume, it may be of interest to readers.

My work on pines as invaders first led me to explore the literature on life-history adaptations of pines – my quest being to discover what traits equipped pines to be such tenacious persisters and aggressive colonizers in South African fynbos, and what could be done to contain their rampant spread. My readings on the subject also led me to examine the conditions that caused range changes within the natural distribution of pines. I also explored the various interactions between disturbance and pine dynamics. Searching for other insights, I scrutinized the palaeoecological literature and was fascinated to find clues to the current range limits of pines in reconstructions of pine migrations following deglaciation. At this time, my colleague William Bond was busy with his PhD studies at the University of California at Los Angeles. One

product of his North American sojourn was his seminal paper entitled 'The tortoise and the hare: ecology of angiosperm dominance and gymnosperm persistence' (Bond 1989). I was intrigued by Bond's findings, as they appeared to explain (or at least provide a framework for understanding) so much of the dynamics of the recent range changes that I had been studying. Soon after his return to South Africa, we collaborated in a study of the determinants of pine distribution based on many published accounts of range changes from numerous localities in the natural range of pines, and in areas where they were invading new habitats (Richardson & Bond 1991). Our main conclusion was that interactions between pine seedlings and various biotic factors in the regeneration niche were fundamentally important in defining range limits. This gave me a new perspective for reflecting on a wide range of topics in the vast literature on pines.

In 1990, I spent a few months in Australia, and was interested to see *Pinus radiata*, which I knew well as an invader of South African fynbos, invading eucalypt forests. Invasions at sites such as Black Mountain in Canberra seemed to be driven by a different suite of factors to those I had identified as being important in pine invasions in South Africa. I started comparing the behaviour of this and other pine species as invaders in different parts of the southern hemisphere. This research forced me to read widely on various aspects of pine ecology. I was totally fascinated by the wealth of information available, but frustrated by the lack of a thorough synthesis. There was a treasure chest of information, but the facts were so scattered that forming clear pictures was difficult.

As I delved deeper into the huge literature on pines, I began corresponding with pine authorities in various fields from many parts of the world. For several years I corresponded with the late Bill Critchfield. His carefully formulated replies to my miscellaneous queries, for example on the evolution of serotiny in pines, reinforced my preoccupation with *Pinus*; it seemed to be the model genus for exploring all sorts of fascinating ecological and evolutionary questions. I was also riveted by the rich information base concerning the impacts of various types of land use on pine dynamics, especially in the Mediterranean Basin and the American Southwest. For the Mediterranean Basin, my interest was fuelled by regular contact with ecologists from various parts of the region at several MEDECOS conferences, and particularly by two visits to the Centre d'Ecologie Fonctionelle et Evolutive in Montpellier in France. For the American Southwest, my curiosity was whetted by accounts in the literature on range management, and by the scientific and popular writings of, and correspondence with, Ronald Lanner, then Professor of Forest Resources at Utah State University.

In several parts of the world, pine taxa are facing extinction as a result of growing human pressure on forests. This is extremely obvious and worrying in parts of Mexico and Central America, and in northern Asia – regions which together store two-thirds of the genetic diversity in *Pinus*. Even very widespread species in more affluent regions have been decimated. This escalating attrition of genetic diversity has untold implications for the future, for example by annulling many options for selective breeding to enhance productivity and disease resistance of pines for human utility. The selective clearing of pines in certain habitats, notably low-altitude sites, has unexplored ramifications for the future of pine forests in the face of global climate change.

The need for a synthesis of the current knowledge on pine ecology and biogeography was obvious. Nicholas Mirov's volume on *The Genus Pinus*, published in 1967, served this function reasonably well, but is now seriously out of date in many aspects. Certainly, Mirov's book did not provide answers to many questions I was asking about pines. There has been enormous progress in virtually every aspect of pine ecology since the 1960s. Given this, it soon became clear that a single-author account that would do justice to emerging perspectives in all fields would be an impossible task. The answer seemed to lie in a carefully planned, multi-author volume. In 1992 I cautiously started compiling an outline for this book. I consulted hundreds of publications and scores of authorities in numerous fields. The final list of chapters and authors (40 contributors from nine countries) was completed after more than three years of correspondence and deliberation.

The book comprises 22 chapters, divided into six parts. The first chapter introduces the volume by placing the genus *Pinus* in perspective. Part Two contains a chapter on phylogeny and systematics, and one on the early evolution of pines. The next part consists of four chapters that detail the historical biogeography of pines in the four major pine regions of the world: northern Asia, Europe, northern North America, and Mexico and Central America. Part Four deals with more recent biogeography for two regions, the Mediterranean Basin and the American Southwest. Also included in this section is an account of the macroecology of pine distribution and abundance. Nine chapters (11–19) are grouped together in a section on Ecological Themes. Although many of the chapters in others sections touch on the role of humans in shaping pine biogeography, three chapters are grouped in the final section on Pines and Humans. This section provides a global review of pines in cultivation, a separate chapter on one remarkable species, *P. radiata*, and an overview of the newest 'pine rise': the rampant spread of invasive pines from sites of cultivation in the southern hemisphere.

A problem confronting the contributors to this volume was the enormous size of the literature on pines. One author complained that reviewing his particular subject was like 'taking hold of an elephant'. Another protested that 'it is like painting a large bridge – by the time you finish, the first part needs painting all over again!' Most authors faced the major problem of deciding what to leave out. It is inevitable that some readers will judge us to have erred in selecting what to include and what to omit. Although the combined reference list for this volume contains about 3000 titles, there are many others that probably deserved mention. I am resigned to the fact that the literature on pines is so huge and rapidly expanding that it will never be possible to tell the whole story in one volume. Nevertheless, I hope that this book will prove to be a useful guide to the most significant literature and a valuable synthesis of current knowledge.

Many people have helped in many ways to bring this book to publication. In the early stages of planning, Phil Rundel (University of California, Los Angeles) and Paul Zedler (San Diego State University) were especially supportive, as was Alan Crowden of Cambridge University Press. During the final stages of editing the book Phil Rundel was a superb 'guide' on a 'pine safari' through the great forests of the Sierra Nevada and the arid South West. During this trip we scrawled the first draft of the introductory chapter and pondered many aspects of pine ecology. His hospitality, enthusiasm, and breadth of knowledge was a great source of inspiration to me. Dick Mack of Washington State University in Pullman is also thanked for his hospitality and for sharing with me his knowledge of the conifer forests of Washington and Idaho. A special word of thanks is due to Richard Cowling who, as Director of the Institute for Plant Conservation at the University of Cape Town, allowed me to take on this time-consuming project when so many other matters required my attention. The late Derek Donald, lecturer and then professor of Silviculture at the University of Stellenbosch, South Africa, from 1960 until his death in 1993 was the source of much information and helpful advice. The following colleagues at the now-defunct South African Forestry Research Institute are thanked for help and support in many ways over the years: Fred Kruger, David Le Maitre, Richard Poynton and Brian van Wilgen.

I am very grateful to the following friends, colleagues and associates for reviewing chapters, providing advice, and helping in various other ways: H. John B. Birks (Botanical Institute, Bergen, Norway; and Environmental Change Research Centre, University College, London, UK), Robert A. Blanchette (Department of Plant Pathology, University of Minnesota, St Paul, Minnesota, USA), Robert A. Boardman (Primary Industries of South Australia, Adelaide, Australia), William J. Bond (Botany Department, University of Cape Town, South Africa), John W. Duffield (Shelton, Washington, USA), William S. Dvorak (CAMCORE, North Carolina State University, Raleigh, North Carolina, USA), Martin V. Fey (Department of Geological Sciences, University of Cape Town, South Africa), Johann G. Goldammer (Fire Ecology and Biomass Burning Research Group, Max Planck Institute for Chemistry, University of Freiburg, Germany), John H. Hoffmann (Department of Zoology, University of Cape Town, South Africa), Henry F. Howe (Department of Biological Sciences, University of Illinois at Chicago, USA), Michael Huston (Environmental Sciences Division, Oak Ridge National Laboratory, Oak Ridge, Tennessee, USA), Harry E. Hutchins (Itasca Community College, Minnesota, USA), Dale Johnson (Biological Sciences Center, Desert Research Institute, Reno, Nevada USA), Fred J. Kruger (CSIR Division of Water, Environment and Forestry Technology, Pretoria, South Africa), Glen M. MacDonald (Department of Geography, UCLA, California, USA), Jeremy J. Midgley (Botany Department, University of Cape Town, South Africa), Jacques Lepart (CNRS, Centre d'Ecologie Fonctionelle et Evolutive, Montpellier, France), Fernando Ojeda (Departamento de Biologia Vegetal y Ecologia, Universidad de Sevilla, Spain), Phil W. Rundel (Department of Biology, UCLA, Los Angeles, California, USA), Costas A. Thanos (Department of Botany, University of Athens, Greece), Geoff D. Tribe (Plant Protection Research Institute, Rosebank, South Africa), Brian W. van Wilgen (CSIR Division of Water, Environment and Forestry Technology, Stellenbosch, South Africa), Robert D. Westfall (USDA Forest Service, Pacific Southwest Research Station, Albany, California, USA), Kathy Willis (Department of Plant Sciences, University of Cambridge, Cambridge, UK) and Costas Zachariades (Institute for Plant Conservation, University of Cape Town, South Africa).

Thanks to the magic of e-mail and the Internet it was possible to communicate regularly with most of the contributors to this volume and with the reviewers. I doubt whether it would have been possible to finish the task of compiling the volume without this tool. I thank all the contributors and reviewers for responding rapidly to my numerous requests and queries.

Thanks are due to the following publishers for granting permission to reproduce material from their publications: Botanical Society of America (*American Journal of Botany*); California Botanical Society (*Madroño*), Columbia University Press; Ecological Society of America (*Ecology*); Duckworth Publishing; Kluwer Academic Press; Macmillan Magazines Ltd (*Nature*); Missouri Botanical Garden (*Annals of the Missouri Botanical Garden*); Sigma XI, The Scientific Research Society (*American Scientist*); South African Association of Botanists (*South African Journal of Botany*) and Timber Press.

Nalini Dickson, who is Librarian at the Institute for Commercial Forestry Research at the University of Natal in Pietermaritzburg, was wonderfully helpful and did numerous literature searches which were extremely helpful in filling various gaps. The staff at the Forestry Faculty and G.S. Gericke libraries at the University of Stellenbosch were also very supportive. My wife, Corlia, helped in countless ways for which I am immensely grateful (I doubt whether she will ever get over the overdose of pines on our 1996 trip to the USA). Craig Benkman, John Birks, Steffan Eggenberg, Diane Erwin, Greg Forsyth, Johann Goldammer, Colin Hughes, Bill Libby, Glen MacDonald, Jose Moreno, Fernando Ojeda, Phil Rundel, Melissa Savage, Diana Tomback and Willy Stock kindly helped in locating photographs to fill various gaps. Jessica Kemper is thanked for her meticulous and cheerful assistance in final checking of the manuscript and proof reading. Thanks to Wendy Paisley, secretary at the Institute for Plant Conservation, for general assistance.

I am extremely grateful to the team at Cambridge University Press, especially Maria Murphy, Jane Bulleid, Ian Sherratt, Alan Crowden and Stephanie Thelwell.

Compiling and editing this book has been a great adventure. I sincerely hope that the insights included here will be of value to students of pine ecology and that they will help in the formulation of strategies for the wise management of these forests in the future.

David M. Richardson

Institute for Plant Conservation, University of Cape Town

October 1996

References

Bond, W.J. (1989). The tortoise and the hare: ecology of angiosperm dominance and gymnosperm persistence. *Biological Journal of the Linnean Society*, **36**, 227–49.

Richardson, D.M. & Bond, W.J. (1991). Determinants of plant distribution: Evidence from pine invasions. *American Naturalist*, **137**, 639–68.

Part one

Introduction

1   Ecology and biogeography of *Pinus*: an introduction

David M. Richardson and Philip W. Rundel

1.1   Introduction

Pines are important, and very often dominant, components of the vegetation over large parts of the northern hemisphere (Fig. 1.1). Besides having major economic value as sources of timber, pulp, nuts, resin and other products, pines also influence ecosystems in many ways. They affect biogeochemical processes, hydrological and fire regimes, and provide food and create habitats for animals. The boreal forest, of which pines are an important component, plays a significant role in determining regional and global climate. For example, the presence of forest in these northern latitudes masks the high reflectance of snow, leading to warmer winter temperatures than would be the case if trees were absent (Bonan, Pollard & Thompson 1992). Pines featured in ancient myths and rituals, have influenced human history, and have been celebrated in visual art, prose, poetry and music (as in Ottorino Respighi's 'The Pines of Rome'). Pines have also been cultivated in many parts of the world, both within and well outside their natural range, and they form the foundation of exotic forestry enterprises in many southern hemisphere countries. *Pinus* is without a doubt the most ecologically and economically significant tree genus in the world.

This chapter provides an introduction to this volume by placing the genus in perspective. We discuss the origin and evolution of pines, the features that distinguish them from other woody taxa, and the position of pines in the landscape in each of the major habitats in which they occur. We consider some of the many interactions between pines and humans, and discuss some recent developments in the study of pines.

1.2   The origin and evolution of pines

The expansion of angiosperms and the concurrent decline of gymnosperms was one of the most important phytogeographic processes in the history of the earth. The earliest-known angiosperms arose in the Early Cretaceous (*c.* 120 million years ago), and there are now between 250 000 and 300 000 extant species. Gymnosperms arose much earlier (Middle Devonian, 365 million years ago), but there have never been more than a few thousand species. Evidence from fossilized cones shows that ancestors of Pinaceae had evolved by the mid-Jurassic, and that *Pinus* had evolved by the Lower Cretaceous. There is some evidence that *Cedrus* and possibly *Larix* appeared before the Tertiary, but the other genera of the family appeared only in the early Tertiary or later (Stewart 1983).

The current diversity of conifers (gymnosperms excluding cycads, *Ephedra*, *Ginkgo*, *Gnetum* and *Welwitschia*) comprises eight families, 68 genera, 629 species and about 176 intraspecific taxa. A large proportion of extant conifers occur in the northern hemisphere: seven families and about 70% of both genera and species (Farjon 1998). More than a third of extant gymnosperm species belong to the Pinaceae, by far the largest family of modern conifers, which is divided into 10 or 11 genera (Chap. 2, this volume). Species of all the large genera in the Pinaceae (notably *Pinus*, *Picea*, *Abies* and *Larix*) are widely distributed throughout the temperate parts of both Old and New Worlds. There are concentrations of species in all these genera in North America and eastern Asia. More than half the species in the Pinaceae (and almost 20% of all gymnosperm species) are included in the remarkable genus *Pinus* which, according to the treatment accepted for this volume (see Chap. 2), contains 111 species (Table 1.1).

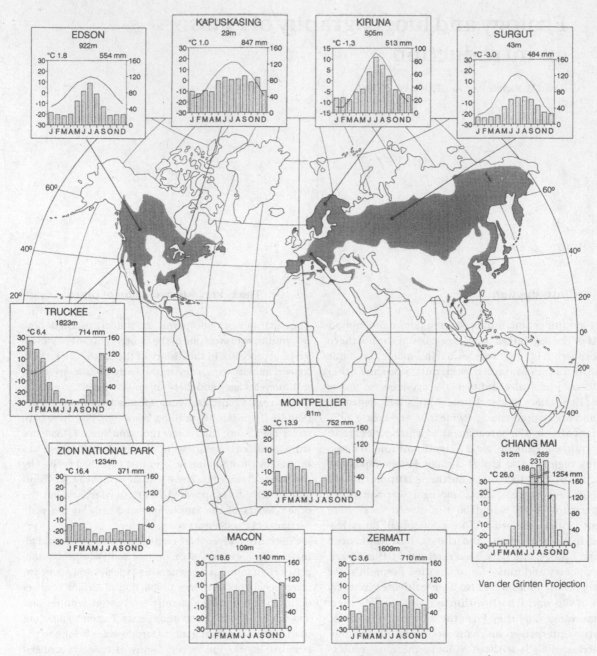

Fig. 1.1. **The distribution of *Pinus* (based on Critchfield & Little 1966) with climographs for representative sites in each of the major habitat categories (see text). Mean monthly temperatures are** indicated by the line and mean precipitation for each month is shown by the bars. Station elevation, mean annual temperature, and mean annual precipitation appear at the top of each climograph.

By the end of the Mesozoic, pines had diversified into two major groups, or subgenera; representatives of both subgenera, *Strobus* (haploxylon or soft pines, with one fibrovascular bundle in the needle) and *Pinus* (diploxylon or hard pines, with two fibrovascular bundles in the needle), survive today. Several early subsections within these subgenera had also evolved by this time, including *Australes*, *Canarienses*, *Cembroides*, *Gerardiana*, *Pineae*, *Pinus*, *Ponderosae* and *Strobi* (Millar & Kinloch 1991). At this stage, pines had migrated throughout the middle latitudes of the northern hemisphere super-continent, Laurasia. Major environmental changes in the Early Cretaceous, between 130 and 90 million years ago, led to the diversification and rapid spread of angiosperms throughout middle latitudes, initiating profound changes in terrestrial ecosystems (Crane, Friis & Pedersen 1995). As the

Pinus taxon[a]	Common name[b]	Needle no.[c]	Needle length (m)[c]	Needle longevity (yrs)[c]	Cone length (cm)[c]	Height (m)[c]	Biogeographic region[d]	Habitat[e]
P. albicaulis	whitebark pine	5	3–7	5–8	4–8	5–10(30)	W North America	subalpine
P. aristata	Colorado, or Rocky Mountain bristlecone pine	5	3–4	10–20	5–6(11)	5–15(30)	Rocky Mts, North America	subalpine
P. armandii	Chinese white, Armand('s), or David's pine	5	8–15(18)	2–3	8–14	20–30	W & C China and Taiwan	temperate montane
P. attenuata	knobcone pine	3	9–15	4–5	8–15	10–20	Baja California, California, SW Oregon,	mediterranean coastal
P. ayacahuite	Mexican white pine	5	8–15(22)	3	10–40	35–50	Mexico, Central America, Arizona, New Mexico	tropical montane
P. balfouriana	foxtail pine	5	(1.5)3–4	10–30	6–9(12)	10–22	California	subalpine
P. banksiana	jack pine	2	2–5	2–4	3–3.5	10–18(20)	Canada, N USA	boreal forest
P. bhutanica	Bhutan white pine	5	12–28	2–3	12–20	25	Himalayas	temperate montane
P. brutia	eastern Mediterranean, or Calabrian pine	2	8–15	?	6–9	10–25	E Medit. Basin	mediterranean coastal
P. bungeana	lacebark pine	3	6–8	3–4	5–6	15(30)	C and N China	temperate montane
P. canariensis	Canary Island pine	3	20–30	2–3	10–20(25)	30	Canary Islands	mediterranean
P. caribaea	Caribbean pine	(2)3(4–5)	15–25	2	5–12	20–30	Caribbean area, Central America	tropical/savanna
P. cembra	Swiss stone, or Arolla pine	5	7–9	3–12	4–10	8–20(25)	C Europe	subalpine
P. cembroides	Mexican pinyon	(2)3(4)	2–6(7)	3–4	1–3.5	5–10(15)	NW Mexico, Texas	arid/montane
P. chiapensis	Chiapas white pine	5	10–12	?	7–16	40	S-C to S Mexico, Guatemala	tropical
P. clausa	sand pine	2	6–9	2–3	3(4–8)	6(10)	SE USA	temperate
P. contorta	lodgepole pine	2	2–8	3–8	2–6	3–46(50)	W USA	temperate
P. c. subsp. bolanderi	Bolander pine	2	2.5–4	(1|2–3)	4–5	6–15	California	montane/subalpine
P. c. subsp. contorta	shore, or beach pine	2	4–6(7)	3–4	4–7	3–10(16)	coastal N California to British Columbia	temperate
P. c. subsp. latifolia	lodgepole pine (Rocky Mountain)	2	(4)5–8	5–18	?	40–46	Rocky Mts, North America	temperate montane
P. c. subsp. murrayana	Sierra (Nevada) lodgepole pine	2	(5–8	?	2–5	15–40(50)	Sierra Nevada, S to Baja California	montane/subalpine
P. cooperi	Cooper pine	5(6–8)	8–10	?	6–10	30–35	W Mexico	tropical montane
P. coulteri	Coulter, or bigcone pine	3	16–30	3–4	20–35	15–25	California, Baja California	mediterranean coastal
P. cubensis	Cuban pine	3	10–14	?	4.5	?	Cuba	tropical/savanna
P. culminicola	Potosi pinyon	(3–4)5(6)	3–5	?	3–5	1–5	NE Mexico	tropical montane/subalpine
P. dabeshenensis	Dabie Shan white pine	5	5–14	?	11–14	20–30	E China	temperate
P. dalatensis	Dalat, or Vietnamese white pine	5	4–10	?	5–10	40	Vietnam	tropical
P. densata	Sikang, or Gaoshan pine	2(3)	8–14	3	4–6	30	China	temperate montane
P. densiflora	Japanese red pine	2	(6)9–12	2–3	3–5	20–30(36)	Japan, Korea, China	temperate
P. devoniana	Michoacán pine	5	20–35	?	20–30	20–30	Mexico, Guatemala	temperate montane
P. discolor	border pinyon	3	2–6	?	2–3	5–10	SW USA, C & NW Mexico	arid/montane
P. donnell-smithii	Donnell Smith pine	5–6(7–8)	15–22	?	10–13	25	Guatemala	tropical/subalpine
P. douglasiana	Douglas pine	5	20–35	?	7–10	20–35	W Mexico	tropical
P. durangensis	Durango pine	6(7–8)	12–20	?	7–10	30–40	N & C Mexico	tropical
P. echinata	shortleaf pine	2(3)	7–11	3–5	4–7	15–30(35)	SE USA	temperate
P. edulis	Colorado pinyon	2	2–4	4–6	3–6	5–15	SW USA	arid
P. elliottii	slash pine	2–3	15(30)	2	8–18	25–30	SE USA	temperate
P. engelmannii	Apache pine	(2)3–4(5)	25–35	?	(10)–15	25(–30)	W Mexico, Arizona, New Mexico	temperate/montane
P. fenzeliana	Fenzel pine	5	4–18	?	6–10	13–20(50)	S China to C Vietnam	temperate

Table 1.1. List of *Pinus* taxa, with common names, selected morphological features, and biogeographic region and habitat. Figures relate to conditions regularly observed in the field (figures in parentheses indicate exceptional dimensions). Only those subspecific taxa that are discussed in detail in this volume are included.

Table 1.1. (cont.)

Pinus taxon [a]	Common name [b]	Needle no. [c]	Needle length (m) [c]	Needle longevity (yrs) [c]	Cone length (cm) [c]	Height (m) [c]	Biogeographic region [d]	Habitat [e]
P. flexilis	limber, or Rocky Mountain white pine	5	3-8	5-6	7-15	7-15(24?)	W North America	subalpine
P. gerardiana	Chilgoza, or Gerard's pine	3	6-10	?	12-20	10-20(25)	Punjab, Afghanistan, Pakistan	temperate montane
P. glabra	spruce pine	2	(4)6-8	2-3	4-9	22-35	SE USA	temperate
P. greggii	Gregg's pine	3	8-15	2-3	8-14	10-15(25)	E Mexico	tropical
P. halepensis	Aleppo pine	2(3)	6-12(15)	2	5-12	10-20(25)	Medit. Basin	mediterranean coastal
P. hartwegii	Hartweg pine	3(4-5)	8-16	?	8-14	20-30	Mexico, Guatemala	tropical/subalpine
P. heldreichii	Heldreich whitebark, or Bosnian pine	2	6-10	2-3(67)	7-8	20(30)	Balkan Peninsula and Italy	temperate
P. herrerae	Herrera pine	3	10(10-25)	?	2-3(4)	20-25(35)	W Mexico	tropical
P. hwangshanensis	Hwangshan (Huangshan) pine	2	5-9	?	4-6	25	C and E China	temperate
P. jaliscana	Jalisco pine	4-5	12-16	?	4-8	20-30	W Mexico	tropical
P. jeffreyi	Jeffrey pine	3	12-15(23)	4-6	15-30	25-50(60)	California, Baja California	temperate montane
P. johannis	Zacatecas pinyon (pine)	3	3-5	?	3-4	2-4	NE Mexico	arid/montane
P. juarezenensis	Sierra Juarez pinyon	5	1.5-4	?	3.5-5	15	Baja California, S California	arid/montane
P. j.×P. monophylla [= P. quadrifolia]	Parry pinyon	4	3-5	?	4-6	5-10	Baja California	arid/montane
P. kesiya	Khasi, or Khasya pine	3	12-20(22)	2	5-7(10)	20-35(45)	SE Asia	tropical
P. koraiensis	Korean stone pine	5	(6)8-13	2	9-20	20-35	Korea, Japan, NE China, Siberia	temperate montane
P. krempfii	Krempf pine	2	3-7	2	7-9	12-30	Vietnam	tropical
P. lambertiana	sugar pine	5	(5)8-10	2-4	25-50(60)	75	Baja California, California, Oregon	temperate montane
P. lawsonii	Lawson's pine	3-5	15-20	?	6-8	25-30	S Mexico	tropical
P. leiophylla	smooth-leaved, or Chihuahuan pine	5	5-9(15)	2	4-6.5(8)	20-25(30)	Mexico, Arizona, New Mexico	temperate montane
P. longaeva	western, Great Basin, or Intermountain bristlecone pine	5	1.5-3	10-33(45)	6-9.5	16	W USA	subalpine
P. luchuensis	Luchu pine	2	15-20	2	<5	<20	Japan, Ryukyu Islands	temperate
P. lumholtzii	Lumholtz pine	3	(15)20-30	?	4-5(7)	10-20	C Mexico	tropical
P. massoniana	Masson, or Chinese red pine	2	15-20	?	5-6	30-40	C and E China, Taiwan	temperate montane
P. maximartinezii	Martinez, or Maxi pinyon	5	7-11	5	15-23	6-10	C Mexico	arid/montane
P. maximinoi	Maximino pine	5	15-28	?	5-8	20-35	S Mexico, Central America	tropical
P. merkusii	Merkus, or Tenasserim pine	(3-4)5(6-8)	17-25	1.5-2	5-9	20-35	SE Asia	tropical/savanna
P. montezumae	Montezuma, or roughbranched pine	(3-4)5	15-25	3	(6)12-15	20-30(35)	Mexico, Guatemala	tropical
P. monophylla	singleleaf pinyon	1(2)	3-6	4-12	5-8	5-10	SW USA to N Baja California	arid
P. monticola	western white pine	5	(4)7-13	3-4	14-25(30)	50-55(70)	W North America	temperate montane
P. morrisonicola	Taiwan white pine	5	4-10	3?	7-11	25(30)	Taiwan	temperate
P. mugo	dwarf mountain pine	2	3-8	5+	3-5(6)	2-6	Europe	subalpine
P. muricata	bishop pine	2	7-15	2-3	4-9	10-15(25)	California, Baja California	mediterranean coastal
P. nelsonii	Nelson pinyon (pine)	3	5-10	?	7-12	5-10	NE Mexico	arid
P. nigra	European black, or Austrian pine	2	8-16	4(8)	3-10	20-40	Europe, Medit. Basin	temperate
P. nubicola	Perry's pine	5-6(7-8)	25-43	?	10-15	25-30	S Mexico, Central America	tropical
P. occidentalis	Hispaniolan pine	(3)4-5	11-18	?	5-7(8)	18	Hispaniola, W Cuba	tropical montane
P. oocarpa	eggcone pine	(3-4)5	20-25	?	6-10	15-30	Mexico & Central America	tropical
P. palustris	longleaf pine	3(5?)	20-45	2	15-25	25-30	SE USA	temperate
P. parviflora	Japanese white pine	5	5-8	3-4	5-10	20-30	Japan	subalpine
P. patula	Mexican weeping pine	3(4-5)	15-25(30)	3-4	7-10	30-35	E Mexico	tropical
P. peuce	Macedonian, or Balkan (white) pine	5	6-12	?	8-15	20-30	Balkan Peninsula	temperate montane

Species	Common name(s)						Distribution	Habitat type
P. pinaster	maritime, or cluster pine	2	(10)15-20(25)	3	10-22	20-35(40)	W Medit. Basin	mediterranean coastal
P. pinceana	weeping, or Pince pinyon	3	6-8(14)	?	5-10	4-10	NE Mexico	arid/montane
P. pinea	Mediterranean stone, or Italian stone, or umbrella pine	2	(8)12-15(20)	2-3	10-15	15-30	Medit. Basin	mediterranean coastal
P. ponderosa	ponderosa, or western yellow pine	(2)3(4-5)	17-25	4-6	5-15	10-50(72)	W USA	temperate montane
P. praetermissa	Styles' pine	5	8-16	?	3-5	15	W Mexico	tropical
P. pringlei	Pringle's pine	3(4-5)	15-25	?	5-8	15-30	S Mexico	tropical
P. pseudostrobus	Mexican false white, or false Weymouth pine	5(6-8)	20-25	?	8-15	30-40	S Mexico, Guatemala	tropical montane
P. pumila	dwarf stone pine	5	4-6	5	3-5(6)	1-4	E Asia	boreal forest, subalpine
P. pungens	Table Mountain pine	2	5-7(9)	3	6-10	15-20	NE USA	temperate
P. radiata	Monterey, or radiata pine	(2)3	9-15	3-4	7-15	10-30(40)	California, Baja California	mediterranean coastal
P. remota	Texas, or paper-shell pinyon	2	3-5	2	2.5-3.5	3-8	Texas, NE Mexico	arid
P. resinosa	red pine	2	12-18	4-5	3.5-6	20-30(40)	NE USA, Canada	temperate
P. rigida	pitch pine	(2)3	5-10(12)	2-3	3-4(5-10)	10-25(30)	NE USA	temperate
P. roxburghii	Chir pine	3	20-30	1-3	10-15(20)	40-50+	Himalayas	temperate montane
P. rzedowskii	Rzedowski pinyon	3(4)5	6-10	5	10-15	15-30	SW Mexico	tropical
P. sabiniana	foothill, or Digger pine	3	15-25(30)	3-4	15-25	15-25	California	mediterranean coastal
P. serotina	pond pine	3	15-20	2-3	5-8	20	SE USA	temperate
P. sibirica	Siberian stone pine	5	(5)10-13	?	6-12	20-35	Central Asia	boreal forest
P. squamata	Qiaojia pine	5	?	?	9	25-30(40)	SW China	subtropical montane
P. strobus	eastern white pine	5	6-10(12)	2	8-20	30(35)	NE USA and Canada	boreal forest
P. sylvestris	Scots pine	2	3-7	2-3	3-6	25(30)	Europe, central Asia	boreal forest, temperate, subalpine
P. tabuliformis	Chinese red pine	2-3	10-12(13-17)	?	4-7	25(30)	N & W-C China	temperate montane
P. taeda	loblolly pine	3	12-22	3-4	6-12(15)	20-30	SE USA	temperate
P. taiwanensis	Taiwan red, or Formosa pine	2	8-12	?	4-8	20-25(35)	Taiwan	tropical montane
P. tecunumanii	Tecun Umán pine	4-5	14-21	?	4-7	50	Central America	tropical
P. teocote	twisted-leaved, or Mexican small-cone, or Aztec pine	(2)3(4-5)	8-15	3	4-7	8-25(30)	Mexico, Guatemala	tropical
P. thunbergii	Japanese black pine	2	7-12	3-4	4-5	30-40	Japan, Korea	temperate
P. torreyana	Torrey pine	5	15-30	3-4	10-15	5-10(15)	California	mediterranean coastal
P. tropicalis	tropical pine	2(3)	15-30	?	?	?	Cuba	tropical/savanna
P. uncinata	Swiss mountain pine	2	(3)5-6	?	4-6	10-20	Europe	temperate montane
P. virginiana	Virginia, or scrub pine	2	4-8	3-4	3-7	8-15(30)	E USA	temperate
P. wallichiana	Himalayan blue pine	5	11-18(20)	3-4	20-30	50+	Himalayas	temperate montane
P. wangii	Wang pine	5	2.5-6	?	4.5-9	20	SW China	temperate
P. washoensis	Washoe pine	3	10-15	4-6	7-10	35(70)	Sierra Nevada	temperate montane
P. yunnanensis	Yunnan (white) pine	2-3	15-20(30)	?	3-7(10)	15-30	China	temperate montane

Notes:

a Nomenclature follows Price, Liston & Strauss (Chap. 2, this volume).

b Criteria used in compiling the list of common names included: (1) the extent and history of common usage in English (some misnomers were discarded); (2) where several names exist we chose the one(s) that describe clearly recognizable features of the pine as first choice, next the geographical range or habitat of the species, next the anglicized version of the latin name. Preference was also given to common names that allude to the position of taxa within super-specific taxa (e.g. ____ white pine for the taxa in section *Strobus*, subsection *Strobi*). In all cases, the first name given is the recommended common name. In a few cases, changes from current common usage are suggested (e.g. *P. brutia* for which we prefer the name East Mediterranean pine; *P. pinea*, commonly known as the Italian stone pine, but here called the Mediterranean stone pine; *P. sabiniana* where foothill pine is used in preference to Digger pine). The new English names for *P. nubicola* and *P. praetermissa* proposed here honour their respective describers, Jesse P. Perry and Brian T. Styles, both prominent scholars of Mexican pines. For a more comprehensive list of names, readers should refer to the Glossary of common names (p. 490).

c Main sources of data: Original species descriptions and Loock (1950); Den Ouden & Boom (1965); Dallimore & Jackson (1966); Mirov (1967); Farjon (1984); Perry (1991); Vidaković (1991); Young & Young (1992); Kindel (1995).

d Main sources of data: Critchfield & Little (1966); Perry (1991).

e Main source of data: Mirov (1967); Perry (1991); Kindel (1995).

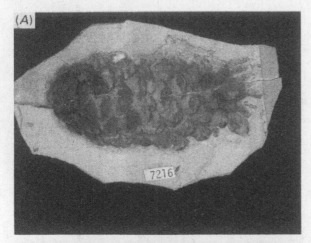

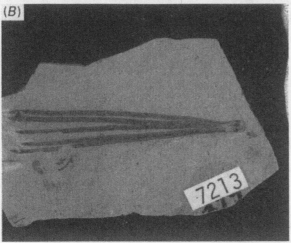

Fig. 1.2. **Evidence reviewed in Chap. 3 (this volume) suggests that pines made their first Tertiary appearance at many middle-latitude locations during the Oligocene, recolonizing areas where they had occurred in the Mesozoic. The figure shows a fossilized cone (A) and needles (B) from the Late Oligocene Creede flora from Colorado, USA (c. 27.2 million years ago; see Table 3.4, p. 79 for palaeo-coordinates). These specimens (UCMP 7213 and 7216) have been assigned to** *Pinus crossii,* **an ancestor of the extant bristlecone and foxtail pines in subsection** *Balfourianae.* **Taxa of this subsection, whose basal position in the phylogeny of the haploxylon pines has been confirmed by recent biochemical studies (Chap. 2, this volume), were probably confined to middle-latitude refugia in the Rocky Mountains during the early Tertiary (photographs kindly supplied by D.M. Erwin, Museum of Paleontology, University of California, Berkeley).**

angiosperms diversified and spread they began to replace the formerly dominant gymnosperms. The latter were deposed to small, cool or dry refugia in polar latitudes and scattered upland refugia at middle latitudes (e.g. the present Rocky Mountains and Japan). These habitats have remained the principal domain of gymnosperms. The widespread displacement of gymnosperms by the rampant angiosperms led to the splitting of several subsections of *Pinus* into northern and southern groups. Kremenetski *et al.* (Chap. 4, this volume) discuss the impor-

tance of the division of subsection *Pinus* into northern refugial populations in western Siberia, mid-latitude populations in eastern Asia, and southern refugial populations in other parts of Asia and Europe for evolution within the genus. Intensive mountain-building events in some areas, with further climate change, created the environmental heterogeneity that drove the radiation of pine taxa in several areas which became secondary centres of diversification in *Pinus* (notably Mexico and north-eastern Asia). Angiosperms which were adapted to tropical conditions declined dramatically throughout middle latitudes following climatic deterioration at the end of the Eocene. This permitted pines to expand their ranges (Fig. 1.2). Radiations of subsections *Contortae, Oocarpae* and *Ponderosae,* and of many species within subsections seem to date to this period. Millar & Kinloch (1991) provide an excellent review of the events described above, and Millar (Chap. 3, this volume) describes in more detail the role of Eocene phenomena in shaping the ecology and bio-geography of *Pinus.*

Like the Eocene, the Pleistocene was also characterized by profound environmental changes. However, whereas events in the Eocene completely reshuffled elements of the genus (Millar 1993), Pleistocene changes caused pine species and populations to shift first south, then north (and to lower, then higher elevations), following the cycle of glacial and interglacial periods. Such migrations had important influences on the genetic diversity of pines. For example, populations of the progenitor of *P. banksiana* and *P. contorta* were separated into disjunct eastern and western populations (Critchfield 1985; see also section 1.6.8 and Chapters 6 and 9, this volume, for further discussion). Whereas North America and northern Europe experienced massive glaciations during the Pleistocene, northern Asia did not. This region was, nonetheless, affected by significant changes in climate associated with the alternation between glacial and non-glacial periods. These fluctuations probably caused the separation of closely related species such as *P. pumila* and *P. sibirica* during glacial periods, and may have played a role in speciation or at least the preservation of distinctive genotypes (Chap. 4, this volume). Although phenomena such as these have been important for the evolution of pines, Pleistocene events seem to have been less pivotal than those of the Eocene (see Chap. 3, this volume). In some areas, such as the Pacific Northwest of North America, pine distributions were not so much split into distinct ranges by glaciations, as fragmented into small, semi-disjunct populations. Such distributions may have served to promote interspecific diversity while not necessarily promoting speciation (see Chap. 6, this volume). Geological history prevented pines from migrating south of the Sahara, south of Nicaragua in the Americas, or from entering the Australian continent; their recent success in the southern hemisphere shows that large parts of these regions are most suitable for pines.

The previous paragraphs have given a very brief (and highly simplified) account of some prominent events that have influenced the evolution and migration of pines since the Cretaceous. To understand the current distribution of pines, however, the changes in abundance and geographic ranges that have occurred since the end of the last glacial period, i.e. during the last 10 000 years, are especially important. The study of fossil pollen from sediments has facilitated the compilation of detailed pollen maps (e.g. Huntley & Birks 1983). Analysis of these has shown that pines generally expanded their ranges more rapidly into deglaciated regions of North America and Europe than other tree taxa in most areas (Chapters 4–6, this volume). Different taxa moved at different rates and different habitats showed different degrees of resistance to invasion by pines and other taxa. Recorded rates of spread of *Pinus* species range from 81–400 m yr^{-1} in North America to 1500 m yr^{-1} in Europe (MacDonald 1993). Species such as *P. banksiana* and *P. contorta* in the western interior of North America probably reached their current distributions relatively recently, whereas others such as *P. sylvestris* migrated rapidly in the early postglacial. Estimates of population growth also vary greatly, with a doubling time of 73 years for *P. sylvestris* to over 1000 years for *P. contorta* subsp. *latifolia* and other undifferentiated pine species (references in MacDonald 1993). Significant variation in population growth, even within a species, has also been observed. For instance, MacDonald & Cwynar (1991) found a high degree of regional variability in population growth rates of *P. contorta* subsp. *latifolia*, suggesting that invading populations spread and/or grew at different rates in different regions. The rates and patterns of these postglacial migrations have left clear imprints on patterns of genetic variation between and within pine taxa (Cwynar & MacDonald 1987; Chap. 13, this volume). Ledig (1993) discusses the examples of *P. jeffreyi* and *P. monticola* which show little genetic differentiation among sites or elevations in the northern parts of their ranges, but substantial differentiation in California. This is presumably because too few generations have passed since they colonized the northern areas to permit genetic adaptation of populations to local environments. Another classic example is the evolution of pygmy-forest edaphic subspecies of *P. contorta* on marine terraces in coastal northern California (Aitken & Libby 1994; see section 1.6.7).

The Quaternary history of the climate and vegetation of southwestern North America was poorly understood until the discovery by Wells & Jorgensen (1964) that middens built by packrats (*Neotoma* spp.) provide abundant plant fossils. Recent analyses of middens preserved for up to 40 000 years at many sites in the Rocky Mountains, the Great Basin, and deserts throughout the western USA and Mexico, have greatly improved our understanding of the history of pines in these areas. Probably the most dramatic floristic change revealed by the study of these 'natural time capsules' was the replacement of pinyon–juniper woodland by desert scrub in the Great Basin between two and three thousand years after a period of rapid warming 13 000–14 000 years ago (Long *et al.* 1990; see Chap. 9, this volume, for further discussion).

The chapters in Part III – *Historical biogeography* describe the events that have shaped pine distribution in four important regions: northern Asia (Chap. 4), Europe (Chap. 5), northern North America (Chap. 6), and Mexico and Central America (Chap. 7). Two chapters in Part IV – *Recent biogeography* describe in more detail the role of events over the past few centuries on the distribution of pines in the Mediterranean Basin (Chap. 8) and the American Southwest (Chap. 9).

1.3 Pines compared with other conifers and broadleaved trees

Pines share certain features with the other genera in the Pinaceae (*Abies, Cathaya, Cedrus, Keteleeria, Larix, Nothotsuga, Picea, Pseudolarix, Pseudotsuga* and *Tsuga*). Price *et al.* (Chap. 2, this volume) show that pines are most similar to *Cathaya* and *Picea* in overall morphology, and to *Cathaya, Larix* and *Picea* if one considers wood anatomy and seed and cone-scale morphology. The immunological comparisons of seed proteins reviewed in Chap. 2 (this volume) suggest that pines and spruces occupy relatively basal positions in the phylogeny of Pinaceae, a finding that is consistent with the fossil record. In terms of ecology, pines are closest to firs (*Abies* spp.) and spruces (*Picea* spp.) with which they frequently co-occur; these three genera are very prominent in the northern hemisphere and often dominate the vegetation in which they occur. It is, however, in the role of aggressive post-disturbance colonizers that pines are most clearly differentiated from firs, spruces, other conifers, and angiosperm trees. An idealized 'pine prototype' would conform with the following profile: a light-demanding, fast-growing, sclerophyllous tree that regenerates as even-aged cohorts following landscape-scale disturbance and retains its position in the landscape by exploiting aspects of its regeneration biology. That this is an oversimplification becomes obvious when one considers the wide range of habitats in which pines occur (see later) and the range of life-history syndromes evident in the genus (Chap. 12, this volume).

Among the factors that have contributed to the rapid migration and population increases of pines in the Holocene are: their abundant output of seeds from an early age; their ability to recruit dense daughter stands on exposed sites soon after disturbance; effective mechanisms for long-distance seed dispersal; a mating system that permits inbreeding and selfing in isolated trees; and various life-history traits that confer resilience at the population level under a wide range of disturbance regimes; and the ability to colonize nutrient-poor sites. These attributes enabled pines to undergo rapid range changes through a combination of neighbourhood diffusion and long-distance dispersal. In such a 'stratified dispersal process', their initial range expansion occurs mainly through the recruitment (in response to disturbance) of large numbers of offspring near parent plants. However, as the range of the founding population expands, new colonies created by long-distance migrants increase in number to cause an accelerating range expansion in the later phase. Pine expansion provides a model case of the stratified diffusion process described by Hengeveld (1989; see also Shigesada, Kawasaki & Takeda 1995). This behaviour also explains the response of pines to recent changes in disturbance regimes (see below).

Abies, Picea and Pinus are all listed as prominent taxa in six of Takhtajan's (1986) 35 floristic regions of the world (Circumboreal; Eastern Asiatic; North American Atlantic; Rocky Mountain; Irano-Turanian; and Madrean). In these regions, pines generally thrive on the poorest soils, whereas firs and spruces require slightly more fertile (and heavier in Picea) soils. Pines are also prominent in Takhtajan's Mediterranean Region, which lacks Picea and has very limited representation of Abies which requires better soils. Regions with the greatest diversity of dominant pine species are the Madrean (29 species listed), North American Atlantic (14 species), Eastern Asiatic (9 species), Rocky Mountain (8 species) and Mediterranean (6 species). Regions where Abies and Picea are prominent but where Pinus is absent as a recognized dominant include Takhtajan's Manchurian Province (Eastern Asiatic Region) and Western Asiatic Subregion (Irano-Turanian Region). Taxa of the other northern hemisphere conifer genera are represented in far fewer floristic regions, and are less abundant and generally less important in ecosystems than pines. Pines also differ from southern hemisphere conifers which, with a few exceptions (e.g. Agathis australis in New Zealand and Araucaria araucana and Fitzroya cupressoides in Chile; Enright et al. 1995), are usually relatively minor components of the vegetation or have highly localized occurrences.

1.4 Morphological traits of pines

1.4.1 Growth form and size

Pines, like many other conifers, have the characteristic of monopodial growth and large size. The largest species of monopodial growth and large size. The largest species of pines in the world are centred in distribution in California and the Pacific Northwest of the USA. Growth conditions in these regions favour immense size in many genera, including Abies, Picea, Pseudotsuga, Sequoia, Sequoiadendron, Thuja and Tsuga (Waring & Franklin 1979). The largest species of pine in both height and girth is P. lambertiana which reaches over 75 m in height and more than 5 m in diameter in the Sierra Nevada of California. Three other pines from the western USA, P. jeffreyi, P. monticola and P. ponderosa, all reach heights of 60 m or more (Table 1.1). Many Mexican pines grow in mountain areas with annual rainfalls of 1200–2000 mm or more (e.g. P. ayacahuite, P. hartwegii, P. maximinoi, P. tecunumanii), but they usually reach heights of only 20–50 m. Similarly, P. caribaea var. hondurensis which grows in Belize, Honduras and Nicaragua, in probably the wettest habitats of any pine where annual rainfall may commonly reach 2000–3000 mm, reaches only 20–30 m in height.

Pines can, however, be quite short in stature in more extreme habitats. The pinyon pines as a group (11 species in section Parrya, subsection Cembroides), growing in habitats with 300–600 mm of rain annually, usually attain heights of 5–10 m when mature (Table 1.1). Timberline pines also may be low-growing, particularly when they occur as krummholz shrubs at the upper limits of tree distribution. Most of these timberline species have the genetic potential for taller growth, and may reach 10–20 m in height under more favourable conditions. At least two species, however, P. mugo in the European Alps, and P. pumila in East Asia and Japan, occur most characteristically with a low shrubby growth form.

The characteristic monopodial growth form of pines is absent in the unusual P. sabiniana in the foothills of California, which lacks apical dominance after the juvenile period and has a branched main trunk like that of hardwood species. Pinus bungeana from China and P. maximartinezii from Mexico have similar habits. Monopodial growth may also be lost in krummholz growth forms of pines which attain a distinctly shrubby canopy.

1.4.2 Whole-tree longevity

Many pines are very long-lived, and the two bristlecone pines, P. aristata and P. longaeva (Fig. 1.3), are the oldest living organisms in the world, with the latter reaching documented ages of nearly 5000 years (Currey 1968). The oldest living P. aristata was aged to 2435 years (Brunstein & Yamaguchi 1992). Nooden (1988) lists two other pine species, P. cembra (1200 years) and P. sylvestris (500 years) among the longest-lived plants in the world, but several other pines could also make this list. According to Schweingruber (1993), P. balfouriana may live as long as the bristlecone pines, P. flexilis can live for more than 2000 years, P. jeffreyi and P. ponderosa for >1000 years, and P. con-

Fig. 1.3. **Western bristlecone pines, *Pinus longaeva*, growing on poor soils on the White Mountains of California, USA, which are characterized by a cold and dry climate. Trees at this site have been aged to nearly 5000 years – the oldest living organisms on Earth. Tree rings from three trees that give a continuous time series from 8000 years ago to the present were analysed for the composition of stable hydrogen isotopes. This record, which shows the presence of a postglacial climate optimum 6800 years ago and a continuous cooling since then, serves as a reference for other climate indicators (Feng & Epstein 1994) (photo: W.D. Stock).**

torta subsp. *murrayana* and *P. monticola* for up to 500 years. *Pinus heldreichii* has been aged to >800 years in Calabria, southern Italy (Serre Bachet 1985).

Pines have played a fundamental role in the development of the modern science of dendrochronology, beginning with the pioneering work of Andrew Douglas in the American Southwest (Fritts 1976). Douglas, an astronomer, became Director of the Lowell Astronomical Observatory in Flagstaff, Arizona, in 1894. With research interests in sunspot activity and possible related impacts on climate, Douglas was drawn to the possibility that tree rings might contain climatic records that would not otherwise be available from existing weather stations. Working with *P. ponderosa* in the Flagstaff area, Douglas developed the concept of cross-dating to compare and extend these tree-ring measures over broad regional areas to identify year-to-year variation in climate. It was this pine research that led him to establish the Laboratory of Tree Ring Research at the University of Arizona in 1906.

Collaborative work beginning in 1914 by Douglas with Clark Wissler, a prominent anthropologist at the American Museum of Natural History, soon led to what were then revolutionary approaches to dating the construction of Indian dwellings in Chaco Canyon and Mesa Verde in the Southwest. These studies allowed the earliest measurement and linkages of floating chronologies to develop long-term records over >2000 years, and had profound impacts in the field of anthropology.

Although the field of dendrochronology has expanded greatly in scope and depth since these early studies and involves work with many tree genera throughout the world (Schweingruber 1993), research with pines still

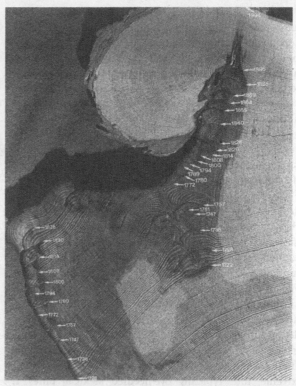

Fig. 1.4. **Cross-section of *Pinus ponderosa* from Kings Canyon National Park, California, showing a record of fire scars from 1722 to 1994. This section was collected by Chris Baisan, Kiyomi Morino, James Risser and Robert Shay of the University of Arizona's Laboratory of Tree-Ring Research.**

forms the heart of this field (see e.g. Cook & Kairiukstis 1990). Pines contain the longest single chronologies available (*P. longaeva* in the western USA; Fig. 1.3) and some of the most sensitive chronologies for evaluating regional patterns of climate (e.g. Brunstein 1996 and references therein). Tree-ring chronologies are also proving to be valuable records of alteration of typical forest growth regimes resulting from atmospheric pollution or other causes. Fire histories over long periods can be inferred from scars on pines (Arno & Sneck 1977; Fig. 1.4), providing useful records of past conditions at a site.

1.4.3 Cone and seed morphology

The form and morphology of pine cones is highly variable, with obvious relationships to the reproductive biology of individual species (Chap. 12, this volume). In terms of size, the greatest length of cone in any pine occurs in *P. lambertiana*, where the elongate cones reach up to 50 cm in length. In terms of weight of fresh cone, the record probably belongs with the large globular cones of *P. coulteri* from California which are 20–35 cm in diameter and may weigh as much as 2.3 kg. Large cones are also present in the Mexican taxa *P. ayacahuite*, *P. devoniana* and *P. maximartinezii*

(Table 1.1). About a third of pine species typically bear cones that are less than 5 cm long. As a broad generalization, it appears that taxa associated with stressful environments have smaller cones. It is intuitive to expect the largest seeds in species with the largest cones, but Keeley & Zedler (Chap. 12, this volume) show that, for all Mexican pine taxa, the correlation between cone size and seed size is poor.

Pinus is far more diverse in the morphology of its seeds than all other Pinaceae combined (see Chap. 14, this volume), a fact that certainly contributes to the wide range of habitats in which pines flourish.

1.4.4 Needle morphology and longevity

Although all pines share the defining morphological trait of possessing pine needles, there is a wide variation in the size and form of display of needles. Needles are arranged in bundles (generally termed fascicles or needle clusters), with the number of needles per fascicle being a reasonably constant and species-specific characteristic in many taxa. Most pine species have two, three or five needles per fascicle, but other numbers are also present (Table 1.1). Only one species has one needle per fascicle: *P. monophylla*, the singleleaf pinyon of the southwestern USA. At the other extreme, four typically five-needled Mexican pines, *P. cooperi*, *P. donnell-smithii*, *P. durangensis* and *P. pseudostrobus*, frequently have six needles per fascicle, and sometimes up to eight. Four-needled fascicles are present in several Latin American pine taxa, although often as a common variant of a three- or five-needled condition. These species include *P. caribaea* (2–5), *P. lawsonii* (3–5), *P. rzedowskii* (3–5), and *P. tecunumanii* (4–5), and the four-needled hybrid between *P. juarezensis* and *P. monophylla* which has been called *P. quadrifolia*.

Adaptive radiation within specific subsections of the genus *Pinus* has taken place, both with and without modifications of needle number per fascicle. For example, almost all of the 19 Old World pines of the subsection *Pinus* have two needles per fascicle – this despite the wide range of habitats occupied by taxa in this group (e.g. *P. resinosa* and *P. sylvestris* in boreal-type forest; *P. nigra* and *P. pinaster* in lower-elevation sites in the Mediterranean Basin; *P. heldreichii*, *P. mugo* and *P. uncinata* at high-elevation sites in the Mediterranean Basin; *P. kesiya*, *P. merkusii* in tropical savannas; and a set of eastern Asian species that occupy a wide range of habitats). At the other extreme are the 11 species of the subsection *Cembroides* which include taxa with 1–5 needles per fascicle, despite the relatively uniform arid environments in which all these species occur. The 16 species in subsection *Strobi* all occupy montane habitats and all typically have five needles per fascicle.

The length and form of pine needles also varies greatly.

The longest needles of any pine species are those of the appropriately named longleaf pine, *P. palustris*, in the southeastern USA (up to 45 cm), and those of *P. nubicola* in Mexico which also reach lengths of over 40 cm. At the other extreme are many pines with very short needles in the 2–8 cm range of maximum lengths (Table 1.1). These short-needled species are almost entirely confined to the arid-adapted pinyon pines and high-elevation, or timberline pines, suggesting a relationship with environmental stress. Some exceptions to this rule are *P. contorta* subsp. *bolanderi* and *P. contorta* subsp. *contorta* along the west coast of the USA, *P. glabra* in the southeastern USA and *P. virginiana* (which grows on very nutrient-poor sites). Variation in needle length in subsection *Strobi* provides further evidence that short needles are associated with stressful environments. There is relatively little difference in needle length among the 16 species in this group (Table 1.1); the two species with shorter needles (*P. flexilis* and *P. parviflora*) are the only taxa in the group that occupy subalpine habitats.

The characteristics of few needles per fascicle and short needle lengths within the pinyon pines of the southwestern USA and Mexico has led to the assumption that these traits are adaptations to arid conditions. Such a correlation, which seems to exist, has yet to be rigorously tested (see Chap. 15, this volume, for discussion). Other pinyons such as *P. juarezensis* in arid areas of Baja California are quite successful with five needles per fascicle. The trait of two needles per fascicle is common among subalpine and timberline pines such as *P. contorta*, the widespread *P. sylvestris* in cool-temperate areas of Eurasia, and many Mediterranean pines. Similarly, some pines of relatively arid environments may also have relatively long needles, as in needles up to 25 cm in length in *P. pinaster* from the Mediterranean Basin. No correlation exists between needle number or needle length within pine species or with either needle length or number of needles per fascicle and mean habitat rainfall for North American pines. Nevertheless, careful studies of ecotypic differentiation and interspecific variation within a single lineage such as the pinyon pines may well show some type of correlation. *Pinus sylvestris* is the ideal taxon for such an investigation within a single species.

There are other interesting needle traits in pines that have not been studied. Several Mexican pine species have drooping, or 'weeping', needles that hang downwards, and there are intermediate morphologies in other species with relatively flexible needles. We have found no suggestions for any possible adaptive correlate for this condition. Such long, fine needles may aid the condensation and drip of fog moisture in tropical mountain areas or in coastal fog zones, as with *P. lumholtzii*, *P. nubicola* and *P. patula* in Mexico, *P. radiata* on Cedros Island (Fig. 21.1; Chap 21, this volume) and the coast of California (see Coffman 1995 for a popular statement of this hypothesis), and *P. canariensis* in the Canary Islands. Such hypotheses remain conjectural, however. Another unusual needle morphology in pine is that of flattened needles which are characteristic of the rare *P. krempfii* from the highlands of Vietnam. Little is known about this unusual species which some authors consider sufficiently distinct to warrant being placed in its own subsection of the section *Pinus* (see Chap. 2, this volume)

One strong environmental correlate of needle traits in pines does exist. Needle longevity is strongly correlated with habitat water- and nutrient relations and/or stress. Tropical pines such as *P. caribaea* and *P. palustris* keep needles for no more than 2–3 years, and *P. roxburghii* usually sheds its needles every year (Singh, Adhikari & Zobel 1994). Temperate forest pines commonly retain their needles for intermediate periods of 4–6 years (Table 1.1). Pinyon pine leaves have relatively greater longevities of up to 10 years. Subalpine pines such as *P. longaeva* retain their leaves for up to 30 years or more, and even 45 years in extreme circumstances – at the timberline in the White Mountains of California (Ewers & Schmid 1981); this is the greatest needle longevity recorded for any conifer. Nebel & Matile (1992) report a cline in needle longevity along an altitudinal gradient (1780–2140 m) in *P. cembra* in Switzerland, with plants at the highest elevations retaining needles the longest (up to 12 years). A similar finding was reported for *P. pumila* in central Japan, where needle longevity was greater at higher altitudes and windward sites than at lower altitudes and leeward sites (Kajimoto 1993).

1.5 Ecophysiological traits of pines in relation to other conifers

Coniferous forest trees characteristically utilize a very different strategy of canopy carbon gain from that of hardwood trees. Conifers have relatively low rates of carbon gain per unit of leaf area compared with deciduous hardwoods, but a far higher leaf area index (LAI). Needles are retained for several to many years, and a clustered arrangement of foliage and regular canopy architecture has evolved to allow maximum irradiance of older foliage. Thus the net primary productivity of conifer forests is typically as great as or greater than that of deciduous hardwood forests in the same climatic regime. However, many years of needle production, resulting in a high LAI, contribute to a high stand-level net primary production in conifers such as *Abies*, *Picea* and *Pseudotsuga*. These species are inherently slow in becoming established in succes-

sional sequences where environmental stress is not extreme. Under these conditions, deciduous hardwood saplings which can attain a full canopy in a single year are much more competitive.

Pines differ from the typical conifer strategy in several respects. Typical ranges of LAI in field populations of pines are only 2–4 $m^2 m^{-2}$, compared with values of 9–11 $m^2 m^{-2}$ in the more shade-tolerant genera *Abies*, *Picea* and *Pseudotsuga* (Jarvis & Leverenz 1983). Nearly twice this LAI would be necessary for the pines to fully absorb incident photosynthetically active radiation (PAR) irradiance, while the shade-tolerant conifers carry far more LAI than can functionally aid in their net primary production. The cost of this high LAI is minimal, however, as individual branch modules have relatively low costs of maintenance respiration and are thus metabolically self-supporting. The value of such extra leaf area lies in increased resource capture, particularly for light, and thus increased competitive strength (Waring 1991). Highly productive plantations of commercial pine species such as *P. radiata* owe much of their productivity to LAI values two or three times those found under field conditions.

The low LAI in pines results largely from the fact that many species carry relatively few years of needles compared with other conifers. Except in pines characteristic of environments of extreme cold or drought stress (see above), 2–5 years of needles in the canopy at any time is typical (Ewers & Schmid 1981; Table 1.1). Thus, despite their relatively low LAI, pines are inherently more effective colonizers than many other conifers because they can attain a full canopy of foliage more rapidly early in succession. Pines do show considerable plasticity in needle retention under conditions of environmental stress, both within a species and between species (Nebel & Matile 1992; Reich *et al.* 1995).

It is interesting to speculate on the potential similarities of the rapid growth and colonizing abilities of many pines and the traits of early-successional hardwood trees. The relatively low LAI of pines and their generally poor shade-tolerance are shared by such hardwoods as temperate *Eucalyptus* and many tropical pioneers. Shade-tolerant conifers such as *Abies*, *Picea*, *Pseudotsuga* and *Sequoia* not only have high LAIs, but are characterized by an architectural form, termed Massart's model. This is characterized by rhythmic orthotropic growth that produces regular whorls of branches at levels determined by the height of the trunk meristem. In contrast to this pattern, most pines and many tropical colonizers (e.g. *Cecropia*, *Macaranga* and *Musanga* spp.) are shade-intolerant and possess a canopy architecture termed Rauh's model (typically expressed in *P. heldreichii*; see Farjon 1984, p. 86) with the rhythmic addition of tiers of branches which are structurally identical to the trunk (Hallé, Oldeman & Tomlinson 1978).

Considering the wide range of ecological habitats in which pines occur (see section 1.6), it is noteworthy that there is relatively little variation in their photosynthetic characteristics. Maximum rates of photosynthetic capacity under field conditions within ecologically plastic species such as *P. contorta* and *P. sylvestris* appear to vary as much as within pines as a group (Chap. 15, this volume). When grown under non-limiting conditions, pines from very different environments exhibit quite similar photosynthetic responses to irradiance (Teskey, Whitehead & Linder 1994), suggesting a considerable degree of phenotypic plasticity. This is equally true both for ecotypes of the same species and for different species. Thus variation in net primary production rates of pines in different environments is less a function of differences in photosynthetic capacity than of climatic factors of cold or drought that limit the period of positive net carbon gain throughout the year.

Seasonal patterns of low temperatures in autumn and winter are clearly important components of the potential net primary production of pine species. Thus pines in cold-temperate environments or timberline habitats have high rates of positive canopy photosynthesis limited to relatively few months of the year. Subtropical and tropical species grow throughout the year, with light and LAI the primary factors controlling net primary production. Although pine species may tolerate habitats over a wide range of annual precipitation, the response of individual species to low tissue-water potentials is surprisingly consistent, as with almost all conifers. Net photosynthesis in pines typically falls to zero at water potentials of -2.0 to -2.8 MPa, thereby limiting the growth potential of pines in semi-arid environments.

Other aspects of the ecophysiology of pines show highly adapted traits between species which adapt them to specific habitats. As logic would suggest, pines of cold habitats commonly have much lower temperature optima for photosynthesis than do species of warm climates. Stomatal responses to environmental water vapour deficit (ΔW) is another trait showing adaptive selection. Species of semi-arid environments such as *P. radiata* are highly sensitive to small changes in ΔW, thereby regulating summer water loss. *Pinus taeda* (a mesic-habitat species) is quite insensitive to changes in ΔW, while *P. sylvestris* shows an intermediate response. Given this range of stomatal sensitivity, it is not surprising, therefore, to find that pines as a group exhibit a broad range of water-use efficiencies in their growth indicated by analyses of stable carbon isotope ratios (Δ) in a broad survey of conifer species. The range of Δ values within the genus *Pinus* is almost as great as that for all conifers, and far broader than that present in other major genera.

1.6 Pines in the landscape

Pines are found in a remarkably wide range of environments, from near the Arctic where winters are very cold and growing seasons are short, to the tropics, where frost never occurs and growth continues through the year (Knight *et al.* 1994). Some pine species form virtually monospecific forests over very large areas whereas others form mixed forests with other conifers (notably *Picea* and *Abies* spp.) and broadleaved trees (notably *Quercus*, *Populus*, *Betula* and *Alnus* spp.), or form savannas or open woodlands. Pines are the dominant trees over large parts of the boreal forest, or taiga, which covers about 12×10^6 km^2 of the northern hemisphere. In boreal-type forests, pines (especially *P. banksiana* and *P. contorta* in North America; *P. pumila*, *P. sibirica* and *P. sylvestris* in Fennoscandia and the former USSR) occur with other conifers (*Abies*, *Larix* and *Picea*) and several broadleaved genera (notably *Betula* and *Populus*). Pines possess a range of specialized mechanisms that enable them to thrive (and usually attain dominance) in these harsh environments. Although the northern coniferous forests contain the greatest area of pine forests, many more species occur in temperate regions. The ranges of the temperate pine species are generally much smaller than those of higher latitudes (Chap. 10, this volume); in temperate regions, and even more so in the tropics, pines are usually associated with acidic, nutrient-poor soils. That pines are not restricted to such sites is clearly shown by their ability to spread into more productive sites, both within and outside their natural ranges, following disturbance that reduces the competitive superiority of vigorous angiosperms (Richardson & Bond 1991). The disturbance regime is thus an important determinant of pine distribution and abundance in the landscape. Fire is the driving force in succession in nearly all pine habitats (Chapters 11 and 12, this volume).

Several authors have defined 'ecological groups' of pines based on their response to disturbance. For example, McCune (1988) described five groups for the pines of North America: (1) thick-barked species tolerant of surface fires; (2) species that become established rapidly from seed after fires; (3) species with moderate tolerance to shade; (4) species found in unusually dry or cold environments, which have wingless, animal-dispersed seed; (5) species of warm, humid environments with rapid growth and short leaf duration. McCune used life-history traits associated with these groups to explain the propensity of different pines to: often form savannas but become dense forest if fires are suppressed (group 1); form seral, even-aged stands (group 2); occur in association with other conifers and broadleaved trees (group 3); form savannas or open woodlands rather than forests (group 4); or form dense, usually seral, forests (group 5). Following similar reasoning,

Fig. 1.5. **Landscape aspect of boreal communities along the Mackenzie River of the Northwest Territories of Canada, south of the town of Norman Wells. The band of paler vegetation behind the lake is mainly *Pinus banksiana* which reaches its northern range limits in this area. As it approaches its northern limits, jack pine becomes increasingly restricted to sandy, well-drained soils such as those associated with glacio-fluvial deposits (photo: G.M. MacDonald).**

Richardson, Cowling & Le Maitre (1990) defined pine life-history syndromes and used these to assess the likelihood of different species invading an alien habitat (South African fynbos) defined in terms of various environmental features or 'filters'. These approaches, although useful abstractions, oversimplify the complex nature of the forces that operate to structure pine forests.

A thorough discussion of the biogeography and ecological habitat selection in pines would require an entire volume for this topic alone. It is, however, important to highlight the broad patterns of geographical and ecological distribution that characterize pines in the northern hemisphere. In the discussion below, we treat major pine habitats, beginning with the coldest and moving to warmer climatic regimes. In the order presented here, these are: boreal forest habitats, subalpine and timberline habitats, temperate forest habitats, mediterranean coastal habitats, arid habitats, and finally lowland and montane tropical habitats.

1.6.1 Boreal forest habitats

Boreal forest or taiga ecosystems are primarily dominated by species of *Abies*, *Larix* and *Picea*, but pines also make important contributions to these coniferous forests. Very cold winters lasting 7–9 months and low-nutrient soils make the boreal forests severe environments for plant growth, and mosaics of boggy soils and arid sandy soils add to this environmental stress.

The dominant pine in the boreal forests of North America is *P. banksiana* which extends from the Rocky Mountain area of western Canada across the continent to Nova Scotia (Fig. 1.5). Unlike the more dominant firs and spruces, *P. banksiana* does not reach the Arctic transition zone for forest vegetation; it is restricted to closed forest

habitats away from this zone. Nevertheless, this species reaches north of latitude 65° N, the northernmost distribution for any North American pine (Burns & Honkala 1990). Pure stands of *P. banksiana* are widespread in the Great Lakes region and central and western Canada on well-drained sand hills, eskers, and rocky ridges; these sites are all too arid for fir and spruce to dominate. *Betula* spp., *Picea mariana* and *Populus* spp. are common associates in mixed stands, as is *P. resinosa* at the southern margin of the boreal forest. *Pinus banksiana* may also regenerate effectively from seed following fires and maintain dominance in successional sequences (St. Pierre, Gagnon & Bellefleur 1992; Gauthier, Gagnon & Bergeron 1993).

The boreal forest or taiga of northern Europe and Asia forms the most extensive area of coniferous forest in the world. Although the number of pine species is relatively low, three species make substantial ecological contributions to this system. These include the two species which reach the northernmost latitude in the world for pines. *Pinus sylvestris* in Norway and *P. pumila* in eastern Siberia each reach above latitude 70° N.

Although *Picea abies* is the most important and widespread climax tree species in the boreal forest of Western Europe, *Pinus sylvestris* is important over large areas (Fig. 1.6). This pine is dominant in drier habitats with coarse rocky and sandy soils, and along the margins of mires. Scots pine also is an effective colonizer of disturbed sites and is resistant to forest fires, thereby giving this species an early successional dominance over *Picea abies*. In the northern areas of Scandinavia where *P. abies* is limited to well-drained soils, and mires are the dominant landscape feature, *Pinus sylvestris* is common on sandy river terraces and around the boggy soils of the mires. There is evidence that *P. sylvestris* is more extensive today in the boreal forest of northern Europe than in the historical past because of human impact from timber harvesting, livestock grazing, and fire (Polunin & Walters 1985; Chap. 5, this volume). *Pinus sylvestris* remains important in the taiga of European Russia, but *P. sibirica* begins to appear to the east.

Pine distributions in northern boreal forest ecosystems of Asia are described in Chap. 4 (this volume). The central Siberian region is dominated by extensive forests of *Larix* to the north, but *P. sibirica* becomes increasingly important towards the south where it may occur in pure stands or mixed with *Betula*, *Larix*, *Picea* and *Populus* (Larson 1980). As the taiga climate becomes drier and more continental in eastern Siberia, the community structure of coniferous forests changes, and *P. pumila* becomes the dominant species on well-drained sites. Forest areas of the Siberian taiga in northern China are dominated by *Larix dahurica*, with forests of *P. pumila* on steep dry slopes (Uemura, Tsuda & Hasegawa 1990).

Fig. 1.7. **Krummholz *Pinus contorta* subsp. *murrayana* on morainal soils near timberline at Nine Lake Basin in the Sierra Nevada of California, USA (photo: P.W. Rundel).**

Fig. 1.6. **Typical open *P. sylvestris* stand in north-central Siberia. Due to frequent surface fires (fire-return intervals ranging from 25 to 100 years) and the germination-inhibiting layer of lichens (predominantly *Cladonia* spp.), understorey vegetation and pine regeneration are sparse. High-intensity, stand-replacing fires provide suitable conditions for germination. Many fire-damaged trees are suitable for dendrochronological (fire history) analysis (photo: J.G. Goldammer).**

1.6.2 Subalpine and timberline habitats

Pines comprise much of the characteristic vegetation at the boundary between subalpine and alpine communities and boreal forests and tundra throughout the northern hemisphere. These timberline pines include subalpine species that may form open forest cover at slightly lower elevations, as with *P. contorta* in the western USA (Fig. 1.7), or *P. sylvestris* in northern Europe. There is also a group of pine species that are limited to this timberline habitat. In the western USA, these timberline pines include two distinctive phylogenetic groups, with *P. aristata*, *P. balfouriana* (Fig. 1.8) and *P. longaeva* from the section *Parrya* subsection *Balfouriana*, and *P. albicaulis* and *P. flexilis* from the section *Strobus* subsection *Cembrae*. Northern and central Europe have fewer timberline pine species. *Pinus cembra*, another member of section *Strobus* subsection *Cembrae*, is largely confined to such habitats in Central Europe, while the ecologically plastic

Fig. 1.8. ***Pinus balfouriana* growing near timberline in the Sierra Nevada of California, USA (photo: P.W. Rundel).**

P. sylvestris extends into such communities in northern Europe. A similar broad ecological plasticity allows two more members of section *Strobus* subsection *Cembrae*, *P. sibirica* in central and east Asia and *P. pumila* in east Asia and Japan, to extend from boreal to treeline habitats. The short growth form of the latter species is apparently an important adaptation in its success at tolerating extreme winter cold (Chap. 15, this volume). The highest occurrence of pines in Asia occurs in the Himalayan Mountains in Tibet, where *P. densata* grows at 3600 m elevation.

Extreme environmental conditions of low winter temperatures, wind abrasion and desiccation, and a short summer growing season all combine to make this an unusually stressful environment in which pines are notably successful compared with other groups of conifers. These conditions lead to characteristic dwarf or krummholz growth forms of trees at the upper limit of tree occurrence above the forest line. Timberline communities, especially those in arid areas, seldom experience fire or other types of disturbance that create openings for pine seedling recruitment in less stressful environments.

Conditions for seedling establishment are thus rare and unpredictable (Chap. 12, this volume). Notable studies of the ecophysiology of timberline pines have been carried out in the Austrian Alps of Central Europe where *P. cembra* is the dominant timberline species.

Timberline pines have several interesting attributes which are associated with their extreme environment. Most notable of these are the remarkable longevities of both whole trees and tissues in this group (see section 1.4). Keeley & Zedler (Chap. 12, this volume) show that variable juvenile mortality has been the driving force favouring longevity in these pines.

1.6.3 Temperate forest habitats

Temperate forests of the northern hemisphere are important habitats for pines, including not only montane coniferous forests but also deciduous forests at lower elevations which are dominated by hardwood trees. We include a brief discussion of pine distributions in these broad habitats, including montane habitats of western North America, eastern North America, Europe and Asia. Montane pines in Mexico are treated below under tropical forest habitats.

Western North America

In western North America, *P. contorta* is the most characteristic subalpine or upper montane pine species. The subspecies *latifolia* assumes single-species stand dominance in the upper montane zone of the Northern and Central Rocky Mountains, subsp. *contorta* does so in the coastal northwestern USA and adjacent Canada, and subsp. *murrayana* in the Sierra Nevada of California in upper montane and subalpine communities. The ecological plasticity shown by this species over its extensive range is remarkable, as discussed below, and no other pine matches it in elevational range (Critchfield 1980). Another widespread, ecologically plastic, and economically important species in the western USA is *P. ponderosa*, with three distinctive varieties in the Rocky Mountains (Fig. 1.9), Sierra Madre Occidental, and California/Pacific Northwest. In the Central and Northern Rocky Mountains, ponderosa pine forests typically occur in relatively open pure stands at the lower margins of montane forests; it is this open forest community that has been romanticized by Hollywood westerns and cigarette commercials. In the Southern Rockies, however, *P. ponderosa* occurs in mixed pine–oak stands with associated species such as *P. cembroides*, *P. leiophylla* and a variety of *Quercus* species.

The contribution of pines to mixed conifer forests in the Sierra Nevada of California is discussed in some detail in section 1.6.7. Besides *P. ponderosa* in the lower montane forests, *P. lambertiana*, *P. jeffreyi* and *P. monticola* are also important forest species, and *P. washoensis* is a rare local endemic (Rundel, Parsons & Gordon 1977). Mixed conifer

Fig. 1.9. *Pinus ponderosa* forest on sandy soils near Spokane, Washington State, USA. The understorey is dominated by alien grasses, including *Bromus tectorum*, which have replaced perennial native grasses such as *Festuca idahoensis* (see text) (photo: D.M. Richardson).

forests here are largely dominated by species of *Abies*, with *Calocedrus* and *Sequoiadendron* as important associates with the pines.

Eastern North America

Despite the predominance of hardwood species in the eastern deciduous forest region of the United States and southeastern Canada, pines play an important ecological role throughout this region. They are particularly important in successional sequences in the northern hardwoods transition to the boreal forest, and on dry and rocky sites further south.

Two pines, *P. resinosa* and *P. strobus*, are widespread and characteristic components of the northern hardwoods ecosystem and the southern margin of the boreal forest. *Pinus resinosa* extends from the Great Lakes region across southern Canada to New England where it may form extensive pure stands, particularly on nutrient-poor sandy soils or after fire on richer soils in a successional sequence leading to replacement by *P. banksiana* or *P. strobus* and eventually hardwood trees. *Pinus strobus* has a similar distribution and habitat preference, but additionally extends southwards down the Appalachian Mountains, and is commonly associated with hardwood species over its range (Fig. 1.10). This species may function as a pioneer in old-field succession (Abrams, Orwig & Demeo 1995), but is a climax species on dry and sandy soils. It is maintained as a dominant over hardwoods with frequent fire or other disturbance (Peterson & Squiers 1995).

The oak–pine–hickory association of the Piedmont region of the eastern deciduous forest has extensive areas of secondary forests dominated by *P. taeda* and *P. echinata* (Braun 1950). Without disturbance, these species are eventually replaced in most sites by mixed hardwood forest (Cain & Shelton 1994). The oak–chestnut association of the

Fig. 1.10. **Pinus strobus community in a pine/hardwood forest region** of Connecticut, USA (photo: P.W. Rundel).

Fig. 1.11. **Salzmann pine, Pinus nigra var. salzmannii, near Pegairolle** de Bueges in the Bueges River Valley, Hérault region, southern France. This variety is endemic to the Mediterranean Basin (photo: D.M. Richardson).

eastern deciduous forest also contains pines as important elements of forest communities in dry and nutrient-poor habitats (Braun 1950). Ridges and peaks in the Appalachian Mountains support pine forests and heaths with an elevational replacement of species from *P. virginiana* at low elevations, to *P. rigida* at intermediate elevations, and *P. pungens* at the highest sites (Whittaker 1956; Racine 1966; Zobel 1969). Pine barrens with a dominance of dwarf *P. rigida* are a prominent feature of fire-prone landscapes in New York and New Jersey on shallow, nutrient-poor soils, and show strong floristic affinities to ridgetop forests of *P. rigida* in the southern Appalachians (Forman 1979). Shrubby oaks are commonly present in this community, and *P. echinata* and *P. virginiana* are occasional associates (Greller 1988). *Pinus rigida* and associated oak species resprout from woody lignotubers following fire, giving them a multi-stemmed, shrubby growth form.

Europe

While the temperate forest region of Europe is characterized by deciduous forest communities, much as in the eastern United States, pines occur as important components of this vegetation in both lowlands and montane habitats. Three areas of temperate deciduous forests can be separated in Europe – the Atlantic, central European, and Mediterranean upland forest regions. All these regions have pines as dominant elements in specific habitats on coarse or nutrient-poor soils, or in successional sequences following disturbance.

Pinus sylvestris, the most widespread European pine (see also 1.6.8), extends from boreal habitats southwards into the deciduous forests of both the Atlantic and central European forest regions. For the Atlantic forest region along the west coast of Europe, *P. sylvestris* is native only in Scandinavia and Scotland, but it has been extensively planted and become naturalized in areas to the south in England and western France. This species was native to this

area in the early and mid-Holocene, and present populations may date back to Roman times (Polunin & Walters 1985). In its natural habitats within the Atlantic region, *P. sylvestris* commonly occurs with hardwoods such as *Alnus*, *Betula* and *Populus*. Another pine, *P. pinaster*, enters this region in southwestern France, northwestern Spain, and Portugal, where it develops open forests on sandy soils and dunes with associated oaks and heathland shrubs.

Pinus sylvestris is again the most dominant species in the forest regions and mountains of central Europe. Scots pine woodlands characteristically occur on sandy soils across the lowlands of northern Germany and Poland, with or without associated hardwoods, and often with ericaceous shrubs in the understorey. This plastic species is successful in montane habitats as well, however, forming woodlands on dry calcareous gravels up to subalpine levels in the Alps. Nutrient-poor acid soils at elevations of 1400–2500 m in the Alps and Carpathians support sparse forests of *P. cembra*, often growing with *Larix decidua* (Tranquillini 1979). Dry dolomitic soils in the mountains of Austria and southwards into Croatia support the typical subspecies of *P. nigra*, the black pine. The dwarf mountain pine, *P. mugo*, forms the highest subalpine community in central Europe, as previously discussed.

Well-developed coniferous forests form notable aspects of the forest cover throughout the montane zone of the Mediterranean region of Europe. The most widely distributed pine species in this region is *P. nigra*, with five distinctive subspecies over its range which extends from western Spain and the Pyrenees across Corsica and Italy to Austria and southwards to Greece and eastwards to the Balkans (Fig. 1.11). *Abies*, *Fagus* and other deciduous hardwoods are common associates.

The Bosnian pine, *P. heldreichii* subsp. *leucodermis*, forms well-developed forest stands with *Abies* and *Fagus* at elevations of 900–1800 m from the central Balkans to Albania

Fig. 1.12. *Pinus koraiensis* forest at Pochungun, Kyunggido, 40 km north of Seoul, South Korea. The main tree species in the forest other than *P. koraienis* are *Abies holophylla, Carpinus cordata, C. laxiflora, Quercus aliena, Q. serrata* and *Acer triflorum*. These forests are conserved for the protection of the tomb of King Sejo, fourth king of the Li Dynasty (photo: Min Hwan Suh).

Fig. 1.13. An old-growth forest in the Xiao Xinghan Mountains, Lesser Xinghan, People's Republic of China, where *Pinus koraiensis* makes up about 60% of the basal area. The remaining basal area comprises hardwood species, especially *Acer mono, Quercus mongolica, Betula costata, Tilia amurensis* and *Populus ussuriensis*. Younger stands have a greater dominance of Korean pine (photo: H. Hutchins).

Fig. 1.14. *Pinus massoniana* in Hong Kong. The pine woodland/savanna is a typical landscape of the subtropical regions of SE China. The ground cover is *Dicranopteris linearis*, a fern (photo: Kam-biu Liu).

and northern Greece. The typical subspecies of this pine has similar ecological requirements, and develops mixed pine forests on limestone soils in the central and western Balkan peninsula and in central Italy (Polunin & Walters 1985). Macedonian pine, *P. peuce*, with *P. cembra* one of only two five-needled pines in Europe, forms pure pine forests above 1700 m from Bulgaria southwards to northern Greece. At lower elevations it may occur mixed with both *Abies* and *Picea*.

Asia

Boreal forest ecosystems of Siberia and Manchuria grade southwards into montane and subalpine communities of conifers with a gradual transition from the widespread forests of *P. pumila, P. sibirica* and *P. sylvestris* to temperate forest communities with different pine species. Such gradients are best seen in northeastern Asia with elevational and latitudinal gradients of pine distribution in the area covered by eastern Siberia, Manchuria, Korea and Japan.

Coastal parts of Japan and southern Korea are dominated by *P. thunbergii* up to elevations of about 1000 m. Montane forests of Japan, Korea and adjacent areas of Manchuria have *P. densiflora* as the dominant pine at low to moderate elevations with *P. koraiensis* increasing in abundance at higher elevations in some areas (Fig. 1.12). *Pinus parviflora* is common in these forests up to 2500 m in Japan. Subalpine forests of the northeastern Asia region are dominated by *P. koraiensis* which extends northwards into eastern Siberia. This species often occurs in mixed-conifer stands with *Abies*. In Siberia and Manchuria it grows at elevations of 600–1000 m, but up to 1050–2600 m in Japan.

Pinus koraiensis/northern hardwood forests occur along montane and subalpine ecological zones from southern Siberia (54° N), the northeastern provinces of the People's Republic of China (PRC; Fig. 1.13), into Northern Korea and the outlying Japanese islands of Honshu and Shikoku. For example, this pine/hardwood mix covers the Lesser Xingan Mountains from 300 to 9200 m, giving way to *Betula ermanii* –*Pinus pumila* forest at higher elevations. Timberline habitats above this limit are dominated by shrubby stands of *P. pumila*, the only haploxylon pine found north of 70° N. Korean pine typically makes up 40–80% of the crown canopy in the mixed montane forests of Heilongjiang Province in China. A few small, outlying stands of *P. sibirica* stands also occur northwest of the Da Xingaling Mountains (52° N, 122° E) in the PRC.

Mixed evergreen forest in the mountains of central and southern China supports a diverse assemblage of pines, including widespread and ecologically plastic species and local endemics (Hou 1983). *Pinus massoniana* is a broad-ranging species across central and eastern China to Taiwan (Fig. 1.14). It is largely restricted to elevations below 1500 m (although it also occurs up to 2000 m) where it may grow in pure stands or more commonly mixed with species of *Cunninghamia*, *Cryptomeria*, *Keteleeria* and *Quercus*. At higher elevations of 1500–3600 m, *P. armandii* is an ecologically important species that occurs in mixed conifer and hardwood forests in Yunnan and Sichuan provinces and again on Taiwan and Hainan. In South Central PRC, this pine grows in clumps like the North American *P. albicaulis* (Chapters 13 and 14, this volume) – probably the result of seed dispersal by the Eurasian nutcracker, *Nucifraga caryocatactes* (H. Hutchins, personal communication). *Pinus tabuliformis* is a third widespread species, ranging in a broad discontinuous band from northeastern Tibet and Yunnan to southern Manchuria and Inner Mongolia. It is reported to occur over a remarkable elevational range from 50 to 3000 m, where it may be associated with *P. armandii* (above 1500 m), and the local endemic *P. bungeana* (now possibly extinct in the wild; H. Hutchins, personal communication) at middle elevations (Mirov 1967). Along its southern margin, *P. tabuliformis* merges with the related *P. yunnanensis* which extends in montane forests from southern Sichuan Province across Young into northern Burma. In southern Young, the needles become shorter and the pine that grows near the Burmese border is known as *P. insularis* by the Chinese, although Price *et al.* (Chap. 2, this volume) treat this taxon as part of *P. kesiya* (see also Turnbull, Armitage & Burley 1980). Related to *P. tabuliformis* is *P. hwangshanensis* which occurs at scattered sites across central and eastern China. The highest occurrence of pines in this region is found with *P. densata* which grows to elevations of 3600 m in western Sichuan and southeastern Tibet (Zheng 1983), making it as much a Himalayan as a Chinese species. Little is known about ecosystem processes related to community structure and dynamics in any of these forests.

Three closely related white pines have localized distributions in lower montane and montane forests across south-central Asia. These are *P. wangii* from limestone hills of Yunnan just north of Burma, *P. fenzeliana* on Hainan Island and scattered sites in southeastern China, and *P. morrisonicola* on Taiwan. All these species occur in mixed hardwood forests from low to moderate elevations up to 2400 m. Higher elevation forests of Taiwan have *P. taiwanensis*, while the related *P. luchuensis* is found to the north on Okinawa, Ryukyu Islands. Mirov (1967) suggests that these species are closely related to *P. parvifolia* from Japan.

Relatively few pine species occur in the Himalayas but several species are important ecological dominants (Ohsawa, Shakya & Numata 1986; Singh & Singh 1987). The ecological significance of pines in the central Himalayas of India has been described in considerable detail (Singh *et al.* 1994). Above 1000 m elevation, subtropical forests of *Shorea robusta* are replaced by *P. roxburghii*, a hardy species

Fig. 1.15. ***Pinus roxburghii* stand near Nainital, Uttar Pradesh, India. This site is typical for the whole Chir pine belt along the Himalayas, showing severe degradation due to the effects of pastoral activities (browsing of *Quercus* spp., trampling and burning of pine seedlings). Since the litter layer consisting of long pine needles is very slippery for cattle, the forests are burned regularly. This prevents regeneration and results in an uneven distribution of age classes. Along the mountain trails, pines are cut to stimulate resinosis for providing torches. Discarded torches cause many wildfires (photo: J.G. Goldammer).**

that colonizes or re-establishes on sites following fire or other disturbance (Fig. 1.15). Climax oak forests which once dominated this zone have been much reduced by cutting, burning, and erosion. Mixed forests of hardwoods and *Abies* become dominant above 1700 m. *Pinus wallichiana*, an upper montane pine, forms pure or mixed pine-hardwood stands between 2300 and 2900 m (Grierson, Long & Page 1980). This species has been reported to grow at elevations up to 3400 m in Bhutan (Farjon 1984) and northwestern Yunnan (Wang 1961). *Pinus densata*, as described above, enters the Himalayas at high elevations in southeastern Tibet. In the western Himalayan regions of Afghanistan and Pakistan, *P. wallichiana* is a common forest species, often mixed with a local endemic, *P. gerardiana*. Another endemic Himalayan pine, *P. bhutanica*, occurs between 1750 and 2400 m elevation, usually in association with *Betula*, *Lithocarpus* and *Quercus* spp. (Grierson *et al.* 1980).

1.6.4 Mediterranean coastal habitats

California

Continuous forests of *P. contorta* subsp. *contorta* and *P. muricata* dominate the northern California coast, with the unusual *P. contorta* subsp. *bolanderi* being a distinctive local endemic on podzolized beach terraces where pines grow with *Cupressus* species. A similar closed-cone pine forest occurred along most of the California coast up through the early Quaternary, but was gradually eliminated by drier conditions of the Holocene (Axelrod 1980). Only relict populations of this coastal pine forest remain today, most prominently with the small native stands of *P. radiata* along the central California coast and on Cedros and Guadalupe Islands off the coast of central Baja California (Chap. 21, this volume). *Pinus muricata* becomes increasingly rare in central California, but persists in small relict stands on Santa Rosa Island of the California Channel Islands, the south-central coast of California, and San Vicente in northwestern Baja California. Finally, *P. torreyana* is a rare endemic known only from Santa Rosa Island and a small coastal area near San Diego in southern California. All these coastal pines are widely planted throughout coastal areas of California, suggesting that dispersal limitations much more than environmental stress restrict their natural distribution today. Relict populations of closed-cone *Cupressus* species also occur along the coasts of California and Baja California, sometimes with closed-cone pines but often with edaphic restriction to serpentine or other unusual soil types (Vogl *et al.* 1977).

Two close relatives of *P. torreyana*, *P. sabiniana* and *P. coulteri*, are common foothill pine species restricted to the mediterranean-climate areas of California. *Pinus sabiniana*, characterized by an unusual branched trunk morphology unlike the typical monopodial growth of pines, and its long grey-green needles, is widespread in woodlands of the coastal ranges and Sierra Nevada foothills. *Pinus coulteri* is confined to the Coast Ranges and Southern California Mountains where it generally occurs in mixed evergreen/hardwood stands at the ecotone between woodland and montane coniferous forest. This species has a variable expression of cone serotiny, with this trait increasing in association with drier stands dominated by chaparral vegetation (Borchert 1985). Another successful foothill pine in southern California is *P. attenuata*, a closed-cone relative of *P. radiata*. Like *P. muricata*, this species is most characteristic of serpentine or other unusual substrata, and is associated with fire-prone chaparral communities (Vogl 1973). It does not resprout after fire, but re-establishes stand dominance from a dormant seed pool. All these foothill species occur in chaparral or open *Quercus* woodlands where fire is a characteristic landscape process.

Mediterranean Basin

Pine forests cover about 13 million hectares – about 5% of the total area of the Mediterranean Basin, and about 25% of the total area under forest. The closely related species in

Fig. 1.16. **Forest dominated by *Pinus halepensis* near Montpellier, southern France. Associated woody taxa are *Cistus monspeliensis*, *Olea europaea*, *Quercus ilex* and *Rosmarinus officinalis* (photo: D.M. Richardson).**

subsection *Halepenses*, *P. brutia* and *P. halepensis* (Fig. 1.16), are by far the most widespread species, and form large forests in the eastern and western parts of the Mediterranean Basin, respectively. These species, and another two, *P. canariensis* (subsection *Canarienses*; confined to the Canary Islands) and *P. pinea* (subsection *Pineae*; with a large, scattered distribution throughout the region except in North Africa), form Klaus's (1989) group of Mediterranean shore and island pines. *Pinus pinaster* (tentatively placed in subsection *Sylvestres*), with a large distribution in the western Mediterranean Basin, covers about 1.3×10^6 ha (Fig. 1.17). Human use and cultivation of pines, especially *P. brutia*, *P. halepensis*, *P. pinaster* and *P. pinea*, and the many changes in land-use practices (notably those affecting the fire regime) since prehistoric times have greatly influenced the distribution of pine forests in the region. Chapter 8 (this volume) gives a detailed account of the complex interaction of ecological factors that govern the distribution of pines in this region.

1.6.5 Arid habitats

The southwestern USA and adjacent areas of Mexico present some of the most arid habitats for pines in the world. Here the drought-adapted pinyon pines occur in mixed stands with species of *Juniperus* to form the distinctive pinyon–juniper woodlands over thousands of square kilometres. In the Zion National Park in southern Utah, with a mean annual rainfall of about 371 mm at

Fig. 1.17. *Pinus pinaster* forest in the Maures–Estérel region of southeastern France. This pine was already well represented in this area in the 18th century, but increased its range considerably as it invaded abandoned fields during the 19th century. Plantations were established in other areas. Outbreaks of the scale insect *Matsucoccus feytaudi*, which spread through the range of *P. pinaster* between 1969 and 1977 (see Chap. 8, this volume) and massive forest fires have caused major problems in these forests (photo: D.M. Richardson).

Fig. 1.18. Pinyon–juniper woodland near Checkerboard Mesa in Zion National Park, Utah, USA. *Pinus edulis* is the dominant pinyon at this elevation (*c.* 2000 m), where it occurs with *P. ponderosa, Juniperus osteosperma* and *Quercus gambelii*. The 2-needled *P. edulis* hybridizes naturally with *P. monophylla* subsp. *fallax* in this area, with pure stands of the latter species dominating at lower elevations (<1600 m) (photo: D.M. Richardson).

1234 m elevation, there are two species of pinyon. *Pinus edulis* is the more widespread species and occurs here at its western margin of distribution (Fig. 1.18). The slightly more drought-adapted *P. monophylla* with a single needle in each fascicle is the common pinyon in the west where winter-rainfall regimes are predominant. Pinyon–juniper woodlands are characterized by an open growth of stout and broad-crowned pinyons and junipers, usually no more than 3–5 m in height, mixed with a variety of evergreen shrub species.

Pinyon pines represent a distinct monophyletic lineage within *Pinus* comprising 11 species (Chap. 2, this volume). Most of these taxa occur in arid montane habitats of

central and northwestern Mexico where mean annual rainfall is 300–600 mm. There is no ecological equivalent of the pinyon pines in the flora of the Mediterranean Basin or its transition to arid lands in North Africa. Similarly, pines are absent from the desert margins and the arid plateau of central Asia. Other drought-adapted conifers may be present in such habitats, as with *Tetraclinis articulata* in Morocco and Tunisia, but there are no pines.

1.6.6 Tropical pine habitats

Neotropics

Although lumped into a single category here, pine habitats in tropical forests of the New and Old Worlds are as varied as those in which pines are found in temperate habitats. Tropical pines may be found from savanna communities at sea level, through dry pine–oak forests or montane hardwood forests at middle elevations, to subalpine conifer stands reaching to more than 4000 m in elevation. Rainfall regimes associated with tropical pines are commonly 1000–1500 mm yr^{-1}, but may be no more than 600 mm in a highly seasonal pattern of rainfall or as high as 2000 mm yr^{-1} in the Caribbean region. The following account was drawn largely from Critchfield & Little (1966), Mirov (1967), Perry (1991) and Vidaković (1991), except where stated otherwise.

Subtropical savannas with pines reach their northern limit of occurrence in the coastal plain of the southeastern USA. The climatic regime of this region is transitional between temperate hardwood forests and subtropical regions with infrequent frosts, and would normally support mixed hardwood forests as a climax community. However, nutrient-poor soils support pine savannas whose dominance over potentially competing hardwood seedlings is maintained by frequent ground fires. The nature of pine adaptation to this fire regime is discussed in more detail in Chapters 11 and 12 (this volume). Dry, sandy soils in the coastal plain of this region generally support mixed stands with such species as *Pinus echinata, P. palustris* and *P. taeda*, with a variety of *Quercus* species. More subtropical sand scrub communities are characterized by an over-storey of *P. clausa* with an understorey of sclerophyllous-leaved shrubs and palmettos (Christensen 1988). More mesic flatlands in these coastal plains support savannas in which single species of pines such as *P. elliottii* (Fig. 1.19), *P. palustris* or *P. serotina* may be dominant as long as frequent fires are present. Scattered shrubs or palmettos often occur in the understorey.

Lowland pine savannas of similar general structure occur in parts of the Caribbean Region and adjacent areas of Central America. These savannas are dominated by the widespread *P. caribaea*, which extends from the Bahama Islands through Cuba to Belize, Honduras and Nicaragua

Fig. 1.19. *Pinus elliottii* var. *densa* recovering after damage caused by Hurricane Andrew, which passed over the Everglades National Park, Florida, USA, on 24 August 1992 (photo: J. Donaldson). Hurricanes are an important disturbance in tropical pine habitats in southern and southeastern USA, and parts of Mexico and Central America. Direct tree mortality is caused by physical damage that either snaps (as shown here) or uproots the trees, damaging the roots (which causes trees to die back). Severed, windthrown and stressed trees are also rapidly colonized by a wide range of insects which cause further mortality.

Fig. 1.20. A managed natural stand of *Pinus oocarpa* near Guaimaca, Department of Francisco Morazan at about 600 m elevation, central Honduras. This pure stand of *P. oocarpa* was selectively thinned for timber production, and subject to regular, periodic prescribed (or accidental) fires – hence the lack of understorey (photo: C.E. Hughes).

Fig. 1.21. A typical small, rounded branchy compact tree of *Pinus cembroides* growing in open, virtually pure stands in the dry mountains of the upper reaches of the Tolantongo valley, north of the town of Ixmiquilapan in the State of Hidalgo in central Mexico at *c.* 1870 m elevation. Edible seeds are harvested and marketed locally (photo: C.E. Hughes).

where it reaches the southernmost point of occurrence of any North American pine at about latitude 12° N. Open savannas of *P. caribaea* typically occur in Belize on sandy soils with a grass understorey that becomes highly flammable during the dry season. Extensive pine savannas in Honduras and Nicaragua occur on infertile gravel or finer quartz sediments (Parsons 1955; Taylor 1962; Alexander 1973). Low mountain forests of relatively pure *P. oocarpa* (Fig. 1.20), or *P. oocarpa* with *P. caribaea*, extend from Belize to Nicaragua in areas where frequent fires similarly act to slow invasions of hardwood competitors on poor soils (Denevan 1961; Kellman 1979). *Pinus cubensis* and *P. tropicalis* are Cuban endemics occurring in low-elevation savanna associations maintained by frequent ground fires.

Mexico and northern Central America, particularly their mountain regions, are home to the greatest diversity of pines in any area of the world. This region, which includes the Caribbean Basin, contains 47 species of pines. The pinyon pines, largely centred in the arid mountains of northwestern Mexico and the adjacent southwestern USA, have been treated above.

The higher Sierra Madre Occidental of northwestern Mexico contains a diverse assemblage of pine species in relatively open montane pine–oak forests. These relatively xeric communities have such taxa as *P. cembroides* (Fig. 1.21), *P. engelmannii*, *P. leiophylla* var. *chihuahuana*, *P. lumholzii* and *P. ponderosa* in mixed-dominance stands that receive about 600–700 mm of rainfall annually. In many respects, these communities are extensions of the montane conifer communities of the southwestern USA (Peet 1988), but they contain more pine species and also new floristic elements in oaks and other hardwood trees. Similar montane communities in central Mexico have a more distinctive tropical composition and receive more rain (600–1000 mm per annum). These areas support mixed-dominance stands of pines and hardwoods, particularly oaks. Fire is an important component of the ecosystem dynamics of these communities, with the grassy understorey burning frequently. *Pinus devoniana* and *P. montezumae* in these communities frequently form relatively pure savanna-like stands and possess a fire-resistent seedling morphology like that of *P. palustris* in the southeastern USA. *Pinus oocarpa* and *P. leiophylla* have the unusual ability to resprout in seedlings 2–4 years of age after light ground fires.

The mesic higher-elevation forests in the Sierra Madre Occidental and local areas of the Sierra Madre Oriental are characterized by the presence of *P. ayacahuite* which extends southwards in similar habitats to central and southern Mexico on moist cool slopes (Veblen 1978). This species grows at elevations of 2000–3500 m in mixed stands of pine and fir. *Pinus hartwegii*, *P. montezumae* and *P. patula* are common associates. There have been few studies of these upper montane forests with respect to the role of fire in structuring community dynamics and species dominance, but fire is certainly important. An interesting associate of this mesic pine–fir community in the northern Sierra Madre Oriental of Nuevo Leon is the bushy local endemic *P. culminicola*, the only pinyon pine to occur outside lower semi-arid communities.

Less seasonal rainfall regimes on the Caribbean slopes of the Mexican volcanoes and in the mountains of Oaxaca in Mexico and Guatemala bring relatively high annual rainfall of 1200–2000 mm to occur at lower elevations. Such conditions allow the development of tall, mixed pine/hardwood communities at intermediate elevations of 1500–2750 m. A notable characteristic of these communities is the strong floristic connections with the eastern deciduous forest association of the eastern USA. Such temperate species as *P. strobus* and *Liquidambar styraciflua* are important community associates here. *Pinus ayacahuite*, *P. devoniana* (= *P. michoacana*; Fig. 1.22) and *P. maximinoi* occur in this area as do several endemic taxa such as *P. nubicola*, *P. patula*, *P. pseudostrobus* var. *apulcensis* and *P. tecunumanii*. Despite the temperate origins of elements of perhumid forest community, frequent mists and high humidity produce a luxuriant growth of epiphytic mosses and vascular plants on exposed forest trees.

Elevations above 2000–2500 m in southern Mexico and Guatemala allow a rich development of conifer forest communities, with nine species of *Pinus*, two species of *Abies*, two species each of *Cupressus*, *Juniperus*, *Taxodium* and *Taxus*, and two species of *Podocarpus* (Veblen 1976; Hartshorn 1988). The upper elevational limits of tropical pines can be seen with *P. hartwegii* which is restricted from about 2800 m to more than 4300 m on the high volcanoes of

Fig. 1.22. **Young densely regenerated natural stand of *Pinus devoniana*, recently burned with no understorey vegetation, in the mountains of the Sierra Madre Occidental, at around 2720 m elevation, 153 km west of Durango, close to the road towards Mazatlan in the State of Durango, north-west Mexico (photo: C.E. Hughes).**

central Mexico and Guatemala. This species frequently forms pure forests of trees 20–30 m in height at 3000–3700 m, and is the only pine to reach timberline in this area. *Juniperus standleyi*, however, reaches above 4000 m in this area (Veblen 1976). *Pinus donnell-smithii* may occur with *P. hartwegii* at subalpine elevations of 3000–3700 m in Guatemala.

Palaeotropics

The ecological relationships of subtropical pines which occur across southern China are poorly known. More characteristically temperate forest pine species such as *P. tabuliformis*, *P. wallichiana* and *P. yunnanensis* reach subtropical forests in the mountains of southern China and northern Burma at the southern end of their distribution. These forests comprise conifer and hardwood genera of mixed temperate forest and other floristic elements of tropical origin (e.g. Jarvis & Helin 1993). The rare local endemic *P. wangii* is restricted to similar habitats in southern China, just north of the Burmese border.

Just four species make up the tropical taxa in the pine flora of Southeast Asia. Two of these are widespread species, and two are rare local endemics. The most widespread tropical Asian species is *P. kesiya* which occurs in a discontinuous belt from the Assam Hills of India across northern Burma and Thailand, and again in the highlands of southern Vietnam and on Luzon Island in the Philippines (Fig. 1.23). Scattered populations are also found throughout Laos and Cambodia and northwards into southern China where this species grades into the closely related *P. yunnanensis* (Critchfield & Little 1966; Turnbull *et al.* 1980). Over most of its range, *P. kesiya* occurs in mixed hardwood forests with *Quercus* and other temperate genera, although it is mixed with Dipterocarpaceae at

Fig. 1.23. *Pinus kesiya* occupies disturbed sites throughout mountainous Southeast Asia and is well adapted to frequent surface fires. The photograph shows a stand near Taunggyi, Shan State, Myanmar (Burma), growing on a site heavily disturbed by erosion (photo: J.G. Goldammer).

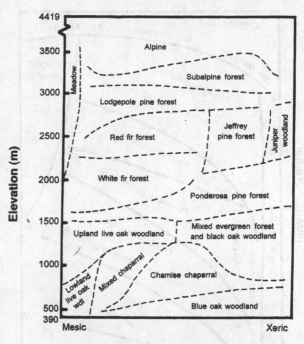

Topographic–moisture gradient

Fig. 1.24. Diagrammatic representation of community positions along elevational, topographic, and moisture gradients for the west slope of the southern Sierra Nevada. Redrawn, with permission, from Vankat (1982).

its lower elevations. Open pure forests of *P. kesiya* are widespread in the highlands of northern Luzon where fire is the critical factor in maintaining open habitats for this species (Kowal 1966). This region has a strongly seasonal monsoon climate, with total annual rainfall of about 1400–1600 mm.

The other widespread pine in Southeast Asia is *P. merkusii* which is common at elevations below 1200 m in northern Burma and Thailand. It also occurs in scattered localities across Laos, Cambodia, Vietnam and on Mindoro and Luzon Islands in the Philippines. Although absent from the Malay Peninsula, *P. merkusii* reappears to the south in the mountains of Sumatra where it reaches to latitude 2° S, the southernmost natural occurrence of pine in the world. The typical communities with *P. merkusii* are open woodlands which are strongly tropical in floristic composition. Species of *Dipterocarpus* and *Shorea* are the common dominants, although the pines may occur in relatively pure stands as well. *Pinus merkusii* and *P. kesiya* occur infrequently in mixed stands at the ecotone between the two species (Werner 1993).

Fire is almost an annual event in the forest communities of both species, although clearly much of this fire cycle has been strongly influenced by humans for centuries. Further studies are urgently needed to determine the relative significance of Tertiary and Quaternary climatic histories and human activities in shaping the markedly scattered patterns of distribution of both species across Southeast Asia. The final two Asian species of tropical pines are poorly known local endemics found only in the highlands of southern Vietnam. These are *P. dalatensis* (which is related to *P. wallichiana* from the Himalayas), and the unusual flat-needled *P. krempfii*. Nothing appears to be known about the ecology of either species.

1.6.7 Topographic gradients in pine distribution

The preceding sections have reviewed the broad-scale patterns of biogeographic distribution of pines within major biogeographic habitats of the northern hemisphere. This discussion, however, obscures the considerable complexity that can occur in pine distribution over complex topographic gradients within a single geographic area. A clear illustration of such patterns can be seen in the complex interactions of elevation and topographic moisture gradients in the ecological distribution of pine species in the southern Sierra Nevada of California (Fig. 1.24). Here, one can encounter as many as 10 species of pines across a landscape gradient from the foothills to the crest of this mountain range (Rundel *et al.* 1977; Vankat 1982).

Blue oak (*Quercus douglasii*) woodlands in the foothills frequently include *P. sabiniana*, while *P. ponderosa* dominates a yellow pine forest extending from 1500 to as much as 2200 m elevation on drier slopes and south-facing exposures. Small, relict populations of *P. monophylla* exist at these elevations on steep canyon slopes. At mesic sites, coniferous forests at these middle elevations are dominated by a *Abies concolor*/mixed conifer forest with *P. lambertiana* as an important associate along with *Sequoiadendron giganteum* and *Calocedrus decurrens*. Drier slopes and exposures at 2200–2800 m are strongly dominated by *P. jeffreyi*, with *P. monticola* as a common associate. Above

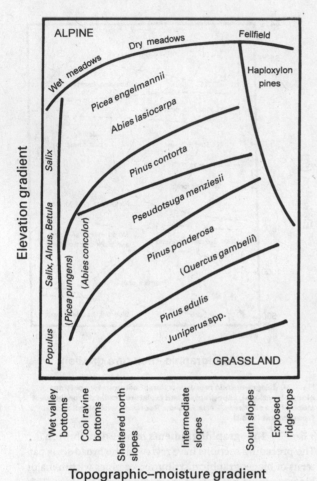

Fig. 1.25. Major vegetation zones of the central Rocky Mountains as related to gradients of elevation and moisture–topography. Species in parentheses are not consistent dominants. Redrawn, with permission, from Peet (1988).

2500–2800 m, *P. contorta* subsp. *murrayana* forms single-species forests. Finally, upper subalpine and timberline habitats above the lodgepole pine zone have *P. balfouriana* as their most important species, with smaller populations of *P. albicaulis* and *P. flexilis*.

A similar but simpler pattern of topographic gradients in pine distribution exists in the central Rocky Mountains of the USA (Fig. 1.25). The interaction of elevation and topographic moisture gradient is particularly strong in this continental region as seen in the steep changes in the elevational distribution of individual communities from mesic to xeric sites. Low pinyon–juniper woodlands co-dominated by *P. edulis* grade into montane woodlands of *P. ponderosa* subsp. *arizonica* at higher elevations. Pines appear again in higher-elevation forests dominated by *P. contorta* subsp. *latifolia*. Although lodgepole pine does not reach upper subalpine or timberline habitats in the Rocky Mountains as it does in the Sierra Nevada, with spruce and

fir taking over such habitats, *P. flexilis* and *P. aristata* are common timberline species.

1.6.8 Widespread versus narrowly-endemic pines

Relatively little direct attention has been given to considerations of the historical, ecological and genetic factors that have interacted to determine the limits of distribution of individual pine species. The extremes of patterns of distribution can be shown by contrasting such widespread and ecologically plastic species as *P. contorta* and *P. sylvestris* with highly localized endemics such as *P. peuce* and *P. radiata*.

Pinus sylvestris has the largest geographic distribution of any pine, ranging in its occurrence from the Scottish Highlands along the Atlantic to the Pacific coast of eastern Siberia, with relict populations throughout the northern Mediterranean Basin. It reaches latitudes from 70° 29' N in Norway to 37° N in Spain, and elevations from sea level to 2600 m. The ecological range of *P. sylvestris* is equally broad. It is common in boreal forests in northern Europe and across Asia where it shows strong dominance on the most xeric slopes and sandy soils. In western Siberia and Mongolia it is a species of the arid steppes and is largely restricted to river courses and the margins of lakes (Mirov 1967).

Populations of *P. sylvestris* in the Mediterranean region and central Europe are thought to be relicts from the Pleistocene (Mirov 1967). Here, *P. sylvestris* occurs in montane and subalpine habitats in the mountains of northern Portugal and central and northern Spain across the Pyrenees to the Alps and Apennines, and then on to the Balkan Peninsula and southwards through the former Yugoslavia to northern Greece. It is also common in the mountains across Turkey and in the Caucasus where it may be found from the coastal to subalpine habitats (Chapters 5 and 8, this volume).

With such a wide range of biogeographic and ecological distribution, it is not surprising to find that *P. sylvestris* is highly plastic and contains considerable genetic diversity (Wang, Szmidt & Lindgren 1991). This genetic diversity is particularly high at the intra-population level, but lower between populations and races, suggesting that there may have been a blending of genetically diverse populations during the Pleistocene when the range of this species was much more restricted (Szmidt & Wang 1993). Much of the ecological success of *P. sylvestris* appears to result from its strong ability to disperse and colonize disturbed sites. Populations of this species in refugia in the Mediterranean Basin were important sources for dispersal and recolonization of glaciated terrain in Europe and northern Asia in the early Holocene.

Pinus contorta is second only to *P. sylvestris* among pines in its wide range of distribution and obvious ecological plasticity. Over its biogeographic range, *P. contorta* grows from latitude 64° N in the Yukon Territory of Canada to 31° N in the Sierra San Pedro Mártir of Baja California, Mexico. It ranges from sea level along the Pacific Coast of the northwestern United States and Canada to elevations of over 3350 m in the Sierra Nevada of California.

Four subspecies of *P. contorta* have been described (Critchfield 1980; Table 1.1); these taxa are sufficiently distinct to lead many authors to recognize them as separate species. *Pinus contorta* subsp. *contorta* grows on low coastal bluffs and sand dunes along the Pacific coast from southeastern Alaska to northern California, whereas *P. contorta* subsp. *bolanderi* is a local endemic on highly podzolized soils along the coast of Mendocino County in northern California. The latter subspecies apparently evolved recently from subsp. *contorta*. Genetic studies have shown that subsp. *bolanderi* has significantly less genetic diversity than its putative parental type, due either to a low number of founding individuals in the extreme edaphic situation in which it grows, or to 'hitch-hiking' of allozyme loci linked to those loci undergoing strong selection for the edaphic stress (Aitken & Libby 1994). As previously described, *P. contorta* subsp. *murrayana* is the dominant species in coniferous forests of the upper montane and subalpine zones of the Sierra Nevada and Cascade Ranges, while subsp. *latifolia* forms the characteristic lodgepole pine forests of the Rocky Mountains from Canada south to Utah and Colorado. Genetic diversity is high, even within individual subspecies of *P. contorta*, and ecotypic differences between populations may be large (Ying & Liang 1994).

Both *P. peuce* and *P. radiata*, with highly localized patterns of distribution, are relictual species which once ranged more widely. The biology and small natural distribution of *P. radiata* along the coast of central California and two islands off the coast of Baja California are discussed in Chap. 21 (this volume). *Pinus peuce* is restricted in its distribution to a few sites in the mountains of southern former Yugoslavia, Albania, Bulgaria and northern Greece. This presumed Tertiary relic is most closely related to *P. wallichiana* of the Himalayas.

There is no simple explanation for the genetic and ecological plasticity in pines. Rapid advances are, however, being made in understanding the nature of the intra- and interpopulation genetic diversity and the nature of those environmental factors which provide the strongest selection regime for pine evolution. Increasing concerns over the consequences of global change make it imperative that there be a better understanding of the role of genetic and phenotypic plasticity in the adaptive responses of pines to changing climatic conditions. This brings us to consider in more detail the many interactions between pines and humans.

1.7 Pines and humans

People have interacted with pines since early hominids first encountered these trees in the Mediterranean Basin, probably about a million years ago. Human activities throughout the Holocene at least, and especially over the past few thousand years, have had a major influence on the behaviour of pines. The present-day distributions of pines over large parts of their range in the Old World are the net result of numerous expansions and contractions caused directly or indirectly by human activities. The consequences of human activities for the vegetation are probably most apparent in the Mediterranean Basin, where marked changes in human population numbers over millennia, numerous wars, and far-reaching changes in land-use practices have exerted major influences (Le Houérou 1981; Thirgood 1981; see also Chapters 8 and 20, this volume). In other parts of Eurasia, such as the northern part of northeast China, which remained sparsely populated until the 20th century, the role of humans in shaping the vegetation is less obvious.

In the North American range of pines, extensive settlements of humans within the range of pines date from about 11 000 years ago (McAndrews 1988). There is evidence of village life supported by agriculture in southern Mexico from 8000 years ago. Humans have, however, probably only been present in sufficient numbers in this region and in the Central American range of pines to have had a marked influence on the biota for about the last 1700 years. Driver (1969) estimated that the population of North America was about 30 million during the 15th century. During the 12–fold increase in the human population of this continent over the last 500 years, much of the natural vegetation has been destroyed or modified. Some pollen diagrams spanning the period of human occupation in North America reflect the influence of human activities on pine forests, e.g. logging of pines and their successional replacement by hardwood species such as *Betula* and *Populus* (McAndrews 1988). The last two centuries have seen massive changes to natural vegetation as human pressures have increased dramatically.

It is beyond the scope of this chapter to describe in detail the multitude of human-induced impacts on pine forests. Rather, we attempt to derive some generalizations from a set of specific examples from localities throughout the range of pines and involving a broad cross-section of taxa (Table 1.2). The examples reviewed here were carefully chosen to show some typical human-induced changes in the extent and condition of pine forests (both positive and negative). This sample includes cases involving more than a third of all pine species from 12 of the 16 subsections. It shows that important human-induced changes can be attributed to one (or more, since interactions between

Agent of change	Pinus taxa	Region	Description
Changes in fire regimes			
Reduced fire frequency			
	P. elliottii var. densa (+); P. clausa (–)	Florida, USA	P. elliottii var. densa increases in importance and replaces P. clausa where fires are suppressed. The former assumes dominance in the sandhill communities where fire-free intervals exceed 50 years (Myers 1985).
	P. elliottii (+); P. palustris (–); P. taeda (+)	SE USA	Pinus palustris covers less than 5% of the area it covered a century ago, partly because of effective fire exclusion and invasion of these lands by hardwoods and more aggressive pine species such as P. elliottii and P. taeda (Landers, van Lear & Boyer 1995).
	P. ayacahuite var. strobiformis (–); P. ponderosa (–)	SE Arizona, USA	Lowered fire frequencies since 1880 in the Pinaleno mts has led to a decline in pines (through reduced seedling recruitment) and an increase in the more shade-tolerant spruce and fir (Grissino-Mayer, Baisan & Swetnam 1995). A similar situation exists on the Paunsaugunt Plateau of southern Utah (Stein 1988).
	P. resinosa (+)	Wisconsin, USA	Red pine invaded sedge meadows protected from fire for 30 years (Vogl 1967).
Increased fire frequency			
	P. caribaea (+/–)	Nicaragua	Burning maintains Caribbean pine savannas, but if fire is too frequent (annually) savannas degrade to grassland (Denevan 1961).
	P. kesiya (–)	Thailand	The regular use of fire to promote early growth of grass kills seedlings before they reach maturity (Cooling 1967; Turnbull et al. 1980). This, and kindling stick harvesting, are severely affecting the structure of P. kesiya forests, e.g. on the southern flanks of Doi Inthanon in northern Thailand (Savage 1994).
	P. merkusii (+/–)	SE Asia	Shifting cultivation practices (increased fire frequency and grazing levels) have major effects on pine regeneration. Where the intensity of cultivation is moderate and the cycles of cutting are sufficiently spaced to allow seedlings to reach maturity, shifting cultivation has favoured P. merkusii. Where cultivation is more intense (shorter cycles of cutting and burning), stands are reduced to scattered trees in various stages of senescence and then eliminated (Cooling 1968).
	P. occidentalis (–)	Haiti	Frequent intense fires through forests disturbed by local farming practices kill seedlings before they acquire fire tolerance (Darrow & Zanoni 1990).
	P. rzedowskii (–)	Michoacán, Mexico	Repeated ground fires have prevented regeneration and killed many large trees (Perry 1991, p. 88).
Grazing/browsing			
	P. contorta (+/–)	W USA	Interactions between climate change, fire history and grazing drive invasions in subalpine meadows. Cessation of sheep grazing sometimes initiates invasions, but halts invasion in other cases. Intense grazing opens vegetation, permitting seedling establishment, but trampling and browsing eliminates seedlings. With less intense grazing, trampling/browsing damage is reduced, and swards are opened up sufficiently to permit establishment and invasion takes place (references in Richardson & Bond 1991).
	P. edulis (+)	SW USA	Pinus edulis and western juniper (Juniperus occidentalis) have formed, or filled in, woodlands in formerly more open western rangelands through effective fire exclusion and heavy grazing that decreased the vigour and cover of herbaceous vegetation (Wright & Bailey 1982).
	P. ponderosa (+)	C Washington, USA	Heavy grazing of herbaceous vegetation is the prime factor in explaining the advance of tree reproduction (Rummell 1951); see Richardson & Bond (1991) for further discussion.
	P. ponderosa var. ponderosa (+)	American Southwest	Working with tree rings from the Navajo Indian Reservation, Savage & Swetnam (1990) showed decreased fire frequencies following increased livestock grazing beginning in 1880. Increases in pine density occurred only in the period 1910–1930 when usually wet years coincided with the fire-free period.
	P. radiata var. binata (–)	Guadalupe Island, Mexico	The population declined from 383 trees in 1964 to 45 in 1988 because of increased browsing of seedlings by introduced goats (Libby 1990).
	P. sylvestris (–)	Scotland	Intensive grazing by red deer and sheep prevented (or at least greatly reduced) natural regeneration between 1850 and 1950 in parts of Scotland (Steven & Carlisle 1959).

Table 1.2. Human-induced changes to the distribution and abundance of *Pinus* taxa. Examples were chosen to illustrate the wide range of human impacts (both recent and in previous centuries) on pine taxa in different parts of the natural and adventive range of *Pinus*. Cases were also selected to emphasize the extent of, and processes causing, different impacts. Many cases evoke more than one agent of change; these are listed under the principal agent. In column 2, (+) and (–) indicate positive and negative effects on the abundance or distribution of particular pine taxa respectively

	Species	Location	
Harvesting activities Harvesting nuts	P. sylvestris (+)	UK	Scots pine rapidly colonized grasslands in parts of lowland Britain after 1954, following the reduction in rabbit numbers caused by myxomatosis (Watt 1971; Crompton 1972; Duffy et al. 1974), implying that rabbit browsing before this curbed pine spread.
	P. gerardiana (–)	Himalayas, India	There is virtually no natural regeneration of Chilgoza pine in the Indian parts of its range (Sehgal & Sharma 1985), and probably in Pakistan as well (see text). The most important factor precluding regeneration is the ruthless harvesting of cones by local inhabitants (100% of cones are usually removed). Inaccessible areas, where harvesting is less severe, comprise only 5% of the total range of P. gerardiana, and prospects for the conservation of this pine are bleak.
	P. koraiensis (–)	China	Human harvesting of cones by knocking off branches and collecting of squirrel-cut cones from the forest floor affects the behaviour of red squirrels (Sciurus vulgaris), which are important seed dispersers, and reduces future seed crops (Hutchins, Hutchins & Liu 1996).
	P. maximartinezii (–)	Sierra de Morones, Mexico	Intense harvesting of nuts of this rare pine over many decades (and changes in land use) has led to reduced regeneration (Martin 1995 p. 200).
Fuelwood gathering	P. edulis (–)	NW New Mexico, USA	Pinus edulis largely disappeared from pinyon–juniper woodlands in Chaco Canyon 1220–520 years ago, when a peak human population (Anasazi) was exerting heavy demands on the fuelwood resource (Chap. 9). Samuels & Betancourt (1982), in a modelling study, showed that a low-density woodland would have been completely depleted within 200 years when subjected to exploitation levels commensurate with 10–12th century estimates of the human population in the canyon.
	P. rigida (–)	NE USA	Pitch pine was eliminated from Nantucket and other islands off the coast of NE USA by fuelwood harvesting, and overgrazing with sheep (references in Ledig 1992).
	P. sylvestris (–)	Pyrenees, France	Cutting of P. sylvestris and other conifers for use in iron smelting in the French Pyrenees in 15–18th centuries (Bonhote & Vernet 1988) probably had a major influence on the forests of this area.
Logging of pines	P. banksiana (+); P. resinosa (–)	Wisconsin Pine Barrens, USA	Pinus resinosa was almost eliminated by initial high-grade logging, followed by severe and repeated fires and more cutting and clearing. Pinus banksiana increased under the conditions that reduced or displaced the longer-lived, infrequent-seeding, and heavily logged P. resinosa (Vogl 1970).
	P. cembra (–/+)	Tyrol, Austria	High-altitude stands of P. cembra in the Viggartal were felled during the Middle Ages to supply timber for mining, resulting in the destruction of the cembran pine belt. Since then the forest has spread back up the slopes, forming a wide transition zone above the forest (Tranquillini 1979; Holtmeier 1990).
	P. chiapensis (–)	Mexico & Guatemala	Nearly all stands of this pine are heavily exploited (the wood is much sought after for use in construction of small homes) and are in danger of being destroyed (Perry 1991). Natural regeneration is scarce/absent in many stands because of heavy grazing and annual fires (Dvorak & Brouard 1987; see also Zamora-Serrano & Velasco-Fiscal 1977).
	P. koraiensis (–)	NE China	A sharp reduction in the abundance of P. koraiensis in NE China over the past 250 years can be ascribed to human disturbance through land clearance (Sun et al. 1991).
	P. monophylla (–)	Great Basin, USA	Mining activities in the 19th century led to cutting of pinyon–juniper woodlands for charcoal production and thus destruction of woodlands in radii of 20–35 km from major mines (Young & Budy 1987).
	P. ponderosa (–)	New Mexico, USA	Anasazi settlements of AD 900–1200 used hundreds of thousands of logs, primarily P. ponderosa, in their houses (Betancourt, Dean & Hull 1986).
	P. ponderosa (–)	California, USA	Low elevations of ponderosa pine, Abies concolor and other species were eliminated a century ago because they were the most accessible sources of timber. Although P. ponderosa is certainly not threatened, Ledig (1993) argues that the genes that enabled this species to occupy droughty, low-elevation sites may have been lost.

Table 1.2. (*cont.*)

Agent of change	*Pinus* taxa	Region	Description
	P. resinosa (–); *P. strobus* (–)	Minnesota, USA	Mixed forests of red and white pine occupied about 1.4 million ha in Minnesota before European settlement. Only about 0.2 million ha (14.3%) remains, and unlogged forests cover only 1.6% of the pre-settlement area (Frelich 1995). Even in the remaining primary forest in State Parks there is almost no pine regeneration, probably because of the combined effect of reduced fire frequency (after 1922) and heavy browsing by deer (Tester, Starfield & Frelich 1997).
	P. sylvestris (–)	Ireland	Human disturbance in Ireland in the first half of the Holocene was the primary cause of forest decline; the landscape was treeless by 2400–1500 years ago (Fossitt 1994).
Logging of broadleaved trees	*P. massoniana* (+)	SE China	The ubiquity of pine woodlands of southeast China (usually dominated by *P. massoniana*) is attributed to intensified human disturbance over the last 1 or 2 millennia, including clearing of forests dominated by *Castanopsis*, *Cryptomeria* and *Quercus* about 1100 years ago (Chap. 4).
	P. yunnanensis (+)	S China	*P. yunnanensis* is generally the first secondary tree to invade following destruction of the original broadleaved forest in southern China (Young & Wang 1989).
Cutting of peat	*P. uncinata* (+)	Frasne (Jura), France	Bégeot & Richard (1996) suggest that *P. uncinata* increased in abundance in the peat bogs of this area following heavy cutting of peat in the 17th century which opened up the vegetation, allowing pine seedlings to become established
Construction/mining activities	*P. halepensis* (+)	Israel	*P. halepensis* spread from nearby plantations and established on the walls of buildings abandoned between 1930 and 1950 at Zar'oniya and Shuni (Karschon, Weinstein & Heth 1976).
	P. muricata (–)	California, USA	Bishop pine groves near Lompoc, Santa Barbara County, have been removed for strip-mining of diatomaceous earth (Barbour et al. 1993).
	P. nigra (+)	Italy	This pine has invaded areas disturbed by mercury mining near Mount Amiata (Bargagli, Barghigiani & Maserti 1986).
Abandonment of agricultural land	*P. cembra* (+)	Alps	*Pinus cembra* regenerates vigorously if not destroyed by red deer or livestock, or if suppressed by vigorous grasses. Abandoned alpine pastures are rapidly invaded (Holtmeier 1990).
	P. densiflora (+)	Japan	The dramatic increase in diploxylon pine pollen registered in almost all regions of Honshu, Shikoku and Kyusho during the late Holocene is due to the colonization of sites cleared for agriculture by *P. densiflora* (Tsukada 1988).
	P. elliottii (+); *P. taeda* (+)	SE USA	These two pines have been planted on millions of hectares in SE USA on old agricultural lands, assisted by government subsidies to convert marginal agricultural lands to forest land (Landers et al. 1995). *Pinus taeda* aggressively colonized abandoned farmland in the second half of the 19th century and was commonly known as the "oldfield pine" (see Ledig 1992).
	P. halepensis (+)	Languedoc, France	The area covered by Aleppo pine in Languedoc increased 3-fold between 1878 and 1904, and another 2.6 times between 1908 and 1978 as the species invaded abandoned lands (Achérar, Lepart & Debussche 1984; Lepart & Debussche 1992). This trend is evident in many parts of western Europe as large areas of agricultural land are abandoned following the "set-aside" policy of the European Union (Chap. 8).
	P. kesiya (+)	SE Asia	On land laid bare by shifting cultivation, *P. kesiya* often forms pure, even-aged stands (Troup 1921; Cooling 1967; Turnbull et al. 1980). Occasional *P. kesiya* in montane forests and lowland broadleaved forests are the result of colonization of man-made clearings (Turnbull et al. 1980).
	P. strobus (+)	S Ontario, Canada	The increase in abundance of this pine in the recent past possibly reflects its spread onto abandoned Indian agricultural fields, although the relative importance of this and climate change are still being debated (see Chap. 6).
	P. sylvestris (+)	Greece	Forests re-established and spread in mountainous parts of northern Greece that had been badly degraded by livestock grazing, following cessation of grazing in 1946 (Zagas 1994).
	P. virginiana (+)	N Virginia, USA	Establishment of many *P. virginiana* stands began in the early 1900s, following the abandonment of areas cleared of hardwood forests and farmed for long periods (Orwig & May 1994).

Purposeful manipulation

Planting

P. wallichiana (+)	E Nepal	This pine regenerates prolifically and rapidly invades grasslands, open areas and abandoned agricultural lands in eastern Nepal, Himalayas (Ohsawa *et al.* 1986).
P. brutia subsp. *eldarica* (+)	Eurasia	Genetic studies (Conkle, Schiller & Grunwald 1988) have shown that this pine occurs naturally only on a single mountain in Azerbaijan, but that planting of seeds along ancient trade routes has extended its range; it now has a scattered distribution from Azerbaijan to Pakistan.
P. elliottii (+); *P. palustris* (−); *P. taeda* (+)	SE USA	Former longleaf pine lands are now planted with *P. elliottii* and *P. taeda*, species native to that region that do not have delayed early growth due to the grass stage of longleaf pine.
P. nigra (+)	Calabria, Italy	More than half the 47 500 ha of pine forests in this region are the result of afforestation projects in the 1950s (Asciuto 1990).
P. pinaster (+); *P. nigra* (+); *P. sylvestris* (+); *P. uncinata* (+)	Cévennes & Mont Aigoual, France	During the 19th and 20th centuries, these species and other conifers were planted over thousands of hectares, replacing a traditional matrix of grazing lands, cultivated fields and heathlands (Lepart & Debussche 1992).
P. pinea (+)	Mediterranean Basin	This species has been planted for its edible seed since ancient times, and it is impossible to tell if its occurrence in all parts of its range is natural, or whether it has been planted there (Mirov 1967).
P. radiata (+)	Australia, Chile, New Zealand, South Africa	This pine occurs naturally in five small populations in California and Mexico, with a total area of less than 12 000 ha. As a plantation species, it currently occupies c. 600 times the current area of the native populations (Chap. 21). It has spread from plantations in all these areas to invade natural and semi-natural vegetation (Chap. 22).
P. taeda (+)	SE USA	From 1952 to 1985 the area of pine plantations in the south and southeast USA has expanded significantly from about 0.75 to 8.5 million ha, mostly *P. taeda* (USDA 1988), making this by far the largest area of pine plantations in the world.
P. torreyana (+)	California, USA	A family of the Kumenay Indians of California who owned a grove of *P. torreyana* managed it by burning it regularly and planting seeds to expand it (Shipek 1989).
P. uncinata (+)	France	Isolated stands of this pine in the Massif Central were regarded as relict populations. However, pollen analysis of the three main stands revealed that the species was introduced to the region in the 19th century (Reille 1989).

Altered biota

Impacts of introduced plants on native pines

P. occidentalis (−)	Haiti	Human modification of disturbance regimes has greatly affected the natural regeneration of this pine in Haitian forests. The invasive alien shrub *Syzygium jambos* forms dense thickets beneath pines which exclude fire (Darrow & Zanoni 1990).
P. ponderosa (−)	Sierra Nevada, California, USA	The alien *Cirsium vulgare* frequently invades clearcuts in the central Sierra Nevada within 2 years of timber harvest, significantly reducing ponderosa pine seedling growth (Randall & Rejmánek 1993).
P. radiata var. *radiata* (−)	California, USA	Threats to the Monterey population involve a variety of urban encroachment issues militating against natural regeneration, including community opposition to the use of prescribed burning, and invasion by alien plants, especially *Genista monspessulana* (Chap. 21).
P. roxburghii (−)	Kumaun, India	Dense thickets of the alien shrub *Lantana camara* have replaced natural forests of *Quercus leucotrichophora* and *P. roxburghii* following over large areas in the Kumaun Himalayas, Uttar Pradesh, following over-utilization of the forests (Bhatt, Rawat & Singh 1994).

Impacts of introduced diseases on native pines

P. albicaulis (−); *P. lambertiana* (−); *P. monticola* (−)	Sierra Nevada, California, USA	In the Sierra Nevada range, several haploxylon pines are threatened by the white pine blister rust (*Cronartium ribicola*), introduced from Europe in 1910. The order of susceptibility is not equal to the order of current status, since spread of inoculum has been from low elevations to high. Current problems are worst in *P. lambertiana*, but *P. albicaulis* and *P. monticola* show signs of severe infection in some areas (C. Millar, unpubl. data).

Table 1.2. (cont.)

Agent of change	Pinus taxa	Region	Description
Impacts of introduced insects on native pines	*P. massoniana* (–)	Guangdong Province, China	The pine needle scale *Hemiberlesia pityophila*, a native of Japan and Taiwan, arrived in mainland China in 1982. Five years later the scale had infested 315 000 ha of pine forest (some of it planted), killing 25% (Wilson 1993).
Air pollution	*P. jeffreyi* (–); *P. ponderosa* (–)	S California, USA	Exposure to high ambient ozone since the early 1950s in the San Bernardino mts has led to a decline in pine growth and seedling establishment and thereby a change in forest composition favouring more ozone-tolerant conifers such as *Abies concolor* and *Calocedrus decurrens* (Miller 1992; see also Barbour *et al.* 1993, p. 109).
	P. maximartinezii (–)	Mexico	Strong oxidant air pollution around Mexico City has a differential but generally negative effect on the growth of conifer species (Miller *et al.* 1994). Plantation populations of *P. maximartinezii* show relatively little growth supression, but needle longevity drops from 5 to 3 years in polluted areas (Hernandez Tejeda & Nieto de Pascual 1996).
	P. strobus (–)	Ontario, Canada	Eastern white pine was the most sensitive tree species to heavy SO_2 pollution around an iron smelter in Ontario. All trees were killed for 27 km downwind of the plant and no seedlings were found within 48 km (Gordon & Gorham 1963).
	P. sylvestris (–)	Europe	Scots pine is one of the most sensitive conifers to the complex of factors of acid deposition and oxidant air pollutants resulting from industrial emissions in Europe (Grodzinski, Weiner & Maycock 1984; Abrahamsen, Stuanes & Tveite 1994).
	P. taeda (–)	SE USA	Ozone and acid precipitation have had complex but negative impacts on the health and productivity of southern pine forests in the USA (Fox & Mickler 1995).
	P. thunbergii (–)	Korea	Impacts of air pollution on pine forests near cities in Korea include needle chlorosis, reduced needle retention, and overall decline in community species diversity (Binkley, Son & Kim 1994).

factors are evident in most cases) of the following categories: changes in fire regimes; changes in grazing/browsing intensity; harvesting and construction activities; the abandonment of agricultural land; purposeful manipulation; the alteration of biotas; and air pollution. We discuss briefly some of the most important aspects of each of these categories, and some implications of these categories.

1.7.1 Changes in fire regime

Fire has played a pivotal role in the evolution and spread of pines since they evolved from their precursor gymnosperms. Human-induced changes to natural fire regimes have had a major impact on vegetation in most fire-prone systems (see e.g. Vale 1982, for a review of impacts in North America), and have had a dramatic effect on pines throughout their range.

The first deliberate use of fire by humans within the range of pines was in Europe, about 400 000 years ago. Intentional ignitions have been an important source of fire in the Mediterranean Basin and elsewhere in Europe for millennia and have had a massive influence on the extent and composition of forests. For example, *P. halepensis*, which is well adapted to frequent intense fires, probably expanded its range into areas formerly covered with forests of less fire-tolerant trees (Chap. 8, this volume). The deliberate use of fire by humans has a much shorter history in the North American range of pines (11 000 years or more), but human-induced changes to fire regimes, in concert with many other impacts, have greatly altered forest dynamics.

In many parts of the natural range of pines, human activities have led to increased fire frequency (Chap. 11, this volume). In Central America, Mexico and Southeast Asia this has often arisen through the agency of slash-and-burn agriculture. This practice has benefited pines, at least temporarily, in some areas (e.g. *P. merkusii* in parts of Southeast Asia), but in most cases this form of land use has devastated forests (notably in Central America; Perry 1991).

Heavy grazing of rangelands has reduced fire frequency in many parts of the American West by reducing fuel loads, and this has had a major impact on vegetation dynamics (e.g. Bahre 1991; McClaran & Brady 1994). Fires have been purposefully excluded from pine forests in several parts of the northern hemisphere. The situation, and the impacts thereof, have been best documented for North America (e.g. Parsons & DeBenedetti 1979; Gruell *et al.* 1982; Agee 1993). Fire exclusion has allowed pines to spread into some areas where the natural fire regime excluded them, and has changed the forest composition in areas where the natural fire regime allowed pines to grow, but where changed fire characteristics have altered processes affecting vegetation dynamics. Vale (1982) provides a succinct

overview, with good examples of the various impacts of fire suppression on pines in North America, and Agee (Chap. 11, this volume) addresses the phenomenon in more detail. Some impacts of fire suppression in pine forests through the disruption of the complex relationships between pines, fire, pathogens and insects are addressed in Chapters 18 and 19 (this volume).

1.7.2 Grazing/browsing

Changes in grazing pressure have triggered changes in pine distribution in many regions, but the phenomenon has been best studied in North America. The following summary was drawn from examples cited by Richardson & Bond (1991). Changes in stocking rates may initiate, sustain or halt range expansions, depending on the regime. Grazing is often cited as one of a suite of disturbance factors that interact to influence conditions affecting seedling establishment. Interactions between grazing and fire regimes (and often alien plants) are complex, and the relative importance of each factor is often not clear from published accounts. The introduction of grazing at moderate to heavy intensities frequently enhances seedling establishment which sometimes leads to range expansions. Such effects have been documented for pine forests adjoining mesic subalpine meadows and mixed grass and brush in more arid regions. Grazing also facilitates pine establishment by opening up vegetation in abandoned fields. In all these cases, grazing reduces the cover of vigorous grasses and thus competition with pine seedlings. Evidence that recruitment decreases, halting invasions, when herbivores are excluded lends additional support to this hypothesis. Areas subjected to heavy grazing may remain susceptible to colonization by pines long after grazing pressure has been greatly reduced or eliminated.

The (likely) expansion of pinyon pines in the arid southwest of North America in the 20th century is particularly interesting. The debate on whether pinyons have indeed invaded millions of hectares of former grassland or shrubland, or whether the 'invasion' constitutes the re-establishment of woodlands on sites from which trees were eliminated or greatly reduced by human activities, has continued for decades. Richardson & Bond (1991) summarize key references favouring the invasion scenario, Lanner (1981) provides a thorough criticism of the rationale behind such views, and West (1988) gives a good general account of human-induced changes to pinyon–juniper woodlands.

Rabbits have influenced pine regeneration in many parts of the natural (e.g. *P. sylvestris* in Britain; Table 1.2) and adventive (Richardson, Williams & Hobbs 1994) ranges of pines. Introduced goats have had severe impacts on pine regeneration in many areas. In at least one case (*P. radiata* var. *binata* on Guadalupe Island) such impacts are threatening the continued survival of a pine taxon (Table 1.2). There are, however, still relatively few published accounts of changes in pine distribution or density as a result of browsing. This is probably because marked increases in the density of browsers are relatively recent phenomena in many areas. Browsers do most damage to young trees rather than mature ones, and most pines are long-lived. The primary effect of changes such as increased moose densities in *P. sylvestris* forests in Sweden are therefore on rates of recruitment to adult growth stages (Edenius, Danell & Nyquist 1995). There are thus substantial time lags between changes in browser density and conspicuous change in forest structure. The literature on the impacts of gazing and browsing on pines is therefore biased in favour of cases that lead to range expansions which are much more obvious.

1.7.3 Harvesting activities

Humans have harvested pines and their products for thousands of years. We consider four categories of harvesting activities that have influenced pine forests: harvesting of nuts; fuelwood gathering; logging of pines; and logging of broadleaved trees.

The harvesting of pine seeds (nuts) for human consumption has a very long history. Kunkel (1984) lists 29 *Pinus* species from which seeds are harvested for human consumption (some only in times of famine). In Europe and Asia, pine seeds, especially those of *P. cembra*, *P. koraiensis*, *P. pinea* and *P. sibirica*, have been harvested from natural forests since prehistoric times. In southwestern North America there is evidence of human use of pine seeds (various pinyons, *P. coulteri*, *P. lambertiana*, *P. sabiniana*) from as long ago as 11 500 years ago. In some societies, pine seeds harvested from natural forests are still an important economic resource. For example, during good years, the harvest of *P. gerardiana* seeds provides income for about 13 000 indigenous people in the Suleiman Mountains of Pakistan (Martin 1995). In the Indian part of its range, such intense harvesting of *P. gerardiana* seeds, and browsing by sheep and goats, prevents regeneration (Sehgal & Sharma 1985; Table 1.2). The endangered *P. maximartinezii* seems destined to be harvested to extinction in its tiny Mexican range where nut collectors often lop off whole branches from trees to collect cones (Styles 1992). Fire, and browsing by goats, burrows and cattle add to the predicament of this pine (Martin 1995 p. 200). Harry Hutchins (personal communication; see also Hutchins, Hutchins & Liu 1996) describes a bizarre annual orgy of exploitation of *P. koraiensis* seeds in Liangshui Forest Preserve in Heilongjiang Province, People's Republic of China. People climb the trees to heights of over 30 m and knock off whole branches to gather cones, breaking the crowns and destroying future cone crops (see also Table 1.2). Although such harvesting is illegal in the reserve, and very dangerous (at least seven people fell to their deaths in 1992 in the Dialing Forest district in 1992), the financial incentives are irresistible. Families can earn between five and seven times their normal annual wages by selling pine nuts for export to Japan and South Korea.

Bark is harvested from at least eight *Pinus* species for human consumption, usually only in times of famine (Kunkel 1984). We could find no evidence that such utilization has had much impact on natural pine forests.

Pines have been cut for fuelwood for many centuries. Le Maitre (Chap. 20, this volume) discusses the history of logging of pine forests of Europe to supply timber for construction and shipbuilding. There is considerable debate among historians concerning the magnitude of the impact of this logging on the extent of the forests over the whole Mediterranean Basin. There can, however, be no doubt that these activities had major impacts in some areas (Thirgood 1981). Scores of publications have documented the impacts of cutting pines for fuel on vegetation structure, and a few examples are given in Table 1.2. In at least one case (*P. rigida* on Nantucket Island), fuelwood harvesting led to the regional extinction of a pine and in many cases harvesting has greatly reduced the abundance of pines. The need for fuelwood in many parts of the natural range of pines still accounts for a large part of the total area of pine forest cleared every year. For example, Buff (1985) reported that 70% of the 3×10^6 m^3 of forest land cleared each year in Greece was to provide fuel. The situation with respect to fuelwood resources in some developing countries is desperate; see, for example, Perry's (1991, p. 211) chilling account of the situation in El Salvador.

The examples in Table 1.2 illustrate a wide range of influences of pine-logging, including a case where selective logging for one pine species (*P. resinosa*) benefited another (*P. banksiana*). There is a massive literature on the history, policies, politics and practices of logging in different parts of the range of pines. A good example of the complexity of the relationship between logging impacts and vegetation structure is the case described for a Californian mixed conifer forest by McDonald (1976) (see Vale 1982 for further discussion).

In many areas, the logging of trees other than pines has had a major influence on pine forests. For example, the clearing of broadleaved forests in parts of Asia has created suitable conditions for pines, allowing species such as *P. massoniana* and *P. yunnanensis* to expand their ranges.

1.7.4 Construction and mining activities

Construction and mining activities are frequently associated with deforestation, either to supply timber to support these endeavours, or to clear vegetation. These activities, which replicate the effects of natural disturbances such as floods, glaciations and landslides, create areas of disturbed ground for recolonization by plants.

1.7.5 Abandonment of agricultural land

Pines have spread into lands following the abandonment of agriculture in many parts of the natural range of *Pinus* for thousands of years. Huntley & Birks (1983) show that the range of *P. sylvestris* expanded greatly in northeastern Europe as it colonized poor, sandy soils following abandonment of cultivation. Many studies have documented the spread of pines and other trees onto abandoned farmlands in the southeastern USA. Pines (notably *P. taeda*) usually invade such sites within 10 years of abandonment, and rapidly form closed stands which are gradually replaced by hardwoods in the absence of major disturbance (e.g. Spring *et al.* 1974; Golley *et al.* 1994 and references therein). Similar trends are evident from nearly all parts of the range of pines (Table 1.2).

1.7.6 Purposeful manipulation

Pines have been widely used and planted by humans in the Mediterranean Basin since prehistoric times. Chapter 20 (this volume) shows how this has influenced the distribution of *P. brutia*, *P. halepensis*, *P. pinaster* and *P. pinea*. The other examples in Table 1.2 range from the local planting of pines by native Americans, which probably had little effect on the distribution of pines, to the huge increase in the total area under pines due to the establishment of plantations of pines both within and well outside the natural range of pines. Widespread planting of pines has led to a major reshuffling of genetic material.

Large-scale afforestation started in the second half of the 19th century in Europe (e.g. the Restauration des Terrains de Montagne in southern France). Sustained, large-scale forestry was, however, not widespread in Europe until the early 20th century, and only expanded to other parts of the world in the second half of this century (Mather 1993; Chap. 20, this volume). The development of forestry with alien species in different parts of the world is well documented by Zobel, van Wyk & Stahl (1987). Some pines have proved highly successful for use in plantations outside their natural ranges where there is a shortage of coniferous species to produce fibres and solid wood products. The main pine species planted in the tropics and subtropics are *P. caribaea*, *P. elliottii*, *P. kesiya*, *P. oocarpa*, *P. patula*, *P. pinaster*, *P. radiata* and *P. taeda* (although the importance of several other Central American/Mexican species is increasing rapidly). Reasons for the widespread use of pines in exotic forestry plantations include: their simple design, with straight trunks and an almost geometrical branching habitat makes them ideal for timber production; as mentioned previously, the fact that LAI values can

be doubled or trebled through silviculture contributes to their high productivity; they grow faster than many other potential species; they are relatively easy to manage in plantations; their seeds are easy to collect, store and germinate; and they are ideally suited for planting in grasslands or scrublands (marginal forest lands) where most afforestation is required.

Large-scale afforestation with pines outside the natural range of *Pinus* has transformed large tracts of former grassland and shrubland to forest. The obvious impact of this form of land transformation on native biodiversity and ecosystem functioning in these areas is causing problems in some areas. For example, large pine plantations in parts of South Africa have greatly reduced streamflow from catchments, with important economic consequences (Dye 1996), and reduced the richness and abundance of native bird, mammal and plant assemblages (Armstrong & van Hensbergen 1996). Similar problems are occurring in the Coast Ranges of South-central Chile where native forests are being cut and replaced with plantations of *P. radiata*.

In the last few decades several pine taxa have spread from plantations outside their natural range, and in several areas invasive pines have major impacts on the functioning of invaded ecosystems (Chap. 22, this volume).

1.7.7 Altered biota

Introduced diseases, insects and plants have had detrimental effects on native pine taxa in many areas. The introduced white pine blister rust (*Cronartium ribicola*) is threatening several haploxylon pines in North America. This disease is expected to become pandemic throughout most of the range of *P. lambertiana* and other species within the next 50–75 years; together with heavy logging, the rust is reducing breeding populations (Table 1.2; see also Millar, Kinloch & Westfall 1996). In parts of Idaho and Montana, the canopy cover of *P. monticola* has declined from as much as 70% to less than 6% as a result of the rust (Hagle & Byler in D'Antonio & Dudley 1995). In Glacier National Park, Kendall & Arno (1990) estimated that 90% of *P. albicaulis* trees had succumbed to the rust by 1990. In Montana, the potential for large seed crops in *P. albicaulis* (an important food resource for bears, squirrels and nutcrackers) has been severely reduced, because the rust kills cone-bearing branches first (Keane, Morgan & Menakis 1994). Lanner (1996) gives a good account of the actual and potential ecosystem-level effects of the elimination of this species. A second locus of introduction of *Cronartium*, in New England, resulted in the spread of white pine blister rust through much of the range of *P. strobus*. Although the disease has not been as severe (except locally) as it has on *P. monticola*, the threat of its presence has limited the dissemination of planting stock of this species (R.D. Westfall, personal communication).

Various human activities have exacerbated the problem of diseases in pine forests. These include the replacement of disease-resistant species with more susceptible species, and fire suppression which favours alternate hosts (e.g. *Quercus* spp.) over disease-resistant and fire-tolerant species (e.g. *P. palustris*). This applies to the situation with fusiform rust in the southern USA, where the relatively disease-susceptible *P. elliottii* and *P. taeda* were planted to replace the more resistant *P. palustris* following the heavy logging of the last-mentioned species in the 19th century (see Chap. 19, this volume). An introduced insect has ravaged native pines in China (Table 1.2). De Groot & Turgeon (Chap. 18, this volume) show that insect outbreaks are' generally more common in human-altered monocultures, although other factors are also implicated (e.g. poor sites conditions and stress caused by drought and/or pollution).

The impacts of invasive alien plants on native pines have probably been much worse than is currently reflected in the literature. In particular, we suspect that the detrimental impacts of invasive alien grasses on pine regeneration in many parts of the northern hemisphere are understated. For example, the impact of the replacement of perennial native grasses such as *Festuca idahoensis* by alien grasses such as *Bromus tectorum* in the understorey of *P. ponderosa* forests in the Pacific Northwest has not been adequately assessed (Fig. 1.9). Such changes certainly change the fire regime (Daubenmire & Daubenmire 1968; Agee 1993) and must have significant impacts on pine regeneration. Similar impacts probably apply to many other taxa in other regions, and the magnitude of such problems is increasing rapidly as invasive alien plants become more widespread and abundant. The widespread movement of insects and disease agents on logs and shipping materials is a very serious threat. Canker fungi and subcortical insects (bark beetles, horntails, etc.) are likely to cause major damage to pine ecosystems in the future (T. Harrington, personal communication).

1.7.8 Air pollution

Large-scale damage to conifer forests caused by acid rain was first observed in Europe in the early 1970s. Symptoms are now evident throughout Central Europe, where damage is most pronounced in Germany and Switzerland, and in other parts of the natural range of pines (Smith 1981; Table 1.2). Acidic deposition, and short rotations which deplete nutrient budgets, are threatening the long-term sustainability of pine forests in the southern USA (Johnson *et al.* 1996) and similar forecasts have also been made in other parts of the northern hemisphere. Air pollution also enhances the likelihood of insect attack (Chap. 18, this volume), and pine mortality due to diseases such as *Armillaria* root rot is increased by pollution and other factors that induce stress in the trees (Chap. 19, this volume). The ecophysiological impacts of atmospheric pollution in pines are reviewed in Chap. 15 (this volume).

1.7.9 Threats to genetic diversity of *Pinus* and ecosystem functioning

The previous paragraphs have described some of the many human influences on pine forests. The selected examples show that moderate disturbance resulting from a wide range of human activities over many centuries has *benefited* pines by reducing competition from more vigorous plants and removing other biotic barriers. Many pine species are extremely tenacious, persisting and even flourishing under fairly severe disturbance regimes. Increasing levels of direct impacts, through logging and planting, and indirect impacts, through the alteration of disturbance regimes and the composition of biotas, and increasing air pollution, are however pushing many pine taxa beyond their tolerance thresholds. This is causing many problems for ecologists and foresters.

The growing pressure on pine forests is rapidly eroding the genetic diversity in the genus. A third of all *Pinus* species are either threatened in their entirety, or have subspecies or varieties that are threatened (Table 1.3). Some taxa listed here were rare before human intervention (e.g. *P. radiata* and *P. torreyana*); others owe their threatened status to human activities. Although many other pine taxa still occupy large ranges, large portions of their genetic diversity have been lost; this may have reduced their ability to respond to changing environmental conditions. Ledig (1993) gives an interesting example for *P. ponderosa*, the widest ranging pine in North America. Low-elevation populations of *P. ponderosa* in California were eliminated a century ago because they were the most accessible timber sources. Ledig suggests that the genes that enabled ponderosa pine to survive on droughty, low-elevation sites may have been lost when those stands were logged (see also Ledig 1992). Low- and middle-elevation populations of many other pine species have also been eliminated as forests have been cut for lumber and firewood and land converted to pasture and crops. In the extremely rugged Yunnan Province of China, all but the most inaccessible stands of *P. yunnanensis* and other pines have been cleared to make way for agriculture. In the mid-1960s 60% of the province was forested, but this had dropped to less than 30% in 1985 (H. Hutchins, unpublished data). Perry (1991) describes the plight of Mexico's forests as follows: 'They . . . continue to disappear, first becoming fragmented then reduced to small patches and parcels and finally to small groups of trees on inaccessible mountain slopes and valleys'. The net effect of these 'secret extinctions' (*sensu* Ledig 1993) on the ability of species to deal with rapidly changing environmental conditions will never be known.

Table 1.3. A preliminary list of threatened *Pinus* taxa, updated from Farjon, Page & Schellevis (1993) by the IUCN-SSC Conifer Specialist Group, applying the new IUCN criteria and categories (IUCN Species Survival Commission 1994) as part of the Conifer Action Plan. Nomenclature for *Pinus* follows Price *et al.* (Chap. 2, this volume). The updated list included *P. kwantungensis* (status: LR); this taxon is not recognized for this volume, and is included in *P. fenzeliana*. The IUCN Status codes are: CR, critically endangered (3 taxa); EN, endangered (8 taxa); VU, vulnerable (16 taxa); LR, lower risk (15 taxa) [LR taxa marked with asterisks are the focus of continuing taxon-specific conservation programmes and would qualify for inclusion in one of the threatened categories if these programmes were stopped]

Pinus taxon	Status
P. armandii var. *amamiana*	VU
P. armandii var. *mastersiana*	LR
P. balfouriana subsp. *balfouriana**	LR
P. balfouriana subsp. *austrina**	LR
P. bhutanica	VU
P. brutia subsp. *eldarica*	VU
P. brutia subsp. *pityusa*	VU
P. bungeana	EN
P. canariensis	LR
P. caribaea var. *caribaea*	VU
P. cembroides subsp. *lagunae*	VU
P. cembroides subsp. *orizabensis*	LR
P. chiapensis	VU
P. cubensis	VU
P. culminicola	EN
P. dabeshanensis	EN
P. dalatensis	CR
P. fenzeliana	VU
P. gerardiana	LR
P. heldreichii var. *heldreichii*	VU
P. heldreichii var. *leucodermis*	LR
P. jaliscana	VU
P. longaeva	VU
P. luchuensis	LR
P. massoniana var. *hainanensis*	CR
P. maximartinezii	EN
P. merkusii	LR
P. muricata	LR
P. nelsonii	VU
P. nigra subsp. *dalmatica*	LR
P. occidentalis	VU
P. peuce	LR
P. pinceana	LR
P. radiata var. *binata*	EN
P. radiata var. *radiata*	LR
P. rzedowskii	EN
P. sylvestris var. *mongolica*	VU
P. sylvestris var. *sylvestriformis*	VU
P. tecunumanii	LR
P. torreyana subsp. *insularis*	EN
P. torreyana subsp. *torreyana*	EN
P. wangii	CR

Besides the problems that the loss of genetic diversity poses for the survival of pine forests in the face of global change, they are also of direct relevance to human survival and prosperity at a global scale. Although tree breeding has already improved productivity (e.g. gains in stem volume of *P. radiata* of about 30% with some selected seed lines in New Zealand; see Chap. 21, this volume) and disease resistance (see Chap. 19, this volume) of pines in exotic plantations, breeders have barely scratched the surface in their efforts to modify pines for human goods

and services (see Chap. 13, this volume, for discussion). Clearly, native gene pools, especially of species with commercial importance, need to be conserved if options are to be kept open for further tree breeding selections. Among the economically important (or potentially so) pines of Mexico and Central America, provenances of at least the following taxa are threatened: *P. ayacahuite*, *P. caribaea* var. *hondurensis*, *P. chiapensis*, *P. greggii*, *P. maximinoi*, *P. oocarpa* and *P. tecunumanii*. For some of these, CAMCORE (Central America and Mexico Coniferous Resources Cooperative) *ex situ* gene conservation banks are the only place where these unique gene pools still exist (Dvorak & Donahue 1992). Judging from the experience in agriculture, species with less obvious commercial value may nevertheless also be valuable in breeding. An example is *P. washoensis*, a narrow endemic in the mountains on the western rim of the Great Basin in California and northwest Nevada which persists in a few populations and is on the verge of extinction largely because of logging in the 19th century. Genetic studies suggest that this pine may be a potentially valuable genetic resource for the yellow pines of North America (Niebling & Conkle 1990).

Human-induced changes to pine forests have also had significant impacts on ecosystem processes, goods and services (*sensu* Christensen *et al.* 1996). Among the important processes affected by changes to pine forests are hydrologic flux and storage, biological productivity, biogeochemical cycling and storage, decomposition and the maintenance of biological diversity. Ecosystems 'goods' influenced include food (nuts), construction materials, medicinal plants, genes, and the aesthetic value of natural pine forests for tourism and recreation. Extracts from pine cones have already shown potential for combatting various tumours and viruses in humans (Sakagami *et al.* 1991), even inhibiting the replication of Human Immunodeficiency Virus (Lai *et al.* 1991; Eberhardt & Young 1996). Among the ecosystem services affected are: the maintenance of hydrological cycles; regulation of climate; the maintenance of the gaseous composition of the atmosphere; the generation and maintenance of soils; the storage and cycling of essential nutrients; and the provision of beauty, inspiration and research opportunities.

Special attention has been given in recent decades to assessing the impacts of various logging strategies on animals associated with pine forests. Two bird species that have been especially well studied in North America are the red-cockaded woodpecker (*Picoides borealis*) and Kirtland's warbler (*Dendroica kirtlandii*). The former species is associated with mature forests of *P. echinata*, *P. elliottii*, *P. palustris* and *P. taeda* in the southern USA, and the latter inhabits young *P. banksiana* forests. The many published studies of the habitat requirements of these species and the measures required to ensure their survival provide models for the study and management of other threatened animals associated with pine forests.

The monumental task of formulating a global strategy to ensure the conservation of genetic diversity in *Pinus* (and the biota associated with natural pine forests) is an urgent priority. The Conifer Action Plan of the Conifer Specialist Group of the IUCN's Species Survival Commission (Farjon 1995) could be the foundation for such a global strategy.

1.8 The study of pines

A huge amount of work has been done on diverse aspects of the ecology and biogeography of *Pinus*, for example as reflected by the 68 522 publications with the word *Pinus* in the title or abstract found in the TREECD data base (1939–95), a CAB Abstracts Forestry Database, published by CAB International, Wallingford, Oxford, UK, and covering the following abstracting journals: *Forestry Abstracts*, *Agroforestry Abstracts* and *Forest Products Abstracts*. The next best covered tree genera in this database were *Picea* (spruces): 31 735; *Abies* (firs): 29 883; *Quercus* (oaks): 21 396; *Populus* (poplars): 15 878; *Eucalyptus* (eucalypts): 12 536; *Pseudotsuga* (Douglas-firs): 10 307; *Larix* (larches): 8556; *Acer* (maples): 8126; and *Acacia* (acacias): 5769). The Science Citation Index (SCI) also lists thousands of pine-related papers. A scan of the SCI for 1990–4 revealed some 2500 primary publications with the words 'Pinus' or 'pine' in the title (list edited to remove spurious records). Most of these dealt with silvicultural aspects, reflecting the huge importance of pines in cultivation. Over 400 papers (24% of those where the subject species could be determined) dealt with one species (*P. sylvestris*). Eight other species were the subject of more than 50 publications (in descending order of importance: *P. taeda*, *P. radiata*, *P. contorta*, *P. ponderosa*, *P. banksiana*, *P. elliottii*, *P. resinosa* and *P. pinaster*); papers on these species comprised 78% of the data base. If we add the other 12 species with more than 10 papers, the cumulative percentage rises to 91%. Sixty-nine *Pinus* species (out of the total of 111) are covered in the data base. This scan of the recent literature shows that many diverse aspects of the ecology and biogeography of pines are being studied. The chapters in this volume cite over 3000 publications in tracing the history of scientific study on many aspects relating to pines.

A wealth of information is included in the series of monographs published by the United States Department of Agriculture (Forest Service) series on 'The Genetics of . . .': *P. banksiana* (Rudolph & Yeatman 1982); *P. contorta* (Critchfield 1980); *P. monticola* (Bingham, Hoff & Steinhoff 1972); *P. palustris* (Snyder, Dinus & Derr 1977); *P. ponderosa* (Wang 1977); *P. resinosa* (Fowler & Lester 1970); *P. strobus*

(Wright 1970); and *P. virginiana* (Kellison & Zobel 1973). Many other USDA Forest Service publications also contain much useful information on *Pinus*, including: *Silvics of Forest Trees of the United States* (Fowells 1965); *The Southwestern Pinyon–Juniper Ecosystem: A Bibliography* (Aldon & Springfield 1973); *Seeds of Woody Plants in the United States* (Schopmeyer 1974); *Ecology, Uses and Management of Pinyon–Juniper Woodlands* (Aldon & Loring 1977); *Proceedings – Pinyon–Juniper Conference* (Everett 1987); *Management of Pinyon–Juniper Woodlands Symposium* (Evans 1988); *Pine–Hardwood Mixtures: A Symposium on Management and Ecology of the Type* (Waldrup 1989); *Silvics of North America. Vol. 1. Conifers* (Burns & Honkala 1990); *Symposium on Whitebark Pine Ecosystems* (Schmidt & McDonald 1990; see also Hoff 1987); *Proceedings – International Workshop on Subalpine Stone Pines and their Environment* (Schmidt & Holtmeier 1994). Important volumes published by other organizations include: *Management of Lodgepole Pine Ecosystems* (Baumgartner 1975); *Lodgepole Pine and its Management* (Baumgartner et al. 1985); and *Symposium on Sugar Pine: Status, Values, and Roles in Ecosystems* (Kinloch, Marosy & Huddleston 1996).

For tropical pine taxa, several monographs (mostly dealing with forestry-related issues, but also containing much useful biogeographic and ecological information) have been published by the Oxford Forestry Institute and the Commonwealth Agricultural Bureau. Notable among these are the monographs on *P. caribaea* (Lamb 1973); *P. kesiya* (Armitage & Burley 1980); *P. merkusii* (Cooling 1968); *P. oocarpa* (Greaves 1982) and *P. patula* (Wormald 1975). There have also been several monographic treatments of *P. radiata* (e.g. Scott 1960; Lewis & Ferguson 1993). Knowledge of the ecophysiology of pines has also progressed rapidly in recent decades (Chap. 15, this volume). Much of this information has been reviewed in recent accounts of the ecophysiology of conifers (Teskey et al. 1994; Smith & Hinckley 1995a, b). All the publications listed above have greatly enhanced our knowledge of pine ecology and biogeography.

Special mention must be made of the classic volume on *The Genus Pinus* by Nicholas Tiho Mirov (1967), until now the only single-volume account of the ecology and biogeography of pines. The publication of this volume was a milestone in the study of pines and it has served as a standard reference for researchers and students for three decades. Although still of considerable interest, Mirov's book is seriously out of date in many respects. The framework for the current volume was prepared giving consideration to the most important advances in studies relating to pines since Mirov's volume. Of the roughly 3000 titles cited in this volume, 85% were published after Mirov's book appeared.

Mirov made important contributions to the study of pine taxonomy and systematics, principally by recognizing the value of biochemicals (terpenes) in determining the relationships of taxa. Two significant advances in plant systematics occurred after Mirov's book was published: the development and acceptance of explicit phylogenetic approaches; and the use of proteins and nucleic acids as sources of comparative data. In particular, comparisons of protein isozymes have been very useful in clarifying relationships among populations within species and between closely related species. Phylogenetic analysis of restriction site and sequence data from chloroplast and nuclear DNA are allowing the explicit reassessment of evolutionary relationships at all levels within *Pinus*. Pinaceae is now the most extensively investigated plant family in allozyme surveys and *Pinus*, partly because of its economic importance, has been the best-studied genus among the conifers (Hamrick & Godt 1996). Mention must be made of the major contribution to forest genetics by William B. Critchfield (1923–89). The many references to his work in this volume attest to Critchfield's legacy in the fields of genetic variation, hybridization, growth and development, biogeography, palaeobotany, systematics and taxonomy (see also Millar 1990). Ledig (Chap. 13, this volume) reviews the advances in genetic studies and the implications for silviculture and conservation.

Systematic revisions of Mesozoic coniferous fossils have shown that Mirov's interpretations of the age, centre of origin, and early spread of *Pinus* were inaccurate. Whereas Mirov postulated a boreal origin of pines, current evidence indicates a temperate beginning for the genus (Chap. 3, this volume). Mirov, in tracing the fossil record of *Pinus*, was guided by Wulff's (1944) history of the world's floras. Among the many limitations of the data available in the 1960s was the lack of any pine fossils from Mexico and Central America. New fossil discoveries have facilitated major progress in describing past migrations of pines. There have also been major advances in the understanding of Quaternary climatic changes and the ways that trees such as pines responded to them. Earlier in the chapter we mentioned the discovery in the early 1960s of the value of packrat middens in reconstructing past vegetation. Major new insights for patterns over the last 40 000 years have emerged recently from studies of these middens (Betancourt, Van Devender & Martin 1990). Besides providing windows on past vegetation structure, packrat middens have also revealed shifts in physiology and morphology of pines in response to climate change (for *P. flexilis* in the Great Basin during the last glacial–interglacial cycle; Van der Water, Leavitt & Betancourt 1994).

Important progress has been made in appreciating the role of fire, and in developing ways of managing it (Chap. 11, this volume). There has been considerable progress in developing better models, and specific information for the generalized statements regarding fire effects that were made in the 1960s. Plant-successional models have moved beyond the simplified notion of monoclimax theory to multiple-pathway models, driven largely by the role of disturbance in succession; fire as a multivariate process has been the most widely studied disturbance. The link between disturbance and diseases of pines (Chap. 19, this volume), and between disturbance and insect outbreaks (Chap. 18, this volume) has been well studied. There is now a much better appreciation for the importance of smaller-scale disturbances (canopy gaps) versus the catastrophic disturbances (e.g. fire) in the dynamics of forest ecosystems.

There have been great advances in understanding with respect to the evolution of life-history traits in pines, e.g. concerning the evolution of serotiny (Chap. 12, this volume), and seed dispersal by birds (Lanner 1996; Chap. 14, this volume), the latter considered by Mirov to be only of local importance. Recent research has shown that conifer cone structure evolves relatively rapidly in response to different assemblages of seed predators, including red squirrels (*Tamiasciurus* spp.; Elliott 1974; Benkman 1995) and crossbills (*Loxia* spp.; Benkman 1993). Research has revealed that pines are obligately mycorrhizal. Axelrod's (1986) view that the evolutionary and ecological success of pines is related to their symbiotic habit has been endorsed by recent work. Access to the differing attributes of a very wide range of compatible heterotrophs is now accepted as a fundamental component in explaining the wide range of habitats occupied by pines throughout their range, and also the striking ecological plasticity within species such as *P. contorta*, *P. ponderosa* and *P. sylvestris*. Understanding of the contribution of mycorrhizas to plant nutrition has benefited considerably from work on pines, as has the knowledge of the roles of mycorrhizas in disease resistance and toxin avoidance (Chap. 16, this volume).

When Mirov published his book in 1967 there had been relatively few detailed studies on the photosynthetic responses, water relations and cold-tolerance of pines. Technological advances have paved the way for analyses that were not possible in the 1960s, and these have greatly improved our understanding of pine biology.

A very large proportion of the huge literature on pines that has accumulated since Mirov's volume has resulted from studies undertaken because of the rapidly changing status of pine forests in many parts of the world. The large changes in the scale of cultivation of pines, both within the natural range of pines, but especially in areas far removed from northern climes, has required major advances in the science of pine silviculture. Escalating threats to pine forests throughout their natural range has motivated research in various disciplines.

The problem of pines as alien invaders in areas well outside their range, not mentioned by Mirov, has assumed major proportions in some regions. It is interesting to note that the rampant spread of *P. pinaster* into natural fynbos in South Africa was cited as an important reason for launching the major SCOPE programme on the ecology of biological invasions in the mid-1980s (Drake *et al.* 1989). These invasions (see Chap. 22, this volume) challenged the prevailing notion that alien invasions were contingent on human disturbance. The many studies that have been done on pine invasions in the past two decades have contributed significantly to the understanding of invasions (e.g. Baskin 1996; Rejmánek & Richardson 1996), and have shed new light on various aspects of the ecology of pines (Chap. 22, this volume).

In as far as is possible in one volume, this book aims to review the current understanding of pine ecology.

Acknowledgements

James Agee, Craig Benkman, John Birks, Tom Harrington, Colin Hughes, Harry Hutchins, Ron Lanner, Aaron Liston, Tom Ledig, Jacques Lepart, Glen MacDonald, Robert Mill, Connie Millar, Bob Price, David Reid, Neil West and Bob Westfall kindly supplied information and/or commented on parts of the chapter. Thanks to Bill Archibold for arranging the preparation of Fig. 1.1. We thank Aljos Farjon and the IUCN-SSC Conifer Specialist Group for supplying the preliminary list of threatened pine taxa. Corlia Richardson helped with data analysis and provided general assistance and useful comment on the manuscript. DMR acknowledges the financial support of the University of Cape Town's University Research Committee.

References

Abrahamsen, G., Stuanes, A.O. & Tveite, B. (eds.) (1994). *Long-Term Experiments with Acid Rain in Norwegian Forest Ecosystems.* Berlin: Springer-Verlag.

Abrams, M.S., Orwig, D.A. & Demeo, T.E. (1995). Dendrochronological analysis of successional dynamics for a presettlement origin of white pine mixed oak forest in the southern Appalachians, USA. *Journal of Ecology,* **83**, 123–33.

Achérar, M., Lepart, J. & Debussche, M. (1984). La colonisation des friches par le pin d'Alep (*Pinus halepensis* Miller) en Languedoc méditerranéen. *Acta Oecologica/Oecologia Plantarum,* **5**, 179–89.

Agee, J.K. (1993). *Fire Ecology of Pacific Northwest Forests.* Washington, DC: Island Press.

Aitken, S.N. & Libby, W.J. (1994). Evolution of the pygmy-forest edaphic subspecies of *Pinus contorta* across an ecological staircase. *Evolution,* **48**, 1009–19.

Aldon, E.F. & Loring, T.T. (1977). *Ecology, Uses and Management of Pinyon–Juniper Woodlands.* USDA Forest Service General Technical Report RM-39.

Aldon, E.F. & Springfield, H.W. (1973). *The Southwestern Pinyon–Juniper Ecosystem: A Bibliography.* USDA Forest Service General Technical Report RM-4.

Alexander, E.B. (1973). A comparison of forest and savanna soils in northeastern Nicaragua. *Turrialba (Costa Rica),* **23**, 181–91.

Armitage, F.B. & Burley, J. (eds.) (1980). *Pinus kesiya.* Tropical Forestry Papers 9. Oxford: Commonwealth Forestry Institute.

Armstrong, A.J. & van Hensbergen, H.J. (1996). Impacts of afforestation with pines on assemblages of native biota in South Africa. *South African Forestry Journal,* **175**, 35–42.

Arno, S.T. & Sneck, K.M. (1977). *A Method for Determining Fire History in Coniferous Forests of the Mountain West.* USDA Forest Service General Technical Report INT-42.

Asciuto, A. (1990). Il pino laricio di Corsica, di Calabria e dell'Etna: aspetti ecologici e produttivi. *Cellulosa e Carta,* **41**, 19–25.

Axelrod, D.I. (1980). History of the maritime closed-cone pines, Alta and Baja California. *University of California Publications in Geological Science,* **120**, 1–143.

Axelrod, D.I. (1986). Cenozoic history of some western American pines. *Annals of the Missouri Botanical Garden,* **73**, 565–641.

Bahre, C.J. (1991). *A Legacy of Change. Historic Human Impacts on Vegetation in the Arizona Borderlands.* Tucson: University of Arizona Press.

Barbour, M., Pavlik, B., Drysdale, F. & Lindstrom, S. (1993). *California's Changing Landscapes. Diversity and Conservation of California Vegetation.* Sacramento: California Native Plant Society.

Bargagli, R., Barghigiani, C. & Maserti, B.E. (1986). Mercury in vegetation of the Mount Amiata area (Italy). *Chemosphere,* **15**, 1035–42.

Baskin, Y. (1996). Curbing undesirable invaders. *BioScience,* **46**, 732–6.

Baumgartner, D.M. (1975). *Management of Lodgepole Pine Ecosystems.* 2 vols. Seattle: Washington State University Cooperative Extension Service.

Baumgartner, D.M., Krebill, R.G., Arnott, J.T. & Weetman, G.F. (eds.) (1985). *Lodgepole Pine: The Species and its Management.* Pullman: Washington State University.

Bégeot, C. & Richard, H. (1996). L'origine récente des peuplements de Pin à crochets (*Pinus uncinata* Miller ex Mirbel) sur la tourbière de Frasne et exploitation de la tourbe dans le Jura. *Acta Botanica Gallica,* **143**, 47–53.

Benkman, C.W. (1993). Adaptation to single resources and the evolution of crossbill (*Loxia*) diversity. *Ecological Monographs,* **63**, 305–25.

Benkman, C.W. (1995). The impact of tree squirrels (*Tamiasciurus*) on limber pine seed dispersal adaptations. *Evolution,* **49**, 585–92.

Betancourt, J.L., Dean, J.S. & Hull, H.M. (1986). Prehistoric long-distance transport of construction beams, Chaco Canyon, New Mexico. *American Antiquity,* **51**, 370–5.

Betancourt, J.L., Van Devender, T.R. & Martin, P.S. (eds.) (1990). *Packrat Middens: The Last 40,000 Years of Biotic Change.* Tucson: University of Arizona Press.

Bhatt, Y.D., Rawat, Y.S. & Singh, S.P. (1994). Changes in ecosystem functioning after replacement of forest by *Lantana* shrubland in Kumaun Himalaya. *Journal of Vegetation Science*, **5**, 67–70.

Bingham, R.T., Hoff, R.J. & Steinhoff, R.J. (1972). *Genetics of Western White Pine*. USDA Forest Service Research Paper WO-12.

Binkley, D., Son, Y. & Kim, Z.S. (1994). Impacts of air pollution on forests: a summary of current situations. *Journal of the Korean Forestry Association*, **83**, 229–38.

Bonan, G.B., Pollard, D. & Thompson, S.L. (1992). Effects of boreal forest vegetation on global climate. *Nature*, **359**, 716–18.

Bonhote, J. & Vernet, J.L. (1988). La memoire des charbonnières. Essai de reconstitution des milieux forestièrs dans une vallée marquée par la metallurgie (Aston, Haute-Ariège). *Revue Forestière Française*, **40**, 197–212.

Borchert, M. (1985). Serotiny and cone-habit variation in populations of *Pinus coulteri* (Pinaceae) in the Southern Coast Ranges of California. *Madroño*, **32**, 29–48.

Braun, E.L. (1950). *Deciduous Forests of Eastern North America*. Philadelphia: Blakiston.

Brunstein, F.C. (1996). Climatic significance of the bristlecone pine latewood frost-ring record at Almagre Mountain, Colorado, USA. *Arctic and Alpine Research*, **28**, 65–76.

Brunstein, F.C. & Yamaguchi, D.K. (1992). The oldest known Rocky Mountain bristlecone pine. *Arctic and Alpine Research*, **24**, 253–6.

Buff, J. (1985). Ein Blick auf Griechenlands Forstwirtschaft. *Holz Zentralblatt*, **111**, 2156.

Burns, R.M. & Honkala, B.H. (eds.) (1990). *Silvics of North America*. Vol. 1. *Conifers*. USDA Agriculture Handbook 654. Washington, DC: USDA Forest Service.

Cain, M.D. & Shelton, M.G. (1994). Indigenous vegetation in a southern Arkansas pine-hardwood forest after half a century without catastrophic disturbance. *Natural Areas Journal*, **14**, 165–74.

Christensen, N.L. (1988). The vegetation of the southeastern coastal plain. In *North American Terrestrial Vegetation*, ed. M.G. Barbour & W.D. Billings, pp. 317–63. Cambridge: Cambridge University Press.

Christensen, N.L., Bartuska, A.M., Brown, J.H., *et al.* (1996). The report of the Ecological Society of America Committee of the Scientific Basis for Ecosystem Management. *Ecological Applications*, **6**, 665–91.

Coffman, T. (1995). *The Cambria Forest. Reflections on its Native Pines and its Eventful Past*. Cambria, Calif.: Coastal Heritage Press.

Conkle, M.T., Schiller, G. & Grunwald, C. (1988). Electrophoretic analysis of diversity and phylogeny of *Pinus brutia* and closely related taxa. *Systematic Botany*, **13**, 411–24.

Cook, E.R. & Kairiukstis, L.A. (eds.) (1990). *Methods of Dendrochronology. Applications in the Environmental Sciences*. Dordrecht: Kluwer.

Cooling, E.N.G. (1967). *Report of a Visit to South East Asia to Obtain Seeds of Tropical Pines*. Forestry Research Bulletin 13. Lusaka: Forestry Department of Zambia.

Cooling, E.N.G. (1968). *Pinus merkusii*. Fast Growing Timber Trees of the Lowland Tropics 4. Oxford: Commonwealth Forestry Institute.

Crane, P.R., Friis, E.M. & Pedersen, K.R. (1995). The origin and early diversification of angiosperms. *Nature*, **374**, 27–33.

Critchfield, W.B. (1980). *Genetics of Lodgepole Pine*. USDA Forest Service Research Paper Wo-37.

Critchfield, W.B. (1985). The late Quaternary history of lodgepole and jack pines. *Canadian Journal of Forestry Research*, **15**, 749–72.

Critchfield, W.B. & Little, E.L. (1966). *Geographic Distribution of Pines of the World*. USDA Forest Service Miscellaneous Publication 991.

Crompton, G. (1972). *History of Lakenheath Warren: An Historical Study for Ecologists*. London: Nature Conservancy.

Currey, D.R. (1968). An ancient bristlecone pine stand. *Ecology*, **44**, 564–6.

Cwynar, L.C. & MacDonald, G.M. (1987). Geographical variation of lodgepole pine in relation to population history. *American Naturalist*, **129**, 463–9.

Dallimore, W. & Jackson, A.B. (1966). *A Handbook of Coniferae and Ginkgoaceae*, 4th edn. London: Edward Arnold.

D'Antonio, C.M. & Dudley, T.L. (1995). Biological invasions as agents of change on islands versus mainlands. In *Islands. Biological Diversity and Ecosystem Function*, ed. P.M. Vitousek, L.L. Loope & H. Adsersen, pp. 103–21. Berlin: Springer-Verlag.

Darrow, W.K. & Zanoni, T. (1990). Hispaniolan pine (*Pinus occidentalis* Swartz). A little known sub-tropical pine of economic potential. *Commonwealth Forestry Review*, **69**, 133–46.

Daubenmire, R. & Daubenmire, J.B. (1968). *Forest Vegetation of Eastern Washington and Northern Idaho*. Washington Agricultural Experiment Station Technical Bulletin 60.

Denevan, W.M. (1961). *The Upland Pine Forests of Nicaragua: A Study in Cultural Plant Geography*. Berkeley: University of California Press.

Den Ouden, P. & Boom, B.K. (1965). *Manual of Cultivated Conifers*. The Hague: Martinus Nijhoff.

Drake, J.A., Mooney, H.A., Di Castri, F., Groves, R.H., Kruger, F.J., Rejmánek, M. & Williamson, M. (eds.) (1989). *Biological Invasions. A Global Perspective*. Chichester: John Wiley.

Driver, H.E. (1969). *Indians of North America*. Chicago: University of Chicago Press.

Duffy, E., Morris, M.G., Sheail, J., Ward, L.K., Wells, D.A. & Wells, T.C.E. (1974). *Grassland Ecology and Wildlife Management*. London: Chapman & Hall.

Dvorak, W.S. & Brouard, J. (1987). An evaluation of *Pinus chiapensis* as a commercial plantation species for the tropics and subtropics. *Commonwealth Forestry Review*, **66**, 165–76.

Dvorak, W.S. & Donahue, J.K. (1992). *CAMCORE Cooperative Research Review 1980–1992*. Raleigh, NC: Department of Forestry, North Carolina State University.

Dye, P.J. (1996). Climate, forest and streamflow relationships in South African forested catchments. *Commonwealth Forestry Review*, **75**, 31–8.

Eberhardt, T.L. & Young, R.A. (1996). Assessment of the anti-HIV activity of a pine cone isolate. *Planta Medica*, **62**, 63–5.

Edenius, L., Danell, K. & Nyquist, H. (1995). Effects of simulated moose browsing on growth, mortality, and fecundity in Scots pine: relations to plant productivity. *Canadian Journal of Forest Resedrch*, **25**, 529–35.

Elliott, P.F. (1974). Evolutionary responses of plants to seed-eaters: pine squirrel predation on lodgepole pine. *Evolution*, **28**, 221–31.

Enright, N.J., Hill, R.S. & Veblen, T.T. (1995). The southern conifers – an introduction. In *Ecology of the Southern Conifers*, ed. N.J. Enright & R.S. Hill, pp. 1–9. Melbourne: Melbourne University Press.

Evans, R.A. (1988). *Management of Pinyon–Juniper Woodlands*. USDA Forest Service General Technical Report INT-249.

Everett, R.L. (ed.) (1987). *Proceedings – Pinyon–Juniper Conference*. USDA Forest Service General Technical Report INT-215.

Ewers, F.W. & Schmid, R. (1981). Longevity of needle fascicles of *Pinus longaeva* (Bristlecone Pine) and other North American pines. *Oecologia*, **51**, 107–15.

Farjon, A. (1984). *Pines. Drawings and Descriptions of the Genus Pinus*. Leiden: E.J. Brill/Dr W. Backhuys.

Farjon, A. (1995). Conifer Specialist Group. *Species (Newsletter of the Species Survival Commission, IUCN – The World Conservation Union)*, **24**, 64–5.

Farjon, A. (1998). *World Checklist and Bibliography of Conifers*. Kew, UK: Royal Botanic Gardens.

Farjon, A., Page, C.N. & Schellevis, N. (1993). A preliminary world list of threatened conifer taxa. *Biodiversity and Conservation*, **2**, 304–26.

Feng, X. & Epstein, S. (1994). Climatic implications of an 8000-year hydrogen isotope time series from Bristlecone pine trees. *Science*, **265**, 1079–81.

Forman, R.T.T. (ed.) (1979). *Pine Barrens: Ecosystem and Landscape*. New York: Academic Press.

Fossitt, J.A. (1994). Late-glacial and Holocene vegetation history of western Donegal, Ireland. *Biology and Environment: Proceedings of the Royal Irish Academy*, B**94**, 1–31.

Fowells, H.A. (1965). *Silvics of Forest Trees of the United States*. US DA Forest Service Agriculture Handbook 271. Washington, DC: USDA Forest Service.

Fowler, D.P. & Lester, D.T. (1970). *Genetics of Red Pine*. USDA Forest Service Research Paper WO-8.

Fox, S. & Mickler, R.A. (1995). *Impact of Air Pollutants on Southern Pine Forests*. New York: Springer-Verlag.

Frelich, L.E. (1995). Old forest in the Lake States today and before European settlement. *Natural Areas Journal*, **15**, 157–67.

Fritts, H.C. (1976). *Tree Rings and Climate*. London: Academic Press.

Gauthier, S., Gagnon, J. & Bergeron, Y. (1993). Population age structure of *Pinus banksiana* at the southern edge of the Canadian boreal forest. *Journal of Vegetation Science*, **4**, 783–90.

Golley, F.B., Pinder, J.E. III, Smallidge, P.J. & Lambert, N.J. (1994). Limited invasion and reproduction of loblolly pines in a large South Carolina old field. *Oikos*, **69**, 21–7.

Gordon, A.G. & Gorham, E. (1963). Ecological aspects of air pollution from an iron-smeltering plant at Wawa, Ontario. *Canadian Journal of Botany*, **41**, 1063–78.

Greaves, A. (1982). *Pinus oocarpa*. Forestry Abstracts, **43**, 503–32.

Greller, A.M. (1988). Deciduous forest. In *North American Terrestrial Vegetation*, ed. M.G. Barbour & W.D. Billings, pp. 287–316. Cambridge: Cambridge University Press.

Grierson, A.J.C., Long, D.G. & Page, C.N. (1980). Notes relating to the flora of Bhutan: III. *Pinus bhutanica*: a new 5-needle pine from Bhutan and India. *Notes from the Royal Botanical Garden, Edinburgh*, **38**, 297–310.

Grissino-Mayer, H.D., Baisan, C.H. & Swetnam, T.W. (1995). Fire history in the Pinaleno Mtns of SE Arizona: effects of human-related disturbances. In *Biodiversity and Management of the Madrean Archipelago: The Sky Islands of Southwestern United States and Northwestern Mexico*, ed. L.F. DeBano, G.J. Gottfried, R.H. Hamre, C.B. Edminster, P.F. Ffolliott & A. Ortega-Rubio, pp. 399–407. USDA Forest Service General Technical Report GTR-264.

Grodzinski, W., Weiner, J. & Maycock, P.F. (eds.) (1984). *Forest Ecosystems in Industrial Regions: Studies on the Cycling of Energy, Nutrients and Pollutants in the Nieopolomice Forest, Southern Poland*. Berlin: Springer-Verlag.

Gruell, G.E., Schmidt, W.C., Arno, S.F. & Reich, W.J. (1982). *Seventy Years of Vegetative Change in a Managed Ponderosa Pine Forest in Western Montana – Implications for Resource Management*. USDA Forest Service General Technical Report INT-130.

Hallé, F., Oldeman, R.A.A. & Tomlinson, P.B. (1978). *Tropical Trees and Forests: An Architectural Analysis*. Berlin: Springer-Verlag.

Hamrick, J.L. & Godt, M.J.W. (1996). Conservation genetics of endemic plant species. In *Conservation Genetics: Case Histories From Nature*, ed. J.C. Avise & J.L. Hamrick, pp. 281–304. New York: Chapman & Hall.

Hartshorn, G.S. (1988). Tropical and subtropical vegetation of Meso-America. In *North American Terrestrial Vegetation*, ed. M.G. Barbour & W.D. Billings, pp. 365–90. Cambridge: Cambridge University Press.

Hengeveld, R. (1989). *Dynamics of Biological Invasions*. London: Chapman & Hall.

Hernandez Tejeda, T. & Nieto de Pascual, C. (1996). Effects of oxidant air pollution on *Pinus maximartinezii* Rzedowski in the Mexico City region. *Environmental Pollution*, **92**, 79–83.

Hoff, R.J. (1987). *Western White Pine: An Annotated Bibliography*. USDA Forest Service General Technical Report SE-58.

Holtmeier, F.-K. (1990). Disturbance and management problems in Larch–Cembra pine forests in Europe. In *Proceedings: Symposium on Whitebark Pine Ecosystems: Ecology and Management of a High-Mountain Resource*, pp. 25–36, USDA (Forest Service) General Technical Report INT-270.

Hou, H.Y. (1983). Vegetation of China with reference to its geographical distribution. *Annals of the Missouri Botanical Garden*, **70**, 509–48.

Huntley, B. & Birks, H.J.B. (1983). *An Atlas of Past and Present Pollen Maps for Europe: 0–13 000 Years Ago*. Cambridge: Cambridge University Press.

Hutchins, H.E., Hutchins, S.A. & Liu, B. (1996). The role of birds and mammals in Korean pine (*Pinus koraiensis*) regeneration dynamics. *Oecologia*, **107**, 120–30.

IUCN Species Survival Commission (1994). *IUCN Red List Categories*. Gland, Switzerland: International Union for the Conservation of Nature.

Jarvis, D.I. & Helin, L. (1993). Vegetation patterns in the *Pinus yunnanensis*–sclerophyllous broadleaved forests, Mianning County, Sichuan Province, China. *Journal of Biogeography*, **20**, 505–24.

Jarvis, P.G. & Leverenz, J.W. (1983). Productivity of temperate, deciduous and evergreen forests. In *Encyclopedia of Plant Physiology*, vol. 12D, ed. O.L. Lange, P.S. Nobel, C.B. Osmond & H. Ziegler, pp. 233–80. Heidelberg: Springer-Verlag.

Johnson, J.D., Chappelka, A.H., Hain, F.P. & Heagle, A.S. (1996). Interactive effects of air pollutants with abiotic and biotic factors on southern pine forests. In *Impact of Air Pollutants on Southern Pine Forests*, ed. S. Fox & R.A. Mickler, pp. 281–312. New York: Springer-Verlag.

Kajimoto, T. (1993). Shoot dynamics of *Pinus pumila* in relation to altitudinal and wind exposure gradients on the Kiso mountain range, central Japan. *Tree Physiology*, **13**, 41–53.

Karschon, R., Weinstein, A. & Heth, D. (1976). *Ecological Studies of the Vegetation on Old Walls*. Tel Aviv, Israel: Division of Forestry, Agricultural Research Organization.

Keane, R.E., Morgan, P. & Menakis, J.P. (1994). Landscape assessment of the decline of Whitebark Pine (*Pinus albicaulis*) in the Bob Marshall Wilderness Complex, Montana, USA. *Northwest Science*, **68**, 213–29.

Kellison, R.C. & Zobel, B.J. (1973). *Genetics of Virginia Pine*. USDA Forest Service Research Paper WO-21.

Kellman, M. (1979). Soil enrichment by neotropical savanna trees. *Journal of Ecology*, **67**, 565–77.

Kendall, K.C. & Arno, S.F. (1990). Whitebark pine: an important but endangered wildlife resource. In *Proceedings – Symposium on Whitebark Pine Ecosystems: Ecology and Management of a High-Mountain Resource*, ed. W.C. Schmidt & K.J. McDonald, pp. 264–73. USDA Forest Service General Technical Report INT-270.

Kim, J.U. & Yim, Y.-J. (1988). Environmental gradient analysis of forest vegetation of Mt. Naejang, southwest Korea. *Korean Journal of Botany*, **31**, 33–40.

Kindel, K.-H. (1995). *Kiefern in Europa*. Stuttgart: Gustav Fischer Verlag.

Kinloch, B.B., Marosy, M. & Huddleston, M.E. (eds.) (1996). *Symposium on Sugar Pine: Status, Values, and Roles in Ecosystems: Proceedings of a Symposium Presented by the California Sugar Pine Management Committee*. Davis: University of California, Division of Agriculture & Natural Resources, Publication 3362.

Klaus, W. (1989). Mediterranean pines and their history. *Plant Systematics and Evolution*, **162**, 133–63.

Knight, D.H., Vose, J.M., Baldwin, V.C., Ewel, K.C & Grodzinska, K. (1994). Contrasting patterns in pine forest ecosystems. In *Environmental Constraints on the Structure and Productivity of Pine Forest Ecosystems: A Comparative Analysis. Ecological Bulletin 43*, ed. H.L. Gholz, S. Linder & R.E. McMurtie, pp. 9–19. Copenhagen: Munksgaard.

Kowal, N.E. (1966). Shifting cultivation, fire, and pine forest in the Cordillera Central, Luzon, Philippines. *Ecological Monographs*, **36**, 389–419.

Kunkel, G. (1984). *Plants for Human Consumption*. Koenigstein, Germany: Koeltz Scientific Books.

Lai, P.K., Tamura, Y., Bradley, W.G., Tanaka, A. & Nonoyama, M. (1991). A novel soluble factor induced by an extract from *Pinus parviflora* Sieb et Zucc can inhibit the replication of Human Immunodeficiency Virus *in vitro*. *Journal of Acquired Immune Deficiency Syndromes*, **4**, 356.

Lamb, A.F.A. (1973). *Pinus caribaea*. Fast Growing Timber Trees of the Lowland Tropics 6. Oxford: Commonwealth Forestry Institute.

Landers, J.L., van Lear, D.H. & Boyer, W.D. (1995). The longleaf pine forests of the southeast: requiem or renaissance? *Journal of Forestry*, **93**, 39–44.

Lanner, R.M. (1981). *The Piñon Pine: A Natural and Cultural History*. Reno: University of Nevada Press.

Lanner, R.M. (1996). *Made For Each Other: A Symbiosis of Birds And Pines*. New York: Oxford University Press.

Larson, J.A. (1980). *The Boreal Ecosystem*. New York: Academic Press.

Ledig, F.T. (1992). Human impacts on genetic diversity in forest ecosystems. *Oikos*, **63**, 87–108.

Ledig, F.T. (1993). Secret extinctions: The loss of genetic diversity in forest ecosystems. In *Our Living Legacy: Proceedings of a Symposium on Biological Diversity*, ed. M.A. Fenger, E.H. Miller, J.A. Johnson & E.J.R. Williams, pp. 127–40. Victoria, Canada: Royal British Columbia Museum.

Le Houérou, H. (1981). Impact of man and his animals on mediterranean vegetation. In *Mediterranean-Type Shrublands. Ecosystems of the World*, Vol. 11, ed. F. di Castri, D.W. Goodall & R.L. Specht, pp. 479–517. Amsterdam: Elsevier.

Lepart, J. & Debussche, M. (1992). Human impact on landscape patterning: Mediterranean examples. In *Landscape Boundaries. Consequences for Biotic Diversity and Ecological Flows*, ed. A.J. Hansen & F. di Castri, pp. 76–106. New York: Springer-Verlag.

Lewis, N.B. & Ferguson, I.S. (1993). *Management of Radiata Pine*. Melbourne: Inkata Press.

Libby, W.J. (1990). Genetic conservation of radiata pine and coast redwood. *Forest Ecology and Management*, **35**, 109–20.

Long, A., Warneke, L.A., Betancourt, J.L. & Thompson, R.S. (1990). Deuterium variations in plant cellulose from fossil packrat middens. In *Packrat Middens: The Last 40,000 Years of Biotic Change*, ed. J.L. Betancourt, T.R. Van Devender & P.S. Martin, pp. 380–96. Tucson: University of Arizona Press.

Loock, E.E.M. (1950). *The Pines of Mexico and British Honduras*. South Africa Department of Forestry Bulletin 34. Pretoria: Department of Forestry.

McAndrews, J.H. (1988). Human disturbance of North American forests and grasslands: The fossil pollen record. In *Vegetation History*, ed. B. Huntley & T. Webb III, pp. 673–97. Dordrecht: Kluwer.

McClaran, M.P. & Brady, W.W. (1994). Arizona's diverse vegetation and contributions to plant ecology. *Rangelunds*, **16**, 208–17.

McCune, B. (1988). Ecological diversity in North American pines. *American Journal of Botany*, **75**, 353–68.

MacDonald, G.M. (1993). Fossil pollen analysis and the reconstruction of plant invasions. *Advances in Ecological Research*, **24**, 67–110.

MacDonald, G.M. & Cwynar, L.C. (1991). Postglacial population history of *Pinus contorta* ssp. *latifolia* in the western interior of Canada. *Journal of Ecology*, **79**, 417–29.

McDonald, P.M. (1976). *Forest Regeneration and Seedling Growth from Five Major Cutting Methods in North Central California*. USDA Forest Service Research Paper PSW-115.

Martin, G.J. (1995). *Ethnobotany*. London: Chapman & Hall.

Mather, A. (1993). Introduction. In *Afforestation. Policies, Planning and Progress*, ed. A. Mather, pp. 1–12. London: Belhaven Press.

Millar, C. (1990). Memorial: William Burke Critchfield (1923–1989). *Madroño*, **37**, 221–3.

Millar, C.I. (1993). Impact of the Eocene on the evolution of *Pinus* L. *Annals of the Missouri Botanical Garden*, **80**, 471–98.

Millar, C.I. & Kinloch, B.B. (1991). Taxonomy, phylogeny, and the coevolution of pines and their stem rusts. In *Rusts of Pines. Proceedings of the 3rd IUFRO Rusts of Pine Working Party Conference*, ed. Y. Hiratsuka, J.K. Samoil, P.V. Blenis, P.E. Crane & B.L. Laishley, pp. 1–38. Edmonton, Alberta: Forestry Canada.

Millar, C.I., Kinloch, B.B. & Westfall, R.D. (1996). Conservation of biodiversity in sugar pine: Effects of the blister rust epidemic on genetic diversity. In *Symposium on Sugar Pine: Status, Values, and Roles in Ecosystems: Proceedings of a Symposium Presented by the California Sugar Pine Management Committee*, ed. B.B. Kinloch, M. Marosy & M.E. Huddleston, pp. 190–9. Davis: University of California, Division of Agriculture & Natural Resources, Publication 3362.

Miller, P.R. (1992). Mixed conifer forests of the San Bernardino Mtns, California. In *The Response of Western Forests to Air Pollution*, ed. R.K. Olson, D. Binkley & M. Bohm, pp. 461–97. New York: Springer-Verlag.

Miller, P.R., DeBauer, M.D.L., Nolasco, A.Q. & Tejeda, T.H. (1994). Comparison of ozone exposure characteristics in forested regions near Mexico City and Los Angeles. *Atmospheric Environment*, **28**, 141–8.

Mirov, N.T. (1967). *The Genus Pinus*. New York: Ronald Press.

Myers, R.L. (1985). Fire and the dynamic relationship between Florida sandhill and sand pine scrub vegetation. *Bulletin of the Torrey Botanical Club*, **112**, 241–52.

Nebel, B. & Matile, P. (1992). Longevity and senescence of needles in *Pinus cembra* L. *Trees: Structure and Function*, **6**, 156–61.

Niebling, C.R. & Conkle, M.T. (1990). Diversity of Washoe pine and comparisons with allozymes of ponderosa pine races. *Canadian Journal of Forest Research*, **20**, 298–308.

Nooden, L.D. (1988). Whole plant senescence. In *Senescence and Aging in Plants*, ed. L.D. Nooden & A.C. Leopold, pp. 391–439. San Diego: Academic Press.

Ohsawa, M., Shakya, P.R. & Numata, M. (1986). Distribution and succession of west Himalayan forest types in the eastern part of the Nepal Himalaya. *Mountain Research and Development*, **6**, 143–57.

Orwig, D.A. & May, M.D. (1994). Land-use history (1720–1992), composition and dynamics of oak–pine forest within the Piedmont and coastal plain of northern Virginia. *Canadian Journal of Forest Research*, **24**, 1216–25.

Parsons, D. & DeBenedetti, S. (1979). Impact of fire suppression on a mixed-conifer forest. *Forest Ecology and Management*, **2**, 21–33.

Parsons, J.J. (1955). The Miskito pine savanna of Nicaragua and Honduras. *Annals of the Association of American Geographers*, **45**, 36–63.

Peet, R.K. (1988). Forests of the Rocky Mountains. In *North American Terrestrial Vegetation*, ed. M.G. Barbour & W.D. Billings, pp. 63–101. Cambridge: Cambridge University Press.

Perry, J.P. (1991). *The Pines of Mexico and Central America*. Portland, Oregon: Timber Press.

Peterson, C.J. & Squiers, E.R. (1995). Competition and succession in an aspen–white pine forest. *Journal of Ecology*, **83**, 449–57.

Polunin, O. & Walters, N. (1985). *A Guide to the Vegetation of Britain and Europe*. Oxford: Oxford University Press.

Racine, C.H. (1966). Pine communities and their site characteristics in the Blue Ridge escarpment. *Journal of the Elisha Mitchell Scientific Society*, **82**, 172–81.

Randall, J.M. & Rejmánek, M. (1993). Interference of bull thistle (*Cirsium vulgare*) with growth of ponderosa pine (*Pinus ponderosa*) seedlings in a forest plantation. *Canadian Journal of Forest Research*, **23**, 1507–13.

Reich, P.B., Kloeppel, B.D., Ellsworth, D.S. & Walters, M.B. (1995). Different photosynthesis–nitrogen relations in deciduous hardwood and evergreen coniferous tree species. *Oecologia*, **104**, 24–30.

Reille, M. (1989). L'origine du pin à crochets dans le Massif Central français. *Bulletin de la Société Botanique de France, Lettres Botaniques*, **136**, 61–70.

Rejmánek, M. & Richardson, D.M. (1996). What attributes make some plant species more invasive? *Ecology*, **77**, 1655–61.

Richardson, D.M. & Bond, W.J. (1991). Determinants of plant distribution: evidence from pine invasions. *American Naturalist*, **137**, 639–68.

Richardson, D.M., Cowling, R.M. & Le Maitre, D.C. (1990). Assessing the risk of invasive success in *Pinus* and *Banksia* in South African mountain fynbos. *Journal of Vegetation Science*, **1**, 629–42.

Richardson, D.M., Williams, P.A. & Hobbs, R.J. (1994). Pine invasions in the Southern Hemisphere: determinants of spread and invadability. *Journal of Biogeography*, **21**, 511–27.

Rudolph, T.D. & Yeatman, C.W. (1982). *Genetics of Jack Pine*. USDA Forest Service Research Paper WO-38.

Rummell, R.S. (1951). Some effects of livestock grazing on ponderosa pine forest and range in Central Washington. *Ecology*, **32**, 594–607.

Rundel, P.W., Parsons, D.J. & Gordon, D.T. (1977). Montane and subalpine forests of the Sierra Nevada and Cascade Ranges. In *Terrestrial Vegetation of California*, ed. M.G. Barbour & J. Major, pp. 559–99. New York: John Wiley.

Sakagami, H., Ohhara, T., Kaiya, T., Kawazoe, Y., Nonoyama, M. & Konno, K. (1991). Molecular species of the antitumor and antiviral fraction from pine cone extract. *Anticancer Research*, **9**, 1593–8.

Samuels, M.L. & Betancourt, J.L. (1982). Modelling the long-term effects of fuelwood harvests on pinyon–juniper woodlands. *Environmental Management*, **6**, 505–15.

Savage, M. (1994). Land-use change and the structural dynamics of *Pinus kesiya* in a hill evergreen forest in Northern Thailand. *Mountain Research and Development*, **14**, 245–50.

Savage, M. & Swetnam, T.W. (1990). Early 19th century fire decline following sheep pasturing in a Navajo ponderosa pine forest. *Ecology*, **71**, 2374–8.

Schmidt, W.C. & Holtmeier, F.-K. (eds.) (1994). *Proceedings – International Workshop on Subalpine Stone Pines and their Environment: The Status of Our Knowledge*. USDA Forest Service General Technical Report INT-309.

Schmidt, W.C. & McDonald, K.J. (compilers) (1990). *Symposium on Whitebark Pine Ecosystems: Ecology and Management of a High Mountain Resource*. USDA Forest Service General Technical Report INT-270.

Schopmeyer, C.S. (tech. co-ord.) (1974). *Seeds of Woody Plants in the United States*. USDA Agriculture Handbook No. 450. Washington, DC: United States Department of Agriculture.

Schweingruber, F.H. (1993). *Trees and Wood in Dendrochronology*. Berlin: Springer-Verlag.

Scott, C.W. (1960). *Pinus radiata*. FAO Forestry and Forestry Products Studies No. 14. Rome: Food and Agriculture Organization of the United Nations.

Sehgal, R.N. & Sharma, P.K. (1985). *Chilgoza the Endangered Social Forestry Pine of Kinnaur*. Nauni, India: Dr Y.S. Parmar University of Horticulture and Forestry.

Serre Bachet, F. (1985). Une chronologie pluriséculaire du sud de l'Italie. *Dendrochronologia*, **3**, 45–66.

Shigesada, N., Kawasaki, K. & Takeda, Y. (1995). Modelling stratified diffusion in biological invasions. *American Naturalist*, **146**, 229–51.

Shipek, F.C. (1989). An example of intensive plant husbandry: the Kumenay of southern California. In *Foraging and Farming. The Evolution of Plant Exploitation*, ed. D.R. Harris & G.C. Hillman, pp. 159–70. London: Unwin & Hyman.

Singh, J.S. & Singh, S.P. (1987). Forest vegetation of the Himalaya. *Botanical Review*, **53**, 80–192.

Singh, S.P., Adhikari, B.S. & Zobel, D.B. (1994). Biomass, productivity and forest structure in the central Himalaya. *Ecological Monographs*, **64**, 401–21.

Smith, W.H. (1981). *Air Pollution and Forests*. New York: Springer-Verlag.

Smith, W.K. & Hinckley, T.M. (eds.) (1995a). *Resource Physiology of Conifers*. San Diego: Academic Press.

Smith, W.K. & Hinckley, T.M. (eds.) (1995b). *Ecophysiology of Coniferous Forests*. San Diego: Academic Press.

Snyder, E.B., Dinus, R.J. & Derr, H.J. (1977). *Genetics of Longleaf Pine*. USDA Forest Service Research Paper WO-33.

Spring, P.E., Brewer, M.L., Brown, J.R. & Fanning, M.E. (1974). Population ecology of loblolly pine *Pinus taeda* in an old field community. *Oikos*, **25**, 1–6.

Stein, S.J. (1988). Fire history of the Paunsaugunt Plateau in southern Utah. *Great Basin Naturalist*, **48**, 58–63.

Steven, H.M. & Carlisle, A. (1959). *The Native Pinewoods of Scotland*. Edinburgh: Oliver and Boyd.

Stewart, W.N. (1983). *Paleobotany and the Evolution of Plants*. Cambridge: Cambridge University Press.

St Pierre, H., Gagnon, R. & Bellefleur, P. (1992). Black spruce (*Picea mariana*) and jack pine (*Pinus banksiana*) regeneration after fire in a boreal forest in Quebec. *Canadian Journal of Forest Research*, **22**, 474–81.

Styles, B.T. (1992). Genus *Pinus*: A Mexican Purview. In *Biological Diversity of Mexico: Origins and Distribution*, ed. T.P. Ramamoorthy, R. Bye, A. Lot & J. Fa, pp. 397–420. New York: Oxford University Press.

Sun, X.-J., Yuan, S.M., Liu, J.-L. & Tang, L.-Y. (1991). The vegetation history of mixed Korean pine and deciduous forests in Chanbai Mt. area, Jilin Province, northeastern China, during the last 13,000 years. *Chinese Journal of Botany*, **3**, 47–61.

Szmidt, A.E. & Wang, X.-R. (1993). Molecular systematics and genetic differentiation of *Pinus sylvestris* L. and *Pinus densiflora* Sieb. et Zucc. *Theoretical and Applied Genetics*, **86**, 159–65.

Takhtajan, A.L. (1986). *Floristic Regions of the World*. Berkeley: University of California Press.

Taylor, B.W. (1962). The status and development of the Nicaraguan pine savannas. *Caribbean Forester*, **23**, 21–6

Teskey, R.O., Whitehead, D. & Linder, S. (1994). Photosynthesis and carbon gain by pines. *Ecological Bulletin (Copenhagen)*, **43**, 35–49.

Tester, J.R., Starfield, A.M. & Frelich, L.E. (1997). Modelling for ecosystem management in Minnesota pine forests. *Biological Conservation*, **80**, 313–24.

Thirgood, J.V. (1981). *Man and the Mediterranean Forest. A History of Resource Depletion*. London: Academic Press.

Tranquillini, W. (1979). *Physiological Ecology of the Alpine Timberline*. Ecological Studies 31. Berlin: Springer-Verlag.

Troup, R.S. (1921). *The Silviculture of Indian Trees*, Vol. 3. Oxford: Clarendon Press.

Tsukada, M. (1988). Japan. In *Vegetation History*, ed. B. Huntley & T. Webb, III, pp. 459–581. Dordrecht: Kluwer.

Turnbull, J.W., Armitage, F.B. & Burley, J. (1980). Distribution and ecology of the *Pinus kesiya* complex. In *Pinus kesiya*, ed. F.B. Armitage & J. Burley, pp. 13–45. Tropical Forestry Paper 9. Oxford: Commonwealth Forestry Institute.

Uemura, S., Tsuda, S. & Hasegawa, S. (1990). Effects of fire on the vegetation of Siberian taiga predominated by *Larix dahurica*. *Canadian Journal of Forest Research*, **20**, 547–53.

USDA (1988). *The South's Fourth Forest: Alternatives for the Future*. Forest Resource Report No. 24. Washington, DC: United States Department of Agriculture Forest Service.

Vale, T.R. (1982). *Plant and People. Vegetation Change in North America*. Washington, DC: Association of American Geographers.

Van der Water, P.K., Leavitt, S.W. & Betancourt, J.L. (1994). Trends in stomatal density and $^{13}C/^{12}C$ ratios of *Pinus flexilis* needles during last glacial-interglacial cycle. *Science*, **264**, 239–43.

Vankat, J.L. (1982). A gradient perspective on the vegetation of Sequoia National Park, California. *Madroño*, **29**, 200–14.

Veblen, T.T. (1976). The urgent need for forest conservation in highland Guatemala. *Biological Conservation*, **9**, 141–54.

Veblen, T.T. (1978). Forest preservation in the western highlands of Guatemala. *Geographical Review*, **68**, 417–34.

Vidaković, M. (1991). *Conifers: Morphology and Variation*, revised English edition. Zagreb: Graficki Zavod Hrvatske.

Vogl, R.J. (1967). Controlled burning for wildlife in Wisconsin. *Tall Timbers Fire Ecology Conference*, **6**, 47–96.

Vogl, R.J. (1970). Fire and the northern Wisconsin pine barrens. *Tall Timbers Fire Ecology Conference*, **10**, 175–209.

Vogl, R.J. (1973). Ecology of knobcone pine in the Santa Ana Mountains, California. *Ecological Monographs*, **43**, 125–43.

Vogl, R.J., Armstrong, W.P., White, K.L. & Cole, K.L. (1977). The closed-cone pines and cypresses. In *Terrestrial Vegetation of California*, ed. M.G. Barbour & J. Major, pp. 295–358. New York: John Wiley.

Waldrup, T.A. (1989). *Pine–Hardwood Mixtures: A Symposium on Management and Ecology of the Type*. USDA Forest Service General Technical Report SE-58.

Wang, C.-W. (1961). *The Forests of China*. Maria Moors Cabot Foundation Publications No. 5. Cambridge, Mass: Harvard University Press.

Wang, C.-W. (1977). *Genetics of Ponderosa Pine*. USDA Forest Service Research Paper WO-34.

Wang, X.-R., Szmidt, A. E. & Lindgren, D. (1991). Allozyme differences among populations of *Pinus sylvestris* L. from Sweden and China. *Hereditas*, **114**, 219–26.

Waring, R.H. (1991). Responses of evergreen trees to multiple stresses. In *Responses of Plants to Multiple Stresses*, ed. H.A. Mooney, W.E. Winner & E.J. Pell, pp. 371–90. San Diego: Academic Press.

Waring, R.H. & Franklin, J.F. (1979). Evergreen coniferous forests of the Pacific Northwest. *Science*, **204**, 1380–6.

Watt, A.S. (1971). Factors controlling the floristic composition of some plant communities in Breckland. *Symposium of the British Ecological Society*, **11**, 137–52.

Wells, P.V. & Jorgensen, C.D. (1964). Pleistocene wood rat middens and climate change in the Mojave Desert – a record of juniper woodlands. *Science*, **143**, 1171–14.

Werner, W. (1993). Pinus *in Thailand*. Stuttgart: Franz Steiner Verlag.

West, N.E. (1988). Intermountain deserts, shrub steppes, and woodlands. In *North American Terrestrial Vegetation*, ed. M.G. Barbour & W.B. Billings, pp. 210–30. Cambridge: Cambridge University Press.

Whittaker, R.H. (1956). Vegetation of the Great Smoky Mountains. *Ecological Monographs*, **26**, 1–80.

Wilson, L.F. (1993). China's Masson pine forests: cure or curse? *Journal of Forestry*, **91**(1), 30–3.

Wormald, T.J. (1975). *Pinus patula*. Tropical Forestry Papers 7. Oxford: Commonwealth Forestry Institute.

Wright, H.A. & Bailey, A.W. (1982). *Fire Ecology: United States and Southern Canada*. New York: John Wiley.

Wright, J.W. (1970). *Genetics of Eastern White Pine*. USDA Forest Service Research Paper WO-9.

Wulff, E.V. (1944). *Isoricheskaia Geografia Rastenii* [Historical Plant Geography]. Moscow: Akad. Nauk SSSR.

Ying, C.C. & Liang, Q. (1994). Geographic pattern of adaptive variation of lodgepole pine (*Pinus contorta* Dougl.) within the species' coastal range: field performance at age 20 years. *Forest Ecology and Management*, **67**, 281–98.

Young, J.A. & Budy, J.D. (1987). Energy crisis in 19th century Great Basin woodlands. In *Proceedings – Pinyon–Juniper Conference*, comp. R.L. Everett, pp. 23–8. USDA Forest Service General Technical Report INT-215.

Young, J.A. & Young, C.G. (1992). *Seeds of Woody Plants in North America*. Revised and enlarged edition. Portland, Oregon: Dioscorides Press.

Young, S.S. & Wang, Z.J. (1989). Comparison of secondary and primary forests in the Ailao region of Yunnan, China. *Forest Ecology and Management*, **28**, 281–300.

Zagas, T. (1994). Die natürliche Bewaldung im Elatia-Gebirge (Griech. Rhodope). *Schweizerische Zeitschrift für Forstwesen*, **145**, 229–40.

Zamora-Serrano, C. & Velasco-Fiscal, V. (1977). *Pinus strobus* var. *chiapensis*, una especie en peligro de extincion en el estado de Chiapas. *Ciencia Forestal*, **2**, 3–23.

Zheng, Wanjun (ed.) (1983). *Sylva Sinica*, Vol. 1. Beijing: Forestry Press.

Zobel, D.B. (1969). Factors affecting the distribution of *Pinus pungens*, an Appalachian endemic. *Ecological Monographs*, **39**, 303–33.

Zobel, B.J., van Wyk, G. & Stahl, P. (1987). *Growing Exotic Forests*. New York: John Wiley.

Part two
Evolution, phylogeny and systematics

2 Phylogeny and systematics of *Pinus*

Robert A. Price, Aaron Liston and Steven H. Strauss

2.1 Introduction

Pinus, with more than 100 species, is the largest genus of conifers and the most widespread genus of trees in the northern hemisphere. Its natural distribution ranges from arctic and subarctic regions of North America and Eurasia south to subtropical and tropical (though usually montane) regions of Central America and Asia, with one species, *P. merkusii*, extending south of the equator in Sumatra (see Critchfield & Little 1966; Mirov 1967; Perry 1991 for detailed treatments of the geography of the genus). The greatest centre of pine species diversity is in North and Central America (c. 70 species, with particular concentrations of species in Mexico, California and the southeastern USA) and in eastern Asia (c. 25 species, with particular concentrations in China). Because of its great economic and ecological importance, the systematics of the genus has received considerable attention from the perspectives of morphology, cytology, crossability, secondary product chemistry, protein electrophoresis, and most recently from restriction site and sequence comparisons of chloroplast and nuclear ribosomal DNA. While the molecular studies have been very useful in developing phylogenetic hypotheses, most studies to date have been restricted to a limited geographic region or include a limited sampling of the diversity within the genus. Thus many questions remain concerning species delimitation and phylogenetic relationships within *Pinus*. Comprehensive studies utilizing both morphological and molecular data must be completed before a definitive worldwide treatment of the genus will be possible.

An outline of our current subgeneric and sectional classification of pines is given in Table 2.1 and a more detailed listing of currently recognized species, sub-

species and varieties, with some more frequently encountered synonyms, is given in Appendix 2.1. Our sectional and subsectional classification largely follows the format of the widely utilized treatment of Little & Critchfield (1969), with several changes in the delimitation of groups following the treatments of Van der Burgh (1973), Rushforth (1987) and Klaus (1989), and the chloroplast DNA restriction site comparisons of Krupkin, Liston & Strauss (1996). Further changes in the number and delimitation of several sectional and subsectional groups are to be anticipated as additional molecular phylogenetic data become available. Based on the results of restriction site and sequence comparisons of the chloroplast genome (Strauss & Doerksen 1990; Wang & Szmidt 1993; R.A. Price & S.H. Strauss, unpublished data) and comparative morphological and biochemical data, we recognize two major lineages of pines: subgenus *Strobus* (haploxylon or soft pines, with one fibrovascular bundle in the needle) and subgenus *Pinus* (diploxylon or hard pines, with two fibrovascular bundles in the needle). Within subgenus *Strobus* we follow Little and Critchfield in recognizing the two sections *Parrya* (with an abaxial ('dorsal') umbo on the cone scale; including the bristlecone, pinyon and lacebark pine groups and the unusual Vietnamese species *P. krempfii*) and *Strobus* (with a terminal umbo on the cone scale; including the white and stone pines). Within subgenus *Pinus* the section *Pinus* appears on the basis of comparisons of chloroplast DNA (Krupkin *et al.* 1996) to be divided into a large North and Central American clade and a group of smaller, largely Eurasian clades. These groups of diploxylon pines only partially correspond to previously recognized sectional groups, and thus two or three sections of diploxylon

Ecology and Biogeography of *Pinus*, ed. D.M. Richardson. Cambridge University Press 1998, pp.49–68.

Table 2.1. Outline of the classification of *Pinus*

Subgenus *Pinus* (Diploxylon or hard pines)
Section *Pinus*
Subsection *Pinus* (Eurasia, N Africa, NE North America, Cuba)
 P. densata, densiflora, heldreichii, hwangshanensis, kesiya, luchuensis, massoniana, merkusii, mugo, nigra, pinaster, resinosa, sylvestris, tabuliformis, taiwanensis, thunbergii, tropicalis, uncinata, yunnanensis
Subsection *Canarienses* (Canary Islands, Himalayas)
 P. canariensis, roxburghii
Subsection *Halepenses* (S Europe, W Asia, N Africa)
 P. brutia, halepensis
Subsection *Pineae* (S Europe)
 P. pinea

New World Diploxylon Pines
Subsection *Contortae* (North America)
 P. banksiana, clausa, contorta, virginiana
Subsection *Australes* (E USA, West Indies and adjacent Central America)
 P. caribaea, cubensis, echinata, elliottii, glabra, occidentalis, palustris, pungens, rigida, serotina, taeda
Subsection *Ponderosae* (W North America, S to Central America)
 P. cooperi, coulteri, donnell-smithii, durangensis, douglasiana, devoniana, engelmannii, hartwegii, jeffreyi, maximinoi, montezumae, nubicola, ponderosa, pseudostrobus, sabiniana, torreyana, washoensis
Subsection *Attenuatae* (W USA and adjacent Mexico)
 P. attenuata, muricata, radiata
Subsection *Oocarpae* (Mexico, S to Central America)
 P. greggii, herrerae, jaliscana, lawsonii, oocarpa, patula, praetermissa, pringlei, tecunumanii, teocote
Subsection *Leiophyllae* (Mexico and adjacent SW USA)
 P. leiophylla, lumholtzii

Subgenus *Strobus* (Haploxylon or soft pines)
Section *Parrya*
Subsection *Balfourianae* (W USA)
 P. aristata, balfouriana, longaeva
Subsection *Krempfianae* (Vietnam)
 P. krempfii
Subsection *Cembroides* (SW USA, Mexico)
 P. cembroides, culminicola, discolor, edulis, johannis, juarezensis, maximartinezii, monophylla, nelsonii, pinceana, remota
Subsection *Rzedowskianae* (Mexico)
 P. rzedowskii
Subsection *Gerardianae* (E Asia, Himalayas)
 P. bungeana, gerardiana

Section *Strobus*
Subsection *Strobi* (North Central America, S & E Asia, SE Europe)
 P. armandii, ayacahuite, bhutanica, chiapensis, dabeshanensis, dalatensis, fenzeliana, flexilis, lambertiana, monticola, morrisonicola, parviflora, peuce, strobus, wallichiana, wangii
Subsection *Cembrae* (NW North America, N Eurasia)
 P. albicaulis, cembra, koraiensis, pumila, sibirica

pines will probably be recognized once they are more thoroughly evaluated with additional molecular phylogenetic data.

Important sources of nomenclatural information utilized in compiling Appendix 2.1 include Rehder (1949), Little & Critchfield (1969), Carvajal & McVaugh (1992) and the listing of 'names in current use' compiled by Farjon (1993). Many questions in the delimitation of species and infraspecific taxa remain unanswered, and thus the list of taxa presented here will be subject to changes as detailed molecular studies are conducted on critical groups of species, especially within the very complex pine floras of

Mexico and eastern Asia. Our classification of the pinyon pines into species and subspecies generally follows the outline of Zavarin (1988) based upon comparisons of morphology and secondary product chemistry. A number of divergent classifications have been recently published for the Mexican and Central American diploxylon pines (see Eguiluz Piedra 1988; Perry 1991; Carvajal & McVaugh 1992; Styles 1993; see also Farjon & Styles 1997 for a revised treatment of the Mexican and Central American species). In both groups, taxonomic novelties continue to be published, and thus our treatments are clearly provisional. In this chapter we present an overview of the history of pine classification, the types of characters that have been used in delimiting groups of species, and the progress towards a phylogenetic classification of the genus from recent molecular systematic studies.

2.2 ***Pinus* in relation to other Pinaceae**

The Pinaceae is a very distinct family of conifers, comprising ten or eleven genera distributed widely throughout the northern hemisphere. The family is supported as monophyletic by a series of shared-derived features that are unique among the conifers. These features include a specialized pattern of proembryogeny (see Doyle 1963; Singh 1978), protein-type sieve cell plastids (Behnke 1974), and the absence of biflavonoids (Geiger & Quinn 1975). Other distinctive morphological features of the family are: the regular occurrence of two ovules per cone scale, each seed usually with a prominent terminal wing derived from the surface of the cone scale (but poorly developed or rudimentary in some pine species); the lack of fusion of the bracts and cone scales in early development; and the presence of two saccae on the pollen grains of most genera. Several different types of molecular studies, including electrophoretic comparisons of phloem proteins (Alosi & Park 1983), immunological comparisons of seed-protein extracts (Price & Lowenstein 1989), and sequence comparisons of the chloroplast gene *rbc*L (Chase *et al.* 1993; Price 1995) and the 18S nuclear ribosomal RNA gene (Chaw *et al.* 1995) have also indicated that the Pinaceae is not closely related to any other extant family of conifers.

Pines form a distinctive natural group within the pine family, supported as monophyletic by a form of shoot dimorphism in which the highly condensed short shoots ('needle clusters' or fascicles) bear one to eight (usually two to five) needle-like leaves sheathed at the base by a series of bud scales. The genus is also distinguished from other members of the family by its often highly woody cone scales with specialized apical regions, the umbo and apophysis, which represent the areas left exposed in the first-year conelet and the mature cone, respectively.

Pines have often been placed in a monogeneric subfamily Pinoideae on the basis of their extreme form of shoot dimorphism (see e.g. Pilger 1926; Krüssmann 1985; Frankis 1988). Pines are most similar in wood anatomy and seed and cone scale morphology to the genera *Cathaya*, *Larix*, *Picea* and *Pseudotsuga*. These five genera regularly exhibit both vertical and horizontal resin canals in their wood, do not have the seed wings partially enfolding the seed, and lack resin vesicles in the seed coat. These features are in contrast to the situation in the genera of subfamily Abietoideae (*Abies*, *Cedrus*, *Keteleeria*, *Nothotsuga*, *Pseudolarix* and *Tsuga*) (see reviews in Frankis 1988; Price 1989; Farjon 1990). *Pinus* is most similar to *Cathaya* and *Picea* in overall morphology, but this is probably due to retention of primitive character states rather than indicating that the three genera form a monophyletic group within the family. For example, pines and spruces both have a pollination drop mechanism, in common with most other families of gymnosperms. In contrast, other mechanisms of pollen reception involving more specialized types of micropylar anatomy are seen in *Larix*, *Pseudotsuga* and most other genera of Pinaceae. The detailed pollination mechanism of *Cathaya* has apparently not yet been studied, but the genus differs from the larches and Douglas firs in having bisaccate pollen.

Results from immunological comparison of seed proteins (Price, Olsen-Stojkovich & Lowenstein 1987) and from *rbc*L sequence comparisons (R.A. Price *et al.*, unpublished data) are consistent with the separation of the abietoid genera (with the possible exception of *Cedrus*) from *Pinus*, *Picea*, and the sister genera *Larix* and *Pseudotsuga*. These results suggest that pines and spruces occupy relatively basal positions in the phylogeny of the family. A recent comparison of restriction sites from six chloroplast-encoded genes (Tsumura *et al.* 1995) placed *Pinus* in a basal position relative to eight of the other genera of Pinaceae, but no bootstrap or decay values were provided to allow assessment of the strength of support for the branches within their tree diagrams. The parsimony and neighbour-joining trees of Tsumura *et al.* both show *Pinus* and the Pinaceae as monophyletic groups, but differ from each other in the arrangement of the other genera of Pinaceae. Sequence comparisons of the *rbc*L gene also strongly support the monophyly of *Pinus* relative to the other genera of the family and suggest that *Picea* is an appropriate outgroup for phylogenetic studies within the genus *Pinus*. *Cathaya* has only very recently become available for inclusion in molecular studies, and may also prove to be a useful outgroup once its phylogenetic position is better established.

2.3 History of pine systematics

Because pines are prominent and morphologically diverse elements in the floras of many different countries of the northern hemisphere, they have been the subject of many, often conflicting, classifications into higher-level infrageneric groups (see e.g. reviews by Mirov 1967; Little & Critchfield 1969; Van der Burgh 1973). The complex nomenclatural history of the genus reflects both changes over time in the conventions and rules of assigning names and ranks to taxa, and differences in the emphases given to particular types of characters and in philosophy of classification. Early classifications of the genus were often highly artificial, emphasizing only a very small number of morphological characters, while classifications from the latter part of the 19th century onwards have attempted to integrate many different types of data from vegetative and reproductive morphology and anatomy, and more recently from secondary product chemistry, crossability, and macromolecular comparisons.

2.3.1 Early classification of the genus

Pinus is a classical Latin name for pines, and was applied by Linnaeus (1753) in his *Species Plantarum* to a group of ten species, including five species of pines (*P. cembra*, *P. pinea*, *P. strobus*, *P. taeda* and *P. sylvestris*) and five species now placed in other genera of the family. Shortly thereafter, Philip Miller (1754) delimited the genus in a narrower fashion closer to our modern circumscription (recognizing firs (*Abies*) and larches (*Larix*) as separate genera, following the pre-Linnaean treatment of Tournefort). Since the separation of the genus *Picea* from *Pinus* by Dietrich (1824), the modern delimitation of the genus has become widely accepted, although treatments as recent as Parlatore (1868) still included the entire family within the genus *Pinus*. The Scots pine, *P. sylvestris*, was chosen from among the original Linnaean species as the lectotype of the genus by Britton (1908).

The first classification of *Pinus* into subdivisions (after the formal starting point of 1753 for our modern system of nomenclature) is that of Duhamel Du Monceau (1755), who recognized three sections based on the number of needles per fascicle: *Bifoliis*, *Trifoliis*, and *Quinquefoliis*. As discussed by Little & Critchfield (1969), these names have come to be replaced under the autonym rule of the International Code of Botanical Nomenclature by sections *Pinus* (for the first two) and *Strobus* (for the latter). If the North American lineage of diploxylon pines is treated as a section separate from section *Pinus*, then *Trifoliis* (in corrected Latin form) needs to be applied as the earliest valid sectional name for this group, with *P. palustris* having been chosen as the lectotype species for this section by Little & Critchfield (1969). An important and often overlooked early classification is that of Loudon (1838), which continued to recognize three largely artificial sections based on the number of needles per fascicle, but also divided the genus into 15 smaller groups on the basis of a larger number of morphological characters from both the cone scales and the needles. Many of the subsectional names published by Loudon have continued to be utilized in the recent classifications of Little & Critchfield (1969) and Van der Burgh (1973). Two other important early classifications of the genus are those of Spach (1842), who divided the genus into four sections (*Cembra, Eupitys, Strobus* and *Taeda*), and Endlicher (1847), who recognized the four sections of Spach (replacing the name *Eupitys* with *Pinaster*) plus two new sections (*Pseudostrobus* and *Pinea*) and five additional sections corresponding to other genera of Pinaceae.

Modern synthetic classifications of *Pinus* largely date to the influential treatment of Engelmann (1880), who utilized characters from many parts of the plant, including the number of needles per fascicle, the position of resin canals in the needle cross-section, degree of denticulation of the needle, persistence or early abscission of the fascicle sheath, degree of development and persistence of the seed wing, form and position on the stem of the ovulate cones, and the detailed morphology of the cone-scale apex. Engelmann recognized two sections within the genus: *Strobus*, divided into subsections *Strobus* and *Cembrae*, and *Pinaster*, divided into six subsections and 16 species groups.

Mayr (1890) presented a synthetic classification which, in addition to the types of features utilized by Engelmann, also considered new characters from wood anatomy which have subsequently also been utilized in many of the more recent treatments of the genus (see e.g. Shaw 1914; Hudson 1960; Van der Burgh 1973). Mayr recognized ten sections within the genus, including the sections recognized by Endlicher with the exception of section *Pinea*, plus the new sections *Balfouria, Banksia, Khasia, Parrya* and *Sula*.

Koehne (1893) was apparently the first worker to divide the genus into two major groups on the basis of the presence of a single fibrovascular bundle in the leaf cross-section (sect. *Haploxylon*, the 'soft pines', equivalent to the current subgenus *Strobus*) versus two fibrovascular bundles in the leaf cross-section (sect. *Diploxylon*, the 'hard pines', equivalent to the current subgenus *Pinus*). In contrast, the earlier classification of Engelmann included the bristlecone, pinyon, and lacebark pines along with the diploxylon pines in section *Pinaster*. Many subsequent classifications have recognized the haploxylon pines and diploxylon pines as the two major subgroups in the genus, beginning with Lemmon (1895), who raised them to the rank of subgenus as subgenus *Strobus* and subgenus *Pinaster*, which has been replaced by subgenus *Pinus* under the autonym rule of the Code of Botanical Nomenclature.

2.3.2 Twentieth century classifications

Many classifications of pines have been proposed since 1900, of which four of the most influential treatments covering the entire genus (Shaw 1914; Pilger 1926; Little & Critchfield 1969; Van der Burgh 1973) are discussed in detail in this section. Their sectional and subsectional treatments of the species of diploxylon pines are compared in Table 2.2, with the species names brought into agreement with those of Appendix 2.1 for ease of comparison. Among other important treatments, Hudson (1960) attempted to correlate morphological characters of pine species with detailed features of their wood anatomy, using the framework of Shaw's classification, and Klaus (1980, 1989) has presented a number of new suggestions concerning the subsectional classification of the genus based on detailed study of cone scale morphology along with other morphological characters. Gaussen (1960) has also reviewed morphology and classification of the genus, and proposed a classification of four subgenera placing particular emphasis on seedling anatomy (see also Ferré 1953, 1965) and size of pollen grains, and introducing a number of new species groups that were not formally published. A considerably more unorthodox classification of the genus has been proposed by Murray (1983; see also Landry 1994), in which both the haploxylon and diploxylon pines are divided into multiple subgenera. This runs counter to the results of recent molecular phylogenetic analyses (see section 2.7), which indicate that the primary subdivision of the genus is between the haploxylon and diploxylon pines. Murray also published a large number of names for species groups at the ranks of series and subseries, which will be available to use for groups below the rank of subsection.

A considerable amount of information on the complex groups of pine species native to Mexico is presented by Martínez (1948), who recognized 39 species in nine sections (some of which were only informally named) and by Perry (1991), who recognized 51 species in ten sections, again using unpublished names for several sectional or subsectional groups. The number of species recognized as distinct among the Mexican pines has varied considerably among different authors (see e.g. Shaw (1909, 1914) who recognized 23 species, and Little & Critchfield (1969) who recognized 33 species). The pine flora of this region is clearly in need of further detailed study utilizing both morphological and molecular approaches. Species delimitation among the pinyon pines (Passini, Cibrian & Eguiluz Piedra 1988) and in the '*Pseudostrobus*' complex (Stead & Styles 1984; Perry 1991) have been particularly controversial. A revised morphological treatment of the Mexican and Central American species was recently completed (Farjon & Styles 1997), and molecular comparisons using the ITS region are in progress (A. Liston, E. Alvarez-Buylla *et al.*, unpublished data).

Table 2.2. Classifications of subgenus *Pinus* according to Shaw (1914, 1924), Pilger (1926), Little & Critchfield (1969) and Van der Burgh (1973). Groupings from the chloroplast DNA restriction site comparisons of Kruplin *et al.* (1996) are indicated by Roman numerals as shown in Fig. 2.1. The names of species used in this table follows the nomenclature and spelling in Appendix 2.1 in cases where synonyms or alternate spellings were used in the original publications

Shaw		Pilger		Little & Critchfield		Van der Burgh	
Subsection *Parapinaster*		Section *Sula*		Section *Pinea*		Section *Leiophylla*	
Group *Leiophyllae*		roxburghii		Subsection *Leiophyllae*		Subsection *Leiophyllae*	
leiophylla	VIII	canariensis	II	leiophylla	VII	chihuahuana	
lumholtzii	VIII			lumholtzii	VII	leiophylla	VIII
Group *Longifoliae*		Section *Eupitys*		Subsection *Canarienses*			
roxburghii		densiflora		canariensis	II	Section *Sula*	
canariensis	II	heldreichii		roxburghii		Subsection *Canariensis*	
Group *Pineae*		luchuensis		Subsection *Pineae*		canariensis	II
pinea	I	massoniana		pinea	I	heldreichii	
		merkusii				roxburghii	
Subsection *Pinaster*		mugo		Section *Pinus*			
Group *Lariciones*		nigra		Subsection *Sylvestres*		Section *Lumholtzii*	
densiflora		pinaster		brutia		Subsection *Lumholtziae*	
kesiya		resinosa	III	densiflora		lumholtzii	VIII
luchuensis		sylvestris	III	halepensis			
massoniana		tabuliformis		heldreichii		Section *Pinea*	
merkusii		thunbergii		hwangshanensis		Subsection *Pineae*	
mugo				kesiya		pinea	I
nigra		Section *Banksia*		luchuensis		Subsection *Halepenses*	
resinosa	III	banksiana		massoniana		halepensis	I
sylvestris	III	clausa		merkusii			
tabuliformis		contorta	IV	mugo		Section *Pinus*	
thunbergii		echinata		nigra		Subsection *Sylvestres*	
tropicalis		glabra		pinaster		densiflora	
Group *Australes*		halepensis		resinosa	III	kesiya	
caribaea		muricata		sylvestris	III	luchuensis	
echinata		pungens		tabuliformis		massoniana	
glabra		virginiana	IV	taiwanensis		merkusii	
lawsonii				thunbergii		mugo	
montezumae		Section *Pinea*		tropicalis		nigra	
occidentalis		pinea		yunnanensis		resinosa	III
palustris				Subsection *Australes*		sylvestris	III
ponderosa	V	Section *Australes*		caribaea		tabuliformis	
pseudostrobus		caribaea		cubensis		taiwanensis	
taeda	VII	lawsonii		echinata		thunbergii	
teocote	VIII	occidentalis		elliottii		tropicalis	
Group *Insignes*		oocarpa	VIII	glabra		uncinata	
attenuata		palustris		occidentalis			
banksiana		pringlei		palustris		Section *Pinaster*	
clausa				pungens		Subsection *Pseudostrobi*	
contorta	IV	Section *Khasia*		rigida		cooperi	
greggii		kesiya		serotina		hartwegii	
halepensis	I			taeda	VII	montezumae	
muricata		Section *Pseudostrobus*		Subsection *Ponderosae*		pseudostrobus	
oocarpa	VIII	engelmannii		cooperi		Subsection *Oocarpae*	
patula	VIII	jeffreyi	V	douglasiana		lawsonii	
pinaster		leiophylla	VIII	durangensis		oocarpa	VIII
pringlei		lumholtzii	VIII	engelmannii		pringlei	
pungens		montezumae		hartwegii		teocote	VIII
radiata	VI	ponderosa	V	jeffreyi	V	Subsection *Australes*	
rigida		pseudostrobus		lawsonii		caribaea	
serotina		teocote	VIII	michoacana		clausa	
virginiana	IV	torreyana	V	montezumae		cubensis	
Group *Macrocarpae*				ponderosa	V	echinata	
coulteri	V	Section *Taeda*		pseudostrobus		elliotti	
sabiniana		attenuata		teocote	VIII	glabra	
torreyana		coulteri	V	washoensis		occidentalis	
		greggii		Subsection *Sabiniana*		palustris	
		patula	VIII	coulteri	V	pinaster	
		radiata	VI	sabiniana		pungens	
		rigida		torreyana		rigida	
		sabiniana		Subsection *Contortae*		serotina	
		serotina		banksiana		taeda	VII
		taeda	VII	clausa		virginiana	IV
				contorta	IV	Subsection *Contortae*	
				virginiana	IV	banksiana	
				Subsection *Oocarpae*		contorta	IV
				attenuata		Subsection *Attenuatae*	
				greggii		attenuata	
				muricata		greggii	
				oocarpa	VIII	muricata	
				patula	VIII	patula	VIII
				pringlei		radiata	VI
				radiata	VI	Subsection *Ponderosae*	
						engelmannii	
						jeffreyi	V
						ponderosa	V
						Subsection *Sabinianae*	
						coulteri	V
						sabiniana	
						Subsection *Torreyanae*	
						torreyana	V

Building upon the classifications of Engelmann, Mayr and Koehne, Shaw (1914) provided an important synthetic classification utilizing characters from throughout the plant body, and attempted to utilize evolutionary trends among the characters. Shaw delimited species in a very conservative fashion, recognizing only 66 species worldwide, some of which are now widely recognized as comprising more than one distinct species. He followed Koehne in dividing the genus into two sections, *Haploxylon* and *Diploxylon*, each composed of two subsections, and used 'groups' as the next major unit of classification below subsection. Shaw divided the haploxylon pines into subsections *Cembra* (with the umbo of the cone scale terminal), and *Paracembra* (with the umbo in a dorsal position on the cone-scale as in the diploxylon pines), and then divided each subsection into three groups, primarily on the basis of characters related to seed dispersal. In his 1914 classification he divided subsection *Cembra* into groups *Cembrae* (stone pines, with indehiscent cones and lacking seed wings), *Flexiles* (white pines with dehiscent cones but only rudimentary seed wings), and *Strobi* (white pines with a well-developed seed wing). He subsequently merged the latter two groups under *Strobi* (Shaw 1924). Subsection *Paracembra* was divided into group *Cembroides* (pinyon pines, with wingless seeds), group *Gerardianae* (the Asian *P. bungeana* and *P. gerardiana*, with very reduced seed wings), and group *Balfourianae* (the bristlecone and foxtail pines, *P. aristata* and *P. balfouriana*, with well-developed seed wings). Shaw (1924) also placed *Pinus krempfii*, described from the highlands of Vietnam in 1921, in group *Balfourianae*. This species of haploxylon pine is unique in the genus in having highly flattened needles which can range up to 4 mm in width (see photograph in Mirov 1967, p. 542).

Shaw (1914) included most diploxylon pine species, which possess both well-developed articulate seed wings and persistent fascicle sheaths, in subsection *Pinaster*. He treated, in subsection *Parapinaster*, a rather disparate group of diploxylon pines which have either a deciduous fascicle sheath (*P. leiophylla* and *P. lumholtzii*), seed wings strongly adnate to the seed body (*P. canariensis* and *P. roxburghii*), or a very reduced seed wing (*P. pinea*). Within subsection *Pinaster*, Shaw recognized group *Lariciones* (equivalent to subsection *Pinus* in the current classification) with large pits on the ray cell walls, and three groups with small pits on the ray cell walls: a very broadly circumscribed group *Australes* with cones dehiscent and abscising from the shoot at maturity, a small group *Macrocarpae* with much thickened cone scales and seed wings (*P. coulteri*, *P. sabiniana*, and *P. torreyana* from western North America), and a large and diverse group *Insignes* with cones persistent on the stems at maturity and sometimes asymmetrical and/or serotinous.

Pilger (1926) provided a rather different classification of the pines in his ground-breaking treatment of the conifer families for Engler's *Die Natürlichen Pflanzenfamilien*. Following the classifications of Mayr and Koehne, he used sections as the primary taxonomic groupings under the subgenera of haploxylon and diploxylon pines (Table 2.2). He divided the haploxylon pines into three sections: *Cembra* (including the stone pines and the additional species of white pines with poorly developed seed wings which were placed in group *Flexiles* by Shaw (1914)), *Strobus* (including the remaining white pines), and *Paracembra* (including *P. bungeana* and *P. gerardiana* and the pinyon pines as two groups under subsection *Gerardianae*, and the foxtail and bristlecone pines, and *Pinus krempfii*, respectively, as two groups under subsection *Balfourianae*).

Within the diploxylon subgenus, Pilger recognized eight sections which only partially correspond in circumscription to the groups of Shaw. In particular, Pilger emphasized needle cross-sectional anatomy and number of needles per fascicle to a much greater degree than Shaw. Pilger included essentially the same group of species in section *Eupitys* that Shaw placed in group *Lariciones*, and treated the species of Shaw's groups *Longifolieae* and *Pineae* as distinct sections, but treated the remaining groups in rather different ways. *Pinus leiophylla* and *P. lumholtzii* were included in section *Pseudostrobus* along with a number of other western North American or Mexican species with uninodal spring-shoots (comprising only a single internode marked at the apex by a cluster of buds), which appears to represent a more appropriate placement than that of Shaw for these two Mexican species (see chloroplast DNA, p. 61). Pilger placed a rather disparate group of species from eastern and western North America (plus *P. halepensis* from Europe) with usually two needles per fascicle, multinodal spring-shoots and often persistent cones in section *Banksia*, while placing some rather similar species with usually three to five needles per fascicle in sections *Australes*, *Pseudostrobus*, or *Taeda* (see Table 2.2). The bigcone pines of Shaw's group *Macrocarpae* were split between sections *Pseudostrobus* and *Taeda* on the basis of having spring-shoots uninodal versus multinodal, and the closely related California closed-cone pines *P. muricata* and *P. radiata* were divided between sections *Banksia* and *Taeda* largely on the basis of needle number per fascicle.

Little & Critchfield (1969) used information from both morphology and crossability studies (see section 2.5 below), and rigorously applied the rules of the Code of Botanical Nomenclature in providing a revised classification of the genus which has been used as the starting point for the classification given in Appendix 2.1. In contrast to Shaw and Pilger, subsections were used as the main unit of classification for groups of species. In addition to the major groups of haploxylon and diploxylon pines placed in subgenera *Strobus* and *Pinus*, respectively, the haploxylon pine *P. krempfii* was placed in a monotypic subgenus *Ducampopinus* (A. Cheval.) de Ferré on the basis of its unusual flattened needles and the putative absence of ray tracheids from the wood. The classification within subgenus *Strobus* is similar to that of Shaw (1924) in that it recognizes two major groups: section *Strobus* (including the stone pines as subsection *Cembrae* and the white pines as subsection *Strobi*) and section *Parrya* (including subsections *Cembroides*, *Gerardianae* and *Balfourianae* with the same circumscriptions as the corresponding groups of Shaw).

Within the diploxylon pines, Little and Critchfield recognized the same three groups treated in Shaw's subsection *Parapinaster* as subsections under section *Pinea*, and placed all other groups under section *Pinus*, equivalent to Shaw's subsection *Pinaster*. Subsection *Sylvestres* in Little & Critchfield's classification (= subsect. *Pinus*) corresponds very closely to Shaw's group *Lariciones* and Pilger's section *Eupitys*, and subsection *Sabinianae* corresponds to Shaw's group *Macrocarpae*, but the circumscriptions of the other subsections of section *Pinus* differ considerably from those of Shaw or Pilger. Subsection *Australes* is redefined as including a group of 11 species from the eastern USA and the Caribbean with multinodal spring shoots, two to three leaves per fascicle, and cones usually opening at maturity. As discussed below, information on interspecific crossability from Duffield (1952) and more recent studies also formed an important basis for Little and Critchfield's circumscription of the section. The serotinous-coned species from Shaw's group *Insignes* were divided between subsection *Australes*, subsection *Oocarpae* (including the California closed-cone pines plus four additional species from Mexico and Central America), and subsection *Contortae* (a group of four North American species with two needles per fascicle and relatively small cones). A large part of the group of western North American and Mexican species treated by Pilger under section *Pseudostrobus* is treated by Little and Critchfield as a broadly defined subsection *Ponderosae*.

The classification of Van der Burgh (1973), also presented in slightly modified form in an appendix in Farjon (1984), differs significantly in structure from that of Little and Critchfield (1969) in that it does not divide the pines into subgenera, and instead recognizes eight sections as the major units within the genus. Van der Burgh placed greater emphasis on characters from wood anatomy than did Little and Critchfield, and recognized somewhat narrower groups as subsections. The classification of Van der Burgh is relatively similar to most other classifications in recognizing two sections of haploxylon pines: section *Strobus*, with subsections *Strobi* and *Cembrae*, and section *Parrya*, with subsections *Balfourianae* (including *P. krempfii*), *Gerardianae*, and *Cembroides*, and two new subsections

Aristatae (including only *P. aristata*) and *Nelsoniae* (including only *P. nelsonii*). Among the diploxylon pines, it recognizes six sections, of which four are small groups: section *Leiophylla* (including *P. leiophylla* and its segregate *P. chihuahuana*), section *Sula* (including *P. canariensis*, *P. leucodermis* (included in *P. heldreichii* in the current treatment) and *P. roxburghii*), section *Lumholtzii* (including only *P. lumholtzii*), and section *Pinea* (including *P. pinea* in a monotypic subsection, and *P. halepensis* in the new subsection *Halepenses*). Section *Pinus* is restricted to the members of subsection *Sylvestres* (= subsect. *Pinus*), while the remaining species of diploxylon pines are placed in section *Pinaster*, including one southern European species (*P. pinaster*) and most of the North or Central American species of diploxylon pines. Eight subsections are recognized within section *Pinaster*, including subsection *Oocarpae* (in a more restricted delimitation than that of Little and Critchfield, with the California closed-cone pines and two additional Mexican species segregated as the new subsection *Attenuatae*), *Australes*, *Contortae* (restricted to only *P. banksiana* and *P. contorta*), and *Ponderosae* (in a much more restricted circumscription than that of Little and Critchfield, with most Mexican species moved to subsection *Pseudostrobi*), and *Sabinianae* (with *P. torreyana* moved to a monotypic subsection *Torreyanae*).

Klaus (1989) proposed several changes to the classification of Van der Burgh on the basis of detailed studies of cone-scale morphology and other morphological characters, some of which are followed in the treatment for this chapter. The unusual Mexican haploxylon pine *P. rzedowskii* was moved into the monotypic subsection *Rzedowskiae* on the basis of having a longer seed wing than those of the pinyon pines. The Mediterranean species *P. brutia* was included with *P. halepensis* in subsection *Halepenses*, which accords well with the results of allozyme comparisons (Conkle, Schiller & Grunwald 1988). Klaus suggested that the Cuban endemic *P. tropicalis* has been misplaced in the almost entirely Eurasian subsection *Pinus* (instead placing it in the North American subsection *Oocarpae*) and also moved the only other New World species of this section (*P. resinosa*) to a new monotypic subsection *Resinosae*.

Some further modifications of the classification of the diploxylon pines were made by Rushforth (1987) in a synoptical table of the genus without discussion of the actual supporting data. Rushforth recognized four sections of diploxylon pines: section *Pinaster* (including *P. brutia*, *P. canariensis*, *P. halepensis*, *P. pinaster*, *P. pinea*, and *P. roxburghii*), section *Leiophyllae* (including only *P. leiophylla* and *P. lumholtzii*), section *Taeda* (including all other North American species except *P. resinosa* and *P. tropicalis*), and section *Pinus* (including all species of subsection *Pinus* except *P. pinaster*). Rushforth's separation of most of the

Mediterranean diploxylon pines and the great majority of the North American diploxylon pines as sections distinct from section *Pinus* is consistent with preliminary data from chloroplast DNA restriction sites (Krupkin *et al.* 1996) and needs to be critically evaluated using a broader sampling of the Mediterranean species. Rushforth's placement of *P. pinaster* in a section with the other Mediterranean hard pines appears to be supported by reports of natural and artificial hybridization of the species with *P. halepensis*, as summarized by Vidaković (1991).

2.4 Morphological characters important to pine classification

Delimitation of species and higher-order groups within *Pinus* has traditionally been primarily based upon certain foliar (needle and needle fascicle) and reproductive (ovulate cone and seed) characters (see treatments in recent floras by Carvajal & McVaugh 1992; Gaussen, Heywood & Chater 1993; Kral 1993; Farjon & Styles 1997; and reviews in Mirov 1967; Farjon 1984; Price 1989).

The number of needles per fascicle is nearly constant within many species of pines, and has frequently been used as a diagnostic character for various sectional or subsectional groups in the genus. On the other hand, some species fairly regularly exhibit variation in the number of needles per fascicle within and among individuals (e.g. two or three needles in *P. echinata* and *P. ponderosa*). Most needle numbers are in the Fibonacci series of one (found only in the pinyon pine *P. monophylla*), two, three, and five (and rarely up to eight, which is infrequently found in the Mexican species *P. durangensis*). Among the haploxylon pines, the number of needles per fascicle is always five in section *Strobus*, but ranges from one to five in section *Parrya* (where it is five in subsection *Balfourianae*, three in subsection *Gerardianae*, two in subsection *Krempfianae*, and one to five in subsection *Cembroides*). Among the diploxylon pines, the number of needles per fascicle is usually two in the Eurasian species (but three in subsection *Canarienses*), two or three in the eastern North American species, and very diverse among and often within the species of western North and Central America, where numbers range from two to eight. Characters of the basal sheath of the fascicle (arising from surrounding bud scales) and the bract subtending the needle cluster have also been important in defining higher-level groups in the genus. The scales of the fascicle sheath abscise as the needles elongate in most species of subgenus *Strobus* (with the exception of *P. nelsonii*) or are retained as a persistent basal sheath in almost all species of subgenus *Pinus* (except for the two Mexican species of subsection *Leiophyllae*). The bracts that subtend the fascicles are decurrent on the stem in subgenus *Pinus*, but not in subgenus *Strobus*, although a somewhat intermediate condition is found in the pinyon pines (Carvajal & McVaugh 1992). Another intriguing character reported by Klaus (1989) is that young plants continue to produce primary leaves rather than needle fascicles at the nodes for an extended period in a number of putatively basal groups in the genus (subsections *Canarienses*, *Halepenses* and *Pineae* among the diploxylon pines and a number of species in section *Parrya* among the haploxylon pines).

Needles can vary substantially in length among and within species, ranging from as little as 2–4 cm in *P. banksiana* and *P. contorta* to as much as 40–45 cm in the longleaf pine, *P. palustris*. Internal and external anatomical characters of the needles can also be very useful in separation of species, and in documentation of interspecific hybridization (see Keng & Little 1961; Little & Righter 1965). The number and position of resin canals in the needle cross-section can vary considerably among species of pines and has often been used in classification (see e.g. Doi & Morikawa 1929; Harlow 1931; Martínez 1948; Jährig 1962). In certain species, however, phenotypic variation in these characters is associated with needle age and environmental factors (see Helmers 1943; White & Beals 1963). The number of layers of cells in the hypodermis of the leaf can also vary significantly among species. In some species the cells are relatively uniform in wall thickness, while others have thicker walls on the inner cell layers. The morphology of the stomatal complexes and patterns of wax deposition on the needles have been surveyed for 51 species of pines by Yoshie & Sakai (1985), who found that the epicuticular wax is restricted to the area of the epistomatal chambers in diploxylon pines and more generally distributed in the stomatal bands in haploxylon pines, and that the stomatal complexes arise abruptly from a much more pronounced depressed band in the latter subgenus.

Comparative wood anatomy of pines has been discussed by several authors (see e.g. Phillips 1941; Greguss 1955; Hudson 1960; Van der Burgh 1973), and features of the cell walls of the ray tracheids and ray parenchyma cells have been utilized as a source of data in many recent higher-order classifications of the genus. The inner walls of the ray tracheids in most species of subgenus *Strobus* are relatively smooth, while those of subgenus *Pinus* usually have evident projections on the inner walls ('dentation') which in some species merge to form a reticulum. The degree of dentation can vary substantially among species and has been scored on a scale of 1 to 14 by Hudson. The greatest depth of dentation among the haploxylon pines is found in a few members of the putatively basal section *Parrya* such as *Pinus nelsonii* and *P. bungeana* (with index scores of 10 and 6, respectively; Hudson 1960), while the smallest amount of dentation in subgenus *Pinus* is found in Old World species such as *Pinus canariensis*, *P. pinea*, and *P. merkusii* (with index scores of 5–6) which also are relatively basal in their subgenus on the basis of molecular comparisons (Wang & Szmidt 1993; Krupkin *et al.* 1996). There is a general tendency to have relatively large fenestriform crossfield pits in section *Strobus* and in subsection *Pinus* of subgenus *Pinus*. Other groups of pines have smaller 'pinoid' crossfield pits similar to those of the other genera of the family (Hudson 1960).

Ovulate cones of pines vary substantially in size and shape among species, ranging from as little as 3–4 cm in length in some diploxylon pine species to as long as 40–50 cm in the white pines *P. lambertiana* and *P. ayacahuite* (Table 1.1, p. 5). Features of the cones related to syndromes of seed dispersal (see Chap. 14, this volume) have frequently been used in sectional classifications of the genus, and details of the morphology of the umbo and apophysis of the cone scale are often very useful in species delimitation in the genus (see e.g. Klaus 1980, 1989). Cone scales in most species of pines become highly woody at maturity, but the degree of lignification is reduced and the cone scales are relatively flexible in section *Strobus* and some species of pinyon pines (Van der Burgh 1973). In addition, the umbo of the cone scale is in an abaxial ('dorsal') position and is often terminated by a spine or prickle in pines of subgenus *Pinus* and section *Parrya* of subgenus *Strobus*, while in section *Strobus* the umbo is in a terminal position on the scale and lacks a prickle. Indehiscent cones from which the cone scales can readily be detached by specialized seed-eating birds constitute an unusual shared-derived feature for the nut pines of subsection *Cembrae* (see Chap. 14, this volume). Another unusual syndrome of features is the production of long persistent serotinous cones in species from several North American groups of diploxylon pines (see McCune 1988), and there has been considerable controversy as to how many origins this condition has had within the subgenus. There is a fairly strong association between presence of cone serotiny in a species and marked asymmetry in cone shape, and these features have been heavily emphasized in some classifications.

Size of the seed body and the degree of development of the seed wing also vary substantially among groups and species of pines, and are associated with changes from wind dispersal to animal dispersal (see Chap. 14, this volume). Well-developed seed wings are present throughout the other genera of the family and thus loss or reduction of the wings appear to be derived states within the genus. Seed wings are well-developed but easily detached ('articulate') in most species of subgenus *Pinus*. However, the seed wings are reduced in length and detachable in the Mediterranean stone pine (*P. pinea*) and in *P. sabiniana* and *P. torreyana* from western North America, while the seed wings are adnate to the seed body in the two species of subsection *Canarienses* (Shaw 1914; Van der Burgh 1973). In subgenus *Strobus* the seed wings are adnate to the seed body in the white pines of subsection *Strobi*, where they are well-developed in several species or very reduced in other species such as *P. armandii* and *P. flexilis* (see Chap. 14, this volume). The seed wings have been almost lost in subsection *Cembrae* and the pinyon pines of subsection *Cembroides*, apparently representing two convergent losses in the history of the subgenus (see Strauss & Doerksen 1990). The seed wings are very reduced and detachable in subsection *Gerardianae* (Shaw 1914) and well-developed in subsections *Balfourianae* and *Krempfianae*, where the wings have been variously described as detachable or adnate (cf. Shaw 1914; Bailey 1970; Van der Burgh 1973) and are in need of further study.

The morphological characters described above should be quite useful in phylogenetic analysis of the genus, particularly in combination with molecular data sets. Several authors have discussed possible evolutionary trends in the genus, usually without conducting outgroup comparisons as a basis for character polarization (see e.g. Shaw 1914; Gaussen 1960; Klaus 1980; Farjon 1984). The only explicit phylogenetic analysis using morphological data that has been published to date for a group within the genus is that of Malusa (1992) for subsection *Cembroides*. A variety of morphological characters useful for comparisons within *Pinus* have also been included in phylogenetic analyses at higher taxonomic levels conducted by Hart (1987) and Nixon *et al.* (1994).

2.5 Contributions from cytology and crossability studies

A number of cases of natural hybridization among pine species have been documented (see Chap. 13, this volume), almost always within the subsectional groups of Little & Critchfield (1969). All of the approximately 90 pine species studied to date are diploids with a chromosome number of $2n=24$ (see Saylor 1964, 1972, 1983) and the genus is relatively conservative in its karyotype. Most species have 11 more or less isobrachial chromosomes and one chromosome with a submedian centromere, while members of subsection *Pinus* (including the North American species *P. resinosa* and *P. tropicalis*) also have a second heterobrachial chromosome (Saylor 1964), as do *P. brutia* and *P. halepensis* which we place in subsection *Halepenses* on the basis of chloroplast DNA comparisons (Krupkin *et al.* 1996). The three species of the Old World subsections *Canarienses* and *Pineae* have only one chromosome with a submedian centromere, and are very similar to one another in detailed karyotype (Saylor 1972). Chromosome sizes and arm-length ratios have been studied for most species of North American diploxylon pines by Saylor (1972) and compared with the subsectional classification of Little and Critchfield. Some caution is needed in interpreting the results, since geographic sampling within species was quite limited and subspecies sometimes differed in details of the karyotype. The results of the relative arm-length comparisons are generally consistent with Little and Critchfield's subsectional classification, but with some notable exceptions that appear to correlate with crossability groups as discussed below.

The detailed karyotypic patterns are consistent with Little and Critchfield's circumscription of subsection *Contortae* rather than that of Van der Burgh (1973), since *P. virginiana* and *P. clausa* are more similar in karyotype to *P. banksiana* and *P. contorta* than to any of the members of subsection *Australes*. Within Little and Critchfield's subsection *Oocarpae*, the three species of California closed-cone pines stand out as being very similar to one another and much less similar to the Mexican taxa *P. greggii*, *P. patula*, *P. pringlei* and *P. oocarpa*. These results, which are consistent with data from crossability and chloroplast DNA restriction site patterns (Krupkin *et al.* 1996) help support placement of the Californian species in a separate subsection *Attenuatae*. Within Little and Critchfield's subsection *Ponderosae*, *P. lawsonii* and *P. teocote* have detailed karyotypic patterns quite different from the other species, which is again consistent with crossability data and chloroplast DNA restriction site comparisons supporting their removal from the group. The karyotypic data tend to favour Little and Critchfield's broad circumscription of the remainder of subsection *Ponderosae*, since several Mexican species of the *Montezumae* group are quite similar in pattern to *P. ponderosa*. *Pinus hartwegii* of the *Montezumae* group and *P. pseudostrobus* and relatives (the *Pseudostrobus* group of Martínez 1948) are somewhat more distinct in karyotype, but are apparently connected to the *Montezumae* group by natural hybridization (Perry 1991; Matos 1995; see discussion below).

Because of the economic importance of pines and the potential utility of hybrid derivatives in programmes of timber-crop improvement, a substantial number of studies of artificial crossability among species have been undertaken. The studies at the Institute of Forest Genetics at Placerville, California (see e.g. Duffield 1952; Critchfield 1963, 1975, 1986) have been particularly extensive and well documented. Although interspecific crossability is a retained primitive state rather than a shared-derived character state, most species of pines appear to be partially or completely reproductively isolated from one another by an accumulation of genetically based barriers (see Chap. 13, this volume). Thus the ability of two species of pines to hybridize relatively freely is generally interpreted as an indication of close relationship.

Artificial crossability was used by Duffield (1952) to help assess relationships within groups of diploxylon pines that were classified in very different ways by Shaw (1914) and Pilger (1926). Duffield's results indicated that the North American species placed by Shaw in group *Australes* form two major crossability groups: a western group of species (including such taxa as *coulteri*, *jeffreyi* and *ponderosa*) and an eastern group of species (including such taxa as *echinata*, *palustris* and *taeda*). Several crosses within these groups yielded successful hybrids, while the two groups were found to be strongly genetically isolated from one another. These and additional studies (see Critchfield 1963; Little & Righter 1965) have shown that crosses can replicably be made in at least some combinations among a broader group of eastern North American species variously placed by Shaw in groups *Australes* and *Insignes*, and by Pilger in sections *Australes*, *Banksia*, and *Taeda*, leading to the revised delimitation of subsection *Australes* used by Little & Critchfield (1969) and followed here.

The placement of the eastern and western groups of 'closed-cone pines' has varied considerably among classifications, and crossability data have allowed their separation into four groups. The closely related and possibly conspecific eastern North American taxa *P. serotina* and *P. rigida* are connected by a number of successful crosses with the other eastern North American species of subsection *Australes*. By contrast, the eastern *P. banksiana* and the western *P. contorta* of subsection *Contortae* are linked by both natural and artificial hybridization, and are strongly isolated from the other species in the genus. The eastern North American species *P. clausa* and *P. virginiana* of subsection *Contortae* can be crossed relatively readily with each other, but are largely isolated from other species (crosses between *P. clausa* and *P. taeda* and between *P. clausa* and *P. elliottii* have yielded small numbers of hybrid seedlings; Critchfield 1963; Saylor & Koenig 1967). Evidence from comparisons of allozymes (Wheeler, Guries & O'Malley 1983; J. Hamrick, personal communication) and chloroplast DNA restriction site patterns (Krupkin *et al.* 1996) discussed below favour the inclusion of *P. clausa* and *P. virginiana* in subsection *Contortae*, rather than subsection *Australes*, where they were placed by Van der Burgh (1973). The three Pacific coastal species of closed-cone pines (*P. attenuata*, *P. muricata* and *P. radiata*), treated here as subsection *Attenuatae*, are linked to one another by successful crosses (although the two 'races' within *P. muricata* are strongly incompatible; see Critchfield 1967). The California closed-cone pines are crossable only with great difficulty with other taxa, including the Mexican group of species with which they were grouped as subsection *Oocarpae* by Little & Critchfield (1969), which is in keeping with their differences in karyotype. In contrast, three Mexican species (*P. greggii*, *P. oocarpa* and *P. patula*) from subsection *Oocarpae* have been found to be readily crossable with one another (Critchfield 1967), which does not provide support for Van der Burgh's (1973) placement of the first and third of these species in subsection *Attenuatae* with the California closed-cone pines.

Crossability studies or observations of apparent natural hybridization also are useful in assessing the delimitation of subsection *Ponderosae*. The Mexican species *P. montezumae* (placed by Van der Burgh in subsection *Pseudostrobi*) and the more northern species *P. engelmannii*, *P. jeffreyi* and *P. ponderosa* (placed by Van der Burgh in subsection *Ponderosae*) are crossable in a number of different combinations (Little & Righter 1965), which tends to support Little & Critchfield's inclusion of *P. montezumae* and relatives in a broader subsection *Ponderosae*. *Pinus montezumae* appears to provide a connection in terms of crossability and morphology between the *P. ponderosa* group and the 'Montezumae' and 'Pseudostrobus' groups of Martínez (1948). Natural hybridization between *P. montezumae* and *P. hartwegii* is indicated by studies of chloroplast DNA and morphology (Matos 1992, 1995). Perry (1991) also notes cases of apparent natural hybridization between *P. montezumae* and *P. devoniana* (of the *Montezumae* group) and *P. pseudostrobus*. Both natural and artificial hybrids have also been documented between *P. coulteri* of subsection *Sabinianae* and *P. jeffreyi* of subsection *Ponderosae* (Zobel 1951; Little & Righter 1965), and suggest a closer phylogenetic relationship between these groups than has been indicated by the classification of Little and Critchfield (see also Krupkin *et al.* 1996). In contrast, *P. lawsonii* and *P. teocote* are crossable with *P. patula* of subsection *Oocarpae* and appear to fall into a separate crossability group from subsection *Ponderosae* (Critchfield 1967).

In some additional cases artificial crossability data have been helpful in supporting morphologically based relationships among disjunct taxa from different continents. The ability of the eastern North American red pine (*P. resinosa*) to be crossed with the European black pine (*P. nigra*) adds support to the placement of the red pine in the almost completely Eurasian subsection *Pinus* (Critchfield 1964). Crossability between the Canary Island endemic *P. canariensis* and the Himalayan *P. roxburghii* (see Keng & Little 1961), helps support their placement in subsection *Canarienses*. Within subgenus *Strobus*, a number of species of white pines (subsection *Strobi*) from both Eurasia and North America have been crossed successfully, while only limited crossability has been found between the putatively closely related subsections *Cembrae* and *Strobi* (Critchfield 1986; Wright 1959).

In summary, a significant number of species within subsections in the classification of Little & Critchfield (1969) can be crossed to yield fertile hybrids, while crosses between members of different subsections only rarely yield viable seedlings. Exceptions to this pattern such as the crossability of the bigcone pines from subsection *Sabinianae* with members of subsection *Ponderosae*, and putative hybridization between *P. caribaea* of subsection *Australes* with *P. oocarpa* (Styles, Stead & Rolph 1982), appear to indicate relatively close phylogenetic relationships between these groups (see also Krupkin *et al.* 1996).

2.6 Contributions from secondary product chemistry and protein comparisons

The taxonomic distribution of phenolic compounds in the heartwood of pines has been reviewed by Erdtman (1959) and Norin (1972). Relatively little variation is seen in the phenolic profiles of diploxylon pines, which generally contain the stilbene compound pinosylvin and its methyl ethers as well as the flavanone compounds pinocembrin and pinobanksin. The profiles of heartwood phenolics in subgenus *Strobus* are much more complex, including *C*-methylated flavonoids such as strobopinin and cryptostrobin and flavone derivatives such as chrysin and tectochrysin. Members of subgenus *Strobus* share the compounds that are widespread in subgenus *Pinus*, leading Erdtman to suggest a loss of enzymatic capabilities within the latter subgenus. Erdtman, Kimland & Norin (1966) found that *P. krempfii* has a complex profile of heartwood phenolics very similar to those of many other haploxylon pines, providing no evidence for its placement in a separate subgenus.

Among pine species, oleoresin profiles from the softwood are much more diverse than those of heartwood phenolics (see reviews in Mirov 1961, 1967). Significant differences are often found among related species, and can be used to help distinguish taxa and document cases of interspecific hybridization. For example, *P. jeffreyi* and *P. sabiniana* have a large percentage of the alkane *n*-heptane in the volatile component of their oleoresins, while *P. coulteri* has several volatile terpenoids in addition to *n*-heptane, and *P. ponderosa* lacks *n*-heptane. The composition of the wood oleoresin can also vary considerably among individuals and populations, however, and compounds of relatively restricted distribution can be found in otherwise unrelated species (e.g. in derived taxa in both subgenera), so considerable caution must be used in inferring relationships from oleoresin profiles alone. A useful review of volatile oil profiles from needle extracts is given by Von Rudloff (1975), and detailed studies of leaf or wood volatile oil profiles from a number of species in subsections *Balfourianae* and *Cembroides* are given by Zavarin *et al.* (1982) and Zavarin & Snajberk (1987).

Population distributions of protein allozymes have been surveyed in many individual species of pines, primarily as a means of assessing levels of within- and among-population variation within species (Hamrick & Godt 1990; see also Chap. 13, this volume). Systematic comparisons of allozyme profiles among species can be particularly valuable in assessing relationships of closely related species and distinctness of taxa. Some important examples of cases where allozymes have been used to study groups of species include Wheeler *et al.* (1983) on subsection *Contortae*, Conkle *et al.* (1988) on *Pinus brutia* and relatives, Millar *et al.* (1988) on the California closed-cone pines, and Goncharenko, Padutov & Silin (1992) and Krutovskii, Politov & Altukhov (1994) on subsection *Cembrae*. Karalamangala & Nickrent (1989) conducted a preliminary study of relationships among a number of Mexican species of diploxylon pines from isozyme profiles, suggesting relatively close relationships between subsections *Leiophyllae* and *Oocarpae* and between *P. ponderosa* and some of the Mexican species (e.g. *P. devoniana* and *P. montezumae*) placed by Little & Critchfield (1969) within subsection *Ponderosae* but in a separate subsection *Pseudostrobi* by Van der Burgh (1973). Niebling & Conkle (1990) used allozyme comparisons to assess the relationships of the narrowly endemic western North American species *P. washoensis* to its widely distributed relative *P. ponderosa*, and provided evidence for the apparent derivation of *P. washoensis* from within the latter species. Additional studies where groups of putatively related species are compared using multiple populations per species will be particularly valuable in conjunction with chloroplast DNA-based phylogenies, since the allozymes are biparentally inherited and often give clear additive patterns in cases of interspecific hybridization. A recent example of this type of combined study is the analysis of the putative hybrid origin of the Chinese species *P. densata* (Wang & Szmidt 1990; Wang *et al.* 1990).

Electrophoretic profiles of seed proteins have been compared for a number of Eurasian pine species of haploxylon and diploxylon pines by Schirone *et al.* (1991) and Piovesan *et al.* (1993). A UPGMA cluster diagram of distances derived from presence or absence of particular bands (Schirone *et al.* 1991) is consistent with results from immunological comparisons of seed proteins (Prager, Fowler & Wilson 1976; Price *et al.* 1987) in separating the two subgenera of pines, and is consistent with chloroplast DNA comparisons (Krupkin *et al.* 1996) in indicating a relatively close relationship between the Mediterranean species *P. brutia*, *P. halepensis*, and *P. pinea*. *Pinus pinaster* clustered with the Mediterranean species in the UPGMA analysis, but none of the American species of diploxylon pines, with which it has been grouped by Van der Burgh (1973) and Klaus (1989), were included for comparison. The results of Piovesan *et al.* (1993), who included a slightly greater number of taxa of pines (17 Eurasian species and one North American species, representing seven subsec-

tions) suggest that the subsections *Cembrae*, *Gerardianae* and *Strobi* of subgenus *Strobus* are quite distinct in their seed protein band patterns. Among the diploxylon pines, six European or Asian species of subsection *Pinus* were relatively similar in band pattern, while considerably greater diversity was seen among five species from subsections *Canarienses*, *Halepenses* and *Pineae*.

2.7 Contributions from DNA analysis

Although the chloroplast genomes of pines and other conifers have a low rate of base substitution relative to many groups of flowering plants (see Bousquet *et al.* 1992; Krupkin *et al.* 1996), restriction site and sequence comparisons of chloroplast DNA have proved useful in providing initial phylogenetic frameworks from which to assess the monophyly of the groups recognized in previous classifications. Because the chloroplast genome is usually paternally inherited in conifers (Neale & Sederoff 1989), its history may differ from that of the species in cases where hybridization has occurred. However, based on experience with artificial crosses, this seems to be generally restricted to relatively closely related groups of species (e.g. within subsections). Another important feature of conifer chloroplast genomes is the relatively high rate of major rearrangements such as inversions, which is apparently associated with the loss of one copy of the inverted repeat (see Strauss *et al.* 1988; Lidholm & Gustaffson 1991). These rearrangements make it difficult to conduct comparisons of restriction fragments of the chloroplast genome between the two subgenera of pines, and between pines and other genera for outgroup comparison, which instead can be conducted using sequence comparisons for relatively slowly evolving genes.

Strauss & Doerksen (1990) compared 19 pine species, representing most major groups within both subgenera, using comparisons of restriction fragments from the chloroplast genome and a smaller number from the mitochondrial and nuclear genomes obtained using six-base cutting enzymes. This analysis provided strong evidence for the distinctness of the subgenera *Pinus* and *Strobus* on the basis of midpoint rooting, although the monophyly of the subgenera could not be addressed directly in the absence of outgroup comparisons. *Pinus longaeva* from subsection *Balfourianae* was placed on the most basal branch in subgenus *Strobus*, with representatives of subsections *Cembroides* and *Gerardianae* branching off successively further from the base of the group, and subsections *Cembrae* and *Strobi* grouping together in a poorly resolved terminal lineage. These results accord well with morphological data in that members of section *Parrya* are much more similar to subgenus *Pinus* than are the members of section *Strobus*. Within subgenus *Pinus*, *P. canariensis* and *P. pinea*, both with relatively little dentation on the ray tracheid walls, were placed on a distinct lineage near the base of the subgenus, while *P. leiophylla*, with a deciduous fascicle sheath but also with a much greater amount of dentation on the ray tracheid walls, was placed well within the major terminal clade of North American diploxylon pines. This change in the taxonomic position of *P. leiophylla* is also consistent with results from the isozyme analyses of Karalamangala & Nickrent (1989), who found that the species grouped relatively closely with *P. oocarpa* and *P. pringlei* of subsection *Oocarpae*.

Wang & Szmidt (1993) conducted a comparison of chloroplast DNA restriction fragment patterns for 18 Asian species of pines using six-base cutting enzymes. They also found strong separation between the two subgenera, with the two species included from section *Parrya* (*P. bungeana* and *P. gerardiana*) branching off near the base of subgenus *Strobus*. The one species of subsection *Cembrae* included in the analysis (*P. sibirica*) is nested well within subsection *Strobi*, grouping with *P. parviflora* with a 95% bootstrap value, suggesting that the former subsection is derived from the latter. Bootstrap support for the pairing of these two subsections (i.e. for section *Strobus*) was 100%, as was also the case in the study of Strauss & Doerksen (1990), who used an almost completely different set of species to represent this group. Within subsection *Pinus* of section *Pinus*, *P. merkusii*, a subtropical to tropical species with relatively little dentation on the ray tracheid walls, came out in a basal position, while the more northerly East Asian species *P. hwangshanensis*, *P. kesiya*, *P. tabuliformis*, *P. taiwanensis*, *P. thunbergii* and *P. yunnanensis* formed a relatively tight terminal cluster.

Krupkin *et al.* (1996) conducted a detailed analysis of chloroplast DNA restriction site mutations from 18 species of subgenus *Pinus* using four-base cutting enzymes in addition to six-base cutting enzymes to give a greater degree of resolution. This study included mostly New World species and had almost no overlap in taxa with the study of Wang & Szmidt (1993). The maximum parsimony tree resulting from this analysis is shown in Fig. 2.1. The results suggest that neither section *Pinea* nor section *Pinus* as delimited by Little & Critchfield (1969) are monophyletic groups, since *P. sylvestris* (the type species of section *Pinus*) groups more closely with *P. canariensis* and *P. pinea* of section *Pinea* than with the other subsections of section *Pinus*. Furthermore, *P. brutia*, placed in subsection *Sylvestres* (= subsect. *Pinus*) by Little and Critchfield, is placed in this analysis as an extremely well-supported sister species of *P. pinea*, in agreement with its treatment by Klaus (1989).

A significant feature of the phylogeny is that there is a long and well-supported branch separating the almost

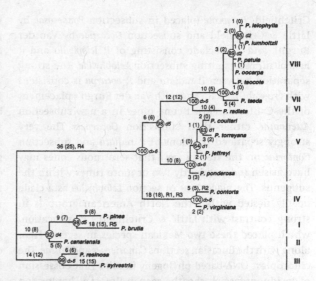

Fig. 2.1. The single minimum-length tree obtained from Wagner parsimony analysis of 204 chloroplast DNA restriction site mutations (reproduced with permission from Krupkin *et al.* 1996). Measures of the strength of support for branches shared by two or more species are provided by bootstrapping and decay of parsimony. Bootstrap values (the percentage out of 2000 random samples from the data set supporting the given branch) are shown within circles at the nodes. Decay values (the number of additional steps required to collapse the branch into a polychotomy) are given by numbers preceded by *d*, with values of 4 or more indicating strong support for the branch. Numbers along the branches indicate the total number of restriction site changes for the group, with the number resulting from four-base cutting enzymes given in parentheses. Numbers preceded by R indicate specific rearrangements found in the chloroplast genomes of certain groups. The Roman numerals at the right side of the figure indicate major clades on the tree, and are also used in the comparisons of classifications in Table 2.2.

completely Eurasian groups representing subsections *Pinus*, *Canarienses*, *Halepenses* and *Pineae*, and the large North American group representing the remaining subsections. *Pinus contorta* and *P. virginiana*, representing the two distinct crossability groups within subsection *Contortae*, form a strongly supported clade at the base of the North American hard pines, providing strong evidence for Little and Critchfield's (1969) circumscription of the subsection, which is adopted here. The western North American species *P. ponderosa*, *P. jeffreyi*, *P. torreyana* and *P. coulteri*, representing subsections *Ponderosae* and *Sabinianae* in the Little and Critchfield classification, group together in a single well-supported clade, in agreement with their crossability relationships and similarities in oleoresin chemistry, and thus are placed together under subsection *Ponderosae* as suggested by Rushforth (1987).

The sister-group to the *Ponderosae* clade is a well-supported group including three major subclades: *P. radiata* (representing the California closed-cone pines), *P. taeda* (representing subsection *Australes* from eastern North America), and a clade of Mexican species. The Mexican group, supported by an 89% bootstrap value, includes *P. oocarpa* and *P. patula* from subsection *Oocarpae sensu* Little &

Critchfield, *P. teocote* (placed in subsection *Ponderosae* by Little & Critchfield and subsection *Oocarpae* by Van der Burgh), and a subclade consisting of *P. leiophylla* and *P. lumholtzii*, representing subsection *Leiophyllae*. The strong separation between *P. radiata* and *P. oocarpa* is consistent with crossability data and with Van der Burgh's placement of the California closed cone pines in a new subsection *Attenuatae* rather than subsection *Oocarpae*. The very strongly separated positions of *P. radiata* and subsection *Contortae* on the tree suggest that serotinous cones may have arisen independently two or more times within the subgenus. The placement of section *Leiophyllae* as a clade deeply nested within the North American group is in strong contrast with Little & Critchfield's classification, which placed these two Mexican species in section *Pinea* along with the Eurasian sections *Canarienses* and *Pineae*. The chloroplast DNA-based phylogeny implies that possession of deciduous fascicle sheaths, seen in almost all haploxylon pines, but only in section *Leiophyllae* among the diploxylon pines, is a homoplasious similarity between these groups.

Perez de la Rosa, Harris & Farjon (1995) examined restriction site variation in PCR-amplified fragments representing *c.* 1500 base pairs of non-coding regions of the chloroplast genome for 12 Mexican pine species (7 from subgenus *Strobus* and 5 from subgenus *Pinus*), with a single *Picea* species as outgroup. Since only a relatively small PCR-amplified region was analysed, few informative mutations were observed, and the resulting strict consensus tree showed only limited resolution. Subgenus *Pinus* and subsection *Strobi* were well-supported as monophyletic groups, but the basal branches within subgenus *Strobus* were largely unresolved. An intriguing finding was the isolated position of the large-coned pinyon species *P. nelsonii*, which accords well with its unusual morphological features such as the high level of dentation in the ray tracheid walls.

Sequence comparisons of the chloroplast *rbcL* gene have proved useful in allowing comparisons of pines to other genera of the family in order to obtain an outgroup rooting for the phylogeny of the genus. The *rbcL* gene has been very widely utilized in systematic comparisons of orders, families and genera of both conifers and flowering plants (see Chase *et al.* 1993), and is the first gene that has been compared among a number of species of pines (Bousquet *et al.* 1992; R.A. Price & S.H. Strauss, unpublished data). These sequence comparisons strongly suggest that the subgenera *Pinus* and *Strobus* are both monophyletic, as opposed to the diploxylon pines being derived from within the extant haploxylon group. *Pinus krempfii* is also placed within subgenus *Strobus*, and appears to be most similar to the Asian subsection *Gerardianae* and the American subsection *Cembroides*, rather than constituting a distinct subgenus as suggested by Ferré (1953) and Little & Critchfield (1969). Results of *rbcL* sequence comparisons are consistent with those from restriction site comparisons in support-

ing a basal split between the Eurasian groups of diploxylon pines plus *P. resinosa* and the North American groups, and are also consistent with a basal position for subsection *Balfourianae* in subgenus *Strobus*.

The recent sequencing of the chloroplast genome of Japanese black pine (*P. thunbergii*) by Wakasugi *et al.* (1994a, b) is of particular value in facilitating design of primers for sequence comparisons of other chloroplast genes that are more rapidly evolving than *rbcL*, and should thus provide a greater level of resolution within the two subgenera. The complete sequence has also provided the intriguing result that the series of *ndh* genes characteristic of the chloroplast genomes of land plants are absent in the Japanese black pine (Wakasugi *et al.* 1994a). Comparative studies of gene content in the chloroplast genomes of pines and other groups of conifers would thus be of considerable interest. Another feature of chloroplast gene evolution is the existence of significant heterogeneity in rates of change among taxa, which has been observed among a number of groups of angiosperms for the *rbcL* gene (see e.g. Bousquet *et al.* 1992). Internal consistency of trees from bootstrap resampling and decay of parsimony suggests that strong support for monophyletic groups can often be obtained using comparisons of chloroplast genes or restriction sites, but differences in rates of change do need to be examined in assessing phylogenies.

Restriction fragment analyses of both the 5S (Moran *et al.* 1992) and 18S–5.8S–26S (Govindaraju, Lewis & Cullis 1992) nuclear ribosomal (rDNA) repeats have been conducted in *Pinus*. Moran *et al.* (1992) partially characterized the 5S rDNA repeats in 30 species representing 14 subsections of the genus. *Pinus radiata*, which has been most fully characterized, has two size classes of the 5S repeats of about 525 and 850 base pairs (bp) in length, the longer of which has an insertion of 330 bp in the downstream spacer region. Representatives of the North American subsections *Australes*, *Contortae*, *Leiophyllae*, *Oocarpae*, *Ponderosae* and *Sabinianae* of subgenus *Pinus* all exhibited both of the size classes found in *P. radiata*, while the European or Asian subsections *Canarienses*, *Pineae* and *Pinus* were found to have only the long repeat units. Representatives of subgenus *Strobus* (including members of subsections *Cembrae*, *Cembroides*, *Gerardianae* and *Strobi*) had only the short repeat units of the type found in *P. radiata*, but some species may have other types of long repeat units with insertions different from the 330 bp insertion characteristic of subgenus *Pinus*.

Govindaraju *et al.* (1992) included 30 pine species representing 11 subsections in their study of the 18S–5.8S–26S region. No restriction fragment variation was found within subsections, and subsection *Leiophyllae* could not be distinguished from subsection *Attenuatae*, in marked contrast to comparisons of chloroplast DNA. The American hard pines formed a well-supported clade, as did the soft pine subgenus. The Eurasian hard pine subsection *Pinus* (represented by five species), however, formed an unresolved trichotomy with the American hard pines and the soft pines in their bootstrap analysis.

Nuclear ribosomal DNA has internal transcribed spacers (ITS) located between the small subunit (18S) and 5.8S rDNA (ITS-1), and between the 5.8S and large subunit (26S) rDNA (ITS-2). The ITS region, including the two spacers and the 5.8S rDNA, has become an important nuclear locus for molecular systematic investigations in flowering plants (see review in Baldwin *et al.* 1995). The popularity of the ITS region can be attributed to the relatively high rate of nucleotide substitutions in the transcribed spacers, permitting comparison of relatively recently diverged species, and to the ease of amplification of the region using highly conserved primers in the rRNA genes. A limitation of the ITS region in angiosperms is that it provides a relatively small amount of sequence data, since the 5.8S rDNA is highly conserved in sequence, and the two ITS spacers total *c.* 400 to 535 bp in length. In the Pinaceae, the ITS region is significantly longer than in flowering plants, varying from 1550 bp in *Pseudotsuga* to 3125 bp in *Picea* (Liston *et al.* 1996). In both subgenera of *Pinus*, the ITS region is approximately 3000 bp in length (Karvonen & Savalonainen 1993; Liston *et al.* 1996). The complete 3037 bp sequence of the ITS region of *Pinus pinea* has recently been published by Marrocco, Gelati & Maggini (1996). The 2631 bp ITS-1 of this species includes five tandem subrepeats of 219–237 bp in length, and restriction site mapping of other species of subgenus *Pinus* suggests the presence of similar subrepeats (Liston *et al.* 1996). The greatly increased length of the ITS-1 and the presence of subrepeats is unknown in other plants. The pattern of sequence evolution in the pine ITS region also differs from that of angiosperms. The ITS-2 shows a relatively low rate of sequence divergence, and can be easily aligned among genera of Pinaceae. The ITS-1, in contrast, is evolving rapidly, and sequences from the two subgenera of pines cannot be reliably aligned (Alvarez-Buylla & Liston 1995). Thus, the ITS-1 has the potential to resolve species-level phylogenies among the pines, while the ITS-2 is useful at the subsectional level and above (Liston *et al.* 1996).

A consideration for phylogenetic analysis of the ITS region in pines is the presence of the rDNA repeats at significantly more chromosomal loci than in diploid angiosperms. In *P. sylvestris*, 10–12 nucleolar organizing regions (NORs) have been observed (Karvonen & Savalonainen 1993). The large number of NORs might slow the process of concerted evolution which tends to homogenize the rDNA repeats within an individual. In fact, non-transcribed spacer heterogeneity has been reported by Karvonen & Savalonainen (1993), who also found that one individual of *P. sylvestris* (out of 97 trees sampled) had a 400 bp deletion in ITS-1 in some of its loci. Although occurring at a low frequency in *P. sylvestris*, a higher rate for this type of mutation would hinder comparative sequence analysis. However, ITS variants can also serve as useful genetic markers. For example, hybrid pines can be recognized by additive RFLP profiles in PCR-amplified ITS regions (Quijada, Liston & Alvarez-Buylla 1995).

2.8 Conclusions

A large amount of comparative systematic data supports the division of the pines into two discrete lineages, subgenus *Pinus* and subgenus *Strobus*. Nucleotide sequence studies using outgroup comparisons are consistent with the monophyly of both subgenera. Several lines of evidence indicate that section *Parrya* forms a basal group within subgenus *Strobus*, but the subsectional relationships among the haploxylon pines are still unclear. No study has incorporated all subsections of haploxylon pines, and the members of subsections *Cembroides* and *Rzedowkianae* are in need of detailed molecular comparisons. Further study is also needed of the unusual Vietnamese species *Pinus krempfii* and the newly discovered Chinese species *P. squamata*.

Among the hard pines, comparisons of chloroplast DNA indicate a major division between the primarily Old World species in subsections *Canarienses*, *Halepenses*, *Pinus* and *Pineae*, and an exclusively New World lineage including all of the North American species except *P. resinosa* and possibly *P. tropicalis*. Evidence from chloroplast restriction sites indicates that section *Pinea* as delimited by Little & Critchfield (1969) comprises two unrelated clades. The Old World members of this group comprise three small and apparently closely related subsections (*Canarienses*, *Halepenses* and *Pineae*). Several lines of evidence argue for a close relationship between the Mexican subsections *Leiophyllae* and *Oocarpae* (including the *teocote* group removed from subsection *Ponderosae*). Two other departures from the subsectional classification of Little and Critchfield are the separation of the California closed-cone pines (subsection *Attenuatae*) from subsection *Oocarpae* and the merger of subsections *Ponderosae* and *Sabinianae* as suggested by evidence from crossability and chloroplast DNA restriction sites. Further research using both nuclear and chloroplast markers is needed to confirm the relationships among the subsections of North American diploxylon pines and the placement of certain Old World species such as *P. heldreichii*, *P. merkusii* and *P. pinaster*.

Acknowledgements

We thank Aljos Farjon, James Hamrick and James Oliphant for helpful comments on the manuscript.

Appendix 2.1. Subdivisions, species, and major infraspecific taxa of *Pinus*. A limited number out of the many published taxonomic or nomenclatural synonyms are given in brackets after the accepted names (see Table 1.1, p. 5 for common names, morphological features, biogeographic region, and habitats of taxa)

Subgenus *Pinus* (Diploxylon or hard pines)
Section *Pinus*
Subsection *Pinus* [subsection *Sylvestres* Loudon]
1 *P. densata* Masters
2 *P. densiflora* Siebold & Zuccarini
   var. *densiflora*
   var. *funebris* (Komarov) Liou & Wang [*P. funebris* Komarov]
3 *P. heldreichii* Christ
   var. *heldreichii*
   var. *leucodermis* (Antoine) Markgraf ex Fitschen [*P. leucodermis* Antoine]
4 *P. hwangshanensis* Hsia
5 *P. kesiya* Royle ex Gordon [*P. khasya* Royle; *P. khasia* Engelmann]
   var. *insularis* (Endlicher) Gaussen [*P. insularis* Endlicher]
   var. *kesiya*
   var. *langbianensis* (Chevalier) Cheng & Fu
6 *P. luchuensis* Mayr
7 *P. massoniana* Lambert
   var. *hainanensis* Cheng & Fu
   var. *henryi* (Masters) Wu [*P. henryi* Masters]
   var. *massoniana*
8 *P. merkusii* Junghuhn & de Vriese [including *P. sumatrana* Junghuhn; *P. latteri* Mason; *P. merkusiana* Cooling & Gaussen; the species perhaps best placed in a separate subsection]
9 *P. mugo* Turra [*P. montana* Miller; *P. mughus* Scopoli]
10 *P. nigra* Arnold
   subsp. *dalmatica* (Visiani) Franco
   subsp. *laricio* (Poiret) Maire [*P. laricio* Poiret]
   subsp. *nigra*
   subsp. *pallasiana* (Lambert) Holmboe [*P. pallasiana* Lambert]
   subsp. *salzmannii* (Dunal) Franco [*P. salzmannii* Dunal]
11 *P. pinaster* Aiton [including *P. maritima* Lamarck; *P. mesogeensis* Fischi & Gaussen; *P. pinaster* subsp. *atlantica* Villar; *P. pinaster* subsp. *hamiltonii* (Tenore) Villar; the species perhaps belonging in a different subsection]
12 *P. resinosa* Aiton
13 *P. sylvestris* Linnaeus [a complex with many often intergrading named varieties and subspecies; some of the major variants are listed as subspecies below; see also Gaussen *et al.* (1993); Vidakovic (1991)]
   subsp. *hamata* (Steven) Fomin
   subsp. *kalundensis* Sukacev
   subsp. *lapponica* Fries
   subsp. *sibirica* Ledebour
   subsp. *sylvestris*
14 *P. tabuliformis* Carrière [spelling corrected from '*P. tabulaeformis*'; cf. Rushforth (1987)]
15 *P. taiwanensis* Hayata
   var. *damingshanensis* Cheng & Fu
   var. *taiwanensis*
16 *P. thunbergii* Parlatore [*P. thunbergiana* Franco; *P. thunbergii* Lambert, nom. illeg.]
17 *P. tropicalis* Morelet
18 *P. uncinata* Miller ex Mirbel
19 *P. yunnanensis* Franchet
   var. *pygmaea* (Xue) Xue
   var. *tenuifolia* Cheng & Law
   var. *yunnanensis*
Subsection *Canarienses* Loudon
20 *P. canariensis* C. Smith
21 *P. roxburghii* Sargent [*P. longifolia* Roxburgh]
Subsection *Halepenses* Van der Burgh
22 *P. brutia* Tenore [*P. halepensis* var. *brutia* (Tenore) Henry]
   subsp. *brutia*
   subsp. *eldarica* (Medwedew) Nahal [*P. eldarica* Medwedew]
   subsp. *pityusa* (Steven) Nahal [*P. pityusa* Steven]
   subsp. *stankewiczii* (Sukaczev) Nahal
23 *P. halepensis* Miller
Subsection *Pineae* Little & Critchfield
24 *P. pinea* Linnaeus

New World Diploxylon Pines
Subsection *Contortae* Little & Critchfield
25 *P. banksiana* Lambert
26 *P. clausa* (Chapman ex Engelmann) Sargent
   var. *clausa*
   var. *immuginata* Ward
27 *P. contorta* Douglas ex Loudon
   subsp. *bolanderi* (Parlatore) Critchfield [*P. bolanderi* Parlatore]
   subsp. *contorta*
   subsp. *latifolia* (Engelmann) Critchfield
   subsp. *murrayana* (Balfour) Critchfield
28 *P. virginiana* Miller
Subsection *Australes* Loudon
29 *P. caribaea* Morelet
   var. *bahamensis* (Grisebach) Barrett & Golfari [*P. bahamensis* Grisebach]
   var. *caribaea*
   var. *hondurensis* (Sénéclauze) Barrett & Golfari
30 *P. cubensis* Grisebach [*P. occidentalis* var. *cubensis* (Grisebach) Silba; including *P. maestrensis* Bisse; *P. occidentalis* var. *maestrensis* (Bisse) Silba]
31 *P. echinata* Miller [*P. mitis* Michaux]
32 *P. elliottii* Engelmann
   var. *elliottii*
   var. *densa* Little & Dorman [*P. densa* (Little & Dorman) de Laubenfels & Silba]
33 *P. glabra* Walter
34 *P. occidentalis* Swartz
35 *P. palustris* Miller [*P. australis* Michaux]
36 *P. pungens* Lambert
37 *P. rigida* Miller
38 *P. serotina* Michaux [*P. rigida* subsp. *serotina* (Michaux) Clausen]
39 *P. taeda* Linnaeus
Subsection *Ponderosae* Loudon
40 *P. cooperi* Blanco [treated as a variety of *P. arizonica* by Farjon & Styles (1997)]
41 *P. durangensis* Martínez [*P. martinezii* Larsen]
42 *P. engelmannii* Carrière [*P. apecheca* Lemmon]
43 *P. jeffreyi* Balfour
44 *P. ponderosa* Douglas ex P. & C. Lawson
   var. *arizonica* (Engelmann) Shaw [*P. arizonica* Engelmann — perhaps representing a distinct species; *P. arizonica* f. *stormiae* Martínez]
   var. *ponderosa*
   var. *scopulorum* Engelmann [subsp. *scopulorum* (Engelmann) Murray]
45 *P. washoensis* Mason & Stockwell [*P. ponderosa* subsp. *washoensis* (Mason & Stockwell) Murray]
'Montezumae Group'
46 *P. devoniana* Lindley [including *P. michoacana* Martínez; *P. wincesteriana* Gordon]
47 *P. donnell-smithii* Masters [needing further study; included in *P. hartwegii* by Farjon & Styles (1997)]
48 *P. hartwegii* Lindley [*P. montezumae* var. *hartwegii* (Lindley) Engelmann; including *P. rudis* Endlicher; *P. montezumae* var. *rudis* (Endlicher) Shaw; see Carvajal & McVaugh (1992); Matos (1995)]
49 *P. montezumae* Lambert [including *P. montezumae* var. *gordonia* (Hartweg ex Gordon) Silba]
'Pseudostrobus Group'
50 *P. douglasiana* Martínez
51 *P. maximinoi* Moore [*P. tenuifolia* Bentham, non Salisbury; *P. pseudostrobus* var. *tenuifolia* (Benth.) Shaw]
52 *P. nubicola* Perry [needing further study; included in *P. pseudostrobus* by Farjon & Styles (1997)]
53 *P. pseudostrobus* Lindley
   var. *apulcensis* (Lindley) Shaw [*P. apulcensis* Lindley; including *P. oaxacana* Mirov]
   var. *estevezii* Martínez [*P. estevezii* (Martínez) Perry]
   var. *pseudostrobus*
'Sabinianae Group' [= subsection *Sabinianae* Loudon]
54 *P. coulteri* D. Don
55 *P. sabiniana* Douglas ex D. Don

Appendix 2.1 (*cont.*)

56 *P. torreyana* Parry ex Carrière
    subsp. *insularis* Haller
    subsp. *torreyana*
Subsection *Attenuatae* Van der Burgh
57 *P. attenuata* Lemmon
58 *P. muricata* D. Don [including var. *borealis* Duffield, *P. remorata* Mason]
59 *P. radiata* D. Don [*P. insignis* Douglas] (see Chap. 21, this volume)
    var. *binata* (Engelmann) Brewer & Watson
    var. *cedrosensis* (Howell) Axelrod
    var. *radiata*
Subsection *Oocarpae* Little & Critchfield
'Oocarpa group'
60 *P. greggii* Engelmann ex Parlatore
61 *P. jaliscana* Perez de la Rosa [*P. macvaughii* Carvajal; *P. patula* var. *jaliscana* (Perez de la Rosa) Silba]
62 *P. oocarpa* Schiede ex Schlechtendal
    var. *oocarpa*
    var. *trifoliata* Martínez
63 *P. patula* Schiede & Deppe ex Schlechtendal & Chamisso
    var. *longipedunculata* Loock ex Martínez
    var. *patula*
64 *P. praetermissa* Styles & McVaugh [*P. oocarpa* var. *microphylla* Shaw]
65 *P. pringlei* Shaw
66 *P. tecunumanii* Eguiluz & Perry [*P. patula* subsp. *tecunumanii* (Eguiluz & Perry) Styles]
'Teocote group'
67 *P. herrerae* Martínez [spelling corrected from '*P. herrarai*' Martínez]
68 *P. lawsonii* Roezl ex Gordon
69 *P. teocote* Schiede & Deppe ex Schlechtendal & Chamisso
Subsection *Leiophyllae* Loudon
70 *P. leiophylla* Schiede & Deppe
    var. *chihuahuana* (Engelmann) Murray [*P. chihuahuana* Engelmann]
    var. *leiophylla*
71 *P. lumholtzii* Robinson & Fernald

Subgenus *Strobus* Lemmon (Haploxylon or soft pines)
Section *Parrya* Mayr
Subsection *Balfourianae* Engelmann
72 *P. aristata* Engelmann
73 *P. balfouriana* Balfour
    subsp. *austrina* R. Mastrogiuseppe & J. Mastrogiuseppe
    subsp. *balfouriana*
74 *P. longaeva* D.K. Bailey
Subsection *Krempfianae* Little & Critchfield
75 *P. krempfii* Lecomte
Subsection *Cembroides* Engelmann
76 *P. cembroides* Zuccarini
    subsp. *cembroides*
    subsp. *lagunae* (Robert-Passini) D.K. Bailey [*P. lagunae* (Robert-Passini) Passini; *P. cembroides* var. *lagunae* Robert-Passini]
    subsp. *orizabensis* D.K. Bailey [*P. orizabensis* (Bailey) Bailey & Hawksworth]
77 *P. culminicola* Andresen & Beaman
78 *P. discolor* D.K. Bailey & Hawksworth
79 *P. edulis* Engelmann

80 *P. johannis* Robert
81 *P. juarezensis* Lanner [*P. quadrifolia* Parlatore ex Sudworth appears to represent a hybrid between this species and *P. monophylla*]
82 *P. maximartinezii* Rzedowski
83 *P. monophylla* Torrey & Fremont
    subsp. *californiarum* (D.K. Bailey) Zavarin [*P. californiarum* Bailey]
    subsp. *fallax* (Little) Zavarin [*P. californiarum* subsp. *fallax* (Little) D.K. Bailey]
    subsp. *monophylla*
84 *P. nelsonii* Shaw
85 *P. pinceana* Gordon
86 *P. remota* (Little) D.K. Bailey & Hawksworth [*P. cembroides* var. *remota* Little; including *P. catarinae* Robert-Passini]
Subsection *Rzedowskianae* Carvajal
87 *P. rzedowskii* Madrigal & Caballero
Subsection *Gerardianae* Loudon
88 *P. bungeana* Zuccarini ex Endlicher
89 *P. gerardiana* Wallich ex D. Don
90 *P. squamata* Li [probably representing a new subsection]
Section *Strobus*
Subsection *Strobi* Loudon
91 *P. armandii* Franchet
    var. *amamiana* (Koidzumi) Hatusima [*P. amamiana* Koidzumi]
    var. *armandii*
    var. *mastersiana* (Hayata) Hayata [*P. mastersiana* Hayata]
92 *P. ayacahuite* Ehrenberg ex Schlechtendal
    var. *ayacahuite*
    var. *strobiformis* (Engelmann) Lemmon [*P. flexilis* var. *macrocarpa* Engelmann; *P. ayacahuite* var. *brachyptera* Shaw; *P. strobiformis* Engelmann; probably representing a distinct species]
    var. *novogaliciana* Carvajal
    var. *veitchii* (Roezl) Shaw
93 *P. bhutanica* Grierson, Long & Page
94 *P. chiapensis* (Martínez) Andresen [*P. strobus* var. *chiapensis* Martínez]
95 *P. dabeshanensis* Cheng & Law
96 *P. dalatensis* de Ferré
97 *P. fenzeliana* Handel-Mazzetti [including *P. kwangtungensis* Chen]
98 *P. flexilis* James
99 *P. lambertiana* Douglas
100 *P. monticola* Douglas ex D. Don
101 *P. morrisonicola* Hayata
102 *P. parviflora* Siebold & Zuccarini
    var. *parviflora* [including *P. himekomatsu* Miyabe & Kudo]
    var. *pentaphylla* (Mayr) Henry [*P. pentaphylla* Mayr]
103 *P. peuce* Grisebach
104 *P. strobus* Linnaeus
105 *P. wallichiana* Jackson [*P. excelsa* Wallich ex D. Don; *P. griffithii* McClellan, nom. illeg.]
106 *P. wangii* Hu & Cheng
Subsection *Cembrae* Loudon
107 *P. albicaulis* Engelmann
108 *P. cembra* Linnaeus
109 *P. koraiensis* Siebold & Zuccarini
110 *P. pumila* (Pallas) Regel
111 *P. sibirica* Du Tour

References

Alosi, M.C. & Park, R.B. (1983). A survey of phloem polypeptides in conifers. *Current Topics in Plant Biochemistry and Physiology*, 2, 250.

Alvarez-Buylla, E. & Liston, A. (1995). The nuclear ribosomal DNA internal transcribed spacer of gymnosperms is significantly longer than angiosperms. Plant Genome III Program, San Diego, CA, January, 1995.

Bailey, D.K. (1970). Phytogeography and taxonomy of *Pinus* subsection *Balfourianae. Annals of the Missouri Botanical Garden*, 57, 210–49.

Baldwin, B.G., Sanderson, M.J., Porter, J.M., Wojciechowski, M.F., Campbell, C.S. & Donoghue, M.J. (1995). The ITS region of nuclear ribosomal DNA: a valuable source of evidence on angiosperm phylogeny. *Annals of the Missouri Botanical Garden*, 82, 247–77.

Behnke, H.-D. (1974). Sieve element plastids of Gymnospermae: their ultrastructure in relation to systematics. *Plant Systematics and Evolution*, **123**, 1–12.

Bousquet, J., Strauss, S.H., Doerksen, A.H. & Price, R.A. (1992). Extensive variation in evolutionary rate of *rbcL* gene sequences among seed plants. *Proceedings of the National Academy of Sciences, USA*, **89**, 7844–8.

Britton, N.L. (1908). *North American Trees*. New York: H. Holt.

Carvajal, S. & McVaugh, R. (1992). *Pinus*. In *Flora Novo-Galiciana: A Descriptive Account of the Vascular Plants of Western Mexico*, Vol. 17, ed. W.R. Anderson, pp. 32–100. Ann Arbor: University of Michigan Herbarium.

Chase, M.D., Soltis, D.E., Olmstead, R.G., *et al.* (1993). Phylogenetics of seed plants: An analysis of nucleotide sequences from the plastid gene *rbcL*. *Annals of the Missouri Botanical Garden*, **80**, 528–80.

Chaw, S.-M., Sung, H.-M., Long, H., Zharkikh, A. & Li, W.-H. (1995). The phylogenetic position of the conifer genera *Amentotaxus*, *Phyllocladus*, and *Nageia* inferred from 18S rRNA sequences. *Journal of Molecular Evolution*, **41**, 224–30.

Conkle, M.T., Schiller, G. & Grunwald, C. (1988). Electrophoretic analysis of diversity and phylogeny of *Pinus brutia* and closely related taxa. *Systematic Botany*, **13**, 411–24.

Critchfield, W.B. (1963). Hybridization of the Southern pines in California. Southern Forest Tree Improvement Commission Publication 22, 1–9.

Critchfield, W.B. (1964). The Austrian × red pine hybrid. *Silvae Genetica*, **12**, 187–92.

Critchfield, W.B. (1967). Crossability and relationships of the closed-cone pines. *Silvae Genetica*, **16**, 91–7.

Critchfield, W.B. (1975). Interspecific hybridization in *Pinus*: A summary review. In *Proceedings of the 14th Meeting of the Canadian Tree Improvement Association. Part 2. Symposium on Interspecific and Interprovenance Hybridization in Forest Trees*, ed. D.P. Fowler & C.W. Yeatman, pp. 99–105. Fredericton, New Brunswick: Canadian Tree Improvement Association.

Critchfield, W.B. (1986). Hybridization and classification of the white pines (*Pinus* section *Strobus*). *Taxon*, **35**, 647–56.

Critchfield, W.B. & Little, E.L. (1966). *Geographic Distribution of the Pines of the World*. USDA Forest Service Miscellaneous Publication 991.

Dietrich, A.G. (1824). *Flora der Gegend um Berlin*. Berlin: G.E. Nauck.

Doi, T. & Morikawa, K.-I. (1929). An anatomical study of the leaves of the genus *Pinus*. *Journal of the Department of Agriculture, Kyushu Imperial University*, **2**, 149–98.

Doyle, J. (1963). Proembryogeny in *Pinus* in relation to that of other conifers — a survery. *Proceedings of the Royal Irish Academy, B*, **62**, 181–216.

Duhamel Du Monceau, H.L. (1755). *Traité des Arbres et Arbustes qui se cultivent en France en pleine terre*, Tome 2. Paris: Guerin & Delatour.

Duffield, J.W. (1952). Relationships and species hybridization in the genus *Pinus*. *Silvae Genetica*, **1**, 93–7.

Eguiluz Piedra, T. (1988). Distribucion natural de los pinos en Mexico. Nota Tecnica 1. Chapingo, Mexico: Centro Genetico Forestal.

Endlicher, I.L. (1847). *Synopsis coniferarum*. Sangalli: Scheitlin & Zollikofer.

Engelmann, G. (1880). Revision of the genus *Pinus*, and description of *Pinus elliottii*. *Transactions of the Saint Louis Academy of Science*, **4**, 161–89.

Erdtman, H. (1959). Conifer chemistry and taxonomy of conifers. In *Proceedings of the Fourth International Congress of Biochemistry*, Vol. II, ed. K. Kratzl & G. Billek, pp. 1–27. Oxford: Pergamon.

Erdtman, H., Kimland, B. & Norin, T. (1966). Wood constituents of *Ducampopinus krempfii* (Lecomte) Chevalier (*Pinus krempfii* Lecomte). *Phytochemistry*, **5**, 927–31.

Farjon, A. (1984). *Pines: Drawings and Descriptions of the Genus Pinus*. Leiden: E.J. Brill & W. Backhuys.

Farjon, A. (1990). *Pinaceae: Drawings and Descriptions of the Genera Abies, Cedrus, Pseudolarix, Keteleeria, Nothotsuga, Tsuga, Cathaya, Pseudotsuga, Larix and Picea*. Konigstein, Germany: Koeltz Scientific Books.

Farjon, A. (1993). Names in current use in the Pinaceae (Gymnospermae) in the ranks of genus to variety. *Regnum Vegetabile*, **128**, 107–46.

Farjon, A. (1995). Typification of *Pinus apulcensis* Lindley (Pinaceae), a misinterpreted name for a Latin American pine. *Novon*, **5**, 252–6.

Farjon, A., & Styles, B. T. (1997). *Pinus*. Flora Neotropica Monograph 75: 1–291.

Ferré, Y. de. (1953). Division du genre *Pinus* en quatres sous-genres. *Compte Rendu de l'Academie des Séances des Sciences, Paris*, **236**, 226–8.

Ferré, Y. de. (1965). Structure des plantules et systématique du genre *Pinus*. *Bulletin de la Société d'Histoire Naturelle de Toulouse*, **100**, 230–80.

Frankis, M.P. (1988). Generic interrelationships in Pinaceae. *Notes from the Royal Botanic Garden, Edinburgh*, **45**, 527–48.

Gaussen, H. (1960). *Pinus* in *Les gymnospermes actuelles et fossiles*. Travaux du Laboratoire Forestier de Toulouse, Tôme II, Section I, Volume I, Chapter XI, Fascicle VI, 1–272.

Gaussen, H., Heywood, V.H. & Chater, A.O. (1993). *Pinus*. In *Flora Europaea*, 2nd edn., Vol. 1, ed. T.G. Tutin, N.A. Burges, A.O. Chater *et al.*, pp. 40–4. Cambridge: Cambridge University Press.

Geiger, H. & Quinn, C. (1975). Biflavonoids In *The Flavonoids*, Part 2, ed. J.B. Harborne, T.J. Mabry & H. Mabry, pp. 692–742. New York: Academic Press.

Goncharenko, G.G., Padutov, V.E. & Silin, A.E. (1992). Population structure, gene diversity, and differentiation in natural populations of cedar pines (*Pinus* subsect. *Cembrae*, Pinaceae) in the USSR. *Plant Systematics and Evolution*, **182**, 121–34.

Govindaraju, D., Lewis, P. & Cullis, C. (1992). Phylogenetic analysis of pines using ribosomal DNA restriction fragment length polymorphisms. *Plant Systematics and Evolution*, **179**, 141–53.

Greguss, P. (1955). *Identification of Living Gymnosperms on the Basis of Xylotomy*. Budapest: Akademia Kiado.

Hamrick, J.L. & Godt, M.J.W. (1990). Allozyme diversity in plant species. In *Plant Population Genetics, Breeding and Genetic Resources*, ed. A.H.D. Brown, M.T. Clegg, A.L. Kahler & B.S. Weir, pp. 43–63. Sunderland, Mass.: Sinauer Associates.

Harlow, W.W. (1931). *The Identification of the Pines of the United States, Native and Introduced, by Needle Structure*. New York State College of Forestry Technical Publication 32.

Hart, J. (1987). A cladistic analysis of conifers: preliminary results. *Journal of the Arnold Arboretum*, **68**, 269–307.

Helmers, A.E. (1943). The ecological anatomy of ponderosa pine needles. *American Midland Naturalist*, **29**, 55–71.

Hudson, R.H. (1960). The anatomy of the genus *Pinus* in relation to its classification. *Journal of the Institute of Wood Science*, **6**, 26–46.

Jährig, M. (1962). Beiträge zur Nadelanatomie und Taxonomie der Gattung *Pinus* L. *Willdenowia*, **3**, 329–66.

Karamangala, R.R. & Nickrent, D.L. (1989). An electrophoretic study of representatives of subgenus *Diploxylon* of *Pinus*. *Canadian Journal of Botany*, **67**, 1750–9.

Karvonen, P. & Savalonainen, O. (1993). Variation and inheritance of ribosomal DNA in *Pinus sylvestris* L. (Scots pine). *Heredity*, **71**, 614–22.

Keng, H. & Little, E.L., Jr (1961). Needle characteristics of hybrid pines. *Silvae Genetica*, **10**, 131–46.

Klaus, W. (1980). Neue Beobachtungen zur Morphologie des Zapfens von *Pinus* und ihre Bedeutung für die Systematik, Fossilbestimmung, Arealgestaltung und Evolution der Gattung. *Plant Systematics and Evolution*, **134**, 137–71.

Klaus, W. (1989). Mediterranean pines and their history. *Plant Systematics and Evolution*, **162**, 133–63.

Koehne, E. (1893). *Deutsche Dendrologie*. Stuttgart: Ferdinand Enke.

Kral, R. (1993). *Pinus*. In *Flora of North America*, Vol. 2, 373–98. New York: Oxford University Press.

Krupkin, A.B., Liston, A. & Strauss, S.H. (1996). Phylogenetic analysis of the hard pines (*Pinus* subgenus *Pinus*, Pinaceae) from chloroplast DNA restriction site analysis. *American Journal of Botany*, **83**, 489–98.

Krüssmann, G. (1985). *Manual of Cultivated Conifers*, 2nd edn. (M. Epp, translator). Portland, Oregon: Timber Press.

Krutovskii, K.V., Politov, D.V. & Altukhov, Y.P. (1994). Genetic differentiation and phylogeny of stone pine species based on isozyme loci. In *Proceedings of the International Workshop on Subalpine Stone Pines and their Environment* ed. W.C. Schmidt & F.-K. Holtmeier, pp. 19–30. Ogden, Utah: USDA Forest Service General Technical Report INT-309.

Landry, P. (1994). A revised synopsis of the pines. 5. The subgenera of *Pinus*, and their morphology and behavior. *Phytologia*, **76**, 73–9.

Lemmon, J.G. (1895). *Handbook of West-American Cone Bearers*, edn. 3. Oakland, California: Pacific Press.

Lidholm, J. & Gustafsson, P. (1991). A three-step model for the rearrangement of the chloroplast *trnK-psbA* region of the gymnosperm *Pinus contorta*. *Nucleic Acids Research*, **19**, 2881–7.

Linnaeus, C. (1753). *Species Plantarum*, edn. 1. Stockholm: Impensis Laurentii Salvii.

Little, E.L., Jr & Critchfield, W.B. (1969). *Subdivisions of the Genus Pinus*. USDA Forest Service Miscellaneous Publication 1144.

Little, E.L., Jr & Righter, F.I. (1965). *Botanical Descriptions of Forty Artificial Pine Hybrids*. USDA Forest Service Technical Bulletin 1345.

Liston, A., Robinson, W.A., Oliphant, J. & Alvarez-Buylla, E. (1996). Length variation in the nuclear ribosomal internal transcribed spacer region of non-flowering seed plants. *Systematic Botany*, **21**, 109–20.

Loudon, J.C. (1838). *Pinus*. In *Arboretum et Fructicetum Britannicum*, Vol. 4, pp. 2152–92. London: Published by the author.

McCune, B. (1988). Ecological diversity in North American pines. *American Journal of Botany*, **75**, 353–68.

Malusa, J. (1992). Phylogeny and biogeography of the pinyon pines (*Pinus* subsect. *Cembroides*). *Systematic Botany*, **17**, 42–66.

Marrocco, R.M., Gelati, M.T. & Maggini, F. (1996). Nucleotide sequence of the internal transcribed spacers and 5.8S region of ribosomal DNA in *Pinus pinea* L. DNA Sequence. *The Journal of Sequencing and Mapping*, **6**, 175–7.

Martínez, M. (1948). *Los Pinos Mexicanos*, edn. 2. Mexico City: Ediciones Botas.

Matos, J.A. (1992). Evolution within the *Pinus montezumae* complex of Mexico: population subdivision, hybridization, and taxonomy. PhD Thesis, Washington University, St Louis, Missouri.

Matos, J.A. (1995). *Pinus hartwegii* and *P. rudis*: A critical assessment. *Systematic Botany*, **20**, 6–21.

Mayr, H. (1890). *Die Waldungen von Nordamerika*. Munich: Rieger.

Millar, C.I., Strauss, S.H., Conkle, M.T. & Westfall, R.D. (1988). Allozyme differentiation and biosystematics of the California closed-cone pines (*Pinus* subsect. *Oocarpae*). *Systematic Botany*, **13**, 351–70.

Miller, P. (1754). *The Gardeners Dictionary*, 4th abridged edn. London: Rivington.

Mirov, N.T. (1961). *Composition of Gum Turpentines of Pines*. USDA Forest Service Technical Bulletin 1239.

Mirov, N.T. (1967). *The Genus Pinus*. New York: Ronald Press.

Moran, G.F., Smith, D., Bell, J.C. & Appels, R. (1992). The 5S RNA genes in *Pinus radiata* and the spacer region as a probe for relationships between *Pinus* species. *Plant Systematics and Evolution*, **183**, 209–21.

Murray, E. (1983). Unum minutum monographum generis pinorum (*Pinus* L.). *Kalmia*, **13**, 11–24.

Neale, D. & Sederoff, R.R. (1989). Paternal inheritance of chloroplast DNA and maternal inheritance of mitochondrial DNA in loblolly pine. *Theoretical and Applied Genetics*, **77**, 212–16.

Niebling, C.R. & Conkle, M.T. (1990). Diversity of Washoe pine and comparisons with allozymes of ponderosa pine races. *Canadian Journal of Botany*, **20**, 1750–9.

Nixon, K.C., Crepet, W.L., Stevenson, D. & Friis, E.M. (1994). A reevaluation of seed plant phylogeny. *Annals of the Missouri Botanical Garden*, **81**, 484–533.

Norin, T. (1972). Some aspects of the chemistry of the order Pinales. *Phytochemistry*, **11**, 1231–42.

Parlatore, P. (1868). Coniferae. In C. De Candolle, *Prodromus Systematis Naturalis Regni Vegetabilis*, Vol. XVI, part 2, pp. 361–521.

Passini, M.F., Cibrian, T. & Eguiluz, Piedra, T. (eds.) (1988). *II Simposio Nacional sobre Pinos Piñoneros*. Chapingo, Mexico: Centro de Genetica Forestal.

Perez de la Rosa, J., Harris, S.A. & Farjon, A. (1995). Noncoding chloroplast DNA variation in Mexican pines. *Theoretical and Applied Genetics*, **91**, 1101–6.

Perry, J.P., Jr (1991). *The Pines of Mexico and Central America*. Portland, Oregon: Timber Press.

Phillips, E.J.W. (1941). *Identification of Softwoods by their Microscopic Structure*. Forest Products Research Bulletin 22. London: Department of Scientific and Industrial Research.

Pilger, R. (1926). *Pinus*. In A. Engler & K. Prantl, *Die Natürlichen Pflanzenfamilien*, Vol. XIII, pp. 331–42. Leipzig: Wilhelm Engelmann.

Piovesan, G., Pelosi, C., Schirone, A. & Schirone, B. (1993). Taxonomic evaluations of the genus *Pinus* (Pinaceae) based on electrophoretic data of salt soluble and insoluble seed storage proteins. *Plant Systematics and Evolution*, **186**, 57–68.

Prager, E.M., Fowler, D.P. & Wilson, A.C. (1976). Rates of evolution in conifers (Pinaceae). *Evolution*, **30**, 637–49.

Price, R.A. (1989). The genera of Pinaceae in the southeastern United States. *Journal of the Arnold Arboretum*, **70**, 247–305.

Price, R.A. (1995). Familial and generic classification of the conifers. *American Journal of Botany*, **82**, Supplement 110.

Price, R.A. & Lowenstein, J.M. (1989). An immunological comparison of the Sciadopityaceae, Taxodiaceae, and Cupressaceae. *Systematic Botany*, **14**, 141–9.

Price, R.A., Olsen-Stojkovich, J. & Lowenstein, J.M. (1987). Relationships among the genera of Pinaceae: an immunological comparison. *Systematic Botany*, **12**, 91–7.

Quijada A., Liston, A. & Alvarez-Buylla, E. (1995). The ITS region of nrDNA as a tool in studies of reticulate evolution. In *VII National and 1st Joint Mexico–US Symposium on Agrobiology, Molecular Physiology and Biotechnology of Crops Important for Mexican Agriculture*, p. 192, Cuernavaca, Morelos, Universidad Autonoma de Mexico.

Rehder, A. (1949). *Bibliography of Cultivated Trees and Shrubs Hardy in the Cooler Temperate tegions of the Northern Hemisphere*. Jamaica Plain, Mass.: Arnold Arboretum.

Rushforth, K. (1987). *Conifers*. London: Helm.

Saylor, L.C. (1964). Karyotype analysis of *Pinus* – group *Lariciones*. *Silvae Genetica*, **13**, 165–70.

Saylor, L.C. (1972). Karyotype analysis of the genus *Pinus* – subgenus *Pinus*. *Silvae Genetica*, **21**, 155–63.

Saylor, L.C. (1983). Karyotype analysis of the genus *Pinus* – subgenus *Strobus*. *Silvae Genetica*, **32**, 119–24.

Saylor, L.C. & Koenig, R.L. (1967). The slash × sand pine hybrid. *Silvae Genetica*, **16**, 134–8.

Schirone, B., Piovesan, G., Bellarosa, R. & Pelosi, C. (1991). A taxonomic analysis of seed proteins in *Pinus* spp. (Pinaceae). *Plant Systematics and Evolution*, **178**, 48–53.

Shaw, G.R. (1909). *The Pines of Mexico*. Arnold Arboretum Publication 1. Boston: J.R. Ruiter.

Shaw, G.R. (1914). *The Genus Pinus*. Arnold Arboretum Publications 5. Forage Village, Mass.: The Murray Printing Co.

Shaw, G.R. (1924). Notes on the genus *Pinus*. *Journal of the Arnold Arboretum*, **5**, 225–7.

Singh, H. (1978). *Embyology of Gymnosperms. Handbuch der Pflanzenanatomie*, edn. 2, Band 10, Teil 2. Berlin: Gebrüder Bornträger.

Spach, E. (1842). *Histoire Naturelle des Végétaux*, Vol. XI. Paris: Roret.

Stead, J.W. & Styles, B.T. (1984). Studies of Central American pines: a revision of the '*pseudostrobus*' group (Pinaceae). *Botanical Journal of the Linnean Society*, **89**, 249–75.

Strauss, S.H. & Doerksen, A.H. (1990). Restriction fragment analysis of pine phylogeny. *Evolution*, **44**, 1081–96.

Strauss, S.H., Palmer, J.D., Howe, G. & Doerksen, A. (1988). Chloroplast genomes of two conifers lack a large inverted repeat and are extensively rearranged. *Proceedings of the National Academy of Sciences, USA*, **85**, 3898–902.

Styles, B.T. (1993). Genus *Pinus*: a Mexican purview. In *Biological Diversity of Mexico: Origins and Distribution*, ed. T.P. Ramamoorthy, R. Bye, A. Lot & J. Fa, pp. 397–420. New York: Oxford University Press.

Styles, B.T., Stead, J.W. & Rolph, K.J. (1982). Studies of variation in Central American pines. 2. Putative hybridization between *Pinus caribaea* and *P. oocarpa*. *Turrialba*, **32**, 229–42.

Tsumura, Y., Yoshimura, K., Tomaru, N. & Ohba, K. (1995). Molecular phylogeny of conifers using RFLP analysis of PCR-amplified specific chloroplast genes. *Theoretical and Applied Genetics*, **91**, 1222–36.

Van der Burgh, J. (1973). Hölzer der niederrheinischen Braunkohlenformation, 2. Hölzer der Braunkohlengruben 'Maria Theresia' zu Herzogenrath, 'Zukunft West' zu Eschweiler und 'Victor' (Zülpich Mitte) zu Zülpich. Nebst einer systematisch-anatomischen Bearbeitung der Gattung *Pinus* L. *Review of Palaeobotany and Palynology*, **15**, 73–275.

Vidaković, M. (1991). *Conifers: Morphology and Variation*, revised English edition. Zagreb: Graficki Zavod Hrvatske.

Von Rudloff, E. (1975). Volatile leaf oil analysis in chemosystematic studies of North American conifers. *Biochemical Systematics and Ecology*, **2**, 131–67.

Wakasugi, T., Tsudzuki, J., Ito, S., Nakashima, K., Tsudzuki, T. & Sugiura, M. (1994a). Loss of all *ndh* genes as determined by sequencing the entire chloroplast genome of the black pine *Pinus thunbergii*. *Proceedings of the National Academy of Sciences, USA*, **91**, 9794–8.

Wakasugi, T., Tsudzuki, J., Ito, S., Shibata, M. & Sugiura, M. (1994b). A physical map and clone bank of the black pine (*Pinus thunbergii*) chloroplast genome. *Plant Molecular Biology Reporter*, **12**, 227–41.

Wang, X.-R. & Szmidt, A.E. (1990). Evolutionary analysis of *Pinus densata* Masters, a putative Tertiary hybrid. 2. A study using species-specific chloroplast DNA markers. *Theoretical and Applied Genetics*, **80**, 641–7.

Wang, X.-R. & Szmidt, A.E. (1993). Chloroplast DNA-based phylogeny of Asian *Pinus* species (Pinaceae). *Plant Systematics and Evolution*, **188**, 197–211.

Wang, X.-R., Szmidt, A.E., Lewandowski, A. & Wang, Z.-R. (1990). Evolutionary analysis of *Pinus densata* Masters, a putative Tertiary hybrid. 1. Allozyme variation. *Theoretical and Applied Genetics*, **80**, 635–40.

Wheeler, N.C., Guries, R.P. & O'Malley, R.P. (1983). Biosystematics of the genus *Pinus*, subsection *Contortae*. *Biochemical Systematics and Ecology*, **11**, 333–40.

White, J.B. & Beals, H.O. (1963). Variation in number of resin canals per needle in pond pine. *Botanical Gazette*, **124**, 251–3.

Wright, J.W. (1959). Species hybridization in the white pines. *Forest Science*, **5**, 210–22.

Yoshie, F. & Sakai, A. (1985). Types of Florin rings, distributional patterns of epicuticular wax, and their relationships in the genus *Pinus*. *Canadian Journal of Botany*, **63**, 2150–8.

Zavarin, E. (1988). Taxonomy of pinyon pines. In *II Simposio Nacional sobre Pinos Piñoneros*, ed. M.F. Passini, T. Cibrian & P. Eguiluz, pp. 29–40. Chapingo, Mexico: Centro de Genetica Forestal.

Zavarin, E. & Snajberk, K. (1987). Monoterpene differentiation in relation to the morphology of *Pinus culminicola*, *Pinus nelsonii*, *Pinus pinceana*, and *Pinus maximartinezii*. *Biochemical Systematics and Ecology*, **15**, 307–12.

Zavarin, E., Snajberk, K., Bailey, D.K. & Rockwell, E.C. (1982). Variability in essential oils and needle resin canals of *Pinus longaeva* from eastern California and western Nevada in relation to other members of subsection *Balfourianae*. *Biochemical Systematics and Ecology*, **10**, 11–20.

Zobel, B. (1951). The natural hybrid between Coulter and Jeffrey pines. *Evolution*, **5**, 405–13.

3 Early evolution of pines

Constance I. Millar

3.1 Introduction

The past two decades have seen an explosion of information on the palaeohistory of the Earth. Evidence from plate tectonics has clarified the position of continents in different ages, continental geomorphology, and the dynamics of inland seaways and changing coastlines. Physical and biological evidence has been used to infer palaeoclimates with increasingly finer resolution in time and space. New fossil discoveries have added to the record of past vegetation, and new diagnostics for identifying taxa have led to systematic revisions of many fossil floras. The improved methods and widespread use of radioisotope dating have added precision to determining the ages of fossil floras.

This information, together with phylogenetic analyses of extant taxa, has contributed new insights and a revised understanding of evolution for many plant groups. In pines, major syntheses have focused on two time periods in the history of the genus. Studies on the Mesozoic history of the pine family, and especially *Pinus*, have significantly changed our understanding of the origin of the genus (Miller 1976, 1977, 1982, 1988; Robison 1977; Blackwell 1984; Stockey & Nishida 1986; Stockey & Ueda 1986). Similarly, studies on the Quaternary history of pines have led to new interpretations about the impact of recent palaeohistoric events on the genetic structure and evolutionary relationships of extant species (Critchfield 1984, 1985).

The broad-scale events that influenced the evolution of the genus between its origins in the Mesozoic (Table 3.1) and its present diversity remain obscure. How did important secondary centres of pine diversity in Mexico, western North America, and eastern Asia originate? How do these areas relate to the primary centres of origin for the genus? What events triggered the diversifications of taxa within the genus, and how have historical events influenced current and fossil distribution? Although there have been important contributions to understanding regional biogeography and evolution of pines in the Tertiary (e.g. Eguiluz Piedra 1985, 1988; Axelrod 1986; Lauria 1991), the impact of Palaeogene (Palaeocene to Oligocene) events, especially the Eocene, on the evolution of the genus as a whole has only begun to be analysed. When recent information on plate tectonics, climate, fossils, and biogeography of pines is brought together, the Eocene emerges as one of the most important phases in pine evolution.

3.2 Mesozoic biogeography

3.2.1 Origin of pines

The prevailing hypothesis until the mid-1970s on the origin of the genus relied on the contemporary interpretation of Mesozoic fossil flora and the prevailing theories of the origin of cool-temperate vegetation (Chaney 1940; Mirov 1967; Mirov & Hasbrouck 1976). Fossil pines had been described from Triassic, Jurassic, and abundant Cretaceous locales, with pines especially abundant and diverse at high northern palaeolatitudes. Mirov's widely cited interpretation dated the genus to the late Palaeozoic or earliest Mesozoic with its origin centred in a far-northern circumpolar continent known as Beringia. According to Mirov, the subsequent evolution of pines unfolded in a steady and progressive migration southwards during the Mesozoic and Tertiary, culminating in a final southward thrust toward the equator during the Pleistocene.

Ecology and Biogeography of Pinus, ed. D.M. Richardson. © Cambridge University Press (1998). pp. 69–91.

Table 3.1. Approximate ages and durations of geological eras from the Mesozoic to present

Era	Period	Epoch	Duration (millions of years)	Millons of years ago
CENOZOIC	Quaternary	Holocene		c. the last 10 000 years
		Pleistocene	2.4	
				2.5
	Tertiary			
	Neogene	Pliocene	4.5	
				7
		Miocene	19	
				26
	Palaeogene	Oligocene	12	
				34[a]
		Eocene	16	
				54
		Palaeocene	11	
				65
MESOZOIC	Cretaceous		71	
				136
	Jurassic		54	
				190
	Triassic		35	
				225

[a] The Oligocene–Eocene boundary is generally accepted to be 34 million years ago, coinciding with the terminal Eocene event. Literature published before the middle 1970s and some current authors accept the boundary as 38 million years ago.

This interpretation was cast into doubt by systematic revisions of Mesozoic coniferous fossils. Alvin, Creber, Miller, Stockey and others compared internal anatomy of fossil and extant pinaceous cones and found four diagnostic traits that characterize *Pinus* (summarized in Miller 1976). When previously described fossils were reanalysed, many that were originally ascribed to *Pinus* were reclassified in the extinct pinaceous genera, *Pityostrobus* and *Pseudoaraucaria*. This applied to all the known pinaceous remains from the Triassic and Jurassic, and many from the Cretaceous. In particular, all of the high-latitude macrofossils, primarily in the collections of Heer (1868–83), originally treated as *Pinus*, were reclassified, in some cases even as angiosperms.

The revisions, combined with new fossil discoveries in the last two decades, result in a Cretaceous fossil record of about 25 species of pines from eight northern hemisphere regions (Table 3.2; Fig. 3.1). About half are known from petrified cones whose internal anatomy has been confirmed as *Pinus*. Others are needle, wood, or pollen fossils, generally accepted to be *Pinus*. Of special note is the distribution of these fossil pines. Although geographic biases may be expected due to unequal distribution of sediments and proximity of fossil locations to active palaeobotanists, fossil pines occur at middle and a few high latitudes, widely spread east and west, with apparent centres in northeastern United States, Japan and western Europe. The earliest known pine, *P. belgica*, from the Early Cretaceous (about 130 million years ago) was found in Belgium. Pollen of an Early Cretaceous pine was also found in amber deposits from Alaska (Langenheim, Smiley &

Gray 1960). All other pine fossils are known from Middle to Late Cretaceous deposits.

The taxonomic diversity of fossil pines in the Cretaceous record is broad; two major subgenera and seven subsections are represented. The origin of the genus is thought to be early-middle Mesozoic, although probably not the Palaeozoic as suggested by Mirov (Miller 1976; Eguiluz Piedra 1985; Axelrod 1986; Millar & Kinloch 1991). A major change from Mirov's thinking concerns the location of the centre of origin of the genus and the paths of subsequent radiation. At the beginning of the Mesozoic, there was one land mass, Pangaea (Smith, Hurley & Briden 1981). By the early Jurassic, a northern super-continent, Laurasia, separated and began to drift from a southern continent, Gondwanaland. Although there may have been more land above sea in northern latitudes than at present (Wolfe 1985), little evidence exists for a circumpolar continent, Beringia, that would have supported the origin of pines (Hickey *et al.* 1983; Eguiluz Piedra 1985; Wolfe 1985).

Most importantly, no fossil evidence exists for a high-latitude Mesozoic centre of origin for pines. Mesozoic pine fossils occur between 31 and 50° N latitude, with only two records from higher latitudes (Table 3.2). Similarly, fossils of six species of *Pseudoaraucaria* and over 20 species of *Pityostrobus*, genera considered most closely related to *Pinus* and most likely to have been the ancestral gene pool to *Pinus*, also occurred exclusively at middle latitudes, concentrated in eastern North America and western Europe (Miller 1976, 1988). Hence, a circumpolar origin for *Pinus* is unsupported, and pine origins in middle latitudes are

Pinus Identification	Affinity	Location	Current Lat/Long	Palaeo Lat/Long	Age (million years)	Reference
P. belgica	subsect. Pinus	Belgium, Wealden	50 N/4 E	39 N/5 E	130	Alvin (1960)
Pinus pollen	genus	Alaska, Kuk River	70 N/160 W	80 N/100 W	Early	Langenheim et al. (1960)
Pinus pollen	genus	Maryland	39 N/79 W	32 N/41 W	Late Early	Brenner (1963)
Pinus pollen	genus	Delaware	39 N/75 W	31 N/39 W	Late Early	Groot et al. (1961)
P. wohlgemuthi	subg. Pinus	France	48 N/4 E	35 N/18 E	Late Early	Alvin (1960)
P. ponderosoides	Ponderosae/Australes	Mississippi, Prentiss Co.	34 N/88 W	31 N/52 W	Early Late	Blackwell (1984)
Pinus pollen	subsect. Pinus	Minnesota	44 N/95 W	40 N/55 W	Early Late	Pierce (1957)
P. clementsii	subsect. Pinus	Minnesota	44 N/95 W	40 N/55 W	Late	Chaney (1954)
Pinus sp.	subsect. Pinus	Delaware	39 N/75 W	32 N/40 W	Late	Penny (1947)
P. triphylla	subsect. Pinus	Massachusetts	41 N/71 W	33 N/35 W	Late	Robison (1977)
	Leiophyllae	New York, Staten Is	41 N/74 W	34 N/37 W	Late	Hollick & Jeffrey (1909)
P. tetraphylla	subsect. Pinus	Japan, Hokkaido	42 N/142 E	48 N/135 E	Late	Stockey & Nishida (1986)
P. bifoliata	subsect. Pinus	Japan, Hokkaido	42 N/142 E	49 N/135 E	Late	Stockey & Nishida (1986)
P. pseudotetraphylla	subsect. Pinus	Japan, Hokkaido	43 N/142 E	49 N/135 E	Late	Stockey & Nishida (1986)
P. flabellifolia	Canariensis	Japan, Hokkaido	43 N/142 E	49 N/135 E	Late	Ogura (1932); Stockey & Nishida (1986)
P. quenstedti	Ponderosae	Kansas	38 N/99 W	34 N/61 W	Late	Lesquereux (1883)
P. quinquefolia	Ponderosae	Massachusetts	41 N/71 W	33 N/35 W	Late	Penny (1947); Robison (1977)
	Ponderosae	New York, Staten Is	41 N/74 W	34 N/37 W	Late	Hollick & Jeffrey (1909)
P. pachydermata	subg. Pinus	Japan, Hokkaido	42 N/142 E	48 N/135 E	Late	Ueda & Nishida (1982)
P. pseudostrobifolia	Ponderosae	Japan, Hokkaido	43 N/142 E	49 N/135 E	Late	Ogura (1932); Stockey & Nishida (1986)
P. pseudoflabellifolia	subg. Pinus	Japan, Hokkaido	44 N/142 E	50 N/136 E	Late	Ueda & Nishida (1982)
P. harborensis	Ponderosae	Japan, Hokkaido	44 N/142 E	50 N/136 E	Late	Stockey & Nishida (1986)
P. hokkaidoensis	Ponderosae/Leiophyllae	Japan, Hokkaido	44 N/142 E	50 N/136 E	Late	Stockey & Ueda (1986)
P. cliffwoodensis	Australes/	New Jersey, Magothy	40 N/74 W	32 N/38 W	Late	Miller & Malinky (1986)
Pinus sp.	subg. Strobus	New York, Staten Is	41 N/74 W	34 N/37 W	Late	Jeffrey (1908)
P. magothensis	Strobi	Delaware	39 N/75 W	31 N/38 W	Late	Penny (1947)
P. yezoensis	subg. Strobus	Japan, Hokkaido	43 N/142 E	49 N/135 E	Late	Stopes & Kershaw (1910)
Pinus pollen	genus	Alaska, Colville River	70 N/151 W	75 N/88 W	Late	Frederiksen et al. (1988)

Table 3.2. Distribution and affinities of fossil pines from Cretaceous deposits listed in approximate order of age (old to young). Palaeo-coordinates from Smith et al. (1981)

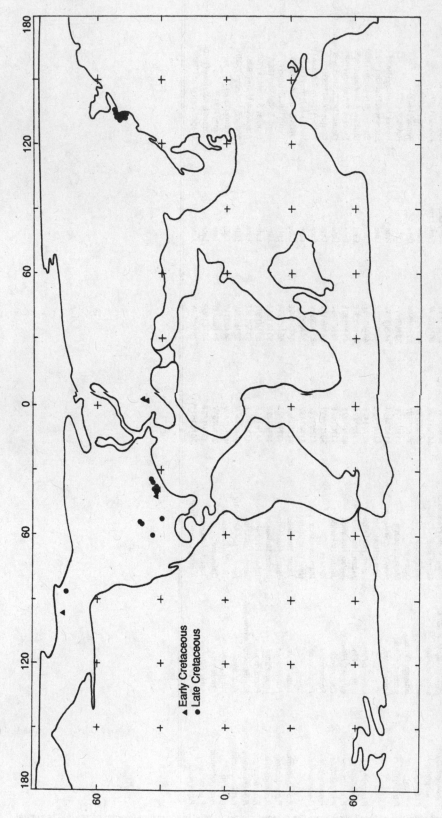

Fig. 3.1. **Distribution of pine fossils from Cretaceous deposits, mapped by estimated palaeo-coordinates on map of the late Cretaceous world. The base map was redrawn from Smith, Hurley & Briden (1981)** *Phanerozoic Palaeocontinental World Maps.*

more likely. The regions of northeastern United States and western Europe, which would have been contiguous in the early and middle Mesozoic (Smith *et al.* 1981), are the current candidates for the centre of origin of the genus (Miller 1976; Eguiluz Piedra 1985; Axelrod 1986; Millar & Kinloch 1991). Alternatively, the diversity of Cretaceous pines and *Pityostrobus* in Japan suggests that pines might have evolved in eastern Asia.

3.2.2 Distribution of pines and climate of the Late Cretaceous

To understand the impact of the early Tertiary on pines, it is important to clarify the known Late Cretaceous distribution of pines and the prevailing climate estimated for the end of the Mesozoic. By the Late Cretaceous, pines had reached eastern and western edges of Laurasia and occurred at middle latitudes in several locations between these extremes (Table 3.2; Fig. 3.1). The widespread distribution of pines in Laurasia by this time indicates that wherever within middle latitudes they originated, their main route of migration was east and west, and not predominately southwards, as Mirov suggested. Migration from eastern North America to western Europe was not impeded until late in the Mesozoic, by which time Laurasia had begun to split into North America and Europe. Laurasia severed first in the south and last in the north. High-latitude connections in the North Atlantic became increasingly reduced towards the end of the Mesozoic, and low seas may have covered the land (Ziegler, Scotese & Barrett 1983; Tiffney 1985a; Parrish 1987). This region would have provided only minor corridors for pine migration. Evidence also exists for land connections at high latitudes in the Bering Sea region between Siberia and Alaska being used as corridors for temperate-adapted flora.

Within the new continents, continuing east–west migration in the Late Cretaceous must have been hindered by seaways that extended the full north–south length of the continents (Kurten 1966; Tiffney 1985a). These seaways divided the continents into separate phytogeographic provinces, creating greater floristic affinities between eastern North America and western Europe, and western North America and eastern Asia, than between the east–west parts of each continent (Wolfe 1975; Tiffney 1985a).

The Late Cretaceous was a time of relative climatic quiescence and equability (Parrish 1987; Upchurch & Wolfe 1987; McGowran 1990). Sea levels were high, and tectonic activity low, creating stable global climates. Although the breakup of Pangaea had commenced, palaeocontinents were still relatively undispersed, resulting in average temperatures in the middle and high latitudes about 10–20 °C warmer than the present (Fig. 3.2) (Savin 1977; Shackleton & Boersma 1981; Miller, Fairbanks & Mountain 1987; Parrish 1987; Upchurch & Wolfe 1987). Evidence on rainfall in the Cretaceous inferred

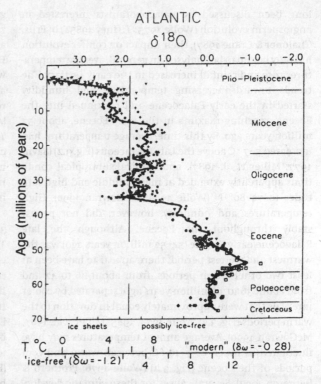

Fig. 3.2. **Palaeotemperatures during the Cenozoic, estimated from deep-sea oxygen isotope records (reprinted by permission from Columbia University Press; Prothero 1994; modified from the original in Miller *et al.* 1987).**

from foliar physiognomy of angiosperms indicates that rainfall patterns were zonal. The northern hemisphere had a humid region around the palaeoequator but extending perhaps 10° more polewards than present (Frakes 1979), a dry zone at low-middle latitudes, and a zone of higher rainfall above latitude 45° N (Parrish 1987). In general, however, latitudinal gradients were shallower, and changes in temperature with latitude were about half to one third of present gradients (Parrish 1987; Upchurch & Wolfe 1987). Temperature and rainfall appear to have been stable annually, with little seasonality at latitudes below 45° N (Upchurch & Wolfe 1987). At higher latitudes, day lengths and precipitation apparently varied seasonally. The major mountain ranges of the northern hemisphere were not developed, or existed only at low elevations. Volcanic activity was minor, so there were few orographic effects on climate and little regional diversity in climate.

3.3 Early Tertiary biogeography

3.3.1 Changing climates

Major changes in climate and vegetation characterized the early Tertiary (Wolfe 1990). Although these events have long been discussed by palaeobotanists interested in angiosperm evolution (Wolfe 1975; Tiffney 1985a, b; Friis, Chaloner & Crane 1987), their impact on conifer evolution has only begun to be analysed. In general, average temperatures rose and rainfall increased in the early Tertiary. The trends toward increasing temperature and humidity started in the early Palaeocene and continued into the Eocene, reaching maxima in the early Eocene, about 52 million years ago. By this time, average temperatures had increased 5–7 °C above the Late Cretaceous (Fig. 3.2) (Savin 1977; Miller *et al.* 1987), and tropical/subtropical conditions apparently extended at many middle and high latitudes to 70–80° N (Wolfe 1985; McGowran 1990). High temperatures and humidity, however, did not persist stably throughout the Eocene. Although the late Palaeocene/early Eocene (54–52 million years ago) was the warmest and wettest period, there appear to have been at least two other warm periods, from about 46 to 42 and from about 36 to 34 million years ago, separated by cooler intervals that were approximately equal in duration to the warm periods (Fig. 3.2; Wolfe 1978, 1985; Miller *et al.* 1987; McGowran 1990). Average annual temperatures may have fluctuated as much as 7–10 °C between warm and cool periods of the Eocene (Fig. 3.2) (Wolfe 1978; Prothero & Berggren 1992). Several causes for these climatic developments have been suggested, including major tectonic events, changes in sea level and ocean circulation, and submarine volcanism resulting in accumulation of atmospheric carbon dioxide and greenhouse heating (Wolfe 1978; Miller *et al.* 1987; Parrish 1987; McGowran 1990; Kerr 1991). Alternatively, large amounts of carbon dioxide may have been produced as a result of ocean–atmosphere interactions following a major extraterrestrial impact at the Cretaceous/Tertiary boundary (O'Keefe & Ahrens 1989), which led to greenhouse warming (Wolfe 1990). Arrangement of continents in large masses and resulting ocean circulations appear to have been the primary cause for generally elevated temperatures of the late Cretaceous and early Tertiary relative to present (Barron 1985a, b; Huntley & Webb 1988).

Although latitudinal gradation was not great, the pattern in the Eocene differed from both the Cretaceous and the later Tertiary and Quaternary: in general, many low-latitude locations were relatively warm and seasonally dry, middle latitudes were warm and wet, and high latitudes were cool and dry (Wolfe 1978; Parrish 1987). Truly arid zones apparently did not exist; there is no evidence for Palaeogene arid deserts or tundra (Axelrod 1979), nor is there evidence for polar ice caps or extensive glaciation (Singh 1988; Prothero & Berggren 1992).

Although these general trends in temperature and humidity existed throughout a broad latitudinal zone worldwide, there appears to have been geographic heterogeneity in the intensity of conditions. The warm humid zone was widest in North America and western Europe, and narrower in central and eastern Asia (Chaney 1940; Hsu 1983; Parrish 1987). In general, continental elevations were low throughout the Palaeogene, and upland areas apparently existed extensively only in one middle-latitude region of western North America and in Antarctica (Axelrod 1966; Wolfe 1985, 1987; Wing 1987; Wolfe & Wehr 1987). In the upland area of western North America, the climate was anomalously temperate compared to other middle-latitude areas. Volcanism and mountain-building in this area during the Eocene also created heterogeneity in local climates and habitats.

3.3.2   The angiosperm boreotropical flora

The changes in climate during the early Tertiary drastically affected global floristics (Wolfe 1975, 1978, 1985; Friis *et al.* 1987). Diverse tropical and subtropical angiosperm floras appeared with increasing geographic representation during the Palaeocene and the warm intervals of the Eocene throughout broad zones at middle latitudes in both northern and southern hemispheres. Originally identified from the London Clay formations of England (Reid & Chandler 1933), similar angiosperm floras have been described from many deposits elsewhere, including western and eastern North America (from the Pacific Coast to Nebraska and Texas; Vermont, Alabama), western and eastern Europe (including England, France, Belgium, Germany, Bulgaria, Ukraine and Russia), northern Egypt, China, and Japan (summaries in Mai 1970; Graham 1972; Wolfe 1975, 1985; Tiffney 1985a, b). Subtropical assemblages occurred north as far as 70° in Alaska, and at other high-latitude locations in Canada, Greenland and Siberia (Wolfe 1977, 1985). In North America, the average zone extended from 30 to 50° N (Wolfe 1985). The origin and major radiation of many angiosperm taxa is documented in these diverse floras.

Plant communities in these Palaeogene floras were adapted to warm, humid and equable conditions. These taxa were similar to those found in modern rainforest vegetation of Malaysia and other extant tropical rainforest regions, and show comparable adaptations. Common genera include *Alangium*, *Engelhardtia*, *Ficus*, *Nypa*, *Nyssa*, *Platycarya*, *Pterocarya* and *Wetherellia*. A few gymnosperms such as *Glyptostrobus*, *Taxodium*, and occasionally *Sequoia* occurred in these floras, but they, too, were apparently adapted to warm, humid conditions.

Recognizing their northern locations and their adaptations to warm, humid conditions, Wolfe (1975, 1977) referred to these widespread angiosperm assemblages as the 'boreotropical flora'. While this assemblage in no way suggests an organic, indivisible unit, it does imply that similarly adapted individual taxa migrated east and west rapidly and unimpeded, at middle latitudes throughout North America and Eurasia in the early Tertiary. The boreotropical flora reached its greatest development in the warm periods of the Eocene. During cool intervals, its latitudinal extent shrank.

3.3.3 Eocene pines

The early Tertiary radiation of many angiosperm lineages and the migration of angiosperm boreotropical taxa east and west at middle latitudes has many parallels with the late Mesozoic radiation and migration patterns of pines. Boreotropical floras occurred in the same locations worldwide in the early Tertiary as pines had in the late Mesozoic. With very few exceptions, pines are not found in boreotropical fossil floras. This prompts the question, where did the pines go?

The Tertiary record of pines begins in the Eocene; no pines are known from the Palaeocene. Pines of the earliest Eocene occur primarily in high (65–80° N) and low (2° N) latitude deposits in North America and Eurasia (Table 3.3; Fig. 3.3). High-latitude locations include central Alaska (Wahrhaftig, Wolfe & Leopold 1969; Dickinson, Cunningham & Ager 1987; Fredericksen, Ager & Edwards 1988), Ellsmere Island, Greenland, Iceland and Spitsbergen (Manum 1962; Schweitzer 1974). Low-latitude pines occurred in Borneo (Muller 1966), and from one low-middle latitude location of the late-early Eocene, near San Diego, California (Axelrod 1986). Pines in these deposits were associated with other pinaceous conifers and with cool-temperate angiosperms such as *Alnus*, *Betula* and *Ulmus*. Ages of these fossils correspond to the first warm humid period of the Eocene (Fig. 3.4).

Fossil pines from the middle Eocene continued to be represented at high and low latitudes, but appeared for the first time in the Tertiary in middle-high latitude locations in North America and Eurasia. Most of the known fossils from the cool period of the earlier middle Eocene are in western North America, from British Columbia (Miller 1973; Stockey 1983, 1984), Washington (Wolfe & Wehr 1987), Idaho (Axelrod 1986), and Colorado (Wodehouse 1933; MacGinite 1953) (Table 3.3; Fig. 3.3). Western North America in general contains some of richest plant-bearing deposits from the Eocene in the world, and dozens of fossil floras have been described. Only a few of these contain pines and other temperate taxa, and these are all concentrated in northern Idaho, central Wyoming, north to central Idaho and British Columbia. This is the region that has been identified as an upland area with average elevations of 1200–1500 m (Axelrod 1965; Axelrod & Raven 1985; Wing 1987; Wolfe 1987). Pine deposits in some of these areas are associated with active volcanoes and high elevations, e.g. Bull Run, Thunder Mountain: (Axelrod 1965, 1986) and Creede (2500 m: Wolfe & Schorn 1989).

The warm humid period of the middle Eocene is represented by few pine fossils (Table 3.3). Pine was present at high latitudes in the Mackenzie Delta of Alaska (80° N; Norris 1982), at low-middle latitudes in Borneo (2° N; Muller 1966) and southern Alabama (36° N; Gray 1960), and also at a few middle latitudes in Nevada (50° N; Axelrod 1966, 1968), within the upland plateau of western North America.

Pine fossils from the subsequent cool period of the later Eocene occurred at middle latitudes in Washington (Miller 1974), Nevada (Axelrod 1966, 1968), Colorado (Leopold & MacGinite 1972; Axelrod 1986), New Mexico (Leopold & MacGinite 1972), Japan (Huzioka & Takahashi 1970; Tanai 1970, interpreted by Wolfe 1985 to be early-late Eocene), and Fushun, China (Hsu 1983), as well as a continuing presence in Borneo (Muller 1966) and at high latitudes (Norris 1982) (Table 3.3).

The latest Eocene marks the final widespread period of tropical conditions at middle latitudes in the Palaeogene (Fig. 3.2). Pine fossils from the late Eocene occur primarily at high latitudes in western Siberia (Dorofeev 1963), Alaska (Norris 1982), and British Columbia (Hopkins, Rutter & Rouse 1972), at low latitudes in Borneo (Muller 1966), and along the Tethys seaway in southeastern Europe (Chiguriaeva 1952). Pines from the southeast coast of China, in the provinces of Jiang-su, Zhejian and Fujian (Hsu 1983), may also be from this period (Table 3.3).

3.3.4 The 'Terminal Eocene Event' and pine expansions

The end of the Eocene/early Oligocene was marked by the most profound climatic event of the Tertiary (Burchardt 1978; Wolfe 1978; Miller *et al.* 1987; Parrish 1987; Singh 1988; McGowran 1990; Prothero & Berggren 1992). The fluctuations from warm to cool periods during the Eocene were minor compared with the drop in regional temperatures at the end of the Eocene. Around 35 million years ago there was a dramatic change in oxygen isotope ratios, indicating a rapid cooling of ocean temperatures (Miller *et al.* 1987). Average annual temperatures dropped *c.* 10–14 °C, in some areas over only one million years (Figs. 3.2, 3.4). The decline in average temperatures was apparently accompanied by a large decrease in rainfall and increase in seasonality. Whereas in the Eocene the mean annual range of temperatures (seasonality) is estimated as only 3–5 °C, in the early Oligocene it is estimated as 25 °C – about twice the range of the present (Wolfe 1971; Prothero & Berggren 1992). In many parts of the world, complex continental climate patterns developed, possibly for the first time, and continental ice sheets first occurred in the Oligocene (Singh 1988; McGowran 1990). Extensive volcanism and mountain-building in the Rocky

Identification	Affinity	Location	Current Lat/Long	Palaeo Lat/Long	Age (million years)	Reference
Pinus pollen	genus	Alaska, Nenana	64 N/148 W	77 N/128 W	54	Wahrhaftig et al. (1969)
Pinus pollen	genus	Alaska, Colville	70 N/151 W	80 N/127 W	Early	Fredericksen et al. (1988)
Pinus? pollen (bisaccate grains)	genus	Alaska, Death Valley	65 N/162 W	80 N/143 W	Early	Dickinson et al. (1987)
Pinus pollen	genus	Canada, Ellsmere Is	77 N/84 W	75 N/18 W	Early	Manum (1962)
Pinus pollen	genus	Greenland	70 N/52 W	60 N/8 W	Early	Manum (1962)
Pinus pollen	genus	Iceland	65 N/21 W	50 N/8 E	Early	Manum (1962)
Pinus pollen	genus	Svalbard	79 N/14 E	70 N/20 E	Early	Manum (1962), Schweitzer (1974)
Pinus pollen	genus	North Dakota, Dunn Co.	47 N/103 W	56 N/70 W	Early	Leopold & MacGinite (1972)
Pinus pollen	genus	NW Borneo	2 N/110 E	2 N/110 E	Late Early	Muller (1966)
P. delmarensis	subg. Strobus	S. California	33 N/117 W	42 N/92 W	47–48	Axelrod (1986)
Pinus pollen	Cembroides	S. California	33 N/117 W	41 N/92 W	47–48	Axelrod & Raven (1985)
Pinus pollen	Ponderosae	S. California	33 N/117 W	41 N/92 W	47–48	Axelrod & Raven (1985)
Pinus pollen	genus	NE China, Fushun	42 N/124 E	45 N/118 E	Mid & Late	Sung & Liu (1976); Hsu (1983)
P. balli	Cembroides	Colorado, Green River	41 N/109 W	50 N/80 W	47	Wodehouse (1933); Brown (1934)
P. driftwoodensis	subsect. Pinus/Ponderosae/Australes	BC, Smithers	55 N/127 W	66 N/97 W	46–47	Stockey (1983)
P. arnoldii	subsect. Pinus	BC, Princeton	50 N/120 W	61 N/90 W	46–47	Miller (1973); Stockey (1984), Phipps et al. (1995)
P. similkameensis	subg. Strobus	BC, Princeton	50 N/120 W	61 N/90 W	46–47	Miller (1973); Phipps et al. (1995)
P. princetonensis	subsect. Pinus	BC, Princeton	50 N/120 W	61 N/90 W	46–47	Stockey (1984)
P. allisonii	subsect. Pinus	BC, Princeton	50 N/120 W	61 N/90 W	46–47	Stockey (1984)
P. andersonii	Ponderosae	BC, Princeton	50 N/120 W	61 N/90 W	46–47	Stockey (1984)
Pinus pollen	genus	Washington, Republic	49 N/118 W	60 N/87 W	46–47	Wolfe & Wehr (1987)
P. balfouroides	Balfourianae	Idaho, Thunder Mt.	45 N/114 W	55 N/84 W	46–47	Brown (1937); Axelrod (1986)
Pinus pollen	genus	Canada, MacKenzie Delta	69 N/134 W	80 N/80 W	Mid–Late	Norris (1982)
Pinus pollen	genus	Alabama, Claiborne Bluffs	31 N/88 W	36 N/62 W	Middle	Gray (1960)
P. wheeleri	Strobi	Nevada, Elko	41 N/116 W	50 N/90 W	42	Axelrod (1968)
P. crossii	Balfourianae	Nevada, Copper Basin	42 N/115 W	51 N/89 W	40–44	Axelrod (1966)
P. wolfei	subsect. Pinus/Contortae	Washington, Little Falls	47 N/123 W	60 N/93 W	Early Late	Miller (1974)
P. alvordensis	Contortae	Nevada, Bull Run	41 N/117 W	50 N/91 W	38	Axelrod (1968)
Pinus pollen	genus	New Mexico, Bernalillo	35 N/107 W	44 N/80 W	38	Leopold & MacGinite (1972)
Pinus sp.	genus	SW Japan, Ube	34 N/130 E	40 N/130 E	Early Late	Huzioka & Takahashi (1972)
Pinus sp.	genus	Japan, Kushiro	43 N/144 W	50 N/135 E	Early Late	Tanai (1972)
Pinus sp.	genus	Colorado, Chaffee	39 N/105 W	49 N/76 W	Early Late	Leopold & MacGinite (1972)
Pinus sp.	genus	BC, Parsnip River	55 N/122 W	66 N/90 W	36–37	Hopkins et al. (1972)
Pinus sp.	genus	Siberia, Tavda River	58 N/65 E	52 N/58 E	35	Dorofeev (1963)
Pinus sp.	genus	Siberia, Ob River	57 N/85 E	54 N/75 E	Late	Dorofeev (1963)
Pinus sp.	genus	Siberia, Tym River	60 N/82 E	55 N/73 E	Late	Dorofeev (1963)
P. spinosa	subsect. Pinus/Ponderosae	Siberia, Irtysch River	57 N/75 E	52 N/67 E	Late	Dorofeev (1963)
Pinus pollen	genus	Ukraine, River Don	47 N/40 E	35 N/38 E	Late	Chiguriaeva (1952)
P. sturgisii	Strobi	Colorado, Florrissant	39 N/105 W	48 N/76 W	34–35	MacGinite (1953); Axelrod (1986)
P. florrissantii	Strobi	Colorado, Florrissant	39 N/105 W	48 N/76 W	34–35	MacGinite (1953); Axelrod (1986)
P. wheeleri	Strobi	Colorado, Florrissant	39 N/105 W	48 N/76 W	34–35	MacGinite (1953); Axelrod (1986)
Pinus pollen	genus	SE China	25–35 N/120 E	28–34 N/118 E	Late	Hsu (1983)

Table 3.3. Distribution and affinities of fossil pines from Eocene deposits listed in approximate order of age (old to young). Palaeo-coordinates from Smith et al. (1981)

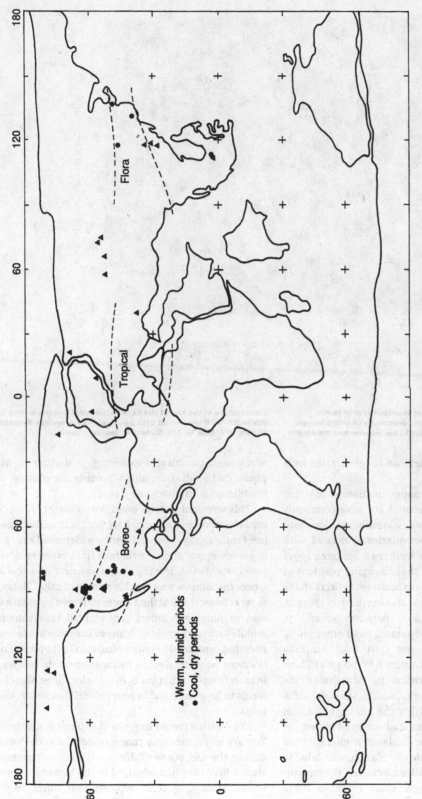

Fig. 3.3. Distribution of pine fossils from Eocene deposits, mapped by estimated palaeo-coordinates on map of the Palaeocene world. The base map was redrawn from Smith *et al.* (1981) *Phanerozoic Palaeocontinental World Maps.*

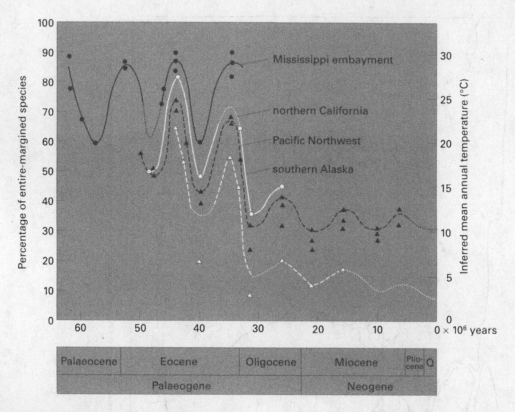

Fig. 3.4. **Estimated temperatures from four latitudes in North America during the Tertiary (based on percentage of entire leaves), indicating the warm and cool episodes in the Eocene and the drop in** temperature at the end of the Eocene (terminal Eocene event), from Wolfe (1978). Reproduced with permission of *American Scientist*, journal of Sigma Xi, The Scientific Research Society.

Mountains, Himalayas, and Mexican ranges created local climatic diversity.

The causes for the major change in climates marking the end of the Eocene are debated but now commonly ascribed to geographic changes in continental locations and critical shifts in plate boundaries associated with these changes (see Singh 1988; Prothero & Berggren 1992). One line of evidence suggests that changing positions of the Earth's continents and oceans had direct effects on the Earth's climate, leading to major cooling (Barron 1985a, b; McGowran 1990; Prothero 1994). Between 40 and 30 million years ago, Antarctica separated from other southern hemisphere continents, and split from Australia around 37 million years ago (Huntley & Webb 1988). These shifts in continents were critical in establishing the Southern Circumpolar Current, which strongly influenced global cooling by inhibiting the mixing of tropical warm and polar cold oceans that had occurred during the Eocene. This appears to have initiated a change from thermospheric to thermohaline circulation, which led to a sharpened temperature gradient between the equator and high latitudes (see Singh 1988; Prothero 1994). Another possibility is that the formation of giant uplifted plateaus in southern Asia and the American West led to

accelerated chemical weathering, a decline in atmospheric carbon dioxide, and a 'greenhouse cooling' effect (Ruddiman & Kutzbach 1989, 1991).

This terminal Eocene event (Wolfe 1978), now considered by improved dating to have been very early Oligocene (see Prothero 1994), was marked by widespread extirpation of boreotropical angiosperm taxa from many middle latitudes worldwide, leaving only remnants in a few areas where the climate presumably remained mild. The extinctions of boreotropical floras were mirrored by great expansion of pines and other cool-adapted taxa into many middle-latitude locations. Many of the same fossil-bearing sites that contained boreotropical taxa in their late Eocene horizons were dominated by cool-temperate conifer taxa in early Oligocene horizons. In some localities, this change seems to have occurred in only one million years (Axelrod 1965).

The existing record suggests that pines made their first Tertiary appearance at many middle-latitude locations during the Oligocene (Table 3.4; Fig. 3.5), recolonizing areas where they had occurred in the Mesozoic. Pines are known from Oligocene deposits that range in North America from the Gulf of Alaska (Wolfe 1972), British Columbia (Banks, Ortiz-Sotomayor & Hartman 1981),

Identification	Affinity	Location	Current Lat/Long	Palaeo Lat/Long	Age (million years)	Reference
Pinus pollen	genus	Oregon, Bridge Cr.	44 N/118 W	47 N/105 W	31	Wolfe (1981); Mason (1927)
Pinus pollen	genus	Oregon, Lyons	45 N/123 W	49 N/110 W	31	Wolfe (1981)
Pinus pollen	genus	North Dakota, Dunn Co.	47 N/103 W	48 N/88 W	Early	Leopold & MacGinite (1972)
Pinus pollen	genus	Montana, Ruby Basin	45 N/112 W	46 N/100 W	32–36	Becker (1961); Wolfe (1977)
P. crossii	Balfourianae	New Mexico, Hillsboro	33 N/108 W	34 N/98 W	32	Axelrod & Bailey (1976)
Pinus pollen	genus	BC, Australian Cr.	56 N/126 W	60 N/110 W	Early	Piel (1971); Rouse & Matthews (1979
Pinus pollen	genus	Kazakhstan, F.S.U.	44–50 N/60–72 E	42–48 N/58–70 E	Early	in Leopold et al. (1992
Pinus pollen	Contortae	Cis-Ohotsk, F.S.U.	44–52 N/132–144 E	50–60 N/128–145 E	Early	in Leopold et al. (1992)
P. escalatensis	Contortae	BC, Escalante Is.	50 N/127 W	56 N/115 W	Early–Late	Banks et al. (1981)
Pinus pollen	genus	NW Borneo	2 N/110 E	5 N/110 E	Early–Late	Muller (1966)
P. buchananii	Ponderosae	Gulf of Alaska	60 N/145 W	68 N/128 W	Late	Wolfe (1972, 1977)
P. avonensis	Ponderosae	Washington, Olympic	48 N/123 W	52 N/110 W	Late?	Underwood & Miller (1980)
P. anthrarivus	Strobi	Montana, Avon	47 N/113 W	50 N/100 W	Late?	Miller (1969)
P. balfouroides	Balfourianae	Idaho, Coal Cr.	45 N/114 W	47 N/103 W	29	Axelrod (1986)
Pinus pollen	genus	Idaho, Coal Cr.	45 N/114 W	47 N/103 W	29	Axelrod (1986)
Pinus pollen	genus	Montana, Beaverhead Basin	45 N/113 W	47 N/102 W	Late	Becker (1961), Leopold & MacGinite (1972)
P. crossii	Balfourianae	Colorado, Chaffee	39 N/105 W	40 N/93 W	Late	Leopold & MacGinite (1972)
P. riogrande	Ponderosae	Colorado, Creede	38 N/105 W	39 N/93 W	27	Schorn & Wolfe (1989)
P. sanjuanensis	Cembroides	Colorado, Creede	38 N/105 W	39 N/93 W	27	Axelrod (1986); Schorn & Wolfe (1989)
Pinus sp. (cf. alvordensis)	Contortae	Colorado, Creede	38 N/105 W	39 N/93 W	27	Axelrod (1986); Schorn & Wolfe (1989)
Pinus pollen (1 grain)	genus	Mexico, Chiapas	16 N/92 W	14 N/83 W	Late?	Langenheim et al. (1967)
Pinus pollen	genus	NW China, Quidam	43 N/963 E	44 N/92 E	Late?	Hsu (1983)
Pinus pollen	genus	NW China, NW Gansu	42 N/95 E	43 N/90 E	Late	Sung (1958)
P. prepityusa	subsect. Pinus	W Transcaucasia	42 N/42 E	35 N/40 E	Late?	Czeczott (1954)
P. maikopiae	subsect. Pinus/Contortae	E Transcaucasia	40 N/50 E	35 N/45 E	Late?	Palibin (1935)

Table 3.4. Distribution and affinities of fossil pines from Oligocene deposits listed in approximate order of age (old to young). Palaeo-coordinates from Smith et al. (1981)

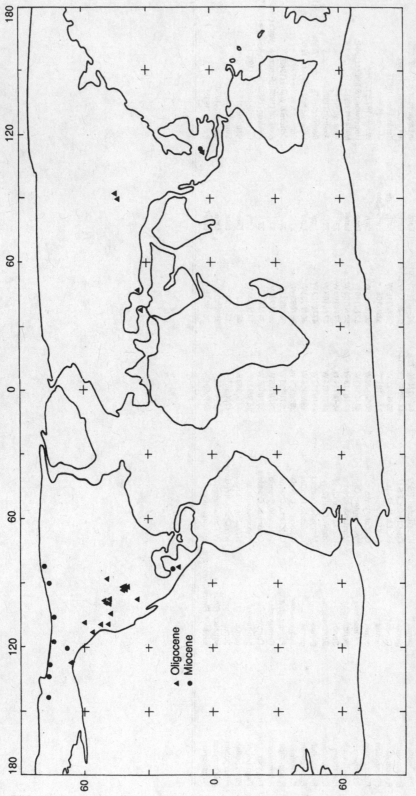

Fig. 3.5. **Distribution of pine fossils from Oligocene and selected high and low latitude Miocene deposits, mapped by estimated palaeo-coordinates on map of the early Miocene world. The base map was redrawn from Smith et al. (1981) Phanerozoic Paleocontinental World Maps.**

Oligocene ▲
Miocene ●

Washington (Underwood & Miller 1980), Montana (Miller 1969; Leopold & MacGinite 1972), Idaho (Axelrod 1986), Oregon (Wolfe 1981) and Colorado (Leopold & MacGinite 1972; Schorn & Wolfe 1989), to a single specimen from Chiapas, Mexico (Langenheim, Hackner & Bartlett 1967) (Table 3.4). In Asia, Oligocene deposits are in eastern and western Transcaucasia (Palibin 1935; Czeczott 1954; Leopold, Liu & Clay-Poole 1992), and in northwest China, in the Quidam Basin, and in the Jinguan Basin of western Gansu (Sung 1958; Hsu 1983). In Japan, pine was present as a minor component of a few floras (Tanai 1972). Pine was abundant in the Oligocene strata of Borneo (Muller 1966). Although some of the pine-bearing deposits were associated with areas of volcanism and mountain-building, others were in lowland and coastal areas where boreotropical flora had flourished in previous epochs.

The climatic deterioration of the early Oligocene was followed by an ameliorating and warming trend during the late Oligocene and into the Miocene (Fig. 3.2). Rainfall, however, stayed moderately low and climatic conditions remained temperate at middle latitudes. The record indicates that pines rose in abundance throughout middle latitudes in North America (Axelrod 1986), Europe (van der Burgh 1973; Klaus 1989), and Asia (Hsu 1983) in the Miocene, and the direct ancestors of many modern pine species can be traced to Miocene pines. The warm conditions of the Miocene supported expanded pine forests at high latitudes, such as Banks Island (78–80° N; Hills, Klovan & Sweet 1974), MacKenzie Delta (72–85° N; Ritchie 1984), and Wrangell Mountains, Alaska (68° N; Wolfe 1969) (Table 3.5). Pine fossils in low-latitude deposits, such as Veracruz, Mexico (17° N; Graham 1976) indicate the continued presence of pine in mountainous low latitudes, despite the fact that lowlands in these latitudes were dominated by boreotropical flora.

3.4 Impact of early Tertiary climate on pine evolution

3.4.1 Eocene pine refugia

The preceding overview of biogeography provides evidence for the hypothesis that pine distributions shifted latitudinally and that pines expanded and contracted in elevational extent several times from the late Mesozoic to the middle Tertiary. In general, the hypothesis states that pine populations (1) occurred throughout middle-latitudes in the Mesozoic, (2) were fragmented and displaced to high and low latitudes and to middle latitude uplands in the Eocene, then (3) reappeared widely throughout middle latitudes in the Oligocene and Miocene, where they remained for the rest of the Tertiary.

The Mesozoic origin and spread of pines east and west throughout Laurasia apparently occurred under a warm temperate climate that was equable and had little latitudinal gradation. Early migration and radiation of pines was favoured in the Mesozoic not only by climatic conditions, but by the absence at high-middle latitudes of competing angiosperm taxa. Although angiosperms apparently originated about the same time as the pines in the Early Cretaceous (Taylor & Hickey 1990), their rise to dominance in middle latitudes did not occur until the Late Cretaceous (Wolfe & Upchurch 1986).

The shift in global climates towards warm humid conditions of middle latitudes in the early Tertiary appears to have favoured the migration and radiation of angiosperms at the expense of pines. Although pines have broad tolerance of climatic and edaphic conditions, they do not survive or grow well in hot and humid climates (Mirov 1967; Bond 1989). More importantly, in these conditions they are poor competitors with angiosperms in seedling establishment, height growth, and reproduction. Conifers dominate in areas where angiosperm competition is reduced, for example by fire, cold, or nutrient shortages (Bond 1989).

Such a situation apparently initiated widespread extirpations of pines over the lowlands at middle latitudes. The remaining areas of suitable pine habitat acted as refugia for pines during the warm, humid periods of the Palaeogene.

In general, the fossil evidence indicates three important refugial zones. The circumpolar high-latitude zone originally thought by Mirov to be the cradle of *Pinus* emerges not as a primary Mesozoic centre of origin but as an important Eocene refugium. Although many Mesozoic fossils from the polar region that had been identified as pine were later discredited, credible new records, mostly pollen, confirm abundant pine at many high-latitude locations in North America and Eurasia during the warm periods of the Eocene (Table 3.3). Pines may have been even more widespread in polar regions during the early Tertiary than indicated by the geography of present land masses. Much of the Arctic Sea is shallow, and tectonic evidence suggests that there may have been more land above sea level in polar regions in the early Tertiary than at present (Smith *et al.* 1981; Wolfe 1985). Land connections between North America and Eurasia in the North Atlantic (Tiffney 1985a, b) would have provided high-latitude corridors for east–west migration of pines during the Eocene (Tiffney 1985a, b).

A major limiting condition for plant growth polewards is light. Tertiary scientists have wondered how, even under warm climates, plants could survive low light conditions of high latitudes. The continents do not appear to have occupied latitudes in this part of the globe significantly

Table 3.5. Distribution and affinities of select high and low latitude fossil pines from Miocene deposits listed in approximate order of age (old to young). Palaeo-coordinates from Smith *et al.* (1981)

Identification	Affinity	Location	Current Lat/Long	Paleo Lat/Long	Age (million yarsa)	Reference
Pinus sp.	*Strobi*	Canada, Banks Is.	72 N/122 W	78 N/92 W	Early	Hills *et al.* (1974)
Pinus sp.	genus	Canada	74 N/122 W	80 N/87 W	Early	Hills *et al.* (1974)
Pinus pollen	genus	MacKenzie Delta/Alaska	70–80 N/130–150 W	72–85 N/94–122 W	Early	Ritchie (1984)
Pinus pollen	genus	NW Borneo	2 N/110 E	5 N/110 E	Early	Muller (1966)
Pinus pollen	genus	Alaska, Nuwok	70 N/160 W	78 N/142 W	22–26	Wolff (1972, 1977)
Pinus pollen	genus	Alaska, Wrangell Mts	62 N/140 W	68 N/122 W	Middle	Wolfe (1969)
Pinus pollen	genus	Mexico, Veracruz	17 N/94 W	17 N/85 W	Middle	Graham (1976)
Pinus sp.	subsect. *Pinus*	Spain		not available		Klaus (1989)
Pinus sp.	subsect. *Pinus*	Greece		not available		Klaus (1989)
Pinus sp.	subsect. *Pinus*	former Yugoslavia		not available		Klaus (1989)
Pinus sp.	subsect. *Pinus*	Romania		not available		Klaus (1989)

different from the present. The suggestion that the earth may have had a lower angle of inclination in the early Tertiary, which would allow more light at high latitudes, has been discredited as untenable from geophysical evidence (Donn 1982; McKenna 1983; Barron 1984; Prothero 1994). Alternatively, geophysical and climatic conditions of the early Tertiary may have increased availability of carbon dioxide and permitted plants to photosynthesize under light regimes that were prohibitive under conditions of lower carbon dioxide concentrations that apparently followed the end of the Eocene (Berner, Lasaga & Carrels 1983; Creber & Chaloner 1984; Ruddiman & Kutzbach 1989, 1991; Kerr 1991). Further, solar radiation may have been greater during the Eocene (McKenna 1980), thus enabling more plant growth at polar latitudes.

Low latitudes in North America and Eurasia also appear to have been refugia for pines during the warm humid periods of the Eocene. There is ample evidence that conditions in general were warmer and drier at low latitudes than at middle latitudes in the Eocene. These conditions have been documented for the southeastern United States, Central America, Taiwan (Wolfe 1975), southern China and southeastern Asia (Guo 1980; Hsu 1983), and Borneo (Muller 1966). During the Palaeogene, the southern boundary of Eurasia was the Tethys seaway. Since the Indian subcontinent had not yet collided with Asia, the area of the present Himalayas marked the Tethys coastline in Asia. Unlike circumpolar regions, the low-latitude dry areas were disjunct and disconnected, even between the southeastern United States and Central America, and refugial populations were unlikely to have been connected by broad migration routes (Wolfe 1975).

There is less abundant fossil evidence for pines at low latitudes than at high latitudes during the Eocene. There are few early Tertiary plant-bearing deposits in the areas of special interest, such as Mexico, Central America and the Mediterranean, mainly because of the lack of depositional sediments (Martin & Harrell 1957; Eguiluz Piedra 1985). Nevertheless, Eocene pine fossils have been found at low latitudes in southeastern Asia, southeastern and southwestern United States, and in areas of southeastern Europe that were at low latitudes during the Eocene (Table 3.3). In Mexico, the single Oligocene specimen of pine at Chiapas (14° N; one of the earliest Tertiary plant records in Mexico), and the abundance of pines in Veracruz (17° N) from the Miocene suggest an earlier pine presence in Mexico and Central America (Table 3.4). Similarly, pines were present in the Oligocene and abundant in Miocene deposits from southern Europe in the region of the Tethys (Palibin 1935; Czeczott 1954; Klipper 1968; van der Burgh 1973; Klaus 1989).

Pines were not absent from middle latitudes during the Eocene. Their presence in select fossil deposits from interior western North America testifies to the presence of mid-latitude refugia. Pines in these regions more commonly date to the cool periods of the Eocene. During these times pine populations apparently expanded, whereas during the warm humid periods they contracted into narrow refugial areas where conditions were favourable. During warm humid periods, pines in interior western North America may have been fragmented into small local refugia and, therefore, poorly represented in fossil deposits.

The interior of western North America was anomalous for the Eocene world in having major upland areas and centres of volcanism. Compressive interactions between the Pacific and North American plates created the force for this broad band of tectonism known as the Laramide orogeny. Laramide deformation continued from the Cretaceous into the Eocene to create a broad uniformly elevated region (summarized by Minckley, Hendrickson & Bond 1986). Several of the pine-bearing deposits occurred at high elevations in Eocene calderas (Thunder Mountain, Idaho; Creede, Colorado) where climates were more temperate than elsewhere at middle latitudes. This region has been documented for angiosperm flora as being one of the first areas at middle latitudes in North America where climatic conditions became temperate during the Palaeogene, and as a centre of conifer diversity, which would serve as a source for migrations into western North America throughout the rest of the Tertiary (Millar 1996). Widespread volcanism along the cordillera added many cubic meters and topographic diversity to the landscape in this region, building a highland that stretched from Arizona to Canada (Leopold & MacGinite 1972; Wing 1987; Wolfe 1987; Ruddiman & Kutzbach 1991).

The juxtaposition of this upland area to lowlands containing boreotropical taxa is suggested by floral compositions of the early Eocene Chalk Bluffs deposit (MacGinite 1953), currently located in the western Sierra Nevada, California. This classic flora is well known from macrofossils for its vast array of tropical-adapted species (71 taxa identified). More recent analysis of pollen from these deposits shows a very small fraction of pine pollen. Pine has light and buoyant pollen, capable of dispersing hundreds of kilometres. Pine probably blew into the Chalk Bluffs region from the uplands of what is now western Nevada (see Millar 1996). Following orogenic activity in the Late Cretaceous/Eocene, western North America was geologically rather quiescent, and orogeny and uplift did not recur until the early Miocene.

Other middle-latitude areas such as eastern Asia may have supported pines during the warm humid periods of the Eocene, although only western North America is known to have had important upland regions. Although the early Tertiary fossil record of eastern Asia is not as complete as in western North America, pines have been found at a few middle-latitude locations in the Eocene. In Japan, the first Tertiary pines date to the last cool period of the Eocene and are not documented for the warm period at the end of the Eocene (Huzioka & Takahashi 1970; Wolfe 1985). These Eocene records may represent marginal populations of species that were centred further north and expanded southwards during the cool periods. Boreotropical angiosperms occurred throughout Asia in the warm humid periods of the Eocene, although they do not appear to have extended as far north as in North America (Chaney 1940; Hsu 1983).

3.4.2 Secondary centres of pine diversity

The widespread extirpation of pine populations from middle latitudes at the end of the Mesozoic is hypothesized to have led to extinction of many pine species and to have greatly depleted genetic diversity in others. Many Cretaceous fossil pines had combinations of traits not known in extant species and represent lineages that went extinct. Of two closely related pinaceous genera of the

Mesozoic, *Pseudoaraucaria* fossils have not been found in rocks younger than the Late Cretaceous, and the youngest *Pityostrobus* fossils date to the early Eocene (Miller 1976). The disappearance in these two genera of many diverse lineages closely related to *Pinus* represents major extinctions in the pine family during the early Tertiary.

Despite this depletion of diversity in pines and taxa related to pines, the tectonic events and consequent migrations of the early Tertiary appear to have culminated in the creation of several new centres of pine diversity and the evolution of new pine lineages. This change appears to have been due in large part to increased tectonic activity. Although the period from the Mesozoic to the Palaeocene had been relatively quiescent and land elevations were generally low across the continents, by the late Eocene several regions were active tectonically. In subsequent epochs, volcanism and mountain-building became increasingly important locally and globally.

The uplift of new mountain ranges and volcanic activity created environmental heterogeneity. Areas where active mountain-building coincided with Palaeogene pine refugia are hypothesized to have become centres of pine radiation. Local climatic diversity was created by elevational differences, and rain shadow and other orographic effects developed. Diversity of soils evolved, with many areas having newly disturbed sites following volcanic activity. Mountain-building created new barriers to migration and gene flow, causing lineages to be fragmented and isolated. All these conditions must have favoured divergence and speciation in pines. Conversely, refugial areas that did not undergo major early Tertiary mountain-building, such as the high-latitude zones, were centres from which pines migrated in the Neogene, but did not become important centres of pine divergence and speciation.

Although pines expanded and contracted from refugial areas during the fluctuating warm and cool periods of the Eocene, it was the major climatic deterioration near the Eocene/Oligocene boundary that seemed to initiate major migrations out of refugia and to coincide with a time when pines appeared more widespread at middle latitudes. These migrations provided further opportunities for divergence. As new environments were encountered, genetic isolation may have occurred, and possibilities for genetic drift by founder effects arose, as well as for hybridization with formerly isolated lineages.

3.4.3 Evolution within *Pinus* in relation to Eocene refugia

By the end of the Palaeogene, all major subsections of *Pinus*, with the possible exception of *Cembrae*, had evolved (Axelrod 1986; Millar & Kinloch 1991). The events of the early Tertiary probably gave rise to at least two subsection-complexes, speciation of lineages within several subsections, and the current biogeography of many subsections and groups within several subsections.

The hypothesis that pines were concentrated into refugial regions during the Eocene explains the current bimodal distribution and pattern of diversity of pines at low and high latitudes (see Axelrod 1986, for maps of pine subsections). In North America three subsections, *Cembroides*, *Leiophyllae* and *Australes*, appear to have been concentrated in southern refugia during the Eocene and radiated from them subsequently (Axelrod 1986). Subsections *Cembroides* and *Leiophyllae* seem to have been limited to western North America and Central American refugia whereas *Australes* may have had refugial areas with a broader southern distribution, including the Gulf Coast (Tables 3.3, 3.4). The current distribution of *Australes* in both the southeastern United States and in Central America/Caribbean suggests that *Australes* may have had refugia in both regions. Although not identifiable to subsection, the abundant pine fossils of the Eocene from southern Alabama (Table 3.3) are from a site that is within the range of several extant species of *Australes*. Small genetic distances between taxa of *Australes* versus *Leiophyllae* (Strauss & Doerksen 1990) suggest that these subsections were in contact or did not diverge from each other until after the Mesozoic.

Four subsections in North America appear to have been fragmented by Eocene events into several refugial regions. A division into northern and southern refugia in subsection *Pinus* is indicated by the extant lineages of *P. resinosa* (latitude 4–52° N) and *P. cubensis* (restricted to Cuba; 22° N). Eocene fossils with affinities to subsection *Pinus* that support a northern refugium have been found in British Columbia and northern Washington (Table 3.3).

Subsection *Contortae* similarly seems to have been divided into northern and southern refugia: the lineages leading to *P. contorta* and *P. banksiana* clearly had northern origins and boreal adaptations whereas those leading to *P. clausa* and *P. virginiana* had southeastern affinities. Eocene fossils with affinities to *Contortae* have been found in northern Washington, and Oligocene fossils in British Columbia, corroborating a northern refugium (Tables 3.3, 3.4).

Subsection *Ponderosae* may have also had both northern and southern refugia. There are 10–17 species endemic or nearly endemic to Mexico and Central America, whereas wide-ranging *P. ponderosa* has a northern distribution and northern ecological affinities in its habitat preference and vegetation associates. Early Tertiary pines most closely allied to *Ponderosae* have been found in British Columbia (Eocene, Table 3.3), northern Washington, and Montana (Oligocene, Table 3.4). Lineages of *Ponderosae* may have been concentrated also in Rocky Mountain refugia as suggested by Oligocene fossils from Colorado with affinities to this subsection (Table 3.3). A recent re-evaluation of evolutionary patterns in *Ponderosae* also indicates northern and southern division in the subsection (Lauria 1991). *Pinus ponderosa* is considered to have originated in northern latitudes, while other distinct phylads likely of the 'Pseudostrobus Group' of subsection *Ponderosae* (see Chap. 2, this volume) had southern origins in Mexico and Central America.

Subsection *Strobi* has northern and southern lineages in North America. Ancestral lineages of *Pinus monticola* and *P. strobus* would have migrated from northern refugia, and those of *P. ayacahuite* and *P. chiapensis* would have migrated from southern refugia. Northern refugia are corroborated by Eocene fossil pines allied to subsection *Strobi* from British Columbia (Table 3.3). Lineages of subsection *Strobi* may also have been fragmented into middle-latitude refugia in the Rocky Mountain region, represented currently by *P. flexilis* and possibly *P. ayacahuite* var. *strobiformis*, and by Oligocene fossils from Idaho and Colorado (Table 3.4).

The small subsection *Balfourianae* is an ancient lineage (Kossack 1989; Strauss & Doerksen 1990; Millar & Kinloch 1991) that appears to have been entirely concentrated in middle-latitude Rocky Mountain refugia during the early Tertiary. Present distribution of the three closely related species is in the Rocky Mountain/Great Basin/Sierra–Cascades Ranges of the western United States between latitude 35 and 41° N. Early Tertiary fossils with probable affinities to this subsection were found in Idaho (Eocene) and Colorado (Oligocene) (Tables 3.3, 3.4).

In Eurasia, sequences of conifer-bearing deposits from the early Tertiary are not as widely distributed as in North America, but evidence for pine evolution in relation to refugia exists. Three ancient subsections, *Canarienses*, *Pineae* and *Krempfianae*, have extant species restricted to regions that were along the Tethys seaway. Fossil evidence exists for Tethys refugial area of pines in southeast Asia, in the region where *P. krempfii* now occurs. Pine was recorded in abundance throughout early Tertiary strata and in coastal southeast China and in northwest Borneo (Tables 3.3, 3.4). Other fossil floras for southeast Asia include temperate taxa, suggesting that pines would have found favourable habitats.

In western Europe, the boreotropical angiosperm flora was recorded abundantly from 45 to 56° N (Chaney 1940). Although there are no fossil records from currently low latitudes in Europe, pine fossils are known from areas of southeast Europe that were along the Tethys seaway (Chiguriaeva 1952; Mirov 1967). Furthermore, there is indication of dry zones in the middle Eocene along the European and north African Tethys, indirectly supporting the occurrence of pines there (Parrish, Ziegler & Scotese 1982; Parrish 1987). Fossil pines allied to *Canariensis* and *Pineae* were widespread along the Tethys seaway in western

Europe during the Miocene (reviewed in Klaus 1989), suggesting more widespread distributions for both subsections in this region in the early Tertiary.

As in North America, in Eurasia several subsections appear to have been divided by Eocene events into northern and southern refugia. In subsection *Pinus*, lineages represented by extant *P. sylvestris* most likely migrated from northern refugia. Eocene pine fossils from high latitudes in western Siberia allied to subsection *Pinus* corroborate a northern refugium in Eurasia. The remaining 14 extant species in subsection *Pinus*, excluding *P. densiflora*, *P. thunbergii*, and possibly *P. yunnanensis*, represent lineages from southern refugia. Although few Eocene deposits from southern Europe are plant-bearing, abundant fossils allied to subsection *Pinus* are known from Oligocene and Miocene deposits in the Tethys region (Tables 3.4, 3.5; Palibin 1935; Czeczott 1954; Klipper 1968; van der Burgh 1973; Klaus 1989). The exceptional three species may represent lineages from middle-latitude refugia, as documented by fossil pollen of unknown affinity from the Eocene and Oligocene of Japan and northern China (Tables 3.3, 3.4). Subsection *Gerardianae* may have been divided between a southern Tethys refugium, as represented by extant *P. gerardiana*, and a middle-latitude refugium in northern China, as represented by extant *P. bungeana*, and by early Tertiary deposits containing pine pollen in China (Tables 3.3, 3.4).

Refugia for subsection *Strobi* in Eurasia were mostly along the Tethys. Of the eight extant Eurasian species in the subsection, only *P. parviflora* seems clearly allied to lineages derived from more northerly refugia. The remaining seven modern species are all distributed along the former Tethys region. No early Tertiary fossils from Eurasia have been specifically allied with subsection *Strobi*, although Miocene fossils of the subsection testify to their presence in that region (Klipper 1968).

The consequence of certain Eocene refugia becoming secondary centres of pine diversity is best described for Mexico and Central America. Although parts of the Mexican and Central American isthmuses were transiently under water during the Mesozoic (Kellum 1944; Eguiluz Piedra 1985), they were elevated during the Eocene and later Tertiary. Tectonic activity was especially great in Mexico during the late Eocene and Oligocene/Miocene. Tertiary volcanic activity significantly reshaped the Sierra Madre Occidental and created the Sierra Madre Oriental and the Transvolcanic Belt. Uplift in the early Eocene rebuilt almost the entire ranges of the Sierra Madre del Sur and Sierra Madre de Chiapas (Eguiluz Piedra 1985).

Mexico and Central America are home to 47 pine species (see Chap. 2, this volume), many of them with several varieties and forms (see also Eguiluz Piedra 1985;

Perry 1991). Within many subsections, extensive radiation and speciation are ongoing which has been attributed to active mountain-building that began in the Eocene (Eguiluz Piedra 1985; Axelrod 1986; Karamangala & Nickrent 1989; Lauria 1991). Active radiation is evidenced by clusters of closely related species, with subsections *Ponderosae* and *Cembroides* being the best examples. There are 10–17 species of *Ponderosae* (probably here including *Pseudostrobus*?) (Eguiluz Piedra 1985) endemic to Mexico and Central America and 10–16 species in *Cembroides* (Zavarin 1988). The difficulty of separating the taxa of *Cembroides*, implying close genetic relationship, has led some botanists to the conclusion that this complex is young. Phylogenies based on DNA-sequence divergence indicate, however, that section *Parrya*, including subsection *Cembroides*, is ancient (Kossack 1989; Strauss & Doerksen 1990). Ancestral *Cembroides* lineages would have been concentrated in Mexican/Central American refugia during the early Tertiary, and the major pulse of radiations now occurring in the subsection may have begun in the early Tertiary.

Two subsection-complexes may have originated in Mexican/Central American refugia (Axelrod 1980, 1986). Unlike other subsections, *Oocarpae* (and the 'Sabinianae Group' of *Ponderosae*; see Chap. 2, this volume) have no Mesozoic fossil record and are not clearly documented until the Miocene. On the basis of specialized and apparently derived morphological adaptations (Shaw 1914, 1924; Little & Critchfield 1969; Klaus 1980; van der Burgh 1984), these groups have long been considered to have originated recently. Phylogenetic analysis of DNA divergence also confirms their relative youth, especially of *Oocarpae* (Strauss & Doerksen 1990). Using fossil and floristic evidence, Axelrod (1980) traced the origins of the northern (California/Baja California) elements of *Oocarpae* to mainland Mexico/Central America prior to the Miocene. The northern lineages appear to have diverged by the time they reached California (Millar *et al.* 1988), indicating that the radiation events occurred further south in the early Tertiary. Although the systematic coherence of the extant Latin American taxa of *Oocarpae* remains uncertain, genetic relationships of some Latin American species link *Oocarpae* with *Australes* and *Ponderosae* (Critchfield 1967), both of which have Mesozoic fossil records, appear to have had Mexican/Central American Palaeogene refugia, and may have been ancestral to *Oocarpae*.

The 'Sabinianae Group' of subsection *Ponderosae* has a limited fossil record consisting of a single taxon allied to *P. sabiniana* and confined to southern California (Axelrod 1986). Fossils are known only from the late Miocene to the Quaternary, and Axelrod (1981) traced the origin of the extant species to Mexican species of *Ponderosae*. Genetic relationships of extant taxa link these subsection pairs on the basis of terpene affinities (Zavarin *et al.* 1967) and crossing evidence (Critchfield 1966; Conkle & Critchfield 1988). Together this evidence points to a Mexican/Central American origin of *Sabinianae/Torreyanae* from early Tertiary *Ponderosae/Pseudostrobus* lines.

3.5 Pleistocene versus Eocene impacts

This chapter emphasizes that tectonic, climatic and biogeographic events of the Eocene had a major impact on pine distributions and evolution. The question may arise whether the effects of the Eocene were so confounded by subsequent events, especially the Quaternary cycles of pluvial and interglacial periods, as to be indecipherable.

The Pleistocene was a time of profound change unprecedented in the history of the earth. In some ways the Pleistocene was analogous to the Eocene. The Pleistocene was also an epoch with fluctuating climates, but the deviations from temperate climates in the Pleistocene were towards glacial conditions, whereas in the Eocene they were towards tropical episodes. The amplitude in average temperature between glacial and interglacial periods of the Pleistocene (5–10 °C; Bowen 1979) was about the same as estimated for the cool and warm periods of the Eocene. The Pleistocene differed in lasting less than 2 million years (cf. 20 million years for the Eocene), in having many more cycles (16–18; Bowen 1979), and in having alternating periods of unequal duration with glacials longer than interglacials.

The events of the Pleistocene had enormous effects on vegetation, including pines. In northern latitudes, pine distributions were displaced by continental ice sheets (e.g. *Pinus contorta/P. banksiana*, Critchfield 1985); in mountainous regions elsewhere, species migrated up or down in elevation (Miki 1957; Van Devender & Spaulding 1979). Along coasts and in other lowlands, pine populations shifted north and south in response to the climate cycles (e.g. *Oocarpae*, Axelrod 1980; Millar 1983). Concomitant to the shifts in distribution of pines were major changes in the genetic structure of species. The flux of population expansion and contraction, coupled with drastically changing selection regimes, affected the structure of genetic variation within species and allowed some species to hybridize (Critchfield 1984, 1985; Kinloch, Westfall & Forrest 1986; Millar 1989).

In general, however, the Pleistocene does not appear to have completely reshuffled the genus in the way the Eocene did, and many of the Tertiary patterns and the evolutionary events that date to that period have been maintained. Notwithstanding the existence of local refugia, Pleistocene events primarily affected *Pinus* in a gradient from north to south, with the effect that species and populations shifted south then north (or down then up in elevation), following the cycle of the glacial and interglacial periods.

The impact of the Eocene, by contrast, was greatest in the latitudinal centre of the genus and had the effect of dissecting the genus and concentrating pines into widely disjunct regions. Furthermore, during the early Tertiary, and unlike the Pleistocene, almost no upland regions (except in interior western North America) could offer local refuge from unfavourable climates. Whereas many pinaceous species appear to have gone extinct in the early Tertiary, no pine extinctions in North America are attributed to the Pleistocene (Critchfield 1984), although western Europe suffered a significant impoverishment of pine flora (Klipper 1968). Some speciation, for example, in *Balfourianae*, *Cembroides* and *Oocarpae*, seems to have been triggered by the Pleistocene, but no major new trends have yet emerged.

Insufficient time has elapsed since the close of the Pleistocene for its full impact to be felt on evolution in *Pinus*. Patterns initiated by the Pleistocene appear minor compared with the effects of the Eocene and are insufficient to erase the evolutionary impacts of the early Tertiary. Thus, many of the major evolutionary patterns of the early Tertiary can still be traced in the biogeography and relationships of extant pines.

3.6 Validation

Like Mirov's earlier reconstructions of pine origins, the arguments developed in this chapter (and see Millar 1993) result in a working hypothesis about the effect of the Eocene on pine evolution. The hypothesis is a reconstruction of pine history based on available information from the fossil record, climatic and tectonic evidence, and biogeographic record of angiosperms and conifers. Many gaps in information exist that, if filled, would corroborate or negate this hypothesis. These gaps fall into several categories. Expanded fossil records are urgently needed, especially in regions hypothesized as Eocene pine refugia that currently have meagre fossil documentation. These include Palaeogene records for Central America and Mexico, low latitudes along the European Tethys, and high latitudes in eastern Asia. Technologies that allow more accurate identification of fossil affinities for macro- and especially microfossils at intra-generic levels would help in tracking the biogeography and evolution of these groups. These technologies need to be applied to published fossil floras, with revised taxonomic lists. Certain pine lineages have especially meagre fossil records, including Cretaceous records of section *Parrya*, especially

Gerardianae and *Balfourianae*, and Cretaceous/early Tertiary records of section *Pinea*. Accurate dating of fossils and correlation with palaeoclimatic and palaeogeological events will also help to track the fine scale paths of these pines. Finally, existing and emerging genetic and molecular technologies should be applied to estimate genetic distances and divergence times among subgeneric groups. Such genetic techniques will allow tests of the hypothesized times of geographic isolation.

References

Alvin, K.L. (1960). Further conifers of the Pinaceae from the Wealden Formation of Belgium. *Mem. Inst. Roy. Sci. Nat. Belgique*, **146**, 1–39.

Axelrod, D.I. (1965). A method for determining the altitudes of Tertiary floras. *Palaeobotanist*, **14**, 144–71.

Axelrod, D.I. (1966). The Eocene Copper Basin Flora of Northeastern Nevada. *University of California Publications in Geological Science*, 59.

Axelrod, D.I. (1968). Tertiary floras and topographic history of the Snake River basin, Idaho. *Bulletin of the Geological Society of America*, **79**, 713–34.

Axelrod, D.I. (1979). Desert vegetation, its age and origin. In *Arid Land Plant Resources*, ed. J.R. Goodin & D.K. Northington, pp. 1–72. Lubbock, Texas: International Center for Arid and Semi-Arid Land Studies.

Axelrod, D.I. (1980). History of the maritime closed-cone pines, Alta and Baja California. *University of California Publications in Geological Sciences*, **120**, 1–143.

Axelrod, D.I. (1981). Holocene climatic changes in relation to vegetation disjunction and speciation. *American Naturalist*, **117**, 847–70.

Axelrod, D.I. (1986). Cenozoic history of some western American pines. *Annals of the Missouri Botanical Garden*, **73**, 565–641.

Axelrod, D.I. & Bailey, H.P. (1976). Tertiary vegetation, climate and altitude of the Rio Grande depression, New Mexico-Colorado. *Paleobiology*, **2**, 235–54.

Axelrod, D.I. & Raven, P.R. (1985). Origins of the Cordilleran flora. *Journal of Biogeography*, **12**, 21–47.

Banks, H.P., Ortiz-Sotomayor, A. & Hartman, C.M. (1981). *Pinus escalantensis*, sp. nov. a new permineralized cone from the Oligocene of British Columbia. *Botanical Gazette (Crawfordsville)*, **142**, 286–93.

Barron, E.J. (1984). Climatic implications of the variable obliquity explanation of Cretaceous–Palaeogene high latitude floras. *Geology*, **12**, 595–7.

Barron, E.J. (1985a). Estimations of the Tertiary global cooling trend. *Palaeogeography, Palaeoclimatology, Palaeoecology*, **50**, 45–61.

Barron, E.J. (1985b). Climate models: applications for the pre-Pleistocene. In *Paleoclimates, Analysis, and Modelling*, ed. A.D. Heclot, pp. 397–421. New York: John Wiley.

Becker, E.W. (1961). Oligocene plants from the upper Ruby River Basin, southwest Montana. *Memoirs of the Geological Society of America*, 82.

Berner, R.A., Lasaga, A. & Carrels, R.M. (1983). The carbonate–silicate geochemical cycle and its effect on atmospheric carbon-dioxide over the past 100 million years. *American Journal of Science*, **283**, 641–83.

Blackwell, W.H. (1984). Fossil ponderosa-like pine wood from the Upper Cretaceous of northeast Mississippi. *Annals of Botany (London)*, **53**, 133–6.

Bond, W.J. (1989). The tortoise and the hare: ecology of angiosperm dominance and gymnosperm persistence. *Journal of the Linnaean Society of Botany*, **36**, 227–49.

Bowen, D.Q. (1979). Geographical perspective on the Quaternary. *Progress in Physical Geography*, **3**, 167–86.

Brenner, G.J. (1963). The spores and pollen of the Potomac Group of Maryland. *Department of Geology, Mines, and Water Resources Bulletin*, 27.

Brown, R.W. (1934). The recognizable species of the Green River flora. *United States Geological Survey Professional Paper*, 185-c, 45–77.

Brown, R.W. (1937). Additions to some fossil floras of the western united States. *US Geological Survey Professional Paper*, 186-j, 163–206.

Burchardt, B. (1978). Oxygen isotope palaeotemperatures from the Tertiary period in the North Sea area. *Nature*, **275**, 121–3.

Chaney, R.W. (1940). Tertiary forests and continental history. *Bulletin of the Geological Survey of America*, **51**, 469–88.

Chaney, R.W. (1954). A new pine from the Cretaceous of Minnesota and its paleoecological significance. *Ecology*, **35**, 145–51.

Chiguriaeva, A.A. (1952). Materials on Eocene vegetation of the right bank of the River Don. *Saratov University Science Proceedings*, **35**, 197–200 [in Russian].

Conkle, M.T. & Critchfield, W.B. (1988). Genetic variation and hybridization of ponderosa pine. In *Ponderosa Pine: The Species and Its Management*, pp. 27–43. Pullman, Washington: Washington State University Cooperative Extension.

Creber, G.T. & Chaloner, W.G. (1984). Climatic indications from tree rings in fossil woods. In *Fossils and Climate*, ed. P.T. Brenchley, pp. 49–74. New York: John Wiley.

Critchfield, W.B. (1966). *Crossability and relationships of the California Big-cone pines*. USDA Forest Service Research Paper, NC-6, 36–44.

Critchfield, W.B. (1967). Crossability and relationships of the California closed cone pines. *Silvae Genetica*, **16**, 91–7.

Critchfield, W.B. (1984). Impact of the Pleistocene on the genetic structure of North American conifers. In *Proceedings of the 8th North American Forest Biology Workshop*, ed. R.M. Lanner, pp. 70–118. Logan, Utah.

Critchfield, W.B. (1985). The late Quaternary history of lodgepole and jack pine. *Canadian Journal of Forest Research*, **15**, 749–72.

Czeczott, H. (1954). The past and present distribution of *Pinus halepensis* and *P. brutia*. Eighth International Botanical Congress, Paris and Nice. 8, 196–7.

Dickinson, K.A., Cunningham, K.D. & Ager, T.A. (1987). Geology and origin of the Death Valley uranium deposit, Seward Peninsula, Alaska. *Economic Geology*, **82**, 1558–74.

Donn, W.L. (1982). The enigma of high-latitude paleoclimate. *Palaeogeography, Palaeoclimatology, Palaeoecology*, **40**, 199–212.

Dorofeev, P.I. (1963). Tertiary floras of western Siberia. *Dok. Akad. Nauk. SSSR*, Moscow [in Russian].

Eguiluz Piedra, T. (1985). Origen y evolucion del genero *Pinus. Dasonomia Mexicana*, 3(6), 5–31.

Eguiluz Piedra, T. (1988). Evolucion de los Pinos Pinoneros Mexicanos. In *Proceedings of II Simposio Nacional Sobre Pinos Pinoneros*, ed. M.F. Passini, D.C. Tovar & T. Eguiluz Piedra, pp. 83–92. Chapingo, Mexico: Centro de Genetica Forestal.

Frakes, L.A. (1979). *Climates Throughout Geological Time*. Amsterdam: Elsevier.

Frederiksen, N.O., Ager, T.A. & Edwards, L.E. (1988). Palynology of Maastrichtian and Paleocene rocks, Lower Colville River, Northslope of Alaska. *Canadian Journal of Earth Sciences*, 25, 512–27.

Friis, E.M., Chaloner, W.G. & Crane, P.R. (eds.) (1987). *The Origin of Angiosperms and Their Biological Consequences*. Cambridge: Cambridge University Press.

Graham, A. (1972). Outline of the origin and historical recognition of floristic affinities between Asia and eastern North America. In *Floristics and Paleofloristics of Asia and Eastern North America*, ed. A. Graham, pp. 1–16. Amsterdam: Elsevier.

Graham, A. (1976). Studies in neotropical paleobotany. II. The Miocene communities of Puerto Rico. *Annals of the Missouri Botanical Garden*, 56, 308–57.

Gray, J. (1960). Temperate pollen genera in the Eocene (Clairborne) flora, Alabama. *Science*, 132, 808–10.

Groot, J.J., Penny, J.S. & Groot, C.R. (1961). Plant microfossils and the age of the Raritan, Tuscaloosa, and Magothy formations of eastern USA. *Palaeontographica*, 108B (3–6), 121–40.

Guo, S. (1980). *Late Cretaceous and Eocene Floral Provinces*. Academia Sinica, Nanjing: Institute for Geology and Paleontology.

Heer, O. (1868–83). *Flora Fossilis Arctica*. 6 vol. Zürich: J. Wurster.

Hickey, L.J., West, R.M., Dawson, M.R. & Choi, D.K. (1983). Arctic terrestrial biota: Paleomagnetic evidence of age disparity with mid-northern latitudes during the late Cretaceous and early Tertiary. *Science*, 221, 1153–6.

Hills, L.V., Klovan, J.E. & Sweet, A.R. (1974).

Juglans eocinera, nov. sp., Beaufort Formation (Tertiary), southwest Banks Island, Arctic Canada. *Canadian Journal of Botany*, 52, 65–90.

Hollick, A. & Jeffrey, E.C. (1909). Studies of Cretaceous coniferous remains from Kreischerville, New York. *Memoirs of the New York Botanical Garden*, 3, 1–137.

Hopkins, W.S., Rutter, N.W. & Rouse, G.E. (1972). Geology, paleoecology, and palynology of some Oligocene rocks in the Rocky Mountain trench of British Columbia. *Canadian Journal of Earth Sciences*, 9, 460–70.

Hsu, J. (1983). Late Cretaceous and Cenozoic vegetation in China, emphasizing their connections with North America. *Annals of the Missouri Botanical Garden*, 70, 490–508.

Huntley, B. & Webb, T. III (1988). *Vegetation History*. Dordrecht: Kluwer.

Huzioka, K. & Takahashi, E. (1970). The Eocene flora of the Ube coal-field, southwest Honshu, Japan. *Journal Minerals College Akita University, Series A*, 4 (5), 1–88.

Jeffrey, E.C. (1908). On the structure of the leaf in Cretaceous pines. *Annals of Botany (London)*, 22, 207–20.

Karamangala, R.R. & Nickrent, D.L. (1989). An electrophoretic study of representatives of subgenus *Diploxylon* of *Pinus. Canadian Journal of Botany*, 67, 1750–9.

Kellum, L. (1944). Geologic history of northern Mexico and its bearing on petroleum exploration. *Bulletin of American Association for Petroleum Geology*, 28, 301–25.

Kerr, R.A. (1991). Did a burst of volcanism overheat ancient earth? *Science*, 251, 746–7.

Kinloch, B.B., Westfall, R.D. & Forrest, G.I. (1986). Caledonian Scots pine: Origins and genetic structure. *New Phytologist*, 104, 703–29.

Klaus, W. (1980). Neue Beobachtungen zur Morphologie des Zapfens von *Pinus* und ihre Bedeutung für die Systematik, Fossilbestimmung, Arealgestaltung, und Evolution der Gattung. *Plant Systematics and Evolution*, 134, 137–71.

Klaus, W. (1989). Mediterranean pines and their history. *Plant Systematics and Evolution*, 162, 133–63.

Klipper, K. (1968). Koniferzapfen aus den Tertiaren Deckschichten des Niederrheinischen Hauptflozes. 1 & 2. *Palaeontographica Abteil B*, 121, 159–68; 123, 213–20.

Kossack, D.S. (1989). The IFG Copia-like element. Characterization of a transposable element present at high copy number in the genus *Pinus* and a history of the pines using IFG as a marker. PhD thesis, University of California, Davis.

Kurten, B. (1966). Holarctic land connections in the early Tertiary. *Commentaries in Biology*, 29(5), 1–5.

Langenheim, R.L., Smiley, C.J. & Gray, J. (1960). Cretaceous amber from the coastal plain of Alaska. *Bulletin of the Geological Society of America*, 71, 1345–66.

Langenheim, J.H., Hackner, B.L. & Bartlett, A. (1967). Mangrove pollen at the depositional site of Oligo–Miocene amber from Chiapas, Mexico. *Botanical Museum Leaflet*, 21, 289–324.

Lauria, F. (1991). Taxonomy, systematics, and phylogeny of *Pinus*, subsection *Ponderosae* London (Pinaceae). Alternative concepts. *Linzer Biologi. Beitr.*, 23(1), 129–202.

Leopold, E.B. & MacGinite, H.D. (1972). Development and affinities of Tertiary floras in the Rocky Mountains. In *Floristics and Paleofloristics of Asia and Eastern North America*, ed. A. Graham, pp. 147–200. London: Elsevier.

Leopold, E.G., Liu, G. & Clay-Poole, S. (1992). In *Eocene–Oligocene Climatic and Biotic Evolution*, ed. D.R. Prothero & W.A. Berggren, pp. 399–420. Princeton, NJ: Princeton University Press.

Lesquereux, L. (1883). Contributions to the fossil flora of the Western Territories III. The Cretaceous and Tertiary flora. *US Geological Survey of the Territories Report*, 8, 1–283.

Little, E.L., Jr. & Critchfield, W.B. (1969). *Subdivisions of the Genus Pinus (Pines)*. USDA Forest Service Miscellaneous Publication 1144.

MacGinite, H.D. (1953). Fossil plants of the Florrissant beds, Colorado. *Publications of the Carnegie Institute of Washington*, 599.

McGowran, B. (1990). Fifty million years ago. *American Scientist*, 78, 30–9.

McKenna, M.C. (1980). Eocene paleolatitudes, climate, and mammals of Ellesmere Island. *Palaeogeography, Palaeoclimatology, Palaeoecology*, 30, 349–62.

McKenna, M.C. (1983). Holocene land mass rearrangement, cosmic events, and Cenozoic terrestrial organisms. *Annals of the Missouri Botanical Garden*, **70**, 459–89.

Mai, D.H. (1970). Subtropische Elemente im Europaischen Tertiar I. *Palaeontol. Abh. Abtteil B*, Palaeobot., **3**, 441–503.

Manum, S. (1962). Studies in the Tertiary flora of Spitsbergen, with notes on Tertiary flora of Ellsmere Island, Greenland, and Iceland. *Norsk Polarinst. Skr*, **125**, S. 1–127.

Martin, P. & Harrell, B.E. (1957). The Pleistocene history of temperate biotas in Mexico and eastern United States. *Ecology*, **38**, 468–80.

Mason, H.L. (1927). Fossil records of some west American conifers. *Publications of the Carnegie Institute of Washington*, **346**, 139–58.

Miki, S. (1957). Pinaceae of Japan, with special reference to its remains. *Journal of the Institute for Polytechnica Osaka City Univer. Ser. D, Biol.*, **8**, 221–72.

Millar, C.I. (1983). A steep cline in *Pinus muricata*. *Evolution*, **37**, 311–19.

Millar, C.I. (1989). Allozyme variation of bishop pine associated with pygmy forest soils in northern California. *Canadian Journal of Forest Research*, **19**, 870–9.

Millar, C.I. (1993). Impact of the Eocene on the evolution of *Pinus* L. *Annals of the Missouri Botanical Garden*, **80**, 471–98.

Millar, C.I. (1996). Tertiary vegetation history. In *Sierra Nevada Ecosystem Project, Final Report to Congress*, Vol. II. *Assessments and Scientific Basis for Management Strategies*, pp. 5.1–5.52. Davis: University of California, Center for Water and Wildland Resources.

Millar, C.I. & Kinloch, B.K. (1991). Taxonomy, phylogeny, and coevolution of pines and their stem rusts. In *Rusts of Pines. Proceedings IUFRO Rusts of Pine Working Party Conference, 1989*, ed. Y. Hiratsuka, J. Samoil, P. Blenis, P. Crane & B. Laishley, pp. 1–38. Banff, Alberta: Forestry Canada, Edmonton, Alberta.

Millar, C.I., Strauss, S.H., Conkle, M.T. & Westfall, R.D. (1988). Allozyme differentiation and biosystematics of the California closed cone pines (*Pinus* subsect. *Oocarpae*). *Systematic Botany*, **13**, 351–70.

Miller, C.N. (1969). *Pinus avonensis*, a new species of petrified cones from the Oligocene of Montana. *American Journal of Botany*, **56**, 972–8.

Miller, C.N. (1973). Silicified cones and vegetative remains of *Pinus* from the Eocene of British Columbia.

Contributions of the University of Michigan Museum of Paleontology, **24**, 101–18.

Miller, C.N. (1974). *Pinus wolfei*, a new petrified cone from the Eocene of Washington. *American Journal of Botany*, **61**, 772–7.

Miller, C.N. (1976). Early evolution in the Pinaceae. *Reviews in Palaeontology and Palynology*, **21**, 101–17.

Miller, C.N. (1977). Mesozoic conifers. *Botanical Review (Lancaster)*, **43**, 217–80.

Miller, C.N. (1982). Current status of Paleozoic and Mesozoic conifers. *Reviews in Palaeontology and Palynology*, **37**, 99–114.

Miller, C.N. (1988). The origin of modern conifer families. In *Origin and Evolution of Gymnosperms*, ed. C.B. Beck, pp. 448–87. New York: Columbia University Press.

Miller, C.N. & Malinky, J.M. (1986). Seed cones of *Pinus* from the Late Cretaceous of New Jersey, USA. *Reviews in Palaeontology and Palynology*, **46**, 257–72.

Miller, K.G., Fairbanks, R.G. & Mountain, G.S. (1987). Tertiary oxygen isotope synthesis, sea level history, and continental margin erosion. *Paleoceanography*, **2**, 1–19.

Minckley, W.L., Hendrickson, D.A. & Bond, C.E. (1986). Geography of western North American fishes: description and relationship to intracontinental tectonism. In *The Zoogeography of North American Fishes*, ed. C.H. Hocutt & E.O. Wiley, pp. 519–613. New York: Wiley Interscience.

Mirov, N.T. (1967). *The Genus Pinus*. New York: Ronald Press.

Mirov, N.T. & Hasbrouck, J. (1976). *The Story of Pines*. Bloomington: Indiana University Press.

Muller, J. (1966). Montane pollen from the Tertiary of northwest Borneo. *Blumea*, **14**, 231–5.

Norris, G. (1982). Spore-pollen evidence for early Oligocene high-latitude cool climatic episode in northern Canada. *Nature*, **297**, 387–9.

Ogura, Y. (1932). On the structure and affinities of some Cretaceous plants from Hokkaido. 2nd Contribution. *Journal of the Faculty of Science, Imp. Univ., Tokyo, Sect. 3 (Bot.)* **2** (7), 455–83.

O'Keefe, J.D. & Ahrens, T.J. (1989). Impact production of CO_2 by the Cretaceous/Tertiary extinction bolide and the resultant heating of the Earth. *Nature*, **338**, 247–9.

Palibin, I.V. (1935). Stages of development of Caspian flora from the Cretaceous

period. *Sovetsk Bot.*, **3**, 10–50 [in Russian].

Parrish, J.T. (1987). Global palaeogeography and palaeoclimate of the late Cretaceous and early Tertiary. In *The Origins of Angiosperms and their Biological Consequences*, ed. F.M. Friis, W.G. Chaloner & P.R. Crane, pp. 51–73. Cambridge: Cambridge University Press.

Parrish, J.T., Ziegler, A.M. & Scotese, C.R. (1982). Rainfall patterns and the distribution of coals and evaporites in the Mesozoic and Cenozoic. *Palaeogeography, Palaeoclimatology, Palaeoecology*, **40**, 67–101.

Penny, J.S. (1947). Studies on the conifers of the Magothy flora. *American Journal of Botany*, **34**, 281–96.

Perry, J.P. (1991). *The Pines of Mexico and Central America*. Portland, Oregon: Timber Press.

Phipps, C.J., Osborn, J.M. & Stockey, R.A. (1995). *Pinus* pollen from the middle Eocene Princeton chert (Allenby Formation) of British Columbia, Canada. *International Journal of Plant Sciences*, **156**(N1), 117–24.

Piel, K.M. (1971). Palynology of Oligocene sediments from central British Columbia. *Canadian Journal of Botany*, **49**, 1885–920.

Pierce, R.L. (1957). Minnesota Cretaceous pine pollen. *Science*, **125**, 26.

Prothero, D.R. (1994). *The Eocene–Oligocene Transition*. New York: Columbia University Press.

Prothero, D.R. & Berggren, W.A. (eds.) (1992). *Eocene–Oligocene Climatic and Biotic Evolution*. Princeton, NJ: Princeton University Press.

Reid, E.M. & Chandler, M.E.J. (1933). *The London Clay Flora*. London: British Museum (Natural History), Vol. 7, 561 pp.

Ritchie, J.C. (1984). *Past and Present Vegetation of the Far Northwest of Canada*. Toronto: University of Toronto Press.

Robison, C.R. (1977). *Pinus triphylla* and *Pinus quinquefolia* from the Upper Cretaceous of Massachusetts. *American Journal of Botany*, **64**, 726–32.

Rouse, G.E. & Matthews, W.H. (1979). Tertiary geology and palynology of the Quesnel area, British Columbia. *Bulletin of Canadian Petroleum Geology*, **27**, 418–45.

Ruddiman, W.F. & Kutzbach, J.E. (1989). Forcing of late Cenozoic Northern

Hemisphere climate by plateau uplift in southern Asia and the American west. *Journal of Geophysics Research*, **91**(D15), 18409–27.

Ruddiman, W.F. & Kutzbach, J.E. (1991). Plateau uplift and climatic change. *Scientific American*, **264**(3), 66–75.

Savin, S.M. (1977). The history of the earth's surface temperature during the past 100 million years. *Annual Review of Earth Planetary Science*, **5**, 319–56.

Schorn, H.E. & Wolfe, J.A. (1989). Taxonomic revision of the *Spermatopsida* of the Oligocene Creede flora, southern Colorado. *US Geological Survey Open-file Report*, 89–433.

Schweitzer, H.J. (1974). Die 'tertiaren' Koniferan Spitsbergen. *Palaeontographica* Abt. B, **149**, 1–89.

Shackleton, N. & Boersma, A. (1981). The climate of the Eocene ocean. *Journal of the Geological Society of London*, **138**, 153–7.

Shaw, G.R. (1914). The genus *Pinus*. *Publications of the Arnold Arboretum*, 5.

Shaw, G.R. (1924). Notes on the genus *Pinus*. *Journal of the Arnold Arboretum*, 5, 225–7.

Singh, G. (1988). History of aridland vegetation and climate: A global perspective. *Biological Review*, **63**, 159–95.

Smith, A.G., Hurley, A.M. & Briden, J.C. (1981). *Phanerozoic Paleocontinental World Maps*. Cambridge: Cambridge University Press.

Stockey, R.A. (1983). *Pinus driftwoodensis* sp. n. from the early Tertiary of British Colombia. *Botanical Gazette (Crawfordsville)*, **144**, 148–56.

Stockey, R.A. (1984). Middle Eocene remains from British Colombia. *Botanical Gazette (Crawfordsville)*, **145**, 262–74.

Stockey, R.A. & Nishida, M. (1986). *Pinus harborensis* sp. nov. and the affinities of permineralized leaves from the Upper Cretaceous of Japan. *Canadian Journal of Botany*, **64**, 1856–66.

Stockey, R.A. & Ueda, Y. (1986). Permineralized pinaceous leaves from the Upper Cretaceous of Hokkaido. *American Journal of Botany*, **73**, 1157–62.

Stopes, M.C. & Kershaw, E.M. (1910). The anatomy of Cretaceous pine leaves. *Annals of Botany (London)*, **24**, 395–402.

Strauss, S.H. & Doerksen, A.H. (1990). Restriction fragment analysis of pine phylogeny. *Evolution*, **44**, 1081–96.

Sung, Z. (1958). Tertiary spore and pollen complexes from the red beds of Chiu-Chuan, Kansu, and their geological and botanical significance. *Acta Palaeontologica Sinica*, **6**, 159–67.

Sung, Z. & Liu, T. (1976). The Paleogene spores and pollen grains from the Fushun coal field, northeast China. *Acta Palaeontologica Sinica*, **15**, 146–62.

Tanai, T. (1970). The Oligocene floras from the Kushiro coal field, Hokkaido, *Japanese Journal of the Faculty of Science, Hokkaido Univ, Ser. 4*, **14**, 383–514.

Tanai, T. (1972). Tertiary history of vegetation in Japan. In *Floristics and Paleofloristics of Asia and Eastern North America*, ed. A. Graham, pp. 235–56. London: Elsevier.

Taylor, D.W. & Hickey, L.J. (1990). An Aptian plant with attached leaves and flowers: Implications for angiosperm origin. *Science*, **247**, 702–4.

Tiffney, B.M. (1985a). The Eocene North Atlantic land bridge: its importance in Tertiary and modern phytogeography of the Northern Hemisphere. *Journal of the Arnold Arboretum*, **66**, 243–73.

Tiffney, B.M. (1985b). Perspectives on the origin of the floristic similarity between eastern Asia and eastern North America. *Journal of the Arnold Arboretum*, **66**, 73–94.

Ueda, Y. & Nishida, M. (1982). On petrified pine leaves from the Upper Cretaceous of Hokkaido. *Japanese Journal of Botany*, **57**, 133–45.

Underwood, J.C. & Miller, C.N. (1980). *Pinus buchananii*, a new species based on a petrified cone from the Oligocene of Washington. *American Journal of Botany*, **67**, 1132–5.

Upchurch, G.R. & Wolfe, J.A. (1987). Mid-Cretaceous to early Tertiary vegetation and climate: evidence from fossil leaves and woods. In *The Origins of Angiosperms and their Biological Consequences*, ed. E.M. Friis, W.G. Chaloner & P.R. Crane, pp. 75–105. Cambridge: Cambridge University Press.

Van Devender, T.R. & Spaulding, W.G. (1979). Development of vegetation and climate in the southwestern United States. *Science*, **204**, 701–10.

van der Burgh, J. (1973). Holzer der Niederrheinischen Braunkohlen formation 2. Holzer der Braunkohlengruben 'Maria Theresia' zu Herzogenrath, 'Zukunft West' zu Eschweiler, und 'Victor' zu Zulpich. Nebst einer systematisch anatomischen Bearbeitung der Gattung *Pinus* L. *Review of Palaeobotany and Palynology*, **15**, 73–275.

van der Burgh, J. (1984). Phylogeny and biogeography of the genus *Pinus*. In *Pines. Drawings and Descriptions of the Genus Pinus*, ed. A. Farjon, pp. 199–206. Leiden: E.J. Brill/Dr W. Backhuys.

Wahrhaftig, C., Wolfe, J.A. & Leopold, E.B. (1969). The coal-bearing group in the Nenana coal field, Alaska. *US Geological Survey Bulletin*, 127–D, 1–30.

Wing, S.L. (1987). Eocene and Oligocene floras and vegetation of the Rocky Mountains. *Annals of the Missouri Botanical Garden*, **74**, 748–84.

Wodehouse, R.P. (1933). Tertiary pollen. II. The oil shales of the Eocene Green River Formation. *Bulletin of the Torrey Botanical Club*, **60**, 479–524.

Wolfe, J.A. (1969). Neogene floristic and vegetational history of the Pacific Northwest. *Madroño*, **20**, 83–110.

Wolfe, J.A. (1971). Tertiary climatic fluctuations and methods of analysis of Tertiary floras. *Palaeogeography, Palaeoclimatology, Palaeoecology*, **9**, 27–57.

Wolfe, J.A. (1972). An interpretation of Alaskan Tertiary floras. In *Floristics and Paleofloristics of Asia and Eastern North America*, ed. A. Graham, pp. 201–33. Amsterdam: Elsevier.

Wolfe, J.A. (1975). Some aspects of plant geography of the Northern Hemisphere during the Late Cretaceous and Tertiary. *Annals of the Missouri Botanical Garden*, **62**, 264–79.

Wolfe, J.A. (1977). Paleogene floras from the Gulf of Alaska region. *US Geological Survey Professional Paper*, 997.

Wolfe, J.A. (1978). A paleobotanical interpretation of Tertiary climates in the Northern Hemisphere. *American Scientist*, **66**, 694–703.

Wolfe, J.A. (1981). A chronologic framework for the Cenozoic megafossil floras of northwestern North America in relation to marine geochronology. In *Pacific Northwest Cenozoic Biostratigraphy*, ed. J.M. Armentrout, pp. 39–47. Special Paper of the Geological Society of America, 184.

Wolfe, J.A. (1985). Distribution of major vegetational types during the Tertiary. *Geophysical Monograph*, **32**, 357–75.

Wolfe, J.A. (1987). An overview of the origins of the modern vegetation and flora of the northern Rocky

Mountains. *Annals of the Missouri Botanical Garden*, **74**, 785–803.

Wolfe, J.A. (1990). Palaeobotanical evidence for a marked temperature increase following the Cretaceous/Tertiary boundary. *Nature*, **343**, 153–6.

Wolfe, J.A. & Schorn, H.E. (1989). Paleoecologic, paleoclimatic, and evolutionary significance of the Oligocene Creede flora, Colorado. *Paleobiology*, **15**, 180–98.

Wolfe, J.A. & Upchurch, G.R. (1986). Vegetation, climatic, and floral changes at the Cretaceous–Tertiary boundary. *Nature*, **324**, 148–52.

Wolfe, J.A. & Wehr, W. (1987). Middle Eocene dicotyledonous plants from Republic, northeastern Washington. *US Geological Society Bulletin*, 1597.

Zavarin, E. (1988). Taxonomy of pinyon pines. In *Proceedings of the II Simposio Nacional Sobre Pinos Pinoneros 6–8 August 1987*, ed. M.F. Passini, D.C. Tovar & T. Eguiluz Piedra, pp. 29–40.

Chapingo, Mexico: Centro de Genetica Forestal.

Zavarin, E. & Hathaway, W., Reichert, Th. & Linhart, Y.B. (1967). Chemo-taxonomic study of *Pinus torreyana* Parry turpentine. *Phytochemistry*, **6**, 1019–23.

Ziegler, A.M., Scotese, C.R. & Barrett, S.R. (1983). Mesozoic and Cenozoic paleogeographic maps. In *Tidal Friction and the Earth's Rotation*, Vol. II, ed. P. Brosche & J. Sunderman, pp. 240–52. Berlin: Springer-Verlag.

Part three
Historical biogeography

4 The late Quaternary dynamics of pines in northern Asia

Constantin V. Kremenetski, Kam-biu Liu and Glen M. MacDonald

4.1 Introduction

Northeastern Asia is home to about 20 *Pinus* species. A diverse fossil flora of pines attests to the presence of the genus in northern Asia since at least the Cretaceous (Millar 1993), making this a potential region of origin of *Pinus* (Mirov 1967; Millar 1993). However, the reclassification of purported Jurassic and Cretaceous Asian pine fossils into extinct pinaceous genera makes the case for an Asian centre of origin less compelling than previously thought (Millar 1993; Chap. 3, this volume). During Eocene climatic shifts northern Asia appears to have served as a refugium for pines displaced from more southerly zones. These displacements occurred because of warm conditions and increased competition from angiosperms (Millar 1993). The division of subsections such as *Pinus* into northern refugial populations in western Siberia, mid-latitude populations in eastern Asia, and southern refugial populations in other parts of Asia and Europe was extremely important for evolution within the genus. *Pinus sylvestris* seems to have its origins in a northern Asian Eocene refuge, while *P. densiflora* and *P. thunbergii* may have arisen from a mid-latitude location near eastern China and Japan (Millar 1993).

Although northern Asia did not experience the massive Pleistocene glaciations of North America and northern Europe, the region was affected by significant changes in climate associated with the alternation between glacial and non-glacial periods (Velichko *et al.* 1984). These climatic variations caused major alterations in vegetation, including the distribution of pines. Although Millar (1993; Chap. 3, this volume) does not believe that these Pleistocene shifts were as significant for the evolution of pines in Asia as were the earlier Eocene events, the separation of closely related species such as *P.*

pumila and *P. sibirica* during glacial periods may have played a role in speciation or at least the preservation of distinctive genotypes. In any case, the modern distribution of pines reflects changes in abundance and geographic ranges that have occurred since the end of the last glacial period. In this chapter we will examine the late Quaternary history of pines in Siberia-Mongolia, China and Japan. Where possible, we will outline the possible locations of refugial populations of *Pinus* during the last glacial maximum. We will attempt to infer the environmental factors that influenced the postglacial distributions of *Pinus* species in northern Asia. Finally, we will discuss some evolutionary implications of the Quaternary history of pines in northern Asia.

Reconstruction of the postglacial history of pine in northern Asia is based mainly upon fossil pollen sampled from peat, lake sediment and loess. Pine macrofossils are also an important source of data. Such palaeoecological data are available from a number of sites in Siberia, China and Japan. Similar data are extremely sparse from southern Asia. *Pinus* pollen cannot be subdivided beyond the section *Strobus* and section *Pinus*. However, macrofossil remains of pines are sometimes encountered and allow for greater floristic precision. As in most parts of the world, pine pollen tends to be over-represented relative to the actual percentages of pine trees in the vegetation. This is because of (1) the high production of pollen by pines, (2) the potential of the grains for long-distance aerial transport, and (3) their high potential for preservation. For example, in Inner Mongolia, eastern Tibet, and southeastern Qinghai, which are near the distributional limits of pines, *Pinus* contributes 10% or more of the total pollen sum (Li 1991; K.-b. Liu, unpublished data). *Pinus* pollen

Ecology and Biogeography of Pinus, ed. D.M. Richardson. © Cambridge University Press (1998), pp. 95–106.

occurs sporadically in surface samples from the interior plateaus and desert basins of Tibet and Xinjiang, hundreds of kilometres beyond the limits of pine (Li 1991). Most strikingly, Wu & Xiao (1995) found 10–40% (mean=26.5%, of total pollen sum) *Pinus* pollen in the Zabuye Salt Lake area of southwestern Tibet, where the nearest pine trees are at least 400 km away. Studies of modern pollen deposition in Zhongtiao Shan & Kunming also demonstrated that *Pinus* pollen occurs at 20–30% in surface sample sites where pine trees are locally absent (Wu & Sun 1987; Yao 1989). The over-representation of pine, particularly in cases where its pollen occurs beyond the range of the genus, argues for caution in attempting to delineate exact range boundaries from the fossil record. The pine pollen record can be considered to provide, at best, an indication of general regions where pine was present in the past.

All ages presented here are radiocarbon years Before Present (BP) with present being taken as AD 1950. Radiocarbon years do not exactly equal calendar years and this difference increases with age (Stuiver & Reimer 1993). For example, a radiocarbon age of 10 000 BP is roughly equivalent to a calendar date of 12 000 years ago.

4.2 Siberia and Mongolia

4.2.1 Introduction

Siberia and Mongolia experienced pronounced changes in vegetation due to climate variations related to glacial cycles. The three common pines that we will discuss are *P. pumila* (dwarf stone pine), *P. sibirica* (Siberian stone pine) and *P. sylvestris* (Scots pine). *Pinus pumila* and *P. sibirica* are members of the section *Strobus* and their pollen cannot be separated. *Pinus sylvestris* belongs to the section *Pinus* and has pollen that is distinctive from the other two species. Sufficient fossil pollen sites exist to produce tentative maps of the postglacial spread of these three species. These maps are based on the assumption that the first continuous appearance of *Pinus* pollen at a site represents the presence of the genus in the regional vegetation (Ritchie & MacDonald 1986). Macrofossil records are present at some sites, but are less common than pollen records.

4.2.2 *Pinus sylvestris*

Pinus sylvestris is extremely widespread in Eurasia (Fig. 4.1) and occupies a number of environmental settings (Fig. 4.2). The northeastern limit of the species in Siberia probably reflects its inability to withstand cool summers and extremely cold winters (Vakovski 1958). The southern limits appear to be determined by the aridity of the steppe zone.

From available Russian studies, a preliminary map of

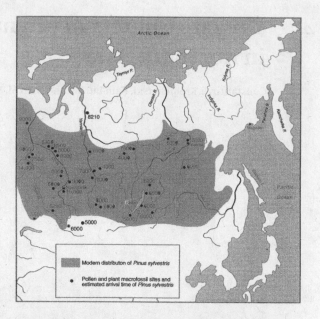

Fig. 4.1. **Modern distribution of *Pinus sylvestris* in Asia and the postglacial spread of the species as evidenced from fossil pollen and plant macrofossil records (based on data from: Neustadt 1967, 1976; Vipper 1968; Kind, Gorshkov & Chernysheva 1969; Levkovskya *et al.* 1970; Neustadt & Selikson 1971; Zubakov 1972; Kutafieva 1973, 1975; Volkov *et al.* 1973; Glebov *et al.* 1974; Kind 1974; Vipper & Golubova 1976; Vipper *et al.* 1976; Khotinsky 1977; Koltsova, Starikov & Zhidovlenko 1979; Arkhipov & Votakh 1980; Koshkarova 1981, 1986, 1989; Zubarev 1981; Firsov *et al.* 1982; Levina *et al.* 1983, 1987; Krivonogov 1988; Glebov 1988; Klimanov & Levina 1989; Chernova *et al.* 1991; Kulagina & Trofimov 1992; Kremenetski, Tarasov & Cherkinski 1994).**

the postglacial spread of *P. sylvestris* can be developed (Fig. 4.1). Most records are from fossil pollen as macrofossils of the species are very rare in Siberia. During the last glacial there were probably only a few isolated locations in Siberia where *P. sylvestris* persisted. A good candidate would be the upper Irtysh River valley where the species has likely been present for the past 14 000 years and perhaps the Ob valley near the city of Novosibirsk where it has been present for at least 10 000 years (Fig. 4.1).

Pinus sylvestris appears to have moved northwards during the early postglacial and grew beyond its present limits along the Yennisy River by 8000 BP. This northward extension of *P. sylvestris* along the Yennisy is convincingly proven by the presence of seeds in a radiocarbon-dated peat deposit (Koshkarova 1986). The species retreated to its present northern limits later in the Holocene, probably as a result of climatic cooling. Expansion southeastward was slower and the species did not reach its northeastern range limits in Siberia until about 6000 BP. It is not clear when it reached its extreme southeastern limits along the Amur River. *Pinus sylvestris* was not present at its extreme southern limits in Siberia and Mongolia until 5000 BP. The relatively slow spread southwards may reflect warm and dry conditions during the early to mid-Holocene, or the

Fig. 4.2. *Pinus sylvestris* occurs in a wide range of habitats in Eurasia, ranging from rocky sites to moist bog margins. The picture shows Scots pine on a rock outcrop on the NW shore of Lake Baikal, Siberia (Russia). Rocky sites such as this provide habitats which are seldom affected by high-intensity, stand-replacing fires. Extremely old pines can be found on such sites (photo: J.G. Goldammer).

Fig. 4.3. **Modern distribution of *Pinus sibirica* in Asia and the postglacial spread of the species as evidenced from fossil pollen and plant macrofossil records (based on data from: Neustadt 1967, 1976; Vipper 1968; Kind *et al.* 1969; Levkovskya *et al.* 1970; Neustadt & Selikson 1971; Volkov *et al.* 1973; Glebov *et al.* 1974; Kind 1974; Vipper & Golubova 1976; Vipper *et al.* 1976; Khotinsky 1977; Koltsova *et al.* 1979; Arkhipov & Votakh 1980; Arkhipov, Levina & Panychev 1980; Koshkarova 1981, 1986, 1989; Zubarev 1981; Levina *et al.* 1983; Krivonogov *et al.* 1985; Krivonogov 1988; Glebov 1988; Klimanov & Levina 1989; Chernova *et al.* 1991; Kulagina & Trofimov 1992).**

relative sparseness of suitable sites for the species in those regions.

4.2.3 *Pinus sibirica*

Pinus sibirica has a more restricted range than *P. sylvestris*, with which it is closely sympatric (Fig. 4.3). In general, its northern and eastern range is controlled by summer and winter cold, while to the south, aridity is the main determinant of distribution.

The pollen of *P. sibirica* cannot be distinguished from that of *P. pumila*. However, the ranges of the two species overlap only in Trans-Baikal region and southern Yakutia (Figs. 4.2, 4.3), where *P. pumila* often grows as understorey in *P. sibirica* forests. Thus, over most of the modern range, pollen of the section *Strobus* can be attributed to *P. sibirica*.

It is likely that the main glacial refuges for *P. sibirica* lay to the southwest in the upper part of the Irtysh River valley where evidence of presence of the species extends back to 12 000 BP (Fig. 4.3). Other refugia include the Yennisy,

Kriaz, Tuva and northern Mongolian intermountain valleys (Vipper *et al.* 1976; Levina *et al.* 1987; Krivonogov 1988). The postglacial spread of the species appears to have been in an easterly direction and occurred slightly more slowly than *P. sylvestris* (Fig. 4.1). *Pinus sibirica* may not have been present at its modern southern range limits in Siberia and Mongolia until 6000–5000 BP (Fig. 4.3). It did not advance to its present eastern limits until 6000–4000 BP. The wide range of arrival dates in most regions suggests that the spread and subsequent population growth of *P. sibirica* occurred by the founding of small populations with variable growth rates. The main period of spread and population growth for *P. sibirica* occurred between roughly 8000 and 4000 BP. This may reflect the predominance of warm conditions in Siberia at that time.

4.2.4 *Pinus pumila*

Pinus pumila occupies the major part of Russian Far East, Medny Island (Komandorski archipelago), Kurily Islands, Sakhalin and adjacent portions of the Japanese Islands, Korea and China (Fig. 4.4). It grows as far west as the Lena and Oleniek River basins and occurs at alpine treeline sites down to sea level. *Pinus pumila* dominates the vegetation over much of the Russian Far East, but may also occur as an understorey tree in *Larix* forests. It is able to extend its

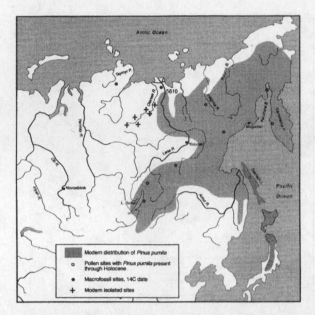

Fig. 4.4. **Modern distribution of *Pinus pumila* in Asia, and: fossil pollen sites with evidence of *P. pumila* presence throughout the Holocene; macrofossil sites presenting evidence of *P. pumila* northern range extension during the Holocene; location of disjunct stands of *P. pumila* along the Oleniek River (based on data from: Tikhomirov 1946, 1949; Korzhuyev & Federova 1962, 1970; Serebrianny 1965; Belorusova *et al.* 1977; Khotinsky 1977; Nikolskaya 1980; Yegorova 1982; Stefanovich *et al.* 1986; Andreev *et al.* 1989; Andreev & Klimanov 1991; Lozkin & Federova 1989).**

range into regions with extremely cold winters due to its prostrate habit during the winter.

Given its ability to withstand cold conditions, during the last glacial maximum *P. pumila* may have been present in many parts of its modern range on isolated sites that possessed suitable soil and microclimatic conditions. If this is the case, the general range of *P. pumila* did not change much from the end of the last glacial through the postglacial and *P. pumila* populations in the mountains of Japan, Korea and China may well be relicts from the period of the last glaciation. Although full glacial records are rare, work from some postglacial sites in Siberia support this contention. In the region of Lake Baikal *P. pumila* was probably present during the whole Holocene. In the Lena River basin of Yakutia (Andreev & Klimanov 1991; Andreev *et al.* 1989) pollen of *P. pumila* occurs without any major fluctuation during the entire Holocene. On Sakhalin Island (Khotinsky 1977) abundant pollen of *P. pumila* is found in deposits from the end of the late glacial. Results of pollen investigations in Kamchatka suggest that thickets of *P. pumila* from the mountains to near sea level have been present since the beginning of the Holocene (Khotinsky 1977). However, near Magadan on the eastern coast of northern Asia the species appears to have been absent until 9000 BP (Lozhkin *et al.* 1993). It is possible that dry-cold conditions during the winter restricted the occur-

rence of the species in the Far East during the last glacial maximum (Lozhkin *et al.* 1993). These small populations, persisting on favourable sites, then expanded during the early postglacial, allowing the rapid appearance of the species in a number of regions.

Detailed analysis shows that there have been changes in *P. pumila* distribution and population density near the edges of its present range. In the southeast there has been an increase in the importance of the species in the late Holocene. At Kradeno Lake near Yakutsk (Khotinsky 1977) there is a rise of *P. pumila* pollen after 5000–4500 BP, which seems representative of an expansion of *P. pumila* thickets in the mountains due to climatic cooling. A similar expansion is reported from near Magadan (Lozhkin *et al.* 1993). A major expansion of *P. pumila* thickets also occurred in Kamchatka after *c.* 5000 BP (Yegorova 1982).

In contrast, northern regions have witnessed a contraction in *P. pumila* range in the late Holocene. Pollen data (Stefanovich *et al.* 1986) show that in the area of Penzhina Bay the species was most widespread between 8000 and 5200 BP. A pollen diagram from Elchikanski Lake demonstrates that the expansion of thickets of *P. pumila* in the Kolyma Mountains began at 8000 BP (Lozhkin & Fedorova 1989). At 5000 BP an expansion of *P. pumila* thickets occurred in the upper part of Indigirka River basin (Belorusova, Lvelius & Ukraintseva 1977). *Pinus pumila* bark was identified in a layer of peat in the first terrace of the Lena River (not in the floodplain) near Chekurovka (Korzhuyev & Fedorova 1962, 1970). The layer of peat is dated 5610 ± 200 BP (Serebrianny 1965). Seeds of *P. pumila* were also identified in non-dated Holocene peat in the eastern part of the Taymyr Peninsula (Nikolskaya 1980). These finds, coupled with the presence today of isolated *P. pumila* in the Oleniek River basin (Fig. 4.4), suggest that between 8000 and 5000 BP the range of *P. pumila* in northeast and north-central Siberia extended further northwest than now. Restriction of *P. pumila* range in this part of Siberia and the isolation of stands in the Oleniek River basin was probably caused by climatic cooling at 5000–4500 BP. At the same time climatic deterioration favoured an expansion of *P. pumila* in the subalpine belt in the southeastern Siberian mountains.

4.3 China

4.3.1 Modern distribution

There are 22 species of *Pinus* in China, including 12 species of the section *Strobus* and 10 species of the section *Pinus* (Zheng 1983). Pines are widely distributed throughout the forested regions of China (Fig. 4.5), from the tropical monsoonal rain forest in Yunnan, Guangdong, Hainan and

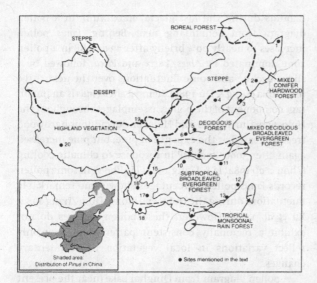

Fig. 4.5. Map of key *Pinus* pollen sites discussed in the text, in relation to the vegetation regions of China: 1. Qindeli, Sanjiang Plain (Xia 1988), 2. Gushantun Bog, Changbei Mountain (Liu 1989), 3. Pulandian, Liaodong Peninsula (Guiyang Institute of Geochemistry 1977), 4. Fenzhuang, Beijing (Kong & Du 1980), 5. Zhongtiao Mountain (Yao 1989), 6. Beizhuangcun, southern loess plateau (Wang & Sun 1994), 7. Qidong, Yangtze River delta (Liu *et al.* 1992), 8. Shennonjia, middle Yangtze River valley (Liu 1990), 9. Jiang-Han Plain, middle Yangtze River valley (Liu 1991), 10. Nanping Bog, Hubei (Gao 1988), 11. Lushan (Li 1985), 12. Daiyun Mountain, Fujian (Liu & Qiu 1994), 13. Han Jiang delta (Zheng 1991), 14. Leizhou Peninsula (Lei & Zheng 1993), 15. Lake Shayema, Sichuan (Jarvis 1993), 16. Xi Hu, Yunnan (Liu *et al.* 1986), 17. Dianchi, Yunnan (Sun *et al.* 1986), 18. Xishuangbanna, Yunnan (Liu *et al.* 1986), 19. Qinghai Lake (Du *et al.* 1989), 20. Zabuye Salt Lake, Tibet (Wu & Xiao 1995). The insert map shows the general distribution of the genus *Pinus* in China. The thick broken lines are the primary vegetation boundaries between the forest, steppe, desert, and highland biomes. The thin broken lines separate different forest regions within the forest biome.

Taiwan in the south to the taiga in Xiao Xingan Ling in the extreme northeast. It extends westwards beyond the forest–steppe ecotone and exists in the grassland regions of southeastern Inner Mongolia, the Loess Plateau, the southeastern part of the Qinghai–Tibetan Plateau, and along the Himalayas. It is absent in the vast desert and plateau regions west of Qinghai Lake, except for the Altai Mountain in the extreme northwestern corner of China (*P. sibirica*). Altitudinally, *Pinus* grows from coastal plains to 3600 m (*P. densata*) in the mountains of western Sichuan and southeastern Tibet (Zheng 1983). The large number of sympatric species means that the following discussion of the history of pine in China must be organized by regions, rather than by species.

Pinus pollen generally accounts for 20–60% of the arboreal pollen (AP) sum in forested regions. In the temperate mixed conifer/hardwood forest of the Northeast, *Pinus* accounts for 15–50% of arboreal pollen and usually co-dominates with *Betula* in the surface samples from the Changbai Mountain (Zhou *et al.* 1984; Sun & Wu 1988a). In the temperate deciduous forest of the Zhongtiao

Mountains *Pinus* co-dominates with *Betula*, *Carpinus* or *Quercus*, and accounts for 30–90% of the arboreal pollen (Yao 1989). In the Shennongjia Mountains of the northern subtropical broadleaved evergreen forest, even where pine trees are locally absent, it still accounts for 10–30% of the modern pollen rain (Liu 1990). In surface samples from Lushan in the northern part of the subtropical broadleaved evergreen forest, *Pinus* occurs at 15–90% and co-dominates with *Castanopsis* and *Quercus* (Li 1985). In surface samples from Yunnan and Sichuan in the southwestern part of the subtropical broadleaved evergreen forest, *Pinus* is present at 20–85% and co-dominates with *Alnus*, *Cyclobalanopsis*, or *Castanopsis/Lithocarpus* (Wu & Sun 1987; Sun & Wu 1988b; Jarvis & Clay-Poole 1992).

4.3.2 Early Quaternary occurrence of pine
Pinus pollen is abundant (40–80% of AP sum) in the classical Early Pleistocene Nihewan Formation in North China (Liu 1988), suggesting that pines were an important component of the glacial-stage vegetation at that time. These pollen assemblages also contain much *Abies* and *Picea*, with a mixture of Tertiary relicts such as *Podocarpus*, *Dacrydium*, *Keteleeria* and *Tsuga*. The data imply that pine existed in an Early Pleistocene forest in North China that was probably without modern equivalent. *Pinus* pollen is present consistently in several long boreholes from the North China Plain and the Lower Yangtze River valley (Liu & Ye 1977; Liu 1988), suggesting that pine persisted in these regions throughout the Quaternary. Nevertheless, fluctuations in *Pinus* pollen percentages in these long records may imply significant changes in pine population size in response to environmental changes in the Quaternary.

4.3.3 Northeast China
Several well-dated pollen records from Northeast China show major changes in the abundance of *Pinus* during the late Pleistocene and Holocene. Pollen assemblages co-dominated by *Pinus* and *Picea/Abies* dating to 23 000–>40 000 BP indicate that pine was an important component of the coniferous forest that prevailed in Northeast China before the last glacial maximum (*c.* 18 000 BP) (Liu 1988). The widespread occurrence of *Coelodonta* and *Mammuthus* fossils and relict periglacial features suggests that late Pleistocene climate was much colder than the present. Pollen data directly dated to the last glacial maximum are rare from this region. Wang & Sun (1994) inferred that much of today's temperate mixed forest and deciduous forest regions of Northeast China and North China were covered by *Artemisia* steppe with some open woodland under a cold and arid climate during the glacial maximum. It is reasonable to expect, however, that *Pinus* was not completely eliminated from the Northeast but

instead survived in small populations in favourable habitats throughout the cold and dry stage. Its populations were probably decimated on the alluvial lowlands and plains, where pine trees are uncommon even today. The mountain ranges (Changbai Mountain and Xiao Xingan Range) may have provided refuge for pine and other conifers (spruce and fir) to survive the increased aridity of the glacial maximum.

Pinus populations in Northeast China were still quite small and restricted during the late glacial and early Holocene. At the Gushantun bog (elevation 500 m) in the Changbai Mountain, *Pinus* pollen is present at 5–10% of a total pollen sum in the basal pollen zone dating to 9500–13 000 BP (Liu 1989). At Qindeli on the Sanjiang alluvial plain near the extreme northeastern 'horn' of Northeast China, *Pinus* pollen is essentially absent in the basal sediments dated to 10 000–12 000 BP (Xia 1988). *Pinus* pollen is also essentially absent in the basal sediments of 8000–10 000 BP age in the Pulandian peatland in the Liaodong Peninsula (Guiyang Institute of Geochemistry 1977). The late glacial to early Holocene pollen spectra from Northeast China are typically dominated by *Betula*, accompanied by relatively high percentages of *Picea/Abies* or *Artemisia*. Pine must have been only a minor component of this *Betula*-dominated forest or woodland.

Pine populations remained small in Northeast China from 10 000 to about 4000 BP. During the mid-Holocene, the climate in Northeast China became warmer and more humid, allowing deciduous hardwoods such as *Corylus*, *Juglans*, *Quercus* and *Ulmus* to increase in abundance (Liu 1988). Since about 5000–4000 BP, *Pinus* populations have increased dramatically throughout Northeast China. In many pollen records *Pinus* increased sharply from <5 to >30%, accompanied by a slight increase in *Picea/Abies* and a remarkable decline in the pollen of deciduous hardwood trees. The expansion of pine in the Northeast during the late Holocene is probably due to climatic cooling, which favoured the temperate pine species like *Pinus koraiensis* over the thermophilous hardwoods. Human disturbance was not the cause for this vegetational change, because the northern part of Northeast China remained sparsely populated until the 20th century.

4.3.4 North China and Northwest China

A pollen record from Beizhuangcun (altitude 570 m) south of the Loess Plateau shows episodic expansions and contractions of *Pinus* populations between 27 000 and 23 000 BP, followed by a decline in forest and expansion of steppe (Wang & Sun 1994). By 18 000 BP forest had virtually disappeared and much of North China was covered by *Artemisia* steppe or open woodland (Wang & Sun 1994). As in the case of Northeast China, pine populations in North China were probably reduced and restricted to favourable

habitats during the last glacial maximum. In a pollen diagram from Fenzhuang near Beijing, *Pinus* pollen increases to nearly 20% briefly after 13 000 BP in a pollen zone dominated by *Abies*, *Larix* and *Picea*, followed by a slight decline and some fluctuations over the next three millennia (Kong & Du 1980). In most sites *Pinus* is an important component of the pollen assemblages throughout the Holocene. During the mid-Holocene deciduous hardwood trees increased at the expense of pine, but pine increased again after 5000–4000 BP in response to climatic cooling (Zhou *et al.* 1984; Li & Liang 1985; Liu 1988). In some pollen records *Pinus* pollen percentages show some remarkable fluctuations during the Holocene (Kong, Du & Zhang 1982; Xu *et al.* 1991). However, these pollen changes do not exhibit a regionally consistent pattern and may only reflect variations in local vegetation or sedimentary changes.

A pollen diagram from Qinghai Lake (near the present western limit of *Pinus* in Northwest China) provides a sensitive record of changes in *Pinus* distribution over the last 11 000 years (Du, Kong & Shan 1989). *Pinus* pollen is essentially absent from the sediments of Qinghai Lake before 8000 BP. During the mid-Holocene (8000–3500 BP), *Pinus* pollen increased to 20–50% of all pollen and spores. This was followed by fluctuating but declining values between 3500 and 1500 BP. *Pinus* pollen has declined to about 5% since 1500 BP. The data therefore suggest that *Pinus* probably extended its range westwards to invade the southern edge of the steppe during the mid-Holocene, in response to a warmer and more humid climate due to a stronger summer monsoon.

4.3.5 Tropical and subtropical China

Pollen data from a long borehole in the Yangtze River delta (Liu & Ye 1977) suggest that southeastern China was forested throughout the Quaternary, and *Pinus* has generally been an important component of the regional vegetation. Superimposed on this general trend are several episodes of *Pinus* population expansion and contraction that have been documented from various sites across the vast territories of the Chinese tropics and subtropics.

Two detailed pollen records from the Hanjiang delta and the Leizhou Peninsula in the tropical monsoonal region of coastal Guangdong document an expansion of *Pinus* populations between 28 000 and 23 000 BP. The increase in *Pinus* pollen is accompanied by peaks in the pollen of subtropical broadleaved deciduous elements (*Carpinus*, *Fagus*, *Quercus*) and tropical alpine conifers (*Dacrydium*, *Podocarpus*, *Taxus*, *Tsuga*), suggesting climatic cooling of about 4 °C in the Chinese tropics (Zheng 1991; Lei & Zheng 1993). By contrast, in Xishuangbanna, Southwest Yunnan, *Pinus* pollen declined abruptly during an interval inferred to be around 30 000 BP when

Podocarpaceae (*Dacrycarpus*, *Dacrydium*) pollen was most abundant, perhaps reflecting a climate with wetter, though not cooler, winters (Liu *et al.* 1986). These pollen changes probably reflect local population changes in the watershed around the lake basin instead of large-scale ecotonal movements or range shifts.

During the last glacial maximum the winter and summer temperatures of the South China Sea were about 6 °C and 2 °C cooler than the present, respectively (Wang & Sun 1994). Significant cooling on land is also evident from the limited pollen data available. *Abies* and *Picea* pollen increased considerably between 21 000 and 14 000 BP in two pollen records from the Hubei Province in the north-central subtropics (Gao 1988; Liu 1991). *Pinus* pollen percentages were also relatively high (10–30%) at that time. At Xi Hu (altitude 1980 m), northwestern Yunnan, *Abies* and *Picea* descended at least 500 m downslope during the period 17 000–15 000 BP when the mean annual temperature was estimated to be 2.5–4.0 °C cooler than the present (Lin, Qiao & Walker 1986). The vegetation of the lower slopes around the lake basin shortly after the last glacial maximum was inferred to be dominated by *Pinus* and evergreen *Quercus* mixed with some *Abies*, *Picea*, *Sabina* and *Tsuga*. *Pinus* pollen accounts for 30–90% of all pollen from trees and shrubs during this period, and remains dominant throughout the entire pollen record spanning the last 17 000 years (Lin *et al.* 1986).

The pollen records from Yunnan suggest that *Pinus* has remained an important or dominant component of the vegetation throughout late-glacial and Holocene times (Lin *et al.* 1986; Sun *et al.* 1986). Some fluctuations do occur in the *Pinus* pollen curve during the last 14 000 years, but precise reconstruction of its history is complicated by problems in dating and the over-representation of *Pinus* pollen. In core DZ18 from Dianchi *Pinus* pollen seems to decrease from 60 to 20% during an unspecified interval around 10 000 BP (Sun *et al.* 1986). *Pinus* pollen becomes more frequent during 10 000–7500 BP in Xi Hu (Lin *et al.* 1986). In another core (DZ13) from Dianchi *Pinus* pollen decreases remarkably between 8000 and 4000 BP, perhaps in response to a more equable climate with warmer winters and reduced rainfall seasonality (i.e. less intense spring drought). *Pinus* expanded again after 4000 BP when the modern climatic pattern with strong rainfall seasonality began to develop (Sun *et al.* 1986).

Elsewhere in the Chinese subtropics *Pinus* also seems to dominate the vegetation during the late Pleistocene and Holocene with only minor population changes. No major changes in *Pinus* abundance were detected in the middle or lower reaches of the Yangtze River valley (Liu 1991; Liu *et al.* 1992). In Lake Shayema (altitude 2400 m), southwestern Sichuan, *Pinus* pollen percentages were relatively low (18% of total pollen sum) between 11 000 and 10 000 BP. It

increased rapidly to 50–60% after 10 000 BP and occurred persistently at about 40% throughout the Holocene (Jarvis 1993).

A pollen study from the Daiyun Mountain of central Fujian provides insight into the origin of the pine woodland in the Chinese subtropics (Liu & Qiu 1994; Qiu 1994). The pine woodland, characteristically dominated by *P. massoniana* with a dense ground cover of *Dicranopteris* ferns and grasses, is a very common secondary vegetation type on the denuded hill slopes of subtropical Southeast China today. A pollen record from a subalpine site (altitude 1360 m) presently surrounded by open pine woodland shows that the basin slopes were originally covered by a dense subtropical mixed conifer–hardwood forest in which pine was absent or rare. The pollen of *Castanopsis*, *Cryptomeria* and *Quercus* declined abruptly about 1100 years ago, apparently as a result of deforestation caused by humans. This was followed by a dramatic increase in *Pinus*, Poaceae, and *Dicranopteris*, indicating the development of the pine woodland (Liu & Qiu 1994; Qiu 1994). The study implies that the ubiquitous pine woodland landscape of Southeast China is a landmark of intensified human disturbance in the Chinese tropics and subtropics during the last one or two millennia.

4.4 Japan

4.4.1 Introduction
Much of the knowledge of the postglacial vegetation history of Japan available in the English-language literature comes from the efforts of Tsukada (1967, 1983, 1985, 1988). This brief account will draw heavily from his published work. Common pine species in Japan include *P. pumila* on alpine sites, *P. koraiensis* and *P. parviflora* in subalpine forests, and *P. densiflora* and *P. thunbergii* in more temperate and coastal forests. *Pinus* pollen is well represented in the modern pollen rain of Japan, ranging from *c.* 6% to 74% of the total sum (Tsukada 1988). The first three species listed above are from the section *Strobus* and not generally differentiated on the basis of pollen. *Pinus densifolia* and *P. thunbergii* are from the section *Pinus* and have so-called diploxylon pollen. They can be differentiated from the other three species, but not from each other. The degree of sympatry between the species makes it difficult to unravel their individual histories except where macrofossils are available.

4.4.2 Glacial distribution
Pine was an important component of the glacial vegetation (Fig. 4.6) of Japan from the northern island of Hokkaido (likely *P. pumila*), through the northeastern

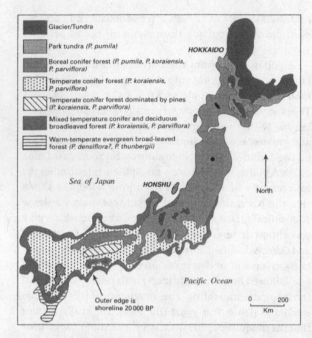

Fig. 4.6. **Glacial vegetation of Japan (c. 20 000 BP) and probable distribution of pine species (vegetation map after Tsukada 1985, 1988).**

portion of the main island, Honshu (probably *P. koraiensis* and *P. parviflora*) and down to lowlands of the south (Tsukada 1985, 1988). The inferred dominance of *P. pumila* in the north reflects the potential of that species to survive the cold conditions that existed during the last glacial maximum. It was probably important in the park tundra of northernmost unglaciated Hokkaido (Fig. 4.6). In Honshu a large number of *P. koraiensis* macrofossil have been recovered from glacial-age sediments (Tsukada 1985). These provide proof of the widespread presence of that species at that time. The general absence of diploxylon pine pollen (Tsukada 1988) argues against the presence, or at least importance, of *P. densiflora* or *P. thunbergii* in the glacial vegetation of the archipelago.

4.4.3 Late-glacial and postglacial history

A decrease in *P. pumila* occurs in central Honshu around 16 000 to 15 000 BP and signals the transition from full-glacial to late-glacial conditions (Tsukada 1988). Further declines in pine pollen occurred between 13 000 and 10 000 BP. As climate warmed, deciduous trees became increasingly important, with *Fagus* reaching northern Honshu and *Quercus* reaching Hokkaido by 10 000 BP. The importance of pine in the temperate regions of the islands continued to decline in response to warm temperatures through the period 7000–4000 BP.

At approximately 2500 to 2000 BP climatic cooling during the late Holocene is manifested in an increase in the importance of *P. pumila* on Hokkaido (Tsukada 1988). A dramatic increase in diploxylon pine pollen is registered in almost all regions of Honshu, Shikoku and Kyusho during the late Holocene. This is probably due to the increased importance of *P. densiflora* as a successional species on sites cleared for agriculture (Tsukada 1988). This rise is time transgressive in response to the spread of agriculture geographically. At some sites in the south the rise of pine occurs as early as 6000–4000 BP, while at northern sites it does not occur until 800–700 BP (Tsukada 1988). Thus, the present abundance of *P. densiflora* in Japan is largely an artefact of human activity.

4.5 Discussion

Pinus is a highly diverse genus in terms of both number of species and intraspecific genetic variability (Chap. 13, this volume). Although Eocene events may be responsible for the main features of pine species diversity in eastern Asia (Millar 1993), the Quaternary history of the region has probably played a significant role in some aspects of both maintaining interspecific differences and also mitigating against speciation events. Three general patterns of pine distributional response to glacial conditions are evident. In the north, *P. sibirica* and *P. sylvestris* were essentially eliminated from their present eastern ranges in Siberia. In northern China, the middle portion of our study area, pine abundances decreased and populations were likely to have been restricted to scattered refugia. In southern China, the abundance of pine was at a maximum during the last glacial in a number of regions. This is also the case for some pines in parts of Japan. These distributional changes could have important impacts on genetic diversity.

In Siberia *P. sibirica* was absent from its modern range during the last glacial. In contrast, *P. pumila* probably persisted in areas of its present range. *Pinus pumila* and *P. sibirica* only became sympatric in the mid-Holocene. There have been as many as 17 glacial periods in the Quaternary, and they have generally lasted longer than interglacials (Bowen 1979). For much of the last 2 million years *P. sibirica* has been isolated from the closely related *P. pumila*. Similarly, *P. sylvestris* has been isolated from other members of the subsection *Sylvestres* such as *P. densiflora* and *P. thunbergii* in eastern Asia during both the last glacial and the Holocene. Such long allopatry would help to foster and maintain genetic differences between these related species.

The persistence of *P. pumila* in small isolated populations in Siberia during glacials and the separation of Japanese and mainland pine populations would serve to enhance genetic variability within the species by promoting and preserving changes brought on by genetic drift and the different selective pressures that might be experienced by different refugial populations. A similar enhancement of genetic diversity through genetic drift and differing selection pressures may have affected pine species such as *P. armandi*, *P. bungeana*, *P. koraiensis* and *P. tabuliformis* that would have experienced population reductions and range fragmentation during glacial periods in northern China and adjacent areas. Conversely, in some southern portions of the study areas, it might be during interglacials, such as the Holocene, when certain pine species experience population declines and range fragmentation.

The high intraspecific genetic variability of pines would impart plasticity that is helpful in the shifting of ranges in response to Quaternary climatic oscillations. During the transitions from glacials to non-glacials and vice versa, pine species would have to accommodate to differing soils, climate, photoperiods and competitive pressures. The shifts in range and changes in abundance could also mitigate against speciation events. Genetically iso-lated populations would be brought back into contact with other populations as ranges and pine abundance increased during favourable periods. In contrast to the Eocene, when climate oscillations occurred over a span of 20 million years, the Quaternary shifts have all occurred over 2 million years. The short duration of isolation and frequent changes in environmental selection pressures during Quaternary glacial and non-glacial episodes might enhance intraspecific variability but preclude the marked evolution of new species that accompanied the Eocene climate variations (Millar 1993).

Acknowledgements

This work was supported by a NSERC Special Collaborative Grant and UCLA support to GMM, and a NOAA Grant (NA56GP0215) and NSF Grants (ATM-9410491, SES-9001343) to KBL. This chapter is PACT (Paleoecological Analysis of Circumpolar Treeline) Contribution 8.

References

Andreev, A.A. & Klimanov, V.A. (1991). Changes of vegetation and climate of the Ungra and Yakokit Rivers interfluvial (Southern Yakutia) in the Holocene. *Botanical Journal*, **76**, 334–51.

Andreev, A.A., Klimanov, V.A., Sulerzhitsky, L.D. & Khotinsky, N.A. (1989). Chronology of landscape and climatic changes of the Central Yakutia in the Holocene. In *Palaeoclimates of the Lateglacial and Holocene*, pp. 116–21. Moscow: Nauka.

Arkhipov, S.A., Levina, T.P. & Panychev, V.A. (1980). Pollen records from two Holocene peat-bogs from the middle and lower Ob Valley. In *Palaeopalynology of Siberia*, pp. 123–7. Moscow: Nauka.

Arkhipov, S.A. & Votakh, M.R. (1980). Pollen records and absolute age of a peat-bog in the Tom River mouth region. In *Palaeopalynology of Siberia*, pp. 118–22. Moscow: Nauka.

Belorusova, Zh.M., Lvelius, N.V. & Ukraintseva, V.V. (1977): Palaeogeography of the Late Pleistocene and Holocene in the region of the Selerikan horse. Quaternary Fauna and Flora of the North-East Siberia. In *Transactions of the Zoological Institute of the USSR Academy of Science*, pp. 265–76. Leningrad: Nauka.

Bowen, D.Q. (1979). Geographical perspective on the Quaternary. *Progress in Physical Geography*, **3**, 167–86.

Chernova, G.M., Mikhaylov, N.N., Denisenko, V.P. & Kozyreva, M.G. (1991). Some problems of the Holocene palaeogeography of South-West Altai. *Izvestia of All-Union Geographical Society*, **2**, 140–6.

Du, Naiqiu, Kong, Zhaochen, & Shan, Fashou (1989). A preliminary investigation on the vegetational and climatic changes since 11 000 years in Qinghai Lake – an analysis based on palynology in core QH85-14C. *Acta Botanica Sinica*, **31**, 803–14.

Firsov, L.V., Volkova, V.S., Levina, T.P., Nikolaeva, I.V., Orlova, L.A., Panychev, V.A. & Volkov, I.A. (1982). Stratigraphy, geochronology and a standard sporo-pollen diagram of the Holocene peat-bog Gladkoye in Novosibirsk City (Pravye Chemy). Problems of the Siberian Pleistocene stratigraphy and palaeogeography. In *Transactions of the Institute of Geology and Geophysics*, pp. 96–107. Novosibirsk: Nauka.

Gao, Fengqi (1988). Analyses of the formation environment of peat and rotted wood in Nanping of Lichuan. *Geographical Research*, **7**, 59–66.

Glebov, F.Z. (1988). *Relations of Forest and Bog in the Taiga Belt*. Novosibirsk: Nauka.

Glebov, F.Z., Toleyko, L.S., Starikov, E.V. & Zhidovlenko, V.A. (1974). Palynology and 14C dating of a peat-bog from the Aleksandrovski district of the Tomsk region (middle taiga sub-belt). In *Types of the USSR Bogs and Principles of Their Classification*, pp. 194–9. Leningrad: Nauka.

Guiyang Institute of Geochemistry (1977). Environmental changes in southern Liaoning Province during the last 10 000 years. *Scientia Sinica, Ser. B*, **22**, 603–14.

Jarvis, D. (1993). Pollen evidence of changing Holocene monsoon climate in Sichuan Province, China. *Quaternary Research*, **39**, 325–37.

Jarvis, D. & Clay-Poole, S. (1992). A comparison of modern pollen rain and vegetation in southwestern Sichuan Province, China. *Review of Palaeobotany and Palynology*, **75**, 239–58.

Khotinsky, N.A. (1977). *The Holocene of Northern Eurasia*. Moscow: Nauka.

Kind, N.V. (1974). Geochronology of the Late Antropogene based on isotope data. In *Transactions of the Geological Institute of the USSR Academy of Sciences 257*. Moscow: Nauka.

Kind, N.V., Gorshkov, S.P. & Chernysheva, M.B. (1969). On the structure and absolute geochronology of the Holocene deposits of the North of Yenisseiski Kriazh. *Bulletin of the Commission for Quaternary Studies*, **36**, 143–8.

Klimanov, V.A. & Levina, T.P. (1989). Climatic changes in the West Siberia in Holocene (based on studies of Yentarny section). In *Forming of Relief, Correlated Deposits and Placers of the North-east of the USSR*, pp. 21–7. Magadan: SVKNII Publishers.

Koltsova, V.G., Starikov, E.V. & Zhidovlenko, V.A. (1979). Evolution of the vegetation and the age of a peat bog in the valley of the Davshe River (Bargusin Natural Reservation). *Bulletin of the Commission for Quaternary Studies*, **49**, 121–4.

Kong, Zhaochen & Du, Naiqiu (1980). Vegetational and climatic changes in the past 30 000–10 000 years in the Beijing region. *Acta Botanica Sinica*, **22**, 330–8.

Kong, Zhaochen, Du, Naiqiu, & Zhang, Zibin (1982). Vegetational development and climatic changes in the last 10,000 years in Beijing. *Acta Botanica Sinica*, **24**, 172–81.

Korzhuyev, S.S. & Fedorova, R.V. (1962). Mammoth from Chekurovka and its environment. *Reports of the USSR Academy of Science*, **143**, 181–3.

Korzhuyev, S.S. & Fedorova, R.V. (1970). About the age of the lower part of the Lena River valley. In *Problems of Geomorphology and Neotectonics of the Siberia Platform Regions*, Vol. 3, pp. 250–63. Novosibirsk: Nauka.

Koshkarova, V.L. (1981). History of taiga forest evolution in the Yenissey River basin (based on macrofossils studies). In *Palaeobotanic Investigations of the Forests of Northern Asia*, pp. 5–30. Novosibirsk: Nauka.

Koshkarova, V.L. (1986). *Seed Flora of Siberian Peat-bogs*. Novosibirsk: Nauka.

Koshkarova, V.L. (1989). Climate changes in the Holocene in the Yenissey River basin (based on palaeocarpological investigations). In *Palaeoclimates of the Lateglacial and Holocene*, pp. 96–8. Moscow: Nauka.

Kremenetski, C.V., Tarasov, P.E. & Cherkinski, A.E. (1994). A history of the Kazakhstan 'island' pine forest in the Holocene. *Botanical Journal*, **79**, 13–29.

Krivonogov, S.K. (1988). Stratigraphy and palaeogeography of the lower part of the Irtysh River valley at the epoch of the last glaciation. In *Transactions of the Institute of Geology and Geophysics 703 Novosibirsk*. Moscow: Nauka.

Krivonogov S.K., Orlova A.A., Panychev V.A. (1985) Seed florae and absolute age of a key section of the I river terrace in the middle part of the Irtysh River valley. In *Palynostratigraphy of the Siberia Mesozoic and Cenozoic*, ed. V.S. Volkova & A.F. Khlonova Novosibirsk, pp. 99–115. Nauka [in Russian].

Kulagina, N.V. & Trofimov, A.G. (1992). Holocene of the Verkhnyaya Angara River valley (northward to Baikal Lake region). *Izvestia of Russian Academy of Sciences Ser. Geology*, **12**, 156–60.

Kutafieva, T.K. (1973). History of forest in southern Evenkia based on pollen analysis of peat bogs. In *Holocene Palynology and Marinopalynology*, pp. 71–5. Moscow: Nauka.

Kutafieva, T.K. (1975). History of forests in the interfluve of the Nizhnaya and Podkamennaya rivers in the Holocene. In *History of Siberian Forests in the Holocene*, pp. 72–95. Krasnoyarsk City: Forest Institute USSR Academy of Sciences Publishers [in Russian].

Lei, Zuoqi & Zheng, Zhuo (1993). Quaternary sporo-pollen flora and paleoclimate of the Tianyang volcanic lake basin, Leizhou Peninsula. *Acta Botanica Sinica*, **35**, 128–38.

Levina, T.P., Orlova, L.A., Panychev, V.A. & Ponomareva, E.A. (1983). Palaeogeography and radiocarbon chronology of Todzhinskaya valley at the lateglacial/early Holocene boundary. In *Glaciations and Palaeoclimate of Siberia in the Pleistocene*, pp. 119–27. Novosibirsk: Institute of Geology Publishers.

Levina, T.P., Orlova, L.A., Panychev, V.A. & Ponomareva, E.A. (1987). Radiochronometry and pollen stratigraphy of the Holocene peat bog Kayakskoye zaimische (Barabinskaya forest-steppe). Regional Geochronology of Siberia and the Far East. In *Transactions of the Institute of Geology and Geophysics 690*, pp. 136–43. Novosibirsk: Nauka.

Levkovskaya, G.M., Kind, N.V., Zavelsky, F.C. & Forova, V.S. (1970). Absolute age of peat-bogs of the Igarka City region and the division of the West Siberia Holocene. *Bulletin of the Commission for Quaternary Studies*, **37**, 94–101.

Li, Wenyi (1985). Pollen analysis of surface samples from Lushan. *Geographical Magazine*, **16**, 91–7.

Li, Wenyi (1991). On dispersal efficiency of *Picea* pollen. *Acta Botanica Sinica*, **33**, 792–800.

Li, Wenyi & Liang, Yulian (1985). Vegetation and environment of the Hypsithermal interval of Holocene in eastern Hebei Plain. *Acta Botanica Sinica*, **27**, 640–51.

Lin, Shaomeng, Qiao, Yulou, & Walker, D. (1986). Late Pleistocene and Holocene vegetation history at Xi Hu, Er Yuan, Yunnan Province, southwest China. *Journal of Biogeography*, **13**, 419–40.

Liu, Guangxiu (1990). Pollen analysis of surface samples from Dajiuhu area, Hubei. *Acta Botanica Boreal.-Occident. Sinica*, **10**, 170–5.

Liu, Guangxiu (1991). Late-glacial and postglacial vegetation and associated environment in Jianghan Plain. *Acta Botanica Sinica*, **33**, 581–8.

Liu, Jinling (1989). Vegetational and climatic changes at Gushantun Bog in Jilin, Northeast China, since 13 000 yr BP. *Acta Palaeontologica Sinica*, **28**, 495–511.

Liu, Jinling & Ye, Pingyi (1977). Quaternary pollen assemblages in Shanghai and Zhejiang Province and their stratigraphic and paleoclimatic significance. *Acta Palaeontologica Sinica*, **16**, 10–33.

Liu, Jinling, Tang, Lingyu, Qiao, Yulou, Head, M.J., & Walker, D. (1986). Late Quaternary vegetation history at Menghai, Yunnan Province, southwest China. *Journal of Biogeography*, **13**, 399–418.

Liu, Kam-biu (1988). Quaternary history of the temperate forests of China. *Quaternary Science Reviews*, **7**, 1–20.

Liu, Kam-biu & Qiu, Hong-lie (1994). Late-Holocene pollen records of vegetational changes in China: climate or human disturbance. *Terrestrial, Atmospheric and Oceanic Sciences*, **5**, 393–410.

Liu, Kam-biu, Sun, Shuncai, & Jiang Xinhe (1992). Environmental change in the Yangtze River delta since 12 000 yr B P. *Quaternary Research*, **38**, 32–45.

Lozhkin, A.V., Anderson, P.M., Eisner, W.R., Ravako, L.G., Hopkins, D.M., Brubaker, L.B., Colinvaux, P.A. & Miller, M.C. (1993). Late Quaternary lacustrine pollen records from southwestern Beringia. *Quaternary Research*, **39**, 314–24.

Lozhkin A.V. & Fedorova I.N. (1989). Vegetation and climate evolution of the USSR North-East in Late Pleistocene and Holocene (based of studies of lakes sediments). In *Forming of Relief, Correlated Deposits and Placers of North-East of the USSR*, ed. V.F. Ivanov & B.F. Palymsky, pp. 3–9. Magadan: SVKNII Publishers.

Millar, C.I. (1993). Impact of the Eocene on the evolution of *Pinus* L. *Annals of the Missouri Botanical Garden*, **80**, 471–98.

Mirov, N.T. (1967). *The Genus Pinus*. New York: Ronald Press.

Neustadt, M.I. (1967). Absolute age of peat bogs of Western Siberia. *Revue Roumaine de Biologie, Serie de Botanique*, **12**, 181–6.

Neustadt, M.I. (1976). Holocene processes in Western Siberia and problems, emerging from these processes. In *Studies and Exploration of the Environment*, pp. 90–9. Moscow: Institute of Geography Publishers.

Neustadt, M.I. & Selikson, E.M. (1971). Neue Angaben zur Stratigraphie der Torfmoore Westsibiriens. *Acta Agralia Fennica*, **123**, 27–32.

Nikolskaya, M.V. (1980). Palaeobotanical characteristics of the late Pleistocene and Holocene deposits on the Taimyr Peninsula. In *Palaeopalynology of Siberia*, pp. 97–111. Moscow: Nauka.

Qiu, Hong-lie (1994). Late-Holocene vegetational history of a subtropical mountain in Southeastern China. PhD Thesis, Louisiana State University.

Ritchie, J.C. & MacDonald, G.M. (1986). The postglacial spread of white spruce. *Journal of Biogeography*, **13**, 527–40.

Serebrianny, L.R. (1965). *The Use of the Radiocarbon Method in Quaternary Geology*. Moscow: Nauka.

Stefanovich, E.I., Klimanov, V.A., Borisova, Z.K. & Vinogradova, S.N. (1986). Holocene palaeogeography on the northern shore of the Penzhina Gulf. *Bulletin of the Commission for Quaternary Studies*, **55**, 97–102.

Stuiver, M. & Reimer, P.J. (1993). Extended ^{14}C data base and revised CALIB3.0 ^{14}C age calibration program. *Radiocarbon*, **35**, 215–30.

Sun, Xiangjun & Wu, Yushu (1988a). Modern pollen rain of mixed conifer forest in Changbai Mt., Northeast China. *Acta Botanica Sinica*, **30**, 549–57.

Sun, Xiangjun & Wu, Yushu (1988b). The distribution of pollen and algae in surface sediments of Dianchi, Yunnan Province, China. *Review of Palaeobotany and Palynology*, **55**, 193–206.

Sun, Xiangjun, Wu, Yushu, Qiao, Yulou & Walker, D. (1986). Late Pleistocene and Holocene vegetation history at Kunming, Yunnan Province, southwest China. *Journal of Biogeography*, **13**, 441–76.

Tikhomirov, B.A. (1946). About origins of associations of dwarf pine (*Pinus pumila* Rgl.). In *Materials of the History of the Flora and Vegetation of the USSR*, 2, pp. 491–537. Moscow-Leningrad: Academic Publishers.

Tikhomirov, B.A. (1949). Dwarf pine (*Pinus pumila* Rgl.), its biology and use. In *Materials for Knowledge of the Fauna and Flora of the USSR*, **6**, 1–106.

Tsukada, M. (1967). Vegetation and climate around 10 000 B P in central Japan. *American Journal of Science*, **265**, 562–85.

Tsukada, M. (1983). Vegetation and climate during the last glacial maximum in Japan. *Quaternary Research*, **19**, 212–35.

Tsukada, M. (1985). Map of vegetation during the last glacial maximum in Japan. *Quaternary Research*, **23**, 369–81.

Tsukada, M. (1988). Japan. In *Vegetation History*, ed. B. Huntley & T. Webb, T. III, pp. 459–581. Dordrecht: Kluwer.

Vaskovski A.P. (1958). New data on limits of trees and shrubs, which dominate in vegetation communities in extreme North-East of the USSR. *Materials on Geology and Minerals of North-East of the USSR*, **13**, 187–204 [in Russian].

Velichko, A.A., Isaeva, L.L., Makeyev, V.M., Matishov, G.G. & Faustova, M.A. (1984). Late Pleistocene glaciation of the Arctic Shelf and the reconstruction of Eurasian ice sheets. In *Late Quaternary Environments of the Soviet Union*, ed. A.A. Velichko, pp. 35–41. Minneapolis: University of Minnesota.

Vipper, P.B. (1968). Relations of forests and steppes in mountain environment of south-west Trans-Baikal region. *Botanical Journal*, **53**, 490–504.

Vipper, P.B., Dorofeiuk, N.I., Meteltseva, E.P. Sokolovskaya, V.T. & Shuliya, K.S. (1976). Vegetation reconstructions of Western and Central Mongolia based on studies of freshwater lake bottom sediments. In *Biological Resources and Environment of the Mongolian People's Republic*, ed. E.M. Lavrenko & E.I. Rachkovskaya, pp. 35–59. Leningrad: Nauka [in Russian].

Vipper, P.B. & Golubova, L.V. (1976). On the history of vegetation of south-western Transbaikal region in the Holocene. *Bulletin of the Commission for Quaternary Studies*, **45**, 45–55.

Volkov, V.S., Gurtovaya, Ye.Ye., Firsov, L.V., Panychev, V.A. & Orlova, L.A. (1973). Structure, age and history of development of the Holocene peat-bog near Gorno- Slinkino village on the Irtysh River. In *Pleistocene of Siberia and Adjacent Regions*, ed. V.N. Sachs, pp. 34–9. Moscow: Nauka.

Wang, Pinxian & Sun, Xiangjun (1994). Last glacial maximum in China: comparison between land and sea. *Catena*, **23**, 341–53.

Wu, Yushu & Sun, Xiangjun (1987). A preliminary study on the relationship between the pollen percentages in forest surface samples and surrounding vegetation on West Mountain of Kunming, Yunnan. *Acta Botanica Sinica*, **29**, 204–11.

Wu, Yushu & Xiao, Jiayi (1995). A preliminary study on modern pollen rain of Zabuye Salt Lake area, Xizang. *Acta Botanica Yunnanica*, **17**, 72–8.

Xia, Yumei (1988). A preliminary study of the vegetation development and climatic changes in the Sanjiang Plain since 12 000 yr BP. *Scientia Geographica Sinica*, **8**, 240–9.

Xu, Qinghai, Wang, Zihui, Wu, Chen, Yu, Shufeng, Li, Yinluo, Lin, Fang, & Zhang, Jing (1991). Vegetational and environmental changes of North Shandong Plain in the past 30 ka B.P. In *Correlation of Onshore and Offshore Quaternary in China*, ed. M.S. Liang & J.L. Zhang, pp. 188–99. Beijing: Science Press.

Yao, Zuju (1989). Surface pollen analysis in Zhongtiao Mountain. *Acta Geographica Sinica*, **44**, 469–77.

Yegorova I.A. (1982). History of the vegetation evolution of Kamchatka peninsula in the Holocene. In *Environmental Evolution of the USSR in the Late Pleistocene and Holocene*, pp. 220–4. Moscow: Nauka.

Zheng, Wanjun (ed.) (1983). *Sylva Sinica*, Vol. 1. Beijing: Forestry Press.

Zheng, Zhuo (1991). Pollen flora and paleoclimate of the Chao-Shan Plain during the last 50 000 years. *Acta Micropalaeontologica Sinica*, **8**, 461–80.

Zhou, Kunshu, Yan, Fuhua, Ye, Yongying, & Liang, Xiulong (1984). Sporo-pollen assemblages in the surface soils of various vegetation zones at the northern slope of Changbaishan Mountain. In *Sporo-pollen Analyses of the Quaternary and Paleoenvironments*, pp. 115–22. Beijing: Science Press.

Zubakov, V.A. (1972). Latest deposits of the West Siberian plain. Leningrad: *Transactions of VSEGEI*, **184**.

Zubarev, A.P. (1981). History of the forest vegetation of middle part of Khamar-Daban mountains in Holocene. In *Palaeobotanic Investigations in Forests of Northern Asia*, pp. 30–44. Novosibirsk: Nauka.

5 The late Quaternary dynamics of pines in Europe

Katherine J. Willis, Keith D. Bennett and H. John B. Birks

5.1 Introduction

Pines occur as native components of the tree flora in most parts of Europe. Jalas & Suominen (1973) and Gaussen, Heywood & Chater (1993) recognize 12 native species (if we take *P. brutia* and *P. halepensis* to be separate species – see below), nine in subgenus *Pinus* (the diploxylon pines) and three in subgenus *Strobus* (the haploxylon pines). *Pinus sylvestris*, the most widespread diploxylon species, occurs from Spain to the Urals and from the Arctic Ocean to northern Greece (Fig. 5.1). It occurs on a variety of soils and in extremely diverse climates (Dallimore & Jackson 1966). It is most abundant as a forest tree in northern Europe as one of the dominant species of the boreal forest (Fig. 5.2; Nikolov & Helmisaari 1992). In central Europe it sometimes forms mixed stands with deciduous species such as *Quercus robur* (Polunin & Walters 1986). The eight other diploxylon pines are all southern European and/or montane in distribution. The most common of these is *P. nigra* which is widespread in southern Europe, and an important forest tree of the Mediterranean Basin and central Europe. Other pines widely distributed in the Mediterranean are *P. brutia* and *P. halepensis*, treated as separate species here (see Chap. 2, this volume, for discussion) but as subspecies of *P. halepensis* in the revision of *Pinus* in Europe by Gaussen *et al.* (1993). *Pinus halepensis* typically occurs above a layer of shrubs such as *Arbutus unedo*, *Erica arborea* and *Myrtus communis* or *Cistus* species (Mirov 1967; Polunin & Walters 1986). Pines that show a distinctive altitudinal zonation include: *P. mugo* which grows at the treeline in the mountains of the Balkans and Central Europe (Mirov 1967; Polunin 1980; Polunin & Walters 1986); *P. heldreichii*, a small tree distributed in the Balkans at mid-altitudes (Mirov 1967; Polunin 1980; Polunin & Walters 1986); and *P. uncinata*, a small tree of the

Pyrenees and the western Alps. Finally, pines that are predominately coastal in their distribution include *P. pinaster*, which occurs in the Iberian Peninsula and eastwards to northern Italy, and *P. pinea* in the Mediterranean Basin. Barbéro *et al.* (Chap. 8, this volume) give a detailed account of the ecology, distribution and recent history of pines in the Mediterranean Basin.

Two of the three haploxylon species are endemic to Europe: *P. cembra*, which forms open woodlands on acidic soils at or near treeline at altitudes of 1400–2500 m in the Alps and Carpathians (Figs. 5.1, 5.3) (Polunin & Walters 1986), and *P. peuce*, which is a rare pine of the Balkan mountains (Fig. 5.1) (Jalas & Suominen 1973; Polunin & Walters 1986). The third species, *P. sibirica*, is a northern Asian species that extends through Russia as far west as 50° E (Fig. 5.1) (Nikolov & Helmisaari 1992; Gaussen *et al.* 1993; Chap. 4, this volume). It can form dense forests, and is a treeline species in the Altai Mountains. In western Siberia it occurs at low elevations, forming forests on swampy plains with *P. sylvestris*, *Abies* and *Picea* (Mirov 1967).

Pines generally occupy marginal habitats in Europe such as montane and boreal environments and dry sandy soils, shallow rendzina-like soils on limestone, ranker soils on rock outcrops, blocky beach ridges, stable screes and boulder fields, and peats. They are not confined to acid soils. *Pinus sylvestris* occurs locally on shallow limestone soils in Scandinavia (Bjørndalen 1980 a, b) and the Alps, and in southern Europe *P. halepensis* and *P. nigra* are frequent on rendzinas. Pines commonly form a surface layer of needle litter up to 5–10 cm thick, even on dry limestone rocks. They avoid waterlogged soils but they can grow on drained or naturally dry or damp peats, for example on bogs in Britain, Fennoscandia and the Pyrenees. Although

Ecology and Biogeography of Pinus, ed. D.M. Richardson. © Cambridge University Press (1998), pp. 107–21.

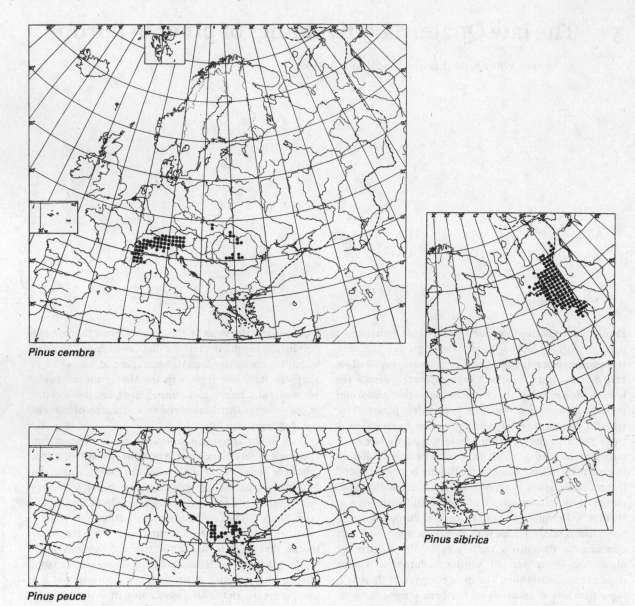

Pinus cembra

Pinus peuce

Pinus sibirica

Fig. 5.1. **The distribution of European *Pinus* species, excluding the species confined to the Mediterranean Basin (see Chap. 8, this volume, for maps for these taxa): *P. cembra*, *P. heldreichii*, *P. mugo*,** *P. peuce*, *P. sibirica*, *P. sylvestris* and *P. uncinata* in Europe. The solid circles show presumed native occurrence of the tree in 50 km² grids of *Atlas Florae Europaeae* (modified from Jalas & Suominen 1973).

pines can grow vigorously in more favourable habitats, they are usually excluded from such areas because of competition from more shade-tolerant trees. They frequently dominate the vegetation in which they occur. The lack of co-dominants is probably a result of environmental extremes. Other trees probably cannot tolerate the dry, nutrient-poor slopes and marginal habitats in which pines achieve dominance.

Pines now also occur as secondary colonizers of abandoned cultivated areas, especially in the Mediterranean Basin (Chap. 8, this volume), and of derelict grasslands and

heaths on shallow soils following changes in land use and reductions in grazing pressure by, for example, deer, rabbits or sheep.

Pine pollen is vesiculate, with two prominent sacci, and can be separated from pollen of other Pinaceae. Pollen from the diploxylon and haploxylon species can often be distinguished (Faegri & Iversen 1989), although the separation has only rarely been made in European pollen-analytical investigations. Pollen of the individual species cannot be recognized. Pine pollen is usually considered to be over-represented in the pollen rain relative

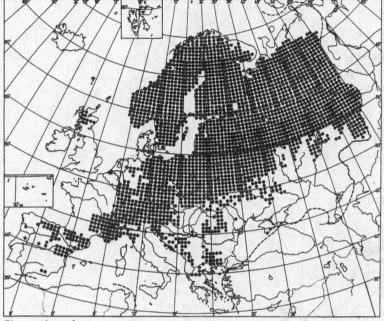

Pinus sylvestris

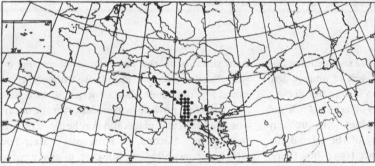

Pinus heldreichii

Fig. 5.1. (cont.)

to other species of forest tree (Andersen 1970; Bradshaw 1993). It is also notoriously well-dispersed (e.g. Prentice 1978). For this reason, palaeoecological records of Pinus pollen have often been disregarded as evidence for presence/absence in a region because of the high production and long-distance dispersal capabilities of the taxon. This is especially the case in regions such as the Arctic that have low local pollen production, resulting in a high representation of pollen transported over long distances, which often includes Pinus (Birks 1973; Hyvärinen 1975, 1976). Unfortunately the assumption that all Pinus pollen must be transported long distances is well established in the literature and other methods of proving the presence of Pinus must be used. These can be divided into macrofossils, including stomatal guard cells, needles, seeds, and

bud-scales, and megafossils, including tree stumps and logs.

Stomatal guard cells of many conifers are lignified and are often preserved along with pollen in suitable sediments. They can be identified at generic level (Trautmann 1953). In a set of 191 surface (0–1 cm) lake-mud samples from throughout Norway, stomatal guard cells occur only when pine is abundant within the catchment of the site. There are, however, some surface samples (c. 10%) that lack stomatal remains even though pine is abundant around the site (H.J.B. Birks and S.M. Peglar, unpublished data). The occurrence of guard cells thus provides unambiguous evidence for the local presence of pine trees around the site (see also Hansen 1994). Little can, however, be inferred from their absence. In eastern England, for example,

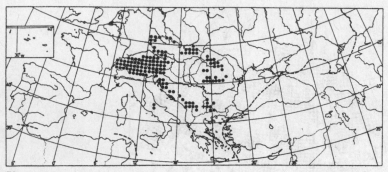

Pinus mugo

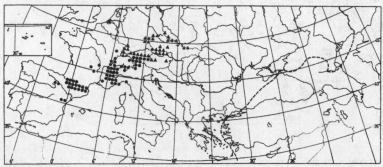

Pinus uncinata

Fig. 5.1. **(cont.)**

Bennett (1983) found pine pollen in the sediments of one lake deposit at frequencies of up to 30% of total pollen, associated with stomatal remains. At a second site, about 3 km away, he found pollen frequencies of less than 10% and no stomata at all, indicating that pines were probably absent from the immediate vicinity of the second site (Bennett 1986). More recently, Fossitt (1994) has found stomata in lake sediments from northwest Ireland with less than 5% pine pollen, suggesting the presence of pine, but in low abundance, and Wick (1995) has used stomatal analysis to help reconstruct Holocene treeline dynamics in the alpine forests of northern Italy (see also Ammann & Wick 1993). Where frequencies of pine pollen are low, therefore, stomatal analyses may help determine whether pine trees were present in low abundance locally or were absent from the surrounding area.

Other fossil remains of pine in Holocene sediments range in size from stumps and logs (e.g. Kullman 1994), often several metres long (megafossils), down to needle fragments that may be only a few millimetres in size. Nearly all parts are recognizable as pine, including wood, leaves, cones, flowers, bark and roots, and may often be identified to species level. Such remains may be discovered *in situ*, principally in mires of various types (e.g. Pilcher *et al.* 1995), but they also occur in any coarse-grained sediment accumulated in a chemical and physical environ-

ment suitable for the preservation of organic materials. Macrofossils, wherever found, are unlikely to have been moved far from their place of origin (Birks & Birks 1980). They therefore provide firm evidence for the occurrence of pine at a particular place, and may be dated directly by the radiocarbon method. Macrofossils rarely occur in sufficient concentration, however, to give any indication of changes in abundance through a stratified sequence. Quantities of trunks of *P. sylvestris* in late-glacial and early Holocene river gravel of south-central European rivers have provided a dendrochronological record that has been used to help calibrate the radiocarbon time scale (Becker, Kromer & Trimborn 1991).

Radiocarbon ages have been converted using the calibration program of Stuiver & Reimer (1993) so that all ages in this paper are given as calendar years before present (AD 1950) unless otherwise stated.

5.2 Full-glacial and late-glacial distribution

Pine was present in Europe throughout the Quaternary in all known glacials and interglacials (van der Hammen, Wijmstra & Zagwijn 1971; West 1980). Full-glacial records for the presence of pine are to be found in the long sedi-

Fig. 5.2. **Southern boreal forest dominated by *Pinus sylvestris* growing on well-drained but moderately fertile soils near Korvas Vaara, Finland (photo: H.J.B. Birks).**

Fig. 5.3. **Stands of *Pinus cembra* growing near the presumed natural treeline at Zirbenwald, Austria (photo: H.J.B. Birks).**

mentary sequences from tectonic lake basins in central and southern Europe (Woillard 1978; de Beaulieu & Reille 1984; Follieri, Magri & Sadori 1988; Pérez-Obiol & Julia 1992). These basins are in regions that were not extensively glaciated and therefore contain continuous sedimentary records covering the last full-glacial period (*c.* 100 000–18 000 years ago). *Pinus* pollen is recorded in all the full-glacial diagrams (e.g. Fig. 5.4), although there is much discussion as to whether this indicates a local presence or long-distance transport. High values of *Pinus* in full-glacial pollen diagrams from France, for example, have often been attributed to long-distance transport of the pollen from southerly refuges (e.g. de Beaulieu & Reille 1984). It is, however, unclear where any 'southerly refuges' might have been located. There is no evidence in the Near East for full-glacial forests nor from central and southern Europe (e.g. van Zeist & Bottema 1991; Bottema, Woldring & Aytuğ 1994). Furthermore, increasing macrofossil evidence (stomata and wood) is being found for the existence of *Pinus* during the full-glacial in regions as far north as Hungary where fossil soils containing charcoal of *P. sylvestris* have been identified from at least 19 sites (Rudner *et al.* 1995). Radiocarbon dating of these layers dates them to between 20 000–30 0000 years ago (Rudner *et al.* 1995; P. Sümegi, personal communication 1995). Charcoal of *P. pinaster* has been found in Portugal from *c.* 33 000 to 22 000 years ago (Figueiral 1995), as well as in the late-glacial. The increasing evidence from the macrofossil record therefore supports the idea of *in situ* populations during the full-glacial. The concept of *Pinus* surviving throughout the full-glacial in central and southern Europe seems even more convincing when the full-glacial environment is considered.

The landscape of central and southern Europe, although not glaciated, would have been undergoing considerable transformation. In the northernmost parts of central Europe, permafrost developed; further south in central Europe, vast swathes of glacial outwash sands (loess) were being deposited by the wind, creating sand dunes, which are still visible in many regions today (Bell & Walker 1992). In southern Europe, relict soils were undergoing intense physical and chemical weathering resulting in nutrient-depleted, acidic soils. Such poor edaphic conditions would have severely limited temperate trees, shrubs and herbs. However, steppe plants appear to have thrived along with one or more species of *Pinus* such as *P. sylvestris*. Soil temperature may have been more restrictive although species such as *P. pumila*, *P. sibirica* and *P. sylvestris* grow reasonably well upon permafrost in the present-day boreal forest (Nikolov & Helmisaari 1992).

Air temperature during the full-glacial would also not have been limiting to the growth of many species of *Pinus*. Computer-simulated modelling for climatic conditions 18 000 years ago (Kutzbach & Guetter 1986; Kutzbach *et al.* 1993) suggests that a steep climatic gradient existed over central and southern Europe. Thus, while some regions had January temperatures as low as −40 °C, others had temperatures of only −5 °C. Temperatures as low as −40 °C

Valle di Castiglione, Rome, Italy

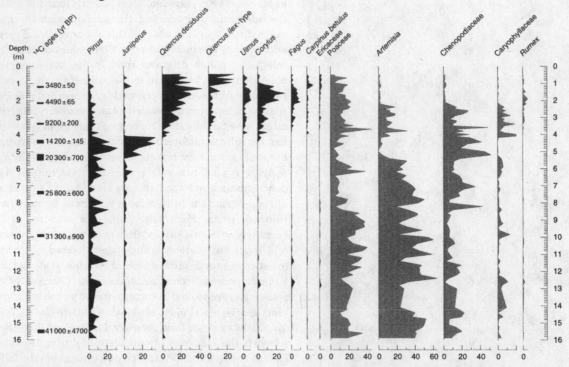

Fig. 5.4. Full-glacial pollen diagram from Valle di Castiglione, Rome, Italy (redrawn, with permission, from Follieri et al. 1988; Magri 1989). Major pollen and spore types are plotted against depth. The scale at the base of the diagram shows percentages of the pollen and spore types. All values are percentages of the total determinable pollen and spores (excluding obligate aquatics and mosses). Dates are given as radiocarbon years before present (BP).

might have been restrictive to all but the hardiest of pines (e.g. *P. cembra*), but many species would have survived winter temperatures of −5 °C.

Prediction of precipitation levels during the last full-glacial is more problematic. In general it is suggested that precipitation was lower than at present, with estimates of 2 mm day^{-1} for both January and July (Kutzbach & Guetter 1986; Kutzbach *et al.* 1993). Limited precipitation would have affected not only the amount of moisture available to plants but also the protective effect of snow-cover against roots and shoots. However, drought-tolerant species such as *P. nigra* and *P. sylvestris* (Ellenberg 1988; Nikolov & Helmisaari 1992) could have survived much drier conditions than those predicted.

In comparison with other temperate trees, shrubs and herbs, therefore, many species of *Pinus* could have survived the full-glacial climatic and environmental conditions. Arguments about long-distance transport or local presence will continue until further macrofossil evidence is found. However, the small amount of evidence currently available supports the notion that more than one *Pinus* species persisted during the full-glacial in central and southern Europe.

5.3 Late-glacial and postglacial (Holocene) distribution change

The onset of the late-glacial in the northern hemisphere with the melting of the ice sheets occurred between *c.* 15 000 and 12 000 years ago (Kutzbach & Guetter 1986; Kutzbach *et al.* 1993). Across central and southern Europe, populations of pine started to increase and numerous pollen sequences across the region record these dynamics. These have been conveniently summarized in map form by Huntley & Birks (1983) (see Fig. 5.5) and Peterson (1983). The date of increase varies greatly between regions although, in general, it would appear that the earliest increases occurred in the lower-altitude sites and then spread to the mountains. In Greece, for example, a large increase in *Pinus* took place in both central and northern regions (sites at 500 m altitude) at *c.* 16 000 years ago (Bottema 1974). On the Iberian Peninsula (173 m) (Pérez-Obiol & Julia 1994) and on the Italian Peninsula (site 40 m) *Pinus* increased from *c.* 16 500 years ago.

In comparison, high-altitude sites appear to indicate an increase up to 3000 years later. In the western Rhodope Mountains in Bulgaria (site 1300 m), for example, a steady increase of diploxylon-type *Pinus* occurred from *c.* 12 700 years ago; this is thought to include *P. mugo*, *P. nigra* and *P. sylvestris* (Huttunen *et al.* 1992). Also present and increasing was *P. peuce*. In northern Italy, high-altitude sites (2250 m) have recorded the increased presence of *P. cembra* as both pollen and macrofossils of needles and stomata (Wick 1995) from only *c.* 10 200 years ago. However, the difference in timing for the increase may relate not only to altitude but also to latitude. Later dates for the increase are found at more northerly sites (Huntley & Birks 1983).

An increase in *Pinus* populations in southern and central Europe at the very beginning of the late-glacial (*c.* 15 000 years ago) supports the notion that many species were already present in the region. An improvement in the climate, such as an increase in temperature and precipitation, especially in southern Europe where soils were not frozen, would have had almost immediate impact upon *Pinus* populations. There were no other trees in significant quantities to outcompete them for light. Increasing solar radiation would have lengthened their growing season and thus productivity. Further north in central Europe and possibly in the high-altitude sites, the situation was slightly different since much of the soil would have been frozen (Bell & Walker 1992). As mentioned above, certain species of *Pinus* can survive on permafrost but it is not ideal for growth and soil temperature can be critical to the plant's metabolic rate (Bonan 1992). Furthermore, in a permafrost region, it has been demonstrated that an increase in solar radiation alone is often not sufficient to melt the permafrost, especially if there is a vegetation/litter layer above (Bonan 1992). In these conditions, a time lag of at least 1000 years (Willis *et al.* 1995) must be expected for the thaw and drainage of soils. This may account for the observed differences in timing between north/south and high-altitude sites for the increase in *Pinus* during the late-glacial.

In northern Europe there is substantial evidence for a late-glacial interstadial between *c.* 15 500 and 13 000 years ago followed by a period of climatic cooling (Younger Dryas), which lasted until *c.* 11 200 years ago when the present Holocene interglacial (= postglacial) commenced (Walker & Lowe 1990). Although the Younger Dryas is recorded in palynological sites throughout northwestern Europe and parts of central Europe (Walker & Lowe 1990; Goslar *et al.* 1993) there is some controversy as to whether its effects can be seen in sites further south (Watts 1980, 1985; Turner & Hannon 1988; Willis 1994). In the pine record for central and southern Europe, the evidence is particularly ambiguous. At some sites a clear decrease in *Pinus* is noted during this time, e.g. in the Rhodope Mountains, Bulgaria (Huttunen *et al.* 1992), Italian Alps (Lowe 1992), and central Europe (Goslar *et al.* 1993); in other regions no obvious change occurs and *Pinus* continues to

dominate, e.g. the Iberian Peninsula (Pérez-Obiol & Julia 1994), and the Hungarian plain (Willis *et al.* 1995). No clear distinction (such as altitude) is apparent between those sites that record a change in *Pinus* and those that do not. Also there is variation in what actually happens to the *Pinus* populations. In the Italian Alps, for example (Lowe 1992), the late-glacial cooling is represented by an increase in *Pinus* whereas in the Rhodope mountains (Huttunen *et al.* 1992) it is represented by a decrease. Again the various species of *Pinus* involved must be considered, with the differences noted probably reflecting the different tolerance levels of the various species of pine in combination with local climatic variability.

Following the effects of the Younger Dryas, widespread increase in all tree types occurred from the beginning of the Holocene at *c.* 11 200 years ago. In southern and central Europe, the expansion of mixed deciduous woodland (*Corylus*, *Fraxinus*, *Quercus*, *Tilia*, *Ulmus*) resulted in the rapid decline of *Pinus*, presumably because of competition for light. At about the same time, there was widespread northerly migration of pine forests across the north European lowlands (Huntley & Birks 1983).

Pine was probably present 11 200 years ago in all but a few parts of Fennoscandia and Britain (Fig. 5.5a). By 10 000 years ago, pine had reached much of the Scandinavian uplands, and had begun to increase in the British Isles (Birks 1989; Fig. 5.6), including the far northwest of Ireland (Fossitt 1994). By 9000 years ago pine had reached Finland and had become locally dominant for the first time in northwest Scotland (Fig. 5.5b). Continued northward expansion at 7800 years ago (Fig. 5.5c) brought pine to northernmost Fennoscandia. By 6800 years ago, pine had declined in southern Fennoscandia, but appears to have been still expanding its range eastwards into the steppe landscapes of eastern Kazakhstan (Kremenetsky, Tarasov & Cherkinsky 1994). By 5700 years ago (Fig. 5.5d) it remained dominant in the extreme north of Finland but occurred only locally further south. The northward extension of pine was almost certainly a response to climatic warming (Iversen 1960), and proceeded most rapidly between 12 400 and 11 200 years ago when spreading rates of at least 1.5 km yr^{-1} were reached. However, only in northeastern Europe and Fennoscandia was pine able to continue its northward migration unchecked. In western Europe its northward spread was checked soon after 11 200 years ago by the expansion of *Corylus* and deciduous *Quercus*. Pine survived longest in areas either with poor soils or far from refugia of deciduous taxa. Ultimately, *P. sylvestris* remained only in the northern boreal forests, sparsely on dry soils in the central European mountains, and in small isolated pockets in the northern European lowlands and northern Britain, where edaphic factors were of primary importance in excluding other taxa (Bennett 1984).

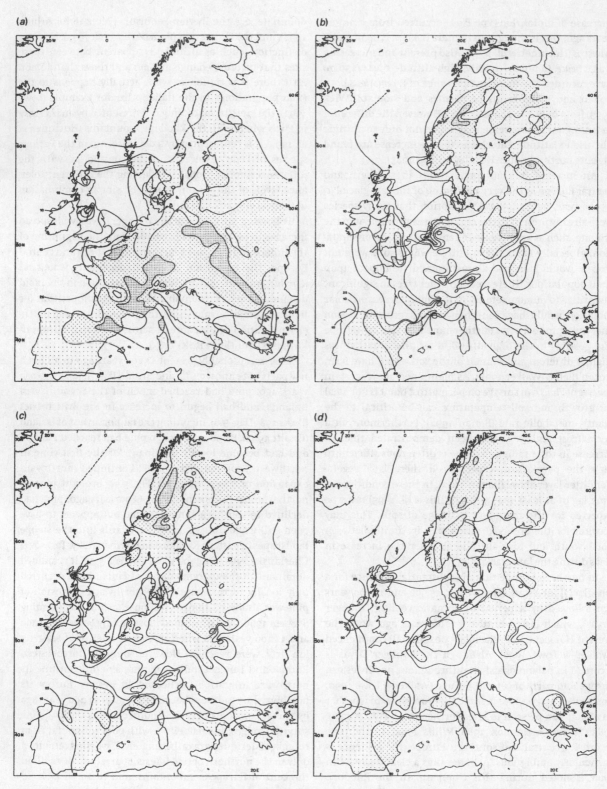

Fig. 5.5. **Isopollen maps of *Pinus* percentages (% total tree and shrub pollen) for (*a*) 11 200 years ago, (*b*) 9000 years ago, (*c*) 7800 years ago** and (*d*) 5700 years ago. The maps are contoured and shaded at 10, 25, 50 and 75% (modified from Huntley & Birks 1983).

Fig. 5.6. Forest of *Pinus sylvestris* near Abernethy, Cairngorms, Scotland, where such woodland has persisted throughout the last 10 000 years (photo: C.E. Hughes).

Fig. 5.7. Stumps of *Pinus sylvestris* c. 5000–4400 years old preserved in blanket peat, Rannoch Moor, Scotland (photo: D.A. Ratcliffe).

When viewed at a finer spatial and temporal scale, the early Holocene spread of pines appears to have been more complex than a progressively expanding range margin. For example, in the British Isles *P. sylvestris* had several seemingly independent centres of early occurrence and subsequent expansion and a diachronous expansion between 10 000 and 6000 years ago (Birks 1989). In northern Scandinavia, *P. sylvestris* appears to have expanded first in areas close to the north Norwegian coast and then south into Finnish Lapland (Hyvärinen 1975, 1976; Seppä 1996). The timing of expansion varies from 9300 to 8000 years ago (Seppä 1996). Further south in Swedish Lappland pine expanded even later (7600–7000 years ago; Seppä 1996; H.J.B. Birks & S.M. Peglar, unpublished data).

5.4 Late Holocene range contraction of *Pinus sylvestris* in northern Europe

Between about 4800 and 4400 years ago there was a marked decline in the abundance of pine in those parts of northern and western British Isles where it still persisted (Birks 1975; Bennett 1984, 1995). This decline appears to have been dramatic, occurring within 100–200 years at some sites, and has left abundant visual evidence in the form of stumps preserved within blanket peat (Fig. 5.7). Ages from many of these have been determined by radiocarbon dating, the overwhelming majority from within a few centuries of 4400 years ago (e.g. Birks 1975; Bridge, Haggart & Lowe 1990). This event was one of the more significant and mysterious shifts in vegetation seen in the British Isles in the course of the Holocene, but some pertinent background details need to be presented before turning to its possible causes. Forest in Scotland (and elsewhere in the British Isles) reached its maximum extent

before about 5700 years ago. Woodland, including pine, developed even on the Western Isles of Scotland (Bennett *et al.* 1990) and in the far northwest of Ireland (Fossitt 1994). A trend away from forest and towards blanket peat, as a direct and indirect consequence of human activities, was established by 4400 years ago, so the last millennium of the expansion of pine was accomplished against a background of overall decline of woodland. The rate of loss of woodland varied from place to place and in time because of clearance and grazing pressure. Some pollen sites near the final range margin of pine (e.g. eastern Skye, Inner Hebrides (Birks & Williams 1983); Fig. 5.8) have high frequencies of pine pollen for a few centuries, possibly indicating that pine was abundant there for only one generation. It is possible that pine was spreading in response to lower competitive pressures in woodlands because of human activity.

The decline of pine in Scotland appears unusually abrupt. Several studies of remains of stumps preserved in peat have shown that only a few generations of trees, at most, were involved. What is not clear is the extent to which this is a true reflection of change in overall pine populations, or a reflection of the pattern of preservation of stumps. Birks (1975) first showed that the ages of some stumps in the peat clustered around 4400 years ago, and suggested that a climatic change had caused the demise of pine over a considerable portion of its range in Scotland at that time. Dubois & Ferguson (1985) determined the ages of many more stumps in the Cairngorms of eastern Scotland, and applied deuterium-isotope analysis of cellulose in the wood. They argued that deuterium abundance was a function of precipitation at the time that the trees were growing, and showed that the cluster of pine stumps from about 4400 years ago had exceptionally low deuterium values, indicating a period of heavy rainfall. On Rannoch Moor in central-west Scotland, Bridge *et al.* (1990) investigated a set of stumps

Loch Ashik, Isle of Skye, Scotland

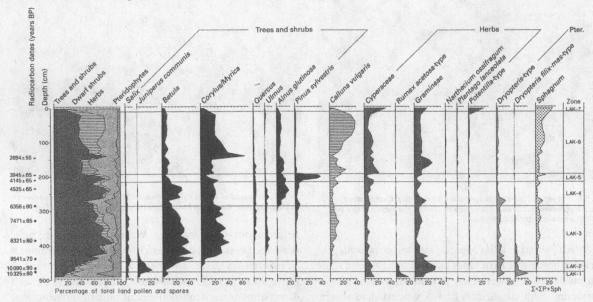

Fig. 5.8. Holocene pollen diagram from Loch Ashik near Kyleakin, isle of Skye. Major pollen and spore types only are plotted against depth. Scale at base of the diagram shows percentages of the silhouettes. All values are percentage of the total determinable pollen and spores (excluding obligate aquatics and mosses). Dates are given as radiocarbon years before present (BP). Pter., pteridophytes; Sph, *Sphagnum* (modified from Birks & Williams 1983).

from the blanket peats there, and carried out pollen analyses of sequences through the peat. They found that pine abundance increased at 7600–7400 years ago, then fluctuated somewhat, before declining in a locally variable manner after 5400 years ago. Combining all available radiocarbon ages for pine stumps in Scotland, they suggested that there was variation in the density and distribution of pine during the Holocene, and they attributed this to variation in precipitation. In particular, there were troughs in pine abundance at 6500–6000, 4200–3800, and after 3500 years ago, with a major decrease between 4400 and 4200 years ago.

Disentangling factors that influence the abundance of pine from factors that might control the preservation of stumps in a peat sequence is not simple. Examination of the pollen record from lake sediments ought to help disentangle the two, by indicating the landscape-scale abundance of pine in forests. Investigations of this type in northwest Ireland (Fossitt 1994) have found little correlation between pollen abundance and tree-stump ages: the latter cluster around 4400 years ago, while the pollen data suggest that the tree was present from c. 10 000 until after 3200 years ago. These data indicate that the problem may be why so many stumps are preserved in peat at 4400 years ago, rather than why pine declined regionally then. Alternatively, the problem may be why pine disappeared from bog habitats at about 4400 years ago, yet survived on better-drained soils in parts of Ireland until medieval times

(Pilcher *et al.* 1995) and in northern Scotland to the present-day.

Elsewhere in northern Europe, much research has been carried out into the Holocene dynamics of trees in Scandinavia. During the course of the Holocene, pollen and macrofossil evidence shows that pine formerly occurred further north than its present limit, and at greater altitudes in the mountains. Eronen & Hyvärinen (1982) and Eronen & Huttunen (1987, 1993) discuss the retreat southwards in terms of a gradual process attributed to a 'deterioration' of climate since c. 5700 years ago. In the central Scandinavian mountains, Kullman (1987a, 1988, 1989, 1995) has obtained ages for 122 stumps of pine, found in lake sediments as well as peat, around its present altitudinal limit. Most of the ages are earlier than 4400 years ago and there are some gaps (e.g. 2200–1400 years ago) with no stumps. He argues that although pine formerly did occur at higher altitudes than today (up to 300 m higher than today about 9000 years ago), it is not possible to say whether clusters of stumps indicate periods favourable for growth of pine or periods favourable for production of dead wood and its preservation. The main period of increased pine limit there (>100 m above present-day limits) was between 9200 and 8000 years ago (Kullman 1995).

In northern Scandinavia, macrofossil remains of pine occur abundantly in lake sediments as well as peats beyond the present latitudinal limit of the tree. *Pinus sylvestris* reached its maximum distribution between *c.* 8000 and 7000 years ago, but between *c.* 5700 and 3200 years ago the northern limit gradually retreated to about its present position (Hyvärinen 1975, 1976; Eronen 1979; Eronen & Huttunen 1987, 1993; Eronen & Hyvärinen 1982). The decline in the altitudinal limit of pine in the mountains of southern Norway occurred between 8000 and 5700 years ago, the period when pine attained its maximum altitudinal and latitudinal extent in the far north (Eronen & Huttunen 1993). The altitudinal limit of pine in the far north shows a marked drop about 5000 years ago (Eronen & Huttunen 1993),after which there is a gradual lowering (Eronen 1979; Eronen & Huttunen 1987). Dendrochronological analyses of the same pine megafossils suggest that the climate became more variable around 5000 years ago (Zetterberg, Eronen & Briffa 1994).

There is also evidence for gradual decline of *P. sylvestris* altitudinal treeline in central Sweden (Kullman 1980, 1987b, 1995). This decline is attributed to long-term, orbitally-forced climatic change, compounded by isostatic uplift (Eronen & Huttunen 1982, 1987, 1993; Eronen & Hyvärinen 1982; Kullman 1980, 1992, 1993, 1995). Recent evidence from Greenland ice cores (e.g. Meese *et al.* 1994) suggests that Holocene climates in the northern hemisphere have been relatively stable since about 8000 years ago, which may make it difficult to sustain a climatic argument for the decline. Although there clearly has been a late-Holocene range reduction in northern and central Scandinavia, both latitudinally and altitudinally, there is no evidence that it has been abrupt. If there was an abrupt decline in the British Isles at 4400 years ago, then it appears to be a phenomenon peculiar to that region. It is, however, possible that the difference in the records from the two areas is due to circumstances of preservation, being permanently available in the case of Scandinavian lake sediments, but only intermittently available in Scottish and Irish blanket peats.

The following hypotheses may be suggested to explain the pine decline in the British Isles (Bennett 1995):

1   Regional climatic change (Birks 1975, 1977; Bennett & Lamb 1984; Birks 1988): can explain the decline and its synchroneity but it is difficult to see why only pine was affected.

2   Volcanic eruption (Blackford *et al.* 1992): not yet established that the decline of pine was exactly synchronous with any eruption (Birks 1994; Hall, Pilcher & McCormac 1994).

3   Anthropogenic pressures: possible, but the decline seems too rapid and over too large an area and there is no evidence for any change in anthropogenic pressures at this time.

4   Change in fire frequency (Bradshaw 1993): possible, if the competitors of pine had been excluded by high fire frequencies and then human effects caused a reduction in fire frequency in the late Holocene.

These hypotheses do not have to be mutually exclusive. One possibility is that pine was expanding over much of its range as a consequence of human activity after 5700 years ago, but onto soils that were becoming marginal because of the general loss of woodland cover, and a period of wetter than average years (as part of the normal variability of climate) then caused the death and preservation of a large number of trees growing on peat. The key test, not yet carried out, is whether the decline of pine, as seen in high-resolution well-dated pollen sequences, is synchronous or variable in time and space.

5.5   Late Holocene expansion of pines in southern Europe

Pine trees are a regular occurrence in the present landscape of central and southern Europe at both high and low altitudes (Bradshaw 1993). As mentioned above, the palaeoecological record from southern Europe suggests that the cover of pine trees decreased rapidly in the early postglacial with the increase of deciduous woodland. However from c. 6000 years ago pines once again started to dominate the southern Europe landscape with a clear distinction between the increase in the lowland regions and that in the mountains. Many low-altitude sites indicate an increase in *Pinus* from *c.* 6000 years ago (e.g. (Bottema 1974; Grieg & Turner 1974; Jahns 1993). Although climatic change such as increased precipitation (Huntley & Prentice 1988) has sometimes been suggested for the rise in pine, most evidence now points to anthropogenic disturbance. The pine increase is accompanied by pollen taxa usually associated with agriculture such as cereals, grasses, and open ground herbaceous species (Behre 1988). In the archaeological record there is evidence of extensive wood use in the Bronze and Iron Ages including wood for charcoal burning (usually *Carpinus betulus/C. orientalis/ Fagus*), building (usually *Fraxinus* and *Quercus*), and fodder (usually *Tilia* and *Ulmus*). However, the clearance of selective deciduous trees would not necessarily have been enough to allow the increase of pine. Species such as *P. sylvestris*, for example, cannot persist in an area of established deciduous forest by invading gaps. Widespread clearance using fire, heavy grazing, and/or depletion of the soil through intensive agriculture must also therefore have been factors in the increase of *Pinus*.

Sedimentological evidence also suggests depletion and erosion of soils from c. 6000 years ago (Willis 1995). Thus, the tolerance of *Pinus* to environmental degradation resulting from Bronze and Iron Age activities probably promoted its re-expansion from the small isolated groups that had established in the early postglacial (e.g. Figueiral 1995). Following this initial clearance, it would appear that in many lowland areas, the soils never recovered and *Pinus* became the established woodland tree as a result of secondary colonization of areas disturbed by cultivation and grazing. Many southern European pine forests are thus of anthropogenic origin, although they have rapidly acquired an associated flora that is generally similar to native woods dominated by the same *Pinus* species.

In the mountainous regions of the Balkan peninsula a rather different picture emerges. A synchronous increase in *Pinus* populations occurred c. 2000 years ago. In the Rhodope mountains (altitude 1200 m, for example, diploxylon *Pinus* increased rapidly to 65% of the total pollen (Huttunen *et al.* 1992); in the Pindus mountains (Willis 1992), *Pinus* (undiff.) increased to 65% total pollen; and in the Rila mountains, *P. peuce*, and *P. sylvestris* (macrofossil evidence) appeared in the vicinity of the lake (Bozilova 1995). No such increase was detected at the low altitudinal sites, nor high altitudinal sites in other regions of southern Europe (e.g. Italian Alps). Again climatic change and anthropogenic activity must be considered. It is difficult to envisage a climatic change that would promote the expansion of *Pinus* in these mountainous regions. An increase in precipitation, for example, would imply that before 2000 years ago the mountains were too dry for *Pinus* growth. Considering the tolerance, in particular, of *P. sylvestris* to drought, this seems unlikely, especially in the more humid mountains. Other climatic variables such as lower January temperatures also seem unlikely to restrict pine growth in the postglacial. Anthropogenic activity must therefore be considered the more likely cause although in contrast to the situation described above, it was probably abandonment that led to the increase. The date of 2000 years ago marks the time of widespread Roman occupation of southern and central Europe. Possibly the Romans instigated a change in farming practices such as the abandonment of summer grazing in the mountains, or caused the reduction in human or animal population sizes. Whatever the cause, it appears that they were probably responsible for the development of the mountain pine forests of the Balkan Peninsula.

5.6 Conclusions

From the palaeoecological evidence it would appear that pine has been present in Europe throughout the Quaternary. In periods of maximum glaciation pines were restricted to southern and central Europe but rapidly expanded northwards during interglacials.

Pinus sylvestris is the only species known to have spread into northern Europe during the interglacials and is thus the only species of pine in Europe to demonstrate invasive behaviour on these broad scales of time and space. In southern Europe there is evidence for an increase of *P. cembra*, *P. mugo*, *P. nigra*, *P. peuce* and *P. pinaster*, but their overall distributions do not seem to have changed markedly. Unfortunately, current levels of taxonomic precision of pine pollen has not yet enabled recognition of the behaviour of many modern species of southern Europe (e.g. *P. halepensis*, *P. pinea* and *P. uncinata*).

Further insights into the biogeographical history of pine in Europe may come from studies on the geographical patterns of genetic and biogeochemical variation within different *Pinus* species, as has done recently for *Quercus petraea* and *Q. robur* by Ferris *et al.* (1993) and for *Abies alba* by Konnert and Bergmann (1995). Early work by Forrest (1980, 1982) and Kinloch, Westfall & Forrest (1986) on monoterpenes and isozymes in *P. sylvestris* demonstrates the potentiality of such studies, particularly when the results are compared with the available palaeoecological record.

The record of *P. sylvestris* in northern Europe is particularly well known because of the abundance of its pollen and macrofossils. It is unusual in this respect relative to other trees of the north temperate zone and provides an opportunity for modelling rates and mechanisms of change which is not available for other species. The behaviour of *P. sylvestris* during the postglacial is unexpectedly complex in time and space and it is likely that other pine species have been similarly dynamic.

In the Holocene it is clear that anthropogenic activity has had a major effect on the behaviour of pines and their present-day distributions in Europe. For example, *P. sylvestris* expanded its range during the early Neolithic in Scotland, and in central Europe one or more pine species increased during the Neolithic and Roman occupations. It is rare to find examples where pines have decreased their range through anthropogenic activity: the *P. sylvestris* decline in northern Europe at 4400 years ago may be the exception.

References

Ammann, B. & Wick, L. (1993). Analysis of fossil stomata of conifers as indicators of the alpine tree line fluctuations during the Holocene. *Paläoklimaforschung*, **9**, 175–85.

Andersen, S.T. (1970). The relative pollen productivity and representation of North European trees, and correction factors for tree pollen spectra. *Danmarks Geologiske Undersogelse II Raekke*, 96.

Becker, B., Kromer, B. & Trimborn, P. (1991). A stable-isotope tree-ring timescale of the late glacial/Holocene boundary. *Nature*, **353**, 647–9.

Behre, K.E. (1988). Some reflections on anthropogenic indicators and the record of prehistoric occupation phases in pollen diagrams from the Near East. In *Vegetation History* ed. B. Huntley & T. Webb III, pp. 633–72. Dordrecht: Kluwer.

Bell, M. & Walker, M.J.C. (1992). *Late Quaternary Environmental Change*. London: Longman .

Bennett, K.D. (1983). Devensian late-glacial and Flandrian vegetational history at Hockham Mere, Norfolk, England. I. Pollen percentages and concentrations. *New Phytologist*, **95**, 457–87.

Bennett, K.D. (1984). The post-glacial history of *Pinus sylvestris* in the British Isles. *Quaternary Science Reviews*, **3**, 133–55.

Bennett, K.D. (1986). Competitive interactions among forest tree populations in Norfolk, England, during the last 10 000 years. *New Phytologist*, **103**, 603–20.

Bennett, K.D. (1995). Postglacial dynamics of pine (*Pinus sylvestris* L.) and pinewoods in Scotland. In *Our Pinewood Heritage*, ed. J.R. Aldhous, pp. 23–39. Farnham, Surrey: Forestry Commission, Royal Society for the Protection of Birds, and Scottish Natural Heritage.

Bennett, K.D., Fossitt, J.A., Sharp, M.J. & Switsur, V.R. (1990). Holocene vegetational and environmental history at Loch Lang, South Uist, Western Isles, Scotland. *New Phytologist*, **114**, 281–98.

Bennett, K.D. & Lamb, H.F. (1984). The history of pine and oak in SW Scotland. *British Ecological Society Bulletin*, **15**, 16–17.

Birks, H.H. (1975). Studies in the vegetational history of Scotland. IV. Pine stumps in Scottish blanket peats. *Philosophical Transactions of the Royal Society of London B*, **270**, 181–226.

Birks, H.J.B. (1973). Modern pollen rain studies in some arctic and alpine environments. In *Quaternary Plant Ecology*, ed. H.J.B. Birks & R.G. West, pp. 143–68. Oxford: Blackwell Scientific Publications.

Birks, H.J.B. (1977). The Flandrian forest history of Scotland: a preliminary synthesis. In *British Quaternary Studies: Recent Advances*, ed. F.W. Shotton, pp. 119–35. Oxford: Clarendon Press.

Birks, H.J.B. (1988). Long-term ecological change in the British uplands. In *Ecological Change in the Uplands*, ed. M.B. Usher & D.B.A. Thompson, pp. 37–56. British Ecological Society Special Publication No. 7 Oxford: Blackwell Scientific Publications.

Birks, H.J.B. (1989). Holocene isochrone maps and patterns of tree-spreading in the British Isles. *Journal of Biogeography*, **16**, 503–40.

Birks, H.J.B. (1994). Did Icelandic volcanic eruptions influence the post-glacial vegetational history of the British Isles? *Trends in Ecology and Evolution*, **9**, 312–14.

Birks, H.J.B. & Birks, H.H. (1980). *Quaternary Palaeoecology*. London: Edward Arnold.

Birks, H.J.B. & Williams, W. (1983). Late-Quaternary vegetational history of the Inner Hebrides. *Proceedings of the Royal Society of Edinburgh B*, **83**, 269–92.

Bjørndalen, J.E. (1980a). Kallstallskogar i Skandinavien-ett förslag till klassificering. *Svensk Botanisk Tidskrift*, **74**, 103–22.

Bjørndalen, J.E. (1980b). Phytosociological studies of basiphilous pine forest in Grenland, Telemark, SE Norway. *Norwegian Journal of Botany*, **27**, 139–61.

Blackford, J.J., Edwards, K.J., Dugmore, A.J., Cook, G.T. & Buckland, P.C. (1992). Icelandic volcanic ash and the mid-Holocene Scots pine (*Pinus sylvestris*) pollen decline in northern Scotland. *The Holocene*, **2**, 260–5.

Bonan, G.B. (1992). Soil temperature as an ecological factor in boreal forests. In *A Systems Analysis of the Global Boreal Forest*, ed. H.H. Shugart, R. Leemans & G.B. Bonan, pp. 126–43. Cambridge: Cambridge University Press.

Bottema, S. (1974). Late Quaternary vegetation history of northwestern Greece. PhD thesis, University of Groningen.

Bottema, S., Woldring, H. & Aytuğ, B. (1994). Late Quaternary vegetation history of northern Turkey. *Palaeohistoria*, **35/36**, 13–72.

Bozilova, E. (1995). The upper forest limit in the Rila mountains from postglacial time – paleoecological evidence from pollen analysis, macrofossil plant remains and ^{14}C dating. In *Advances in Holocene Palaeoecology in Bulgaria*, ed. E. Bozilova & S. Tonkov, pp. 1–9. Sofia: Pensoft Publishers.

Bradshaw, R.H.W. (1993). Forest response to Holocene climatic change: equilibrium or non-equilibrium. In *Climate Change and Human Impact on the Landscape*, ed. F.M. Chambers, pp. 57–65. London: Chapman & Hall.

Bridge, M.C., Haggart, B.A. & Lowe, J.J. (1990). The history and palaeoclimatic significance of subfossil remains of *Pinus sylvestris* in blanket peats from Scotland. *Journal of Ecology*, **78**, 77–99.

Dallimore, W. & Jackson, A.B. (1966). *A Handbook of Coniferae and Ginkgoaceae*. London: Edward Arnold.

de Beaulieu, J.-L. & Reille, M. (1984). A long Upper Pleistocene pollen record from Les Echets, near Lyons, France. *Boreas*, **13**, 111–32.

Dubois, A.D. & Ferguson, D.K. (1985). The climatic history of pine in the Cairngorms based on radiocarbon dates and stable isotope analysis, with an account of the events leading up to its colonization. *Review of Palaeobotany and Palynology*, **46**, 55–80.

Ellenberg, H. (1988). *Vegetation Ecology of Central Europe*. Cambridge: Cambridge University Press.

Eronen, M. (1979). The retreat of pine forest in Finnish Lapland since the Holocene climatic optimum: a general discussion with radiocarbon evidence from subfossil pines. *Fennia*, **157**(2), 93–114.

Eronen, M. & Huttunen, P. (1987). Radiocarbon-dated subfossil pines from Finnish Lapland. *Geografiska Annaler*, **69**A, 297–304.

Eronen, M. & Huttunen, P. (1993). Pine megafossils as indicators of Holocene climatic changes in Fennoscandia. *Paläoklimaforschung*, **9**, 29–40.

Eronen, M. & Hyvärinen, H. (1982). Subfossil pine dates and pollen diagrams from northern Fennoscandia. *Geologiska Föreningens i Stockholm Förhandlingar*, **103**, 437–45.

Faegri, K. & Iversen, J. (1989). *Textbook of Pollen Analysis*, 4th edn. Chichester: John Wiley.

Ferris, C., Oliver, R.P., Davy, A.J. & Hewitt, G.M. (1993). Native oak chloroplasts reveal an ancient divide across Europe. *Molecular Ecology*, **2**, 337–44.

Figueiral, I. (1995). Charcoal analysis and the history of *Pinus pinaster* (cluster pine) in Portugal. *Review of Palaeobotany and Palynology*, **89**, 441–54.

Forrest, G.I. (1980). Genotypic variation among native Scots Pine populations in Scotland based on monoterpene analysis. *Forestry*, **53**, 101–28.

Forrest, G.I. (1982). Relationship of some European Scots Pine populations to native Scottish woodlands based on monoterpene analysis. *Forestry*, **55**, 19–37.

Follieri, M. Magri, D. & Sadori, L. (1988). 250 000 year pollen diagram from Valle di Castiglione (Roma). *Pollen et Spores*, **30**, 329–56.

Fossitt, J.A. (1994). Late-glacial and Holocene vegetation history of western Donegal, Ireland. *Biology and Environment: Proceedings of the Royal Irish Academy*, **94**B, 1–31.

Gaussen, H., Heywood, V.H. & Chater, A.O. (1993). *Pinus*. In *Flora Europaea*, Vol. 1. ed. T.G. Tutin, N.A. Burges, A.O. Chater et al., pp. 40–4. Cambridge: Cambridge University Press.

Goslar, T., Kuc, T., Ralska-Jasiewiczowa, M., et al. (1993). High-resolution lacustrine record of the late glacial/Holocene transition in central Europe. *Quaternary Science Reviews*, **12**, 287–94.

Grieg, J.R.A. & Turner, J. (1974). Some pollen diagrams from Greece and their archaeological significance. *Journal of Archaeological Science*, **1**, 177–94.

Hall, V.A., Pilcher, J.R. & McCormac, F.G. (1994). Icelandic volcanic ash and the mid-Holocene Scots pine (*Pinus sylvestris*) decline in the north of Ireland: no correlation. *The Holocene*, **4**, 79–83.

Hansen, B.C.S. (1994). Refining the paleoecological record by analysis of conifer stomata. *Abstracts of 13th Biennial Meeting of the American Quaternary Association (AMQUA)*, p. 91.

Huntley, B. & Birks, H.J.B. (1983). *An Atlas of Past and Present Pollen Maps for Europe 0–13 000 Years Ago*. Cambridge: Cambridge University Press.

Huntley, B. & Prentice, I.C. (1988). July temperatures in Europe from pollen data, 6000 years before present. *Science*, **241**, 687–90.

Huttunen, A., Huttunen, R., Vasari, R., Panovska, H. & Bozilova, E. (1992). Late-glacial and Holocene history of flora and vegetation in the western Rhodopes mountains, Bulgaria. *Acta Botanica Fennica*, **144**, 63–80.

Hyvärinen, H. (1975). Absolute and relative pollen diagrams from northernmost Fennoscandia. *Fennia*, **142**, 1–23.

Hyvärinen, H. (1976). Flandrian pollen deposition rates and tree-line history in northern Fennoscandia. *Boreas*, **5**, 163–75.

Iversen, J. (1960). Problems of the Early Post-Glacial Forest Development in Denmark. *Danmarks Geologiske Undersogelse IV Raekke*, **4**(3), 1–32.

Jahns, S. (1993). On the Holocene vegetation history of the Argive Plain (Peloponnese, southern Greece). *Vegetation History and Archaeobotany*, **2**, 187–203.

Jalas, J. & Suominen, J. (1973). *Atlas Florae Europaeae, 2. Gymnospermae*. Cambridge: Cambridge University Press.

Kinloch, B.B., Westfall, R.D. & Forrest, G.I. (1986). Caledonian Scots pine: origins and genetic structure. *New Phytologist*, **104**, 703–29.

Konnert, M. & Bergman, F. (1995). The geographical distribution of genetic variation of silver fir (*Abies alba*, Pinaceae) in relation to its migration history. *Plant Systematics and Evolution*, **196**, 19–30.

Kremenetsky, K.V., Tarasov, P.E. & Cherkinsky, A.E. (1994). Holocene history of the Kazakhstan "island" pine forests. *Botanichesky Jhurnal*, **79**, 13–29.

Kullman, L. (1980). Radiocarbon dating of subfossil Scots pine (*Pinus sylvestris*) in the southern Swedish Scandes. *Boreas*, **9**, 101–6.

Kullman, L. (1987a). Long-term dynamics of high-altitude populations of *Pinus sylvestris* in the Swedish Scandes. *Journal of Biogeography*, **14**, 1–8.

Kullman, L. (1987b). Little ice age decline of a cold marginal *Pinus sylvestris* forest in the Swedish Scandes. *New Phytologist*, **106**, 567–84.

Kullman, L. (1988). Holocene history of the forest–alpine tundra ecotone in the Scandes Mountains (central Sweden). *New Phytologist*, **108**, 101–10.

Kullman, L. (1989). Tree-limit history during the Holocene in the Scandes mountains, Sweden, inferred from subfossil wood. *Review of Palaeobotany and Palynology*, **58**, 163–71.

Kullman, L. (1992). Orbital forcing and tree-limit history: hypothesis and interpretation of evidence from Swedish Lappland. *The Holocene*, **2**, 131–7.

Kullman, L. (1993). Holocene thermal trend inferred from tree-limit history in the Scandes Mountains. *Global Ecology and Biogeography Letters*, **2**, 181–8.

Kullman, L. (1994). Palaeoecology of pine (*Pinus sylvestris*) in the Swedish Scandes and a review of the analysis of subfossil wood. *Geografiska Annaler*, **76**A, 247–59.

Kullman, L. (1995). Holocene tree-limit and climate history from the Scandes Mountains, Sweden. *Ecology*, **76**, 2490–502.

Kutzbach, J.E., Guetter, P.J., Behling, P.J. & Sehling, R. (1993). Simulated climate changes: results of the COHMAP climate-model experiments. In *Global Climates Since the Last Glacial Maximum*, ed. H.E.J. Wright, J.E. Kutzbach, T. Webb III, W.F. Ruddimann, F.A. Street-Perrott & P.J. Bartlein, pp. 24–93. Minneapolis: University of Minnesota Press.

Kutzbach, J.F. & Guetter, P.J. (1986). The influence of changing orbital parameters and surface boundary conditions on climate simulations for the past 18 000 years. *Journal of Atmospheric Science*, **43**, 1726–59.

Lowe, J.J. (1992). Lateglacial and early Holocene lake sediments from the northern Apennines, Italy – pollen stratigraphy and radiocarbon dating. *Boreas*, **21**, 193–208.

Magri, D. (1989). Interpreting long-term exponential growth of plant populations in a 250 000-year pollen record from Valle di Castiglione (Roma). *New Phytologist*, **112**, 123–8.

Meese, D.A., Gow, A.J., Grootes, P., Mayewski, P.A., Ram, M., Stuiver, M., Taylor, K.C., Waddington, E.D. & Zielinski, G.A. (1994). The accumulation record from the GISP2 core as an indicator of climate change throughout the Holocene. *Science*, **66**, 1680–2.

Mirov, N.T. (1967). *The Genus Pinus*. New York: Ronald Press.

Nikolov, N. & Helmisaari, H. (1992). Silvics of the circumpolar boreal forest tree species. In *A Systems Analysis of the Global Boreal Forest*, ed. H.H. Shugart, R. Leemans & G.B. Bonan, pp. 13–85. Cambridge: Cambridge University Press.

Pérez-Obiol, R. & Julia, R. (1994). Climatic change on the Iberian peninsula recorded in a 30 000-yr pollen record from Lake Banyoles. *Quaternary Research*, **41**, 91–8.

Peterson, G.M. (1983). Recent pollen spectra and zonal vegetation in the western USSR. *Quaternary Science Reviews*, **2**, 281–321.

Pilcher, J.R., Baillie, M.G.L., Brown, D.M., McCormac, F.G., MacSweeney, P.B. & McLawrence, A.S. (1995). Dendrochronology of subfossil pine in the north of Ireland. *Journal of Ecology*, **83**, 665–71.

Polunin, O. (1980). *Flowers of Greece and the Balkans – a Field Guide*. Oxford: Oxford University Press.

Polunin, O. & Walters, M. (1986). *A Guide to the Vegetation of Britain and Europe*. Oxford: Oxford University Press.

Prentice, I.C. (1978). Modern pollen spectra from lake sediments in Finland and Finnmark, north Norway. *Boreas*, **7**, 131–53.

Rudner, E., Sümegi, P., Toth, I., Beszeda, I. & Hertelendi, E. (1995). The vegetation of the upper-Weichselian in the central and southern part of the Great Hungarian Plain. *7th European Ecological Congress*, 108 (abstract).

Seppä, H. (1996). Post-glacial dynamics of vegetation and tree-lines in the far north of Fennoscandia. *Fennia* **174**, 1–96.

Stuiver, M. & Reimer, P.J. (1993). Extended ^{14}C database and revised CALIB radiocarbon calibration program. *Radiocarbon*, **35**, 215–30.

Trautmann, W. (1953). Zur Unterscheidung fossiler Spaltoffnungen der mitteleuropaischen Coniferen. *Flora*, **140**, 523–33.

Turner, C. & Hannon, G. (1988). Vegetational evidence for late Quaternary climatic changes in southwest Europe in relation to the influence of the north Atlantic Ocean. *Philosophical Transactions of the Royal Society of London B*, **318**, 451–85.

van der Hammen, T., Wijmstra, T.A. & Zagwijn, W.H. (1971). The floral record of the late Cenozoic of Europe. In *The Late Cenozoic Glacial Ages* ed. K.K. Turekian, pp. 391–424. New Haven, Conn: Yale University Press.

van Zeist, W. & Bottema, S. (1991). *Late Quaternary Vegetation of the Near East*. Wiesbaden: Dr. Ludwig Reichert Verlag.

Walker, M.J.C. & Lowe, J.J. (1990). Reconstructing the environmental history of the last glacial-interglacial transition: evidence from the Isle of Skye, Inner Hebrides, Scotland. *Quaternary Science Reviews*, **9**, 15–49.

Watts, W.A. (1980). Regional variation in response of vegetation to lateglacial climatic events in Europe. In *Studies in the Lateglacial of Northwest Europe*. ed. J.J. Lowe, J.M. Gray & J.E. Robinson, pp. 1–21. Oxford: Pergamon.

Watts, W.A. (1985). A long pollen record from Laghi di Monticchio, southern Italy: a preliminary account. *Journal of the Geological Society of London*, **142**, 491–9.

West, R.G. (1980). Pleistocene forest history in East Anglia. *New Phytologist*, **85**, 571–622.

Wick, L. (1995). Early Holocene reforestation and vegetation change at a lake near the alpine forest limit: Lago Basso (2250m asl), northern Italy. *Dissertationes Botanicae*, **234**, 555–63.

Willis, K.J. (1992). The late Quaternary vegetational history of northwest Greece. II. Rezina marsh. *New Phytologist*, **121**, 119–38.

Willis, K.J. (1994). The vegetational history of the Balkans. *Quaternary Science Reviews*, **13**, 769–88.

Willis, K.J. (1995). Land degradation in the Balkans: variations in time and space. In *L'homme et la Dégradations de l'Environnement*, ed. S. van der Leeuw, pp. 161–74. Juan-les-Pins: Editions APDCA.

Willis, K.J., Sümegi, P., Braun, M. & Tóth, A. (1995). The late Quaternary environmental history of Bàtorliget, N.E. Hungary. *Palaeogeography, Palaeoclimatology, Palaeoecology*, **118**, 25–47.

Woillard, G.M. (1978). Grande Pile Peat Bog: a continuous pollen record for the last 140 000 years. *Quaternary Research*, **9**, 1–21.

Zetterberg, P., Eronen, M. & Briffa, K.R. (1994). Evidence on climatic variability and prehistoric human activities between 165 BC and AD 1400 derived from subfossil Scots pine (*Pinus sylvestris* L.) found in a lake in Utsjoki, northernmost Finland. *Bulletin of the Geological Society of Finland*, **66**, 107–24.

6 The late Quaternary dynamics of pines in northern North America

Glen M. MacDonald, Les C. Cwynar and Cathy Whitlock

Ecology and Biogeography of Pinus, ed. D.M. Richardson. © Cambridge University Press (1998). pp. 122–36.

6.1 Introduction

The pine flora of northern North America contains species of the subsections *Cembrae, Strobi, Pinus, Ponderosae* and *Contortae*. Much of the phylogenetic diversity in North American pines can be traced back to fragmentation and isolation of species into northern and southern populations during the Eocene and major differentiations within and between subsections probably occurred then (Millar 1993; Chap. 3, this volume). For example, Eocene to Oligocene aged fossils of ancestral forms of *Pinus banksiana/contorta* of subsection *Contortae* and *P. monticola/strobus* of subsection *Strobi* have been found in British Columbia, Washington, Idaho and Montana (Millar 1993). Such finds point out the antiquity of these species in northern North America and suggest the potential for early differentiation from southern species. Mirov (1967) speculated upon the potential importance of subsequent Quaternary events in shaping the distributions of pines, but was hampered by the lack of chronological control for most of the palaeoecological studies that he had available. However, the latter work of Critchfield (1980, 1984, 1985), based upon more recent palaeoecological studies with reliable chronological control, is important in highlighting that much of the additional genetic diversity in North American pines, including some speciations, may be explained by the changes in geographic distributions caused by repeated glacial and non-glacial cycles during the last 2 million years. In addition, the migration of pines into their modern ranges following the last glaciation, may have affected the distribution of genetic variation within the species (Cwynar & MacDonald 1987; Chap. 13, this volume).

A relatively large body of evidence exists regarding the late Quaternary history of pine in northern North America. Many of these data come from fossil pollen studies. Geneticists such as Critchfield have made use of pollen records to interpret the biogeographic and genetic history of North American pines. Typically, pine is more strongly represented in the fossil pollen record than any other tree genus. The abundance of pine pollen reflects three principal factors. First, pine produces large amounts of pollen. An individual *P. contorta* (lodgepole pine) tree may produce 25 billion grains in a single year (Critchfield 1985). Second, the outer wall of the grains contains relatively high amounts of the polymer sporopollenin, compared with many other genera of plants, which is resistant to corrosion and oxidation (Havinga 1984). Third, much of northern North America was glaciated (Dyke & Prest 1987) and contains numerous glacially derived lakes and peatlands in which pine and other pollen types are deposited and readily preserved. A similar situation occurs in northern Europe and portions of northern Asia. Intensive efforts have been made in both North America and western Europe to use networks of fossil pollen records from lakes and peatlands to reconstruct the late Quaternary migrational histories of pines and other trees (Davis 1983; Huntley & Birks 1983; Delcourt & Delcourt 1987; Webb et al. 1993; Chap. 5, this volume).

Palaeoecologists retrieve cores of lake sediments and peat to reconstruct past vegetation. The pollen is concentrated from the sediment by use of chemical processing. Chronological control for the fossil records is generally provided by the radiocarbon dating of organic material from the lake sediment. Radiocarbon dates are reported as years BP (Before Present with present taken as AD 1950). It should be noted that years BP do not exactly equal calendar years before present. The difference between calender and radiocarbon years generally decreases over the postglacial so that a radiocarbon age of 1000 BP is about equal to 1000 years while a radiocarbon age of 10 000 BP is equivalent to almost 12 000 calendar years BP (Stuiver & Reimer 1993). Thus calculations of migration rates, etc. based on radiocarbon dates are only approximations and are prone to bias.

Two important problems limit the use of fossil pollen analysis to reconstruct the postglacial history of pine. First, it is only possible to identify pine pollen grains to the taxonomic level of section. Pollen from species of the section *Strobus* are marked by small verrucae on the distal portion of the central body of the grains (called haploxylon pollen by palynologists). In northeastern North America it is possible to separate *Pinus banksiana* (jack pine) and *P. resinosa* (red pine) of the section *Pinus* (diploxylon pollen – lacking distal verrucae) from *P. strobus* (eastern white pine), but not from each other. Second, large amounts of pine pollen are transported long distances and deposited at sites far removed from the trees. For example, up to 25% of the pollen found in lakes on the grasslands of Canada comes from trees in the Rocky Mountains some 200 km or more to the west (MacDonald & Ritchie 1986). Although pine trees generally produce large amounts of pollen, macrofossils of *P. contorta* have been found in deposits where very little pine pollen is present. This suggests that in some cases, pine species may initially invade at population densities so low that they are difficult to detect using pollen analysis (Peteet 1991). It is not possible to reconstruct the **exact** range limits of pine in the past from the pollen record. The pollen records are likely to provide good evidence of the establishment of pine as an important element in the regional vegetation rather than evidence of the first occurrence of the genus at a very specific locale.

In this chapter, we use fossil pollen records to reconstruct the general history of pine distribution in northern North America over the late Quaternary. We pay particular attention to three questions: What is known about glacial refugia for these conifers? What were the geographic patterns and rates of range adjustments since deglaciation? And, what were the consequences of changes in climate on Holocene distributions of pine? To organize our presentation we divide the northern portion of the continent into three broad regions: Northeastern North America, the Northern Continental Interior, and the Pacific Northwest (including adjacent Rocky Mountains) and Alaska. We conclude with a discussion of the potential impact of late Quaternary history on the genetic variability of the northern North American pines.

6.2 Northeastern North America

The northeastern pines which we consider here are *P. banksiana*, *P. resinosa* and *P. strobus* (Fig. 6.1). *Pinus banksiana* has the northernmost range limits of the three species and

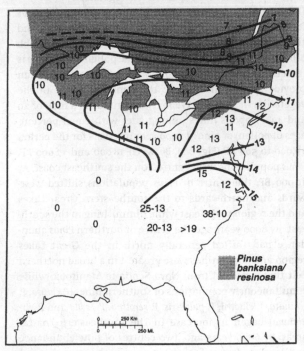

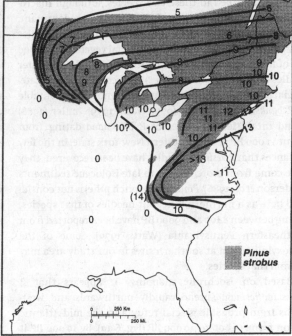

Fig. 6.1. **Range extension of *P. banksiana/resinosa* and *Pinus strobus* following the retreat of glacial ice. Lines are isochrones that connect points of similar arrival times, indicating the position of the** expanding frontier at 1000-year intervals. The shaded area indicates the modern range limits (after Davis 1983).

is a shade intolerant tree found on well-drained, recently burned sites in the boreal forest. *Pinus resinosa* is found along the northern edge of the eastern United States and southern edge of eastern Canada. It is generally associated with rocky and open habitat. *Pinus strobus* has the southernmost distribution of the three species. It occurs on well-drained to mesic soils and may be found as a canopy dominant in mixed deciduous–coniferous forests. The northward migration of these pines during the postglacial period has been summarized by Davis (1983). She assumed that initial arrival of pine in a region would be represented by a sharp and large increase in pine pollen deposition. She identified the first sharp increase in pine pollen in the fossil record for a network of radiocarbon-dated sites and then produced maps of pine isochrones which are lines joining sites with equivalent arrival times. These isochrones are interpreted as the leading edge of an expanding population, i.e. the migration front. Similar maps, together with reconstructions have been produced by Delcourt & Delcourt (1987). The rate of northward expansion of the front for northeastern pines was spatially and temporally variable, but generally ranged between 80 and 400 m yr^{-1} (Davis 1983; Delcourt & Delcourt 1987).

6.2.1 *Pinus banksiana/P. resinosa*
The most common northern pines in eastern North America are *P. banksiana* and *P. resinosa* and diploxylon pollen from northeastern North America is therefore generally ascribed to these two species, although macrofossil studies suggest that *P. banksiana* was dominant. Watts (1970) showed from cross-sections of fossil needles and pollen evidence that the full-glacial vegetation (20 000 BP) of northwestern Georgia contained *P. banksiana*. Later studies of macrofossils (needles) have confirmed that lateglacial pollen of the diploxylon type is mostly attributable to *P. banksiana* (Delcourt 1979; Watts 1979). Miller (1988) found intact female cones of *P. banksiana* dating from about 11 000 BP in northwestern New York state. In the few instances that *P. resinosa* needles have been recovered, they have come from more middle to late-Holocene sediments (Anderson *et al.* 1986). *Pinus rigida* (pitch pine) is not considered here as a northern pine, but needles of that species, dating between 8400 and 9000 BP have been reported from northeastern Pennsylvania (Watts 1979). Some of the diploxylon pollen at southern sites in our study area may be from that species.

Based on isochrone mapping, it appears that *P. banksiana/resinosa* spread rapidly northwards and westwards from probable glacial refugia in the mid-Atlantic region from 14 000 to 10 000 BP (Fig. 6.1*a*). By 10 000 BP it was a significant component of the vegetation from the Atlantic Coast of eastern Canada to well west of the Great Lakes. After 10 000 BP, expansion continued to the north-

east and *P. banksiana* reached its modern northern limits in eastern Canada at 7000 BP. Considering the distance of the glacial refuge for jack pine relative to its modern northern limits, the spread of this species was one of the fastest for any tree in eastern North America (Davis 1983).

6.2.2 *Pinus strobus*
The haploxylon pollen of *P. strobus* is readily separated from the diploxylon grains of the two other northern pines. In addition, macrofossil needles of *P. strobus* are frequently found (Watts 1979; Davis, Spear & Shane 1980; Anderson *et al.* 1986; Peteet *et al.* 1993). The species initially spread from a likely refugium on either the mid-Atlantic Coastal Plain or the adjacent area of exposed continental shelf, migrating northwards and westwards (Fig. 6.1*b*). The northward migration of *P. strobus* appears to have been slower than *P. banksiana* and at many sites the former species subsequently replaced the latter as the dominant pine in the vegetation (Watts 1973; Brubaker 1975; Szeicz & MacDonald 1991). By 5000 years ago, *P. strobus* had moved beyond its present range limit in northeastern Canada and subsequently retreated southwards during the last few millennia (Terasmae & Anderson 1970). In the western part of its range, the spread of *P. strobus* was relatively slow and it did not reach Minnesota until the last thousand years (Jacobson 1979).

6.2.3 Abundance of pine in the vegetation
Delcourt & Delcourt (1987) and Webb and his colleagues have mapped pollen abundances for various species and times (Webb, Bartlein & Kutzbach 1987; Webb 1988; Webb & Bartlein 1988). If we assume that the pollen abundance is directly related to tree abundance, then these maps can provide some indication of the density of pine in the vegetation in the past. Webb (1988) produced a map of 20 and 40% isopolls (lines joining sites with equal percentages of pollen) of pine at 2000-year intervals for the period 18,000 to 500 BP (Fig. 6.2). Between 18 000 and 12 000 BP, *Pinus* populations were centred on the southeast coast. At 10 000 BP, the centre of *Pinus* populations shifted westwards and northwards to the southeastern Great Lakes and the region to the east while diminishing in the southeast. By 8000 years ago the centre of northern *Pinus* abundance had shifted generally north in the Great Lakes region and populations were greatest in a broad northern belt that stretched from Nova Scotia to Manitoba, while simultaneously populations of southern pines (*P. clausa*, *P. echinata*, *P. elliottii*, *P. palustris*, *P. serotina*, *P. taeda* and *P. virginiana*) began to increase in the southeastern coastal areas. By 6000 years ago, two centres of pine abundance had developed: one of southern pines in the southeast and another of northern pines whose centre of distribution had shifted from the east to the west around the northern

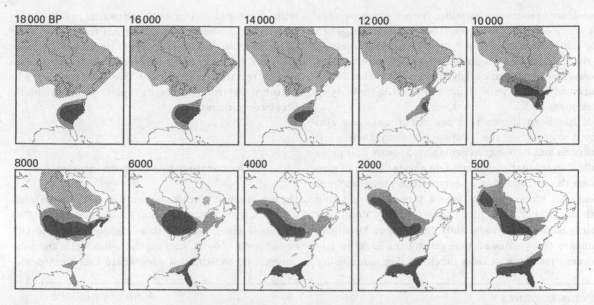

Fig. 6.2. **Map of 20% (light shading) and 40% (dark shading) isopolls (contours of equal pollen percentages) for *Pinus* at 2000-year** intervals. The map includes the position of the Laurentide ice-sheet (cross-hatching) (after Webb 1988).

Great Lakes. Over the next 2000 years, northern pines showed little change whereas southern pines expanded their population areas and sizes to their present status. However, from more detailed studies, there is evidence of a significant peak in the abundance of *P. strobus* at many sites in southern Ontario just before European colonization about 200 years ago (Campbell & McAndrews 1991).

6.2.4 Controls on postglacial distributions

It has been speculated that climate variation is directly responsible for many features of the postglacial history of pine and other trees in northeastern North America (Webb *et al.* 1987, 1993; Prentice, Bartlein & Webb 1991). The main control on the general northward spread of pines in northeastern North America was certainly warming at the close of the last glaciation. The relative rates of spread and patterns of spread for different tree species may also reflect the individualistic response of the species to subtle variations in climate over the Holocene (Prentice *et al.* 1991). However, Davis (1983) has suggested that variations in seed dispersal may have also played a role.

Both climatic control and seed dispersibility may be reflected in the postglacial histories of *P. banksiana* and *P. strobus*. The two species possess winged seeds that are readily transported by wind. The seeds of *P. banksiana* are the smallest of any pines (Mirov 1967) and this might provide the species with a slight advantage over *P. strobus* for rapid migration. In addition, jack pine is a boreal species and its modern range limits lie far north of those for eastern white pine. It is likely that jack pine could have tolerated cold conditions near the glacial margins and

invaded the recently deglaciated landscapes more quickly due to this tolerance. It is worth noting that *Picea* (spruce) dominated the recently deglaciated terrain of northeastern North America in advance of the pines (Davis 1983). The genus has winged seeds that are even smaller than those of pines, coupled with an ability to withstand even more extreme boreal climates than jack pine. In contrast, the more southerly ranging species *Tsuga canadensis* (eastern hemlock) possesses small winged seeds, but did not appear at most northern sites until 1000 years after *P. strobus*. In this case, both climate and soil requirements may have been more important than seed dispersibility in controlling the rate of northward migration (Davis 1983).

In a recent study, Webb *et al.* (1993) compared changes in pine abundance with estimates of past climate derived from a community climate model and from pollen-climate response surfaces. They concluded that the decline in pine abundance in the southern Great Lakes region after 9000 BP was probably the result of increasing moisture and decreasing summer temperatures. This prompted an increase in deciduous trees at the expense of pine. The slow rate of expansion of *P. strobus* westwards in the Great Lakes region has been attributed to arid conditions during the prairie period between 8000 and 4000 BP (Jacobson 1979). The southward retreat of *P. strobus* in Quebec and portions of northern Ontario has been related to cooling after 5000 BP (Terasmae & Anderson 1970; Campbell & McAndrews 1991). At a finer spatial and temporal scale, Campbell & McAndrews (1993) linked forest-stand models with inferred climate changes during the Little Ice Age (AD 1450–1850) and concluded that the increase in *P. strobus* abundance in southern Ontario in the recent past was due to cooling of 1–2° C. However, it has also been suggested that this recent increase in *P. strobus* reflected succession by that species on abandoned Indian agricultural fields (McAndrews 1988; Clark & Royall 1995) and heated debate continues on this issue (Campbell & McAndrews 1995; Clark 1995).

It has been shown that soil conditions have also played a role in the postglacial abundance of *P. banksiana* and *P. strobus*. In general, it appears that pines could dominate the vegetation on a wide variety of sites before the introduction of competitive pressures from deciduous trees. In areas where they are competing with deciduous trees, pines have remained dominant mainly on areas of coarse soils. In northern Michigan, Brubaker (1975) showed that during the early to late Holocene jack pine was able to remain dominant on dry

soils developed on glacial outwash, while being replaced by *P. strobus* and deciduous trees on moister sites. Szeicz & MacDonald (1991) presented evidence that, during the early to mid-Holocene, *P. strobus* was able to remain the dominant tree on coarse soils in southern Ontario while being replaced on more mesic sites by deciduous trees.

6.3 The continental northern interior

The two common pines in the continental northern interior (Fig. 6.3) are *P. banksiana* and *P. contorta* subsp. *latifolia* (lodgepole pine). Both of these closely related species extend north of 60° N. They are shade-intolerant and most common on well-drained sites where they form dense,

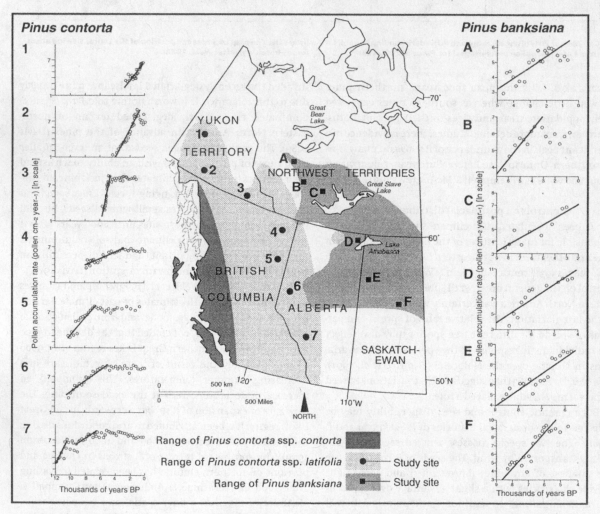

Fig. 6.3. Modern distributions of *Pinus banksiana* and *P. contorta* in the northern interior of North America and the location of sediment coring sites. Arranged along the side of the map are pollen accumulation diagrams for the postglacial spread of *P. banksiana* and *P. contorta* (after MacDonald & Cwynar 1985, 1991; McLeod 1991; K.T. McLeod & G.M. MacDonald, unpublished data).

even-aged stands following fires. *Pinus flexilis* (limber pine) occurs along the western edge of the northern interior plains. It does not grow as far north as the other two species. On the plains, *P. flexilis* occurs in open stands along rocky ridges and stony slopes. The main source of information on the postglacial history of the species is fossil pollen from lake sediments. *Pinus flexilis* is a member of the section *Strobus* and its pollen is easy to separate from that of jack and lodgepole pine. Differentiation of *P. banksiana* and *P. contorta* pollen is not possible. In general, *P. banksiana* is restricted to the eastern portion of the northern interior and *P. contorta* to the west (Fig. 6.3). It is assumed that diploxylon pollen from eastern sites represents *P. banksiana* and *P. contorta* in the west. The two species are sympatric and hybridize in central Alberta making the tracing of their history there difficult. In contrast to eastern North America, fossil pollen sites in the northern interior are too sparse to allow the construction of isochrone maps.

The Quaternary history of *P. contorta* has been used to explain the current distribution of genetic diversity in the species (Wheeler & Guries 1982; Critchfield 1985; Cwynar & MacDonald 1987). The divergence and speciation of *P. banksiana* and *P. contorta* probably resulted from allopatric processes set in motion by the range disruption caused by continental glaciation (Dancik & Yeh 1983; Critchfield 1985). In addition, the lower genetic diversity of Yukon populations of *P. contorta* relative to those from further south has been cited as evidence that refugial populations of the species persisted in ice-free portions of the Yukon during glacial periods (Wheeler & Guries 1982).

6.3.1 *Pinus contorta* subsp. *latifolia*

The postglacial history of *P. contorta* subsp. *latifolia* in the northern interior was reconstructed by MacDonald & Cwynar (1985, 1991; Cwynar & MacDonald 1987), based on the premise that *P. contorta* was present at a site when diploxylon *Pinus* exceeded 15% of the pollen assemblage at a site (Fig. 6.3). Even if a more liberal reconstruction is used, for example the premise that the first continuous occurrence of pine pollen represents the arrival of the species, the geographic pattern remains generally similar, although the estimated arrival times of lodgepole pine become 1000–2000 years earlier (Fig. 6.3). *Pinus contorta* was probably present in mountain valleys in the southwestern corner of the northern interior by 12 200 BP (MacDonald & Cwynar 1985). It migrated northwards along the eastern slope of the Rocky Mountains, reaching central eastern British Columbia by 8000 BP. Lodgepole pine reached the Yukon border region by 5000 BP and extended to its present northern limits in the Yukon in the last thousand years (MacDonald & Cwynar 1985, 1991). There is no

evidence of a refugial population existing in the unglaciated portions of the Yukon during the last glacial maximum and the lodgepole pines in the northern interior probably originated from glacial refugia in the Rocky Mountains (MacDonald & Cwynar 1985). The average rate of northward expansion for *P. contorta* was roughly 150–200 m yr^{-1}.

MacDonald & Cwynar (1991) used pollen accumulation rates (PARs: pollen grains cm^{-2} yr^{-1}) to reconstruct changes in population density as *P. contorta* expanded northwards. They found that the time between the first continuous accumulation of pine pollen at a site and the time when pollen accumulation reached modern rates ranged from c. 800 years to more than 5000 years. The population doubling times for lodgepole pine calculated by MacDonald & Cwynar (1991) ranged from 80 years to over 1000 years. They suggested that the long interval between initial invasion and the growth of the populations to modern densities demonstrated that lodgepole pine advanced by the establishment of small founding populations, well in advance of the main population of the trees. These advance populations appear to have served as seed sources for further advance. These small founding populations grew to modern densities at relatively slow rates. Slowest population growth rates appear to have occurred in areas dominated by damp organic soils, unsuitable for lodgepole pine.

6.3.2 *Pinus banksiana*

Pinus banksiana probably migrated into the northern interior from the southeast and entered from near the Great Lakes at about 10 000 BP (see preceding section). The subsequent northwestward spread of jack pine has been reconstructed from a series of pollen records from western Canada (MacDonald 1987; McLeod 1991; K.T. McLeod and G.M. MacDonald, unpublished data). The pollen evidence suggests a progressive northward movement similar to that of *P. contorta* (Fig. 6.3). By 8500 BP, *P. banksiana* was in central Saskatchewan and Alberta. It reached northern Alberta by 8000 BP and the central Northwest Territories by 4500 BP. It probably did not reach its extreme northern limits in the Mackenzie River valley until the last 2000 years. The rate of movement for this expansion was roughly 200 to 300 m yr^{-1}, but decreased northwards.

The long time between the start of pine pollen accumulation and when it reached modern accumulation rates suggests that the initial invading populations of *P. banksiana* were small and took a considerable time to reach modern densities. The time between the first continuous appearance of *P. banksiana* pollen and when it reached modern abundance ranges from 2000 to 5000 years. Population doubling times ranged from 200 to 1300 years

(K.T. McLeod and G.M. MacDonald, unpublished data). Doubling times were generally slowest in the north.

6.3.3 *Pinus flexilis*

The postglacial record of *P. flexilis* is sparser than that of *P. banksiana* or *P. contorta*. A site on the western edge of the interior plains in southwestern Alberta shows moderately high amounts of haploxylon pine pollen at 10 000 BP, which has been attributed to *P. flexilis* (MacDonald 1989). That site lies approximately 300 km south of the northern range limits of limber pine today. The same site shows a marked increase in the amount of haploxylon pine pollen at the expense of diploxylon pollen between 8000 and 5000 BP. This is accompanied by a marked decrease in *Picea* pollen and an increase in herb pollen. These changes may indicate a displacement of *Picea glauca* (white spruce) and *P. contorta* forest by *P. flexilis*–grassland vegetation, typical of dry and rocky ridges along the eastern margins of the northern plains today (MacDonald 1989).

6.3.4 Controls on postglacial distributions

Warming at the close of glaciation allowed the northward spread of pine into the northern interior. Clearly all three species had glacial refugia south of the ice-sheets. The refugia for *P. contorta* and *P. flexilis* were probably in unglaciated portions of the Rocky Mountains and adjacent Great Plains (see next section). In contrast, the refuge for *P. banksiana* was to the southeast in the mid-Atlantic region (see preceding section).

The northward extension of pine into the northern interior continued into the last few millennia and contrasts with the early Holocene spread of the genus in the northeastern portion of the continent. The spread of *P. banksiana* across Saskatchewan, Alberta and the Northwest Territories was not that much slower than its northward spread in the east. Its late assumption of its modern range boundaries probably reflects the great distance of the western Northwest Territories from the glacial refugia of the species. *Pinus contorta* spread into the northern interior from closer refugia, but proceeded at a slower rate. It may be that lodgepole pine, known for a high degree of shade intolerance, had difficulty penetrating the closed spruce forest that established in western Canada soon after deglaciation (MacDonald 1987; MacDonald & McLeod 1996). Alternatively, the timing and rate of the northward spread of lodgepole pine may reflect some unknown pattern of climatic change over the course of the Holocene. The increased importance of *P. flexilis* at lower elevations along the eastern flanks of the Rocky Mountains between 8000 and 5500 BP is probably due to increased aridity at that time (MacDonald 1989).

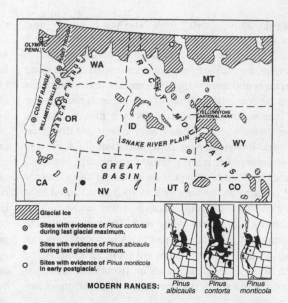

Fig. 6.4. **Areas mentioned in text, modern distributions and known glacial refuges for pines in the Pacific Northwest and adjacent northern Rocky Mountains.**

6.4 The Pacific Northwest and Alaska

Our knowledge of the Quaternary history of pine in the Pacific Northwest and Alaska (Fig. 6.4) is based primarily on fossil pollen records and wherever possible by the association of pollen with plant macrofossils. Pollen is identified to the level of section *Pinus* (diploxylon pollen group) and section *Strobus* (haploxylon pollen group) and the assignment of species often rests on modern phytogeography. *Pinus contorta* and *P. ponderosa* (ponderosa pine) are both diploxylon pines. *Pinus albicaulis* (whitebark pine), *P. flexilis* (limber pine), *P. lambertiana* (sugar pine) and *P. monticola* (western white pine) are haploxylon types. In this section, we will examine the Quaternary history of *P. albicaulis*, *P. contorta* and *P. monticola*. *Pinus flexilis* and *P. ponderosa* are not considered in this discussion because their palaeoecologic record is sparse in the region. Although ponderosa pine is the most widely distributed pine in North America today, little is known of its distribution before the late Holocene. Notable exceptions, however, are the occurrence of *P. ponderosa* in Yellowstone National Park during the last interglaciation (Baker 1986) and its early appearance in the central Sierra Nevada at *c.* 12 000 BP (Smith & Anderson 1992). In comparison with the northeast, fossil pollen sites in this region are spatially sparse and the construction of isochrone maps for any of the pine species is not possible.

6.4.1 *Pinus contorta*

In the Pacific Northwest *P. contorta* occurs as two subspecies. *Pinus contorta* subsp. *contorta* (shore pine) occurs on

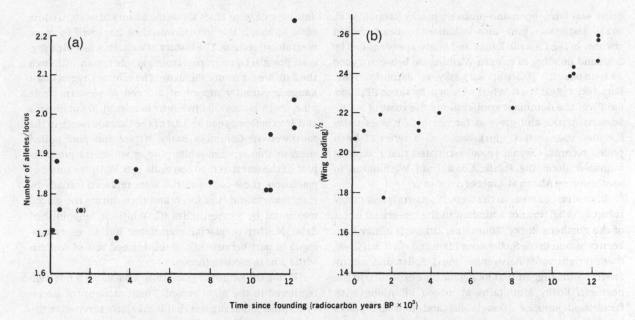

Fig. 6.5. **Relationships between the time of invasion by *Pinus contorta* subsp. *latifolia* and the genetic diversity of the modern population as evidenced by allozyme number (*a*) and modern seed** dispersibility as evidenced by the ratio of seed wing area to seed mass (*b*) (from Cwynar & MacDonald 1987).

a variety of sites along the coastline. *Pinus contorta* subsp. *latifolia* occurs on well-drained sites at higher elevations inland and is a common invader following fires. Both subspecies are shade-intolerant. The glacial range of *P. contorta* was nearly as great as its modern distribution (Fig. 6.4). Needles of lodgepole pine and diploxylon pine pollen have been recovered at sites in the Coast Range before and during the full-glacial period, *c.* 25 000–13 000 BP (Fig. 6.5; Heusser 1977; Worona & Whitlock 1995). During the glacial maximum, *P. contorta* was part of a forest of pine, spruce (probably *Picea engelmannii*, Engelmann spruce), *Abies* (fir), and *Tsuga mertensiana* (mountain hemlock) that extended from south of the Cordilleran ice sheet on the Olympic Peninsula in northwestern Washington to western Oregon. The species was also present in subalpine parkland/tundra east of the Coast Range from the Willamette Valley of Oregon to within 30 km of the Cordilleran ice sheet in the Puget Trough of western Washington (Barnosky 1981, 1985a; Tsukada, Sugita & Hibbert 1981; P.K. Schoonmaker and M.A. Worona, unpublished data). In the Columbia Basin of eastern Washington, *P. contorta* was common in forests before 25 000 BP, but absent when full-glacial conditions at low elevations were cold, dry, and probably windy (Barnosky 1985b). It may have survived in unglaciated parts of the eastern Cascade Range at this time. Further east in the Rocky Mountains, lodgepole pine was present at middle elevations during the glacial maximum, where it was apparently restricted from the valley floor by cold dry conditions and from high

elevations by glacial ice. In a record from the eastern Snake River Plain of Idaho, diploxylon pine pollen in full-glacial sediments is attributed to *P. contorta* stands in nearby foothills (Beiswenger 1991).

The nature of the glacial refugia for lodgepole pine probably varied across the range. The harshness of the climate in the interior Pacific Northwest and northern Rocky Mountains may have supported ancestors of *P. contorta* subsp. *latifolia*. This subspecies also may have survived in protected habitats at middle elevations in the Cascade Range and eastern Olympic Mountains, enabling it to colonize the Puget Lowland and Okanogan region by migrating downslope (Whitlock 1992). In contrast, *P. contorta* subsp. *contorta* may have grown in the unglaciated Coast Range and even on the continental shelf, where a more moderate climate probably prevailed. Although there is no direct evidence, the presence of populations on the continental shelf would explain the rapid colonization of coastal British Columbia and southeastern Alaska (Hebda & Whitlock 1997).

Pinus contorta was well adapted to take advantage of the unstable landscapes created during deglaciation, and its maximum abundance in the Pacific Northwest was during the late glacial. Beginning at about 14 000 BP in the south and ending at *c.* 10 000 BP in the north, lodgepole pine forest (presumably dominated by *P. contorta* subsp. *contorta*) followed the retreat of Cordilleran ice from Washington to southeastern Alaska. Lignin analysis of late-glacial sediments in the central Puget Lowland suggests that the pine forest was fairly open and probably patchy (Leopold *et al.* 1982). Lodgepole pine also colonized outwash-choked streams in the Cascade Range and landscapes denuded by Scabland flooding in eastern Washington between 13 000 and 12 000 BP (Martin, Barnosky & Barnosky 1982; Barnosky 1985a; Sea & Whitlock 1995). By 10 000 BP, *P. contorta* was the dominant conifer along the coast of southwestern Alaska and grew as far north as it does today (Cwynar 1990; Peteet 1991). Based on a series of fossil pollen records, Cwynar (1990) estimated that *P. contorta* migrated along the Pacific Coast from Washington to southwestern Alaska at a rate of *c.* 670 m yr^{-1}.

In contrast to areas to the west, *P. contorta* (presumably subsp. *latifolia*) was not abundant in the late-glacial forest of the northern Rocky Mountains, although it may have been common in the Snake River Plain and northern Great Basin (Bright 1966; Beiswenger 1991). Pollen and macrofossil records suggest that lodgepole pine expanded in the northern Rocky Mountains at 10 000 BP. Along with *Pseudotsuga menziesii* (Douglas-fir) and in some places *Populus tremuloides* (aspen), lodgepole pine grew above its present range between 10 000 and 4000 BP at the expense of spruce–fir–whitebark pine forest. In Yellowstone National Park, the infertile rhyolitic plateaus today are covered almost exclusively by lodgepole pine forest. The fossil record suggests that these plateaus were treeless in the late-glacial period even though subalpine conifers, including spruce, fir, and whitebark pine, were growing nearby on andesitic and other substrates (Whitlock 1993; Whitlock & Bartlein 1993; Whitlock, Bartlein & Van Norman 1995). Lodgepole pine invaded all of the Yellowstone region at about 10 000 BP, and on rhyolitic substrates, pine forests have persisted to the present with little modification. The present dominance of lodgepole pine in the low-elevation forests of the Rocky Mountains has been attributed to its ability to grow well on infertile soils and in areas of frequent disturbance (Despain 1983; Whitlock 1993). On non-rhyolitic regions, more mesophytic taxa were favoured in the late Holocene.

6.4.2 *Pinus monticola*

The range of *P. monticola* extends throughout the Pacific Northwest, but it seldom constitutes more than 5% of the trees in a stand (Arno & Hammerly 1977). Western white pine occurs from sea level to about 1500 m elevation in southwestern British Columbia, 600 to 1800 m in the eastern Cascades and 1800 to 2300 m in central California. It is abundant on poor sites and competes with Douglas fir and *Larix occidentalis* (western larch) as a seral dominant (Fowells 1965). Based on the genetic uniformity of populations north of 45° N, Critchfield (1984) speculated that the glacial refugia of western white pine were some distance from the ice, perhaps in western Oregon. The fossil pollen

data provide some clues about the history of western white pine, although the picture would be improved by more macrofossil records. The history of western white pine suggests postglacial migration from a single region, although the data are far from conclusive. The central Oregon Coast Range apparently supported a forest of western white pine, *Tsuga heterophylla* (western hemlock), *T. mertensiana* and *Abies* before 25 000 BP. East of the Cascade Range in the southwestern Columbia Basin, haploxylon pine pollen suggests that western white pine or whitebark pine was part of the forest at *c.* 90 000 years ago. With the onset of glaciation, these communities were replaced initially by *Pinus contorta* and *Picea* forest and then during the glacial maximum by steppe/tundra (C. Whitlock, unpublished data). Neither preglacial assemblage has an exact analogue in part because of the diminished role of western white pine in modern forests.

During the height of glaciation, *P. monticola* is not well-registered in the fossil record. Small amounts of haploxylon pine pollen suggest that it may have survived in the Coast Range when the forest consisted of Engelmann spruce, lodgepole pine, mountain hemlock, and fir (Fig. 6.4). Western white pine reappeared in the Coast Range as well as the southern Puget Trough at the close of the glacial period, when it grew in a mixed forest of high- and low-elevation species. It spread northwards from Oregon and was present in the southern Puget Lowland at about 7500 BP (Barnosky 1981), in the northern Puget Lowland at *c.* 7000 BP (Cwynar 1987), and in the Fraser Lowland at 4700 BP (Wainman & Mathewes 1987) (although the pollen record from the Fraser Lowland places its appearance as early as 7000 BP; Mathewes 1973). *Pinus monticola* was present in the central Sierra Nevada of California by 6300 BP, but its southern distribution before that time is unknown (Anderson 1990).

In the northern Rocky Mountains, needles of *P. monticola* are found at a low-elevation site in western Glacier National Park as early as *c.* 10 500 BP and pollen data suggest that the species was present there throughout the Holocene (Whitlock 1995). Haploxylon pine is present in pollen records from the Okanogan region at about 5800 BP (Mack, Rutter & Valastro 1978) and in the eastern interior Plateau of British Columbia at *c.* 4000 BP, as a component of the interior Douglas fir forest (Hebda 1995).

6.4.3 *Pinus albicaulis*

Pinus albicaulis is a common constituent of subalpine vegetation in the xeric areas of the Cascade Range, Coast Range, northern Rocky Mountains and Sierra Nevada. Its range extends from low elevations in the Rocky Mountains, where it occurs as a minor element, to the alpine zone, where it can form dense stands of *krummholz* (Arno & Hammerly 1977).

Macrofossils of whitebark pine from a site in the central Great Basin of Nevada (Fig. 6.4) suggest that the species survived the last glaciation south of its present range (Nowak *et al.* 1994). These populations may have been the source for the postglacial colonization of *P. albicaulis* in the Sierra Nevada (Anderson 1990). Sites in the Snake River Plain and northern Rocky Mountains contain negligible amounts of haploxylon pine pollen before 12 000 BP (Anderson 1990; Beiswenger 1991; Whitlock *et al.* 1995), so little is known of the glacial refugia for populations that invaded the Rocky Mountains. The actual timing of the earliest postglacial appearance of *P. albicaulis* varies from site to site and shows no obvious spatial pattern to suggest migration from a single source. In the Yellowstone region, *P. albicaulis* first appears at the end of the late glacial period when it accounts for as much as 25% of the pollen. The pollen is attributed to whitebark pine, but *P. flexilis* may also have been a contributor (Bright 1966; Davis, Sheppard & Robertson 1986; Whitlock 1993).

With onset of warmer and drier conditions at 10 000 BP, whitebark pine, spruce, and subalpine fir were restricted to high elevations in the northern Cascade Range, Coast Range, Rocky Mountains and Great Basin (Mehringer, Arno & Peterson 1977; Whitlock 1993). Subfossil wood from the southern Canadian Cordillera and northern Rocky Mountains indicates that *P. albicaulis* grew above its present altitude in the early Holocene (Luckman 1988; Clague & Mathewes 1989). In the late Holocene, whitebark pine increased in abundance in subalpine and montane forests and the position of the treeline shifted downslope (Mehringer *et al.* 1977).

6.4.4   Controls on postglacial distributions

Due to the topographic and environmental heterogeneity of the Pacific Northwest, the glacial distribution of *P. contorta* was probably spatially extensive, although populations were probably small and restricted to favourable sites. The near synchronous appearance of lodgepole pine at many sites in the Rocky Mountains at 10 000 BP argues against long-distance migration from any one refuge. Rather, scattered refugial populations could probably expand quickly when the climate became suitable. The fossil pollen evidence does, however, suggest that *P. contorta* did not have glacial refugia in Alaska, but migrated northwards from Washington in the late-glacial period.

Postglacial changes in the abundance of *P. contorta* may reflect both changing climate and edaphic conditions. In the Pacific Northwest, *P. contorta* was most abundant during the late glacial period when it colonized new areas created by ice recession. Open forests of pine extended from Washington to coastal Alaska with little change for several centuries. With progressive soil development and possibly a shift to cooler conditions *c.* 11 000–10 000 BP

(Mathewes 1993), other conifers expanded their range and formed a mixed forest of lodgepole pine, spruce and hemlock. Conditions became warmer and drier than at present after 10 000 BP, and these taxa were replaced by Douglas-fir and red alder (*Alnus rubra*). During most of the Holocene, lodgepole pine has been confined to stressed habitats and higher elevations in the Pacific Northwest, although its abundance has increased in the late Holocene, probably in response to regional cooling.

In the northern Rocky Mountains, lodgepole pine was most abundant in the early Holocene when warm dry conditions and frequent fires maintained widespread seral communities. Charcoal records from the northern Rocky Mountains suggest that fires were more frequent than today, and lodgepole pine probably dominated early-successional forests (Mehringer *et al.* 1977; Millspaugh 1997; S.H. Millspaugh, unpublished data). As the climate became cooler and moister in the last 4000 years and fire return intervals lengthened, spruce, fir and whitebark pine became more important in montane forests (Whitlock 1993). Today, lodgepole pine is dominant only in areas with poor soils and frequent disturbance.

Pinus monticola was apparently abundant in the Coast Range before the last glaciation but was reduced to small populations as the climate became cooler and drier in the glacial period. Its migration northwards is traced from western Oregon in the late-glacial period to western Washington and southwestern British Columbia in the early and middle Holocene. Low-elevation forests in the Pacific Northwest featured an expansion of western white pine and other mesophytic taxa with the establishment of cool humid conditions in the late Holocene. Western white pine may also have survived the last glaciation in small populations in northwestern Montana in as much as it was present in Glacier National Park during the late-glacial period. There is general agreement that the expansion of *P. monticola* to its present range in the interior valleys of British Columbia reflects the cooler and wetter climate that has developed over the late Holocene.

The record of postglacial expansion of *P. albicaulis* is too sparse and asynchronous to draw many conclusions regarding its environmental controls. However, the growth of whitebark pine above its modern range during the period 10 000 to 4000 BP suggests the influence of warmer conditions during the early to mid-Holocene.

6.5   Discussion

The postglacial history of pines in northern North America highlights a great amount of inter- and intra-specific diversity in the timing and rates at which species

Table 6.1. Dates at which selected North American pine species reached their modern northern range limits following deglaciation, and approximate rates of northward migration

Pinus species	Time when northern limits reached (BP)	Migration rate (m yr^{-1})
P. banksiana		
Eastern North America	7000	400
Western North America	2000	250
P. contorta		
Interior Canada (subsp. *latifolia*)	<1000	170
Coastal Canada and Alaska (subsp. *contorta*)	10 000	670
P. strobus		
Northeastern North America	6000	320
Western Great Lakes Region	1000	130

reached their modern ranges and abundances in the vegetation (Table 6.1). For example, in eastern North America *P. banksiana* had one of the fastest migration rates of any species and reached its northern range limits in the early Holocene. In contrast, the northward spread of that species in the west was slower and it did not reach its modern range limits until the late Holocene. The contrast between the western coastal migration of *P. contorta* (probably subsp. *contorta*) and the interior migration of the species (probably subsp. *latifolia*) is even more pronounced. *Pinus contorta* reached its range limits in southern Alaska over 10 000 years ago, but approached its northern limits in the Yukon only over the last millennium. In addition, *P. contorta* reached its maximum abundance in the Pacific Northwest rapidly at the end of the glacial period and then declined, while the species experienced slower population growth rates in most of the interior of Canada and reached maximum abundance in the late Holocene.

Pines generally have a high degree of genetic variability (Chap. 13, this volume), and it might be tempting to attribute the wide variability in timing and rates of postglacial migration to this. However, quite comparable variability in migration rates and timings is evident in other North American tree species (Davis 1983; Delcourt & Delcourt 1987; MacDonald 1993). For example, *Tsuga canadensis* was able to reach its northeastern range limits by 6000 BP but, like *P. strobus*, did not reach its range limits in the western Great Lakes region until the last 2000 years (Davis 1983). An opposite situation occurred for *Larix laricina* (tamarack) which arrived in the western Great Lakes by 12 000–10 000 BP and did not reach its northeastern range limits until 7000 BP (Davis 1983). *Picea glauca* migrated northwards incredibly rapidly in western Canada and reached its range limits by 9000–10 000 BP. In the northeast, that species did not reach its northern limits along the coast of Labrador until the mid- to late Holocene (Ritchie & MacDonald 1986).

Given the wide variability in the rates and timing of migration by most North American trees it can only be concluded that physical factors such as climate, soils, barriers such as the Great Lakes and distance from glacial refugia, in combination with biological factors such as fecundity, seed dispersal characteristics, and competitive abilities impart a high degree of variability to the postglacial histories of pines and most other tree genera. No one feature such as genetic variability or fecundity etc. appears to explain all of the variability in postglacial histories.

Some aspects of the genetic variability in the northern North American pines may probably be explained by their glacial and postglacial histories. Allozyme analysis suggests that the speciation of *P. banksiana* and *P. contorta* probably took place during the Pleistocene as a result of continental glaciation which split the ancestral species into disjunct eastern and western populations (Dancik & Yeh 1983). These species do have a sympatric distribution and produce viable hybrids in central Alberta. However, during the postglacial, the species have probably only been in contact for the past 6000 to 8000 years. It is probable that glacial conditions were more the 'norm' of the Quaternary than are conditions such as the present postglacial. In fact, the last time that global temperatures were as warm as the postglacial was probably over 120 000 years ago (Sancetta, Imbie & Kipp 1973; Dansgaard *et al.* 1982). Thus, the sympatry between the two species could be considered unusual in the Quaternary and distinct speciation is maintained by the long periods of isolation of *P. banksiana* and *P. contorta* during glacial periods.

Glacial distributions may also explain the high degree of genetic variability in *P. contorta* relative to *P. banksiana* and many other pines (Critchfield 1980; Dancik & Yeh 1983). During the long glacial periods *P. contorta* is likely to have persisted in several disjunct refugia in the mountainous portion of the western United States. Not only would these small populations be subjected to differing selective pressures due to differences in climate, soils and competition, but they would also be more susceptible to genetic drift. In contrast, *P. banksiana* probably persisted in a more compact and potentially less environmentally variable refuge in the southeastern United States.

The postglacial fragmentation of pine populations may have played a role in maintaining and promoting genetic variability. In northeastern North America *P. strobus* was more abundant or even dominant on a wide variety of soils in the early postglacial. Upon the arrival of deciduous species, competitive pressures appear to have fragmented its distribution and restricted its dominance. Despite the potential for pine pollen to be carried long distances, and the winged nature of many pine seeds, the genetic neighbourhoods for pines are often small and pollination is most likely to occur from nearby trees (Chap. 13, this volume). Thus, fragmentation of populations, even on a landscape scale, might promote limited genetic drift and maintain genetic variability. A similar situation could be envisioned for *P. contorta* in portions of the Pacific Northwest where it was extremely widespread and reached maximum dominance in the late glacial, becoming more restricted in the postglacial.

The process of postglacial migration may have itself had an appreciable impact on the distribution of genetic diversity within the pines. Genotypic differences have been detected in *P. banksiana* populations in the Great Lakes region that appear to relate to the impact of the lakes as barriers to initial migration and subsequent gene flow (Critchfield 1985; Chap. 13, this volume). The fossil pollen data do not support the contention that *P. contorta* persisted in the Yukon through the last glaciation. Thus, the low allelic diversity of northern populations cannot be attributed to isolation in the Yukon during the last glacial maximum. Cwynar & MacDonald (1987) compared the history of lodgepole pine migration reconstructed from fossil pollen records with data on genetic diversity. They found that populations that were established during the first expansion of lodgepole pine into Canada during the early postglacial had a higher number of alleles per locus than the more recently established populations (Fig. 6.5a). They hypothesized that the genetic differences between Yukon and southern populations of lodgepole pine were caused by stochastic losses of diversity during postglacial migration. The northward migration of lodgepole pine occurred by the establishment of small advance populations (MacDonald & Cwynar 1991). Each of these populations would contain only a subset of the total genetic variation in the originating population. The establishment of these small founding populations produces population bottlenecks and associated genetic founder effects in which the genetic diversity of the advance population is less than that found in the main population. These advance populations then serve as sources for the further founding of advance populations, thereby compounding the potential loss of genetic diversity relative to the original population. Ledig (Chap. 13, this volume) refers to such migrations as saltational events and discusses the potential loss of genetic diversity due to this

phenomenon. The small genetic neighbourhoods of pines (Chap. 13, this volume) could work to maintain these geographic differences in genetic diversity over long periods, even if spatially continuous pine distributions were subsequently established between the originating populations and the advance populations. Cwynar & MacDonald (1987) also found evidence that selection occurred for small dispersible seeds as pine advanced northwards (Fig. 6.5b). Small-seeded genotypes would be selected for as migration proceeds as these progeny have the highest potential for establishment at sites distant from the originating population. Such selection pressures may be very small, but could be significant as species such as lodgepole pine migrate thousands of kilometres over periods of thousands of years.

Although the discussion above has focused on the role of Quaternary environmental change in promoting variability, environmental changes may also work against speciation. With migration to new ranges during periods of climatic change, mixing must occur between different genotypes. This would be particularly true during contractions to spatially compact glacial refugia. In addition, environmental conditions and associated selection pressures will change as species shift their boundaries. These factors will operate against allopatric speciation caused by genetic drift or selection. In this context it is interesting to note the far higher number of pine species in the mountainous regions of Mexico (Chap. 7, this volume) compared to the Pacific Northwest and Rocky Mountains which have probably experienced far greater disruptions of pine distributions during the Quaternary.

Acknowledgements

This chapter is dedicated to the memory of Bill Critchfield, who did so much to generate hybrid vigour between pine geneticists and paleoecologists. This research was supported by a NSERC Operating Grant and UCLA funding to Glen MacDonald, NSERC Operating Grants to Les Cwynar and a grant to Cathy Whitlock from the National Science Foundation (ATM-9307201).

References

Anderson, R.S. (1990). Holocene forest development and paleoclimates within the central Sierra Nevada, California. *Journal of Ecology*, **78**, 470–89.

Anderson, R.S., Davis, R.B., Miller, N.G. &

Stuckenrath, R. (1986). History of late- and post-glacial vegetation and disturbance around Upper South Branch Pond, northern Maine. *Canadian Journal of Botany*, **64**, 1977–86.

Arno, S.F. & Hammerly, R.P. (1977). *Northwest Trees*. Seattle: The Mountaineers.

Baker, R.G. (1986). Sangamonian (?) and Eisconsinan paleoenvironments in Yellowstone National Park. *Geological Society of America Bulletin*, **97**, 717–36.

Barnosky, C.W. (1981). A record of late Quaternary vegetation from Davis Lake, southern Puget Lowland, Washington. *Quaternary Research*, **16**, 221–39.

Barnosky, C.W. (1985a). Late Quaternary vegetation near Battle Ground Lake, southern Puget Trough, Washington. *Geological Society of America Bulletin*, **96**, 263–71.

Barnosky, C.W. (1985b). Late Quaternary vegetation in the southwestern Columbia Basin, Washington. *Quaternary Research*, **23**, 109–22.

Beiswenger, J.M. (1991). Late Quaternary vegetation history of Grays Lake, Idaho. *Ecological Monographs*, **61**, 165–82.

Bright, R.C. (1966). Pollen and seed stratigraphy of Swan Lake, southeastern Idaho; its relation to regional vegetational history and to Lake Bonneville history. *Tebiwa*, **9**, 1–47.

Brubaker, L.E. (1975). Postglacial forest patterns associated with till and outwash in north central upper Michigan. *Quaternary Research*, **5**, 499–528.

Campbell, I.D. & McAndrews, J.H. (1991). Cluster analysis of late Holocene pollen trends in Ontario. *Canadian Journal of Botany*, **69**, 1719–30.

Campbell, I.D. & McAndrews, J.H. (1993). Forest disequilibrium caused by rapid Little Ice Age cooling. *Nature*, **366**, 336–8.

Campbell, I.D. & McAndrews, J.H. (1995). Charcoal as evidence for Indian-set fires: a comment on Clark and Royall. *Holocene*, **5**, 369–70.

Clague, J.J. & Mathewes, R.W. (1989). Early Holocene thermal maximum in western North America: new evidence from Castle Peak, British Columbia. *Geology*, **17**, 277–80.

Clark, J.S. (1995). Climate and Indian effects on southern Ontario forest: a reply to Campbell and McAndrews. *Holocene*, **5**, 371–9.

Clark, J.S. & Royall, P.D. (1995). Transformation of a northern hardwood forest by aboriginal (Iroquois) fire: charcoal evidence from Crawford Lake. *Holocene*, **5**, 1–9.

Critchfield, W.B. (1980). *Genetics of Lodgepole Pine*. USDA Forest Service Research Paper WO-37.

Critchfield, W.B. (1984). Impact of the Pleistocene on the genetic structure of North American conifers. In

Proceedings of the Eighth North American Forest Biology Workshop, ed. R.M. Lanner, pp. 70–118. Logan: Utah State University.

Critchfield, W.B. (1985). The late Quaternary history of lodgepole and jack pines. *Canadian Journal of Forest Research*, **17**, 749–72.

Cwynar, L.C. (1987). Fire and forest history of the northern Cascade Range. *Ecology*, **68**, 791–802.

Cwynar, L.C. (1990). A late Quaternary vegetation history from Lily Lake, Chilkat Peninsula, southeast Alaska. *Canadian Journal of Botany*, **68**, 1106–12.

Cwynar, L.C. & MacDonald, G.M. (1987). Geographical variation of lodgepole pine in relation to population history. *American Naturalist*, **129**, 463–9.

Dancik, B.P. & Yeh, F.C. (1983). Allozyme variability and evolution of lodgepole pine (*Pinus contorta* var. *latifolia*) and jack pine (*P. banksiana*) in Alberta. *Canadian Journal of Genetics and Cytology*, **25**, 57–64.

Dansgaard, W., Clausen, H.B., Gundenstrup, N., Hammer, C.U., Johnsen, S.F., Kristinsdottir, P.M. & Reeh, N. (1982). A new Greenland deep ice core. *Science*, **218**, 1273–7.

Davis, M.B. (1983). Quaternary history of deciduous forests of eastern North America and Europe. *Annals of the Missouri Botanical Garden*, **70**, 550–63.

Davis, M.B., Spear, R.W. & Shane, L.C.K. (1980). Holocene climate of New England. *Quaternary Research*, **14**, 240–50.

Davis, O.K., Sheppard, J.C. & Robertson, S. (1986). Contrasting climatic histories for the Snake River Plain, Idaho, resulting from multiple thermal maximum. *Quaternary Research*, **26**, 321–39.

Delcourt, H.R. (1979). Late Quaternary vegetation history of the eastern Highland Rim and adjacent Cumberland Plateau of Tennessee. *Ecological Monographs*, **49**, 255–80.

Delcourt, P.A. & Delcourt, H.R. (1987). *Long-Term Forest Dynamics of the Temperate Zone*. New York: Springer-Verlag.

Despain, D.G. (1983). Nonpyrogenous climax lodgepole pine communities in Yellowstone National Park. *Ecology*, **64**, 231–4.

Dyke, A.S. & Prest, V.K. (1987). Late Wisconsin and Holocene history of the Laurentide Ice Sheet. *Geographie Physique et Quaternaire*, **41**, 237–64.

Fowells, H.A. (1965). *Sylvics of Forest Trees of the United States*. USDA Forest Service Agriculture Handbook 271.

Havinga, A.J. (1984). A 20-year experimental study investigation into the differential corrosion susceptibility of pollen and spores in various soil types. *Pollen et spores*, **26**, 541–58.

Hebda, R.J. (1995). British Columbia vegetation and climate history with focus on 6 Ka BP. *Geographie Physique et Quaternaire*, **49**, 55–79.

Hebda, R.J. & Whitlock, C. (1997). Environmental history. In *The Rain Forests of Home: Profile of a North American Bioregion*, ed. P.K. Schoonmaker, B. von Hagen & E.C. Wolf, pp. 227–54. Washington, DC: Island Press.

Heusser, C.J. (1977). Quaternary paleoecology of the Pacific slope of Washington. *Quaternary Research*, **8**, 282–306.

Huntley, B. & Birks, H.J.B. (1983). *An Atlas of Past and Present Pollen Maps for Europe 0–13 000 Years Ago*. Cambridge: Cambridge University Press.

Jacobson, G.L., Jr (1979). The paleoecology of white pine (*Pinus strobus*) in Minnesota. *Journal of Ecology*, **67**, 697–726.

Leopold, E.B., Nickmann, R.J., Hedges, J.I. & Ertel, J.R. (1982). Pollen and lignin records of late Quaternary vegetation, Lake Washington. *Science*, **218**, 1305–7.

Luckman, B.H. (1988). 8000 year old wood from the Athabasca Glacier, Alberta. *Canadian Journal of Earth Sciences*, **25**, 148–51.

McAndrews, J.H. (1988). Human disturbance of North American forests and grassland: the fossil pollen record. In *Vegetation History*, ed. B. Huntley & T. Webb III, pp. 673–97. Dordrecht: Kluwer.

MacDonald, G.M. (1987). Postglacial vegetation history of the Mackenzie River basin. *Quaternary Research*, **28**, 245–62.

MacDonald, G.M. (1989). Postglacial palaeoecology of the subalpine forest-grassland ecotone of southwestern Alberta: new insights on vegetation and climatic change in the Canadian Rocky Mountains and adjacent foothills. *Paleogeography, Paleoclimatology, Paleoecology*, **73**, 155–173.

MacDonald, G.M. (1993). Fossil pollen

analysis and the reconstruction of plant invasions. *Advances in Ecological Research*, **24**, 67–110

MacDonald, G.M. & Cwynar, L.C. (1985). A fossil pollen based reconstruction of the late Quaternary history of lodgepole pine (*Pinus contorta* ssp. *latifolia*) in the western interior of Canada. *Canadian Journal of Forestry Research*, **15**, 1039–44.

MacDonald, G.M. & Cwynar, L.C. (1991). Postglacial population growth rates of *Pinus contorta* ssp. *latifolia* in western Canada. *Journal of Ecology*, **79**, 417–29.

MacDonald, G.M. & McLeod, K.T. (1996). The Holocene closing of the 'Ice-Free' Corridor: A biogeographical perspective. *Quaternary International*, **32**, 87–95.

MacDonald, G.M. & Ritchie, J.C. (1986). Modern pollen surface samples and the interpretation of postglacial vegetation development in the western interior of Canada. *New Phytologist*, **103**, 245–68.

Mack, R.N., Rutter, N.W. & Valastro, S. (1978). Late Quaternary pollen record from the Sanpoil River valley, Washington. *Canadian Journal of Botany*, **56**, 1642–50.

McLeod, K.T. (1991). The postglacial population spread of *Picea mariana*, *Picea glauca* and *Pinus banksiana* across the Western Interior of Canada. MSc thesis, McMaster University, Hamilton, Canada.

Martin, J.E., Barnosky, A.D. & Barnosky, C.W. (1982). Fauna and flora associated with the West Richland mammoth from the Pleistocene Touchet formation in south-central Washington. *Research Reports of the Burke Memorial Museum*, **3**, 1–61.

Mathewes, R.W. (1973). A palynological study of postglacial vegetation changes in the University Research Forest, southwestern British Columbia. *Canadian Journal of Botany*, **51**, 2085–103.

Mathewes, R.W. (1993). Evidence for Younger Dryas-Age cooling on the north Pacific coast of America. *Quaternary Science Reviews*, **12**, 321–31.

Mehringer, P.J. Jr, Arno, S.F. & Peterson, K.L. (1977). Postglacial history of Lost Trail Pass Bog, Bitterroot Mountains, Montana. *Arctic and Alpine Research*, **9**, 345–68.

Millar, C.I. (1993). Impact of the Eocene on the evolution of *Pinus* L. *Annals of the Missouri Botanical Garden*, **80**, 471–98.

Miller, N.G. (1988). The late Quaternary Hiscock site, Geneseo County, New York: paleoecological studies based on pollen and plant macrofossils. *Bulletin of the Buffalo Society of Natural Sciences*, **33**, 83–93.

Millspaugh, S.H. (1997). *Late-glacial and Holocene variations in fire frequency in the Central Plateau and Yellowstone-Lamar Provinces of Yellowstone National Park.* Ph.D. dissertation, University of Oregon, Eugene, OR, USA.

Mirov, N.T. (1967). *The Genus Pinus.* New York: Ronald Press.

Nowak, C.L., Nowak, R.S., Tausch, R.S. & Wigand, P.E. (1994). A 30 000 year record of vegetation dynamics at a semi-arid locale in the Great Basin. *Journal of Vegetation Science*, **5**, 579–90.

Peteet, D.M. (1991). Postglacial migration history of lodgepole pine near Yakutat, Alaska. *Canadian Journal of Botany*, **69**, 786–96.

Peteet, D.M., Daniels, R.A., Heusser, L.E., Vogel, J.S., Southon, J.R. & Nelson, D.E. (1993). Late-glacial pollen, macrofossils, and fish remains in northeastern USA – the Younger Dryas Oscillation. *Quaternary Science Reviews*, **12**, 597–612.

Prentice, I.C., Bartlein, P.J. & Webb, T. III (1991). Vegetation and climate change in eastern North America since the last glacial maximum. *Ecology*, **72**, 2038–56.

Ritchie, J.C. & MacDonald, G.M. (1986). The postglacial spread of white spruce. *Journal of Biogeography*, **13**, 527–40.

Sancetta, C., Imbrie, J. & Kipp, N.G. (1973). Climatic record of the past 130 000 years in the North Atlantic deep-sea core V23–82: correlation with the terrestrial record. *Quaternary Research*, **3**, 110–16.

Sea, D.S. & Whitlock, C. (1995). Postglacial vegetation and climate of the Cascade Range, central Oregon. *Quaternary Research*, **43**, 370–81.

Smith, S.J. & Anderson, R.S. (1992). Late Wisconsin paleoecologic record from Swamp Lake, Yosemite National Park, California. *Quaternary Research*, **38**, 91–102.

Stuiver, M. & Reimer, P.J. (1993). Extended ^{14}C data base and revised CALIB3.0 14C age calibration program. *Radiocarbon*, **35**, 215–30.

Szeicz, J.M. & MacDonald, G.M. (1991). Postglacial vegetation history of oak savanna in southern Ontario. *Canadian Journal of Botany*, **69**, 1507–19.

Terasmae, J. & Anderson, T.W. (1970). Hypsithermal range extension of white pine (*Pinus strobus* L.) in Quebec, Canada. *Canadian Journal of Earth Sciences*, **7**, 406–13.

Tsukada, M., Sugita, S. & Hibbert, D.M. (1981). Paleoecology of the Pacific Northwest I. Late Quaternary vegetation and climate. *Verhandlungen der Internationalen Vereingung für Theoretische und Angewandte Limnologie*, **21**, 730–7.

Wainman, N. & Mathewes, R.W. (1987). Forest history of the last 12 000 based on plant macrofossil analysis of sediment from Marion Lake, southwestern British Columbia. *Canadian Journal of Botany*, **65**, 2179–87.

Watts, W.A. (1970). The full-glacial vegetation of northwestern Georgia. *Ecology*, **51**, 17–33.

Watts, W.A. (1973). Rates of change and stability in vegetation in the perspective of long periods of time. In *Quaternary Plant Ecology*, ed. H.J.B. Birks & R.G. West, pp. 195–206. Oxford: Blackwells.

Watts, W.A. (1979). Late Quaternary vegetation of central Appalachia and the New Jersey coastal plain. *Ecological Monographs*, **49**, 427–69.

Webb, T., III (1988). Eastern North America. In *Vegetation History*, ed. B. Huntley & T. Webb III, pp. 385–414. Dordrecht: Kluwer.

Webb, T. III & Bartlein, P.J. (1988). Late Quaternary climate change in eastern North America: The role of modelling experiments and empirical studies. In *Late Pleistocene and Early Holocene Paleoecology and Archeology of the Eastern Great Lakes Region*, ed. R.S. Laub, N.G. Miller & D.W. Steadman. *Bulletin of the Buffalo Society of Natural Sciences*, **33**, 3–13.

Webb, T., III, Bartlein, P.J., Harrison, S.P. & Anderson, K.H. (1993). Vegetation, lake levels, and climate in eastern North America for the past 18 000 years. In *Global Climates Since the Last Glacial Maximum*, ed. H.E. Wright, Jr, J.E. Kutzbach, T. Webb III, W.F. Ruddiman, F.A. Street-Parrott & P.J. Bartlein, pp. 415–67. Minneapolis: University of Minnesota Press.

Webb, T. III, Bartlein, P.J. & Kutzbach, J.E. (1987). Climate change in eastern North America during the last 18 000 years: comparisons of pollen data with model results. In *North America and Adjacent Oceans during the Last Deglaciation, The*

Geology of North America, vol. K-3, ed. W.F. Ruddiman & H.E. Wright, Jr, pp. 447–62. Boulder, Colorado: Geological Society of America.

Wheeler, N.C. & Guries, R.P. (1982). Biogeography of lodgepole pine. *Canadian Journal of Botany*, **60**, 1805–14.

Whitlock, C. (1992). Vegetational and climatic history of the Pacific Northwest during the last 20 000 years: implications for understanding present-day biodiversity. *The Northwest Environmental Journal*, **8**, 5–28.

Whitlock, C. (1993). Postglacial vegetation history and climate of Grand Teton and southern Yellowstone National Parks. *Ecological Monographs*, **63**, 173–98.

Whitlock, C. (1995). *The History of* Larix occidentalis *During the Last 20,000 Years of Environmental Change in Ecology and Management of* Larix *Forests*. USDA Forest Service General Technical Report INT-319, 83–90. Ogden, Utah: USDA Forest Service Intermountain Research Station.

Whitlock, C. & Bartlein, P.J. (1993). Spatial variations of Holocene climatic change in the Yellowstone region. *Quaternary Research*, **39**, 231–8.

Whitlock, C., Bartlein, P.J. & Van Norman, K.J. (1995). Stability of Holocene climate regimes in the Yellowstone region. *Quaternary Research*, **43**, 433–6.

Worona, M.A. & Whitlock, C. (1995). Late Quaternary vegetation and climate history near Little Lake, central Coast Range, Oregon. *Geological Society of America Bulletin*, **107**, 867–76.

7 The history of pines in Mexico and Central America

Jesse P. Perry, Jr, Alan Graham and David M. Richardson

7.1 Introduction

Mexico and Central America have the greatest number of pine species of any region of similar size in the world. Shaw (1909) reported 18 species and 17 varieties from Mexico; Standley (1920–6) listed 28 species and two varieties from Mexico; Martínez (1948) cited 39 species, 18 varieties and nine forms from Mexico; Loock (1950) reported 55 species, varieties and forms from Mexico and Honduras; Mirov (1967) listed 29 'effectively published' pine species in Mexico; Eguiluz Piedra (1988) recognized 69 taxa (species, subspecies and varieties) in Mexico; and Perry (1991) recognized 51 species (and many varieties and forms) from Mexico and Central America. Of the 111 *Pinus* species accepted for inclusion in this book (Chap. 2, this volume), 47 occur in Mexico and Central America. Aljos Farjon and the late Brian Styles, in an independent assessment of taxa for their forthcoming monograph of *Pinus* of Mexico and Central America for *Flora Neotropica*, also recognize 47 species (Oxford Forestry Institute 1996).

This amazing pine flora includes 31 diploxylon species (44% of the species in section *Pinus*, with representatives in seven of the 11 subsections), and 16 haploxylon pines – 12 species from section *Parrya* (67% of the total, with representatives in two of the five subsections) and four species from section *Strobus* (19% of the total, with representatives in one of the two subsections). Most of the species belong to three groups that are largely endemic to Mexico: the pinyon pines (12 species in subsections *Cembroides* and *Rzedowskianae*); and two major groups within the North American clade of diploxylon pines recognized by Price *et al.* (Chap. 2, this volume) on the basis of (very limited) data from chloroplast DNA plus crossability and morphology (subsection *Ponderosae sensu lato*, including the *Montezumae*, *Pseudostrobus* and *Sabinianae* groups; and sub-

section *Oocarpae sensu lato*, here taken to include the *Teocote* group and very closely allied to subsection *Leiophyllae*; see Chap. 2, this volume). Controversy over the delimitation of many species and the fact that hybridization between species is common suggests that these largely Mexican groups have speciated relatively recently. Some species in the pinyon pine group (e.g. *P. culminicola*, *P. maximartinezii*, *P. nelsonii*, *P. pinceana* and *P. rzedowski*) are highly localized, relictual taxa that are clearly defined taxonomically, whereas other species of pinyons are often very difficult to delimit unambiguously. Thirty-five of the 47 pine species (74%) are endemic to the region (or virtually so; some have marginal range extensions into the USA). The endemic species are concentrated in the subsections *Ponderosae* (11 of the 14 species in this taxon are endemic to the region), *Oocarpae* (10 out of 10) and *Cembroides* (9 out of 11). Despite the great richness of pines in the region, and although pines have been better studied than other tree genera, many Mexican and Central American pine taxa have been poorly studied. The palaeoecology of pines in the region is also less well understood than in other regions with rich pine floras.

The large and diverse pine flora of Mexico and Central America owes its existence to the migration, over millions of years, of pines southwards from northern North America. Many plants from eastern North America (north of Mexico) migrated southwards around the Gulf of Mexico during the Tertiary, eventually reaching Mexico and Central America (Steyermark 1950; Graham 1973a, b, 1976). Miller (1977) suggests that the ancestors of modern diploxylon pines in Mexico probably reached that region no later than the mid-Tertiary, and by the mid-Cenozoic at the latest. During their migration, many diploxylon pines,

Ecology and Biogeography of Pinus, ed. D.M. Richardson. © Cambridge University Press (1998), pp. 137–49.

through introgression and hybridization, developed forms, varieties, subspecies and species that were adapted to the different environmental conditions of the Mexican and Central American highlands (Shaw 1914; Martínez 1948; Loock 1950; Mejorada & Huguet 1959; Mirov 1967; Eguiluz Piedra 1985; Perry 1991). The surviving pines, over geological ages and vast continental changes, formed a secondary centre of pine evolution (Mirov 1967; Chap. 3, this volume). The pine forests of Mexico and Central America are now an amazing mixture of species, varieties and hybrids (Shaw 1914; Martínez 1948; Loock 1950; Mirov 1967; Rzedowski 1978; Eguiluz Piedra 1985; Perry 1991).

In Mexico and Central America, coniferous forests dominated by pines covered 22.6 million hectares at the end of 1980, a third of the total area under natural forests in the region (Lanly 1995). Pines occur in a range of forest types, sometimes forming monospecific stands (e.g. *P. durangensis*, *P. engelmannii*, *P. hartwegii*, *P. leiophylla*, *P. oocarpa*, *P. patula*, *P. ponderosa* var. *arizonica*), assemblages of up to seven *Pinus* species, or mixed stands with other conifers (notably *Abies*, *Juniperus* and *Pseudotsuga* spp.) or *Quercus* spp. (Perry 1991; Styles 1993). Pine habitats in the region range from the driest in the world for pine growth (e.g. on the central plateau of Mexico where many taxa in subsection *Cembroides* occur, i.e. *P. johannis*, *P. maximartinezii*, *P. nelsonii* and *P. pinceana*; and Baja California, where *P. cembroides* subsp. *lagunae*, *P. juarezensis* and hybrids between the last-mentioned species and *P. monophylla* occur) to among the wettest (e.g. in Belize, Honduras and Nicaragua where *P. caribaea* var. *hondurensis* occurs). In many areas, there is a clear altitudinal zonation, one pine taxon replacing another with an increase in elevation (e.g. Yeaton 1982).

Martínez (1948) described three great pine groups of Mexico; (1) a **Ponderosa** group – mostly three-needled pines: *P. arizonica* and varieties (now *P. ponderosa* var. *arizonica*), *P. engelmannii*, *P. jeffreyi*, and *P. durangensis* which has six needles (Perry 1991); (2) a **Montezumae** group – mostly five-needled pines: *P. cooperi*, *P. devoniana* (formerly *P. michoacana* with its varieties *cornuta* and *quevedoi* and forma *nayaritana* and *procera*), *P. donnell-smithii*, *P. douglasiana*, *P. hartwegii*, *P. montezumae* and *P. montezumae* var. *lindleyi* (Perry 1991); and (3) a **Pseudostrobus** group – mostly five-needled pines: *P. maximinoi*, *P. nubicola*, *P. pseudostrobus* (forma *coatepecensis*, *megacarpa* and *protuberans*) (Perry 1991). Mirov (1967), following Martínez's scheme, described three great pine complexes of America: a *Ponderosa* complex, a *Montezumae* complex, and a *Pseudostrobus* complex.

Martínez's **Ponderosa** group occurs mostly in north-western Mexico in the Sierra Madre Occidental (Fig 7.1). The **Montezumae** group occurs mostly along the Transvolcanic Belt, with some species occurring along the Sierra Madre Oriental and southwards into Guatemala. The **Pseudostrobus** group also occurs primarily along the Transvolcanic Belt, the Sierra Madre Oriental and southwards into Guatemala, El Salvador and Honduras.

This chapter deals very briefly with the preglacial spread of pines in continental North America, the post-glacial migration of pines in Mexico and Central America, and the impacts of these and other factors on the diversification and expansion of *Pinus* in the region. Discussion is virtually confined to the diploxylon pines of the subsection *Ponderosae* which is represented in the region by 13 species, 11 of them endemic; in further discussion we follow the terminology of Price *et al.* (Chap. 2, this volume) for groups within this subsection. More is known of the history of this subsection than of the others, and the basic patterns described for this group probably also apply to the other diploxylon groups. Subsection *Cembroides* is not discussed here (but see Lanner 1981; Passini 1982a, b; Styles 1993; Chap. 9, this volume for recent reviews).

7.2 Preglacial spread of pines in continental North America

The palaeobotanical literature on the evolution and spread of pines in North America is extensive and fascinating (Mason 1927, 1932; Axelrod 1976, 1980, 1983, 1986, 1988, 1990; Axelrod & Hill 1988; Axelrod & Cota 1993; Chap. 3, this volume). As pines migrated southwards from northern North America during the Cretaceous and Tertiary, they were evolving under changing ecological conditions. Fossils from Canada and the northern USA show that diploxylon and haploxylon pines were already differentiated during these periods (Chap. 3, this volume). Adaptations to enormous physical and climatic fluctuations on the North American continent had a profound impact on the diversification of pine species in Mexico and Central America (Critchfield 1984).

The rise of the Cretacean Sea, extending from the Arctic Ocean to the Gulf of Mexico (Fig. 7.2), effectively divided and isolated the pine forests of western and eastern North America (Mirov 1967). Most of Mexico, the Isthmus of Tehuantepec, the Yucatán Peninsula and most of Guatemala were covered by this sea. Pines already established in northeastern America and in the ancient Appalachian Mountains and Ozark Plateau were isolated for millions of years from the pines of western America.

Lesquereux (1883) published a report on the Cretaceous and Tertiary floras in which he described leaves and a cone of *P. quenstedti* Heer from the Cretaceous of Kansas (see Chap. 3, this volume for details). In 1895, he also described a pine from Cretaceous fossil plants of Minnesota, which

Fig. 7.1. **Looking west by north across Mexico's Sierra Madre Occidental near El Salto, Durango. Typical pines in this area belong** to Martínez's (1948) Ponderosa and Montezumae groups (photo: J.P. Perry). Reproduced, with permission, from Perry (1991).

he considered to be the same pine he had described from the Cretaceous of Kansas, i.e. *P. quenstedti* (Lesquereux 1895). Chaney (1954), in describing a new pine from the Cretaceous of Minnesota (*P. clementsii* Chaney), compared Lesquereux's description of *P. quenstedti* from the Cretaceous of Kansas with *P. quenstedti* from the Cretaceous of Minnesota; he found the specimens from the two areas to be completely different. Both Chaney and Lesquereux commented on the similarity of these specimens, with their leaves in fives and their elongate and asymmetrical cones, to the extant pines of Mexico. The presence of a five-needled, diploxylon pine in the Cretaceous of Kansas that closely resembles the living Mexican pines, particularly a form of *P. devoniana* (formerly *P. michoacana* var. *cornuta*), is extremely important for our understanding of the migration of pines southwards to Mexico. It is even more important for understanding the widespread development and evolution of five-needled diploxylon pines in Mexico and

Central America. These developments will be discussed later in this chapter.

Tectonic plate movements in the late Cretaceous and early Tertiary formed the crumpled mountains of Mexico's Sierra Madre Oriental which consist primarily of limestone (de Cserna 1989). Tremendous volcanic activity in Mexico during this period formed the great volcanic peaks and enormous layers of lava and ash (see also Chap. 3, this volume). Mexico's Transvolcanic Belt, extending from the Pacific Coast to the Gulf of Mexico, linked the Sierra Madre Occidental with the Sierra Madre Oriental (Eguiluz Piedra 1985; Perry 1991). Volcanic activity along the Transvolcanic Belt and southwards into Guatemala and Nicaragua continues to this day. The isthmus of Tehuantepec and the Guatemalan highlands emerged as the Cretacean Sea subsided, opening a pathway for the southward migration of pines from eastern and western North America through Mexico into Central America.

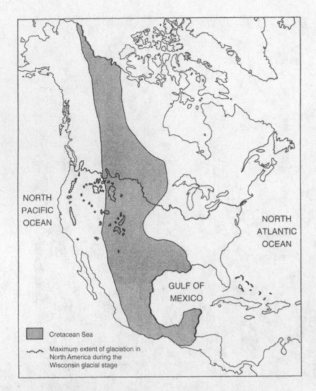

NORTH
PACIFIC
OCEAN

NORTH
ATLANTIC
OCEAN

GULF OF
MEXICO

▨ Cretacean Sea

⌒⌒ Maximum extent of glaciation in
    North America during the
    Wisconsin glacial stage

Fig. 7.2. **The Cretacean Sea and the southern limit of glaciation in North America. Slightly modified from Mirov (1967) and Perry (1991).**

It seems likely that pines migrated in Mexico along two routes: the Sierra Madre Occidental (Fig. 7.1) in the west; and the Sierra Madre Oriental in the east (Mirov 1967). Several authors have suggested that northern temperate elements migrated southwards into Mexico during major periods of glaciation in the Pleistocene (e.g. Deevey 1949; Dressler 1954). Others have postulated that these elements began migrating into Mexico and Central America during the early and middle Tertiary (Chaney 1936; Steyermark 1950; Braun 1955; Martin & Harrell 1957). Of particular interest to some writers has been the migration of broadleaved species from the eastern USA to Mexico and Central America (Steyermark 1950; Miranda & Sharp 1950; Hernandez *et al.* 1951; Graham 1973a, b, 1976; Wendt 1993). A classic example is the occurrence of *Liquidambar styraciflua* along the eastern escarpment of Mexico into Guatemala, Honduras and El Salvador. Harrell (1951) listed 16 tree genera shared between the eastern USA and *Liquidambar/Quercus* forest of Rancho del Cielo in the Mexican state of Tamaulipas (among the trees listed in the forest were *P. montezumae* and *P. patula*). Graham (1973b) listed 53 genera in common between the eastern USA and northeastern Mexico. Steyermark (1950) postulated that pines and deciduous trees migrated southwards from eastern USA and the Ozark uplands into eastern Mexico,

Guatemala and Central America during the early Tertiary. During this period, pines from the southwestern USA and Rocky Mountains were probably moving southwards into Mexico and Central America (Steyermark 1950). Martin & Harrell (1957), in summarizing the views of Steyermark and others, also concluded that interpreting the eastern temperate (Arcto-Tertiary) element in the Mexican montane biota exclusively in terms of Pleistocene dispersal raised more distributional problems than it solved. They suggested that an earlier (pre-Pliocene) arrival of temperate plants was more in accord with the available evidence. They did not rule out the possibility that temperate plants and animals from eastern North America migrated to Mexico during the Pleistocene, since they state 'In the Pleistocene the arid zone (southern Texas) ameliorated only enough to allow passage of those temperate animals and plants that could exist under savanna conditions.'

Dressler (1954) reached a different conclusion from that of Martin & Harrell (1957). He argued that it was in the Pleistocene that climatic and physiographic conditions conducive to a floristic exchange between eastern Mexico and the eastern USA seem to have existed. He also noted that 'Evidence from pollen analyses indicates that east Texas had a cooler and more mesic climate in the past . . . and shows a long period of climatic fluctuation for Mexico . . .'. Dressler (1954) also pointed out that 'The main vegetational continuity in the east probably occurred not through the most direct route now available but through the Big Bend region, being determined by physiography.' It seems likely that Dressler had in mind a 'vegetational continuity' extending from the southern Appalachian Mountains (altitude 1153 m north of Atlanta), to the Ozark Plateau's Boston Mountains (781 m), the Ouachita Mountains (811 m), the Wichita Mountains of central Texas (alt. 756 m), Edwards Plateau (758 m), and the Chisos Mountains (2400 m) of the Big Bend region in western Texas – a southern extension of the Rocky Mountains. From the Big Bend area, Mexico's Sierra Madre Oriental extends southeastwards, rising gradually from 1900 m to 2200 and 2500 m until near the cities of Monterrey and Saltillo, where they rise abruptly to 3100–3700 m. This is a migration route quite different from one along the coastal plain of Texas and one that, in Dressler's view, probably served as a 'vegetational continuity' from the Appalachian Mountains to northeastern Mexico during the Pleistocene. If it did, this route could also have served as a southward migration path for pines.

Regarding the question of Pleistocene versus Tertiary arrival of Pinaceae in Mexico, there is evidence that *Pinus* was already present in southern Mexico by the beginning of the Miocene. *Picea* was also present in the Miocene, and *Abies* is known from the Pliocene.

There are few Cenozoic floras from Mexico and Central America (Graham 1994, figs. 1 and 2), and only some of these contain Pinaceae. Thus, the data base is meagre and new discoveries could significantly change the presently known stratigraphic and palaeogeographic range of the family. Nonetheless, a general picture is emerging that is internally consistent, and in accord with concepts based on the distribution and systematics of the extant species. This picture reveals the relatively recent introduction of Pinaceae into northern Latin America compared with their occurrence in the north.

The only report of macrofossils of Pinaceae from northern Latin America is of *Pinus* from the Pleistocene of Cuba (Berry 1934). Three fossil cones are identified as *P. caribaea*.

The existing Tertiary microfossil record for the family begins in central and southern Mexico in the early to middle Miocene (no records of Tertiary age are known from northern Mexico). Langenheim, Hackner & Bartlett (1967, p. 319) refer to 'a few poorly preserved grains probably of *Pinus*' from the Oligo-Miocene (probably early Miocene) La Quinta Formation at Simojovel, Chiapas, Mexico. Pollen of *Pinus* has also been reported from the early to middle Miocene Méndez Formation (Palacios Chavez & Rzedowski 1993), from the middle to late Miocene Ixtapa Formation of Chiapas (Martínez-Hernández 1992), and from the middle Pliocene Paraje Solo Formation of Veracruz, Mexico (Graham 1976; Fig. 7.3). The 'probable' nature of the specimens from the La Quinta Formation, and the age range cited for the Méndez and Ixtapa formations suggest the early/middle Miocene transition as a conservative estimate for the initial appearance of *Pinus* in southern Mexico. *Picea* is first recorded in the region in the early to middle Miocene Méndez flora, and *Abies* and *Picea* are known from the middle Pliocene Paraje Solo assemblage.

Plant microfossils of Tertiary age are also known from Central America. These are from the late Eocene Gatuncillo flora of Panama (Graham 1985; Graham, Stewart & Stewart 1985), the early Miocene Uscari sequence from Costa Rica (Graham 1987), the early Miocene Culebra, Cucaracha, La Boca floras of Panama (Graham 1988a, b, 1989), and the Mio-Pliocene Gatun flora of Panama (Graham 1991a, b, c). There are no Pinaceae present in these floras. In the Quaternary, fossil pollen of *Pinus* appears in sediments from Deep Sea Drilling Project Site 565 off western Costa Rica (Horn 1985). A small amount of *Pinus* pollen is found in modern surface samples, and it is difficult to estimate how much of the fossil pollen was transported by wind or marine currents and how much was the result of possible range extension southwards during glacial intervals. Pollen of *Pinus* was not recovered from either modern surface samples, or from Quaternary sediments in Panama (Bartlett & Barghoorn 1973).

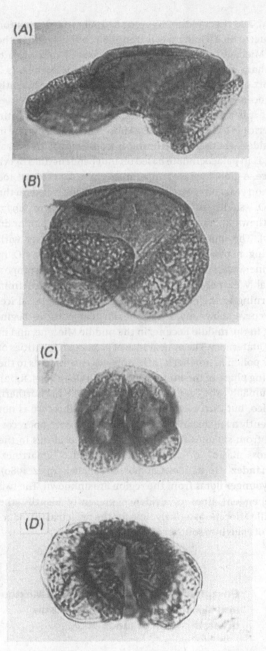

Fig. 7.3. **Fossil Pinaceae/Podocarpaceae from the middle Pliocene Paraje Solo Formation, state of Veracruz, southeastern Mexico. *A, Abies; B, Picea; C, Pinus; D, Podocarpus.***

The distributional history of *Pinus* revealed by these floras, and others in continental North America, suggests its earliest appearance in the northern temperate zones, introduction into southern Mexico by the early to middle Miocene, meagre representation in northern Central America in the Quaternary, and absence from southern Central America throughout the Cenozoic. This record is paralleled in the Antilles in that *Pinus* is absent from the middle Eocene Chapelton Formation of Jamaica (Graham

1993), and the middle Oligocene San Sebastian Formation of Puerto Rico (Graham & Jarzen 1969), but is found in the late Miocene to middle Pliocene Artibonite Group of Haiti (Graham 1990). The only exception to this pattern is a report of pollen resembling *Pinus sylvestris* of 'apparently Oligocene age' from Cuba (Areces 1987).

If it is assumed that *Pinus* appears in northern Latin America in the early to middle Miocene, it is appropriate to consider whether this timing is consistent with global trends in palaeoclimate based on independent lines of evidence. A global palaeotemperature curve for the past 100 million years has been developed from $^{18}O/^{16}O$ ratios in the $CaCO_3$ shells of marine invertebrates (Savin 1977; Matthews & Poore 1980; Miller, Fairbanks & Mountain 1987). The amount of ^{18}O in the shells increases with cooling marine waters. The relative amount of ^{18}O in marine waters also increases as the lighter ^{16}O disproportionably evaporates, but temporarily is prevented from returning to the ocean by incorporation into glacial ice. The curve shows major declines in temperatures beginning in the middle Eocene, in the middle Miocene, and in the Quaternary. The appearance of sustained quantities of *Pinus* pollen in northern Latin America corresponds to the cooling phase of the middle Miocene (Graham 1995, fig. 2). Presumably the genus appeared earlier in northern Mexico, but direct evidence from the fossil record is not presently available. For example, Pinaceae were not recovered from samples from various Eocene localities in the Burgos Basin of northeastern Mexico (Martínez-Hernández, Hernández-Campos & Sánchez-López 1980), but younger floras from the region are unknown. The two independent lines of evidence presently identify the middle Miocene as a likely time for the principal introduction of *Pinus* into southern Mexico.

7.3 Postglacial migration of pines in Mexico and Central America; its impact on diversification of pines

The great changes in North America towards the end of the Cretaceous, the tremendous geographic uplifts in western North America, and the lowered temperatures towards the end of the Tertiary followed by major glaciations during the Pleistocene, all had a huge impact on the evolution of pines in North America (Critchfield 1984). We have already mentioned the isolation of eastern and western pines by the Cretacean Sea and the southward migration of pines into Mexico during the Tertiary, and possibly also in the Pleistocene. During the Pleistocene, the southeastern USA became a refugium for the northeastern pines (e.g. *P. banksiana*) and, as shown previously, pines had already

migrated southwards into Mexico during the Tertiary (see also Steyermark 1950). It is possible that some eastern pines and broadleaved species also bridged the Texas 'barrier' during the Pleistocene to become established in the Sierra Madre Oriental of eastern Mexico and extended their range southwards (Dressler 1954).

A similar pattern of migration is assumed to have been taking place in western North America and Mexico during the Tertiary (Chap. 6, this volume). The changes in climate accompanying glaciation probably added impetus to the southward movement of pines along Mexico's Sierra Madre Occidental. The current distribution (Fig. 7.4) and composition of pine forests in Mexico and Central America was probably established in the early Holocene, but they would have changed slightly with each reshuffling during the numerous glacial/interglacial intervals. The composition of the forests was also probably changing as the northern species adapted to new environments in Mexico and Central America. The climates of Mexico and the southern USA fluctuated (Graham & Heimsch 1960; Delcourt & Delcourt 1993), temperatures along Mexico's high mountain ranges decreased (Lorenzo 1969), and many pine species along these ranges evolved in response to the new environment.

7.4 The diversification of pines in Mexico and Central America

Many of the pine assemblages south of the Rio Grande are dramatically different from those in the USA and Canada. An obvious difference is the great number of five-needled diploxylon taxa in Mexico and Central America, notably taxa in the *Montezumae* and *Pseudostrobus* groups of subsection *Ponderosae* and the *Oocarpa* group in subsection *Oocarpae*, and their scarcity in the pine forests north of the Rio Grande. The predominance of five-needled pines south of the Rio Grande is puzzling since these pines probably originated in the central USA during the Cretaceous: at least, early records are from this region (Lesquereux 1883; Chaney 1954). How did diploxylon five-needled pines spread to Mexico and Central America and become an important component of pine forests in this region? What follows is an outline of what may have occurred.

Lesquereux's (1883) five-needled pine from the Cretaceous of Kansas (*P. quenstedti*) evolved before the rise of the Cretacean Sea. It was already present in the western pine forests and in the Ozarks and the Appalachian Mountains when the rise of the Cretacean Sea divided the western and eastern pine forests of North America (Fig. 7.2). As mentioned previously, both Lesquereux (1883) and Chaney (1954) pointed out the similarity of *P. quenstedti*

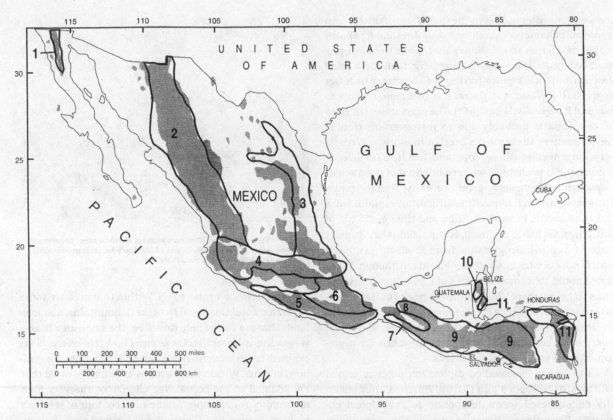

Fig. 7.4. **The distribution of pine forests along the principal mountain ranges of Mexico and Central America (based on information in: Leopold 1950; Loock 1950; Rzedowski 1978; Eguiluz Piedra 1985; Perry 1991; Styles 1993). 1. Sierra de Juarez and San Pedro Martir; 2. Sierra Madre Occidental (see Fig. 7.1); 3. Sierra Madre Oriental; 4. Eje Volcanico (The Transvolcanic Belt); 5. Sierra Madre del Sur; 6. Macizo** de Oaxaca; 7. Sierra Madre de Chiapas; 8. Macizo Central de Chiapas; 9. Highlands of Guatemala, El Salvador, Honduras and Nicaragua; 10. Maya Mountains of Belize; 11. Coastal plain forests of *Pinus caribaea* var. *hondurensis* in Belize, Honduras and Nicaragua. Modified from Perry (1991).

from the Cretaceous of Kansas, with the living Mexican pines, '. . . of which *P. michoacana* var. *cornuta* Mart. [now *P. devoniana*] has cones most similar to the Kansas fossil . . .'. With the diploxylon pines divided into eastern and western groups by the Cretacean Sea, their evolutionary development diverged as they adapted to different environments. *Pinus quenstedti* is probably the early ancestor of the present *P. palustris* and *P. devoniana* – the latter being variable and widespread in Mexico and Central America. The ancestor of yet another five-needled diploxylon group of pines which is now widespread in Mexico and Central America (the *Montezumae* group in subsection *Ponderosae*), probably developed at the same time and along with *P. quenstedti*. Like *P. devoniana*, *P. montezumae* is highly variable (Shaw 1914; Martínez 1948; Mirov 1967) and occurs widely in Mexico and Central America. It is also closely related to *P. devoniana* and the two species hybridize readily in the field (Martínez 1948; Perry 1991). Both these groups of five-needled pines may have migrated to Mexico during the Tertiary or the early Pleistocene. The early members of the *Montezumae* group probably moved south-

wards along Mexico's eastern mountains, the Sierra Madre Oriental (Area 3 in Fig. 7.4). However, a few taxa in this group may also have migrated southwards along Mexico's Sierra Madre Occidental (Area 2 in Fig. 7.4), mingling with pines of other groups in subsection *Ponderosae*. The present *P. montezumae* has been successfully crossed with *P. ponderosa* (Duffield 1952), leading Price *et al.* (Chap. 2, this volume) to suggest that *P. montemuzae* appears to provide a connection in terms of crossability and morphology between Martínez's (1948) three groups (**Montezumae**, **Ponderosae** and **Pseudostrobus**).

Considerable evidence supports the relationships outlined above. First is the description of *P. quenstedti* by Lesquereux and Chaney discussed previously. Perry (1991) noted the similarity between *P. devoniana* and related species (subsection *Ponderosae*) and *P. palustris* (subsection *Australes*) in the southeastern USA. The very long needles – mostly in groups of five (in *P. devoniana*), the large, long cones and growth form of the young and mature trees are strikingly similar. Although *P. palustris* is primarily a three-needled pine, Harrar & George (1962) found many trees with needles in fives in the Gulf states of the USA. An important character shared by *P. devoniana* and *P. palustris* is that of a 'grass stage'. *Pinus palustris* has the most pronounced grass stage (see Wahlenberg 1946 for a good description); other pines having it to a greater or lesser degree are *P. elliottii* var. *densa*, *P. engelmannii*, *P. montezumae* and *P. tropicalis* (Mirov 1967). The grass stage in *P. elliottii* var. *densa* is probably due to introgression from *P. palustris* where their ranges overlap in Florida. With its very long needles (three, four and five in a fascicle), *P. engelmannii* is probably a western offshoot of the ancient *P. quenstedti–Montezumae* group of early Tertiary pines. The grass stage in *P. tropicalis* is difficult to explain. Since this pine occurs in western Cuba and the nearby Isle of Youth (formerly Isle of Pines), is it possible that its grass stage is due to introgression from *P. elliottii* var. *densa* which now occurs in southern Florida. Although Mirov (1967) noted that some varieties of *P. pseudostrobus* have a grass stage, extensive field studies by one of us (J.P.P.) indicate that is not so. Since *P. montezumae* and *P. pseudostrobus* are often found growing together, the grass stage of *P. montezumae* could easily have been mistaken for a presumed grass stage in *P. pseudostrobus*.

As the early five-needled diploxylon pines migrated southwards in Mexico and Central America, they generally encountered warm temperate to subtropical climates. Exceptions, however, were the very high mountain ranges where freezing temperatures locally occurred during some months of the year. Mexico's Transvolcanic Belt appears to have provided a particularly favourable environment for the development of new forms and varieties within the *Montezumae* group of subsection *Ponderosae*. It is here, and possibly further south in Guatemala, El Salvador and Honduras, that another branch of the five-needled diploxylon pines may have evolved, i.e. the *Pseudostrobus* group. Several authors, among them Shaw (1914), Martínez (1948), Mirov (1967) and Perry (1991), have noted the striking differences among the three groups of pines. Of particular interest is the absence of a grass stage in the *Pseudostrobus* group, the non-decurrent bases of the leaf bracts on the branchlets of *P. pseudostrobus*, and the smooth bark on the stems and branches of young *P. pseudostrobus* trees (Shaw 1914; Loock 1950; Perry 1991). Needles of *P. pseudostrobus* are generally more slender than those of the *Montezumae* group.

Despite the differences noted here, *P. montezumae* and *P. pseudostrobus* are closely related and hybridize readily in the field (Martínez 1948; Mirov 1967; Perry 1991). The great number of forms, varieties and species that have evolved within and between these related groups indicates a high degree of genetic affinity. Ranges of these groups overlap consistently and both occur (with a few minor variations)

Fig. 7.5. ***Pinus hartwegii*, Mexico's highest altitude pine, growing at the timberline (3750 m) below one of Mexico's famous mountains, Ixtaccihuatl (5300 m) (photo: J.P. Perry).**

between latitude 13 and 22° N. Within that area are some of Mexico's and Guatemala's highest mountains, and lowlands that are completely frost-free. The same area is also very active volcanically. These pines have, therefore, been exposed to an extremely variable environment for millions of years. Within these groups, taxa have evolved that are adapted to the boreal-type climate of the very high mountains. For example, on the slopes of four of Mexico's highest mountains (Cerro Potosí, Nevada de Toluca, Ixtaccihuatl and Popocatépetl), the pine taxa vary with changes in altitude: *P. pseudostrobus* occurs at lower altitudes followed by *P. montezumae* at slightly higher elevations, followed by different forms of the variable *P. hartwegii* (which previously included the taxon *P. rudis*; Matos 1995) which grows up to the timberline (Leopold 1950; Loock 1950; Mirov 1967; Favela 1991; Perry 1991; Fig. 7.5). These altitudinal differences hold true throughout the range of these three species: a classic example of altitudinal zonation.

The previous sections have dealt almost exclusively with the *Montezumae* and *Pseudostrobus* groups of five-needled pines in subsection *Ponderosae*, but the closed-cone groups of Mexico and Central America, *P. oocarpa* (five needles) and *P. patula* (three needles), have probably followed a similar evolutionary process. The *P. oocarpa* group (*sensu* Perry 1991) has three species and three varieties. *Pinus oocarpa* has the greatest north–south range of all the Mexican and Central American pines, extending from northwestern Mexico in southern Sonora, southwards through Mexico and Guatemala, Belize, El Salvador, Honduras and northwestern Nicaragua. Over its 3000 km range the species has adapted to very different environmental conditions, ranging from dry-temperate in Sonora, to humid-tropical in Honduras and Belize. The *P. patula* group (*sensu* Perry 1991) with four species and two varieties also has a wide north–south range which extends southwards along the Sierra Madre Oriental of eastern Mexico into Chiapas State. At their southern limits, the ranges of *P. oocarpa* and *P. patula* (and their related species and varieties) overlap. Here in southern Mexico, Guatemala, Belize and Honduras, these taxa are hybridizing to form new varieties, subspecies and species. Recent evidence (Styles, Stead & Rolph 1982; Squillance & Perry 1992) also indicates that natural hybridization is occurring between *P. caribaea* var. *hondurensis* and *P. oocarpa*. Hybridization also appears to be occurring between *P. oocarpa* and *P. tecunumanii*. Further genetic studies are required in this region.

Styles (1993) designated six principal areas of pine distribution in Mexico, based on information from Eguiluz Piedra (1985). Perry (1991) showed the pine forests extending further southwards from Mexico along the mountain ranges of Guatemala, Honduras, El Salvador and Nicaragua. Figure 7.4 shows the distribution of pine forests along the principal mountain ranges of Mexico and Central America. Styles (1993) noted six centres of species diversity of pines in Mexico. One of these, which corresponds to Area 4 in Fig. 7.4, was a particularly important centre of evolution and speciation of the genus *Pinus* because it connects the Sierra Madre Occidental and Sierra Madre Oriental ranges. Topographical (volcanic) disturbance has provided many microhabitats, allowing hybridization, adaptive radiation, and speciation to occur.

The greatest diversity of pine species in Mexico and Central America, occurs in those areas with the greatest and most abrupt changes in topography. The Transvolcanic Belt (Area 4 in Fig. 7.4) is an excellent example. Here are found Mexico's great snow-capped volcanic peaks scattered along an east-west range of mountains with innumerable, small, fertile valleys. Rainfall along this volcanic range averages about 1000 mm yr^{-1} (800–1500 mm), and climate varies from boreal to tropical. Some of Mexico's finest pine forests and greatest *Pinus* diversity (43 species, subspecies, varieties and forms) occur here.

Another area with great and abrupt topographic changes occurs in Guatemala, Honduras and El Salvador (Area 9 in Fig. 7.4). In Guatemala there are several very high, active volcanos dominating fertile and tropical valleys (Fig. 7.6) with temperate mountain ranges extending into Honduras and El Salvador. Physiographically and climatically, this area is quite similar to Area 4. Pine diversity here (as in area 4) is very great, with about 20 species, subspecies, varieties and forms (Perry 1991). Finally, 65% of these taxa are classified in the five-needled *Montezumae* and *Pseudostrobus* groups of diploxylon pines in subsection *Ponderosae*.

Fig. 7.6. **The highlands of Guatemala have for thousands of years provided an extremely diverse environment that has favoured the diversification of *Pinus* taxa. Here active volcanoes provide a background for thousands of small man-made clearings almost always surrounded by pine forests (photo: J.P. Perry).**

7.5 Human influences

Extensive settlements of humans within the natural range of pines in the New World date from at least as early as *c.* 11 000 years ago (Chap. 20, this volume). There is, however, little information available on the impact of humans specifically on pine forests in Mexico and Central America. The record is also complicated by the fact that wet and dry intervals have occurred during the period of human occupation. This makes it difficult to disentangle the relative roles of climate and human disturbance in the waxing and waning of pine forests. Also, hard water contamination of lake sediment renders many ^{14}C dates unreliable. Early agriculture in the region is evident by the presence of plants such as *Zea* (corn), *Phaseolus* (bean), and *Cucurbita* (gourds and squash), increases in pollen of weeds typically associated with agriculture (chenoams (Chenopodiaceae–Amaranthaceae)), ragweed (*Ambrosia*), other Asteraceae, grasses), and charcoal. According to reviews by Brown (1985) and McAndrews (1988), agriculture widespread to the extent that it affected the natural vegetation (i.e. farming) is evident by *c.* 4000 years ago. A pollen diagram from Lake Patzcuaro at 2044 m elevation in Mexico shows a 44 000-year record dominated by *Pinus* and *Quercus* (Fig. 7.7; Watts & Bradbury 1982). At this site, human disturbance due to agriculture becomes evident *c.* 3700 years ago. Neither this event, nor ensuing disturbances (which lead to a decline in *Alnus* pollen) had

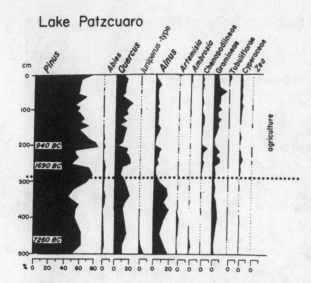

Lake Patzcuaro

Fig. 7.7. Pollen diagram from Lake Patzcuaro, Mexico. Note the decline in *Alnus* (alder) pollen after the introduction of agriculture, and the general decline of *Pinus* pollen with the appearance or increase of ragweed (*Ambrosia*), chenoams and Graminaeae (including *Zea*, corn). Reproduced, with permission, from McAndrews (1988), after Watts & Bradbury (1982).

much effect on the dominance of *Pinus* and *Quercus* (see McAndrews 1988 for further discussion). However, a similar pollen record, from the Valley of Mexico (Clisby & Sears 1955), shows a marked decline in *Pinus* and *Quercus*, suggesting the destruction of upland forest.

Humans have certainly been present in Mexico and Central America in sufficient numbers to have had a marked influence on the biota since AD 300, the beginning of the Classic Period in Mesoamerica. The population at Tikal, a lowland Maya centre in Guatemala, is estimated at about 45 000 in AD 550 and it covered an area of 123 km² (Weaver 1972). Mayapán in the Yucatán Peninsula had a population of 11 000–12 000 in AD 1263–83 over an area of about 5 km². By the time of Cortez's landing near Veracruz, Mexico in 1519, beginning the Spanish conquest of Mexico, the population of the capital city of Tenochtitlán numbered about 300 000. Thus, the vegetation of Mexico and Central America, including the pine forest, has experienced some impact from farming and other kinds of human disturbance for at least 4000 years. This intensified beginning about AD 300, and is presently proceeding virtually unchecked at unprecented rates.

7.6   Conclusions

Among the pines that migrated slowly southwards into Mexico were the ancient five-needled diploxylon pines first described by Lesquereux (1883). Both he and Chaney (1954)

noted the resemblance of Lesquereux's specimens to the living Mexican pines, particularly to *P. michoacana* var. *cornuta* (now *P. devoniana*) which has cones most similar to the Kansas fossil. It seems reasonable to conclude that Lesquereux's five-needled diploxylon pines were indeed the ancestors of the *Montezumae* and *Pseudostrobus* groups of pines that are presently widespread in Mexico and Central America. These groups share many important morphological and chemical features, and appear to stem from a central source which has been termed a *P. montezumae* complex (Martínez 1948; Mirov 1967). As noted earlier, *P. montezumae* crosses with *P. devoniana* and its varieties, *P. ponderosa*, and with *P. pseudostrobus*. *Pinus devoniana* and *P. pseudostrobus* apparently do not hybridize. These pines are clearly closely related. Hybrids resulting from the crosses noted here are vigorous and fertile. Within the highly variable environment found along the mountain ranges and valleys in Mexico and Central America, many of these hybrids survived in 'ecological niches' quite different from those of their parents. Their progeny, over millions of years, have through the process of evolution formed new races, varieties, subspecies and species of pines.

Shaw (1909), almost ninety years ago wrote,

In Mexico and Central America are found extremes of climate within small areas and easily within the range of dissemination from a single tree. The cause of the bewildering host of varietal forms, connecting widely contrasted extremes, seems to lie in the facile adaptability of these pines, which are able to spread from the tropical base of a mountain to a less or greater distance towards its snow-capped summit.

Mirov (1967), commenting on the amazing diversity of the Mexican pines, wrote,

The abundance of pine species in Mexico, their widespread interspecific hybridization, and their pronounced intraspecific variability all bespeak that in the tropical highlands of Mexico and Central America, there has developed a secondary centre of evolution and speciation of the genus *Pinus*.

Although we now have a better understanding of how this amazing pine flora originated, a great deal of work remains to be done before a complete picture emerges.

It is important to note that growing human pressure is severely threatening many *Pinus* taxa in Mexico and Central America (Perry 1991). Besides jeopardizing opportunities for testing and developing new races of pines for use in afforestation worldwide, the rampant destruction of forests is also obliterating large sections of the irreplaceable natural laboratory in which lie answers to many questions concerning the evolution and genetic structure of pines. Although it is encouraging to note the progress made by the Central America and Mexico Coniferous Resources Cooperative (CAMCORE) which is striving for the *ex situ* preservation of pines and other tree taxa from this region (Dvorak & Donahue 1992), every effort must be made to conserve at least representative tracts of each remaining major forest type in the region.

References

Areces, A. (1987). Consideraciones sobre la supuesta presencia de *Pinus sylvestris* L. en el Oligoceno de Cuba. *Publicacion Centro Investigaciones y Desarrollo del Petroleo, Serie Geologica*, **2**, 27–40.

Axelrod, D.I. (1976). History of the coniferous forests, California and Nevada. *University of California Publications in Botany*, **70**, 1–62.

Axelrod, D.I. (1980). History of the maritime closed-cone pines, Alta and Baja California. *University of California Publications in Geological Science*, **120**, 1–143.

Axelrod, D.I. (1983). New Pleistocene conifer records, coastal California. *University of California Publications in Geological Science*, **127**, 1–108.

Axelrod, D.I. (1986). Cenozoic history of some western American pines. *Annals of the Missouri Botanical Garden*, **73**, 565–641.

Axelrod, D.I. (1988). Paleoecology of a late Pleistocene Monterey pine at Laguna Niguel, southern California. *Botanical Gazette*, **149**, 458–64.

Axelrod, D.I. (1990). Ecologic differences have separated *Pinus remorata* and *P. muricata* since the early Pleistocene. *American Journal of Botany*, **77**, 289–94.

Axelrod, D.I. & Cota, J. (1993). A further contribution to closed-cone pine (Oocarpae) history. *American Journal of Botany*, **80**, 743–51.

Axelrod, D.I. & Hill, T.G. (1988). *Pinus* × *critchfieldii*, a late Pleistocene hybrid pine from coastal southern California. *American Journal of Botany*, **75**, 558–69.

Bartlett, A.S. & Barghoorn, E.S. (1973). Phytogeographic history of the Isthmus of Panama during the past 12 000 years (a history of vegetation, climate, and sea-level change). In *Vegetation and Vegetational History of Northern Latin America*, ed. A. Graham, pp. 203–99. Amsterdam: Elsevier.

Berry, E.W. (1934). Pleistocene plants from Cuba. *Torrey Botanical Club Bulletin*, **61**, 237–40.

Braun, E.L. (1955). The phytogeography of unglaciated eastern United States and its interpretation. *Botanical Review*, **21**, 297–335.

Brown, R.B. (1985). A summary of Late-Quaternary pollen records from Mexico west of the Isthmus of Tehuantepec. In *Pollen Records of Late Quaternary North American Sediments*, ed. V.M. Bryant, Jr & R.G. Holloway, pp. 71–93. Dallas: American Association of Stratigraphic Palynologists.

Chaney, R.W. (1936). Plant distribution as a guide to age determination. *Washington Academy Science Journal*, **26**, 313–24.

Chaney, R.W. (1954). A new pine from the Cretaceous of Minnesota and its paleoecological significance. *Ecology*, **35**, 145–51.

Clisby, K.H. & Sears, P.B. (1955). Palynology in southern North America. Part 3. Pleistocene climate in Mexico. *Bulletin of the Geological Society of America*, **66**, 511–20.

Critchfield, W.B. (1984). Impact of the Pleistocene on the genetic structure of North American conifers. In *Proceedings of the Eighth North American Forest Biology Workshop*, ed. R.M. Lanner, pp. 70–118. Logan: Utah State University.

de Cserna, Z. (1989). An outline of the geology of Mexico. In *The Geology of North America – An Overview*, Vol. A, ed. A.W. Bally & A.R. Palmer, pp. 233–64. Boulder, Colorado: Geological Society of America.

Deevey, E.S. (1949). Biogeography of the Pleistocene. *Bulletin of the Geological Society of America*, **60**, 1315–1416.

Delcourt, P.A. & Delcourt, H.R. (1993). Paleoclimates, paleovegetation, and paleofloras during the late Quaternary. In *Flora of North America North of Mexico*, ed. Flora of North America Editorial Committee, pp. 71–94. Oxford: Oxford University Press.

Dressler, R.L. (1954). Some floristic relationships between Mexico and the United States. *Rhodora*, **56**, 81–96.

Duffield, J.W. (1952). Relationships and species hybridization in the genus *Pinus*. *Zeitschrift für Forstgenetik*, **1**, 93–100.

Dvorak, W.S. & Donahue, J.K. (1992). *CAMCORE Cooperative Research Review 1980–1992*. Raleigh, NC: CAMCORE, North Carolina State University.

Eguiluz Piedra, T. (1985). Origin y evolucion del genero *Pinus* (con referencia especial a los pinos mexicanos). *Dasonomia Mex.*, **3** (6), 5–31.

Eguiluz Piedra, T. (1988). Distribucion natural de los pinos en Mexico. *Nota Tecnica Centro de Genetica Forestal A.C.*, **1**, 1–6.

Favela, L.S. (1991). Taxonomia de *Pinus pseudostrobus* Lindl., *Pinus montezumae* Lamb. y *Pinus hartwegii* Endl. *Reporte Cientifico Facultad de Ciências Forestales, Universidad Autonoma de Nuevo León*, **26**, 1–30.

Graham, A. (ed.) (1973a). *Vegetation and Vegetational History of Northern Latin America*. Amsterdam: Elsevier.

Graham, A. (1973b). History of the arborescent temperate element in the northern Latin American biota. In *Vegetation and Vegetational History of Northern Latin America*, ed. A. Graham, pp. 301–14. Amsterdam: Elsevier.

Graham, A. (1976). Studies in neotropical paleobotany. II. The Miocene communities of Veracruz, Mexico. *Annals of the Missouri Botanical Garden*, **63**, 787–842.

Graham, A. (1985). Studies in neotropical paleobotany. IV. The Eocene communities of Panama. *Annals of the Missouri Botanical Garden*, **72**, 504–34.

Graham, A. (1987). Miocene communities and paleoenvironments of southern Costa Rica. *American Journal of Botany*, **74**, 1501–18.

Graham, A. (1988a). Studies in neotropical paleobotany. V. The lower Miocene communities of Panama – the Culebra Formation. *Annals of the Missouri Botanical Garden*, **75**, 1440–66.

Graham, A. (1988b). Studies in neotropical paleobotany. VI. The lower Miocene communities of Panama – the Cucaracha Formation. *Annals of the Missouri Botanical Garden*, **75**, 1467–79.

Graham, A. (1989). Studies in neotropical paleobotany. VII. The lower Miocene communities of Panama – the La Boca Formation. *Annals of the Missouri Botanical Garden*, **76**, 50–66.

Graham, A. (1990). Late Tertiary microfossil flora from the Republic of Haiti. *American Journal of Botany*, **77**, 911–26.

Graham, A. (1991a). Studies in neotropical paleobotany. VIII. The Pliocene communities of Panama – introduction and ferns, gymnosperms, angiosperms (monocots). *Annals of the Missouri Botanical Garden*, **78**, 190–200.

Graham, A. (1991b). Studies in neotropical paleobotany. IX. The Pliocene communities of Panama – angiosperms (dicots). *Annals of the Missouri Botanical Garden*, **78**, 201–23.

Graham, A. (1991c). Studies in neotropical paleobotany. X. The Pliocene communities of Panama – composition, numerical representations, and paleocommunity paleoenvironmental reconstructions. *Annals of the Missouri Botanical Garden*, **78**, 465–75.

Graham, A. (1993). Contribution toward a palynostratigraphy for Jamaica: the status of Tertiary paleobotanical studies in northern Latin America and preliminary analysis of the Guys Hill Member (Chapelton Formation, middle Eocene) of Jamaica. In *Biostratigraphy of Jamaica*, ed. R.M. Wright & E. Robinson, pp. 443–61. Geological Society of America Memoir 182. Boulder, Colorado: Geological Society of America.

Graham, A. (1994). Neogene palynofloras and terrestrial paleoenvironments in northern Latin America. In *Pliocene Terrestrial Environments and Data/Model Comparisons*, ed. R.S. Thompson, pp. 23–30. US Geological Survey Open-File Report 94–23.

Graham, A. (1995). Development of affinities between Mexican/Central American and northern South American lowland and lower montane vegetation during the Tertiary. In *Biodiversity and Conservation of Neotropical Montane Forests*, ed. S.P. Churchill, H. Balslev, E. Forero & J. Luteyn, pp. 11–22. Bronx: The New York Botanical Garden.

Graham, A. & Heimsch, C. (1960). Pollen studies of some Texas peat deposits. *Ecology*, **41**, 785–90.

Graham, A. & Jarzen, D.M. (1969). Studies in neotropical paleobotany. I. The Oligocene communities of Puerto Rico. *Annals of the Missouri Botanical Garden*, **56**, 308–57.

Graham, A., Stewart, R.H. & Stewart, J.L. (1985). Studies in neotropical paleobotany. III. The Tertiary communities of Panama – geology of the pollen-bearing deposits. *Annals of the Missouri Botanical Garden*, **72**, 485–503.

Harrar, E.S. & George, J. (1962). *Guide to Southern Trees*. New York: Dover.

Harrell, B.E. (1951). The birds of Rancho del Cielo. An ecological investigation in the oak–sweet gum forests of Tamaulipas, Mexico. MA thesis, University of Minnesota.

Hernandez, E.X., Crum, H. Fox, W.B. & Sharp, A.J. (1951). A unique vegetational area in Tamaulipas. *Bulletin of the Torrey Botanical Club*, **78**, 458–63.

Horn, S.P. (1985). Preliminary pollen analysis of Quaternary sediments from Deep Sea Drilling Project site 565, western Costa Rica. Initial Reports of the Deep Sea Drilling Project LXXXIV: 533–47.

Langenheim, J.H., Hackner, B.L. & Bartlett, A. (1967). Mangrove pollen at the depositional site of Oligo-Miocene amber from Chiapas, Mexico. *Botanical Museum Leaflets*, **21**, 289–324.

Lanner, R.M. (1981). *The Piñon Pine: A Natural and Cultural History*. Reno: University of Nevada Press.

Lanly, J.P. (1995). The status of tropical forests. In *Tropical Forests: Management and Ecology*, ed. A.E. Lugo & C. Lowe, pp. 18–32. New York: Springer-Verlag.

Leopold, A.S. (1950). Vegetation Zones of Mexico. *Ecology*, **31**, 507–18.

Lesquereux, L. (1883). Contributions of the fossil flora of the western Territories. III. The Cretaceous and Tertiary floras. *US Geological Survey of the Territories Report*, **8**, 1–283.

Lesquereux, L. (1895). Cretaceous fossil plants from Minnesota. *Minnesota Geological and Natural History Survey Final Report*, **3**(1), 1–22.

Loock, E.E.M. (1950). *The Pines of Mexico and British Honduras*. South Africa Department of Forestry Bulletin 34. Pretoria: Department of Forestry.

Lorenzo, J.L. (1969). *Condiciones Periglaciares de las Altas montanas de Mexico*. Mexico City: Departamento De Prehistoria. Instituto Nacional De Anthropologia e Historia.

McAndrews, J.H. (1988). Human disturbance of North American forests and grasslands: the fossil pollen record. In *Vegetation History*, ed. B. Huntley & T. Webb III, pp. 673–97. Dordrecht: Kluwer.

Martin, P.S. & Harrell, B.E. (1957). The Pleistocene History of Temperate Biotas in Mexico and Eastern United States. *Ecology*, **38**, 468–80.

Martínez, M. (1948). *Los Pinos Mexicanos*, 2nd edn. Mexico City: Ediciones Botas.

Martínez-Hernández, E. (1992). Caracterización ambiental del Terciario de la región de Ixtapa, estado de Chiapas – un enfoque palinoestratigráfico. *Univ. Nacional Autonoma México, Inst. Geologia, Revista*, **10**, 54–64.

Martínez-Hernández, E., Hernández-Campos, H. & Sánchez-López, M. (1980). Palinologia del Eoceno en el noreste de Mexico. *Univ. Nacional Autonoma México, Inst. Geologia, Revista*, **4**, 155–66.

Mason, H.L. (1927). Fossil records of some west American conifers. *Carnegie Institute of Washington Publication*, **16**, 261–82.

Mason, H.L. (1932). A phylogenetic series of California closed-cone pines suggested by fossil record. *Madroño*, **2**, 49–55.

Matos, J.A. (1995). *Pinus hartwegii* and *P. rudis*: a critical assessment. *Systematic Botany*, **20**, 6–21.

Matthews, R.K. & Poore, R.Z. (1980). Tertiary ^{18}O record and glacio-eustatic sea-level fluctuations. *Geology*, **8**, 501–4.

Mejorada, N.S. & Huguet, L. (1959). Conifers of Mexico. *Unasylva*, **1**, 24–35.

Miller, C.N., Jr (1977). Mesozoic conifers. *Botanical Review*, **43**, 217–80.

Miller, K.G., Fairbanks, R.G. & Mountain, G.S. (1987). Tertiary oxygen isotope synthesis, sea level history, and continental margin erosion. *Paleoceanography*, **2**, 1–19.

Miranda, F. & Sharp, A.J. (1950). Characteristics of the vegetation in certain temperate regions of eastern Mexico. *Ecology*, **31**, 313–33.

Mirov, N.T. (1967). *The Genus Pinus*. New York: Ronald Press.

Oxford Forestry Institute (1996). *Seventy-First Annual Report. 1995*. Oxford: Oxford Forestry Institute.

Palacios Chavez, R. & Rzedowski, J. (1993). Estudio palinologico de las floras fosiles del Miocene inferior y principios del Mioceno medio de la region de Pichucalco, Chiapas, Mexico. *Actas Botánica Mexicana*, **24**, 1–96.

Passini, M.F. (1982a). *Les Forêts de Pinus cembroides s.l. au Mexique. Etude Phytogeographique et Ecologique*. Mission Archeologique et Ethnologique Française au Mexique. Etudes Mesoamericaines II-5. Paris, France: Editions Recherche sur les civilisations.

Passini, M.F. (1982b). The Mexican stone pines (pinyons) of the *Cembroides* group. *Forest Genetic Resources Information*, **11**, 29–33.

Perry, J.P. (1991). *The Pines of Mexico and Central America*. Portland, Oregon: Timber Press.

Rzedowski, J. (1978). *Vegetacion de Mexico*. Mexico City: Editorial Limusa.

Savin, S.M. (1977). The history of the earth's surface temperature during the past 100 million years. *Annual Review of Earth and Planetary Science*, **5**, 319–55.

Shaw, G.R. (1909). The Pines of Mexico. *Journal of the Arnold Arboretum*, **1**, 1–30.

Shaw, G.R. (1914). The genus *Pinus*. *Journal of the Arnold Arboretum*, **5**, 1–96.

Squillance, A.E. & Perry, J.P. (1992). *Classification of* Pinus patula, P. tecunumanii, P. oocarpa, P. caribaea var. hondurensis, *and related taxonomic entities*. USDA Forest Service Research Paper SE-285.

Standley, P.C. (1920–6). Trees and shrubs of Mexico. *Contributions of the United States National Herbarium*, **23**, 1–848.

Steyermark, J.A. (1950). Flora of Guatemala. *Ecology*, **31**, 368–72.

Styles, B.T. (1993). Genus *Pinus*: a Mexican purview. In *Biological Diversity of Mexico: Origins and Distribution*, ed. T.P. Ramamoorthy, R. Bye, A. Lot & J. Fa, pp. 397–420. New York: Oxford University Press.

Styles, B.T., Stead, J.W. & Rolph, K.J. (1982). Studies of variation in Central American pines. 2. Putative hybridization between *P. caribaea* var. *hondurensis* and *P. oocarpa*. *Turrialba*, **32**, 229–42.

Wahlenberg, W.G. (1946). *Longleaf Pine*. Washington, DC: Charles Lathrop Pack Forestry Foundation.

Watts, W.A. & Bradbury, J.P. (1982). Paleoecological studies at Lake Patzcuaro on the west-central Mexican Plateau and at Chalco in the Basin of Mexico. *Quaternary Research*, **17**, 56–70.

Weaver, M.P. (1972). *The Aztecs, Maya, and Their Predecessors*. New York: Seminar Press.

Wendt, T. (1993). Composition, floristic affinities, and origins of the canopy tree flora of the Mexican Atlantic slope rain forests. In *Biological Diversity of Mexico: Origins and Distribution*, ed. T.P. Ramamoorthy, R. Bye, A. Lot & J. Fa, pp. 595–680. New York: Oxford University Press.

Yeaton, R.I. (1982). The altitudinal distribution of the genus *Pinus* in the western United States and Mexico. *Boletin de la Societa Botanica de Mexico*, **42**, 55–71.

Part four
Macroecology and recent biogeography

8   Pines of the Mediterranean Basin

Marcel Barbéro, Roger Loisel, Pierre Quézel, David M. Richardson and François Romane

8.1   Introduction

We define the Mediterranean Basin as the area around the Mediterranean Sea that experiences a true mediterranean-type bioclimate (Daget 1977a, b; Nahal 1981; Quézel 1985; Quézel & Barbéro 1985). This region, which covers about 2.3×10^6 km², has an unusual geographical and topographical diversity, with high mountains, peninsulas, and one of the largest archipelagos in the world (46 000 km of coastline). There are several hundred islands, with a total area of about 103 000 km² (4% of the area), and about 18 000 km of coastline (39% of the total for the region) (Kolodny 1974).

The region can be divided into four parts (northwest,

northeast, southwest and southeast) by lines running east to west (southern Anatolia–Malta–Strait of Gibraltar) and north to south (Adriatic Sea–Malta–Tunisia) (Fig. 8.1). These lines which take into account the mean annual potential evapotranspiration (Blondel & Aronson 1995), separate the Mediterranean Basin into zones that are reasonably discrete in terms of at least the following factors, which must be considered in any account of the vegetation of the region: features of soil and climate; impacts of human cultures; skills in resource management; obstacles of poverty; deep-rooted traditions; economic forces; and political events (Thirgood 1981). It is beyond the scope of

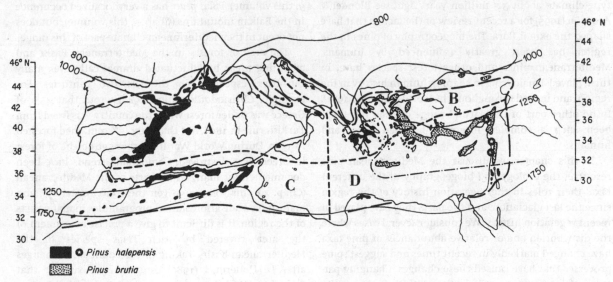

Fig. 8.1. **The Mediterranean Basin (———————), as defined for this chapter, showing the distribution of *Pinus brutia* and *P. halepensis* (modified from Quézel 1980a), and potential evapotranspiration (in** mm; Blondel & Aronson 1995). The short-broken lines (– – –) divide the region into four sectors: North-West (A), North-East (B), South-West (C) and South-East (D).

Ecology and Biogeography of Pinus, ed. D.M. Richardson. © Cambridge University Press (1998). pp. 153–70.

this chapter to review the current understanding of how these factors have interacted to shape vegetation in different parts of the region (but see Pons & Quézel 1985). One fundamental aspect must, however, be mentioned. Conditions in the northwest of the region favour a more settled human society, with more stable traditions, individual title to property and inheritance of land. In the east, however, the arid environment, together with exposure to waves of immigration from Arabia and Asia, has mitigated against settled life, and a pastoral society still predominates over considerable areas. Such contrasting forms of land use over millennia (humans have occupied the region for almost a million years) have played a dramatic role in shaping the region's vegetation. The marked cultural differences between west and east continue to play a major role in defining contrasts in land use, with the former being increasingly characterized by the development of industry, urbanization and tourism, and a concurrent decrease of agriculture and pastoralism. On the other hand, the main concern in Afro-Asiatic Mediterranean countries at present is the rapidly growing human population, which represents an obstacle to sustainable socioeconomic development. It is against this template that we must consider the recent history and contemporary dynamics of pines in the Mediterranean Basin.

Pines are important components of many landscapes in this region, and have played major roles in the origin of its flora and vegetation (Quézel 1985). *Pinus* spp., along with other Eurasian and Holarctic elements such as *Acer*, *Betula*, *Cercis*, *Corylus*, *Fagus*, *Fraxinus*, *Quercus*, *Tilia* and *Ulmus* that are now prominent components of the flora, invaded the Mediterranean Basin after the advent of a mediterranean-type climate about 3.5 million years ago; see Blondel & Aronson (1995) for a recent review of the factors that have shaped the extant flora. The biogeography of pines in the region has been greatly influenced by humans. Mediterranean pines and other forest species have, in turn, played a major role in shaping human history in the region, and indeed throughout the world (Thirgood 1981). In no other part of the natural range of *Pinus* has there been such a complex interplay between pines and humans.

In this chapter we discuss the Mediterranean pine resource, the ecology and biogeography of the different taxa, their roles in the vegetation history of the region since the last glaciation, and the parts they have played in recent vegetation history. We consider several cases where the distribution and/or relative abundances of pine taxa have changed markedly in recent times and suggest some processes that have caused these changes. Changing patterns of land use are still having major effects on the dynamics of the Mediterranean pine forest; we explore salient features of the changing Mediterranean environment and discuss the implications for the Mediterranean pine forests.

8.2 The Mediterranean pine resource

The extant pine flora of the Mediterranean Basin (hereafter 'the Mediterranean pines') comprises 10 *Pinus* species (Klaus 1989). Included in this list is *P. sylvestris* which, although widespread in the region, is more characteristic of extra-mediterranean Europe (Chap. 5, this volume). Chloroplast DNA studies (reviewed in Chap. 2, this volume) show the Mediterranean pines to be divided between a small group of species in subsections *Canarienses* (*P. canariensis*), *Halepenses* (*P. brutia* and *P. halepensis*) and *Pineae* (*P. pinea*), which correspond to almost all of Klaus's (1989) group of Mediterranean shore and island pines; and a separate, larger group comprising part of subsection *Pinus* (*P. heldreichii*, *P. mugo*, *P. nigra*, *P. pinaster*, *P. sylvestris* and *P. uncinata*) which fall into Klaus' group of mountain pines from areas surrounding the Mediterranean. The chloroplast DNA restriction site tree (see Fig. 2.1, p. 61) suggests that the subsections *Canarienses*, *Halepenses* and *Pineae* form a well-supported clade separate from subsection *Pinus* (the exact position of *P. pinaster* is still questionable until it is subjected to molecular comparisons; Chap. 2, this volume). Pines of the subgenus *Strobus*, which comprise more than a third of species in the genus, are not represented in the region, although taxa in this group (*P. juarezensis*, *P. monophylla*) do occupy mediterranean-type climates in North America (Quézel & Barbéro 1989; Chap. 9, this volume). *Pinus peuce* has a very localized occurence in the Balkan mountains (Chap. 5, this volume), but does not occur in the mediterranean-climate part of this range.

The area of forests in the Mediterranean Basin and adjoining areas has fluctuated dramatically over many centuries, especially in the last two centuries. For example, Christodoulopoulos (1947) wrote that 45% of Greece was under forest when the country was freed from Turkish rule in 1829, but that only 15% remained forested in 1940. During World War II another 500 000 ha of forest in Greece was destroyed. Comparable trends have been documented for many other parts of the Mediterranean (Chap. 20, this volume). Given these circumstances, and many large-scale afforestation programmes in many parts of the region, it is difficult to give a precise assessment of the area covered by each *Pinus* species in the Mediterranean Basin. Taking into account some changes after Le Houérou's (1980) estimates, we suggest that *P. brutia* and *P. halepensis* now cover 6.8×10^6 ha, *P. nigra* (*sensu lato*) 3.5×10^6 ha, *P. pinaster* (*s.l.*) 1.3×10^6 ha, *P. pinea* 320 000 ha and other *Pinus* species 300 000 ha, giving a total of almost 13×10^6 ha. Although pine forests cover only 5% of the total area of the Mediterranean Basin, they comprise about 25% of the forested area. In North Africa and in Anatolia 75% of the forested area comprises pine forests, and the proportion is probably higher in central and northern Taurus. In the northwestern part of the Basin, although pines are less dominant, pine forests are nonetheless important features of the landscapes, particularly in Provence, eastern Spain, Corsica, and southern Greece.

The Mediterranean pines can be divided into several groups on the basis of their distribution and general ecology (Quézel 1980a). The two most common species, *P. brutia* and *P. halepensis* (Fig. 8.2), occupy large parts of the eastern and western part of the Mediterranean Basin, respectively (Fig. 8.1; Nahal 1962), *P. halepensis* being the only pine with a considerable part of its natural range in North Africa (*P. nigra* subsp. *mauretanica* and the subspecies *maghrebiana* and *renoui* of *P. pinaster* also occur in North Africa). *Pinus brutia* and *P. halepensis*, which are very similar ecologically and genetically, are vicariant taxa that rarely co-occur (Akman, Barbéro & Quézel 1978); see Biger & Liphschitz (1991) for a concise review of differences between these taxa. Where the two taxa do coexist (in two small districts in Greece, in southeastern Anatolia and in Lebanon), they form natural hybrids (Panetsos 1975; see Chap. 13, this volume, for discussion). Isolated populations of *P. halepensis* in Asia Minor and the Near East, at the eastern limit of the Mediterranean Basin, pose intriguing questions regarding past distribution patterns and migration pathways (Schiller, Conkle & Grunwald 1986).

Among the other Mediterranean pines, two species, *P. pinea* and *P. pinaster*, occur at low and medium altitudes.

Pinus pinea, whose natural range is difficult to define because it has been planted so widely for so long (Mirov 1967, p. 245), now occurs in scattered populations throughout the Mediterranean Basin except in North Africa (Figs. 8.3, 8.4). *Pinus pinaster* is restricted to the western part of the Mediterranean (Figs. 8.3, 8.5), where many subspecific taxa have been recognized but which are difficult to distinguish (Del Villar 1933; Baradat & Marpeau-Bezard 1988; Chap. 2, this volume). The most important subspecies are *pinaster*, which occurs mainly in the Atlantic region but also in southern France, *hamiltonii* (= *P. mesogeensis* Fieschi & Gaussen), including subsp. *renoui* H. del Villar (in Algeria and Tunisia), and subsp. *maghrebiana* (in Morocco and southern Spain) which is ecologically, but not taxonomically, very well differentiated. We include *P. canariensis*, which is found only in the

Fig. 8.2. *Pinus halepensis* (Aleppo pine) in the foothills of Euboia, Greece. High-elevation sites are occupied by *P. nigra* and *Abies cephalonica*, with *Platanus orientalis* occurring on wet sites (photo: J.G. Goldammer).

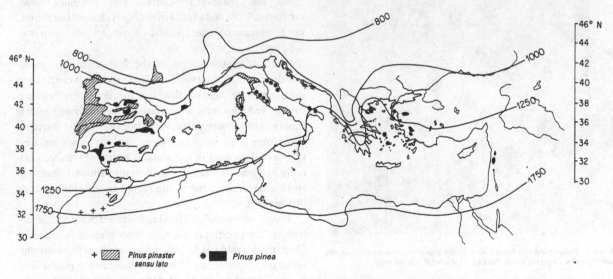

Fig. 8.3. The distribution of *Pinus pinaster* and *P. pinea* in the Mediterranean Basin (+ isolated localities of *P. pinaster*; *, human-introduced *P. pinea*; modified from Quézel 1980a), and potential evapotranspiration (in mm; Blondel & Aronson 1995).

+ *Pinus pinaster sensu lato*   * *Pinus pinea*

Fig. 8.4. *Pinus pinea* on the Mediterranean coast north of Gerona, Catalonia, northeastern Spain. Seeds of this pine have been used as food by humans since prehistoric times and were widely traded. This species was also widely planted, making it impossible to determine its natural range (photo: D.M. Richardson).

Fig. 8.6. *Pinus nigra* subsp. *pallasiana* occurs naturally in Greece, Turkey and Cyprus. The picture shows a forest of this pine in eastern Taurus, Turkey after degradation by humans and overgrazing (photo: P. Quézel).

Fig. 8.5. *Pinus pinaster* forest with *Stipa gigantea* understorey on dry sandy soils, Lago da Casa, Portugal (photo: H.J.B. Birks).

central and eastern Canary Islands (Wildprett della Torre & Aguilar 1987), in this ecological set.

Another group, found especially in the mountains, is represented principally by *P. nigra* whose populations, from the Rif Mountains of Morocco to the eastern Taurus mountains of Turkey (Fig. 8.6), are divided into several subspecies and numerous varieties (Debazac 1971; Chap. 2, this volume; Figs. 8.6, 8.7). Its maximum distribution is, however, in the southern Balkans and Anatolia (subspecies *nigra* and *pallasiana*). Also in this group is *P. sylvestris* which, although not truly a Mediterranean pine, has some populations that are very well adapted to the Mediterranean mountain environment, particularly in Spain (var. *nevadensis* Christ, var. *pyrenaica* Svob, var. *iberica* Svob) and southern France (var. *aquitana* Schott, var. *brigantiaca* Gaussen) (Castroviejo *et al.* 1986; Tutin *et al.* 1964–93).

The last group comprises the high-mountain pines: *P. heldreichii* (including var. *leucodermis*) which is confined to southern Italy, the northwestern Balkans and eastern Greece, and *P. uncinata*, a vicariant species of *P. mugo*, which occurs in the western Alps, the Pyrenees and the Sierra de Javalambre (Figs. 8.8, 8.9). Pollen analysis of the isolated stands of *P. uncinata* in the Massif Central of France, originally regarded as relict populations, showed that the species was introduced to the region in the 19th century (Reille 1989).

Pines have been widely used and planted by human inhabitants of the Mediterranean Basin since prehistoric times. In Chap. 20 (this volume), Le Maitre discusses how this has influenced the distribution of *P. brutia*, *P. halepensis*, *P. pinaster* and *P. pinea*. Since the second half of the 19th century, reafforestation programmes have been undertaken in most parts of the

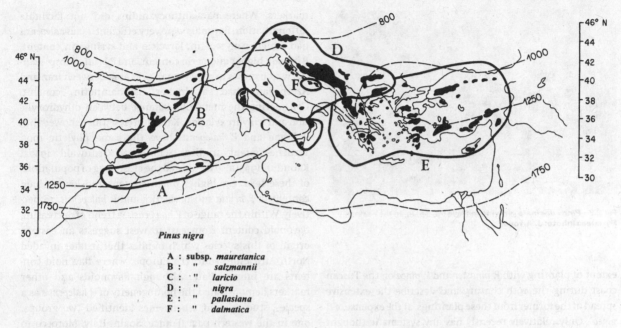

Fig. 8.7. The distribution of five subspecies of *Pinus nigra* in the Mediterranean Basin (modified from Quézel 1980a), and potential evapotranspiration (in mm; Blondel & Aronson 1995).

Pinus nigra

A : subsp. *mauretanica*
B : " *salzmannii*
C : " *laricio*
D : " *nigra*
E : " *pallasiana*
F : " *dalmatica*

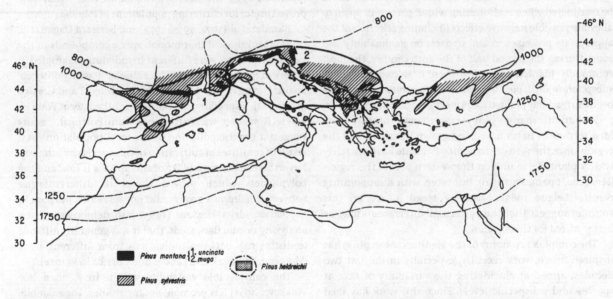

Fig. 8.8. The distribution of *Pinus heldreichii*, *P. mugo*, *P. sylvestris* and *P. uncinata* in the Mediterranean Basin (modified from Quézel 1980a), and potential evapotranspiration (in mm; Blondel & Aronson 1995). The solid line indicates the southern limit of *P. sylvestris* and **the dashed line shows the total area where *P. heldreichii* is present. *Pinus mugo* and *P. uncinata* are mapped together, as they were previously mapped as one species (*P. montana*).**

region (notably in Italy, France, Spain and Turkey), often to control erosion. Such programmes have used mainly species native to the region (especially *P. nigra* subsp. *nigra*, *P. pinaster*, *P. pinea*, *P. sylvestris* and *P. uncinata*). Such plantings have undoubtedly had a major influence on the distribution of these species in the region. For example, Paci & Romoli (1992)

describe the 'natural diffusion' of *P. nigra* from plantations into *Brachypodium pinnatum*-dominated grasslands in the Arezzo Province of Tuscany. More than 50% of 47 500 ha of pine forests in Calabria in southern Italy are the result of afforestation (largely with *P. nigra*) dating from the 1950s (Asciuto 1990). Gatteschi & Milanese (1988) document the

Fig. 8.9. *Pinus uncinata* growing in Navarra, Spain, in the eastern Pyrenees (photo: J. Arroyo).

extent of planting with *P. pinaster* and *P. pinea* on the Tuscan coast during the 19th century and describe the extensive spread of the former from these plantings 'at the expense of *P. pinea*'. Only relatively recently has any systematic thought been given to ecology and genetics when selecting species for use in reafforestation (Quézel, Barbéro & Loisel 1990). Gil *et al.* (1990) give a detailed account of the many factors that need to be considered when reafforesting with *P. pinaster* in Spain to alleviate possible negative effects (including selection of the appropriate provenances and impacts on flammability and soils). During the second half of the 20th century, the most commonly planted taxa have been *P. halepensis*, *P. nigra* subsp. *laricio* and *P. pinaster sensu lato*. *Pinus brutia sensu lato* has been fairly widely planted for protection against erosion.

Although many pine taxa from outside the Mediterranean Basin have been grown in arboreta in the region since the 19th century, only *P. radiata* has been fairly widely planted – mainly in the western part of the region (Morocco, Spain, Portugal), but often with disappointing results (Seigue 1985). Lavery & Mead (Chap. 21, this volume) suggest that the eastern Mediterranean may be better suited for this species.

The complex taxonomy of the Mediterranean pines has inspired much work recently, especially in the last two decades, aimed at elucidating the variability of taxa at species- and subspecific levels. Since this work has shed some light on the recent history of Mediterranean pines (and has posed many more questions), we review some important findings briefly here.

Much taxonomical work has focused on *P. brutia* and *P. halepensis* (Iconomou, Valkanas & Bucchi 1964; Panetsos 1975; Riva & Vendramin 1983; Schiller *et al.* 1986; Schiller & Grunwald 1987a, b; Conkle, Schiller & Grunwald 1988; Weinstein 1989a, b; Bariteau 1992; Kaundun 1995). Kaundun (1995) was not able to differentiate between these species using morphological features and used chemical

markers. Where para-anthocyanidins did not facilitate differentiation, flavonals were very efficient. *Pinus halepensis* had a higher myrecetin, larycitin and syringetin content, *P. brutia* a higher quercetin content, and *P. brutia* subsp. *eldarica*, which grows in a very small arid area between Iran and Azerbaijan, had a higher isorhamnetin content. Considering the metabolic balance between dihydroxylation and trihydroxylation, Kaundun (1995) also showed that *P. brutia* and *P. halepensis* have the same phyletic root. Terpene-content studies (Schiller & Grunwald 1987a; Kaundun 1995) also facilitated the clustering of populations of these taxa, e.g. higher pinen content and lower pinen content in *P. brutia* subsp. *eldarica* and *P. halepensis*, respectively. Within the range of *P. halepensis*, a trend of decreasing flavonoid content from east to west suggests an eastern origin of this species which implies that it later invaded North Africa and southern Europe, where flavonoid contents are intermediate. Although flavonoids and other markers demonstrated the homogeneity of *P. halepensis* as a species, studies using isoenzymes identified two groups, one in the western part (France, Spain, Italy, Morocco to Libya) and one in the eastern part of the Mediterranean Basin (Schiller *et al.* 1986). Neither protoanthocyanidin (Kaundun 1995) nor isoenzymes (Conkle *et al.* 1988) have proved useful for clustering populations of *P. brutia*.

Baradat *et al.* (1972, 1975, 1978) and Bernard-Dagan *et al.* (1971, 1982) showed that monoterpene compounds in the wood of *P. pinaster* were different in individuals sampled in different parts of the Mediterranean Basin (Morocco, Spain, Portugal, and Landes, Maures, Esterel and Corsica in France). Especially in Landes (southeastern France), where *P. pinaster* was introduced from Portugal, results suggest a genetic pollution of the local populations as a result of centuries of cultivation of this species by humans. Nevertheless, the variability of this species is low, and the polyphenols which proved useful in differentiating between *P. halepensis* groups, did not serve this purpose in *P. pinaster*. Idrissi-Hassani (1985) also demonstrated, by analysing needle flavonoids, that it was generally difficult to distinguish between individuals from different areas (Morocco, Portugal, and Landes and Corsica in France).

The considerable variability within in *P. nigra* (see Vidaković 1991) has prompted many studies. For example, Bikay-Bikay (1977) and Bonnet-Masimbert (1979), established that the allelic map of the glutamatoxaloacetate transminase loci of 40 provenances delineated four subspecies (*clusiana*; *laricio*; *nigricans*; *pallasiana*). They were, however, unable to distinguish between *P. nigra* from Corsica and Calabria, or even between the subspecies *nigricans* and *pallasiana* from Greece and the former Yugoslavia. More recently, Fineschi (1984), who analysed shikinate deshydrogenase in provenances from Corsica, Calabria, Central Italy, the Alps and former Yugoslavia, found geo-

Fig. 8.10. **The Corsican pine, *Pinus nigra* subsp. *laricio*, dominates about 45 000 ha of forest on Corsica, and is an important resource for the French timber industry. Between 900 and 1500 m elevation this pine co-occurs with *P. pinaster*. At higher elevations (e.g. at La Restonica shown here) it forms pure stands. (photo: F. Ojeda).**

graphic differences, especially between Corsican and Calabrian provenances. Taking into account a larger area from Morocco to Crimea and using five enzymatic systems, Nikolic & Tucic (1983) could describe three main groups. Terpenes have also been used to assess variability within *P. nigra* (Arbez & Millier 1971; Arbez, Bernard-Dagan & Fillon 1974; Zinkel, Maggee & Walter 1985; Arbez 1988). These authors, who used individual trees from arboreta, differentiated four groups. More recently, Bojovic (1995), using samples collected directly from the different regions, confirmed these results and could distinguish between the following subspecies: (i) *nigra* (Slovenia, northern Italy, Austria); (ii) *pallasiana* (Greece; central and eastern part of the species area); (iii) *salzmannii* including *clusiana* (north-western part); and (iv) *laricio* (Corsica (Fig. 8.10), Calabria, Sicily). Bojovic (1995) also differentiated the Corsican populations (high limned content). This terpene analysis largely supported the results of Lauranson (1989) and Lauranson & Lebreton (1991) who used flavonic compounds.

For *P. sylvestris*, Laracine (1984) and Laracine-Pittet & Lebreton (1988) distinguished two chemomorphs by studying polyphenolic compounds: (i) with taxifolin (T+) and (ii) without taxifolin (T−). The populations with T− were found to occur mainly at higher altitudes in western Europe, particularly in the Mediterranean Basin, where Scots pine generally grows at medium altitudes but can reach elevations of 2 000 m. The chemomorph T+ was found to occur mainly at lower altitudes in central Europe, including southern Scandinavia. Some chemical similarities have been reported between *P. sylvestris* and *P. uncinata* (e.g. the absence of taxifolin in *P. uncinata* everywhere and in *P. sylvestris* growing at the highest altitude). Dissimilarities include low and high quercetin contents in *P. uncinata* and *P. sylvestris* respectively. Otherwise, the quercetin variability in *P. uncinata* indicated a high intrapopulation variability which facilitates the colonization of sites with climatic constraints (Lauranson 1989). The similarities between the two species in terms of chemistry, but also aspects of their life cycles (e.g. concomitant pollination and ovule maturation), facilitated hybridization between the two species; hybrids are very variable.

Pinus mugo and *P. sylvestris* also hybridize in France (Alpes Maritimes), Italy (Bergama Province), the former Yugoslavia and in the Carpathians (Barbéro 1972; Szweykowski & Bobowicz 1982). The similarities of the polyphenolic composition of the *P. sylvestris* in the high altitude (chemotype T−), and the two other species (*P. mugo* and *P. uncinata*) has been mentioned by several authors (e.g. Barbéro 1979).

The preceding paragraphs point to an intricate tapestry of pine taxonomy within the Mediterranean Basin – much of it the result of the long history of human occupation and manipulation of genotypes in the course of cultivation and translocation within the region. Much work remains to be done before we will have a clear understanding of the origins of, and relationships between, pine taxa in different parts of the Mediterranean Basin.

8.3 The present ecological status of pines and their biogeography in the Mediterranean Basin

A schematic representation of the role of substrata, bioclimate and altitude in shaping the distribution of pines in the Mediterranean Basin (Fig. 8.11) shows that:

- *P. brutia*, *P. halepensis* and *P. sylvestris* generally grow on marls, limestone and dolomites, although Akman *et al.* (1978) report that *P. brutia* also occurs on volcanic soils ('green rocks').

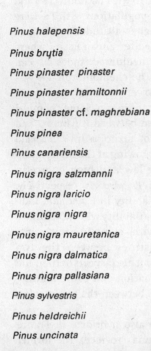
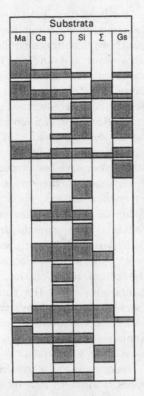
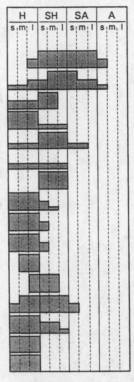
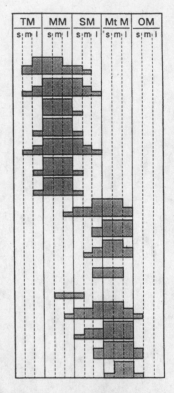

Fig. 8.11. **Ecological features of the Mediterranean pines (the taxa listed in Greuter, Burdet & Long 1984) with respect to substratum (Ma, marls; Ca, limestone; D, dolomites; Si, acid sands; Σ, volcanic (ultrabasic) rocks; Gs, sandstone), bioclimate (H, humid; SH, subhumid; SA, semi-arid; A, arid) and altitude *étages* (TM, thermo-Mediterranean; MM, meso-Med.; SM, supra-Med.; MtM, mountain-Med.; OM, oro-Med.). Marls, limestone, dolomites, volcanic rocks and sandstone are irregularly widespread in the Mediterranean Basin, but all are generally quite abundant. Limestone and dolomite occur mainly in the western, and ultrabasic rocks in the eastern parts of the Mediterranean Basin. The substratum, bioclimate and altitude *étages* (see text for a definition) facilitated the estimation of temperatures and rainfall limits (Quézel 1985). Three levels of importance (very abundant, abundant and rare) are described for each species (represented by the thickness of the bars). The symbols s, m and l denote superior, intermediate (*moyen*) and low (*inférieur*). The figure integrates published work and the authors' unpublished data.**

- Two subspecies of *P. pinaster* (*pinaster* and *hamiltonii*), prefer the acid soils, including those derived from sandstone, and tolerate the dolomites. On the other hand, the races of the *maghrebiana* group grow on various substrata (Fig. 8.12).
- *P. pinea* grows best on sandstone and sandy substrata (Fig. 8.4).
- *P. canariensis* is confined to volcanic rocks (Fig. 8.13).
- The subspecies of *P. nigra* grow on various substrata, depending on the local conditions of the stratigraphy. Among these subspecies, *pallasiana* has a very high ecological plasticity.
- The distribution of the high-mountain pines (*P. heldreichii*, *P. mugo* and *P. uncinata*) is also more dependent on the local stratigraphy than the parent rock itself.

The Mediterranean pines grow mainly in humid and subhumid types (if we follow Emberger's (1939) classification of Mediterranean climates) where the annual rainfall is greater than 600 mm (Le Houérou 1971; Quézel 1985). The species growing in the driest areas are undoubtedly *P. brutia* and *P. halepensis* which occur throughout the semi-arid bioclimate and even in the arid bioclimate (low part). *Pinus halepensis* is especially tolerant of dry conditions, and rarely occurs in humid areas. The only other pines that occur in semi-arid conditions are *P. nigra* subsp. *pallasiana* and *P. pinaster* subsp. *maghrebiana*. All the Mediterranean pines, including *P. halepensis*, grow in the subhumid climate but the high-mountain pines and some subspecies of *P. nigra* grow best in humid bioclimates.

The *étage* concept, although not well established in the English-language ecological literature, is widely used in the Mediterranean Basin. Since many important publications dealing with Mediterranean pines have used the concept, we use it here to illustrate the broad-scale distribution of pines. *Etages* are essentially belts or zones of vegetation described on the basis of topography, altitude and regional climate (Flahault & Schröter 1910; Emberger 1971). We use the concept *sensu* Quézel (1974) and Ozenda

Fig. 8.12. **Remnant forests of *Pinus pinaster* subsp. *maghrebiana* on marls in the semi-arid High Atlas mountains of Morocco (photo: P. Quézel).**

Fig. 8.14. **On Mount Olympus in eastern Greece, *P. nigra* var. *pallasiana* forms large clumps between 600 and 1700 m elevation. At higher elevations it is gradually replaced by *P. heldreichii* which forms open forest, and eventually grows (as shown here) as isolated trees up to the treeline (photo: G.G. Forsyth).**

Fig. 8.13. ***Pinus canariensis* colonizing recent volcanic debris on the island of Tenerife, Canary Islands, Spain. This pine forms distinct belts of sparse woodland on the main islands, at elevations of between 400 and 2200 m, depending on the aspect (photo: J. Arroyo).**

(1975), who distinguish the following *étages* in the Mediterranean Basin: infra-Mediterranean; thermo-Med.; meso-Med.; supra-Med.; mountain-Med.; oro-Med.; and alti-Med. Pines occur in all these zones where they exhibit a wide range of behaviours.

One group, the 'thermophile' pines (*P. brutia* and *P. halepensis*), is found (as the name implies) in the thermo-Mediterranean *étage*, but also often in the meso-Mediterranean, and even in the supra-Mediterranean. The group of *P. canariensis*, *P. pinaster* and *P. pinea* occurs mainly in the meso-Mediterranean *étage* with a very high plasticity of subspecific taxa of *P. pinaster* in the *maghrebiana* group. The ecological status of the subspecies of *P. nigra* is quite complex (Vidaković 1991 provides a thorough review). Most subspecies are found in the mountain-Mediterranean *étage*, but there are some exceptions: subsp. *dalmatica* is mainly found in the meso-Mediterranean, whereas subsp. *salzmannii* (Dunal) Franco occurs from the meso-Mediterranean (French races) up to the mountain-Mediterranean, as does subsp. *pallasiana* which also penetrates the oro-Mediterranean *étage* in the Taurus mountains (Akman *et al.* 1978; Fig. 8.6). *Pinus heldreichii* (Fig. 8.14) and *P. uncinata*, which occur mainly in the mountain-Mediterranean *étage*, also occur in the oro-Mediterranean *étage* (Quézel & Barbéro 1985).

8.4 The role of pines in the vegetation history of the Mediterranean Basin after the last glaciation

8.4.1 Pollen analysis: results and problems

Pine pollen is very well represented in pollen spectra in the Mediterranean area from as far back as the Miocene

(Suc 1980, 1984). Unfortunately, it is not possible to differentiate pine pollen to the species level within sections of the genus; it is thus difficult to use pollen data to elucidate changes in abundance and distribution of Mediterranean pines; see Baruch (1986) for a practical example. The fact that pollen of diploxylon pines can easily be distinguished from that of haploxylon pines is of no use here, since the latter is not represented in the region. Another complicating factor is that *Pinus* pollen is very widely dispersed by wind (Chap. 13, this volume). Studies on modern pollen rain have indicated that a very low percentage of pine pollen (about 5%) is carried for great distances and that increases of *c.* 20–30% of pine pollen in spectra can be taken to indicate local growth. It has, nonetheless, been difficult to infer changes in the relative abundance of different pine species from pollen spectra in the region. For example, pollen data has been of no value for determining the changing relative abundance of *P. brutia* and *P. halepensis* in Israel (Biger & Liphschitz 1991).

Pine pollen has been recovered from sites throughout the Mediterranean Basin dating back as far as Wuermian time until the end of the glaciation period, 10 000 to 15 000 BP (Bazile-Robert, Suc & Vernet 1980). Biger & Liphschitz (1991) give a concise review of the occurrence of pine pollen in cores in the eastern Mediterranean (see also Chap. 5, this volume).

In the northern part of the Mediterranean Basin, where a steppe flora (rich in *Artemisia*, Chenopodiaceae, *Ephedra* and *Helianthemum* spp.) dominated until 10 000 BP, *Pinus* pollen is very abundant. This pollen (probably mainly *P. sylvestris*, but possibly other Mediterranean pines in some places), occurred with *Juniperus* pollen. This suggests that the vegetation was then of a pre-steppe type (i.e. a steppe with some trees which occurs before or after a steppe in the vegetation succession), or even a steppe type (Abi-Saleh *et al.* 1976; Barbéro & Romane 1992). The climate was dry and very cold; the annual rainfall probably ranged between 300 and 400 mm and the mean annual temperature between 8 and 10 °C (Quézel 1989; see also the independent computer-simulated models of past climates by Kutzbach *et al.* 1993). The rapid increase in the temperature after 10 000 BP, with some fluctuations, led to the spread of conifers, particularly pines, and of broadleaved trees which increased in abundance after the end of the Boreal time (7500 BP), even if most of the conifers, especially pine, began to increase slowly before 10 000 BP (Reille 1975; Triat-Laval 1978). As a rule, these trends also apply in the eastern Mediterranean Basin, although some differences are apparent (Horowitz 1971; van Zeist & Bottema 1977; Chap. 5, this volume).

South of the Mediterranean, where pollen data are scarce, the vegetation was probably a pre-steppe type during the Wuermian time and pines were associated with *Juniperus* and sclerophyllous *Quercus* spp. After 8000 BP, when the climate in the Mediterranean Basin was probably very similar to the present climate, the amount of *Pinus* pollen decreased, in places drastically. Later, the amount of pine pollen increased again, as the level of human disturbance increased (Pons & Quézel 1985; Reille 1992). In some cases it is possible, for the species recognizable by their pollen, to identify the species involved, e.g. *P. pinaster* in the High Atlas Mountains of Morocco (Reille 1976) and Corsica (Reille 1975), *P. halepensis* in Provence, France (Triat-Laval 1978), and *P. nigra* in Anatolia, Turkey (van Zeist, Woldrig & Stapert 1975). Finally, since 1500 BP, the amount of pine pollen in the pollen spectra increases rapidly everywhere around the Mediterranean except in some parts of Morocco where the percentages of pine pollens remain generally very low (Reille 1977) as in the present vegetation (Quézel 1980a). For some sites this trend probably reflects a true increase in the extent of pine forests; in others it is more likely due to a decrease in the other woody vegetation types, especially the deciduous, malacophyllous or sclerophyllous, broadleaved vegetation types (Triat-Laval 1978; Blondel & Aronson 1995; see Chap. 5, this volume for further discussion).

8.4.2 Other data

The remains of human activities are the best source of information for the study of vegetation change in the Mediterranean Basin, even though the choice of material used by humans creates a biased data set that must be interpreted with care. Despite some uncertainty because of the patchy data, the remains of prehistoric fireplaces studied by archaeologists generally support the results of pollen studies (Pons & Thinon 1987; M. Thinon, unpublished data). Pine remains in charcoal have yielded much useful information, e.g. the presence at 10 000 BP of *P. nigra* subsp. *salzmannii* in Provence (France), where it no longer occurs (Vernet 1986).

The study of charcoal remains is beginning to yield important results. For example, it was shown that the maximum elevation of *P. mugo* and *P. uncinata* in the high Mediterranean mountains moved from 2700–2800 m to 2200–2400 m over the last millennium (M. Thinon, personal communication). Charcoal data also confirmed that most of the steppes in North Africa were created by the removal of *P. halepensis* or *P. halepensis*–*Juniperus turbinata* woodlands (Mikesell 1960). Many other studies have been done recently, but the results have yet to be published (Pons & Thinon 1987).

Dendroarchaeology (the analysis of wood remains from archaeological sites and historical buildings) showed that *P. brutia* and *P. halepensis*, now both widespread in Israel, were absent from this region until introduced by humans between 6000 and 1000 BP. Furthermore, the absence of timber of these species from monumental buildings points not only to their absence from Israel, but also to a more restricted distribution in nearby areas from which other tree species were imported (both pines are very suitable for building purposes). This suggests substantial changes in vegetation composition in the region during the past few centuries as a result of human activities (Biger & Liphschitz 1991). Schiller & Genizi (1993) arrived at the same conclusion after attempting to identify the origin of *P. brutia* plantations in Israel by analysing needle-resin composition.

8.5 The role of pines in recent vegetation history

The role of pines in the vegetation dynamics of the Mediterranean Basin has been debated for decades. Many authors (notably Braun-Blanquet 1936; Molinier 1937) rejected the notion of stable, pine-dominated vegetation types and therefore of true pine climaxes in the region. In many places around the Mediterranean Sea, the pine-dominated vegetation is undeniably an intermediate step in succession to a climax state dominated by broadleaved trees. However, as knowledge of the Mediterranean vegetation has accrued, it has became clear that pine-dominated 'climax' communities, or stable, pine-dominated vegetation arising from 'pine climax' formations that have been irreversibly modified by long periods of human influence ('paraclimaxes'), are very common.

The pine paraclimaxes are very frequent, particularly in the northern part of the Mediterranean Basin, where the 'pine stage' is an essential phase in the succession leading to the climax. Thus, it is possible to associate a pine with each broadleaved tree of a climax formation, at least for the vegetation of low or medium altitude. For example, *Quercus ilex* and *Q. rotundifolia* occur with *P. halepensis*; *Q. calliprinos* with *P. brutia*; *Q. suber* with *P. pinaster* subsp. *hamiltonii*; and *Q. pubescens* with *P. sylvestris*. This kind of association is not evident throughout the natural range of each pine species, but occurs mainly in humid, subhumid, or even in low-lying, semi-arid bioclimates, i.e. where succession can lead to a climax formation that can be classified as a forest.

Where the climate is very limiting, as in North Africa (low precipitation, particularly in summer), successional stages are incomplete, i.e. broadleaved trees are missing from the last stage. In such cases, the climax is often either a pre-steppe vegetation, or a stage just before the forest where the conifers, mainly pines, are very abundant (Quézel *et al.* 1980b). In the arid and upper and middle semi-arid bioclimates of the Near East and North Africa, the dynamics of *P. brutia* and *P. halepensis* fit this pattern.

This type of process, which leads to a stable vegetation formation dominated by pines, also occurs on some substratum types regardless of bioclimate and altitude. *Pinus brutia* growing on the marls in the Near East (Akman *et al.* 1978) illustrates this process which is also found in Syria and in Lebanon (Barbéro *et al.* 1976). The same process exists on the ultrabasic rocks, again for *P. brutia* in the same area, and for *P. pinaster* subsp. *maghrebiana* in the Malaga area of southern Spain (Asensi & Diez-Garretas 1987). On dolomites in the Languedoc of southern France, the *P. nigra* subsp. *salzmannii* forests are also good examples of this process (Barbéro & Quézel 1988).

The mountain pines (*P. nigra* and *P. sylvestris*) generally form climax vegetation types in the mountain-Mediterranean *étage*; for example, *P. nigra* subsp. *pallasiana* (Quézel 1980a) in southern Anatolia and *P. sylvestris* in northern Anatolia (Quézel, Barbéro & Akman 1980a). On the other hand, these species only act as paraclimax of broadleaved tree climax (*Fagus sylvatica* and deciduous *Quercus* spp.) in the supra-Mediterranean *étage*. The high-mountain pines belong to climax vegetation in the oro-Mediterranean *étage* and sometimes in the sub-alpine *étage* of the European mountains.

8.6 Vegetation change scenarios

Although it is relatively easy to assess the general features of the pine vegetation, and the role of each species in the Mediterranean Basin forest (Barbéro, Quézel & Loisel 1990b), it is much more complicated to determine the processes that drive succession in each case. Nevertheless, we describe some examples from southern France which highlight the importance of, and interactions between, factors such as fire, land abandonment and pest dynamics.

Land abandonment is probably the main feature affecting vegetation dynamics in southern France, where *P. halepensis* plays an important role as a pioneer (Fig. 8.15). Most rural areas in this region have been transformed by the exodus of humans to the cities which started at the end of the nineteenth century and accelerated after World War I and especially after World War II. The area covered by *P. halepensis* increased threefold in some regions following land abandonment (Achérar, Lepart & Debussche 1984; Barbéro *et al.* 1990b; Lepart & Debussche 1991). These authors attribute this expansion to life-history traits of *P. halepensis*; it is a short-lived tree which bears cones when it is about 6 years old and produces many seeds that can quickly invade exposed ground. It is, however, a transient species which, being unable to reproduce in its own understorey, is replaced by *Quercus ilex* or *Q. pubescens* in the absence of fire.

Fig. 8.15. ***Pinus halepensis* invading abandoned fields near Montpellier, southern France (photo: D.M. Richardson).**

The same kind of process probably occurred in the Cévennes mountains with *P. pinaster* a century ago, but in this case the pine was introduced by humans. This species rapidly invaded abandoned fields and became an important feature of the landscape. Whereas the expansion of *P. halepensis* described above was caused by natural seed dispersal, the spread of *P. pinaster* in the Cévennes was enhanced by landowners who disseminated the seeds widely (Marek 1994; Marek & Romane 1995). Extensive planting of *P. pinaster* in France and Portugal for dune sand reclamation and subsequent spread from these plantings has also contributed to the increase in the distribution of this species. The rate of expansion of *P. pinaster* appears to have declined in the last few decades, probably mainly because most areas affected by human disturbance have already been colonized, in many cases by shrubs such as *Calluna vulgaris* and *Cytisus scoparius* (Loisel 1976a; Marek 1994; Marek & Romane 1995).

A very different scenario occurred in the *P. pinaster* forests of the Maures hills where pest dynamics plays an important role. In this part of southern France, it appears that *P. pinaster* invaded and replaced the original vegetation dominated by *Quercus suber* several centuries ago (Loisel 1976a). For some decades, these pines have been attacked by insects (notably the Maritime pine bast scale *Matsucoccus feytaudi*; Homoptera, Margarodidae) which killed all the trees older than 10–15 years. Although *P. pinaster* continues to disperse seeds which germinate, the present vegetation is often dominated by other species with only young pines. Studies have failed to detect any differences in resistance between pine genotypes in specific environments (Mazurek *et al.* 1982; Mazurek & Romane 1986). However, a study was done of the susceptibility of seven *P. pinaster* provenances to *M. feytaudi* in a plantation established in 1965 at Maures. This showed Cuenca (Spain) and Tamjoute (Morocco) provenances to be symptom-free, while other provenances were affected to different degrees, with the Genova (Italy) and Maures provenances being very heavily affected (Schvester & Ughetto 1986).

Fire is a major factor in the dynamics of Mediterranean pine forests, notably in *P. halepensis* forests (Barbéro, Loisel & Quézel 1988; Chap. 11, this volume). Humans started using fire about 400 000 years ago in the region (Prodon, Fons & Athias-Binche 1987), but human-modified fire regimes probably only started having significant impacts on the region's vegetation about 10 000 years ago (Thirgood 1981; Trabaud 1984). Humans intentionally used fire intensively for agriculture, to improve grazing for livestock and to protect themselves and their herds from predators. Fire frequency has increased dramatically in the 20th century. During the 1980s, about 0.56×10^6 ha of woodland and shrubland burned every year in the Mediterranean region (ECE/FAO 1990). About a third of all landscapes in mediterranean-climate France have burned at least once in the past 25 years (Ramade 1990). The average areas burned annually in Spain, France, Italy and Algeria between 1980 and 1985 were 2460, 406, 1638 and 489 km^2 respectively (Ramade 1990). In some areas, the average fire-return interval may be as little as 5 years (Trabaud, Christensen & Gill 1993). Such short fire cycles have massive impacts on the vegetation, in many cases leading to depauperate communities dominated by species that resprout or germinate quickly after fire. A feature of ecosystems affected by frequent fires is the lack of true successional processes whereby different species assemblages replace one another over time (see Blondel & Aronson 1995 for a recent review). Vegetation recovery following fire in the Mediterranean region involves endogenous processes of local plant species which progressively return the burned system to a state very similar to that which prevails before fire, with very few pioneer species colonizing such systems (Trabaud 1987).

Pinus halepensis readily replaces itself in areas that are frequently burnt. It can also '. . . enlarge its area after fire on outskirts around sites in which it is (or it was) present owing to its wind-disseminated seeds' (Trabaud 1990). This invasion by pine after fire is a general trend in the Mediterranean Basin (*P. halepensis* and *P. pinaster* in the western part, *Pinus brutia* in the eastern part), and contributes to the widespread occurrence of pine forests around the Mediterranean (see also Chap. 11, this volume). This dominance corresponds to a paraclimax in humid and subhumid climate and to a climax in semi-arid, and sometimes arid climate. This trend, which seems peculiar to the Mediterranean Basin, is quite different from the situation in California where pines, of which there are many more species, remain confined to a smaller part of the landscape in the corresponding climates (Quézel 1979, 1980a; Quézel & Barbéro 1989). The pattern observed in the

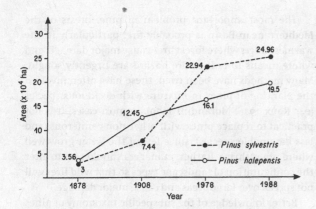

Fig. 8.16. Changes in surface area (×10⁴ha) of *Pinus halepensis* and *P. sylvestris* since 1878 in the Provence–Alpes–Côte d'Azur region of southern France. Modified from Barbéro *et al.* (1990b).

Mediterranean Basin is, however, also evident at the south-western tip of Africa where *P. halepensis* readily invades the fire-prone fynbos (Richardson 1988; Chap. 22, this volume).

8.7 The present: models of dynamics

Although, as mentioned previously, it is difficult to delineate the areas occupied by the Mediterranean pines, it is possible to describe some trends and their causes. The trends differ markedly between the northern and the southern parts of the Mediterranean Basin. For example, around the western Mediterranean Sea, the area under pines has decreased drastically in the Maghreb area (particularly *P. halepensis*), whereas in Europe the area of pine forests has generally increased. More precisely, in Algeria, Morocco and Tunisia, *P. halepensis* currently occupies only half of its potential range (Barbéro *et al.* 1990a; Quézel 1980a, b). The reason for this is the large increase in the extent of human activities, including clearing of the forests for cultivation, cutting for various purposes (to feed livestock, build houses, obtain charcoal), overgrazing and frequent fires.

Trends in southwestern Europe are totally different, and the area of pine forests is increasing rapidly. For example, the extent of two of the most common conifers in southern France (*P. halepensis* and *P. sylvestris*) has increased rapidly in recent decades (Achérar *et al.* 1984; Barbéro *et al.* 1990a, b; Lepart & Debussche 1991; Fig. 8.16). The same pattern is evident, albeit less obvious, for *P. pinaster* and *P. pinea*, but not for all other pine taxa in the region. *Pinus nigra* subsp. *salzmannii* has decreased in many places because of the expansion of *P. pinaster* after fire (Barbéro & Quézel 1988).

The overall increase in the area under pines is essentially attributable to aspects of the biology of pines and changes in land use. The reafforestation policy in some countries (e.g. Algeria and Spain) explains this extension but no more than 5 or 10%. The biology and the ecology of (some) pine species allows them to become 'weedy' within their natural range (Barbéro & Quézel 1989; Barbéro *et al.* 1990a; Richardson & Bond 1991; Chap. 22, this volume). The pines, which often behave like invaders, produce large numbers of seeds at an early age (less than 10 years in some cases). These seeds can germinate in various vegetation types like abandoned fields, grasslands, matorral and burned areas (Loisel 1976b; Trabaud 1980; Abbas, Barbéro & Loisel 1984; Barbéro *et al.* 1987).

Marked changes in land use throughout the Mediterranean Basin are also very important for understanding the increase in the area under pines. The area of abandoned fields, grasslands and matorral has generally increased over the past few decades, particularly in the European Union where a policy of setting aside land has prevailed (Barbéro *et al.* 1988; Barbéro & Quézel 1989; Bourdeau 1992). These trends are also increasing the fires and the burned areas that leads to vegetation types where pine germination easily occurs (Trabaud 1980, 1990; Naveh 1993).

8.8 Conclusions

The pines that are now prominent components of the flora invaded the Mediterranean Basin about 3.5 million years ago. The ecological plasticity of these pine species, which are found in all the altitude *étages* of the Mediterranean Basin, is quite remarkable. They often occur with the other conifers under very harsh environmental conditions, e.g. on the border of the Sahara Desert in North Africa (*P. halepensis* with *Juniperus turbinata*), or at high altitudes (*P. nigra* with *Abies* and *Cedrus* spp.). This high plasticity probably explains the very dynamic behaviour of most of these pines and the very important roles that they play in vegetation dynamics around the Mediterranean Sea wherever human-induced disturbances are not too severe.

The role of human-induced changes to disturbance regimes in structuring pine forests is quite different in northern (mainly Europe) and southern parts of the Basin (North Africa). In western Europe (France and Spain), pines are often invaders occupying the abandoned fields induced by the 'set-aside' policy of the European Union. On the contrary, in many areas of the southern part of the Basin, where the shortage of fuel and timber is most severe (e.g. in Algeria and Morocco), the area under pines is often decreasing drastically.

These two very different scenarios create many current problems in both cases. There is the problem of genetic pollution when pines are moved within the region. Some provenances of widespread species are not adapted to conditions in areas away from their natural ranges. Such maldaptation is manifested in many ways. For example, as described previously for *P. pinaster* with regard to *Matsucoccus feytaudi*, some provenances are more susceptible to devastating outbreaks of insect pests than others. Different provenances of *P. halepensis* also display different levels of resistance. Indeed, most of the devastating outbreaks of insect pests in the Mediterranean Basin have occurred in planted, rather than natural, pine stands, since the former often occur in suboptimum climatic and edaphic conditions (e.g. Questienne 1979 for Morocco). Similarly, provences of *P. halepensis* display different levels of resistance to frost and low water potentials (Schiller & Brunori 1992). Thus, the plasticity must be considered in all reafforestation plans to avoid such problems.

The most important problem in pine forests of the Mediterranean Basin is probably fire, particularly in the warmest *étages* where forest fires cause major damage, and where means to reduce fire hazard are urgently sought. Many methods have been tried; these have often involved the planting of pines in mixture with deciduous species (e.g. Roux 1946; Morandini 1979). In most cases, it is not practical to replace pines with other non-coniferous (and less flammable species), since few of the latter grow well where pines thrive. Rather, managers should aim to alter the configuration of landscape types so that wild fires will not spread over large areas and cause major damage.

Better knowledge of the subspecific taxonomy of pines is also required to facilitate improved taxon–site matching, and better use of the diverse genetic resource of Mediterranean pines for production (Belghazi & Romane 1993, 1994).

References

Abbas, H., Barbéro, M. & Loisel, R. (1984). Réflexions sur le dynamisme actuel de la régénération du pin d'Alep en région méditerranéenne dans les pinèdes incendiées en Provence calcaire (de 1973 à 1979). *Ecologia Mediterranea*, **10**, 85–104.

Abi-Saleh, B., Barbéro, M., Nahal, I. & Quézel, P. (1976). Les séries forestières de végétation au Liban. Essai d'interprétation schématique. *Bulletin de la Société Botanique de France, Lettres Botaniques*, **123**, 541–60.

Achérar, M., Lepart, J. & Debussche, M. (1984). La colonisation des friches par le pin d'Alep (*Pinus halepensis* Miller) en Languedoc méditerranéen. *Acta Oecologica–Oecologia Plantarum*, **5**, 179–89.

Akman, Y., Barbéro, M. & Quézel, P. (1978). Contribution à l'étude de la végétation forestière d'Anatolie méditerranéenne. *Phytocoenologia (Stuttgart)*, **5**, 1–79.

Arbez, M., (1988). Méthodes biochimiques de caractérisation variable des arbres forestiers. *Revue Forestière Française*, **40**, 71–6.

Arbez, M., Bernard-Dagan, C. & Fillon, C. (1974). Variabilité intraspécifique des monoterpènes de *Pinus nigra* Arn. Bilan des premiers résultats. *Annales des Sciences Forestières*, **31**, 57–70.

Arbez, M. & Millier, C. (1971). Contribution à l'étude de la variabilité géographique de *Pinus nigra* Arn.

Annales des Sciences Forestières, **28**, 23–49.

Asciuto, A. (1990). Il pino laricio di Corsica, di Calabria e dell'Etna: aspetti ecologici e produttivi. *Cellulosa e Carta*, **41** (1), 19–25.

Asensi, A. & Diez-Garretas, B. (1987). Andalusia occidental. In *La Vegetacion de España*. Publicación del Universita de Alcala, 197–229.

Baradat, P., Bernard-Dagan, C., Fillon, C., Marpeau, A. & Pauly, G. (1972). Les terpènes du pin maritime: aspects biologiques et génétiques. Hérédité de la teneur en monoterpènes. *Annales des Sciences Forestières*, **29**, 307–34.

Baradat, P., Bernard-Dagan, C. & Marpeau, A. (1978). Variations des terpènes à l'intérieur et entre populations de pin maritime. *Proceedings of Conference of Biochemical Genetics of Forest Trees (Umea)*, 151–69.

Baradat, P., Bernard-Dagan, C., Pauly, G., Zimmerman, W. & Fillon, C. (1975). Les terpènes du pin maritime : aspects biologiques et génétiques. Hérédité de la teneur en myrcènes. *Annales des Sciences Forestières*, **32**, 29–54.

Baradat P. & Marpeau-Bezard, A. (1988). *Le pin maritime. Biologie et génétique des terpènes pour la connaissance et l'amélioration de l'espèce*. PhD thesis, University of Bordeaux I.

Barbéro, M. (1972). *Etudes phytosociologiques et écologiques comparées des végétations orophiles alpines, subalpines et*

mésogéenne des Alpes Maritimes. PhD thesis, University of Provence, Marseille.

Barbéro, M. (1979). Les remontées méditerranéennes sur le versant italien des Alpes. *Ecologia Mediterranea*, **4**, 109–32.

Barbéro, M., Bonin, G., Loisel, R., Miglioretti, F. & Quézel, P. (1987). Incidence of exogenous factors on the regeneration of *Pinus halepensis* after fire. *Ecologia Mediterranea*, **13**, 51–6.

Barbéro, M., Bonin, G., Loisel, R. & Quézel, P. (1990a). Changes and disturbances of forest ecosystems caused by human activities in the western part of the Mediterranean Basin. *Vegetatio*, **87**, 151–73.

Barbéro, M., Chalabi, N., Nahal, I. & Quézel, P. (1976). Les formations à conifères méditerranéens en Syrie littorale. *Ecologia Mediterranea*, **2**, 87–100.

Barbéro, M., Loisel, R. & Quézel, P. (1988). Perturbations et incendies en région méditerranéenne française. *Instituto Estudios Pyrenaicos (Jaca y Huesca)*, 409–19.

Barbéro, M. & Quézel, P. (1988). Signification phytoécologique et phytosociologique des peuplements naturels de Pin de Salzmann en France. *Ecologia Mediterranea*, **14**, 41–63.

Barbéro, M. & Quézel, P. (1989). Structures, architectures forestières à sclérophylles et prévention des incendies. *Bulletin d'Ecologie*, **20**, 25–35.

Barbéro, M., Quézel, P. & Loisel, R. (1990b). Les apports de la phytoécologie dans l'interprétation des changements et perturbations induits par l'homme sur les écosystèmes forestiers méditerranéens. *Forêt Méditerranéenne*, **12**, 194–215.

Barbéro, M. & Romane, F. (1992). Evolution actuelle de la forêt méditerranéenne. In *Les Recherches en France sur les écosystèmes Forestiers*, ed. G. Landmann, pp. 19–20. Paris: Ministère de l'Agriculture et de la Forêt.

Bariteau, M. (1992). Variabilité géographique et adaptation aux contraintes du milieu méditerranéen des pins de la section *Halepenses*: résultats (provisoires) d'un essai en plantations comparatives en France. *Annales des Sciences Forestières*, **49**, 261–76.

Baruch, U. (1986). The late Holocene vegetational history of Lake Kinneret (Sea of Galilee). *Israel Paleorient*, **12**, 37–48.

Bazile-Robert, E., Suc, J.P. & Vernet, J.L. (1980). Les flores méditerranéennes et l'histoire climatique depuis le Miocène. *Naturalia Monspeliensia*, n° h.s., 33–40.

Belghazi, B. & Romane, F. (1993). Relations entre caractéristiques morphométriques des arbres et milieu. Cas du pin maritime (*Pinus pinaster* Sol. var. *maghrebiana*) dans le Maroc septentrional. *Cahiers Agriculture*, **2**, 338–42.

Belghazi, B. & Romane, F. (1994). Productivité du pin maritime (*Pinus pinaster* Sol. var. *magh.*) en peuplements artificiels dans le Nord du Maroc. *Forêt Méditerranéenne*, **15**, 391–6.

Bernard-Dagan, C., Fillon, C., Pauly, G. & Baradat, P. (1971). Les terpènes du pin maritime: aspects biologiques et génétiques. I – Variabilité de la composition monoterpénique dans un individu, entre individus et entre provenances. *Annales des Sciences Forestières*, **28**, 223–58.

Bernard-Dagan, C., Pauly, G., Marpeau, A., Gleizes, M., Garde, J.P. & Baradat, P. (1982). Control and compartmentation of terpen biosynthesis in leaves of *Pinus pinaster*. *Physiologie végétale*, **20**, 775–95.

Biger, G. & Liphschitz, N. (1991). The recent distribution of *Pinus brutia*: a reassessment based on dendroarchaeological and dendrohistorical evidence from Israel. *The Holocene*, **1**, 157–61.

Bikay-Bikay, V. (1977). *Mise en évidence des isoenzymes par la technique de microélectrophorèse: application à l'étude de la taxonomie expérimentale chez* Pinus nigra *Arn*. PhD thesis, Nancy I University, France.

Blondel, J. & Aronson, J. (1995). Biodiversity and ecosystem function in the Mediterranean Basin: human and non-human determinants. In *Mediterranean-Type Ecosystems. The Function of Biodiversity*, ed. G.W. Davis & D.M. Richardson, pp. 43–119. Berlin: Springer-Verlag.

Bojovic, S. (1995). *Biodiversité du pin noir en région Méditerranéenne*. PhD thesis, Aix-Marseille III University, Marseille.

Bonnet-Masimbert, A. (1979). Populations et isoenzymes chez *Pinus nigra*. *Bulletin de la Société Botanique de France, Actualités Botaniques*, **126**, 67–78.

Bourdeau, P. (1992). General conclusions and perpectives for future national and international research programmes on forest ecosystems. In *Responses of Forest Ecosystems to Environmental Changes*, ed. A. Teller, P. Mathy & J.N.R. Jeffers, pp. 560–3. London: Elsevier.

Braun-Blanquet, J. (1936). II. La lande à romarin et bruyère (Rosmarino-Ericion) en Languedoc. *Communication SIGMA (Montpellier)*, **48**, 8–23.

Castroviejo, S., Laínz, M., Lopez González, G., Montserrat, P., Muñoz Garmendia, F., Paiva, J. & Villar, L. (1986). *Flora Iberica, Tome 1*. Madrid: Real Jardin Botanico.

Christodoulopoulos, A. (1947). Anadasoseis. *Dasos*, **4**, 3–10.

Conkle, M.T., Schiller, G. & Grunwald, C. (1988). Electrophoretic analysis of diversity and phylogeny of *Pinus brutia* and closely related taxa. *Systematic Botany*, **13**, 411–24.

Daget, P. (1977a). Le bioclimat méditerranéen : caractères généraux, modes de caractérisation. *Vegetatio*, **34**, 1–20.

Daget, P. (1977b). Le bioclimat méditerranéen: analyse des formes climatiques par le système d'Emberger. *Vegetatio*, **34**, 87–103.

Debazac, E. (1971). Contribution à la connaissance de l'écologie et de la répartition de *Pinus nigra* dans le sud est de l'Europe. *Annales des Sciences Forestières*, **28**, 91–139.

Del Villar, E.H. (1933). Sobre el habitat calizo de *Pinus pinaster*. *Boletin de la Sociedad Española de Historia Natural*, **33**.

ECE/FAO (1990). *Forest Fire Statistics*. Geneva: ECE/TIM/51.

Emberger, L. (1939). Aperçu général sur la végétation du Maroc. Commentaire de la carte phytogéographique du Maroc au 1/1 500 000ᵉ. *Veröffentlichungen des Geobotanisches Rübel Institut (Zürich)*, **14**, 40–157.

Emberger, L. (1971). *Travaux de Botanique et d'écologie*. Paris: Masson.

Fineschi, S. (1984). Determination of the origin of an isolated group of trees of *Pinus nigra* through enzyme markers. *Silvae Genetica*, **33**, 169–72.

Flahault, C. & Schröter, C. (1910). *Nomenclature Phytogéographique. Rapports et Propositions*. Bruxelles: IIIᵉ Congrès International de Botanique.

Gatteschi, P. & Milanese, B. (1988). Condizioni della vegetazione del litorale toscano a sud di Livorno. *Monti e Boschi*, **39**(2), 5–10.

Gil, L., Gordo, J., Alia, R., Catalan, G. & Pardos, J.A. (1990). *Pinus pinaster* Aiton en el paisaje vegetal de la Peninsula Iberica. *Ecologia Madrid Fuera de Serie*, **1**, 469–96.

Greuter, W., Burdet, H.M. & Long, G. (eds.) (1984). *Med-Checklist. A Critical Inventory of the Circum-Mediterranean Countries*. Vol. 1. *Pteridophyta* (ed. 2), *Gymnospermae, Dycotyledones*. Geneva: Edition des Conservatoire et Jardin botaniques.

Horowitz, A. (1971). Climatic and vegetational developments in Northeastern Israel during Upper Pleistocene–Holocene times. *Pollen et Spores*, **13**, 255–78.

Iconomou, N., Valkanas, C. & Bucchi, J. (1964). Composition of gum terpentines of *Pinus halepensis* and *P. brutia* grown in Greece. *Journal of Chromatography*, **16**, 29–33.

Idrissi-Hassani, M. (1985). *Etude de la variabilité flavonique chez deux conifères Méditerranéens: le pin maritime et le genévrier thurifère*. PhD thesis, Lyon I University, France.

Kaundun, S. (1995). *Contribution biochimique à la connaissance systématique du pin d'Alep (*Pinus halepensis *Mill.)*. PhD thesis, Lyon I University, France.

Klaus, W. (1989). Mediterranean pines and their history. *Plant Systematics and Evolution*, **162**, 133–63.

Kolodny, E.Y. (1974). *La Population des Iles de la Grèce, Essai de Géographie Insulaire en*

Méditerranée Orientale. Aix-en-Provence: Edisud.

Kutzbach, J.E., Guetter, P.J., Behling, P.J. & Sehling, R. (1993). Simulated climate changes: results of the COHMAP climate-model experiments. In *Global Climates Since the Last Glacial Maximum*, ed. H.E.J. Wright, J.E. Kutzbach, T. Webb III, W.F. Ruddimann, F.A. Street-Perrott & P.J. Bartleïn pp. 24–93. Minnesota: University of Minneapolis Press.

Laracine, C. (1984). *Etude de la variabilité flavonique infraspécifique chez deux conifères: le pin sylvestre et le genévrier commun*. PhD thesis, Lyon I University, France.

Laracine-Pittet, C. & Lebreton, P. (1988). Flavonoid variability within *Pinus sylvestris*. *Phytochemistry*, **27**, 2663–6.

Lauranson, J. (1989). *Exploration de la biodiversité biochimique chez les conifères. Contribution à l'étude de l'hybridation Pinus uncinata Ram. Pinus silvestris L. et à la connaissance du complexe spécifique Pinus nigra Arn*. PhD thesis, Claude Bernard University, Lyon, France.

Lauranson, J. & Lebreton, P. (1991). Flavonoid variability within and between natural populations of *Pinus uncinata*. *Biochemical Systematics and Ecology*, **19**, 659–64.

Le Houérou, H. (1971). *L'écologie Végétale dans la Région Méditerranéenne*. Montpellier: Centre International des Hautes Etudes Agronomiques Méditerranéennes, Institut Agronomique Méditerranéen.

Le Houérou, H. (1980). L'impact de l'homme et de ses animaux sur la forêt méditerranéenne. *Forêt Méditerranéenne*, **2**, 31–4.

Lepart, J. & Debussche, M. (1991). Invasion processes as related to succession and disturbance. In *Biogeography of Mediterranean Invasions*, ed. R.H. Groves & F. di Castri, pp. 159–77, Cambridge: Cambridge University Press.

Loisel, R. (1976a). *La végétation de l'étage Méditerranéen dans le sud-est Français*. PhD thesis, Aix-Marseille III University, France.

Loisel R. (1976b). Place et rôle du genre *Pinus* dans la végétation du sud-est méditerranéen français. *Ecologia Mediterranea*, **2**, 131–52.

Marek, U. (1994). *La régénération du pin maritime (Pinus pinaster Ait.) dans les châtaigneraies abandonnées des Cévennes,* France. Diplomarbeit, Universität des Saarlandes, Saarbrücken.

Marek, U. & Romane, F. (1995). Cluster pine regeneration processes and their effects on landscape changes in southern France. In *The Importance of Spatial and Temporal Perspectives for Understanding Vegetation Pattern and Process*. Proceedings of the 38th Symposium of the International Association for Vegetation Science, p. 51. Houston, Texas: Rice University.

Mazurek, H., Godron, M., Romane, F. & Schvester, D. (1982). Structure and dynamic variations of Cluster Pine (*Pinus pinaster* Ait. subsp. *pinaster*) vegetation along a climatic gradient of the Maures Hills (Acid Provence, Var, France). In *Struktur und Dynamik von Wäldern*, ed. H. Dierschke, Berichte der Internationalen Symposien der Internationalen Vereinigung für Vegetationskunde, pp. 149–73. Vaduz: J. Cramer.

Mazurek, H. & Romane, F. (1986). Dynamics of young *Pinus pinaster* vegetation in a Mediterranean area: diversity and niche-strategy. *Vegetatio*, **66**, 27–40.

Mikesell, M.W. (1960). Deforestation in northern Morocco. Burning, cutting, and browsing are changing a naturally wooded area into a land of scrub. *Nature*, **132**, 441–8.

Mirov, N.T. (1967). *The Genus Pinus*. New York: Ronald Press.

Molinier, R. (1937). Carte des associations végétales des massifs de Carpiagne et Marseilleveyre. *Communication SIGMA, Montpellier*, n° 58.

Morandini, R. (1979). Sylviculture et incendies. *Revue Forestière Française*, **1**, 5–9.

Nahal, I. (1962). Le pin d'Alep. *Annales des Sciences des Eaux et Forêts*, **19**, 472–686.

Nahal, I. (1981). The Mediterranean climate from biological viewpoint. In *Mediterranean-Type Shrublands*, ed. F. Di Castri, D.W. Goodall & H.A. Mooney, pp. 63–86. *Ecosystems of the World*, Vol. 11. Amsterdam: Elsevier.

Naveh, Z. (1993). Trends in Mediterranean landscapes. In *Agricultural Landscapes in Europe*. Summaries of Oral Presentation and Posters of International Association for Landscape Ecology Congress, Rennes (France), 1993, p. 17.

Nikolic, D. & Tucic, N. (1983). Isoenzyme variation within and among populations of European black pine (*Pinus nigra* Arn.). *Silvae Genetica*, **32**, 82–9.

Ozenda, P. (1975). Sur les étages de végétation dans les montagnes du bassin méditerranéen. *Documents Cartographiques Ecologiques (Grenoble)*, **16**, 1–32.

Paci, M. & Romoli, G. (1992). Studio sulla diffusione spontanea del pino nero sui pascoli del Passo dello Spino. *Annali Accademia Italiana di Scienze Forestali*, **41**, 191–226.

Panetsos, C.P. (1975). Natural hybridation between *Pinus halepensis* and *Pinus brutia* in Greece. *Silvae Genetica*, **24**, 129–200.

Pons, A. & Quézel, P. (1985). The history of the flora and vegetation and past and present human disturbance in the Mediterranean area. In *Plant Conservation in the Mediterranean Area*, ed. C. Gómez-Campo, pp. 25–43. Dordrecht: Junk.

Pons, A. & Thinon, M. (1987). The role of fire from paleoecological data. *Ecologia Mediterranea*, **13**, 3–11.

Prodon, R., Fons, R. & Athias-Binche, F. (1987). The impact of fire on animal communities in mediterranean area. In *The Role of Fire in Ecological Systems*, ed. L. Trabaud, pp. 121–57. The Hague: SPB Academic Publishers.

Questienne, P. (1979). Notes sur quelques insectes nuisibles aux pins au Maroc. *Annales de Gembloux*, **85**, 113–30.

Quézel, P. (1974). *Les forêts du pourtour méditerranéen*. Notes techniques M.A.B. 2, pp. 9–34. Paris: UNESCO.

Quézel, P. (1979). Matorrals méditerranéens et Chaparrals californiens. Quelques aspects comparatifs de leur dynamique, de leurs structures et de leur signification écologique. *Annales des Sciences Forestières*, **36**, 1–12.

Quézel, P. (1980a). Biogéographie et écologie des conifères sur le pourtour méditerranéen. In *Actualités d'Ecologie Forestière. Sol, Flore, Faune*, ed. P. Pesson, pp. 205–55. Paris: Gauthier-Villars.

Quézel, P. (1980b). L'homme et la dégradation récente des forêts au Maghreb et au Proche–Orient. *Naturalia Monspeliensia*, n° h.s., 147–52.

Quézel, P. (1985). Definition of the Mediterranean region and the origin of its flora. In *Plant Conservation in the Mediterranean Area*, ed. C. Gómez-Campo, pp. 9–24. Dordrecht: Junk.

Quézel, P. (1989). Mise en place des structures de végétation circum-méditerranéennes naturelles. Landscape Ecology Proceedings of the Man & Biosphere. XVIth International Grasslands Congress, 16–32.

Quézel, P. & Barbéro, M. (1985). *Carte de la Végétation Potentielle de la Région Méditerranéenne. Feuille n° 1: Méditerrannée Orientale.* Paris: CNRS éditeur.

Quézel, P. & Barbéro, M. (1989). Zonation altitudinale des structures forestières de végétation en Californie méditerranéenne: leur interprétation en fonction des méthodes utilisées sur le pourtour méditerranéen. *Annales des Sciences Forestières*, **46**, 233–50.

Quézel, P., Barbéro, M. & Akman, Y. (1980a). Contribution à l'étude de la végétation forestière d'Anatolie septentrionale. *Phytocoenologia (Stuttgart)*, **8**, 365–519.

Quézel, P., Barbéro, M., Bonin, G. & Loisel, R. (1980b). Essai de corrélations phytosociologiques et bioclimatiques entre quelques structures actuelles et passées de la végétation méditerranéenne. *Naturalia Monspeliensia*, n° h.s., 89–100.

Quézel, P., Barbéro, M. & Loisel, R. (1990). Les reboisements en région méditerranéenne. Incidences biologiques et économiques. *Forêt Méditerranéenne*, **12**, 103–14.

Ramade, F. (1990). *Conservation des Ecosystèmes Méditerranéens.* Plan Bleu 3. Paris: Economica.

Reille, M. (1975). *Contribution pollenanalytique à l'histoire tardiglaciaire et Holocène de la végétation de la Montagne Corse.* PhD thesis, Aix-Marseille III University, France.

Reille, M. (1976). Analyse pollinique de sédiments postglaciaires dans le Moyen-Atlas et le Haut-Atlas marocains: premiers résultats. *Ecologia Mediterranea*, **2**, 153–70.

Reille, M. (1977). Contribution pollenanalytique à l'histoire holocène de la végétation des montagnes du Rif (Maroc septentrional). Recherches françaises sur le Quaternaire. *INQUA 1977. Supplément au Bulletin de l'Association Française pour l'Etude du Quaternaire*, 1977-1, **50**, 53–76.

Reille, M. (1989). L'origine du pin à crochets dans le Massif Central français. *Bulletin de la Société Botanique de France, Lettres Botaniques*, **136**, 61–70.

Reille, M. (1992). New pollen analytical researches in Corsica: the problem of *Quercus ilex* L. and *Erica arborea* L.; the origin of *Pinus halepensis* Miller forests. *New Phytologist*, **122**, 359–78.

Richardson, D.M. (1988). Age structure and regeneration after fire in a self-sown *Pinus halepensis* forest on the Cape Peninsula, South Africa. *South African Journal of Botany*, **54**, 140–4.

Richardson, D.M. & Bond, W.J. (1991). Determinants of plant distribution: evidence from pine invasion. *American Naturalist*, **137**, 639–68.

Riva, L. & Vendramin, G. (1983). Prime observazioni sull'ibrido artificiale *Pinus brutia* × *Pinus halepensis*. *Italiano Forestal Montane*, **38**, 234–48.

Roux, G. (1946). Le reboisement dans la région landaise. *Revue des Eaux et Forêts*, **84**, 473–91.

Schiller, G. & Brunori, A. (1992). Aleppo pine (*Pinus halepensis* Mill.) in Umbria (Italy) and its relation to native Israeli populations. *Israel Journal of Botany*, **41**, 123–7.

Schiller, G., Conkle, M.T. & Grunwald, C. (1986). Local differentiation among Mediterranean populations of Aleppo pine in their isoenzymes. *Silvae Genetica*, **35**, 11–19.

Schiller, G. & Genizi, A. (1993). An attempt to identify the origin of *Pinus brutia* Ten. plantations in Israel by needle resin composition. *Silvae Genetica*, **42**, 63–8.

Schiller, G. & Grunwald, C. (1987a). Resin monoterpene in range-wide provenance trials of *Pinus halepensis* Mill. in Israel. *Silvae Genetica*, **36**, 109–14.

Schiller, G. & Grunwald, C. (1987b). Cortex resin monoterpene composition in *Pinus brutia* provenances grown in Israel. *Biochemical Systematics and Ecology*, **15**, 389–94.

Schvester, D. & Ughetto, F. (1986). Différences de sensibilité a *Matsucoccus feytaudi* DUC (Homoptera: Margarodidae) selon les provenances de pin maritime (*Pinus pinaster* Ait). *Annales des Sciences Forestières*, **43**(4), 459–74.

Seigue, A. (1985). *La Forêt Méditerranéenne et ses Problèmes.* Paris: Editions Maisoneuve et Larose.

Suc, J.P. (1980). *Contribution à la connaissance du Pliocène et du Pleistocène Inférieur des régions Méditerranéennes d'Europe Occidentale par l'analyse palynologique des dépôts du Languedoc-Roussillon (sud de la France) et de la Catalogne (nord-est de l'Espagne).* PhD thesis, Sciences et Techniques du Languedoc University, Montpellier; France.

Suc, J.P. (1984). Origin and evolution of the Mediterranean vegetation and climate in Europe. *Nature*, **307**, 429–32.

Szweykowski, J. & Bobowicz, M.A. (1982). Variation in *Pinus sylvestris*, *Pinus mugo* and relative-hybrid populations in Central Europe. *Bulletin de la Société des Amis des Sciences et des Lettres (Poznam)*, série D, **22**(0), 43–50.

Thirgood, J.V. (1981). *Man and the Mediterranean Forest.* New York: Academic Press.

Trabaud, L. (1980). *Impact biologique et écologique des feux de végétation sur l'organisation, l'évolution et la structure de la végétation des zones de Garrigues du Bas-Languedoc.* PhD thesis, Sciences et Techniques du Languedoc University, Montpellier; France.

Trabaud, L. (1984). Man and fire: impacts on mediterranean vegetation. In *Mediterranean-Type Shrublands.* ed. F. di Castri, D.W. Goodall & R.L. Specht, pp. 523–37. *Ecosystems of the World*, Vol 11. Amsterdam: Elsevier.

Trabaud, L.V. (1987). Dynamics after fire of sclerophyllous communities in the Mediterranean Basin. *Ecologia Mediterranea*, **13**, 25–37.

Trabaud, L. (1990). Fire as an agent of plant invasion? A case study in the French Mediterranean vegetation. In *Biological Invasions in Europe and the Mediterranean Basin*, ed. F. di Castri, A.J. Hansen & M. Debussche, pp. 417–37. Dordrecht: Kluwer.

Trabaud, L.V., Christensen, N.L. & Gill, A.M. (1993). Historical biogeography of fire in temperate and mediterranean ecosystems. In *Fire in the Environment: Its Ecological and Atmospheric Importance*, ed. P.J. Crutzen & J.G. Goldammer, pp. 277–95. New York: John Wiley.

Triat-Laval, H. (1978). *Contribution pollenanalytique à l'histoire tardi- et post-Glaciaire de la végétation de la Basse Vallée du Rhône.* PhD thesis, Aix-Marseille III University, France.

Tutin, T.G., Heywood, V.H., Burges, N.A. *et al.* (1964–93). *Flora Europaea*, Vols. 1–5.

Cambridge: Cambridge University Press.

van Zeist, W., Woldrig, H. & Stapert, D. (1975). Late Quaternary vegetation and climate of Southwestern Turkey. *Palaeohistoria*, **17**, 55–143.

van Zeist, W. & Bottema, S. (1977). Palynological investigations in western Iran. *Palaeohistoria*, **19**, 19–87.

Vernet, J.L. (1986). Changements de végétation, climats et action de l'homme au Quaternaire en Méditerranée occidentale. In *Proceedings of the Symposium on Climatic Fluctuations During the Quaternary in the Western Mediterranean*, ed. A. Lopez-Ver, pp. 535–47. Madrid: Universidad Autónoma de Madrid.

Vidaković, M. (1991). *Conifers: Morphology and Variation*, revised English edition. Zagreb: Graficki Zavod Hrvatske.

Weinstein, A. (1989a). Provenance evaluation of *Pinus halepensis*, *Pinus brutia* and *Pinus eldarica* in Israel. *Forest Ecology and Management*, **26**, 215–25.

Weinstein, A. (1989b). Geographic variation and phenology of *Pinus halepensis*, *Pinus brutia* and *Pinus eldarica* in Israel. *Forest Ecology and Management*, **27**, 99–108.

Wildprett della Torre, W. & Aguilar, M. (1987). España insular. II. Las Canarias. In *La Vegetacion de España. Publicación del Universita de Alcala*, 515–44.

Zinkel, D.F., Maggee, T.V. & Walter, J. (1985). Major resin acids of *Pinus nigra* needles. *Phytochemistry*, **24**, 1273–7.

9 The recent history of pinyon pines in the American Southwest

Ronald M. Lanner and Thomas R. Van Devender

9.1 Introduction

The pinyon pines are a distinctive group of North American species forming the subsections *Cembroides* and *Rzedowskianae* within section *Parrya* in subgenus *Strobus* in the genus *Pinus* (Chap. 2, this volume). They occur as far north as southern Idaho, USA (Lat. 42° 16' N) in the case of *Pinus monophylla*, and as far south as southern Puebla, Mexico (Lat. 18° 27' N) in the case of *P. cembroides*, thus spanning about 26° 49' of latitude. Since the discovery of *P. cembroides* by Karwinski in 1831, fresh discoveries and an irregularly evolving species concept have periodically increased both the number of recognized species and subspecific taxa, and the level of controversy regarding them. Biosystematic disharmony and nomenclatural chaos continue to engulf these most unprepossessing of pines.

In one area of research, however, the pinyon pines have been characterized in recent years by a clarity exceeding that of other groups of North American conifers – the unfolding story of these species' distributional changes during the most recent glacial (the Wisconsin) and the present interglacial (the Holocene). Our purpose in this chapter is to synthesize our current understanding of these changes among the pinyon species.

9.1.1 Pinyon pine characteristics

Morphology
The pinyons receive their common name from the Spanish *piñón*, which refers to the large, wingless, edible seed that reminded early explorers of those of *P. pinea*. These nut-like seeds have played an important historical and cultural role as a human foodstuff in the American Southwest and Mexico (Lanner 1981). All the pinyons found in the USA (*P. cembroides*, *P. discolor*, *P. edulis*, *P. juarezensis*, *P. monophylla* and *P. remota*) are low, broad-crowned, short-needled trees with thick, furrowed bark on mature trunks and smooth grey bark on young stems and branches (Figs. 9.1, 9.2). They tend to be highly resinous, and often bleed pitch profusely from wounds. The number of needles in the fascicle ranges from one (unique to this group) to five. Pinyon cone scales have pseudo-dorsal umbos, and deep recesses on the seed-bearing surface (Fig. 9.3). In contrast to other pines, the surface tissue of pinyon cone scales (spermoderm) forms an irregular rim around the cone-scale recess, instead of forming a wing. This rim holds the seed within the recess longer than it would otherwise remain in the cone. Several of the pinyon pines are known to form summer shoots spontaneously on vigorous shoots, after the spring shoot has expanded (Lanner 1970).

Ecology
The pinyon pines inhabit semi-arid areas where rainfall may be summer-monsoonal (*P. discolor*, *P. edulis*, *P. remota*) or near-mediterranean in distribution (*P. juarezensis*, *P. monophylla*). *Pinus edulis* reaches high elevations (Lanner 1981) in the Rocky Mountains; *P. monophylla* experiences a continental climate in the Great Basin. Both species are very cold-tolerant, as well as drought resistant (Chap. 15, this volume). Pinyon pines usually form woodlands of widely to closely spaced trees in association with species of *Juniperus* (Fig. 9.4). This belt is often intermediate in elevation between desert scrub below, and montane coniferous forest above. Seeds of *P. edulis* and *P. monophylla* are known to be dispersed by birds of the family Corvidae, which cache them in the soil as a food store, thereby effecting regeneration. The role of rodents in regeneration is unclear. All the pinyons probably depend on vertebrates for seed dispersal and establishment of their seedlings.

Ecology and Biogeography of *Pinus*, ed. D.M. Richardson. © Cambridge University Press (1998), pp. 171–82.

Fig. 9.1. **Young *Pinus juarezensis* at La Rumorosa, Baja California, Mexico (photo: R.M. Lanner).**

Fig. 9.2. ***Pinus monophylla* subsp. *fallax* (Arizona singleleaf pinyon) in the Florida Mountains of New Mexico, USA (photo: R.M. Lanner).**

Evolution

Based on the fossil record and geological evidence, Millar (Chap. 2, this volume) argues that the subsection *Cembroides* is one of several that originated in Mexico or Central America during the Eocene. Lanner (1981) suggested a secondary centre of pinyon evolution in Mexico based on the diversity of relict species there, and the occurrence of summer shoots even in northern species. He has also argued (Lanner 1996) that the natural selective pressures leading to speciation of the pinyons from a 'conventional white pine' ancestor have been mediated by the jays that harvest, cache and eat pinyon seeds (pine nuts). The earliest fossils of such birds date to the upper Miocene (Lanner 1981). Natural hybridization followed by introgression may also have been an evolutionary factor in this group, at least in the north, where the subspecies of *P. monophylla* hybridize readily with *P. edulis* (Lanner & Hutchison 1972; Lanner 1974a; Gafney & Lanner 1987; Lanner & Phillips 1992) and *P. juarezensis* (Lanner 1974b).

9.2 Data sources for Pleistocene/Holocene studies

The major data sources for these syntheses are packrat middens. Packrats (*Neotoma* spp.) are rodents indigenous to western North America (Fig. 9.5). In arid areas of the American Southwest, their middens may be preserved for tens of thousands of years and are a unique source of Late Pleistocene perishable materials. They build large 'houses' from local materials collected in their home range (usually within 30 m). These collections generally represent a large percentage of the local plant species. Middens are the portions of a packrat house that become compacted, and hardened with encrusted urine. Middens may be as much as 2 m in depth, and provide well-preserved material that can be identified (often) to the species level, and can be radiocarbon-dated (Finley 1990).

Other sources of data cited in this chapter are palynological studies, which have been of limited value in our area until a couple of decades ago, because of their earlier inability to distinguish species of *Pinus* pollen; rare

Fig. 9.3. **Open cone of Great Basin singleleaf pinyon, *Pinus monophylla* subsp. *monophylla*, showing the large nut-like seeds retained in their cone-scale recesses by rims of spermoderm (photo: R.M. Lanner).**

Fig. 9.4. **An extensive woodland of Great Basin singleleaf pinyon, *Pinus monophylla* subsp. *monophylla*, and Utah juniper, *Juniperus osteosperma*, in eastern Nevada, USA (photo: R.M. Lanner).**

Fig. 9.5. **The packrat, *Neotoma* sp. (sketch: H. Wilson).**

macrofossils found in rockshelters or human-occupied caves; and hybrid zones that allow inferences to be made regarding past species occurrences.

9.3 Species case studies

There are five pinyon pine species with a fossil background sufficiently rich to shed some light on their Wisconsin occurrence, and how they have responded to the warmer conditions of the Holocene. These will be taken in order by progressing from east to west, mainly by citing the carbon-dated remains from packrat middens recovered from areas that are now within deserts. For simplification, dates are give in years BP (Before Present; =before 1950), without the error term (±) being cited.

9.3.1 *Pinus remota*, Texas pinyon

Pinus remota currently has a relict distribution in the mountains of northeastern Mexico and adjacent Texas (Fig. 9.6). It grows at elevations of 460–1600 m on limestone substrates from the southwestern edge of the Edwards Plateau west to the Glass Mountains in Texas. But during the mid- to late Wisconsin, this species was widely distributed in the limestone ranges of the Chihuahuan Desert region (Van Devender 1990a). Needles from six middens place it in the

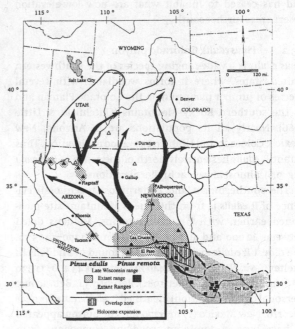

Fig. 9.6. **Late Wisconsin and extant ranges of Colorado pinyon, *Pinus edulis*, and Texas pinyon, *P. remota*, and presumed routes of their Holocene migrations. ▲, △, macrofossil sites mentioned in the text.**

Sierra de la Misericordia, Durango (1250 m) and at Cañon de la Fragua, Coahuila (900–930 m) at 11 730 and 13 590 BP respectively. Today woodland species are absent from most of the lower ranges in this area (Van Devender 1990a). Further north, where the Chihuahuan Desert reaches its lowest elevations in the Big Bend area of the Rio Grande, middens have documented the widespread presence of Ice Age woodlands. *Pinus remota* remains have been found at Maravillas Canyon (610 m) from 27 820 to 11 240 BP (Wells 1966); in the Rio Grande Village area (600–835 m) from 45 600 to 26 430 BP (at which time it was uncommon), and from 21 830 to 11 470 BP (when it was dominant; Van Devender 1990a); as well as in Wisconsin-aged woodland midden assemblages from Dagger Mountain, Burro Mesa, Santa Elena Canyon, Terlingua, Shafter, Bennett Ranch, Steeruwitz Hills, and the Quitman Mountains (670–1430 m; Lanner & Van Devender 1981). Only in the Hueco Mountains (1270–1495 m) east of El Paso, Texas, has *P. remota* been found together with *P. edulis*, in middens dated between 42 000 and 10 750 BP (Lanner & Van Devender 1981). Evidence from climatically sensitive isotopes in continuous sediment cores from the ocean floors (Imbrie & Imbrie 1979; Porter 1989) suggests that the climatic conditions of about 12 000 BP were about average for the 2.4 million years of the Pleistocene. Thus we can hypothesize that widespread Ice Age woodlands dominated by *P. remota* were the normal situation during most of the Pleistocene throughout the Chihuahuan Desert. Since then, *P. remota* has contracted its range about 300 km from the northwest, and has ceased to inhabit what are now low-elevation desert areas further south.

9.3.2 *Pinus edulis*, Colorado pinyon

Pinus edulis is a wide-ranging species of the southwestern United States, where it forms woodlands with several species of juniper (*Juniperus*) on the Colorado Plateau and in the southern Rocky Mountains (Critchfield & Little 1966). Nearly all its populations are in Arizona, New Mexico, Colorado and Utah *c.* 1290–2900 m elevation. Thus extant *P. edulis* lies entirely north of the range of *P. remota* (Fig. 9.6), almost always at higher elevations.

The southernmost and easternmost macrofossil evidence of *P. edulis* is from the Hueco Mountains site mentioned earlier, with *P. remota* remains. Middens dated between 42 000 and 10 750 BP, were from 1270–1495 m. The oldest *P. edulis* macrofossils were found in middens in Shelter Cave, New Mexico (1440 m), dated at 43 400–11 330 BP (Lanner & Van Devender 1981). Recently J.L. Betancourt (personal communication) identified *P. edulis* remains near Socorro, New Mexico (1710–1770 m). The pinyon appeared to have been a dominant woodland component from >38 000 to 10 975 BP, but became occasional to rare during 10 040–9515 BP. Another apparent replacement of *P. edulis*,

in this case by grasses, occurred in the Coyote Hills of southwestern New Mexico (1505 m) where *P. edulis*, which was present in a midden from 13 830 BP, had disappeared by 10 635 BP. According to Harris (1985), mid-Wisconsin *P. edulis* needles were found among vertebrate fossils and bat guano in U-Bar Cave, also in southwestern New Mexico. Further New Mexican *P. edulis* locations are in the Sacramento Mountains (1555 m) at 18 300–16 260 BP, and the Sierra de los Pinos (1710 m) at 11 560 BP (J.L. Betancourt, personal communication).

The fossil record provides two especially interesting case studies of *P. edulis* in New Mexico. (1) On the Sevilleta Long Term Ecological Reserve, near Socorro, *P. edulis* responded to climatic warming after 11 560 BP by moving upslope. Samples from 1770–1800 m were dominated by *P. edulis* from 9200 BP until a few centuries ago, when they declined, possibly due to drought (J.L. Betancourt, personal communication). Essentially, the late Wisconsin upper elevational limit of *P. edulis* (*c.* 1780 m) became its Holocene lower elevational limit. (2) In Chaco Canyon, northwestern New Mexico, *P. edulis–Juniperus* woodland replaced late Wisconsin–early Holocene montane conifer forest between 9460 and 8300 BP (Betancourt & Van Devender 1981). Then *P. edulis* largely disappeared during 1220–520 BP, when a peak human population (Anasazi) was exerting heavy demands on the fuelwood resource.

One datum only is available for Utah. *Pinus edulis* did not reach Allen Canyon Cave (2195 m) in the Abajo Mountains of southeastern Utah until 3400 BP. At present, *P. edulis* in Utah extends about 475 km further north (Lanner & Hutchison 1972).

The only other middens found to contain *P. edulis* remains are from Arizona. Cinnamon & Hevly (1988) have reported late Wisconsin pinyon remains from Wupatki National Monument (1530 m), 13 670–10 890 BP. This population may have provided the founders for late Wisconsin–Holocene *P. edulis* at other nearby sites in the Grand Canyon to the north (Cole 1990). Van Devender (1986) has reported two-needled pinyon fascicles from Peach Springs Wash (1300 m) dating to 16,580 BP. Peach Springs Wash is near the westernmost present-day location of *P. edulis* (Lanner & Phillips 1992).

Several studies have been undertaken of natural hybridization and introgression involving *P. edulis* and *P. monophylla* (Lanner & Hutchison 1972; Lanner 1974a; Gafney & Lanner 1987; Lanner & Phillips 1992) (Fig. 9.7). Materials similar to those found in extant zones of hybridization have been found in middens at Rampart Cave in Arizona dated at 12 650 BP, and at Desert Almond Canyon, a lower Grand Canyon site dated at 12 600 BP (Phillips 1977). One- and two-needled fascicles have also appeared in six samples from the Santa Catalina Mountains (1555 m) near Tucson, Arizona dated 17 950–13 670 BP. By 12 360 BP

Fig. 9.7. **A zone of pinyon hybridization in Zion National Park, Utah, USA. *Pinus monophylla* subsp. *fallax* grows in the valley bottom and *P. edulis* on the mesa tops. Hybrid populations are found on the mid-slopes (photo: D.J. Gafney).**

only *P. monophylla* needles were found, suggesting that *P. edulis* and the hybrid zone had moved further upslope. Neither species now occurs there, but *P. discolor* is present in the same mountain range.

Taken together, the data suggest that during the Wisconsin in south-central New Mexico and adjacent Texas, *P. edulis* occurred below about 1770 m at 31–34° N (Van Devender 1986; J.L. Betancourt, personal communication). The dramatic Holocene expansion, during which *P. edulis* spread across the Colorado Plateau and northward up both sides of the Rocky Mountains to *c.* 40° N, was probably due largely to the activity of the Pinyon jay (*Gymnorhinus cyanocephalus*) and Clark's nutcracker (*Nucifraga columbiana*), corvids known to disperse and establish pinyons and other wingless-seeded pines over substantial distances (Vander Wall & Balda 1981). Avian dispersal may also have been responsible for establishment of the northeastern outlier of *P. edulis* at Owl Canyon, Colorado, 25 km northwest of Fort Collins. Alternatively, it may have resulted from native Americans dropping or caching the edible seeds along an established trade route. Midden and tree-ring evidence suggest that the small population at Owl Canyon was established between 1290 and 420 BP (Betancourt *et al.* 1991). Lanner & Hutchison (1972) reported several relictual pinyon pine stands in northeastern Utah that included *P. edulis* × *monophylla* hybrids and introgressants. They suggested that the *P. edulis* were the remains of more extensive populations that had expanded northwards from the Uinta Basin during a warmer period of the present interglacial, and then contracted during a cooler period, after having interbred with *P. monophylla*. However, there is no fossil evidence to support this view. The northernmost *P. edulis* relict was in the Crawford Mountains (2060–2200 m) at about latitude 41° 38′ N.

While new materials continue to be unearthed – literally – that document the Wisconsin disposition of *P. edulis*

in the south, our ignorance of its northerly migration remains nearly complete. Perhaps the most important contribution that can now be made, using palaeoecological tools, is a clear understanding of when Colorado pinyon established its present range. Like *P. monophylla*, this species has been an important food plant to native Americans. Knowledge of its prehistoric presence should shed light on the habitation of the American Southwest by human populations.

9.3.3   *Pinus discolor*, border pinyon

Pinus discolor is found in southeastern Arizona and southwestern New Mexico, south into the Mexican states of Tamaulipas and San Luis Potosí. The only fossil records of this little-known species were from late Wisconsin (17 950–12 360 BP) midden samples from Pontatoc Ridge in the Santa Catalina Mountains (1460 m) near Tucson. The assemblages were dominated by *Cupressus arizonica*, *Pseudotsuga menziesii* and *Pinus ponderosa*. This is now a desert grassland–desert scrub ecotone, and extant *P. discolor* in the Santa Catalinas is at 1220–2440 m; but there is none near the midden site. As discussed above, there is evidence that *P. edulis* and *P. monophylla* were in the Santa Catalinas at the same time as *P. discolor*. The presence of three pinyon species in the same mountain range is most unusual in the American Southwest.

9.3.4   *Pinus juarezensis*, Sierra Juárez pinyon

This is a 5-needled pinyon which crosses readily with *P. monophylla* subsp. *californiarum* to produce hybrid swarms of trees bearing 1–5 needles per fascicle (Lanner 1974b) (Fig. 9.8). It has a narrowly linear distribution area from the San Jacinto Mountains of southern California south into the Sierra San Pedro Mártir of Baja California. Wells (1987) reported abundant remains of this species (as *P. quadrifolia*) in a full-glacial midden from the granite boulder fields of La Cataviña in central Baja California (*c.* 18 000 BP). A second sample from the site, dated at 21 000 BP, contained only 5-needled fascicles, indicating a pure stand. This location appears to be 80–100 km south of the southernmost extant *P. juarezensis* (Critchfield & Little 1966; Gerhard & Gulick 1967). By 10 130 BP *P. juarezensis* had disappeared from La Cataviña as pinyon–juniper woodland was replaced by chaparral similar to that now found between Ensenada and San Diego. A midden from Joshua Tree National Monument, California (1280 m) dated at 12 790 BP contained both two- and three-needled fascicles, numbers commonly found in hybridized *P. juarezensis* stands (Lanner 1974b) and not elsewhere in the region. This site is at least 50 km northeast of the closest extant *P. juarezensis*, in the San Jacinto Mountains, where hybrids and introgressants occurred at 1600 m (Lanner 1974b). These data indicate a modest contraction of the range of *P. juarezensis* during the Holocene.

Fig. 9.8. **A hybrid swarm of *Pinus monophylla* subsp. *californiarum* and *P. juarezensis* at La Rumorosa, Baja California, Mexico. In the background is the Sierra Juárez (photo: R.M. Lanner).**

9.3.5 *Pinus monophylla* Torr. & Frém., singleleaf pinyon, comprising:
subsp. *fallax*, Arizona singleleaf pinyon
subsp. *monophylla*, Great Basin singleleaf pinyon
subsp. *californiarum*, California singleleaf pinyon

Singleleaf pinyon is the most northerly species of the *Cembroides*. It is the only single-needled pine, a condition brought about by the failure to develop a second needle primordium in each of its dwarf shoots (Gabilo and Mogensen 1973). The solitary needle primordium develops into a stout needle with a circular cross-section. The taxonomy of this species became more complex when Bailey (1987, 1988) first split *Pinus monophylla* into *P. monophylla* and *P. californiarum*; and then split the latter species into subspecies *fallax* and *californiarum*. 'Fallax' had been regarded by Little (1968) as a single-needled variety of *P. edulis*, and later by Lanner (1974a) as a geographic race of *P. monophylla*. Bailey's 'new species', *P. californiarum*, was separated from *P. monophylla* mainly on the basis of minor differences in metric needle characters. Some of Bailey's data were derived from subjectively chosen samples that were analysed by non-standardized statistical techniques. His methods are therefore not strictly repeatable, and the results cannot be compared with previously published data. It is therefore more appropriate to view Bailey's three taxa as subspecies or varieties of a single geographically varying species (Chap. 2, this volume). These will be discussed progressing from east to west.

Arizona singleleaf pinyon (subsp. *fallax*)
Today, at elevations of 550–1500 m, the Sonoran Desert is dominated by a Sonoran desertscrub with giant saguaro cactus (*Carnegia gigantea*) and foothills paloverde (*Cercidium microphyllum*) among the distinctive species. But

during the middle and late Wisconsin, this zone bore a growth of woodland trees. The principal species was the Arizona singleleaf pinyon, which was associated with junipers and oaks (*Quercus*) (Van Devender 1990b). The Sonoran Desert midden records for Wisconsin woodland that included *fallax* are very numerous in Arizona: from east to west, Santa Catalina Mountains (altitude 1460 m), Tucson Mountains (890 m), Waterman Mountains (795 m), Picacho Peak (655 m), Ragged Top (860 m), Castle (790 m), Ajo Mountains (915–975 m), New Water Mountains (725 m), Kofa Mountains (860 m), and Tinajas Altas Mountains (460–550 m). These samples were dated between >43 300 and 10 880 BP. Site-specific details can be found in Anderson & Van Devender (1991), Lanner & Van Devender 1974, Van Devender, Mead & Rea (1991), and Van Devender & Wiens (1993). Additional unpublished records of Arizona *fallax* remains are from near Superior (>42 000 and 16 625 BP; J.L. Betancourt, personal communication), Harcuvar Mountains, 730 m (26 600 BP; A. Dickey and T.R. Van Devender, unpublished data), and Harquahala Mountains, 795 m (22 140 BP; J. McAuliffe and T.R. Van Devender, unpublished data). At present, *fallax* docs not grow in any of the locations where it has been found in packrat middens. Its present distribution is in mountains south of the Mogollon Rim, on the southern edge of the Colorado Plateau, from southwestern New Mexico to northwestern Arizona at elevations of *c.* 1250–1800 m. Thus its Wisconsin range was further south and at lower elevations than now.

Great Basin singleleaf pinyon (subsp. *monophylla*)
The recent history of the typical *P. monophylla*, the Great Basin singleleaf pinyon, is a classical example of a Holocene expansion from southern refuges occupied during the Wisconsin (Fig. 9.9). In the southern portions of the Mojave Desert there are several middle- to late Wisconsin midden records of pinyon–juniper woodland. These include the Turtle Mountains (700–850 m) dated at 19 500 and at 13 900–12 600 BP (Wells & Berger 1967) and the Newberry Mountains (Leskinen 1975). Further north in the Mojave, remains from the Specter Range (1100–1190 m) show pinyon appearing during the full-glacial, about 19 000 BP, and at Robber's Roost in the Scodie Mountains, dated at 13 800–12 800 BP (Van Devender & Spaulding 1979). The lowest fossil occurrence of pinyon in this region was in the Skeleton Hills (925 m) at 17 900 BP, which was also the northernmost full-glacial occurrence of the pinyon (36° 38' N). Thus, during the last glacial, pinyon–juniper woodlands were widespread in the ranges of the southern Mojave Desert, poised to expand northwards upon the collapse of the latest Wisconsin–early Holocene pluvial, and the accompanying climatic warming episode (Spaulding 1990). Grayson (1993) has summarized what happened in the Great Basin during the

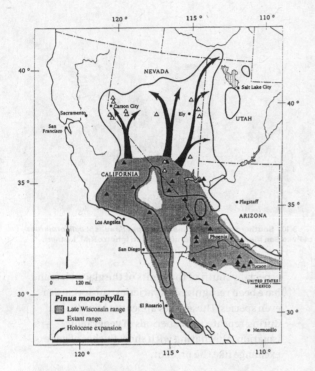

Fig. 9.9. The Late Wisconsin and extant ranges of singleleaf pinyon, *Pinus monophylla*, and its expansion during the Holocene. ▲, △, macrofossil sites mentioned in the text.

Holocene. As Grayson emphasizes, despite the considerable number of midden and pollen analyses that have been made throughout the Great Basin, none has shown any evidence whatever of the presence of singleleaf pinyon earlier than 6590 BP.

The oldest such pinyon remains found so far – in Meadow Valley Wash in southern Nevada – have been dated to 6590 BP (Madsen 1972). Progressing northwards, pinyon macrofossils appear progressively later: in the Schell Creek Range at 6250 BP (but not at >6300 BP), and in the Snake Range at 6120 BP (Thompson 1990). At Gatecliff Shelter, in the Toquima Range of central Nevada, Rhode & Thomas (1983) found pinyon wood charcoal and burnt needles dated to 5350 BP. Thompson & Hattori (1983) analysed middens from Mill Canyon, in the vicinity of Gatecliff, and found pinyon remains dating to 4790 BP, but absent in samples dated 9000–9500 BP. The pollen record from Gatecliff Shelter supports these findings, demonstrating a significant increase in pinyon and juniper pollens during the period 6000–5600 BP (Thompson & Kautz 1983). Thus singleleaf pinyon clearly spread northwards at a fairly rapid rate in the eastern Great Basin. Spread was apparently slower in the western Great Basin, as Nowak *et al.* (1994) did not detect *P. monophylla* in middens older than 3000 BP in the Reno, Nevada region. Lanner (1983) has shown that if Clark's nutcracker is

assumed to have been its major disperser/establisher, singleleaf pinyon could well have arrived at Gatecliff Shelter even well before 6000 BP, given the life-history traits of the pinyon pine, and the availability of feasible migration routes from the northern Mojave.

One enigma must be mentioned. In Danger Cave, Utah, north of all of the locations previously mentioned, Madsen & Rhode (1990) found singleleaf pinyon nutshells dated at 7920 and 7410 BP that had apparently been carried into the cave. But from where? As Madsen & Rhode (1990) point out, it is unlikely that pinyon had arrived that early from the south without leaving traces in the middens discussed above. Another possibility, though speculative, is that it arrived from the east or even the north, originating in stands that resulted from a more rapid northerly spread of pinyon up the Wasatch Range and eastern edge of the Great Basin. Singleleaf pinyon relicts are now scattered in several northeast Utah mountain ranges that lie to the east of the nearby Wasatch Range (Lanner & Hutchison 1972), as well as at one location actually on the Wasatch (Wheatgrass Canyon in Weber County; S. Clark, personal communication). In addition, there are singleleaf pinyon populations in a number of places south of the Great Salt Lake (Stansbury Mountains, West Tintic Mountains, Ophir Canyon, Mercur) (Fig. 9.9). The present-day occurrence of singleleaf pinyon in an arc formed to the north by the Black Pine (Idaho), Raft River, and Grouse Creek Mountains (Utah) suggests that Danger Cave may have been surrounded by pinyon-bearing areas at moderate distances.

Has the Holocene expansion of Great Basin singleleaf pinyon from a few Mojave Desert refugia resulted in a genetically depauperate taxon? Apparently not. Allozyme analyses by Hamrick, Schnabel & Wells (1994) showed that *P. monophylla* maintains a high degree of genetic diversity both within and among its populations. These conclusions agree with those of Smith & Preisler (1988), who found a considerable degree of variability in oleoresin composition in Great Basin singleleaf pinyon from several locations. Data from some of the northernmost (thus presumably youngest) pinyon populations in the Great Basin show marked variability in number of leaf resin canals (Lanner 1974a). Means of 10-tree plots, with 100 needles per tree sampled, are: Black Pine Mountains, Idaho, 5.9 leaf resin canals; Raft River Mountains, Utah, 4.4; Grouse Creek Mountains, Utah, 4.8; Wells, Nevada, 3.3. In a group of less intensively sampled plots further south in Utah, mean resin canal number varied from 2.0 (Richfield) to 7.7 (Wah Wah Mountains) (R.M. Lanner, unpublished data). Further studies are needed to analyse genetic variability and architecture in this migrating taxon, but these data indicate that Great Basin singleleaf pinyon is by no means genetically homogeneous.

Variability may have been conserved during the Wisconsin by occupation of numerous refugia that contained much of the species' genetic diversity, and may have been enhanced since then by hybridization with *P. edulis*. This taxon continues to spread northwards, at least in southern Idaho, where the northernmost populations grow vigorously and regenerate normally to form all-aged stands.

Another possible migration route by which singleleaf pinyon might have entered the Great Basin following the Wisconsin, is from the Tehachapi Mountains northeastwards around the south spur of the Sierra Nevada, or possibly across that range. Needles and seed coats recovered from the Robber's Roost middens in the extreme northern Mojave Desert, were consistent in resin canal number, stomatal row number, seed-coat thickness, and seed length with extant Great Basin singleleaf pinyon (R.M. Lanner, unpublished data). Needles recently examined from several extant singleleaf pinyon stands along a 'Tehachapi route' in California, are more similar in resin canal and stomatal row number to extant stands further west (north of Santa Barbara) and at the Sierra Nevada–Great Basin interface (Mono Lake) than to Mojavean stands to the south and east (R.M. Lanner, unpublished data). The occurrence of *P. monophylla* macrofossils in packrat middens from Kings Canyon on the west slope of the Sierra Nevada, and its presence there today, indicates this species has been there for at least 35 000 years (Cole 1983). Finally, pinyon remains from the McKittrick tar pits in the San Joaquin Valley (180 km southwest of Kings Canyon) suggest strongly that Pleistocene woodlands of singleleaf pinyon were quite extensive in California northwest of the Mojave Desert.

In discussing impacts of the Pleistocene on the genetic structure of North American conifers, Critchfield (1984, p. 73) speculated that most of them

> ... have responded to glacial cycles in a generally similar way: (a) at the onset of glaciation, a shifting, contraction, and fragmentation of the geographic range, (b) during glaciation, a long period of geographic isolation and possibly genetic differentiation among refugial populations, and (c) at the end of a glacial episode, a shifting and expansion of the range, and reestablishment of genetic contact with populations (and species) from other refugia.

One possible consequence of this sequence of events, according to Critchfield (1984 p. 75), is the redistribution of their genetic variation. Thus they

> ... illustrate an advanced stage of subspecies evolution. Their geographic races differ conspicuously,

Fig. 9.10. **Southern outlier of *Pinus monophylla* subsp. *californiarum* at Paso San Matías, Baja California, Mexico (photo: R.M. Lanner).**

especially south of the limits of the glacial ice, and have been recognized as varieties, subspecies, or even species. These races appear to have evolved during repeated long periods of geographic isolation, alternating with short flushes of gene exchange like the present.

Critchfield's statements, written prior to the publication of much of the currently available evidence on singleleaf pinyon, sounds remarkably like a blueprint for that species' recent evolution.

California singleleaf pinyon (subsp. *californiarum*)

The California singleleaf pinyon, *P. monophylla* subsp. *californiarum*, which is differentiated from the typical subspecies mainly by its higher number of leaf resin canals (Lanner 1974a) and length of fascicle-sheath scales (Bailey 1987), is distributed around the Mojave Desert in southern Nevada, southeastern California, and northern Baja California. A series of five midden samples dated 30 950–11 190 BP from near Misión San Fernando Velicata (700 m) is presumed to have contained subsp. *californiarum* and *Juniperus californica* remains. This area now supports Sonoran desert scrub with boojum tree (*Fouquieria columnaris*). The nearest mapped location of extant presumed California singleleaf pinyon is the species' southernmost outlier, about 120 km to the south (Critchfield & Little 1966; Gerhard & Gulick 1967) (Fig. 9.10).

9.4 The historical period

Because of their value to native Americans as a staple food plant, pinyon pines were noted early in the Mexican

Fig. 9.11. A 'chained' area of singleleaf pinyon–Utah juniper woodland in central Utah, USA, about 1984 (photo: R.M. Lanner).

colonial period, and in exploration of the American Southwest. The first Spaniards to 'discover' *P. edulis*, for example, were Alvar Núñez Cabeza de Vaca and other survivors of a shipwreck in the Gulf of Mexico in 1528. For eight years this party wandered across Texas, New Mexico, Arizona and Sonora, and Cabeza de Vaca wrote of his encounters with *P. edulis* nuts (Lanner 1981). Coronado's journalist Castañeda also made note of pinyon woodland in early New Mexico ('. . . the country is all wilderness, covered with pine forests. There are great quantities of the pine nuts'), and of pine nuts being eaten by the Zuñi, in the 1540s (Lanner 1981). Many other historic figures called attention to pinyon pine trees, forests, and seeds, though few were in any way descriptive. Lieutenants Joseph Ives and A.W. Whipple were exceptions, in mentioning the denseness of some of the woodland they traversed.

The settling of Spaniards in the Southwest resulted in heavy use of woodlands for grazing, nut gathering, fuelwood, fencing, and building materials (Randles 1949). These impacts increased greatly upon American settlement of the area. In the Great Basin, starting in the 1850s, woodlands were also heavily impacted by burning, extremely heavy grazing (especially by sheep), and use of pinyon pine for industrial production of charcoal, fuelwood, fencing and lumber (Lanner 1981).

During the 1950s, public lands management agencies in the West began a programme of woodland eradication aimed at converting forest to pasture (Fig. 9.11). The 'chaining' or 'cabling' of several million acres of woodland was later justified as a control programme to protect rangelands from encroaching trees (USDA Forest Service 1973, 1974). The woodland was portrayed as an invader that has been actively spreading since settlement, especially in the Great Basin (Lanner 1981). While the appearance of junipers in lower-elevation areas where they were formerly absent is uncontested, the spread of pinyon pines is

less obvious. There are two reasons for this. First, vast areas of woodland were clearcut for charcoal manufacture in the Great Basin, during the 19th century. Therefore, the large-scale reappearance of the species is in many cases a reoccupation of its former sites effected by avian seed dispersal (Lanner 1981). Second, no surveys have been made to determine the history of present-day pinyon pine stands. The occurrence of trees as much as 200–400 years old in 'treated' areas indicates that while tree removal may be formally justified by a hypothesis of 'invasion', no evidence of such an invasion was required to implement the programme (Lanner 1981).

Despite the decades-old alarm occasioned by juniper encroachment on rangelands, very little research has been undertaken to establish its cause or causes. Hypotheses that have been advanced are lack of natural control by fire due to fire suppression efforts, or reduction of light fuels due to grazing, spread of juniper seed by livestock, reduction of grass competitiveness due to overgrazing, and climatic change (Johnsen 1962; Richardson & Bond 1991; Miller & Wigand 1994). Palaeoecological study methods are now being applied to an understanding of juniper dynamics (Miller & Wigand 1994; Nowak et al. 1994), promising rapid progress in the near future. The lumping of pinyon pines with junipers, despite their very different biologies, is a further confusing factor in understanding the behaviour of these coniferous woodlands in which the species may occur together, or singly. For example, while the above-cited findings of Johnsen (1962) dealt specifically with a juniper, they are often extrapolated to pinyon. Though Miller & Wigand (1994) dealt almost entirely with juniper species, the title of their paper implies otherwise. The most egregious sign of confusion of these taxa is the frequent use of the term 'pinyon–junipers', referring to trees growing in pinyon–juniper woodland. Even in the compiling of a very carefully prepared pine atlas (Critchfield & Little 1966), the occurrence of *P. monophylla* near its northern border is in error because foresters reporting this species' supposed occurrence were actually reporting the occurrence of *Juniperus osteosperma* growing alone (W.B. Critchfield, personal communication *c.* 1970).

Notwithstanding any merits that are perceived to follow from woodland eradication, no persuasive evidence has yet been brought forth to support the idea that pinyon pines have been engaged in a regional invasion, or migration, into historic grasslands or shrublands. The political polarization surrounding this issue (Sauer 1988) probably guarantees that little if any disinterested research is likely to illuminate it. At present, woodland eradication in the pinyon–juniper zone is nearly an abandoned practice, partly because it is controversial, but mainly because it is almost always a prohibitively expensive way to grow grass. A major expense is the need to re-treat areas by hand after the trees start to be brought in again by their seed dispersers. So it seems likely that woodland area will increase gradually in the future, as old, uneconomic clearings again fill up with trees. Whether impending global climate change will set in motion a major new migration, like that of the Holocene, remains to be seen.

9.5   Concluding note

Virtually all of the scientific findings cited in this chapter in regard to pinyon pine systematics, hybridization and migration have entered the literature during the past quarter-century. When Mirov was engaged in writing his *The Genus Pinus* during the early to mid-1960s, many references still recognized only one pinyon species in the United States (*P. cembroides*), with four varieties (*cembroides, edulis, monophylla* and *quadrifolia*). The epithets *remota, discolor, fallax, juarezensis* and *californiarum* were as yet unuttered. Mirov, having had no experience of natural pinyon pine hybrids, cited a personal communication from E.L. Little Jr stating that he (Little) had been unable to detect intermediate forms. Nevertheless, Mirov (1967, p. 334) called for biosystematic and genetic studies of this group.

Nor was anything known in 1967 of pinyon migrations during the Holocene. Packrat middens had just been shown to be a source of identifiable prehistoric plant remains, but few results had been published, and the method had not yet been taken up by new investigators. Today, pinyon biosystematics and hybridization are being pursued by molecular as well as traditional means, and many investigators have joined in the search for palaeoecological data. Mirov, who pioneered the use of chemical methods to study pine biosystematics, took an active interest in the exciting developments that occurred between publication of his *magnum opus* and his death in 1980, and he encouraged others to focus on the pinyons, which had captured his imagination as a mysterious and exotic group of pines (N.T. Mirov, personal communications 1967–80).

Acknowledgements

We thank Julio L. Betancourt for generous sharing of unpublished research results, and Bill Clark for guiding us to full-glacial sites in Baja California.

References

Anderson, R.S. & Van Devender, T.R. (1991). Comparison of pollen and macrofossils in (*Neotoma*) middens: a chronological sequence from the Waterman Mountains, southern Arizona, USA. *Review of Paleobotany and Palynology*, **69**, 1–28.

Bailey, D.K. (1987). A study of *Pinus* subsection *Cembroides*.I. The single-needle pinyons of the Californias and the Great Basin. *Notes Royal Botanical Garden Edinburgh*, **44**, 275–310.

Bailey, D.K. (1988). The single-needle pinyons – one taxon or three? In *Plant Biology of Eastern California. Natural History of the White-Inyo Range Symposium*, Vol. 2, ed. C.A. Hall & V. Doyle-Jones, pp. 69–91. Los Angeles: University of California White Mountain Research Station.

Betancourt, J.L., Schuster, W.S., Mitton, J.B. & Anderson, R.S. (1991). Fossil and genetic history of a pinyon pine (*Pinus edulis*) isolate. *Ecology*, **72**, 1685–97.

Betancourt, J.L. & Van Devender, T.R. (1981). Holocene vegetation in Chaco Canyon. *Science*, **214**, 656–8.

Cinnamon, S.K. & Hevly, R.H. (1988). Late Wisconsin macroscopic remains of pinyon pine on the southern Colorado Plateau. *Current Research in the Pleistocene*, **5**, 47–8.

Cole, K. (1983). Late Pleistocene vegetation of Kings Canyon, Sierra Nevada, California. *Quaternary Research*, **19**, 117–29.

Cole, K.L. (1990). Late Quaternary vegetation gradients through the Grand Canyon. In *Packrat Middens: The Last 40,000 Years of Biotic Change*, ed. J.L. Betancourt, T.R. Van Devender & P.S. Martin, pp. 240–58. Tucson: University of Arizona Press.

Critchfield, W.B. (1984). Impact of the Pleistocene on the genetic structure of North American conifers. In *Eighth North American Forest Biology Workshop Proceedings*, ed. R.M. Lanner, pp. 70–118. Logan: Utah State University.

Critchfield, W.B. & Little, E.L. (1966). *Geographic Distribution of Pines of the World*. USDA Forest Service Miscellaneous Publication 991.

Finley, R.B. Jr. (1990). Woodrat ecology and behavior. In *Packrat Middens: The Last 40,000 Years of Biotic Change*, ed. J.L. Betancourt, T.R. Van Devender & P.S. Martin, pp. 28–42. Tucson: University of Arizona Press.

Gabilo, E.M. & Mogensen, H.L. (1973). Foliar initiation and fate of the dwarf shoot apex in *Pinus monophylla*. *American Journal of Botany*, **60**, 671–7.

Gafney, D.J. & Lanner, R.M. (1987). Evolutionary sorting of pinyon pine taxa in Zion National Park, Utah. In *Proceedings – Pinyon–Juniper Conference*, comp. R.L. Everett, pp. 288–92. Ogden, Utah: USDA Forest Service.

Gerhard, P. & Gulick, H.E. (1967). *Lower California Guidebook*. Glendale: Arthur H. Clark.

Grayson, D.K. (1993). *The Desert's Past: A Natural Prehistory of the Great Basin*. Washington, DC: Smithsonian Institution Press.

Harris, A.H. (1985). Preliminary report on the vertebrate fauna of U-Bar Cave, Hidalgo County, New Mexico. *New Mexico Geology*, **74**, 77–84.

Hamrick, J.L., Schnabel, A.F., & Wells, P.V. (1994). Distribution of genetic diversity within and among populations of Great Basin conifers. In *Natural History of the Colorado Plateau and Great Basin*, ed. K.T. Harper, L.L. St Clair, K.H. Thorne & W.M. Hess, pp. 147–61. Niwot: University Press of Colorado.

Imbrie, J. & Imbrie, K.P. (1979). *Ice Ages: Solving the Mystery*. Hillside: Enslow.

Johnsen, T.N. (1962). One-seed juniper invasion of northern Arizona grasslands. *Ecological Monographs*, **32**, 187–207.

Lanner, R.M. (1970). Origin of the summer shoot of pinyon pines. *Canadian Journal of Botany*, **48**, 1759–65.

Lanner, R.M. (1974a). Natural hybridization between *Pinus edulis* and *Pinus monophylla* in the American Southwest. *Silvae Genetica*, **23**, 108–16.

Lanner, R.M. (1974b). A new pine from Baja California and the hybrid origin of *Pinus quadrifolia*. *The Southwestern Naturalist*, **19**, 75–95.

Lanner, R.M. (1981). *The Piñon Pine: A Natural and Cultural History*. Reno: University of Nevada Press.

Lanner, R.M. (1983). The expansion of singleleaf piñon in the Great Basin. In *The Archaeology of Monitor Valley 2. Gatecliff Shelter*, ed. D.H. Thomas, pp. 167–71. New York: American Museum of Natural History.

Lanner, R.M. (1996). *Made For Each Other: A Symbiosis of Birds And Pines*. New York: Oxford University Press.

Lanner, R.M. & Hutchison, E.R. (1972). Relict stands of pinyon hybrids in northern Utah. *Great Basin Naturalist*, **32**, 171–5.

Lanner, R.M. & Phillips, A.M. III (1992). Natural hybridization and introgression of pinyon pines in northwestern Arizona. *International Journal of Plant Sciences*, **153**, 250–7.

Lanner, R.M. & Van Devender, T.R. (1974). Morphology of pinyon pine needles from fossil packrat middens in Arizona. *Forest Science*, **20**, 207–11.

Lanner, R.M. & Van Devender, T.R. (1981). Late Pleistocene piñon pines in the Chihuahuan Desert. *Quaternary Research*, **15**, 278–90.

Leskinen, P.H. (1975). Occurrence of oaks in Late Pleistocene vegetation in the Mohave Desert of Nevada. *Madroño*, **23**, 234–5.

Little, E.L. Jr. (1968). Two new pinyon varieties from Arizona. *Phytologia*, **17**, 329–42.

Madsen, D.B. (1972). Paleoecological investigations in Meadow Valley Wash, Nevada. In *Great Basin Cultural Ecology: A Symposium*, ed. D.D. Fowler, pp. 57–66. Reno: Desert Research Institute.

Madsen, D.B. & Rhode, D. (1990). Early Holocene piñon (*Pinus monophylla*) in the northeastern Great Basin. *Quaternary Research*, **33**, 94–101.

Miller, R.F. & Wigand, P.E. (1994). Holocene changes in semi-arid pinyon–juniper woodlands. *BioScience*, **44**, 465–74.

Mirov, N.T. (1967). *The Genus Pinus*. New York: Ronald Press.

Nowak, C.L., Nowak, R.S., Tausch, R.J. & Wigand, P.E. (1994). Tree and shrub dynamics in northwestern Great Basin woodland and shrub steppe during the Late-Pleistocene and Holocene. *American Journal of Botany*, **81**, 265–77.

Phillips, A.M. III (1977). Packrats, plants, and the Pleistocene in the Lower Grand Canyon. PhD thesis, University of Arizona, Tucson.

Porter, S.C. (1989). Some geological implications of average Quaternary glacial conditions. *Quaternary Research*, **32**, 245–61.

Randles, Q. (1949). Pinyon–juniper in the Southwest. In *Trees, Yearbook of Agriculture*, ed. A. Stefferud, pp. 342–7. Washington, DC: United States Department of Agriculture,

Rhode, D. & Thomas, D.H. (1983). Flotation analysis of selected hearths. In *The Archaeology of Monitor Valley. 2. Gatecliff Shelter*, ed. D.H. Thomas, pp. 151–7. New York: American Museum of Natural History.

Richardson, D.M. & Bond, W.J. (1991). Determinants of plant distribution: evidence from pine invasions. *American Naturalist*, **137**, 639–68.

Sauer, J.D. (1988). *Plant Migration: The Dynamics of Geographic Patterning in Seed Plant Species*. Berkeley: University of California Press.

Smith, R.H. & Preisler, H.K. (1988). Xylem monoterpenes of *Pinus monophylla* in California and Nevada. *The Southwestern Naturalist*, **33**, 205–14.

Spaulding, W.G. (1990). Vegetational and climatic development of the Mohave Desert: the last glacial maximum to the present. In *Packrat Middens: The Last 40,000 Years of Biotic Change*, ed. J.L. Betancourt, T.R. Van Devender & P.S. Martin, pp. 166–99. Tucson: University of Arizona Press.

Thompson, R.S. (1990). Late Quaternary vegetation and climate in the Great Basin. In *Packrat Middens: The Last 40,000 Years of Biotic Change*, ed. J.L. Betancourt, T.R. Van Devender & P.S. Martin, pp. 200–39. Tucson: University of Arizona Press.

Thompson, R.S. & Hattori, E.M. (1983). Packrat (*Neotoma*) middens from Gatecliff Shelter and Holocene migrations of woodland plants. In *The Archaeology of Monitor Valley 2. Gatecliff Shelter*, ed. D.H. Thomas, pp. 157–67. New York: American Museum of Natural History.

Thompson, R.S. & Kautz, R.R. (1983). Pollen analysis. In *The Archaeology of Monitor Valley 2. Gatecliff Shelter*, ed. D.H. Thomas, pp. 136–51. New York: American Museum of Natural History.

USDA Forest Service (1973). *Pinyon–Juniper Chaining Program on National Forest Lands in the State of Utah*. Ogden, Utah: USDA Forest Service.

USDA Forest Service (1974). *Pinyon–Juniper Chaining Programs on National Forest Lands in the State of Nevada*. Ogden, Utah: USDA Forest Service.

Vander Wall, S.B. & Balda, R.P. (1981). Ecology and evolution of food storage behavior in conifer-seed-caching corvids. *Zeitschrift für Tierpsychologie*, **56**, 217–42.

Van Devender, T.R. (1986). Late Quaternary history of pinyon–juniper–oak woodlands dominated by *Pinus remota* and *Pinus edulis*. In *Proceedings – Pinyon–Juniper Conference*, comp. R.L. Everett, pp. 99–103. Ogden, Utah: USDA Forest Service.

Van Devender, T.R. (1990a). Late Quaternary vegetation and climate of the Chihuahuan Desert, United States and Mexico. In *Packrat Middens: The Last 40,000 Years of Biotic Change*, ed. J.L. Betancourt, T.R. Van Devender & P.S. Martin, pp. 104–33. Tucson: University of Arizona Press.

Van Devender, T.R. (1990b). Late Quaternary vegetation and climate of the Sonoran Desert, United States and Mexico. In *Packrat Middens: The Last 40,000 Years of Biotic Change*, ed. J.L. Betancourt, T.R. Van Devender & P.S. Martin, pp. 134–65. Tucson: University of Arizona Press.

Van Devender, T.R., Mead, J.I. & Rea, A.M. (1991). Late Quaternary plants and vertebrates from Picacho Peak, Arizona, with emphasis on *Scaphiopus hammondi* (western spadefoot). *The Southwestern Naturalist*, **36**, 302–14.

Van Devender, T.R. & Spaulding, W.G. (1979). Development of vegetation and climate in the southwestern United States. *Science*, **204**, 701–10.

Van Devender, T.R. & Wiens, J.F. (1993). Holocene changes in the flora of Ragged Top, south-central Arizona. *Madroño*, **40**, 246–64.

Wells, P.V. (1966). Late Pleistocene vegetation and degree of pluvial climatic change in the Chihuahuan Desert. *Science*, **153**, 970–5.

Wells, P.V. (1987). Systematics and distribution of pinyons in the late Quaternary. In *Proceedings – Pinyon–Juniper Conference*, ed. R.E. Everett, pp. 104–8. USDA Forest Service General Technical Report INT-215.

Wells, P.V. & Berger, R. (1967). Late Pleistocene history of coniferous woodland in the Mohave Desert. *Science*, **155**, 1640–7.

10 Macroecological limits to the abundance and distribution of *Pinus*

George C. Stevens and Brian J. Enquist

10.1 Introduction

Pines provide the biogeographer with valuable clues for understanding the ecological processes at work in fragmenting geographical ranges, in setting the limits to geographical distributions, and in generating the great diversity of species on the planet. This is somewhat ironic given that the genus *Pinus* is not particularly speciose (comprising about 111 species) and that pines, for the most part, are not a part of super-rich tropical forest assemblages (Chap. 1, this volume). However, it is precisely these aspects of pines (their moderate richness and their extratropical distribution) that allow one to test biogeographical hypotheses previously untestable. This chapter is motivated by an attempt to provide insights into the 'macroecological factors' (*sensu* Brown 1995) that influence the distribution of pines and the success of their introduction or invasion into extra-limital areas.

By 'macroecological' we mean the central tendencies or repeatable generalizations that emerge when patterns of distribution are compared across *Pinus* species. The macroecological approach is not new. In fact, one of the great pioneers in the study of pines, Nicholas Mirov, illustrated the macroecological approach when he overlaid latitudinal distributions of pines to show the mid-latitude peak in species richness (reproduced here using data from Critchfield & Little (1966) in Fig. 10.1). He used that information, together with morphological and ecological perspectives, to propose a north temperate origin for the genus (see Chap. 3, this volume for discussion). We extend this approach by applying it to a broader range of phenomena.

For example, Mirov (1967, p. 313) states that some pines of the western USA (*P. aristata*, *P. sabiniana*, *P. torreyana* and others) have 'lost their capacity to expand' geographically.

There are at least two broad categories of forces limiting the expansion of the geographical range of a species: highly restricted tolerances of the species itself, and a patchy distribution of habitat types over which the species can potentially spread. The limited 'capacity' may be a characteristic of the species or a feature of the regional landscape. As will be shown shortly, the latitude (i.e. a feature of the landscape) at which these three species occur supports other *Pinus* species, many of which are patchily distributed. Given the observation that other pines at these latitudes show highly fragmented distributions it is probably not a characteristic of the species, but rather a consequence of the land they cover that accounts for the lack of 'capacity to expand' in the three species named by Mirov.

The foregoing test of alternative hypotheses is a macro-

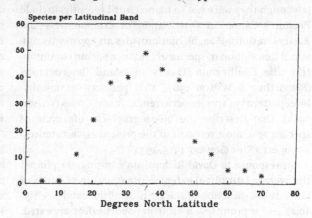

Fig. 10.1. **The latitudinal gradient in species richness for the genus *Pinus* (data from Critchfield & Little 1966). Points represent the number of species of pines found within each latitudinal band (of 5° width).**

Ecology and Biogeography of *Pinus*, e. D.M. Richardson. © Cambridge University Press (1998), pp. 183–90.

ecological approach to Mirov's implicit question of what limits the distribution of particular pines. If due entirely to the species themselves and not to latitude, then other species should not be so restricted at the chosen latitudes. If due entirely to latitudinal effects and not species then other species at the same latitude should be just as restricted in distribution. While the either/or dichotomy poorly reflects the potential for multiple causation, the results of the test indicate that latitudinal effects are a stronger determinant of range fragmentation than species identity. An entirely different and equally valuable approach would be to test the capacity of particular species to expand by attempting to introduce them to new areas.

Before developing the macroecological analysis of pines in more detail we need to note that this analysis would not be possible were it not for pioneers like Mirov (1967) whose devotion to gathering and summarizing the diverse and large literature on pines remains unrivalled. In addition, the distribution maps of Critchfield & Little (1966) are wonderfully complete, detailed, and virtually untapped as a resource for biogeographical analysis. For no other genus of woody plants are such extensive and high-quality data available. This makes, on the one hand, the analysis presented here difficult to replicate with other taxa. On the other hand, the analysis of these data becomes all the more valuable given the uniqueness of the data set.

10.2 Rapoport's Rule in pines

In the broadest sense, macroecology is the study of supra-species patterns, of multiple site patterns, or of patterns of multiple eras. Familiar macroecological patterns (although they were not so named until recently) include Preston's Canonical Distribution of Commonness and Rarity (Preston 1962a, b), that provides an apparently universal description of species abundance within communities; the Equilibrium Theory of Island Biogeography (MacArthur & Wilson 1967), that predicts community-level patterns of species occurrence; Taxon Cycles (Wilson 1961), that describe the biogeographical life cycle of species; and, most relevant to the present circumstances, Rapoport's Rule (Stevens 1989, 1992).

In response to David Richardson's urging, we plotted the average latitudinal extent of pines within 5° bands of latitude as a function of the latitude of the band (see Fig. 10.2). As he promised, a clear Rapoport effect appeared. Rapoport's Rule is the tendency for the size of the geographical range of a species (here measured by the number of 5° bands of latitude through which the range passes) to

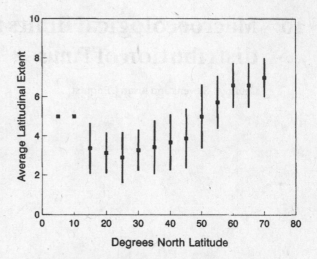

Fig. 10.2. **Rapoport's latitudinal rule for the genus *Pinus* (data from Critchfield & Little 1966). The average latitudinal extent of species decreases from high latitude to low. Note that 'latitudinal extent' is the count of 5° latitudinal bands through which the geographical range of the species passes. Error bars represent one Standard Deviation of the mean. The two left-most points are based on one species (*P. merkusii*). Their apparent deflection upward is probably an artefact of small sample size.**

decrease with decreasing latitude (Stevens 1989). It is also seen in a wide range of taxa other than pines (e.g. marine invertebrates, birds, fish, and mammals). Consequently, Rapoport's Rule qualifies as a robust macroecological pattern – one that transcends taxonomic boundaries.

Since the ranges of most pine species extend across more than one 5° band of latitude, the points in Fig. 10.2 are not independent. As has been discussed elsewhere (Stevens 1989, 1992), this violates assumptions necessary for detailed regression analyses. As a result, here and throughout this chapter, no meaningful comparison can be made between curves with regard to the 'goodness of fit' of points to regression lines. However, this limitation of the statistical analysis of the data does not invalidate the observation that geographical range size and latitude show a consistent relationship. That taxa as different as trees and tetrapods both exhibit Rapoport phenomena provides its own form of statistical improbability against which random associations between latitude and range size can be compared (Colwell & Hurtt 1994).

So, despite the statistical inconvenience of the pattern, Rapoport phenomena are real. Our challenge is to tease apart the factors that produce them. How much of the pattern is due to differences between species in how they use the environment? Are temperate and subtropical species much more specialized than polar ones? How much of the pattern is due to subtle changes in the 'lay of the land' at different latitudes? Are low latitudes more heterogeneous in the suitability of habitats? These are the questions for which pines are an excellent choice of study organism.

Rapoport's Rule has been explained as the result of interspecific differences in the ecological tolerances of the individuals that make up each species (Stevens 1989). For a given tree to survive in a high-latitude site it must be able to tolerate a wide range of environmental conditions. As the seasons progress, individual plants are exposed to high temperatures then low, high soil moisture levels then low, etc. Plants found at lower latitudes experience a narrower range of environmental conditions. This is not to say that conditions at lower-latitude sites are more benign or less challenging (biological stress is relative, not absolute; see Janzen 1967); there is just a narrower range of conditions over the course of the year. Natural selection acting on individual plants will create a situation in which pines that reside at high latitudes will have broad tolerances. A species composed of broadly tolerant individuals will itself be broadly distributed. Rapoport's Rule is the result.

An alternative explanation, on equal footing with the Climatic Variability Hypothesis mentioned above, is the Competitive Pressure Hypothesis (Pianka 1989). According to this view, the range boundaries of low-latitude plants are more limited by competition with other plant species than are the range boundaries of high-latitude plants. The implicit assumption is that the more species that potentially interact with each other, the higher the competitive pressure on any given species. Since low latitudes support more species than high, the effects of competition are thought to be higher there. This competition is what limits geographical range size in low latitudes.

Common sense might suggest that, given the history of glaciation experienced by nearly all northern temperate species, differential rates of dispersal from glacial refugia might explain Rapoport's Rule (Hultén (1937) being the most visible, but certainly not the only proponent of this view). Under this hypothesis, Rapoport's Rule appears because those species of high latitudes are those that disperse well. While this makes intuitive sense, it does not fit well with two observations:

- There is no correlation between seed size (a simple measure of dispersal potential) and size of current geographical range in *Pinus* (data from Burns & Honkala 1990; $n = 27$; Spearman Rank Correlation Coefficient $= 0.18$, $p \gg 0.05$). Those species with large ranges at present may simply have been those that, for whatever historical accident, ended up in high latitudes. Once established at high latitudes, they spread in a manner consistent with Rapoport's Rule.
- Marine fishes show Rapoport's Rule (Stevens 1989). While ocean temperatures changed during glacial periods, there was no barrier to dispersal in fishes as there was for terrestrial organisms. The rate of dispersal away from a past glacial barrier could not play a role in the distribution of marine fishes – yet they show Rapoport's Rule. This suggests that Rapoport's Rule is not solely an outcome of terrestrial glaciation.

Shortly we will make a distinction between the interspecific Rapoport's Rule that applies to multiple-species assemblages or community-level patterns (the only Rapoport's Rule currently being discussed in the literature: Pagel, May, & Collie 1991; France 1992; Rohde 1992; Gage & May 1993; Rohde, Heap & Heap 1993; Colwell & Hurtt 1994; Letcher & Harvey 1994; Macpherson & Duarte 1994; Roy, Jablonkski, & Valentine 1994; Smith, May, & Harvey 1994; Taylor & Gotelli 1994), and a new intraspecific Rapoport phenomenon that applies to the sizes of range fragments within the geographical range of a single species. It is this new intraspecific Rapoport's phenomenon that will be the main emphasis of this chapter (see section 10.4) as a means of better understanding the interspecific Rapoport's Rule. The analysis of range fragments within a species will shed light on the role of both climate and competition in the distribution of pines.

The analysis will follow a dichotomous series of tests, much like the simple one mentioned earlier for Mirov's question of the limits in the capacity of pines to expand their ranges. By looking solely within rather than between species, the potential effect of species replacements with latitude in producing Rapoport's Rule can be eliminated. If range fragments themselves show a kind of Rapoport's effect (in essence an intraspecific Rapoport's Rule for range fragment size) then it is likely that some feature of the landscape produces the interspecific Rapoport's Rule seen in pines (Fig. 10.2) rather than a species effect *per se*.

But what feature of the landscape – climate or interspecific competition? For this analysis one needs look at the sizes of range fragments as a function of latitude. If they follow a monotonic pattern of decline in size with decreasing latitude, then climate is likely to be the cause. If the extratropical peak in *Pinus* species richness deflects the sizes of range fragments away from simple linear relationships with latitude, then competition is likely a cause of range fragmentation.

Two hypotheses for the interspecific Rapoport's Rule in pines will be considered: the Climatic Variability hypothesis and the Competitive Pressure hypothesis. The results of this analysis will have broad applicability to other taxa whose data are not as well maintained or publicized as they are for the genus *Pinus*. Before continuing, some methodological notes are in order.

10.3 Statistical methods and visualization protocol

Given the non-independence of average range size along latitudinal gradients mentioned earlier, the statistical limitation that applies to interspecific Rapoport's Rules also applies to intraspecific patterns. Least-squares regression will be used to fit lines to data, but no attempt will be made to justify the fit using probabilistic arguments. When there is no relationship between range fragment size and latitude, a horizontal line will be drawn. If a regression line drawn through the points would be statistically significant had the points been independent, then a fitted line is drawn. The statistical inference being applied is only with regard to the shapes of the curves (using partial F-tests to determine when a higher-order term should be included in the regression equation), not to the improbability of such fitted lines arising by chance alone. Average Fragment Extent (measured in units of degrees latitude) is a 'running average' and produces quite smooth curves. The smoothness and lack of scatter is entirely a consequence of the extension of many range fragments over multiple latitudes. The finer one divides the latitudinal gradient, the smoother the curves become. No importance should be placed on the smoothness of the transitions.

A 'range fragment' will be defined as a cohesive cluster of sightings that is represented as a continuous blob on a distribution map. Some observer bias is thus introduced into the study, but this is unavoidable no matter what means of identifying range fragments were used (e.g. kriging, dithering). However, the amount of detail in the range maps of Critchfield & Little (1966) should dissuade anyone from the conclusion that the patterns being described here are artefactual. There is simply no tendency for pines with broad distributions to be represented with less detail on range maps. Critchfield & Little's maps of species with large distributions have isolated populations (even single points) identified. The large scale of the maps has not led the cartographers to become more conservative in their definition of isolates. In fact, there is a positive correlation between the scale of the map and the number of range fragments identified ($n=94$, $r^2=0.23$, $p < 0.05$). Moreover, the relative positions of identified fragments (the central component of this analysis) would be uniformly affected by any biases that may be introduced by these scale or observer biases. While the problem of scale is unavoidable it is not fatal to this analysis.

All the data for this study were taken from the maps in Critchfield & Little (1966). Two classes of species were not included in the detailed biogeographical analysis (all were included in the drawing of Figs. 10.1, 10.2). Mexican species were excluded because their distribution is rela-

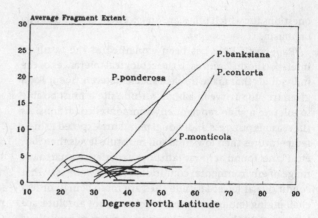

Fig. 10.3. **Rapoport's latitudinal rule for range fragments of western North American *Pinus* spp., including *P. albicaulis, P. attenuata, P. balfouriana, P. banksiana, P. cembroides, P. contorta, P. edulis, P. flexilis, P. jeffreyi, P. lambertiana, P. leiophylla, P. monophylla, P. monticola, P. ponderosa, P. sabiniana* and *P. ayacahuite* var. *strobiformis*. Five species were excluded to minimize the overlapping of lines on the figure (*P. aristata, P. coulteri, P. muricata, P. radiata* and *P. torreyana*). All of these excluded species showed no relationship between average fragment size and latitude, fell between 31 and 41° N, and had average fragment sizes of less than 1.**

tively poorly known. Other species were excluded because their known ranges consist of single points or small collections of single points and so no latitudinal trends in range fragment size could be determined. Rather than list all of the excluded species, the 62 species (56% of the species in the genus) included are given in the legends of Figs. 10.3–10.10.

10.4 Rapoport's intraspecific rule for range fragments

The interspecific Rapoport's Rule can be broken down into its component parts by considering the patterns of range fragment size as a function of latitude for pines of western North America (Fig. 10.3). This graph was produced by measuring the latitudinal range in degrees of each range fragment for a given species at a given latitude and then averaging them over each degree of latitude. Each average is the average of all range fragments found at that latitude. Species with more northern distributions tend to have larger total latitudinal extent to their geographical ranges (seen as well in Fig. 10.2) and larger average range fragments at any given latitude (Fig. 10.3). The average sizes of range fragments within a species are correlated with latitude. Rapoport (1982) found a similar pattern for the ranges of mammalian subspecies within a species. Here we are seeing the pattern for range fragments not differentiated into subspecies rank.

The existence of an intraspecific Rapoport's effect (i.e.

seeing that average range fragment size is correlated with latitude) discounts the role of species replacements with latitude as driving the interspecific Rapoport's Rule (Fig. 10.2). The differences between tropical and extratropical species do not completely account for the differences in average geographical range size at different latitudes. Even within a single species, where ecological tolerances, competitive abilities, etc. are held nearly constant, range fragments show a Rapoport effect. This means that the biotic and abiotic features of the landscape are changing with decreasing latitude. Of course the dichotomy is not as clear as 'landscape effects' versus 'species effects'. Ecotypic variation within pines is likely to exist. Our point is that both must be considered.

It appears (Fig. 10.3) that the congestion of species between 30 and 45° N in western North America is associated with smaller range fragments in those latitudes; this applies even for species whose ranges extend outside the congested zone to the north (i.e. *P. banksiana*, *P. contorta*, *P. monticola* and *P. ponderosa*) or to the south (unlabelled convex curves of *P. cembroides*, *P. leiophylla* and *P. ayacahuite* var. *strobiformis*). Many of these small range fragments occur within the range of potential competitors, and in areas where mixed-species stands occur. While we do not have direct evidence of competition limiting the distribution of these pines, the circumstantial evidence is quite strong. The sizes of range fragments over broad latitudinal gradients change in a way consistent with the argument that range boundaries are set by competitive interactions with other species. While this may not be a revelation to the reader, one of us (G.C.S.) has been quite vocal in his reluctance to accept the idea that competition plays any role in setting range boundaries or generating the interspecific Rapoport's Rule (see Stevens 1992). This is a public admission of error or, perhaps, stated in a more charitable fashion, public evidence of learning.

We have been rather careful in our wording of the expected correlation between range fragment size and latitude. When previously discussing interspecific Rapoport's Rule (Stevens 1989, 1992) only positive correlations between range and latitude were considered. However, Fig. 10.3 shows that predictive relationships between range fragment size and latitude for intraspecific comparisons can take a variety of shapes. This too is evidence for the Competitive Pressure hypothesis. If the competitive pressures to which these pines are exposed are not monotonic, neither should be the average sizes of range fragments. It is worthwhile looking more closely at the fit and the shape of the intraspecific Rapoport's rule for particular species. Of particular interest are the existence of maxima and minima at intermediate latitudes and how they correspond to latitudes of high species richness for pines.

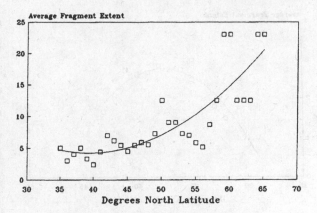

Fig. 10.4. **Rapoport's latitudinal rule for range fragments of *Pinus contorta*. The J-shaped curve is also shown in Fig. 10.3.**

For *P. contorta* (Fig. 10.4) the relationship between latitude and average range fragment size is positive. Isolated range fragments tend to be larger at higher latitudes than at lower latitudes. The relationship appears bowed (meaning that were the points independent, the addition of a higher-order term, here x^2, would result in a significant partial F), but not markedly so. Many of the concave curves (those with midpoint minima) are J-shaped rather than U-shaped (the exceptions will be discussed shortly). For the most part, the largest range fragments are located toward the northern latitudinal extreme for pines (the 10 exceptions among the 62 species studied here each will be explained in turn). Finding that most pines have relatively large range fragments at high latitude is consistent with the Rapoport's Rule found for the entire genus *Pinus* (Fig. 10.2). It doubtless can be explained in the same way by invoking some combination of the Climatic Variability and Competitive Pressure hypotheses.

Consider now the exceptions to the general case of pines having larger average range fragments at high latitude. As will be seen, when a strongly bowed curve is studied more closely, only the Competitive Pressure hypothesis makes sense. *Pinus monophylla* (seen in Fig. 10.3 unlabelled, shown in Fig. 10.5 isolated and scaled up) exhibits a convex relationship between the average sizes of range fragments and latitude. It is one of eight species worldwide that show such a curve. We propose that all eight arise from similar underlying processes. On the southern end of *P. monophylla*'s distribution the range fragments are relatively smaller than at mid-latitudes (consistent with both the Climatic Variability and Competitive Pressure hypotheses) but on the northern end a reverse Rapoport pattern appears. This reversal can be explained easily by the Competitive Pressure hypothesis but not at all by the Climatic Variability hypothesis.

The inverted U-shaped curves for *P. monophylla* and three other species in Fig. 10.3 (*P. cembroides*, *P. leiophylla*,

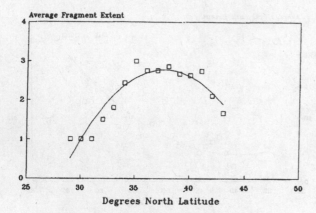

Fig. 10.5. **Rapoport's latitudinal rule for range fragments of *Pinus monophylla*. The inverted U-shaped curve is also shown in Fig. 10.3.**

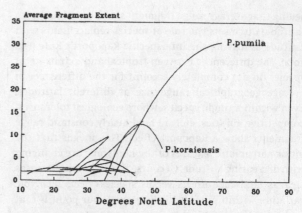

Fig. 10.6. **Rapoport's latitudinal rule for range fragments of Asian *Pinus* spp., including *P. armandii*, *P. densiflora*, *P. fenzeliana*, *P. gerardiana*, *P. kesiya*, *P. koraiensis*, *P. massoniana*, *P. merkusii*, *P. parviflora*, *P. pumila*, *P. roxburghii*, *P. tabuliformis*, *P. thunbergii* and *P. wallichiana*. Four species were excluded to minimize the overlapping of lines on the figure (*P. luchuensis*, *P. morrisonicola*, *P. taiwanensis* and *P. yunnanensis*). All these excluded species showed no relationship between average fragment size and latitude, fell between 23 and 29° N, and had average fragment sizes of less than 2.**

P. ayacahuite var. *strobiformis*) and four other species in subsequent figures have convinced us that interspecific interactions influence the sizes of range fragments and must be considered when explaining Rapoport's Rule. The downward bow on the range fragment size toward the north is likely to be a consequence of the congestion of species at mid-latitudes for western North America (Fig. 10.3).

To save some face in light of past error, we hasten to mention that, despite the long-standing hypothesis (Darwin 1859; Matthew 1915; reviewed and expanded by Darlington 1958 and many others) that northern temperate species are limited on their northern geographical boundary by abiotic conditions (e.g. frost-free days, extreme cold, winter desiccation) and on their southern geographical boundary by biotic conditions (e.g. competitors, parasites, predators), we have presented evidence that biotic interactions reduce range fragment size on either northern or southern edges depending on the latitude at which the potential competitors reside. An advantage of working with pines is that their peak in species richness is decidedly extratropical (see Fig. 10.1). It is the large number of species at mid-latitudes that deflects downward the sizes of range fragments of species that fall to the south of the latitudinal peak in *Pinus* species richness. It takes a taxon that has such an extratropical peak in species richness (Fig. 10.1) to separate cleanly the effects of latitude from the effects of species richness. Again, the revelation is that, apart from latitude (and the associated climatic effects), the number of potentially competing species can influence range size. There are at least two forces generating the *interspecific* Rapoport's Rule: climate and competition.

Everything that has been said thus far about the pines of western North America (Fig. 10.3) also applies to other regions of the world (Figs. 10.6–10.8). Pine species of high latitudes tend to have large geographical ranges with fragments larger than those of other species. Where there is a

buildup of pines at mid-latitudes there is a reduction in range fragment size that produces strong curvilinear effects in the relationship between range fragment size and latitude. Despite the unanimity of results, there are some interesting species whose curves demand explanation.

In Asia (Fig. 10.6), *P. koraiensis* shows an unusual mid-range maximum for a northern pine species. On closer inspection it is clear from Critchfield & Little (1966) that *P. koraiensis* and *P. pumila* have mutually exclusive distributions. The downward bend of the *P. koraiensis* curve at either latitudinal extreme is further evidence that competition can limit the sizes of range fragments and the northward distribution of a species. Were *P. koraiensis* to be introduced to higher latitudes where *P. pumila* is absent, we would expect it to thrive and its range fragments in higher latitudes to increase monotonically (after some equilibration). This is a wonderful example of how current distributions are unlikely to predict future distributions on allopatric continents. Because the relationship between range fragment size and latitude is an inverted U-shape we deduce that competition must be limiting the distribution of the species on both northern and southern extremes. Climatic variability alone would not produce this pattern.

In Europe (Fig. 10.7) the tendency for high-latitude pines to show reduced range fragment size in areas of high species richness is startlingly clear. *Pinus sylvestris* shows a clearly asymptotic, J-shaped distribution of range fragment sizes. *Pinus sibirica* may seem slightly unusual, but that line is based on a single large range fragment extending from the 40s to the 60s degrees latitude. It has no sec-

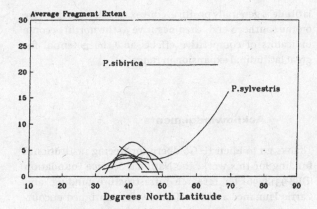

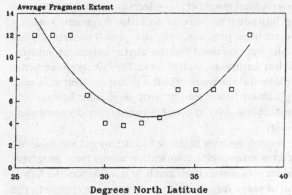

Fig. 10.7. Rapoport's latitudinal rule for range fragments for European *Pinus* spp. including *P. brutia, P. cembra, P. halepensis, P. mugo, P. nigra, P. pinaster, P. pinea, P. sibirica* and *P. sylvestris.* Two species were excluded to minimize the overlapping of lines on the figure (*P. heldreichii* and *P. peuce*). Both these excluded species showed no relationship between average fragment size and latitude, fell between 40 and 44° N, and had average fragment sizes of less than 0.1.

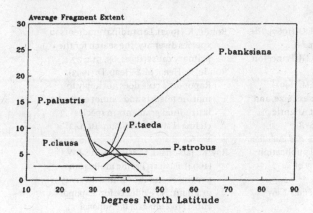

Fig. 10.8. Rapoport's latitudinal rule for range fragments for eastern North American *Pinus* spp., including *P. banksiana* (also seen in Fig. 10.3.), *P. clausa, P. echinata, P. elliottii, P. glabra, P. palustris, P. pungens, P. rigida, P. serotina, P. strobus, P. taeda* and *P. virginiana.* One species (*P. resinosa*) was excluded to minimize the overlapping of lines on the figure. It showed no relationship between average fragment size and latitude, fell between 40 and 50° N, and had an average fragment size of 0.43.

ondary fragments (this perhaps being due to poor data quality in the former Soviet Union).

Finally, eastern North America (Fig. 10.8) has some interesting patterns given the other regions of the globe previously discussed. *Pinus palustris* and *P. taeda* both have an unusual U-shaped relationship between latitude and range fragment size. They are the only species that show larger range fragments at their latitudinal extremes than in their mid-ranges. *Pinus palustris* and *P. taeda* are the two remaining exceptions to the statement 'range fragments tend to be larger toward the northern edge of the range' made earlier.

Closer inspection of the intraspecific Rapoport effect

Fig. 10.9. Rapoport's latitudinal rule for range fragments of *Pinus palustris.* The U-shaped curve is also shown in Fig. 10.8.

for *P. palustris* (Fig. 10.9; the curve for *P. taeda* is similar) shows it to be an exception that proves the rule. Both *P. palustris* and *P. taeda* have geographical ranges that extend into peninsular Florida. This is an area of low pine species richness. The rebound of the sizes of their range fragments re-emphasizes how interspecific competition influences the sizes of ranges. Once these two species enter the peninsula, there is little stopping their ranges from expanding to lower latitudes. From two lines of evidence (competitive release as told here and competitive exclusion as seen earlier in the downward deflection of range fragment size caused by areas of congestion on Figs. 10.3, 10.6, 10.7, 10.8), we have found support for the idea that range size in pines is influenced by intrageneric competition.

The potential for these two pines to expand northward were they freed from competition to the north has been confirmed by analogy with other pines introduced to new settings. Of the eight species listed by Richardson *et al.* (1994) as being introduced and then spreading vigorously over large areas in the southern hemisphere, half (*P. contorta, P. nigra, P. ponderosa* and *P. sylvestris*) are species showing the inverted U-shaped correlations which indicate competitive limits to distribution. This finding is remarkable in that the aforementioned authors were hard pressed to come up with species-specific factors that accounted for colonization success (but see Chap. 22, this volume). The analysis of average range fragment size appears to be a good predictor of potential spread of introduced species – certainly better than range size alone.

10.5   Conclusions

We have described a new way of considering at geographical distributions that reveals several macroecological features of pines that biologists have overlooked. First, the geographical ranges of pines decrease in size with decreasing latitude (Fig. 10.2), as do range fragments within a species (Figs. 10.3, 10.6, 10.7, 10.8). Small pockets of pines might be explained by the juxtaposition of unusual habitat features in isolated areas. They are, however, better explained as the result of both a global gradient of increasing habitat heterogeneity with decreasing latitude and the nibbling away at suitable habitat sites by competing species.

Second, we have presented an argument and evidence for how interspecific competition across large geographical areas expresses itself and how it influences the formation of range fragments. By looking at the change in the sizes of range fragments with latitude we can make predictions about the success or failure of potential invasions of pines. While the relation between range fragment size and latitude is generally positive, convex correlations (positive on the southern end, then negative to the north) become indicators of competitive effects and the potential for great latitudinal expansion in range.

Acknowledgements

Thanks go to Ellen H. Goldberg for freeing institutional funding for this work, the National Science Foundation (DEB-9318096) for research infrastructure funding, and Carrie Finnance and Kit Matthew for continued encouragement. The manuscript was improved by comments from James H. Brown, Dave Richardson, and two anonymous reviewers.

References

Brown, J.H. (1995). *Macroecology*. Chicago: University of Chicago Press.

Burns, R.M. & Honkala, B.H. (1990). *Silvics of North America*. Vol. 1. *Conifers*. Washington DC: USDA Forest Service.

Colwell, R.K. & Hurtt, G.C. (1994). Nonbiological gradients in species richness and a spurious Rapoport effect. *American Naturalist*, 114, 570–95.

Critchfield, W.B. & Little, E.L. (1966). *Geographic Distribution of Pines of the World*. USDA Forest Service Miscellaneous Publication 991.

Darlington, P.J., Jr (1958). Area, climate, and evolution. *Evolution*, 13, 488–510.

Darwin, C. (1859). *On the Origin of Species by Means of Natural Selection*. London: J. Murray (reprinted 1979 New York: Avenel Books).

France, R. (1992). The North American latitudinal gradient in species richness and geographical range of freshwater crayfish and amphipods. *American Naturalist*, 139, 342–54.

Gage, J.D. & May, R.M. (1993). A dip into the deep seas. *Nature*, 365, 609–10.

Hultén, E. (1937). *Outline of the History of Arctic and Boreal Biota During the Quaternary Period*. Stockholm: Aktiebolaget Thule.

Janzen, D.H. (1967). Why mountain passes are higher in the tropics. *American Naturalist*, 101, 233–49.

Letcher, A.J. & Harvey, P.H. (1994). Variation in geographical range size among mammals of the Palearctic. *American Naturalist*, 144, 30–42.

MacArthur, R.H. & Wilson, E.O. (1967). *The Equilibrium Theory of Island Biogeography*. Princeton, NJ: Princeton University Press.

Macpherson, E. & Duarte, C.M. (1994). Patterns in species richness, size, and latitudinal range of East Atlantic fishes. *Ecography*, 17, 242–8.

Matthew, W.D. (1915). *Climate and Evolution* [reprinted 1939]. Special publication of the New York Academy of Science 1.

Mirov, N.T. (1967). *The Genus Pinus*. New York: Ronald Press.

Pagel, M.D., May, R.M., & Collie, A.R. (1991). Ecological aspects of the geographical distribution and diversity of mammalian species. *American Naturalist*, 137, 791–815.

Pianka, E.R. (1989). Latitudinal gradient in species diversity. *Trends in Ecology and Evolution*, 4, 223.

Preston, F.W. (1962a). The canonical distribution of commonness and rarity: Part I. *Ecology*, 43, 185–215.

Preston, F.W. (1962b). The canonical distribution of commonness and rarity: Part II. *Ecology*, 43, 410–32.

Rapoport, E.H. (1982). *Areography: Geographical Strategies of Species*. Oxford: Pergamon.

Richardson, D.M., Williams, P.A. & Hobbs, R.J. (1994). Pine invasions in the Southern Hemisphere: determinants of spread and invadability. *Journal of Biogeography*, 21, 511–27.

Rohde, K. (1992). Latitudinal gradients in species diversity: the search for the primary cause. *Oikos*, 65, 514–27.

Rohde, K., Heap, M. & Heap, D. (1993). Rapoport's rule does not apply to marine teleosts and cannot explain latitudinal gradients in species richness. *American Naturalist*, 142, 1–16.

Roy, K., Jablonski, D. & Valentine, J.W. (1994). Eastern Pacific molluscan provinces and latitudinal diversity gradient: No evidence for 'Rapoport's rule'. *Proceedings of the National Academy of Sciences, USA*, 91, 8871–4.

Smith, F.D.M., May, R.M., & Harvey, P.H. (1994). Geographical ranges of Australian mammals. *Journal of Animal Ecology*, 63, 441–50.

Stevens, G.C. (1989). The latitudinal gradient in geographical range: how so many species can co-exist in the tropics. *American Naturalist*, 133, 240–56.

Stevens, G.C. (1992). The elevational gradient in altitudinal range: an extension of Rapoport's latitudinal rule to elevation. *American Naturalist*, 140, 893–911.

Taylor, C.M. & Gotelli, N.J. (1994). The macroecology of *Cyprinella*: correlates of phylogeny, body size, and geographical range. *American Naturalist*, 144, 549–69.

Wilson, E.O. (1961). The nature of the taxon cycle in the Melanesian ant fauna. *American Naturalist*, 95, 169–93.

Part five
Ecological themes

11 Fire and pine ecosystems

James K. Agee

11.1 Introduction

Wildland fires and trees of the genus *Pinus* are inextricably linked over space and time. Fire has been responsible for specialized adaptations possessed by the pines and for much of the wide distribution of *Pinus* across its native range in the northern hemisphere and an expanding range as an alien species in the southern hemisphere. Understanding why pines are so widespread requires an understanding of the evolutionary relationship between pines and fire, and the many ways in which fire has interacted with this dominant genus of the Pinaceae.

Fire has long been recognized as an important ecological factor associated with pines; it was given more space than any other physiological or ecological factor in Mirov's (1967) classic *The Genus Pinus*. Since then the literature on fire and pines has expanded considerably, essentially providing more conceptual ideas and specific information for the generalized statements made in the 1960s. Plant-successional models have moved beyond the simplified notion of monoclimax theory to multiple-pathway models (Noble & Slatyer 1980; Christensen 1988), driven largely by the role of disturbance in succession, and fire as a multivariate process has been the most widely studied disturbance.

This chapter focuses on the role of fire at the scale of communities and ecosystems. Although some attention is also given to describing life-history traits and strategies that are important for understanding higher-level responses, readers should consult Chap. 12 (this volume) for a detailed account of the role of fire in shaping pine life histories.

11.2 The fire environment of *Pinus*

Fire, part of the development of *Pinus* since it evolved from its precursor gymnosperms, has been on the earth for hundreds of millions of years, although its source has varied over space and time. The autecology and synecology of *Pinus* are in part a result of this long history of interaction with fire. Although pines are found over a very wide range of northern hemisphere environments, those environments have had fire as a common disturbance factor.

11.2.1 Natural ignitions

The most common natural ignition factor is lightning. Tens of thousands of thunderstorms occur over the earth's surface each day; as many as 1800 may be in progress at any moment (Trewartha 1968; Taylor 1974). Many lightning strikes are cloud-to-cloud, and although some of these may not start fires, enough strike combustible fuels with sufficient energy to ignite them. A meteorological basis for natural fires has existed for thousands of millennia (Komarek 1967, 1968). Where volcanoes exist, both lightning and lava-related ignitions may occur. Natural ignitions have not been constant over the earth's surface. They are usually most significant in relatively dry and exposed (upper slope) locations, but may spread significantly from there to other places on the landscape. Natural ignitions can account for much of the fire activity in many ecosystems of the world.

11.2.2 Cultural ignitions

There is considerable debate about how long humans have controlled fire, with early evidence from Africa suggesting that campfires were used more than a million years ago (Brain & Sillen 1988). There is little doubt that humans had major impacts on vegetation once they began to use fire at will. In Europe, humans first used fire about 400 000 years ago, while in North America this occurred less than 30 000

Ecology and Biogeography of *Pinus*, ed. D.M. Richardson. © Cambridge University Press (1998), pp. 193–218.

Table 11.1. Characteristics of disturbance regimes (adapted from White & Pickett 1985)

Descriptor	Definition
Type	The agent of disturbance: fire, wind, etc.
Frequency	Mean number of events per time period, expressed in several ways:
Probability	Decimal fraction of events per year
Return interval	The inverse of probability; years between events
Rotation/cycle	Time needed to disturb an area equal in size to the study area
Predictability	A scaled inverse function of variation in the frequency
Extent	Area disturbed per time period or event
Magnitude	Described as *intensity* (physical force, such as energy released per unit area and time for a fire, or windspeed for a hurricane), or *severity*, a measure of the effect on the organism or ecosystem)
Synergism	Effect on the occurrence of the same or other disturbances in the future
Timing	The seasonality of the disturbance, linked to differential susceptibility of organisms to damage based on phenology

years ago (Pyne 1982; Prodon, Fons & Athias-Binche 1987; Trabaud, Christensen & Gill 1993). Humans used fire to herd game, in warfare, to signal via smoke, to cook, and to favour plant species used in their diet or daily life. Each culture used fire in different ways and, like natural ignitions, human-caused fires were more important in some locations, such as valleys, than others. The recent spread of pines through their natural range has been associated with human-based increases in disturbance (Richardson & Bond 1991). In Japan, pine forests spread rapidly since about 1500 years ago, when slash-and-burn agriculture expanded (Kamada, Nakagoshi & Nehira 1991). Even today, more than 99% of ignitions in Japanese forests are of human origin (Nakagoshi, Nehira & Takahashi 1987).

11.2.3 Fire as a disturbance factor

Fire is an important disturbance factor that has generally favoured *Pinus* throughout its natural range in the northern hemisphere. The recognition of fire as a complex agent of change in ecosystems is not new, but its incorporation into successional theory is more recent. Disturbance was once considered an exogenous, catastrophic event that affected ecosystems (White 1979). It is now embraced as a much broader concept that has multivariate character (Table 11.1). White & Pickett (1985) define disturbance as 'any relatively discrete event in time that disrupts ecosystem, community, or population structure and changes resources, substrate availability, or the physical environment.' Fire fits quite well into this definition, as it is discrete (although a single fire may burn for months), and changes ecosystem, community, and population structure (Agee 1993). It may favour certain species by selectively removing small or fire-sensitive species, and changing the physical environment simultaneously. Effects on canopy

cover and altered surface albedo can increase surface temperature ranges (Fowler & Helvey 1978). If the role of fire in an ecosystem can be described in terms of the characters in Table 11.1, a good start can be made towards describing the effects of fire.

Fire frequency is expressed as either a point or area frequency (Agee 1993). Where trees are repeatedly scarred by low-intensity fires, a point fire-return interval is calculated by measuring the length of fire-free intervals recorded by annual growth rings on single trees (Fig. 11.1) or the combined record from a cluster of adjacent trees. Where fires have historically been intense, evidence of past fires is collected from age classes of trees that regenerated after past events. Fire frequency is then expressed as area or landscape fire-return intervals, such as natural fire rotation (Heinselman 1973) or negative exponential or Weibull distributions (Johnson & Van Wagner 1985). Disturbance intensity is usually measured by fireline intensity, the rate of energy release per unit length of fireline (Byram 1959; Rothermel & Deeming 1980). Fireline intensity is empirically related to flame length. Variation in frequency and intensity, as well as mean values, are important to describe the ecological outcomes of fire (Wright & Bailey 1982). Fire extent is important for large fires where heavy-seeded species may be killed, so that substantial time may pass before they may again colonize interior portions of the fire. Fire timing (e.g. spring, summer or autumn) may be critical for some species that may be very vulnerable at certain phenological stages but resistant at other times. Synergism, the interaction of fire with other disturbance agents such as insects, disease or wind, is difficult to predict but can have important secondary effects on ecosystem response after fire.

Fire has an important role in the nutrient functioning of pine ecosystems. Nutrient cycling effects relate in part to the nature of the fire, as described above, and in part to where the nutrients are stored in the ecosystem (such as foliage, forest floor, etc.), and what capacity the system has to retain the nutrients affected by fire. At its minimum impact level, fire will simply heat the system, and water loss may be the only chemical change. At higher temperatures carbon begins to volatilize, then nitrogen (at 175 °C) and sulphur (at 375 °C) (DeBell & Ralston 1970; Tiedemann 1987). The latter two elements are often limiting nutrients for plant growth. Elements with boiling points above 600 °C are typically left as ash oxides: potassium, magnesium, calcium and sodium. Subsequently, in the presence of water, they will hydrolyse and as they leach into the soil create more basic soils (higher pH values). In very low intensity fires, only the forest floor nutrient reservoir will be affected (Agee 1993). As fire intensities and duration increase, forest canopy and soil nutrients may be directly affected by the fire. Soil nutrients and soil biota are con-

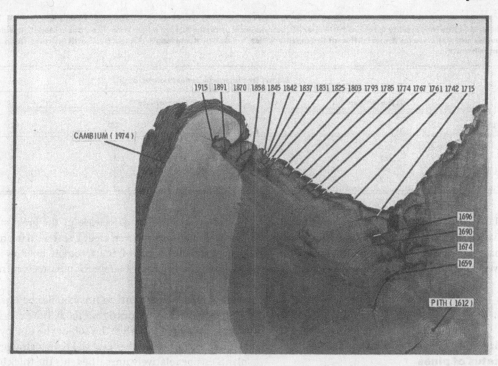

1915 1891 1870 1858 1845 1842 1837 1831 1825 1803 1793 1785 1774 1767 1761 1742 1715

CAMBIUM (1974)

1696
1690
1674
1659

PITH (1612)

Fig. 11.1. **Fire scars on a section from a *Pinus ponderosa* tree show 21 fire scars between 1659 and 1915. Each fire killed only a small** portion of the cambium, allowing regrowth and providing a record of the date of each fire (photo: S.F. Arno, USDA Forest Service).

centrated in the surface horizons, so that the extent of heating of this layer is critical in determining the chemical and biological impact of fire.

The proportion of nutrients above and below ground will vary by location. In temperate pine forests, most of the nitrogen is found below ground (e.g. Grier 1975), and much of it is therefore insulated from direct fire impact. Intense fires tend to volatilize or mobilize larger quantities of all nutrients while reducing nutrient sinks, such as vegetation, at least temporarily. Other nutrient sinks, such as the soil, may be critical in preserving nutrient stores. Coarse-textured soils with low nutrient-holding capacity may be more significantly affected by a fire than a more fertile soil, in that liberated nutrients may be leached below the rooting zone more rapidly (McNabb & Cromack 1990).

The longer-term impacts of fire may depend on the balancing of losses with gains of nutrients. Where nitrogen-fixing sources are present (legumes, *Alnus* spp., *Ceanothus* spp.) and are favoured by disturbance, losses of nutrients like nitrogen may be replaced relatively quickly (Youngberg & Wollum 1976). Nutrients that have limited geological sources, like sulphur, depend on atmospheric inputs, and losses of sulphur from fire may take longer to be replaced from such inputs.

Soil biota will usually be altered by soil heating. Fire has direct effects by killing or injuring organisms, and

indirect effects in altering plant succession, affecting soil organic matter transformations, and changing microclimates (Borchers & Perry 1990). On less fertile sites where pines often dominate, mycorrhizal associations may be very important in facilitating water and nutrient dynamics for the pines (Chap. 16, this volume).

11.2.4 The fire regime

Although myriad combinations of the fire characteristics listed above may occur, simplified systems can help to categorize fire effects in different ecosystems. One system is based on the physical characteristics of fire (Heinselman 1981). Six categories of fire are defined, ranging from short-interval, low-intensity fires to long-interval, intense fires. Another system is based on vegetation classification units, with potential vegetation (*sensu* Daubenmire 1968) integrating general environmental conditions and the successional status of the various plant species (Davis, Clayton & Fischer 1980). A third system (Agee 1993) uses a broader classification based on fire severity, or the effect of fire on the dominant vegetation. The high-severity fire regime includes those vegetation types where fire kills most of the above-ground vegetation, while the low-severity fire regimes are those where historical fires had little effect on mortality of the dominant vegetation. Moderate-severity fire regimes include the range between low and high severity. This system will be used to describe the ecological role

Table 11.2. The probability of mortality for three *Pinus* species, as computed using the FOFEM (First Order Fire Effects Model). Results are ordered by species and size class for fires of different fireline intensities, indexed by flame length (Keane, Reinhardt & Brown 1993). Original output is in English units

| | Size of Tree (diameter breast height, cm) | | | | | | | |
| | *P. ponderosa* | | | *P. lambertiana* | | | *P. contorta* | |
Flame length (m)	12	50	100	12	50	100	12	50
0.60	0.6	0.1	0	0.7	0.3	0.1	0.8	0.6
1.21	0.8	0.1	0	0.8	0.3	0.1	0.8	0.6
1.82	1.0	0.1	0	1.0	0.3	0.1	1.0	0.6
2.42	1.0	0.3	0	1.0	0.3	0.1	1.0	0.8

of fire in the genus *Pinus*. Obviously, the severity of fire within a given ecosystem will depend not only upon the physical character of the fire, but upon the degree to which the vegetation is adapted to survive or regenerate after fire.

11.3 Fire adaptations and successional status of pines

The important morphological adaptations to fire in pines are not simply the result of the presence or absence of fire, but rather to specific combinations of fire frequency and intensity. The major fire adaptations possessed by pines include thick bark, serotiny, rapid development, and sprouting (see Chap. 12, this volume). Thick bark helps pines survive low-intensity fire by insulating the cambium against lethal temperatures. Many pines with thick bark also have thick twigs (McCune 1988) that have high heat capacity and therefore better protection for buds. Cone serotiny is an adaptation that makes many *Pinus* species resilient to fire. After high-intensity fires, the long-closed cones that have stored seeds open and spread abundant seed across the ash-covered landscape. Most species are polymorphic for cone serotiny, allowing establishment after disturbances other than fire, such as insect attack (Muir & Lotan 1985). Some non-serotinous species, such as *P. coulteri*, *P. sabiniana*, and *P. torreyana* retain seed for years within persistent cones on the trees (Borchert 1985).

Sprouting from the root collar or stem occurs in some species, but may be restricted to juvenile stages (Stone & Stone 1954; McCune 1988). Rapid development in its broadest interpretation is characteristic of many *Pinus* species but is applied here to those members of the genus that have a specialized 'grass stage' which is an adaptation to frequent fire. The tree has the appearance of a perennial bunchgrass for several years while it develops a deep root system (Chap. 12, this volume). Fires during the grass stage rarely kill the meristem because it is protected by the thick tuft of green needles. At release of the grass stage, the meristem then grows several feet a year until it is above the typical scorch height of the frequent but low-intensity fires, and the thick bark adaptation protects it from then on.

The response of plants to fire can also be classified by life-history characteristics (Rowe 1983). The *evader* strategy, with seeds stored in either the soil or the canopy, fits the serotinous trait of some pines; the *resister* strategy, where plants escape relatively unscathed, fits the thick bark trait of many pines; and the *endurer* strategy, where plants resprout, fits the sprouting trait of some pines. The *avoider* strategy, applied to plants without any adaptations to fire, includes few if any of the genus *Pinus*.

Tree mortality from fire can be predicted, and can be cumulatively assessed at the stand level. Fire can injure trees in several ways: foliage or bud scorch, bole damage, or root damage (Agee 1993), and tissue death is assumed to occur at 60 °C for one minute. Height (h) of crown scorch is a function of fireline intensity (I) (h_s [m]$=0.148$ $I^{2/3}$; Van Wagner 1973). Lethal temperatures at the cambium can be predicted from bark thickness (BT) as t [min]$=2.9$ $[BT]^2$ (Peterson & Ryan 1986). These two equations, along with tree morphology parameters, have been integrated into equations that predict a probability of mortality from volume of crown scorch and bark thickness (Peterson & Ryan 1986; Ryan & Reinhardt 1988). The computer program FOFEM (First Order Fire Effects Model; Keane, Reinhardt & Brown 1993) predicts mortality by species and size class, given inputs of flame length or scorch height. Several runs of the model showed that *P. ponderosa* is more fire tolerant than *P. lambertiana*, which is more fire tolerant than *P. contorta* (Table 11.2).

The role of fire in plant succession has been studied for much of the 20th century, yet it is only recently that fire has been incorporated into successional theory. Pines were at the centre of early fire studies, with Clements (1910) studying *P. contorta* var. *latifolia* in the Rocky Mountains, and Chapman (1912) studying *P. palustris* in the southeastern USA. Yet so little progress was made in those early

Table 11.3. *Physical and ecological characteristics of three fire regimes. Most pines have ecological amplitudes such that they are found in forest communities in more than one fire regime*

Character	Fire regime		
	Low-severity	Moderate-severity	High-severity
Fire-return interval	Short (<20 years)	Variable (10–70 years)	Variable but often>30 yrs
Fire intensity	Low; flame length (FL)<1 m	Often low to high in same fire; FL variable	High; FL >3 m
Tree life history strategy favoured	*Resisters*	Diverse: *resisters, endurers, evaders, invaders*	*Endurers, evaders, invaders*
Landscape character			
Patch size[a]	Small	Medium	Large
Ecological edge[b]	Low	High	Moderate

Notes:
[a] Patch size refers to a landscape element created by fire that is different from unburned landscape.
[b] Ecological edge refers to the total perimeter of edge between landscape patches differentiated on the basis of species composition and/or structure.

years that Clements (1935), seemingly oblivious of his earlier work, stated: 'Under primitive conditions, the great [vegetation] climaxes of the globe must have remained essentially intact, since fires from natural causes must have been both relatively infrequent and localized.' When ecologists finally abandoned attempts at a grand unifying theory for plant ecology (Christensen 1988), multiple-pathway models emerged that could incorporate disturbance in its multivariate forms (e.g. Noble & Slatyer 1980). By focusing on mechanisms of vegetation change, community-level responses of the various species can be tracked in relation to the permutations of the disturbance regime. Some *Pinus* species are pioneering, early-successional species that are replaced later in succession; others constitute the climax or potential vegetation on a site. Because some pines are widespread across many ecosystems, they may play either role in different ecosystems.

11.4   The fire regimes of pines

All over the world, fire and the dominance of pines are positively correlated. As noted earlier in this chapter, the relation is not between pines and simply the presence of fire, but between pines and particular historical or natural fire regimes. Within these generalized fire regimes, common characters are present (Table 11.3). While the particular species of *Pinus* and its associated trees and understorey components will change within a fire regime, these basic principles apply. Some *Pinus* species may be found in more than one fire regime where they have broad geographical ranges, so most species are not confined to only one fire regime. In the low-severity fire regime, fire-return intervals are shortest, and the negative-feedback relationship between fire occurrence and fuel energy results in low-intensity fires. Mature pines are rarely killed. As fire-return intervals lengthen, typical fireline intensities increase,

and fire severity increases eventually to stand-replacement levels. In areas where historical fires were uncommon, such as Japan, fires tend to be of high severity, with few *resister* and many *avoider* species. In *P. densiflora* forests, almost all species are top-killed or totally killed, but *evader* (called B strategy), *endurer* (called C strategy), and *invader* (called D strategy) life-history strategies are common for both shrubs and trees (*P. densiflora* has a D strategy) (Nakagoshi *et al.* 1987).

In the low-severity fire regime, *resister* species of *Pinus* are usually dominant, and most short understorey dominants have *endurer* or *evader* strategies. Few *avoider* species persist over time. In the high-severity fire regime, even the overstorey species have *endurer* or *evader* life-history strategies, while the moderate-severity fire regime has a mix of *resister*, *endurer* and *evader* species dominating different landscape patches.

The landscape ecology of fire regimes also has a common character. In the low-severity fire regimes, landscape patches tend to be small (often < 0.5 ha) and as most fires have little impact on the overstorey, patch similarity is high and edge is low: the only dissimilar patches are those senescent patches in the transition to regenerating forest. The high-severity fire regime typically has large patches (e.g. Eberhart & Woodard 1987) and although edge may increase with size of fire, the maximum edge tends to be in the moderate-severity fire regime, with highly interspersed patches of variable species composition and structure depending on the severity of the last fire.

The fire regime concept is useful in explaining many of the nutrient effects of fire in pine ecosystems. In the low-severity fire regimes, typical fires consume only forest floors, often with little impact on the overstorey pines. Many of the understorey species will sprout, so that minor amounts of liberated nutrients are retained by the system. Losses of nitrogen may be replaced by legumes or actinorrhizal hardwoods (Youngberg & Wollum 1976) and erosional losses are minimal (Agee 1993). Low to moderate soil heating results in minor effects on soil organisms, but short-term decreases in soil invertebrates usually occur (Metz & Farrier 1973; Borchers & Perry 1990). Stands of *P. taeda* with frequent, low-intensity burns had more diverse populations of Collembola (common insects that frequent organic matter) than unburned stands (Metz & Dindal 1975).

Effects on ecosystem function are more significant in the moderate- and high-severity fire regimes, because the stability of the system is upset by more severe disturbance. Forest floor consumption may be higher, soil heating more severe, and more nutrients liberated while nutrient sinks (live vegetation, organic matter) are reduced. Many of the nutrients are unavailable to plants before burning, but some will be lost by volatilization or leaching as the ecosystem recovers. In moderate-severity fire regimes, residual trees may be important in retaining sources of mycorrhizal fungi for the regenerating cohort of pines (Borchers & Perry 1990). Hardwood shrubs and trees may harbour many of the same fungal species as pines, and their ability to sprout can stabilize populations of mycorrhizae until pines can recolonize the site (Amaranthus & Perry 1987). In high-severity fire regimes, up to 30–40% of ecosystem nitrogen may be volatilized (McNabb & Cromack 1990), and in seral *P. contorta* stands, microbial biomass declines after burning (Entry, Stark & Lowenstein 1986). Such infrequent but intense fires may have more damaging effects on site productivity than fires in low-severity fire regimes (Waring & Schlesinger 1985).

The case histories of six temperate species of *Pinus*, presented below, each exhibit sets of multiple-successional pathways. Two species representing pine forests in each of three natural fire regimes were chosen to represent the spectrum of ecological interactions of pines and fire (Fig. 11.2). In the low-severity fire regime, *P. palustris* and *P. ponderosa* were chosen. In these ecosystems, historical fires were a factor maintaining ecosystem stability, with stable landscape structure over time. In the moderate-severity fire regime, *P. sylvestris* and climax forests of *P. contorta* were selected. Heterogeneity in species composition and structure at a landscape scale was favoured by historical fires of variable severity. *Pinus banksiana* and *P. halepensis* represent the high-severity fire regime, where burned patches may be large and generally of high intensity, and differ in structure from the surrounding landscape for many decades. These examples are but some of the many ways in which fire interacts with pine forests.

11.5 Low-severity fire regimes

11.5.1 *Pinus ponderosa*

Pinus ponderosa (ponderosa pine) is the widest ranging pine in North America, occurring from the west coast of the continent east to South Dakota. As a drought-tolerant species, it persists in stressed environments and is often the first tree species encountered as one moves from grassland or steppe towards more mesic forests. In the driest environments it forms savannas, while in more mesic forests it occurs with a wide range of other conifers, including *Abies concolor*, *A. grandis*, *Calocedrus decurrens*, *Pinus lambertiana* and *Pseudotsuga menziesii*. Its environment is characterized by cool to cold winters and warm, dry summers with prolonged drought of several months' duration. This drought may occur in different seasons across the range of *P. ponderosa*. In California and the Pacific Northwest, the drought occurs in summer, while in the southwestern USA, a two-season (spring–autumn) drought is separated by a typically moist summer.

Fire-return intervals are generally very short in forests where *P. ponderosa* is dominant. Historically, most of these forests had a substantial herbaceous component which, with the long needles of the pine, maintained a continuous fine fuelbed across the forest floor. Some of the shortest fire-return intervals, such as the 1.8 years reported for the Chimney Springs area of Arizona (Dieterich 1980), are partly the result of data colection methods. The fire scars over a 40 ha area were treated as a single sample, which may overestimate the point frequency of fire (e.g. Agee 1993). Other studies have used records from individual trees alone, and since fires do not scar every tree, these may underestimate the actual point frequency: 3–36 years (Weaver 1959; Soeriaatmadja 1966). Bork (1985) reported fire-return intervals using a range of area clustering in two climax *P. ponderosa* forests: 4 to 7 years (165 ha areas), 7–20 to 16–38 years (16 ha areas), and 4–100 to 13–74 years (individual trees). Fire-return intervals have been noted to lengthen where heavy grazing removed most of the grassy understorey fuels (Weaver 1959; Savage & Swetnam 1990).

The predictability of fire-return intervals may play an important role in forests where *P. ponderosa* is but one of several mixed-conifer dominants. Most of the associated species are resistant to fire when mature, but have thin bark as saplings. Occasional long fire-free intervals allows cohorts of those more fire-sensitive species to develop thicker bark before fire burns through the forest, allowing partial survival of the cohort. Substantial numbers of these shade-tolerant individuals surviving that fire will also survive subsequent fires (Thomas & Agee 1986).

These frequent fires provided a negative feedback on fire intensity. Fuels were removed often enough to reduce potential fire intensity. Early accounts of fires in these types, and the pattern of scarring rather than killing trees, suggest that fires of low intensity were the norm in *P. ponderosa* forests. One tree in the eastern Cascades of Washington, suppressed because of its proximity to a large pine, survived five fires while between basal diameters of

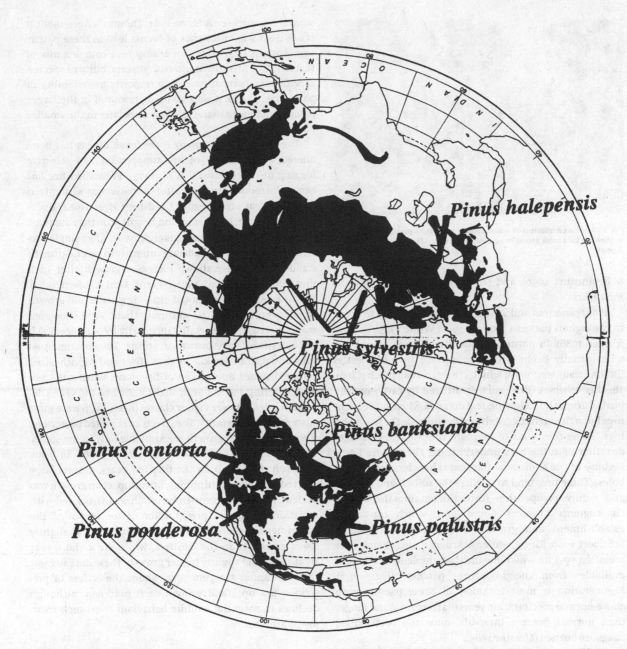

Pinus halepensis

Pinus sylvestris

Pinus banksiana

Pinus contorta

Pinus ponderosa

Pinus palustris

Fig. 11.2. **The ranges of the six *Pinus* species described in this chapter (for *P. contorta* only that part of the range of *P. contorta* where it forms climax forests is shown. Background map modified from Critchfield & Little (1966).**

2–15 cm and heights of 2–5 m, suggesting fires with scorch heights below 5 m and residence times of less than a minute. Simulation models in mixed-conifer forests suggest that at 10-year fire-return intervals, typical fire intensities will range from 50–100 kW m^{-1}, or flame lengths < 0.5 m (Keane, Arno & Brown 1990).

These light fires burned extensively across the landscape. In the eastern Cascades of Washington, some fires burned across many square kilometres, stopping at natural firebreaks such as rivers or at the edge of areas burned in the previous several years that may have been fuel-limited (J. Barnes, unpublished data, Wenatchee National Forest, Entiat, Washington). In the southwestern USA, fire-free and widespread fire years were found to be synchronous with the low (*El Niño*) and high (*La Niña*) phases of the Southern Oscillation. During *El Niño* years, little fire activity occurred because of above-normal precipitation carried northwards from the tropics (Swetnam

Fig. 11.3. **The clustering of small patches of *Pinus ponderosa* is evident in this pure stand in southern Oregon, USA (photo: J.K. Agee).**

& Betancourt 1990), and the pattern was reversed in *La Niña* years.

Pinus ponderosa and its associated species tend to occur in contagious patterns (Cooper 1960; White 1985; Thomas & Agee 1986). In natural stands, stand development patterns usually begin with a group kill by bark beetles (*Dendroctonus brevicomis*) which creates a snag patch (Chap. 18, this volume). This patch is utilized by cavity-nesting birds during the following few decades when subsequent fires burn through the patch and decompose the snags and logs. These fires tend to be more intense and of longer duration than the fires underburning the mature forest because of the additional fuels and their large size (>1 m boles). These fires tend to sterilize the soil around the logs and reduce competition from rhizomatous herbs (e.g. *Calamagrostis rubescens*, *Carex* spp.), which can affect establishment and growth of *P. ponderosa* (Larson & Schubert 1969; Riegel, Miller & Krueger 1992). Patch sizes are often <0.3 ha which means that ample seed source is available from neighbouring patches (Fig. 11.3). Regeneration is most common in these patches with above-normal precipitation years (Barrett 1979) and longer than normal fire-free intervals, once the beetle-killed snags are burned (Morrow 1985).

Subsequent fires thin sapling and pole-sized stands and create another even-sized patch on the landscape. Hypotheses about even-aged patches (Cooper 1960) comprising the entire forest have been found somewhat simplistic, as there are large age ranges within patches (White 1985). Nevertheless, at the landscape level, there are numerous clusters of generally different ages (Morrow 1985) representing an all-age/all-size landscape comprising small clusters of more even distributions of age and size. In climax *P. ponderosa* forests, all the clusters are *P. ponderosa*, whereas in mixed conifer forests, there are clusters of *P. ponderosa* and other clusters of *P. lambertiana* and *Abies*

concolor (Bonnicksen & Stone 1981; Thomas & Agee 1986). It is not clear what sequence of events lead to these patchy aggregations of species, but it may be a complex mix of variability of fire-return interval, susceptibility to species-specific insects or disease, or a response to variability in patch size, with *P. ponderosa* being favoured in the larger patches and more shade-tolerant associates in the smaller patches.

The fire regime in many *P. ponderosa* forests has been altered by a century of fire suppression and selective logging of pines (Biswell *et al.* 1973). Prescribed fire has been reintroduced to these altered ecosystems with mixed results. In eastern Oregon, Landsberg *et al.* (1984) documented volume, basal area, and height growth reductions compared with control stands eight years after single prescribed fires of low to moderate intensity in second-growth stands. Others have shown increased growth after such fires (Morris & Mowat 1958; Wyant, Omi & Laven 1983). Swezy & Agee (1991) noted little impact if old-growth stands were burned in autumn; there were, however, major increases in mortality (primarily by bark beetles) if the older trees were burned in spring. They documented substantial fine-root kill by these fires and hypothesized vigour declines due to loss of fine roots when the trees were entering the major drought period of the year. Old-growth trees in lower vigour classes (Keen 1943) were most susceptible to mortality. The effects of repeated prescribed fire on *P. ponderosa* growth are available only in the southwestern USA for rotations of 1, 2, 4, 6, 8 and 10 years (Peterson *et al.* 1994). In the first few years, despite large fuel reductions and thinnings, little impact on growth was evident. Most of the growth variation was associated with variation in annual precipitation. After year 8 of the experiment, the 1, 2, 8 and 10-year treatments had slightly lower growth than the controls, while the 4 and 6 year treatments had slightly higher growth. These data suggest that we cannot yet generalize about the effects of prescribed fire on stand growth of *P. ponderosa*, although declines in potential wildfire behaviour from such treatments are clear.

11.5.2 *Pinus palustris*

Pinus palustris (longleaf pine) occurs across 25 million ha of the Coastal Plain region of southeastern North America. Its range includes long growing seasons (300–350 days) and 1000–1500 mm of evenly distributed annual precipitation, with 2–4 week droughts during which fires may occur (Christensen 1981, 1991). High lightning frequencies are typical of this region (Komarek 1968), but because the thunderstorms are usually associated with substantial precipitation, fires may not be ignited by most cloud-to-ground strikes. Burning by Native Americans was probably an important ignition source before the 1800s (Komarek

1974). *Pinus palustris* occurs in two major vegetation types: the xeric sandhill type, along with *Quercus laevis*, and the more widespread pine savanna type, where *P. taeda* and *P. elliottii* also occur. *Aristida stricta* is a common graminoid associate in both types, and *Andropogon* spp. in the pine savanna type. These grasses, with fuel from other herbs and the long needles of *P. palustris*, can provide ample fuel for fire spread within a year or two of a previous fire.

Pinus palustris has a low-severity fire regime. Fires are frequent but of low intensity, with fire frequency exhibiting a negative feedback effect on fire intensity. In the xeric sandhills, minimum fire-return intervals are probably 3–5 years, because of limitations on continuous fuels (Christensen 1981). Fire-return intervals may increase to 30–40 years where *Aristida* is absent or where *P. palustris* has been locally extirpated by logging. 'Natural' fire frequencies have been described as from 2–10 years over most of the range (Chapman 1932; Heyward 1939; Wells 1942; Wahlenberg 1946). This has been supported by three lines of other evidence (Platt, Glitzenstein & Streng 1991): the rapid invasion of pine savannas by hardwoods if fire is excluded for a few years, the high frequency of lightning, and a large number of herbaceous and woody species that resprout vigorously and can survive fire. Localized disturbance (lightning) can spread widely across the landscape through understorey grass and combustible leaf litter of the pines.

At such high fire frequencies, limited biomass can accumulate between fires, so that fire intensity is limited by the available energy. In savanna landscapes, fires must have spread widely across the landscape, but historical fire extent has not been studied in this type. With possible conditions for fire spread occurring year-round, fires in any season are possible. Prescribed fire studies have included comparisons of autumn, winter, spring, and summer fires (Greene 1931; Boyer 1991).

Pinus palustris has an unusual adaptation to a regime of high-frequency, low-intensity fires: the 'grass stage'. It has large seeds that prefer mineral soil seedbeds, so an immediate post-fire environment is considered favourable (Chapman 1932). Early growth is concentrated in root development, particularly a long taproot. The apical bud is protected near ground level by a bunchgrass-like clump of needle-like leaves, and the bud has fire-resistant scales (Christensen 1981). While in the grass stage, the needles are subject to infection by *Scirrhia acicola*, called brown spot needle blight, which is controlled by fires scorching the needles (Siggers 1944). Most pines in the grass stage will survive these low-intensity surface fires, while other tree species will be killed. After 3–5 years in the grass stage (Allen 1964), the apical bud begins to grow rapidly, and in another 3–4 years will have reached a height where it is protected against scorch from an average fire. The zone

Fig. 11.4. **The savanna-like nature of *Pinus palustris* forest is maintained by frequent, low-intensity fires (photo: W.D. Boyer, USDA Forest Service).**

between 0.3 and 1.5 m is the most critical for the bud; below that level the grass stage protects the bud and above that level lethal temperatures rarely occur.

Fire is an essential process in the growth and development of *P. palustris* stands (Chapman 1932). The herbaceous layer comprising the understorey of *P. palustris* savannas, in the absence of fire, will be replaced quickly by hardwoods on productive sites (Christensen 1981). In the northern coastal plain, *P. taeda* may replace *P. palustris*, while further south *P. elliottii* may become dominant with longer fire-return intervals. Frequent fires control the hardwoods and other conifers and maintain the savanna (Fig. 11.4). In its development from a seedling, through sapling and pole sizes over the first 25 years of its life, *P. palustris* may survive as many as 10 fires (Mattoon 1925).

Once in the pole stage, a frequent fire regime will have little effect on *P. palustris* density, as the thick bark developed by young trees will protect cambial tissues. Effects of fire on subsequent tree growth have been shown to be both positive and negative in comparison to control stands where fire has been excluded. Plots in young *P. palustris* burned biennially in the spring had better height and diameter growth than unburned stands (Grelen 1975). In contrast, Boyer (1991) found both height growth and volume growth were less in biennially burned stands (burned in winter, spring and summer) compared with controls. Effects were less apparent after stands were 30 years old. Maintenance prescribed burns to reduce competition for pines and reduce fire hazard have been suggested at 3-year intervals (Sackett 1975), well within the natural range of disturbance variability in these ecosystems.

Some *P. palustris* stands are 'islands' or 'hammocks' that are isolated from more continuous savanna landscapes. These areas are cut off from other savanna areas by broadleaved forests that occupy moister areas. These hammocks probably burn less frequently (Harper 1911;

Christensen 1981), but may occupy sites dry enough that *P. palustris* can still maintain landscape dominance. When these areas do burn, the accumulated fuel energy release kills competing vegetation and leaves *P. palustris* as a dominant, preventing conversion to other vegetation types (Myers 1985).

A mature forest patch can be maintained for centuries without further regeneration, but there is little information available on historical patterns of stand development. Most of the natural stands were cleared as a result of logging and agriculture beginning in the early 1800s (Platt *et al.* 1988). Adult tree density is negatively associated with pregrass stage *P. palustris* (Grace & Platt 1995), and juveniles in areas of low needle litter accumulations are larger and have higher post-fire survival than juveniles near mature trees. This negative feedback between mature trees and regeneration results in a strongly aggregated pattern of regeneration in openings (Platt *et al.* 1988). The openings may be associated with patches where trees have been attacked by southern pine beetles (Chap. 18, this volume). Windthrow may occur in stands where fusiform rust (*Cronartium fusiforme*) has weakened trees (Chap. 19, this volume), although this is more likely to occur on loblolly pine (*P. taeda*) than *P. palustris*. This pattern is less evident as stands develop over time. In a Georgia forest relatively undisturbed except by frequent fire, tree recruitment at the landscape level (60 ha) was almost continuous over a two-century period, suggesting a small patch size and very stable landscape structure over time in the presence of fire (Platt *et al.* 1988).

11.6 Moderate-severity fire regimes

11.6.1 *Pinus sylvestris*

Pinus sylvestris (Scots pine) has the widest distribution of any pine, with a range encompassing over 10 800 km^2 (Nikolov & Helmisaari 1992). It occurs north of 37° N and generally above 50° N east of the Caspian Sea. The natural range of *P. sylvestris* spans the longitudes from Scotland (10° W) to northeastern Asia (150° E). It grows on dry, nutrient-poor soils to wet nutrient-rich sites, and generally exhibits poor growth on peat and rocky soils. In the forest zones, it is a xerophyte, while in the grasslands of southern Siberia it is a mesophyte (Mirov 1967). Across this wide range of environments *P. sylvestris* is found with most of the boreal species of Europe and Asia.

Pinus sylvestris has a moderate-severity fire regime. The fire regime of *P. sylvestris* appears to include a wider range of fire frequencies and less intense burning than for the boreal pines of North America. The boreal North American pines are estimated to have fire-return intervals for gener-

ally intense fires at *c.* 60+ years and generally have the shortest fire-return intervals of the vegetation types present (Wein & MacLean 1983). In contrast, the fire regime of *P. sylvestris* forests ranges from those fire frequencies to conditions where fire appears to play an inconsequential role in forest dynamics. Near the treeline, for example, fire is not a significant ecological factor in *P. sylvestris* forests. Kullman (1986) found no evidence of past fire at the tree limit in the Scandes Mountains of central Sweden. Humid climate and extensive mires in the area have caused this area to be fire-free. No fire scars were observed on trunks of live trees or on logs, and no charcoal was found in the humus. However, most stands have burned at least once in the last half-millennium (Siren 1974; Zackrisson 1980; Gromtsev 1993), and many sites have burned at much more frequent intervals. A mire in northern Sweden (64° N) had six charcoal layers in the peat profiles, with a 50-year fire-return interval between 1653 and 1758 and no fire since then (Agren, Isaakson & Zackrisson 1983). Fire frequency around a *P. sylvestris*-dominated lake basin in eastern Finland was about 75 years, using charcoal influx in a varved lake (Tolonen 1983). Fire scars on one *P. sylvestris* in Sweden showed fires at intervals of 42, 41, 49, 47, 27, 17 and 47 years, with a mean of 39 years (Zackrisson 1977). Sannikov (1985) developed a succession hypothesis for *P. sylvestris* forest in Siberia using a fire-return interval of 20 years. In northeastern China, Goldammer & Di (1990) estimated a fire-return interval of 25 years (range 6–54). This wide range of mean fire-return intervals encompasses the expected range of variation for a species with such a wide boreal range.

The implication of multiple fire scars is that many repeated fires are of low to moderate intensity. While some fires can be stand-replacing events (Fig. 11.5), many occur at intensities low enough to allow some tree survival. Thick bark is the primary fire adaptation possessed by *P. sylvestris* (Nikolov & Helmisaari 1992). Multiple-aged stands, representing past fires of moderate severity, and low-intensity fires are common: in northeastern China (Goldammer & Di 1990), in the southern taiga of Siberia (Kurbatskii & Ivanova 1980), in Russia (Kotov 1980), and in Sweden (Zackrisson 1977, 1980). Artsybashev (1983) identifies surface fires in grassy forests and carried by lichens, as well as those which consume considerable forest floor and understorey (*Vaccinium* spp.). However, even-aged stands also exist, and suggest past fires of stand-replacement intensity (Artsybashev 1967; Kurbatskii & Ivanova 1980). These fires occur in pure pine forests, often with coniferous understorey and either *Calluna* or *Ledum* spp. (Artsybashev 1983). Crown fires are less common in the boreal forests of Asia (Shcherbakov 1977), particularly if *Betula* is intermixed with *P. sylvestris*, and account for about 25% of the fires within the range of *P. sylvestris* in

Fig. 11.5. **Fire can be a stand-replacing event even in _Pinus sylvestris_ forest. This forest is part of the Bor Forest Island Fire Experiment in central Siberia. _A._ Before burning, the overstorey is _P. sylvestris_ and the understorey _Cladonia_-dominated lichen cover. _B._ One day after the experimental burn, the overstorey appears intact, with the dominant trees showing little crown scorch. Coarse woody debris and lichen cover have been consumed. _C._ One year later, bark beetles (_Dendroctonus_ spp.) have infested the stand, with the synergistic effect of insects resulting in almost total mortality of the overstorey (FIRESCAN 1994) (photos: J.G. Goldammer).**

Russia (Artsybashev 1967). Estimates of severely burned and non-regenerating _Pinus_ forests in Russia showed only about 0.5% in that category, compared with 4.5% for more easterly larch forests (Dixon & Krankina 1993). This variable-intensity regime is characteristic of the moderate-severity fire regimes (Agee 1993).

Some fires in _P. sylvestris_ forests can be of great extent. Fires over 150 000 km^2 burned in Siberia in 1915 (Shostakovitch 1925). In 1987, major fires were occurring in Siberia while the Black Dragon fire was burning in northeast China (Goldammer & Di 1990). In Russia, significant amounts of land burn each year: $1.4–2.4 \times 10^6$ ha (1971–85) to 7×10^6 ha (1987–91) (Dixon & Krankina 1993). Broadleaved forest can act as a natural firebreak (Otto 1980), as the leaves of deciduous trees have higher moisture content. Fire extent is influenced by season of burn, as understorey plants will typically dampen fire spread early in the season and accelerate it later (Kurbatskii & Ivanova 1980).

Pinus sylvestris is the most fire-tolerant tree in the boreal forests of Europe and Asia (Uggla 1958, 1974). Its thick bark (Pobedinsky 1979) allows it to remain undamaged after low-intensity fires, but even if 90% of the cambium is killed the tree can survive (Zackrisson 1977). Forest fires create conditions for successful regeneration of *P. sylvestris* by consuming the humus layer and exposing mineral soil (Drackenberg 1981; Sannikov 1983). Once established, seedlings can grow 10–15 cm yr^{-1} (maximum *c.* 70–100 cm yr^{-1}) to maximum heights of *c.* 45–50 m in the southern range to *c.* 35 m in the northern range (Nikolov & Helmisaari 1992). *Pinus sylvestris* can begin seeding at 8–20 years of age (Nikolov & Helmisaari 1992) which allows it to expand its own seed source where seed is limiting, and also possibly to maintain a site presence when two fires occur in a 20-year period. In the *Calluna*-type, seedlings grow better on burned sites, but in the more mesic *Myrtillus*- or *Oxalis–Myrtillus* types, seedlings start with higher height growth on burned areas but height growth drops below unburned areas after a decade (Braathe 1974). Height growth of *P. sylvestris* is about twice that of *Picea abies* in open environments. Seedling and sapling mortality from moose and vole grazing, drought, and snow blight (*Phacidium infestans*) can slow tree recovery on burned sites (Uggla 1958).

A variety of stand architectures can exist in this variable-severity fire regime. On the driest, most frequently burned sites, very uneven-aged stands exist (Zackrisson 1980), and *P. sylvestris* can be a climax species (Nikolov & Helmisaari 1992). As fire-return intervals lengthen (and presumably fire severity increases) stands with two or three discrete age classes are found, and single-aged stands from high-severity fires also are found (Zackrisson 1980). Sites with longer fire-return intervals are usually more mesic sites, where *P. sylvestris* is seral to *Picea abies* (Gromtsev 1993), and fire acts to favour *Pinus sylvestris*, as it is more fire tolerant than *Picea abies*. Near the treeline, where fire is less significant, age structure of *P. sylvestris* stands may represent long-term climatic trends (Kullman 1986). Pines are found on warmer, south-facing slopes, where their presence appears to be the result of the recolonization of forests that existed on these sites before the Little Ice Age.

Most of the understorey species in *P. sylvestris* forests are well-adapted to fire (Sannikov 1983). The widely distributed *Calluna vulgaris* reproduces from seed, while *Vaccinium vitis-idaea* and *V. myrtillus* sprout. In the Transurals of Russia, about two-thirds of the herbs and shrubs are highly resistant to fires (Chizhov & Sannikov 1978). Vegetation change in 40 to 70-year-old post-fire forests are minor, with increases in humus-rooted herbs such as *Linnaea borealis* and some mosses (Vanha-Majamaa & Lahde 1991).

Pinus sylvestris forests occur across wide landscape gradients, and fire frequency and effects are likely to differ across the gradient. South-facing slopes will burn more frequently (Zackrisson 1977) and uneven-aged forests are therefore more likely on those slopes. North-facing slopes are more likely have spruce as an understorey component that will increase over time, although the long life of *P. sylvestris* is likely to allow it to maintain co-dominance even if fire is absent for centuries. Some stand-replacement fires will favour birches (*Betula pubescens* and *B. verrucosa*), which will have subsequent dampening effects on landscape flammability (Otto 1980). The moderate-severity fire regime in *P. sylvestris* forests has greatly affected landscape diversity in the boreal forests of Europe and Asia.

11.6.2 *Pinus contorta* [climax]

Pinus contorta (lodgepole pine) is widely distributed across western North America, and is found from sea level to subalpine environments. Some varieties are more serotinous (var. *latifolia*) than others (var. *murryana*), and *P. contorta* serves a variety of ecological roles in these diverse forest ecosystems. In the Rocky Mountains, it is primarily an early-seral species favoured in high-severity fire regimes over *Pseudotsuga menziesii* or *Abies lasiocarpa* by infrequent but high-intensity fires. However, there are other areas where *P. contorta* is the only tree species that can reproduce with or without fire, so it functions in both early- and late-seral roles. These climax *P. contorta* forests are found in Colorado (Moir 1969), in Yellowstone National Park (Despain 1983), and widely across south-central Oregon (Zeigler 1980; Stuart 1983). They exist in areas with severe limitations on productivity: either severe microclimate or very infertile soils. The remainder of this discussion focuses on these forests where *P. contorta* is dominant in early and late succession.

These *P. contorta* forests have a moderate-severity fire regime, and disturbance by insects (*Dendroctonus ponderosae*) and diseases (Chapters 18 and 19, this volume) are closely tied to the fire regime, with strong synergistic interactions. Fire-return intervals are available from both age class and fire scar records. Where these forests are found in large patches, the average fire-return interval is probably 60–80 years (Agee 1981; Chappell 1991), although individual trees may survive up to 350 years on these sites (Stuart 1983). Where surrounded by other forest types and found in only small patches, the fire-return intervals may mimic those of the surrounding forest types. Under most conditions, these are considered fuel-limited ecosystems, so that fires in adjacent forests may not spread into the *P. contorta* stands. An unusual characteristic of these forests is that productivity is so low and phloem so thin so *Dendroctonus ponderosae* attacks may result in only strip kills of cambium along the lower trunk, which can easily

Fig. 11.6. **Formation of basal (A), tapering basal (B, C), and ellipsoidal (D) fire wounds on *Pinus contorta* after a 1980 fire at Crater Lake, Oregon. These wounds are compared with fire scars (E–H) formed by** an 1898 fire on the neighbouring Fremont National Forest (from Gara *et al.* 1986).

Fig. 11.7. **The 'cigarette burns' in climax *Pinus contorta* forest smoulder along log corridors and continue as long as there are continuous logs in contact with each other (photo: J.K. Agee).**

be mistaken for fire scars (Stuart *et al.* 1983; Gara *et al.* 1986; Fig. 11.6).

Fires may range from crown fires to slow spreading, log-to-log smouldering fires (Fig. 11.7), from intensities exceeding 5000 kW m⁻¹ to those so low they cannot be measured with the usual flame-length criteria. The logs that burn are usually the most continuous fuel in the landscape, even though they may have among the highest dead fuel moisture (Agee 1993). Litter fuels and the associated flora are too sparse to carry surface fires under most conditions. Fire extent has not been quantified over large areas. To some degree it may depend on the continuity of logs that were the product of previous disturbances, or the continuity of crown fuels in the event of the unusual wind-driven crown fire event.

Synergism with other disturbances is the most unique attribute of the disturbance regime in these *P. contorta* forests. It is perhaps more obvious in this forest type because of its exceptionally low productivity (1–2 m³ ha⁻¹ yr⁻¹; Volland 1976). Trees killed by fire or bark beetles, when partially decayed, are the vectors for the smouldering 'cigarette burns' of later decades, which slowly smoulder along the matrix of fallen logs (Fig. 11.7) and result in tree death either directly by fire or by bark beetles. Those trees that survive may be wounded, allowing pathogens to create decay columns attractive decades later to further bark beetle attack. Where fires have burned along log corridors, tree roots passing under the log may be scarred on one side and create an infection court for fungi, primarily *Poria asiatica* (Littke & Gara 1986). The subsequent decay column moves only a few centimetres per year, but may be 1–2 m up the central bole after a century (Gara *et al.* 1985). If a tree has had several logs burn near its base, its roots may be substantially damaged, and it is likely to be immediately attacked by *Dendroctonus ponderosae*. If the tree survives with incipient decay, it will grow more slowly than undecayed trees and be preferentially selected for attack by *D. ponderosae* in later decades (Geiszler, Gara & Littke 1984). Fire, insects and disease are tightly linked in the stand dynamics of these *P. contorta* forests.

Pinus contorta is not well-adapted to survive fire at its base, because of very thin bark. It often has most of its roots near the surface, particularly where the soils are derived from pumice, so substantial root damage is possible if fire burns near the tree base. Most of the climax *P. contorta* forests have non-serotinous cones, which may be due to the fire regime favouring this character. Even in *P. contorta* forests where the serotinous character is expressed, cone polymorphism exists and allows regeneration after non-fire disturbances such as bark beetle attack (Muir & Lotan 1985). Annual seed production in the non-serotinous forests may exceed 1.5 million seeds ha^{-1}, with high viability (Stuart, Agee & Gara 1989). Regeneration after fire or beetle attack is rapid and abundant, with most of the regeneration near snags and logs. These areas provide microclimate buffering (against extreme hot and cold temperature) and are areas with the most available soil moisture. A clear association exists between seedling groups and insect- or fire-killed trees (Stuart 1983). Seedling densities may exceed 5000 ha^{-1} by year 10 (Agee 1993).

The trees killed by disturbance begin to fall within a year or two after death, and the process is largely complete within a decade. Most logs are by then in contact with the soil, and a slow decay process begins. The extent and severity of the previous disturbance will affect the distribution and density of new logs, and therefore the probability of future log-smouldering events, which depend on continuous intersections of logs to carry the slow-moving fires. Litter layers are too thin and discontinuous to carry fire even after a century.

If the disturbance is not a crown fire, the stand is usually thinned from above (larger size classes removed).

Dendroctonus ponderosae will attack up to two-thirds of the trees, focusing on large size classes with low vigour due to drought, fire damage or decay (Chap. 18, this volume). Although fire typically thins from below, in these forests with such patchy fire spread fire-killed trees many range over a wide size range, but subsequent beetle attack on damaged trees will result in a high thin. The combination of fire and beetle on one site reduced average tree age from 124 to 70 years, with average diameter of trees removed (18 cm) about twice that of the residual stand (Agee 1993).

Multiple-aged stands with age classes separated by 60–80 years typically result from this complex disturbance regime. This interval is caused by two developmental sequences of logs and trees. The residual trees from previous disturbances require about that length of time to grow to a size where they are both susceptible to and capable of supporting a brood of *D. ponderosae*. The logs require about that length of time to decay internally so that they have an exterior hard rind of sapwood that conserves energy and allows moisture ahead of the smouldering front to be evaporated (Agee 1993). Intervening age classes need not be alternating beetle–fire–beetle episodes, but three-aged stands are not unusual. Every few episodes a stand-destroying fire is possible, so that the multiple age classes usually have an initiating age class that is widespread and even-aged (Gara *et al.* 1985).

The coupling of fire, insects, and disease disturbances is crucial to an understanding of the dynamics of climax *P. contorta* ecosystems. In these relatively simple ecosystems (one tree species, and fewer than 10 other vascular plants), these dynamics can be unravelled much more easily than in more highly structured ecosystems. It is likely, however, that the synergism of disturbances in *P. contorta* forests exists in more complex and as yet less understood form in other forest ecosystems.

11.7 High-severity fire regimes

11.7.1 *Pinus halepensis*

Pinus halepensis (Aleppo pine) is a serotinous-coned pine found around the rim of the Mediterranean Basin, comprising about 10% of the shrubland/forest complex (Le Houérou 1974; Chap. 8, this volume). Its distribution follows the 1.5 °C isotherm of mean minimum temperature for the coldest months of the year (Trabaud, Michels & Grosman 1985), with annual precipitation in the range 300–900 mm (Le Houérou 1974). Summer temperatures average 20–25 °C and a prolonged three-month drought is typical. Drier environments may be dominated by *Quercus suber* (Summers 1939), higher elevations often have other pine species (*P. pinaster, P. pinea, P. nigra*), and calcareous soils within the range of *P. halepensis* are often dominated by *Quercus* spp. Generally, *P. halepensis* is restricted to the poorer soils with fewer nutrients (Kutiel & Naveh 1987).

The Mediterranean Basin has such a long history of intense human use of wildlands that the natural fire regime is difficult to define. The climate is such that each year conditions for widespread wildfire exist, and the *P. halepensis* zone is one of the most flammable and fire-prone within the Mediterranean Basin. During the 1950s to 1970s, over 200 000 ha yr^{-1} burned in Mediterranean forest and shrublands, and in the 1980s up to 550,000 ha yr^{-1} burned (Trabaud *et al.* 1993), with about one-third of the total in *P. halepensis* forest (Le Houérou 1974). In Greece, about 60% of burned area (1956–71) was in *P. halepensis* and *P. brutia* forest (Liacos 1974). Lightning ignitions account for only 2% of the area burned (Trabaud *et al.* 1993); intentional ignitions have been a powerful source of fire for millennia (Naveh 1974; Chap. 8, this volume).

The maximum forest fire-return interval is estimated to be about 30–50 years. In the Riviera 'red zone', so named because of its propensity to burn, the probability of a pine living 20 years without being burned is about 60% (Naveh 1990), making the average fire-return interval about 25 years. Due to the relatively dense cover of shrubs in these forests, the fireline intensity of a typical fire is sufficient to kill the pines. At about 20 m height for a mature tree (Trabaud, Grosman & Walter 1985), a flame length of 2.2 m would be sufficient to scorch 100% of the crown. In areas with light canopies of pines, a 30–70% shrub cover, with shrubs 0.5–1.2 m tall (May 1990) would be sufficient under summer weather conditions to create such fireline intensities.

Pinus halepensis is considered 'fire-sensitive' (Zohar *et al.* 1988) because it is usually killed by wildfires in semi-natural landscapes. However, low-intensity prescribed fires have been used with success in dense stands with little shrub understorey (Naveh 1974). After forests are about 30 years old, the bark is thick enough to protect the trees from heat from low-intensity surface fires (Liacos 1974). The bark thickness of *P. halepensis* and its resistance to low-intensity fires should be comparable to several fire-resistant western North American conifers, including *P. ponderosa* (Ryan, Rigolot & Botelho 1993). Because *P. halepensis* possesses an evader strategy with its serotinous cones, it is quite resilient, even in the face of intense fires. It is a precocious seed producer, at 7 years (Trabaud *et al.* 1985a), which serves two ecological functions: the species can survive two fires in a decade, and early colonizers may produce seedlings for the same cohort of post-fire colonizers, as colonization may sometimes continue for a decade or two (Trabaud *et al.* 1985a).

The immediate post-fire environment is often ash-rich. This can provide benefits particularly to herbaceous species (Naveh 1990). Amounts of available nitrogen, phosphorus, calcium and iron may be 3–10 times higher than on unburned areas. Seed from serotinous cones of *P. halepensis* rains down into the ash. Near the trunks of large trees, seedling survival can be low, as the thick ash produced from the litter causes osmotic stress to emerging seedlings (Ne'eman, Lahav & Izhaki 1992). Growth of the few seedlings that become established near the trunk is high, however, as there is less competition due to the previous occupation of the growing space by the tree killed by the fire. Highest seedling densities occur away from the trunks of large trees, and establishment may be as high as 50 000–200 000 seedlings per hectare (Ne'eman *et al.* 1992; see also Chap. 12, this volume). Growth in the open is slower, and competition from shrubs and herbs may kill up to 50% of the seedlings (Kutiel, Naveh & Kutiel 1990). An unusual drought year, or grazing, if it occurs on the area, may cause additional seedling mortality (Naveh 1990; Trabaud *et al.* 1993). Seed density on the ground is not associated with distance from an unburned seed source, for several reasons (Trabaud *et al.* 1985a): there may be seed from cones in the burned area that opened immediately after the fire; additional seeds may rain down from later opening cones on the burned trees; and polymorphic open cones from either individual residual trees or unburned adjacent forest may provide seed.

An initial-floristics model (Egler 1954) is typical of post-fire succession. Pines establish immediately after fire, although after 10 years they may be only 1 m tall (Trabaud *et al.* 1985a). The fire tends to kill some of the shrub understorey, and species like *Cistus clusii*, *C. incanus*, *C. salvifolius*, *Rosmarinus officinalis* and *Ulex parviflorus* recover from seed banks in the soil (May 1990). Root sprouters like *Arbutus andrachne* and *Pistachia lenticus* will generally recover their pre-fire cover faster than the species that are obligate seeders. Phenology may be accelerated in burned areas (Trabaud & deChanterac 1985). A larger amount of available nitrogen after fire (even though total nitrogen decreases) accelerates the understorey recovery (Kutiel *et al.* 1990). Within a decade, the understorey shrub biomass stabilizes at about 10–15 t ha^{-1}, and the herbaceous layer at 1–2 t ha^{-1}. The tree layer continues to increase in biomass over time, ranging from 30 to 175 t ha^{-1} and producing a litter layer of 12–16 t ha^{-1} (Trabaud *et al.* 1985b; Fig. 11.8).

The growth of individual *P. halepensis* trees slows down as they age, with height growth of 15 cm yr^{-1} (Trabaud *et al.* 1985a). Because the genetic potential for height growth is limited, and there is a high probability fire will return to the site before the tree has reached maximum height, the chance of trees surviving fire is limited. If the stand is open, the shrubs, primarily sclerophyllous, will burn with enough intensity to kill the trees. If the stand is closed and

Fig. 11.8. **A very dense *Pinus halepensis* forest, about 15 years after a stand-replacing fire, near Montpellier, France (photo: D.M. Richardson).**

dense, the forest may burn with a crown fire. An evader strategy, given these outcomes, is a prescription for success for *P. halepensis*.

The landscape ecology of *P. halepensis* cannot be separated from its human history. Because it is so well adapted to frequent, intense fire, it may have expanded its range into areas that were once forests of less fire-tolerant conifers (Trabaud *et al.* 1993). It is clearly more successful, for example, than *P. pinaster* (May 1990). However, in areas where repeated fire may occur within a decade, the seed sources for *P. halepensis* may be exhausted, and shrublands may result (Le Houérou 1977). Although *P. halepensis* might disappear locally from the landscape if it were protected from fire for 100 years or more (Naveh 1990), this is a remote possibility for most Mediterranean landscapes given current fire patterns. The biodiversity of today's Mediterranean landscape reflects a long history of fire, and that biodiversity is managed largely by accident as wildfires burn and reburn the landscape. If fire is ultimately controlled in these warm, dry environments, there will still be a place for intelligent use of fire over space and time to maintain this important element of Mediterranean landscapes.

11.7.2 *Pinus banksiana*

Across the northern part of North America is a widespread forest region known as the boreal forest. Although there is considerable variability in the composition and structure of the boreal forest from east to west and north to south, it is characterized by long, cold winters, and short, warm summers. Closed boreal forest in more southerly areas, where *P. banksiana* (jack pine) is a dominant species on coarse-textured and nutrient-poor soils, gradually changes to the north to more open forest stands that are interspersed with bogs and some closed-canopy forest, a mosaic called the taiga. The major tree species in the southerly boreal forest along with *P. banksiana* are *Picea*

mariana, Picea glauca, Betula papyrifera and *Populus tremuloides*.

Very long days during June and July and temperatures exceeding 33 °C (Viereck 1983) can create severe fire weather. Such weather encouraging intense fires (Fig. 11.9) generally begins with several weeks without rain, occasional days of low humidity and high day temperatures, and lightning (Van Wagner 1983). Lightning is the cause of about 90% of the area burned in the North American boreal forest (Stocks & Street 1983). Large crown fires may occur in spring months when conifer foliar moisture dips just before new buds flush, making the crown more flammable (Chrosciewicz 1986). This depression of foliar moisture may be due to an influx of carbohydrates in the leaves, increasing dry weight, rather than a loss of water (Little 1970).

The predominance of crown fire in the boreal forest is due to the continuous surface fuel layer, composed of dead foliage, moss and lichens, and fine shrubs with dead twigs, along with aerial foliage of adequate mass, bulk density, resin and wax content, and moisture content to sustain combustion (Van Wagner 1983). Low-intensity fires occur, but are usually few and small, and have little effect on the large, high-intensity crown fires (Johnson 1992). Deep organic layers may smoulder after the passage of the fire front. This may have ecological significance, but is relatively unimportant to the rate of fire spread (Van Wagner 1983). Fires will burn through an ericaceous shrub layer (which often have high heats of combustion in their foliage; Sylvester & Wein 1981) and ignite lichens on lower branches of trees, which will then ignite the tree crown.

Heinselman (1981) estimates an average fire-return interval between 50 and 150 years, with more fire in the southern boreal forest (Johnson & Rowe 1975). In the MacKenzie River valley of Northwest Territory, fire frequencies were 30–70 years for jack pine, 80–90 years for *Picea mariana*, and over 300 years for *P. glauca* forests (Rowe *et al.* 1974). Fires in the boreal forest can be very large, ranging from thousands to hundreds of thousands of hectares (Heinselman 1981); often large fires are clustered in particular years of severe fire weather (Rowe *et al.* 1974).

Pinus banksiana stands are the most likely to sustain surface or crown fires over the season. *Picea mariana* is generally less flammable but under severe fire weather will carry fire similar to the pine stands (Van Wagner 1983). Sustained rates of spread of 100 m min^{-1} and fireline intensities of 50 000–150 000 kW/m^{-1} have been measured in northern Alberta (Kiil & Grigel 1969).

The influence of stand age on flammability continues to be debated. One of the assumptions used in calculating fire cycles using the negative exponential distribution is equal flammability with age (Van Wagner 1978). There is little quantitative evidence to test this assumption,

Fig. 11.9. *A, B.* Fires in boreal forest are often very intense (photos: B. Stocks, Canadian Forestry Service).

although Van Wagner (1983) suggests some variation in surface fire and crown fire potential over time. Recent research in *P. banksiana* stands suggests that fire inensity can be severe in both immature and mature pine (Stocks 1987, 1989). In immature stands, fires do not spread well under light wind and high fuel moisture, but develop crown fire behaviour with frontal intensities exceeding 40 000 kW m^{-1} if a vigorous surface fire is generated. Fires are more likely to spread under light wind in the mature stands because of the open understorey, with moderate rates of spread creating crown fire conditions. Intense surface fire behaviour is the key to generating crown fire behaviour in *P. banksiana* stands. In passive crown fires and surface fires, some *P. resinosa*, *P. strobus* and *P. banksiana* may survive, as they have thicker bark than *Populus tremuloides*, *Picea glauca* or *Abies balsamea* (Johnson 1992). Heterogeneous terrain, caused by fragmentation of the forest by water bodies, may be associated with variation in fire extent (Dansereau & Bergeron 1993).

After burning, the serotinous cones of jack pine open and spread seed over the competition-free forest floor. Cones subject to 450 °C for 30 seconds or 350 °C for 3 minutes show no decrease in seed germination (Beaufait 1960). The rapid growth of jack pine seedlings allows them to overtop competing vegetation (Ahlgren 1960, 1976), and they can produce viable seed within 5–7 years (Ahlgren 1959). On the Mack Lake fire in Michigan, seed density averaged 800 000 per hectare, and seedling density ranged from 5000 to 43 000 ha^{-1} after one growing season (Simard et al. 1983). Younger trees may exhibit substantial polymorphism and produce many non-serotinous cones, which may aid recolonization; older trees may have old cones open, which may extend the species presence past one life cycle of *P. banksiana* if there is open ground nearby (Gauthier, Bergeron & Simon 1993).

Vegetation in the southern boreal forest is well-adapted to burning. *Pinus banksiana* has serotinous cones. *Populus tremuloides* and *P. balsamifera*, as *fire endurers*, reproduce prolifically from root suckers, while *Betula papyrifera*, even though it will sprout from the stump, generally uses the *fire invader* strategy of reproducing from seed (Viereck & Schandelmeier 1980). *Picea glauca* and *Abies balsamea* are *fire avoiders*, with little adaptation to fire, and require refugia to contribute seed to the burned areas (Rowe & Scotter 1973). If fires occur more than 10 years apart, *P. banksiana* will usually dominate the tree component of the post-fire stand. Two crown fires in quick succession may allow aspen and birch to dominate the post-fire stand, while long fire-free intervals will encourage replacement of hardwoods by *Picea* spp. (Alexander & Euler 1981). Of the major species in the boreal forests, the pines (*P. banksiana* and *P. contorta*) are most likely to regenerate after fire, based on seed retention on tree, early-season seed production, frost

hardiness, unpalatability, seedling growth rate, and ability to tolerate full exposure (Rowe & Scotter 1973). *Picea mariana* has a *fire evader* semi-serotinous cone habit. Fire-killed spruce can retain viable seed in its canopy for several years (LeBarron 1939), and substantial seed may also lie on the forest floor.

Deep forest floors are typical of boreal forests (Van Wagner 1983). Fires in duff with moderate fuel moisture may consume 5 cm of the forest floor, while extended smouldering associated with low duff moisture may allow most of the forest floor to burn (Viereck et al. 1979). Species relying on underground roots or rhizomes or buried seed will be detrimentally affected with increasing forest floor consumption (Flinn & Wein 1977; Moore & Wein 1977). *Fire endurer* shrubs such as species of *Vaccinium* and *Ledum*, which rely on rhizomes, sprout best after moderate-intensity fires, while deep burns will encourage *fire invaders* like *Epilobium* spp. to seed in.

Successional relationships after fire in the boreal forest are complex; a typical sere in the more southerly boreal forest where *P. banksiana* is dominant is described below to illustrate typical post-fire succession. *Pinus banksiana*, at the southerly edge of the boreal forest, is found in a landscape mosaic of *P. resinosa* and *P. strobus*, generally on the poorest sites. On the associated pine sites that burn with two closely spaced crown fires, *P. resinosa* and *P. strobus* may be eliminated and *P. banksiana* and *Betula papyrifera* will capture the growing space (Eyre & LeBarron 1944; Kilburn 1960; Heinselman 1973, 1981). If the area is burned repeatedly at short intervals, the pine will disappear and sprouting hardwoods and brush, with the *endurer* strategy, will dominate the site (Frissell 1973). Multiple-successional pathways (*Carex* meadows, stratified canopies of shrubs, or early dominance by *P. banksiana*) are common in the southern boreal forest (Abrams, Sprugel & Dickmann 1985).

A unique attribute of young jack pine stands is their low branching habit above typically well-drained soil. Such sites are excellent habitat for the endangered Kirtland's warbler (*Dendroica kirtlandii*), which nests on the ground and uses the low branches for perches (Miller 1963). Patches of 30 ha or more are the minimum size for Kirtland's warbler breeding sites; such patches are well within the range of crown fires for jack pine.

At the landscape level, it is tempting to think of *P. banksiana* ecosystems being in some sort of dynamic equilibrium, so while there is major instability and change at the individual stand level, there is a more stable set of stand ages at the landscape level. This is unlikely to occur for several reasons. First, *P. banksiana* stands are but one element of a southerly boreal patch mosaic that includes other stand types with different disturbance histories. Second, the patches that burn are not burned uniformly.

Small fires tend to be uniform crown fires with no unburned islands, while larger fires are characterized by more unburned islands, larger islands, and more edge (Eberhart & Woodard 1987). Stringers of unburned forest may be associated with slight depressions or higher soil moisture (Quirk & Sykes 1971). Finally, as is true for most fairly long-lived conifer species, changing climate in the southern boreal forest has changed disturbance regimes so that no equilibrium has existed in the last several centuries (Baker 1992).

11.8   Fire and tropical pines

The community examples used above focus primarily on temperate pine species, yet fire has also been important in the spread, persistence and demise of pines in the tropics. About a third of pine species range into the tropics, and the 10–15% that are endemic to the tropics generally have very limited ranges (Critchfield & Little 1966; Chap. 10, this volume). Fire in these pine forests has become more common as population pressures have increased slash-and-burn agriculture, fuelwood cutting, and grazing (Goldammer 1990). As fires spread more frequently across the landscape, pines are favoured over the associated hardwood species (Denevan 1961; Kowal 1966), but as fire-return intervals become shorter and shorter, the 'fire-hardened' pine forests become degraded, and grasses such as *Imperata* spp. may replace the pine forests.

Annual burned area in Thailand and Burma in the 1980s constituted 14–21% of the forested areas of these countries (Goldammer & Peñafiel 1990), and much of this is in pine forests. Statistics on burned areas in other tropical countries are not available but are probably comparable. The presence of fire favours the pines, as they are shade-intolerant species and without disturbance would be succeeded by tropical hardwoods (except at high elevation with frost, where pines may coexist without fire; Denevan 1961). In central America, *P. caribaea* and *P. oocarpa*, and in Southeast Asia, *P. kesiya* and *P. roxburghii*, have been favoured by the presence of fire (Denevan 1961; Kowal 1966; Goldammer & Peñafiel 1990). When fire-return intervals are less than 3 years, pine savannas are maintained but tree regeneration does not occur. If the pines are cut or otherwise die and the frequent fires continue, grasslands replace the savanna. Clearly, fire exclusion and uncontrolled fires will be detrimental to tropical pine forests; an integrated fire management strategy, including prescribed fire, will provide the best way to maximize net benefits from fire in tropical pine forests (Goldammer & Peñafiel 1990).

11.9   The role of fire in pine invasions and plantation management in the southern hemisphere

Pines have been widely planted around the world (Chap. 20, this volume). Although most have not spread widely from their native ranges, some have become important weeds in these new environments, and fire management problems have been created by large blocks of this relatively flammable vegetation.

Pine invasions
Some *Pinus* species have invaded large tracts of natural vegetation in the southern hemisphere where they decrease native biodiversity, alter fire regimes, reduce water yields, change nutrient cycling processes, and alter biomass accumulation processes (Richardson & Bond 1991; Richardson, Williams & Hobbs 1994; van Wilgen, Richardson & Seydack 1994). The invasive spread of *Pinus* spp. is closely linked to fire and their adaptations to these new fire regimes. A few *Pinus* species appear to comprise most of the problems (Richardson *et al.* 1994): *P. contorta* in New Zealand, *P. halepensis* in South Africa, *P. nigra* in New Zealand, *P. patula* in Madagascar, Malawi, and South Africa, *P. pinaster* in South Africa, New Zealand, and Uruguay, and *P. ponderosa* and *P. sylvestris* in New Zealand. *Pinus radiata* has escaped plantation culture in Chile and Australia, and is a much larger problem than *P. pinaster*, the next most invasive pine. Pines like *P. halepensis*, which do not appear to be invasive threats at the margins of their home ranges (Trabaud 1991) can be serious invaders elsewhere (Richardson 1988). Several factors, including fire, appear to be responsible for the spread of pines into wildlands of the southern hemisphere (Chap. 22, this volume).

Fire is a natural disturbance factor in most southern hemisphere countries where pine invasions are a problem, and some but not all introduced *Pinus* spp. can exploit the environmental conditions induced by wildfires. These pines are fire-resilient species with small seeds, low seed-wing loadings, short juvenile periods, moderate to high degrees of serotiny, and relatively poor fire tolerance as adults (Richardson, Cowling & Le Maitre 1990). Species with these attributes were identified through multivariate analyses, including the invasive species of *Pinus* noted above plus the following additional species that are considered high risk: *P. attentuata*, *P. banksiana*, *P. clausa*, *P. contorta* (all three varieties), *P. muricata*, *P. pungens*, *P. serotina*, and *P. virginiana*. Most of these species are strongly serotinous (e.g, the *evader* strategy of Rowe 1983) and adapted to high-severity fire regimes as described earlier in this chapter, but have not been widely planted because of generally poor growth form for timber. In areas where fires can be frequent and intense, such as the fynbos of South

Africa, these species are well-adapted to colonize the post-fire environment.

Fires allow the serotinous cones to open, and there is little competition from native plants during the invasion window (Johnstone 1986). With a short juvenile period, another generation of seed can be available within a decade. In an area of fynbos burned with an average fire-return interval of 13 years (range 9–14), *P. halepensis* could maintain densities above 7000 stems per hectare (Richardson 1988). Most of the trees established after the last fire 14 years before sampling occurred. In the area of the disastrous Tunnel Fire in Oakland, California, *P. radiata*, outside its small native range in California, was regenerating well near large trees killed by the fire (J.K. Agee, personal observation). *Pinus radiata* also spreads rapidly after fire in South Africa. Initial colonizers spread up to 3 km from an established plantation over a several decade period, and then increased density rapidly after wildfire burned through the area (Richardson & Brown 1986).

Management options to reduce or eliminate these invasive pines are limited. Trees can be cut, and then burned within a few months after seed germination to kill the regenerating cohort before it produces seed (van Wilgen *et al.* 1994). Alternatively, two fires can be introduced in an interval shorter than the minimum time to maturity for a species, a strategy widely applied where seed banks in the soil exist (Agee 1993). Other mechanical and chemical means exist, too. Evaluation of techniques requires not only the knowledge of effects on the *Pinus* spp. but effects on the native flora. Inevitable fires in the southern hemisphere countries are likely to continue a pattern of spread for the *Pinus* spp. with invasive characteristics.

Pine plantation management and fire

The major pine species used in tropical plantations include *P. caribaea*, *P. elliottii*, *P. patula*, *P. pinaster*, *P. radiata* and *P. taeda* (de Ronde *et al.* 1990). Few plantations have been established with fire potential in mind. Given the generally close spacing of trees, and the slow decomposition rates of detritus, fuel loading and the potential for stand-replacing, high-severity fires are highly probable (Goldammer 1993). Prescribed fire has been applied in many tropical and subtropical countries: Bahamas, Belize, Brazil, Costa Rica, Fiji, Honduras, Nicaragua, Panama, Spain and Venezuela. However, there are few places where prescribed fire has been routinely applied as a management tool in pine plantations (de Ronde *et al.* 1990).

Management objectives for the use of fire include fire hazard reduction, site preparation, enhancing nutrient availability, and encouraging herbaceous understorey growth for grazing. Lack of experience in prescribed fire can result in unwanted effects, such as higher than desir-able fireline intensity, and discourage further experimentation. Initial fires in *P. caribaea* plantations, for example, must avoid torching caused by extended retention of dead needles (de Ronde *et al.* 1990). Backing fires, where the fire spreads slowly through the plantation into the wind (as contrasted to a heading fire of higher intensity spreading in the same direction as the wind), are recommended (Goldammer 1993). To date, the potential for using prescribed fire in pine plantations has not been realized. It can be most effectively applied in plantations where pines with thick bark have been planted, so that fuels can be consumed and lower branches pruned without excessive cambial heating.

11.10 Conclusions

Pines have a long evolutionary history of fire and have adapted to natural fire regimes in remarkable ways to be tolerant of, or resilient to, disturbance. The fire regime concept is an appropriate way of organizing information on fire effects, as it allows categorization of effects from various historical frequencies, intensities, and extents. However, the natural fire regimes of the past are not the regimes of the present, nor will they be the regimes of the future. Adaptations of pines to fire may, in some cases, be relictual, and place them at risk in new fire environments.

Fire has been responsible for much of the spread and persistence of tropical pines. Without disturbance, these pines are replaced by hardwoods. Many tropical areas have been degraded to pine savannas by almost annual burning, and are now much less productive than they would be with less frequent fires that might still favour pines. In Nicaragua's Miskito Coast region, for example, annual burning (almost 90% of pine lands burn each year) has led to reduced vigour and a decline in wood production of *P. caribaea* var. *hondurensis* (Koonce, Paysen & Corcoran 1995). Koonce *et al.* describe this burning pattern as 'too much, too often'.

Human interaction with burning patterns has had the opposite effect in Scandinavia. Sweden has been so successful in eradicating fire from its landscape that wildfire statistics are no longer collected (Pyne 1995). In the process, the biodiversity that was once produced by the mosaic of patchy burning has also declined, and a number of species are threatened with extinction. In North America, attempts at excluding fire have been less successful, but have brought major changes to fire regimes. High-severity fire regimes have replaced low-severity regimes, leading to mortality of thick-barked species such as *P. ponderosa* in crown-scorching or crown-consuming fires (Agee 1993). *Pinus contorta*, with its serotinous or semi-serotinous cones, is better adapted to the new regime.

Where pines have been introduced into fire-prone environments outside their natural range, the fire-resilient pines, already capable of escaping cultivation, have a high potential for continued spread across these new landscapes.

The genus *Pinus* seems set to survive and prosper in a changing world, but the changes will affect species differently. Successful taxa will be those best adapted to current and future fire regimes, whereas those with adaptations that can be considered anachronistic in changing environments will fail.

The application of fire to the management of pines requires a much deeper understanding of the effects of fire regimes. Pines are but one taxon in the complex ecosystems of which they form a part, and we currently have limited tools for evaluating the effects of more than one characteristic of fire regimes (e.g. frequency interactions with seasonality) even on the *Pinus* component of ecosystems. Long-term research installations, particularly in low-severity fire regimes where prescribed fires are most likely to find management applications, will be needed to evaluate these interactions.

Across the ecosystems of the world, fire has and will always occur. The many adaptations of pines to fire suggest that *Pinus* taxa will also be present. The challenge to humankind is to manage natural and cultural fire ignitions in pine-dominated landscapes to meet the needs of society. In some cases, this will mean less fire, and in others it will mean more fire through the application of prescribed burns. Creation and maintenance of appropriate fire regimes is the key to the successful conservation of pines.

References

Abrams, M.D., Sprugel, D.G. & Dickmann, D.I. (1985). Multiple successional pathways on recently disturbed jack pine sites in Michigan. *Forest Ecology and Management*, **10**, 31–48.

Agee, J.K. (1981). *Initial Effects of Prescribed Fire in a Climax* Pinus contorta *Forest: Crater Lake National Park*. National Park Service Report CPSU/UW 81–4. Seattle: College of Forest Resources, University of Washington.

Agee, J.K. (1993). *Fire Ecology of Pacific Northwest Forests*. Washington, DC: Island Press.

Agren, J., Isaakson, L. & Zackrisson, O. (1983). Natural age and size of *Pinus sylvestris* and *Picea abies* on a mire in the inland part of northern Sweden. *Holarctic Ecology*, **6**, 228–37.

Ahlgren, C.E. (1959). Some effects of fire on forest reproduction in northeastern Minnesota. *Journal of Forestry*, **57**, 194–200.

Ahlgren, C.E. (1960). Some effects of fire on reproduction and growth of vegetation in northeastern Minnesota. *Ecology*, **41**, 431–45.

Ahlgren, C.E. (1976). Regeneration of red pine and white pine following wildfire and logging in northeastern Minnesota. *Journal of Forestry*, **74**, 135–40.

Alexander, M.E. & Euler, D.L. (1981). Ecological role of fire in the uncut boreal mixedwood forest. In *Canadian Forest Service, Canada–Ontario Joint Forestry Research Committee Symposium Proceedings O-P-9*, pp. 42–64. Sault Ste. Marie, Ontario, Canada: Great Lakes Forest Research Centre.

Allen, R.M. (1964). Contributions of roots, stems, and leaves to height growth of longleaf pine. *Forest Science*, **10**, 14–16.

Amaranthus, M.P. & Perry, D.A. (1987). Effects of soil transfer on ectomycorrhiza formation and the survival and growth of conifer seedlings on old, nonforested clearcuts. *Canadian Journal of Forest Research*, **17**, 944–50.

Artsybashev, E.S. (1967). Achievements of the USSR in the protection of forests from fire. *LenNIILKH* 1967 (May), 1–16.

Artsybashev, E.S. (1983). *Forest Fires and Their Control*. New Delhi, India: Amerind Pub. Co. [Translation of Russian version of 1974].

Baker, W.L. (1992). Effects of settlement and fire suppression in landscape structure. *Ecology*, **73**, 1879–87.

Barrett, J.W. (1979). *Silviculture of Ponderosa Pine in the Pacific Northwest: The State of Our Knowledge*. USDA Forest Service General Technical Report PNW-97.

Beaufait, W.R. (1960). Some effects of high temperatures on the cones and seeds of jack pine. *Forest Science*, **6**, 194–9.

Biswell, H.H., Kallander, H.R., Komarek, R., Vogl, R.J. & Weaver, H. (1973). *Ponderosa Fire Management*. Tall Timbers Research Station Miscellaneous Publication 2. Tallahassee, Florida: Tall Timbers Research Station.

Bonnicksen, T.M. & Stone, E.C. (1981). The giant sequoia–mixed conifer forest community characterized through pattern analysis as a mosaic of aggregations. *Forest Ecology and Management*, **3**, 307–28.

Borchers, J.G. & Perry, D.A. (1990). Effects of prescribed fire on soil organisms. In *Natural and Prescribed Fire in the Pacific Northwest*, ed. J. Walstad, S.R. Radoesvich & D.V. Sandberg, pp. 143–57. Corvallis: Oregon State University Press.

Borchert, M. (1985). Serotiny and cone-habit variation in populations of *Pinus coulteri* (Pinaceae) in the southern Coast Ranges of California. *Madroño*, **32**, 29–48.

Bork, J. (1985). Fire history in three vegetation types on the east side of the Oregon Cascades. PhD thesis, Oregon State University, Corvallis.

Boyer, W.D. (1991). Eighteen years of seasonal burning in longleaf pine: Effects on overstory growth. In *Proceedings of the 12th International Conference on Fire and Forest Meteorology*, pp. 602–10. Bethesda, Md: Society of American Foresters.

Braathe, P. (1974). Prescribed burning in Norway – effects on soil and regeneration. *Tall Timbers Fire Ecology Conference*, **13**, 211–22.

Brain, C.K. & Sillen, A. (1988). Evidence from the Swartkrans cave for the earliest use of fire. *Nature*, **336**, 464–6.

Byram, G.M. (1959). Combustion of forest fuels. In *Forest Fire: Control and Use*, ed. K.P. Davis, pp. 155–82. New York: McGraw-Hill.

Chapman, H.H. (1912). Forest fires and forestry in the southern states. *American Forests*, **18**, 510–17.

Chapman, H.H. (1932). Is the longleaf pine a climax? *Ecology*, **13**, 328–34.

Chappell, C.B. (1991). Fire ecology and seedling establishment in Shasta red fir (*Abies magnifica* var. *shastensis*) forests of Crater Lake National Park, Oregon. MS thesis, University of Washington, Seattle.

Chizhov, B.F. & Sannikov, N.S. (1978). Fire resistance of plants in the herbaceous–shrubby tier in pine forests of the Transurals region. *Lesovedenie*, **5**, 67–76.

Christensen, N.L. (1981). Fire regimes in southeastern ecosystems. In *Fire Regimes and Ecosystem Properties: Proceedings of the Conference*, ed. H.A. Mooney, T.M. Bonnicksen, N.L. Christensen, J.E. Lotan & W.A. Reiners, pp. 112–36. USDA Forest Service General Technical Report WO-26.

Christensen, N.L. (1988). Succession and natural disturbance: paradigms, problems, and preservation of natural systems. In *Ecosystem Management for Parks and Wilderness*, ed. J.K. Agee & D.R. Johnson, pp. 62–86. Seattle: University of Washington Press.

Christensen, N.L. (1991). Vegetation of the southeastern Coastal Plain. In *North American Terrestrial Vegetation*, ed. M.G. Barbour & W.D. Billings, pp. 317–64. New York: Cambridge University Press.

Chrosciewicz, Z. (1986). Foliar moisture content variations in four coniferous tree species of central Alberta. *Canadian Journal of Forest Research*, **16**, 157–62.

Clements, F.E. (1910). *The Life History of Lodgepole Burn Forests*. USDA Forest Service Bulletin 79.

Clements, F.E. (1935). Experimental ecology in the public service. *Ecology*, **16**, 342–63.

Cooper, C.F. (1960). Changes in vegetation, structure, and growth of southwestern ponderosa pine forests since white settlement. *Ecological Monographs*, **30**, 129–64.

Critchfield, W.B. & Little, E.L. (1966). *Geographic Distribution of Pines of the World*. USDA Forest Service Miscellaneous Publication 991.

Dansereau, P.-R. & Bergeron, Y. (1993). Fire history in the southern boreal forest of northwestern Quebec. *Canadian Journal of Forest Research*, **23**, 25–32.

Daubenmire, R. (1968). *Plant Communities*. New York: Harper & Row.

Davis, K.M., Clayton, B.D. & Fischer, W.C. (1980). *Fire Ecology of Lolo National Forest Habitat Types*. USDA Forest Service General Technical Report INT-79.

DeBell, D.S. & Ralston, C.W. (1970). Release of nitrogen by burning light forest fuels. *Soil Science Society of America Proceedings*, **34**, 936–8.

Denevan, W.M. (1961). *The Upland Pine Forests of Nicaragua: A Study in Cultural Plant Geography*. Berkeley: University of California Press.

de Ronde, C., Goldammer, J.G., Wade, D.D. & Soares, R.V. (1990). Prescribed fire in industrial pine plantations. In *Fire in the Tropical Biota: Ecosystem Processes and Global Challenges*, ed. J.G. Goldammer, pp. 216–72. Berlin: Springer-Verlag.

Despain, D.G. (1983). Nonpyrogenous climax lodgepole pine communities in Yellowstone National Park. *Ecology*, **64**, 231–4.

Dieterich, J.H. (1980). *Chimney Springs Forest Fire History*. USDA Forest Service General Technical Report RM-220.

Dixon, R.K. & Krankina, O.N. (1993). Forest fires in Russia: carbon dioxide emissions to the atmosphere. *Canadian Journal of Forest Research*, **23**, 700–5.

Drackenberg, B. (1981). *Kompendium i allman dendrologie samt barrtrads och barrvirkesegenskaper*. Umea, Sweden: SLU Institut for Skoglig Standortslara [In Swedish].

Eberhart, K.E. & Woodard, P.M. (1987). Distribution of residual vegetation associated with large fires in Alberta. *Canadian Journal of Forest Research*, **17**, 1207–12.

Egler, F. (1954). Vegetation science concepts, I: Initial floristic composition – a factor in old field successional development. *Vegetatio*, **4**, 412–7.

Entry, J.A., Stark, N.M. & Lowenstein, H. (1986). Effect of timber harvesting on microbial biomass fluxes in a northern Rocky Mountain forest soil. *Canadian Journal of Forest Research*, **16**, 1076–81.

Eyre, F.H. & LeBarron, R.K. (1944). *Management of Jack Pine Stands in the Lake States*. USDA Forest Service Technical Bulletin 863.

FIRESCAN (1994). Fire in boreal ecosystems of Eurasia: First results of the Bor Forest Island Fire Experiment, Fire Research Campaign Asia-North (FIRESCAN). *World Resource Review*, **6**, 499–523.

Flinn, M.A. & Wein, R.W. (1977). Depth of underground plant organs and theoretical survival during fire. *Canadian Journal of Botany*, **55**, 2550–4.

Fowler, W.B. & Helvey, J.D. (1978). *Changes in the Thermal Regime After Prescribed Burning and Select Tree Removal (Grass Camp, 1975)*. USDA Forest Service Research Paper PNW-234.

Frissell, S.S. (1973). The importance of fire as a natural ecological factor in Itasca State Park, Minnesota. *Quaternary Research*, **3**, 397–407.

Gara, R.I., Agee, J.K., Littke, W.R. & Geiszler, D.R. (1986). Fire wounds and beetle scars: Distinguishing between the two can help reconstruct past disturbances. *Journal of Forestry*, **84**, 47–50.

Gara, R.I., Littke, W.R., Agee, J.K., Geiszler, D.R., Stuart, J.D. & Driver, C.H. (1985). Influence of fires, fungi, and mountain pine beetles on development of a lodgepole pine forest in south-central Oregon. In *Lodgepole Pine: The Species and its Management*, ed. D.M. Baumgartner, R.G. Krebill, J.T. Arnott & G.F. Weetman, pp. 153–62. Pullman: Washington State University.

Gauthier, S., Bergeron, Y. & Simon, J.-P. (1993). Cone serotiny in jack pine: ontogenetic, positional, and environmental effects. *Canadian Journal of Forest Research*, **23**, 394–401.

Geiszler, D.R., Gara, R.I. & Littke, W.R. (1984). Bark beetle infestations of lodgepole pine following a fire in south central Oregon. *Zeitschrift für Angewandte Entomologie*, **98**, 389–94.

Goldammer, J.G. (1990). Fire in the pine–grassland biomes of tropical and subtropical Asia. In *Fire in the Tropical Biota: Ecosystem Processes and Global Challenges*, ed. J.G. Goldammer, pp. 45–64. Berlin: Springer-Verlag.

Goldammer, J.G. (1993). Fire management. In *Tropical Forestry Handbook*, ed. L. Pancel, pp. 1221–68. Berlin: Springer-Verlag.

Goldammer, J.G. & Di, X. (1990). Fire and forest development in the Daxinganling montane-boreal coniferous forests, Heilongjiang, northeast China – a preliminary model. In *Fire in Ecosystem Dynamics: Mediterranean and Northern Perspectives*, ed. G.J. Goldammer & M.J. Jenkins, pp. 175–84. The Hague: SPB Academic Publishing.

Goldammer, J.G. & Peñafiel, S.R. (1990). Fire in the pine–grassland biomes of tropical and subtropical Asia. In *Fire in the Tropical Biota: Ecosystem Processes and Global Challenges*, ed. J.G. Goldammer, pp. 45–62. Ecological Studies 84. Berlin: Springer-Verlag.

Grace, S.L. & Platt, W.J. (1995). Effects of adult tree density and fire on the demography of pregrass stage juvenile longleaf pine (*Pinus palustris* Mill.). *Journal of Ecology*, **83**, 75–86.

Greene, S.W. (1931). The forest that fire made. *American Forests*, **37**, 583–4.

Grelen, H.E. (1975). *Vegetative Response to Twelve Years of Seasonal Burning on a Louisiana Longleaf Pine Site*. USDA Forest Service Research Note SO-192.

Grier, C.C. (1975). Wildfire effects on nutrient distribution and leaching in a coniferous ecosystem. *Canadian Journal of Forest Research*, **5**, 599–608.

Gromtsev, A.N. (1993). Pattern of occurrence of fire in spontaneous forests of northwestern taiga landscapes. *Soviet Journal of Ecology*, **24**(3), 161–4.

Harper, R.M. (1911). The relation of climax vegetation to islands and peninsulas. *Bulletin of the Torrey Botanical Club*, **38**, 515–25.

Heinselman, M.L. (1973). Fire in the virgin forests of the Boundary Waters Canoe Area, Minnesota. *Quaternary Research*, **3**, 329–82.

Heinselman, M.L. (1981). Fire intensity and frequency as factors in the distribution and structure of northern ecosystems. In *Fire Regimes and Ecosystem Properties: Proceedings of the Conference*, pp. 7–57. USDA Forest Service General Technical Report WO-26.

Heyward, F. (1939). The relation of fire to stand composition of longleaf pine forests. *Ecology*, **20**, 287–304.

Johnson, E.A. (1992). *Fire and Vegetation Dynamics: Studies from the North American Boreal Forest*. Cambridge: Cambridge University Press.

Johnson, E.A. & Rowe, J.S. (1975). Fire in the subarctic wintering ground of the Beverley caribou herd. *American Midland Naturalist*, **94**, 1–14.

Johnson, E.A. & Van Wagner, C.E. (1985). The theory and use of two fire history models. *Canadian Journal of Forest Research*, **15**, 214–20.

Johnstone, I.M. (1986). Plant invasion windows: a time-based classification of invasion potential. *Biological Review*, **61**, 369–94.

Kamada, M., Nakagoshi, N. & Nehira, K. (1991). Pine forest ecology and landscape management: A comparative study in Japan and Korea. In *Coniferous Forest Ecology From an International Perspective*, ed. N. Nakagoshi & F. Golley, pp. 43–62. The Hague: SPB Academic Publishing.

Keane, R.E., Arno, S.F. & Brown, J.K. (1990). Simulating cumulative fire effects in ponderosa pine/Douglas-fir forests. *Ecology*, **71**, 189–203.

Keane, R.E., Reinhardt, E.D. & Brown, J.K. (1993). FOFEM: a first order fire effects model for predicting the immediate consequences of wildland fire in the United States. In *Proceedings of the 12th International Conference on Fire and Forest Meteorology, Jekyll Island, GA*, pp. 628–31. Bethesda, Md: Society of American Foresters.

Keen, F.P. (1943). Ponderosa pine tree classes redefined. *Journal of Forestry*, **41**, 249–53.

Kiil, A.D. & Grigel, J.E. (1969). *The May, 1968 Forest Conflagrations in Central Alberta*. Alberta. Canadian Forest Service Information Report A-X-24.

Kilburn, P.D. (1960). Effects of logging and fire on xerophytic forests in northern Michigan. *Bulletin of the Torrey Botanical Club*, **87**, 402–5.

Koonce, A.L., Paysen, T.E. & Corcoran B.M. (1995). Fire in a tropical savanna – a double-edged sword. In *The Biswell Symposium: Fire Issues and Solutions in Urban Interface and Wildland Ecosystems*, ed. D.R. Weise & R.E. Martin. USDA Forest Service General Technical Report PSW-158.

Komarek, E.V. (1967). The nature of lightning fires. *Tall Timbers Fire Ecology Conference*, **7**, 5–41.

Komarek, E.V. (1968). Lightning and lightning fires as ecological forces. *Tall Timbers Fire Ecology Conference*, **8**, 169–97.

Komarek, E.V. (1974). Effects of fire on temperate forests and related ecosystems: Southeastern United States. In *Fire and Ecosystems*, ed. T.T. Kozlowski & C.E. Ahlgren, pp. 251–77. New York: Academic Press.

Kotov, M.M. (1980). [Comparison of the morphological characters of two generations of *P. sylvestris*.] *Lesovodstvo-Lesnye-Kul'tury-i-Pochvovedenie*, **9**, 107–11.

Kowal, N.E. (1966). Shifting cultivation, fire, and pine forest in the Cordillera Central, Luzon, Philippines. *Ecological Monographs*, **36**, 389–419.

Kullman, L. (1986). Late Holocene reproductional patterns of *Pinus sylvestris* and *Picea abies* at the forest limit in central Sweden. *Canadian Journal of Botany*, **64**, 1682–90.

Kurbatskii, N.P. & Ivanova, G.A. (1980). [Effect of herbaceous plants and *Vaccinium vitis-idaea* on ground fires in Scots pine forests]. *Lesnoe Khoyastvo*, **5**, 48–50.

Kutiel, P. & Naveh, Z. (1987). Soil properties beneath *Pinus halepensis* and *Quercus calliprinos* trees on burned and unburned mixed forest on Mt Carmel, Israel. *Forest Ecology and Management*, **20**, 11–24.

Kutiel, P., Naveh, Z. & Kutiel, H. (1990). The effect of wildfire on soil nutrients and vegetation in an Aleppo pine forest on Mount Carmel, Israel. In *Fire in Ecosystem Dynamics: Mediterranean and Northern Perspectives*, ed. J.G. Goldammer & M.J. Jenkins, pp. 85–94. The Hague: SPB Academic Publishing.

Landsberg, J.D., Cochran, P.D., Frink, M.M. & Martin, R.E. (1984). *Foliar Nitrogen Content and Tree Growth after Prescribed Fire in Ponderosa Pine*. USDA Forest Service Research Note PNW-412.

Larson, M.M. & Schubert, G.H. (1969). *Root Competition between Ponderosa Pine Seedlings and Grass*. USDA Forest Service Research Paper RM-54.

LeBarron, R.K. (1939). The role of forest fires in the reproduction of black spruce. *Proceedings of the Minnesota Academy of Science*, **7**, 10–14.

Le Houérou, H.N. (1974). Fire and vegetation in the Mediterranean basin. *Tall Timbers Fire Ecology Conference*, **13**, 237–77.

Le Houérou, H.N. (1977). Fire and vegetation in North Africa. In *Proceedings of the Symposium on the Environmental Consequences of Fire and Fuel Management in Mediterranean Ecosystems*, ed. H.A. Mooney & C.E. Conrad, pp. 334–41. USDA Forest Service General Technical Report WO-3.

Liacos, L.G. (1974). Present studies and history of burning in Greece. *Tall Timbers Fire Ecology Conference*, **13**, 65–95.

Littke, W.R. & Gara, R.I. (1986). Decay of fire-damaged lodgepole pine in south-central Oregon. *Forest Ecology and Management*, **17**, 279–87.

Little, C.H.A. (1970). Seasonal changes in carbohydrate and moisture content in needles of balsam fir (*Abies balsamea*). *Canadian Journal of Botany*, **48**, 2021–8.

Mattoon, W.R. (1925). *Longleaf Pine*. USDA Forest Service Bulletin 1061.

May, T. (1990). Vegetation development and surface runoff after fire in a catchment of southern Spain. In *Fire in Ecosystem Dynamics: Mediterranean and Northern Perspectives*, ed. J.G. Goldammer & M.J. Jenkins, pp. 117–26. The Hague: SPB Academic Publishing.

McCune B. (1988). Ecological diversity in North American pines. *American Journal of Botany*, **75**, 353–68.

McNabb, D.H. & Cromack, K. Jr (1990). Effects of prescribed fire on nutrients and soil productivity. In *Natural and Prescribed Fire in the Pacific Northwest*, ed. J. Walstad, S.R. Radosevich & D.V. Sandberg, pp. 125–42. Corvallis: Oregon State University Press.

Metz, L.J. & Dindal, D.A. (1975). Collembola populations and prescribed burning. *Environmental Entomology*, **4**, 583–7.

Metz, L.J. & Farrier, M.H. (1973). Prescribed burning and populations of soil mesofauna. *Environmental Entomology*, **2**, 433–40.

Miller, H.A. (1963). Use of fire in wildlife management. *Tall Timbers Fire Ecology Conference*, **2**, 18–30.

Mirov, N.T. (1967). *The Genus Pinus*. New York: Ronald Press.

Moir, W.H. (1969). The lodgepole pine zone in Colorado. *American Midland Naturalist*, **81**, 87–98.

Moore, J.M. & Wein, R.W. (1977). Viable seed populations by soil depth and potential site recolonization after disturbance. *Canadian Journal of Botany*, **55**, 2408–12.

Morris, W.G. & Mowat, E.L. (1958). Some effects of thinning a ponderosa pine thicket with a prescribed fire. *Journal of Forestry*, **56**, 203–9.

Morrow, R.J. (1985). Age structure and spatial pattern of old-growth ponderosa pine in Pringle Falls Experimental Forest, central Oregon. MS thesis, Oregon State University, Corvallis.

Muir, P.S. & Lotan, J.E. (1985). Disturbance history and serotiny of *Pinus contorta* in western Montana. *Ecology*, **66**, 1658–68.

Myers, R.L. (1985). Fire and the dynamic relationship between Florida sandhill and sand pine scrub vegetation. *Bulletin of the Torrey Botanical Club*, **112**, 241–52.

Nakagoshi, N., Nehira, K. & Takahashi, F. (1987). The role of fire in pine forests in Japan. In *The Role of Fire in Ecological Systems*, pp. 91–119. The Hague: SPB Academic Publishing.

Naveh, Z. (1974). Effects of fire in the Mediterranean region. In *Fire and Ecosystems*, ed. T.T. Kozlowski & C.E. Ahlgren, pp. 401–34. New York: Academic Press.

Naveh, Z. (1990). Fire in the Mediterranean – a landscape ecological perspective. In *Fire in Ecosystem Dynamics: Mediterranean and Northern Perspectives*, ed. J.G. Goldammer & M.J. Jenkins, pp. 1–20. The Hague: SPB Academic Publishing.

Ne'eman, G., Lahav, H. & Izhaki, I. (1992). Spatial pattern of seedlings 1 year after fire in a Mediterranean pine forest. *Oecologia*, **91**, 365–70.

Nikolov, N. & Helmisaari, H. (1992). Silvics of the circumpolar boreal forest tree species. In *A Systems Analysis of the Global Boreal Forest*, ed. H. Shugart, R. Leemans & G. Bonan, pp. 13–84. New York: Cambridge University Press.

Noble, I.R. & Slatyer, R.O. (1980). The use of vital attributes to predict successional changes in plant communities subject to recurrent disturbance. *Vegetatio*, **43**, 5–21.

Otto, H.J. (1980). [Silvicultural considerations and measures following the forest fires of 1975 and 1976 in Lower Saxony, West Germany.] *Forstwissenschaftliches Centralblatt*, **99**, 385–6.

Peterson, D.L. & Ryan, K.C. (1986). Modeling post-fire conifer mortality for long-range planning. *Environmental Management*, **10**, 797–808.

Peterson, D.L., Sackett, S.S., Robinson, L.J. & Haase, S.M. (1994). The effects of repeated prescribed burning on *Pinus ponderosa* growth. *International Journal of Wildland Fire*, **4**, 239–47.

Platt, W.J., Evans, G.W. & Rathbun, S.L. (1988). The population dynamics of a long-lived conifer (*Pinus palustris*). *American Naturalist*, **131**, 491–525.

Platt, W.J., Glitzenstein, J.S. & Streng, D.R. (1991). Evaluating pyrogenicity and its effects on vegetation in longleaf pine savannas. *Tall Timbers Fire Ecology Conference*, **17**, 143–61.

Pobedinsky, A.B. (1979). *The Scotch Pine*. Lesnaya promyshlenost [in Russian].

Prodon, R., Fons, R. & Athias-Binche, F. (1987). The impact of fire on animal communities in mediterranean area. In *The Role of Fire in Ecological Systems*, ed. L. Trabaud, pp. 121–57. The Hague: SPB Academic Publishers.

Pyne, S.J. (1982). *Fire in America: A Cultural History of Wildland and Rural Fire*. Princeton, NJ: Princeton University Press.

Pyne, S.J. (1995). *World Fire*. New York: Henry Holt and Co.

Quirk, W.A. & Sykes, D.J. (1971). White spruce stringers in a fire-patterned landscape in interior Alaska. In *Fire in the Northern Environment*, pp. 179–98. USDA Forest Service, Pacific Northwest Forest and Range Experiment Station, Portland, Oregon.

Richardson, D.M. (1988). Age structure and regeneration after fire in a self-sown *Pinus halepensis* forest on the Cape Peninsula, South Africa. *South Africa Journal of Botany*, **54**, 140–4.

Richardson, D.M. & Bond, W.J. (1991). Determinants of plant distribution: evidence from pine invasions. *American Naturalist*, **137**, 639–68.

Richardson, D.M. & Brown, P.J. (1986). Invasion of mesic mountain fynbos by *Pinus radiata. South Africa Journal of Botany*, **52**, 529–36.

Richardson, D.M., Cowling, R.M. & Le Maitre, D.C. (1990). Assessing the risk of invasive success in *Pinus* and *Banksia* in South African mountain fynbos. *Journal of Vegetation Science*, **1**, 629–42.

Richardson, D.M., Williams, P.A. & Hobbs, R.J. (1994). Pine invasions in the Southern Hemisphere: determinants of spread and invadability. *Journal of Biogeography*, **21**, 511–27.

Riegel, G.M., Miller, R.F. & Krueger, W.C. (1992). Competition for resources between understory vegetation and overstory *Pinus ponderosa* in northeast Oregon. *Ecological Applications*, **2**, 71–85.

Rothermel, R.C. & Deeming, J.E. (1980). *Measuring and Interpreting Fire Behavior for Correlation with Fire Effects*. USDA Forest Service General Technical Report INT-142.

Rowe, J.S. (1983). Concepts of fire effects on plant individuals and species. In *The Role of Fire in Northern Circumpolar Ecosystems*, ed. R.W. Wein & D.A. MacLean, pp. 135–54. New York: John Wiley.

Rowe, J.S., Bergsteinsson, J.L., Padbury, G.A. & Hermesh, R. (1974). *Fire Studies in the Mackenzie Valley*. Canadian Department of Indian Affairs and Northern Development. Arctic Land Use Research Program Report 73-74-61.

Rowe, J.S. & Scotter, G.W. (1973). Fire in the boreal forest. *Quaternary Research*, **3**, 444–64.

Ryan, K.C. & Reinhardt, E.D. (1988). Predicting postfire mortality of seven western conifers. *Canadian Journal of Forest Research*, **18**, 1291–7.

Ryan, K.C., Rigolot, E. & Botelho, H. (1993). Comparative analysis of fire resistance and survival of Mediterranean and western North American conifers. In *Proceedings of the 12th International Conference on Fire and Forest Meteorology*, pp. 701–8. Jekyll Island, GA. Bethesda, Md: Society of American Foresters.

Sackett, S.S. (1975). Scheduling prescribed burns for hazard reduction in the Southeast. *Journal of Forestry*, **73**, 143–7.

Sannikov, S.N. (1983). Cyclical erosional–pyrogenic theory of common-pine natural renewal. *Soviet Journal of Ecology*, **14**, 7–16.

Sannikov, S.N. (1985). Hypothesis of pulsed pyrogenic stability of pine forests. *Soviet Journal of Ecology*, **16**, 69–75.

Savage, M. & Swetnam, T.W. (1990). Early nineteenth-century fire decline following sheep pasturing in a Navajo ponderosa pine forest. *Ecology*, **71**, 2374–8.

Shcherbakov, I.P. (1977). Forest vegetation in burned and logged areas of Yakutsk. In *North American Forest Lands at Latitudes North of 60 Degrees – Symposium Proceedings*, pp. 68–84. Fairbanks: University of Alaska.

Shostakovitch, V.B. (1925). Forest conflagrations in Siberia. *Journal of Forestry*, **23**, 365–71.

Siggers, P.V. (1944). *The Brown Spot Needle Disease of Pine Seedlings*. USDA Forest Service Technical Bulletin 870.

Simard, A.J., Haines, D.A., Blank, R.W. & Frost, J.S. (1983). *The Mack Lake fire*. USDA Forest Service General Technical Report NC-83.

Siren, E. (1974). Some remarks on fire ecology in Finnish forests. *Tall Timbers Fire Ecology Conference*, **13**, 191–209.

Soeriaatmajda, R.E. (1966). Fire history of the ponderosa pine forests of the Warm Springs Indian Reservation, Oregon. PhD thesis, Oregon State University, Corvallis.

Stocks, B.J. (1987). Fire behavior in immature jack pine. *Canadian Journal of Forest Research*, **17**, 80–6.

Stocks, B.J. (1989). Fire behavior in mature jack pine. *Canadian Journal of Forest Research*, **19**, 783–90.

Stocks, B.J. & Street, R.B. (1983). Forest fire weather and wildfire occurrence in the boreal forest of northwestern Ontario. In *Resources and Dynamics of the Boreal Zone*, ed. R.R. Riewe & I.R. Methven, pp. 249–65. Ottawa: Association of Canadian Universities Northern Studies.

Stone, E.L. & Stone, M.H. (1954). Root collar sprouts in pine. *Journal of Forestry*, **52**, 487–91.

Stuart, J.D. (1983). Stand structure and development of a climax Lodgepole Pine forest in South-Central Oregon. PhD thesis, University of Washington, Seattle.

Stuart, J.D., Agee, J.K. & Gara, R.I. (1989). Lodgepole pine regeneration in an old, self-perpetuating forest in south-central Oregon. *Canadian Journal of Forest Research*, **19**, 1096–104.

Stuart, J.D., Geiszler, D.R., Gara, R.I. & Agee, J.K. (1983). Mountain pine beetle scarring of lodgepole pine in south-central Oregon. *Forest Ecology and Management*, **5**, 207–14.

Summers, T.W. (1939). Some impressions of Algerian forestry. *Empire Forestry Journal*, **18**, 235–43.

Swetnam, T.W. & Betancourt, J.L. (1990). Fire–Southern Oscillation relations in the southwestern United States. *Science*, **249**, 1017–20.

Swezy, D.M. & Agee, J.K. (1991). Prescribed fire effects on fine root and tree mortality in old growth ponderosa pine. *Canadian Journal of Forest Research*, **21**, 626–34.

Sylvester T.W. & Wein, R.W. (1981). Fuel characteristics of arctic plant species and simulated plant community flammability by Rothermel's model. *Canadian Journal of Botany*, **59**, 898–907.

Taylor, A.R. (1974). Ecological aspects of lightning in forests. *Tall Timbers Fire Ecology Conference*, **13**, 455–82.

Thomas, T.L. & Agee, J.K. (1986). Prescribed fire effects on mixed conifer forest structure at Crater Lake, Oregon. *Canadian Journal of Forest Research*, **16**, 1082–7.

Tiedemann, A.R. (1987). Combustion losses of sulfur from forest foliage and litter. *Forest Science*, **33**, 216–33.

Tolonen, K. (1983). The post-glacial fire record. In *The Role of Fire in Northern Circumpolar Ecosystems*, ed. R.W. Wein & D.A. MacLean, pp. 21–44. New York: John Wiley.

Trabaud, L. (1991). Is fire an agent favouring plant invasions? In *Biogeography of Mediterranean Invasions*, ed. R.H. Groves & F. di Castri, pp. 179–91. Cambridge: Cambridge University Press.

Trabaud, L.V., Christensen, N.L. & Gill, A.M. (1993). Historical biogeography of fire in temperate and Mediterranean ecosystems. In *Fire in the Environment: The Ecological, Atmospheric, and Climatic Importance of Vegetation Fires*, ed. P. Crutzen & J.G. Goldammer, pp. 277–95. New York: John Wiley.

Trabaud, L. & deChanterac, B. (1985). The influence of fire on the phenological behavior of Mediterranean plant species in Bas-Languedoc (southern France). *Vegetatio*, **60**, 119–30.

Trabaud, L., Grosman, J. & Walter, T. (1985a). Recovery of burnt *Pinus halepensis* Mill. forest. I. Understorey and litter phytomass development after wildfire. *Forest Ecology and Management*, **12**, 269–77.

Trabaud, L., Michels, C. & Grosman, J. (1985b). Recovery of burnt *Pinus halepensis* Mill. forest. II. Pine reconstitution after wildfire. *Forest Ecology and Management*, **13**, 167–79.

Trewartha, G.T. (1968). *An Introduction to Climate*, 4th edn. New York: McGraw-Hill.

Uggla, E. (1958). *Ecological Effects of Fire on North Swedish Forests*. Uppsala, Sweden: Almquist and Wiksells Boktryckeri AB.

Uggla, E. (1974). Fire ecology in Swedish forests. *Tall Timbers Fire Ecology Conference*, **13**, 171–90.

Vanha-Majamaa, I. & Lahde, E. (1991). Vegetation changes in a burned area planted by *Pinus sylvestris* in northern Finland. *Annales Botanici Fennici*, **28**, 161–70.

Van Wagner, C.E. (1973). Height of crown scorch in forest fires. *Canadian Journal of Forest Research*, **3**, 373–8.

Van Wagner, C.E. (1978). Age-class distribution and the forest fire cycle. *Canadian Journal of Forest Research*, **8**, 220–7.

Van Wagner, C.E. (1983). Fire behavior in northern conifer forests and shrublands. In *The Role of Fire in Northern Circumpolar Ecosystems*, ed. R.W. Wein & D.A. MacLean, pp. 65–80. New York: John Wiley.

van Wilgen, B.W., Richardson, D.M. & Seydack, A.H.W. (1994). Managing fynbos for biodiversity: Constraints and options in a fire-prone environment. *South Africa Journal of Science*, **90**, 322–9.

Viereck, L.A. (1983). The effects of fire in black spruce ecosystems of Alaska and northern Canada. In *The Role of Fire in Northern Circumpolar Ecosystems*, ed. R.W. Wein & D.A. MacLean, pp. 210–19. New York: John Wiley.

Viereck, L.A. & Schandelmeier, L. (1980). *Effects of Fire in Alaska and Adjacent Canada – A Literature Review*. US Bureau of Land Management Technical Report BLM-Alaska 6.

Viereck, L.A., Foote, M.J., Dyrness, C.T., Van Cleve, K., Kane, D. & Seifert, R. (1979). *Preliminary Results of Experimental Fires in the Black Spruce Type of Interior Alaska*. USDA Forest Service Research Note PNW-332.

Volland, L.A. (1976). *Plant Communities of the Central Oregon Pumice Zone*. R-6 Area Guide 4–2. USDA Forest Service, Pacific Northwest Region, Portland, Oregon.

Wahlenberg, W.G. (1946). *Longleaf Pine*. Washington, DC: Charles Lathrop Pack Forestry Foundation.

Waring, R.H. & Schlesinger, W.H. (1985). *Forest Ecosystems. Concepts and Management*. New York: Academic Press.

Weaver, H. (1959). Ecological changes in the ponderosa pine forests of the Warm Springs Indian Reservation in Oregon. *Journal of Forestry*, **57**, 15–20.

Wein, R.W. & MacLean, D.A. (1983). An overview of fire in northern ecosystems. In *The Role of Fire in Northern Circumpolar Ecosystems*, ed. R.W. Wein & D.A. MacLean, pp. 1–20. New York: John Wiley.

Wells, B.W. (1942). Ecological problems of the southeastern United States coastal plain. *Botanical Review*, **8**, 533–61.

White, A.S. (1985). Presettlement regeneration patterns in a southwestern ponderosa pine stand. *Ecology*, **66**, 589–94.

White, P.S. (1979). Pattern, process, and natural disturbance in vegetation. *Botanical Review*, **45**, 229–99.

White, P.S. & Pickett, S.T.A. (1985). Natural disturbance and patch dynamics: an introduction. In *The Ecology of Natural Disturbance and Patch Dynamics*, ed. S.T.A. Pickett & P.S. White, pp. 3–13. New York: Academic Press.

Wright, H.A. & Bailey, A.W. (1982). *Fire Ecology: United States and Southern Canada*. New York: John Wiley.

Wyant, J.G., Omi, P.N. & Laven, R.D. (1983). Fire effects on shoot growth characteristics of ponderosa pine in Colorado. *Canadian Journal of Forest Research*, **13**, 620–35.

Youngberg, C.T. & Wollum, A.G. II (1976). Nitrogen accretion in developing *Ceanothus velutinus* stands. *Soil Science Society of America Proceedings*, **40**, 109–12.

Zackrisson, O. (1977). Influences of forest fires on the northern Swedish boreal forest. *Oikos*, **29**, 22–32.

Zackrisson, O. (1980). Forest fire history: Ecological significance and dating problems in the north Swedish boreal forest. In *Proceedings of the Fire History Workshop*, ed. M. Stokes & J. Dieterich, pp. 120–5. USDA Forest Service General Technical Report RM-181.

Zeigler, R. (1980). The vegetation dynamics of *Pinus contorta* forest, Crater Lake National Park, Oregon. MSc thesis, Oregon State University, Corvallis.

Zohar, Y., Weinstein, A., Goldman, A. & Genizi, A. (1988). Fire behavior in conifer plantations in Israel. *Forest-Mediterranéenne*, **10**, 423–6.

12 Evolution of life histories in *Pinus*

Jon E. Keeley and Paul H. Zedler

'Where soils have been used for some time, have become depleted, and have been given up, pines are known to establish themselves where few other trees can survive' Parsons (1955).

12.1 Introduction

Two extremes may be identified in the study of life histories. One is the traditional approach in which the attributes and patterns of growth and development characteristic of individual species are described and classified. The other is quantitative and reduces life histories to a small set of demographic descriptors, which may be empirically difficult to measure but are logically and mathematically tractable, allowing for generalizations which are hopefully applicable to large groups of organisms (e.g. Stearns 1992). Our approach is mid-way between these extremes, and focuses on synthesizing the multitude of life-history characteristics by describing life-history strategies, which we define as sets of attributes that determine the ecological role of groups of species. Strategies represent an evolutionary compromise to the biotic and abiotic factors that affect the survival and fecundity of individuals and the times and places in which a species can be successful. Grime (1979) proposed a scheme for understanding strategies that emphasizes the trade-offs that are necessary to balance the effect of factors that can be grouped into three major categories: competition, disturbance and stress. Stress, as he defined it, refers to abiotic conditions that limit photosynthetic production and include predominantly nutrient and temperature limitations. Competition, which could be thought of as biotic stress, largely represents limitations to photosynthetic production arising from interference with other species.

Disturbance is the sum of all environmental factors that maintain communities in a state of disequilibrium, e.g. fires, floods, wind, etc. Grime's scheme, though simple relative to the complexity of life histories to be explained, is useful for distinguishing the unique qualities that separate pines from other groups of woody plants. The first obvious generality is that pines, as with most other conifers, are more tolerant of abiotically stressful sites than angiosperms. Throughout their extensive range they are most abundant where photosynthetic productivity, and therefore competition from other plants, is limited by a lack of resources or by disturbance, or both. This proclivity is shown not only in their natural range, but also by their capacity to prosper and invade in nutrient-limited or disturbed sites far beyond their natural range in the southern hemisphere (Richardson, Williams & Hobbs 1994).

The association of pines and poor soils is almost certainly an ancestral trait, but it may have been reinforced by competition with the emerging angiosperms which were better suited to exploit high-nutrient conditions. Among conifers, including other genera in the Pinaceae (*Abies, Picea, Tsuga*) as well as more distantly related groups (*Actinostrobus, Cupressus, Fitzroya, Juniperus, Taxodium, Thuja*), tolerance of poor soils and extreme conditions is the rule. Thus, when the pines radiated in parallel with the angiosperms (Chapters 1 and 3, this volume) – they specialized along ancestral lines. As forest ecosystems developed through the Tertiary, angiosperms came to dominate the most fertile and favourable sites. There seem to be many reasons why this was so. Bond (1989) has pointed out several physiological and developmental advantages of most gymnosperms on infertile sites. As site fertility increases, the net photosynthetic capacity of sclerophyllous-leaved conifers falls significantly below that of broadleaved trees (Reich *et al.* 1995; see also Chap. 15, this volume). Additionally, Hickey & Doyle (1977) point out the potential advantage accorded the newly evolved angiosperms because of the unique highly reticulate venation

Ecology and Biogeography of Pinus, ed. D.M. Richardson. © Cambridge University Press (1998), pp. 219–51.

pattern and laminate structure, giving them the potential for more rapid growth rates on fertile sites, initially as understorey shrubs in low-light environments but later as canopy trees. It is also likely that there was a positive feedback between leaf anatomy/physiology and soil fertility, with angiosperm litter favouring more rapid decomposition, better soil structure, and richer soil biota: a contrast that can be observed today between forests dominated by broadleaved trees and those dominated by pines (Nakane 1994; Aerts 1995).

Although tolerance of low-fertility stress is general among conifers, a capacity to invade and exploit open, disturbed, or early to mid-successional stages of forest development is not; it is in the role of aggressive post-disturbance invaders that pines are most clearly differentiated from other conifers. Many conifers have a considerable capacity to endure in conditions of low light (e.g. *Abies* spp.). Others require the conditions present in old-growth forests for seedling establishment (e.g. *Tsuga* spp.). In contrast, pines have a high light requirement and seedling establishment requires mineral soil or only light litter, and open or slightly shaded conditions (Fowells 1965). Growth rates are generally relatively high, and for some species very high. Conifers like *Cupressus* spp. share with pines the traits of disturbance-related establishment and a high light requirement, but in other respects have a conservative stress-tolerant life history with slower growth rates, and a correspondingly limited invasive capacity with relictual distributions to match.

Fire is the only pre-human disturbance factor that occurs with sufficient frequency and intensity in nearly all climatic zones to be a consistent and strong selective pressure affecting the radiation of pines. We believe that fire and fire effects account for much of the diversity in pine life histories, both for persistence of adults and recruitment of seedlings. Pines have exploited fire-prone environments by modifying their life cycle at different stages, e.g. dispersal, seedling development, adult tolerance and vegetative regeneration (Knight *et al.* 1994). Therefore to understand the diversity of pine life histories it is necessary to explore how fire acts in different regions and different plant communities.

The most fundamental aspects of a fire regime are fire frequency and fire intensity (see also Chap. 11, this volume). Fire frequency, the number of fires per unit time, is largely determined by the rate of fuel accumulation and the number and seasonal distribution of lightning ignitions. Fire frequency becomes a potent selective force when it is at or above levels that produce recurrence intervals shorter than the lifespan of the adult trees. Under these conditions, species that cannot survive the fire either by resisting or by establishing seedlings after fire will quickly be pushed to extinction, and individuals that

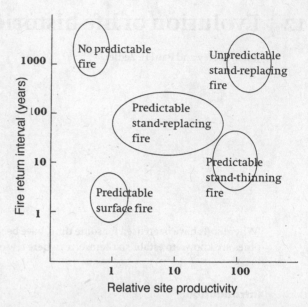

Fig. 12.1. **Fire regimes generated by patterns of site productivity and fire recurrence interval.**

develop traits that benefit most from fire will leave more offspring. In landscapes of the present and recent past, anthropogenic fire has been common and while it has certainly affected the distribution and abundance of pines in many places (e.g. Denevan 1961; Wang 1961; Chap. 1, this volume), we maintain that it is too recent to have been a major factor in life-history evolution.

Fire intensity, the heat release per unit time and per unit length of a fire, is important because it determines the degree of damage to tissues exposed to the fire and also the likelihood of spread, most importantly the probability that fires will carry into the canopies and propagate through the crowns. It is affected by quantity and structure of fuels and therefore indirectly by fire frequency. It is also sensitive to changes in fuel moisture which generally vary significantly with the seasons. Fire intensity is primarily important in life-history evolution because fire severity limits the adaptive options.

12.2 Ecological strategies in pines

The ecological role of pines may be described as trees specialized for moderate to low-fertility habitats and the exploitation of open conditions imposed by severe limitations on growth, because of aridity and/or cold, or resulting from disturbance. Variations on this basic theme are attributable to evolutionary histories that differed with respect to site productivity, disturbance frequency (fire in particular), and the interactions between them (Fig. 12.1).

Fig. 12.2. *Pinus aristata* on dolomite in the White Mountains, California, USA (photo: J.E. Keeley).

On extreme sites where growing seasons are cut short by low temperatures and aridity, fuel accumulation will rarely be sufficient to sustain fire spread, regardless of ignition frequency, and fire is either absent or too rare to be a significant evolutionary influence. In such situations competitive stress imposed by other woody plants is low but abiotic stress is high. With increased productivity competitive stress would be expected to increase, but this can be nullified or reduced by an increase in the frequency and intensity of fires, which are to be expected with the more rapid and spatially uniform accumulation of biomass. Each landscape will have a unique set of properties that determine rate of fuel accumulation, its size and distribution, the probability of ignition, and the moisture conditions at the time of ignition. These will determine the probability of a site being subjected to (1) stand-replacing crown fires; (2) ground fires that consume understoreys; or (3) stand-thinning fires that burn in a mosaic of ground and crown fire. This will have a profound influence on the evolutionarily stable strategies possible for a given landscape. Here we will relate these patterns of life history to fire regimes across the natural productivity gradient.

12.2.1 Very low site productivity [no predictable fires]

Timberline pines

In western North America, half a dozen pine species occur in timberline forests above altitudes of 2500 m. The most stressful timberline sites are those on the desertic side of the main mountain ranges. Here, biomass production is severely limited by the combination of very short, cool growing seasons and rocky poorly developed soils (Fig. 12.2). Productivity is very low, and therefore fire almost unknown (Fig. 12.1). High winds may be locally important, but are not generally a cause of significant mortality, and thus more of a stress than a disturbance. Species characteristic of these very extreme sites include *P. aristata*

(Colorado bristlecone pine), *P. flexilis* (limber pine) and *P. longaeva* (western bristlecone pine). Growth rates are extremely slow. Wright (1963) reported ages up to 40 years for 'seedlings' less than 15 cm tall. Mature trees are noted for their great longevity. Many individuals are more than 1000 years old, and for *P. longaeva* maximum ages can exceed 4000 years. This remarkable tenacity is thought to be a result of the dramatic reduction in respirational demand, arising from branch and bark die-back, leaving old plants with only a narrow strip of living tissue (Wright 1963; LaMarche 1969). Because of their longevity, seedling establishment in these species can be very low and sporadic without imperilling the population. At one site Billings & Thompson (1957) reported no *P. longaeva* seedlings or saplings less than 10 cm basal diameter and very few less than 30 cm. Wright & Mooney (1965) found that *P. longaeva* seedling establishment is restricted by *Artemisia tridentata* (sagebrush) which shades them out. In the White Mountains of eastern California, *P. longaeva* establishes largely on phosphorus-deficient dolomite soils, which apparently exclude sagebrush but allow the sclerophyllous pines to persist. On the adjacent quartzite soils the pines are eliminated due to competition with sagebrush and lack of tolerance of the more xeric conditions created by the better-drained substrate, a critical factor on these rain shadow sites that receive <350 mm precipitation per year. Thus, lacking periodic disturbances that open up sites, seedling recruitment is restricted to fairly exacting conditions of nutrient-deficient soils in arid ranges; conditions that make seedling establishment precarious and unpredictable.

We interpret the life history of these long-lived trees to be the expected outcome of selection in an environment in which opportunities for establishment are rare, but the capacity to endure is rewarded by the opportunity to establish offspring in circumstances where there is minimal threat of competitive displacement. If theoretical models beginning with Murphy (1968) are to be believed, it is variable juvenile mortality that has been the driving force favouring longevity.

At timberline in more mesic ranges (with annual precipitation 500–1000 mm), enough biomass may accumulate to carry wildfires occasionally and thus some species such as *P. albicaulis* (whitebark pine) and *P. flexilis* may have the opportunity to disperse into burned sites, previously dominated by other conifers (Shankman 1984; Habeck 1987; Tomback 1986; Tomback, Hoffman & Sund 1990). On these sites the pines play a seral role, with higher growth rates and curtailed life spans (Veblen 1986; Morgan & Bunting 1990; Keane & Morgan 1993). As shade-tolerant conifers re-invade such sites, *P. albicaulis* seedling establishment ceases (Fig. 12.3) and the populations survive only in more extreme subalpine sites. On rare occasions, typical

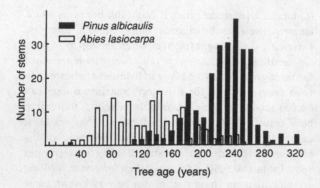

Fig. 12.3. **Age distribution for whitebark pine (*Pinus albicaulis*) and subalpine fir (*Abies lasiocarpa*) (redrawn from Morgan & Bunting 1990).**

timberline species such as *P. flexilis* occur below 2000 m, but due to greater fire-return intervals, never reach the longevity typical of higher subalpine sites (Schuster *et al.* 1995).

Other timberline species such as the European alpine *P. cembra* (Swiss stone pine), Eurasian boreal *P. sibirica* (Siberian stone pine), Korean *P. koraiensis* (Korean stone pine), and East Asian alpine *P. pumila* (dwarf stone pine) are similar in their persistence on extreme high-elevation/latitude sites, and ability to colonize burned forests (Paryski 1971; Contini & LaVarelo 1982; Lanner 1990). As discussed later, both these Old World and New World subalpine pines depend upon a remarkable coevolution with birds for seed dispersal (see also Chap. 14, this volume), which play a prominent role in the colonization of burned forests.

Desertic pines

The growing season for timberline species is largely limited by low temperatures, but for lower-elevation western North American desert or pinyon pines, the growing season is cut short by both winter cold and summer heat, as well as by severe water deficits. Above-ground seedling growth rates are extremely low, but survival of established trees is high with individuals reaching ages of 400 to > 900 years (Fowells 1965). Seedling establishment is episodic and population age structure is markedly affected by drought (St Andre, Mooney & Wright 1965), which differentially reduces seedling and sapling recruitment more than other age classes (Fig. 12.4). Seedlings survive best in the shade and moisture provided by 'nurse plants' and establishment in the open is rare (Fowells 1965). Consequently, the desertic pines are often relatively shade-tolerant, and can apparently colonize higher elevation closed-canopy pine communities (Barton 1993). Because of their thin bark and lack of self-pruning, however, they are intolerant of fire, which is probably the primary control on their upper elevational distribution.

Within their natural range, biomass accumulation is limited and thus disturbance from fires is infrequent and

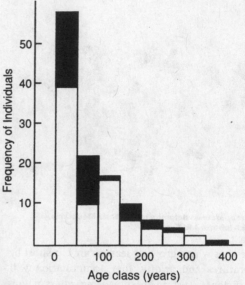

Fig. 12.4. **Age distribution for singleleaf pinyon pine (*Pinus monophylla*) affected by severe drought (redrawn from St Andre *et al.* 1965) ■, seedling recruitment; □, sapling recruitment.**

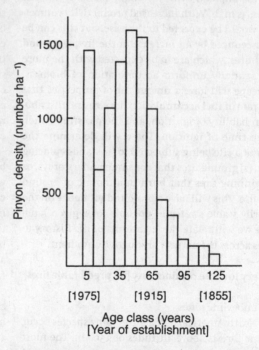

Fig. 12.5. **Age distribution for singleleaf pinyon pine (*Pinus monophylla*) recolonizing a site burned in the mid-19th century (redrawn from Tausch & West 1988).**

unpredictable, being dependent upon exceptional rainfall years that lead to herbaceous growth sufficient to carry fire. If stands are burned, pinyon pines such as *P. mono-phylla* (single-needle pinyon) are typically eliminated from the site and invade very slowly (Fig. 12.5) (Barney &

Table 12.1. Distribution of characteristics interpreted as evolutionary adaptations to fire in selected species of pines

Pinus taxa	Grass Stage	Resprouting	Serotiny	Thick Bark	Self-pruning
attenuata	–	–	++++	–	–
banksiana	–	–	++++	–	–
brutia	–	–	++	+	+
canariensis	–	++++	–	+++(+)	–
clausa	–	++++	++++/–	–	–
contorta subsp.					
latifolia	–	–	++++/–	–	–
coulteri	–	–	–/+	++	–
devoniana	++++	–	–	++	++
echinata	–	++++	–	+	+
elliottii var.					
densa	–/+	–	–	++	++
greggii	–	–	++++	+	–
halepensis	–	–	++/–	+++	+
jaliscana	–	–	++++	+++	+++
jeffreyi	–	–	–	++++	++++
lambertiana	–	–	–	++++	++++
leiophylla	–	++++	–/+	+++	+++
merkusii	++++/–	–	–	++	++
montezumae	++++	–	–	+++	+++
muricata	–	–	+++	+++	+
oocarpa	–	++++	++++/–	+	+
palustris	++++	–	–	++	++++
patula	–	–	+++	++	+
ponderosa	–	–	–	++++	++++
pringlei	–	–	–/+	+++	+++
pseudostrobus	–	–	–	+++	++++
pungens	–	+	++++	+	+
radiata	–	–	++	++	+
resinosa	–	–	–	++	+++
rigida	–	++++/–	++/–	++	–
serotina	–	++++	++++	+	–
torreyana	–	–	–/+	+	+
virginiana	–	+	++++	+	+

responses to fire (Table 12.1). Most of the species in this conspicuously fire-adapted group are members of the subgenus *Pinus* (hard pines). Ranking of characteristics in Table 12.1 is relative and should be loosely interpreted to indicate importance. Some traits such as serotiny are well developed in some subspecific forms and not in others. Quantitative comparative data on characters such as bark thickness and degree of self-pruning are lacking and these rankings are based on personal observations or inferred from descriptions and photographs in the literature. Limitations are manifold. For example, bark thickness is only one measure of degree of insulation and this character, as well as degree of self-pruning, may have adaptive origins unrelated to fire. The 'grass stage', resprouting, and serotiny characteristics are considered in all instances to be directly selected for by fire. Here we will relate these to patterns of fire regime and site productivity, but see Chap. 11 (this volume) for coverage of community-level responses to fire.

Low to moderately productive sites with high fire frequency [predictable ground fires]

One of the most remarkable life-history traits of any woody plant is exhibited by species of pines at low latitudes in unproductive sites typically on excessively drained sandy soils in warm humid climates with frequent low-intensity ground fires fuelled by a herbaceous understorey. The most studied species in this category is *P. palustris* (longleaf pine), which is widely distributed on relatively unproductive nutrient-poor sandy sites in the coastal plain of the southeastern USA (Christensen 1977). These conditions lead to a sparse canopy cover and establishment of herbaceous wiregrass (*Aristida* spp.) understorey. Despite the high rainfall (>1000 mm yr^{-1}), the sparse canopy results in high soil albedo and rapid drying of fuels. Summer thunderstorms generate the highest frequency of lightning strikes in North America (Komarek 1974), and nearly 60% of the wildfires in the USA occur in this small region (Christensen 1981). On longleaf pine sites the fuel load of the herbaceous understorey is insufficient to generate fires that are much threat to the adult pines, although temperatures are sufficient to kill seedlings and saplings of most species (Chapman 1936; Garren 1943; Landers 1991). In response to these conditions, *P. palustris* has evolved delayed seedling development, whereby internode elongation is depressed for the first five to 20 years of development (Fowells 1965), resulting in seedlings persisting with a bunchgrass growth form (Fig. 12.6). During this grass stage, needles with high fire resistance are produced soon after germination and protect the terminal bud from fire, thus making seedlings extremely resistant to fire (Croker & Boyer 1975). During the grass stage an extensive root system and carbohydrate stores accumulate

Fischknecht 1974; Tausch & West 1988). Some pinyons, e.g. *P. cembroides* (Mexican pinyon) appear to be somewhat resistant to occasional fires that may occur once a century or more (Moir 1982).

In summary, the pines of both stressful high-elevation and desert habitats seem to have evolved towards longevity, most probably because the stress of the environment falls most heavily on the youngest and smallest individuals, thus making establishment rare and highly variable. Fire or other disturbances have little impact on mature individuals and the extreme and stressful sites minimize competition from other woody species. Pines adapted to such extreme sites are in the subgenus *Strobus* (soft pines), suggesting that members of this group have an inherent capacity to endure such abiotic stresses.

12.2.2 Low to moderate site productivity [predictable fires]

On sites with sufficient moisture and nutrients to produce a continuous fuel bed, but no more than relatively moderate plant productivity, all pines exhibit life-history characteristics that we interpret as evolutionary

Fig. 12.6. **Longleaf pine (*Pinus palustris*) grass stage (photo: J.E. Keeley).**

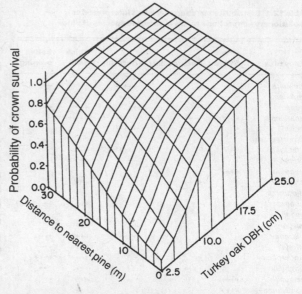

Fig. 12.7. **Effect of proximity to longleaf pine on postfire crown survival of oaks (from Rebertus *et al.* 1989).**

(Pessin 1938; Brown 1964). After many years in the grass stage, seedlings 'bolt' and exhibit rapid internode elongation, resulting in very high growth rates, with as much as 1.5 m growth in three years (Mattoon 1922). During this stage the saplings are susceptible to injury from wildfires (Garren 1943), but become increasingly resistant to fires due to development of moderately thick bark and self-pruning of lower branches. After several decades, trees are quite resistant to fire, and may live for several hundred years (Platt, Evans & Rathbun 1988), unless an absence of fire allows fuels to accumulate and generate fires that can scorch the crowns (Glitzenstein, Platt & Streng 1995).

There can be little doubt that the regime of predictable ground fires has selected for the unusual pattern of seedling/sapling growth, and also permitted and possibly selected for the extreme shade intolerance which makes seedling establishment rare in the vicinity of established adults (Garren 1943; Grace & Platt 1995a). The result is a species dependent on fire and therefore sensitive to changes in fire regime. The greatest threat to the persistence of *P. palustris* is competition from oaks and other hardwoods and fire plays a critical role in reducing oak canopies (Moser 1989; Rebertus, Williamson & Moser 1989; Glitzenstein *et al.* 1995). It has been hypothesized that the needle attributes that produce higher temperatures in pine leaf litter relative to oak litter, have been selected because they enhance sites for pine recruitment (Williamson & Black 1981). This is supported by the observed relationship between proximity to longleaf pines and crown survival of oaks after fire (Fig. 12.7).

Undisturbed systems exist in a dynamic balance between pine density, understorey fuels, competing hardwoods, and fire recurrence. In undisturbed systems, as pine tree density increases, shading and pine leaf litter accumulation increase so that conditions for seedling growth are less favourable and fires are hotter, killing both pine seedlings and hardwoods (Grace & Platt 1995b).

Patches that appear because of the death of trees are most suitable for recruitment because they will have less intense fires and ample light, leading to recruitment and the ultimate disappearance of the patch. In general high fire frequency stimulates understorey herbaceous fuels which reduce oak and other hardwoods, thus favouring pines (Platt, Glitzenstein & Streng 1991). This balance can be upset by human interference. It has also been hypothesized that anthropogenic removal of longleaf pines has eliminated pyrogenic fuels necessary to sustain frequent light fires (Christensen 1981). When fires are suppressed, recruitment is halted (Fig. 12.8), and stands stagnate (Hartnett & Krofta 1989).

The grass-stage attribute is uncommon in *Pinus*, but it occurs in widely disjunct and distantly related taxa and has undoubtedly evolved more than once. It is a prominent part of the life cycle of *P. merkusii* in Southeast Asia; this is the only species that occurs naturally in the southern hemisphere, ranging from approximately latitude 20° N to 2° S (Mirov 1967). As with *P. palustris*, it is distributed on nutrient-poor and well-drained sandy soils in association with xerophytic dipterocarps and a herbaceous understorey (Turakka, Luukkanen & Bhumibhanon 1982). The relationship of *P. merkusii* to fire is also very similar to that of *P. palustris*. Despite the relatively high annual rainfall (1000–2000 mm), these xeric poorly vegetated sites are susceptible to frequent fires, especially during the hot dry season which may average 37 °C (Koskela, Kuusipalo & Sirikul 1995). Adult trees survive these low-intensity fires due to thick bark and moderate self-pruning (Werner 1993; Koskela *et al.* 1995). Annual fires are often lethal to

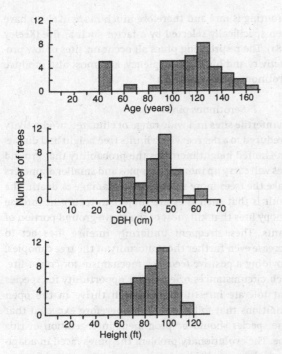

Fig. 12.8. Lack of longleaf pine recruitment on sites protected from fire (redrawn from Hartnett & Krofta 1989).

first-year seedlings, but by the second year of growth *P. merkusii* seedlings develop fire-resistant needles and remain in a grass stage for five years or more. Duration of the grass stage in *P. merkusii* was found to increase with degree of competition (Koskela *et al.* 1995) and the same has been found for *P. palustris* (Brown 1964). Frequent fires and herbaceous fuels maintain fire temperatures at levels tolerable for adult trees as well as the grass-stage seedlings, and reduce competitive inhibition of seedling growth. Koskela *et al.* (1995) proposed that the grass stage was not selected by frequent fires alone, but rather the slow growth rates attainable on xeric sites put seedlings in a state vulnerable to fire for an exceptionally long period. Accumulating carbohydrates during the grass stage allowed rapid growth rates and the potential for rapidly reaching a size that made the sapling immune to fire.

This grass stage also has been reported from several other species (Table 12.1) from similar subtropical oligotrophic sites with high fire frequency (Mirov 1967; Perry 1991). These include a southern Florida variety of *P. elliottii*, the Cuban *P. tropicalis*, and southern Mexican pines (*P. devoniana* and *P. montezumae*). Several of these species, such as *P. merkusii* and *P. elliottii*, exhibit regional variation in presence of the grass stage (Ketcham & Bethune 1963; Mirov 1967; Perry 1991; Koskela *et al.* 1995). Although not well studied, it appears that this grass stage has been selected out of these species in more productive environments, perhaps due to greater fire intensities that make survival of this stage more

tenuous. Although second-year seedlings of *P. pseudostrobus* form a grass-like mound, they do not remain in that stage (Perry 1991; see also Chap. 7, this volume).

Not all subtropical pine species with a regime of predictable ground fires have evolved a grass stage; for example, *P. caribaea* which is common in the llanos of eastern Nicaragua and Honduras (Mirov 1967), lacks the grass stage. One explanation is that the pines are invaders into an anthropogenic landscape from what might have been small original refuges (Parsons 1955) and therefore may not have had time to adapt to the new conditions. But it seems equally likely that the conditions for the evolution of the grass stage may be dependent on a combination of circumstances absent for some species. Some of these may be subtle. For example, the pattern of distribution of fuels may be such that patches exist in which seedling establishment is possible and sapling survival great enough that there is no fitness gain to individuals that have tendencies towards reduced height growth. A constant is, however, that all pines that occur under this regime of predictable low-intensity ground fires have thick bark and extensive self-pruning that ensure adult survival.

Low to moderately productive sites with high-intensity fires [predictable stand-replacing fires]

With better site conditions, community productivity increases and alters the environment in several important ways. Herbaceous fuels are replaced by woody fuels, and greater water availability means fires are dependent upon unique weather conditions that generate droughts of sufficient duration to dry fuels; thus, fire-return intervals increase (which allows for greater fuel accumulation). These changes lead to an increase in fire intensity sufficient to generate stand-replacing crown fires (Fig. 12.1). Also, as site productivity increases, increased post-fire competition selects for rapid re-establishment, through resprouting and/or serotiny (Table 12.1).

Resprouting pines

In the southeastern USA, as fire intensity increases the grass-stage strategy (as in *P. palustris*) is replaced by the capacity for resprouting from the stem or root collar (Fig. 12.9). Basal resprouting after fire is common in three pines of the eastern USA: *P. echinata* (shortleaf pine), which is best developed on well-drained soils; *P. serotina* (pond pine), which forms extensive stands in poorly-drained depressions; and *P. rigida* (pitch pine), which spans the range of infertile conditions from excessively drained to poorly drained sites (Fowells 1965; Stone & Stone 1954). Seedlings of these three species are vigorous resprouters from axillary buds produced at the base of the stem. Basal sprouting is, however, largely restricted to seedlings and small

Fig. 12.9. **Resprouting shortleaf pine (*Pinus echinata*) sapling from southeastern USA (photo: J.E. Keeley).**

saplings. Limited ability for basal resprouting is known from a few other species (Table 12.1), including the Mexican–Central American *P. oocarpa* (Ponce 1985). In many of these species, as well as others, e.g. *P. clausa* (sand pine) and *P. canariensis* (Canary Island pine), older saplings and young trees resprout epicormically (Stone & Stone 1943). These preformed buds are well insulated by bark and survive fire in trees up to several decades old (Ledig & Little 1979); they also respond to other disturbances that open the canopy.

It appears that most basal resprouting species are from summer-rainfall climates, which may reflect the increased likelihood of rootstock survival. Another factor could be competition. These sites generally lack post-fire seeding species, but have an abundance of vigorous resprouting broadleaved herbs and shrubs, which may make the sites less favourable for post-fire seeding (Webber 1935; Boerner 1981; Ostertag & Menges 1994). In many woody plants, basal resprouting is commonly interpreted as an adaptation to fire. This assertion has, however, been questioned because sprouting is almost universal in woody angiosperms. It is not, however, the case for conifers where basal

sprouting is rare and therefore much more likely to have been specifically selected by a factor such as fire (Keeley 1981). The resprouting pines all occur on sites of low productivity and high fire frequency, and most also produce serotinous cones (Table 12.1).

Serotinous pines
On infertile sites in a wide range of climates, productivity is reduced to a degree which limits tree height and diameter. Limited height increases the probability that ground fires will carry up into the canopies and smaller diameters make the trees more susceptible to damage or death. The result is that such sites are subject to recurrent intense canopy fires that kill most of the above-ground portion of plants. These frequent uniformly intense fires act to increase even further the uniformity of the tree canopies, providing a positive feedback mechanism for crown fire. Such circumstances offer a clear opportunity for species that tolerate infertility and which thrive in the open conditions that follow fire. It is therefore expected that pine species should be prominent in vegetation of this type. The evolutionary problem the pines faced in adapting to these conditions was how to equip a population of small trees to survive intense fire. The solution was cone serotiny.

In North America, the working definition of cone serotiny is that of Critchfield (1957). He defined a serotinous cone as one that remained closed and on the tree for one or more years after the seeds mature, but could open rapidly when high temperatures melted the resin that seals the cone scales. Lamont *et al.* (1991) considered serotiny more generally and proposed that it be defined as 'canopy seed storage' where at least part of the previous seed crop is retained when the current year's crop of seeds is mature. As will be discussed below, the more general definition of Lamont *et al.* is necessary to encompass the full range of variation in this important life-history trait in pines. The essential feature of serotiny is captured by both definitions: i.e. serotinous species can accumulate seeds, and in the serotinous pines this results in a mass-release of seeds within hours or days after fire. In some species, release can occur at other times. Resin is reported to melt between 45 and 60 °C; on open, xeric sites cones may open on hot summer days without fire (Cameron 1953; Beaufait 1960). Serotinous cones commonly have other features that suggest fire adaptation. Smith (1970) noted that serotinous pines allocate more resources to protective cone structure than to seeds; as a consequence, seeds can survive cone temperatures >200 °C (Beaufait 1960; Knapp & Anderson 1980). In the Californian closed-cone species *P. attenuata* (knobcone pine) and *P. muricata* (bishop pine), and several other serotinous species, the cones not only have thick apophyses but also are strongly reflexed to lie

Fig. 12.10. **Serotinous cones of knobcone pine (*Pinus attenuata*)** (photo: J.E. Keeley).

along the branch. Across serotinous and non-serotinous populations of *P. contorta* var. *latifolia* (lodgepole pine), these traits are indicative of serotiny (Tinker *et al.* 1994). Such traits may minimize radiative and convective heating during fire (however, see the discussion on coevolution).

Serotiny in pines varies among species. At one extreme are the strongly serotinous species such as the Californian *P. attenuata* (Fig. 12.10), in which most of the cone crop remains closed for 15–20 years (Badron 1949; Vogl 1973). At the other extreme are species such as *P. torreyana* (Torrey pine) and *P. coulteri* (Coulter pine) (McMaster & Zedler 1981; Borchert 1985) in which the cones remain on the trees for years but they open at maturity without a heat stimulus. A large proportion of the seeds drop in the first weeks, but some are retained in the proximal portion of the cone and fall over an extended period, possibly in response to scale movement caused by wetting and drying. Similar patterns of cone retention and gradual release of seeds have also been noted for *P. leiophylla* (smooth-leaved pine) and a few other Mexican pines (Perry 1991). Other species such as *P. radiata* (Monterey pine) are intermediate. Their cones

remain closed at maturity but do open without fire after a few years in response to normal temperature extremes or other causes. As a result, their canopies often contain large numbers of open cones.

Serotiny also varies within species. In *P. banksiana* (jack pine), *P. contorta* and *P. rigida* (pitch pine) there is strong circumstantial evidence that serotiny is under relatively simple genetic control, involving perhaps just a single gene that regulates the melting temperature of cone resin (Teich 1970). Ledig & Fryer (1972) concluded that serotinous *P. rigida* were homozygous for the serotiny gene, and heterozygotes at this locus had a mix of serotinous and non-serotinous cones. In light of such a relatively simple genetic control, it is to be expected that this polymorphism could be spatially and temporally quite dynamic.

Indeed, most serotinous pine species are polymorphic for this feature, having populations or subspecific taxa with no closed cones, or polymorphic populations or individuals with both open and closed cones. For example, *P. contorta* is widely distributed throughout the western USA. In the Cascade Mountains of Oregon and Washington, the Sierra Nevada Range of California and Transverse and Peninsular ranges of southern California and adjacent Baja California, var. *murrayana* is uniformly non-serotinous, whereas in the Rocky Mountains var. *latifolia* is often serotinous (Critchfield 1957). These patterns are clearly correlated with differences in site productivity and fire frequency. In the Sierra Nevada, *P. contorta* var. *murrayana* is a subalpine species distributed up to altitudes of 3500 m, in environments characterized by a short growing season, low fuel accumulation, infrequent droughts and low lightning fire incidence (Parker 1986; Sheppard & Lassoie 1985). As is common in other high-elevation pines, *P. contorta* in the Sierra Nevada is quite long-lived (>600 years; Fowells 1965). Also, stands are uneven-aged with seedling recruitment restricted to gaps created by windthrown trees, or to meadows experiencing a drop in water table (Helms & Ratliff 1987) or seedlings may initiate primary succession on granitic outcrops (Rundel 1975; Parker 1986). These lodgepole forests are not replaced by more shade-tolerant species, perhaps due to the shallow granitic soils (Rundel, Parsons & Gordon 1977; Parker & Peet 1984). Where this species co-occurs with more stress-tolerant taxa, e.g. *P. albicaulis*, or less stress-tolerant taxa, e.g. *P. jeffreyi* (Jeffrey pine), forests appear to result from a lottery-type coexistence which derives from infrequent, species-specific pulses of seedling recruitment in response to differing sets of environmental conditions (Carpenter 1991).

In the Rocky Mountains, lodgepole pine may form similar uneven-aged populations on stressful subalpine sites (Despain 1983; Habeck 1987; Johnson & Fryer 1989; Jakubos & Romme 1993). These trees are likewise non-serotinous, and the infrequent fires are patchy enough to

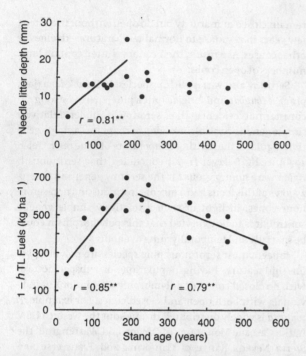

Fig. 12.11. **Chronosequence of changes in small fuels capable of supporting ignition and initial fire spread in Rocky Mountain lodgepole pine (*Pinus contorta* subsp. *latifolia*) forests (redrawn from Romme 1982).**

leave sufficient seed trees for restocking (Stahelin 1943). More commonly, however, this pine forms dense even-aged stands at lower elevations, down to 1800 m (Arno 1980; Peet 1988). In these situations the longer growing season contributes to a higher growth potential (Rehfeldt 1983, 1987). With higher productivity comes greater fuel accumulation, and with lower elevation a greater probability of summer drought coincident with lightning ignition, a condition that occurs infrequently enough (perhaps only once every few hundred years) that when fires do come they are stand-replacing crown fires (Romme 1982; Muir 1993). On these sites lodgepole pine is generally serotinous, has a shorter lifespan, and in the long absence of fire, can be replaced by more mesic, shade-tolerant *Abies* (fir) and *Picea* (spruce) species (Whipple & Dix 1979; Peet 1988; Tait, Cieszewski & Bella 1988) – successional changes that reduce the potential for stand-replacing fires (Fig. 12.11).

There is evidence that small-scale differences in fire frequency select for genetic differences in populations, and thus frequency of serotiny may reflect local predictability of fire (Lotan 1967; Perry & Lotan 1979). Strongly serotinous species have little to no seed dispersal in the absence of fire and thus are clearly fire-dependent. Serotinous species typically generate dense even-aged stands: for example, Lowery (1984) reported over 400 000 eight-year-old trees per hectare and Fowells (1965) reported nearly 250 000

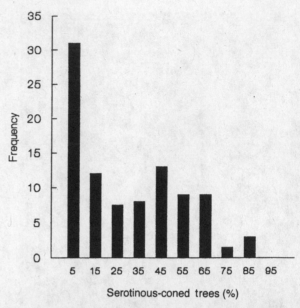

Fig. 12.12. **Frequency distribution of serotinous and non-serotinous-coned trees in a Rocky Mountain lodgepole pine stand (redrawn from Perry & Lotan 1979).**

trees per hectare at 70 years. Lodgepole pine retains many dead branches, making stands particularly susceptible to stand-replacing crown fires, and raising the possibility that the lack of self-pruning may have evolved to increase pyrogenecity *vis-à-vis* the Mutch Hypothesis (Mutch 1970; Bond & Midgely 1995). While stand-replacing fires are conducive to cone opening (Johnson & Gutsell 1993), and lead to reduced competition from more shade-tolerant species, very intense fires may reduce post-fire seedling recruitment (Ellis *et al.* 1994).

In the Rocky Mountains *P. contorta* var. *latifolia* is typified by mixed populations of serotinous and non-serotinous trees (Fig. 12.12). Despite the widespread occurrence of serotiny throughout its range, the open-cone phenotype remains high, suggesting fires are unpredictable and variable in time and space (Perry & Lotan 1979). Detailed studies of geographical variation in the Rocky Mountains showed that percentage serotiny in a population was not correlated with any environmental variables such as elevation, slope aspect or incline, and topographic position (Muir & Lotan 1985). Following fire, percentage serotiny has been shown to be a more important predictor of seedling density than aspect, inclination, or soil type (Tinker *et al.* 1994).

A similar pattern of serotinous and non-serotinous populations is observed in *P. rigida* in the eastern USA. On the well-drained sandy New Jersey Pine Barrens, populations are uniformly serotinous, whereas elsewhere this character declines along a cline of increasing forest productivity. Ledig & Fryer (1972) maintained that this cline

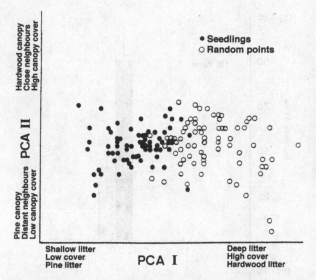

Fig. 12.14. **Pinus coulteri, a chaparral–oak woodland pine that does not self-prune** (photo: J.E. Keeley).

Fig. 12.13. **Principal components affecting Table Mountain pine (Pinus pungens) recruitment compared with distribution of random points (from Williams & Johnson 1992).**

was established and maintained by fire occurrence and gene flow. Givnish (1981) doubted the presence of such a cline and thus the importance of gene flow; rather, he provided evidence that serotinous populations were largely maintained by active selection from frequent stand-replacing fires. As site productivity increases, fuel moisture conditions reduce the likelihood of fires occurring before serotinous species such as *P. pungens* (Table Mountain pine) and *P. rigida* are outcompeted by shade-tolerant broadleaved trees, thus selecting against serotinous phenotypes (Zobel 1969; Givinish 1981; Williams & Johnson 1990, 1992). On these sites, in the absence of fire, limited seedling recruitment may occur on sites with low canopy coverage and shallow litter (Fig. 12.13). Likewise, as site productivity decreases, for example on sparsely vegetated rock outcrops, fire frequency declines and *P. rigida* forms uneven-aged non-serotinous populations that persist indefinitely with little threat of being displaced (Abrams & Orwig 1995); the same has been observed for *P. pungens* (Barden 1977a, b). Laessle (1965) has suggested that landscape patterns control serotiny polymorphism in *P. clausa*. In the peninsula of Florida this species forms large contiguous stands that are subject to periodic stand-replacing crown fires, leading to even-aged serotinous populations. In the Florida panhandle, on the other hand, it occurs in small, uneven-aged populations, in a mosaic of more mesic vegetation types and has a fire regime of low-intensity ground fires that generate uneven-aged, non-serotinous populations.

The importance of active selection maintaining serotiny is also suggested by the persistence of serotinous *P. contorta* subsp. *bolanderi* (Bolander pine) on the sandy pygmy forests of northern California, within dispersal distance of non-serotinous populations of *P. contorta* subsp. *contorta* (Critchfield 1957; Westman 1975). Another example would be the Californian *P. coulteri* which has non-serotinous cones in most woodland/forest populations, but an increasing percentage of serotinous cones on trees juxtaposed with chaparral brushlands (Borchert 1985). Chaparral populations of this pine have an even-aged structure resulting from synchronized seed dispersal following stand-replacing crown fires (Vale 1979; Borchert 1985). Woodland/forest populations, however, have an uneven-aged population structure, due to continuous seed dispersal and a fire regime of stand-thinning ground fires (Borchert 1985). Coincident with this pattern is a polymorphic branching habitat (Zobel 1953); on chaparral landscapes it branches near the ground (Fig. 12.14), making canopy ignition almost certain, whereas on forested sites it self-prunes for a considerable length of the bole, increasing the chances of surviving low-intensity ground fires.

Pinus banksiana is a serotinous species closely related to *P. contorta* subsp. *latifolia* that occupies relatively unproductive sites in the boreal forest of North America, being a relatively recent colonizer of glacial sands (Payette 1993; Desponts & Payette 1993; Béland & Bergeron 1993). Due to the very short growing season, fuel accumulation is slow and the fire-return interval relatively long – about 100 years – and of sufficient intensity to generate stand-replacing crown fires. High fire intensity appears necessary to release seeds and to destroy competing vegetation, thus both seedling density and growth are positively correlated with fire intensity (Weber, Hummel & van Wagner 1987). Over much of its range it forms even-aged stands with recruitment restricted to the first decade after fire (Cayford & McRae 1983; St-Pierre & Gagnon 1992; Gauthier, Bergeron & Simon 1993a). Despite natural thinning of smaller trees in the first few decades, decomposition is slow and stands are generally very dense with much standing dead wood (Yarranton & Yarranton 1975; Govindaraju 1984; Stocks 1987; Kenkel 1988), making the site increasingly susceptible to fire. This pattern, however, is affected by site productivity: on less productive xeric sites, populations tend to be uneven-aged as recruitment occurs in gaps (Abrams 1984; Gauthier, Gagnon & Bergeron 1993b) from cones opened by heat generated during the summer on these more exposed sites (Sterrett 1920; Fowells 1965). *Pinus banksiana* is a relatively short-lived species (100–200 years) and if sites escape fires for more than 200 years this species will be displaced by less fire-dependent species such as *Picea mariana* (Gauthier *et al.* 1993b).

Five serotinous pine species are distributed on relatively unproductive sandy or rocky sites in the eastern USA, including several that are vigorous resprouters such as *P. clausa*, *P. rigida* and *P. serotina*, as well as *P. pungens* and *P. virginiana* (Virginia pine). None of these are very long lived but experience fires, typically at 10–50 year intervals, which result in stand-replacing crown fires (Crutchfield & Trew 1961; Zobel 1969; Forman & Boerner 1981). Less intense ground fires are not as effective at inducing pine recruitment for most of these serotinous species (Barden & Woods 1976; Groeschl, Johnson & Smith 1992). Rapid regeneration by resprouting broadleaved species (Webber 1935; Boerner 1981; Buchholz 1983) puts a premium on rapid seedling recruitment from serotinous cones. The southern Mexican and Central American highlands, with a summer-rain climate, also have numerous serotinous species, including *P. greggii*, *P. jaliscana*, *P. leiophylla*, *P. oocarpa*, *P. patula*, and *P. pringlei*.

Serotinous pines also are frequent in summer-droughted mediterranean-climate regions; *P. brutia* (Eastern Mediterranean pine) and *P. halepensis* (Aleppo pine) of the Mediterranean Basin, and *P. attenuata*, *P. contorta* subsp. *bolanderi*, *P. muricata*, and *P. radiata* of Oregon, California and Baja California. These are moderately productive environments with rapid fuel accumulation and annual summer droughts conducive to fires.

Ignition sources are not nearly as ubiquitous as they are in regions such as the southeastern USA; thus fire-return intervals are 10 to 100 years or more (Keeley 1982; Abbas, Barbéro & Loisel 1984; Trabaud, Christensen & Gill 1993). This fire-return interval, coupled with the juxtaposition of these forests with highly flammable shrublands and the lack of self-pruning in most of these pines, nearly always results in stand-replacing crown fires (Vogl *et al.* 1977; Trabaud, Michels & Grosman 1985; Thanos *et al.* 1989; Moravec 1990; Ne'eman, Lahav & Izhaki 1992a). These post-fire environments are different from serotinous pine habitats in other regions in the abundance and diversity of broadleaved seeding species; many of these are annuals capable of rapid growth and vigorous competition. Bond

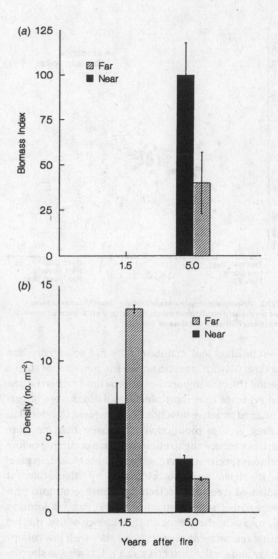

Fig. 12.15. **Postfire *Pinus halepensis* seedling density (*a*) and relative biomass (*b*) near and far from burned tree skeletons (from Ne'eman 1996).**

(1989) hypothesized that such competition should be particularly severe for obligate-seeding species (e.g. serotinous pines) and recruitment would be most successful on sites with reduced establishment of other species. There is evidence that this may be the case for *P. halepensis*, where post-fire competition is markedly reduced near the pre-fire pine skeleton and in this microhabitat pine seedlings are most likely to succeed (Fig. 12.15) (Ne'eman *et al.* 1992a, b; Ne'eman 1997). It is not known how much of this pattern can be explained by the more intense heat likely to be generated within the canopy shadow or by the inhibitory effect of ash on germination (Ne'eman, Meir & Ne'eman 1993a, b). A similar pattern is seen in the Californian serotinous pine forests; a post-fire inhibition of herbaceous growth within the canopy shadow of *P. attenuata*

Fig. 12.16. **Postfire herb growth in chaparral (left) compared with relatively depauperate regrowth beneath burned knobcone pines (right) (photo: J.E. Keeley).**

Fig. 12.17. **Thick bark and self-pruning typical of *Pinus ponderosa* in California's Sierra Nevada (photo: J.E. Keeley).**

skeletons (Fig. 12.16), where pine seedling recruitment is concentrated (G. Ne'eman, J.E. Keeley & C.J. Fotheringham, unpublished data).

Pinus attenuata may also avoid intense post-fire competition by its tendency for establishing on nutrient-poor serpentine soils, a common characteristic of most of the closed-cone conifers in California (Vogl 1973; Vogl *et al.* 1977; Zedler, Gautier & Jacks 1984).

In summary, environments with predictable, high-intensity stand-replacing fires select for resprouting and serotinous adaptations in pines. Most of these pines are relatively short-lived (100–200 years) and become reproductive at an early age (<10 years), and most occur in dense even-aged stands. Nearly all species are polymorphic for scrotiny and this appears to be under simple genetic control that can be altered by changes in the fire regime over one to several generations.

12.2.3   High site productivity

One obvious conclusion from Fig. 12.1 is that fire-return interval is not a simple function of site productivity. Pines are most susceptible to displacement by other species on productive sites, and therefore they are, or were before human intervention, dependent on fire disturbance to create opportunities for establishment and early survival. In an oxygen-rich atmosphere, plant biomass is always susceptible to fire when drought or drying conditions coincide with ignition. At one extreme, on some productive sites recurrent drought and frequent lightning ignition can cause high fire frequency which reduces fuel loads and favours understorey species that will be readily burned in less than the most extreme drying conditions. These factors reduce intensity and favour stand-thinning fires rather than stand-replacing fires. In moister situations with less frequent or less extreme episodes of drought, fuel accumulation will be greater, stands denser, and the

conditions necessary to propagate fires more extreme, with the result that fires will be rare, but stand-replacing when they do occur. Which position along this continuum of fire conditions a pine species has occupied will determine the evolutionary options that have been open to it.

Productive sites with high fire frequency [predictable stand-thinning fires]

Pinus ponderosa persists in environments with high fire-return intervals such as the mixed coniferous belt of the western USA. Wet winter and spring growing seasons provide sufficient deep moisture to allow large trees to grow and shade out much of the understorey, but woody debris and other fuels accumulate rapidly in the understorey. Summer convection storms ignite dry fuels and have historically led to frequent fires, at intervals of 2 to 20 years, except during the recent fire suppression era (Show & Kotok 1924; Cooper 1960; Lunan & Habeck 1973; McBride & Laven 1976; Kilgore & Taylor 1979; Dieterich & Swetnam 1984; Habeck 1990; Arno, Scott & Hartwell 1995). Limited understorey reduces the chances that fire will enter the canopy and destroy adult trees, and these conditions have selected for attributes that enhance survival of frequent moderate-intensity fires: self-pruning of dead branches and thick bark (Fig. 12.17). Fires in these forests burn in a mosaic pattern of low-intensity ground fire interspersed with flare-ups that are lethal to (1) patches of more shade-tolerant species, (2) younger denser stands of trees with sufficient 'ladder fuels' to carry fire into the canopy, or (3) susceptible diseased or damaged trees (Agee, Wakimoto & Biswell 1978; Harrington 1987a; Kalabokidis & Wakimoto 1992; Chap. 11, this volume). Post-fire recruitment is largely restricted to these highly localized sites of crown fire, from seed sources in adjacent sites subjected to lighter ground fire.

A mosaic burning pattern with highly variable local intensity at a small scale makes invasible patches available to the fire-resistant mature trees whose cone crops are high in the tree and can survive even severe crown damage (Rietveld 1976). *Pinus ponderosa* seeds are winged and disperse within a 100 m radius, allowing them to saturate small patches suitable for seedling establishment. This recruitment pattern leads to even-aged patches of *P. ponderosa* on the scale of <0.5 ha (Cooper 1960; West 1969; Agee 1993). Within these patches certain trees will dominate, resulting in a very uneven-sized distribution (Knowles & Grant 1983), and the suppressed trees are likely to be thinned out in a subsequent fire. On some sites successful recruitment may be uncommon (White 1985); however, due to their longevity (>700 years; Swetnam & Brown 1992), forest equilibrium will persist if successful establishment occurs roughly every 500 years for each individual in the community. Serotiny is clearly not essential or necessarily even advantageous for a species which can survive multiple fires and is presented with many opportunities for establishment during its lifetime.

Though *P. ponderosa* thrives with frequent fire, under many circumstances it can survive without it. For example, on arid margins of its range, *P. ponderosa* recruitment occurs in gaps created by natural mortality or in brushfields or grasslands (Mirov 1967; Vale 1977; McTague & Stansfield 1994). On these arid margins fires are less frequent or rare and recruitment is more continuous and less restricted to temporarily available patches (West 1969). On some nutrient-deficient arid sites, *P. ponderosa* may have a marked competitive advantage over broadleaved shrubs and herbs (DeLucia, Schlesinger & Billings 1989). As site aridity increases seedlings of this species cannot maintain adequate water balance (Barton 1993). On the high-elevation edge of its distribution, *P. ponderosa* stands lack recruitment, possibly due to historical climate change (Stein 1988), or replacement by the more cold-hardy *P. jeffreyi* (Haller 1959) and both are replaced on moister sites by *P. lambertiana* (sugar pine) (Yeaton 1981). All three of these montane pines have a nearly identical mode of resisting fire through thick bark, self-pruning, and crown heights of 40–60 m. If the fire regime is altered by fire suppression, fuel accumulation results in local extirpation (Kilgore 1973; Talley & Griffin 1980; White & Vankat 1993; Habeck 1994).

Some highly productive sites in the mid-western and eastern parts of Canada and adjacent USA also experience high-frequency, low-intensity stand-thinning fires. Using charcoal stratigraphic analysis and fire scars on *P. resinosa* (red pine), Clark (1990) estimated the mixed conifer/hardwood forests of Minnesota had experienced fires at 8–13 year intervals over the last 750 years, and similar fire-return intervals have been reported from red pine fire scars in northern Vermont (Engstrom & Mann 1991). Mature *P. resinosa* survive these fires due to the unusually

fire-resistant thick bark (van Wagner 1970) and extensive self-pruning that eliminates ladder fuels that would lead to stand-replacing crown fires; Fowells (1965) noted that this pine self-prunes better than any other conifer. Seedling recruitment for *P. resinosa* shares an important feature with the western *P. ponderosa*; low-intensity ground fires that burn surface fuels are insufficient for recruitment; rather, seedling, establishment requires local flare-ups that remove patches of vegetation, creating an even-aged patchwork mosaic (Butson, Knowles & Farmer 1987; Roberts & Mallik 1987; Bergeron & Brisson 1990; Engstrom & Mann 1991; Flannigan 1993).

Highly productive sites with unpredictable fires [unpredictable stand-replacing fires]

Highly productive sites in climates lacking an annual summer drought produce vegetation with high fuel moisture and therefore a low probability of fires propagating. This does not preclude fires, but it means that they are rare and unpredictable, and when they do occur are high-intensity and stand-replacing fires (Whitney 1986; Clark 1993). Other disturbances, both large- and small-scale, such as windthrow, insect outbreaks, and extreme drought therefore play a more prominent role in community dynamics. These circumstances require species that have some tolerance of closed-canopy conditions while also favouring species that can exploit the rare opportunities for invasion of disturbances. Highly specialized traits favouring fire survival or immediate post-fire seedling establishment at the cost of establishment at other times will tend to be selected against. Pines of subsection *Strobi* are the group that seems best equipped to deal with these constraints and opportunities. For example, *P. strobus* (eastern white pine) of the eastern and midwestern USA does not necessarily require large disturbances, burned sites or mineral soil to survive. It can exploit small gaps and small disturbances (Abrams, Orwig & Demeo 1995; Ziegler 1995). But it also is a vigorous invader of sites disturbed by wildfires or, more commonly, by hurricanes through wind dispersal of light winged seeds from surviving trees. Colonization can occur either in the first years after fire, or into early-successional forests 15–20 years after the fire (Goff 1967). Rapid growth and tall stature are essential for these species to gain a foothold before the sites are closed to invasion by more shade-tolerant broadleaved trees (Goff & Zedler 1968; Ahlgren 1976; Leak, Cullen & Frieswyk 1995).

Pinus strobus is one of the tallest pines, often towering over the surrounding hardwood forest. Because of its moderately long lifespan, it may persist long after other pines have perished (Fig. 12.18). Height is also of selective value because it increases the chances of seed falling on new disturbed sites.

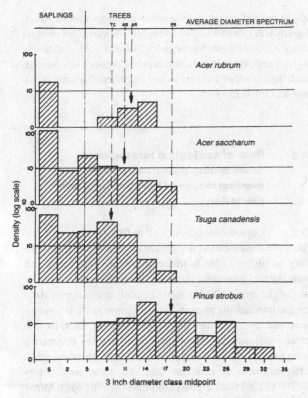

Fig. 12.18. **Size distribution of dominant trees in upland forests in the Great Lakes area of the Midwestern USA (from Goff & Zedler 1968).**

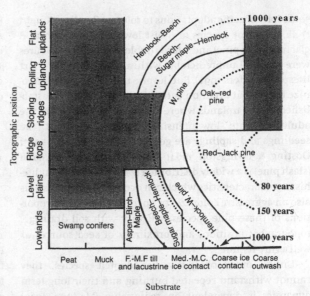

Fig. 12.19. **Distribution of jack pine (*Pinus banksiana*), red pine (*P. resinosa*), and white pine (*P. strobus*) along topographic and substrate gradients in Michigan (from Whitney 1986).**

This life history has provided little selective pressure for fire-resistant morphology or growth patterns. Thick bark and self-pruning of lower branches are of minimal value (Landers 1991), because a given site may not burn for a thousand years; however, local extinction is a possibility (Whitney 1986). But within a landscape with suitable soils, extinction of *P. strobus* at larger scales is extremely unlikely. Over long periods of time, the regional population of *P. strobus* probably shifts considerably in its pattern and overall abundance, replacing or being replaced by other pines such as *P. resinosa* primarily in response to changes in fire pattern and temporal regime (Fig. 12.19).

Other species in subsection *Strobi* have life-history characteristics that are broadly comparable to the well-studied *P. strobus*. In the western USA, *P. monticola* (western white pine), like the eastern white pine, is a long-lived forest species of intermediate tolerance and limited fire resistance that occurs primarily on mesic, and therefore in the west, higher-elevation habitats (Fowells 1965; Cwynar 1987). It readily invades burned sites (Agee & Smith 1984; Morrison & Swanson 1990), but on mesic sites, in the long absence of fire, it is displaced by more shade-tolerant conifers (Huberman 1935; Parker 1988). On extreme sites it can play a role in primary succession (Jackson & Faller 1973) where it may intergrade with timberline species dis-

cussed above. In light of this broad ecological amplitude it is consistent that Rehfeldt (1984) described this species as having a generalist genotype.

Another Californian species in the white pine group, *P. lambertiana*, is a characteristic component of montane forest over a considerable latitudinal range, but is rarely dominant. Where it occurs, it is often among the largest trees; it can also achieve great age (Fowells 1965). Its large size confers some degree of fire resistance on the oldest individuals, and its distribution therefore overlaps those of the more fire-demanding and fire-tolerant *P. jeffreyi* and *P. ponderosa*. Like white pine, *P. lambertiana* can invade recent burns (Talley & Griffin 1980), but its shade tolerance also permits saplings to persist beneath the canopy of other trees or in small openings, providing it with an ability to exploit small gaps. Its seedlings germinate in either mineral soil or litter (Rundel *et al.* 1977).

Mexican members of subsection *Strobi*, e.g. *P. ayacahuite* (including var. *strobiformis*) and *P. chiapensis*, occur in montane habitats where rainfall is higher and more reliable, generally as subdominants or in small groves (Perry 1991). These are sites where productivity is higher, competition from other conifers and broadleaved trees is more a factor, and fires, at least until recent centuries, are rare.

In the southeastern USA, a group of pines in subsection *Australes* lacks obvious adaptations to fire beyond a marked ability to invade disturbed areas, including burns. These species have high growth rates, relatively short lifespans, and produce an abundance of readily dispersed seeds, and some have specific adaptations to tolerate flooding (Knight et al. 1994). For species of moist lowland sites, such as *P. glabra* (spruce pine) and *P. taeda* (loblolly pine), natural fires were rare and unpredictable within their relatively short lifespan (Landers 1991). But despite this, both species colonize burns that occur on better-drained soils. When established on the uplands where fires are more common, the adults tolerate low-intensity ground fires, but the seedlings and saplings are generally killed (Garren 1943; Oosting & Livingston 1964; Landers 1991). *Pinus elliottii* (slash pine) is a widely distributed species with similar life-history characteristics which can occur on moist sites. It is also an aggressive colonizer of burned forests. The proportions of these pine species will vary with soil drainage, topographic position, and proximity of seed sources as well as fire regime.

Unlike more obviously fire-adapted species, they cannot withstand repeated burning and their long-term dynamics is dependent on re-invasion (Mattoon 1922; Borman 1953; Cooper 1957; Clewell 1976; Gibson & Good 1987; Landers 1991, Doren, Platt & Whiteaker 1993). Under natural conditions it is likely that on moist sites, fires were not the most important disturbance. The southeast is subject to hurricanes and tornadoes and the invasion of pines into the devastated areas has produced large, even-aged stands of these species so often that the names 'hurricane forests' (Turner 1935) or 'harricans' (Doren et al. 1993) have entered into the local language.

Several pines in other parts of the world have a similar ecology and lack of specific adaptation to fire but clear dependence upon a successional role on disturbed sites, e.g. the European *P. nigra* (Trabaud & Campant 1991; Regato-Pajares & Elena-Rosselló 1995) or the Japanese *P. densiflora* (Nakagoshi, Nehira & Takahashi 1987). *Pinus sylvestris* (Scots pine), which because of its vast range and occurrence in highly varied habitats (Walter 1968), must also be considered one of the great successes of the genus. Like other pines, Scots pine has a high-light requirement (Kreeb 1983) and it is therefore never a climax species where a lack of disturbance or suitably favourable conditions permit shade-tolerant species to be strong competitors. Its present wide distribution may, in fact, be considered as the premier example of a pine exploiting disturbance, in this case the vast continent-wide 'disturbance' of glaciation (Pravdin 1969; Chap. 5, this volume). Like many of the species discussed above, it is a generalist colonizer in more heavily forested regions (Steven & Carlisle 1959), and is dependent on fire or, more recently, humans for its persistence as a dominant (Uggla 1974). In marginal situations such as in the far north, fire is not necessary for *P. sylvestris* to persist, and it exists in permanent stands with mixed-age populations sustained by rare waves of regeneration related to climatic variation (Zackrisson et al. 1995). Like widespread species of North America, *P. sylvestris* shows considerable variation over its range (Carlisle 1958; Mirov 1967; Pravdin 1969; Vidaković 1991), suggesting that there has probably been selection in each region which has fine-tuned the physiology, morphology, and life history.

12.3 Role of ecological strategies in determining patterns of cone production, seed germination, and predation

12.3.1 Cone development and pollination

Pines are monoecious, with male strobili typically on tertiary or higher-order branches lower in the crown and female strobili on more vigorous main shoots in the upper part of the crown (Owens 1991). Strobili or cone primordia are initiated in the growing season prior to their appearance and 9–12 months before pollination; thus, in seasonal environments they undergo a winter dormancy prior to opening. There is evidence that haploxylon pines differentiate primordia later and have a shorter developmental period than diploxylon pines (Fowells 1965; Mirov 1967; Owens 1991).

The timing of the release of the wind-dispersed pollen appears to be controlled by local conditions, beginning in January at low latitudes and low elevations and extending into late summer at high latitudes and elevations (Young & Young 1992). Self-fertilization is possible, but generally has been observed to lead to reduced seed set (Chap. 14, this volume). The spatial separation of the two sexes of strobili is therefore probably a means of decreasing self-pollination (Forshell 1974; Smith, Hamrick & Kramer 1988; Sorensen 1994). Once in place, the growth of the pollen tube from the pollen grains is slow, and typically undergoes a winter dormancy period prior to fertilization, usually one year after pollination.

Seeds develop rapidly and mature and disperse about six months later, generally in the second autumn after pollination. In a few species the seed maturation period is longer, e.g. *P. leiophylla*, *P. maximartinezii* and *P. pinea* (Donahue & Lopez 1995). In some species there is a delay in seed dispersal, which varies from a few months to many years in the case of strongly serotinous species. Three Californian pines (*P. coulteri*, *P. muricata* and *P. radiata*) illustrate further complexities of seed release. All are polymorphic for serotiny but the non-serotinous cones delay dispersal until mid- to late winter, despite the fact that seed maturation is completed by early autumn (Sudworth 1908; Burns & Honkala 1990; Young & Young 1992). Thus, even these 'open cones' would be functionally serotinous, since dispersal would be after the natural fire season, which is typically during autumn, at the end of the summer drought, when Santa Ana *föhn*-type winds result in the most extensive wildfires. Delaying dispersal could have another selective basis since it may reduce predation by limiting the time seeds are on the ground prior to predictable rainfall.

The 2+ year reproductive cycle in pines (Chap. 13, this volume) may limit life-history evolution, an obvious constraint being the impossibility of an annual growth habit. It is unknown whether this long reproductive cycle is an effectively irreversible trait rooted deeply in ontogeny, or if it is a trait maintained by current selection. Certainly the glacial pace of pine pollen tube growth is amazing in comparison to the seemingly instantaneous rate of angiosperms. As Bond (1989) has observed, *Zea mays* pollen tubes can traverse the centimetres-long styles of the maize tassel and effect fertilization in hours, whereas pine pollen requires nearly a year to complete the same task. A key difference that may point to the selective value of delay is that angiosperms are required to commit resources to ovule development before 'knowing' whether or not pollination was successful, whereas in pines unsuccessful pollination has lower costs, since the megaspore does not initiate megagametophyte development until after pollination is completed (Mirov 1967; Chap. 13, this volume). While this slow pace of pollen tube development appears to be unique to gymnosperms it may be a response to winter dormancy as it apparently is not characteristic of tropical pines (Mirov 1962). Other features of the long reproductive cycle are not unique to gymnosperms. Numerous angiosperms initiate floral primordia 9–12 months in advance of pollination and there are *Quercus* species throughout the Old and New Worlds that require up to 18 months following pollination for seed maturation.

Cone size varies by an order of magnitude within the genus, from 2 to 50 cm. This variation is only partly related to seed size (Fig. 12.20). For example, an analysis of log-transformed cone length and seed length data for the Mexican pines from Perry (1991) shows no overall relationship between cone size and seed size ($r^2 = 0.004$, $p > 0.60$, d.f.=67). But when the species are grouped as 'arid' and 'mesic', a significant trend is found within the more numerous smaller-seeded mesic group ($r^2 = 0.46$, $p < 0.0001$, d.f.=53) but little improvement within the larger-seeded arid group ($r^2 = 0.01$, $p > 0.69$, d.f.=13). The other pattern that emerges is that arid (mostly pinyon-type) species have on average smaller cones (6.4 vs 11.7 cm). The large seed size of arid pines is explained by the combination of an animal dispersal syndrome with the need for seedlings in arid regions to get a strong start in the first season. It is less clear why cones should be smaller;

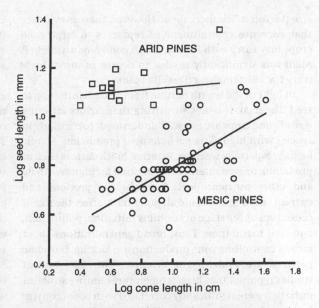

Fig. 12.20. **Relationship between seed-wing length and cone length for arid-adapted and mesic-adapted Mexican pines (data from Perry 1991).**

however, these pines have a relatively high ratio of seeds to cone biomass, something of potential selective value in these unproductive environments.

In addition to the larger between-species variation, there is often large intraspecific variation. One example is *P. sabiniana* from the interior of California which produces very small cones in the northern portion of its range and cones more than twice as large in the southern part of its range (Stockwell 1939).

12.3.2 Patterns of cone production

Pine species vary by more than an order of magnitude in their age of first reproduction, ranging roughly from 5 to 50 years (Strauss & Ledig 1985). As a general rule, species in high fire frequency habitats begin reproduction earlier than others. Timberline and desertic species tend to delay initiation of reproduction, but some species in moderately productive environments (e.g. *P. lambertiana*) also delay reproduction for many decades.

Periodicity and magnitude of cone crops varies greatly between pine species. Many are noted for mast years at intervals of 3–10 years, whereas others produce similar-sized crops each year (Fowells 1965). Mast fruiting of course is of little value for serotinous species and as will be shown in the concluding section is not characteristic of such pines. In other species there is evidence that mast fruiting may play a role in predator satiation. Annual *P. ponderosa* cone production over a 10 year period ranged from about 800 to > 11 000 per hectare and percentage cone predation by squirrels was inversely related to crop size (Larson & Schubert 1970). However, there is evidence that excessive commitment of resources to large cone crops may carry with it a cost; high cone production in P. edulis was significantly related to degree of insect herbivory the following year (Forcella 1980).

In light of the length of time from strobili initiation to seed dispersal, it is not surprising that factors affecting size of cone crops are not well understood, particularly in species with high variance in annual production. Abiotic factors, especially weather, external biotic factors such as predation, or even internal competition for photosynthate and other nutrient pools between the previous and current year's cones, could all potentially affect the size of cone crops at the stage of strobilus initiation, pollination, and seed maturation. Thus, broad generalizations about factors controlling cone production are lacking, but some patterns are evident. For example, two independent studies reported that high temperatures during strobilus initiation were significantly correlated with cone crop size 2+ years later in P. ponderosa (Maguire 1956; Daubenmire 1960) and the same has been observed for P. resinosa (Lester 1967). Reduced water stress around the time of strobilus initiation has likewise been correlated with high cone crops in P. elliottii (Varnell 1976), P. monticola (Rehfeldt, Stage & Bingham 1971) and P. taeda (Dewers & Moehring 1970). For P. monticola, early-summer water stress in the year following pollination was correlated with strobilus abortion. Although such a conclusion may be premature, it appears that environmental factors early in the reproductive cycle are more critical than later in the cycle.

12.3.3 Seed germination

Given adequate moisture and temperatures, seeds of most pines will germinate readily following release from the cone. In fact, pine seeds germinate so readily that sometimes this happens before they are dispersed. In a form of P. heldreichii var. leucodermis some seeds germinate and start seedling development while still in the cones (Tucovic & Stilnovic 1972). Some species, especially those of high latitudes or high elevations, require a cold (<5 °C) stratification treatment, or such treatment speeds the rate of germination (Young & Young 1992). However, generalizations are dangerous as seeds of some high-elevation species, e.g. P. aristata, will germinate without pre-chilling. Nor is germination behaviour uniform within species. Some differences have been attributed to ecotypic response to incubation temperature (Haasis & Thrupp 1931), and many environmental factors have been shown to affect germination including pretreatment scarification of the seed coat, light, and temperature during germination. In some pines there is an interaction between light and stratification; light eliminated the dependence on cold stratification for P. ponderosa (Li, Burton & Leadem 1994), whereas in the

European P. heldreichii no such interaction was observed (Borghetti et al. 1989). Such complexity is commonplace in seeds of many species and illustrates the need for field studies combined with factorial laboratory experiments.

In most pine habitats, conditions suitable for pine seed germination are met within months of dispersal, thus it is not surprising that there is little or no annual seed carryover in the soil (Quick 1956; Pratt, Black & Zamora 1984). Seeds of some pines have been shown to germinate poorly under closed canopies or heavy litter (Harrington & Kelsey 1979) and allelopathy may play a role (Harrington 1987b). But seeds on the ground that do not germinate in the first growing season are not likely to survive long due to predation and those that do will lose viability within a year or two (Wahlenberg 1960; Fowells 1965; Johnson & Fryer 1989).

Seed viability can be maintained over many years when seeds are retained within serotinous cones. Germination will be little impaired for five years in closed cones and, although viability does decline with age, some viable seeds may remain after a decade or more (Badron 1949; Andresen 1963; Barden 1979; McMaster & Zedler 1981; Borchert 1985). Seeds that are dispersed after fire may be affected by the pattern of fire intensity within the site. Germination is inhibited by ash in P. halepensis (Ne'eman 1993a, b) and P. banksiana seeds (Thomas & Wein 1994) and for the former species this appears to have important implications for the spatial distribution of seedlings (Fig. 12.15).

12.3.4 Life history evolution and coevolution

Herbivory and parasitism

Animals have assuredly had a long evolutionary association with pines, both as predators and dispersers. But despite a respectable collection of studies that have explored evidence of coevolution in both vegetative and reproductive traits, no attempts have been made to explain animal–plant interactions for the genus as a whole.

Pine resin, though probably performing many functions, is certainly a key element in plant–animal interactions. Pine resins contain toxic terpenes and there is much species-specific variation in their monoterpene composition. Associated with this are species-specific parasites and predators that have evolved tolerance to these compounds. The level of specificity in the interaction is illustrated by the clonal scale insect (Nuculaspis californica) which occurs on P. ponderosa. It has differentiated into separate demes on individual trees in response to differences in xylem and phloem monoterpenes (Edmunds & Alstad 1978).

For mammalian predators that can move among trees, the response to individual pine variation is behavioural

(Pederson & Welch 1985; Zhang & States 1991). Thus, tassel-eared squirrels can select *P. ponderosa* trees with smaller amounts of monoterpenes than in trees of the same species not used as food sources (Thomas 1979; Farentinos *et al.* 1981), and selected trees suffer significant reductions in several fitness components (Snyder 1992, 1993). Individual differences in susceptibility to insect predation on seeds have also been noted for this pine species (Schmid, Mata & Mitchell 1986). Evolving resistance to such parasitism/predation is possibly complicated by the fact that resistance to one predator, such as the bark beetle (*Dendroctonus*), apparently does not involve the same phloem chemistry as resistance to a parasite such as dwarf mistletoe (*Arceuthobium*) (Linhart, Snyder & Gibson 1994).

It is curious that conifers appear to be more susceptible to tree-killing insect predators than angiosperms; for example, bark beetles (Scolytidae) are responsible for over half of all natural deaths in over-mature conifers and are known to decimate entire forests, particularly if trees have been weakened by drought or other stresses (Ohmart 1989; Chap. 18, this volume). While phylogenetic and energy allocation constraints may explain this pattern, there may be indirect explanations, such as the adaptive value of the consequent increased probability of stand-replacing wildfires in these over-mature forests (Mutch 1970; Bond, Maze & Desmet 1995). Another view would be to attribute this problem to the tradeoffs that are necessary in life-history evolution. Pines generally, and especially those that face competition from other tree species, are adapted to catching disturbance-generated waves and riding them – staying ahead of more shade-tolerant competitors by virtue of rapid growth and an ability to exploit open conditions. It is reasonable that species with this life-history mode might evolve more general defences that are dependent on vigour. Thus bark beetles are fended off in healthy trees by 'pitching out', a defence lacking in subtlety and very effective while trees are healthy, but subject to failure when trees are stressed (Speight & Wainhouse 1989). Further, the correlation of pine establishment and disturbance would mean that pine populations would tend to have narrower age distributions than many angiosperms. This too might make them susceptible to outbreak. Finally, as with senescence in animals, there may historically have been a correlation between the onset of susceptibility of pines to insects and a significant decline in the prospect of future offspring. If so, selection for alternative means of resistance may have been weak. Pines can tolerate such episodes of mortality so long as the ecosystem continues to serve up large disturbances at the required frequency.

We can expand the realm of coevolutionary scenarios by considering the hypothesis of Rice & Westoby (1982), which argues that complex life cycles have evolved in heteroecious rusts as a means of making them agents of interference competition by angiosperms against conifers in a kind of primaeval biological warfare. For example, some rust fungi, such as white pine blister rust, are potentially lethal to the teliospore host pine but much less damaging when present in the aecial stage on host *Ribes* shrubs (see also Chap. 19, this volume). This asymmetry suggests it is possible that the fitness of infected *Ribes* might be greater than non-infected individuals because they inoculate the overstorey pines, causing mortality which could open the canopy and create opportunities for *Ribes* colonization. It is not clear, however, that the selection coefficients would be consistent enough to promote such a complex balance. Among other questions is whether pines can use such a system to their advantage in eliminating hardwoods.

Certain life-history characters may have unexpected impacts on other aspects of the life history. For example, Smith & Balda (1979) noted that five orders of insects, six families of birds, and two orders of mammals in various combinations can exploit the cones and seeds of most pine species. However, *P. contorta* is an exception to this pattern because only one squirrel and one coreid bug attack its serotinous cones. They hypothesized that the ever-present cone crop in this species, coupled with the low annual variance in cone crop size, allowed for competitive exclusion of most predators. In other pines, very high variance in magnitude of annual cone crops prevented exclusion by maintaining predator populations in perpetual disequilibrium.

Herbivory can significantly alter reproductive patterns. *Pinus edulis* trees attacked by the moth *Dioryctria albovittella* over a period of years produced over 40% less wood and, due to the pattern of herbivory, buds that would differentiate into female strobili were destroyed, producing functionally male plants (Whitham & Mopper 1985).

Granivory and seed dispersal

Smith (1970) hypothesized that discriminatory feeding behaviour of squirrels (*Tamiasciurus* spp.) could act as selective agents and alter pine life-history characteristics. He compared the serotinous *P. contorta* var. *latifolia* with the open-cone *P. contorta* var. *murrayana*. Serotinous cones serve as a year-round food source and Smith assumed that for serotinous species there must be intense selection for cones that are harder, more asymmetrical, and shorter-stalked than cones from the non-serotinous variety. Increased allocation of resources to cone protection may explain why var. *latifolia* has fewer seeds per cone; but it is also possible that squirrels select against high seed numbers because they choose cones with more seeds in an effort to maximize their feeding rate (Elliott 1974).

These arguments of animal-driven selection are convincing, but determining the degree to which cone characteristics are predator-driven is complicated by the role of fire. High-intensity fires could select for harder, asymmetric, sessile cones which would reduce the chances of ignition and thus better protect seeds. Since the open-cone variety *murrayana* is a subalpine tree not subjected to the same fire regime (see above), it is not possible to distinguish between the importance of fire and seed predation. Linhart (1978) encountered the same problem in distinguishing between the selective influence of fire and squirrels on cone morphology in Californian closed-cone pines. Axelrod (1980) contends that interactions with present-day faunas cannot explain traits that originated more than 10 million years ago, although he does not rule out a role for ancestral seed predators. He also is sceptical that fire has been a major factor in cone evolution, a remarkable assertion that runs counter to our view that pines and fire have probably been inextricably linked since the genus originated.

Sorting out the role of extinct faunas is highly problematic, but for a genus of such great antiquity and apparent uniformity over time, it may be necessary if we are to have a clear understanding of the evolution of some pine characteristics. For example, the massive cones of *P. coulteri*, which are the heaviest (up to 2.3 kg) and hardest of all pines and with massive hooked, spine-like apophyses (>5 cm), seem spectacularly over-designed for protection from any extant animal (it has been humorously suggested that they were designed to maim anyone attempting to cut down the tree). It is possible that some elements of the megafauna that disappeared at the end of the Pleistocene (Diamond 1992) may have preyed on these cones. The predation, to be selectively important, would have to be on cones still on the tree. This suggests an arboreal animal or, as Coulter pines are often relatively small trees, perhaps a very large animal that could push over or uproot trees in the manner of elephants.

The Pleistocene extinctions may make it difficult to understand other pine life-history traits. These massive mammals would have generated disturbances of a type and on a scale that cannot be equalled by the historic fauna, and which may have provided opportunities for pine establishment. Quite possibly some pines, that today only recruit after fire, may have evolved under a very different disturbance regime. This is suggested by contemporary studies in Africa, which show that anthropogenic-driven extinction of modern elephants, the major disturbance-creating megaherbivores, leads to the expansion of shrublands and woodlands and a switch from low-intensity stand-thinning fires to more catastrophic stand-replacing wildfires (Owen-Smith 1988).

Cone and seed predation is not restricted to serotinous pines. For example, Tevis (1953) described the near complete decimation of one year's *P. lambertiana* seed crop;

prior to cone opening squirrels cut cones from the tree and woodpeckers slashed them, and seeds were extracted by jays and numerous perching birds. Seeds that reach the ground are probably even more at risk, and are subject to intense predation from mice, chipmunks and squirrels (Martell 1979; Vander Wall 1994). For *P. jeffreyi* and *P. ponderosa* seeds, Vander Wall (1994) estimated a half-life on the ground of 120 hours and predicted that <<1% of the seed crop would go undetected by seed predators. It has been suggested that this predation may favour crypsis and different colour morphs to match substrates (Ager & Stettler 1983) and may account for why the seed wing is readily detachable in most species.

Not all seed removal, however, represents predation because seeds are often not recovered from caches. In fact, Vander Wall (1992) estimated that in one year over 40% of successful seedling recruitment in *P. jeffreyi* came from seed caches, and the mean distance of animal dispersed seeds 13 to 25 m. Further, based on simulated seed fall, he proposed that animals were quantitatively more effective in the dispersal of winged pine seeds than wind. However, maximum dispersal distances are potentially greater by wind than by ground-dwelling rodents, and therefore a combination of wind and animal dispersal may be optimal. The importance of rodent seed predators to the dispersal and planting of seeds is likely to be widespread in pines and has been documented for others such as *P. strobus* in the eastern USA (Abbott & Quink 1970) and the Asian *P. koraiensis* (Hayashida 1989).

The North American timberline *P. albicaulis* and Euro-Asian timberline stone pines (subsection *Cembrae*) share cone and seed characteristics that reflect a long evolutionary symbiosis, with birds in the genus *Nucifraga* (nutcrackers) playing a near-obligate role as dispersal agents (Paryski 1971; Tomback 1981; Lanner 1982, 1990; Chap. 14, this volume). This coevolved mutualism leads to a greatly enhanced chance of seeds reaching safe sites within the subalpine environment and is mandatory for colonizing occasional burn sites. In addition, it makes it possible for these pines to occasionally colonize sites tens of kilometres distant (Tomback 1982; Tomback *et al.* 1990).

There seems little doubt that this is a coevolved system, because of the otherwise inexplicable cone and seed characteristics. These pines – both New World and Old World species in the subsection *Cembrae* – have truly indehiscent cones; unlike serotinous cones that open with heat, these cones, even after falling from the tree, require mechanical opening by snapping off the cone scales along a unique fracture zone that exposes the seed. In addition, these cones lack the armature characteristic of many others, particularly serotinous pines. The timberline *P. flexilis*, in a different subsection to *P. albicaulis*, illustrates remarkable convergence in its large wingless seeds

...ispersed by Clark's nutcrackers (Woodmansee 1977; Lanner, Hutchins & Lanner 1980). *Pinus aristata*, on the other hand has smaller winged seeds, yet it too is dispersed to some extent by nutcrackers (Lanner, Hutchins & Lanner 1984).

Desertic pinyon pines are similar to timberline pines in their large wingless seeds that are dispersed by pinyon jays (*Gymnorhinus cyanocephalus*) and nutcrackers (Vander Wall & Balda 1977; Ligon 1978; Styles 1993; see also Chap. 14, this volume) and, as seems to be the case with timberline species, caching behaviour of pinyon jays leads to planting of seeds in seemingly more favourable sites for seedling establishment (Ligon 1978).

12.4   Conclusions on life-history evolution in pines

We view life-history evolution in pines to be driven by the interaction of site productivity and fire frequency, and these factors are correlated with particular structural and functional syndromes. To explore the extent to which fire regimes outlined in Fig. 12.1 can drive pine life-history evolution we have done a principal components analysis similar to the one presented by McCune (1988). Our analysis differs in the inclusion of European and Asian species and a slightly different set of variables. Data came from Fowells (1965), Strauss & Ledig (1985), Loehle (1988), McCune (1988), Landers (1991), and papers referred to earlier in the text. The decision as to which species to include was based on availability of data for all variables, although some species were not included because we lacked a clear understanding of their natural fire regime. For example, *P. contorta* var. *murrayana* possibly does not regularly experience any one of the fire regimes outlined in Fig. 12.1; depending on elevation it may experience no fires, or ground fires or even crown fires. Other pines difficult to assign to a particular fire regime include *P. coulteri* (evidence presented above suggests that it may persist under various fire regimes) and *P. torreyana* (its restriction to coastal bluffs and its relatively short lifespan suggest it may have evolved under an unpredictable fire regime, but this is debatable).

For the 38 species (see Appendix 12.1) used in the principal components analysis (Fig. 12.21), fire regime explained 29% of the variance in the 13 life-history characters (Table 12.2). Life-history characteristics clearly separate species in some fire-regime environments and the component loadings (Table 12.3) indicate that a different set of characters separates species on axis 1 and 2. Here we will discuss the life-history attributes that are correlated with these five fire regimes.

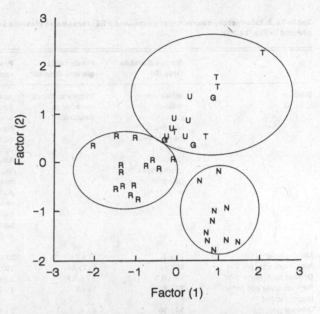

Fig. 12.21. **Principal components analysis for 38 pines from the five fire regimes illustrated in Fig. 12.1. N, no predictable fires; G, predictable ground fires; R, stand-replacing fires; T, stand-thinning fires; U, unpredictable stand-replacing fires. Based on 13 characters listed in Appendix 12.1; data from Sudworth (1908); Fowells (1965); Peterson (1967); Strauss & Ledig (1985); Elias (1987); McCune (1988); Loehle (1988); and Perry (1991).**

12.4.1   Environments with no predictable fires

These are timberline and deserptic pines adapted to extreme site conditions, which are largely fire-free unproductive sites. Pines from these habitats separated clearly along both axis 1 and 2 (Fig. 12.21). Extreme temperatures and short growing season are likely to have selected for their slow growth rates and short stature. Low growth rates may have selected for the long juvenile period and extremely long intervals between mast reproductive years (Table 12.2), and these, coupled with poor recruitment success, may have selected for the great longevity observed in these pines. The lack of predictable fire in these environments has resulted in little self-pruning (indeed, extreme temperatures may select against it) and thin bark (Table 12.3). The fact that these pines are found on abiotically stressful sites is a powerful argument against extreme temperatures and aridity playing a role in the evolution of bark thickness in pines and for fire as an important selective agent.

The dominant factor separating these pines from fire-adapted species is in the measure of dispersability, specifically the ratio of seed-wing length/seed weight (Table 12.3), reflecting the importance of bird dispersal in timberline and desertic species. These pines have substantially larger seeds (Table 12.2) and a different allometric relationship between seed weight and wing size/shape (Benkman 1995b). Considering the whole

Table 12.2. Life-history characteristics (means ± SE) for species classified by fire regime (letter designations for fire regimes are same as those plotted in Fig. 12.21)

	No predictable fires (N)	Predictable ground-fires (G)	Predictable stand-replacing fires (R)	Predictable stand-thinning fires (T)	Unpredictable stand-replacing fires (U)
Species	albicaulis aristata balfouriana cembra cembroides edulis flexilis monophylla quadrifolia[a] sibirica	elliotti var. densa merkusii palustris	attenuata banksiana brutia clausa contorta subsp. latifolia echinata halepensis leiophylla muricata pungens radiata rigida serotina virginiana	engelmannii jeffreyi lambertiana ponderosa resinosa	elliottii var. elliottii monticola pinaster strobus sylvestris taeda
Mature height (m)	20 ± 3	31 ± 3	29 ± 3	53 ± 8	45 ± 6
Maximum age (yr)	457 ± 67	217 ± 44	120 ± 17	405 ± 63	285 ± 35
Growth rate (scale: 1–3)	1.0 ± 0.0	1.7 ± 0.7	2.0 ± 0.2	1.4 ± 0.2	2.3 ± 0.2
Reproductive age (yr)	27 ± 3	15 ± 3	9 ± 2	28 ± 9	10 ± 1
Reproductive interval (yr)	14 ± 10	4 ± 1	2 ± 1	4 ± 1	5 ± 1
Cone length (cm)	7 ± 1	13 ± 4	7 ± 1	20 ± 7	16 ± 3
Seed weight (mg)	221 ± 48	55 ± 22	13 ± 2	82 ± 38	22 ± 3
Wing length (mm)	0.5 ± 0.2	2.8 ± 0.6	1.7 ± 0.2	2.4 ± 0.2	2.2 ± 0.2
Ratio of wing-length/seed wt	0.01 ± 0.01	0.06 ± 0.01	0.17 ± 0.02	0.07 ± 0.03	0.11 ± 0.01
Bark thickness (cm)	1.7 ± 0.2	3.2 ± 0.3	2.9 ± 0.5	6.2 ± 0.7	3.2 ± 0.1
Self-pruning (scale: 1–10)	1.9 ± 0.6	6.7 ± 1.7	2.4 ± 0.4	8.8 ± 0.7	6.5 ± 1.0
Grass stage (% of species)	0	100	0	0	0
Serotiny (% of species)	0	0	92	0	0
Resprouting (% of species)	0	0	46	0	0

[a] According to the taxonomic treatment accepted for this volume (Chap. 2, this volume), P. quadrifolia Parlatore ex Sudworth is a hybrid between P. juarezensis and P. monophylla.

Table 12.3. Component loadings for principal components analysis of 13 life-history characters (see Appendix 12.1; WLXSW = wing length/seed weight) for the 38 Pinus taxa presented graphically in Fig. 12.21. Components 1 and 2 explained 28.8% and 24.2% of variance, respectively

Life history characters	Component 1	Component 2
WLXSW	−0.830	0.148
SEROTIN	−0.823	−0.148
MAXAGE	0.770	−0.044
MINAGE	0.769	−0.081
SEEDWT	0.618	−0.490
RESPROUT	−0.602	−0.030
GRATE	−0.508	0.384
PRUNE	0.346	0.814
HEIGHT	0.142	0.770
WINGL	−0.268	0.764
BARK	0.181	0.741
CONELENG	0.417	0.704
GRASS	0.146	0.199
INTERVAL	0.195	−0.174

genus, wing size increases and wind-dispersal capacity increases with seed mass, up to about 90 mg, after which further increases in mass are correlated with a decline in wind dispersal capacity. This reflects the qualitative shift from wind to bird dispersal, and these seed character-istics may have been selected for by nutcrackers and pinyon jays. However, evolution of the very large seed could have other origins, i.e. establishment under extreme temperatures and drought might select for large seeds (e.g. Baker 1972; Fig. 12.20). If true, this would explain the apparent 'loss' of the seed wing, because wings sufficient to carry these large seeds would require extraordinarily large cones (a conclusion drawn from the fact that the two largest seed wings are found in P. coulteri and P. lambertiana, which happen to have the largest cones in the genus). Alternatively, wind dispersal might have been selected against on these very windy sites where wind dispersal might result in much of the seed crop being lost to unfavourable sites. If aridity and/or wind selected for large wingless seeds, this of course would put a high premium on the coevolution of a bird-dispersal mutualism.

Squirrels may also have affected cone structure in these otherwise bird-dispersed pines. In mountain ranges where squirrels are present, P. albicaulis allocates a far greater proportion of energy to protection in the form of a thicker seed coat, and greater cone and resin mass (Benkman 1995a).

12.4.2 Environments with predictable stand-replacing fires

Species in these environments group separately from all other species on axis 1 (Fig. 12.21). The preponderance of serotinous and resprouting species is not surprising, although the lack of these adaptations in other fire regimes is noteworthy (Table 12.2). Thin bark in these fire-type pines reflects the low probability of surviving high-intensity fires. The very short juvenile period and short lifespan were probably selected for by the low to moderate fire-return interval. It may be that these life-history traits are correlated with allometric patterns of biomass allocation to foliage versus structural organs (Strauss & Ledig 1985). Alternatively, features such as short lifespan and poor self-pruning may have been selected to enhance flammability and increase probability of stand-replacing fires (Bond & Midgley 1995). Low seed weight would contribute to greater numbers of seeds, which could be adaptive in light of the fact that seeds are exposed to predators for many years, plus these smaller prey items would reduce feeding efficiency. Other attributes of low seed mass, such as higher wind dispersability, seem less important in relation to the dispersal strategy which is temporal rather than spatial. On the other hand, these pines do have the highest ratio of wing length/seed weight, suggesting high wind dispersability. This may have been selected because stand-replacing fires are likely to generate a vast area of suitable habitat at the time of seed dispersal. Moderate armature and numerous small cones reflects the selective pressure of predation on the constantly available canopy seed source. Predation may also account for annual to biennial cone production.

12.4.3 Environments with predictable ground fires or stand-thinning fires

Species adapted to light ground fires can survive with moderately thick bark and self-pruning (Table 12.2). These environments present a potential threat to seedling survival and thus the evolution of the grass stage in several species is noteworthy; however, some pines persist in such pine savannas without this adaptation.

As site productivity increases, the closed canopy reduces the herbaceous understorey – as well as pine recruitment – and increases the woody fuels that lead to sporadic flare-ups and localized stand-thinning. Increased fire intensity has selected for thicker bark, greater self-pruning, taller stature, larger denser cones, and longer lifespans (Table 12.2). Most pines in these fire regimes are on relatively arid sites and this may account for the significantly greater seed weight than in pines subjected to predictable stand-replacing fires. High seed mass in turn would select for large cones; however, high seed-predator populations are likely to be maintained in these productive environments, selecting for well-armed cones (McCune 1988). On average, the potential wind dispersability of these pines is not great; however, rodent seed predators appear to play a major role in dispersing and planting these seeds. Additionally, localized flare-ups potentially generate suitable seed beds in proximity to seed source trees, minimizing pressure for long-distance dispersal.

Moderately frequent fires require that saplings reach sufficient size to tolerate repeat fires at a relatively young age. The long juvenile period is consistent with allocation of resources to vegetative growth early in development. These allocation patterns are reflected in far less foliage/structure ratios in these species, as noted by Strauss & Ledig (1985) for *P. jeffreyi* and *P. lambertiana*. The high allocation of resources to protection from fire, and the low probability of crown fires, have selected for long lifespans.

12.4.4 Environments with unpredictable stand-replacing fires

Productive sites where fires are infrequent produce a fire regime of unpredictable high-intensity stand-replacing crown fire (Fig. 12.1). Life-history attributes are most similar to pines adapted to predictable stand-thinning fires (Fig. 12.21; Table 12.2). This suggests that moderately thick bark and self-pruning may occasionally be rewarded. Since the fire-return interval may exceed the lifespan of species such as *P. strobus*, and other more localized types of disturbance may be more frequent than fire, these pines are largely dependent upon wide spatial dispersal of seeds. Tall stature and high wing length/seed weight ratio would be of selective value (Table 12.2). Lack of cone armature on relatively large cones is surprising in these highly productive environments with high seed-predator populations. However, predation may be reduced by long intervals between mast cone years. High growth rate and short juvenile period reflects selection in the face of the rapid colonization by more shade-tolerant broadleaved competitors and the better growth conditions on highly productive sites. Shade tolerance (relative to other pines) may reflect the strategy of maintaining a 'seedling bank', which could take advantage of disturbances other than fire, e.g. windthrows, snowfall, hurricanes etc.

12.4.5 Environments not considered

Not all pine species will fall clearly into one of these life-history modes, but in accordance with niche packing theory, we should expect species intermediate to one or the other of these groups. Two other factors will come into play. Some species are very widespread and encounter more than one fire regime. Also, over time, fire regimes change (e.g. Clark 1989) and adaptations at one point in time may not reflect the selective environment at the time of origin. Changes in distribution will occur far more rapidly than changes in life history and thus sorting out the life history evolution for narrow endemics, which were once widespread under a different climatic regime (e.g. *P. torreyana*) may be impossible. It is to be expected that over time, species distribution will expand and contract as its suite of traits fits the frequency of adaptive peaks and valleys in the landscape.

In conclusion, we argue that life-history evolution in pines is best understood by relating structural and functional characteristics to site productivity and fire-return interval and these parameters have more explanatory power than models based on fire alone (Rowe 1983;

Landers 1991), or on *r*- and *K*-selection (Caswell 1982, Turner 1985, Strauss & Ledig 1985), competition for light (Govindaraju 1984), moisture requirements (Yeaton 1978, 1981), or plant defence (Loehle 1988).

Acknowledgements

We thank Drs Ronald Lanner, Thomas Ledig and Dave Richardson for comments on an earlier draft of the manuscript. Thanks also to Drs G. Whitney and C. Williams for providing original figures.

References

Abbas, H., Barbéro, M. & Loisel, R. (1984). Réflexions sur le dynamisme actuel de la régéneration naturelle du pin d'Alep (*Pinus halepensis* Mill.) dans les pinèdes incendiées en Provence Calcaire (de 1973 à 1979). *Ecologia Mediterranea*, **10**, 85–104.

Abbott, H.G. & Quink, T.F. (1970). Ecology of eastern white pine seed caches made by small forest mammals. *Ecology*, **51**, 271–8.

Abrams, M.D. (1984). Uneven-aged jack pine in Michigan. *Journal of Forestry*, **82**, 306–7.

Abrams, M.D. & Orwig, D.A. (1995). Structure, radial growth dynamics and recent climatic variations of a 320-year-old *Pinus rigida* rock outcrop community. *Oecologia*, **101**, 353–60.

Abrams, M.D., Orwig, D.A. & Demeo, T.E. (1995). Dendroecological analysis of successional dynamics for a presettlement-origin white-pine–mixed-oak forest in the southern Appalachians, USA. *Journal of Ecology*, **83**, 123–33.

Aerts, R. (1995). The advantages of being evergreen. *Trends in Ecology and Evolution*, **10**, 402–7.

Agee, J.K. (1993). *Fire Ecology of Pacific Northwest Forests*. Washington, DC: Island Press.

Agee, J.K. & Smith, L. (1984). Subalpine tree reestablishment after fire in the Olympic Mountains, Washington. *Ecology*, **65**, 810–19.

Agee, J.K., Wakimoto, R.H. & Biswell, H.H. (1978). Fire and fuel dynamics of Sierra Nevada conifers. *Forest Ecology and Management*, **1**, 255–65.

Ager, A.A. & Stettler, R.F. (1983). Local variation in seeds of ponderosa pine. *Canadian Journal of Botany*, **61**, 1337–44.

Ahlgren, C.E. (1976). Regeneration of red pine and white pine following wildfire and logging in northeastern Minnesota. *Journal of Forestry*, **74**, 135–40.

Andresen, J.W. (1963). Germination characteristics of *Pinus rigida* seeds borne in serotinous cones. *Brotéria, Séie de Ciências Naturais*, **32**, 153–78.

Arno, S.F. (1980). Forest fire history in the northern Rockies. *Journal of Forestry*, **78**, 460–5.

Arno, S.F., Scott, J.H. & Hartwell, M.G. (1995). *Age-class Structure of Old Growth Ponderosa Pine/Douglas-fir Stands and its Relationship to Fire history*. USDA Forest Service Research Paper INT-481.

Axelrod, D.I. (1980). History of the maritime closed-cone pines, Alta and Baja California. *University of California Publications in Geological Sciences*, **120**, 1–143.

Baker, H.G. (1972). Seed weight in relation to environmental conditions in California. *Ecology*, **53**, 997–1010.

Badron, O.A. (1949). Maintenance of seed viability in closed-cone pines. MS thesis, University of California, Berkeley.

Barden, L.S. (1977a). Drought and survival in a self-perpetuating *Pinus pungens* population: equilibrium or nonequilibrium. *American Midland Naturalist*, **119**, 253–7.

Barden, L.S. (1977b). Self-maintaining populations of *Pinus pungens* Lam. in the southern Appalachians. *Castanea*, **42**, 316–23.

Barden, L.S. (1979). Serotiny and seed viability of *Pinus pungens* in the southern Appalachians. *Castanea*, **44**, 44–7.

Barden, L.S. & Woods, F.W. (1976). Effects of fire on pine and pine–hardwood forests in the southern Appalachians. *Forest Science*, **22**, 399–403.

Barney, M.A. & Fischknecht, N.C. (1974). Vegetation changes following fire in the pinyon–juniper type of west-central Utah. *Journal of Range Management*, **27**, 91–6.

Barton, A.M. (1993). Factors controlling plant distributions: drought, competition, and fire in montane pines in Arizona. *Ecological Monographs*, **63**, 367–97.

Beaufait, W.R. (1960). Some effects of high temperatures on the cones and seeds of jack pine. *Forest Science*, **6**, 194–9.

Benkman, C.W. (1995a). The impact of tree squirrels (*Tamiasciurus*) on limber pine seed dispersal adaptations. *Evolution*, **49**, 585–92.

Benkman, C.W. (1995b). Wind dispersal capacity of pine seeds and the evolution of different seed dispersal modes in pines. *Oikos*, **73**, 221–4.

Bergeron, Y. & Brisson, J. (1990). Fire regime in red pine stands at the northern limit of the species range. *Ecology*, **71**, 1352–64.

Béland, M. & Bergeron Y. (1993). Ecological factors affecting abundance of advanced growth in jack pine (*Pinus banksiana* Lamb.) stands of the boreal forest of northwestern Québec. *Forestry Chronicle*, **69**, 561–8.

Billings, W.D. & Thompson, J.H. (1957). Composition of a stand of old bristlecone pines in the White Mountains of California. *Ecology*, **38**, 158–60.

Boerner, R.E.J. (1981). Forest structure dynamics following wildfire and prescribed burning in the New Jersey pine barrens. *American Midland Naturalist*, **105**, 321–33.

Bond, W.J. (1989). The tortoise and the hare: ecology of angiosperm dominance and gymnosperm persistence. *Biological Journal of the Linnean Society*, **36**, 227–49.

Bond, W.J., Maze, K. & Desmet, P. (1995). Fire life histories and the seeds of chaos. *Ecoscience*, **2**, 252–60.

Bond, W.J. & Midgley, J.J. (1995). Kill thy neighbour: an individualistic argument for the evolution of flammability. *Oikos*, **73**, 79–85.

Borchert, M. (1985). Serotiny and cone-habit variation in populations of *Pinus coulteri* (Pinaceae) in the southern Coast Ranges of California. *Madroño*, **32**, 29–48.

Borghetti, M., Vendramin, G.G., Giannini, R. & Schettino, A. (1989). Effects of stratification, temperature and light on germination of *Pinus leucodermis*. *Acta Oecologica/Oecologia Plantarum*, **10**, 45–56.

Bormann, F.H. (1953). Factors determining the role of loblolly pine and sweetgum in early old-field succession in the Piedmont of North Carolina. *Ecological Monographs*, **23**, 339–58.

Brown, C.L. (1964). *The Seedling Habit of Longleaf Pine*. Georgia Forest Research Council and School of Forestry, University of Georgia.

Buchholz, K. (1983). Initial responses of pine and oak to wildfire in the New Jersey pine barren plains. *Bulletin of the Torrey Botanical Club*, **110**, 91–6.

Burns, R.M. & Honkala, B.H. (1990). *Silvics of North America*, Vol. 1. *Conifers*. USDA Forest Service, Agriculture Handbook 654.

Butson, R.G., Knowles, P. & Farmer R.E.Jr. (1987). Age and size structure of marginal, disjunct populations of *Pinus resinosa*. *Journal of Ecology*, **75**, 685–92.

Cameron, H. (1953). Melting point of the bonding material in lodgepole and jack pine cones. *Canada Department of Research and Development, Forestry Branch, Silvicultural Leaflet*, 86, 1.

Carlisle, A. (1958). A guide to the named variants of Scots pine (*Pinus sylvestris* Linnaeus). *Forestry*, **31**, 203–40.

Carpenter, C.C. (1991). The environmental control of seedling establishment in a subalpine forest Association of Red Fir and Lodgepole Pine at Deadman Creek, Mono County, California. PhD thesis, University of California, Davis.

Caswell, H. (1982). Life history theory and the equilibrium status of populations. *American Naturalist*, **120**, 317–39.

Cayford, J.H. & McRae, D.J. (1983). The role of fire in jack pine forests. In *The Role of Fire in Northern Circumpolar Ecosystems*, ed. R.S. Wein & D.A. McLean, pp. 183–99. New York: John Wiley.

Chapman, H.H. (1936). Effect of fire in preparation of seedbed for longleaf pine seedlings. *Journal of Forestry*, **34**, 852–4.

Christensen, N.L. (1977). Fire and soil-plant nutrient relations in a pine–wiregrass savanna on the coastal plain of North Carolina. *Oecologia*, **31**, 27–44.

Christensen, N.L. (1981). Fire regimes in southeastern ecosystems. In *Proceedings of the Conference on Fire Regimes and Ecosystem Properties*, ed. H.A. Mooney, T.M. Bonnicksen, N.L. Christensen, J.E. Lotan & W.A. Reiners, pp. 112–36. USDA Forest Service General Technical Report WO-26.

Clark, J.S. (1989). Ecological disturbance as a renewal process: theory and application to fire history. *Oikos*, **56**, 17–30.

Clark, J.S. (1990). Fire and climate change during the last 750 yr in northwestern Minnesota. *Ecological Monographs*, **60**, 135–59.

Clark, J.S. (1993). Fire, climate change, and forest processes during the past 2000 years. In *Elk Lake, Minnesota: Evidence for Rapid Climate Change in the North-central United States*, ed. J.P. Bradbury & W.E. Dean, pp. 295–308. US Geological Survey Special Paper 276. Denver, Colorado: USGS.

Clewell, A.F. (1976). A remnant stand of old-growth slash pine in the Florida panhandle. *Bulletin of the Torrey Botanical Club*, **103**, 1–9.

Contini, L. & LaVarelo Y. (1982). *Le pin cembro* Pinus cembra L. *Répartition, écologie, Sylviculture et Production*. Paris: Institute National de la Recherche Agronomique.

Cooper, C.F. (1960). Changes in vegetation, structure, and growth of southwestern ponderosa pine forest since white settlement. *Ecological Monographs*, **30**, 129–64.

Cooper, R.W. (1957). Silvicultural characteristics of slash pine (*Pinus elliottii*, Egelm. var. *elliottii*). USDA Forest Service Research Paper SE-81.1.

Critchfield, W.B. (1957). *Geographic Variation in* Pinus contorta. Harvard University, Cambridge, Mass.: Maria Moors Cabot Foundation Publication No. 3.1.

Croker, T.C. Jr. & Boyer, W.D. (1975). *Regenerating Longleaf Pine Naturally*. USDA Forest Service Research Paper SO-105.

Crutchfield, D.M. & Trew, I.F. (1961). Investigation of natural regeneration of pond pine. *Journal of Forestry*, **59**, 264–66.

Cwynar, L.C. (1987). Fire and the forest history of the Cascade Range. *Ecology*, **68**, 791–802.

Daubenmire, R.F. (1960). A seven-year study of cone production as related to xylem layers and temperature in *Pinus ponderosa*. *American Midland Naturalist*, **64**, 189–93.

DeLucia, E.H., Schlesinger, W.H. & Billings, W.D. (1989). Edaphic limitations to growth and photosynthesis in Sierran and Great Basin vegetation. *Oecologia*, **78**, 184–90.

Denevan, W.M. (1961). The upland pine forests of Nicaragua. A study in cultural plant geography. *University of California Publications in Geography*, **12**, 251–320.

Despain, D.G. (1983). Nonpyrogenous climax lodgepole pine communities in Yellowstone National Park. *Ecology*, **64**, 231–4.

Desponts, M. & Payette, S. (1993). The Holocene dynamics of jack pine at its northern range limit in Québec. *Journal of Ecology*, **81**, 719–27.

Dewers, R.S. & Moehring, D.M. (1970). Effect of soil water stress on initiation of ovulate primordia in loblolly pine. *Forest Science*, **16**, 219–21.

Diamond, J. (1992). *The Third Chimpanzee*. Los Angeles: University of California Press.

Dieterich, J.H. & Swetnam, T.W. (1984). Dendrochronology of a fire-scarred ponderosa pine. *Forest Science*, **30**, 238–47.

Donahue, J.K. & Lopez, C.M. (1995). Observations of *Pinus maximartinezii* Rzed. *Madroño*, **42**, 19–25.

Doren, R.F., Platt, W.J. & Whiteaker, L.D. (1993). Density and size structure of slash pine stands in the everglades region of south Florida. *Forest Ecology and Management*, **59**, 295–311.

Edmunds, G.F. Jr & Alstad, D.N. (1978). Coevolution in insect herbivores and conifers. *Science*, **199**, 943–5.

Elias, T.S. (1987). *Conservation and Management of Rare and Endangered Plants*. Sacramento: California Native Plant Society.

Elliott, P.F. (1974). Evolutionary responses of plants to seed-eaters: pine squirrel predation on lodgepole pine. *Evolution*, **28**, 221–31.

Ellis, M., von Dohlen, C.D., Anderson, J.E. & Romme, W.H. (1994). Some important factors affecting density of lodgepole pine seedlings following the 1988 Yellowstone fires. In *Plants and their Environments: Proceedings of the First Biennial Scientific Conference on the Greater Yellowstone Ecosystem*, ed. D.G. Despain, pp. 139–50. USDI National Park Service, Natural Resources Publication Office, Technical Report NPS/NRYELL/NRTR-93/XX.

Engstrom, F.B. & Mann, D.H. (1991). Fire ecology of red pine (*Pinus resinosa*) in northern Vermont, USA. *Canadian Journal of Forest Research*, **21**, 882–9.

Farentinos, R.C., Capretta, P.J., Keener, R.E. & Littlefield, V.M. (1981). Selective herbivory in tassel-eared squirrels: role of monoterpenes in ponderosa pines chosen as feeding trees. *Science*, **213**, 1273–5.

Flannigan, M.D. (1993). Fire regime and the abundance of red pine. *International Journal of Wildland Fire*, **3**, 241–7.

Forcella, F. (1980). Cone predation by pinyon cone beetle (*Conopthorus edulis*; Scolytidae): dependence on frequency and magnitude of cone production. *American Naturalist*, **116**, 594–8.

Forman, R.T.T. & Boerner, R.E. (1981). Fire frequency and the pine barrens of New Jersey. *Bulletin of the Torrey Botanical Club*, **108**, 34–50.

Forshell, C.P. (1974). Seed development after self-pollination and cross-pollination of Scots pine, *Pinus sylvestris* L. Swedish University of Agricultural Sciences, Faculty of Forestry, Uppsala, Sweden, Studia Forestalia Suecica No. 118.1.

Fowells, H.A. (1965). *Silvics of Forest Trees of the United States*. USDA Forestry Service Agriculture Handbook No. 271. Washington, DC; USDA.

Garren, K.H. (1943). Effects of fire on vegetation of the southeastern United States. *Botanical Review*, **9**, 617–54.

Gauthier, S., Bergeron, Y. & Simon, J.P. (1993a). Cone serotiny in jack pine: ontogenetic, positional, and environmental effects. *Canadian Journal of Forest Research*, **23**, 394–401.

Gauthier, S., Gagnon, J. & Bergeron, Y. (1993b). Population age structure of *Pinus banksiana* at the southern edge of the Canadian boreal forest. *Journal of Vegetation Science*, **4**, 783–90.

Gibson, D.J. & Good, R.E. (1987). The seedling habitat of *Pinus echinata* and *Melampyrum lineare* in oak–pine forest of the New Jersey pinelands. *Oikos*, **49**, 91–100.

Givnish, T.J. (1981). Serotiny, geography, and fire in the pine barrens of New Jersey. *Evolution*, **35**, 101–23.

Glitzenstein, J.S., Platt, W.J. & Streng, D.R. (1995). Effects of fire regime and habitat on tree dynamics in North Florida longleaf pine savannas. *Ecological Monographs*, **65**, 441–76.

Goff, F.G. (1967). Upland vegetation. In *Soil Resources and Forest Ecology of Menominee County, Wisconsin*, ed. G.W. Olsen, F.D. Hole & C.J. Milford, pp. 60–89. Madison, Wisconsin: Bulletin 85. Soil Series No. 60. University of Wisconsin, Geological and Natural History Survey, Soil Survey Division.

Goff, F.G. & Zedler, P.H. (1968). Structural gradient analysis of upland forests in the western Great Lakes area. *Ecological Monographs*, **38**, 65–86.

Govindaraju, D.R. (1984). Mode of colonization and patterns of life history in some North American conifers. *Oikos*, **43**, 271–6.

Grace, S.L. & Platt, W.J. (1995a). Neighborhood effects on juveniles in an old-growth stand of longleaf pine, *Pinus palustris*. *Oikos*, **72**, 99–105.

Grace, S.L. & Platt, W.J. (1995b). Effects of adult tree density and fire on the demography of pregrass stage juvenile longleaf pine (*Pinus palustris* Mill.). *Journal of Ecology*, **83**, 75–86.

Grime, J.P. (1979). *Plant Strategies and Vegetation Processes*. New York: John Wiley.

Groeschl, D.A., Johnson, J.E. & Smith, D.W. (1992). Early vegetative response to wildfire in a Table Mountain–pitch pine forest. *International Journal of Wildland Fire*, **2**, 177–84.

Haasis, F.W. & Thrupp, A.C. (1931). Temperature relations of lodgepole-pine seed germination. *Ecology*, **12**, 728–44.

Habeck, J.R. (1987). Present-day vegetation in the northern Rocky Mountains. *Annals of the Missouri Botanical Garden*, **74**, 804–40.

Habeck, J.R. (1990). Old-growth ponderosa pine–western larch forests in western Montana: ecology and management. *Northwest Environmental Journal*, **6**, 271–92.

Habeck, J.R. (1994). Using general land office records to assess forest succession in ponderosa pine/Douglas-fir forests in western Montana. *Northwest Science*, **68**, 69–78.

Haller, J.R. (1959). Factors affecting the distribution of ponderosa and Jeffrey pines in California. *Madroño*, **15**, 65–71.

Harrington, M.G. (1987a). Ponderosa pine mortality from spring, summer, and fall crown scorching. *Western Journal of Applied Forestry*, **2**, 14–16.

Harrington, M.G. (1987b). *Phytotoxic Potential of Gambel Oak on Ponderosa Pine Seed Germination and Initial Growth*. USDA Forest Service, Research Paper RM-277.

Harrington, M.G. & Kelsey, R.G. (1979). *Influence of some Environmental Factors on Initial Establishment and Growth of Ponderosa Pine Seedlings*. USDA Forest Service Research Paper INT-230.1.

Hartnett, D.C. & Krofta D.M (1989). Fifty-five years of post-fire succession in a southern mixed hardwood forest. *Bulletin of the Torrey Botanical Club*, **116**, 107–13.

Hayashida, M. (1989). Seed dispersal by red squirrels and subsequent establishment of Korean pine. *Forest Ecology and Management*, **28**, 115–29.

Helms, J.A. & Ratliff, R.D. (1987). Germination and establishment of *Pinus contorta* var. *murrayana* (Pinaceae) in mountain meadows of Yosemite National Park, California. *Madroño*, **34**, 77–90.

Hickey, L.J. & Doyle J.A. (1977). Early Cretaceous fossil evidence for angiosperm evolution. *Botanical Review*, **43**, 3–104.

Huberman, M.A. (1935). The role of western white pine in forest succession in northern Idaho. *Ecology*, **16**, 137–51.

Jackson, M.T. & Faller, A. (1973). Structural analysis and dynamics of the plant communities of Wizard Island, Crater Lake National Park. *Ecological Monographs*, **43**, 441–61.

Jakubos, B. & Romme, W.H. (1993). Invasion of subalpine meadows by lodgepole pine in Yellowstone National Park, Wyoming. *Arctic and Alpine Research*, **25**, 382–90.

Johnson, E.A. & Fryer, G.I. (1989). Population dynamics in lodgepole pine–engelmann spruce forests. *Ecology*, **70**, 1335–45.

Johnson, E.A. & Gutsell, S.L. (1993). Heat budget and fire behaviour associated with the opening of serotinous cones in two *Pinus* species. *Journal of Vegetation Science*, **4**, 745–50.

Kalabokidis, K.D. & Wakimoto, R.H. (1992). Prescribed burning in uneven-aged stand management of ponderosa pine/Douglas fir forests. *Journal of Environmental Management*, **34**, 221–35.

Keane, R.E. & Morgan, P. (1993). Landscape processes affecting the decline of whitebark pine (*Pinus albicaulis*) in the Bob Marshall Wilderness Complex, Montana, USA. In *12th Conference on Fire and Forest Meteorology*, pp. 195–208. Bethesda, Md: Society of American Foresters.

Keeley, J.E. (1981). Reproductive cycles and fire regimes. In *Proceedings of the Conference on Fire Regimes and Ecosystem Properties*, ed. H.A. Mooney, T.M. Bonnicksen, N.L. Christensen, J.E. Lotan & W.A. Reiners, pp. 231–77. USDA Forest Service General Technical Report WO-26.

Keeley, J.E. (1982). Distribution of lightning and man-caused wildfires in California. In *Proceedings of the Symposium on Dynamics and Management of Mediterranean-Type Ecosystems* ed. C.E. Conrad & W.C. Oechel, pp. 431–7. USDA Forest Service General Technical Report PSW-58.

Kenkel, N.C. (1988). Pattern of self-thinning in jack pine: testing the random mortality hypothesis. *Ecology*, **69**, 1017–24.

Ketcham, D.E. & Bethune, J.E. (1963). Fire resistance of South Florida slash pine. *Journal of Forestry*, **61**, 529–30.

Kilgore, B.M. (1973). The ecological role of fire in Sierran conifer forests: its application to national park management. *Quaternary Research*, **3**, 496–513.

Kilgore, B.M. & Taylor, D. (1979). Fire history of a sequoia–mixed conifer forest. *Ecology*, **60**, 129–42.

Knapp, A.K. & Anderson, J.E. (1980). Effect of heat on germination of seeds from serotinous lodgepole pine cones. *American Midland Naturalist*, **104**, 370–2.

Knight, D.H., Vose, J.M., Baldwin, V.C.,

Ewel, K.C. & Grodzinska, K. (1994). Contrasting patterns in pine forest ecosystems. In *Environmental Constraints on the Structure and Productivity of Pine Forest Ecosystems: A Comparative Analysis*. Ecological Bulletin 43, ed. H.L. Gholz, S. Linder & R.E. McMurtie, pp. 9–19. Copenhagen: Munksgaard.

Knowles, P. & Grant, M.C. (1983). Age and size structure analyses of Engelmann spruce, ponderosa pine, lodgepole pine, and limber pine in Colorado. *Ecology*, **64**, 1–9.

Komarek, E.V. (1974). Effects of fire on temperate forests and related ecosystems: southeastern United States. In *Fire and Ecosystems*, ed. T.T. Kozlowski & C.E. Ahlgren, pp. 251–77. New York: Academic Press.

Koskela, J., Kuusipalo, J. & Sirikul, W. (1995). Natural regeneration dynamics of *Pinus merkusii* in northern Thailand. *Forest Ecology and Management*, **77**, 169–79.

Kreeb, K.H. (1983). *Vegetationskunde*. Stuttgart: Verlag Eugen Ulmer.

Laessle, A.M. (1965). Spacing and competition in natural stands of sand pine. *Ecology*, **46**, 65–72.

LaMarche, V.C., Jr (1969). Environment in relation to age of bristlecone pines. *Ecology*, **50**, 53–9.

Lamont, B.B., Le Maitre, D.C., Cowling, R.M & Enright, N.J. (1991). Canopy seed storage in woody plants. *Botanical Review*, **57**, 277–317.

Landers, J.L. (1991). Disturbance influences on pine traits in the southeastern United States. *Proceedings of the Tall Timbers Fire Ecology Conference*, **17**, 61–98.

Lanner, R.M. (1982). Adaptations of whitebark pine for seed dispersal by Clark's nutcracker. *Canadian Journal of Forest Research*, **12**, 391–402.

Lanner, R.M. (1990). Biology, taxonomy, evolution, and geography of stone pines of the world. In *Proceedings — Symposium on Whitebark Pine Ecosystems: Ecology and Management of a High-Mountain Resource*, pp. 14–24. USDA Forest Service General Technical Report INT-270.

Lanner, R.M., Hutchins, H.E. & Lanner, H.A. (1980). Dispersal of limber pine seed by Clark's Nutcracker. *Journal of Forestry*, **78**, 637–9.

Lanner, R.M., Hutchins, H.E. & Lanner, H.A. (1984). Bristlecone pine and

Clark's nutcracker: probable interaction in the White Mountains, California. *Great Basin Naturalist*, **44**, 357–60.

Larson, M.K. & Schubert, G.H. (1970). *Cone Crops of Ponderosa Pine in Central Arizona, Including the Influence of Albert Squirrels*. USDA Forest Service Research Paper RM-58.1.

Leak, W.B., Cullen, J.B. & Frieswyk, T.S. (1995). *Dynamics of White Pine in New England*. USDA Forest Service Paper NE-699.

Ledig, F.T. & Fryer, J.H. (1972). A pocket of variability in *Pinus rigida*. *Evolution*, **26**, 259–66.

Ledig, F.T. & Little, S. (1979). Pitch pine (*Pinus rigida* Milli): ecology, physiology, and genetics. In *Pine Barrens. Ecosystem and Landscape*, ed. R.T.T. Forman, pp. 347–71. New York: Academic Press.

Lester, D.T. (1967). Variation in cone production of red pine in relation to weather. *Canadian Journal of Botany*, **45**, 1683–91.

Li, X.J., Burton, P.J. & Leadem, C.L. (1994). Interactive effects of light and stratification on the germination of some British Columbia conifers. *Canadian Journal of Botany*, **72**, 1635–46.

Ligon, J.D. (1978). Reproductive interdependence of piñon jays and piñon pines. *Ecological Monographs*, **48**, 111–26.

Linhart, Y.B. (1978). Maintenance of variation in cone morphology in California closed-cone pines: the roles of fire, squirrels and seed output. *Southwestern Naturalist*, **23**, 29–40.

Linhart, Y.B., Snyder, M.A. & Gibson, J.P. (1994). Differential host utilization by two parasites in a population of ponderosa pine. *Oecologia*, **98**, 117–20.

Loehle, C. (1988). Tree life history strategies: the role of defenses. *Canadian Journal of Forest Research*, **18**, 209–22.

Lotan, J.E. (1967). Cone serotiny of lodgepole pine near West Yellowstone, Montana. *Forest Science*, **13**, 55–9.

Lowery, D.P. (1984). *Lodgepole Pine (Pinus contorta Dougl. ex Loud.)*. USDA Forest Service, FS-253.

Lunan, J.S. & Habeck, J.R. (1973). The effects of fire exclusion on ponderosa pine communities in Glacier National Park, Montana. *Canadian Journal of Forest Research*, **3**, 574–9.

Maguire, W.P. (1956). Are ponderosa pine crops predictable? *Journal of Forestry*, **54**, 778–9.

Martell, A.M. (1979). Selection of conifer seeds by deer mice and red-backed voles. *Canadian Journal of Forest Research*, **9**, 201–4.

Mattoon, W.R. (1922). *Longleaf Pine*. USDA Forest Service Bulletin 1061.

Mattoon, W.R. (1922). *Slash Pine*. USDA Forest Service Bulletin 1256

McBride, J.R. & Laven, R.D. (1976). Scars as an indicator of fire frequency in the San Bernardino Mountains, California. *Journal of Forestry*, **74**, 439–42.

McCune, B. (1988). Ecological diversity in North American pines. *American Journal of Botany*, **75**, 353–68.

McMaster, G.S. & Zedler, P.H. (1981). Delayed seed dispersal in *Pinus torreyana* (Torrey pine). *Oecologia*, **51**, 62–6.

McTague, J.P. & Stansfield, W.F. (1994). Stand and tree dynamics of uneven-aged ponderosa pine. *Forest Science*, **40**, 289–302.

Mirov, N.T. (1962). Phenology of tropical pines. *Arnold Arboretum Journal*, **43**, 218–19.

Mirov, N.T. (1967). *The Genus Pinus*. New York: Ronald Press.

Moir, W.H. (1982). A fire history of the High Chisos, Big Bend National Park, Texas. *Southwestern Naturalist*, **27**, 87–98.

Moravec, J. (1990). Regeneration of northwest African *Pinus halepensis* forests following fire. *Vegetatio*, **87**, 29–36.

Morgan, P. & Bunting, S.C. (1990). Fire effects in whitebark pine forests. In *Proceedings — Symposium on Whitebark Pine Ecosystems: Ecology and Management of a High-Mountain Resource*, pp. 166–70. USDA Forest Service General Technical Report INT-270.

Morrison, P.H. & Swanson, F.J. (1990). *Fire History and Pattern in a Cascade Range Landscape*. USDA Forest Service General Technical Report, PNW-254.1.

Moser, E.B. (1989). Longleaf pine pyrogenicity and turkey oak mortality in Florida xeric sandhills. *Ecology*, **70**, 60–70.

Muir, P.S. & Lotan, J.E. (1985). Disturbance history and serotiny of *Pinus contorta* in western Montana. *Ecology*, **66**, 1658–68.

Muir, P.S. (1993). Disturbance effects on structure and tree species composition of *Pinus contorta* forests in western Montana. *Canadian Journal of Forest Research*, **23**, 1617–25.

Murphy, G.I. (1968). Pattern in life history and the environment. *American Naturalist*, **102**, 390–404.

Mutch, R.W. (1970). Wildland fires and ecosystems – a hypothesis. *Ecology*, **51**, 1046–51.

Nakagoshi, N., Nehira, K. & Takahashi, F. (1987). The role of fire in pine forests of Japan. In *The Role of Fire in Ecological Systems*, ed. L. Trabaud & R. Proden, p. 157. Commission of European Communities, Ecosystems Research Reports.

Nakane, K. (1994). Modelling the soil carbon cycle of pine ecosystems. In *Environmental Constraints on the Structure and Productivity of Pine Forest Ecosystems: A Comparative Analysis. Ecological Bulletins 43*, ed. H.L. Gholz, S. Linder & R.E. McMurtie, pp. 161–72. Copenhagen: Munksgaard.

Ne'eman, G. (1997). Regeneration of natural pine forest – A review of the work done after the 1989 fire in Mount Carmel Israel. *International Journal of Wildland Fire*, **7**, 295–306.

Ne'eman, G., Lahav, H. & Izhaki, I. (1992a). The resilience of vegetation to fire in an east-Mediterranean pine forest on Mount Carmel, Israel: the effects of post-fire management regimes. In *Proceedings of the International Workshop on the Role of Fire in Mediterranean Ecosystems*, ed. L. Trabaud & R. Prodon, pp. 7–14. Banyuls-Sur-Mer, France.

Ne'eman, G., Lahav, H. & Izhaki, I. (1992b). Spatial pattern of seedlings 1 year after fire in a Mediterranean pine forest. *Oecologia*, **91**, 365–70.

Ne'eman, G., Meir, I. & Ne'eman, R. (1993a). The effect of ash on the germination and early growth of shoots and roots of *Pinus, Cistus* and annuals. *Seed Science and Technology*, **21**, 339–49.

Ne'eman, G., Meir, I & Ne'eman, R. (1993b). The influence of pine ash on the germination and early growth of *Pinus halepensis* Mill. and *Cistus salviifolius* L. *Water Science and Technology*, **27**, 525–32.

Ohmart, C.P. (1989). Why are there so few tree-killing bark beetles associated with angiosperms? *Oikos*, **54**, 242–63.

Oosting, H.J. & Livingston, R.B. (1964). A resurvey of a loblolly pine community twenty-nine years after ground and crown fire. *Bulletin of the Torrey Botanical Club*, **91**, 387–95.

Ostertag, R. & Menges, E.S. (1994). Patterns of reproductive effort with time since last fire in Florida scrub plants. *Journal of Vegetation Science*, **5**, 303–10.

Owens, J.N. (1991). Flowering and seed set. In *Physiology of Trees*, ed. A.S. Raghavendra, pp. 247–71. New York: John Wiley.

Owen-Smith, N. (1988). *Megaherbivores. The Influence of Very Large Body Size on Ecology*. Cambridge: Cambridge University Press.

Parker, A.J. (1986). Persistence of lodgepole pine forests in the central Sierra Nevada. *Ecology*, **67**, 1560–7.

Parker, A.J. (1988). Stand structure in subalpine forests of Yosemite National Park, California. *Forest Science*, **34**, 1047–58.

Parker, A.J. & Peet, R.K. (1984). Size and age structure of conifer forests. *Ecology*, **63**, 1685–9.

Parsons, J.J. (1955). The Miskito pine savanna of Nicaragua and Honduras. *Association of the American Geographers Association*, **45**, 36–63.

Paryski, W.H. (1971). [Translation of] Stone-pine *Pinus cembra* L. *Polish Academy of Sciences Institute of Dendrology and Kornik Arboretum, Popular Scientific Monographs 'Our Forest Trees'*, 1.

Payette, S. (1993). The range limit of boreal tree species in Québec–Labrador: an ecological and palaeoecological interpretation. *Review of Palaeobotany and Palynology*, **79**, 7–30.

Pederson, J.C. & Welch, B.L. (1985). Comparison of ponderosa pines as feed and nonfeed trees for Abert squirrels. *Journal of Chemical Ecology*, **11**, 149–57.

Peet, R.K. (1988). Forests of the Rocky Mountains. In *North American Terrestrial Vegetation*, ed. M.G. Barbour & W.D. Billings, pp. 63–101. Cambridge: Cambridge University Press.

Perry, D.A. & Lotan, J.E. (1979). A model of fire selection for serotiny in lodgepole pine. *Evolution*, **33**, 958–68.

Perry, J.P. (1991). *The Pines of Mexico and Central America*. Portland, Oregon: Timber Press.

Pessin, L.J. (1938). The effect of vegetation on the growth of longleaf pine seedlings. *Ecological Monographs*, **8**, 115–49.

Peterson, R. (1967). *The Pine Tree Book*. New York: Brandywine Press.

Platt, W.J., Evans, G.W. & Rathbun, S.L. (1988). The population of dynamics of a long-lived conifer (*Pinus palustris*). *American Naturalist*, **131**, 491–525.

Platt, W.J., Glitzenstein, J.S. & Streng, D.R. (1991). Evaluating pyrogenicity and its effects on vegetation in longleaf pine savannas. *Proceedings of the Tall Timbers Fire Ecology Conference*, **17**, 143–61.

Pratt, D.W., Black, R.A & Zamora, B.A. (1984). Buried viable seed in a ponderosa pine community. *Canadian Journal of Botany*, **62**, 44–52.

Pravdin, L.F. (1969). Scots pine; variation, intraspecific taxonomy and selection [translated from Russian]. Israel Program for Scientific Translations, Jerusalem.

Ponce, D.H.E. (1985). Basal sprouting in *Pinus oocarpa*. *Turrialba*, **35**(1), 96–101.

Quick, C.R. (1956). Viable seeds form the duff and soil of sugar pine forests. *Forest Science*, **2**, 36–42.

Rebertus, A.J., Williamson, G.B. & Moser, E.B. (1989). Longleaf pine pyrogenicity and turkey oak mortality in Florida xeric sandhills. *Ecology*, **70**, 60–70.

Regato-Pajares, P. & Elena-Rosselló, R. (1995). Natural black pine (*Pinus nigra* subsp. *salzmannii*) forests of the Iberian eastern mountains: development of the phytoecologcial basis for their site evaluation. *Annales des Sciences Forestières*, **52**, 589–606.

Rehfeldt, G.E. (1983). Adaptation of *Pinus contorta* populations to heterogeneous environments in northern Idaho. *Canadian Journal of Forest Research*, **13**, 405–11.

Rehfeldt, G.E. (1987). Components of adaptive variation in *Pinus contorta* from the inland northwest. USDA Forest Service Research Paper INT-375.1.

Rehfeldt, G.E., Stage, A.R. & Bingham, R.T. (1971). Strobili development in western white pine: periodicity, prediction, and association with weather. *Forest Service*, **17**, 454–61.

Rehfeldt, J. (1984). Microevolution of conifers in the northern Rocky Mountains: a view from common gardens. *Proceedings Eighth North American Forest Biology Workshop*. Logan, Utah.

Reich, P.B., Kloeppel, B.D., Ellsworth, D.S. & Walters, M.B. (1995). Different photosynthesis–nitrogen relations in deciduous hardwood and evergreen coniferous tree species. *Oecologia*, **104**, 24–30.

Rice, B. & Westoby, M. (1982). Heteroecious rusts as agents of interference competition. *Evolutionary Theory*, **6**, 43–52.

Richardson, D.M., Williams, P.A. & Hobbs, R.J. (1994). Pine invasions in the Southern Hemisphere: determinants of spread and invadability. *Journal of Biogeography*, **21**, 511–27.

Rietveld, W.J. (1976). Cone maturation in ponderosa pine foliage scorched by wildfire. USDA Forest Service Research Note RM-317.

Roberts, B.A. & Mallik, A.U. (1987). Responses of red pine to wildfire in Newfoundland. Canadian Botanical Association, Bulletin 20.1.

Romme, W.H. (1982). Fire and landscape diversity in subalpine forests of Yellowstone National Park. *Ecological Monographs*, **52**, 199–221.

Rowe, J.S. (1983). Concepts of fire effects on plant individuals and species. In *The Role of Fire in Northern Circumpolar Ecosystems*, ed. R.W. Wein & D.A. MacLean, pp. 135–54. New York: John Wiley.

Rundel, P.W. (1975). Primary succession on granite outcrops in the montane southern Sierra Nevada. *Madroño*, **23**, 209–19.

Rundel, P.W., Parsons, D.J. & Gordon, D.T. (1977). Montane and subalpine vegetation of the Sierra Nevada and Cascade Ranges. In *Terrestrial Vegetation of California*, ed. M.G. Barbour & J. Major, pp. 559–99. New York: John Wiley.

Schmid, J.M., Mata, S.A. & Mitchell, J.C. (1986). Number and condition of seeds in ponderosa pine cones in central Arizona. *Great Basin Naturalist*, **46**, 449–51.

Schuster, W.S.F., Mitton, J.B., Yamaguchi, D.K. & Woodhouse, C.A. (1995). A comparison of limber pine (*Pinus flexilis*) ages at lower and upper treeline sites east of the continental divide in Colorado. *American Midland Naturalist*, **133**, 101–11.

Shankman, D. (1984). Tree regeneration following fire as evidence of timberline stability in the Colorado Front Range, USA. *Arctic and Alpine Research*, **16**, 413–17.

Sheppard, P.R. & Lassoie, J.P. (1985). Fire regime of the lodgepole pine communities of the San Jacinto Mountains, California. In *Proceedings of the Symposium and Workshop on Wilderness Fire*, ed. J.E. Lotan, B.M. Kilgore, W.C. Fischer & R.W. Mutch, p. 376. USDA Forest Service General Technical Report INT-182.

Show, S.B. & Kotok, E.I. (1924). *The Role of Fire in the California Pine Forests*. USDA, Department Bulletin 1294, Washington, DC: USDA.

Smith, C.C. (1970). The coevolution of pine squirrels (*Tamiasciurus*) and conifers. *Ecological Monographs*, **40**, 349–71.

Smith, C.C. & Balda, R.P. (1979). Competition among insects, birds and mammals for conifer seeds. *American Zoologist*, **19**, 1065–83.

Smith, C.C., Hamrick, J.L. & Kramer, C.L. (1988). The effects of stand density on frequency of filled seeds and fecundity in lodgepole pine (*Pinus contorta* Dougl.). *Canadian Journal of Forest Research*, **18**, 453–60.

Snyder, M.A. (1992). Selective herbivory by Abert's squirrel mediated by chemical variability in ponderosa pine. *Ecology*, **73**, 1730–41.

Snyder, M.A. (1993). Interactions between Abert's squirrel and ponderosa pine: the relationship between selective herbivory and host plant fitness. *American Naturalist*, **141**, 866–79.

Sorensen, F.C. (1994). Frequency of seedlings from natural self-fertilization in Pacific Northwest ponderosa pine (*Pinus ponderosa* Dougl. ex. Laws). *Silvae Genetica*, **43**, 100–7.

Speight, M.R. & Wainhouse, D. (1989). *Ecology and Management of Forest Insects*. Oxford: Clarendon Press.

Stahelin, R. (1943). Factors influencing the natural restocking of high altitude burns by coniferous trees in the central Rocky Mountains. *Ecology*, **24**, 19–30.

St. Andre, G., Mooney, H.A. & Wright, R.D. (1965). The pinyon woodland zone in the White Mountains of California. *American Midland Naturalist*, **73**, 225–39.

Stearns, S.C. (1992). *The Evolution of Life Histories*. Oxford: Oxford University Press.

Stein, S.J. (1988). Explanations of the imbalanced age structure and scattered distribution of ponderosa pine within a high-elevation mixed coniferous forest. *Forest Ecology and Management*, **25**, 139–53.

Sterrett, W.D. (1920). *Jack Pine*. USDA Bulletin 829. Washington, DC: USDA.

St-Pierre, H. & Gagnon, R. (1992). Régénération après feu de l'épinette noire (*Picea mariana*) et du pin gris (*Pinus banksiana*) dans la forêt boréale, Québec. *Canadian Journal of Forest Research*, **22**, 474–81.

Steven, H.M. & Carlisle, A. (1959). *The Native Pinewoods of Scotland*. Edinburgh: Oliver and Boyd.

Stockwell, W.P. (1939). Cone variation in digger pine. *Madroño*, 5, 72–3.

Stocks, B.J. (1987). Fire behavior in immature jack pine. *Canadian Journal of Forest Research*, 17, 80–6.

Stone, E.L. Jr & Stone, M.H. (1943). Dormant buds in certain species of *Pinus*. *American Journal of Botany*, 30, 346–51.

Stone, E.L. Jr & Stone, M.H. (1954). Root collar sprouts in pine. *Journal of Forestry*, 52, 487–91.

Strauss, S.H. & Ledig, T. (1985). Seedling architecture and life history evolution in pines. *American Naturalist*, 125, 702–15.

Styles, B.T. (1993). Genus *Pinus*: a Mexican purview. In *Biological Diversity of Mexico: Origins and Distribution*, ed. T.P. Ramamoorthy, R. Bye, A. Lot & J. Fa, pp. 397–420. New York: Oxford University Press.

Sudworth, G.B. (1908). *Forest Trees of the Pacific Slope*. Washington, DC: USDA Forest Service.

Swetnam, T.W. & Brown, P.M. (1992). Oldest known conifers in the southwestern United States: temporal and spatial patterns of maximum age. In *Old-growth forests in the southwest and Rocky Mountain regions. Proceedings of a workshop*, ed. M.R. Kaufmann, W.H. Moir & R.L. Bassett, pp. 24–38. USDA Forest Service General Technical Report RM-213.

Tait, D.E., Cieszewski, C.J. & Bella, I.E. (1988). The stand dynamics of lodgepole pine. *Canadian Journal of Forest Research*, 18, 1255–60.

Talley, S.N. & Griffin, J.R. (1980). Fire ecology of a montane pine forest, Junipero Sierra Peak, California. *Madroño*, 27, 49–60.

Tausch, R.J. & West, N.E. (1988). Differential establishment of pinyon and juniper following fire. *American Midland Naturalist*, 119, 174–84.

Teich, A.H. (1970). Cone serotiny and inbreeding in natural populations of *Pinus banksia* and *Pinus contorta*. *Canadian Journal of Botany*, 48, 1805–9.

Tevis, L. Jr (1953). Effect of vertebrate animals on seed crop of sugar pine. *Journal of Wildlife Management*, 17, 128–31.

Thanos, C.A., Marcou, S., Christodoulakis, D. & Yannitsaros, A. (1989). Early post-fire regeneration in *Pinus brutia* forest ecosystems of Samos island (Greece). *Acta Oecologica/Oecologia Plantarum*, 19, 79–94.

Thomas, G.R. (1979). The role of phloem sugars in the selection of ponderosa pine by the Kaibab squirrel. MA thesis, San Francisco State University.

Thomas, P.A. & Wein, W.R. (1994). Amelioration of wood ash toxicity and jack pine establishment. *Canadian Journal of Forest Research*, 24, 748–55.

Tinker, D.B., Romme, W.H., Hargrove, W.W., Gardner, R.H. & Turner, M.G. (1994). Landscape-scale heterogeneity in lodgepole pine serotiny. *Canadian Journal of Forest Research*, 24, 897–903.

Tomback, D.F. (1981). Notes on cones and vertebrate-mediated seed dispersal of *Pinus albicaulis* (Pinaceae). *Madroño*, 28, 91–4.

Tomback, D.F. (1982). Dispersal of whitebark pine seeds by Clark's nutcracker: a mutualism hypothesis. *Journal of Animal Ecology*, 51, 451–67.

Tomback, D.F. (1986). Post-fire regeneration of krummholz whitebark pine: a consequence of nutcracker seed caching. *Madroño*, 33, 100–10.

Tomback, D.F., Hoffmann, L.A. & Sund, S.K. (1990). Coevolution of whitebark pine and nutcrackers: implications for forest regeneration. In *Proceedings – Symposium on Whitebark Pine Ecosystems: Ecology and Management of a High-Mountain Resource*, pp. 118–29. USDA Forest Service General Technical Report INT-270.

Trabaud, L. & Campant, C. (1991). Difficulté de recolonisation naturelle du pin de Salzmann *Pinus nigra* Arn. ssp. *salzmannii* (Dunal) Franco après incendie. *Biological Conservation*, 58, 329–43.

Trabaud, L.V., Christensen, N.L. & Gill, A.M. (1993). Historical biogeography of fire in temperate and Mediterranean ecosystems. In *Fire in the Environment: The Ecological, Atmospheric, and Climatic Importance of Vegetation Fires*, ed. P.J. Crutzen & J.G. Goldammer, pp. 277–95. New York: John Wiley.

Trabaud, L., Michels, C. & Grosman, J. (1985). Recovery of burnt *Pinus halepensis* Mill. II. Pine reconstitution after wildfire. *Forest Ecology and Management*, 13, 167–79.

Tucovic, A. & Stilnovic, S. (1972). Viviparousness in *Pinus heldreichii* Christ. *Genetika-Yugoslavia*, 4, 193–200.

Turakka, A., Luukkanen, O. & Bhumibhamon, S. (1982). Notes on *Pinus kesiya* and *P. merkusii* and their natural regeneration in watershed areas of northern Thailand. *Acta Forestalia Fennica*, 178, 1–33.

Turner, L.M. (1935). Catastrophes and pure stands of southern shortleaf pine. *Ecology*, 16, 213–15.

Turner, D.P. (1985). Successional relationships and a comparison of biological characteristics among six northwestern conifers. *Bulletin of the Torrey Botanical Club*, 112, 421–8.

Uggla, E. (1974). Fire ecology in Swedish forests. *Proceedings of the Tall Timbers Fire Ecology Conference*, 13, 171–90.

Vale, T.R. (1977). Forest changes in the Warner Mountains, California. *Annals of the Association of American Geographers*, 67, 28–45.

Vale, T.R. (1979). *Pinus coulteri* and wildfire on Mount Diablo, California. *Madroño*, 26, 135–40.

Van Wagner, C.E. (1970). Fire and red pine. *Proceedings of the Tall Timbers Fire Ecology Conference*, 10, 211–19.

Vander Wall, S.B. (1992). The role of animals in dispersing a 'wind-dispersed' pine. *Ecology*, 73, 614–21.

Vander Wall, S.B. (1994). Removal of wind-dispersed pine seeds by ground-foraging vertebrates. *Oikos*, 69, 125–32.

Vander Wall, S.B. & Balda, R.P. (1977). Coadaptations of the Clark's Nutcracker and the piñon pine for efficient seed harvest and dispersal. *Ecological Monographs*, 43, 125–43.

Varnell, R.J. (1976). *Cone and Seed Production in Slash Pine: Effects of Tree Dimensions and Climatic Factors*. USDA Forest Service Research Paper SE-145.1.

Veblen, T.T. (1986). Age and size structure of subalpine forests in the Colorado Front Range. *Bulletin of the Torrey Botanical Club*, 113, 225–40.

Vidaković, M. (1991). *Conifers: Morphology and Variation*, revised English edition. Zagreb: Graficki Zavod Hrvatske.

Vogl, R.J. (1973). Ecology of knobcone pine in the Santa Ana Mountains, California. *Ecological Monographs*, 43, 125–43.

Vogl, R.J., Armstrong, W.P., White, K.L. & Cole, K.L. (1977). The closed-cone pines and cypresses. In *Terrestrial Vegetation of California*, ed. M.G. Barbour & J.

Major, pp. 295–358. New York: John Wiley.

Wahlenberg, W.G. (1960). *Loblolly Pine. Its Use, Ecology, Regeneration, Protection, Growth and Management.* Durham, NC: School of Forestry, Duke University.

Walter, H. (1968). *Die Vegetation der Erde. Band II: Die Gemasessigten und Arktischen Zonen.* Stuttgart: Gustav Fischer Verlag.

Wang, C. (1961). *The Forests of China.* Cambridge, Mass.: Maria Moors Cabot Foundation Publication, Harvard University, Series No. 5.

Webber, H.J. (1935). The Florida scrub, a fire-fighting association. *American Journal of Botany,* **22**, 344–61.

Weber, M.G., Hummel, M. & van Wagner, C.E. (1987). Selected parameters of fire behavior and *Pinus banksiana* Lamb. regeneration in eastern Ontario. *Forestry Chronicle* (October), 340–6.

Werner, W.L. (1993). Pinus in Thailand. Stuttgart: Franz Steiner Verlag.

West, N.E. (1969). Tree patterns in central Oregon ponderosa pine forests. *American Midland Naturalist,* **81**, 584–90.

Westman, W.E. (1975). Edaphic climax pattern of the pygmy forest region of California. *Ecological Monographs,* **45**, 109–35.

Whipple, S.A. & Dix, R.L. (1979). Age structure and successional dynamics of a Colorado subalpine forest. *American Midland Naturalist,* **101**, 142–58.

White, A.S. (1985). Presettlement regeneration patterns in a southwestern ponderosa pine stand. *Ecology,* **66**, 589–94.

White, A.S. & Vankat, J.L. (1993). Middle and high elevation coniferous forest communities of the North Rim region of Grand Canyon National Park, Arizona, USA. *Vegetatio,* **109**, 161–74.

Whitham, T.G. & Mopper, S. (1985). Chronic herbivory: impacts on architecture and sex expression of pinyon pine. *Science,* **228**, 1089–91.

Whitney, G.G. (1986). Relation of Michigan's presettlement pine forests to substrate and disturbance history. *Ecology,* **67**, 1548–59.

Williams, C.E. & Johnson, W.C. (1990). Age structure and the maintenace of *Pinus pungens* in pine-oak forests of southwestern Virginia. *American Midland Naturalist,* **124**, 130–41.

Williams, C.E. & Johnson, W.C. (1992). Factors affecting recruitment of *Pinus pungens* in the southern Appalachian Mountains. *Canadian Journal of Forest Research,* **22**, 878–87.

Williamson, G.B. & Black, E.M. (1981). High temperature of forest fires under pines as a selective advantage over oaks. *Nature,* **293**, 643–4.

Woodmansee, R.G. (1977). Clusters of limber pine trees: a hypothesis of plant–animal coaction. *Southwestern Naturalist,* **21**, 511–17.

Wright, R.D. (1963). Some ecological studies on bristlecone pine in the White Mountains of California. PhD thesis, Los Angeles: University of California.

Wright, R.D. & Mooney, H.A. (1965). Substrate-oriented distribution of bristlecone pine in the White Mountains of California. *American Midland Naturalist,* **73**, 257–84.

Yarrranton, M. & Yarranton, G.A. (1975). Demography of a jack pine stand. *Canadian Journal of Botany,* **53**, 310–14.

Yeaton, R.I. (1978). Some ecological aspects of reproduction in the genus *Pinus* L. *Bulletin of the Torrey Botanical Club,* **105**, 306–11.

Yeaton, R.I. (1981). Seedling characteristics and elevational distributions of pines (Pinaceae) in the Sierra Nevada of central California: a hypothesis. *Madroño,* **28**, 67–77.

Young, J.A. & Young, C.G. (1992). *Seeds of Woody Plants in North America.* Portland, Oregon: Dioscorides Press.

Zackrisson, O., Nilsson, M., Steijlen, I. & Hoernberg, G. (1995). Regeneration pulses and climate–vegetation interactions in nonpyrogenic boreal Scots pine stands. *Journal of Ecology,* **83**, 469–83.

Zedler, P.H., Gautier, C.R. & Jacks, P. (1984). Edaphic restriction of *Cupressus forbesii* (Tecate cypress) in southern California, USA – a hypothesis. In *Being Alive on Land,* ed. N.S. Margaris, M. Arianoutsou-Faraggitaki & W.C. Oechel, pp. 237–43. The Hague: Dr W. Junk.

Zhang, X. & States, J.S. (1991). Selective herbivory of ponderosa pine by Abert Squirrels: a reexamination of the role of terpenes. *Biochemical Systematics and Ecology,* **19**, 111–15.

Ziegler, S.W. (1995). Relict eastern white pine (*Pinus strobus* L.) stands in southwestern Wisconsin. *American Midland Naturalist,* **133**, 88–100.

Zobel, B. (1953). Geographic range and intraspecific variation of Coulter pine. *Madroño,* **12**, 1–7.

Zobel, D.B. (1969). Factors affecting the distribution of *Pinus pungens,* an Appalachian endemic. *Ecological Monographs,* **39**, 303–33.

Pinus taxa	Firereg	Height	Grate	Maxage	Minage	Interval	Coneleng	Seedwt	Wingl	Grass	Resprout	Serotin	Bark	Prun
albicaulis	N	33	1.0	500	25.0	4.0	5.715	127	0.0		1	1	1.270	1
aristata	N	15	1.0	723	20.0	102.0	8.255	20	1.524		1	1	1.575	1
balfouriana	N	19	1.0	700	20.0	5.5	10.160	27	2.540		1	1	1.575	2
cembra	N	23	1.0	700	27.0	8.0	8.0	413	0.0		1	1	1.0	1
cembroides	N	8	1.0	300	25.0	6.5	4.445	350	0.0		1	1	1.270	1
edulis	N	12	1.0	350	50.0	3.5	5.080	239	0.0		1	1	1.575	1
flexilis	N	25	1.0	300	30.0	3.0	16.510	103	0.0		1	1	3.810	7
monophylla	N	15	1.0	150	22.5	1.5	5.080	379	0.726		1	1	1.905	1
quadrifolia[a]	N	9	1.0	250	15.0	3.0	5.080	379	0.254		1	1	1.524	2
sibirica	N	40	1.0	600	30.0	5.5	7.620	175	0.0		1	1	1.651	2
elliotti var. *densa*	G	26	3.0	150	10.0	3.0	10.795	31	2.540	10	1	1	3.048	5
merkusii	G	31	1.0	200	15.0	1.5	7.620	34	1.900	10	1	1	3.810	5
palustris	G	37	1.0	300	20.0	6.0	20.320	99	3.810	10	1	1	2.794	10
attenuata	R	15	3.0	87	6.5	1.0	11.430	16	3.175		1	10	0.965	1
banksiana	R	31	3.0	80	9.0	3.4	4.445	4	0.838		1	10	1.270	1
brutia	R	31	2.0	150	8.5	1.0	8.890	28	1.800		1	10	4.445	2
clausa	R	25	2.0	60	5.0	1.5	6.858	6	1.524		1	10	1.016	1
contorta subsp. *latifolia*	R	46	1.0	120	7.5	1.0	3.505	5	1.270		1	10	1.016	1
echinata	R	31	3.0	200	12.5	6.5	5.080	10	1.270		10	10	2.083	4
halepensis	R	25	2.0	150	17.5	1.0	8.890	22	1.400		1	10	3.810	3
leiophylla	R	25	1.0	300	29.0	3.0	4.445	11	0.838		10	7	6.985	6
muricata	R	28	3.0	84	5.5	2.5	7.620	9	2.032		7	10	6.350	3
pungens	R	19	1.0	80	5.0	2.0	7.620	13	2.540		1	10	1.905	4
radiata	R	46	3.0	85	7.5	1.0	12.700	28	1.905		7	10	4.445	2
rigida	R	31	3.0	100	3.5	6.5	6.350	7	2.540		10	10	3.048	3
serotina	R	25	1.0	87	7.0	1.0	5.715	8	1.905		10	10	1.575	1
virginiana	R	31	1.5	100	5.0	1.0	5.715	9	0.762		10	10	1.270	2
engelmanni	T	25	1.0	400	29.0	3.5	14.986	31	2.540		1	1	4.445	7
jeffreyi	T	55	1.0	400	8.0	3.0	25.400	114	2.540		1	1	7.620	10
lambertiana	T	69	2.0	425	60.0	4.0	45.720	217	25.40		1	1	6.985	10
ponderosa	T	71	1.0	600	18.0	3.5	10.795	38	1.702		1	1	7.620	10
resinosa	T	46	1.0	200	22.5	5.0	4.826	9	2.540		1	1	4.445	7
elliottii	U	31	3.0	200	8.5	3.0	10.795	31	2.540		1	1	3.048	5
monticola	U	62	2.0	400	13.5	5.0	24.130	17	2.540		1	1	3.175	7
pinaster	U	37	2.0	250	12.5	4.0	25.400	30	2.300		1	1	3.175	6
strobus	U	67	2.0	325	7.5	6.5	15.240	17	2.540		1	1	3.175	3
sylvestris	U	40	2.0	350	10.0	5.0	7.620	10	1.100		1	1	3.175	10
taeda	U	34	3.0	187	7.5	8.0	10.795	25	1.905		1	1	3.505	8

[a] According to the taxonomic treatment accepted for this volume (Chap. 2, this volume), *P. quadrifolia* Parlatore ex Sudworth is a hybrid between *P. juarezensis* and *P. monophylla*.

Appendix 12.1. **Raw data for 38 pine species used in principal components analysis shown in Fig. 12.21.** Firereg = fire regime (codes defined in Table 12.2); Height = metres; Grate = relative growth rate (1=slow, 3 = rapid); Maxage = maximum age (years); Minage = minimum age (years); Interval = period between significant cone crops (years); Coneleng = cone length (cm); Seedwt = seed weight (g); Wingl = seed wing length (cm); Grass = seedling grass stage development (1 = absent, 10 = well-developed); Resprout = relative resprouting capacity (1 = absent, 10 = well-developed); Serotin = degree of cone serotiny (1 = absent, 10 = well-developed); Bark = bark thickness (cm); Prun = relative self-pruning capacity (1 = absent, 10 = well-developed).

13 Genetic variation in *Pinus*

F. Thomas Ledig

13.1 Introduction

Pines are, on average, among the most variable of organisms, both among and within populations, as revealed by measures of quantitative genetic variation (Cornelius 1994), and by diversity at isozyme loci (Hamrick & Godt 1990). The high level of diversity within populations is a result of a genetic system that effectively provides for the creation, storage and release of genetic variation. The observed level and pattern of variation within species of the genus *Pinus*, and therefore their ecological role, depend to a great extent on their genetic system. Wind-borne pollen provides cohesiveness among populations and guarantees that most patterns of geographic variation will be clinal. On the other hand, systems of seed dispersal are conducive to rare colonization events that provide opportunities for the action of genetic drift and protect the genetic integrity of founding populations during race formation. Another paradox is that ecotypes separated by less than a kilometre can coexist because the tremendous seed output of pines results in net fitness despite high selection intensities. The genetic system that allows for the storage and release of such high levels of genetic variation makes pines ideal material for breeders and facilitates their domestication, with the result that several species have entered world commerce because of their adaptability to plantation culture (Chap. 20, this volume). High levels of genetic variation, tremendous reproductive capacity, and facultative selfing also contribute to the success of pines as invaders when introduced into the southern hemisphere where they are not native (see Chap. 22, this volume).

13.2 The genetic system

13.2.1 Reproductive system

The genetic system includes everything that provides for the creation, storage and release of variation: the recombination system, the reproductive system, the mating system, the dispersal system, and the mutation system, among others. Of course, these are not mutually exclusive systems. Several aspects of the genetic system will be discussed in this chapter, and this first section will deal with some physical features of reproduction.

The reproductive system of pines is exclusively sexual. A few species will sprout from the stump after harvest or from the root collar when the top is killed by fire. Examples are *P. canariensis*, *P. echinata* and *P. rigida*. However, this is not a method of multiplication. The branches of a few species may layer (form roots) if brought into contact with the soil while still attached to the tree (e.g. Lutz 1939), but this is very rare and of negligible importance in the reproduction of pines. Neither is there any evidence of apomixis. When pollination is prevented by isolating the megasporangiate strobili with suitable barriers, no seeds are set and cones fail to mature.

I will refer to the reproductive structures as megasporangiate strobilus and microsporangiate strobilus. These are often referred to as female and male flowers, respectively. However, they are not morphologically flowers, which is why the gymnosperms in general and the pines in particular are distinguished from the angiosperms, the flowering plants.

The interval between initiation of strobili and maturation of seeds is two years in most species. In brief: the primordia are initiated within the buds in late summer; the strobili emerge from the buds the following spring; within

Ecology and Biogeography of Pinus, ed. D.M. Richardson. © Cambridge University Press, 1998, pp. 251–80

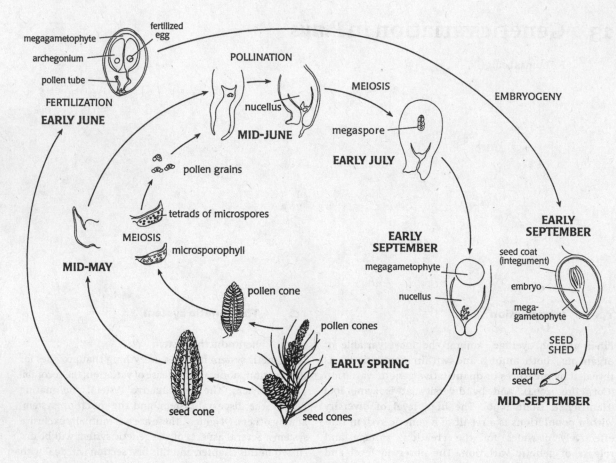

Fig. 13.1. **The reproductive cycle of pine.**

a period of days or weeks, anthesis occurs in the male or pollen strobili and the female strobili reach a receptive stage and are pollinated; fertilization occurs one year after pollination, in the second spring; after fertilization, seeds develop and mature over a period of several months and are shed from the cone in the autumn, about two years after the primordia were first visible within the bud (Fig. 13.1). A few species, such as *P. torreyana* and *P. maximartinezii*, seem to require three years to complete the cycle (McMaster & Zedler 1981; Donahue & Mar 1995), but the exact timing is unclear.

A more detailed description of the reproductive cycle in the megasporangiate strobilus is necessary to understand some of the unique features of pine genetics. The description will be taken largely from Ferguson (1904), McWilliam & Mergen (1958), and Owens & Molder (1977), and are most accurate for species of the subgenus *Pinus*. Details for species of subgenus *Strobus* differ in some particulars. The pines are like lower plant orders that have alternation of generations (sporophyte and gametophyte) because they have multicellular gametophytes, and they

differ from primitive orders because the gametophytes are not free-living but are parasitic within the sporophyte. The primordium of the megasporangiate strobilus in *P. elliottii* is initiated in summer, and by late summer, bracts are evident. The cone scales arise in the axils of the bracts and the ovules are borne on the abaxial surface of the scale, two ovules per scale. The ovules may not be initiated until the following spring, depending on species. The ovule consists of an integument surrounding the nucellus. An opening in the integument, the micropyle, is directed towards the cone axis. The megasporocyte, or spore mother cell, in the centre of the nucellus is apparent from its large size. The megasporocyte undergoes meiosis at about the time of anthesis (i.e. the time that pollen is shed from the microsporangiate strobili). Meiosis results in a linear file of three or four cells, depending on whether both of the initial daughter cells undergo the second meiotic division. In either case, all cells degenerate except the one furthest from the micropyle, which becomes the megaspore. This development differs from that in angiosperms where several of the meiotic daughter cells survive and participate in further development, as in the maize and lily models.

At about the time of meiosis, the ovule becomes receptive to pollination. The cone scales open until they are at nearly right angles to the cone axis, exposing the ovules. The nucellus secretes a pollination droplet between the arms of the micropyle, and pollen lodges on the sticky surface of the droplet, which is fairly high in glucose, fructose and sucrose (McWilliam 1958). The droplet is withdrawn, apparently reabsorbed, drawing any trapped pollen into a depression on the surface of the nucellus called the nucellar or pollen chamber. The middle cell layer in the micropylar neck enlarges, closing the neck. If pollen is plentiful, two to four pollen grains may be lodged in each pollen chamber, a very significant factor in the genetics of pine.

The pollen grain(s) germinates and begins to penetrate the nucellar tissue. At the time of anthesis the microspore has already undergone two mitotic divisions, followed by degeneration of two nuclei (known as the prothallial cells), and contains a generative nucleus and a tube nucleus. After germination, the generative nucleus divides to form the so-called stalk and body cells. Further development proceeds slowly over the summer. Meanwhile the nucellus grows, and at the end of the season the pollen grain is further from the megaspore than it was at the time of pollination. Within the nucellus, the megaspore begins gametogenesis, undergoing five free-nuclear divisions (meaning that no cell walls are formed) to form 32 nuclei. In *P. elliottii* overwintering occurs in this stage, but the sequence varies depending upon species and climate.

Development in the following spring (one year after pollination) is rapid. More mitotic divisions in the megagametophyte result in a free-nuclear state of about 1024 or 2048 haploid nuclei before wall formation begins. Archegonia, similar to those in ferns and mosses, are initiated on the surface of the megagametophyte. The archegonia are composed of neck cells and a large central cell. The central cell divides to produce a small ventral canal cell and a large egg. The egg nucleus enlarges and descends to the centre of the egg cell. Although a particular ovule may produce only one archegonium, the number is usually higher, and all species studied produce multiple archegonia as a rule. The maximum number varies from two to seven among the 15 species tabulated by Lill (1976). Multiple archegonia are an important feature of the mating system of pines and explain how species of the genus maintain high levels of variation.

While the archegonia are developing, the pollen tubes grow rapidly in the direction of the megagametophyte and the body cell divides to form two unequal, unflagellated sperm. The tip of the pollen tube forces through the neck

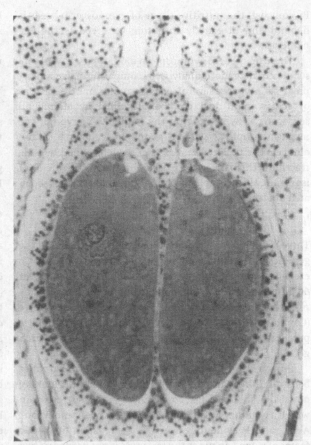

Fig. 13.2. **Fertilization of the egg in *Pinus nigra* (previously published in McWilliam & Mergen 1958). Two archegonia are visible. In the one on the left, the sperm nucleus and egg nucleus have merged. In the one on the right, a pollen tube has just entered the archegonium.**

cells of the archegonium and discharges its contents into a receptive vacuole in the egg (Fig. 13.2). The larger sperm approaches the egg nucleus and their outlines merge, although the two sets of chromosomes remain distinct throughout the first mitotic division of the zygote. Because several pollen grains may have been trapped in the pollen chamber, each of the several archegonia formed on the megagametophyte may be fertilized, and multiple embryos may develop; this is called polyembryony. Competition occurs among embryos and only rarely does more than one survive to germinate. This has great significance for the mating system, as will be explained later.

During embryogenesis, the megagametophyte continues to grow and provides the nutritive tissue for the embryo and eventually for the germinating seedling. Though the nutritive tissue of the mature seed is often called endosperm, this is not, strictly speaking, correct. The angiosperms have a true endosperm, a polyploid tissue resulting from the fusion of pollen and seed nuclei.

By contrast, the nutritive tissue of gymnosperm seeds is haploid and of exactly the same genotype as the egg nucleus.

These features of gymnosperm biology have useful consequences for genetic analysis. By comparing megagametophyte genotypes of several seeds from the same tree, alleles of co-dominant loci can be identified by Mendelian segregation. By comparing the haploid genotype of the megagametophyte with the diploid genotype of the embryo, the pollen genotype can be determined by 'subtraction'. This is routinely done for allozymes, alleles of isozymes, the multiple molecular forms of enzymes separated by electrophoresis.

Segregation

The use of electrophoresis has provided more data on the genetics of pines in the last 15 years than had been gathered in the preceding 75 years (Hamrick & Godt 1996). Over 60 different gene loci (in about 40 enzyme systems) can be revealed by electrophoresis in the megagametophyte (Conkle *et al.* 1982). Rothe (1994) tabulates electrophoretic procedures and stains for 78 enzymes separated on starch gels.

Pines are ideal material for genetic studies of isozymes because inheritance can be inferred without making crosses. A standard test for allelism is the test cross, in which a heterozygote is crossed to a recessive homozygote. A 1:1 ratio is evidence of meiotic segregation for allelic forms of a gene. In pines the test cross is not needed; if two isozymes are allelic, this can be demonstrated by a 1:1 segregation ratio among the haploid megagametophytes of a single tree. In most cases, normal segregation is the rule, and studies of genotypic-frequency distributions within populations show no significant departures from Hardy–Weinberg equilibrium.

Nevertheless, segregation distortion, or deviation from the expected 1:1 ratio, is frequently observed (Strauss & Conkle 1986). In some cases of segregation distortion, alleles may have aberrant segregation ratios across the entire population (up to 15.6% deviation from expectations has been observed in *P. attenuata*), and in other cases distortion is apparent only in single trees. 'Selfish' alleles, or segregation distorters, are one explanation. These are alleles that are linked to the allozyme loci and affect meiosis, increasing the frequency with which they are transmitted. Another explanation is selection against one of the alleles during gametogenesis or embryogenesis. The long period of gametophytic development and the complexity of the process, as explained above, together with the possibility of competition among embryos during embryogenesis, make selection an especially viable explanation. The fact that the less common allele is almost always the one which is under-represented (Rudin 1977;

Millar 1985; Strauss & Conkle 1986) may be evidence that selection is responsible for the observed distortion of segregation ratios, since uncommon alleles may represent mutants with impaired function.

Linkage

The pine megagametophyte system, which makes the study of segregation simple, also makes it possible to quantify linkage among allozyme loci without the necessity of crosses. Loci with allele combinations that tend to segregate together can be identified and the recombination fraction used to map linkage groups.

Taken together, the many studies of allozyme inheritance reveal a striking conservatism of linkage in pines and in the Pinaceae in general. Genes coding for glutamate-oxaloacetate transaminase (GOT) and phosphoglucose isomerase (PGI) were one of the first pairs of linked loci detected in pine, with a recombination frequency of 4% (Guries, Friedman & Ledig 1978). The same linkage has been detected in *P. attenuata*, *P. contorta*, *P. jeffreyi*, *P. ponderosa*, *P. strobus*, and *P. taeda*, as well as in several other Pinaceae in the genera *Abies* and *Pseudotsuga* (see Strauss & Conkle 1986). Conkle (1981) identified five linkage groups in *P. jeffreyi* and *P. taeda* and these were quite similar between species, a further indication that the pine genome has been relatively stable. These highly conserved linkages parallel the highly conserved karyotype of pines, discussed below. Recently, isozyme loci have been mapped with regard to other biochemical markers, such as restriction fragment length polymorphisms (RFLPs) and random amplified DNA polymorphisms (RAPDs), on the *P. taeda* genome (M.M. Sewell, personal communication 1995).

The rate of recombination varies among species and even among trees within species (Strauss & Conkle 1986). Sexual differences are also known: the rate of recombination is 43% greater in male gametes than in female gametes of *P. radiata* (Moran, Bell & Hilliker 1983). These differences are probably genetic.

Organelle inheritance

Inheritance of organellar DNA in pines is unlike that in angiosperms. In angiosperms, both chloroplasts and mitochondria are inherited through the maternal side, or sometimes chloroplast inheritance is biparental. In pines, and in all conifers studied to date, chloroplasts are inherited through the pollen parent (Neale & Sederoff 1989). The mitochondria, as in almost all other plants and animals, are inherited maternally. This suggests the existence of a very complex system in the egg, whereby maternal proplastids are recognized and destroyed or fail to multiply, leaving only paternal chloroplasts. Conversely, paternal mitochondria from the pollen tube must be destroyed as the contents of the pollen tube are released into the egg cytoplasm. The paternal inheritance of chloroplasts is all the more surprising given studies on the cytology of fertilization in *P. nigra* which indicated that both the mitochondria and the plastids contributed by the pollen degenerate after fertilization (Camefort 1966).

The paternal inheritance of chloroplast DNA, the maternal inheritance of mitochondrial DNA, and the biparental inheritance of nuclear DNA suggest ways to track the descent and mating of pines that is not equalled in any other organism. The system has not yet been fully exploited for evolutionary and ecological studies.

Summary of the pine reproductive system

In summary, the reproductive cycle of pines has several advantages for genetic analysis. The large, primitive megagametophyte provides a mass of tissue for isozyme electrophoresis. Haplogenetics is possible because the megagametophyte develops from a single meiotic product, the megaspore, and the egg has the same genotype as the nutritive tissue surrounding it. Therefore, segregation and linkage can be quantified by analysing multiple seeds from the same tree without the need for test crosses. Linkage analysis has confirmed the results of karyotype analysis: the pine genome is highly conserved. Furthermore, the pollen contribution to the embryo can be determined by 'subtracting' the contribution of the egg, a feature quite useful in mating system analysis, which will be described below.

Linkage analysis suggests that recombination rates are higher in pollen sporogenesis than in the meiotic divisions leading to the female gametophyte (Moran *et al.* 1983; Groover *et al.* 1995). This makes adaptive sense because pollen is broadcast widely whereas seeds tend to stay closer to the seed parent. Therefore, there is less correlation between the environment in which the male parent grew and that of its progeny than between the environment of the seed parent and its progeny. Expressed another way, the seed parent and its progeny are more likely to share the same environment 'on average' than the pollen parent and its progeny. Because adjacent trees may be related and inbred progeny are inferior to outcrosses, selection does not favour pollination between near neighbours. Widely dispersed pollen may be more effective in producing offspring than locally-dispersed pollen (e.g. Coles & Fowler 1976). That places a premium on recombination in the pollen parent because at least some of its recombinant offspring may, by chance, be preadapted to the distant environment in which they find themselves.

Each ovule has multiple archegonia and may be pollinated by a number of pollen grains in a good year. Both male and female gametophytes have a long development period (at least one year) and the male must function in a foreign environment, providing an opportunity for competition among pollen tubes and selection among gametes, both male and female. Multiple archegonia and multiple pollination events also provide an opportunity for competition and selection among embryos within the ovule, since only one usually survives to germinate. Segregation distortion is evidence that selection among gametes or embryos operates within the ovule (e.g. Fowler 1964a). Competition among embryos may be the most important mechanism by which pines maintain a highly outcrossing mating system; the embryos within an ovule that result from selfing and are homozygous for deleterious recessive alleles can be eliminated without a reduction in seed production. That is, competition among embryos does not reduce reproductive output (i.e. fitness) of the parent, and a high level of diversity can be maintained at little reproductive cost. On the other hand, the lack of self-incompatibility mechanisms in pines provides flexibility in the mating system. In years when pollination is poor, little competition occurs among embryos, which provides for some seed production even from selfing.

13.2.2   Mating system

The mating system usually refers to the levels of inbreeding (selfing and consanguineous mating) and outcrossing. The mating system is affected by mechanisms of self-incompatibility and self-sterility, and also by population structure (e.g. density) and the temporal and spatial distribution of relatives.

Spatial and temporal separation of the sexes should favour outcrossing and, therefore, recombination. Pines are monoecious, meaning that the male and female reproductive structures are borne on the same tree but separated. The megasporangiate strobili occur predominantly high in the crown and near the ends of branches in the mid- or lower crown, and the microsporangiate strobili are borne predominantly on the interior and lower portions of the crown (Fowler 1964b). However, substantial overlap occurs, creating the opportunity for geitonogamy.

Temporal separation of the sexes, or dichogamy, is observed in some species but not in others. *Pinus palustris* and *P. ponderosa* are protandrous, meaning that peak pollen shed occurs a few days before peak female receptivity on the same tree (Snyder, Dinus & Derr 1977; Wang 1977). *Pinus nigra* is weakly protandrous on average, but some trees may even be protogynous (Matziris 1994), meaning that females are receptive before pollen is shed. *Pinus contorta* in British Columbia is reported as slightly protandrous (Owens, Simpson & Molder 1981). However, in coastal California it is generally protogynous (although individuals can be synacmous or even protandrous), and inland the situation varies from year to year (Critchfield 1980). *Pinus banksiana*, *P. echinata*, *P. monticola*, *P. resinosa*, *P. strobus*, *P. taeda* and *P. virginiana* are all reported as synacmous (Wasser 1967; Fowler & Lester 1970; Wright 1970; Bingham, Hoff & Steinhoff 1972; Kellison & Zobel 1973; Rudolph & Yeatman 1982). Temporal separation between male and female gametes seems, in general, to be weak.

Studies in controlled pollination have failed to reveal any incompatibility system. When intraspecific crosses fail, the cause seems to be post-fertilization. Thus, prezygotic mechanisms to prevent selfing seem to be absent or poorly developed.

The rate of selfing has been estimated using morphological markers, such as albino and virescent mutants. These markers are recessive and lethal or deleterious, and are assumed to be in low frequency in the population. Therefore, the fraction of marker phenotypes in the progeny array should be equal to one-quarter of the selfed fraction. Selfing in seed orchards of *P. elliottii* was estimated at 2.5% and in natural stands at *c.* 6% (Squillace & Goddard 1982), although a wide variability was evident (Squillace & Kraus 1963). For nine carriers of mutant phenotypes, the rate of natural selfing was 0–5%, but for two others it was 23% and 27%. The higher estimates of selfing for some trees in natural stands may result from crossing among relatives as well as from variation in self-fertility. Most estimates of self-fertility in pines, obtained from morphological markers, indicate considerable self-sterility (Table 13.1). An exception is the genetically depauperate *P. resinosa*, but even in this species, outcrossing is the dominant mode; the rate of outcrossing in natural stands was 79 to 98%, depending on year and stand (Fowler 1965a, b). However, an isolated tree apparently produced 100% selfed progeny.

The use of isozyme electrophoresis has provided a much more rigorous method of measuring the outcrossing rate and one that can be used in a variety of situations. When single-locus estimates of outcrossing are calculated, they vary widely from locus to locus, although generally the average is greater than 90% in pines (Table 13.1). Multilocus estimates are preferred to single-locus estimates because the former are less sensitive to violations of the assumptions inherent in the calculation of outcrossing (Shaw, Kahler & Allard 1981). Multilocus estimates, like the mean of single-locus estimates, usually exceed 90%. Outcrossing, whether measured by morphological or isozyme markers, seems to be between 0.91 and 0.98 for pines (Schemske & Lande 1985).

Multilocus estimates of outcrossing tend to be equal to or higher than the mean of single-locus estimates (e.g. Furnier & Adams 1986a; El-Kassaby *et al.* 1987). For example, based on isozymes the mean of single-locus estimates in *P. banksiana* was 88% and the multilocus estimate was 91%, although the difference was not statistically significant (Cheliak *et al.* 1985). Differences between multilocus and single-locus estimates of out-

Table 13.1. Outcrossing (%) estimates in *Pinus*

Pinus taxon	t_s[a]	t_m[b]	Marker	Reference
P. banksiana	81	–	morphol.	Fowler (1965c)
P. banksiana	90	–	morphol.	Sittman & Tyson (1971)
P. banksiana	93	–	morphol.	Rudolph (1979)
P. banksiana	88	91	isozyme	Cheliak *et al.* (1985)
P. banksiana	88	–	isozyme	Snyder, Stewart & Strickler (1985)
P. banksiana	85	89	isozyme	Xie & Knowles (1991)
P. caribaea	–	89.8	isozyme	Matheson, Bell & Barnes (1989)
P. cembra	70.7	68.6	isozyme	Krutovskii, Politov & Altukov (1995)
P. contorta	94.4	–	isozyme	Danzmann & Buchert (1983)
P. contorta	100	99	isozyme	Epperson & Allard (1984)
P. contorta	97.4	94.8	isozyme	Perry & Dancik (1986)
P. contorta	93.3	–	morphol.	Sorensen (1987)
P. engelmannii	72.9	72.4	isozyme	Bermejo (1993)
P. heldreichii var. *leucodermis*	78.8	80.3	isozyme	Morgante, Vendramin & Olivieri (1991)
P. jeffreyi	88.9	93.5	isozyme	Furnier & Adams (1986a)
P. koraiensis	93.6	97.4	isozyme	Krutovskii *et al.* (1995)
P. maximinoi	–	65	isozyme	Matheson *et al.* (1989)
P. monticola	95.2	97.7	isozyme	El-Kassaby *et al.* (1987)
P. oocarpa	–	87.4	isozyme	Matheson *et al.* (1989)
P. ponderosa	95.8	–	isozyme	Mitton *et al.* (1977)
P. ponderosa	97.6	–	morphol.	Mitton *et al.* (1981)
P. ponderosa	84.1	96.0	isozyme	Mitton *et al.* (1981)
P. pungens	–	112[c]	isozyme	Gibson & Hamrick (1991b)
P. sibirica	86.2	89.4	isozyme	Krutovskii *et al.* (1995)
P. sylvestris	91.0	94.4	isozyme	Muona & Harju (1989)
P. sylvestris	93.6	–	isozyme	Yazdani *et al.* (1989)

[a] Mean of single-locus estimates of outcrossing.
[b] Multi-locus estimate of outcrossing.
[c] Although the maximum possible rate of outcrossing is 100%, estimates may exceed that because of sampling error and violations of assumptions.

crossing could occur as a result of inbreeding other than selfing (i.e. consanguineous mating), which would depress the single-locus estimates of outcrossing but have less affect on the multilocus estimate. If crosses among relatives are influencing the single-locus estimates, that may suggest that related individuals are clustered to some degree. An important topic for further research is whether positive assortative or negative assortative mating occurs in pines. Isozymes have provided evidence for partial assortative mating in Sitka spruce (*Picea sitchensis* (Bong.) Carr.) seed orchards (Chaisurisri, Mitton & El-Kassaby 1994), though no examples have yet been reported in pines.

Given that neither spatial nor temporal separation of sexes is strong, that relatives tend to be clustered (e.g. Coles & Fowler 1976; Tigerstedt *et al.* 1982), and that self-incompatibility seems to be lacking in most species of pines, a high level of outcrossing must be maintained by some other mechanism. Partial self-sterility resulting from inbreeding depression seems to be a major part of the explanation. Seed set to controlled self-pollination in pines is about 66% of the value for controlled outcrosses on average (Franklin 1970; Chap. 14, this volume). When pollen mixes of self and outcross pollen are used in controlled pollination, the ratios of self to outcross progeny are much less than 50% except in completely self-fertile trees (Barnes, Bingham & Squillace 1962; Kraus & Squillace 1964), which is evidence for selection in the embryonic stage. Even in the highly self-fertile *P. resinosa*, embryonic selection occurs against homozygotes for a chlorophyll allele (Fowler 1964a). The multiple archegonia of pine megagametophytes provide a mechanism that allows for selection among embryos without sacrificing seed yields (Buchholz 1929; Sarvas 1962; Dogra 1967).

Selection against selfed embryos and trees is inferred from virtually all isozyme studies that compare embryo genotypes with those of seedlings, saplings or mature trees. For example, homozygosity is highest at the embryo stage, less at the sapling stage, and least in mature trees of *P. radiata* (Plessas & Strauss 1986). The lower fitness of selfed embryos is also suggested by studies of seed from seroti-nous cones. Serotinous cones remain closed after seed maturity until they are opened by the heat generated in forest fires. Homozygosity is highest in the youngest cones of *P. banksiana*, which represent the most recent mating season, and decreases with increasing cone age (Cheliak *et al.* 1985; Snyder, Stewart & Strickler 1985). This parallels a decrease in seed germination with increasing age of cone, which is observed in *P. contorta* and *P. rigida* (e.g. Clements 1910; Andresen 1963). The decrease in germination with age indicates an opportunity for selection. However, such a correlation was not obvious in *P. banksiana* (Snyder *et al.* 1985).

Even if selfed zygotes pass all the hurdles of embryonic development, inbreeding depression continues to act, reducing the chances that they will survive and reproduce. In general, germination of seeds from self-pollination is only 89% that of seeds from cross-pollination; seedling mortality is 31% greater for selfs than for outcrosses; and height growth (which affects competitive ability, crown size and, therefore, reproductive output) of selfs is 10% less than that of outcrosses (Franklin 1970).

Nevertheless, the pine mating system is flexible enough to permit inbreeding and selfing in isolated trees. This flexibility is important for colonizing species and facilitates invasion of new habitat. Pines in exotic locales often escape, establishing colonies at some distance from the seed source (Chap. 22, this volume). Seed dispersants of *P. radiata* in New Zealand established colonies over 3 km distant from the parent (Fig. 13.3), and *P. contorta* founded colonies 16–18 km from their parents. These advance colonists were surrounded by a second, progeny generation. Bannister (1965) argued that the progeny would largely be the products of selfing, and any successive generation would result largely from sib matings.

The predominant pattern for pines is one of out-

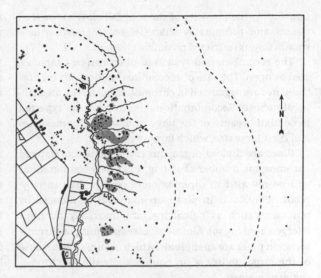

Fig. 13.3. **Map showing location of advance colonists of *Pinus radiata* (dots) originating by seed dispersal from planted trees at A, B, and C. The arcs indicate distances of 2 and 4 km from the sources. After Bannister (1965).**

crossing. Outcrossing promotes recombination, and it maintains variation because new recessive alleles are shielded from exposure to selection. The barriers to selfing are largely postzygotic and are fully effective only when the opportunity exists for competition with outcrossed siblings. The barrier does not preclude selfing and, therefore, pines can be successful colonizers. The ability to self may have enabled pines to migrate successfully to new habitat as glaciers waxed and waned.

13.2.3 Recombination system

Organization of the genetic material must strike a balance between immediate fitness and long-term flexibility. The greater the recombination of genetic elements during reproduction, the greater the chances that at least some progeny will find a match in the range of environments to which they may be exposed. Conversely, in a spatially and temporally homogeneous environment, restricted recombination would be favoured because it increases the likelihood of offspring that resemble their parents and therefore will be adapted to the environment in which their parents were obviously successful. A late-successional species that reproduced under the canopy of the parental generation might benefit by restricted recombination (i.e. a 'closed' recombination system). A colonizing species might require an 'open' recombination system that favoured the generation of new combinations of genes. It is noteworthy that recombination in the pollen parent may be 26 to 43% greater than in the seed parent (Moran *et al.* 1983; Groover *et al.* 1995). Because pollen dispersal distance is greater than seed dispersal, the pollen parent is analogous to a colonizer of new habitat.

Conversely, the majority of the seed fall close to the seed parent and presumably share its environment, which should favour restricted recombination.

The recombination system of pines can be characterized as open. The rate of recombination depends on how the genes are organized in chromosomes and the mechanics of meiosis. Recombination rates in pines are expected to be high because of the large number of chromosomes and their large size, which favours crossover.

Pines are diploid organisms ($2n = 24$) with a haploid chromosome number of 12 (Fig. 13.4). New combinations will be revealed in diploids more readily than in polyploids. Polyploids in pines, spontaneous or induced by mutagens such as colchicine, are invariably aberrant (Mergen 1958, 1959). Almost all cases of multiple chromosomes in pines are aneuploids, which are unstable because of the impossibility of an equitable division of chromosomes at meiosis.

The haploid karyotype of pines is remarkably uniform (it is difficult to tell one chromosome from another), and the karyotype is highly conserved in the genus (Sax & Sax 1933). Most pine species have a karyotype of 11 long, metacentric to sub-metacentric chromosomes and one short, heterobrachial chromosome. The exceptions are the species of subsection *Pinus*. Pines of that subsection have 10 metacentrics and two short, heterobrachial chromosomes. In a cytological study of nine species of pines, the longest metacentric chromosome was only 9–15% larger than the median, and gradation from the longest to the shortest chromosome was so gradual that they could be distinguished with great difficulty, if at all (Pederick 1967a). The ratio of the long arm to the short arm was between 1.01 and 1.20. Some successful attempts have been made to distinguish the metacentrics by the pattern of secondary constrictions (Pederick 1967b) and by fluorescent banding using base-specific fluorochromes (Hizume, Ohgiku & Tanaka 1983). However, recognition seems to depend very greatly on the skills of the observer

and techniques are not always successfully transferred among laboratories.

Chromosome rearrangements can restrict recombination in heterozygous combination. Inversions have been detected by 'bridge and fragment' configurations following meiosis (Saylor & Smith 1966; Pederick 1968; Runquist 1968). Bridge-fragment configurations had a frequency of 0.5% in meiotic cells of 12 *P. radiata* trees (Pederick 1968). The frequency of bridge-fragment configurations in pines is greater than that reported for other taxa (Saylor & Smith 1966), but the inversions are invariably small, as indicated by the size of the fragment, and it is not clear that they are important in holding adapted gene complexes together. Inversions might be expected in a genome characterized by highly repeated elements (see below) because of the increased opportunity for misalignment during pairing. No translocations have been reported in pines.

The conservation of the karyotype has significance for hybridization in pines. Because of the low frequency of rearrangements, meiosis in hybrids is likely to be normal and, therefore, hybrids are fully or nearly fully fertile. Hybridization will be discussed below, but obviously the ability to share genes among species extends the concept of open recombination to the genus level.

The amount of DNA in the pine karyotype is remarkable. In one study, the DNA content in four pines ranged from c. 30 to 88 pg per diploid genome (Rake *et al.* 1980). The 34.7 pg of DNA per diploid nucleus in *P. lambertiana* (Dhillon 1980) is in sharp contrast to the 6 pg in maize, 3.2 pg in humans, 0.7–1.2 pg in eucalypts, or 0.2 pg in fruit flies (Hinegardner 1976; Grattapaglia 1994).

Much of the pine DNA is in highly repeated, multi-copy sequences. The single-copy fraction of the *P. strobus* genome was estimated as only 24% of the total DNA (Kriebel 1989). Much of the 'extra' DNA in pines seems to be non-coding. Probably only 0.1% is expressed in mRNA. A highly repeated element of about 6000 base pairs (bp) in length, called the Institute of Forest Genetics element (IFG element), was isolated from *P. lambertiana* DNA (Kossack 1989). Its base pair sequence shows that it is related to transposons, or 'jumping genes', in maize and other well-studied plants. The IFG element is present in the genomes of all pine species studied and the number of copies is 10^4. However, it does not account for all the 'extra' DNA in the pine genome; the total length of the genome of *P. strobus*, for example, is about 2×10^{10} bp (Kriebel 1989), and copies of the IFG element represent only 3×10^3 bp.

Although the karyotype is remarkably uniform, the amount of DNA varies greatly from species to species in the pines. *Pinus lambertiana* has 1.6 times the nuclear DNA content of *P. radiata* and 2.0 times the DNA of *P. rigida* (Dhillon 1980). It is tempting to speculate that the greater amounts of DNA in pines, in general, may favour recombi-

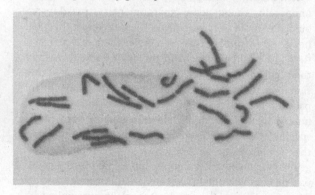

Fig. 13.4. **Chromosomes from a mitotic cell of *Pinus radiata*, showing the diploid number of 24 (11 pairs of metacentric to sub-metacentric chromosomes and one pair of short, heterobrachial chromosomes).**

nation. It is also tempting to speculate that species with large amounts of nuclear DNA may have the most open recombination systems because chromosome length is related to DNA content. On the other hand, pines with low nuclear DNA contents may have shorter generations and may occupy earlier seral stages in succession and may be the most invasive species in the genus (Rejmánek 1996). Dispersal and invasiveness may be facilitated by small seed size (or at least, by low wing loading, the ratio of seed mass to surface area of the seed wing), and seed size is correlated with cellular DNA content (Wakamiya et al. 1993). Northern species that have migrated back and forth during swings from glacial to interglacial periods may have been subjected to selection for small seeds and, therefore, reduced DNA content. Even within conifer species, DNA content has been reported to decrease with increasing latitude (e.g. Miksche 1968), although the data are suspect because procedures lacked internal standards. Nevertheless, intraspecific correlations of seed size and latitude are known in several species of pine (Chap. 14, this volume).

Unfortunately, very few species have been characterized as to their DNA content. Nuclear volume is closely correlated with DNA content, but comparisons of nuclear volume among different studies are apparently not possible; for example, nuclear volume for the same or related species differs greatly among studies (Miksche 1967; Mergen & Thielges 1967). Comparison of DNA content or nuclear volume and life-history characteristics in a range of pine species would be a fruitful area of research.

Hybridization and introgression

Many naturally-occurring, interspecific hybrids are known in the pines, leading some taxonomists to consider it a taxonomically 'critical' group (i.e. one in which species are intrinsically difficult to classify; see Chap. 2, this volume). However, as in other taxa, hybridization is usually local and is often associated with disturbance.

Because genetic barriers to hybridization are frequently ecological in pines, isolation breaks down when habitats are disrupted. Pinus palustris and P. taeda in Mississippi and Louisiana are morphologically distinct species and were ecologically isolated prior to a period of intensive and destructive logging between 1910 and 1930 (Chapman 1922). Hybrids were able to survive in the relative absence of competition on the logged-over lands, resulting in the appearance of a new pine known as 'Sonderegger pine'. Hybridization was followed by a generation of backcrossing (Namkoong 1966) but, over time, separation between the two species seems to have been re-established. Hybridization when it occurs is generally local; i.e. it is observed in some areas of sympatry but not in others. The Sonderegger hybrids, common in Mississippi and Louisiana, were apparently rare in many other portions of the sympatric range of P. palustris and P. taeda.

If ecological or phenological barriers to hybridization are overcome, other barriers may come into play. In the subgenus Strobus, failure usually occurs post-fertilization, during embryogenesis (Kriebel 1972). In the subgenus Pinus, barriers may occur at many stages, from the failure of alien pollen to germinate on the nucellus, to failure of the pollen tube to grow to the egg, to failures during embryogenesis (McWilliam 1959). Despite the presence of well-defined isolating mechanisms, few are complete, permitting some gene exchange even among species that belong to different subsections.

Extensive trials in pine hybridization were carried out at the Institute of Forest Genetics, Placerville, California (Critchfield 1975). With few exceptions there were no marked barriers to crossing between races of a species in pines. Most species were partially or completely isolated from each other by genetic barriers. Seed set in most compatible combinations was below 40% of that obtained in intraspecific crosses. Hybrid inviability was rare or absent and hybrid fertility was high. Crossing among subsections (Little & Critchfield 1969) was very rare and the two subgenera, Pinus and Strobus, were completely isolated.

Hybridization in pines seems to be significant as a mechanism that can provide for an influx of variation, a further broadening of the potential for recombination (Stebbins 1959). The lack of double fertilization in pines, a contrast to the situation in angiosperms, is important in a discussion of hybridization. In angiosperms the second sperm from the pollen tube fertilizes a fusion nucleus derived from the maternal side (from the megaspore or megaspore mother-cell). Thus, in angiosperms not only must the sperm and egg function in the development of a viable embryo, a sperm and fusion nucleus must function in development of the nutritive tissue, the endosperm. Double fertilization creates a double barrier to hybridization in angiosperms that does not exist in the pines.

A trickle of genes across species barriers could easily have been an important event in the survival and long-term evolution of pines. Pinus brutia and P. halepensis provide an example of cryptic introgression. The two species hybridize extensively in two small districts in Greece, but only a few kilometres away no hybrids or backcrosses are obvious (Stebbins 1950). However, isozyme electrophoresis reveals alleles of P. brutia in populations of P. halepensis in western Italy, Albania, and Libya as well as in Greece (Schiller, Conkle & Grunwald 1986).

Hybridization may even have led to the origin of new taxa in pines. Although P. densata is sometimes accepted as a variety of P. tabuliformis, it combines the allozyme and cpDNA genomes of the latter and P. yunnanensis, and may be a stabilized Tertiary hybrid of these species (Wang et al.

1990; Wang & Szmidt 1990; see also Chap. 2, this volume). However, speciation by hybridization should be uncommon because hybrids in pine are not normally isolated from their parents. Pines lack major morphological or physiological incompatibility systems. And unlike angiosperms, speciation by amphidiploidy is virtually impossible because polyploidy is so developmentally disruptive in pines (Mergen 1958, 1959). Therefore, hybridization in pines is more likely to lead to introgression than to speciation.

Seed dispersal and migration

Vagility is important to coherence within species. The greater the capacity for dispersal, the greater the possibility to exchange genes among populations and the more open the recombination system. Dispersal also affects the mating system.

Pine seeds are adapted for dispersal by wind, birds, and mammals, including humans. Wings facilitate dispersal away from the parent tree by wind. Of the 71 species within the subgenus *Pinus*, all but three are effectively winged. Of these three, only *P. pinea* is wingless. Two others, *P. sabiniana* and *P. torreyana* (both in subsection *Sabinianae*) have only rudimentary wings. By contrast, within the subgenus *Strobus*, only 13 of 40 species are effectively winged.

The mean dispersal distance of winged seeds is so low that gene flow would have little power to counterbalance selection over large geographic distances. For example, *P. palustris* seed-fall at a distance of 30 m from the source was only about 10% of seed-fall within the stand and was virtually zero at 90 m (Fig. 13.5; Boyer 1958). However, mean dispersal distance may be less important than the extreme event in preventing differentiation by genetic drift. Extreme events are, by their nature, virtually impossible to detect except, perhaps, in the case of non-native species in an alien environment. For alien species, the occurrence of reproduction at a distance from the nearest seed source can be used to estimate the maximum range of dispersants. Winged seeds of *P. contorta* and *P. radiata* have apparently dispersed to distances of 16–18 km and 3 km, respectively, from their seed trees in New Zealand (Bannister 1965).

Despite the possibility of transport over moderately long distances, many lines of evidence suggest that limited seed dispersal results in the clustering of relatives within stands. Because the distribution of wind-dispersed seeds around a parent is leptokurtic (Levin 1981), siblings should cluster around their seed parent. Contagious distribution of isozyme alleles is one evidence of family structure in pines (Linhart *et al.* 1981; Tigerstedt *et al.* 1982; Yazdani, Lindgren & Stewart 1989). The mean of single-locus estimates of mating system parameters are frequently higher

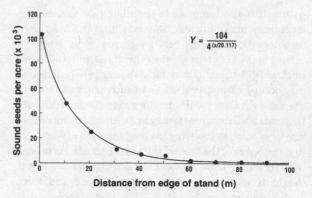

$$Y = \frac{104}{4^{(x/20.117)}}$$

Fig. 13.5. **Number of sound seeds deposited per acre in relation to distance from stand edge in *Pinus palustris*. After Boyer (1958).**

than multilocus estimates (Table 13.1), although the difference is probably not statistically significant. If the differences are real, they may indicate that mating among relatives is occurring, which would be likely if relatives tended to be clustered. A reduction in seed set from crosses among neighbours, relative to seed set in crosses among distant trees, also suggests the existence of family structure in pines (E.B. Snyder, personal communication reviewed in Ledig 1974) and other conifers (Coles & Fowler 1976). Results from studies of quantitative genetic variation are in agreement: in progeny of old-field *P. strobus* stands, much of the variation in height could be explained by crossing among relatives (Ledig & Smith 1981; Ledig 1992).

Many species in the subgenus *Strobus* are wingless (or the wing is left attached to the cone scale) and, presumably, coevolved with an animal disperser (Chap. 14, this volume). Of 40 species, 22 are wingless and three more have only rudimentary wings. Another species, *P. ayacahuite*, is mixed, i.e. seeds of var. *brachyptera* are wingless and varieties *veitchii* and *ayacahuite* have winged seeds. Note, however, that var. *brachyptera* is considered by some to be synonymous with the wingless-seeded *P. strobiformis* (Chap. 2, this volume). In general, species with bird-dispersed seeds are shorter in stature than those with wind-dispersed seeds, perhaps because dispersal distance depends on the height at which seeds are released in anemophilous species (Niklas 1994) but not in animal-dispersed species.

Dispersal by birds offers greater opportunity to spread seeds far from the parent than dispersal by wind. Clark's nutcracker can carry up to 95 seeds of *P. edulis* at one time and transport them as far as 22 km from the seed parent (Vander Wall & Balda 1977). *Pinus albicaulis* and *P. flexilis* also depend on Clark's nutcracker for dispersal (Lanner & Vander Wall 1980; Chap. 14, this volume). Bird-dispersed seeds have a better chance of survival than wind-dispersed

Fig. 13.6. *Pinus albicaulis* clumps in the Sierra Nevada of California originate from seed caches. Several germinants may fuse at the base as they grow. Some evidence suggests that stems in a clump may be related, on average, as closely as half-siblings (photo: F.T. Ledig).

seeds because birds bury seeds in suitable habitats, whereas the destination of wind-dispersed seeds is completely up to chance.

Caching by avian dispersers may also have an effect on population structure. *Pinus albicaulis* is often a multi-stemmed tree, and the different stems are genetically distinct but apparently related (Fig. 13.6; Furnier *et al.* 1987). It seems likely that the Clark's nutcracker, the disperser of *P. albicaulis* seed, would fill its sublingual pouch at a single tree or a small number of adjacent trees. Within multi-stemmed clusters of *P. flexilis*, individual stems are related as slightly less than half-sibs on the average (Schuster & Mitton 1991). No-one has yet investigated the effect that clumping of siblings has on the mating system. Not all bird-cached species of pine are multi-stemmed, however: *P. edulis* is single-stemmed, suggesting that only one seed in a cache survives.

Mammals disperse both winged and wingless seeds. Squirrels cache cones of *P. contorta* (Smith 1970) and the dis-

tance from the parent tree must of necessity be short. However, even pines with winged seeds may depend on secondary dispersal by animals. Chipmunks cached *P. jeffreyi* seeds, placed to simulate wind dispersal, an average of at least 13–25 m further from the source (Vander Wall 1992). Although small mammals may increase the mean dispersal distance, the rare, extreme long-distance dispersal of pine seeds must depend on wind or birds.

The most important long-distance dispersers may be humans. *Pinus brutia* subsp. *eldarica* occurs in scattered locations from Azerbaijan to Pakistan (Conkle, Schiller & Grunwald 1988). It may be native on a single mountain, Eliar-Ugi in Azerbaijan, but elsewhere it seems to have been spread and managed as a semi-domesticated land-race, desirable for its large seeds. Its known locations follow ancient trade routes along the path taken by Alexander the Great. Stands of *P. halepensis* in Israel seem to be derived, at least in part, from populations in western Europe or North Africa (Grunwald, Schiller & Conkle 1986). In North America, edible pine seeds were one of the many items of trade among Native Americans (Davis 1974). No evidence suggests that these seeds were ever planted, but they were deliberately transported, creating the possibility of accidental sowing. Native Americans may have established an isolated population of *P. edulis* in Colorado, 200 km north of the more-or-less continuous range of the species, 450 years ago, according to Betancourt *et al.* (1991). Intercontinental transfers through human intervention have culminated in the naturalization of many pine species and the invasive spread of at least 19 species in the southern hemisphere (Richardson, Williams & Hobbs 1994; Chap. 22, this volume).

Because the reproductive output of pines is tremendous, even a rare event – long-distance transport – may occur. Based on a mean annual cone production of 1692 per hectare (Fielding 1960) and tabulated seed yields per cone (Krugman & Jenkinson 1970), a hectare of *P. radiata* would produce nearly 2×10^6 seeds in its first 40 years. The mean annual seed crop in 25 m tall *P. sylvestris* in Finland is about 800 000 seeds per hectare (Sarvas 1962). *Pinus palustris*, about 17 m in mean height and 40–60 years old, produced about 257 000 seeds ha^{-1} yr^{-1} (Boyer 1958). For North American pines, seed production generally runs from 400 000 to over 4 430 000 seeds ha^{-1} yr^{-1} in good seed years (Burns & Honkala 1990). During a 10-year period, the lowest annual production for *P. echinata* was 4900 and the highest was 1 845 000 seeds ha^{-1}.

The tremendous seed output of trees in general, and pines in particular, is important to evolutionary survival. Despite changing environments (such as the swings that accompany glaciation and deglaciation), the high reproductive output ensures that most pine species can produce a number of variants, which increases the chance that some genotypes will be preadapted to a changed environment. In order to replace themselves, pine stands must necessarily undergo a huge reduction in numbers; because accidental deaths must occur, pines can easily accommodate intense selection pressures. If selection occurs before density-dependent mortality takes over (i.e. at the seedling stage or earlier), then it will have no affect on reproduction.

Migration rates in pines (i.e. the rates at which the ranges of pines expanded) after the end of the last glacial period have been inferred from the dates at which pine pollen began to accumulate in bogs and lakes at various latitudes in Europe and North America (MacDonald 1993). In eastern North America, the estimated migration rate was 81–400 m yr^{-1}. For *P. monophylla*, Lanner (1983) estimated a rate of 240 m yr^{-1}. In Britain the rates were <100 to 700 m yr^{-1}. In both cases, the range of migration rates for pines includes the lowest and the highest rates recorded for tree species. In continental Europe the estimate was 1500 m yr^{-1}; only alder (*Alnus* spp.) spread faster in the postglacial period, and the difference is marginal. Thus, palynology suggests that pines were, perhaps, the most highly mobile of tree species.

Spread can occur as either a wave or as a series of saltational events, which has important consequences for the genetic structure of pines. If species spread as a wave, then clinal patterns of variation should be the rule. If saltational events produced colonies far in advance of the species' front, then the founder effect may have left an imprint on the patterns of genetic variation.

Patterns of variation in several species suggests that saltation played a role. In *P. coulteri*, diversity is twice as great in small, relict, southern populations as in northern ones (Ledig 1987). Heterozygosity at isozyme loci decreased in a stepwise fashion along the mountain ranges from Baja California to Mount Diablo near San Francisco (Fig. 13.7). The decrease in expected heterozygosity reflects, in part, a decrease in the number of alleles. The major differences in diversity are coincident with major disjunctions in the range of *P. coulteri*. Some alleles disappear between Baja California and the Laguna Mountains east of San Diego and never reappear further north. More drop out across the even wider gaps between California's Peninsular Ranges and the Transverse Ranges, still more between the Transverse Ranges and the Coast Ranges, and others between the Santa Lucia Mountains and the Diablo Range.

The pattern is especially strong in *P. coulteri* because of its fragmented range and, partly, because its large seed would dictate that dispersal from one mountain range to another was a rare event indeed. Therefore, the number of founders would be correspondingly smaller than for a small-seeded species. However, a similar pattern occurs in

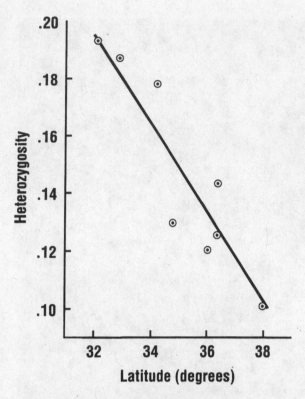

Fig. 13.7. **The decrease in heterozygosity with latitude in *Pinus coulteri* may reflect the action of genetic drift, acting through the founder effect. As the species expanded its range northward by long distance transport at the end of the Pleistocene, successive colonizing events resulted in the loss of alleles.**

other conifers. In *P. contorta* the founding date of northern populations has been established by palynological records, and the number of alleles per locus is related to the time since founding (Cwynar & MacDonald 1987). That is, loss of alleles occurred as *P. contorta* migrated northward in the current interglacial, apparently by long-distance dispersal events that involved at least 70 km jumps. Other examples are giant sequoia (*Sequoiadendron giganteum* (Lindl.) Buchholz), *P. jeffreyi* and *P. monticola* (Fins & Libby 1982; Steinhoff, Joyce & Fins 1983; Furnier & Adams 1986b), and additional cases will probably be found upon further investigation.

In summary, both wind and animals can disperse pine seeds great distances. This contributes to open recombination and prevents differentiation on the one hand, but on the other provides for the rare colonization event that can lead to changes in gene frequency through genetic drift. Drift in pines operates primarily through the founder effect. On balance, seed dispersal in pines would probably favour ecotypic patterns of variation, and heavy-seeded, wind-dispersed species would be more likely to show disjunct patterns of variation than light-seeded species. However, the distribution of suitable habitat may be of as much significance as the dispersal mechanism. Clustering of sibs or of parents and progeny in wind-distributed seeds and clumping of sibs in bird caches are also a result of seed dispersal in pines (see Fig. 14.5 in Chap. 14, this volume). The resulting family structure within populations will affect the mating system and perhaps lead to the release of variation because of the more frequent exposure of recessive alleles due to inbreeding (i.e. crossing among relatives).

Pollen dispersal and gene flow

During development, pine pollen develops two 'wings', air bladders that form between the intine and exine of the pollen grain. The bisaccate pollen of pine is a significant adaptation for anemophily, dispersal by wind. Terminal velocity of bisaccate pine pollen is much less than would be predicted by their size (Jackson 1994). The potential for transport by wind is tremendous; pollen has been deposited in great quantities at distances up to 58 km from forests (reviewed by Lanner 1966 and Koski 1970). Conversely, studies of pollen dispersal from point sources show a highly leptokurtic distribution (Wright 1952). Thus, the effect of pollen transport on gene flow among populations has been controversial:

> ... the distance of pollen dispersal is short and gene flow among subpopulations within a forest seldom occurs
>
> *Sakai & Park (1971)*

> It [gene flow] can be stated to be so great ... that the existence in the same subpopulation of neighborhoods that differ from one another cannot be regarded as possible
>
> *Koski (1970)*

Pollen production is obviously tremendous, but few measurements are available. *Pinus sylvestris* in Finland produced an average of 50×10^6 grains ha^{-1} yr^{-1} on poor sites and 190×10^6 ha^{-1} yr^{-1} on fertile sites (Sarvas 1962). Pohl (1937) estimated 346×10^6 grains were produced by a 10-year-old Scots pine.

Pollen distribution is frequently measured by catching pollen on sticky traps at various distances from an isolated source (Colwell 1951; Wright 1952; Wang, Perry & Johnson 1960). Using this technique, most tree pollen is recaptured close to the source. The standard deviation for dispersal distance (± 1 SD includes 91% of the pollen) for two pines was 67 m for *P. elliottii* (Wang *et al.* 1960) and 17 m for *P. edulis* (Wright 1952). There are several problems with this technique. No independent measure of the pollen released is available; therefore, significant amounts might travel long distances and be uncounted. Neither are the effects of

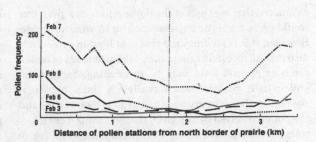

Fig. 13.8. **Pollen frequency of *Pinus elliottii* remains high even in the centre of large treeless areas, which suggests the possibility for substantial gene flow in pines. After Wang *et al.* (1960).**

turbulent wind pockets or thermal shells measured, nor are they measurable (Buell 1947; Lanner 1966). In addition, point sources differ from plane sources. The standard deviation of pollen fall from a 2 ha stand of *P. wallichiana* was 469 m (Siddiqui & Khattak 1975), much greater than estimates from point sources of other pines.

Pollen counts at substantial distances (1.6–2.4 km) from forest fronts are very high and do not depend on distance beyond a certain point in pines (Fig. 13.8; Wang *et al.* 1960) and other conifers (Silen 1962). Pine pollen is one of the major components of the pollen catch in ponds on Block Island, a treeless island 19 km off the northeastern coast of the United States (Jackson & Dunwiddie 1992). Although long-distance transport of pollen is possible, the possibility exists that it is not effective in pollination by the time it reaches the target area (Strand 1957). For example, phenological differences in flowering among areas may ensure temporal isolation. A separation of 400 m in elevation ensures a complete lack of overlap in the timing of pollen shed in *P. flexilis* in the Rocky Mountains of Colorado (Schuster, Alles & Mitton 1989).

A 'neighbourhood' in the terminology of Wright (1969) is the area of a subpopulation within which mating is essentially random; i.e. an area 'from which the parents of an individual at the center may be treated as if drawn at random'. This may be more intuitive if a neighbourhood is thought of as a group homogeneous within itself and beyond which, in a continuum, increasing opportunity exists for differentiation because of isolation by distance (relative to gene flow). The neighbourhood size is the number of trees within a neighbourhood, and depends on population density as well as dispersal ability. Neighbourhood sizes calculated from the standard deviation of pollen dispersal distance from point sources, assuming no seed dispersal, range from 11 trees for *Pinus edulis* to 349 for *P. elliottii* (Wright 1952; Wang *et al.* 1960).

The view of a leptokurtic pollen distribution leads to the improbable corollary that every tree is likely to be pollinated predominantly by its nearest neighbour (Wright 1952; Wang *et al.* 1960). For example, Colwell (1951) calculated that over 90% of the pollinations on a given tree would come from the adjacent tree on its windward side. However, this is far from the truth, at least in extensively distributed, forest-forming pines. Measurements of pollen catch at ground level may be misleading; background pollen in *P. sylvestris* was actually 1.5 times higher at 100–150 m above the ground than at 20 m (Koski 1970), and the pollen received by any given tree was a mix from many trees. In a definitive series of experiments, Koski (1970) found that a single *P. sylvestris* tree would account for no more than 10% of the pollen received by a close neighbour. Beyond an area of about 60–70 m², pollen from trees labelled with ³²P represented less than 1% of the total pollen catch. A high level of background pollen results from the overlapping of pollen shadows from a great many trees, any one of which contributes only a small proportion to the total. In *P. radiata* plantations also, less than 2% of the pollen received by a given tree would come from any other single neighbour (Bannister 1965). Although half of the pollen received by a given tree in *P. sylvestris* stands probably originates from trees within a 50 m radius, that area would circumscribe at least 25–30 trees (Koski 1970). The effective population size in *P. sylvestris* spans distances of tens to hundreds of kilometres (Koski 1970).

The distribution of isozyme alleles has verified the high rate of gene flow among populations, and most species maintain a high proportion of their total variation within populations. G_{st}, a measure of the proportion of the total genic diversity that exists among populations, is usually less than 10% and often less than 5%. That means that more than 90% of genic diversity, and often above 95%, is within populations ($1 - G_{st}$; Table 13.2). Even in species with insular or disjunct populations that have been separated for millions of years or where distinct races have evolved because of selection or drift, variation due to differences among populations is surprisingly low; for example, *c.* 13% for *P. radiata* and 22% for *P. muricata*. The number of migrants per generation (N_m), calculated from isozyme data, is generally greater than 5, even among widely separated populations (Table 13.2), suggesting that differentiation by drift is unlikely.

Alleles may be clustered (e.g. Tigerstedt *et al.* 1982), yet gene flow is so great that even widely-separated populations of extensively distributed pines cannot be considered isolated. The paradox can be explained by reference to seed distribution and, perhaps, by local selection pressures. The relatively low vagility of seeds leads to a population substructure whereby some relatives occur in close proximity. On the other hand, pollen vagility promotes an extremely open recombination system and reduces the possibility of differentiation by genetic drift. Despite the low vagility of seeds relative to pollen, colonization of remote habitat through seed dispersal does occur; when it occurs, it provides instant isolation for the founders.

In pine species that occur in small, fragmented populations so that background levels of pollen are not as high as in widely distributed species like *P. elliottii*, *P. ponderosa* or *P. sylvestris*, the situation may be different and the conclusions of Colwell (1951), Wright (1952), and Wang *et al.* (1960) may be more nearly valid. For example, it seems clear that drift has operated in *P. coulteri*, *P. jeffreyi*, *P. contorta* and *P. monticola* during range expansion, as described above, and gene flow by pollen flight has not resulted in a homogeneous gene pool. Drift has also occurred in *P. torreyana*, where its two small populations are both fixed at all 59 isozyme loci studied, but fixed for different alleles at two of the loci (Ledig & Conkle 1983). The small size of the populations and, therefore, low output of pollen, and their separation (280 km), undoubtedly preclude gene flow between them.

On a more local scale, Hamrick, Blanton & Hamrick (1989) ascribed differences among fragmented populations of *P. ponderosa*, and among patches of cohorts within populations, to genetic drift. Populations were distributed across a roughly 120 km gradient on the geographic margin of the species range. Age classes in *P. taeda* were also differentiated on the basis of allele frequency (Roberds & Conkle 1984). However, these patterns are transient in time and gene flow will erase or rearrange them in the next generation.

In summary, gene flow through pollen exchange in many pines is so great that recombination is maximized within populations and random drift, which leads to differentiation among populations, is retarded. High levels of recombination provide for the release of variation necessary to adapt to changing conditions. Nevertheless, for narrowly distributed endemics and other species with fragmented distributions, gene flow will be restricted or entirely eliminated and genetic drift becomes a possibility. In some species, the tracks of dispersal during postglacial migrations are still obvious, suggesting that gene flow via pollen has not bridged all gaps, as in *P. coulteri*.

13.2.4 Mutation system

The use of isozymes revealed that pines were among the most genically diverse plants (Hamrick, Linhart & Mitton 1979). Although recent surveys of gene diversity have tended to lower earlier estimates of heterozygosity (e.g. Ledig 1986; Hamrick & Godt 1990), heterozygosities in trees, in general, exceed those in herbaceous plants by twofold. Expected heterozygosity under the Hardy–Weinberg equilibrium ranges from 0 to *c.* 0.33 for the 48 pine species for which results are based on 10 or more loci in Table 13.2. The modal value lies between 0.13 and 0.16. This means that most individuals are expected to be heterozygous at about 13–16% of their loci. Most of the

Table 13.2. Genic diversity in *Pinus* populations[a].

Pinus taxon	Loci	Polymorphic[b] >0%	>5%	H_o[c]	$1-G_{st}$[d]	Nm[e]	Sample	Reference
P. albicaulis	16	–	56.3	0.204	–	–	1 pop.; Alberta	Krutovskii *et al.* (1995)
P. aristata	22	46.4	–	0.139	–	–	5 pop.; rangewide	R.D. Hiebert & J.L. Hamrick (unpubl. data), cited in Hamrick, Mitton & Linhart (1981)
P. attenuata	22	73	27	0.125	88	–	10 pop.; rangewide	Conkle (1981)
P. attenuata	43	58	49	0.087	88	–	10 pop.; rangewide	Strauss & Conkle (1986)
P. attenuata	32	55	–	0.131	77.8	–	31 pop.; rangewide	Millar *et al.* (1988)
P. ayacahuite	23	59.3	–	0.154	95.3	–	14 pop., including 3 varieties; rangewide	F.T. Ledig *et al.*, unpubl. data
P. ayacahuite var. *strobiformis*	23	46.1	–	0.154		–	2 pop.; Nuevo León, Coahuila	F.T. Ledig *et al.*, unpubl. data
P. balfouriana	23	57.6	–	0.208	–	–	4 pop.; rangewide	R.D. Hiebert & J.L. Hamrick (unpubl. data), cited in Hamrick *et al.* (1981)
P. banksiana	21	81	52	0.115	98.0	–	3 pop.; Alberta	Dancik & Yeh (1983)
P. banksiana	27	47	–	0.141	93.4	–	32 pop.; central Ontario	Danzmann & Buchert (1983)
P. banksiana	20	–	53.3	0.192	97.9	–	3 pop.; 10 km apart in Manitoba	Ross & Hawkins (1986)
P. banksiana	35	–	44	0.104	–	–	5 pop.; Alberta, Ontario	Wheeler & Guries (1987)
P. banksiana	29	44.1	–	0.148	93	–	4 pop.; New Hampshire, New York, Ontario	Misenti & DeHayes (1989)
P. banksiana	22	–	60.2	0.158	98.2	–	4 pop.; 15 km apart in Quebec	Gauthier, Simon & Bergeron (1992)
P. brutia	31	35.3	–	0.111	–	–	14 pop., including 4 subspecies; rangewide	Conkle *et al.* (1988)
P. caribaea	18	72	48	0.212	–	–	7 pop.; Central America	P.D. Hodgkiss, unpubl. data
P. caribaea	16	75	–	0.268	86.9[f]	–	4 pop., including 2 varieties; Bahamas, Belize	Matheson *et al.* (1989)
P. cembra	8	65.9	–	0.260	68[h]	–	11 isolated pop.; Europe, Asia	Szmidt (1982)
P. cembra	15	33.3	–	0.118	–	–	1 pop.; Ukraine	Goncharenko, Padutov & Silin (1992)
P. cembra	19	59	26.3	0.109	96.0	–	1 pop.; Ukraine	Krutovskii *et al.* (1995)
P. contorta	25	59	45	0.160	–	–	9 pop.; British Columbia, Yukon	Yeh & Layton (1979)
P. contorta	39	90	44	0.185	93.9	–	1 pop.; California	Conkle (1981)
P. contorta	42	68[g]	–	0.116	98.0	–	32 pop.; rangewide	Wheeler & Guries (1982)
P. contorta	21	86	71	0.184	99.0	–	5 pop.; Alberta	Dancik & Yeh (1983)
P. contorta	9	44	44	0.135	–	–	4 pop.; within 2 km in Colorado	Knowles (1984)
P. contorta	23	64	–	0.165	–	–	17 pop.; British Columbia, Yukon	Yeh *et al.* (1985)
P. contorta	35	–	69	0.118	98.4	36.4	23 pop.; Canada, NW USA	Wheeler & Guries (1987)
P. contorta	–	–	–	0.17	94.3	–	17 pop., within a single watershed; Colorado	J.L. Hamrick & C.C. Smith (unpubl. data), cited in Hamrick (1987)
P. coulteri	33	65.8	45	0.148	–	–	66 pop.; rangewide	Yang & Yeh (1993)
P. densata	13	49	61.5	0.210	98.7	11.1	8 pop.; rangewide	F.T. Ledig, unpubl. data
P. densiflora	14	–	64.3	0.275	94.2	–	6 pop.; Japan, China	Szmidt & Wang (1993)
P. echinata	22	53	–	0.113	97.4	10.0	18 pop.; rangewide	Edwards & Hamrick (1995)
P. edulis	17	–	58.8[g]	0.145	98.8	–	1 subdivided pop., Colorado	Betancourt *et al.* (1991)
P. edulis	10	–	54.4	0.213	97.1	3.4	5 pop.; Colorado	Premoli, Chischilly & Mitton (1994)
P. engelmannii	26	45.0	36	0.100	87	–	23 pop.; rangewide	Bermejo (1993)
P. flexilis	10	100	–	0.320[i]	–	–	2 pop.; Colorado	Schuster *et al.* (1989)
P. halepensis	17	58.8	43.1	0.179	98.7	–	3 pop.; Greece	Loukas, Vergini & Krimbas (1983)
P. halepensis	30	15.3	–	0.040	–	–	19 pop.; rangewide	Schiller *et al.* (1986)
P. halepensis	30	13.3	–	0.045	–	–	8 pop.; Israel	Conkle *et al.* (1986)
P. heldreichii var. *leucodermis*	23	24.2	–	0.123	94.6	–	7 pop.; Italy, Greece	Boscherini *et al.* (1994)
P. jeffreyi	43	86	39.7	0.261	–	–	4 pop.; central California	Conkle (1981)
P. jeffreyi	20	–	67	0.22	86.2	–	14 pop.; Oregon, California, Baja California	Furnier & Adams (1986b)
P. koraiensis	15	66.7	–	0.273	–	–	1 pop.; Russia	Goncharenko *et al.* (1992)
P. koraiensis	17	–	43.2	0.131	96.0	–	3 pop.; Russia	Krutovskii *et al.* (1995)
P. lambertiana	19	79	58	0.275	–	–	58 individuals; California	Conkle (1981)
P. longaeva	14	79[g]	–	0.327	96.2	–	5 pop.; Nevada, Utah	Hiebert & Hamrick (1983)
P. maximartinezii	33	30.3	–	0.122	–	–	1 pop.; rangewide	F.T. Ledig, unpubl. data

Table 13.2. (*cont.*)

Pinus taxon	Loci	Polymorphic[b] >0%	Polymorphic[b] >5%	H_e[c]	$1-G_{st}$[d]	Nm[e]	Sample[a]	Reference
P. maximinoi	16	62	–	0.17	–	–	1 pop.; Honduras	Matheson et al. (1989)
P. monticola	12	65	51	0.180	85.2	–	28 pop.; rangewide	Steinhoff et al. (1983)
P. monticola	14	64	–	0.18	–	–	1 pop.; British Columbia	El-Kassaby et al. (1987)
P. mugo	11	100	–	0.263	–	–	1 pop.; Switzerland	Neet-Saqueda (1994)
P. muricata	26	–	–	–	85.9	–	19 pop.; northern California	Millar (1983)
P. muricata	32	47	–	0.118	78	–	7 pop.; rangewide	Millar et al. (1988)
P. muricata	45	24.9	–	0.077	–	–	18 pop.; northern California	Millar (1989)
P. nigra	4	78.8	–	0.336	–	–	40 pop.; rangewide	Bonnet-Masimbert & Bikay-Bikay (1978)
P. nigra	4	100	75	0.272	86.5	–	28 pop.; Mediterranean	Nicolic & Tucic (1983)
P. nigra	16	70	–	0.208	94.0	–	5 pop.; Austria, Bulgaria, Greece, Italy, Corsica	Scaltsoyiannes et al. (1994)
P. oocarpa	18	72	55	0.183	–	–	8 pop.; Central America	P.D. Hodgkiss, unpubl. data
P. oocarpa	32	65	–	0.27	–	–	2 pop.; Mexico, El Salvador	Millar et al. (1988)
P. oocarpa[f]	16	74	–	0.192	89.6[k]	–	5 pop.; Nicaragua, Belize	Matheson et al. (1989)
P. palustris	19	100	84	0.150	–	–	24 pop.; rangewide	Duba (1985)
P. pinceana	27	59.3	–	0.161	82.1	–	3 pop.; rangewide	M.T. Conkle et al., unpubl. data
P. ponderosa	21	62	38	0.123	88	–	10 pop.; pooled; Washington, Idaho, Montana	O'Malley, Allendorf & Blake (1979)
P. ponderosa	7[f]	100	–	0.289	95.8	–	1 pop.; subdivided into 6 groups	Linhart et al. (1981)
P. ponderosa	29	90	52	0.186	98.5	–	400 trees; Washington, Idaho, Montana	Allendorf, Knudsen & Blake (1982)
P. ponderosa	23	74	35	0.124	98.5	–	6 small, isolated pop.; Montana	Woods, Blake & Allendorf (1983)
P. ponderosa	12[l]	100	–	0.29	–	13.5	11 pop.; Colorado	Hamrick et al. (1989)
P. ponderosa var. ponderosa	23	68	–	0.155	–	–	14 pop.; California, NW USA	Niebling & Conkle (1990)
P. ponderosa var. scopulorum	23	75	–	0.164	–	–	20 pop.; Rocky Mountains	Niebling & Conkle (1990)
P. pumila	15	80.0	–	0.334	95.6[f]	5.6	4 pop.; eastern Siberia, Sakhalin Island	Goncharenko et al. (1992)
P. pumila	22	67[g]	65	0.257	95.7	–	5 pop.; eastern Siberia, Sakhalin Island	Goncharenko, Padutov & Silin (1993a)
P. pumila	20	–	60.0	0.249	97.9	–	3 pop.; north of Kamchatka Peninsula	Krutovskii et al. (1995)
P. pungens	21	58.3	–	0.204	86.5	4.6	20 pop.; rangewide	Gibson & Hamrick (1991a)
P. radiata	37	–	33.2	0.127	96.5	–	3 pop.; California	Plessas & Strauss (1986)
P. radiata	32	58	–	0.141	87	–	5 pop.; rangewide	Millar et al. (1988)
P. radiata	31	46.4[g]	–	0.098	83.8	–	5 pop.; rangewide	Moran, Bell & Eldridge (1988)
P. resinosa	9	0	–	0	–	–	5 pop.; Ontario, New Brunswick, Nova Scotia	Fowler & Morris (1977)
P. resinosa	27	15	4	0.007	–	–	2 pop.; Minnesota	Allendorf et al. (1982)
P. resinosa	46	0	0	0.002	–	–	50 trees; Wisconsin	R.P. Guries, unpubl. data
P. resinosa	>4	0	–	0	–	–	6 pop.; NW margin of species	Simon, Bergeron & Gagnon (1986)
P. rigida	21	100	76	0.146	97.7	17.2	11 pop.; rangewide	Guries & Ledig (1982)
P. rigida	33	64	–	0.228	93	–	6 pop.; Massachusetts, New Jersey, Pennsylvania, Vermont, Quebec	Misenti & DeHayes (1989)
P. sabiniana	29	93	62	0.128	–	–	8 pop.; rangewide	F.T. Ledig, unpubl. data
P. sibirica	19	57.9[g]	42.1	0.147	–	–	1 pop.; Russia	Krutovskii, Politov & Altukhov (1988)
P. sibirica	20	50	–	0.171	–	–	1 pop.; Russia	Goncharenko et al. (1988)
P. sibirica	15	56.7	–	0.185	97.3[f]	–	6 pop.; central Russia	Goncharenko et al. (1992)
P. sibirica	20	44	41	0.161	95.9	6.2	8 pop.; central Russia	Goncharenko, Padutov & Silin (1993b)
P. sibirica	19	–	46.2	0.158	97.5	–	9 pop.; Siberia	Krutovskii et al. (1995)
P. strobus	17	53	–	0.330	92	–	35 selected clones; Maine, New Hampshire	Eckert, Joly & Neale (1981)
P. strobus	12	69	–	0.236	98.1	8.6	27 pop. sampled in a provenance test; rangewide	Ryu & Eckert (1983)
P. strobus	18	50.6	–	0.180	98.1	–	10 pop.; Quebec	Beaulieu & Simon (1994)
P. sylvestris	3	100	–	0.365	98	–	3 pop.; Quebec	Rudin et al. (1974)
P. sylvestris	12	91.7	–	0.321	–	–	22 selected trees; Poland	Krzakowa, Szweykowski & Korczyk (1977)
P. sylvestris	3	98.3	–	0.313	84	–	19 pop.; Poland, Germany, Hungary, Turkey	Mejnartowicz (1979)
P. sylvestris	9[j]	100	–	0.303	99	–	3 pop.; Sweden	Gullberg, Yazdani & Rudin (1982)
P. sylvestris	11[i]	100	91	0.310	98.3	–	9 pop.; Sweden	Gullberg et al. (1985)

Species	Loci	%P[b]		N_m[e]	$1-G_{st}$[d]	G_{st}[c]	Location	Reference
P. sylvestris	16	84	—		97.2	0.309	14 pop.; Scotland	Kinloch, Westfall & Forrest (1986)
P. sylvestris	13[i]	100	—		98.3	0.283	3 pop.; Sweden	Muona & Harju (1989)
P. sylvestris	14	—	66.3		92.2	0.211	7 pop., including 3 varieties; Sweden, China	Wang, Szmidt & Lindgren (1991)
P. sylvestris	14	—	65.3	8.7	98.9[n]	0.235[m]	16 pop., including 4 varieties; Sweden, Turkey, China	Szmidt & Wang (1993)
P. sylvestris	11	90.9	—		—	0.271	3 pop.; Switzerland	Neet-Sarqueda (1994)
P. sylvestris	21	82.5[g]	69.6		96.7	0.273	18 pop.; Latvia, Ukraine, Russia	Goncharenko, Silin & Padutov (1994)
P. sylvestris	8	95	—		92.4	0.356	13 pop.; eastern Europe, Turkey	Prus-Glowacki & Bernard (1994)
P. sylvestris	7	97	—		98.3	0.344	5 pop.; 4 sampled in provenance tests and 1 natural stand; Poland	Szweykowski, Prus-Glowacki & Hrynkiewicz (1994)
P. tabuliformis	13	—	53.8		—	0.195	1 pop.; China	Wang et al. (1990)
P. taeda	10	100	90		—	0.362	1 pop.; North Carolina	Conkle (1981)
P. taeda	25	96	80		—	0.282	90 selected clones; southeastern USA	Conkle (1981)
P. torreyana	59	0	0		0	—	2 pop.; rangewide	Ledig & Conkle (1983)
P. uncinata	11	84.4	—		—	0.266	7 pop.; Switzerland	Neet-Sarqueda (1994)
P. virginiana	2	50	—		97	0.590	2 pop.; Virginia	Witter & Feret (1978)
P. washoensis	23	58.7	—		98.4	0.148	3 pop.; Nevada, California	Niebling & Conkle (1990)
P. yunnanensis	13	—	46.2		—	0.169	1 pop.; China	Wang et al. (1990)

Notes:

[a] Although this table was intended to be complete, some references may have been missed, particularly those in Russian. Occasionally, authors present preliminary data in review publications and values differ slightly from those in the final research article because of rounding or a slight change in interpretation. Decisions were made to exclude some reports when it seemed apparent that data were redundant, i.e. that the same data set was repeated. Nevertheless, some items in the table are undoubtedly redundant in part. Published works based on theses were preferred to the theses themselves. Missing values, indicated by a dash, could sometimes be calculated from data in the reference cited, but this was not always attempted for this table.

[b] Mean percentage polymorphic loci within populations, based on two criteria, unless otherwise noted: (1) >0% = a locus was considered polymorphic if more than one allele was detected; 2) >5% = a locus was considered polymorphic if the most common allele was present in a frequency less than 95%.

[c] Mean expected heterozygosity within populations. Presentation in some reports was less clear than in others, and it was not always possible to decide whether H_e was the mean of within-population diversity or total diversity.

[d] Proportion of total genic diversity within populations, $1 - G_{st}$. G_{st} is the coefficient of gene differentiation (Nei 1975) and is almost identical to Wright's (1969) F_{st}. If the reference cited did not report G_{st}, $1 - F_{st}$ was tabulated.

[e] N_m is the number of migrants per generation (Slatkin 1985).

[f] If var. bahamensis is separated from var. hondurensis, $1 - G_{st} = 99$.

[g] A locus was considered polymorphic if the most common allele was present in a frequency less than 99%.

[h] From El-Kassaby (1991).

[i] Only polymorphic loci analysed.

[j] Possibly includes P. patula subsp. tecunumannii (now P. tecunumannii; Chap. 2, this volume).

[k] With possible P. patula subsp. tecunumannii omitted, $1 - G_{st} = 97.5$.

[l] Based on 19 loci.

[m] Calculated from varietal averages.

[n] Within varieties.

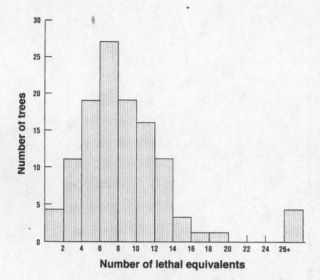

Fig. 13.9. Frequency distribution for number of embryonic lethal equivalents in *Pinus taeda*. After Franklin (1972).

allozyme variants are probably neutral or nearly so with respect to selection. However, they are good indicators of the level of variation throughout the pine genome.

Mutants that are obviously deleterious (i.e. recessive lethals such as pigment deficiencies) occur in high frequencies in pines; 133 of 712 *P. elliottii* were carriers (Snyder, Squillace & Hamaker 1966). In *P. taeda*, 22 different morphological mutants were found in the progeny of 30 of 119 trees by selfing (Franklin 1969). Conifers in general have a high genetic load (Ledig 1986, see table 2), measured as embryonic lethal equivalents. An embryonic lethal equivalent is defined as a true recessive lethal allele or recessive deleterious genes of such number that if dispersed in different embryos would cause, on average, one selective death (Morton, Crow & Muller 1956). An embryonic lethal equivalents can be calculated from the reduction in seed yield following self-pollination. *Pinus taeda*, which has one of the highest average heterozygosities reported in pines (Conkle 1981), has about 8.5 embryonic lethal equivalents per tree up to a maximum of nearly 26 in a sample of 116 trees (Fig. 13.9; Franklin 1972). These estimates are conservative because they fail to take into account polyembryony (Bramlett & Popham 1971). Each ovule produces several egg cells in conifers, and if pollination is adequate, all may be fertilized. Therefore, not every embryonic death results in loss of reproductive capacity. Moreover, the total genetic load is much higher than is indicated by embryonic lethal equivalents because additional mortality occurs after germination (Franklin 1970).

The true genetic loads for pines are, in general, much higher than those reported for animals or annual plants, in agreement with the high level of allozyme heterozygosity in pines. Embryonic lethal equivalents in pines average from

0.3 to 9.4 (generally 3–9) per zygote (Ledig 1986, table 2). In one review, Levin (1984) reported a range in lethal equivalents from 1.2 to 5.2 in agronomic plants and a similar spread, from 1 to 4, in humans, fruit flies and flour beetles.

Several factors may explain the high levels of variation in pines, including the mating system, dispersal and gene flow, spatial diversity of the habitat, temporal heterogeneity in conjunction with a long life cycle and, perhaps most importantly, mutation rate. The effect of life-history characteristics on mutation rate and the role of mutation in maintaining genetic variation have been given little recognition. In animals, the germ line is cut off at an early stage of embryogeny and undergoes few divisions before the sexual division that results in gamete production (Brewbaker 1964). Plants have no germ line; spore mother cells derive from vegetative cell lines. Therefore, many more cell divisions and many more opportunities for DNA-copying error intervene in plants than in animals. The more cell divisions, the higher the mutation rate, as observed in human males. Mutation rate in males increases linearly as a function of age because with ageing, more cell divisions (30 divisions at puberty, 380 divisions at age 28, 540 divisions at age 35) intervene in sperm formation. Conversely, mutation rate is relatively constant in human females where divisions leading to the eggs are completed early in development.

Mutations may accumulate with age even in the absence of cell division; the viability load transmitted to progeny was greater in old fruit flies than in young ones (Andjelkovic *et al.* 1979). Mutations accumulate in stored seed even in the absence of division and at low rates of metabolic activity (Harrington 1970). The great longevity of pines (centuries to millennia) provides the opportunity for the accumulation of many new mutations during their lifetime.

Unless there are offsetting differences in the rate of mutation per cell division, more mutations might accumulate per generation in pines than in herbaceous perennials, and more in herbaceous perennials than in annuals. Mutation rate per year may be the same in annuals and trees, but the impact on individual fitness or reproductive capacity will be much greater in trees because each tree will have accumulated more mutations than an annual. The number of divisions between zygote and gamete can be estimated in pines by assuming that four cells result from two divisions of each cell cut off by the apical meristem, and that the length of the wood cells, the primary tracheids, is 2 mm. Thus, a gamete produced in a cone at the top of a 12 m pine represents 1500 divisions. Assuming a mutation rate of 10^{-8} per cell division, 1.5 new mutants would be expected in every 10^5 gene loci. For mammals, the same rate per cell division would result in only five mutations in 10^7 gene loci: that is, the rate is 30 times greater in pine. For a 24 m pine, the rate should be double that in a 12 m tree. A sizeable proportion of new mutations will be deleterious, so long-lived pines will have a tremendous genetic load compared with mammals.

Several types of evidence suggest that mutational load may explain the high genic diversity in pines (and other forest trees) and may be an important force in shaping the mating system. One line of evidence is a relationship recognized by M.T. Conkle (personal communication 1986) between genic diversity and self-fertility. Conifers with the highest heterozygosity, such as *Pseudotsuga menziesii*, suffer the most depression in seed yield due to embryo abortion when selfed. At the opposite extreme is *P. resinosa*, which has almost no detectable variation (Fowler & Morris 1977), and can be selfed with no reduction in seed set (Fowler 1965a). A second line of evidence is Wiens's (1984) observation that ovule abortion was highest in woody perennials, somewhat lower in herbaceous perennials, and lowest in annuals, which suggests that the genetic load increases with longevity or size.

Though direct evidence is scant, it supports the hypothesis that in large, long-lived, woody species mutation rates are indeed higher than those in small, annual plants. In two populations of red mangrove (*Rhizophora mangle* L.), the rate of mutation to chlorophyll-deficient lethal recessives was 5.9×10^{-3} and 7.4×10^{-3} per haploid genome per generation (Klekowski & Godfrey 1989). By comparison, analogous mutation rates for chlorophyll-deficient lethals in annual, herbaceous plants range from 3.1×10^{-4} to 5.6×10^{-5}, one to two orders of magnitude lower.

In summary, genic diversity in pines is, on the average, high, as revealed by allozymes and morphological variation. However, the existence of genetically depauperate species, like *P. resinosa* and *P. torreyana*, proves that genic diversity is not necessary for survival. It is significant that the true genetic load is also high in pines. The several mechanisms that promote outbreeding and favour gene flow in tree species might explain the maintenance of genic diversity within populations, but they are not the 'cause' of the diversity. Selection in spatially and temporally heterogeneous habitats, followed by wide dispersal, may be a contributing factor. However, for many loci the explanation may not be selection-dependent; because of the nature of plant growth, germ cells descend from vegetative cell lines, providing many more opportunities for accumulation of mutations in plants than in animals. Opportunities for the accumulation of mutations are greatest in large, long-lived, tree species, such as pines, and mutational load will drive the mating system towards outcrossing, which will protect all forms of genetic variation, which in turn magnifies the effect of inbreeding, driving the system further towards outcrossing, and so on.

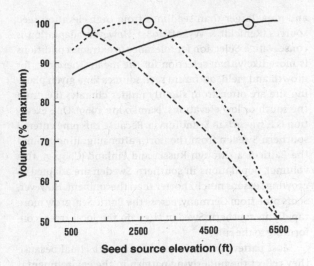

Fig. 13.10. Mean volume per tree (as a percentage of maximum) of 50-year-old *Pinus ponderosa* from a range of elevations when grown in uniform gardens at 960 ft (293 m; broken line), 2 730 ft (832 m; dotted line), and 5 650 ft (1722m; solid line) in the Sierra Nevada of California (M.T. Conkle, unpublished data).

13.3 Patterns of variation in pines

Common garden studies (usually called provenance tests or seed source tests in the forestry literature) have shown pronounced patterns of geographic variation in most pine species that have been investigated. However, many species have not been studied, particularly those of minor commercial value. Information is particularly scarce on the pines of Mexico, which represent half the world's species.

In common garden studies, the local provenance usually seems to be among the best adapted (i.e. has the fastest growth rate and often the best survival). For example, volume at 12 years of age was negatively correlated with distance from the planting site in *P. rigida* (Kuser & Ledig 1987). In the classic study of *P. ponderosa* along an elevational gradient (Conkle 1973), provenances from a 105 km transect that ranged in elevation from 38 m to 2109 m performed best when planted at an elevation close to that of their origin (Fig. 13.10).

Another common trend is that trees from southern sources grow more rapidly than those from northern ones (e.g. Wells & Wakeley 1966; Ledig, Lambeth & Linzer 1976; Williams & Funk 1978) and those from an elevation slightly lower than the planting site perform as well as the local provenance (Conkle 1973). Populations from southern latitudes or low elevations are usually adapted to longer growing seasons than those from northern latitudes or high elevations. The effects of elevation are especially striking. In a common environment, seedlings from low-elevation seed sources of *Pinus contorta* grew for a longer period and grew larger than seedlings from high-elevation seed sources (Rehfeldt & Wykoff 1981). However, adaptation is conservative. Selection for tolerance to extreme conditions is more likely than selection for the mean. Therefore, for growth and yield, optimum seed sources for a given planting site are often from slightly milder climates (i.e. from the south or low elevations; Namkoong 1969). One exception is *P. sylvestris* in Scandinavia. Because this pine entered southern Sweden from the north after migrating around the Baltic Sea through Russia and Finland (Chap. 5, this volume), populations in southern Sweden are adapted to growing seasons much shorter than the ambient. However, Scots pine from Germany, across the Baltic Sea, grow more rapidly in southern Sweden than do the local trees, conforming to the rule.

Most patterns of variation in pines are clinal because they reflect the underlying pattern in the environmental factors that exert selection pressure (Langlet 1959). Temperature, precipitation, length of the growing season, etc. are continuously distributed variables. Dry-matter content of needles in *P. sylvestris* provenances is correlated with photoperiod on the first day of the year with an average normal temperature of 6 °C at their origin (Langlet 1959). As mentioned above, many growth traits (height, diameter, volume) vary continuously with latitude or elevation, surrogates for length of the growing season and correlated climatic variables (Wells & Wakeley 1966; Kuser & Ledig 1987).

Clines in gene frequency are also known. For example, in *P. lambertiana* the frequency for a major dominant gene for resistance to white pine blister rust is near zero in northwestern California, increases gradually to a frequency of 8.2% in the southern Sierra Nevada, and then declines again to zero in Baja California (Kinloch 1992).

Clinal patterns of variation may also arise because of gene flow resulting from pollen dispersal. Pollen flow can create clines between populations in discrete pockets of selection or between previously isolated populations following hybridization at contact points. High fire frequency in the Pine Plains (pockets of pygmy forest in New Jersey) has selected for *P. rigida* that are serotinous-coned, an obvious adaptation to fire (Ledig & Little 1979; Chap. 12, this volume). A cline in cone serotiny, resulting in part from pollen flow, extends out in all directions from the Pine Plains into the surrounding Pine Barrens (Ledig & Fryer 1972), but especially downwind (Givnish 1981). The blue and green races of *P. muricata* differ in gene frequency for a locus that controls allozymes of glutamate-oxaloacetate transaminase (GOT). Where the two races hybridize along an interface at Sea Ranch, California, a steep cline less than 3 km wide results (Millar 1983).

Ecotypic patterns of variation are rarer than clinal patterns, but have been noted where the underlying environ-

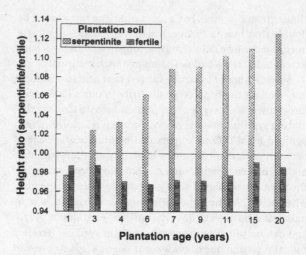

Fig. 13.11. **When grown on serpentinite, the height of progeny from *P. ponderosa* trees native to infertile, serpentinite soil increases relative to the height of progeny from trees native to fertile soil for at least the first twenty years after planting. Conversely, when grown on fertile soil, progeny from trees native to fertile soil maintain a slight advantage in height. Relative performance is shown by histograms of the ratio for height of 'serpentinite' progeny/height of 'fertile' progeny (J.L. Jenkinson, unpubl. data).**

mental factors are sharply discontinuous. Soil type or substrate is the variable that most often shows abrupt discontinuity. *Pinus ponderosa* has adapted to pockets of serpentine soil in California. Progeny from parents growing on infertile serpentinite were compared with those from fertile soils only 3 kilometres away. When grown on infertile serpentinite, the progeny of serpentinite origin had 50% greater volume at 20 years than the progeny from parents native to fertile soils (J.L. Jenkinson, personal communication 1995). The difference in height and volume increased gradually from year 1 to year 20; the difference in height increased by about 0.5% each year (Fig. 13.11). When the two edaphic ecotypes were compared on fertile soil of metasedimentary origin, progeny originating from that soil exceeded in height those of serpentinite origin, although the difference was not great.

Genetically distinct populations of 'blue' *P. muricata* in northern California have evolved on the soils of ancient Pleistocene beaches. These soils have extremely low pH, podzolized iron hardpans, and both mineral toxicities and mineral deficiencies. Genetic integrity is maintained even though the blue *P. muricata* are embedded in a matrix of populations adapted to more fertile soils (Millar 1989). Like the interfaces between serpentine and non-serpentine soils, the interfaces between fertile and infertile soils on the ancient beach terraces are very abrupt.

Both clinal patterns and ecotypic patterns of variation may occur within the same species. For example, north–south clines are obvious in *P. taeda*, but a break also occurs between populations east and west of the

Mississippi River. The east–west discontinuity was probably established by selection or drift during the Pleistocene, when the species' distribution was split by the DeSoto Canyon (Wells, Switzer & Schmidtling 1991). More complex patterns are also known. The latitudinal cline in growth of *Pinus monticola* was very steep in the central and southern Cascade Mountains of Oregon and California, USA, but little differentiation was obvious north or south of this region (Rehfeldt, Hoff & Steinhoff 1984). Explanations for such complex patterns will probably depend on better knowledge of migration patterns and refugial history.

Migration history as well as selection may influence the patterns of variation within pine species, e.g. in *P. contorta*, referred to above (Cwynar & MacDonald 1987). In *P. banksiana*, abrupt boundaries between regions seem to have been established because of differences in migration history. After five years in a common garden, *P. banksiana* trees from upper Michigan were 20% shorter than those from lower Michigan even though the two ecotypes were separated only by the 7 km wide Straits of Mackinac (Arend *et al*. 1961). Apparently, this species migrated from two glacial refugia. Trees from the eastern refugium dispersed into lower Michigan, between Lakes Michigan and Huron, while the western refugium was the source of the trees that dispersed into Upper Michigan northwest of Lake Michigan and south of Lake Superior. Other sharp boundaries in *P. banksiana* were detected in Ontario, Canada, and interpreted with respect to postglacial migration (Skeates 1979).

Genetic variation within populations is generally high. Usually, above 90% of the variation detected at isozyme loci is within populations. Variation for quantitative traits is also high. A recent review for forest trees found that median estimates of heritability (the ratio of additive genetic variance to total variance) were 0.18 for tree volume, 0.19 for diameter, and 0.25 for height (Cornelius 1994). These medians include estimates from a variety of conifers and hardwoods, but 48% (height) to 83% (volume) of the studies reviewed were based on pines and, therefore, the results are highly indicative of heritabilities in this group. Wood specific gravity had an even higher heritability; a median of 0.48. Stem straightness, branching, morphological and structural traits had median heritabilities slightly higher than those for growth traits, with medians ranging from 0.23 to 0.26.

In summary, geographic patterns of variation in pines can be either clinal or ecotypic. Clinal patterns predominate because environmental variables are usually continuously distributed and because dispersal mechanisms and gene flow through pollen flight are highly effective in pines. However, discontinuous, or ecotypic, patterns are known and are maintained by intensive selection or because insufficient time has elapsed to erase the effects of past isolation.

13.4 Domestication

Some pine species are in an incipient stage of domestication. Even in the short time since attempts at domestication began (usually less than 50 years), breeders have made impressive changes in the pine phenotypes. *Pinus radiata* is native to about 6000 hectares in California, USA and Cedros and Guadalupe Islands off Baja California, Mexico, but has already been planted on *c.* 4×10^6 ha outside its range, more than 95% of this in New Zealand, Chile, Australia and Spain (Rogers & Ledig 1996; Chap. 21, this volume). Realized gains from breeding this species are large. The volume of improved *P. radiata* in Australia was 9–29% greater than that of controls, depending on test conditions, 10–12 years after planting (Eldridge & Matheson 1983). Gains were also apparent in straightness and branching characteristics (see Chap. 21, this volume).

Landraces have developed in pine species introduced to new, exotic environments. Even casual introductions have resulted in landraces after one or two generations. For example, in Australia, progeny of early introductions of *P. ponderosa* and *P. radiata* were superior to any provenances introduced later (Eldridge 1974), presumably because of natural selection for adaptation to Australian conditions. Several other examples, including *P. elliottii*, were reported in South Africa (Marsh 1970). Natural selection has improved adaptation, relative to the initial introductions, in only one generation. The restriction to gene flow created when a species is introduced makes it simple to establish and maintain genetic integrity in an evolving landrace.

Much of the increase in yield of agricultural crops in historical times has been due to selection for adaptation to the new environments into which they were transported (Evans 1980). The major adjustment was to modify phenological responses. It will be very easy to modify phenological responses such as date of initiation or cessation of terminal growth in pines; heritabilities for phenological characteristics are moderate to high in pines and other conifers (Smith, White & Hodge 1993; Kaya & Temerit 1994), and date of terminal growth cessation or terminal bud formation is often correlated with latitude of origin in common garden studies, e.g. in *P. strobus* (Santamour 1960; Genys 1968) and *P. elliottii* (Bengston, McGregor & Squillace 1967). However, similar correlations are difficult to detect in species like *P. ponderosa*, native to complex topography that distorts regional climatic patterns (Squillace & Silen 1962).

Domestication will result in a loss of diversity within

breeding populations. The action of sampling error (drift), expressed as a loss of some of the rare alleles found in wild populations, is already observed in breeding populations (Yow *et al.* 1992). Moreover, gains through breeding are achieved by changing gene frequencies, increasing the frequency of 'favourable' alleles and reducing the frequency of alternative alleles or eliminating them entirely. However, the among-population component of genetic variation will increase as breeding populations are pushed in different directions, and total diversity will increase.

13.5 Conservation

Conservation of genetic resources is, or should be, an integral part of any breeding programme. For commercially important pine species, genetic resources are conserved in several forms: seed banks, clone banks, progeny tests, and seed orchards. For example, the North Carolina State-Industry Cooperative Tree Improvement programme had preserved over 8000 selections of *P. taeda* within its first two decades of operation (McConnell 1980). The expansion of some species through plantation programmes has itself probably enriched the genetic resources of pines.

Nevertheless, three areas of concern are obvious: (1) non-commercial species are not always adequately protected; (2) populations, even of commercial species, are often subject to loss; (3) in some countries, deforestation and lack of adequate programmes for reforestation threaten extensive loss of genetic resources.

Even in a highly developed country like the United States, no programme for the conservation of forest genetic resources exists. Non-commercial species may or may not be adequately protected in national parks, national forests, and other reserved areas. Furthermore, the US Endangered Species Act of 1973 protects plant taxa, but not populations. A species could conceivably be reduced to a single population before it was classed as endangered and received protection under the Act. Any genetic diversity among populations would be lost. Low-elevation populations of several species (e.g. *P. ponderosa* in California) were lost as a result of destructive logging a century ago (Shoup 1981). The same scenario has been repeated through the millennia and continues today (Ledig 1992). These are the 'secret extinctions' (Ledig 1993).

Mexico should be an especially important focus for conservation. It is a secondary centre of speciation for the pines. Almost half of the 111 species of pine recognized by Price *et al.* (Chap. 2, this volume) are native to Mexico (Chap. 7, this volume). Yet Mexico is ranked fourth in the world in rate of deforestation; current net annual deforestation rate conservatively averages 1.3% (Cairns, Dirzo & Zadroga

1995). The danger to low-elevation populations of pines has been recognized for at least three decades (Jasso 1970). Fortunately, two organizations are active in collection and *ex situ* conservation of forest genetic resources in Mexico – the Central America and Mexico Coniferous Resources Cooperative (CAMCORE) and the Oxford Forestry Institute (Barnes & Burley 1990; Dvorak 1990). However, the job is immense, and for Mexico, pines are only one element of an endangered flora that includes 2000–3000 tree species (J. Rzedowski, personal communication 1989).

13.6 Conclusions

The genetic system of pines, in general, favours the creation and recombination of genetic variation. The creation of genetic variation, or mutation, is probably higher in long-lived woody perennials than in any other organisms because of the opportunity for the accumulation of somatic mutations in cell lines that give rise to reproductive structures.

The release and recombination of this variation is favoured by a high chromosome number ($n = 12$) and chiasma frequency (2.4). The modal chromosome number in diploid annual plants is apparently about 7 (Darlington & Wylie 1955). Furthermore, these chromosomes are relatively short (with several exceptions such as *Tradescantia* spp. and *Trillium* spp.), suggesting that the number of chiasmata and, therefore, the frequency of crossover is less than that in pines. Monoecy and separation of sexes within the crown of pines also favour recombination by promoting outcrossing. The lack of complete barriers to hybridization facilitates recombination and replenishment of genetic variation via interspecific gene flow.

Thus, it is not surprising that pines are among the most genetically diverse of organisms. Diversity has permitted pines to evolve in concert with environmental change over the 200 million years since they diverged from ancestral forms (Millar 1993). However, diversity has its negative aspects; most mutations are deleterious and recombinants may be so unlike their parents that many are maladapted, especially in a constant environment. These effects are offset by the high fecundity of pines.

Seed dispersal varies among species, as can be anticipated by the wide range in seed weights – from about 900 to 300 000 seeds kg^{-1}, a difference of over 300-fold. Furthermore, some species are wind-dispersed and others animal-dispersed, mainly by birds. Both wind- and bird-dispersed species are capable of being transported long distances. The best direct evidence is (1) the occurrence of colonists at some distance from planted stands in countries where pines have been introduced, and (2) observations of birds that cache seeds many kilometres from the source. Indirect evidence is provided by the rapid rates at which pines migrated northwards during the current interglacial and by patterns of genetic diversity that are best explained as 'foot prints' of founder effects left by long-distance colonists. Because pines lack self-incompatibility systems, facultative selfing is possible and they can colonize new habitat after long-distance dispersal. Genetic drift is likely to have influenced patterns of genetic variation in new colonies.

Pollen output in pines can be tremendous and wind-dispersed pollen can cover great distances. Therefore, gene flow can counteract drift in all but the most extreme colonization events and maintain genetic variation at levels characteristic of the species. The homogenizing effect of gene flow can be seen in the distribution of isozyme alleles in pines, the high estimates of migration, and the low values for population differentiation.

Because of effective gene flow, it should be anticipated that most variation in pines will be clinal. However, ecotypic variation is also known, usually a response to edaphic factors. Extreme selection pressures can be effective in establishing and maintaining ecotypic differentiation because of the high fecundity of pines. Differences among populations have evolved over distances as short as a few kilometres.

Because of their high diversity, many species of pine respond rapidly to selection and are easily domesticated (and may, incidentally, be aggressive invaders). Tree breeders have barely scratched the surface in their efforts to modify pines for human goods and services. Great gains are possible, but only if genetic diversity is conserved. Conservation of commercially important species will be nearly automatic, but that is not the case for non-commercial species. Yet, judging from the agricultural model, non-commercial species will eventually be valuable in breeding. They are already valuable in their own right for ecological and aesthetic reasons. Conservation efforts are sorely needed for many Asian pines and for the especially rich genetic resources of the Mexican pines.

References

Allendorf, F.W., Knudsen, K.W. & Blake, G.M. (1982). Frequencies of null alleles at enzyme loci in natural populations of ponderosa and red pine. *Genetics*, **100**, 497–504.

Andjelkovic, M., Marinkovic, D., Tucic, N. & Tosic, M. (1979). Age-affected changes in viability and longevity loads of *Drosophila melanogaster*. *American Naturalist*, **114**, 915–20.

Andresen, J.W. (1963). Germination characteristics of *Pinus rigida* seed borne in serotinous cones. *Brotéria, Serie de Ciências Naturais*, **32**, 151–78.

Arend, J.L., Smith, N.F., Spurr, S.H. & Wright, J.W. (1961). Jack pine geographic variation – five-year results from lower Michigan tests. *Michigan Academy of Science, Arts, and Letters*, **46**, 219–38.

Bannister, M.H. (1965). Variation in the breeding system of *Pinus radiata*. In *The Genetics of Colonizing Species*, ed. H.G. Baker & G.L. Stebbins, pp. 353–72. New York: Academic Press.

Barnes, B.V., Bingham, R.T. & Squillace, A.E. (1962). Selective fertilization in *Pinus monticola* Dougl. II. Results of additional tests. *Silvae Genetica*, **11**, 103–11.

Barnes, R.D. & Burley, J. (1990). Tropical forest genetics at the Oxford Forestry Institute: exploration, evaluation, utilization and conservation of genetic resources. *Forest Ecology and Management*, **35**, 159–69.

Beaulieu, J. & Simon, J.-P. (1994). Genetic structure and variability in *Pinus strobus* in Quebec. *Canadian Journal of Forest Research*, **24**, 1726–33.

Bengston, G.W., McGregor, W.H.D. & Squillace, A.E. (1967). Phenology of terminal growth in slash pine: some differences related to geographic seed source. *Forest Science*, **13**, 402–12.

Bermejo, B.V. (1993). Genetic diversity and the mating system in *Pinus engelmannii* Carr. PhD thesis, Madison: University of Wisconsin, Madison.

Betancourt, J.L., Schuster, W.S., Mitton, J.B. & Anderson, R.S. (1991). Fossil and genetic history of a pinyon pine (*Pinus edulis*) isolate. *Ecology*, **72**, 1685–97.

Bingham, R.T., Hoff, R.J. & Steinhoff, R.J. (1972). *Genetics of Western White Pine*. USDA Forest Service Research Paper WO-12.

Bonnet-Masimbert, M. & Bikay-Bikay, V. (1978). Variabilité intraspécifique des isozymes de la glutamate-oxaloacetate-transaminase chez *Pinus nigra* Arnold intérêt pour la taxonomie des sous espèces. *Silvae Genetica*, **27**, 71–9.

Boscherini, G., Morgante, M., Rossi, P. & Vendramin, G.G. (1994). Allozyme and chloroplast DNA variation in Italian and Greek populations of *Pinus leucodermis*. *Heredity*, **73**, 284–90.

Boyer, W.D. (1958). Longleaf pine seed dispersal in south Alabama. *Journal of Forestry*, **56**, 265–8.

Bramlett, D.L. & Popham, T.W. (1971). Model relating unsound seed and embryonic lethal alleles in self-pollinated pines. *Silvae Genetica*, **20**, 192–3.

Brewbaker, J.L. (1964). *Agricultural Genetics*. Englewood Cliffs, NJ: Prentice-Hall.

Buchholz, J.T. (1929). The embryogeny of the conifers. *Proceedings of the International Congress of Plant Sciences*, **1**, 359–92.

Buell, M.F. (1947). Mass dissemination of pine pollen. *Journal of the Elisha Mitchell Scientific Society*, **63**, 163–7.

Burns, R.M. & Honkala, B.H. (eds.) (1990). *Silvics of North America*. Vol. 1. *Conifers*. USDA Forest Service. Agriculture Handbook 654. Washington, DC: USDA Forest Service.

Cairns, M.A., Dirzo, R. & Zadroga, F. (1995). Forests of Mexico: a diminishing resource? *Journal of Forestry*, **93**(7), 21–4.

Camefort, H. (1966). Observations sur les mitochondries et les plastes d'origine pollinique après leur entrée dans une oosphère chez le Pin noir (*Pinus laricio* Poir. var. *austriaca=Pinus nigra* Arn.). *Comptes Rendus de l'Académie des Science de Paris, Série D*, **263**, 959–62.

Chaisurisri, K., Mitton, J.B. & El-Kassaby, Y.A. (1994). Variation in the mating system of Sitka spruce (*Picea sitchensis*): evidence for partial assortative mating. *American Journal of Botany*, **81**, 1410–15.

Chapman, H.H. (1922). A new hybrid pine (*Pinus palustris* × *Pinus taeda*). *Journal of Forestry*, **20**, 729–34.

Cheliak, W.M., Dancik, B.P., Morgan, K., Yeh, F.C.H. & Strobeck, C. (1985). Temporal variation of the mating system in a natural population of jack pine. *Genetics*, **109**, 569–84.

Clements, F.E. (1910). *The Life History of Lodgepole Burn Forests*. USDA Forest Service Bulletin 79. Washington, DC: USDA Forest Service.

Coles, J.F. & Fowler, D.P. (1976). Inbreeding in neighboring trees in two white spruce populations. *Silvae Genetica*, **25**, 29–34.

Colwell, R.N. (1951). The use of radioactive isotopes in determining spore distribution patterns. *American Journal of Botany*, **38**, 511–23.

Conkle, M.T. (1973). Growth data for 29 years from the California elevational transect study of ponderosa pine. *Forest Science*, **19**, 31–9.

Conkle, M.T. (1981). Isozyme variation and linkage in six conifer species. In *Proceedings of a Symposium on Isozymes of North American Forest Trees and Forest Insects*, ed. M.T. Conkle, pp. 11–17. USDA Forest Service General Technical Report PSW-48.

Conkle, M.T., Hodgskiss, P.D., Nunnally, L.B. & Hunter, S.C. (1982). *Starch Gel Electrophoresis of Conifer Seeds: A Laboratory Manual*. USDA Forest Service General Technical Report PSW-64.

Conkle, M.T., Schiller, G. & Grunwald, C. (1988). Electrophoretic analysis of diversity and phylogeny of *Pinus brutia* and closely related taxa. *Systematic Botany*, **13**, 411–24.

Cornelius, J. (1994). Heritabilities and additive genetic coefficients of variation in forest trees.

Canadian Journal of Forest Research, **24**, 372–9.

Critchfield, W.B. (1975). Interspecific hybridization in *Pinus*: a summary review. In *Proceedings of the Fourteenth Meeting of the Canadian Tree Improvement Association: Part 2. Symposium on Interspecific and interprovenance Hybridization in Forest Trees*, ed. D.P. Fowler & C.W. Yeatman, pp. 99–105. Fredericton, New Brunswick: Canadian Tree Improvement Association.

Critchfield, W.B. (1980). *Genetics of Lodgepole Pine*. USDA Forest Service Research Paper WO-37.

Cwynar, L.C. & MacDonald, G.M. (1987). Geographical variation of lodgepole pine in relation to population history. *American Naturalist*, **129**, 463–9.

Dancik, B.P. & Yeh, F.C. (1983). Allozyme variability and evolution of lodgepole pine (*Pinus contorta* var. *latifolia*) and jack pine (*P. banksiana*) in Alberta. *Canadian Journal of Genetics and Cytology*, **25**, 57–64.

Danzmann, R.G. & Buchert, G.P. (1983). Isozyme variability in Central Ontario jack pine. In *Proceedings of the Twenty-eighth Northeastern Forest Tree Improvement Conference*, ed. R.T. Eckert, pp. 232–48. Durham, NH: Institute of Natural and Environmental Resources, University of New Hampshire.

Darlington, C.D. & Wylie, A.P. (1955). *Chromosome Atlas of Flowering Plants*, 2nd edn. London: George Allen and Unwin.

Davis, J.T. (1974). *Trade Routes and Economic Exchange Among the Indians of California*. Ramona, California: Ballena Press.

Dhillon, S.S. (1980). Nuclear volume, chromosome size and DNA content relationships in three species of *Pinus*. *Cytologia*, **45**, 555–60.

Dogra, P.D. (1967). Seed sterility and disturbances in embryogeny in conifers with particular reference to seed testing and tree breeding in Pinaceae. *Studia Forestalia Suecica* 45. Stockholm, Sweden: Royal College of Forestry.

Donahue, J.K. & Mar L.C. (1995). Observations on *Pinus maximartinezii* Rzed. *Madroño*, **42**, 19–25.

Duba, S.E. (1985). Polymorphic isozymes from megagametophytes and pollen of longleaf pine: characterization, inheritance, and use in analyses of genetic variation and genotype

verification. In *Proceedings of the Eighteenth Southern Forest Tree Improvement Conference*, pp. 88–98. Long Beach, Mississippi: Southern Forest Tree Improvement Committee.

Dvorak, W.S. (1990). CAMCORE: industry and governments' efforts to conserve threatened forest species in Guatemala, Honduras and Mexico. *Forest Ecology and Management*, **35**, 151–7.

Eckert, R.T., Joly, R.J. & Neale, D.B. (1981). Genetics of isozyme variants and linkage relationships among allozyme loci in 35 eastern white pine clones. *Canadian Journal of Forest Research*, **11**, 573–9.

Edwards, M.A. & Hamrick, J.L. (1995). Genetic variation in shortleaf pine, *Pinus echinata* Mill. (Pinaceae). *Forest Genetics*, **2**, 21–8.

Eldridge, K.G. (1974). Forest tree improvement in Australia. In *Forest Tree Breeding in the World*, ed. R. Toda, pp. 170–80. Tokyo, Japan: Government Forest Experiment Station of Japan.

Eldridge, K.G. & Matheson, A.C. (1983). Assessing gain. In *Radiata Pine Breeding Manual*, ed. A.C. Matheson & A.G. Brown, pp. 10.1–10.6. Canberra: CSIRO Division of Forest Research.

El-Kassaby, Y.A. (1991). Genetic variation within and among conifer populations: review and evaluation of methods. In *Biochemical Markers in the Population Genetics of Forest Trees*, ed. S. Fineschi, M.E. Malvolti, F. Cannata & H.H. Hattemer, pp. 61–76. The Hague: SPB Academic Publishing.

El-Kassaby, Y.A., Meagher, M.D., Parkinson, J. & Portlock, F.T. (1987). Allozyme inheritance, heterozygosity and outcrossing rate among *Pinus monticola* near Ladysmith, British Columbia. *Heredity*, **58**, 173–81.

Epperson, B.K. & Allard, R.W. (1984). Allozyme analysis of the mating system in lodgepole pine populations. *Journal of Heredity*, **75**, 212–4.

Evans, L.T. (1980). The natural history of crop yield. *American Scientist*, **68**, 388–97.

Ferguson, M.C. (1904). Contributions to the knowledge of the life history of *Pinus* with special reference to sporogenesis, the development of the gametophytes and fertilization. *Proceedings of the Washington Academy of Sciences*, **6**, 1–202.

Fielding, J.M. (1960). Branching and flowering characteristics of Monterey pine. Forestry and Timber Bureau

Bulletin No. 37. Canberra: Forestry and Timber Bureau,

Fins, L. & Libby, W.J. (1982). Population variation in *Sequoiadendron*: seed and seedling studies, vegetative propagation, and isozyme variation. *Silvae Genetica*, **31**, 102–10.

Fowler, D.P. (1964a). Pre-germination selection against a deleterious mutant in red pine. *Forest Science*, **10**, 335–6.

Fowler, D.P. (1964b). Effects of inbreeding in red pine, *Pinus resinosa* Ait. *Silvae Genetica*, **13**, 170–7.

Fowler, D.P. (1965a). Effects of inbreeding in red pine, *Pinus resinosa* Ait. II. Pollination studies. *Silvae Genetica*, **14**, 12–23.

Fowler, D.P. (1965b). Effects of inbreeding in red pine, *Pinus resinosa* Ait. III. Factors affecting natural selfing. *Silvae Genetica*, **14**, 37–46.

Fowler, D.P. (1965c). Natural self-fertilization in three jack pines and its implications in seed orchard management. *Forest Science*, **11**, 56–8.

Fowler, D.P. & Lester, D.T. (1970). *Genetics of Red Pine*. USDA Forest Service Research Paper WO-8.

Fowler, D.P. & Morris, R.W. (1977). Genetic diversity in red pine: evidence for low genic heterozygosity. *Canadian Journal of Forest Research*, **7**, 343–7.

Franklin, E.C. (1969). Mutant forms found by self-pollination in loblolly pine. *Journal of Heredity*, **60**, 315–20.

Franklin, E.C. (1970). *Survey of Mutant Forms and Inbreeding Depression in Species of the Family Pinaceae*. USDA Forest Service Research Paper SE-61.

Franklin, E.C. (1972). Genetic load in loblolly pine. *American Naturalist*, **106**, 262–5.

Furnier, G.R. & Adams, W.T. (1986a). Mating system in natural populations of Jeffrey pine. *American Journal of Botany*, **73**, 1002–8.

Furnier, G.R. & Adams, W.T. (1986b). Geographic patterns of allozyme variation in Jeffrey pine. *American Journal of Botany*, **73**, 1009–15.

Furnier, G.R., Knowles, P., Clyde, M.A. & Dancik, B.P. (1987). Effects of avian seed dispersal on the genetic structure of whitebark pine populations. *Evolution*, **41**, 607–12.

Gauthier, S., Simon, J.-P. & Bergeron, Y. (1992). Genetic structure and variability in jack pine populations: effects of insularity. *Canadian Journal of Forest Research*, **22**, 1958–65.

Genys, J.B. (1968). Geographic variation in eastern white pine. *Silvae Genetica*, **17**, 6–12.

Gibson, J.P. & Hamrick, J.L. (1991a). Genetic diversity and structure in *Pinus pungens* (Table Mountain pine) populations. *Canadian Journal of Forest Research*, **21**, 635–42.

Gibson, J.P. & Hamrick, J.L. (1991b). Heterogeneity in pollen allele frequencies among cones, whorls, and trees of Table Mountain pine (*Pinus pungens*). *American Journal of Botany*, **78**, 1244–51.

Givnish, T.J. (1981). Serotiny, geography, and fire in the Pine Barrens of New Jersey. *Evolution*, **35**, 101–23.

Goncharenko, G.G., Padutov, V.E., Krutovskii, K.V., Podzharova, Z.S., Kirgizov, N.Y. & Politov, D.V. (1988). Level of genetic variability in *Pinus sibirica* in Altai. [Translated from Doklady Biological Sciences, Proceedings of the Academy of Sciences of the USSR, **299**, 139–41.]

Goncharenko, G.G., Padutov, V.E. & Silin, A.E. (1992). Population structure, gene diversity, and differentiation in natural populations of cedar pines (*Pinus* subsect. *Cembrae*, *Pinaceae*) in the USSR. *Plant Systematics and Evolution*, **182**, 121–34.

Goncharenko, G.G., Padutov, V.E. & Silin, A.E. (1993a). Allozyme variation in natural populations of Eurasian pines. I. Population structure, genetic variation, and differentiation in *Pinus pumila* (Pall.) Regel from Chukotsk and Sakhalin. *Silvae Genetica*, **42**, 237–46.

Goncharenko, G.G., Padutov, V.E. & Silin, A.E. (1993b). Allozyme variation in natural populations of Eurasian pines. II. Genetic variation, diversity, differentiation, and gene flow in *Pinus sibirica* Du Tour in some lowland and mountain populations. *Silvae Genetica*, **42**, 246–53.

Goncharenko, G.G., Silin, A.E. & Padutov, V.E. (1994). Allozyme variation in natural populations of Eurasian pines. III. Population structure, diversity, differentiation and gene flow in central and isolated populations of *Pinus sylvestris* L. in eastern Europe and Siberia. *Silvae Genetica*, **43**, 119–32.

Grattapaglia, D. (1994). Eucalypts catch-up with the genome research wave. *Dendrome: Forest Tree Genome Research Updates*, **1**(1), 5–7.

Groover, A.T., Williams, C.G., Devey, M.E.,

Lee, J.M. & Neale, D.B. (1995). Sex-related differences in meiotic recombination frequency in *Pinus taeda*. *Journal of Heredity*, **86**, 157–8.

Grunwald, C., Schiller, G. & Conkle, M.T. (1986). Isozyme variation among native stands and plantations of Aleppo pine in Israel. *Israel Journal of Botany*, **35**, 161–74.

Gullberg, U., Yazdani, R. & Rudin, D. (1982). Genetic differentiation between adjacent populations of *Pinus sylvestris*. *Silva Fennica*, **16**, 205–14.

Gullberg, U., Yazdani, R., Rudin, D. & Ryman, N. (1985). Allozyme variation in Scots pine (*Pinus sylvestris* L.) in Sweden. *Silvae Genetica*, **34**, 193–201.

Guries, R.P., Friedman, S.T. & Ledig, F.T. (1978). A megagametophyte analysis of genetic linkage in pitch pine (*Pinus rigida* Mill.). *Heredity*, **40**, 309–14.

Guries, R.P. & Ledig, F.T. (1982). Genetic diversity and population structure in pitch pine (*Pinus rigida* Mill.). *Evolution*, **36**, 387–402.

Hamrick, J. L. (1987). Gene flow and distribution of genetic variation in plant populations. In *Differentiation Patterns in Higher Plants*, ed. K. M. Urbanska, pp. 53–67. New York: Academic Press.

Hamrick, J. L., Blanton, H. M. & Hamrick, K. J. (1989). Genetic structure of geographically marginal populations of ponderosa pine. *American Journal of Botany*, **76**, 1559–68.

Hamrick, J.L. & Godt, M.J.W. (1990). Allozyme diversity in plant species. In *Plant Population Genetics, Breeding and Genetic Resources*, ed. A.H.D. Brown, M.T. Clegg, A.L. Kahler, & B.S. Weir, pp. 43–63. Sunderland, Mass.: Sinauer Associates.

Hamrick, J.L. & Godt, M.J.W. (1996). Conservation genetics of endemic plant species. In *Conservation Genetics: Case Histories From Nature*, ed. J.C. Avise & J.L. Hamrick, pp. 281–304. New York: Chapman & Hall.

Hamrick, J.L., Linhart, Y.B. & Mitton, J.B. (1979). Relationships between life history characteristics and electrophoretically detectable genetic variation in plants. *Annual Review of Ecology and Systematics*, **10**, 173–200.

Hamrick, J.L., Mitton, J.B. & Linhart, Y.B. (1981). Levels of genetic variation in trees: influence of life history characteristics. In *Proceedings of a Symposium on Isozymes of North American Forest Trees and Forest Insects*, ed. M.T.

Conkle, pp. 35–41. USDA Forest Service General Technical Report PSW-48.

Harrington, J.F. (1970). Seed and pollen storage for conservation of plant gene resources. In *Genetic Resources in Plants – Their Exploration and Conservation*. IBP handbook No. 11, ed. O.H. Frankel & E. Bennett, pp. 501–21. Oxford: Blackwell Scientific Publishers.

Hiebert, R.D. & Hamrick, J.L. (1983). Patterns and levels of genetic variation in Great Basin bristlecone pine, *Pinus longaeva*. *Evolution*, **37**, 302–10.

Hinegardner, R. (1976). Evolution of genome size. In *Molecular Evolution*, ed. F. J. Ayala, pp. 179–99. Sunderland, Mass.: Sinauer Associates.

Hizume, M., Ohgiku, A. & Tanaka, A. (1983). Chromosome banding in the genus *Pinus*. I. Identification of chromosomes in *P. nigra* by fluorescent banding method. *Botanical Magazine(Tokyo)*, **96**, 273–6.

Jackson, S.T. (1994). Pollen and spores in Quaternary lake sediments as sensors of vegetation composition: theoretical models and empirical evidence. In *Sedimentation of Organic Particles*, ed. A. Traverse, pp. 253–86. Cambridge: Cambridge University Press.

Jackson, S.T. & Dunwiddie, P.W. (1992). Pollen dispersal and representation on an offshore island. *New Phytologist*, **122**, 187–202.

Jasso, M.J. (1970). Impact of silviculture on forest gene resources. *Unasylva*, **24**, 70–5.

Kaya, Z. & Temerit, A. (1994). Genetic structure of marginally located *Pinus nigra* var *pallasiana* populations in central Turkey. *Silvae Genetica*, **43**, 272–7.

Kellison, R.C. & Zobel, B.J. (1973). *Genetics of Virginia Pine*. USDA Forest Service Research Paper WO-21.

Kinloch, B.B., Jr (1992). Distribution and frequency of a gene for resistance to white pine blister rust in natural populations of sugar pine. *Canadian Journal of Botany*, **70**, 1319–23.

Kinloch, B.B., Westfall, R.D. & Forrest, G.I. (1986). Caledonian Scots pine: origins and genetic structure. *New Phytologist*, **104**, 703–29.

Klekowski, E.J. & Godfrey, P.J. (1989). Ageing and mutation in plants. *Nature*, **340**, 389–91.

Knowles, P. (1984). Genetic variability among and within closely spaced populations of lodgepole pine.

Canadian Journal of Genetics and Cytology, **26**, 177–84.

Koski, V. (1970). A study of pollen dispersal as a mechanism of gene flow in conifers. *Communicationes Instituti Forestalis Fenniae*, **70**, 1–78.

Kossack, D.S. (1989). The IFG copia-like element. Characterization of a transposable element present at high copy number in the genus *Pinus* and a history of the pines using IFG as a marker. PhD thesis., University of California, Davis.

Kraus, J.F. & Squillace, A.E. (1964). Inheritance of yellow oleoresin and virescent foliage in slash pine. *Silvae Genetica*, **13**, 114–6.

Kriebel, H.B. (1972). Embryo development and hybridity barriers in the white pines (section *Strobus*). *Silvae Genetica*, **16**, 89–97.

Kriebel, H.B. (1989). DNA sequence components of the *Pinus strobus* nuclear genome. *Canadian Journal of Forest Research*, **15**, 1–4.

Krugman, S.L. & Jenkinson, J.L. (1970). *Pinus* L. pine. In *Seeds of Woody Plants in the United States*, tech. coord. C.S. Schopmeyer, pp. 598–638. USDA Forest Service Agricultural Handbook No. 450. Washington, DC: USDA Forest Service.

Krutovskii, K.V., Politov, D.V. & Altukhov, Y.P. (1988). Genetic variability of Siberian stone pine, *Pinus sibirica* Du Tour. II. Level of allozyme variability in a natural population in Western Sayan. [Translated from *Genetika*, **24**, 118–24.]

Krutovskii, K.V., Politov, D.V. & Altukhov, Y.P. (1995). Izozyme study of population genetic structure, mating system and phylogenetic relationships of the five stone pine species (subsection *Cembrae*, section *Strobi*, subgenus *Strobus*). In *Population Genetics and Genetic Conservation of Forest Trees*, ed. P. Baradat, W.T. Adams & G. Müller-Starck, pp. 279–304. Amsterdam: SPB Academic Publishing.

Krzakowa, M., Szweykowski, J. & Korczyk, A. (1977). Population genetics of Scots pine (*Pinus sylvestris* L.) forests. Genetic structure of plus-trees in Bolewice near Poznań (West Poland). *Bulletin de l'Académie Polonaise des Sciences, Série des Sciences Biologiques*, **25**, 583–90.

Kuser, J.E. & Ledig, F.T. (1987). Provenance and progeny variation in pitch pine from the Atlantic Coastal Plain. *Forest Science*, **33**, 558–64.

Langlet, O. (1959). A cline or not a cline – a question of Scots pine. *Silvae Genetica*, **8**, 13–22.

Lanner, R.M. (1966). Needed: a new approach to the study of pollen dispersion. *Silvae Genetica*, **15**, 50–2.

Lanner, R.M. (1983). The expansion of singleleaf piñon in the Great Basin. In *The Archaeology of Monitor Valley 2. Gatecliff Shelter*, ed. D.H. Thomas, pp. 167–71. New York: American Museum of Natural History.

Lanner, R.M., & Vander Wall, S.B. (1980). Dispersal of limber pine seed by Clark's nutcracker. *Journal of Forestry*, **78**, 637–9.

Ledig, F.T. (1974). An analysis of methods for the selection of trees from wild stands. *Forest Science*, **20**, 2–16.

Ledig, F.T. (1986). Heterozygosity, heterosis, and fitness in outbreeding plants. In *Conservation Biology: The Science of Scarcity and Diversity*, ed. M.E. Soulé, pp. 77–104. Sunderland, Mass.: Sinauer Associates.

Ledig, F.T. (1987). Genetic structure and the conservation of California's endemic and near-endemic conifers. In *Conservation and Management of Rare and Endangered Plants*. Proceedings from a conference of the California Native Plant Society, ed. T. S. Elias, pp. 587–94. Sacramento: California Native Plant Society.

Ledig, F.T. (1992). Human impacts on genetic diversity in forest ecosystems. *Oikos*, **63**, 87–108.

Ledig, F.T. (1993). Secret extinctions: the loss of genetic diversity in forest ecosystems. In *Our Living Legacy: Proceedings of a Symposium on Biological Diversity*, ed. M.A. Fenger, E.H. Miller, J.F. Johnson & E.J.R. Williams, pp. 127–40. Victoria: Royal British Columbia Museum.

Ledig, F.T. & Conkle, M.T. (1983). Gene diversity and genetic structure in a narrow endemic, Torrey pine (*Pinus torreyana* Parry ex Carr.). *Evolution*, **37**, 79–85.

Ledig, F.T. & Fryer, J.H. (1972). A pocket of variability in *Pinus rigida*. *Evolution*, **26**, 259–66.

Ledig, F.T., Lambeth, C.C. & Linzer, D.I.H. (1976). Nursery evaluation of a pitch pine provenance trial. In *Proceedings of the Twenty-third Northeastern Forest Tree Improvement Conference*, pp. 93–108. New Brunswick, New Jersey.

Ledig, F.T. & Little, S. (1979). Pitch pine (*Pinus rigida* Mill.): ecology, physiology, and genetics. In *Pine Barrens: Ecosystem and Landscape*, ed. R.T.T. Forman, pp. 347–71. New York: Academic Press.

Ledig, F.T. & Smith, D.M. (1981). The influence of silvicultural practices on genetic improvement: height growth and weevil resistance in eastern white pine. *Silvae Genetica*, **30**, 30–6.

Levin, D.A. (1981). Dispersal versus gene flow in plants. *Annals of the Missouri Botanical Garden*, **68**, 233–53.

Levin, D.A. (1984). Inbreeding depression and proximity-dependent crossing success in *Phlox drummondii*. *Evolution*, **38**, 116–27.

Lill, B.S. (1976). Ovule and seed development in *Pinus radiata*: postmeiotic development, fertilization, and embryogeny. *Canadian Journal of Botany*, **54**, 2141–54.

Linhart, Y.B., Mitton, J.B., Sturgeon, K.B. & Davis, M.L. (1981). Genetic variation in space and time in a population of ponderosa pine. *Heredity*, **46**, 407–26.

Little, E.L., Jr & Critchfield, W.B. (1969). *Subdivisions of the Genus Pinus (Pines)*. USDA Forest Service Miscellaneous Publication 1144. Washington, DC: USDA Forest Service.

Loukas, M., Vergini, Y. & Krimbas, C.B. (1983). Isozyme variation and heterozygosity in *Pinus halepensis*. *Biochemical Genetics*, **21**, 497–509.

Lutz, H.J. (1939). Layering in eastern white pine. *Botanical Gazette*, **101**, 505–7.

McConnell, J.L. (1980). The Southern forest – past, present and future. In *Proceedings of the Servicewide Workshop on Gene Resource Management*, pp. 17–26. Sacramento, California: Timber Management Staff, USDA Forest Service.

MacDonald, G.M. (1993). Fossil pollen analysis and the reconstruction of plant invasions. *Advances in Ecological Research*, **24**, 67–110.

McMaster, G.S. & Zedler, P.H. (1981). Delayed seed dispersal in *Pinus torreyana* (Torrey pine). *Oecologia*, **51**, 62–6.

McWilliam, J.R. (1958). The role of the micropyle in the pollination of *Pinus*. *Botanical Gazette*, **120**, 109–17.

McWilliam, J.R. (1959). Interspecific incompatibility in *Pinus*. *American Journal of Botany*, **46**, 425–33.

McWilliam, J.R. & Mergen, F. (1958). Cytology of fertilization in *Pinus*. *Botanical Gazette*, **119**, 247–9.

Marsh, E.K. (1970). Selecting adapted races of introduced species. In *Second World Consultation on Forest Tree Breeding*, vol. 2, pp. 1249–61. Rome: Food and Agriculture Organization of the United Nations.

Matheson, A.C., Bell, J.C., & Barnes, R.D. (1989). Breeding systems and genetic structure in some Central American pine populations. *Silvae Genetica*, **38**, 107–13.

Matziris, D.I. (1994). Genetic variation in the phenology of flowering in black pine. *Silvae Genetica*, **43**, 321–8.

Mejnartowicz, L. (1979). Genetic variation in some isoenzyme loci in Scots pine (*Pinus silvestris* L.) populations. *Arboretum Kornickie*, **24**, 91–104.

Mergen, F. (1958). Natural polyploidy in slash pine. *Forest Science*, **4**, 283–95.

Mergen, F. (1959). Colchicine-induced polyploidy in pines. *Journal of Forestry*, **57**, 180–90.

Mergen, F. & Thielges, B.A. (1967). Intraspecific variation in nuclear volume in four conifers. *Evolution*, **21**, 720–4.

Miksche, J.P. (1967). Variation in DNA content of several gymnosperms. *Canadian Journal of Genetics and Cytology*, **9**, 717–22.

Miksche, J.P. (1968). Quantitative study of intraspecific variation of DNA per cell in *Picea glauca* and *Pinus banksiana*. *Canadian Journal of Genetics and Cytology*, **10**, 590–600.

Millar, C.I. (1983). A steep cline in *Pinus muricata*. *Evolution*, **37**, 311–19.

Millar, C.I. (1985). Inheritance of allozyme variants in bishop pine (*Pinus muricata* D. Don). *Biochemical Genetics*, **23**, 933–46.

Millar, C.I. (1989). Allozyme variation of bishop pine associated with pygmy-forest soils in northern California. *Canadian Journal of Forest Research*, **19**, 870–9.

Millar, C.I. (1993). Impact of the Eocene on the evolution of *Pinus* L. *Annals of the Missouri Botanical Garden*, **80**, 471–98.

Millar, C.I., Strauss, S.H., Conkle, M.T. & Westfall, R.D. (1988). Allozyme differentiation and biosystematics of the Californian closed-cone pines (*Pinus* subsect. *Oocarpae*). *Systematic Botany*, **13**, 351–70.

Misenti, T.L. & DeHayes, D.H. (1989). Genetic diversity of marginal vs. central populations of pitch pine and jack pine. In *Proceedings of the Thirty-first Northeastern Forest Tree Improvement Conference and the Sixth North Central Tree Improvement Association*, ed. M.E. Demeritt, Jr, pp. 63–72. University Park: Pennsylvania State University.

Mitton, J.B., Linhart, Y.B., Davis, M.L. & Sturgeon, K.B. (1981). Estimation of outcrossing in ponderosa pine, *Pinus ponderosa* Laws., from patterns of segregation of protein polymorphisms and from frequencies of albino seedlings. *Silvae Genetica*, **30**, 117–21.

Mitton, J.B., Linhart, Y.B., Hamrick, J.L. & Beckman, J.S. (1977). Observations on the genetic structure and mating system of ponderosa pine in the Colorado Front Range. *Theoretical and Applied Genetics*, **51**, 5–13.

Moran, G.F., Bell, J.C. & Eldridge, K.G. (1988). The genetic structure and the conservation of the five natural populations of *Pinus radiata*. *Canadian Journal of Forest Research*, **18**, 506–14.

Moran, G.F., Bell, J.C. & Hilliker, A.J. (1983). Greater meiotic recombination in male vs. female gametes in *Pinus radiata*. *Journal of Heredity*, **74**, 62.

Morgante, M., Vendramin, G.G. & Olivieri, A.M. (1991). Mating system analysis in *Pinus leucodermis* Ant.: detection of self-fertilization in natural populations. *Heredity*, **67**, 197–203.

Morton, N.E., Crow, J.F. & Muller, H.J. (1956). An estimate of the mutational damage in man from data on consanguineous marriages. *Proceedings of the National Academy of Sciences, USA*, **42**, 855–63.

Muona, O. & Harju, A. (1989). Effective population sizes, genetic variability, and mating system in natural stands and seed orchards of *Pinus sylvestris*. *Silvae Genetica*, **38**, 221–8.

Namkoong, G. (1966). Statistical analysis of introgression. *Biometrics*, **22**, 488–502.

Namkoong, G. (1969). Nonoptimality of local races. In *Proceedings of the Tenth Southern Conference on Forest Tree Improvement*, pp. 149–53. Houston, Texas: Texas A & M University, College Station.

Neale, D.B. & Sederoff, R.R. (1989). Paternal inheritance of chloroplast DNA and maternal inheritance of mitochondrial DNA in loblolly pine. *Theoretical and Applied Genetics*, **77**, 212–16.

Neet-Sarqueda, C. (1994). Genetic differentiation of *Pinus sylvestris* L. and *Pinus mugo* aggr. populations in Switzerland. *Silvae Genetica*, **43**, 207–15.

Nei, M. (1975). *Molecular Population Genetics and Evolution.* Amsterdam: North-Holland Publishing Co.

Nicolic, D. & Tucic, N. (1983). Isoenzyme variation within and among populations of European black pine (*Pinus nigra* Arnold). *Silvae Genetica,* 32, 80–8.

Niebling, C.R. & Conkle, M.T. (1990). Diversity of Washoe pine and comparisons with allozymes of ponderosa pine races. *Canadian Journal of Forest Research,* 20, 298–308.

Niklas, K.J. (1994). *Plant Allometry: The Scaling of Form and Process.* Chicago: University of Chicago Press.

O'Malley, D.M., Allendorf, F.W. & Blake, G.M. (1979). Inheritance of isozyme variation and heterozygosity in *Pinus ponderosa. Biochemical Genetics,* 17, 233–50.

Owens, J.N. & Molder, M. (1977). Seed cone differentiation and sexual reproduction in western white pine (*Pinus monticola*). *Canadian Journal of Botany,* 55, 2574–90.

Owens, J.N., Simpson, S.J. & Molder, M. (1981). Sexual reproduction of *Pinus contorta.* I. Pollen development, the pollination mechanism, and early ovule development. *Canadian Journal of Botany,* 59, 1828–43.

Pederick, L.A. (1967a). Cytogenetic studies in *Pinus radiata* D. Don. PhD thesis, University of Melbourne.

Pederick, L.A. (1967b). The structure and identification of the chromosomes of *Pinus radiata* D. Don. *Silvae Genetica,* 16, 69–77.

Pederick, L.A. (1968). Chromosome inversions in *Pinus radiata. Silvae Genetica,* 17, 22–6.

Perry, D.J. & Dancik, B.P. (1986). Mating system dynamics of lodgepole pine in Alberta, Canada. *Silvae Genetica,* 35, 190–5.

Plessas, M.E. & Strauss, S.H. (1986). Allozyme differentiation among populations, stands, and cohorts in Monterey pine. *Canadian Journal of Forest Research,* 16, 1155–64.

Pohl, F. (1937). Die Pollenkorngewichte einiger windblütiger Pflanzen und ihre ökologische Bedeutung. *Beihefte zum Botanischen Centralblatt,* 57a, 112–72.

Premoli, A.C., Chischilly, S. & Mitton, J.B. (1994). Levels of genetic variation captured by four descendant populations of pinyon pine (*Pinus edulis* Engelm.). *Biodiversity and Conservation,* 3, 331–340.

Prus-Glowacki, W. & Bernard, E. (1994). Allozyme variation in populations of *Pinus sylvestris* L. from a 1912 provenance trial in Pulawy (Poland). *Silvae Genetica,* 43, 132–8.

Rake, A.V., Miksche, J.P., Hall, R.B. & Hansen, K.M. (1980). DNA reassociation kinetics of four conifers. *Canadian Journal of Genetics and Cytology,* 22, 69–79.

Rehfeldt, G.E., Hoff, R.J. & Steinhoff, R.J. (1984). Geographic patterns of genetic variation in *Pinus monticola. Botanical Gazette,* 145, 229–39.

Rehfeldt, G.E. & Wykoff, W.R. (1981). Periodicity in shoot elongation among populations of *Pinus contorta* from the northern Rocky Mountains. *Annals of Botany,* 48, 371–7.

Rejmánek, M. (1996). A theory of seed plant invasiveness: the first sketch. *Biological Conservation* (in press).

Richardson, D.M., Williams, P.A. & Hobbs, R.J. (1994). Pine invasions in the Southern Hemisphere: determinants of spread and invadability. *Journal of Biogeography,* 21, 511–27.

Roberds, J.H. & Conkle, M.T. (1984). Genetic structure in loblolly pine stands: allozyme variation in parents and progeny. *Forest Science,* 30, 319–29.

Rogers, D.L. & Ledig, F.T. (eds.) (1996). *The Status of Temperate North American Forest Genetic Resources.* Genetic Resources Conservation Program, Report No. 16. Davis: University of California.

Ross, H.A. & Hawkins, J.L. (1986). Genetic variation among local populations of jack pine (*Pinus banksiana*). *Canadian Journal of Genetics and Cytology,* 28, 453–8.

Rothe, G.M. (1994). *Electrophoresis of Enzymes: Laboratory Methods.* Berlin: Springer-Verlag.

Rudin, D. (1977). Leucine-amino-peptidases (LAP) from needles and macrogametophytes of *Pinus sylvestris* L. *Hereditas,* 85, 219–26.

Rudin, D., Eriksson, G., Ekberg, I. & Rasmuson, M. (1974). Studies of allele frequencies and inbreeding in Scots pine populations by the aid of the isozyme technique. *Silvae Genetica,* 23, 10–13.

Rudolph, T.D. (1979). Seed production in the first eight years and frequency of natural selfing in a jack pine seed orchard. In *Proceedings of the Thirteenth Lake States Forest Tree Improvement Conference,* pp. 33–47. USDA Forest Service General Technical Report NC-50.

Rudolph, T.D. & Yeatman, C.W. (1982). *Genetics of Jack Pine.* USDA Forest Service Research Paper WO-38.

Runquist, E.W. (1968). Meiotic investigations in *Pinus sylvestris* L. *Hereditas,* 60, 77–128.

Ryu, J.B. & Eckert, R.T. (1983). Foliar isozyme variation in twenty seven provenances of *Pinus strobus* L.: genetic diversity and population structure. In *Proceedings of the Twenty-eighth Northeastern Forest Tree Improvement Conference,* ed. R.T. Eckert, pp. 249–61. Durham, NH: Institute of Natural and Environmental Resources, University of New Hampshire.

Sakai, K.-I. & Park, Y.-G. (1971). Genetic studies in natural populations of forest trees. III. Genetic differentiation within a forest of *Cryptomeria japonica. Theoretical and Applied Genetics,* 41, 13–17.

Santamour, F.S., Jr (1960). *Seasonal Growth in White Pine Seedlings from Different Provenances.* Forest Research Note 105. Upper Darby, Pennsylvania: Northeastern Forest Experiment Station, USDA Forest Service.

Sarvas, R. (1962). Investigations on the flowering and seed crop of *Pinus silvestris. Communicationes Instituti Forestalis Fenniae,* 53, 1–198.

Sax, K. & Sax, H.J. (1933). Chromosome number and morphology in the conifers. *Journal of the Arnold Arboretum,* 14, 356–74.

Saylor, L.C. & Smith, B.W. (1966). Meiotic irregularity in species and interspecific hybrids of *Pinus. American Journal of Botany,* 53, 453–68.

Scaltsoyiannes, A., Rohr, R., Panetsos, K.P. & Tsaktsira, M. (1994). Allozyme frequency distributions in five European populations of black pine (*Pinus nigra* Arnold). I. Estimation of genetic variation within and among populations. II. Contribution of isozyme analysis to the taxonomic status of the species. *Silvae Genetica,* 43, 20–30.

Schemske, D.W. & Lande, R. (1985). The evolution of self-fertilization and inbreeding depression in plants. II. Empirical observations. *Evolution,* 39, 41–52.

Schiller, G., Conkle, M.T. & Grunwald, C. (1986). Local differentiation among Mediterranean populations of Aleppo pine in their isoenzymes. *Silvae Genetica*, **35**, 11–19.

Schuster, W.S., Alles, D.L. & Mitton, J.B. (1989). Gene flow in limber pine: evidence from pollination phenology and genetic differentiation along an elevational transect. *American Journal of Botany*, **76**, 1395–403.

Schuster, W.S.F. & Mitton, J.B. (1991). Relatedness within clusters of a bird-dispersed pine and the potential for kin interactions. *Heredity*, **67**, 41–8.

Shaw, D.V., Kahler, A.L. & Allard, R.W. (1981). A multilocus estimator of mating system parameters in plant populations. *Proceedings of the National Academy of Sciences USA*, **78**, 1298–1302.

Shoup, L.H. with Baker, S. (1981). Speed power, production, and profit: railroad logging in the Goosenest District, Klamath National Forest, 1900–1956. Prepared in fulfilment of USDA Forest Service Contract No. 00–91W8-0-1911. Yreka, California: Klamath National Forest.

Siddiqui, K.M. & Khattak, M.S. (1975). Pollen dispersion in Himalayan blue pine. *Pakistan Journal of Forestry*, **25**, 153–8.

Silen, R.R. (1962). Pollen dispersal considerations for Douglas-fir. *Journal of Forestry*, **60**, 790–5.

Simon, J.-P., Bergeron, Y. & Gagnon, D. (1986). Isozyme uniformity in populations of red pine (*Pinus resinosa*) in the Abitibi Region, Quebec. *Canadian Journal of Forest Research*, **16**, 1133–5.

Sittman, K. & Tyson, H. (1971). Estimates of inbreeding in *Pinus banksiana*. *Canadian Journal of Botany*, **49**, 1241–5.

Skeates, D.A. (1979). Discontinuity in growth potential of jack pine in Ontario: a 20-year provenance assessment. *Forestry Chronicle*, **55**, 137–41.

Slatkin, M. (1985). Rare alleles as indicators of gene flow. *Evolution*, **39**, 53–65.

Smith, C.C. (1970). The coevolution of pine squirrels (*Tamiasciurus*) and conifers. *Ecological Monographs*, **40**, 349–71.

Smith, C.K., White, T.L. & Hodge, G.R. (1993). Genetic variation in second-year slash pine shoot traits and their relationship to 5- and 15-year volume in the field. *Silvae Genetica*, **42**, 266–75.

Snyder, E.B., Dinus, R.J. & Derr, H.J. (1977). *Genetics of Longleaf Pine*. USDA Forest Service Research Paper WO-33.

Snyder, E.B., Squillace, A.E. & Hamaker, J.M. (1966). Pigment inheritance in slash pine seedlings. In *Proceedings of the Eighth Southern Conference on Forest Tree Improvement*, pp. 77–85. Savannah, Georgia: Southern Forest Tree Improvement Committee.

Snyder, T.P., Stewart, D.A. & Strickler, A.F. (1985). Temporal analysis of breeding structure in jack pine (*Pinus banksiana* Lamb.). *Canadian Journal of Forest Research*, **15**, 1159–66.

Sorensen, F.C. (1987). Estimated frequency of natural selfing in lodgepole pine. *Silvae Genetica*, **36**, 215–16.

Squillace, A.E. & Goddard, R.E. (1982). Selfing in clonal seed orchards of slash pine. *Forest Science*, **28**, 71–8.

Squillace, A.E. & Kraus, J.F. (1963). The degree of natural selfing in slash pine as estimated from albino frequencies. *Silvae Genetica*, **12**, 46–50.

Squillace, A.E. & Silen, R.R. (1962). Racial variation in ponderosa pine. *Forest Science Monograph*, **2**, 1–27.

Stebbins, G.L. (1950). *Variation and Evolution in Plants*. New York: Columbia University Press.

Stebbins, G.L. (1959). The role of hybridization in evolution. *Proceedings of the American Philosophical Society*, **103**, 231–51.

Steinhoff, R.J., Joyce, D.G. & Fins, L. (1983). Isozyme variation in *Pinus monticola*. *Canadian Journal of Forest Research*, **13**, 1122–32.

Strand, L. (1957). Pollen dispersal. *Silvae Genetica*, **6**, 129–36.

Strauss, S.H. & Conkle, M.T. (1986). Segregation, linkage, and diversity of allozymes in knobcone pine. *Theoretical and Applied Genetics*, **72**, 483–93.

Szmidt, A.E. (1982). Genetic variation in isolated populations of stone pine (*Pinus cembra* L.). *Silva Fennica*, **16**, 196–200.

Szmidt, A.E., & Wang, X.-R. (1993). Molecular systematics and genetic differentiation of *Pinus sylvestris* (L.) and *P. densiflora* (Sieb. et Zucc.). *Theoretical and Applied Genetics*, **86**, 159–65.

Szweykowski, J., Prus-Glowacki, W. & Hrynkiewicz, J. (1994). The genetic structure of Scots pine (*Pinus sylvestris* L.) population from the top of Szczeliniec Wielki Mt., Central Sudetes. *Acta Societatis Botanicorum Poloniae*, **63**, 315–24.

Tigerstedt, P.M.A., Rudin, D., Niemela, T. &

Tammisola, J. (1982). Competition and neighboring effect in a naturally regenerating population of Scots pine. *Silva Fennica*, **16**, 122–9.

Vander Wall, S.B. (1992). The role of animals in dispersing a 'wind-dispersed' pine. *Ecology*, **73**, 614–21.

Vander Wall, S.B. & Balda, R.P. (1977). Coadaptations of the Clark's nutcracker and the piñon pine for efficient seed harvest and dispersal. *Ecological Monographs*, **47**, 89–111.

Wakamiya, I., Newton, R.J., Johnston, J.S. & Price, H.J. (1993). Genome size and environmental factors in the genus *Pinus. American Journal of Botany*, **80**, 1235–41.

Wang, C.-W. (1977). *Genetics of Ponderosa Pine*. USDA Forest Service Research Paper WO-34.

Wang, C.-W., Perry, T.O. & Johnson, A.G. (1960). Pollen dispersion of slash pine (*Pinus elliottii* Engelm.) with special reference to seed orchard management. *Silvae Genetica*, **9**, 78–86.

Wang, X.-R. & Szmidt, A.E. (1990). Evolutionary analysis of *Pinus densata* Masters, a putative Tertiary hybrid. 2. A study using species-specific chloroplast DNA markers. *Theoretical and Applied Genetics*, **80**, 641–7.

Wang, X.-R., Szmidt, A.E., Lewandowski, A. & Wang, Z.-R. (1990). Evolutionary analysis of *Pinus densata* Masters, a putative Tertiary hybrid. 1. Allozyme variation. *Theoretical and Applied Genetics*, **80**, 635–40.

Wang, X.-R., Szmidt, A.E. & Lindgren, D. (1991). Allozyme differentiation among populations of *Pinus sylvestris* (L.) from Sweden and China. *Hereditas*, **114**, 219–26.

Wasser, R.G. (1967). *A Shortleaf and Loblolly Pine Flowering Phenology Study*. Virginia Division of Forestry, Occasional Report 28.

Wells, O.O., Switzer, G.L. & Schmidtling, R.C. (1991). Geographic variation in Mississippi loblolly pine and sweetgum. *Silvae Genetica*, **40**, 105–19.

Wells, O.O. & Wakeley, P.C. (1966). Geographic variation in survival, growth, and fusiform-rust infection of planted loblolly pine. *Forest Science Monograph*, **11**, 1–40.

Wheeler, N.C. & Guries, R.P. (1982). Population structure, genic diversity, and morphological variation in *Pinus contorta* Dougl. *Canadian Journal of Forest Research*, **12**, 595–606.

Wheeler, N.C. & Guries, R.P. (1987). A

quantitative measure of introgression between lodgepole and jack pines. *Canadian Journal of Botany*, **65**, 1876–85.

Wiens, D. (1984). Ovule survivorship, brood size, life history, breeding systems, and reproductive success in plants. *Oecologia*, **64**, 47–53.

Williams, R.D. & Funk, D.T. (1978). Eighteen-year performance of an eastern white pine genetic test plantation in southern Indiana. *Proceedings of the Indiana Academy of Science*, **87**, 116–19.

Witter, M.S. & Feret, P.P. (1978). Inheritance of glutamate oxalo-acetate transaminase isozymes in Virginia pine megagametophytes. *Silvae Genetica*, **27**, 129–34.

Woods, J.H., Blake, G.M. & Allendorf, F.W. (1983). Amount and distribution of isozyme variation in ponderosa pine from eastern Montana. *Silvae Genetica*, **32**, 151–6.

Wright, J.W. (1952). *Pollen Dispersion of Some Forest Trees*. USDA Forest Service, Northeastern Forest Experiment Station Paper 46.

Wright, J.W. (1970). *Genetics of Eastern White Pine*. USDA Forest Service Research Paper WO-9.

Wright, S. (1969). *Evolution and the Genetics of Populations.*, vol. 2. *The Theory of Gene Frequencies*. Chicago: University of Chicago Press.

Xie, C.Y. & Knowles, P. (1991). Spatial genetic substructure within natural populations of jack pine (*Pinus banksiana*). *Canadian Journal of Botany*, **69**, 547–51.

Yang, R.-C. & Yeh, F.C. (1993). Multilocus structure in *Pinus contorta* Dougl. *Theoretical and Applied Genetics*, **87**, 568–76.

Yazdani, R., Lindgren, D. & Stewart, S. (1989). Gene dispersion within a population of *Pinus sylvestris*.

Scandinavian Journal of Forest Research, **4**, 295–306.

Yeh, F.C., Cheliak, W.M., Dancik, B.P., Illingworth, K., Trust, D.C. & Pryhitka, B.A. (1985). Population differentiation in lodgepole pine, *Pinus contorta* spp. [sic] *latifolia*: a discriminant analysis of isozyme variation. *Canadian Journal of Genetics and Cytology*, **27**, 210–18.

Yeh, F.C. & Layton, C. (1979). The organization of genetic variability in central and marginal populations of lodgepole pine *Pinus contorta* spp. [sic] *latifolia*. *Canadian Journal of Genetics and Cytology*, **21**, 487–503.

Yow, T.H., Wagner, M.R., Wommack, D.E. & Tuskan, G.A. (1992). Influence of selection for volume growth on the genetic variability of southwestern ponderosa pine. *Silvae Genetica*, **41**, 326–33.

14 Seed dispersal in *Pinus*

Ronald M. Lanner

14.1 Introduction

The biogeography of a taxon essentially comes down to its success in colonizing parcels of the Earth's surface, and to do that it must first get there. Theoretically, a pine can 'get there' in one of two ways: by slow, incremental migration of vegetative parts that become rooted and independent along the way; or through transport of diaspores by biotic or abiotic agents. At its present state of evolutionary development, use of the vegetative option by *Pinus* is almost nil. The only taxa reported to regenerate vegetatively are *P. mugo* and *P. pumila*. The latter is typically shrubby, even mat-like in form, and branches held in contact with moist earth by the snowpack produce adventitious roots (Saito & Kawabe 1990; Khomentovsky 1994), resulting in dense thickets. Decay in the basal parts of adventitiously rooted prostrate stems eventually severs the rooted branches (layers) from their parent trunks, making them independent trees of the same genotype, i.e. a clone (Khomentovsky 1994). The development of adventitious roots on *P. mugo* is described by Mirov (1967) as 'occasional'. According to Arno & Hoff (1989), *P. albicaulis* has in at least one instance layered, but the habit appears to be too rare to have significance in this species. The unlikelihood of other pines reproducing vegetatively *in situ* is highlighted by the difficulty of rooting pine cuttings or layers in tree improvement programmes (Wright 1976). Despite its vegetative reproductive abilities, *P. pumila* has an effective system of sexual reproduction, which will be discussed below.

The dispersal of diaspores has had a profound effect on the biogeography, indeed on all aspects of the population biology of pines. Close-in dispersal allows for the conservative strategy of establishing new recruits under proven conditions that are suitable for growth to the reproductive stage, assuming establishment occurs in open or disturbed areas not beneath the parent stand's canopy. Dispersal at a distance allows diaspores to sample a much wider range of conditions than those experienced by their parents, thus being exposed to a nexus of natural selective factors new to the line. Once outlying populations have become established, and have begun to differentiate genetically, long-distance dispersal makes gene flow possible between central and marginal stands. Thus seed dispersal is a major determinant not only of a species' geographic occurrence and successional behaviour, but of its genetic structure as well (Chap. 13, this volume).

Seed dispersal is customarily discussed in relation to morphological adaptation of diaspores, or the mechanics of factors effecting dispersal. But the process of dispersal is only one of many life history processes, and it is ultimately related to what goes on before and after, as well as elsewhere. Thus dispersal has a context.

14.2 The context of pine seed dispersal

The dispersal of pine seeds has both temporal and spatial contexts. In this chapter dispersal to a 'safe site' where the seed will germinate will be considered the final act in a chain of events that begins with cone initiation. While early events will not be discussed in great detail (see Chap. 12, this volume, for further discussion), they will be viewed from the standpoint of their bearing on dispersal.

14.2.1 Readiness to produce seed

The minimum age at which seeds are borne is quite variable among pines. On rare occasions, pines may bear cones

Ecology and Biogeography of Pinus, ed. D.M. Richardson. © Cambridge University Press, Cambridge (1998), pp. 281–95.

Table 14.1. *Pinus* species classified by observation of minimum ages at which seed were borne. Data from Krugman & Jenkinson (1974)

Less than 10 years	10 to 20 years	Over 20 years
P. attenuata	P. albicaulis	P. cembra
P. banksiana	P. aristata	P. edulis
P. brutia	P. armandii	P. engelmannii
P. clausa	P. balfouriana	P. lambertiana
P. contorta	P. canariensis	P. leiophylla var.
P. coulteri	P. densiflora	chihuahuana
P. echinata	P. flexilis	P. sibirica
P. elliottii	P. halepensis	
P. jeffreyi	P. koraiensis	
P. kesiya	P. monophylla	
P. monticola	P. nigra	
P. muricata	P. peuce	
P. ponderosa var.	P. ponderosa var.	
scopulorum	ponderosa	
P. pungens	P. roxburghii	
P. radiata	P. torreyana	
P. rigida	P. wallichiana	
P. serotina		
P. strobus		
P. sylvestris		
P. taeda		
P. thunbergii		
P. virginiana		

Table 14.2. Yield of filled seed in several pine species following self- and open- or cross-pollination

Pinus taxon	Self-pollinated filled seed (%)	Open- or cross-pollinated filled seed (%)	Reference
P. ponderosa	23.7	66.5	Sorensen (1970)
P. sylvestris	13.4	71.4	Johnsson (1976)
P. virginiana	16.4	90.7	Bramlett & Pepper (1974)
P. edulis	14.4	90.5	Lanner (1980)
P. contorta	14.4	86.7	Smith, Hamrick & Kramer (1988)
P. radiata	34.2	78.9	Griffin & Lindgren (1985)
Mean	19.4	80.8	

even in their second year, before producing secondary needles (Mergen 1961), but this is an aberrant event of no biological significance. According to Krugman & Jenkinson (1974), about half the pines for which data were available have produced seeds before attaining 10 years of age, and only a few species are not known to have borne seed by age 20 years (Table 14.1). Although the age of bearing is largely determined by environmental factors (Matthews 1963), growth rate, and – perhaps – tree size, age is probably an important correlate. Thus seed crops from young populations are likely to be sparse, and little dispersal will occur from them.

14.2.2 Initiation of cone crops

Once readiness-to-bear has been achieved, the initiation of cone primordia becomes the next critical step. According to Owens (1991), pines of subgenus *Pinus* initiate both pollen cones and seed cones prior to bud dormancy, while those of subgenus *Strobus* initiate pollen cones prior to dormancy, but seed cones after dormancy, only a few weeks prior to pollination. Initiation of pollen cones has received far less attention in the literature than has seed cone initiation, and there have been far fewer attempts to stimulate it. Just the same, no theory has yet been fashioned that can explain or unify the disparate correlations to be found in a copious literature that has accumulated over a century of research. For example, a large body of evidence suggests a cause-and-effect relationship between physiological stress conditions in pines and initiation of cone crops. But equally persuasive evidence relates cone pro-

duction to favourable growing conditions and high vigour (Matthews 1963). It is not an overstatement to say that the initiation conditions that will eventually lead to a dispersal event are not yet understood and cannot be predicted.

14.2.3 Cone and seed development

Even excluding the depredations of insect pests and pathogens, there are numerous post-initiation factors that can constrain the magnitude of future seed dispersal. For example, Sarvas (1962) has shown that a lack of pollination of *P. sylvestris* conelets leads to conelet abscission. In a cone in which many of the ovules have been pollinated, those that have not received pollen fail to develop as seeds, producing wings only. Low levels of pollination may occur in years when the crop is small, as between mast years, or because of pollen washout in a rainy spring (McDonald 1962). Seed yield can also be reduced by genetic factors, such as the collapse of embryos following self-fertilization due to homozygosity of lethal alleles. Most pines are moderately self-fertile but produce fewer filled seed per cone when selfed (Table 14.2). Isolated trees self-pollinate at a higher frequency than stand-grown trees, and thus bear many empty seeds. In fact, Fowler (1965) reported an isolated *P. resinosa* that produced 100% of selfed seeds. In *P. elliottii*, selfing reduced mean sound seeds per cone from 34 to nine (Squillace & Kraus, cited in Stern & Roche 1974). It is therefore apparent that, other things being equal, seed dispersal from a closed stand of pines is biologically more meaningful than dispersal from a woodland of widely scattered trees, because the latter would produce a much larger proportion of selfed progeny, leading to greater mortality.

14.2.4 Timing of seed dispersal

Masting and non-masting species
The temporal occurrence of dispersal may be an important factor in dispersal success. For example, in pines that have a strong masting habit, with 'boom-and-bust' periodicity, a

great many seeds are dispersed during a few weeks every several years. Among non-serotinous western American species this pattern typifies *P. flexilis*, *P. jeffreyi*, *P. monticola* and *P. ponderosa*. Several other species are less clearly periodic, and may produce seed crops in most years. These include *P. longaeva* (Lanner 1988) and *P. monophylla* (Lanner 1983). This group of pines disperses fewer seeds in a single crop, but does so more frequently. Open-cone (non-serotinous) populations of *P. contorta* var. *latifolia* in northern Utah (R.M.L., personal observations) and Alberta (Hellum 1983) disperse seeds virtually every year. More frequent seeders are able to 'sample' more years, and therefore more temporal variability in ecological conditions, while masting species can sample spatial heterogeneity more intensively within a mast year. The consequences of this difference remain to be explored, including the question of which bearing schedule leads to the greatest overall production of seed and establishment of seedlings.

Phenology of cone opening

The commonly held stereotype is that pine cones mature in late summer to early autumn, that they dry and open almost immediately, and that they then shed their winged seeds on the wind. The cones then cease to function. This pattern is perhaps typified by *P. strobus*, reported to have a period of seed-fall extending from early September to late November, most of which occurred during a one-week period in late September to early October (Graber 1970). The actual situation is far more complex. For example, serotinous pines can retain their seeds in closed cones for several years.

Cone serotiny

Among the world's numerous serotinous or partially serotinous pines are *P. attenuata*, *P. banksiana*, *P. brutia*, *P. clausa*, *P. contorta*, *P. greggii*, *P. halepensis*, *P. jaliscana*, *P. muricata*, *P. oocarpa*, *P. patula*, *P. pinaster*, *P. pringlei*, *P. pungens*, *P. radiata*, *P. rigida*, *P. serotina* and *P. virginiana* (Krugman & Jenkinson 1974; McMaster & Zedler 1981; Perry 1991; Frankis 1992; R.M.L., personal observations). Serotiny probably evolved in these taxa as an adaptation to fire (Chap. 12, this volume). For example, *P. banksiana* and *P. contorta* have cones that open and shed seeds accumulated over many years when fire-heat softens the resin bonding the cone scales (Krugman & Jenkinson 1974). Trees bearing serotinous cones will disperse seeds in whatever year a facilitating fire occurs. Thus dispersal is determined not by plant characteristics so much as by unpredictable environmental events.

Cone serotiny and climate

Is cone serotiny in pines limited to or characteristic of certain types of climates? In other words, has serotiny

Table 14.3. Types of climate (after Köppen) characterizing ranges of serotinous North American and Euro-Mediterranean pines. Aw, tropical wet/dry savanna; Bs, subtropical steppe; Ca, humid subtropical, warm summer; Cb, humid mesothermal marine cool summer; Cs, dry summer subtropical (Mediterranean); Db, humid continental, cool summer

subsection *Pinus*	
Pinus pinaster	Cs
subsection *Halepenses*	
P. brutia	Cs
P. halepensis	Cs
subsection *Contortae*	
P. banksiana	Db
P. clausa	Ca
P. contorta	Cb, undiff. highlands
P. virginiana	Ca, Cb
subsection *Australes*	
P. pungens	Ca, Da
P. rigida	Cb, Db
P. serotina	Ca
subsection *Attenuatae*	
P. attenuata	Cb, Cs
P. muricata	Cs
P. radiata	Cs
subsection *Oocarpae*	
P. greggii	Bs
P. oocarpa	Aw
P. patula	Aw
P. pringlei	Aw, undiff. highlands

evolved as a response to a particular climate? When the natural ranges of serotinous pines of subsections *Attenuatae*, *Australes*, *Contortae*, *Halepenses*, *Oocarpae* and *Pinus* are superimposed on a world map showing climates of the earth following Köppen, a fairly consistent pattern emerges (Table 14.3). Of the 16 species, 13 are found in humid mesothermal (C) types, and nine of these are in dry-summer Mediterranean (Cs) or warm-summer (Ca) subtypes. Three are found in tropical wet and dry savanna (Aw), and three in humid continental types (Da, Db). Two species are relegated to 'undifferentiated highlands', and the range of *P. greggii* is erroneously classified as desert. The Cs species are concentrated in California and along the Mediterranean littoral, with most of the others from the eastern USA and Mexico. None are from the vast area between Asia Minor and the Far East, including southeast Asia, though numerous pines of subgenus *Pinus* grow there. *Pinus banksiana* is unusual among serotinous pines in having a humid, continental, cool-summer climate (Db), though the serotinous spruce *Picea mariana* inhabits the subarctic belt immediately to the north of *P. banksiana* (Dc). Species numbers may appear not to add properly because some occur in two climatic zones. It appears from these data that most serotinous pines inhabit areas where summers are hot and/or dry, but it is also apparent that there are large areas where the summers are indeed hot and dry, but serotinous pines are lacking.

According to Saracino & Leone (1991), 81% of *P. halepensis* seeds that were disseminated within two months after a fire fell in the first month. These were found to be larger and of greater mass than the seeds shed during the second month.

Some subtleties of seed dispersal from serotinous species include non-serotiny in young trees (Gauthier, Bergeron & Simon 1992) and partial opening by sun-drying (Frankis 1992). It is possible that when closed cones are induced to open and shed seeds by fire heat, seeds could be dispersed by updraughts and winds generated by the fire itself. However, the heat of such fires may consume such seeds during transport (S. Arno & J. Brown, personal communication), or markedly reduce viability if the seeds are exposed to temperatures above 75 °C (Knapp & Anderson 1980).

Several pines (P. coulteri, P. sabiniana, P. torreyana) are reported to disperse seeds gradually from slowly-opening cones (Krugman & Jenkinson 1974; Borchert 1985). McMaster & Zedler (1981) have suggested this is an evolutionary compromise where intense crown fires alternate with long, fire-free intervals. Post-fire seed dispersal may be facilitated by canopy reduction, and the resulting freer flow of air currents, and by updraughts forming over the blackened soil surface (Lamont et al. 1991).

Variation in non-serotinous species

Even among pines whose cones apparently open promptly on maturity to disperse winged seeds, there is a wide variety of dispersal dates. Cone ripening and seed shed in P. caribaea var. hondurensis in Belize occurs during the rainy months of July and August (Loock 1977). Yet P. maximinoi, which occurs in similarly tropical conditions in nearby Mexico and Honduras, disperses seeds in the dry months of December and January (Perry 1991). Indeed, many Mexican pines (P. devoniana, P. douglasiana, P. durangensis, P. hartwegii, P. montezumae) also disperse seeds in December and January or 'the winter months' (Perry 1991). The southernmost of all pines, P. merkusii, sheds its seeds during March–May in Cambodia, April–June in Thailand, and June–November in Sumatra (Cooling 1968). Cones of P. nigra planted in northern Utah open and shed seeds in January, February or March, depending on when there is sufficient sunshine and lowered humidity to cause cone scales to 'crack open' audibly. This occurs earlier in cones on the south-facing side of the crown than on the north-facing side. They frequently close up again when humidity increases. In the same vein, Geiger (1965) cites observations by Kohlermann in pine forests of Hesse (presumably P. sylvestris) in which the optimum range of relative humidities for seed-fall was 55–65%. When humidity increased further, the cones closed up. P. Dreyfus (personal communication) reports similar behaviour of P. nigra in southern France. According to Büsgen & Münch (1929), both P. nigra and P. sylvestris disperse seeds in February and March, presumably in Germany. Seed dispersal of associated pines in the Himalayas of northeast Pakistan appears to come just before, or at the beginning of, the summer monsoon (April–May) in the case of P. roxburghii, and just after the monsoon (September–October) for P. wallichiana (Siddiqui & Pervez 1978; Das 1987). Thus, although most pines disperse their seeds in the autumn months of September and October (Krugman & Jenkinson 1974), there is greater variability than is usually acknowledged. The ecological consequences of varying phenology of seed dispersal in pines awaits serious investigation.

Another area of inquiry might be the mechanics of cone-scale retention and opening in serotinous and slowly-opening cones. Harlow, Coté & Day (1964) attributed the reflexing of pine cone-scales to differential shrinkage of tissues, but the range of cone behaviour patterns in Pinus suggests the subject might well be expanded.

14.2.5 Pine seed morphologies

Pinus is far more diverse in the morphology of its seeds than all other Pinaceae combined. Pine seeds may be fully winged, equipped with rudimentary wings, or have no attached structure (Fig 14.1). Wings may be removable (articulate) or firmly integrated (adnate) (Shaw 1914). Benkman (1995b) speculated that the upper limit of wing length in large pine seeds may be determined by the energetic costs of constructing cones whose scales would be long enough to contain them. He suggests that this physical constraint might explain why such large-seeded species as P. sabiniana and P. torreyana have relatively short wings (see Fig. 14.1). Seed coats can vary in thickness over a wide range. Seed length can vary by a factor of 10, and seed mass by a factor of 270 among species of the genus (Krugman & Jenkinson 1974). Shapes vary from ellipsoid through pear-shaped and cylindrical to almost triangular. Coats may be reddish, purplish, greyish, brown, black, or mottled (Krugman & Jenkinson 1974). Coat thickness is seldom reported, but among the pinyon pines (subsection Cembroides) those of P. remota may measure 0.15 mm (Lanner & Van Devender 1981) and those of P. cembroides as much as 1.0 mm (Little 1968). Seed coats of P. sabiniana attain a thickness of 1.25 mm (Vander Wall & Balda 1977). Several of these factors have a direct bearing on seed dispersal. The winged condition is viewed as an adaptation to wind dispersal, and winglessness as an adaptation to dispersal by jays and nutcrackers (Aves: Corvidae). Implications of this dichotomy are discussed later in this chapter. It is possible that seed coat colour and mottling may serve as camouflage when viewed against certain substrates, thus conferring protection from predators that use visual rather than olfactory cues. This does not appear to have been investigated, however. Countering this idea is Vander Wall's (1993) report that chipmunks were more

Table 14.4. Reduction in length of the longest seed of a cone and filled-seed percentage with increasing elevation in *Pinus albicaulis* in Western Wyoming, USA. Collections were made on 22–24 August 1978. Figures in parentheses are sample sizes

Elevation (m)	Length of longest seed (cm)	Filled seed (%)
Squaw Basin and Wind River Lake		
2848	0.86 (22)	96.0 (82)
2879	0.95 (13)	87.2 (109)
2897	0.73 (3)	16.7 (20)
Rendezvous Mountain (9 locations)		
3056–3109	0.59 (29)	0 (128)

efficient in gathering winged seeds of *P. jeffreyi* than experimentally de-winged seeds. This implies the winged seeds were more visible to the rodents, in effect acting as 'flags'. Since the chipmunks are agents of regeneration (see below), this raises the possibility that the wing has evolved as more than just a device to facilitate wind dispersal. *Pinus edulis* seeds containing collapsed embryos, like those that often result from selfing, are tan in colour because they lack the pigmentation and waxy surface of viable seeds. Thus pigmentation may require the presence of a viable embryo within (R.M.L., personal observations).

14.2.6 Latitudinal effects on seed mass
Numerous reports indicate that the seed mass of pines is negatively correlated with latitude. This relationship has been demonstrated in *P. contorta* (Maschning 1971), *P. strobus* (Genys 1968), and especially *P. sylvestris* (Wright & Bull 1963; Reich, Oleksyn & Tjoelker 1994). According to the last-mentioned authors, seed mass has been shown also to vary according to collection year, age of the seed tree, cone size, location of the seed within the cone and the cone upon the tree, stand density, nutrient availability, and even the seed tree's resin-tapping history. Correlations of seed mass and seedling vigour, while interesting and important (Reich *et al.* 1994), are not relevant here, but what is relevant is the relative wind dispersability of seeds of different mass. This will be discussed below.

14.2.7 Altitudinal effects on seed soundness
Surprisingly little work seems to have been done on altitudinal effects on conifer seed characters. Tranquillini (1979) reported work of Tschermak that showed *Picea abies* (Norway spruce) cone crops to be far less frequent at the timberline than in closed forest downslope. Reich *et al.* (1994) reported that *P. sylvestris* seed mass was unrelated to altitude of the seed stands, which ranged in elevation from 40 to 1420 m. That represents only the lower half of Scots pine's elevational range, however (Vidaković 1991).

Perhaps samples taken closer to the upper limit would show a reduction in seed mass similar to that at higher latitudes. According to Tranquillini (1979), Holzer found a reduction in mean mass from 900 to 500 mg per thousand seeds of *Picea abies* over 700 m from a closed forest stand to the timberline in the Austrian Alps. An unpublished study conducted by Karen Snethen shows a pronounced reduction of filled seed percentage of *P. albicaulis* with increasing elevation, and a corresponding decrease in length of the longest seed of a cone (Table 14.4). Thus seed dispersal could only have occurred from the lower-elevation stands during that year (1978), as the upper-elevation stands yielded no sound seed at all. Since empty pine seeds are known to result from lack of pollination (Sarvas 1962), this may have been a contributory factor. The upper-slope trees were of 'flag-form', growing in *krummholz* thickets, and bore very few seed cones. It is likely that they also bore too few staminate cones to ensure pollination. Another likely factor is that the short growing season at upper elevations did not permit normal maturation of cones and seeds. Lengths of cones collected on 22–24 August 1978 were negatively correlated with elevation. Thus, between 2840 and 2900 m (Squaw Basin, Wyoming) mean lengths of five collections totalling 56 cones were 4.8–6.6 cm. The mean lengths of nine collections totalling 38 cones collected nearby at 3056–3109 m (Rendezvous Mountain, Wyoming) were 3.0–5.6 cm (Fig. 14.2).

The near absence of viable seeds at the treeline may be of general occurrence among North American conifers. Personal communications received by Karen Snethen in 1979 indicated that *krummholz* conifers at the treeline rarely, at best, have been observed to produce viable seeds (*Picea engelmannii* [M. Grant, J. Marr, P. Wardle, M. Baig], *Picea mariana* [P. Marchand], *Abies lasiocarpa* [M. Baig], *Pinus albicaulis* [M. Baig], and *Larix lyallii* [M. Baig]). Baig tested seeds of all the above species except *Picea mariana* with tetrazolium chloride and found <1% viability. Dahms (1984) reported that *krummholz* thickets of *Picea engelmannii* and *Abies lasiocarpa* in Colorado produced only about 2% as much seed as in the montane forest below. The treeline seeds were often empty, selfed, or insect-damaged. Thus, genetically differentiated populations like those described by Grant & Mitton (1977) of *A. lasiocarpa* and *P. engelmannii* may consist of site-selected non-breeding individuals arising from seed blown upslope from closed forest stands. Such populations would exist only through continued seed dispersal from below, but could become self-perpetuating if climatic warming lengthened the growing season, allowing them to become sexually reproductive.

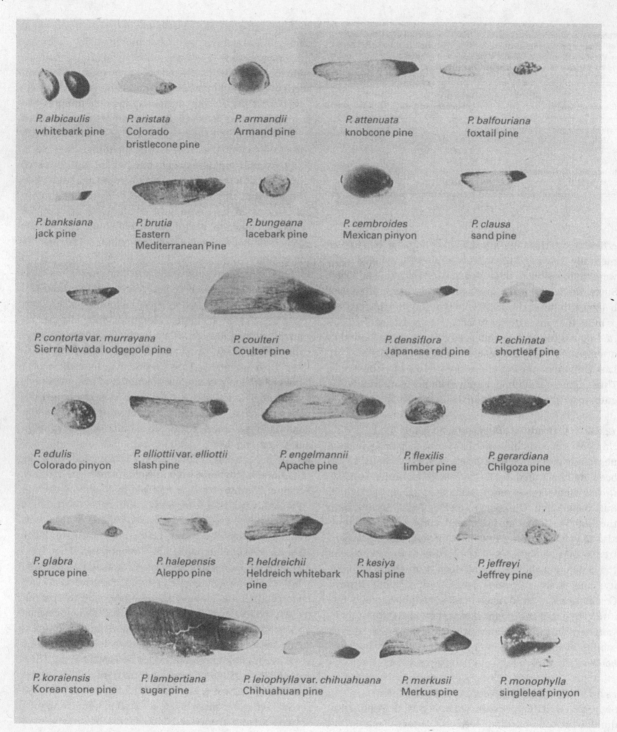

Fig. 14.1. *Pinus* seeds as shed naturally from their cones.
Reproduced, with permission, from Krugman & Jenkinson (1974).

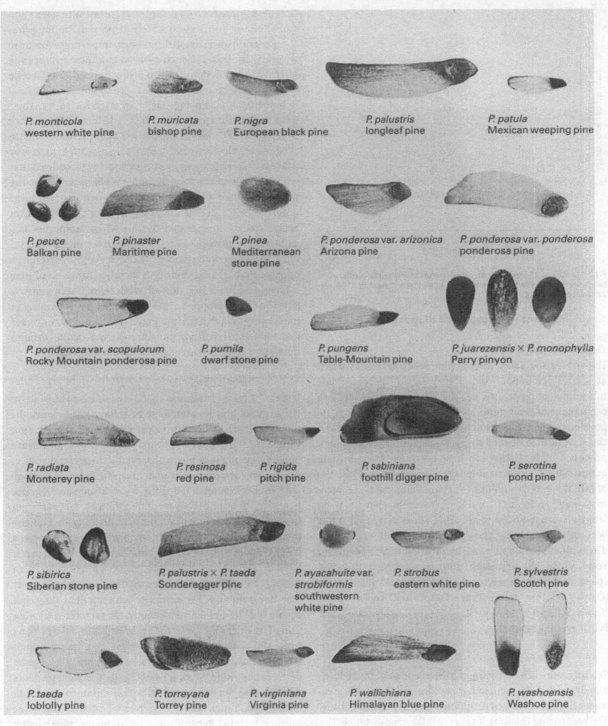

P. monticola
western white pine

P. muricata
bishop pine

P. nigra
European black pine

P. palustris
longleaf pine

P. patula
Mexican weeping pine

P. peuce
Balkan pine

P. pinaster
Maritime pine

P. pinea
Mediterranean
stone pine

P. ponderosa var. arizonica
Arizona pine

P. ponderosa var. ponderosa
ponderosa pine

P. ponderosa var. scopulorum
Rocky Mountain ponderosa pine

P. pumila
dwarf stone pine

P. pungens
Table-Mountain pine

P. juarezensis × P. monophylla
Parry pinyon

P. radiata
Monterey pine

P. resinosa
red pine

P. rigida
pitch pine

P. sabiniana
foothill digger pine

P. serotina
pond pine

P. sibirica
Siberian stone pine

P. palustris × P. taeda
Sonderegger pine

P. ayacahuite var.
strobiformis
southwestern
white pine

P. strobus
eastern white pine

P. sylvestris
Scotch pine

P. taeda
loblolly pine

P. torreyana
Torrey pine

P. virginiana
Virginia pine

P. wallichiana
Himalayan blue pine

P. washoensis
Washoe pine

Fig. 14.1. (cont.)

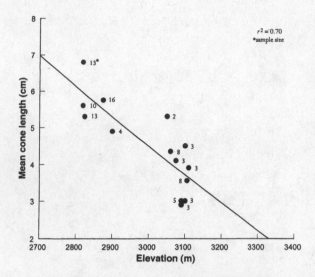

Fig. 14.2. **Effect of elevation on mean length of *Pinus albicaulis*** (whitebark pine) seed cones collected in western Wyoming, USA, on 22–24 August 1978.

14.3 Agents of pine seed dispersal

14.3.1 Wind

The wind dispersal of winged pine seeds has been studied at least since 1860, when Henry David Thoreau made numerous observations of the dispersal and seedling occurrence of *Pinus rigida* and *P. strobus* (Thoreau 1993). He pointed out the importance of the seed wing, and showed the significance of seed dispersal as a forest-successional process. Dingler (1889) may have been the first to initiate quantitative process studies when he determined the rate of fall of *P. sylvestris* seeds of varying wing length to range from 0.43 to 0.83 m s^{-1}, and to be inversely related to wing length (Stern & Roche 1974). Since those early days it has become apparent that the major factors influencing wind dispersal of seeds (and other biota) are rate of fall (settling velocity), height of release into the atmosphere, wind speed and turbulence, and morphological adaptations (Okubo & Levin 1989). These factors will be examined briefly. Seed morphology will not be treated separately, but is integrated elsewhere.

Rate of fall

Settling velocities of winged pine seeds are affected primarily by seed mass and wing size. Most studies of this parameter have consisted of dropping seeds of pines in still air, and measuring the time it takes to fall a fixed distance. The (usually unstated) assumption is that the more slowly a seed settles, the longer it is exposed to lateral air movement, and thus the greater is its potential to be carried a long distance. When samaras (including winged pine seeds) are dropped, they quickly attain a terminal velocity which is a function of their size. Therefore, according to

Greene & Johnson (1993), larger samaras fall faster than smaller ones, and are less widely dispersed. Among pines this generalization has little practical importance, because wing size increases in such a way as to almost balance the effects of increased mass. Thus, using data cited by Greene & Johnson (1993), while seed mass of *P. contorta* (7.9 mg) and *P. lambertiana* (180.0 mg) differ by a factor of 23, their descent velocities (0.81 vs 1.21 m s^{-1}) differ only by a factor of 1.5. Seeds of *P. contorta* and *P. radiata* have virtually identical descent velocities (0.81, 0.82 m s^{-1}), yet differ in mass by a factor of seven (7.9 vs 55.2 mg). Lanner (1985) showed that winged seeds of the long-winged *P. wallichiana* fell at the rate of 1.12 m s^{-1} but at 4.06 m s^{-1} when the wing was removed. Seeds of *P. flexilis* with the short, stubby wings occasionally found in that usually wingless species (Fig. 14.1) fell at 4.18 m s^{-1} with the wing intact, and 4.61 m s^{-1} after the wing was removed (Lanner 1985). It therefore appears that settling velocity is largely a function of *wing loading*, the mass per unit of wing area. Augspurger & Franson (1987) dropped artificial winged fruits varying in mass, wing area and morphology, from a 40 m tower under different wind conditions, and found wing loading to be an important determinant of mean dispersal distance.

Height of release

It seems intuitively obvious that a seed released high above the ground, and its attendant boundary layer, would be blown further than one released closer to the ground. Okubo & Levin (1989) accounted for the effects of release height by using it to normalize modal dispersal distance, which was then plotted against mean wind speed normalized by rate of seed fall, in a wind dispersal diagram. However, the studies they modelled show far less variation in seed release height than occurs naturally in an aerodynamically rough forest canopy, where trees of various heights and crowns of various lengths are the seed sources. Further, air behaves differently inside a forest than in an opening in which seed releases are simulated from a tower or other structure. For example, the turbulent flow that occurs within and above a canopy, characterized by ejection of warm, moist air from the forest, followed by strong sweeps of cool, dry air into the forest, is of major importance in transporting fluxes of heat, momentum and mass in the canopy (Gao, Shaw & Paw U 1989). Thus experiments done in the open do not realistically simulate forest conditions. Also, much wind dispersal of seed occurs over varying terrain. Recent work (Achtemeier 1994) using smoke-plumes to simulate atmospheric plankton, shows that plumes break up into chaotic patterns, leading to the transport of biota to 'unexpected' places. Empirical data on the airborne behaviour of spruce seeds dispersed naturally from cones about 15 m above the ground likewise suggest unexpected results (Zasada & Lovig 1983). One seed travelled 121 m, but landed only 13 m from the source tree. While airborne, it ascended twice and descended three times. Seeds followed circuitous paths, both vertically and horizontally. Some, lofted upwards, disappeared. In studying settling velocities of *P. contorta* seeds by dropping them from a seven-storey building, Cremer (1971) found that '...occasional slight updraughts caused some...seeds to disappear above the roof of the building ...' These observations are reminiscent of Sarvas' (1962) observations of pollen clouds that moved with a prevailing wind, but also irregularly upwards and downwards in air turbulence.

Another empirical observation that suggests past consideration of release height is oversimplified, is that of so-called 'take-off sites' on hills and ridges, from which enhanced-distance dispersal of seeds has been said to occur (Richardson, Williams & Hobbs 1994).

Wind speed and turbulence

These factors overlap with those just discussed, but will be considered as occurring on a larger spatial scale. Little experimental work has been done to explore their effects. In a seed-release trial, Augspurger & Franson (1987) found a group of artificial samaras to be updraughted far beyond the majority of the fruits released. Twelve of them landed 172–277 m from the release point. The continuously distributed fruits had landed no further than 107 m. They concluded that 'normal methods' of study usually miss such events, and that more study of 'extreme tails' is needed. But it is difficult for this viewpoint to receive serious attention because the weight of evidence that has been gathered is consistent in showing winged-seed dispersal to be mainly a local affair. For example, Buttrick (1914) reported that *c.* 80% of *P. palustris* seeds were caught in seed traps within 40 m of the seed tree, and the highest seed concentration was within 10 m of the tree. Yocom (1968) found that half of the *P. echinata* seeds trapped in a forest opening fell within 20 m of the forest edge, and 85% within 50 m. Seedfall was densest at the edge and within 10 m of the edge, inside the forest. Fowells (1950) reported seedfall in concentric rings around isolated *P. lambertiana* seed trees. Seedfall density was 53 500 per hectare within 32 m of the tree, but only 9125 ha^{-1} in a ring extending from 55 to 64 m. Similar studies have been reviewed by McCaughey, Schmidt & Shearer (1986), and are legion in the forestry literature. Thus one concludes that seeds are dispersed along a negative exponential curve whose peak frequency is at or just downwind from the forest edge, and the tail descends to a value close to zero within a few tens of metres.

Dissatisfied with the state of knowledge of long-distance dispersal, Greene & Johnson (1995) recently tested two models that predicted no discernible reduction in seed deposition under high wind speeds at 300–1600 m from the source. They point out that empirical observation of seed deposition many kilometres from the source is very difficult methodologically, and suggest the use of islands in a large lake. But they point out that at a distance of 10 km, a seed-trap area of 100 m^2 might be required to catch a single seed, even during a mast year.

The situation is made even more complex by unstable meteorological conditions. According to McCaughey *et al.* (1986), Shearer observed that early-ripening seeds of *Pseudotsuga menziesii* in the Rocky Mountains were dispersed largely by upslope thermal winds, while late-ripening seeds were dispersed by winds associated with storm fronts. The modern concept of thunderstorm development has implications for long-distance seed transport. According to Moran & Morgan (1995), a thunderstorm begins with a cumulus stage consisting of an upward surge of air to 8000–10 000 m, as saturated air updraughts strongly enough to suspend water droplets and ice crystals. In the mature stage, updraughting ceases, and precipitation causes a downdraught which spreads ahead of the thundercell. Gust fronts form along the downdraught front, triggering secondary thunderstorm cells tens of kilometres ahead of the parent cell. Studies of seed dispersal during such unsettled, turbulent conditions do not seem to have been made. The parallel literature of pollen dispersal also shows theory supporting a negative exponential curve of short reach, while empirical data provide strong evidence of frequent long-distance dispersal (Lanner 1966). I suggest that relatively long-distance dispersal of pine seeds will be found to be biologically significant if the methodological problems inherent in such studies can be solved. But this cannot be done in the typically short transects of seed traps lined out from a forest edge during fair weather.

Another wind phenomenon is the gliding of seeds on a snow or ice surface. Holmboe noticed this as early as 1898 in *P. sylvestris* (Stern & Roche 1974). McCaughey *et al.* (1986) report such movement occurring over 'considerable distances' for numerous Pinaceae, but cite no figures. To be effective, this dispersal means would require that seeds be released onto the snow in the winter, an unlikely phenology in pines of cold climates. Further, as Greene & Johnson (1995) suggest, many such seeds would probably be trapped in surface depressions or buried by later snowfalls. The importance of this phenomenon is undetermined.

14.3.2 Birds [Family Corvidae]

The dispersal of pine seeds by birds is of special biological significance because it usually occurs in the context of a mutualism. The interacting members of the mutualism are pines with wingless or nearly wingless seeds, and birds of several genera of Corvidae, the crow family. Seed winglessness (including near-winglessness) is the majority

Fig. 14.3. **Clark's Nutcracker (*Nucifraga columbiana*) in a *Pinus contorta* subsp. *latifolia* tree in Rocky Mountain National Park, Colorado, USA (photo: D.F. Tomback).**

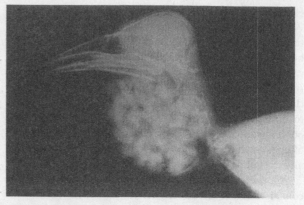

Fig. 14.4. **X-ray image of Clark's Nutcracker (*Nucifraga columbiana*) with 28 seeds of singleleaf pinyon (*Pinus monophylla*) in its sublingual pouch. The seeds weigh 31 grams and the bird weighs 141 g (photo: S.B. Vander Wall).**

condition among pines of subgenus *Strobus*, and the habit is disseminated throughout the North Temperate Zone among species of subsections *Cembrae*, *Cembroides* and *Strobi*. Subgenus *Pinus* has no strictly wingless-seeded species, but does include four species whose somewhat reduced or rudimentary wings are probably of little use in wind dispersal. None of these pines – *P. coulteri*, *P. pinea*, *P. sabiniana* and *P. torreyana* – has yet been sufficiently studied to identify their dispersers with confidence. In the case of *P. pinea* it is premature to conclude, as Blondel & Aronson (1995) have done, that there are no close bird–plant interactions in Mediterranean habitats. According to Turček & Kelso (1968), the Azure-winged Magpie (*Cyanopicus cyanus*) stores *P. pinea* seeds in the ground. Additionally, Frankis (1992) has presented intriguing evidence of a possible evolving mutualism involving the nuthatch *Sitta krueperi* and *P. brutia*. However, the rest of this discussion will deal with subgenus *Strobus*, the soft pines, and their corvid dispersers, because that is where the available evidence is concentrated. Since this topic has recently been covered in great detail in several books (Crocq 1990; Vander Wall 1990; Marzluff & Balda 1992; Lanner 1996), the treatment here will be cursory.

Avian dispersal in subsection *Cembroides*

The only pinyon pines that have received sufficient study to conclude they are mutualistic with corvids are *P. edulis* and *P. monophylla*. Four corvids – Scrub Jay (*Aphelocoma coerulescens*), Steller's Jay (*Cyanocitta stelleri*), Pinyon Jay (*Gymnorhinus cyanocephalus*) and Clark's Nutcracker (*Nucifraga columbiana*; Fig. 14.3) – disperse pinyon pine seeds. All four species extract seeds from open cones: Clark's Nutcracker and the Pinyon Jay also remove them from still-closed cones by pecking away the scale tissues to make them accessible, and both bird species perform

'tests' on seeds in order to discard those that are empty or unsound (Lanner 1996). Seeds can be carried in the mouth (Scrub Jay), in an extensible oesophagus (Steller's and Pinyon Jays), or a sublingual pouch (Clark's Nutcracker; Vander Wall & Balda 1981; Fig. 14.4). Approximate pinyon seed-carrying capacities of these four corvids are, respectively, 5, 18, 50, and 90 seeds. Flight distances range from local (Scrub and Steller's Jays), to 1–5 km (Pinyon Jay), to 22 km (usually 5–15 km) (Clark's Nutcracker) (Vander Wall 1990). Much, if not all, regeneration of pinyon pines stems from seeds cached in the soil (typical depth, 2–3 cm), and not recovered by the birds. Numerous coadaptations of the pines and birds strongly suggest that their relationship is a coevolved mutualism (Marzluff & Balda 1992). Similar morphologies and behaviours among other pinyons, and their coexistence with the same or related corvids, suggests the mutualism extends to all members of *Cembroides*: *P. maximartinezii*, *P. nelsonii* and *P. pinceana*.

Avian dispersal in subsection *Cembrae*

It has long been known that seeds of the stone pines *P. cembra*, *P. pumila* and *P. sibirica* are harvested from their 'indehiscent' cones (Shaw 1914), transported, and cached in the soil as a winter food by Eurasian Nutcrackers (*Nucifraga caryocatactes*); and that seeds not recovered by the birds frequently germinate. Recent work has shown in detail that a similar relationship exists in North America between *P. albicaulis* and *N. columbiana* (Tomback 1978; Fig. 14.5). While this may not be quite a 'one-to-one' mutualism, it does appear that the pine is completely dependent on its avian disperser for its regeneration (Hutchins & Lanner 1982), and the nutcracker is, at least, heavily dependent on the pine for its major year-round food source, during the juvenile stage as well as the adult (Vander Wall & Hutchins 1983). A recent study (Hutchins, Hutchins & Liu of *P. sibirica* about 1000 km from the natural ranges of both participants (Lanner & Nikkanen 1990).

Avian dispersal in other *Strobus* subsections
Among pines of subsection *Strobi*, dispersal of seeds and establishment of seedlings is well established in *P. flexilis* (Lanner & Vander Wall 1980) and *P. ayacahuite* var. *strobiformis* (Benkman, Balda & Smith 1984). The Eurasian Nutcracker has been identified as a disperser of *P. parviflora* (Hayashida 1989). Nutcrackers are native to the mountains of Taiwan, where *P. armandii* occurs, and they are likely to be mutualists. Regeneration of *P. peuce* outside Finnish plantations has been attributed to the Eurasian Nutcracker (Lanner & Nikkanen 1990). Clark's Nutcracker has been implicated in the regeneration of *P. longaeva* (subsection *Balfourianae*) (Lanner, Hutchins & Lanner 1984; Lanner 1988). Both pines of subsection *Gerardianae*, *P. gerardiana* and *P. bungeana*, have functionally wingless seeds, and occur within the ranges of nutcracker subspecies. They may be considered presumed mutualists, though data on them are lacking. Clark's Nutcrackers have been observed harvesting seeds of *P. contorta* (D.F. Tomback, personal communication), *P. jeffreyi* (Tomback 1978), and *P. ponderosa* (Giuntoli & Mewaldt 1978). In these cases, however, the pines have winged seeds, and no mutualistic relationship has been postulated. There is no longer any question that dispersal of wingless pine seeds by corvids is a major forest-ecological, biogeographic and evolutionary phenomenon in the northern hemisphere (Lanner 1996).

Fascinating and unexpected versions of avian dispersal of pine seed involving two cockatoo species (*Calyptorhynchus* spp.; Psittacidae) have been reported from Australia. The spread of *P. elliottii* from plantations into *Eucalyptus* forests in Queensland was attributed to glossy black-cockatoos (*C. lathami*) which carry off whole cones, allowing some seeds to fall free and eventually germinate. A similar phenomenon, involving *P. pinaster* and Carnaby's cockatoo (*C. funereus latirostrus*) has been described from Western Australia (see Chap. 22, this volume).

14.3.3 Mammals
Small mammals, especially rodents, are well known as consumers of conifer seeds, and are usually regarded by foresters as destructive pests. But rodents of many kinds store seeds, or seed-filled cones, in or on the soil after having transported them some distance. Therefore they are potential agents of dispersal and establishment. As with corvids, dispersal and establishment may occur together. But they may also be independent of each other. An example of a mammal causing seedling establishment in the absence of dispersal has been described by Alexander, Larson & Olson (1986). They found regeneration of *P. strobus* in a New Hampshire mixed conifer–hardwood

Fig. 14.5. **Whitebark pine (*Pinus albicaulis*) seedling cluster, arising from a nutcracker cache, Mammoth Mountain, Inyo National Forest, eastern Sierra Nevada (photo: D.F. Tomback).**

1996) indicates that the same may be true of *P. koraiensis*, the least known of the *Cembrae*. The *Cembrae* pines exhibit a remarkable suite of morphological, anatomical and physiological adaptations that facilitate the location, harvest, transport and preservation of their seeds in a way that appears beneficial to the needs of nutcrackers. These include such diverse traits as branching habit, cone colour, cone-scale anatomy, seed winglessness, cone placement (Lanner 1982), and variable seed dormancy (Lanner & Gilbert 1994). Numerous morphological, anatomical and behavioural traits of nutcrackers also serve to facilitate nutcracker utilization of pine seeds, and therefore increase the nutcracker's ability to regenerate the *Cembrae* pines. As a result, the relationship of these pines to nutcrackers has been frequently cited as an example of a coevolved mutualism. There is considerable evidence that both Clark's Nutcracker and *P. albicaulis* have speciated from Asian forbears that arrived in North America as trans-Beringian immigrants (Lanner 1996). A mutualism has become established in Finland since the late 1960s between Siberian nutcracker immigrants and plantations forest to be enhanced by the soil disturbance created by Grey Squirrels (*Sciurus carolinensis*) while burying acorns and hickory nuts. The versatility of animals as dispersers and establishers is further highlighted by the report of this same eastern North American squirrel transporting seeds and establishing seedlings of a pine from the Mediterranean Basin (*P. pinea*) in South Africa (Richardson, Williams & Hobbs 1994).

North American Red (Pine) Squirrel, *Tamiasciurus hudsonicus*

This animal larder-hoards great numbers of unopened cones of pines and other Pinaceae (Finley 1969), and scatter-hoards some seeds in soil caches (Kendall 1980). It does not disperse seeds widely, due to its small territories (0.25–1.0 ha) (Smith 1970). It has been found not to be an agent of *P. albicaulis* regeneration (Hutchins & Lanner 1982). Benkman (1995a) suggested that *Tamiasciurus* has acted to constrain the evolution of some cone traits that aid seed dispersal by birds.

Old World Red Squirrel, *Sciurus vulgaris*

This non-territorial squirrel is primarily a scatter-hoarder, and does not defend its food caches, which are numerous and relatively small. An effective dispersal role for this species has been reported by Miyaki (1987) and Hayashida (1989) in regard to seeds of *P. koraiensis*. They reported scatter-hoarding of seeds as much as 1.8 km from source in studies conducted in plantations in Hokkaido, Japan. The pine plantations were surrounded by deciduous broadleaved forests at low elevations (140–240 m; Miyaki 1987), and no nutcrackers were present. Thus the squirrel was the only disperser. Similar studies made in a natural old-growth *P. koraiensis* forest in northeastern China, however, found the Eurasian Nutcracker to be the major disperser of Korean stone pine (Hutchins *et. al.* 1996).

White-footed Mouse, *Peromyscus leucopus*

This small North American rodent disperses seeds of *P. strobus* less than 50 m into soil caches (Abbott & Quink 1970).

Chipmunks, *Eutamias* and *Tamias* spp.

Eutamias sibirica is known to scatter-hoard large quantities of *P. sibirica* seed (Shtil'mark 1963). Dispersal behaviour has not been clearly documented. *Tamias amoenus*, the Yellow Pine Chipmunk, and *T. speciosus*, the Lodgepole Pine Chipmunk, have been shown by Vander Wall (1992, 1993, 1994) to play a critically important role in the regeneration of *P. jeffreyi* in the Sierra Nevada range of California and Nevada by placing seeds into small scatter-hoarded soil caches from which many germinate. In one set of experi-

ments, Vander Wall (1993) created, around two trees, artificial seed shadows simulating the negative exponential distribution of wind-dispersed seeds around seed trees. All of these seeds were placed within 12 m of their supposed source trees. Chipmunks were allowed to gather up the seeds, and their caches were later found to have been made 2–69 m from the source trees. Redistribution of the pine seeds was largely *against* the prevailing wind. Thus Vander Wall demonstrated that the dispersal of *P. jeffreyi* seeds by wind, at least in the near vicinity of the seed trees, was one of two distinct dispersal stages. The second stage, mediated by the chipmunks, tended to redistribute the seeds into a more uniform spacing regime, and moved them further from the influence of other *P. jeffreyi* trees into more open patches. Vander Wall (1993) contrasts wind dispersal with animal dispersal, and concludes that the former is random with regard to the necessities for seedling establishment, while the latter represents a higher quality of dispersal. The possibility should be considered that redistribution of wind-dispersed seeds by rodents may be an integral part of the dispersal/establishment process in some other pines with effectively winged seeds. However, Benkman (1995b) suggests that small-seeded conifers with winged seeds that are effectively dispersed by wind appear not to benefit from the activities of ground-foraging rodents. Clearly, this is an area that needs more attention from researchers. The traditional – one might say dogmatic – idea among foresters that wind-dispersed seeds, however large, somehow find safe sites by falling into soil cracks and crevices, appears to lack the benefit of evidence. It may have come about as a rationalization based on the unlikelihood that seeds could lie on the soil surface from seedfall to germination without succumbing to predation or dehydration. Perhaps more species of *Pinus* are dependent on animals for their regeneration than have been acknowledged.

14.4 Conclusions

Some years ago, in discussing pine shoot development, I concluded that it was not '. . . the stereotyped affair it is often represented to be, but is richly diverse' (Lanner 1976). The same can be said of seed dispersal in this most complex of conifer genera. The data reviewed here suggest that dispersal of *winged pine seeds* can be a multi-level process. Initial dispersal of seeds that are moved *laterally by normal winds* leads to the negative exponential distribution. The result is a high concentration of seeds on the ground close to their source, diminishing exponentially with distance. These seeds can then be redistributed by rodents attracted to the concentration of nutritious food, and if properly cached and left uneaten and unrecovered, can become germinants. This process results in augmentation and perpetuation of existing pine populations, and establishment of pines in nearby patches of proven environment (Vander Wall 1993).

The unknown fraction of seeds *carried long distances in turbulent atmospheric structures* may have an important genetic impact: as Hamrick (1987) put it, '. . . these relatively rare, long-distance events are the most effective in increasing neighborhood sizes and decreasing population differentiation'. Long-distance dispersal is riskier than local dispersal, but can lead to new populations in new places, giving evolution a spatial as well as a temporal dimension.

Among *wingless-seeded pines*, avian seed dispersal allows the construction of large seeds filled with large quantities of nutrient, surely an advantage to species growing in stressful habitats. Dispersal is both local and long-distance, and much of it is into habitats chosen for their openness. Seeds are placed in good germination sites, and over-caching, normal bird mortality, and forgetfulness guarantee some will always be available to sprout. This is a relatively recent development in *Pinus*, and selection of winged-seeded species in the direction of winglessness

continues (Lanner 1996). Our understanding and appreciation of avian dispersal in *Pinus* has indeed come a long way since Mirov (1967) concluded that it was of only local importance.

Like all conclusions, these are based on an interpretation of available evidence, but not all the needed evidence is yet available. For example, the ecology of dispersal phenology is little understood. Cone-scale mechanics is overdue for a second look, as it is now receiving in *Cupressus* (Lev-Yadun 1994). The issue of seed camouflage and 'flags' needs attention. The question of whether high-elevation 'races' are self-perpetuating, or merely the temporary detritus of dispersal from below, needs to be addressed. Realistic studies of what happens to real seeds in strong weather should replace the seemingly endless efforts to model idealistic data to keep arriving at the same answers. More of the 'Bird Pines' and their dispersers need to be studied, to deepen our understanding of this phenomenon. And finally, the positive role of rodents needs far more attention than it has received in the past. How important are rodents to pines? Are there unsuspected mutualisms awaiting discovery? Careful seed-by-seed investigation is needed in order to forge a stronger understanding of the regenerative process of which seed dispersal is but one essential step.

References

Abbott, H.G. & Quink, T.F. (1970). Ecology of eastern white pine seed caches made by small forest mammals. *Ecology*, **51**, 271–8.

Achtemeier, G.L. (1994). Evidence of chaos in model-simulated nocturnal movement of atmospheric plankton. In *21st Conference on Agricultural and Forest Meteorology and 11th Conference on Biometeorology and Aerobiology*, pp. 389–92. Boston, Mass.: American Meteorological Society.

Alexander, L., Larson, B.C. & Olson, D.P. (1986). The influence of wildlife on eastern white pine regeneration in mixed hardwood–conifer forests. In *Eastern White Pine: Today and Tomorrow*, ed. D.T. Funk, pp. 40–4. USDA Forest Service General Technical Report WO-51.

Arno, S.F. & Hoff, R.J. (1989). *Silvics of Whitebark Pine* (Pinus albicaulis). USDA Forest Service General Technical Report INT-253.

Augspurger, C.K. & Franson, S.E. (1987). Wind dispersal of artificial fruits varying in mass, area, and morphology. *Ecology*, **68**, 27–42.

Benkman, C.W. (1995a). The impact of tree squirrels (*Tamiasciurus*) on limber pine seed dispersal adaptations. *Evolution*, **49**, 585–92.

Benkman, C.W. (1995b). Wind dispersal capacity of pine seeds and the evolution of different seed dispersal modes in pines. *Oikos*, **73**, 221–4.

Benkman, C.W., Balda, R.P. & Smith, C.C. (1984). Adaptations for seed dispersal and the compromises due to seed predation in limber pine. *Ecology*, **65**, 632–42.

Blondel, J. & Aronson, J. (1995). Biodiversity and ecosystem function in the Mediterranean Basin: human and non-human determinants. In *Mediterranean-Type Ecosystems. The Functions of Biodiversity*, ed. G.D. Davis & D.M. Richardson, pp. 43–119. Ecological Studies 109. Berlin: Springer-Verlag.

Borchert, M. (1985). Serotiny and cone-habit variation in populations of *Pinus coulteri* (Pinaceae) in the southern Coast Ranges of California. *Madroño*, **32**, 29–48.

Bramlett, D.L. & Pepper, W.D. (1974). Seed yield from a diallel cross in Virginia pine. In *Seed Yield From Southern Pine Seed Orchards Colloquium*, ed. J. Kraus, pp. 49–55.

Büsgen, M. & Münch, E. (1929). *The Structure and Life of Forest Trees*. London: Chapman & Hall.

Buttrick, P.L. (1914). Notes on germination and reproduction of longleaf pine in southern Mississippi. *Forestry Quarterly*, **12**, 532–7.

Cooling, E.N.G. (1968). *Pinus merkusii*. Fast Growing Timber Trees of the Lowland Tropics No. 4. Oxford: Commonwealth Forestry Institute.

Cremer, K.W. (1971). Speeds of falling & dispersal of seed of *Pinus radiata* and *P. contorta*. *Australian Forest Research*, **5**, 29–32.

Crocq, C. (1990). *Le Casse-Noix Moucheté (Nucifraga caryocatactes)*. Lechevalier-Chabaud.

Dahms, A. (1984). The natural reproduction of some selected tree species in the forest-tundra ecotone of the Colorado Front Range in an ecological view. Abstract of thesis, Institut für Geographie, Münster.

Das, P.K. (1987). Short- and long-range monsoon predictions in India. In *Monsoons*, ed. J.S. Fein & P.L. Stephens, pp. 549–78. New York: Wiley Interscience.

Dingler, H. (1889). *Die Bewegung der Pflanzlichen Flugorgane*. Munich.

Finley, R.B. (1969). Cone caches and middens of *Tamiasciurus* in the Rocky Mountain region. *University of Kansas Museum of Natural History Miscellaneous Publication*, **51**, 223–73.

Fowells, H.A. (1950). Some observations on the seedfall of sugar pine. USDA Forest Service, California Forest and Range Experiment Station Research Note 70.

Fowler, D.P. (1965). Effects of inbreeding in red pine, *Pinus resinosa* Ait. II. Pollination studies. *Silvae Genetica*, **14**, 12–23.

Frankis, M.P. (1992). Krüper's Nuthatch *Sitta krueperi* and Turkish pine *Pinus brutia*: an evolving association? *Sandgrouse*, **13**, 92–7.

Gao, W., Shaw, R.H. & Paw U, K.T. (1989). Observation of organized structure in turbulent flow within and above a forest canopy. *Boundary-Layer Meteorology*, **47**, 349–77.

Gauthier, S., Bergeron, Y. & Simon, J.-P. (1992). Cone serotiny in jack pine: ontogenetic, positional and environmental effects. *Canadian Journal of Forest Research*, **23**, 394–401.

Geiger, R. (1965). *The Climate Near the Ground*. Cambridge, Mass.: Harvard University Press.

Genys, J.B. (1968). Geographic variation in eastern white pine: two-year results of testing range-wide collections in Maryland. *Silvae Genetica*, **17**, 6–12.

Giuntoli, M. & Mewaldt, L.R. (1978). Stomach contents of Clark's nutcracker collected in western Montana. *Auk*, **95**, 595–8.

Graber, R.E. (1970). *Natural Seedfall in White Pine (Pinus strobus L.) stands of varying density*. USDA Forest Service Research Note NE-119.

Grant, M.C. & Mitton, J.B. (1977). Genetic differentiation among growth forms of Engelmann spruce and subalpine fir at tree line. *Arctic and Alpine Research*, **9**, 259–63.

Greene, D.F. & Johnson, E.A. (1993). Seed mass and dispersal capacity in wind-dispersed diaspores. *Oikos*, **67**, 69–74.

Greene, D.F. & Johnson, E.A. (1995). Long-distance wind dispersal of tree seeds. *Canadian Journal of Botany*, **73**, 1036–45.

Griffin, A.R. & Lindgren, D. (1985). Effect of inbreeding on production of filled seed in *Pinus radiata* – experimental results and a model of gene action. *Theoretical and Applied Genetics*, **71**, 334–43.

Hamrick, J.L. (1987). Gene flow and distribution of genetic variation in plant populations. In *Differentiation Patterns in Higher Plants*, ed. K.M. Urbanska, pp. 53–67. London: Academic Press.

Harlow, W.M., Coté, W.A., Jr & Day, A.C. (1964). The opening mechanism of pine cone scales. *Journal of Forestry*, **62**, 538–40.

Hayashida, M. (1989). Seed dispersal and regeneration patterns of *Pinus parviflora* var. *pentaphylla* on Mt. Apoi in Hokkaido. *Research Bulletins of the College Experimental Forests (Hokkaido University)*, **46**, 177–90.

Hellum, A.K. (1983). Seed production in serotinous cones of lodgepole pine. In *Lodgepole Pine: Regeneration and Management*, ed. M. Murray, pp. 23–7. USDA Forest Service, General Technical Report PNW-157.

Hutchins, H.E., Hutchins, S.A. & Liu, B. (1996). The role of birds and mammals in Korean pine (*Pinus koraiensis*) regeneration dynamics. *Oecologia*, **107**, 120–30.

Hutchins, H.E. & Lanner, R.M. (1982). The central role of Clark's nutcracker in the dispersal and establishment of whitebark pine. *Oecologia (Berlin)*, **55**, 192–201.

Johnsson, H. (1976). Contribution to the genetics of empty grains in the seed of pine (*Pinus sylvestris*). *Silvae Genetica*, **25**, 10–4.

Kendall, K.C. (1980). Bear–squirrel–pine nut interactions. In *Yellowstone Grizzly Bear Investigation. Annual Report 1978–79*, ed. R.R. Knight, B.M. Blanchard, K.C. Kendall & L.E. Oldenburg, pp. 51–60. Washington: National Park Service.

Khomentovsky, P.A. (1994). A pattern of *Pinus pumila* seed production ecology in the mountains of central Kamtchatka. In *Proceedings – International Workshop on Subalpine Stone Pines and their Environment: The Status of Our Knowledge*, ed. W.C. Schmidt & F.K. Holtmeier, pp. 67–77. USDA Forest Service General Technical Report INT-309.

Knapp, A.K. & Anderson, J.E. (1980). Effect of heat on germination of seeds from serotinous lodgepole pine cones. *American Midland Naturalist*, **104**, 370–2.

Krugman, S.L. & Jenkinson, J.L. (1974). *Pinus* L. Pine. In *Seeds of Woody Plants in the United States*, ed. S.C. Schopmeyer, pp. 598–638. USDA Forest Service Agriculture Handbook 450. Washington, DC: USDA.

Lamont, B.B., Le Maitre, D.C., Cowling, R.M. & Enright, N.J. (1991). Canopy seed storage in woody plants. *The Botanical Review*, **57**, 277–317.

Lanner, R.M. (1966). Needed: a new approach to the study of pollen dispersion. *Silvae Genetica*, **15**, 50–2.

Lanner, R.M. (1976). Patterns of shoot development in *Pinus* and their relationship to growth potential. In *Tree Physiology and Yield Improvement*, ed. M.G.R. Cannell & F.T. Last, pp. 223–43. London: Academic Press.

Lanner, R.M. (1980). A self-pollination experiment in *Pinus edulis*. *Great Basin Naturalist*, **40**, 265–7.

Lanner, R.M. (1982). Adaptations of whitebark pine for seed dispersal by Clark's Nutcracker. *Canadian Journal of Forest Research*, **12**, 391–402.

Lanner, R.M. (1983). The expansion of singleleaf piñon in the Great Basin. In *The Archaeology of Monitor Valley, 2. Gatecliff Shelter*, ed. D.H. Thomas, *Anthropological Papers of the American Museum of Natural History*, **59**, 167–71.

Lanner, R.M. (1985). Effectiveness of the seed wing of *Pinus flexilis* in wind dispersal. *Great Basin Naturalist*, **45**, 318–20.

Lanner, R.M. (1988). Dependence of Great Basin bristlecone pine on Clark's Nutcracker for regeneration at high elevations. *Arctic and Alpine Research*, **20**, 358–62.

Lanner, R.M. (1996). *Made For Each Other. A Symbiosis of Birds and Pines*. New York: Oxford University Press.

Lanner, R.M. & Gilbert, B. (1994). Nutritive value of whitebark pine seeds, and the question of their variable dormancy. In *Proceedings – International Workshop on Subalpine Stone Pines and their Environment: The Status of Our Knowledge*, ed. W.C. Schmidt & F.-K. Holtmeier, pp. 206–11. USDA Forest Service General Technical Report INT-309.

Lanner, R.M., Hutchins, H.E. & Lanner, H.A. (1984). Bristlecone pine and Clark's Nutcracker: probable interaction in the White Mountains, California. *Great Basin Naturalist*, **44**, 357–60.

Lanner, R.M. & Nikkanen, T. (1990). Establishment of a *Nucifraga–Pinus* mutualism in Finland. *Ornis Fennica*, **67**, 24–7.

Lanner, R.M. & Vander Wall, S. B. (1980). Dispersal of limber pine seed by Clark's Nutcracker. *Journal of Forestry*, **78**, 637–9.

Lanner, R.M. & Van Devender, T.R. (1981). Late Pleistocene piñon pines in the Chihuahuan Desert. *Quaternary Research*, **15**, 278–90.

Lev-Yadun, S. (1994). Living serotinous cones in *Cupressus sempervirens*. *International Journal of Plant Sciences*, **156**, 50–4.

Little, E.L., Jr. (1968). Two new pinyon varieties from Arizona. *Phytologia*, **17**, 329–42.

Loock, E.E.M. (1977). *The Pines of Mexico and British Honduras*. Pretoria: Republic of South Africa Department of Forestry.

McCaughey, W.W., Schmidt, W.C. & Shearer, R.C. (1986). Seed dispersal characteristics of conifers in the inland mountain West. In *Proceedings – Conifer Tree Seed in the Inland Mountain West Symposium*, ed. R.C. Shearer, pp. 50–62. Ogden, Utah: Intermountain Research Station.

McDonald, J.E. (1962). Collection and washout of airborne pollen and spores by raindrops. *Science*, **135**, 435–7.

McMaster, G.D. & Zedler, P.H. (1981). Delayed seed dispersal in *Pinus torreyana* (Torrey pine). *Oecologia (Berlin)*, **51**, 62–6.

Marzluff, J.N. & Balda, R.P. (1992). *The Pinyon Jay: Behavioral Ecology of a Colonial and Cooperative Corvid*. London: Poyser.

Maschning, E. (1971). The variations in number of cotyledons in some provenances of *Pinus contorta*. *Silvae Genetica*, **20**, 10–4.

Matthews, J.D. (1963). Factors affecting the production of seed by forest trees. *Forestry Abstracts*, **24**, i–xiii.

Mergen, F. (1961). Natural and induced flowering in young pine trees. In *Sixth Southern Conference on Forest Tree Improvement Proceedings*, pp. 129–35. Gainesville: University of Florida.

Mirov, N.T. (1967). *The Genus Pinus*. New York: Ronald Press.

Miyaki, M. (1987). Seed dispersal of the Korean Pine, *Pinus koraiensis*, by the Red Squirrel, *Sciurus vulgaris*. *Ecological Research*, **2**, 147–57.

Moran, J.M. & Morgan, M.D. (1995). *Essentials of Weather*. Englewood Cliffs, NJ: Prentice Hall.

Okubo, A. & Levin, S.A. (1989). A theoretical framework for data analysis of wind dispersal of seeds and pollen. *Ecology*, **70**, 329–38.

Owens, J.N. (1991). Flowering and seed set. In *Physiology of Trees*, ed. A.S. Raghavendra, pp. 247–71. New York: John Wiley.

Perry, J.P., Jr. (1991). *The Pines of Mexico and Central America*. Portland, Oregon: Timber Press.

Reich, P.B., Oleksyn, J. & Tjoelker, M.G. (1994). Seed mass effects on germination and growth of diverse European Scots pine populations. *Canadian Journal of Forest Research*, **24**, 306–20.

Richardson, D.M., Williams, P.A. & Hobbs, R.J. (1994). Pine invasions in the Southern Hemisphere: determinants of spread and invadability. *Journal of Biogeography*, **21**, 511–27.

Saito, S. & Kawabe, M. (1990). On the forest vegetation of Mt Higashi-Nupukaushinupuri, Tokachi, Hokkaido. (2) On two thickets of *Pinus pumila*. *Bulletin of the Higashi Taisetsu Museum of Natural History*, **12**, 17–29.

Saracino, A. & Leone, V. (1991). Osservazioni sulla renovacione de Pino d'Aleppo (*Pinus halepensis* Mill.) in soprauoli percorsi dal fuoco. I. La disseminazione. *Monti e Boschi*, **42**, 39–46.

Sarvas, R. (1962). Investigations on the flowering and seed crop of *Pinus sylvestris*. *Communicationes Instituti Forestalis Fenniae*, **53**.4, 1–198.

Shaw, G.R. (1914). *The Genus Pinus*. Cambridge, Mass: The Riverside Press.

Shtil'mark, F.R. (1963). Ecology of the chipmunk (*Eutamias sibiricus* Laxm.) in cedar forests of the Western Sayan Mountains. *Zoologicheskii Zhurnal*, **42**, 92–102 [translated by L. Kelso].

Siddiqui, K.M. & Pervez, M. (1978). *Collections of Seed of Different Seed Sources and Establishment of Provenance Trials of Himalayan Blue Pine, Pinus wallichiana A.B. Jacks, syn: Pinus griffithii*. Annual Research Report. Peshawar: Pakistan Forest Research Institute.

Smith, C.C. (1970). The coevolution of pine squirrels (*Tamiasciurus*) and conifers. *Ecological Monographs*, **40**, 349–71.

Smith, C.C., Hamrick, J.L. & Kramer, C.L. (1988). The effects of stand density on frequency of filled seeds and fecundity in lodgepole pine (*Pinus contorta* Dougl.). *Canadian Journal of Forest Research*, **18**, 453–60.

Sorenson, F.C. (1970). *Self-fertility of a Central Oregon Source of Ponderosa Pine*. USDA Forest Service Research Paper PNW-109.

Stern, K. & Roche, L. (1974). *Genetics of Forest Ecosystems*. New York: Springer-Verlag.

Thoreau, H.D. (1993). *Faith in a Seed*. Washington: Island Press.

Tomback, D.F. (1978). Foraging strategies of Clark's nutcrackers. *Living Bird*, **16**, 123–61.

Tranquillini, W. (1979). *Physiological Ecology of the Alpine Timberline*. New York: Springer-Verlag.

Turček, F.J. & Kelso, L. (1968). Ecological aspects of food transportation and storage in the Corvidae. *Commun. Behav. Biol. Pt. A*, **1**, 277–97.

Vander Wall, S.B. (1990). *Food Hoarding in Animals*. Chicago: University of Chicago Press.

Vander Wall, S.B. (1992). Establishment of Jeffrey pine seedlings from animal caches. *Western Journal of Applied Forestry*, **7**, 14–20.

Vander Wall, S.B. (1993). Cache site selection by chipmunks (*Tamias* spp.) and its influence on the effectiveness of seed dispersal in Jeffrey pine (*Pinus jeffreyi*). *Oecologia (Berlin)*, **96**, 246–52.

Vander Wall, S.B., (1994). Removal of wind-dispersed pine seeds by ground-foraging vertebrates. *Oikos*, **69**, 125–32.

Vander Wall, S.B. & Balda, R.P. (1977). Coadaptations of the Clark's nutcracker and the piñon pine for efficient seed harvest and dispersal. *Ecological Monographs*, **47**, 89–111.

Vander Wall, S.B. & Balda, R.P. (1981). Ecology and evolution of food-storage behavior in conifer-seed-caching corvids. *Zeitschrift für Tierpsychologie*, **56**, 217–42.

Vander Wall, S.B. & Hutchins, H.E. (1983). Dependence of Clark's nutcracker, *Nucifraga columbiana*, on conifer seeds during the post-fledging period. *The Canadian Field-Naturalist*, **97**, 208–14.

Vidaković, M. (1991). *Conifers: Morphology and Variation*, revised English edition. Zagreb: Graficki Zavod Hrvatske.

Wright, J.W. (1976). *Introduction to Forest Genetics*. New York: Academic Press.

Wright, J.W. & Bull, W.I. (1963). Geographic variation in Scotch pine: Results of a 3-year Michigan study. *Silvae Genetica*, **12**, 1–25.

Yocom, H.A. (1968). Shortleaf pine seed dispersal. *Journal of Forestry*, **66**, 422.

Zasada, J.C. & Lovig, D. (1983). Observations of primary dispersal of white spruce, *Picea glauca*, seed. *The Canadian Field-Naturalist*, **97**, 104–6.

15 Ecophysiology of *Pinus*

Philip W. Rundel and Barbara J. Yoder

Ecology and Biogeography of Pinus, ed. D.M. Richardson. © Cambridge University Press, Cambridge (1998), pp. 296–323.

15.1 Introduction

Were Nicholas Mirov to review the literature today, he would be amazed at the progress that has been made over the past three decades in the study of the ecophysiology of pines. The relatively few and scattered papers that existed on the photosynthetic response, water relations and cold tolerance of pines when Mirov's (1967) book was published have been superseded by literally hundreds of papers on these subjects. Research on pine ecophysiology in the broad sense, including studies of responses to air pollution and increased concentrations of atmospheric carbon dioxide, continue to show an almost logarithmic increase, and will thus no doubt continue to make major strides in the future.

This chapter is not the first to attempt a broad review of the ecophysiology of pines. The physiological ecology of pines and other conifers in the western USA was reviewed a decade ago by Lassoie, Hinckley & Grier (1985) and Smith (1985). More recently, a major review of conifer ecophysiology has been presented in two volumes by Smith & Hinckley (1995a, b), and a broad review of pine ecology by Gholz, Linder & McMurtrie (1994). These references should be consulted for broader perspectives of the ecology and physiology of conifers.

Our approach in writing this chapter has been to focus first on the physiological processes of photosynthesis, water relations, and respiration that are fundamental in the response of pines to their physical environment. A large literature on pine nutrient relations and fertilization studies is not reviewed here, except as it relates directly to photosynthetic capacity. We then take an ecological perspective on the adaptive traits of pines in their response to the complex environmental stresses which they face in a series of differing habitats. Finally, we briefly review the ecophysiological responses of pines to atmospheric pollutants, particularly ozone and other oxidant pollutants, and to changing atmospheric concentrations of CO_2.

15.2 Physiological processes

15.2.1 Photosynthesis

There is a wealth of information on leaf photosynthetic rates from greenhouse and field studies of pines, in part due to the availability of field-portable infrared gas analysers at relatively low cost. However, it is often difficult to compare results among reports because the gas exchange rates are expressed on different bases (total leaf area, projected leaf area, or leaf dry weight) and because of differences in methods and instrumentation. For example, Gower, Reich & Son (1993) showed that photosynthetic capacity across several species varied by 300% when expressed per unit weight, but by only 50% when expressed per unit area. The difference between using an area versus weight basis for CO_2 assimilation is particularly important when considering photosynthetic rates of pines, because the leaf area per unit dry weight (specific leaf area) of pine needles is highly variable and can be unusually low. For example, the specific leaf area for *Pinus contorta* and *P. ponderosa* can be 30 $cm^2 g^{-1}$ or less (Cregg 1994; Yoder et al. 1994), for 30–40 year-old *P. strobus* in the study of Gower et al. (1993) it was 74, and for shade needles of *P. radiata* it can approach 300 $cm^2 g^{-1}$ (Rook, Bollmann & Hong 1987).

Further confusion can arise from expressing net photosynthesis in terms of total leaf surface area or one-sided, projected leaf area. Again, the variability of pines, in leaf geometry in this case, can lead to very different interpretations depending on the measurement basis. For pines with three-needled fascicles, like *P. ponderosa*, the ratio between total and projected leaf area is 2.36 if the needles are assumed to be thirds of a cylinder; for pines with two-needled fascicles, like *P. contorta*, it is 2.57; and for *P. monophylla* (the only pine with one-needled, cylindrical fascicles) it is 3.14 (π). Stenberg *et al.* (1995b) argued that it is more appropriate to define gas exchange of conifer needles in terms of the total, rather than projected, needle surface. Because needles function as optically 'black' surfaces, essentially no visible light impending on one surface is transmitted to the other side of the needle. Also, stomata occur on all surfaces of pine needles. On the other hand, in a direct beam of radiation, light is captured only by the projected surface. The choice depends on the application; however, published information on gas exchange of pine needles is more commonly presented on a projected area basis.

Shoot structure also has a strong influence on photosynthetic rates, due to leaf orientation with respect to the light source and shading among needles (Smith, Schoettle & Cui 1991; Stenberg *et al.* 1995a; Sprugel, Brooks & Hinckley 1996). Some authors have suggested that the best measure of photosynthetic capacity is on the basis of the silhouette area of whole shoots, with leaves in their natural orientation with respect to the sun (Carter & Smith 1985). Smith *et al.* (1991) found that net photosynthetic rates of *P. contorta* and *Abies lasiocarpa* were 5–10 times higher when expressed per unit shoot-silhouette area compared with a total leaf area basis. In fact, net photosynthetic rates of these conifers expressed relative to the silhouette area of shoots were among the highest recorded for any species. Comparisons between *A. lasiocarpa* and *P. contorta* revealed that the shade-tolerant fir had more structural differentiation between sun and shade shoots compared with the less tolerant pine, leading to greater variability in silhouette leaf area as sun angles changed over the course of a day or season. The shade shoots of the fir intercepted the greatest amount of radiation at the solar zenith, whereas sun shoots of fir and both sun and shade shoots of the pine showed a strong midday depression in solar interception. These results emphasize the strong interdependence between shoot structure and leaf biochemistry in determining whole-plant CO_2 uptake. Unfortunately, few gas exchange studies include information on shoot structure and leaf orientation, so comprehensive comparisons of photosynthetic activity within and among species must be restricted primarily to unit leaf area or leaf weight bases.

15.2.2 Differences in photosynthetic capacity among and within species

Photosynthetic capacity of individual leaves (defined as the net photosynthetic activity in saturating light, ambient CO_2, and otherwise 'optimal' conditions), is much lower for gymnosperms than for angiosperms, although there is considerable variation within these groups. More distinct is the relatively low photosynthetic capacity of evergreens (including all pines) compared with deciduous species. Across species, higher nitrogen concentration and higher maximum net photosynthetic rates are inversely correlated with leaf longevity (Gower *et al.* 1993).

When photosynthetic capacity is evaluated per unit projected leaf area, pines often have high rates compared with other conifers. In a comparison of photosynthetic capacities of over 100 tree species by Ceulemans & Saugier (1991), the highest reported rates for *P. sylvestris* and *P. radiata* were approximately double those for other conifers (*c.* 16 μmol m^{-2} s^{-1} compared with 2–10 μmol m^{-2} s^{-1}). In part, the differences may result from the greater intensity of study for these two commercially important species compared with many other conifers. *Populus* sp., in turn, had approximately twice the photosynthetic capacity of *Pinus sylvestris* and *P. radiata*. (20–27 μmol m^{-2} s^{-1}). Because of the low specific leaf area of pine needles, their photosynthetic capacity tends to be quite low relative to other species on a leaf weight basis.

The maximum net photosynthetic rates reported for pines ranges from less than 5 to more than 20 μmol m^{-2} s^{-1} when expressed on a projected needle area basis (Table 15.1). The highest reported rates occur among the most economically important species, including *P. radiata*, *P. sylvestris* and *P. taeda*, although this may result partly from the intensity of study of these species. Teskey, Whitehead & Linder (1994b) noted that, in general, the differences in photosynthetic capacity between pine species are small compared to the variation within species.

Genetic variations in photosynthetic capability have been reported within races and subspecies of pines, although results are conflicting. In *P. ponderosa*, which occupies a broad diversity of sites, Monson & Grant (1989) noted significant differences between progeny from coastal versus interior trees. Trees resulting from a cross between the coastal and interior varieties (*ponderosa* × *scopulorum*) had lower maximum photosynthetic rates, lower needle nitrogen, lower stomatal conductance and higher water-use efficiency compared with trees with two coastal parents (*ponderosa* × *ponderosa*). On the other hand, Cregg (1994) studied 27 open-pollinated families of *P. ponderosa* with known differences in drought response, and found no significant differences in needle gas exchange, the relationships between photosynthesis and stomatal

Table 15.1. Maximum net photosynthesis (A_{max}) and stomatal conductance ($g_{s\,max}$) at ambient CO_2 and saturating light of current age-class needles for several pine species. Leaf area is presented on a projected area basis. Where the original reference was in terms of the total leaf surface, conversion factors of 2.36 and 2.57 total surface/projected surface were used for 3-needled pines and 2-needled pines, respectively

Pinus taxon	A_{max} (μmol m^{-2} s^{-1})	$g_{s\,max}$ (mmol m^{-2} s^{-1})	Reference
P. aristata	4.8	127.9	Schoettle (1994)
P. banksiana	9.5		Reich et al. (1995)
P. brutia subsp. eldarica	25	52	Garcia et al. (1994[a])
P. contorta	8.2		Dick, Jarvis & Leakey (1991)
P. contorta	5.1		Dykstra (1974)
P. contorta	6.9		Higginbotham et al. (1985)
P. contorta	9		Smith (1980)
P. elliottii	7.6	94	Teskey et al. (1994a)
P. jeffreyi	6.5	105	Patterson & Rundel (1989)
P. ponderosa	3–4		Cregg (1994)
P. ponderosa	12.8	188	B.J. Yoder et al. unpubl. data
P. radiata	11.3		Benecke (1980)
P. radiata	9.4	71	Conroy et al. (1988)
P. radiata	9.0		Hollinger (1987)
P. radiata	7.8		Rook & Corson (1978)
P. radiata	14.2	378	Thompson & Wheeler (1992)
P. resinosa	11.1		Gower et al. (1993)
P. resinosa	5.0/7.6		Reich et al. (1995)
P. strobus	8.5		Gower et al. (1993)
P. strobus	8	80	Maier & Teskey (1992)
P. strobus	8.1/8.5		Reich et al. (1995)
P. sylvestris	10.3	228	Beadle et al. (1985)
P. sylvestris	9.5	159	DeLucia et al. (1991)
P. sylvestris	7.7	159	Küppers & Schulze (1985)
P. sylvestris	17.2		Smolander & Oker-Blom (1989)
P. sylvestris	9.8		Smolander et al. (1987)
P. sylvestris	10.3		Strand & Öquist (1985)
P. sylvestris	14.4	113	Troeng & Linder (1982b)
P. sylvestris	24.4		James, Grace & Hoad (1994)
P. sylvestris	12.5		Reich et al. (1995)
P. taeda	14.2	420	Fites & Teskey (1988)
P. taeda	12.3	360	Thomas, Lewis & Strain (1994)
P. taeda	11.8	198	Teskey et al. (1986)

[a] Measurements are from gas exchange of entire canopies.

conductance, or leaf water potential in response to drought. Similarly, Grulke, Hom & Roberts (1993) found no significant differences in gas exchange between two P. ponderosa families with different growth rates, at either ambient or elevated CO_2 levels.

15.2.3 Seasonal variation in photosynthesis

The photosynthetic activity of evergreen conifers generally extends well beyond the growing season. For P. elliottii in Florida, Teskey, Gholz & Cropper (1994a) estimated that total carbon gain during the winter months accounted for about 19% of the annual total. This carbon is translocated to the roots, and results in significant accumulation of starch. Starch concentrations reached a maximum 70 mg g^{-1} in coarse roots of mature P. elliottii trees (Gholz & Cropper 1991). Even pines in regions with cold winters, such as P. sylvestris in Sweden, fix considerable amounts of carbon during autumn after stem growth ceases, and store significant amounts of starch in the roots (Ågren et al. 1980). Still, there is pronounced seasonal variation in photosynthesis in most pine ecosystems. Net photosynthesis during the winter may fall to zero for several months in the coldest regions (Troeng & Linder 1982a; Jurik, Briggs & Gates 1988).

Seasonal variation in net photosynthesis results from shifts in leaf-level capacity for photosynthesis and total leaf area as well as annual cycles of climate and radiation. Among the earliest ecophysiological studies of conifers were reports of depressed photosynthetic capacity during winter months (e.g. Bourdeau 1959; McGregor & Kramer 1963; Schulze, Mooney & Dunn 1967). Photoinhibition, resulting from an excess of absorbed light energy relative to consumption by the carbon reduction processes, is a primary cause of reduced photosynthetic capacity in winter. Cold air temperature and high light are a particularly lethal combination for photoinhibition. In addition, cold soil temperatures decrease the conductivity of roots to water and increase the viscosity of water, which can result in severe desiccation of foliage (see below). This condition can also result in photoinhibition, or may reduce photosynthesis directly through lower stomatal conductance. In P. strobus, Jurik et al. (1988) found that photosynthetic depressions during winter were more strongly correlated with low soil temperature than with low air temperature.

Studies of P. cembra, P. radiata, P. strobus and P. sylvestris show that photosynthetic capacity rises quickly in the spring and reaches a maximum early in the growth season (Jurik et al. 1988; Teskey et al. 1994b). However, in other cases the post-winter rise is more gradual, peaking in September for P. taeda (McGregor & Kramer 1963) and P. ponderosa (Hadley 1969) seedlings. Teskey et al. (1994a) pointed out that spring recovery tends to be most rapid in pines from cold climates. Very high photosynthetic capacity in spring may result from a high demand for carbohydrates during this period of rapid growth (Maier & Teskey 1992).

15.2.4 Photosynthetic response to light

Over 70 years ago, Bates (1925) demonstrated that growth in low-light conditions resulted in a dramatic reduction in root growth for eight species of conifer seedlings, including six pines, and concluded that low-light environments make seedlings more vulnerable to other stresses, such as desiccation or low nutrient availability. Indeed, later studies have shown that pines grown in shaded environments are more likely to suffer injury or death due to

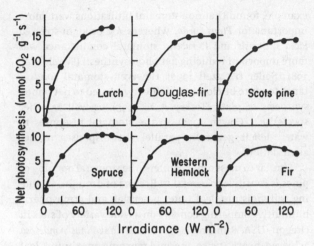

Fig. 15.1. **Net photosynthesis in relation to irradiance for shade-intolerant conifers:** *Larix decidua* **(larch),** *Pinus sylvestris* **(Scots pine) and** *Pseudotsuga menziesii* **(Douglas-fir); and shade-tolerant conifers:** *Abies alba* **(fir),** *Picea abies* **(spruce) and** *Tsuga hetrophylla* **(western hemlock). Adapted from Szaniawski & Wierzbicki (1978).**

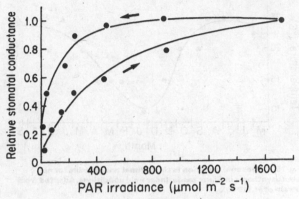

Fig. 15.2. **Hysteresis of stomatal conductance in response to increasing and decreasing PAR irradiance. Adapted from Ng & Jarvis (1980).**

drought (e.g. Vance & Zaerr 1991). As Kozlowski, Kramer & Pallardy (1991) point out, the immediate cause of death in such a situation may be desiccation, but the ultimate cause is reduced photosynthesis due to low light.

Pines generally have low tolerance to shade (Franklin & Dyrness 1973). Compared with shade-tolerant plants, intolerant plants tend to have higher light-saturated photosynthetic capacity, higher dark respiration rates, higher saturation irradiance for photosynthesis, and higher light compensation levels for photosynthetic activity. Although photosynthetic capacity of pines tends to be low compared with angiosperms, Szaniawski & Wierzbicki (1978; Fig. 15.1) showed that shade-intolerant conifer species (*Larix decidua*, *Pinus sylvestris*, *Pseudotsuga menziesii*) showed precisely these responses when compared with three tolerant species (*Abies alba*, *Picea abies* and *Tsuga heterophylla*). There was nearly a two-fold difference in light-saturated photosynthesis and dark respiration rates between the groups.

The shape of photosynthetic light response curves is quite similar among pine species (*Pinus elliottii*, *P. radiata*, *P. sylvestris*, *P. taeda*) growing in non-stressed conditions (Teskey *et al.* 1994b). For all of these species, photosynthesis was 90% saturated at an irradiance of 1000 μmol m^{-2} s^{-1}. Other publications show lower photosynthetic light-saturation levels. For example, a compilation of hundreds of measurements of photosynthesis for *P. ponderosa* over a growing season shows photosynthetic light saturation occurring around 750 μmol m^{-2} s^{-1} (Hadley 1969). Light response curves are quite sensitive to pre-conditioning on both short and long time scales. Even fully mature needles show reversible acclimation to the current light environment with large shifts in photosynthetic capacity

(Przykorska-Zelawska & Zelawski 1981). As the maximum rate of photosynthesis increases, the photosynthetic photon flux density (PPFD) required to saturate photosynthesis increases as well. Over shorter time scales, Ng & Jarvis (1980) found that the responses of stomatal conductance of *P. sylvestris* to varying PPFD was strongly dependent on whether the light was increasing or decreasing (Fig. 15.2). The initial slope of the light response curve was much steeper under decreasing light than under increasing light, even with ¾ to 1 hour's acclimation to each new light level, but the differences disappeared with 2 hours' acclimation and at high temperatures. These results have important implications in experimental determinations of photosynthetic responses to light.

Photosynthetic rates and light response curves are strongly impacted by the direction and coherence of the light environment. Net photosynthesis may be 20% higher or more with diffuse radiation compared with an equal flux density of direct radiation (Ludlow & Jarvis 1971; Oker-Blom, Lahti & Smolander 1992), which is more commonly employed in experimental measurements. Direct irradiation to only one surface of a conifer needle produces self-shading among chloroplasts, resulting in less sharpness, or convexity, in the light response curve (Leverenz 1987) (see Stenberg *et al.* 1995a, for a thorough discussion of this topic).

15.2.5 Photosynthetic responses to temperature

Variations in temperature affect photosynthesis through a variety of avenues. Increased leaf temperatures may profoundly affect leaf transpiration rates which, in turn, may have impacts on stomatal conductance and photosynthesis. Elevated leaf temperatures also increase rates of photorespiration and dark respiration, reducing net photosynthesis. Cold soil temperatures inhibit water uptake by roots, which may result in desiccation of foliage

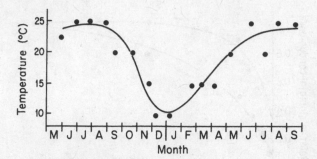

Fig. 15.3. Seasonal variation in the optimal temperature for net photosynthesis in *Pinus taeda* under field conditions. Adapted from Strain *et al.* (1976).

and reduction of potential photosynthesis. These topics are discussed in more detail later in this chapter.

For conifers in general, the photosynthetic response to temperature is gradual, with optimal temperatures ranging between 15 and 30 °C. Like most plants (Berry & Bjorkman 1980), pines show a profound ability to adapt and acclimate photosynthetic temperature responses. Optimal temperatures for photosynthesis are higher for pines from warmer climates (Teskey *et al.* 1994a), and optimal temperatures for a given tree change seasonally as the environmental temperatures shift (Fig. 15.3).

At extreme temperatures, reversible or irreversible damage to photosynthetic apparatus reduces photosynthetic rates. Night frosts can significantly reduce photosynthetic activity for the next day or so (e.g. Tranquillini 1959; Troeng & Linder 1982a; Teskey *et al.* 1986). Most pines acquire some tolerance to freezing temperatures through cold hardening, a complex process that confers the ability for supercooling, or ability to avoid ice crystallization at sub-freezing temperatures (see Havranek & Tranquillini 1994, for a summary of cold-hardiness in conifers). The ability to acclimate to cold temperatures varies among pine species. Analyses of chlorophyll *a* fluorescence and visual assessment showed that *P. contorta* needles were more hardy and acclimated to low temperatures earlier than *P. sylvestris* after both were subjected to freezing temperatures in field conditions (Lindgren & Hallgren 1992). However, cold hardening does not appear to protect pines from photoinhibition at low, but above-freezing temperatures (Strand & Öquist 1985).

15.2.6 Responses of photosynthesis to soil and atmospheric water deficits

As the amount of water in the soil or the atmosphere decreases, photosynthetic rates decrease as well. The reduction may be due both to stomatal closure and to direct inhibitory effects on the processes of CO_2 fixation, and the relative limitations of each process to total photosynthesis varies among species. Teskey *et al.* (1986), for

example, found that non-stomatal limitations were more important for *Pinus taeda*, whereas for drought-stressed *Picea sitchensis* and *P. rubens*, stomatal conductance was more important reducing net photosynthesis (Beadle *et al.* 1981; Seiler & Cazell 1990). However, stomatal conductance tends to be closely and linearly related to net photosynthesis of pines (Teskey *et al.* 1995), as well as other species (e.g. Cowan 1977), so the stomatal responses to water deficit generally parallel the responses of net assimilation.

Similar to most other conifers, the stomatal opening of pines is sensitive to several indicators of drought, including soil moisture, leaf water potential, and atmospheric humidity. Comparing conifers on a wide variety of sites in Oregon, USA, Running (1976) found that *Pinus ponderosa* achieved nearly twice the mid-morning maximum leaf conductance as *Psuedotsuga menziesii* or *Abies grandis* at comparable pre-dawn water potentials. The ponderosa pine had a lower threshold of leaf water potential for stomatal closure than did Douglas-fir (-1.8 MPa for *P. ponderosa* compared with -2.0 MPa for *P. menziesii*), yet they were able to maintain high stomatal conductance longer through the morning because that threshold was not reached. Running (1976) suggested that differences in diurnal responses may be due, in part, to differences in xylem water flow resistances among the species, as described in more detail below.

Sandford & Jarvis (1986) compared the stomatal responses to atmospheric humidity of two pine species, *Pinus contorta* and *P. sylvestris*, and two other conifers, *Picea sitchensis* and *Larix ×eurolepis*. In all of the species, stomatal closure occurred as vapour pressure deficit increased from 0.4 to 2.0 kPa. However, the two pine species maintained significantly higher stomatal conductance at low vapour pressure compared with the other species. Among pine species, the stomatal response to humidity deficit appears greatest in pines native to drought-prone regions (Teskey *et al.* 1994b). Stomatal responses to vapour pressure may also vary as a function of tree age and hydraulic capacity to supply water to needles. Recent experiments revealed that stomatal conductance in the upper crown of old *Pinus ponderosa* (c. 200 years) was much more sensitive to vapour pressure compared with foliage of younger (c. 50 years old) trees in identical environments. When hydraulic capacity was increased by removing half of the needles on the branches of old trees, their response to vapour pressure was similar to younger trees (R.M. Hubbard *et al.*, unpublished data).

The stomatal response to humidity deficit can be strongly dependent on soil water stress. Maier & Teskey (1992) found that light-saturated photosynthesis and stomatal conductance of mature *P. strobus* did not respond to humidity deficit when pre-dawn water potential was greater than -1.0 MPa.

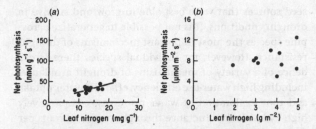

Fig. 15.4. **Relationships between mass-based (a) and area-based (b) net photosynthesis in relation to needle nitrogen content for nine evergreen conifer species measured in Wisconsin (Juniperus virginiana, Picea abies, P. glauca, Pinus banksiana, P. resinosa, P. strobus, P. sylvestris, Thuja occidentalis). Adapted from Reich et al. (1995).**

15.2.7 Photosynthesis and leaf nutrition

Across widely varying vegetation types, photosynthetic capacity is closely and positively correlated with nitrogen concentration of foliage (Field & Mooney 1986). Several studies have also reported positive correlations between photosynthetic capacity and total nitrogen concentration solely among conifers (Gower et al. 1993) and among pine species (Linder & Troeng 1980; Smolander & Oker-Blom 1989). However, in other studies with conifers, the correlation between leaf nitrogen and photosynthetic capacity was weak (e.g. P. strobus; Reich & Schoettle 1988), or there was no correlation. In fertilization studies with P. radiata, Sheriff, Nambiar & Fife (1986) found no photosynthetic response to increased leaf nitrogen, although there was a response to increased leaf phosphorus and an even greater response to increases in nitrogen and phosphorus together. For the same species, Thompson & Wheeler (1992) reported significant correlations between CO_2 assimilation and nitrogen only when trees were irrigated, and stomatal conductance was unusually high. Teskey et al. (1994a) found no significant increase in photosynthesis through the growing season of fertilized, 23-year-old P. elliottii compared with control trees despite a large increase in leaf nitrogen through at least part of the year. Similarly, Reich et al. (1995) found no significant correlation for nine evergreen conifers in Wisconsin between leaf nitrogen and photosynthetic capacity on a leaf area basis (Fig. 15.4b), although there was good correlation for broadleaved deciduous species. However, there was a significant correlation ($r^2 = 0.59$) for conifers between nitrogen and photosynthetic capacity on a leaf mass basis (Fig. 15.4a). This study also found statistically different relationships between net photosynthesis and leaf nitrogen in hardwoods and conifers, which was attributed to leaf longevity.

On the strength of the many reports of positive relationships between leaf nitrogen and photosynthetic capacity, many process-level models of canopy gas exchange use leaf nitrogen concentration as a surrogate

for maximum photosynthetic capacity, so it is important to resolve disparate reports in the literature. Positive correlations between leaf nitrogen and photosynthetic capacity are expected when light-saturated photosynthesis is limited by the activity of the carboxylating enzyme, Rubisco, and when the distribution of nitrogen to Rubisco relative to other pools is constant. These conditions are often met, but there are important exceptions.

Photosynthesis may be more limited by the supply of inorganic phosphate (P_i) than by Rubisco activity when the supply rate of P_i via the phosphate translocator in the chloroplast envelope is low relative to the CO_2 fixation rate. This may occur when ambient CO_2 concentrations are high (e.g. Sage, Sharkey & Seemann 1989), when the export rate of sucrose is low (e.g. when growth is slow), or when availability of phosphorus is low (e.g. Lewis et al. 1994), which is true in many pine ecosystems. Also, species that form dense canopies or individuals grown in the shade typically allocate a greater proportion of nitrogen to non-photosynthetic pools (Kull & Jarvis 1995), although these conditions are not as likely for shade-intolerant pines. Finally, nitrogen allocation among pools can vary greatly according to environmental conditions, and evergreen species show strong seasonal shifts in nitrogen distribution among pools. Free amino acids may account for as much as 50% of total leaf nitrogen for fertilized conifers at certain times of the year (Billow, Matson & Yoder 1994), with arginine alone constituting up to 27% of the total nitrogen of fertilized P. sylvestris in early spring (Näsholm & Ericsson 1990). The foliar protein concentration was similar in fertilized and unfertilized P. sylvestris (Näsholm & Ericsson 1990), despite a near doubling of total nitrogen in response to fertilization. In these situations, a poor correlation between photosynthetic capacity and total nitrogen is expected.

15.2.8 Effects of leaf and tree age on photosynthesis

Net photosynthesis as well as dark respiration rates are generally reduced in older cohorts of conifer needles, both with sequential flushes within a year (e.g. Naidu et al. 1993) and with sequential years of needle growth (e.g. Freeland 1952; Chabot & Hicks 1982; Teskey, Grier & Hinckley 1984; Schoettle 1994). Typically, photosynthetic capacity declines at a rate of about 30–50% per year (e.g. Freeland 1952; Linder & Troeng 1980; Chabot & Hicks 1982), although the capacity of immature, current-year needles is sometimes lower than needles formed in the previous year (e.g. Linder & Troeng 1980). In one comparison, the rate of photosynthetic decline with needle age for a true fir species (Abies concolor) greatly exceeded the rate of decline for three pine species (P. ponderosa, P. strobus, P. sylvestris; Freeland 1952). To some extent, the reductions may be related to retranslocation of nitrogen from older to newer foliage, because older needles also tend to have reduced tissue nitrogen concentrations. At least for *P. elliottii*, older age classes of needles also have higher starch concentrations (Gholz & Cropper 1991), and it is possible that photosynthesis is suppressed by starch accumulation. From studies with stable isotopes of carbon, Naidu *et al.* (1993) found that the average CO_2 concentration in the leaf mesophyll (c_i) of *P. taeda* increased in older flushes, suggesting that the enzymatic processes of photosynthesis decreased more than stomatal conductance as the foliage aged.

Average net photosynthesis also decreases with increasing size and age of trees (Mooney, Wright & Strain 1964; Grulke & Miller 1994; Schoettle 1994; Yoder *et al.* 1994). For *P. contorta* and *P. ponderosa*, Yoder *et al.* (1994) found that light-saturated, one-year-old foliage of old trees (>200 years) averaged 14–30% lower photosynthetic rates compared with younger, mature trees (40–60 years). Foliar nitrogen concentration and maximum photosynthetic rates were similar for young and old trees of these species in comparable growing conditions (Schoettle 1994; Yoder *et al.* 1994). The difference in average net assimilation was associated with different diurnal patterns in the two age classes: stomata of foliage on large, old trees began closing earlier in the day (Yoder *et al.* 1994). In a study comparing large and small *P. aristata*, on the other hand, large old trees had lower foliar nitrogen and phosphorus concentrations as well as reduced photosynthetic capacity compared with younger trees (Schoettle 1994).

15.3 Water relations

Many pine species occupy sites with low rainfall, i.e. they are drought resistant. Resistance to drought may come either through tolerance of tissue desiccation during drought periods, or through mechanisms that allow avoidance of tissue desiccation despite low water availability in the environment. One component of drought tolerance is maintenance of turgor in leaf cells through lowering the osmotic potential. In *P. ponderosa*, low irradiance significantly reduced the ability of seedlings to adjust osmotically, and shaded seedlings showed more tissue damage at low leaf water contents compared with unshaded seedlings (Vance & Zaerr 1991).

As a genus, pines tend to be drought avoiders compared with co-occurring species. For example, *P. ponderosa* is able to maintain leaf water potentials an average 0.2 MPa higher than *Abies concolor* in the same environmental conditions (Barker 1973). Within *P. ponderosa*, Cregg (1994) concluded that there was no significant relationship between drought tolerance of seedlings and needle gas exchange, but there was an optimum shoot/root ratio for

seed sources that were best able to grow and survive in drought conditions. It is not possible to generalize across pine species the most important mechanisms of drought resistance. However, for individual species there is evidence of a variety of mechanisms of drought avoidance, including high water-use efficiency, efficient water uptake and root exploration of water resources, comparatively high hydraulic conductance through stems and roots per unit leaf area, and high capacitance, or water storage. These are discussed in more detail below. There is also evidence from old and very recent studies that pines, and perhaps other conifers, may absorb atmospheric moisture through needles, thereby improving leaf water relations. Stone, Went & Young (1950), Stone & Fowells (1955) and Stone (1957) determined that *P. coulteri* and *P. ponderosa* were able to absorb atmospheric water vapour and that artificial dew applied at night improved survival in very dry soil. The reliability of these data to establish foliar uptake of water at ecologically significant levels has been questioned, however (Rundel 1982). More recently, Boucher, Munson & Bernier (1995) reported that artificial dew significantly improved shoot water potential and increased stomatal conductance and root growth of *P. strobus* seedlings. They concluded that high water potential values in field-grown *P. strobus* after nights with heavy dew resulted from direct foliar absorption of water from dew.

Plant responses to drought conditions are strongly dependent on other environmental conditions. High irradiance, for example, often amplifies the instantaneous responses of photosynthesis and stomatal conductance to soil humidity deficits. However, survival in drought conditions may also depend on adequate light for photosynthesis. Vance & Zaerr (1991) found that shaded *P. ponderosa* seedlings were much more likely to die from drought stress than unshaded seedlings.

15.3.1 Water-use efficiency

Above, we discussed some of the ways that pines limit water loss through stomatal closure. Water-use efficiency (WUE; mol CO_2 fixed per mol H_2O transpired) tends to be high for pines compared with other species. For example, Carter & Smith (1988) reported that the instantaneous WUE (measured in a cuvette over short periods) of open-grown *P. contorta* was more than double that of *Abies lasiocarpa* and *Picea engelmannii*, and Kaufmann (1985) estimated a similar difference in stand-level WUE for the same species over a 120-year period. On the other hand, Smit & Van den Driessche (1992) found much higher stomatal conductance (nearly double early in the growth season) for *P. contorta* than for *Pseudotsuga menziesii* seedlings; their rooting characteristics also differed (see below).

High water-use efficiency of pines may be at least partially due to efficiency in heat dissipation. The narrow needles of pines are at least five times as efficient as a broad leaf in dissipating heat (Waring & Schlesinger 1985), although the geometric arrangement of needle packing can reduce this efficiency (Smith & Carter 1988). As leaf temperatures increase, the leaf-to-air vapour pressure gradient also increases, affecting transpiration rates and potentially leaf water relations and stomatal conductance. For example, for an air temperature of 30 °C and relative humidity of 25%, the leaf-to-air vapour pressure gradient is 45% higher when leaf temperature is 5 °C above air temperature, than when leaf temperature equals air temperature. If stomatal conductance remains constant, transpiration increases proportionately to the vapour pressure gradient. Thus, lower leaf temperatures can significantly reduce water loss. In addition, alleviated from the need to use evaporative cooling to regulate leaf temperatures, pines may close stomata on hot, dry days with moderate wind, but this could result in lethal leaf temperatures for a broadleaved species in the same environment.

The very low stomatal conductance of many pines may also account for their high water-use efficiency in some cases (Teskey et al. 1994b). However, in other cases the comparatively high WUE of pines appears to result more from relatively high photosynthetic rates than from low stomatal conductance. For example, typical net photosynthetic rates in the early summer for mature P. ponderosa in the Pacific Northwest of the USA are higher than for two other important conifers in the region, Pseudotsuga menziesii and Tsuga heterophylla (Fig. 15.5). The stomatal conductance of the pines is also higher, but not by as great a margin. As a result, the pines have lower internal CO_2 and greater WUE, especially during the morning hours, compared with the Douglas-fir and hemlock. The trees for these measurements were on different sites characteristic of the range for each species, with the pine site the most xeric and the hemlock site the most mesic, but soil moisture was abundant for all sites during these measurements.

The high water-use efficiency of the pines is also evident from studies of stable carbon isotopes, ^{12}C and ^{13}C. With certain exceptions, discrimination (Δ) against the heavier isotope decreases as water-use efficiency increases (e.g. Farquhar et al. 1988). (The efficient heat dissipation of conifer needles eliminates one important exception to this rule.) For P. ponderosa in Fig. 15.5, Δ values of cellulose extracted from sun foliage averaged 15.6, compared with 16.6 for Pseudotsuga menziesii and 17.3 for Tsuga heterophylla (B.J. Yoder et al., unpublished data). A summary of 11 reports of Δ values for conifers (Marshall & Zhang 1993) showed the lowest values of pines were for P. edulis (12.9), P.

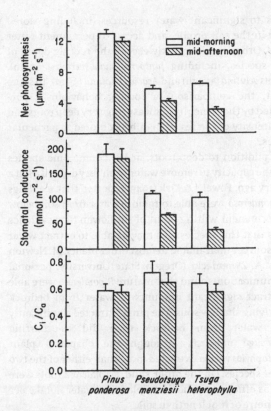

Fig. 15.5. **Gas exchange characteristics of net photosynthesis, stomatal conductance, and ratio of internal to ambient CO_2 concentration for *Pinus ponderosa* in eastern Oregon and *Pseudotsuga menziesii* and *Tsuga heterophylla* in western Oregon.** Adapted from B.J. Yoder et al. unpublished data.

jeffreyi (15.9), P. monophylla (12.5) and P. ponderosa (15.1); these values were similar to the range for Juniperus species (12.4–16.9). Pinus albicaulis and P. edulis had intermediate values, both 16.6. Pinus contorta, P. echinata, P. massioniana, P. ponderosa and P. taeda had the highest reported Δ values for the pines, with values ranging from 17.7 to 19.5. Larix, Lagarostrobos franklinii and Picea abies also had values in this range. Local environmental conditions as well as genetic predisposition strongly influence WUE as well as isotope discrimination, so these values cannot be taken as a strict species ranking. However, the Δ values for the conifer species in general is low compared with crop plants, indicating their comparatively high WUE.

15.3.2 Rooting behaviour

Rooting behaviour depends as much on site characteristics as it does on genetic predisposition. In a comprehensive review of root depth and radius for 211 species, Stone & Kalisz (1991) reported that the typical rooting depth of pines (2–5 m) was similar to the range for other species. However, their summary revealed particulary deep roots for Pinus edulis (c. 20 m), P. flexilis (10 m), P. ponderosa (24 m), P. radiata (8 m) and P. sylvestris (8 m). Deep roots can provide access to significant water resources, including stored water in the soil profile, and access to permanent water tables. Other species greatly exceed the rooting depth of *Pinus* species, including *Juniperus monosperma* (>60 m), *Prosopis velutina* (>53 m) and *Acacia raddiana* (35 m). To some extent, the comparison in rooting behaviour is constrained by the difficulty of measuring very deep roots and the intensity of the research efforts devoted to particular species.

In addition to deep roots, at least some pine species have the capacity to remove water from very dry soil. Half a century ago, Fowells & Kirk (1945) reported that seedlings of *P. ponderosa* were able to remove water from soil below the permanent wilting point of sunflower. New evidence shows that this species also may be able to extract water from sources unavailable to most other plants. M. Newton and M.A. Zwieniecki (Oregon State University, personal communication) found that mature *P. ponderosa* were able to extract significant amounts of water from bedrock underlying shallow soils. The pines extracted significantly more water from bedrock than did co-occurring *Psuedotsuga menziesii*, although the chaparral plant *Arctostaphylos viscida* extracted more than either of the two conifer species. On the other hand, *P. ponderosa* roots were twice as efficient as chaparral plants in water uptake per centimetre of root length in soil.

Rooting behaviour can have a significant impact on the response of seedlings to drying soils, and deeply penetrating roots may be more valuable than high water-use efficiency. Smit & van den Driessche (1992) compared the response to soil drought in one-year-old seedlings of *Pinus contorta* and *Pseudotsuga menziesii* planted in deep pots. The Douglas-fir seedlings had higher water-use efficiency; however, the pine seedlings had greater dry matter production, water use, stomatal conductance and new root length than the Douglas-fir seedlings even in dry treatments, and they extracted significantly more water below 40 cm soil depth.

15.3.3 Water storage in sapwood

Mature pines of most species can achieve very large sizes, and most of the volume is in the main stem. The combination of relatively low wood density, high ratio between cross-sectional area of sapwood and leaf area (see below) and large stem volume results in a very large amount of water storage, or capacitance, in the stems of large pine trees. Waring, Whitehead & Jarvis (1979) estimated the amount of stored water that is potentially 'available' for transpiration in 40-year-old *P. sylvestris*, defining 'available' water as the volume of water in fully hydrated tissue multiplied by the observed changes in relative water content of the tissue. They concluded that a densely stocked stand had a water storage capacity of over 20 mm, and that 1–1.5

mm per day, or *c.* ⅓–½ the total daily transpirational flux, could be removed from stem sapwood via xylem cavitation.

For stored water to be an important buffering system against seasonal drought events over many years, it is necessary for trees to restore the lost water, i.e. the cavitated tracheids must be refilled. Although circumstantial evidence indicates that refilling occurs in gymnosperms (e.g. Waring & Running 1978), there has been some controversy over the reversibility of cavitation, especially considering that the xylem is never in a state of positive pressure. Recent experiments, however, demonstrated that cavitated tracheids in branchwood specimens of *P. sylvestris* were able refill under negative water potentials (Edwards *et al.* 1994). Edwards *et al.* postulate that the air bubbles within cavitated tracheids re-dissolve into the water in surrounding tissue because the partial pressures of gases in the bubble exceed their concentration in the aqueous solution owing to pressure differences across the meniscus.

15.3.4 Tree hydraulics

The driving force for water movement from the soil, through a plant, and to the atmosphere is a gradient of decreasing water potential. As soils dry out, soil water potential falls, and leaf water potential must decrease by the same amount in order to maintain a constant flux of water from roots to foliage. Also, for a given water potential difference between soil and foliage, the rate of water flow will be lower in a tree that has high resistance to flow in roots and stems compared with one that has low resistance. The total resistance along the hydraulic pathway depends on the combined influences of the permeability, the cross-sectional area, and the path length of water-conducting tissue in the roots, stems and branches, and leaves. Zimmermann (1983) coined the term 'hydraulic architecture' to describe hydraulic conductivity (analogous to 'permeability') throughout a plant and its relationship to leaf area. Comprehensive analyses of hydraulic architecture have been published for only a few tree species (e.g. Ewers & Zimmermann 1984a, b), and to our knowledge no thorough studies of hydraulic architecture have been completed for pine species.

15.3.5 Differences in hydraulic resistance among species

Reviewing reports of total hydraulic resistance for a variety of species, Hellkvist, Richards & Jarvis (1974) concluded that conifers have low hydraulic resistance compared with herbaceous angiosperms. In addition, the hydraulic resistance of two pine species, *P. resinosa* and *P. sylvestris*, was 20% or less than that of *Picea sitchensis* (Hellkvist *et al.* 1974; Jarvis 1976). This means that for a given leaf area, leaf and soil water potential, and vapour pressure deficit, the pines can sustain higher stomatal conductance and transpiration rates compared with the spruce (Jarvis 1976) or with the herbaceous angiosperms.

Several experimental and theoretical analyses conclude that the resistance in the main stem of trees is generally small compared with the resistance in small-diameter branches (Richter 1973; Ewers & Zimmermann 1984a, b; Tyree 1988; Yang & Tyree 1993). On the other hand, Roberts (1977) estimated that the total resistance in branches of *P. sylvestris* was about one-third of the resistance in stems; and Sellin (1994) noted that the distribution of resistance in roots, stems and branches is strongly influenced by growing conditions.

Resistance in the shoot (stem + branches + leaves) may account for anywhere between 20 and 60% of the total plant hydraulic resistance (Sperry 1995), and the relative resistances between roots and shoots varies with species and environment. Roberts (1977) and Running (1980) reported that more than half of the total hydraulic resistance was in the roots of *P. sylvestris* and *P. contorta*, respectively. In contrast, the shoot offered more resistance than the roots in *Picea abies* (Roberts 1978) and *Pseudotsuga menziesii* (Nnyamah, Black & Tan 1978).

Because membrane permeability decreases at low temperatures, the proportion of the total resistance in roots increases in cold soils. In *P. contorta*, roots accounted for 67% of total resistance when soil temperature was 7 °C and 93% at 0 °C (Running & Reid 1980). The viscosity of water is also increased at low temperatures, which further limits the flux of water from soil to leaves. Together, low root permeability and high viscosity of water can reduce water flow to the extent that foliage becomes severely desiccated in cold temperatures. The sensitivity to cold soils differs among pine species, and in general species from cold climates are less sensitive than species from warmer climates. For example, in cold soil temperatures *P. taeda*, which is native to the southeastern USA, shows a bigger decrease in leaf water potential than *P. contorta*, which grows in colder montane regions of the western USA (Day, Heckathorn & DeLucia 1991; DeLucia, Day & Oquist 1991). Tranquillini (1976) also found that drought-tolerant pines (*P. cembra* and *P. mugo*) survived at high altitudes better than drought-sensitive *Picea abies* because the pines were more able to maintain a favourable leaf water content at low temperatures.

15.3.6 Xylem anatomy, wood permeability, and vulnerability to embolism

The sapwood of conifers is generally less permeable to water than that of angiosperms (Wang, Ives & Lechowicz 1992; Sperry 1995). The variations in permeability are due to different xylem anatomies. In conifers, water moves through tracheids, which are connected mainly through pit-pairs, whereas perforated vessels conduct water in the angiosperms. Vessels are generally larger than tracheids, and the vessel perforations offer much less resistance to water flow than do bordered pits.

Permeability also varies within conifer species in different environmental conditions, within different parts of a tree, and among different conifer species. In favourable environments, tracheids tend to be larger and have thinner walls (Carlquist 1975), conferring higher xylem permeability. Permeability increases with stem size (Edwards & Jarvis 1982; Cochard 1992) and tree age (Pothier, Margolis & Waring 1989), and is generally higher at the tops of trees than near the base (Pothier *et al.* 1989). Comparing the permeabilities of a broad array of species, Siau (1984) concluded that the permeability of sapwood of *Pinus* species is among the highest of all gymnosperms, similar to that of *Pseudotsuga menziesii* and one or more orders of magnitude greater than that of *Cedrus* and *Picea* species. On the other hand, Cochard (1992) found that wood from *Pinus sylvestris* and *Picea abies* had similar permeabilities, but that permeability was lower in these two species than in *Cedrus atlantica*, *C. deodara*, *Pseudotsuga menziesii*, *Abies alba* and *A. bornmulleriana*. Whitehead, Edwards & Jarvis (1984) also reported comparatively low sapwood permeability in *Pinus contorta*:, i.e. 50% less than that of *Picea sitchensis*.

Because the water movement through plants is driven along a negative hydrostatic gradient, and the water in water-conducting elements is under tension, these elements are susceptible to failure through cavitation and embolism: vaporization of the liquid phase and diffusion of air into the conduit. Sperry (1995) pointed out that the range of water potentials over which a plant is vulnerable to cavitation unambiguously limits its ability to tolerate desiccation, and Tyree & Ewers (1991) maintained that vulnerability to cavitation may be the most important parameter determining drought resistance in trees. Typically, the range of water potentials 'normally' encountered by plants in their native environments coincides with the range of vulnerability (e.g. Milburn 1991). Cold temperatures, especially repeated freeze–thaw cycles, can also induce cavitation, so vulnerability to cavitation may also affect cold tolerance and species distribution along temperature gradients (Tyree, Davis & Cochard 1994; Sperry 1995; Woodward 1995). This may be one reason why drought-tolerant species are often also cold-tolerant.

Conifers are generally less vulnerable to cavitation than forest angiosperms (Tyree *et al.* 1994), and tracheids are better able to refill after cavitation (Edwards *et al.* 1994). Conifers typically experience a 50% loss in hydraulic conductivity at water potentials ranging between −2.5 MPa and −7 MPa (Tyree *et al.* 1994). Unfortunately, there is little information available on the vulnerability of pine species to cavitation. Losing half of its hydraulic conductivity at about -3.0 MPa, *Pinus sylvestris* is a comparatively vulnerable conifer (Cochard 1992; Tyree *et al.* 1994). This is consistent with a 'drought-avoidance' strategy of rapid stomatal closure in response to water stress and deep rooting habits (Cochard 1992).

15.3.7 Leaf area/sapwood area ratios

For a particular site and species, the ratio between leaf area and the cross-sectional area of supporting sapwood tends to be constant (Grier & Waring 1974; Waring *et al.* 1977; Waring, Schoeder & Oren 1982), and this relationship has been used widely to predict leaf area from a more easily measured dimension. For conifers, the leaf area/sapwood area ratio ranges between 0.08 and 0.75 m^2 cm^{-2} (Margolis *et al.* 1995). Pine species fall in the lower end of this range. For *P. contorta* and *P. taeda* the ratio has been reported between 0.11 and 0.30, and for all other pine species that have been measured, the ratio falls between these values (Margolis *et al.* 1995). Within a species, the amount of leaf area relative to sapwood is highest where site conditions are mild. For *P. ponderosa* in mild, wet forests of western Oregon, the ratio is 0.25 (Waring *et al.* 1982), and it decreases to 0.201 in high elevation, montane sites in the eastern Sierra Nevada and further to 0.104 in the Great Basin desert (DeLucia, Callaway & Schlesinger 1994). The ratio decreases with evaporative demand and increases with sapwood permeability (Whitehead *et al.* 1984; Oren, Werk & Schulze 1986; Mencuccini & Grace 1995). For *P. sylvestris* growing in sites with widely varying evaporative demand, the leaf area/sapwood area ratio changed so that trees maintained similar water potential gradients despite the climatic differences (Mencuccini & Grace 1995).

15.4 Respiration

Respiratory losses of carbon can be categorized into two groups. *Growth* respiration releases the energy necessary for constructing new tissue; its rate depends almost exclusively on the type and amount of tissue that is produced. On the other hand, *maintenance* respiration, the cost of maintaining existing tissue, is sensitive to environmental influences, especially temperature.

15.4.1 Respiration in foliage

Gymnosperms generally have lower foliar respiration rates than angiosperms (Sprugel *et al.* 1995). This is not surprising because maintenance respiration and photosynthetic capacity are positively correlated (Ceulemans & Saugier 1991). Foliar respiration rates also vary among and within conifer species. In a recent review, Sprugel *et al.* (1995) showed that *P. elliottii* had the lowest measured rates of foliar respiration among a group of over a dozen conifers, possibly because the species is adapted to a warm subtropical climate. For conifers in general, adaptation to high elevation or acclimation to cold temperatures is associated with relatively high rates of foliar respiration (Sprugel *et al.* 1995)

15.4.2 Respiration in woody tissue

For large trees, maintenance respiration of roots and stems can consume a significant portion of fixed carbon (Sprugel & Benecke 1991; Ryan *et al.* 1994). Maintenance respiration of stems is fairly consistent across pine species (including *P. contorta*, *P. elliottii*, *P. ponderosa* and *P. resinosa*), averaging about 23 μmol CO_2 s^{-1} g^{-1} sapwood at 10 °C (Ryan 1990; Ryan *et al.* 1995). In contrast, the rate for *Tsuga heterophylla* stems is about 40% lower (Ryan *et al.* 1995), and for *Abies amabilis* it is about 50% higher (Sprugel 1990). However, the species differences are relatively unimportant to maintenance respiration on a stand level compared to the effects of temperature and sapwood volume (Ryan *et al.* 1995). Because of the complexities inherent in measuring root respiration, it is not possible at this time to conclude meaningful differences among species or forest types (Sprugel *et al.* 1995).

15.5 Ecophysiological response to environmental stress

15.5.1 Pines in dry habitats

Dwarf conifer or pinyon–juniper woodlands are widespread in the American Southwest (Chap. 9, this volume) and northwestern Mexico (Chap. 7, this volume); these represent some of the driest habitats in the world for pine growth. Mean annual rainfall in the pinyon-juniper zone, forming a transition from desert communities below and conifer forests above, typically ranges from 250 to 500 mm. Lower limits of pine distribution in these communities are commonly assumed to be related to drought stress. While pinyon pines are able to tolerate extended periods of drought, seasonal carbon gain drops to unsustainable rates with increasing aridity.

The pinyon pines show much less adaptation for drought resistance than do the juniper species in their pinyon–juniper community. Minimum midday water potentials of *Pinus edulis* at a low-elevation site at 1960–1980 m in northern New Mexico fell to only −2.1 MPa in midsummer (Lajtha & Getz 1993). In contrast, *Juniperus monosperma* in the same community remained metabolically active for much of the day in midsummer, allowing midday water potentials to fall to −4.2 MPa. Similar patterns of modest minimum water potentials in *P. edulis* have been reported in other field studies (Barnes & Cunningham 1987; Wilkins & Klopatek 1987; Malusa 1992).

The midsummer pattern of diurnal gas exchange of pinyon and juniper provides an explanation for these patterns of water relations. While both *P. edulis* and *J. monosperma* exhibit low rates of net carbon assimilation in the early morning hours, strong stomatal control by the pinyon pines causes gas exchange to close down in mid-morning, thereby preventing strongly negative water potentials from occurring (Lajtha & Barnes 1991). Experimental growth studies with seedlings of these two species have found that net photosynthesis dropped to zero in *P. edulis* at a water potential of −1.8 MPa, but at a far lower −4.6 MPa in *J. monosperma* (Barnes & Cunningham 1987).

Gas exchange measurements have suggested that juniper has higher water-use efficiencies than pinyon pines during the dry season. However, integrated measurements of WUE using carbon isotope fractionation show that the pinyon pine may actually have marginally higher WUE than juniper integrated over the period of needle formation (Lajtha & Getz 1993). Integrated WUE as indicated by tissue carbon isotope ratios was lowest at the arid end of its range of occurrence in northern New Mexico, but levelled off at less arid (higher elevation) sites. Since stand canopy cover varied significantly across this gradient of sites, biotic factors of below-ground competition for water may be equally important as climatic factors in influencing physiological adaptation in drought response (Lajtha & Getz 1993).

Rooting architecture and components of tissue water relations have much to do with the competitive interactions of pinyon pines and sagebrush (*Artemisia tridentata*) in the Great Basin area of the western USA. While sagebrush is much more drought-tolerant than competing pinyon species such as *P. monophylla*, its high water-use efficiency forces a tradeoff in having a low nitrogen-use efficiency for photosynthesis. Thus pinyon pines, as well as *P. jeffreyi* and *P. ponderosa*, are able to outcompete *A. tridentata* in Great Basin sites on strongly nutrient-deficient soils where the slow growth rate of the conifers is advantageous (DeLucia, Schlesinger & Billings 1988; Schlesinger, DeLucia & Billings 1989; DeLucia & Schlesinger 1991). Deep root systems in the pines release them from competition for water with the associated desert shrubs.

Some pinyon pines of arid areas have relatively small numbers of needles in each fascicle (Chap. 9, this volume). It has been hypothesized that reduction in needle number might represent an adaptation to drought by reducing needle surface-to-volume ratio (Haller 1965). Malusa (1992) partially tested this hypothesis by comparing tissue water

relations in a hybrid complex between the one-needled *P. monophylla* subsp. *californiarum* and the two-needled *P. edulis*. He found no significant difference in either diurnal or seasonal patterns of water potential between these one- and two-needled forms over a two-year study.

Although aridity would seem to be the strongest environmental factor influencing pinyon pines in the semi-arid pinyon–juniper habitat, there is a strong interaction between water and nitrogen availability in controlling photosynthetic capacity (Lajtha & Getz 1993). Nitrogen fertilization of *P. edulis* under field conditions significantly increased maximum rates of photosynthesis in both wet and dry seasons. Within single trees, not only maximum rates of photosynthesis but also nitrogen-use efficiency declined with needle age, suggesting that older needles were not as efficient as younger tissues in fixing carbon. The maximum rates of net photosynthesis in 6-year needles of *P. edulis* were found to be only about 55% of the rate in first-year needles.

There has been little experimental study of factors controlling the upper elevational limits of pinyon pines in the southwestern USA. Community gradients suggest that increased resource competition and shading from *P. ponderosa* at higher and more mesic habitats is the critical factor. Daubenmire (1943) has suggested that cold tolerance may also be a factor in the upper elevational limits of pinyon pines in the Rocky Mountains, but this hypothesis has not been tested.

15.5.2 Pines in timberline environments

Pines frequently occur at the timberline or in high subalpine communities throughout the northern hemisphere at or near the limits of tree growth. Survival of trees under the extreme environmental conditions presented by such habitats requires physiological adaptations to maintain a positive net carbon balance over annual cycles with short growing seasons as well as resistances to low winter temperatures and potential desiccation or needle erosion from strong winds. Annual carbon balance near the upper or northern limit of tree growth is strongly limited by a short growing season, low air and soil temperatures, and low atmospheric concentrations of CO_2. Frozen soils also restrict water uptake, subjecting trees to potential problems of desiccation in evergreens exposed to clear skies and high-velocity winds. Buffering against these environmental stresses, however, are a number of factors which favour higher levels of net photosynthetic production. These include high levels of summer irradiance, heating of photosynthetic tissues well above air temperatures, water availability from snow melt and summer rains, and cool nights and low soil temperatures that restrict respiratory losses of carbon. Much of our knowledge of the ecophysiology of timberline pines and other conifers comes from classic studies during the 1950s and 1960s in the Austrian Alps by researchers at the Institute for Subalpine Forest Research.

The severe environmental stresses put on the growth of pines and other tree species near the timberline can be readily seen in the steady decline of tree height with elevation. Trees eventually occur as dwarfed krummholz individuals at the treeline (Tranquillini 1979). Morphological effects other than growth form alone have been documented with increasing elevation in timberline pines. In *P. cembra*, mean needle length shortens by about 30% (Tranquillini 1965), and cuticle depth and epidermal and hypodermal layers of needles become thinner as populations increase in elevation from 1300 to 2000 m at the timberline (Baig & Tranquillini 1976). Studies with other timberline conifers have shown that mean annual wood production drops steadily with lower temperatures associated with increasing elevation, and that the lignin content of the wood itself declines as well (Tranquillini 1979).

Respiratory carbon losses represent a strong limiting factor for pine growth near the timberline because of the limited amount of carbon fixation possible during the short growing season. Unlike photosynthesis, respiration rates increase rapidly when warm temperatures occur during periods of winter dormancy. There are also data to suggest that high-elevation trees have inherently higher rates of tissue respiration at lower temperatures compared with lower-elevation species or genotypes, perhaps as a physiological adaptation to aid in photosynthetic activity at such temperatures (Tranquillini 1979). The data supporting this hypothesis are equivocal, however.

The photosynthetic responses of pines near the timberline show strong physiological adaptations to low growth temperatures. Maximum rates of photosynthesis in *P. cembra* seedlings at the timberline were at 12 °C at low light intensities and 17 °C at moderate light intensities (Tranquillini & Turner 1961). Median needle temperatures at the beginning of the growing season in May were well below these levels at 7 °C, but approached these peaks in the warmest month of July. *Pinus longaeva* at the timberline in the White Mountains of California had an optimal temperature for photosynthesis at 10–15 °C (Mooney, West & Brayton 1966). There is strong acclimation, however, in these temperature responses, and trees planted at lower elevations or grown under warmer conditions exhibit higher temperatures for maximal photosynthesis (Mooney *et al.* 1964). Even in these cold-adapted pines, however, photosynthetic rates drop off rapidly at air temperatures below 10 °C.

Under conditions of high solar irradiance and low wind speeds, temperatures of growth meristems and needles of timberline pines may increase well above surrounding cool air temperatures, thereby increasing net photosyn-

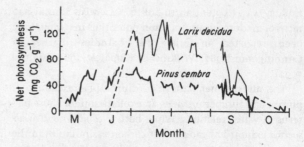

Fig. 15.6. **Seasonal cycles of net daily photosynthesis for field seedlings of *Larix decidua* and *Pinus cembra* at timberline (2000 m) in the Austrian Alps. Adapted from Tranquillini (1979).**

thesis. This effect is particularly strong in low krummholz pines where mean meristem–air temperature difference was found to be 4.3 °C and differences as much as 10 °C were recorded (Grace, Allen & Wilson 1989). Needle temperatures in *P. cembra* in the Austrian Alps have been reported to reach as much as 20 °C above air temperatures.

The most detailed data on seasonal patterns of net photosynthesis over an annual cycle for pines at the timberline remain those made with *P. cembra* more than 30 years ago (Tranquillini 1957, 1959, 1964). Seedlings of this species exhibited positive rates of net photosynthesis as soon as they emerged from under snow cover in the spring, usually in May at 2070 m elevation in the Austrian Alps (Fig. 15.6). Low soil temperatures initially restricted maximum photosynthetic rates, but rising air and soil temperatures within a few weeks increased daily maximum rates of photosynthesis three-fold to about 60 mg CO_2 g^{-1} d^{-1}. Daily photosynthetic production remained high until early July, when these rates declined significantly at time that new vegetative growth was initiated. Although solar irradiance, air temperatures, and soil moisture availability were not limiting, daily photosynthetic production generally remained at this lower rate through September. After September, lowered levels of solar irradiance and lower environmental temperatures steadily decreased photosynthetic production. Increasingly severe frosts in mid-November reached to the root zone of *P. cembra*, sharply restricting water uptake and quickly leading to stomatal closure and an end to any CO_2 uptake. Seedlings were generally covered by deep snow by early snows.

This seasonal pattern of daily photosynthesis in seedlings of *P. cembra* differs notably from that of another conifer in the same community at the timberline, *Larix decidua* (Fig. 15.6). The deciduous larch initiated photosynthesis approximately one month later than *P. cembra*, and ceased activity one month earlier, but these temporal differences were compensated for by maximum rates of daily photosynthetic production that were twice as high (Tranquillini 1979).

Mature trees of *P. cembra* which stand above snow cover near the timberline throughout the winter have been found to have roughly the same annual pattern of photosynthetic production as the seedlings described above. Reduced solar irradiance and lowered environmental temperatures reduce photosynthetic assimilation after September, and soil frost brings on a true dormancy in November (Tranquillini 1979). Winter dormancy in *P. cembra* appears to be long and fixed, as brief periods of favourable weather conditions associated with *föhn* winds during the winter do not result in any photosynthetic activity (Pisek & Winkler 1958). Winter dormancy ceases only in late April as ambient temperatures rise and solar irradiance increases. The breaking of this dormancy is primarily induced by temperature, although Tranquillini (1979) suggested that endogenous physiological rhythms and day length may also be involved. Overall, *P. cembra* experiences a period of continuous negative carbon balance for five months (Tranquillini 1957, 1964). Low respiratory rates associated with winter dormancy, however, reduce the carbohydrate costs of maintenance respiration during this period.

Severe frosts in spring or autumn reduce net photosynthesis to near zero for a day or more after such an event, presenting a significant limiting factor during the growing season in *P. cembra* (Tranquillini 1959). A similar effect has been noted in other cold-adapted pine species such as *P. contorta* (Fahey 1979) and *P. sylvestris* (Pelkonen, Hari & Luukanen 1977; Troeng & Linder 1982a).

Seasonal studies of photosynthetic production of *P. longaeva* near the timberline in the White Mountains of California have shown similar patterns to those previously described for *P. cembra*. Severe cold temperatures in autumn sharply reduced rates of net photosynthesis, and gross levels of photosynthesis remained at or close to zero through April (Schulze *et al.* 1967). There was a striking difference, however, in seasonal patterns of respiration between the two species. While *P. cembra* and other timberline conifers in Europe show reduced rates of winter respiration associated with complete dormancy, *P. longaeva* was found to have high tissue respiration rates until late winter, with these dropping sharply in spring as air temperatures increase and soils thaw. Schulze *et al.* (1967) have suggested that negative carbon balance through the long winter period may deplete carbohydrate reserves and thus bring on this spring depression of rates.

The high winter rates of respiration in *P. longaeva* present severe problems of carbon balance for this species. Even ignoring respiration by trunk and root tissues during winter, which were unmeasured, Schulze *et al.* (1967) calculated that photosynthetic production from at least half of the growing season in *P. longaeva* was necessary to provide for carbohydrate reserves to survive a normal

winter. It is therefore not surprising that this species exhibits a slow rate of vegetative growth. In contrast, the higher rates of photosynthetic production and lower winter respiratory rates in *P. cembra* allow this species quickly to restore carbohydrate reserves for the following winter.

The widespread *P. sylvestris* in northern Europe also limits its photosynthetic production to the warmer months of the year. Carbon gain for this species in Sweden was zero from late November until early April because of frozen soils and low air temperatures (Linder & Lohammar 1981; Troeng & Linder 1982a). More than 90% of the annual carbon fixation occurs in just six months of the year.

Climatic conditions of timberline habitats can lead to severe stresses resulting from frost, wind damage, and desiccation. Timberline pines appear to exhibit ecophysiological adaptations to minimize all of these potential limiting factors. Both *P. cembra* and *P. mugo* in the European Alps exhibit greater resistance to low winter temperatures than do lower-elevation trees, and thus survive cold winters without harm (Tranquillini 1979). Drought tolerance accounts for a large part of this resistance. Young needle tissues of *P. cembra* which are more sensitive to frost damage are not initiated until summer when frost events that could produce damage are rare.

Wind is an important environmental factor for plant growth near the timberline (Tranquillini 1979; Hadley & Smith 1983). Timberline conifers in Europe, including *P. cembra*, are much less sensitive to this factor than are broadleaved shrubs and trees in these habitats (Tranquillini 1979). This species can maintain open stomata even in strong winds, with only recently formed needles showing stomatal sensitivity (Caldwell 1970a). Observed declines in net photosynthesis under high wind speeds were attributed largely to shading as clusters of needles are pressed closely together (Caldwell 1970b). Even with such self-shading, photosynthesis in *P. cembra* was reduced by only 15–40% at high wind velocities of 15 m s^{-1}, compared with complete stomatal closure in *Rhododendron ferrugineum* under such conditions.

Visible damage to pines and other tree species at the timberline is largely the result of desiccation effects resulting from a complex of environmental and biotic interactions which Tranquillini (1979) has termed *frost-desiccation* (Fig. 15.7). Low temperatures, nutrient deficiencies, and late growth initiation may all lead to incomplete cuticle development, and thus decreased resistance to cuticular transpiration under severe winter conditions. Wind abrasion on needles increases this effect. Such higher winter levels of potential cuticular water loss under conditions of frozen soils with low hydraulic conductivity promotes desiccation damage as tissue water stress results.

This model explains the observed differences in tree

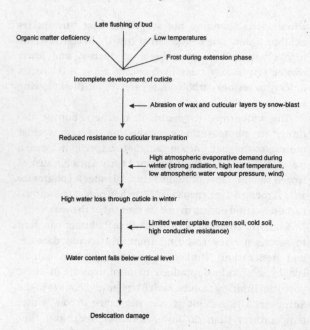

Fig. 15.7. Empirical model of the causal chain of relationships leading to desiccation of terminal shoots of timberline pines and other conifers during winter (Tranquillini 1979).

growth-form and desiccation tolerance near the timberline. Tree forms of growth become increasingly restricted with elevation to scattered individuals on the most favourable microsites, with dwarfed krummholz individuals at the limits of growth. Considerable interspecific difference can be observed in the elevational distribution of tree- and krummholz forms of growth. Drought-resistant *P. aristata* and *P. flexilis* in the Rocky Mountains of the western USA occur as trees well above elevations where other conifers occur only as krummholz (Wardle 1965), and *P. cembra* and *P. mugo* reach higher elevations in central Europe than *Picea abies* (Tranquillini 1979). These severe timberline habitats are often associated with long needle retention. *Pinus aristata* and *P. longaeva* in the western USA commonly retain physiologically active needles for 10–15 years, and occasionally as long as 45 years (Ewers & Schmid 1981).

15.5.3 Pines in montane habitats
Ecophysiological studies of montane and lower subalpine conifers have focused more on *P. ponderosa* in the western USA than on any other species. Ponderosa pine is a dominant species in open forest stands in lower and less mesic montane environments throughout this region. Within its range, ponderosa pine occurs from sea level to >3000 m elevation. Some care must be taken in interpreting research studies, therefore, because of the strong genetic differences and phenotypic plasticity that exist within Rocky Mountain populations of this species as well as in

populations from the summer-dry climatic regime of the Sierra Nevada and Cascade Mountains.

Pinus ponderosa shows numerous adaptations to drought conditions. Rocky Mountain populations of ponderosa pine generally have higher rates of stomatal conductance than co-occurring conifers in other genera (Running 1976; Jackson & Spomer 1979; Smith *et al.* 1984). However, ponderosa pine exhibits a lower osmotic potential at the point of zero turgor and a greater range of osmotic potential during water stress than competing fir species. Rooting architecture with a deep taproot contributes further to this drought resistance and allows access to soil moisture at depth (Jackson & Spomer 1979). Diurnal patterns of net photosynthesis for ponderosa pine in the Rocky Mountains frequently show maximum rates in mid-morning followed by a midday depression as stomatal control limits water loss (Hadley 1969). Such midday stomatal closure was not observed on cloudy days.

Diurnal and seasonal patterns of photosynthesis in Sierra Nevada populations of *P. ponderosa* have shown that solar irradiance limits photosynthetic capacity during early summer growth (Helms 1970, 1971). By late summer, however, water stress and high vapour-pressure gradients became the strongest limiting factors in this summer-dry environment, and midday stomatal closure is typical. Rates of incremental wood growth for *P. ponderosa* in the Rocky Mountains are significantly correlated with amount of spring rainfall (Peterson, Arbaugh & Robinson 1993).

The tissue water relations of *P. ponderosa* have not been studied in detail over annual cycles. Experimental studies of components of tissue water relations of seedlings at various degrees of moisture depletion found little effect of water treatment on measured parameters, with the exception of apoplastic water fraction, suggesting that osmotic adjustment was occurring in response to soil moisture availability (Anderson & Helms 1994). The osmotic potential at full and zero turgor averaged −1.7 and −2.5 MPa, respectively, in autumn. Ponderosa pines with their deep roots are able to survive in areas with only about 250 mm of annual precipitation on isolated outcrops of hydrothermally altered rock in the Great Basin west of the Sierra Nevada where low nutrient conditions prevent competition from desert shrubs (DeLucia *et al.* 1988; DeLucia & Schlesinger 1991). Water-use efficiencies in these desert populations of ponderosa pine, as shown by carbon isotope fractionation, were not different from that found in populations from the Sierra Nevada or in associated pinyon pines ($\Delta = 14$–15‰), and significantly higher than those found in associated desert shrubs ($\Delta = 16$–17‰).

This same pattern of seasonal changes in maximum photosynthetic capacity found in ponderosa pine has been shown in the related *P. jeffreyi* in somewhat more mesic

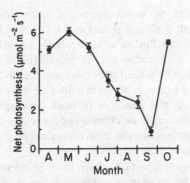

Fig. 15.8. **Seasonal cycle of net photosynthesis in *Pinus jeffreyi* in the Sierra Nevada of California at 2100 m elevation in Sequoia National Park. Adapted from Patterson & Rundel (1993).**

montane habitats in the Sierra Nevada. Needle photosynthetic rates were highest at the beginning of the growing season in May and June, and dropped steadily over the course of the summer until lowest seasonal rates occurred in September (Patterson & Rundel 1993; Fig. 15.8). Seasonal drought stress manifested with low water potentials and increasing vapour pressure gradients were associated with this decline. Photosynthetic capacity increased markedly, however, with the first autumn rains, even though air temperatures became less than optimal. *Pinus jeffreyi* trees maintained positive rates of net photosynthesis throughout the winter months, with solar irradiance as a limiting factor. These rates of photosynthesis on clear winter days were higher than those in September at the end of the summer drought period.

Older needles of *P. jeffreyi* decline in photosynthetic capacity by about 10% with each year of age (Patterson & Rundel 1993). This change was associated with an increase in carbon isotope discrimination (Δ). These changes in Δ could result from a decline in water-use efficiency of these needles due to a decline in mean intercellular CO_2. However, it may also reflect a change in needle composition with the accumulation of isotopically lighter compounds such as cellulose and the reduction in content of heavier compounds such as starch (Patterson & Rundel 1993).

Pinus contorta is another widespread montane to subalpine species that has been the focus of many field and experimental studies. Like ponderosa pine, *P. contorta* exhibits considerable genetic and ecotypic differentiation over its range throughout the western USA. *Pinus contorta* var. *latifolia* is a widespread ecological dominant at middle elevations in open, early-successional conditions throughout much of the Rocky Mountains.

Early ecophysiological studies of *P. contorta* were reviewed by Lopushinsky (1975) and Bassman (1985). Both drought and cold temperatures are major limiting factors for growth of lodgepole pine. Seedling studies have

reported an optimum temperature for photosynthesis at *c.* 20 °C, with a four-fold decline in rates at 2 °C (Dykstra 1974). Freezing night temperatures have been found to inhibit stomatal opening of *P. contorta* on the following day (Fahey 1979; Smith *et al.* 1984). Some of this effect may be the result of reduced water uptake by roots in cold soils (Running & Reid 1980).

Photosynthesis and stomatal conductance of *P. contorta* are strongly influenced by snow cover in the early growing season. Day, DeLucia & Smith (1989) found that individuals growing in cold soil (<1 °C) had 25–40% lower rates of photosynthesis than trees on warm soils (>10 °C). Low intercellular concentrations of CO_2 were present, providing a stomatal limitation on photosynthesis. Competing *Picea engelmannii* at the sites studied had a more extensive root system, and were thus able to utilize nearby warmer soils outside the snow packs. *Pinus contorta* lacked this ability, although deeper roots provided more access to soil moisture at depths where temperatures were less limiting. These differences may well explain an important aspect of the higher elevational limits of occurrence for *Picea engelmannii*.

Stomatal conductance, and thus presumably net photosynthetic uptake of carbon, is highest in *Pinus contorta* early in the growing season, and there is closure at midday (Fetcher 1976). By late summer, however, maximum conductance rates decline sharply at midday when water potentials decline to −1.7 MPa or below. Seasonal patterns of decline in maximum leaf conductance and predawn water potential have also been recorded (Running 1980). Such stomatal sensitivity, however, may result from a complex of factors including vapour pressure gradient and solar irradiance (Fetcher 1976; Knapp & Smith 1981).

Aspects of the water relations of *P. contorta* aid this species in its success as a colonizing species at low to middle elevations. Field studies have consistently found higher rates of stomatal conductance in lodgepole pine than other conifers at the same location in the central Rocky Mountains (Kaufmann 1982; Smith *et al.* 1984). High leaf conductance occurred at higher water potentials than in other species, suggesting greater water availability. Greater rooting depth or extent may be involved with this adaptation, but there is evidence that the thick sapwood of *P. contorta* may provide comparatively larger pools of available stored water (Lassen & Okkonen 1969).

Unlike many conifers, *P. contorta* is relatively tolerant of flooded soils, and thus is an important colonizer at the edge of wet meadows. The relative insensitivity of lodgepole pine roots to poorly oxygenated soils may be due to an increased capacity for O_2 transport in the root system (Coutts & Philipson 1978 a, b; Philipson & Coutts 1978).

15.5.4 Pines in cool temperate habitats

Pines species are frequently ecologically dominant on nutrient-poor soils in cool temperate regions which would otherwise support deciduous forest communities. Good examples of such species are *P. strobus* in the northeastern USA and *P. taeda* in the coastal plain and piedmont forests of southeastern USA. *Pinus sylvestris* in central Europe is another example, although it extends to much colder habitats at high latitudes (Chap. 5, this volume).

Seasonal and diurnal variation in photosynthetic assimilation in *P. strobus* are controlled by a variety of not only external factors, but also internal regulation. Light and temperature are the principal external controls affecting photosynthesis through the year. Although cold winter temperatures may prevent net photosynthesis in *P. strobus* over much of its range, low rates of net assimilation occur throughout the year near its southern extent (McGregor & Kramer 1963). Predawn water potentials in this species only rarely drop below -1.0 MPa, suggesting that water stress may not normally be a strong limiting factor for growth. However, low rates of stomatal conductance under drought conditions have been associated with lowered predawn water potentials of -1.1 to -1.4 MPa (Maier & Teskey 1992).

Developing foliage of the single annual growth flush in *P. strobus* presents a strong carbon sink in the spring, when foliage area doubles over a 10-week period (Maier & Teskey 1992). This sink appears to stimulate photosynthetic production in the older foliage. Net photosynthesis in existing foliage of *P. strobus* peaked sharply in spring to 6–7 μmol m^{-2} s^{-1} as new flushing was initiated, and then dropped sharply in July to 4–5 μmol m^{-2} s^{-1} as the young needles matured. Young developing foliage during its spring growth thus depended upon carbohydrates imported from other parts of the tree, particularly older foliage. Experimental removal of developing foliage in *P. resinosa* has been shown to significantly reduce photosynthetic rates in one-year-old foliage (Gordon & Larson 1968).

Pinus rigida is widespread on nutrient-poor soils in the eastern USA, and is most notable in its dominance on the oligotrophic Pine Barrens of New York and New Jersey. Although tolerant of nutrient conditions, this pine grows across a broad range of moisture availability from dry sites to poorly drained sands to swampy sites. Fertilization increases growth significantly (Ledig & Clark 1977), suggesting that its restriction to nutrient-poor sites is due to poor competitive ability. The optimal temperature for net photosynthesis in a latitudinal range of populations of *P. rigida* was found to be very consistent at about 25 °C (Ledig, Clark & Drew 1977). Day temperatures as low as 17 °C or as high as 32 °C substantially lowered rates of seedling growth (Good & Good 1976).

The ecophysiology of *P. taeda* has been well studied. In many respects, this species is transitional between cool temperate habitats to the north and warm subtropical habitats to the south. Unlike *P. strobus* and *P. sylvestris*, *P. taeda* exhibits multiple flushes of new needle growth during the year, a characteristic of tropical pines. Furthermore, the northern limit of occurrence of *P. taeda* is determined by the presence of severe winter ice storms which may decimate tree canopies through breakage. Even in the relatively moderate climate regime where it grows, low winter temperatures or frosts reduce net photosynthesis to zero (Teskey *et al.* 1986).

The ecophysiology of *P. taeda* has been broadly investigated in relation to both moisture and temperature stress (Teskey *et al.* 1986, 1987). Reductions of photosynthesis in *P. taeda* at low soil temperatures have been shown to result from stomatal limitations, but carbohydrate feedbacks causing non-stomatal limitations may also be involved (Day *et al.* 1991). Seasonal patterns of net photosynthesis vary significantly in genetic populations of *P. taeda*. (Boltz, Bongarten & Teskey 1986).

15.5.5 Pines in tropical habitats

Pines are dominant and co-dominant trees in many warm tropical habitats in Central America, the Caribbean region, and Southeast Asia. These species generally exhibit multiple flushes of needle production through the growing system in seasonal climate regimes, and may grow continuously in non-seasonal climates (Mirov 1967). Needle retention is relatively short in these species, averaging only 2–4 years. Comparatively little work has been done on the ecophysiology of these species.

Many tropical pines grow in wetlands with seasonally flooded soils. Such species have root intercellular spaces that enable them to tolerate extended periods of anoxia (Fisher & Stone 1991; Eissenstat & Van Rees 1994).

Seasonal studies of carbon cycling and photosynthetic uptake have been carried out with *P. elliottii*, in coastal areas of the southeastern USA where frosts are rare and rainfall occurs throughout the year. Plantations of this species in Florida showed a strong pattern of seasonal change in canopy photosynthesis with a peak in mean daily rate from April to June (Cropper & Gholz 1993, 1994). This peak corresponded to the period of maximum leaf expansion and growth, rather than the period of maximum canopy leaf area which occurred several months later. Ontogenetic effects appear to be involved as well, however, with older tissue showing reduced photosynthetic capacity (Beyers, Reichers & Temple 1992). The seasonal pattern of photosynthetic assimilation suggests that day length may be the most significant environmental factor in production so long as water is not limiting. Appreciable levels of photosynthetic assimilation occurred throughout the year so long as severe winter frosts did not occur (Teskey *et al.* 1994a).

Hurricane force winds can be a major factor in the large-scale mortality of pines in the Caribbean region and southeastern USA (Gresham, Williams & Lipscomb 1991). Conifers are generally more likely than hardwood trees to snap in strong winds because of lower wood strength and density, while the hardwoods with less flexible stems are more prone to uprooting (Putz *et al.* 1983).

15.6 The atmospheric environment

15.6.1 Ecophysiological impacts of atmospheric pollution

As ambient concentrations of oxidant air pollutants and acidic deposition have increased in many parts of the world over the past century, forest trees have been subject to significant impacts on their physiology and growth. Indeed, so pervasive are pollutant effects in many parts of the world that many, if not most, field forest studies realistically involve such pollutants as an environmental stress whether or not this fact is acknowledged. The literature on pollutant impacts on the growth and ecophysiology of forest trees is immense, and several studies have focused on pine species.

Recent texts have broadly reviewed the responses of forest trees to SO_2 (Winner, Mooney & Goldstein 1985), ozone (Guderian 1985; Olson, Binkley & Böhm 1992), and acid deposition (Schulze, Lange & Oren 1989), while a broad review of pollutant interactions with forests has been presented by Smith (1990). The evolutionary impacts of air pollution as a selective force influencing plant productivity and fitness have been considered by Taylor, Pitelka & Clegg (1991). The physiological impacts of air pollutants on plant processes have been reviewed by Darrall (1989).

Understanding the direct ecophysiological impacts of air pollutants on pine species is often a complex task. Visible evidence of needle necrosis or early needle senescence is often pointed to as evidence of pollutant impacts, but physiological processes may be significantly affected long before such morphological evidence is seen. Interpretation of the degree of physiological impact is further complicated by such variables as effective pollutant dose, phenological stage or age of tissues affected, seed source, nutritional status or water balance of trees, tree age, and the integrated effects of multiple stresses. Furthermore, chronic pollutant stress may increase the sensitivity of pines to other environmental stresses or pathogens through secondary impacts.

The point of first impact of gaseous air pollutants on pine tissues is largely the substomatal spaces within the needles. The further transfer of these pollutants to other

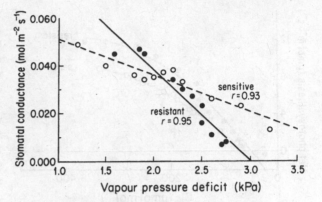

Fig. 15.9. **Stomatal conductance in relation to changes in vapour pressure deficit (VPD) in ozone-resistant and ozone-sensitive individuals of *Pinus jeffreyi* at Sequoia National Park in the Sierra Nevada of California. Adapted from Patterson & Rundel (1989).**

tissues follows pathways similar to those of CO_2, but transfer rates are affected by the diffusivity, solubility and chemical properties which affect the rate of movement of each type of molecule. Because stomatal conductance is generally correlated with level of pollutant exposure internally, species with slow growth rates and low stomatal conductance are typically less sensitive than are rapidly growing species with high mean conductance rates.

Foliar chlorosis due to ozone in pines is associated with the degradation of photosynthetic membranes and cell lysis (Karenlampi 1986; Evans & Fitzgerald 1993). Early senescence of needle cohorts is correlated with reductions in leaf metabolites (Miller, McCutchan & Milligan 1972) or oxidative damage to plasmalemma (Evans & Ting 1973). Photosynthetic rates were found to decline with degree of chlorotic mottle in ozone-sensitive *P. jeffreyi*, with a sharper decline up to about 30% needle mottle, and a more gentle change with greater amounts of damage (Patterson & Rundel 1995). These data suggest that there may be compensatory changes in photosynthetic capacity as larger areas of the needle surface are damaged. Such foliar chlorosis did not appear in the first season of needle growth in *P. jeffreyi*, but increased in area linearly through the second and third seasons of growth (Patterson & Rundel 1995).

Reduced growth rates in pine populations impacted by ozone appear to result from both stomatal and non-stomatal effects on net photosynthesis. Stomatal limitations on photosynthesis in *P. jeffreyi* have been demonstrated with controlled experiments of stomatal response to vapour pressure gradients under field conditions. Sensitive trees showing chlorotic leaf mottle maintained significantly higher conductance than resistant individuals in the same populations at vapour pressure deficit (VPD) levels below 2.5 kPa (Patterson & Rundel 1989; Fig. 15.9).

Ozone damage has also been shown to affect photosynthetic capacity of *P. jeffreyi* through non-stomatal

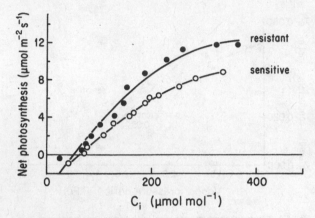

Fig. 15.10. **Net photosynthesis in relation to internal CO₂ concentration (Cᵢ) in ozone-resistant and ozone-sensitive individuals of *Pinus jeffreyi* at Sequoia National Park in the Sierra Nevada of California. Adapted from Patterson & Rundel (1989).**

mechanisms acting on the photosynthetic machinery. Ozone-sensitive trees were found to have a lower carboxylation efficiency in two-year-old needles than resistant trees, as shown by a 25% slower initial increase in photosynthesis in response to increasing concentrations of internal CO_2 (Patterson & Rundel 1989; Fig. 15.10). Furthermore, CO_2-saturated photosynthetic rates were 20% lower for sensitive trees, suggesting damage to the electron transport or RuBP regeneration systems of the photosynthetic apparatus. Similar effects of ozone on carboxylation efficiency and rates of RuBP carboxylation efficiency have been reported for *P. taeda* (Teskey *et al.* 1986; Sasek & Richardson 1989).

Experimental studies using controlled ozone fumigation of pine seedlings have not shown a consistent pattern of response. Although ozone exposure generally reduces photosynthetic capacity, exceptions have been reported. Exposures of four western conifers, including *P. ponderosa*, to moderate levels of ozone over two months produced no change in rates of net photosynthesis compared with controls in clean air (Bytnerowicz & Grulke 1992). Similarly, net photosynthesis in seedlings of *P. serotina* was unaffected by exposure to elevated ozone, while rates in two seed sources of *P. strobus* were enhanced by such treatment (Barnes 1972). These results are difficult to interpret, but may relate to the conditions of exposure or other environmental stress factors. Drought stress in seedlings of *P. ponderosa* may protect physiological processes from damage by reducing exposure through stomatal closure (Beyers *et al.* 1992; Temple *et al.* 1994).

The metabolic activity of *P. jeffreyi* and *P. ponderosa* throughout the year may make these species more sensitive than many other conifers to ozone damage. Premature autumn loss of ozone-damaged needles in these species sharply reduces canopy leaf area in the winter months, thereby leading to reduced starch reserves and restricted growth potential in the following spring (Patterson & Rundel 1989). This growth restriction may occur not only above ground, but below ground as well, with decreased spring root growth and lower root carbohydrate storage (Anderson *et al.* 1991).

Significant impacts of ozone on the growth and survival of pines in conifer forests are best described for the San Bernardino Mountains of southern California where long-term studies have been carried out (Miller 1992). Field studies over several decades have established that there is a range in susceptibility to ozone among the seven native conifers present. Pine species show great variability, with *Pinus jeffreyi* and *P. ponderosa* the most susceptible species, while *P. lambertiana* was considered the least susceptible and *P. coulteri* intermediate (Miller 1992).

Ozone impacts on growth of forest pines have also been documented for the Sierra Nevada (Peterson *et al.* 1987, 1991), and suggested for the southeastern USA (Sheffield *et al.* 1985; Knight 1987). Although local sites in the Rocky Mountains have moderately high ambient levels of atmospheric ozone, visible pollutant injury in *P. ponderosa* and growth reduction attributable to ozone have not been found (Peterson, Arbaugh & Robinson 1993).

Plant exposure to sulphur dioxide causes many biochemical and physiological impacts (Winner *et al.* 1985). SO_2 directly affects the photophosphorylation of ATP as sulphate competes for phosphate sites during ATP synthesis. Resultant shortage of ATP restricts carbon fixation, and thus pool sizes of carbohydrates and other metabolites (Bytnerowicz & Grulke 1992). Protein metabolism may also be affected. Experimental seedling studies have shown significant reduction of rates of net photosynthesis in seedlings of *P. ponderosa* (Houpis 1989) and *P. contorta* × *P. banksiana* hybrids (Amundson *et al.* 1986). In the latter site, high ambient exposure to SO_2 is associated with lowered photosynthetic capacity, foliar chlorosis, and shortened needle retention in mature trees.

Acid mists with a pH of 2.5 to 4.0 have been found to have a fertilization effect in enhancing net photosynthesis in a number of conifers, including *P. taeda* (Reich *et al.* 1987; Hanson, McLaughlin & Edwards 1988). Little or no interaction of a combination of ozone and acid mist was found in studies with *P. taeda* (Taylor *et al.* 1991; Reich *et al.* 1987) or *P. sylvestris* (Skeffington & Roberts 1985). Highly acidic fogs with a pH below 2.5 act differently and reduce net photosynthetic capacity in seedlings of *P. ponderosa* (Temple 1988) and *P. jeffreyi* (Takemoto & Bytnerowicz 1992).

Air pollution may also impact the ecophysiology of pines, both directly and indirectly, through effects on mineral nutrient status. Since ozone exposure typically reduces the relative allocation of carbon to tree roots, total rates of nutrient uptake in exposed plants may be reduced

(Tingey *et al.* 1986; Edwards *et al.* 1991). Acid rain may also have such an effect by leaching magnesium and other critical elements from forest soils (Schulze *et al.* 1989). Nitrate deposition from gaseous or particulate forms of nitrogen is another such interaction. This nitrate can result in stand fertilization and thus altered conditions for growth, but in many cases such nitrate additions may have a phytotoxic effect (Wellburn 1990).

15.6.2 Global change and elevated CO$_2$

Questions of forest response to elevated ambient CO$_2$ concentrations are critically important in understanding the potential role of forest systems as a buffer in global carbon budgets (Bazzaz & Fajer 1992; Gucinski, Vance & Reiners 1995). Since growth rate in conifers is a function of photosynthetic rate and allocation of fixed carbon between foliage and non-photosynthetic tissues, there has been an emphasis in recent years on ecophysiological studies of the photosynthetic responses and resource allocation in pine species exposed to elevated CO$_2$ concentrations. These studies, involving single-factor growth experiments as well as studies of multiple stress interactions, have focused heavily on three species – *P. ponderosa*, *P. radiata* and *P. taeda*.

Rates of photosynthesis in plant species generally increase with CO$_2$ fertilization, so long as irradiance, water and nutrients are not limiting (Eamus & Jarvis 1989; Mooney *et al.* 1991). The magnitude of increased rates of photosynthesis in plant species following exposure to 500–700 μmol mol^{-1} concentrations of ambient CO$_2$, however, have been found to be quite variable. Even within a single pine species, *P. taeda*, photosynthetic rates have been reported to increase anywhere from insignificant amounts to double control levels (Tolley & Strain 1984a, b; Fetcher *et al.* 1988; Liu & Teskey 1994).

The nature of possible acclimation in pine species after long-term exposure to elevated levels of CO$_2$ remains unclear. While most studies have not shown such down-regulation of photosynthetic capacity (Teskey *et al.* 1994b; Teskey 1995), seven-year-old *P. ponderosa* exposed to twice ambient CO$_2$ levels for 18 months were found to have lower photosynthetic rates than control plants (Surano *et al.* 1986). High concentrations of CO$_2$ (>1000 μmol mol^{-1}) have been found to lower photosynthetic capacity in both *P. contorta* and *P. taeda* (Tolley & Strain 1984b; Higginbotham *et al.* 1985).

Stomatal conductance in pines, like photosynthesis, shows variable responses to exposure to elevated concentrations of CO$_2$. A four-month exposure of *P. radiata* seedlings to elevated CO$_2$ decreased stomatal conductance by 35% (Hollinger 1987), but other studies found no difference in conductance rates after 22 months exposure in the same species (Conroy, Barlow & Beverage 1986a; Conroy *et*

al. 1986b, 1988). This difference in response is likely to reflect both differences in the genetic origin of the seedlings used and the nature of experimental treatments. Drought and nutrient limitations can clearly interact to form complex interactions with CO$_2$ fertilization.

15.6.3 Modelling atmospheric effects

Despite the large literature on the ecophysiological impacts of atmospheric pollutants and increased atmospheric concentrations of CO$_2$ on the growth of pines and other trees, a thorough mechanistic understanding of this process is still lacking (Pell *et al.* 1994; Winner 1994). The difficulty in reaching this goal results in part from the complexity of multiple stress interactions to which forest trees are exposed. Inherent genetic variation in species sensitivity, together with the poor understanding of phenotypic plasticity in physiological response, adds to this difficulty.

Most existing data on air pollutant effects and CO$_2$ interactions have come from controlled exposure experiments with seedlings, but relatively little progress has been made in effectively scaling from seedling to tree studies. This question of scaling is a broad one that is of critical importance in extrapolating not only from seedling experiments, but also from branch measurements to canopy studies. A number of physiological models have been developed in recent years to attempt to bridge these gaps in our understanding (Reich *et al.* 1990; Jarvis *et al.* 1991; Weinstein, Beloin & Yanai 1991). Although these models are still imperfect, they provide an important first step in progress that is badly needed. Tree ecophysiology, demography, and ecosystem studies must all come together if this progress is to occur.

15.7 Conclusions

Despite remarkable progress over the past two decades in developing and expanding our understanding of the ecophysiology of pines, comments made 30 years ago by Mirov (1967) remain true. As he wrote, 'Our knowledge of the genus *Pinus* is rather uneven.' This statement remains notably true.

Much of what has been learned about the ecophysiology of pines has come from studies of a relatively small number of species. *Pinus ponderosa* and *P. taeda* have been the subject of intensive studies both under natural circumstances and in their role as model systems for understanding the impacts of atmospheric pollutants and changing atmospheric CO$_2$ concentrations on forest ecosystems. *Pinus sylvestris* has served a similar role for European research. Despite these studies, our understanding of the relative significance of genetic variation and phenotypic plasticity in mediating pine response to environmental stress remains relatively poor. The economic importance of *P. radiata* in the southern hemisphere has made this species another focus of ecophysiological study, although relatively little of this research has been carried out in its natural habitat in California.

Although improved instrumentation has made new and exciting research on pine ecophysiology possible, the inherent problems in such research remain the same as they did when Mirov wrote his text. Experimental material for pine research has come largely from wild forest plants with varied and unknown genetic heterogeneity. Thus intraspecific variation in physiological traits is often as great as or greater than interspecific variation. Long life histories in pines make for inherent difficulties for breeding experiments to understand the inheritance of physiological traits. Added to this is the large size of pines and complex architecture of pine canopies. It is not surprising that most controlled growth experiments with pines have been done with seedlings.

Studies of photosynthesis and its controls in pine species have made great strides in recent years. While these studies have largely been focused at a needle level of scale, work with scaling from needle to branch and branch to canopy show real promise for advancement. Similarly, there has been significant progress in investigations of water relations in pine species, and the mechanisms through which pines gain much of their relative drought resistance compared with many other conifers. The hydraulic architecture of pines remains poorly studied, but is likely to be a significant area for research in the future.

Ecophysiological components of adaptations to specific environmental conditions are relatively well understood for timberline and montane pines, and for selected temperate forest pines. An equivalent level of understanding, however, is still lacking for pines in other selective environments. Much of our knowledge of the ecophysiology of mediterranean-climate pines comes from plantation studies of *P. radiata*, but there are few equivalent studies of pine species from the Mediterranean Basin. There are only very limited data available on the ecophysiology of tropical pine species despite their economic importance. What information is available is largely from research on the subtropical *P. elliottii* and subtropical traits existing in *P. taeda*. Major opportunities for field-orientated studies of pine ecophysiology exist in Mexico where widespread speciation has produced sharp gradients of pine distribution from lowland tropical and subtropical habitats to high montane and subalpine ecosystems.

References

Ågren, G.I., Axelsson, B., Flower-Ellis, J.G.K., Linder, S., Persson, H., Staaf, H. & Troeng, E. (1980). Annual carbon budget for a young Scots pine. *Ecological Bulletin (Copenhagen)*, **32**, 307–14.

Amundson, R.G., Walker, R.B. & Legge, A.H. (1986). Sulfur gas emissions in the boreal forest: the West Whitecourt case study. VII. Pine tree physiology. *Water, Air and Soil Pollution*, **29**, 129–47.

Anderson, C.P., Hogsett, W.E., Wessling, R. & Plocher, M. (1991). Ozone decreases spring root growth and root carbohydrate content in ponderosa pine the year following exposure. *Canadian Journal of Forest Research*, **21**, 1288–91.

Anderson, P.D. & Helms, J.A. (1994). Tissue water relations of *Pinus ponderosa* and *Arctostaphylos patula* exposed to various levels of soil moisture depletion. *Canadian Journal of Forest Research*, **24**, 1495–502.

Baig, M.N. & Tranquillini, W. (1976). Studies on upper timberline: morphology and anatomy of Norway spruce (*Picea abies*) and stone pine (*Pinus cembra*) needles from various habitat conditions. *Canadian Journal of Botany*, **54**, 1622–32.

Barker, J.E. (1973). Diurnal patterns of water potential in *Abies concolor* and *Pinus ponderosa*. *Canadian Journal of Forest Research*, **3**, 556–64.

Barnes, F.J. & Cunningham, G. L. (1987). Water relations and productivity in pinyon–juniper habitat types. In *Proceedings – Pinyon–Juniper Conference*, pp. 406–11. USDA Forest Service General Technical Report INT-215.

Barnes, R.L. (1972). Effects of chronic exposure to ozone on photosynthetic and respiration of pines. *Environmental Pollution*, **3**, 133–8.

Bassman, J.H. (1985). Selected physiological characteristics of lodgepole pine. In *Lodgepole Pine: The Species and its Management*, ed. D.M. Baumgartner, R.G. Krebill, J.T. Arnott & G.F. Weetman, pp. 27–43. Pullman: Washington State University Cooperative Extension.

Bates, C.G. (1925). The relative light requirement of some coniferous seedlings. *Journal of Forestry*, **23**, 869–79.

Bazzaz, F.A. & Fajer, E.D. (1992). Plant life in a CO_2-rich world. *Scientific American*, **264**, 68–74.

Beadle, C.L., Neilsen, R.E., Jarvis, P.G. & Talbot, H. (1981). Photosynthesis as related to xylem water potential and carbon dioxide concentration in Sitka spruce. *Physiologia Plantarum*, **52**, 391–400.

Beadle, C.L., Neilson, R.E., Talbot, H. & Jarvis, P.G. (1985). Stomatal conductance and photosynthesis in a mature Scots pine forest. I. Diurnal, seasonal and spatial variation in shoots. *Journal of Applied Ecology*, **22**, 557–71.

Benecke, U. (1980). Photosynthesis and transpiration of *Pinus radiata* under natural conditions in a forest stand. *Oecologia*, **44**, 192–8.

Berry, J. & Björkman, O. (1980). Photosynthetic response and adaptation to temperature in higher plants. *Annual Review of Plant Physiology*, **31**, 491–543.

Beyers, J.L. Riechers, G.H. & Temple, P.J. (1992). Effects of long-term ozone exposure and drought on the photosynthetic capacity of ponderosa pine (*Pinus ponderosa* Laws). *New Phytologist*, **122**, 81–90.

Billow, C., Matson, P. & Yoder, B. (1994). Seasonal biochemical changes in coniferous canopies and their response to fertilization. *Tree Physiology*, **14**, 563–74.

Boltz, B.A., Bongarten, B.C. & Teskey, R.O. (1986). Seasonal patterns of net photosynthesis of loblolly pine from diverse origins. *Canadian Journal of Forest Research*, **16**, 1063–8.

Boucher, J.-F., Munson, A.D. & Bernier, P.Y. (1995). Foliar absorption of dew influences shoot water potential and root growth in *Pinus strobus* seedlings. *Tree Physiology*, **15**, 819–23.

Bourdeau, P.F. (1959). Seasonal variations of the photosynthetic efficiency of evergreen conifers. *Ecology*, **40**, 63–7.

Bytnerowicz, A. & Grulke, N. (1992). Physiological effects of air pollutants on western trees. In *The Response of Western Forests to Air Pollution*, ed. R.K. Olson, D. Binkley & M. Böhm, pp. 183–233. New York: Springer-Verlag.

Caldwell, M.M. (1970a). The effect of wind on stomatal aperture, photosynthesis and transpiration of *Rhododendron ferrugineum* L. and *Pinus cembra* L. *Zentralbl. Gesamte Forstwes.*, **87**, 193–201.

Caldwell, M.M. (1970b). Plant gas exchange at high wind speeds. *Plant Physiology*, **46**, 535–7.

Carlquist, S. (1975). *Ecological Strategies of Xylem Evolution*. Berkeley: University of California Press.

Carter, G.A. & Smith, W.K. (1985). Influence of shoot structure on light interception and photosynthesis of conifers. *Plant Physiology*, **79**, 1038–43.

Carter, G.A. & Smith, W.K. (1988). Microhabitat comparisons of transpiration and photosynthesis in three subalpine conifers. *Canadian Journal of Botany*, **66**, 963–9.

Ceulemans, R.J. & Saugier, B. (1991). Photosynthesis. In *Physiology of Trees*, ed. A.S. Raghavendra, pp. 21–50. New York: John Wiley.

Chabot, B.F. & Hicks, D.J. (1982). The ecology of leaf life spans. *Annual Review of Ecology and Systematics*, **13**, 229–59.

Cochard, H. (1992). Vulnerability of several conifers to air embolism. *Tree Physiology*, **11**, 73–83.

Conroy, J.P., Barlow, E.W.R. & Beverage, D.I. (1986a). Response of *Pinus radiata* seedlings to carbon dioxide enrichment at different levels of water and phosphorus: growth, morphology and anatomy. *Annals of Botany*, **57**, 165–77.

Conroy, J.P, Küppers, M., Küppers, B., Virgona, J. & Barlow, E.W.R. (1988). The influence of CO_2 enrichment, phosphorus deficiency and water stress on the growth, conductance, and water use of *Pinus radiata* D. Don. *Plant, Cell and Environment*, **11**, 91–8.

Conroy, J.P., Smillie, R.M., Küppers, M., Virgona, J. & Barlow, E.W.R. (1986b). Chlorophyll *a* fluorescence and photosynthetic and growth responses of *Pinus radiata* to phosphorus deficiency, drought stress, and high CO_2. *Plant Physiology*, **81**, 423–9.

Coutts, M.P. & Philipson, J.J. (1978a). Tolerance of tree roots to waterlogging. I. Survival of sitka spruce and lodgepole pine. *New Phytologist*, **80**, 63–9.

Coutts, M.P. & Philipson, J.J. (1978b). Tolerance of tree roots to waterlogging. II. Adaptations of sitka spruce and lodgepole pine to waterlogged soils. *New Phytologist*, **80**, 71–7.

Cowan, I.R. (1977). Stomatal behavior and the environment. *Advances in Botanical Research*, **4**, 117–228.

Cregg, B.M. (1994). Carbon allocation, gas exchange and needle morphology of *Pinus ponderosa* genotypes known to differ in growth and survival under imposed drought. *Tree Physiology*, **14**, 883–98.

Cropper, W.P. & Gholz, H.L. (1993). Simulation of the carbon dynamics of a Florida slash pine plantation. *Ecological Modelling*, **66**, 231–49.

Cropper, W.P. & Gholz, H.L. (1994). Evaluating potential response mechanisms of a forest stand to fertilization and night temperatures: a case study using *Pinus elliottii*. *Ecological Bulletin (Copenhagen)*, **43**, 154–60.

Darrall, N.M. (1989). The effect of air pollutants on physiological processes in plants. *Plant, Cell and Environment*, **12**, 1–30.

Daubenmire, R.F. (1943). Vegetational zonation in the Rocky Mountains. *Botanical Review*, **9**, 326–93.

Day, T.A., DeLucia, E.H. & Smith, W.K. (1989). Influence of cold soil and snowcover on photosynthesis and leaf conductance in two Rocky Mountain conifers. *Oecologia*, **80**, 546–52.

Day, T.A., Heckathorn, S.A. & DeLucia, E.L. (1991). Limitations of photosynthesis in *Pinus taeda* (loblolly pine) at low soil temperatures. *Plant Physiology*, **96**, 1246–54.

DeLucia, E.H., Callaway, R.M. & Schlesinger, W.H. (1994). Offsetting changes in biomass allocation and photosynthesis in ponderosa pine (*Pinus ponderosa*) in response to climate change. *Tree Physiology*, **14**, 669–77.

DeLucia, E.H., Day, T.A. & Oquist, G. (1991). The potential for photoinhibition of *Pinus sylvestris* L. seedlings exposed to high light and low soil temperature. *Journal of Experimental Botany*, **42**, 611–17.

DeLucia, E.H. & Schlesinger, W.H. (1991). Resource-use efficiency and drought tolerance in adjacent Great Basin and Sierran plants. *Ecology*, **72**, 51–8.

DeLucia, E.H., Schlesinger, W.H. & Billings, W.D. (1988). Water relations and the maintenance of Sierran conifers on hydrothermally altered rock. *Ecology*, **69**, 303–11.

Dick, J.M., Jarvis, P.G. & Leakey, R.R.B. (1991). Influence of male and female cones on needles CO_2 exchange rates of field-grown *Pinus contorta* Doug. trees. *Functional Ecology*, **5**, 422–32.

Dykstra, G.F. (1974). Photosynthesis and carbon dioxide transfer resistance of lodgepole pine seedlings in relation to irradiance, temperature, and water potential. *Canadian Journal of Forest Research*, **4**, 201–6.

Eamus, D. & Jarvis, P.G. (1989). The direct effects of increase in the global atmospheric CO_2 concentration on natural and commercial temperate trees and forests. *Advances in Ecological Research*, **19**, 1–55.

Edwards, G.S., Edwards, N.T., Kelly, J.M. & Mays, P.A. (1991). Ozone, acidic precipitation, and soil Mg effects on growth and nutrition of loblolly pine seedlings. *Environmental and Experimental Botany*, **31**, 67–78.

Edwards, W.R.N. & Jarvis, P.G. (1982). Relations between water content, potential, and permeability in stems of conifers. *Plant, Cell and Environment*, **5**, 217–77.

Edwards, W.R.N., Jarvis, P.G., Grace, J. & Moncrieff, J.B. (1994). Reversing cavitation in tracheids of *Pinus sylvestris* L. under negative water potentials. *Plant, Cell and Environment*, **17**, 389–97.

Eissenstat, D.M. & Van Rees, K.C.J. (1994). The growth and function of pine roots. *Ecological Bulletin (Copenhagen)*, **43**, 76–91.

Evans, E.H. & Fitzgerald, G.A. (1993). Histological effects of ozone on slash pine (*Pinus elliottii* var. *densa*). *Environmental and Experimental Botany*, **33**, 505–13.

Evans, L. & Ting, I. (1973). Ozone-induced membrane permeability changes. *American Journal of Botany*, **60**, 155–62.

Ewers, F.W. & Schmid, R. (1981). Longevity of needle fascicles of *Pinus longaeva* (bristlecone pine) and other North American pines. *Oecologia*, **51**, 107–15.

Ewers, F.W. & Zimmermann, M.H. (1984a). The hydraulic architecture of balsam fir (*Abies balsamea*). *Physiologia Plantarum*, **60**, 453–8.

Ewers, F.W. & Zimmermann, M.H. (1984b). The hydraulic architecture of eastern hemlock (*Tsuga canadensis*). *Canadian Journal of Botany*, **2**, 940–6.

Fahey, T.D. (1979). The effect of night frost on the transpiration of *Pinus contorta* ssp. *latifolia*. *Oecologia Plantarum*, **14**, 483–90.

Farquhar, G.D., Hubick, K.T., Condon, A.G. & Richards, R.A. (1988). Carbon isotope fractionation and plant water-use efficiency. In *Stable Isotopes in Ecological Research*, ed. P.W. Rundel, J.R. Ehleringer & K.A. Nagy, pp. 21–40. New York: Springer-Verlag.

Fetcher, N. (1976). Patterns of leaf resistance to lodgepole pine transpiration in Wyoming. *Ecology*, **57**, 339–45.

Fetcher, N., Jaeger, C.H., Strain, B.R. & Sionit, N. (1988). Long-term elevation of atmospheric CO_2 concentration and the carbon exchange rates of saplings of *Pinus taeda* L. and *Liquidamber styraciflua* L. *Tree Physiology*, **4**, 255–62.

Field, C. & Mooney, H.A. (1986). The photosynthesis–nitrogen relationship in wild plants. In *On the Economy of Plant Form and Function*, ed. T.J. Givnish, pp. 25–55. New York: Cambridge University Press.

Fisher, H.M. & Stone, E.L. (1991). Iron oxidation at the surface of slash pine roots from saturated soils. *Soil Science Society of America Journal*, **55**, 1123–9.

Fites, J.A. & Teskey, R.O. (1988). CO_2 and water vapor exchange of *Pinus taeda* in relation to stomatal behavior, test of an optimization hypothesis. *Canadian Journal of Forest Research*, **18**, 150–7.

Fowells, H.A. & Kirk, B.M. (1945). Availability of soil moisture to ponderosa pine. *Journal of Forestry*, **43**, 601–4.

Franklin, J.F. & Dyrness, C.T. (1973). *Natural Vegetation of Oregon and Washington*. USDA Forest Service General Technical Report PNW-8.

Freeland, R.O. (1952). Effect of age of leaves upon the rate of photosynthesis in some conifers. *Plant Physiology*, **27**, 685–90.

Garcia, R.L., Idso, S.B., Wall, G.W. & Kimball, B.A. (1994). Changes in net photosynthesis and growth of *Pinus eldarica* seedlings in response to atmospheric CO_2 enrichment. *Plant, Cell and Environment*, **17**, 971–8.

Gholz, H.L. & Cropper, W.P. (1991). Carbohydrate dynamics in mature *Pinus elliottii* var. *elliottii* trees. *Canadian Journal of Forest Research*, **21**, 1742–7.

Gholz, H.L., Linder, S. & McMurtrie, R.E. (eds.) (1994). *Pine Ecosystems*. Ecological Bulletin, Vol. 43, Copenhagen.

Good, R.E. & Good, N.F. (1976). Growth analysis of pitch pine seedlings under three temperature regimes. *Forest Science*, **22**, 445–8.

Gordon, J.C. & Larson, P.R. (1968). Seasonal course of photosynthesis, respiration, and distribution of ^{14}C in young *Pinus resinosa* trees as related to wood formation. *Plant Physiology*, **43**, 1617–24.

Gower, S.T., Reich, P.B. & Son, Y. (1993). Canopy dynamics and aboveground production of five tree species with different leaf longevities. *Tree Physiology*, **12**, 327–45.

Grace, J., Allen, S.J. & Wilson, C. (1989). Climate and the meristem temperatures of plant communities near the tree-line. *Oecologia*, **79**, 198–204.

Gresham, C.A., Williams, T.M. & Lipscomb, D.J. (1991). Hurricane Hugo damage to southeastern coastal forest tree species. *Biotropica*, **23**, 420–6.

Grier, C.C. & Waring, R.H. (1974). Coniferous foliage mass related to sapwood area. *Forest Science*, **20**, 205–6.

Grulke, N.E., Hom, J.L. & Roberts, S.W. (1993). Physiological adjustment of two half-sib families of ponderosa pine to elevated CO_2. *Tree Physiology*, **12**, 391–401.

Grulke, N.E. & Miller, P.R. (1994). Changes in gas exchange characteristics during the life span of giant sequoia: implications for response to current

and future concentrations of atmospheric ozone. *Tree Physiology*, **14**, 659–68.

Gucinski, H., Vance, E. & Reiners, W.A. (1995). Potential effects of global climate change. In *Ecophysiology of Coniferous Forests*. ed. W.K. Smith & T.M. Hinckley, pp. 309–31. San Diego: Academic Press.

Guderian, R. (ed.) (1985). *Air Pollution by Photochemical Oxidants*. Berlin: Springer-Verlag.

Hadley, E.B. (1969). Physiological ecology of *Pinus ponderosa* in southwestern North Dakota. *American Midland Naturalist*, **81**, 289–314.

Hadley, E.B. & Smith, W.K. (1983). Influence of wind exposure on needle desiccation and mortality for timberline conifers in Wyoming, USA. *Arctic and Alpine Research*, **15**, 127–35.

Haller, J.R. (1965). The role of 2-needle fascicles in the adaptation and evolution of ponderosa pine. *Brittonia*, **17**, 354–82.

Hanson, P.J., McLaughlin, S.B. & Edwards, N.T. (1988). Net CO_2 exchange of *Pinus taeda* shoots exposed to variable ozone levels and rain chemistries in field and laboratory settings. *Physiologia Plantarum*, **74**, 635–42.

Havranek, W.M. & Tranquillini, W. (1994). Physiological processes during winter dormancy. In *Ecophysiology of Coniferous Forests*, ed. W.K. Smith & T.M. Hinckley, pp. 95–124. San Diego: Academic Press.

Hellkvist, J., Richards, G.P. & Jarvis, P.G. (1974). Vertical gradients of water potential in Sitka spruce trees measured with the pressure chamber. *Journal of Applied Ecology*, **7**, 637–67.

Helms, J.A. (1970). Summer net photosynthesis of ponderosa pine in its natural environment. *Photosynthetica*, **4**, 243–53.

Helms, J.A. (1971). Environmental control of net photosynthesis in naturally growing *Pinus ponderosa* Laws. *Ecology*, **53**, 92–101.

Higginbotham, K.O., Mays, J.M., L'Hirondelle, S. & Krystofiak, D.K. (1985). Physiological ecology of lodgepole pine (*Pinus contorta*) in an enriched CO_2 environment. *Canadian Journal of Forest Research*, **15**, 417–21.

Hollinger, D.Y. (1987). Gas exchange and dry matter allocation responses to elevation of atmospheric CO_2 concentration in seedlings of three tree species. *Tree Physiology*, **3**, 193–202.

Houpis, J.L. (1989). Seasonal effects of sulfur dioxide on the physiology and morphology of *Pinus ponderosa* seedlings. PhD thesis, Berkeley: University of California.

Jackson, P.A. & Spomer, G.G. (1979). Biophysical adaptations of four western conifers to habitat water conditions. *Botanical Gazette*, **140**, 428–32.

James, J., Grace, J. & Hoad, S.P. (1994). Growth and photosynthesis of *Pinus sylvestris* at its altitudinal limit in Scotland. *Journal of Ecology*, **82**, 297–306.

Jarvis, P.G. (1976). The interpretation of the variations in leaf water potential and stomatal conductance found in canopies in the field. *Philosophical Transactions of the Royal Society, London B*, **273**, 593–610.

Jarvis, P.G., Barton, C.V.M., Dougherty, P.M., Teskey, R.O. & Masshedev, J.M. (1991). MAESTRO. In *Acidic Deposition: State of Science and Technology.*, Vol. 3, ed. P.M. Irving, pp. 167–78. Washington, DC: US National Acid Precipitation Assessment Program.

Jurik, T.W., Briggs, G.B. & Gates, D.M. (1988). Springtime recovery of photosynthetic activity of white pine in Michigan. *Canadian Journal of Botany*, **66**, 138–41.

Karenlampi, L. (1986). Relationship between macroscopic symptoms of injury and cell structural changes in needles of ponderosa pine exposed to air pollution in California. *Annales of Botanici Fennici*, **23**, 255–64.

Kaufmann, M.R. (1982). Leaf conductance as a function of photosynthetic photon flux density and absolute humidity difference from leaf to air. *Plant Physiology*, **69**, 1018–22.

Kaufmann, M.R. (1985). Species differences in stomatal behavior, transpiration and water use efficiency in subalpine forests. In *Crop Physiology of Forest Trees*, ed. P.M.A. Tigenstedt, P. Puttonen & V. Koski, pp. 39–52. Helsinki: University of Helsinki.

Knapp, A.K. & Smith, W.K. (1981). Water relations and succession in subalpine conifers in southeastern Wyoming. *Botanical Gazette*, **142**, 502–11.

Knight, H.A. (1987). The pine decline. *Journal of Forestry*, **85**, 25–8.

Kozlowski, T.T., Kramer, P.J. & Pallardy, S.G. (1991). *The Physiological Ecology of Woody Plants*. San Diego: Academic Press.

Kull, O. & Jarvis, P.G. (1995). The role of nitrogen in scaling up photosynthesis. *Plant, Cell and Environment*, **18**, 1174–82.

Küppers, M. & Schulze, E.-D. (1985). An empirical model of net photosynthesis and leaf conductance for the simulation of diurnal courses of CO_2 and H_2O exchange. *Australian Journal of Plant Physiology*, **12**, 513–26.

Lajtha, K. & Barnes, F.J. (1991). Carbon gain and water use in pinyon pine–juniper woodlands of northern New Mexico: field versus phytotron chamber measurements. *Tree Physiology*, **9**, 59–67.

Lajtha, K. & Getz, J. (1993). Photosynthesis and water-use efficiency in pinyon–juniper communities along an elevational gradient in northern New Mexico. *Oecologia*, **94**, 95–101.

Lassen, L.E. & Okkonen, R. (1969). *Sapwood Thickness of Douglas-fir and five other Western Softwoods*. USDA Forest Service Research Paper FPL-124.

Lassoie, J.P., Hinckley, T.M. & Grier, C.C. (1985). Coniferous forests of the Pacific Northwest. In *Physiological Ecology of North American Plant Communities*, ed. B.F. Chabot & H.A. Mooney, pp. 127–61. New York: Chapman & Hall.

Ledig, F.T. & Clark, J.G. (1977). Photosynthesis in a half-sib family experiment in pitch pine. *Canadian Journal of Forest Research*, **7**, 510–14.

Ledig, F.T., Clark, J.G. & Drew, A.P. (1977). The effects of temperature on photosynthesis of pitch pine from northern and southern latitudes. *Botanical Gazette*, **138**, 7–12.

Leverenz, J.W. (1987). Chlorophyll content and the light response curve of shade-adapted conifer needles. *Physiologica Plantarum*, **71**, 20–9.

Lewis, J.D., Griffin, K.L., Thomas, R.B. & Strain, B.R. (1994). Phosphorus supply affects the photosynthetic capacity of loblolly pine grown in elevated carbon dioxide. *Tree Physiology*, **14**, 1229–44.

Linder, S. & Lohammar, T. (1981). Amount and quality of information on CO_2 required for estimating annual carbon balance of coniferous trees. *Stud. For. Suec.*, **160**, 73–87.

Linder, S. & Troeng, E. (1980). Photosynthesis and transpiration of 20-year-old Scots pine. *Ecological Bulletin (Copenhagen)*, **32**, 165–81.

Lindgren, K. & Hallgren, J.-E. (1992). Cold acclimation of *Pinus contorta* and *Pinus sylvestris* assessed by chlorophyll fluorescence. *Tree Physiology*, **13**, 97–106.

Liu, S. & Teskey, R. O. (1994). Responses of foliar gas exchange to long-term elevated CO_2 concentrations in mature loblolly pine trees. *Tree Physiology*, **15**, 351–9.

Lopushinsky, W. (1975). Water relations and photosynthesis in lodgepole pine. In *Management of Lodgepole Pine Ecosystems*, Vol. 1, ed. D.M.B. Baumgartner, pp. 135–53. Pullman: Washington State University Press.

Ludlow, M.M. & Jarvis, P.G. (1971). Photosynthesis in sitka spruce (*Picea sitchensis* (Bong.) Carr.). I. General characteristics. *Journal of Applied Ecology*, **8**, 925–53.

McGregor, W.H.D. & Kramer, P.J. (1963). Seasonal trends in rates of photosynthesis and respiration of loblolly pine and white pine seedlings. *American Journal of Botany*, **50**, 760–5.

Maier, C.A. & Teskey, R.O. (1992). Internal and external control of net photosynthesis and stomatal conductance of mature eastern white pine (*Pinus strobus*). *Canadian Journal of Forest Research*, **22**, 1387–94.

Malusa, J. (1992). Xylem pressure potentials of single- and double-needled pinyon pines. *Southwestern Naturalist*, **37**, 43–8.

Margolis, H.A., Oren, R., Whitehead, D. & Kaufmann, M.R. (1995). Leaf area dynamics of conifer forests. In *Ecophysiology of Coniferous Forests*, ed. W.K. Smith & T.M. Hinckley, pp. 181–223. San Diego: Academic Press.

Marshall, J.D. & Zhang, J. (1993). Altitudinal variation in carbon isotope discrimination by conifers. In *Stable Isotopes and Plant Carbon–Water Relations*, ed. J.R. Ehleringer, A.E. Hall & G.D. Farquhar, pp. 187–200. San Diego: Academic Press.

Mencuccini, M. & Grace, J. (1995). Climate influences the leaf area/sapwood area ratio in Scots pine. *Tree Physiology*, **15**, 1–10.

Milburn, J.A. (1991). Cavitation and embolisms in xylem conduits. In *Physiology of Trees*, ed. A.S. Raghavendra, pp. 163–74. New York: John Wiley.

Miller, P.R. (1992). Mixed conifer forests of the San Bernardino Mountains, California. In *The Response of Western Forests to Air Pollution*. ed. R. K. Olson, D. Binkley & M. Böhm, pp. 461–500. New York: Springer-Verlag.

Miller, P.R. McCutchan, M.H. & Milligan, H.P. (1972). Oxidant air pollution in the Central Valley, Sierra Nevada foothills, and Mineral King Valley of California. *Atmospheric Environment*, **6**, 623–33.

Mirov, N.T. (1967). *The Genus Pinus*. New York: Ronald Press.

Monson, R.K. & Grant, M.C. (1989). Experimental studies of ponderosa pine. III. Differences in photosynthesis, stomatal conductance, and water-use efficiency between two genetic lines. *American Journal of Botany*, **76**, 1041–7.

Mooney, H.A., Drake, B.G., Luxmoore, R.J., Oechel, W.C. & Pitelka, L.F. (1991). Predicting ecosystem responses to elevated CO_2 concentrations. *BioScience*, **41**, 96–104.

Mooney, H.A., West, M. & Brayton, R. (1966). Field measurements of the metabolic responses of bristlecone pine and big sagebrush in the White Mountains of California. *Botanical Gazette*, **127**, 105–13.

Mooney, H.A., Wright, R.D. & Strain, B.R. (1964). The gas exchange capacity of plants in relation to vegetation zonation in the White Mountains of California. *American Midland Naturalist*, **72**, 281–97.

Naidu, S.L., Sullivan, J.H., Teramura, A.H. & DeLucia, E.H. (1993). The effects of ultraviolet-B radiation on photosynthesis of different-aged needles in field-grown loblolly pine. *Tree Physiology*, **12**, 151–62.

Näsholm, T. & Ericsson, A. (1990). Seasonal changes in amino acids, protein and total nitrogen in needles of fertilized Scots pine trees. *Tree Physiology*, **6**, 267–81.

Ng, P.A.P. & Jarvis, P.G. (1980). Hysteresis in the response of stomatal conductance in *Pinus sylvestris* L. needles to light: observations and a hypotheses. *Plant, Cell and Environment*, **3**, 207–16.

Nnyamah, J.U., Black, T.A. & Tan, C.S. (1978). Resistance to water uptake in a Douglas-fir forest. *Soil Science*, **126**, 63–76.

Oker-Blom, P., Lahti, T. & Smolander, H. (1992). Photosynthesis of a Scots pine shoot: A comparison of two models of shoot photosynthesis in direct and diffuse radiation fields. *Tree Physiology*, **10**, 111–25.

Olson, R. K., Binkley, D. & Böhm, M. (eds.) (1992) *The Response of Western Forests to Air Pollution*. New York: Springer-Verlag.

Oren, R., Werk, K.S. & Schulze, E.-D. (1986). Relationships between foliage and conducting xylem in *Picea abies* (L.) Karst. *Trees*, **1**, 61–9.

Patterson, M.T. & Rundel, P.W. (1989). Seasonal physiological responses of ozone stressed Jeffrey pine in Sequoia National Park, California. In *Effects of Air Pollution on Western Forests*, ed. R.K. Olson & A.S. Lefon, pp. 419–28. Pittsburgh, Pa: Air and Waste Management Association.

Patterson, M.T. & Rundel, P.W. (1993). Carbon isotope discrimination and gas exchange in ozone-sensitive and -resistant populations of Jeffrey pine. In *Stable Isotopes and Carbon–Water Relations*, ed. J.R. Ehleringer, A. Hall & G.D. Farquhar, pp. 213–25. San Diego: Academic Press.

Patterson, M.T. & Rundel, P.W. (1995). Stand characteristics of ozone-stressed populations of *Pinus jeffreyi* (Pinaceae): extent, development, and physiological consequences of visible injury. *American Journal of Botany*, **82**, 150–8.

Pelkonen, P., Hari, P. & Luukanen, O. (1977). Decrease of CO_2 exchange in Scots pine after naturally occurring or artificial low temperatures. *Canadian Journal of Forest Research*, **7**, 462–86.

Pell, E.J., Temple, P.J., Friend, A.C., Mooney, H.A. & Winner, W.E. (1994). Compensation as a plant response to ozone and associated stresses: an analysis of the ROPIS experiments. *Journal of Environmental Quality*, **23**, 429–36.

Peterson, D.L., Arbaugh, M.J. & Robinson, L.J. (1991). Regional growth trends of ozone-injured ponderosa pine (*Pinus ponderosa*) in the Sierra Nevada, California, USA. *Holocene*, **1**, 50–61.

Peterson, D.L., Arbaugh, M.J. & Robinson, L.J. (1993). Effects of ozone and climate on ponderosa pine (*Pinus ponderosa*) growth in the Colorado Rocky Mountains. *Canadian Journal of Forest Research*, **23**, 1750–9.

Peterson, D.L., Arbaugh, M.J., Wakefield, V.A. & Miller, P.R. (1987). Evidence of growth reduction in ozone-stressed Jeffrey pine (*Pinus jeffreyi* Grev. and Balf.) in Sequoia and Kings Canyon National Parks. *Journal of the Air Pollution Control Association*, **37**, 906–12.

Philipson, J.J. & Coutts, M.P. (1978). Tolerance of tree roots to waterlogging. III. Oxygen transport in lodgepole pine and sitka spruce roots

of primary structure. *New Phytologist*, **80**, 341–9.

Pisek, A. & Winkler, E. (1958). Assimilationsvermögen und Respiration der Fichte (*Picea excelsa* Link) in verschiedener Höhenlage und der Zirbe (*Pinus cembra* L.) und Sonnenblume (*Helianthus annuus* L.). *Planta*, **53**, 532–50.

Pothier, D., Margolis, H.A. & Waring, R.H. (1989). Changes in saturated sapwood permeability and total sapwood conductance with stand development. *Canadian Journal of Forest Research*, **19**, 432–9.

Przykorska-Zelawska, T. & Zelawski, W. (1981). Changes in photosynthetic capacity of non-growing needles of conifers during their acclimation to shade. *Acta Physiologiae Plantarum*, **3**(1), 33–41.

Putz, F.E., Coley, P.D., Lu, K., Montalvo, A. & Aiello, A. (1983). Uprooting and snapping of trees: structural determinants and ecological consequences. *Canadian Journal of Forest Research*, **13**, 1011–20.

Reich, P.B., Ellsworth, D.S., Kloeppel, B.D., Fownes, J.H. & Gower, S.T. (1990). Vertical variation in canopy structure and CO_2 exchange of oak–maple forests: influence of ozone, nitrogen, and other factors on simulated canopy carbon gain. *Tree Physiology*, **7**, 329–45.

Reich, P.B., Kloeppel, B.D., Ellsworth, D.S. & Walters, M.B. (1995). Different photosynthesis–nitrogen relations in deciduous hardwood and evergreen coniferous tree species. *Oecologia*, **104**, 24–30.

Reich, P.B. & Schoettle, A.W. (1988). Role of phosphorus and nitrogen in photosynthetic and whole-plant carbon gain and nutrient use efficiency in eastern white pine. *Oecologia*, **77**, 25–33.

Reich, P.B., Schoettle, A.W., Stroo, H.F., Troiano, J. & Amundson, R.G. (1987). Influence of O_3 and acid rain on white pine seedlings grown in five soils. I. Net photosynthesis and growth. *Canadian Journal of Botany*, **65**, 977–87.

Richter, H. (1973). Frictional potential losses and total water potential in plants: a re-evaluation. *Journal of Experimental Botany*, **31**, 983–94.

Roberts, J. (1977). The use of tree-cutting techniques in the study of water relations of mature *Pinus sylvestris* L. *Journal of Experimental Botany*, **28**, 751–67.

Roberts, J. (1978). The use of the 'tree cutting' technique in the study of the water relations of Norway spruce, *Picea abies* (L.) Karst. *Journal of Experimental Botany*, **29**, 465–71.

Rook, D.A., M.P. Bollmann & Hong, S.O. (1987. Foliage development within the crowns of *Pinus radiata* trees at two spacings. *New Zealand Journal of Forest Science*, **17**, 297–314.

Rook, D.A. & Corson, M.J. (1978). Temperature and irradiance and the total daily photosynthetic production of a crown of a *Pinus radiata* tree. *Oecologia*, **36**, 371–82.

Rundel, P.W. (1982). Water uptake by organs other than roots, In *Water Relations and Carbon Assimilation*. ed. O.L. Lange, P.S. Nobel, C.B. Osmond & H. Ziegler, pp. 111–34. Berlin: Springer-Verlag.

Running, S.W. (1976). Environmental control of leaf water conductance in conifers. *Canadian Journal of Forest Research*, **6**, 104–12.

Running, S.W. (1980). Field estimates of root and xylem resistance in *Pinus contorta* using root excision. *Journal of Experimental Botany*, **31**, 555–69.

Running, S.W. & Reid, C.P. (1980). Soil temperature influences on root resistance of *Pinus contorta* seedlings. *Plant Physiology*, **65**, 635–40.

Ryan, M.G. (1990). Growth and maintenance respiration in stems of *Pinus contorta* and *Picea engelmannii*. *Canadian Journal of Forest Research*, **20**, 48–57.

Ryan, M.G., Gower, S.T., Hubbard, R.M., Waring, R.H., Gholz, H.L. Cropper, W.P. & Running, S.W. (1995). Woody tissue maintenance respiration of four conifers in contrasting climates. *Oecologia*, **101**, 133–40.

Ryan, M.G., Linder, S., Vose, J.M. & Hubbard, R.M. (1994). Dark respiration in pines. *Ecological Bulletin (Copenhagen)*, **43**, 50–63.

Sage, R.F., Sharkey T.D. & Seemann, J.R. (1989). Acclimation of photosynthesis to elevated CO_2 in five C_3 species. *Plant Physiology*, **89**, 590–6.

Sandford, A.P. & Jarvis, P.G. (1986). Stomatal responses to humidity in selected conifers. *Tree Physiology*, **2**, 89–103.

Sasek, T.W. & Richardson, C.J. (1989). Effects of chronic doses of ozone on loblolly pine: photosynthetic characteristics in the third growing season. *Forest Science*, **35**, 745–55.

Schlesinger, W.H., DeLucia, E.H. & Billings, W.D. (1989). Nutrient-use efficiency of woody plants on contrasting soils in the western Great Basin, Nevada. *Ecology*, **70**, 105–13.

Schoettle, A.W. (1994). Influence of tree size on shoot structure and physiology of *Pinus contorta* and *Pinus aristata*. *Tree Physiology*, **14**, 1055–68.

Schulze, E.-D., Lange, O.L. & Oren, R. (eds.) (1989). *Forest Decline and Air Pollution*. Berlin: Springer-Verlag.

Schulze, E.-D., Mooney, H.A. & Dunn, E.L. (1967). Wintertime photosynthesis of bristlecone pine (*Pinus aristata*) in the White Mountains of California. *Ecology*, **48**, 1044–7.

Seiler, J.R. & Cazell, B.H. (1990). Influence of water stress on the physiology and growth of red spruce seedlings. *Tree Physiology*, **6**, 69–77.

Sellin, A. (1994). Variation in hydraulic architecture of *Picea abies* (L.) Karst. Trees grown under different environmental conditions. PhD thesis, Tartu University, Estonia.

Sheffield, R.M., Cost, N.D., Bechtold, W.A. & McClure, J.P. (1985). *Pine Growth Reductions in the Southeast*. USDA Forest Service Research Bulletin SE-83.

Sherriff, D.W., Nambiar, E.K.S. & Fife, D.N. (1986). Relationships between nutrient status, carbon assimilation and water use efficiency in *Pinus radiata* (D. Don) needles. *Tree Physiology*, **2**, 73–88.

Siau, J.F. (1984). *Transport Processes in Wood*. Berlin: Springer-Verlag.

Skeffington, R.A. & Roberts, T.M. (1985). The effects of ozone and acid mist on Scots pine saplings. *Oecologia*, **65**, 201–6.

Smit, J. & Van den Driessche, R. (1994). Root growth and water use efficiency of Douglas-fir (*Pseudotsuga menziesii* (Mirb.) Franco) and lodgepole pine (*Pinus contorta* Dougl.) seedlings. *Tree Physiology*, **11**, 401–10.

Smith, W.H. (1990). *Air Pollution and Forests: Interactions between Air Contaminants and Forest Ecosystems*. New York: Springer-Verlag.

Smith, W.K. (1980). Importance of aerodynamic resistance to water use efficiency in three conifers under field conditions. *Plant Physiology*, **65**, 132–5.

Smith, W.K. (1985). Western montane forests. In *Physiological Ecology of North American Plant Communities*, ed. B.F. Chabot & H.A. Mooney, pp. 95–126. New York: Chapman & Hall.

Smith, W.K. & Carter, G.A. (1988). Shoot structural effects on needle temperatures and photosynthesis in conifers. *American Journal of Botany*, **75**, 496–500.

Smith, W.K. & Hinckley, T.M. (eds.) (1995a). *Resource Physiology of Conifers*. San Diego: Academic Press.

Smith, W.K. & Hinckley, T.M. (eds.) (1995b). *Ecophysiology of Coniferous Forests*. San Diego: Academic Press.

Smith, W.K., Schoettle, A.W. & Cui, M. (1991). Importance of the method of leaf area measurement to the interpretation of gas exchange of complex shoots. *Tree Physiology*, **8**, 121–7.

Smith, W.K., Young, D.R., Carter, G.A., Hadley, J.L. & McNaughton, G.M. (1984). Autumn stomatal closure in six conifer species of the central Rocky Mountains. *Oecologia*, **63**, 237–42.

Smolander, H. & Oker-Blom, P. (1989). The effect of nitrogen content on the photosynthesis of Scots pine needles and shoots. *Annales de Sciences Forestières*, **46**, 473–5.

Smolander, H., Oker-Blom, P., Ross, J., Kellomäki, S. & Lahti, T. (1987). Photosynthesis of a Scots pine shoot: test of a shoot photosynthesis model in direct radiation field. *Agricultural and Forest Meteorology*, **39**, 67–80.

Sperry, J.S. (1995). Limitations on stem water transport and their consequences. In *Plant Stems: Physiology and Functional Morphology*, ed. B.L. Gartner, pp. 105–24. San Diego: Academic Press.

Sprugel, D.G. (1990). Components of woody-tissue respiration in young *Abies amabilis* trees. *Trees*, **4**, 88–98.

Sprugel, D.G. & Benecke, U. (1991). Measuring woody-tissue respiration and photosynthesis. In *Techniques and Approaches in Forest Tree Ecophysiology*, ed. J. P. Lassioe & T.M. Hinckley, pp. 329–55. Boca Raton, Florida: CRC Press.

Sprugel, D.G., Brooks, J.R. & Hinckley, T.M. (1996). Effects of light on shoot geometry and needle morphology in *Abies amabilis*. *Tree Physiology* **16**, 91–8.

Sprugel, D.G., Ryan, M.G., Brooks, J.R., Vogt, K.A. & Martin, T.A. (1995). Respiration from the organ level to the stand. In *Resource Physiology of Conifers*, ed. W.K. Smith & T.M. Hinckley, pp. 255–300. San Diego: Academic Press.

Stenberg, P., DeLucia, E.H., Schoettle, A.W. & Smolander, H. (1995a). Photosynthetic light capture from cell to canopy. In *Resource Physiology of Conifers*, ed. W.K. Smith & T.M. Hinckley, pp. 3–38. San Diego: Academic Press.

Stenberg, P., Linder, S. & Smolander, H. (1995b). Variation in the ratio of shoot silhouette area to needle area in fertilized and unfertilized Norway spruce trees. *Tree Physiology*, 15, 705–12.

Stone, E.C. (1957). Dew as an ecological factor. II. The effect of artificial dew on the survival of *Pinus ponderosa* and associated species. *Ecology*, 38, 414–22.

Stone, E.C. & Fowells, H.A. (1955). The survival value of dew as determined under laboratory conditions. I. *Pinus ponderosa*. *Forest Science*, 1, 183–8.

Stone, E.C. & Kalisz, P.J. (1991). On the maximum extent of tree roots. *Forest Ecology and Management*, 46, 59–102.

Stone, E.C., Went, F.W. & Young, C.L. (1950). Water absorption from the atmosphere by plants growing in dry soil. *Science*, 111, 546–8.

Strain, B.R., Higginbotham, K.O. & Mulroy, J.C. (1976). Temperature preconditioning and photosynthetic capacity of *Pinus taeda* L. *Photosynthetica*, 10, 47–53.

Strand, M. & Öquist, G. (1985). Inhibition of photosynthesis by freezing temperatures and high light levels in cold acclimated seedlings of Scots pine (*Pinus sylvestris*). I. Effects on the light-limited and light-saturated rates of CO_2 assimilation. *Physiologia Plantarum*, 64, 425–30.

Surano, K.A., Daley, P.F., Houpis, J.L., Shinn, J.H., Helms, J.A., Palassou, R.J. & Costella, M.P. (1986). Growth and physiological responses of *Pinus ponderosa* Dougl. ex P. Laws. to long-term elevated CO_2 concentrations. *Tree Physiology*, 2, 243–59.

Szaniawski, R.K. & Wierzbicki, B. (1978). Net photosynthetic rate of some coniferous species at diffuse high irradiance. *Photosynthetica*, 12, 412–17.

Takemoto, B. & Bytnerowicz, A. (1992). Effects of acid fog on growth, physiology and biochemistry of ponderosa pine (*Pinus ponderosa*) and white fir (*Abies concolor*) seedlings. *Environmental Pollution*, 79, 235–41.

Taylor, G.E., Pitelka, L.F. & Clegg, M.K. (eds.) (1991). *Plant Ecological Genetics and Air Pollution*. New York: Springer-Verlag.

Temple, P.J. (1988). Injury and growth of Jeffrey pine and giant sequoia in response to ozone and acidic mist. *Environmental and Experimental Botany*, 28, 323–33.

Temple, P.J., Reichers, G.H., Miller, P.R. & Lennox, R.W. (1994). Growth response of ponderosa pine to long-term exposure to ozone, wet and dry acidic deposition and drought. *Canadian Journal of Forest Research*, 23, 59–66.

Teskey, R.O. (1995). A field study of the effects of elevated CO_2 on the carbon assimilation, stomatal conductance and leaf and branch growth of *Pinus taeda* trees. *Plant, Cell and Environment*, 18, 565–73.

Teskey, R.O., Bongarten, B.C., Cregg, B.M., Dougherty, P.M. & Hennessey, T. C. (1987). Physiology and genetics of tree growth response to moisture and temperature stress: an examination of of the characteristics of loblolly pine (*Pinus taeda* L.). *Tree Physiology*, 3, 41–61.

Teskey, R.O., Fites, J.A., Samulson, J.J. & Bongarten, B.C. (1986). Stomatal and nonstomatal limitations to net photosynthesis in *Pinus taeda* L. under different environmental conditions. *Tree Physiology*, 2, 131–42.

Teskey, R.O., Gholz, H.L. & Cropper, W.P. (1994a). Influence of climate and fertilization on net photosynthesis of mature slash pine. *Tree Physiology*, 14, 1215–27.

Teskey, R.O., Grier, C.C. & Hinckley, T.M. (1984). Change in photosynthesis and water relation with age and season in *Abies amabilis*. *Canadian Journal of Forest Research*, 14, 77–84.

Teskey, R.O., Sheriff, D.W., Hollinger, D.Y. & Thomas, R.B. (1995). External and internal factors regulating photosynthesis. In *Resource Physiology of Conifers*, ed. W.K. Smith & T.M. Hinckley, pp. 105–40. San Diego: Academic Press.

Teskey, R.O., Whitehead, D. & Linder, S. (1994b). Photosynthesis and carbon gain by pines. *Ecological Bulletin (Copenhagen)*, 43, 35–49.

Thomas, R.B., Lewis, J.D. & Strain, B.R. (1994). Effects of leaf nutrient status on photosynthetic capacity in loblolly pine (*Pinus taeda* L.) seedlings grown in elevated CO_2. *Tree Physiology*, 14, 947–60.

Thompson, W.A. & Wheeler, A.M. (1992). Photosynthesis by mature needles of field-grown *Pinus radiata*. *Forest Ecology and Management*, 52, 225–42.

Tingey, D.T., Rodecap, K.D., Lee, E.II., Moser, T.J. & Hogsett, W.E. (1986). Ozone alters the concentrations of nutrients in bean plants. *Angewandte Botanik*, 60, 481–93.

Tolley, L.C. & Strain, B.R. (1984a). Effects of CO_2 enrichment on growth of *Liquidambar styraciflua* and *Pinus taeda* seedlings under different irradiance levels. *Canadian Journal of Forest Research*, 14, 343–50.

Tolley, R.B. & Strain, B.R. (1984b). Effects of CO_2 enrichment and water stress on growth of *Liquidambar styraciflua* and *Pinus taeda* seedlings. *Canadian Journal of Botany*, 62, 135–9.

Tranquillini, W. (1957). Standortsklima, Wasserbilanz und CO_2-Gaswechsel junger Zirben (*Pinus cembra* L.) an der alpine Waldgrenze. *Planta*, 49, 612–61.

Tranquillini, W. (1959). Die Stoffproduktion der Zirbe (*Pinus cembra* L.) an der Waldgrenze während eines Jahres. *Planta*, 54, 107–51.

Tranquillini, W. (1964). Photosynthesis and dry matter production of trees at high altitudes. In *The Formation of Wood in Forest Trees*, ed. M.H. Zimmerman, pp. 505–18. Academic Press, New York.

Tranquillini, W. (1965). Über den Zusammenhang zwischen Entwicklungszustand und Dürrenresistenz junger Zirben (*Pinus cembra* L.) im Pflanzengarten. *Mitteilungen der Forstlichen Bundesversuchsanstalt Wien*, 75, 457–87.

Tranquillini, W. (1976). Water relations and alpine timberline. In *Water and Plant Life*, ed. O.L. Lange, L. Kappen & E.-D. Schulze, pp. 473–91. Berlin: Springer-Verlag.

Tranquillini, W. (1979). *Physiological Ecology of the Alpine Timberline*. Berlin: Springer-Verlag.

Tranquillini, W. & Turner, H. (1961). Untersuchungen über die Pflanzentemperaturen in der subalpinen Stufe mit besonderer Berücksichtigung der Nadeltemperaturen der Zirbe. *Mitteilungen der Forstlichen Bundesversuchsanstalt Mariabrunn*, 59, 127–51.

Troeng, E. & Linder, S. (1982a). Gas exchange in a 20-year-old stand of Scots pine. I. Net photosynthesis of current and one-year-old shoots within and between seasons. *Physiologia Plantarum*, 54, 7–14.

Troeng, E. & Linder, S. (1982b). Gas exchange in a 20-year-old stand of Scots pine. II. Variation in net photosynthesis and transpiration within and between trees. *Physiologia Plantarum*, **54**, 15–23.

Tyree, M.T. (1988). A dynamic model for water flow in a single tree: evidence that models must account for hydraulic architecture. *Tree Physiology*, **4**, 195–217.

Tyree, M.T., Davis, S.D. & Cochard, H. (1994). Biophysical perspectives of xylem evolution: is there a tradeoff of hydraulic efficiency or vulnerability to dysfunction? *International Association of Wood Anatomists Journal*, **15**, 355–60.

Tyree, M.T. & Ewers, F.W. (1991). The hydraulic architecture of trees and other woody plants. *New Phytologist*, **119**, 345–60.

Vance, N.C. & Zaerr, J.B. (1991). Influence of drought stress and low irradiance on plant water relations and structural constituents in needles of *Pinus ponderosa* seedlings. *Tree Physiology*, **8**, 175–84.

Wang, J., Ives, N.E & Lechowicz, M.J. (1992). The relation of foliar phenology to xylem embolism in trees. *Functional Ecology*, **6**, 469–75.

Wardle, P. (1965). A comparison of alpine timberlines in New Zealand and North America. *New Zealand Journal of Botany*, **3**, 113–35.

Waring, R.H., Gholz, H.L., Grier, C.C. & Plummer, M.L. (1977). Evaluating stem conducting tissue as an estimator of leaf area in four woody angiosperms. *Canadian Journal of Botany*, **55**, 1474–7.

Waring, R.H. & Running, S.W. (1978). Sapwood water storage: its contribution to transpiration and effect upon water conductance through the stems of old-growth Douglas-fir. *Plant, Cell and Environment*, **1**, 131–40.

Waring, R.H. & Schlesinger, W.H. (1985). *Forest Ecosystems: Concepts and Management*. San Diego: Academic Press.

Waring, R.H., Schroeder, P.E. & Oren, R. (1982). Application of the pipe model theory to predict canopy leaf area. *Canadian Journal of Forestry Research*, **12**, 556–60.

Waring, R.H., Whitehead, D. & Jarvis, P.G. (1979). The contribution of stored water to transpiration in Scots pine. *Plant, Cell and Environment*, **2**, 309–17.

Weinstein, D.A., Beloin, R.M. & Yanai, R.D. (1991). Modeling changes in red spruce carbon balance and allocation in response to interacting ozone and nutrient stresses. *Tree Physiology*, **9**, 127–46.

Wellburn, A.R. (1990). Why are atmospheric oxides of nitrogen usually phytotoxic and not alternative fertilizers? *New Phytologist*, **115**, 395–429.

Whitehead, D., Edwards, W.R.N. & Jarvis, P.G. (1984). Conducting sapwood area, foliage area, and permeability in mature trees of *Picea sitchensis* and *Pinus contorta*. *Canadian Journal of Forest Research*, **14**, 940–7.

Wilkins, S.D. & Klopatek, J.M. (1987). Plant water relations in ecotonal areas of pinyon–juniper and semi-arid shrub ecosystems. In *Proceedings – Pinyon–Juniper Conference*, pp. 412–17. USDA Forest Service General Technical Report INT-215.

Winner, W.E. (1994). Mechanistic analysis of plant responses to air pollution. *Ecological Applications*, **4**, 651–61.

Winner, W.E., Mooney, H.A. & Goldstein, R.A. (eds.) (1985). *Sulfur Dioxide and Vegetation: Physiology, Ecology, and Policy Issues*. Stanford, Calif.: Stanford University Press.

Woodward, F.I. (1995). Ecophysiological controls of conifer distributions. In *Ecophysiology of Coniferous Forests*. ed. W.K. Smith & T.M. Hinckley, pp. 79–94. San Diego: Academic Press.

Yang, S. & Tyree, M.T. (1993). Hydraulic resistance in *Acer saccharum* shoots and its influence on leaf water potential and transpiration. *Tree Physiology*, **12**, 231–42.

Yoder, B.J., Ryan, M.G., Waring, R.H., Schoettle, A.W. & Kaufmann, M.R. (1994). Evidence of reduced photosynthetic rates in old trees. *Forest Science*, **40**, 513–27.

Zimmermann, M.H. (1983). *Xylem Structure and the Ascent of Sap*. New York: Springer-Verlag.

16 The mycorrhizal status of *Pinus*

David J. Read

16.1 Introduction

In *Pinus*, the apices of the distal feeding roots, together with those regions proximal to the apex which would otherwise bear root hairs, become ensheathed early in their development by a pseudoparenchymatous mantle of fungal tissue. The fungi forming the mantle thus occupy the interface between the soil and the root surface in that critical part of the system which is involved in nutrient transfer processes between the plant and its environments. From the mantle, hyphae grow inwards between the cortical cells to form a labyrinthine structure, the Hartig net, which is intimately associated with the radial walls of these cells. Hyphae also grow outwards from the mantle to form an extraradical mycelial system which explores the soil and provides connections with sexual structures of the fungus that develop above or below ground. Collectively the mantle, Hartig net and extraradical mycelium form what is referred to as a mycorrhiza or 'fungus-root'. Because the hyphae do not penetrate the cortical cells, this type of symbiosis is called an *ecto*mycorrhiza to distinguish it from those *endo*mycorrhizal types seen in most herbaceous and many woody plants where intracellular penetration is the norm (Smith & Read 1997). The ectomycorrhizal condition is found in most members of the Pinaceae as well as in several angiospermous families of woody plants, notably the Dipterocarpaceae, Fagaceae and Myrtaceae, but certain features of the morphology of ectomycorrhiza of *Pinus* are distinctive, most notably the dichotomization of the lateral roots. Although it is not uncommon, particularly in nursery situations, to observe intracellular penetration of root cortical cells from the Hartig net or mantle, producing a so-called ectendomycorrhiza (Mikola 1965), this is essentially a transient phase rarely seen in healthy adult trees.

Recent studies (Le Page *et al.* 1997) suggest that the ectomycorrhizal condition developed early in the evolution of *Pinus*, since exquisitely preserved dichotomous roots characteristic of the genus, with Hartig net, mantle and extraradical mycelium all intact, have been recovered as fossils of early Eocene age (*c.* 50 million years ago) from the Princeton chert of British Columbia. Such findings lend support to the hypothesis developed by Axelrod (1986) that pines coevolved with their fungal symbionts.

The very first experimental studies of 'mycorrhiza' were carried out on a member of the genus *Pinus* by Frank (1894). Following his earlier (1885) recognition that the fine roots of *Pinus*, along with those of many other genera of woody plants, were ensheathed by fungal hyphae, Frank (1894) made the critical observation that this type of association, now known as an ectomycorrhiza, is predominantly formed in the acidic surface organic horizons of the forest soil profile where nitrification is inhibited. When he grew plants of *P. sylvestris* in this type of substrate he found (Fig. 16.1) that only when they were mycorrhizal did they grow normally. While hitherto it had been widely believed that any relationship between roots and fungi was necessarily of a pathogenic kind, Frank's work with *P. sylvestris* showed, for the first time, that the association had the potential to be beneficial. It also enabled him to develop a far-sighted theory of the function of the ectomycorrhizal association.

In what became known as his 'Organic nitrogen theory', Frank postulated in the same seminal paper of 1894 that mycorrhizal associations proliferated in the superficial layers because the fungi were able to mobilize the nitrogen (N) contained in the accumulating organic residues. He went further to suggest that a significant proportion of the carbon requirement of the fungus might

Ecology and Biogeography of *Pinus*, ed. D.M. Richardson. © Cambridge University Press, Cambridge (1998), pp. 324–40.

Fig. 16.1. The original demonstration by Frank (1894) of the effect of removal, by sterilization (centre) of mycorrhizal fungi, on the growth of *Pinus sylvestris*. Spontaneous recolonization of soil by saprophytes (left) provides modest growth enhancement, but healthy growth is observed only in the mycorrhizal condition (right).

regulation of photosynthetic activity by determination of sink-strength have all been demonstrated.

This chapter will first analyse the structural and functional attributes of mycorrhizal associations in *Pinus* so as to provide the background for a holistic appreciation of the possible contribution of the symbiosis to the success of pines in their natural environment and as invaders of environments far removed from the natural range of the genus.

16.2 Structural and epidemiological aspects of *Pinus* mycorrhiza

16.2.1 The colonization process

be satisfied in the process of assimilation of organic N, its demands upon the plant being thereby reduced.

In the century since his experiments with *P. sylvestris*, the thrust of research on the function of the symbiosis moved away from considerations of natural substrates as sources of plant nutrients. Hatch (1937) established that the extent of mycorrhizal colonization of *Pinus* was determined by soil fertility, it being greatest under the conditions of low or unbalanced nutrient availability in which, Axelrod (1986) suggests, members of the genus have evolved, and in which they occur to this day in nature. Colonization is much reduced in fertile soil.

Measurement of mineral concentration of tissues of *Pinus* after application of nutrients to soil led Hatch to the view, still largely accepted, that the internal nutrient status of the plant, particularly of phosphorus and nitrogen, was the factor that determined the extent of mycorrhizal colonization.

Observations such as these provided impetus for studies of the ability of mycorrhizal fungi to capture, transport and store nutrients in the simple mineral form. Studies of phosphate capture have predominated. As an inevitable consequence of this shift in emphasis, mycorrhizal fungi of plants such as *Pinus* have increasingly been seen simply as scavengers of mineral nutrients, the assumption being that the major processes of mobilization of these were carried out by a distinct population of non-symbiotic decomposer organisms. However, Frank's original view that ectomycorrhizal fungi may themselves be directly involved in the processes of nutrient mobilization has been re-asserted (Read 1982), again largely on the basis of work using *Pinus* spp. as test plants.

While studies of nutrient relations of the host plant have occupied the central ground in research on mycorrhiza of the genus *Pinus*, there is increasing evidence that the symbiosis may have a broader range of functions. Amongst these, exclusion of pathogenic fungi, enhancement of resistance to potentially toxic metal ions, and

While all ectomycorrhizal trees have heterorhizic root systems which are differentiated into long roots of potentially indefinite growth and short roots of restricted growth, the demarcation between the two types of root is particularly well defined in *Pinus* spp. In nature, short roots become converted by fungal colonization into mycorrhizas, whereas the long roots are incompletely colonized, the apex usually being free of fungal mycelium. The long roots have no sheath but behind the apex the fungus grows between the outer epidermal layers of the root to form a Hartig net. In studies of *P. nigra*, Aldrich-Blake (1930) emphasized that intrinsic anatomical differences between long and short roots occurred even in the absence of fungal colonization. He showed that long roots have a stele of significantly greater diameter than that of short roots, and that they alone had a root cap and could show secondary growth. The main morphological feature of the short root is that even in the absence of colonization, it develops a dichotomous structure which is a characteristic of *Pinus* spp. Aldrich-Blake also made the important observation that short roots which failed to become colonized by a mycorrhizal fungus aborted soon after emergence from the long root. As a result, virtually all of the dichotomous short roots seen on the long root of *Pinus* spp. are mycorrhizal (Fig. 16.2).

Analysis of *P. strobus* by Hatch & Doak (1933) and Hatch (1937) confirmed that the main impact of fungal colonization on development of the short roots was to enable continuation of their growth. Further and repeated dichotomization of single short roots can lead to a complex structure with up to 80 individual tips, the whole of which can be invested in an outer mantle of fungal tissue to form a complex semi-globose structure called a tuberculate mycorrhiza. The overall effect of colonization on the structure of the root system is thus to increase the longevity, branching and surface area of the short root system (Fig. 16.3).

Fig. 16.2. **Cluster of ectomycorrhizal roots of pine formed by repeated dichotomies of a short root colonized and ensheathed by a basidiomycetous fungus (*Russula*) (photo: Dr A.F.S. Taylor).**

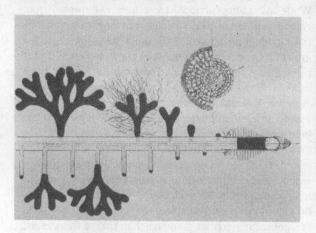

Fig. 16.3. **The diagrammatic representation by Hatch (1937) of the sequential development of mycorrhizal short roots along the axis of a long root. Root cap and piliferous zone are shown on the long root. While short roots that are not colonized abort, those that are colonized either from the Hartig net of the long root or from the soil become ensheathed by a fungal mantle and may continue to divide dichotomously until clusters are produced. A transverse section through a short root (inset) shows the investing mantle and intercellular penetration of the fungus to form the Hartig net.**

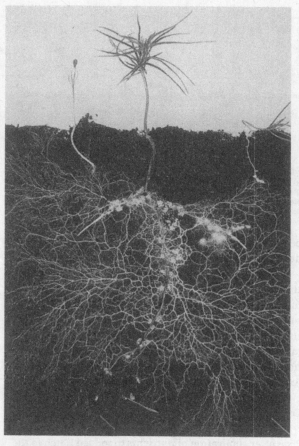

Fig. 16.4. **Emerging radicle of newly germinated seedling of *Pinus sylvestris* becoming incorporated (right) into the mycelial network of the mycorrhizal fungus *Suillus bovinus* growing from and supported by the established central plant. Depending on the degree of host specificity shown by the fungus in question, such infection processes lead to incorporation, at intra- or interspecific levels, of individual plants, into guilds which are interconnected by their mycorrhizal fungi.**

The pattern of colonization of short roots, as they emerge through the axis of the long root, is determined by the age of the plant. In adult plants of *P. sylvestris*, Robertson (1954) showed that colonization occurred primarily from the fungal tissues of the Hartig net that was already established in the long root. Under exceptional circumstances, for example in *P. radiata* growing in Australian soils of extremely low P status, it has been reported that no such colonization of the long root occurs (Lamb & Richards 1974). In this case, short roots even on mature root-systems must be colonized by inoculum resident in the soil.

The radicle growing from newly germinating seed, likewise, carries no fungal colonization and the short roots emerging through it can only be colonized by fungal mycelia or spores occupying the soil in their immediate vicinity. In those most frequent of cases where a seed germinates close to or under the canopy of an established tree of the same species or even genus, the lack of host specificity shown by most ectomycorrhizal fungi will ensure that colonization will be by mycelia already established on these mature individuals (Fig. 16.4).

Where seed is dispersed into areas containing no other ectomycorrhizal host plants, colonization can only occur if spores of an appropriate fungus are present in the soil. This can be an important factor determining whether successful establishment in alien environments takes place and is further discussed below.

16.2.2 Epidemiology and the notion of fungal successions in ectomycorrhiza of pine

Because some species of ectomycorrhizal fungi have a greater ability than others to colonize emerging short roots from spores, a distinct suite of fungi characteristically appears as the first mycorrhizal associates when *Pinus* seedlings invade disturbed sites or are grown in nurseries. Chu-Chou (1979) recognized a succession of species associated with developing stands of *P. radiata* in New Zealand, fungi such as *Hebeloma crustuliniforme*, *Laccaria laccata* and *Thelephora terrestris* occurring, probably as a result of their superior ability to colonize roots from spores, on the youngest trees, but being replaced as stands aged by other species, notably *Amanita muscaria* and *Scleroderma verrucosum*. These produced abundant fruit bodies in stands greater than 15 years old. The observations of Chu-Chou (1979) have since been repeated with other tree genera and have given rise to the concept of 'succession' of fungal species (Mason *et al.* 1982), those characteristic of young stands being referred to as 'early-stage' and those of older plantations as 'late-stage' fungi. The same types of sequence have been recorded in other species of *Pinus*, notably *P. banksiana* growing in its natural range (Danielson 1984; Visser 1995), and *P. patula* planted as an exotic in India (Natarajan, Mohan & Ingleby 1992).

Attractive though the simple notion of succession appears, it must be viewed with some caution. Some fungi, notably species of *Rhizopogon*, have the ability to colonize young and old pines and to persist throughout the life of a stand. Danielson (1991) has suggested a separate 'multi-stage' category for this type of fungus. Further, it is now recognized that patterns of succession which are based purely on records of appearance of fruit bodies above ground may not reflect, in either qualitative or quantitative terms, the structure of the population of fungi occurring on the roots themselves. This point has been made forcibly in a recent study of the mycorrhizal fungi associated with a mature stand of a native pine, *P. muricata*, in California (Gardes & Bruns 1996). In this work, the species composition of fungi appearing above ground as fruit bodies was compared with that occurring as mycorrhizal symbionts on roots of *P. muricata*, these being identified by molecular methods based upon the polymerase chain reaction (PCR). This amplifies DNA sequences obtained from vegetative mycelium so that they can be matched with those from fruit bodies of known identity. Some of the species occurring most commonly as fruit bodies, for example *Suillus pungens*, were rarely seen below ground; others which formed large numbers of mycorrhizal roots, such as *Russula amoenolens* and some 'thelephoroid' fungi, were poorly represented or did not appear at all in the above-ground fruiting record. It is evident that analysis of fruit body production may not in itself provide an accurate picture of the mycorrhizal population below ground.

Fig. 16.5. **Seedling of *Pinus* infected with the ectomycorrhizal fungus *Suillus bovinus* and grown in an observation chamber on natural (non-sterile) forest soil. The effectiveness of the exploitation of the environment by the mycelial fan is clear. The mycelium is at an early stage of development but differentiation between the advancing hyphal front (lower arrow) and mycelial rhizomorphs (upper arrow) is evident. The ratio of the length of absorptive mycelium to that of the root is around 10^4:1 at this stage.**

16.3 The extramatrical mycelial system of pine ectomycorrhiza

While they recognized the possibility that increased branching of short roots induced by mycorrhizal colonization might yield greater absorptive surface area and so help to explain the growth promotion seen in colonized plants, few of the earlier workers recognized that far more significant increases of surface area were provided by the vegetative mycelia of the fungi growing from the colonized roots. Study of the extent of this mycelium has been facilitated by the use of transparent observation chambers which enable analysis of its growth from colonized host plants over natural substrates. In these, mycelia of typical mycorrhizal associates of *Pinus* spp., such as *Suillus bovinus*, can be seen to extend from the host plant (Fig. 16.5) across and through natural non-sterile substrates at rates of 2–3 mm day^{-1}. The mycelium advances in a fan-shaped front made up of individual hyphae which effectively explore

Table 16.1. Relative rates of uptake of phosphate (expressed as μmol P per unit dry matter per day) and efficiency (expressed as P uptake/gram carbon invested in dry matter) of the uptake by roots and by ectomycorrhizal hyphae of the dimensions typically seen in *Pinus*. See text for details. From Yanai *et al.* (1995)

Fertility (μmol P l^{-1})	Uptake (μmol P day^{-1})		Efficiency (μmol P g^{-1})	
	Roots	Hyphae	Roots	Hyphae
190	6.7	323	150	4310
100	6.3	302	140	4010
50	5.4	264	122	3520

the medium. Behind the advancing front, individual hyphae aggregate to form root-like linear organs or rhizomorphs within which, in many of the fungi forming these structures, some tissue differentiation occurs. Narrow outer hyphae *c.* 3 μm in diameter, with cytoplasmic contents, surround central wide hyphae of *c.* 20 μm diameter which lack content and in which transverse septa collapse to produce tubular lengths of uninterrupted space analogous to that seen in xylem vessels. The rhizomorphs connect the advancing hyphal front to the fungal mantle of the colonized root. In some cases, for example that of *Pisolithus tinctorius* associated with *Pinus virginiana* growing on anthracite coal wastes (Schramm 1966), the rhizomorphs have been shown to extend over several metres from the roots. Finlay & Read (1986a) demonstrated that carbon fixed as $^{14}CO_2$ by pine seedlings was translocated rapidly to the vegetative mycelium exploring the soil.

The effectiveness of the fungal hypha as a structure for exploration of substrate can be judged in terms of both its dimensions and its physiological properties. Concerning dimension, an individual hypha is a cylinder of diameter of 3–6 μm compared with that of a root hair which in the case of, for example, *P. radiata* (Bowen 1968) is 15–20 μm, and of a lateral root of pine which is typically of the order of 500 μm (Kelly, Barber & Edwards 1992). Using these typical dimensions of the components of the ectomycorrhizal root and with the assumption that the carbon (C) costs per unit mass of absorbing tissue in roots and hyphae are the same, Yanai, Fahey & Miller (1995) employed a modelling approach to quantify the likely effectiveness of the two types of structure in absorption of the nutrient phosphorus (P) (Table 16.1). Because of the much greater length and surface area achieved by hyphae per unit of mass, both the rates of uptake and the efficiency of the process of P uptake achieved by the fungal structure are orders of magnitude greater than for the root both at low and relatively high levels of P supply.

The narrower hyphal elements have, as individuals, a much greater ability to penetrate micro-pores and a surface/volume ratio *c..* 100 times that of a root hair. Collectively, they provide an enormous increase of exploratory potential. Read & Boyd (1986) measured lengths of mycelia of *Suillus bovinus* growing from *P. sylvestris* in observation chambers of the kind shown in Fig. 16.5, and compared them with root lengths to achieve a mycelial length/root length (M:R) ratio. As mycelia spread from the plant, the mean hyphal lengths increased with time from 10 to 80 m per cm of root length giving a M:R range of 1000–8000:1.

When converted to lengths of mycelium per unit weight or volume of soil, values of the order of 200 m g^{-1} dry soil or 2000 m cm^{-3} of fresh soil are obtained (Read 1991). These must be compared with published estimates of root length per unit volume of soil in plantations of *Pinus*, in order to see their significance. Roberts (1976) in an intensive study of root distribution in plantations of *P. sylvestris*, found between 1 and 5 cm of root length per cm^3 of forest soil. Taking the highest of these values and an associated mycelial length of 2000 m cm^{-3} the M:R ratio on a soil volume basis is 200 000:1. The seemingly low density of the 'coarse' root system of pine has been contrasted by physiologists (Kramer & Bullock 1966; Newman 1969) with the situation seen in plants such as grasses with 'fibrous' root systems and its apparent ineffectiveness highlighted. Such comparisons fail to appreciate the differences of strategy employed by the two types of plant. Whereas pine roots are simply food bases which nourish an extremely dense mycelial system, the roots of grasses themselves constitute the major absorptive component of this type of plant, notwithstanding the fact that they may be colonized, in a facultative manner, by fungi of the vesicular-arbuscular kind.

Values of mycelial length of the order 200 m g^{-1} dry soil obtained in microcosms are the same as those reported by Finlay & Söderström (1992) for a *P. sylvestris* forest in Sweden in which the dominant ectomycorrhizal fungus was *Lactarius rufus*. From such values it is possible to calculate the annual carbon demand imposed by the fungal symbiont on the tree. Taking 200 m g^{-1} dry soil as being representative of ectomycorrhizal hyphal length for a pine stand and using conversion factors, Finlay and Söderström estimated this length to be equivalent to 3.5 kg of live mycelial dry weight per hectare of standing crop and an annual production of 70 kg mycelial biomass per hectare. When the fungal content in the mycorrhizal sheath, *c.* 730 kg ha^{-1}, and that of fruit bodies, *c.* 30 kg ha^{-1}, were added, a total fungal biomass of 830 kg $ha^{-1} y^{-1}$ was obtained (Finlay & Söderström 1992). Assuming that 40% of this biomass is in the form of carbon, and that the conversion efficiency is 60%, the carbon demand of the ectomycorrhizal fungi in this forest will be 830 kg C $ha^{-1} y^{-1}$. Since photosynthetic fixation of carbon in the same forest is known to be of the order of 5800 kg C $ha^{-1} y^{-1}$, it becomes evident that the ectomycorrhizal fungi consume 14–15% of the assimilated carbon. This figure is comparable with those obtained by Vogt *et al.* (1982) and Fogel & Hunt (1979) for ectomycorrhizal conifer forests in North America. It is somewhat lower than the 30% of total assimilate production observed by direct measurement of mycelial respiration in microcosms containing *Pinus* plants colonized by a number of mycorrhizal fungi (Söderström & Read 1987). However, this is to be expected in view of the fact that the former system consists of mature trees which themselves have a large component of maintenance respiration, and the latter employed seedlings with little storage capability.

In a study of carbon allocation using seedlings of *Pinus ponderosa* colonized by *Hebeloma crustuliniforme*, Rygiewicz & Anderson (1994) showed that though the fungus constituted only a small proportion, *c.* 5%, of total plant dry weight its presence caused allocation of carbon, fed as $^{14}CO_2$ to the plant, to be increased by about 23% relative to that seen in non-mycorrhizal plants. The fungus received *c.* 7% of total ^{14}C fixed and 60% of this carbon was respired. In these experiments hyphal respiration represented 19.4% of total below-ground respiration.

Söderström & Read (1987) showed, by measurement of respiratory output of the mycelium of a number of fungal species before and after their connection with the plant was cut, that activity of the fungi was almost completely dependent on the supply of current assimilate from the host plant. The same appears also to be true of production of fungal fruit bodies. Numbers and sizes of basidiomes of *Laccaria bicolor* developing in association with *P. strobus* have been shown to be directly proportional to photon flux densities (PFD) and rates of photosynthetic assimilation of the host plant (Lamhamedi, Godbout & Fortin 1994). Basidiome development ceased at low PFD values.

While it has been the convention to regard the allocation of carbon to the fungal symbiont in ectomycorrhiza as a 'cost', it is increasingly recognized that photosynthesis is a sink-regulated process in which unloading of assimilate from source tissues in the leaves can stimulate assimilation of carbon (Luxmoore *et al.* 1996). Significant increases in rates of carbon fixation by pine plants have been observed to follow colonization by mycorrhizal fungi even when no enhancement of mineral nutrient input is involved. Such stimulation has been recorded in *P. contorta* (Reid, Kidd & Ekwebelam 1983), *P. pinaster* (Conjeaud, Scheromm & Mousain 1996), *P. ponderosa* (Miller, Durall & Rygiewicz 1989) and *P. taeda* (Rousseau & Reid 1990, 1991). If the size of the sink for assimilate is reduced, for example by removal of fungal fruit bodies developing under the tree, both stomatal conductance and photosynthesis have been shown (Lamhamedi *et al.* 1994) to decrease instantaneously.

In view of these responses, it is evident that the notion of 'cost' of carbon allocation to the mycorrhizal symbiont may be inappropriate. Even if sink-related increases of assimilation do not fully compensate for the large allocation of C to the fungus, it is preferable to regard carbon transfer from autotroph to heterotroph as an investment, and then to consider what the returns, in addition to stimulation of photosynthetic activity, might be.

16.4 Returns on carbon investment in mycorrhizal *Pinus* plants

16.4.1 Enhancement of nutrient capture

Faced with the complexities of the soil environment, some authors have concluded that progress towards understanding of mycorrhizal biology can only be achieved under simplified and strictly controlled conditions in the laboratory. This approach was exemplified by the pioneering studies of Melin (1925), who first examined in culture the nutritional requirements of mycorrhizal fungi then synthesized mycorrhiza on *P. sylvestris* in sterilized sand and demonstrated that the mycelia of these fungi could capture phosphate (Melin & Nilsson 1950), as well as nitrogen in both mineral (Melin & Nilsson 1952), and amino acid (Melin & Nilsson 1953) forms and facilitate their transfer to the host plant. Bowen (1973) and his group in Australia were among the first to examine the nutrient relations of mycorrhizal and non-mycorrhizal *Pinus* plants in natural substrates. Working first with a P-deficient soil in the field it was shown (Theodorou & Bowen 1970) that positive responses of *P. radiata* to mycorrhizal colonization were greatest in those fungi, particularly *Rhizopogon* spp., which produced extensive mycelial rhizomorph systems. This confirmed earlier work on *P. radiata* by Stone (1950). It was considered that the benefits were attributable to increased access to poorly mobile phosphate ions, and evidence was provided that these were captured and subsequently stored in the hyphal mantle surrounding colonized roots (Bowen & Theodorou 1967). Skinner & Bowen (1974) subsequently showed that rhizomorphs of *R. luteolus* growing from roots of *P. radiata* into sterile forest soil were able to absorb and transport phosphorus over distances up to 12 cm.

This role of mycorrhizal rhizomorphs has been confirmed (Finlay & Read 1986b) under non-sterile conditions using *P. contorta* and *Suillus bovinus* as symbionts. Phosphate, applied to soil as ^{32}P, can be captured at considerable distance from the root and transferred by the fungus to the mantle of all roots that it has colonized (Fig. 16.6A, B). Onward transfer to shoots of plants is also detectable.

Interesting though these observations of enhanced P capture are, it must be recalled that in boreal forest

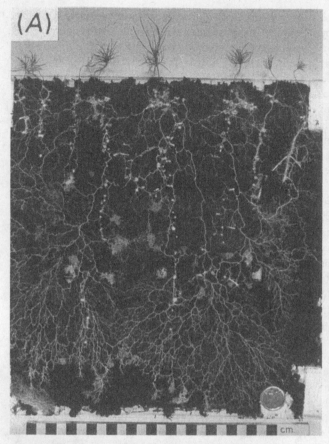

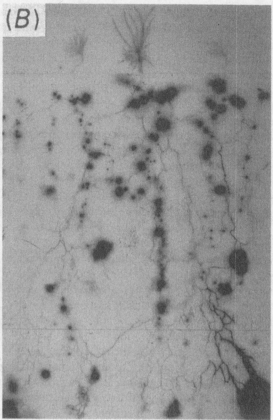

Fig. 16.6 A. **A group of *Pinus* plants growing on non-sterile soil in a transparent observation chamber showing colonization and interconnection by mycelium of *Suillus bovinus*. The larger central plant (*Pinus sylvestris*) was planted in the mycorrhizal condition and surrounding plants (*P. contorta*) were added as uncolonized individuals. The fungus has spread from the central plant to colonize others in the chamber (cf. Fig. 16.4) and so integrate them into a common mycelial network. By feeding phosphorus, as $^{32}PO_4$, to the mycelium (bottom right) the processes of uptake and transport of the element can be followed.**

B. **Autoradiograph of A after feeding of $^{32}PO_4$ to the mycelium of S. bovinus. P is shown to be transferred through the mycelium, to accumulate in roots of interconnected plants as well as in the shoots of 'receiver' plants. Such studies reveal that in ectomycorrhizal plants it is the mycorrhizal mycelium rather than the roots which is primarily responsible for absorption and transport of nutrients and that the mycelium has the ability to integrate the distribution of nutrients from resource-rich 'patches' to interconnected plants (from Finlay & Read 1986b).**

ecosystems of the type dominated by *Pinus* spp., nitrogen rather than phosphorus is not only the nutrient required in greatest amount but also that most limiting to plant growth (Tamm 1991; Chap. 17, this volume). In view of this, selection favouring mechanisms which enhance access to this element is to be expected, and Frank's hypothesis that mycorrhizal roots of pine accumulate in the organic residues in order to mobilize N, appears to be not unreasonable, particularly because we now know that the fermentation horizon (FH) of the soil profile in which their roots selectively proliferate are those in which N mobilization is greatest (Staaf & Berg 1977). The difficulty arises when we come to test the hypothesis.

Complex organic polymers, when presented as potential substrates to mycorrhizal fungi in sterile form, are unlikely to be qualitatively representative of those which will be encountered by the fungi in nature. The sterilization process itself produces structural changes. Alternatively, natural substrates presented in the non-sterile condition contain a mixed microbial population of potential decomposers which makes quantification of activity, specifically of the ectomycorrhizal fungi, impossible. It is in this situation that a combination of the simplified approach in which the biochemical potential of the fungi can be examined, with that using complex natural substrates, can yield advances.

Laboratory studies using chemically defined soluble proteins as high molecular weight nitrogenous substrates under aseptic conditions (Abuzinadah & Read 1986a) have revealed that some ectomycorrhizal fungal associates of *Pinus*, referred to as being 'protein-fungi', can readily use such polymers as sole source of nitrogen. Importantly, when these fungi are grown in mycorrhizal association with pine plants and supplied with protein as the sole N source, a significant proportion of the nitrogen, which they assimilate largely in the form of amino compounds (Read, Leake & Langdale 1989), is transferred to the host plant (Abuzinadah & Read 1986b). In pure culture, when grown with protein as sole source of N, and supplied with carbon as they would be by their host plants in nature, the 'protein fungi' release ammonium only when exogenous carbon sources are exhausted (Read *et al.* 1989). It seems inappropriate in view of these observations to regard mineralized forms as being the only, or even the most important, sources of 'plant available' N in the soil around ectomycorrhizal roots of pine.

It has been confirmed that 'protein fungi' produce an acid carboxyproteinase, and that those fungi which are typical colonists of *Pinus* spp. in northern boreal-forest environments with thick organic horizons, e.g. *S. bovinus* and *Paxillus involutus*, show more proteolytic activity than those such as *Pisolithus tinctorius* that occur most characteristically in warmer regions with a greater propensity for mineralization and nitrification of N (Read 1991). By adding freshly collected organic matter of the FH of a pine-forest soil to microcosms supporting *Pinus–Suillus* associations, Bending & Read (1995) recently showed not only that the hyphae of the fungus show a particular affinity for such substrates, in which they proliferate intensively to form dense patches of mycelium, but also that these fungi are involved in mobilization and export of nutrients from the exploited organic matter. Selective allocation by the autotroph, of carbon to the fungus forming the patch, was revealed by autoradiographic analysis of ^{14}C distribution.

When the concentrations of the major nutrients N, P and K were determined before and after the main period of carbon allocation to the introduced FH material it was found (Bending & Read 1995) that exploitation by *S. bovinus* did indeed lead to export of these elements, the concentrations, of which declined by 23, 22 and 30% respectively over the *c.* 40-day period during which the patches were selectively importing carbon. Equivalent values for substrates colonized by *Thelephora terrestris* in a parallel series of chambers were 13 and 21% for N and K respectively, while no change in P concentration was observed with this fungus.

The values of N and P release by *S. bovinus* are comparable with the estimates, 32 and 33% respectively, reported by Entry, Rose & Cromack (1991) to occur over a growing season in *Pseudotsuga* litter colonized by naturally occurring ectomycorrhizal mat-forming fungi. They convert to fluxes of 78, 4.1 and 6.1 μg per grown g FH per day for N, P and K respectively over the 40 days of active occupation of the introduced material. Measurements of rates of N mineralization were made in uncolonized samples of the FH substrate incubated under identical conditions. These

revealed that microbial ammonification was insufficient to account for the observed pattern of N depletion, and add support to the view that the ectomycorrhizal fungi are directly involved in exploitation of the organic residues.

One cautionary note must, however, be sounded. Since the natural substrates employed in these studies contain a normal microflora and fauna, such results do not provide conclusive proof that mycorrhizal fungi are directly involved in mobilization. A process of facilitated attack by saprotrophs could be occurring, carbon leaked from mycorrhizal fungi being used to sustain detritivores. While some facilitation of this kind may occur, Ingham *et al.* (1991) could find no evidence of consistent increases of bacterial over fungal numbers in soils colonized by dense mats of ectomycorrhizal mycelium. The density of the mycelium formed by the mycorrhizal fungus itself in 'patches' and 'mats' would appear physically to preclude a large increase in biomass of fungal saprotrophs. These observations, coupled with the knowledge that some ectomycorrhizal fungi have the ability to produce a suite of enzymes capable of initiating decomposition processes, e.g. lignase (Haider & Hutterman 1984; Haselwandter, Bobleter & Read 1990; Trojanowski, Griffiths & Caldwell 1992), polyphenol oxidase (Giltrap 1982; Griffiths & Caldwell 1992) peroxidase (Griffiths & Caldwell 1992) and fatty acid esterase (Hutchinson 1990; Caldwell, Castellano & Griffiths 1991), as well as those proteases and phosphatases which enable direct attack on N- and P-containing polymers, all suggest that they are directly rather than secondarily involved in mobilization as well as export of essential nutrients from colonized litter.

The reduction of quality of these substrates, that has been detected most readily as an increase of C:N ratio in patches (Bending & Read 1995) and mats (Griffiths *et al.* 1990) as exploitation proceeds, may significantly contribute to the slowing in rate of decomposition observed by Gadgil & Gadgil (1975) in litter occupied by mycorrhizal roots of pine.

The capture of phosphorus from organic residues of the FH, facilitated by ectomycorrhizal associates of pine, may under some circumstances be as important as that of nitrogen. The dynamics of the turnover of P in forest soil are thought to be controlled by the same energy-dependent microbial activities as are those of N (Berg & McClaugherty 1989). The residues of membranes, mostly phospholipid, and of the nucleic acids DNA and RNA, are likely to be among the primary products of cell turnover in soil. The ability of the non-carbon-limited ectomycorrhizal mycelium to express phospho-mono- and di-esterase activity in close association with the main sites of turnover in localized pockets of organic matter could be of greater biological significance in P-limited ecosystems than is their ability to access organic N. In the less acid mineral soils of warmer climates which are now widely occupied by *Pinus* spp. there may be a population of nitrifying bacteria in the rooting environment, so that P, still largely present in organic form, may become the major growth limiting element. Here, and increasingly in the boreal forests of the northern hemisphere where N saturation is occurring as a result of pollutant inputs (Aber *et al.* 1989), phosphatase activities may be key determinants of ecosystem productivity, and the need for further study of the deployment of these enzymes by ectomycorrhizal fungi as well as for improved characterization of the substrates which they attack has been emphasized (Griffiths & Caldwell 1992).

Observation of activities of ectomycorrhizal fungi in laboratory microcosms helps to provide a functional basis for the selective exploitation by ectomycorrhizal pine roots of the superficial organic horizons of the soil, first described by Frank. These are the substrates in which mobilization of N and P is most active. In pine forest soil, it has been shown that periods of net immobilization associated with freshly fallen litter (Berg & Söderström 1979), are followed by a phase of mineralization in the FH layer. The conventional explanation of this process is that it arises when the saprotrophs become carbon-limited (Berg & Staaf 1981; Berg & McClaugherty 1989). Most considerations of the nutrient dynamics of the FH horizon have failed to recognize that its substrates are selectively colonized by ectomycorrhizal fungi which, because of their attachment to the root, are relatively free of carbon limitation. Under these circumstances it is likely that the mycorrhizal fungi are able to express their biochemical capability to mobilize the N and P from the substrates at a particular stage during which, because of their carbon limitation, the remainder of the microbial population is inactive.

16.4.2 Enhancement of resistance to toxicity

The organic horizons in which ectomycorrhizal roots of *Pinus* spp. characteristically proliferate have a natural acidity. Low pH itself poses a threat to root growth of pine but it also leads to increased solubility of metal ions. Resistance both to low pH and to metal toxicity has been shown to be dramatically increased by ectomycorrhizal colonization (Marx 1991; Marx, Maul & Cordell 1992). In the most extreme sites inoculation of *Pinus* spp. with *Pisolithus tinctorius* has been shown to be particularly beneficial, enabling highly significant increases of survivorship and growth (Table 16.2). These effects have led to the use of *P. tinctorius* as an inoculant in many afforestation programmes (see below).

It appears that in such adverse sites selection operates rapidly to produce ectomycorrhizal symbionts of

Table 16.2. Percentage increase in survival and volume growth of pine seedlings after 2 to 4 years with *Pisolithus tinctorius* ectomycorrhiza over controls with naturally occurring ectomycorrhiza on various adverse sites (from Marx 1975b, Marx *et al.* 1992)

Pinus taxon	Site	Adversity	% increase in seedling Survival	% increase in seedling Volume
P. resinosa	Coal spoil	pH 3.0	214	60
P. echinata	Coal spoil	pH 4.1	5	400
P. virginiana	Coal spoil	pH 3.1	87	444
P. virginiana	Coal spoil	pH 3.8	480	422
P. rigida	Coal spoil	pH 3.8	0	420
P. rigida	Coal spoil	pH 3.4	57	215
P. rigida	Coal spoil	pH 4.3	8	180
P. taeda	Coal spoil	pH 3.3	20	415
P. taeda	Coal spoil	pH 3.4	14	750
P. taeda	Coal spoil	pH 4.1	41	400
P. taeda	Coal spoil	pH 3.4	96	800
P. taeda	Coal spoil	pH 4.3	16	380
P. taeda	Kaolin spoil	low fertility	0	1100
P. taeda	Fullers' earth	low fertility	0	47
P. taeda	Copper basin	eroded	0	45
P. virginiana	Copper basin	eroded	0	88
P. taeda	Borrow pit	droughty	17	412

high resistance to the prevalent contaminant. Colpaert & van Assche (1987) isolated strains of *Suillus luteus* from zinc-contaminated soil that were able to grow in the presence of 1000 $\mu g\ g^{-1}$ of the element. Strains of the same fungus obtained from carpophores growing on uncontaminated soil showed little or no growth above 100 $\mu g\ g^{-1}$ Zn. It was subsequently shown (Colpaert & van Assche 1992) that Zn-resistant strains of *S. bovinus* conferred significantly more Zn tolerance upon plants of *Pinus sylvestris* than did non-resistant strains, and that metal resistance was most probably attributable to binding of the metal in the extramatrical mycelium of the fungus. This enables the plant to avoid the effects of the metal, and so to maintain the flow of assimilatory compounds to the roots, on which the fungus is dependent for its functions.

16.4.3 Enhancement of resistance to fungal pathogens

Ectomycorrhizal fungi of pines can provide their hosts with enhanced resistance to attack by fungal pathogens. Both *Pisolithus tinctorius* and *Thelephora terrestris* have been shown to reduce the impacts of the root pathogen *Phytophthora cinnamomi* on *Pinus* spp. (Marx 1969, 1973, 1975a). Similarly, inoculation with *Laccaria laccata* reduced the incidence of disease caused by the pathogen *Fusarium oxysporum* in *P. sylvestris* (Chakravarty & Unestam 1987). *Paxillus involutus* reduced 'damping-off' disease caused by *Pythium* spp. in *Pinus resinosa* (Chakravarty & Unestam 1991). It has been hypothesized that protection against fungal pathogens is achieved as a result of the physical barriers imposed by the hyphal mantle (Marx 1973) or by the production of phenolic compounds in the host tissues in response to the presence of the mycorrhizal fungus (Sylvia & Sinclair 1983).

While both of these effects may indeed be involved in contributing to defence in the adult plant, there is evidence that ectomycorrhizal fungi exert direct antibiotic effects upon would-be pathogens. Duchesne et al. (1988a, b) observed that inoculation of seedlings of P. resinosa with the fungus Paxillus involutus significantly reduced pathogenicity of F. oxysporum before mycorrhizal colonization took place. Increases of seedling survival were associated with a six-fold decrease in sporulation of F. oxysporum in the rhizosphere of the host plant (Duchesne, Peterson & Ellis 1987). Ethanol extracts of the rhizosphere (Duchesne, Peterson & Ellis 1989) indicated that fungitoxic effects were present within three days of inoculation of pine seedlings with P. involutus. Disease suppression at this critical stage of plant development before formation of the ectomycorrhizal symbiosis may be of particular significance in the regeneration niche.

16.5 Ecological aspects of the mycorrhizal symbiosis in *Pinus*

16.5.1 The requirement for mycorrhizal colonization

The ecological dependence of pine species on mycorrhizal colonization was strikingly demonstrated when many early attempts to introduce them into exotic environments, particularly in Africa and Central America, failed (Kessell 1923, 1927; Roeloffs 1930; Clements 1941; Pearson 1950; Mikola 1953, 1969; van Suchtelen 1962). It was learned from these failures that if soils lacked appropriate inoculum, the trees would die, and methods were developed to ensure that fungal symbionts were introduced. In afforestation practices these routinely involved introduction of forest humus or nursery soil from native pine areas (Rayner 1938; Hacskaylo 1970). In general, both the failures of the initial plantings and the responses to inoculation with mycorrhizal fungi were greatest in areas that were most remote from ectomycorrhizal forests, for example in Trinidad (Lamb 1956), Puerto Rico (Hackskaylo & Vozzo 1971), Nigeria and Zambia (Mikola 1970). As a result of these experiences, instruction for establishment of nurseries for growth of *Pinus* spp. in such areas normally routinely recommended addition of ectomycorrhizal inoculum. Such approaches were even recommended for establishment of new nurseries in treeless areas of the temperate zone in the northern hemisphere (Stoeckeler & Slabauch 1965).

In the more temperate areas of Africa as well as in New Zealand and Australia it appears that relatively few difficulties were encountered when introducing exotic pine species (Mikola 1973). The implication of this observation is that mycorrhizal inoculum was already present in these environments. The first introductions of *Pinus* spp. to exotic locations probably preceded afforestation programmes and occurred as growing plants which would have brought with them their established mycorrhizal flora (Mikola 1969). *Pinus radiata* was first introduced in this form to Australia in 1857 but from Kew Gardens in Britain (Fielding 1957).

16.5.2 The fungi of exotic pine plantations

Because most of the human immigrants to the southern hemisphere, as well as the tree seedlings, came from Europe, the introduced mycorrhizal fungi were also mainly European species despite the fact that many of the tree species, e.g. P. elliottii, P. patula and P. radiata, were of North American origin. This is reflected in lists (Table 16.3) of records of occurrence of mycorrhizal fungi in exotic pine plantations. Only *Suillus brevipes*, among the 28 species listed, is from North America. It is evident from these lists that the diversity of ectomycorrhizal fungi of exotic *Pinus* stands is extremely low compared with that seen in native stands, numbers of fungal species recorded being probably two orders of magnitude lower than would be expected in the natural habitat of a given pine species. Malajczuk (1987) estimates that there are fewer than 20 ectomycorrhizal fungal species in pine plantations of Australia, which cover more than 750 000 ha. This contrasts with calculations suggesting as many as 2000 spp. as possible associates of pine in its natural habitats (Molina, Massicotte & Trappe 1992).

Synthesis experiments carried out under standardized aseptic conditions (Malajczuk, Molina & Trappe 1982) have confirmed that fungal species other than those typically found in association with P. radiata in exotic plantations have the potential to form mycorrhiza with this host (Table 16.4). In nature, however, the particular soil conditions and climate of the exotic environment appear, in combination, to favour the selection of a relatively small number of fungal symbionts.

In exotic plantations the dominant fungal symbionts are *Rhizopogon luteolus* and *R. roseolus* (= *R. rubescens*), *Suillus luteus* and *S. granulatus*, *Hebeloma crustuliniforme*, *Paxillus involutus*, *Amanita muscaria* and *Lactarius deliciosus*. *Rhizopogon* species occur globally as dominant mycorrhizal fungi in exotic pine plantations and characteristically colonize plants at a very early age. They have been shown to be effective as colonists of *Pinus radiata* seedlings from spores (Theodorou 1971) and to be particularly efficient in stimulating growth of young plants in P-deficient soils

Table 16.3. Records of the occurrence of mycorrhizal fungi as fruit bodies in exotic pine plantations

Fungus	Country and reference
Amanita excelsa	South Africa: Lundquist (1986); Van der Westhuizen & Eicher (1987)
A. muscaria	Australia: Purnell (1957) India: Natarajan et al. (1992) New Zealand: Birch (1937); Rawlings (1950, 1951, 1960); Chu-Chou (1979) South Africa: Stephens & Kidd (1953b); Mikola (1969); Marais & Kotzé (1975)
A. pantherina	South Africa: Lundquist (1986, 1987); Van der Westhuizen & Eicher (1989)
A. rubescens	South Africa: Lundquist (1986, 1987); Van der Westhuizen & Eicher (1987)
Boletus edulis	South Africa: Stephens & Kidd (1953a); Mikola (1969); Marais & Kotzé (1975) Uruguay: Singer (1964)
B. piperatus	New Zealand: Rawlings (1950, 1960); McNabb (1968); Chu-Chou (1979) South Africa: Mikola (1969) Trinidad: Singer (1964)
Cenococcum geophilum	Australia: Lamb & Richards (1970, 1971)
Hebeloma crustuliniforme	South Africa: Stephens & Kidd (1953b); Mikola (1969) Kenya: Gibson (1963) Argentina: Takacs (1961a) New Zealand: Chu-Chou (1979)
Laccaria laccata	India: Natarajan et al. (1992) New Zealand: Rawlings (1951, 1960); Chu-Chou (1979) South Africa: Lundquist (1986); Van der Westhuizen & Eicher (1987)
Lactarius deliciosus	Australia: Purnell (1957); Mikola (1969) Chile: Mikola (1969) South Africa: Stephens & Kidd (1953a); Lundquist (1987); Van der Westhuizen & Eicher (1987)
Paxillus involutus	Australia: Malajczuk (1987)
P. panuoides	South Africa: Lundquist (1987)
Pisolithus tinctorius	Puerto Rico: Mikola (1969) Surinam: van Suchtelen (1962)
Rhizopogon luteolus	Australia: Purnell (1957); Chilvers (1973); Lamb & Richards (1971); Skinner & Bowen (1974) New Zealand: Rawlings (1950); Chu-Chou (1979) South Africa: Donald (1975)
R. roseolus	Australia: Young (1940); Lamb & Richards (1970, 1971) New Zealand: Birch (1937); Rawlings (1950) Nigeria: Mikola (1969) South Africa: Mikola (1969) Trinidad: Mikola (1969)
Russula caerulea	South Africa: Van der Westhuizen & Eicher (1987)
R. capensis	South Africa: Van der Westhuizen & Eicher (1987)
R. cyanoxantha	South Africa: Lundquist (1987)
R. sardonia	South Africa: Van der Westhuizen & Eicher (1987)
Scleroderma bovista	Argentina, Uruguay: Takacs (1961a, b) New Zealand: Birch (1937); Rawlings (1950, 1960) Kenya: Gibson (1963)
S. citrinum	India: Natarajan et al. (1992) South Africa: Van der Westhuizen & Eicher (1987)
S. verrucosum	New Zealand: Chu-Chou (1979) South Africa: Van der Westhuizen & Eicher (1987)
Suillus bovinus	Australia: Mikola (1969) South Africa: Stephens & Kidd (1953a); Lundquist (1987); Van der Westhuizen & Eicher (1987)
S. brevipes	Argentina: Singer (1963) India: Natarajan et al. (1992) New Zealand: McNabb (1968)

Table 16.3. (cont.)

Fungus	Country and reference
S. granulatus	Argentina: Singer (1964); Takacs (1961a) Australia: Young (1940); Purnell (1957); Mikola (1969); Lamb & Richards (1971) Brazil: Mikola (1969) Chile: Singer (1963); Mikola (1969) New Zealand: McNabb (1968); Chu-Chou (1979) South Africa: Mikola (1969)
S. luteus	Argentina: Singer (1963, 1964) Australia: Purnell (1957); Mikola (1969) Chile: Mikola (1969) Kenya: Gibson (1963) New Zealand: Birch (1937); Rawlings (1950); McNabb (1968); Chu-Chou (1979) South Africa: Mikola (1969); Van der Westhuizen & Eicher (1987) Uganda: Brown (1963) Uruguay: Takacs (1961b)
Thelephora terrestris	New Zealand: Chu-Chou (1979) South Africa: Lundquist (1986, 1987); Van der Westhuizen & Eicher (1987)
Tuber rapaeodurum	South Africa: Marais & Kotzé (1975)

Table 16.4. Results of monoxenic synthesis experiments in which Pinus radiata was grown with pure cultures of a range of mycorrhizal fungi (from Malajczuk et al. 1982)

Fungus	Extent of colonization
Amanita muscaria	+ + +
Astraeus pteridis	+ + +
Boletus edulis	+ +
Cenococcum geophilum	+
Hebeloma crustuliniforme	+ + +
Hydnangium carneum	–
Hymenogaster albellus	–
Laccaria laccata	+ +
Lactarius deliciosus	+
Melanogaster intermedius	+ + +
Paxillus involutus	+ + +
Pisolithus tinctorius	+
Rhizopogon occidentalis	+
R. vinicolor	+
R. vulgaris	+ +
Scleroderma laeve	+
Suillus brevipes	+ + +
S. grevillei	+
S. lakei	+ + +
S. tomentosus	+
Thelephora terrestris	+ + +
Truncocolumella citrina	+ + +

Note:
Mycorrhizal intensity: + + +, 70–100% short roots colonized; + +, 30–69% short roots colonized; +, 1–29% short roots colonized; –, no formation.

(Theodorou & Bowen 1970). *Suillus granulatus* is the most widespread *Suillus* species on pines in warmer regions globally, while *S. luteus* becomes more important in cooler climates such as the East African Highlands (Gibson 1963), southern Chile and Argentina (Singer 1963).

The inoculum first imported to enable establishment of pine in exotic locations, whether already present on introduced plants or in soil collected in natural pine forest, will have contained a greater diversity of fungal species. It must therefore be concluded that the depauperate but regionally distinctive mycoflora that subsequently developed is a product of selection processes in which certain species best suited to local conditions were able to extend their occupancy. Their dominance will frequently have been further facilitated by the lack of competition from any native population of ectomycorrhizal fungi.

In southern Africa, the introductions of other genera of trees that are hosts to a broad spectrum of ectomycorrhizal fungi, most notably *Eucalyptus* spp. from Australia and *Quercus* spp. from Europe, are likely to have provided some sources of inoculum for the pines. However, it is a notable feature of pine plantations that their mycorrhizal flora is largely dominated by pine-specific fungi. The extensive planting of *Pinus* spp. throughout the warm temperate and subtropical regions of both hemispheres coupled with the implementation of inoculation programmes, will in all likelihood have led to a situation in which, as hypothesized by Richardson, Williamson & Hobbs (1994), virtually no region of the globe is free from exposure to mycorrhizal inoculum in the form of spores. What is more, as increasingly aggressive strains of fungi such as *Pisolithus tinctorius* are employed in inoculation programmes, the potential for colonization of isolated trees in alien ecosystems should increase.

The use of *P. tinctorius* as an inoculant fungus is of interest because, as noted above, it is not one of those fungi normally invading natural or alien plantations of pine. In particular, its selection may have repercussions for natural ecosystems worldwide. The original isolate of *P. tinctorius* was obtained from a sporophore found under loblolly pine (*Pinus taeda*) in Georgia, USA around 1970. Since that time it has been repeatedly inoculated onto and re-isolated from pine roots and in sequential isolates has shown increasing aggressiveness (Marx 1981, 1991). It is now referred to as a 'super-strain' and in recent times has been introduced as an inoculant of pine species to Brazil, Canada, China, Congo, France, Ghana, India, Liberia, Malawi, Mexico, Nigeria, South Korea, Thailand and Venezuela (Marx 1991), many millions of seedlings being grown in association with it every year. While the commercial advantages of vigorous early growth of exotic plantations is obvious, little thought has been given to the ecological consequences of distributing aggressive genotypes of the super-strain category around the world. It is possible that it will enhance invasiveness of pine and it will be of considerable interest to see whether this fungus becomes established in countries such as those listed in Table 16.3, in which *P. tinctorius* has not hitherto been recorded.

16.5.3 Sources and types of fungal inoculum

Spores of ectomycorrhizal fungi do not germinate readily under any circumstance. Theodorou & Bowen (1973)

obtained only 46–69% germination in *R. luteolus*, 34% in *S. luteus* and 31% in *S. granulatus* when these were applied to the surfaces of *P. radiata* roots. It is partly for this reason that spores are poor sources of inoculum in forestry. It is known that very high densities, for example of *Pisolithus tinctorius* spores, are necessary to ensure extensive colonization of pine seedlings (Marx et al. 1989). Vegetative inoculum is normally preferred in nursery practices if only because colonization is obtained so much more rapidly by this means than by spores which germinate very slowly.

In the absence of roots of a potential host, spores of such fungi normally fail to germinate at all and, as far as is known, none of their mycelia have a free-living existence in the soil. They are dependent on the host root not only to stimulate germination but also for the supply of carbon necessary to express their well-developed potential to exploit (extensively and intensively) the resources of their environment.

As a result of these biological attributes, the chances of successful colonization of pine seedlings and the amounts of colonization obtained, both of which will contribute to the determination of survivorship and early growth of an isolated plant after germination in an alien environment, are likely to be heavily dependent not simply on *occurrence* of spore deposition but on its *frequency*. This inoculum potential will be greatest close to a source of fungal sporophores. It would be high near an established stand of ectomycorrhizal hosts and diminish with distance from the edge of the stand, though at a rate much slower than that seen in propagules, such as seeds, of greater weight.

Richardson et al. (1994) comment that the largest stands of self-sown pine in all regions are close to plantations of the tree and that a large density of immigrating plant propagules clearly increases the likelihood of invasion. The same arguments apply to availability of inoculum. There is thus much to suggest that the processes of invasion by host and fungus must occur with some simultaneity to ensure the development of the symbiosis which, in turn, is a prerequisite for successful establishment in the new environment.

Once a single tree is colonized by mycorrhizal fungi and established in an alien environment, the vegetative mycelia growing from its roots provide, as described earlier (Fig. 16.4), a much more vigorous and effective source of inoculum. Seeds arriving from these established individuals, or from elsewhere, germinating in the vicinity will be rapidly colonized by the fungi occupying the soil domain in which they germinate. This may be a primary factor determining the clumping of seedlings around established pines that produces a two-phase pattern of invasion in systems such as fynbos (Richardson & Brown 1986).

The ability of the colonized roots of trees to act as refuges for ectomycorrhizal fungi, during and after fire, has been recognized (Amaranthus, Molina & Perry 1990). The deeper mycorrhizal roots of these trees retain viability and themselves provide foci for colonization of newly germinating seedlings, hence facilitating regeneration in the post-fire environment.

16.6 Summary and conclusions

The natural dominance of pines on the most impoverished acidic soils in mid- to high latitudes of the northern hemisphere is evidence of the ability of members of this genus to compete effectively for nutrients with would-be competitors. This ability led Axelrod (1986), after an extensive review of the origin, evolution and speciation of the genus *Pinus* to speculate that 'mycorrhizae have given Pinaceae, and *Pinus* in particular, an adaptive edge as environments have become more extreme whether in cold regions, the seasonal tropics or in sandy sites with acid soils'. From what has been said earlier, it is clear that his speculation is largely justified by the experimental evidence. This confirms that a number of features of its mycorrhizal biology may indeed make significant contribution to the success of *Pinus* in impoverished environments. These can be summarized as follows.

- The ectomycorrhizal mycelial network provides an extremely effective method of scavenging for nutrients in infertile environments and one that is several orders of magnitude more efficient in terms of carbon 'cost' than a non-mycorrhizal system, while being more extensive than that of vesicular-arbuscular mycorrhiza.
- The ectomycorrhizal fungus has the ability to exploit the major nutrients nitrogen and phosphorus contained in organic substrates, where they are unlikely to be available to non-mycorrhizal plants or those colonized by vesicular-arbuscular mycorrhiza.
- The antibiotic and physical defences gained by roots through the presence of the ensheathing fungal mantle reduce the susceptibility of the plant to attack by pathogens.
- The greater resistance to environmental toxins such as metals shown by many ectomycorrhizal associates of pine enables continued growth of the mycelium and sequestration of metals in its biomass, and provides avoidance of toxicity in the otherwise susceptible tissues of the plant.

When these attributes are introduced into nutrient-poor alien environments, particularly those that are disturbed or are dominated by plants which lack any or all such attributes, they are likely to provide potent advantage to the alien. Studies of invasion by pines (Richardson & Bond 1991) suggest that they are most successful in grasslands, old-fields or shrublands of the 'fynbos' type. It is a feature of such systems that the plants which dominate them before invasion by pine, being non-ectomycorrhizal, may be at a particular disadvantage when faced with competition for nutrients by the alien species. These disadvantages will be accentuated when the distinctive growth form above ground also enables superior competition for light. Success above ground, however, is dependent on capture and allocation of nutrients from the below-ground environment. To the extent that open woodlands and forests appear to be relatively resistant to invasion by pines (Richardson & Bond 1991), there is some support for the view that invasions are encouraged by niche dissimilarity. It can be hypothesized that forest ecosystems already with a preponderance of ectomycorrhizal trees would be more resistant to invasion because some of the most important of the symbiotic attributes of pine, in particular the ability to scavenge effectively for nutrients, were already deployed. Invasion of systems of this kind, especially if, as appears likely, they contain a high proportion of fungal species not readily compatible with *Pinus*, should be more heavily dependent on disturbance which will release substrates for attack by symbionts of the invader.

There is much scope for research in this area. We lack even the most basic information concerning the early symbiotic events involved in the successful establishment of pine in the regeneration niche of an alien environment. As a result, we are poorly placed to evaluate the processes whereby such plants come to dominate these environments, often to the exclusion of the native flora. Studies of mycorrhizal biology of pines in their natural range and in laboratory environments suggest that their success is attributable in large part to a superior ability to compete for minerals and to exploit nutrients contained in organic residues. However, without direct analysis of the events occurring below ground as establishment of pine takes place, it is unlikely that the mechanisms underlying its success as an invasive alien will be fully understood.

Acknowledgements

I thank Dr J.M. Theron for his critical review of the manuscript and Jayne Young for her assistance in its preparation.

References

Aber, D.J., Nadelhoffer, K.J., Steudler, P. & Melillo, J.M. (1989). Nitrogen saturation in northern forest ecosystems. *BioScience*, **39**, 378–86.

Abuzinadah, R.A. & Read, D.J. (1986a). The role of proteins in the nitrogen nutrition of ectomycorrhizal plants. I. Utilization of peptides and proteins by ectomycorrhizal fungi. *New Phytologist*, **103**, 481–93.

Abuzinadah, R.A. & Read, D.J. (1986b). The role of proteins in the nitrogen nutrition of ectomycorrhizal plants. III. Protein utilization by *Betula, Picea* and *Pinus* in mycorrhizal association with *Hebeloma crustuliniforme*. *New Phytologist*, **103**, 507–14.

Aldrich-Blake, R.N. (1930). The root system of the Corsican Pine. *Oxford Forest Memoirs*, **12**, 1–64.

Amaranthus, M.P., Molina, R. & Perry, D.A. (1990). Soil organisms, root growth and forest regeneration. *Proceedings of the Society of American Foresters, National Convention*, Spokane, Washington, pp. 89–93.

Axelrod, D.I. (1986). Cenozoic history of some western American pines. *Annals of the Missouri Botanical Garden*, **73**, 565–641.

Bending, G.D. & Read, D.J. (1995). The structure and function of the vegetative mycelium of ectomycorrhizal plants. V. The foraging behaviour of ectomycorrhizal mycelium and the translocation of nutrients from exploited organic matter. *New Phytologist*, **130**, 401–9.

Berg, B. & McClaugherty, C.A. (1989). Nitrogen and phosphorus release from decomposing litter in relation to the disappearance of lignin. *Canadian Journal of Botany*, **67**, 1148–56.

Berg B. & Söderström, B. (1979). Fungal biomass and nitrogen in decomposing Scots pine needle litter. *Soil Biology and Biochemistry*, **11**, 339–41.

Berg, B. & Staaf, H. (1981). Leaching, accumulation and release of nitrogen in decomposing forest litter. In *Terrestrial Nitrogen Cycles*, ed. F.E. Clark & T. Rosswall, pp. 163–78. Ecological Bulletin 33. Stockholm: Swedish Natural Science Research Council.

Birch, T.T.C. (1937). A synopsis of forest fungi of significance in New Zealand. *New Zealand Journal of Forestry*, **6**, 109–25.

Bowen, G.D. (1968). Phosphate uptake by mycorrhizas and uninfected roots of *Pinus radiata* in relation to root distribution. *Ninth International Congress of Soil Science Transactions*, II, 219–28.

Bowen, G.D. (1973). Mineral nutrition of ectomycorrhizae. In *Ectomycorrhizae – Their Ecology and Physiology* ed. G.C. Marks & T.T. Kozlowski, pp. 151–97. New York: Academic Press.

Bowen, G.D. & Theodorou, C. (1967). Studies on phosphate uptake by mycorrhiza. Proceedings of the 14th IUFRO Congress, Munich, Germany, **5**, 116–38.

Brown, J.L. (1963). *Working Plan for Mafuga Central Forest Reserve*. Uganda: Forestry Department.

Caldwell, B.A., Castellano, M.A., Griffiths, R.P. (1991). Fatty acid esterase production by ectomycorrhizal fungi. *Mycologia*, **83**, 233–6.

Chakravarty, P. & Unestam, T. (1987). Differential influence of ectomycorrhizae on plant growth and disease resistance of *Pinus sylvestris* seedlings. *Journal of Phytopathology*, **120**, 104–20.

Chakravarty, P. & Unestam, T. (1991). Mycorrhizal fungi prevent disease in stressed pine seedlings. *Journal of Phytopathology*, **188**, 335–40.

Chilvers, G.A. (1973). Host range of some eucalypt mycorrhizal fungi. *Australian Journal of Botany*, **21**, 103–11.

Chu-Chou, M. (1979). Mycorrhizal fungi of *Pinus radiata* in New Zealand. *Soil Biology and Biochemistry*, **11**, 557–62.

Clements, J.B. (1941). The introduction of pines into Nyasaland. *Nyasaland Agriculture Quarterly Journal*, **1**(4), 5–15.

Colpaert, J.V. & van Assche, J.A. (1987). Heavy metal tolerance in some ectomycorrhizal fungi. *Functional Ecology*, **1**, 415–21.

Colpaert, J.V. & Van Assche, J.A. (1992). Zinc toxicity in ectomycorrhizal *Pinus sylvestris*. *Plant and Soil*, **143**, 201–11.

Conjeaud, C., Scheromm, P. & Mousain, D. (1996). Effects of phosphorus fertilisation and ectomycorrhizal infection on the carbon balance in maritime pine seedlings (*Pinus pinaster* Soland. in Ait.). *New Phytologist*, **133**, 345–51.

Danielson, R.M. (1984). Ectomycorrhizal associations in Jack pine stands in north-eastern Alberta. *Canadian Journal of Botany*, **62**, 932–9.

Danielson, R.M. (1991). Temporal changes and effects of amendments on the occurrence of sheathing (ecto-) mycorrhizal fungi of conifers growing in oil sands tailings and coal spoil. *Agricultural Ecosystems and Environment*, **35**, 261–81.

Donald, D.G.M. (1975). Mycorrhizal inoculation for pines. *South African Forestry Journal*, **92**, 27–9.

Duchesne, L.C., Peterson, R.L. & Ellis, B.E. (1987). The accumulation of plant-produced antimicrobial compounds in response to ectomycorrhizal fungi: a review. *New Phytologist*, **68**, 17–27.

Duchesne, L.C., Peterson, R.L. & Ellis, B.E. (1988a). Interaction between the ectomycorrhizal fungus *Paxillus involutus* and *Pinus resinosa* induces resistance to *Fusarium oxysporum*. *Canadian Journal of Botany*, **66**, 558–62.

Duchesne, L.C., Peterson, R.L. & Ellis, B.E. (1988b). Pine root exudate stimulates antibiotic synthesis by the ectomycorrhizal fungus *Paxillus involutus*. *New Phytologist*, **108**, 470–6.

Duchesne, L.C., Peterson, R.L. & Ellis, B.E. (1989). The time-course of disease suppression and antibiosis by the ectomycorrhizal fungi *Paxillus involutus*. *New Phytologist*, **111**, 693–8.

Entry, J.A., Rose, C.L. & Cromack, K., Jr (1991). Litter decomposition and nutrient release in ectomycorrhizal mat soils of a Douglas fir ecosystem. *Soil Biology and Biochemistry*, **23**, 285–90.

Fielding, J.M. (1957). Introduction of *Pinus radiata* to Australia. *Australian Forester*, **21**, 15–16.

Finlay, R.D. & Read, D.J. (1986a). The structure and function of the vegetative mycelium of ectomycorrhizal plants. I. Translocation of ^{14}C-labelled carbon between plants interconnected by a common mycelium. *New Phytologist*, **103**, 143–56.

Finlay, R.D. & Read, D.J. (1986b). The structure and function of the vegetative mycelium of ectomycorrhizal plants. II. The uptake and distribution of phosphorus by mycelial strands inter-connecting host plants. *New Phytologist*, **103**, 157–65.

Finlay, R.D. & Söderström, B. (1992). Mycorrhiza and carbon flow to the soil. In *Mycorrhiza Functioning*, ed. M.F. Allen, pp. 134–62. London: Chapman & Hall.

Fogel, R. & Hunt, G. (1979). Fungal and arboreal biomass in a western Oregon Douglas fir ecosystem: distribution patterns and turnover. *Canadian Journal of Forestry Research*, **9**, 245–56.

Frank, A.B. (1885). Ueber die Wurzelsymbiose beruhende Ernährung gewisser Bäume durch unterirdische Pilze. *Berichte der Deutschen Botanische Gesellschaft*, **3**, 128–45.

Frank, A.B. (1894). Die Bedeutung der Mykorrhizapilze für die gemeine Kiefer. *Forstwissenschaftliches Centralblatt*, **16**, 1852–90.

Gadgil, R.L. & Gadgil, P.D. (1975). Suppression of litter decomposition by mycorrhizal roots of *Pinus radiata*. *New Zealand Journal of Forest Science*, **5**, 35–41.

Gardes, M. & Bruns, T. (1996). Community structure of ectomyccorhizal fungi in a *Pinus muricata* forest: above and below ground views. *Canadian Journal of Botany* **74**, 1572–83.

Gibson, I.A.S. (1963). Eine Mitteilung über die Kiefernmykorrhiza in den Waldern Kenias. In *Mykorrhiza – Internationales Mykorrhizasymposium 25–30 April 1960, Weimar*, ed. W. Rawald & A. Lyr, pp. 49–51. Jena: Gustav Fischer.

Giltrap, N.J. (1982). Production of polyphenol oxidases by ectomycorrhizal fungi with special reterence to *Lactarius* species. *Transactions of the British Mycological Society*, **78**, 75–81.

Griffiths, R.P. & Caldwell, B. (1992). Mycorrhizal mat communities in forest soils. In *Mycorrhizas in Ecosystems*, ed. D.J. Read, D.H. Lewis, A.H. Fitter & I.J. Alexander, pp. 98–105. Wallingford, UK: Commonwealth Agricultural Bureau.

Griffiths, R.P., Caldwell, B.A., Cromack, K. & Morita, R.Y. (1990). Microbial dynamics and chemistry in Douglas fir forest soils colonised by ectomycorrhizal mats: I. Seasonal variation in nitrogen chemistry and nitrogen cycle transformation rates. *Canadian Journal of Forestry Research*, **20**, 211–18.

Hacskaylo, E. (1970). Biological amendments to improve forest soils. *Journal of Forestry*, **68**, 332–4.

Hacskaylo, E. & Vozzo, J.A. (1971). Inoculation of *Pinus caribaea* with ectomycorrhizal fungi in Puerto Rico. *Forest Science*, **17**, 239–41.

Hatch, A.B. (1937). The physical basis of mycotrophy in the genus *Pinus*. *Black Rock Forest Bulletin* 6. Cambridge, Mass. Harvard University Press.

Hatch, A.B. & Doak, K.D. (1933). Mycorrhizal and other features of the root system of *Pinus*. *Journal of the Arnold Arboretum*, **14**, 324–34.

Haselwandter, K., Bobleter, O. & Read, D.J. (1990). Utilisation of lignin by ericoid and ectomycorrhizal fungi. *Achive für Mikrobiologie*, **153**, 352–4.

Hutchinson, L.J. (1990). Studies on the systematics of ectomycorrhizal fungi in axenic culture. II. The enzymatic degradation of selected carbon and nitrogen compounds. *Canadian Journal of Botany*, **68**, 1522–30.

Ingham, E.R., Griffiths, R.P., Cromack, K. & Entry, J.A. (1991). Comparison of direct versus fumigation-flux microbial biomass estimates from ectomycorrhizal mat and non-mat soils. *Soil Biology and Biochemistry*, **23**, 465–71.

Kelly, J.M., Barber, S.A. & Edwards, G.S. (1992). Modelling magnesium, phosphorus and potassium uptake by loblolly pine seedlings using a Barber–Cushman approach. *Plant and Soil*, **135**, 209–18.

Kessell, S.L. (1923). Some observations on the establishment of pine nurseries in Western Australia. *Australia–New Zealand Association for Advancement of Science Proceedings*, 1923, 749.

Kessell, S.L. (1927). Soil organisms. The dependence of certain pine species on a biological soil factor. *Empire Forestry Journal*, **6**, 70–4.

Kramer, P. & Bullock, H.C. (1966). Seasonal variations in the proportions of suberised and unsuberised roots of trees in relation to the absorption of water. *American Journal of Botany*, **53**, 200–4.

Lamb, A.F.A. (1956). *Exotic Forest Trees in Trinidad and Tobago*. Trinidad: Government Printing Office.

Lamb, R.J. & Richards, B.N. (1970). Some mycorrhizal fungi of *Pinus radiata* and *Pinus elliottii* var. *elliottii* in Australia. *Transactions of the British Mycological Society*, **54**, 371–8.

Lamb, R.J. & Richards, B.N. (1971). Effect of mycorrhizal fungi on the growth and nutrient status of slash and radiata pine seedlings. *Australian Forestry*, **35**, 1–7.

Lamb, R.J. & Richards, B.N. (1974). Inoculation of pines with mycorrhizal fungi in natural soils. II. Effects of density and time of application of inoculum and phosphorus amendment on seedling yield. *Soil Biology and Biochemistry*, **6**, 173–7.

Lamhamedi, M.S., Godbout, G. & Fortin, J.A. (1994). Dependence of *Laccaria bicolor* basidiome development on current photosynthesis of *Pinus strobus* seedlings. *Canadian Journal of Forest Research*, **24**, 1797–804.

Le Page, B.A., Currah, R., Stockey, R. & Rothwell, G.W. (1997). Fossil ectomycorrhizae from the Middle Eocene. *Americam Journal of Botany*, **84**, 410–12.

Lundquist, J.E. (1986). Fungi associated with *Pinus* in South Africa. Part I. The Transuaal. *South African Forestry Journal*, **138**, 1–14.

Lundquist, J.E. (1987). Fungi associated with *Pinus* in South Africa. Part II. The Cape. *South African Forestry Journal*, **140**, 4–15.

Luxmoore, R.J., Oren, R., Sherriff, D.W. & Thomas, R.B. (1995). Source sink storage relationships of conifers. In *Resource Physiology of Conifers – Acquisition, Allocation and Utilization*, ed. W.K. Smith & T.M. Hinckley, pp. 75–103. New York: Academic Press.

McNabb, R.F.R. (1968). The Boletaceae of New Zealand. *New Zealand Journal of Botany*, **6**, 137–76.

Malajczuk, N. (1987). Ecology and management of ectomycorrhizal fungi in regenerating forest ecosystems in Australia. In *Mycorrhizae in the Next Decade – Practical Applications and Research Priorities, 7th NACOM*, ed. D.M. Sylvia, L.L. Hung, & J.H. Graham, pp. 118–20. Gainesville: University of Florida.

Malajczuk, N., Molina, R. & Trappe, J.M. (1982). Ectomycorrhiza formation in *Eucalyptus*. I. Pure culture synthesis, host specificity and mycorrhizal compatibility with *Pinus radiata*. *New Phytologist*, **91**, 467–82.

Marais, L.J. & Kotzé, J.M. (1975). Mycorrhizal associates of *Pinus patula* in South Africa. *South African Forestry Journal*, **92**, 13–6.

Mason, P.A., Last, F.T., Pelham, J. & Ingleby, K. (1982). Ecology of some fungi associated with an ageing stand of birches (*Betula pendula* and *B. pubescens*). *Forest Ecology and Management*, **4**, 19–39.

Marx, D.H. (1969). The influence of ectotrophic ectomycorrhizal fungi on the resistance of pine roots to pathogenic infections. I. Antagonism of mycorrhizal fungi to pathogenic fungi and soil bacteria. *Phytopathology*, **59**, 153–63.

Marx, D.H. (1973). Mycorrhizae and feeder root diseases. In *Ectomycorrhizae: Their Ecology and Physiology*, ed. G.C. Marks & T.T. Kozlowski, pp. 351–82. New York: Academic Press.

Marx, D.H. (1975a). Role of ectomycorrhizae in the protection of pine from root infection by *Phytophthora cinnamomi*. In *Biology and Control of Soil-borne Plant Pathogens*, ed. J.W. Bruehl, pp. 119–30. St Paul, Minnesota: American Phytopathology Society.

Marx, D.H. (1975b). Mycorrhiza and establishment of trees on strip-mined land. *Ohio Journal of Science*, **75**, 288–97.

Marx, D.H. (1981). Variability in ectomycorrhizal development and growth among isolates of *Pisolithus tinctorius* as affected by source age and re-isolation. *Canadian Journal of Forest Research*, **11**, 168–74.

Marx, D.H. (1991). The practical significance of ectomycorrhizae in forest establishment. In *Ecophysiology of ectomycorrhizae of forest trees. Marcus Wallenberg Foundation Symposia Proceedings*, **7**, 54–90.

Marx, D.H., Cordell, C.E., Maul, S.B. & Ruehle, J.L. (1989). Ectomycorrhizal development on pine by *Pisolithus tinctorius* in bare-root and container seedlings nurseries. II. Efficacy of various vegetative and sport inocula. *New Forests*, **3**, 57–66.

Marx, D.H., Maul, S.B. & Cordell, C.E. (1992). Application of specific ectomycorrhizal fungi in world forestry. In *Frontiers in Industrial Mycology*. ed. G.F. Leatham, pp. 78–98. New York: Chapman & Hall.

Melin, E. (1925). *Untersuchungen über die Bedeutung der Baummykorriza*. p. 152. Jena: Gustav Fischer.

Melin, E. & Nilsson, H. (1950). Transfer of radioactive phosphorus to pine seedlings by means of mycorrhizal hyphae. *Physiologia Plantarum*, **3**, 88–92.

Melin, E. & Nilsson, H. (1952). Transport of labelled nitrogen from an ammonium source to pine seedlings through mycorrhizal mycelium. *Svensk Botanisk Tidskrift*, **46**, 281–5.

Melin, E. & Nilsson, H. (1953). Transfer of labelled nitrogen from glutamic acid pine seedlings through the mycelium of *Boletus variegatus* (S.W.) Fr. *Nature*, **171**, 434.

Mikola, P. (1953). An experiment on the invasion of mycorrhizal fungi into soil. *Karstenia*, **2**, 33–4.

Mikola, P. (1965). Studies on the ectendotrophic mycorrhiza of pine. *Acta Forestalia Fennica*, **79**, 1–56.

Mikola, P. (1969). Mycorrhizal fungi of exotic forest plantations. *Karstenia*, **10**, 169–76.

Mikola, P. (1970). Mycorrhizal inoculation in afforestation. *International Review of Forest Research*, **3**, 123–46.

Mikola, P. (1973). Application of mycorrhizal symbiosis in forestry practice. In *Ectomycorrhizae*, ed. G.C. Marks & T.T. Kozlowski, pp. 383–411. New York: Academic Press.

Miller, S.L., Durall, D.M. & Rygiewicz, P.T. (1989). Temporal allocation of ^{14}C to extramatrical hyphae of ectomycorrhizal ponderosa pine seedlings. *Tree Physiology*, **5**, 239–50.

Molina, R., Massicotte, H. & Trappe, J.M. (1992). Specificity phenomena in mycorrhizal symbiosis: Community ecological consequences and practical applications. In *Mycorrhizal Functioning*, ed. M.F. Allen, pp. 357–423. New York: Chapman & Hall.

Natarajan, K., Mohan, V. & Ingleby, K. (1992). Correlation between basidiomata production and ectomycorrhizal formation in *Pinus patula* plantations. *Soil Biology and Biochemistry*, **24**, 279–80.

Newman, E.I. (1969). Resistance to water flow in soil and plants. *Journal of Applied Ecology*, **6**, 1–12.

Pearson, A.A. (1950). Cape Agarics and Boleti. *Transactions of the British Mycological Society*, **33**, 276–316.

Purnell, H. (1957). Notes on fungi in Victorian Plantations. III. The mycorrhizal fungi. *Plantation Technical Papers, Forests Commission of Victoria*, **3**, 9–13.

Rawlings, G.B. (1950). The mycorrhizas of New Zealand forests. *Forest Research of New Zealand Forest Service*, **1**, 15–17.

Rawlings, G.B (1951). The mycorrhizas of trees in New Zealand forests. *New Zealand Forest Research Notes*, **1**(3), 15–17.

Rawlings, G.B. (1960). Some practical aspects of forest mycotrophy. *New Zealand Society for Soil Science Proceedings*, **3**, 41–8.

Rayner, M.C. (1938). The use of soil or humus inocula in nurseries and plantations. *Empire Forestry Journal*, **17**, 236–43.

Read, D.J. (1982). In support of Franks organic nitrogen theory. *Angewandte Botanik*, **61**, 25–37.

Read, D. J. (1991). Mycorrhizas in ecosystems. *Experientia*, **47**, 376–91.

Read, D.J. & Boyd, R. (1986). Water relations of mycorrhizal fungi and their host plants. In *Water, Fungi and Plants*, ed. P. Ayres & L. Boddy, pp. 287–303. Cambridge: Cambridge University Press.

Read, D.J., Leake, J.R. & Langdale, A.R. (1989). The nitrogen nutrition of mycorrhizal fungi and their host plants. In *Nitrogen, Phosphorus and Sulphur Utilization by Fungi*, ed. L. Boddy, R. Marchant & D.J. Read, pp. 181–204. London: Academic Press.

Reid, C.P.P., Kidd, F.A. & Ekwebelam, S.A. (1983). Nitrogen nutrition, photosynthesis and carbon allocation in ectomycorrhizal pine. *Plant and Soil*, **71**, 415–32.

Richardson, D.M. & Bond, W.J. (1991). Determinants of plant distribution: Evidence from pine invasions. *American Naturalist*, **137**, 639–68.

Richardson, D.M. & Brown, P.J. (1986). Invasion of mesic mountain fynbos by *Pinus radiata*. *South African Journal of Botany*, **52**, 529–36.

Richardson, D.M., Williams, P.A. & Hobbs, R.J. (1994). Pine invasions in the Southern Hemisphere: determinants of spread and invadability. *Journal of Biogeography*, **21**, 511–27.

Roberts, J. (1976). A study of root distribution and growth in a *Pinus sylvestris* L. (Scots pine) plantation in Thetford Chase, East Anglia. *Plant and Soil*, **44**, 607–21.

Roeloffs, J.W. (1930). Ovr kunstmatige Verjonging van *Pinus merkusii* en *Pinus khasya*. *Tectona*, **23**, 874.

Robertson, N.F, (1954). Studies of the mycorrhiza of *Pinus silvestris*. I. The pattern of development of mycorrhizal roots and its significance for experimental studies. *New Phytologist*, **53**, 253–83.

Rousseau, J.V.D. & Reid, C.P.P. (1990). Effects of phosphorus and ectomycorrhizas on the carbon balance of loblolly pine seedlings. *Forest Science*, **36**, 101–12.

Rousseau, J.V.D. & Reid, C.P.P. (1991). Effects of phosphorus fertilization and mycorrhizal development on phosphorus nutrition and carbon balance of loblolly pine. *New Phytologist*, **117**, 319–26.

Rygiewicz, P.T. & Anderson, C.P. (1994). Mycorrhizae alter quality and quantity of carbon allocated below ground. *Nature*, **369**, 58–60.

Schramm, J.R. (1966). Plant colonization studies on black wastes from anthracite mining in Pennsylvania. *Transactions of the American Philosophical Society*, **56**(1), 1–194.

Singer, R. (1963). Der Ektotroph, seine Definition, geographische Verbreitung und Bedeutung in der Forstökologie. *Mykorrhiza*, Intern. Mykorrhizasymposium, Weimar 1960, 223–31.

Singer, R. (1964). Boletes and related groups in South America. *Nova Hedwigia*, **7**, 93–132.

Skinner, M.F. & Bowen, G.D. (1974). The uptake and translocation of phosphate by mycelial strands of pine mycorrhizas. *Soil Biology and Biochemistry*, **6**, 53–6.

Smith, S.E. & Read, D.J. (1997). *Mycorrhizal Symbiosis*. London: Academic Press.

Staaf, H. & Berg, B. (1977). Mobilisation of plant nutrients in a Scots pine forest mor in central Sweden. *Silva Fennica*, **11**, 210–17.

Stephens, E.L. & Kidd, M.M. (1953a). *Some South African Edible Fungi*. Cape Town: Longmans, Green & Co.

Stephens, E.L. & Kidd, M.M. (1953b). *Some South African Poisonous Fungi*. Cape Town: Longmans, Green & Co.

Stoeckeler, J.H. & Slabauch, P.E. (1965). *Conifer Nursery Practice in the Prairie Plains*. US Department of Agriculture Handbook 279.

Stone, E.L. (1950). Some effects of mycorrhiza on the phosphorus nutrition of Monterey pine seedlings. *Soil Science Society of America Proceedings*, **14**, 340–5.

Söderström, B. & Read, D.J. (1987). The respiratory activity of intact and excised ectomycorrhizal mycelial systems growing in unsterilised soil. *Soil Biology and Biochemistry*, **19**, 231–6.

Sylvia, D.M. & Sinclair, W.A. (1983). Phenolic compounds and resistance to fungal pathogens induced in primary roots of Douglas-fir seedlings by the ectomycorrhizal fungus *Laccaria laccata*. *Phytopathology*, **73**, 390–7.

Takacs, E.A. (1961a). Inoculation de especies de pinos con hongos formadores de micorizas. *Silvicultura (Uruguay)*, **15**, 5–17.

Takacs, E.A. (1961b). Algunas especies de hongos formadores de micorizas en arboles forestales cultivados en la Argentina. *Revista Forestal Argentina*, **5**(3), 80–2.

Tamm, C.O. (1991). *Nitrogen in Terrestrial Ecosystems*. Berlin: Springer-Verlag.

Theodorou, C. (1971). Introduction of mycorrhizal fungi into soil by spore inoculation of seed. *Australian Forestry*, **35**, 23–6.

Theodorou, C. & Bowen, G.D. (1970). Mycorrhizal responses of radiata pine in experiments with different fungi. *Australian Forestry*, **34**, 183–91.

Theodorou, C. & Bowen, G.D. (1973). Inoculation of seeds and soil with basidiospores of mycorrhizal fungi. *Soil Biology and Biochemistry*, **5**, 765–71.

Trojanowksi, J., Haider, K. & Hutterman, A. (1984). Decomposition of ^{14}C-labelled lignin, holocellulose and lignocellulose by mycorrhizal fungi. *Archives of Microbiology*, **139**, 202–6.

Van Suchtelen, M.J. (1962). Mykorrhiza bij *Pinus* spp. in de tropen. *Meddelanden Landbouw Hogeschod Gent*, **27**, 1104–6.

Van der Westhuizen, G.C.A. & Eicher, A. (1987). Some fungal symbionts of ectotrophic mycorrhizae of pines in South Africa. *South African Forestry Journal*, **143**, 20–4.

Visser, S. (1995). Ectomycorrhizal fungal succession in Jack pine stands following wildfire. *New Phytologist*, **129**, 389–401.

Vogt, K.A., Grier, C.C., Meier, C.E. & Edmunds, R.L. (1982). Mycorrhizal role in net primary production and nutrient cycling in *Abies amabilis* ecosystems in Western Washington. *Ecology*, **63**, 370–80.

Yanai, R.D., Fahey, T.J. & Miller, S.L. (1995). Efficiency of nutrient acquisition by fine roots and mycorrhizae. In *Resource Physiology of Conifers – Acquisition, Allocation and Utilization*, ed. W.K. Smith & T.M. Hinckley, pp. 75–103. New York: Academic Press.

Young, H.E. (1940). Mycorrhiza and growth of *Pinus* and *Araucaria*. The influence of different species of mycorrhiza forming fungi on seedling growth. *Journal of the Australian Institute of Agricultural Science*, **6**, 21–5.

17 Effects of pines on soil properties and processes

Mary C. Scholes and Thomas E. Nowicki

17.1 Introduction

Vegetation and the environment are coupled, dynamic systems. Dokuchaev was one of the first researchers to recognize the strong interdependence of vegetation and soils (Joffe 1938). This is a critical aspect of the widely accepted model, refined by Jenny (1941), of pedogenesis as an interplay of climate, organisms, relief, parent material and time. Many studies have shown how the environment markedly influences vegetation structure and productivity (Clements 1916; Gleason 1939; Whittaker 1978; Shugart 1984; Roberts 1987), but there are fewer documented cases in which it can be shown how vegetation influences the environment. Pines are perhaps an exception to this, as the dynamic nature of the pine–soil relationship has been highlighted by interest in the factors accounting for their widespread distribution, their economic importance, and concern regarding the effects of natural as well as human-induced encroachment of pines into sensitive ecosystems.

In their natural range, pines appear to be particularly well adapted to marginal habitats where a combination of interrelated factors enables them to compete successfully with alternative vegetation types (Delcourt & Delcourt 1987; Richardson & Bond 1991; Chap. 1, this volume). A particularly prominent aspect of pine biogeography is the common association between pines and acidic nutrient-poor soils, a relationship exemplified in the vast boreal forests of the northern hemisphere (Wilde 1958; Fitzpatrick 1980; Delcourt & Delcourt 1987). However, pines are not restricted to these soil types. In many parts of the world, and particularly in the southern hemisphere, pine expansion and commercial forestry activities have led to the development of extensive pine forests in areas underlain by a wide variety of soils. Pine species are known to thrive in soils developed on a variety of parent materials

(including loess, fluvial sediments, dune sands, sandstones, granites, calcareous lithologies and ultramafic rocks) and with textures ranging from sandy to clay-rich (Mirov 1967).

This chapter reviews some of the important dynamic links between soils and pines. The first section describes the characteristic features of soils and nutrient-cycling processes typically associated with the old boreal forests of the northern hemisphere, an environment in which podzolization plays a dominant role. The second section deals with the impacts of pine afforestation in areas previously covered by other vegetation types. Such impacts are evident in areas recently colonized by pine (e.g. parts of southwestern USA, Central America, and eastern Asia), and where pine plantations have been established in environments which differ greatly from habitats in the natural range of pines.

17.2 Characteristic soil properties and processes associated with pines

17.2.1 Soil types and processes

In the northern latitudes, pines typically exist in dynamic equilibrium with podzolic soil types which are commonly associated with heath-type vegetation and coniferous forests in temperate climatic zones (Tamm 1950; Wilde 1958; Buurman 1984). Although these species are not restricted to areas underlain by soils which can be defined as podzols *sensu stricto* (Fig. 17.1), the characteristic features of podzols and the podzolization process itself epitomise pedogenic processes associated with pine forests in

Ecology and Biogeography of Pinus, ed. D.M. Richardson. © Cambridge University Press, Cambridge (1998), pp. 341–53.

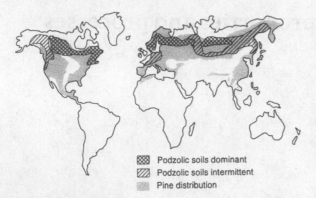

Podzolic soils dominant
Podzolic soils intermittent
Pine distribution

Fig. 17.1. **Schematic map showing the global distribution of podzols (from Fenwick & Knapp 1982) in relation to the natural distribution of pines (from Critchfield & Little 1966).**

temperate as well as Mediterranean climates. It is therefore worthwhile to review some of the main aspects of podzols and podzolization processes, and their relationship with pines.

On world soil maps, podzols are shown as extending in a circumpolar belt approximately from the Arctic Circle southwards to 50° N in Eurasia, and slightly further south, to the latitude of the Great Lakes, in North America (Fig. 17.1). South of this main circumpolar zone, the distribution of podzols is far more restricted, and they occur only in areas with a combination of parent material, climate and vegetation favourable for their formation (Buurman 1984). Latitudes north of the Arctic Circle are dominated by tundra soils which typically support mostly mosses, lichens and a few grass species, but very few trees (Fitzpatrick 1986). In contrast to podzolic soils, very cold temperatures and restricted drainage due to permafrost significantly retard pedogenic processes in tundra soils, resulting in immature organic-rich soils and limited horizonation (Fenwick & Knapp 1982).

Podzolic soils are characterized by the presence, just below the organic upper soil horizon, of a bleached quartz-rich horizon which is severely depleted of clay minerals and the oxides and hydroxides of iron and aluminium (sesquioxides). In podzols *sensu stricto*, this bleached E horizon is underlain by a dark, often highly indurated, blackish-brown B horizon characterized by an accumulation of sesquioxides in intimate association with high concentrations of humic substances (Fitzpatrick 1980). Slow litter decomposition and restricted faunal activity in the organic upper soil layers lead to the formation of *mor* humus, characterized by distinct horizonation into an upper undisturbed litter layer, a partly decomposed fermentation horizon, and a lower humus horizon (Fenwick & Knapp 1982). The characteristic E and B horizons of podzols are formed primarily as a

result of the downward percolation of dissolved humic substances (humic and fulvic acids) generated by slow decomposition processes in this mor humus layer. The humic substances form mobile complexes with iron and aluminium, as well as base cations (primarily calcium and magnesium), and together with natural carbonic acid facilitate the dissolution and removal of clay minerals (Bloomfield 1955; Ugolini *et al.* 1977). The mobilized metals are leached out of the upper mineral soil and are deposited lower down in the soil profile, along with the associated humic substances, resulting in the formation of the ferrihumic B horizon. When prevalent over long time periods, this process results in substantial acidification and depletion of nutrients in the upper soil horizons.

Podzols typically develop in environments which favour the generation and mobility of dissolved humic substances. They occur primarily in temperate and Mediterranean climates and to a large extent are restricted to quartzitic parent materials. In soils with a moderate to high clay content, the effects of podzolization are limited due to the formation of clay–humus complexes that severely restrict humic substance mobility. Although podzols are by no means restricted to areas underlying coniferous forests, they are particularly prevalent in these areas, and it is likely that in these environments the vegetation type, and in particular the nature of its litter, plays an important role in podzolization processes (e.g. Buurman 1984 and references therein). Several studies have demonstrated the podzolizing effect of pines (Ball & Williams 1974; Fenwick & Knapp 1982; Miles 1985; James & Wharfe 1989; Wardenaar & Sevink 1992). This probably results from the refractory polyphenol-rich nature of pine litter which decomposes slowly and promotes the generation of mor humus as well as humic and fulvic acids (Kononova 1966; Schnitzer & Kahn 1979). It should be emphasised, however, that podzolization is by no means uniquely associated with pines. Other coniferous species, as well as heath vegetation, are also closely associated with podzols, both spatially and genetically.

Although the natural distribution of pines generally reflects their adaptation to acidic, nutrient-poor soil conditions, a comparison of pine and soil global distribution patterns (Fig. 17.1) clearly indicates that large areas of pine forest are not underlain by podzols. These include western and southeast USA, Central America, portions of southern Europe and Southeast Asia. Even in the presence of pines, the effects of podzolization are overshadowed in these areas, mainly due to the influence of climatic factors and/or parent materials. In the dry climates of the western USA and parts of southern Europe, low precipitation limits leaching and weathering processes. As a result, soils in these regions typically display limited horizonation and their properties are to a large extent determined by the underlying parent material. In general, they are neutral or alkaline and have relatively high base status. In contrast to this, regions with warm humid clim-ates (e.g. southeast USA, Central America and Southeast Asia) are generally dominated by deep, highly weathered and leached soils which are typically acidic and poor in nutrients. Rapid decomposition and mineralization of organic matter severely limits the generation of mobile humic sub-stances, and leaching is dominated by inorg-anic, non-pod-zolizing processes.

17.2.2 Nutrient cycling

Tree growth in the cold conditions prevailing in the boreal forests of the northern latitudes (i.e. the taiga) is slow. Long-term fertilization trials suggest that nutrients in these systems are transferred mainly between the decom-posing litter, the endomycorrhizal roots and the trees, without ever passing through the mineral soil (Ovington 1962; Switzer & Nelson 1972). In addition, it appears that pines are able to conserve nutrients effectively by invest-ing low amounts of nutrients per unit of leaf and bole-wood production, and by returning low amounts of nutrients in litterfall relative to above-ground tissue (Bockheim & Leide 1991). Therefore, in the absence of significant logging, nutrient losses in pine ecosystems are small.

Fire is a crucial aspect of pine ecosystems and in the nutrient-poor systems of the *taiga* it plays a major role in nutrient-cycling processes (Chap. 11, this volume). Combustion of forest litter removes acids (primarily in the form of CO_2 and H_2O) and liberates mineral bases and nutrients for uptake by the regenerated vegetation. However, an increasing amount of evidence suggests that pines, via their mycorrhizal fungi, are capable of efficient nutrient uptake directly from soil organic matter, i.e. without intermediate mineralization by independent soil microbes, the activity of which is often severely suppressed in acid soils (Chap. 16, this volume). Furthermore, a recent study by Northup et al. (1995) sug-gests that high concentrations of polyphenolics, devel-oped in pine litter in acid environments, enhance concentrations of dissolved organic nitrogen at the expense of mineral nitrogen forms (Fig. 17.2), thereby enhancing their competitive edge over other species which are unable to utilize organic nitrogen effectively. As discussed in the previous section, by reducing rates of forest litter decomposition, polyphenolics reinforce the podzolization process, and in this way may further limit competition from species which are less tolerant of acid, nutrient-poor soils.

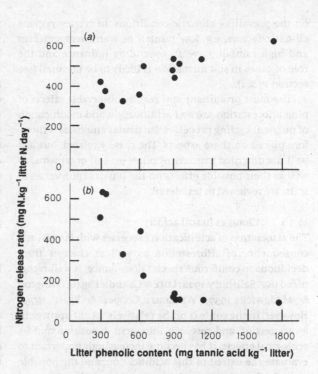

Fig. 17.2. *Pinus muricata* Oa horizon (moderate to highly decomposed organic matter) litter nitrogen release versus concentration of total phenolics. (*a*) dissolved organic nitrogen; (*b*) mineral nitrogen ($NH_4^+ + NO_3^-$). The litter was sampled from under monospecific cluster pine in three contrasting soil acidity/fertility conditions on the Ecological Staircase, near Mendocino, California. Reproduced, with permission, from Northup et al. (1995).

17.3 Soil changes associated with pine afforestation

This section describes some of the important changes in soil properties and processes which occur when pines colo-nize or are planted in areas previously covered by other vegetation types. This is particularly relevant in the south-ern hemisphere where pine expansion, or the develop-ment of commercial pine plantations, occurs on soils which are often very different from the podzolic soils characteristic of the natural range of pines in temperate climates. Certain processes, such as soil maturation during interglacials, may result in the evolution of soils from conditions favouring coniferous vegetation to those in which angiosperms become dominant (Kershaw & McGlone 1995). However, in general, the establishment of coniferous vegetation tends to enhance acidity and reduce nutrient availability (Miles 1985; Ogden & Stewart 1995), and in this sense tends to drive soils towards conditions prevailing in the boreal forests of the northern hemi-sphere. The extent to which this occurs specifically as a result of pine afforestation will depend on the relative importance of other soil-forming factors, and in particular on the prevailing climatic conditions. In certain regions climatic factors, e.g. low rainfall or warm temperatures and high rainfall, have an overriding influence and the role of pines in soil formation is likely to be minimal (see section 17.2.1).

The most prominent and readily observable effects of pine afforestation are soil acidification and modification of nutrient-cycling processes. Particular emphasis is therefore placed on these aspects. The closely related, but less well documented impacts of pines on soil organisms, as well as their possible effects on the physical properties of soils, are reviewed in less detail.

17.3.1 Changes in soil acidity

The association of acidification processes with forests, as a consequence of afforestation as well as changes from deciduous to coniferous species dominance, is well recognized (e.g. Salisbury 1922; Lutz & Chandler 1946; Ovington & Madgwick 1957; Williams, Cooper & Pyatt 1979). However, in the context of the relatively recent upsurge in awareness of, and research into, acid deposition and its ecological effects, it has become increasingly important to evaluate the extent of this acidification, and the possible contribution that it makes to the observed widespread acidification of surface waters in northern Europe and America. The naturally acidic, poorly buffered soils that typically occur under pine forests are particularly sensitive to further acid inputs. It is therefore perhaps not surprising that the effects of acid deposition, on both terrestrial and aquatic ecosystems are most prominent in the forested regions of northern Europe and the northeastern United States. However, due to the spatial, and in many cases temporal, correlation between air pollution impacts and afforestation in these regions, it is difficult to determine to what extent acidification processes associated with afforestation *per se* directly contribute to the observed ecological effects.

Soil acidification has been intensively researched, particularly with respect to its effect on plant growth and the ecological impacts of acid deposition (e.g. Kennedy 1986; Reuss & Johnson 1986; Ulrich & Sumner 1991). It is a naturally occurring process associated with soil formation, and is caused primarily by the removal from the soil of base cations, liberated by hydrolytic exchange and dissolution reactions, and their replacement with the acidic cations H^+, Al^{3+} and to a lesser extent Fe^{3+} and Mn^{2+} (Sposito 1989; Fey, Manson & Schutte 1990). The main driving force of soil acidification is the addition of protons which displace base cations from the soil exchange complex, and drive hydrolytic dissolution reactions. In the absence of atmospheric pollution, the dominant sources of protons are carbonic and natural organic acid dissociation, soil nitrogen transformations (mineraliza-

tion, nitrification), and excess cation uptake by biomass with accompanying release of protons (Bredemeier, Matzner & Ulrich 1990). In stable ecosystems, biomass uptake and nitrogen transformations are balanced by acid-consuming or neutralizing processes, i.e. the return of dead biomass with accompanying base cations to the soil, denitrification, and nitrogen uptake by plants (Binkley & Richter 1987; Ulrich 1991). Permanent acidification of soils occurs only when there is a net export of basic components, i.e. the solubilized basic cations, out of the soil. Thus, net acidification can only occur by cation leaching, or if natural biogeochemical cycles are disrupted in such a way that proton generating processes are uncoupled from those resulting in proton consumption (Ulrich 1986; van Breemen 1991).

Soil acidity is measured in terms of intensity and capacity factors. Soil pH is the most important intensity factor, and reflects the concentration of H^+ in aqueous solutions that are in equilibrium with the soil solid phase. Capacity factors represent stored acidity or alkalinity and reflect the extent to which the soil can accommodate added acids or bases without associated changes in pH, i.e. they reflect the soil's buffer capacity. Commonly measured capacity factors include acid neutralizing capacity, exchangeable acidity, and base saturation (the ratio of base cations to acid cations on the soil exchange complex). Strictly speaking, acid soils are those with pH values < 7, but problems associated with acidification generally occur only when the soil pH drops below $c.$ 5.5 (Sumner, Fey & Noble 1991) and/or when total exchangeable acidity exceeds $c.$ 15% of the soil cation exchange capacity (Sposito 1989). The ecological effects of soil acidification arise primarily because under very acidic conditions (i.e. at pH below 5) Al^{3+} is liberated by mineral (e.g. gibbsite or kaolinite) dissolution and dissociation of Al–organic matter complexes (Neal *et al.* 1987; Mulder & Stein 1994), resulting in significantly enhanced exchangeable aluminium concentrations. In the presence of accompanying strong acid anions (e.g. Cl^-, NO_3^-, SO_4^{2-}), high concentrations of exchangeable Al^{3+} lead to significantly elevated and possibly toxic aluminium concentrations in the soil solution (Reuss & Johnson 1986; Adams, Ali & Lewis 1990). In addition, under these conditions the availability of Ca^{2+} and Mg^{2+} to plants is severely limited due to the replacement of exchangeable bases by H^+ and Al^{3+} (Sumner *et al.* 1991; Sverdrup, Warfvinge & Jönsson 1993). Other potentially phytotoxic elements such as manganese and boron may be solubilized under acidic conditions and may contribute to the plant stresses already mentioned (Sumner *et al.* 1991). In addition to these effects on terrestrial ecosystems, soil acidity can have a major impact on the chemistry and consequently the quality of associated surface waters (Reuss & Johnson 1986; Cresser & Edwards 1987).

Table 17.1. *Potential sources of acidity in forest ecosystems (adapted from Bredemeier et al. 1990). The first five sources may arise out of natural processes, whereas the last three are associated with atmospheric deposition of acidifying pollutants*

Proton source	Generalized reactions
Assimilation of surplus inorganic cations over anions in organic matter	$R\text{-}OH + M^+ \rightarrow R\text{-}OM + H^+$ (M^+ = nutrient cation)
Generation and protolysis of organic acids	$R\text{-}OH \rightarrow R\text{-}O^- + H^+$ (pK_a ~2.5–5.5)
Preferential uptake of NH_4^+ over NO_3^-	$R\text{-}OH + NH_4^+ \rightarrow R\text{-}NH_2 + H_2O + H^+$ (H^+ source) $R\text{-}OH + H^+ + NO_3^- \rightarrow R\text{-}NH_2 + H_2O$ (H^+ sink)
Carbonic acid dissociation	$CO_2 + H_2O \rightarrow H^+ + HCO_3^-$ (pK_a ~ 6.3)
Disruption of ecosystem-internal nitrogen cycle, net nitrification and nitrate leaching	$R\text{-}NH_2 + O_2 \rightarrow R\text{-}OH + H^+ + NO_3^-$
Buffering of deposited H^+ in the forest canopy with subsequent restoration of foliar buffer capacity and H^+ excretion to the soil	
Nitrification of deposited NH_4^+	$NH_4^+ + 2O_2 \rightarrow 2H^+ + NO_3^- + H_2O$ (H^+ source)
Direct flux of free H^+ to the soil in throughfall	$H_2SO_4 \rightarrow 2H^+ + SO_4^{2-}$ $HNO_3 \rightarrow H^+ + NO_3^-$

The most important potential sources of acidity in forest ecosystems in general and pine forests in particular, are listed in Table 17.1. It is clear that, in the absence of acid deposition, the acidifying effect of pine forests is primarily associated with leaching of organic acids generated in slowly decomposing pine litter, and with the phytocycling of nutrients (and accompanying protons), those of particular importance being nitrogen, sulphur, and the base cations (primarily Ca^{2+} and Mg^{2+}). More specifically, the disruption of these nutrient cycles by vegetational changes and forest management practices are the prime causes of soil acidification. The relative importance of these two sources depends on factors which may limit the generation or mobility of organic acids (e.g. high clay content) and on those which influence the degree to which nutrient cycles are disrupted (e.g. forest management practices).

A distinction can be made between acidification associated with 'natural' colonization of areas previously covered by other vegetation types, and that associated with commercial forestry. A number of acidification processes may operate in the former case. Net accumulation of base cations in biomass occurs temporarily during the aggradation phase of forest development and may result from retarded litter decay processes and resultant net litter accumulation on the forest floor (Krug & Frink 1983; Richter 1986). Depending on the nature of the original vegetation, the addition of pine litter may result in enhanced leaching of organic acids. Preference for ammo-

nium-nitrogen (i.e. high NH_4^+/NO_3^- uptake ratios) may contribute to acidification by significantly enhancing the net uptake of positively charged ionic species with associated release of protons into the soil (Arnold 1992). The influence of these processes is particularly critical when pines colonize areas underlain by naturally acidic soils which support fragile, low-biomass ecosystems (e.g. grassland and heath). In the case of commercial pine plantations, export of nutrient cations in harvested biomass and rapid oxidation of forest litter following clearfelling (with associated nitrate leaching; Bormann & Likens 1979; Mitchell *et al.* 1989), intensifies acidification processes significantly. After several rotations, even originally well-buffered soils can be acidified dramatically and depleted of mineral nutrients as a result of these processes. The nutrient depletion and acidification processes impact primarily on the upper mineral soil horizons (to depths of 10 to *c.* 60 cm), with a significant proportion of the removed nutrients ultimately accumulating in thick surface litter layers (Billett, Fitzpatrick & Cresser 1993).

Recent studies provide substantial evidence of extensive acidification by pines (and other forest species). These studies can generally be classified into two broad groups: those based on historical evidence, and those based on the measurement and calculation of proton budgets.

The historic approach involves evaluating changes in soil properties with time, and is based either on comparisons of soil chemical data obtained for single forested areas over lengthy time periods, or on a comparison of data for adjacent forested and unforested areas of similar climate, parent material, and topography (the 'modified historic approach'). Numerous historical studies in northern Europe and North America (e.g. Johnson *et al.* 1991 and references cited therein), as well as in the southern hemisphere (e.g. Morris 1986; Du Toit & Fey 1996), have demonstrated significant decreases in the pH and base status of topsoil (with associated increases in exchangeable acidity) as a result of forestry (Table 17.2). The highly variable nature of results obtained from these studies stems from a number of factors, including differences in tree species, climate, topography, parent materials and pre-afforestation acidity status. Examples of data generated by proton budget studies are given in Table 17.3. These show that in acid soils, proton additions associated with net biomass uptake of nutrient cations can constitute a significant proportion of the total acid load, even in areas which are subject to very high rates of acid deposition (e.g. north-western Germany, northeastern USA).

The data in Tables 17.2 and 17.3 suggest that, in terms of changes in pH and base status, and in terms of proton budgets, acidification associated with pine species is not significantly more intense than that associated with many other forest species, and in some cases other species can

Table 17.2. Examples of changes in pH and exchangeable base cation concentration (EBC) as determined by historical studies on forest soils

Locality	Species	Years	ΔpH	ΔpH yr⁻¹ (×10)	pH$_{final}$	ΔEBC(%)	Reference
W USA	*Pinus radiata*	c. 100	1.5	0.15	c. 5.5[a]	–	Amundson & Tremback (1989)
USA	*Eucalyptus globulus*	6	1.0	1.67	4.9[a]	–	Johnson et al. (1991)
Canada	*Pinus resinosa*	45	0.7	0.15	5.3[a]	–	Brand, Kehoe & Connors (1986)
SE USA	*Pinus taeda*	20	0.3–0.8	0.15–0.4	–	30–80	Binkley et al. (1989)
South Africa	*Pinus patula*	>45	0.34	0.07	4.02[b]	76	du Toit (1993)
South Africa	*Acacia mearnsii*	>45	0.54	0.12	3.78[b]	42	du Toit (1993)
South Africa	*Eucalyptus* spp.	>45	0.29	0.06	4.25[b]	78	du Toit (1993)
Swaziland	*Pinus patula*	30	0.15	0.05	3.79[b]	~50	Morris (1986)

Notes:
[a] Medium for pH measurement unspecified.
[b] pH measured in 1M KCL.

Table 17.3. Examples of estimated rates of net annual proton inputs (in kmol ha⁻¹yr⁻¹) based on proton budget and flux calculations. Rates are given for proton production associated with: biomass cation uptake (int-H⁺); bicarbonate leaching (HCO₃⁻); and atmospheric deposition of acidity (ext-H⁺). The latter includes indirect proton additions associated with canopy leaching induced by acid deposition, and excess uptake of deposited NH₄⁺ relative to NO₃⁻

Locality	Species	int-H⁺	HCO₃⁻	ext-H⁺	Reference
Netherlands	mixed deciduous	1.5	–	3.0	Verstraten et al. (1990)
NW Germany	*Fagus* spp.	1.05	–	2.61	Bredemeier (1987)
NW Germany	*Picea* spp.	1.14	–	5.56	Bredemeier (1987)
NW Germany	*Pinus* spp.	0.68	–	1.80	Bredemeier (1987)
NW Germany	*Quercus* spp.	1.48	–	2.56	Bredemeier (1987)
NE USA	mixed hardwoods	1.0	0.1	1.4	van Breemen, Driscoll & Mulder (1984)
Sweden	*Pinus* spp.	1.0	0.7	1.7	van Breemen et al. (1984)
S USA	various	0.83–1.44	0.3–0.75	1.0	Richter (1986)

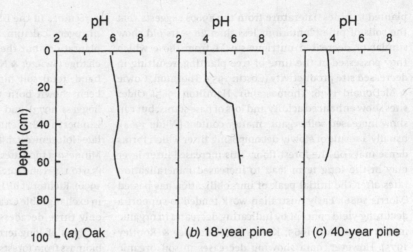

Fig. 17.3. **pH profiles for acid brown soils developed under oak woodland which has been partially replaced by pine at Simon's Copse, Surrey, England. (a) Undisturbed oak woodland; (b) 18-year-old pine; (c) 40-year-old pine. Reproduced, with permission, from Fenwick & Knapp (1982).**

have very dramatic effects (e.g. *Eucalyptus globulus*, Table 17.2). The relative effect of different tree species will depend on factors such as tree growth rate, relative accumulation rates of base cations, degree of nitrogen fixation, the ash alkalinity of tree litter, and the spatial distribution of roots. These factors, particularly growth rate, are in turn dependent to varying degrees on 'external' variables such as climate. Nonetheless, all other factors being equal, tree species which accumulate base cations (particularly calcium) to a greater extent than pines (e.g. *Populus* and *Quercus* species) (Alban 1982; Johnson & Todd 1987), and species which are capable of fixing large amounts of nitrogen (e.g. certain *Eucalyptus* and *Acacia* species, and *Alnus rubra*) (Van Miegroet & Cole 1985; Johnson *et al.* 1991) will have a more prominent effect than pine species on the acidity and nutrient status of underlying mineral soil. This may not always be the case, however. Fenwick & Knapp (1982) report a dramatic decrease in soil pH together with incipient podzolization following partial replacement of a *Quercus* woodland, by *Pinus* and *Larix* (Fig. 17.3). The oak woodland is underlain by sandy acid brown soils (original pH of 6). It appears, therefore, that in sandy soils with limited buffering capacity, the nature of the pine litter may play a prominent role, with the leaching of organic acids generated in the mor horizon leading to podzolization and acidification of the underlying mineral soil. These processes are likely to be less prominent under broadleaved forest species such as oaks.

It has been demonstrated that the addition of organic matter can ameliorate soil acidity, and that the extent to which this occurs is proportional to the ash alkalinity of the tree litter (Noble & Randall 1996). Thus, the low ash alkalinity of pine litter (Noble & Randall 1996), as well as the general lack of bioturbation in the acidic forest floor (Fenwick & Knapp 1982), may further contribute to soil acidity by limiting the extent to which the acidifying effect of net biomass uptake of base cations is neutralized by the reincorporation of alkalinity in the form of organic matter.

17.3.2 Changes in nutrient availability and cycling

One of the key aspects distinguishing forests from other vegetation types is the nature of their nutrient-cycling processes. Forests in general, and particularly pine forests in marginal environments, rely heavily on internal translocation and tight cycling between the forest floor and the trees (section 17.2). Thus, a prominent effect of the introduction of pine forests is a significant modification of the distribution, fluxes and to some extent the chemical forms of the major nutrient elements, i.e. nitrogen, sulphur and the base cations (primarily Ca, Mg and K). It has been suggested that changes in soil nitrogen forms and the various soil organic matter pools may be some of the first soil alterations following afforestation (Fisher & Eastburn 1974; Heng & Goh 1984; Turner & Lambert 1988).

Several studies have addressed changes in soil carbon and other nutrient stocks after afforestation of grasslands and the conversion of a range of vegetation types to pine plantations. Extensive studies have been undertaken in Australia and New Zealand where large areas have been planted to pines. Literature from the 1960s suggests that the soils of pine plantations less than 20 years old show similar or decreased nutrient stocks from those which they possessed at the time of tree planting, resulting in decreased site productivity (Costin 1953; Thornton, Cowie & McDonald 1956; Thomas 1961; Hamilton 1965). Older sites show enhanced acidity and loss of base status, but can show increased soil organic matter content (Wilde 1964), usually a result of slowly decomposing litter which forms dense mats on the forest floor. This increased litter layer may in the long term lead to increased mineralization rates after the initial peak of immobilization has passed (Morris 1986). Early Australian work tended to support a declining-yield concept by indicating decreases in organic matter (Hamilton 1965; Keeves 1966; Routley & Routley 1975). However, data showing decreases in soil organic matter are usually from soil samples taken shortly after planting where large losses have occurred as a result of enhanced microbial respiration. These initial losses may not markedly affect the cycling of nutrients over long rotation times (Turner & Kelly 1985). Extensive studies in the 1980s in Australia and New Zealand showed that the planting of pines resulted in a redistribution of nutrients within the ecosystem, including increased CO_2 evolution rates and changed proportions of nitrate and ammonium availability. If harvest losses are accounted for, however, then the changes in total quantities of nutrients in the system are small (Switzer & Nelson 1972; Alban 1982; Turner & Kelly 1985).

Recent studies undertaken in the high country of South Island, New Zealand, show that planting of conifers in denuded grasslands can lead to increased productivity. Although the effects are variable, topsoils (10 cm depth) under conifer plantations in this region are generally enriched in inorganic and plant-available phosphorus (total P was not significantly affected) relative to equivalent grassland soils (Davis & Lang 1991; Belton, O'Conner & Robson 1995). Mineralizable nitrogen concentrations are significantly enhanced under young ($\leq$ 20 years old) pine plantations in relatively wet environments (1000–1300 mm rainfall per annum), whereas the reverse was observed for similar soil types under older pine stands in drier environments (500–600 mm rainfall yr^{-1}) (Davis & Lang 1991). Enhanced mineralizable N is generally accompanied by moderate decreases in organic carbon concentration, and may be caused by enhanced mineralization of residual grassland organic matter under the young pine stands (Fisher & Stone 1969; Davis & Lang 1991). However, plant-available phosphorus concentrations were found to be consistently elevated in older ($\geq$20-year-old) plantations, suggesting that other processes, such as nutrient transfer from the subsoil, may be involved (Davis & Lang 1991; Belton et al. 1995).

Studies in the USA indicate that the redistribution of nitrogen, calcium and magnesium occurs under pine plantations but the total amounts in the system do not change (Switzer & Nelson 1972; Alban 1982). On the other hand, total soil nitrogen and exchangeable bases have been shown both to increase (Richards 1962; Rolfe & Boggess 1973) and to decrease under pines (Turner & Lambert 1988; Richter et al. 1994). A study of the effects of late Holocene establishment of jack pine on prairie soils in Minnesota indicates significant decreases in soil organic C (5.9 to 1.4%) and total soil N (0.220 to 0.034%) (Almendinger 1990). Richter et al. (1994) document significant decreases in exchangeable Ca and Mg (down to depths of 60 cm) after only three decades under loblolly pine (Pinus taeda). An analysis of long-term (120 years) data on soil and plant biomass from forests in the eastern USA (Federer et al. 1989) has shown that atmospheric inputs balance the outputs with respect to nitrogen dynamics. For potassium, magnesium and phosphorus the total pool may decrease by 2–10% in 120 years, depending on the site and the harvest intensity. Calcium losses by leaching and harvest could reduce total soil and biomass Ca by 20–60%. It is suggested that although calcium deficiency does not generally occur now in acid forest soils, it seems likely in future if anthropogenic leaching and intensive harvest removal continue (Federer et al. 1989).

Most of the studies referred to above deal with the effects of replacing grassland or other relatively low biomass plant communities by pines. The relative effect of different forest tree species on soil nutrient status is less well documented, but several studies indicate that the effect of pines on exchangeable Ca and to a lesser extent Mg concentrations in the mineral soil is significantly less dramatic than that of trees which accumulate calcium to greater degrees, e.g. certain species of Picea, Populus and Quercus (Alban 1982; Johnson & Todd 1987). However, as pointed out by Alban (1982), the relative effect of different tree species is unlikely to be constant as it depends to a large extent on their growth rates (and other factors – see last paragraph, section 17.3.1), which in turn are determined by factors such as climate, soil properties and forest management practices.

Although these data indicate that the effects of pines on nutrient pools and fluxes are highly variable, some generalizations can be suggested:

- The introduction of pines typically results in the depletion of upper mineral soil horizons of nutrients, and their accumulation in overlying organic horizons. Litter accumulation is related to a number of factors and climate appears to exert a particularly important influence. The chemical nature of the litter itself, however, and particularly high concentrations of lignin and other phenolic compounds, are likely to play a critical role. In addition to the immobilizing effect on nutrients, enhanced generation of humic and fulvic acids in thick, slowly decomposing pine litter increases the potential for podzolization under newly established pine forests. The ultimate effect of these processes is to reinforce the importance of the unique nutrient-cycling processes utilized by pines (section 17.2).

- Because nitrogen is typically the limiting nutrient in pine ecosystems, the effects of pines on soil nitrogen are likely to be particularly significant. Nitrogen mineralization and nitrification is generally reduced under pines owing to low pH conditions, poorer substrate quality and possibly raised polyphenol concentrations (section 17.2.2). Thus one would expect the introduction of pines to cause a shift in dominant available nitrogen form from nitrate to ammonium, and from mineral to predominantly organic.

- In commercial pine plantations, high growth rates and export of harvested biomass lead to significant reductions in the total size of nutrient pools.

17.3.3 Changes in soil organisms

Decomposition of organic matter is regulated by inter-actions between the substrate quality, the biota and the microclimate. Temperature and soil moisture have a major impact on soil microflora activity (Swift, Heal & Anderson 1979). Mutual relationships among the biota, especially between the microflora and their invertebrate grazers, have also been reported to be extremely important in affecting the degradation process. In addition, rates of decomposition may be influenced by interspecific rela-tionships within the faunal component (Anderson & Ineson 1984). Several decomposition stages have been rec-ognized on fallen pine needles, each being associated with the activity of animal and microbial organisms. Some studies have been carried out on the characterization of the faunal communities under coniferous forests.

To the well-known fungal successions that have been so far described by mycologists must be added nematodes, amoebae, enchytraeids, sciarid larvae, oribatid mites and earthworms (Ponge 1991). Coniferous forests tend to be dominated by Collembola, mites and woodlice and to be low in those organisms requiring large amounts of calcium (Hagvar 1984a, b; Vilkamaa & Huhta 1986).

The range of organisms found is influenced by the methodology used. Nylon mesh bags with a mesh size of less than 1 mm will lead to a marked underestimation in numbers and diversity (Ponge 1991). The soil faunal verti-cal distribution and the age structure of the population are mainly influenced by soil moisture. In dry periods higher frequencies of Collembola can be expected in the deeper soil layers (Faber & Verhoef 1991). The effect of acid-ification of forest soils on soil arthropod communities has been frequently studied. Mild acidification causes a posi-tive reaction in the density of Collembola, mostly due to an increase in a few acidophilic species (Hagvar 1984a), while acid-sensitive species decrease in numbers. Other workers have shown that acidification decreased the number of Collembola (Heungens & Van Daele 1984). Among the mites, species react differently to acidification with both positive and negative reactions (Hagvar & Amundsen 1981; Heungens & Van Daele 1984).

The balance between the rates of mineralization and immobilization determines the amounts of inorganic nutrients that go into soil solution. These may then be taken up by the plant roots or lost to leaching. Soil fauna, via enhanced nutrient mobilization and favourable changes in the structural soil properties, exert a positive influence on plant growth (Setala & Huhta 1991). Microbial growth in pine litter is restricted mainly by the nitrogen availability in each of the decomposing layers whereas in *Populus* forests nitrogen deficiency declines in the later stages of decay with phosphorus deficiency increasing (Parmelee, Ehrenfeld & Tate 1993). Nutrient cycling in soils and litter is influenced mainly by the chemical composi-tion of the litter and by the ecological characteristics of the dominant species, and not simply as a function of trophic group or biomass.

It has been suggested that some soil fauna may be used as indicator species to monitor the vitality of a forest. In the mineral soil horizons, pine roots enhanced the activity of the microbial component, and nematode and micro-arthropod abundances increased at higher root densities (Scheu & Parkinson 1995). Experiments in Europe have addressed changes in arthropod communities due to enhanced nitrogen inputs from pollution and subsequent soil acidification. An increase in the number of mites was observed with a decrease in the number of Collembola (Faber & Verhoef 1991; Hogervorst, Verhoef & van Stralen 1993). In Sweden microbial biomass has been shown to decrease with an increase in soil nitrogen. Fungal fruit-body production and mycorrhizal infection frequency and composition have been shown both to increase and to decrease after forest fertilization, the response being species-specific (Arnebrant, Baath & Sonderstrom 1990). It is difficult to separate cause and effect; it is unclear whether it is the nitrogen fertilizer which is affecting the microbial populations directly, or whether this is due to secondary effects such as changes in pH or patterns of root growth.

The search for the universal indicator species may be a fruitless one. This is partly because of the lack of taxonomic information and the very limited data base of macro- and microfauna in forested soils. It has been suggested that the approach should be to collect as much information as possible about soil organisms in a range of forest types and ecosystems. This information should not be solely a taxonomic exercise. The knowledge of the species name should be linked to its functionality in the system. It may be the relative proportions of functional groups in various systems that will allow us to quantify change and to make predictions as to whether that degree of change would lead to uncoupling of vital processes.

17.3.4 Changes in soil physical properties

Numerous studies have documented the physical changes associated with the establishment of a plantation (e.g. Zwolinski, Donald & van Laar 1993). Less information is available on the changes to soil physical properties as a direct result of pine growth. Available data do not show distinct trends and appear to be influenced mostly by management practices (Page 1968; Rolfe & Boggess 1973; Turner & Lambert 1988; Ohta 1990). This is especially true for data on soil erosion and runoff. Bulk densities and hydraulic conductivities have been shown both to improve and to deteriorate under pine forests (Rolfe & Boggess 1973; Ohta 1990). Most of these effects were evident only in the top 5 cm of the soil. In areas influenced by fire, pine afforestation may have an indirect effect on soil physical properties due to the enhanced fire intensity associated with significant biomass increases. In particular, the very hot fires associated with burning pine plantations can greatly reduce soil wettability, resulting in increased rates of surface runoff and soil erosion (Scott & Van Wyk 1990).

17.4 Conclusions

It is clear from the above account that soils play a prominent role in the ecology of pines. The adaptation of pines to acidic and nutrient-poor soil conditions suggests that soil properties, particularly soil acidity and nutrient status, are major determinants of pine distribution. However, a number of processes associated with pine growth and nutrient cycling have substantial impacts on their host soils. These generally lead to acidification and nutrient depletion, which in turn provide pines with a competitive advantage over alternative vegetation types. Consequently, it is in many cases very difficult to quantify the extent to which pine distribution is determined by soil properties (or covariant, closely related factors such as climate and parent materials), as opposed to soil properties being determined by processes specifically related to pine growth. This is particularly relevant with respect to the argument that current vegetation theory and analyses overemphasize environmental determination of vegetation composition and neglect the effects of vegetation on the environment (Roberts 1987).

References

Adams, W.A., Ali, A.Y. & Lewis, P.J. (1990). Release of cationic Al from acidic soils into drainage water and relationship to land use. *Journal of Soil Science*, **41**, 255–68.

Alban, D.H. (1982). Effects of nutrient accumulation by aspen, spruce, and pine on soil properties. *Soil Science Society of America Journal*, **46**, 853–60.

Almendinger, J.C. (1990). The decline of soil organic-matter, total-N, and available water capacity following late-Holocene establishment of jack pine on sandy mollisols, north-central Minnesota. *Soil Science*, **150**, 680–94.

Amundson, R.G. & Tremback, B. (1989). Soil development on stabilized dunes in Golden Gate Park, San Francisco. *Soil Science Society of America Journal*, **53**, 1798–806.

Anderson, J.M. & Ineson, P. (1984). Interactions between microorganisms and soil invertebrates in nutrient flux pathways of forest ecosystems. In *Invertebrate–microbial Interactions*, ed. J.M. Anderson, A.D.M. Rayner & D.W.H. Walton, pp. 59–88. Cambridge: Cambridge University Press.

Arnebrant, K., Baath, E. & Sonderstrom, B. (1990). Changes in microfungal community structure after fertilization of Scots pine forest soil with ammonium nitrate or urea. *Soil Biology and Biochemistry*, **22**, 309–12.

Arnold, G. (1992). Soil acidification as caused by nitrogen uptake pattern of Scots pine (*Pinus sylvestris*). *Plant and Soil*, **142**, 41–51.

Ball, D.F. & Williams, W.M. (1974). Soil development on coastal dunes at Holkham, Norfolk, England. *Transactions of the 10th International Congress of Soil Science*, Vol. **6**, 380–6.

Belton, M.C., O'Conner, K.F. & Robson, A.B. (1995). Phosphorus levels in topsoils under conifer plantations in Canterbury high country grasslands. *New Zealand Journal of Forestry Science* **25**, 265–82.

Billett, M.F., Fitzpatrick, E.A. & Cresser, M.S. (1993). Long-term changes in the nutrient pools of forest soil organic horizons between 1949/50 and 1987, Alltcailleach Forest, Scotland. *Applied Geochemistry*, Supplementary Issue No. 2, 179–83.

Binkley, D. & Richter, D. (1987). Nutrient cycles and H+ budgets of forest ecosystems. *Advances in Ecological Research*, **16**, 1–51.

Binkley, D., Valentine, D., Wells, C. & Valentine, U. (1989). An empirical model of the factors contributing to a 20-yr decrease in soil pH in an old-field plantation of loblolly pine. *Biogeochemistry*, **8**, 39–54.

Bloomfield, C. (1955). Leaf leachates as a factor in pedogenesis. *Journal of the Science of Food and Agriculture*, **6**, 641–51.

Bockheim, J.G. & Leide, J.E. (1991). Foliar nutrient dynamics and nutrient-use efficiency of oak and pine on a low-fertility soil in Wisconsin. *Canadian Journal of Forest Research*, **21**, 925–34.

Bormann, F.H. & Likens, G.E. (1979). *Pattern and Process in a Forested Ecosystem.* New York: Springer-Verlag.

Brand, D.G, Kehoe, P & Connors, M. (1986). Coniferous afforestation leads to soil acidification in central Ontario. *Canadian Journal of Forestry Research*, **16**, 1389–91.

Bredemeier, M. (1987). Quantification of ecosystem-internal proton production from the ion balance of the soil. *Plant and Soil*, **101**, 273–80.

Bredemeier, M., Matzner, E. & Ulrich, B. (1990). Internal and external proton load to forest soils in Northern Germany. *Journal of Environmental Quality*, **19**, 469–77.

Buurman, P. (1984). *Podzols.* Van Nostrand Reinhold Soil Science Series. New York: Van Nostrand Reinhold.

Clements, F.E. (1916). *Plant Succession. An Analysis of the Development of Vegetation.* Carnegie Institute of Washington Publication 242.

Costin, A.B. (1953). On coniferous forests in Australia. *Australian Forestry*, **17**, 21–5.

Cresser, M. & Edwards, A. (1987). *Acidification of Fresh Waters.* Cambridge: Cambridge University Press.

Critchfield, W.B. & Little, E.L. (1966). *Geographic Distribution of Pines of the World.* USDA Forest Service Miscellaneous Publication 991. Washington, DC: USDA Forest Service.

Davis, M.R. & Lang, M.H. (1991). Increased nutrient availability in topsoils under conifers in the South Island high country. *New Zealand Journal of Forestry Science*, **21**, 165–79.

Delcourt, P.A. & Delcourt, H.R. (1987). *Long-Term Forest Dynamics of the Temperate Zone: A Case Study of Late-Quaternary Forests of Eastern North America.* New York: Springer-Verlag.

du Toit, B. (1993). Soil acidification under forest plantations and the determination of the acid neutralizing capacity of soils. MSc thesis, University of Natal, Pietermaritzburg.

du Toit, B. & Fey, M.V. (1996). Soil acidification under forest plantations in South Africa. In *Abstracts of the 4th International Symposium on Plant–Soil interactions at low pH*, Belo Horizonte, Minas Gerais, Brazil.

Faber, J.H. & Verhoef, H.A. (1991). Functional differences between closely-related soil arthropods with respect to decomposition processes in the presence or absence of pine tree roots. *Soil Biology and Biochemistry*, **23**, 15–23.

Federer, C.A., Hornbeck, J.W., Tritton, L.M., Martin, C.W., Pierce, R.S. & Smith, C.T. (1989). Long-term depletion of calcium and other nutrients in Eastern US forests. *Environmental Management*, **13**, 593–601.

Fenwick, J.M. & Knapp, B.J. (1982). *Soils. Process and Response.* London: Duckworth.

Fey, M.V., Manson, A.D. & Schutte, R. (1990). Acidification of the pedosphere. *South African Journal of Science*, **86**, 403–6.

Fisher, R.F. & Eastburn, R.P. (1974). Afforestation alters prairie soil nitrogen status. *Soil Science Society of America Proceedings*, **38**, 366–72.

Fisher, R.F. & Stone, E.L. (1969). Increased availability of nitrogen and phosphorus in the root zone of conifers. *Soil Science Society of America Proceedings*, **33**, 955–61.

Fitzpatrick, E.A. (1980). *Soils. Their Formation, Classification and Distribution.* London: Longman.

Fitzpatrick, E.A. (1986). *An Introduction to Soil Science*, 2nd edn. Essex: Longman.

Gleason, H.A. (1939). The individualistic concept of the plant association. *American Midland Naturalist*, **21**, 92–110.

Hagvar, S. (1984a). Effects of liming and artificial acid rain on Collembola and Protura in coniferous forest. *Pedobiologia*, **24**, 255–96.

Hagvar, S. (1984b). Six common mite species (Acari) in Norwegian coniferous forest soils: Relations to vegetation types and soil characteristics. *Pedobiologia*, **27**, 355–64.

Hagvar, S. & Amundsen, T. (1981). Effects of liming and artificial acid rain on the mite (Acari) fauna in coniferous forest. *Oikos*, **37**, 7–20.

Hamilton, C.D. (1965). Changes in the soil under *Pinus radiata*. *Australian Forestry*, **29**, 275–89.

Heng, S. & Goh, K. (1984). Organic matter in forest soils and the mineralization of soil carbon and nitrogen. *Soil Biology and Biochemistry*, **16**, 201–2.

Heungens, A. & Van Daele, E. (1984). The influence of some acids, bases and salts on the mite and Collembola population of a pine litter substrate. *Pedobiologia*, **27**, 299–311.

Hogervorst, R.F., Verhoef, H.A. & van Stralen, N.M. (1993). Five-year trends in soil arthropod densities in pine forests with various levels of vitality. *Biology and Fertility of Soils*, **15**, 189–95.

James, P.A. & Wharfe, A.J. (1989). Time scales in soil development in a coastal sand dune system, Ainsdale, North-West England. In *Perspectives in Coastal Dune Management*, ed. F. van der Meulen, P.D. Jungerius & J. Visser, pp. 287–96. The Hague: SPB Academic Publishers.

Jenny, H. (1941). *The Factors of Soil Formation.* New York: McGraw-Hill.

Joffe, J.S. (1938). *Pedology.* New Brunswick.

Johnson, D.W & Todd, D.E. (1987). Nutrient export by leaching and whole-tree harvesting in a loblolly pine and mixed oak forest. *Plant and Soil*, **102**, 99–109.

Johnson, D.W., Cresser, M.S., Nilsson, S.I., Turner, J., Ulrich, B., Binkley, D. & Cole, D.W. (1991). Soil changes in forest ecosystems: evidence for and probable causes. *Proceedings of the Royal Society of Edinburgh B*, **97**, 81–116.

Keeves, A. (1966). Some evidence of loss of productivity with excessive rotations of *Pinus radiata* in the south-east of South Australia. *Australian Forestry*, **30**, 51–63.

Kennedy, I.R. (1986). *Acid Soil and Acid Rain.* Letchworth, Hertfordshire: Research Studies.

Kershaw, A.P. & McGlone, M.S. (1995). The quaternary history of southern conifers. In *The Ecology of the Southern Conifers*, ed. N.J. Enright & R.S. Hill, pp. 30–63. Carlton, Victoria: Melbourne University Press.

Kononova, M.M. (1966). *Soil Organic Matter*, 2nd edn. London: Pergamon Press.

Krug, E.C. & Frink, C.R. (1983). Acid rain on acid soil: a new perspective. *Science*, **221**, 520–5.

Lutz, H.J. & Chandler, R.F. (1946). *Forest Soils*. New York: John Wiley.

Miles, J. (1985). The pedogenic effects of different species and vegetation types and the implications of succession. *Journal of Soil Science*, **36**, 571–84.

Mirov, N.T. (1967). *The Genus Pinus*. New York: Ronald Press.

Mitchell, M.J., Driscoll, C.T., Fuller, R.D., David, M.B. & Likens, G.E. (1989). Effect of whole-tree harvesting on the sulfur dynamics of a forest soil. *Soil Science Society of America Journal*, **53**, 933–40.

Morris, A.R. (1986). Soil fertility and long term productivity of *Pinus patula* plantations in Swaziland. PhD thesis, University of Reading.

Mulder, J. & Stein, A. (1994). The solubility of aluminium in acidic forest soils: Long-term changes due to acid deposition. *Geochimica et Cosmochimica Acta*, **58**, 85–94.

Neal, C., Skeffington, R.A., Williams, R. & Roberts, D.J. (1987). Aluminium solubility controls in acid waters: the need for a reappraisal. *Earth Planetary Science Letters*, **86**, 105–12.

Noble, A.D. & Randall, P.J. (1996). Effects of tree leaf litter properties on soil acidity. In *Abstracts of the 4th International Symposium on Plant–Soil interactions at low pH*, Belo Horizonte, Minas Gerais, Brazil.

Northup, R.R., Yu, Z., Dahlgren, R.A. & Vogt, K.A. (1995). Polyphenol control of nitrogen release from pine litter. *Nature*, **377**, 227–9.

Ogden, J. & Stewart, G.H. (1995). Community dynamics of the New Zealand conifers. In *The Ecology of the Southern Conifers*, ed. N.J. Enright & R.S. Hill, pp. 81–119. Carlton, Victoria: Melbourne University Press.

Ohta, S. (1990). Initial soil changes associated with afforestation with *Acacia auriculiformis* and *Pinus kesiya* on denuded grasslands of the Pantabangan Area, Central Luzon, the Philippines. *Soil Science Plant Nutrition*, **36**, 633–43.

Ovington, J.D. (1962). Quantitative ecology and the woodland ecosystem concept. *Advances in Ecological Research*, **1**, 103–98.

Ovington, J.D. & Madgwick, H.A.I. (1957). Afforestation and soil reaction. *Journal of Soil Science*, **8**, 141–9.

Page, G. (1968). Some effects of conifer crops on soil properties. *Commonwealth Forestry Review*, **47**, 52–62.

Parmelee, R.W., Ehrenfeld, J.G. & Tate, R.L. III (1993). Effects of pine roots on microorganisms, fauna, and nitrogen availability in two soil horizons of a coniferous forest spodozol. *Biology and Fertility of Soils*, **15**, 113–19.

Ponge, J.F. (1991). Succession of fungi and fauna during decomposition of needles in a small area of Scots pine litter. *Plant and Soil*, **138**, 99–113.

Reuss, J.O. & Johnson, D.W. (1986). *Acid Deposition and the Acidification of Soils and Waters*. New York: Springer-Verlag.

Richards, B.N. (1962). Increased supply of soil nitrogen brought about by *Pinus*. *Ecology*, **43**, 538–41.

Richardson, D.M. & Bond, W.J. (1991). Determinants of plant distribution: evidence from pine invasions. *American Naturalist*, **137**, 639–68.

Richter, D.D. (1986). Sources of acidity in some forested Udults. *Soil Science Society of America Journal*, **50**, 1584–9.

Richter, D.D., Markewitz, D., Wells, C.G., Allen, H.L., April, R., Heine, P.R. & Urrego, B. (1994). Soil chemical change during three decades in an old-field loblolly pine (*Pinus taeda* L.) ecosystem. *Ecology*, **75**, 1463–73.

Roberts, D.W. (1987). A dynamical systems perspective on vegetation theory. *Vegetatio*, **69**, 27–33.

Rolfe, G.L. & Boggess, W.R. (1973). Soil conditions under old field and forest cover in Southern Illinois. *Soil Science*, **56**, 1945–50.

Routley, R. & Routley, V. (1975). *The Fight for the Forests. The Takeover of Australian Forests for Pinewood Chips and Intensive Forestry*. Canberra: Research School of Social Services, Australian National University.

Salisbury, E.J. (1922). Stratification and hydrogen-ion concentration of the soil in relation to leaching and plant succession with special reference to woodlands. *Journal of Ecology*, **9**, 220–40.

Scheu, S. & Parkinson, D. (1995). Successional changes in microbial biomass, respiration and nutrient status during litter decomposition in an aspen and pine forest. *Biology and Fertility of Soils*, **19**, 327–32.

Schnitzer, M. & Kahn, S.U. (1979). *Soil Organic Matter*. New York: Elsevier.

Scott, D.F. & van Wyk, D.B. (1990). The effects of wildfire on soil wettability and hydrological behaviour of an afforested catchment. *Journal of Hydrology*, **121**, 239–56.

Setala, H. & Huhta, V. (1991). Soil fauna increase *Betula pendula* growth: laboratory experiments with coniferous forest floor. *Ecology*, **72**, 665–71.

Shugart, H.H. (1984). *A Theory of Forest Dynamics*. New York: Springer-Verlag.

Sposito, G. (1989). *The Chemistry of Soils*. Oxford: Oxford University Press.

Sumner, M.E., Fey, M.V.F. & Noble, A.D. (1991). Nutrient status and toxicity problems in acid soils. In *Soil Acidity*, ed. B. Ulrich & M.E. Sumner, pp. 149–82. Berlin: Springer-Verlag.

Sverdrup, H., Warfvinge, P. & Jönsson, C. (1993). Critical loads of acidity for forest soils groundwater and first-order streams in Sweden. In *Critical Loads: Concept and Applications*, ed. M. Hornung & R.A. Skeffington, pp. 54–67. ITE symposium No. 28. London: Institute of Terrestrial Ecology.

Swift, M.J., Heal, O.W. & Anderson, J.M. (1979). *Decomposition in Terrestrial Ecosystems*. Oxford: Blackwell Scientific Publications.

Switzer, G.L. & Nelson, L.E. (1972). Nutrient accumulation and cycling in loblolly pine (*Pinus taeda*) plantation ecosystems: the first twenty years. *Soil Science Society of America Proceedings*, **36**, 143–7.

Tamm, O. (1950). *Northern Coniferous Forest Soils*. Oxford: Scrivener Press.

Thomas, J. (1961). Two rotations of *Pinus radiata*. *Institute Foresters Australian Newsletter*, December 1961.

Thornton, R.H., Cowie, J.D. & McDonald, D.C. (1956). Mycelial aggregation of sandy soil under *Pinus radiata*. *Nature*, **177**, 231–2.

Turner, J. & Kelly, J. (1985). Effect of radiata pine on soil chemical characteristics. *Forest Ecology and Management*, **11**, 257–70.

Turner, J. & Lambert, M.J. (1988). Soil properties as affected by *Pinus radiata* plantations. *New Zealand Journal of Forestry Science*, **18**, 77–91.

Ugolini, F.C., Minden, R., Dawson, H. & Zachara, J. (1977). An example of soil processes in the *Abies amabilis* zone of central Cascades, Washington. *Soil Science*, **124**, 291–302.

Ulrich, B. (1986). Natural and anthropogenic components of soil acidification. *Zeitschrift für Pflanzenernährung und Bodenkunde*, **149**, 702–17.

Ulrich, B. (1991). An ecosystem approach to soil acidification. In *Soil Acidity*, ed. B. Ulrich & M.E. Sumner, pp. 28–79. Berlin: Springer-Verlag.

Ulrich, B. & Sumner, M.E. (1991). *Soil Acidity*. Berlin: Springer-Verlag.

van Breemen, N. (1991). Soil acidification and alkalinization. In *Soil Acidity*, ed. B. Ulrich & M.E. Sumner, pp. 1–7. Berlin: Springer-Verlag.

van Breeman, N., Driscoll, C.T. & Mulder, J. (1984). Acidic deposition and internal proton sources in acidification of soils and waters. *Nature*, **307**, 599–604.

Van Miegroet, H. & Cole, D.W. (1985). Acidification sources in red alder and douglas-fir soils – Importance of nitrification. *Soil Science Society of America Journal*, **49**, 1274–9.

Verstraten, J.M., Dopheide, J.C.R., Duysings, J.J.H.M., Tietema, A. & Bouten, W. (1990). The proton cycle of a deciduous forest ecosystem in the Netherlands and its implications for soil acidification. *Plant and Soil*, **127**, 61–9.

Vilkamaa, P. & Huhta, V. (1986). Effects of fertilization and pH on communities of Collembola in pine forest soil. *Annali Zoologici Fennici*, **23**, 167–74.

Wardenaar, E.C.P. & Sevink, J. (1992). A comparative study of soil formation in primary stands of Scots pine (planted) and poplar (natural) on calcareous dune sands in the Netherlands. *Plant and Soil*, **140**, 109–20.

Whittaker, R.H. (1978). Direct gradient analysis. In *Ordination of Plant Communities*, ed. R.H. Whittaker, pp. 7–50. The Hague: Junk.

Wilde, S.A. (1958). *Forest Soils*. New York: Ronald Press.

Wilde, S.A. (1964). Changes in soil productivity produced by pine plantations. *Soil Science*, **97**, 276–8.

Williams, B.L., Cooper, J.M. & Pyatt, D.G. (1979). Effects of afforestation with *Pinus contorta* on nutrient content, acidity, and exchangeable cations in peat. *Forestry*, **51**, 29–35.

Zwolinski, J.B., Donald, D.G.M. & van Laar, A. (1993). Regeneration procedures of *Pinus radiata* in the Southern Cape Province. Part I: Modification of soil physical properties. *South African Forestry Journal*, **167**, 1–8.

18 Insect–pine interactions

Peter de Groot and Jean J. Turgeon

Ecology and Biogeography of Pinus, ed. D.M. Richardson. © Cambridge University Press, Cambridge (1998). pp. 354–80.

18.1 Introduction

Ecological patterns are the outcome of a complex set of historical and contemporary factors that interact with each other. As integral components of pine ecosystems, herbivorous insects occur under a wide range of abiotic conditions and interact with natural enemies, inter- and intraspecific competitors, and the host tree. Insects, by influencing individual trees or their components, are an intrinsic part of ecosystem functioning by, among other activities, increasing nutrient cycling, reversing stand stagnation and re-initiating forest succession. Insects generally have stable relationships with their hosts and are maintained at relatively low population densities by their interactions with a wide range of factors. Nevertheless, some insects occasionally reach extreme population densities (outbreaks), which may cause severe tree stress, mortality and significant economic impact. Pines can adapt to the many consequences of insect feeding and have evolved various defences to survive and reproduce; insects, in turn, have adapted to overcome these defences.

In the first part of this chapter, we examine the literature and other sources to develop an ecological profile of insect communities exploiting pines. For this, we have chosen the geographic area of Canada and the USA, primarily because the insects here have been relatively well identified and host record data were available to us. In the second part of the chapter we examine insect–plant interactions, emphasizing the plant perspective. We begin by exploring how insects contribute to ecosystem processes, how pines defend against insect herbivory and the counter-adaptations of insects to these defences. Next we examine how certain forestry practices and environmental stresses can affect insect–tree interactions.

The literature on pine insects is huge (a very conservative estimate would be 8000 references between 1939 and 1994 [CAB International – Tree CD]): we have therefore concentrated on more recent literature and reviews. We hope that this broad overview will stimulate additional interest and appreciation of the role of insects in pine ecology.

18.2 Insect fauna of pines in Canada and the United States

There are 39 native and at least 25 introduced species of pines in Canada and the USA (Little & Critchfield 1969; Young & Young 1992; Chap. 2, this volume); the ratio of hard to soft pines in the native and introduced pine flora is 24:15 and 16:9, respectively. Some native pines are widely distributed (e.g. Pinus banksiana, P. contorta, P. echinata, P. ponderosa, P. strobus, P. taeda and P. virginiana), whereas others have restricted ranges (e.g. P. aristata, P. engelmannii and P. radiata) (Critchfield & Little 1966). Pines characteristically grow on sandy, well-drained soil, but can be found on very harsh and impoverished soils to nutrient-rich soils. Most natural pine communities are long-lived ($\geq$80 years) and can occur in monotypic or polytypic stands. Against a mosaic of host range, site quality, community structure, and spatial, temporal and genetic heterogeneity, insects are provided with a diverse spectrum of potential niches.

Insects exploit all pine components (e.g. roots, foliage) throughout the tree's entire lifespan, and can play an even greater role in recycling the tree's nutrients after death. For this reason, a study of insect communities associated with pines cannot be restricted only to phytophagous species, which feed on living tissues (e.g. phloem, twigs, seed cones) but must also include those living and feeding in xylem, which is considered to be dead tissue (Strong, Lawton & Southwood 1984).

The objective of this first section is to provide a community profile of the known indigenous and introduced insects exploiting native and introduced pines. This profile was gleaned from the following sources: Rose & Lindquist (1973), Furniss & Carolin (1977), Hedlin et al. (1980), Martineau (1984), Drooz (1985), Ives & Wong (1988), Johnson & Lyon (1988), Gagné (1989), Turgeon & de Groot (1992), Wood & Bright (1992), Mattson et al. (1994), Nystrom & Britnell (1994), Turgeon (1994), and databases from the Canadian forest insect and disease surveys (FIDS) and from the USA. For each insect we recorded, if possible: (1) the tree component(s) exploited (foliage; twig, shoot or small branch; bole or large branch; root or root collar; pollen cone; seed cone); (2) its feeding habit (boring, grazing, mining, sap-sucking, gall-forming); (3) whether it was native or introduced; (4) the status of its host when colonized (healthy, injured, stressed, dying, or recently dead, etc.); (5) host specificity (mono-, oligo-, or polyphagous); (6) the length of its life cycle; and (7) overwintering site and stage. Insects were considered monophagous if feeding was apparently restricted to the genus Pinus, oligophagous if apparently restricted to Pinaceae, and polyphagous otherwise. Insects that obtain only shelter from pines (e.g. carpenter ants), or those that feed exclusively on dead or dry seasoned wood (e.g. powderpost beetles and termites) were excluded from our analysis. Because of space limitations, the data for this analysis will be published separately.

We begin by estimating species richness, a simple indicator of overall community diversity. Next, we examine the structure of the fauna by focusing specifically on patterns in host specificity and in associations with different host components and follow with a presentation of patterns in life-history adaptations. We also briefly compare these patterns with those from other regions of the world.

18.2.1 Insect diversity

Native pines of Canada and the USA are hosts to at least 1111 species of insects (Fig. 18.1). These insects represent eight orders, 58 families and 340 genera. Most of this entomofauna is either Lepidoptera (42%) or Coleoptera (36%). The remaining species belong to the Homoptera (9%), Hymenoptera (8%), Diptera (3%), Hemiptera (1%), Orthoptera (1%) and Thysanoptera (1%). The number of families in each order is ≤ 12, except for the Lepidoptera which has 20. Most families are represented by fewer than a dozen genera (Fig. 18.2). The only notable exceptions are two families of Coleoptera (Cerambycidae and Scolytidae) and three families of Lepidoptera (Geometridae, Noctuidae and Tortricidae). In these families, diversity

exceeds 24 genera. For more than 90% of the genera, diversity is restricted to seven species or fewer (Fig. 18.3). One genus of Homoptera (Cinara (Aphididae)) and two genera each of Hymenoptera (Neodiprion (Diprionidae) and Acantholyda (Pamphiliidae)) and of Lepidoptera (Eupithecia (Geometridae) and Dioryctria (Pyralidae)) have more than 20 species. The most diverse genus feeding on native pines is Pityophthorus (Coleoptera: Scolytidae), with 84 species.

The species richness of the insect fauna feeding on introduced pines is much lower than that feeding on native pines. Only 211 species of insects, from seven orders, 41 families and 118 genera, are recorded as exploiting introduced pines (Fig. 18.2). The overall general pattern of distribution of insect species among taxa higher than species, however, is similar for introduced pines as it is for native pines. The Lepidoptera and Coleoptera are again the dominant orders, representing 52 and 19% of the total fauna, respectively. The richness of Homoptera (14%) and Hymenoptera (12%) is slightly higher than for native pines. Diptera, Hemiptera and Orthoptera each represent c. 1% of the fauna. The number of families in each order is below 10, except for Lepidoptera which has 16. Most families contain fewer than six genera (Fig. 18.2). Among the exceptions are three families of Lepidoptera (Geometridae, Noctuidae and Tortricidae), which all have more than 10 genera. All genera have eight species or less, except Neodiprion, which has 11 (Fig. 18.3).

Most of the 211 species exploiting introduced pines also feed on native pines. Indeed, only 14 species, 10 of which are Lepidoptera, are recorded exclusively on introduced pines (J.J.T., unpublished data). Most of these, however, are polyphagous insects and pines are not common hosts (Drooz 1985).

For most individual native pine species, insect diversity is less than 50 species (Table 18.1). Only a few species of pine, namely P. banksiana, P. contorta, P. monticola, P. ponderosa, P. resinosa and P. strobus, are exploited by more than 200 species of insects. Richness of introduced pine species does not exceed 10 species of insects, except for P. sylvestris which hosts almost 200 species (Table 18.2). Furthermore, this pattern of limited diversity on each pine species is also reflected at taxonomic levels higher than species (i.e. genus, family and order).

Thirty-one species of introduced insects are recorded as feeding on native pines. Most of these species were introduced from Europe during this century (see Mattson et al. 1994). Introduced species like the European pine shoot moth, Rhyacionia buoliana (Lepidoptera: Tortricidae) and the European pine sawfly, Neodiprion sertifer, have become serious pests primarily in young plantations (Drooz 1985). Only 16 of these 31 introduced species of insects were recorded on introduced pines. Virtually all introduced species of insects adopted a new plant species in the same

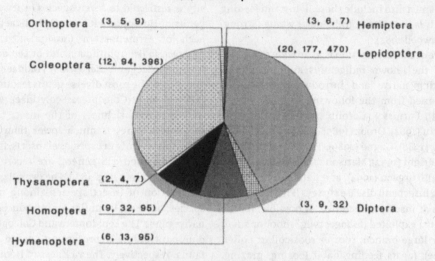

(a) **Native pines**

Orthoptera (3, 5, 9)   (3, 6, 7) Hemiptera

Coleoptera (12, 94, 396)   (20, 177, 470) Lepidoptera

Thysanoptera (2, 4, 7)

Homoptera (9, 32, 95)   (3, 9, 32) Diptera

Hymenoptera (6, 13, 95)

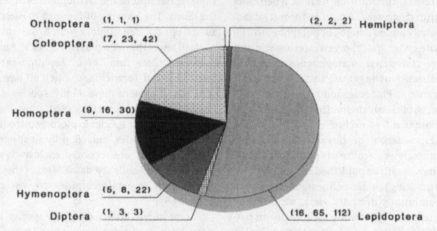

(b) **Introduced pines**

Orthoptera (1, 1, 1)   (2, 2, 2) Hemiptera

Coleoptera (7, 23, 42)

Homoptera (9, 16, 30)

Hymenoptera (5, 8, 22)

Diptera (1, 3, 3)   (16, 65, 112) Lepidoptera

Fig. 18.1. **Composition of the insect fauna exploiting native (*a*) and introduced (*b*) pines of Canada and the United States. Parentheses** include the number of taxa (families, genera, species) in each insect order.

genus as their host in the region of origin, or the same pine species which has been introduced in North America (Mattson *et al.* 1994).

The accuracy of our estimates of insect diversity on pines is difficult to assess, for at least three reasons. First, we could not establish, either from the literature citations or from the FIDS database, the number of host records for all insect species. Thus, we could not differentiate accidental host records from true host records. Second, insect-host records are incomplete. Records exist for 35 of the 39 native pines and for 12 of the 25 introduced pines (Tables 18.1, 18.2), suggesting that survey intensity was inconsistent. Pine species of little economic importance or of limited distribution, except *P. radiata*, undoubtedly received less attention. Third, there is a lack of biological knowledge about several insects,

thus making it difficult to ascertain their association with pines. For example, FIDS database includes records of at least 50 species of click beetles (Coleoptera: Elateridae). Joly (1975) reports that in Europe, larvae of this family can be either phytophagous (seriously damaging roots of pines in plantations), saprophagous, or entomophagous (feeding on Coleoptera larvae living in dead wood). In North America, the role of most Elateridae living in the soil is largely unknown, although some species are suspected of confining their feeding to dead and well-decayed wood (Furniss & Carolin 1977; Drooz 1985).

18.2.2 Structure

The first difficulty encountered while examining the structure of the pine entomofauna was to assess accurately the

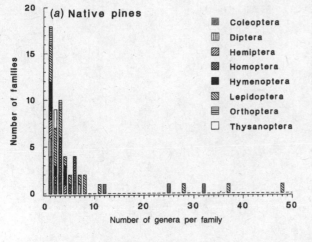

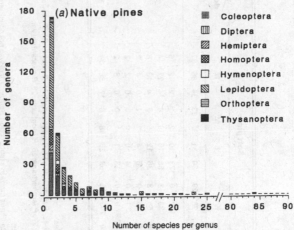

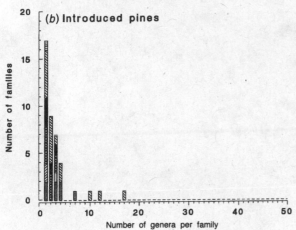

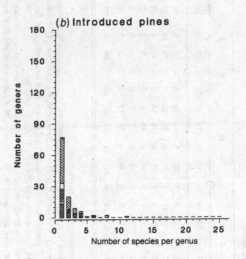

Fig. 18.2. **Distribution of the number of genera per family of insects exploiting native (a) and introduced (b) pines of Canada and the United States.**

Fig. 18.3. **Distribution of the number of species per genus of insects exploiting native (a) and introduced (b) pines of Canada and the United States.**

richness associated with each tree component. Part of this problem stemmed from the fact that feeding by several species is not limited to a single component. In some instances, adults and larvae of the same species feed on different components (e.g. adults of *Buprestis* spp. (Coleoptera: Buprestidae) feed on foliage whereas larvae exploit the main stem). In others, one stage exploits several components based on their relative availability. For example, larvae of *Dioryctria* and *Rhyacionia* spp. can feed on foliage, twigs, seed and pollen cones (Drooz 1985). This problem was further compounded by the lack of biological knowledge indicating whether congeneric species behave similarly. Another difficulty was determining whether insects exploiting several components did so on all pine species recorded as hosts. Finally, clear patterns of community structure were difficult to obtain because host records were analysed on a qualitative basis, making it difficult if not impossible to distinguish common from occasional

hosts or to discriminate between records representing absence of feeding from those suggesting lack of examination. Notwithstanding these limitations, some interesting patterns emerged from this analysis.

Host specificity

The host specificity was determined for 82 and 88% of the insects exploiting native and introduced pines, respectively. Of the insects with known host specificity exploiting native pines, 51% were apparently monophagous, 28% were oligophagous, and 21% were polyphagous (Table 18.3). For insects feeding on introduced pines, these proportions were 46, 31 and 23%, respectively (Table 18.4). This pattern of host specificity, where insects are predominantly monophagous, was observed on most native and introduced pine species. Among the exceptions are species such as *Pinus monticola*, *P. ponderosa* and *P. strobus*

Table 18.1. Number of insect species associated with each component of *Pinus* taxa native to Canada and the USA

Component	Hard pines[a]																								Soft pines											Total richness
	A[b]	B								C							D				E			F	K					L	M		N			
	1[c]	2	3	4	5	6	7	8	9	10	11	12	13	14	15	16	17	18	19	20	21	22	23	25	26	27	28	29	30	31	32	33	34	36	37	
Foliage	193	27	27	3	17	6	46	7	37	0	4	172	0	4	8	1	236	8	274	28	5	3	30	0	28	12	132	0	173	10	2	2	2	12	6	607
Twigs	62	23	18	5	16	9	27	6	30	8	25	90	8	1	10	11	2	77	9	86	33	5	9	23	16	28	10	33	13	56	15	10	4	22	14	258
Boles	41	19	10	2	8	2	14	2	18	5	7	106	0	0	7	0	0	55	0	119	0	9	9	17	9	25	19	55	0	36	12	5	3	13	8	265
Roots	17	6	2	0	5	1	7	0	6	1	0	37	0	0	0	0	0	26	0	44	4	3	6	0	0	3	0	18	0	26	0	1	1	0	2	112
Pollen cones	4	6	11	7	0	0	1	1	5	0	0	9	0	0	4	0	0	7	1	7	7	2	1	2	0	0	0	1	0	4	0	0	0	2	2	36
Seed cones	15	11	17	2	18	3	8	2	19	1	12	35	0	0	4	0	0	18	4	24	13	5	1	6	1	6	4	0	0	13	1	1	0	8	2	90
Species richness[d]	275	65	61	8	52	14	81	12	80	13	72	367	1	23	25	4	345	16	451	67	20	21	66	28	70	40	203	16	259	28	18	8	17	56	34	1111

Notes:

[a] Hard pines, subgenus *Pinus* (section *Pinus*); Soft pines, subgenus *Strobus* (sections *Strobus* and *Parrya*).

[b] Subsections: A, *Pinus*; B, *Australes*; C, *Ponderosae*; D, *Contortae*; E, *Attenuatae*; F, *Leiophyllae*; K, *Strobi*; L, *Cembrae*; M, *Balfourianae*; N, *Cembroides*.

[c] 1, *P. resinosa*; 2, *P. echinata*; 3, *P. elliottii*; 4, *P. glabra*; 5, *P. palustris*; 6, *P. pungens*; 7, *P. rigida*; 8, *P. serotina*; 9, *P. taeda*; 10, *P. engelmannii*; 11, *P. jeffreyi*; 12, *P. ponderosa*; 13, *P. washoensis*; 14, *P. coulteri*; 15, *P. sabiniana*; 16, *P. torreyana*; 17, *P. banksiana*; 18, *P. clausa*; 19, *P. contorta*; 20, *P. virginiana*; 21, *P. attenuata*; 22, *P. muricata*; 23, *P. radiata*; 25, *P. leiophylla*; 26, *P. flexilis*; 27, *P. lambertiana*; 28, *P. monticola*; 29, *P. ayacahuite* var. *strobiformis*; 30, *P. strobus*; 31, *P. albicaulis*; 32, *P. aristata*; 33, *P. balfouriana*; 34, *P. cembroides*; 36, *P. edulis*; 37, *P. monophylla*.

[d] Species richness is often smaller than the sum of insects reported on each tree component because many species exploit more than one component.

No insect records for the following native taxa: 24, *P. leiophylla* var. *chihuahuana* (= *P. chihuahuana*); 35, *P. discolor*; 38, *P. juarezensis*; and 39, *P. remota*. *Pinus longaeva* (subsection *Balfourianae*) was not considered in this analysis. Taxon 24 was deleted in proof as it is treated as a variety of another species.

Table 18.2. Number of insect species associated with each component of *Pinus* taxa introduced to Canada and the USA

Component	Hard pines[a]											Soft pines		Total richness
	A[b]							B	G	I	J	K	L	
	41[c]	42	46	47	48	49	50	51	52	53	55	57	60	
Foliage	6	3	34	20	0	130	3	3	0	1	2	0	1	155
Twigs	3	1	16	15	0	49	3	0	1	1	2	1	0	56
Boles	1	1	5	10	1	24	2	0	1	0	0	0	0	32
Roots	0	0	4	4	0	13	0	0	1	0	0	0	0	15
Pollen cones	0	0	0	0	0	3	0	0	0	0	0	0	0	3
Seed cones	1	0	7	6	0	10	1	0	0	0	0	0	0	12
Species richness[d]	10	4	49	39	1	186	9	4	2	1	2	1	1	211

Notes:

[a] Hard pines, subgenus *Pinus* (section *Pinus*); Soft pines, subgenus *Strobus* (sections *Strobus* and *Parrya*).

[b] Subsections: A, *Pinus*; B, *Australes*; G, *Oocarpae*; I, *Canarienses*; J, *Pineae*; K, *Strobi*; L, *Cembrae*.

[c] 41, *P. densiflora*; 42, *P. halepensis*; 46, *P. mugo*; 47, *P. nigra*; 48, *P. pinaster*; 49, *P. sylvestris*; 50, *P. thunbergii*; 51, *P. caribaea*; 52, *P. patula*; 53, *P. canariensis*; 55, *P. pinea*; 57, *P. parviflora*; 60, *P. cembra*.

[d] Species richness is often smaller than the sum of insects reported on each tree component because many species exploit more than one component. No insect records for the following introduced species: 40, *P. brutia*; 43, *P. heldreichii*; 44, *P. kesiya*; 45, *P. merkusii*; 54, *P. roxburghii*; 56, *P. armandii*; 58, *P. peuce*; 59, *P. wallichiana*; 61, *P. koraiensis*; 62, *P. pumila*; 63, *P. sibirica*; and 64, *P. gerardiana*.

where oligophagous species predominate. Whether these variations in host specialization result from inconsistencies in survey intensities or relate to specific characteristics of the various tree components remains to be determined.

Host plant specialization by insects varied greatly among tree components (Fig. 18.4). Insects exploiting twigs, stems, and pollen and seed cones were mainly monophagous. Foliage feeders have equal proportions of mono- and polyphagous species whereas root feeders have similar proportions of mono- and oligophagous species.

Host specificity also differed between insect orders (Fig. 18.5). Among species with known host specialization, Hymenoptera, Homoptera, Hemiptera, Diptera and Coleoptera were predominantly monophagous whereas Lepidoptera and Orthoptera were mainly polyphagous; Thysanoptera were mostly oligophagous. Whether the degree of host specialization also varies among feeding groups was not investigated. But Björkman & Larsson (1991a) reported significant differences in host specialization between mining (predominantly monophagous) and chewing (mostly polyphagous) insects exploiting the foliage of three conifer species native to Sweden, including *P. sylvestris*. Many hypotheses have been proposed to explain variations in host specificity among phytophagous insects (see references in Björkman & Larsson 1991a).

Foliage

More than half of the insect species recorded on native or introduced pines are, at one stage of development or another, phyllophagous (Tables 18.1, 18.2). This fauna, although comprising predominantly Lepidoptera, was represented by all eight orders. Foliage was consumed either by immature stages (Lepidoptera, Hymenoptera and Diptera), adults (Coleoptera), or both (Hemiptera, Homoptera, Orthoptera and Thysanoptera). The principal feeding behaviour was chewing or grazing (most Lepidoptera; all Coleoptera, Hymenoptera and Orthoptera). Other feeding habits included sap-sucking (Hemiptera, Homoptera and Thysanoptera), galling (all Diptera), and a few families of needle-mining Lepidoptera (i.e. all Argyresthiidae, most Gelechiidae and Yponomeutidae, and some Tortricidae).

Foliage feeding by coleopteran adults is of little consequence for tree health, usually causing only needle-browning or light defoliation (Furniss & Carolin 1977; Drooz 1985). Although most dipterans make their galls in the current year's foliage, causing needles to swell or to turn brown (Drooz 1985), some can kill young trees (Barbosa & Wagner 1989). Because sucking of foliage by Hemiptera occurs predominantly during the early nymph stages, which eat considerably less than older nymphs, its impact is considered unimportant. Homopterans such as *Elatobium* spp., *Pineus* spp. and *Toumeyella* spp., (Aphididae, Phylloxeridae and Coccidae, respectively) can kill trees or branches, although the most significant impact of most species is to reduce vigour and growth (Berryman 1986; Barbosa & Wagner 1989) and increase susceptibility to attack by other pests such as bark beetles (Furniss & Carolin 1977). Defoliation by most pamphiliid sawflies is usually not severe, except possibly for the introduced *Acantholyda erythrocephala*, which is causing mortality of *P. resinosa* in eastern North America (Drooz 1985). Diprionid sawflies, on the other hand, are much more destructive (Fig. 18.6A), as defoliation not only causes loss of height and radial growth but often results in weak, stunted or dead trees (Barbosa &

Table 18.3. Host specificity for insects exploiting native pines[a]

Specificity	A		B					C										D					E	F	K				L		M		N			Overall proportion
	1	2	3	4	5	6	7	8	9	10	11	12	13	14	15	16	17	18	19	20	21	22	23	25	26	27	28	29	30	31	32	33	34	36	37	
Monophagous[b]	45	77	86	88	74	82	65	91	75	85	72	56	100	91	82	25	40	100	37	76	74	68	55	86	60	55	33	81	32	67	83	88	65	79	87	51
Oligophagous[b]	30	20	14	12	22	18	28	9	16	15	25	33	0	9	18	50	33	0	41	21	26	32	32	14	35	32	48	19	40	26	17	12	35	19	13	28
Polyphagous[b]	25	3	0	0	4	0	7	0	9	0	3	11	0	0	0	25	27	0	22	3	0	0	13	0	5	13	19	0	28	7	0	0	0	2	0	21
Unknown[c]	13	5	5	0	6	21	7	8	6	9	0	7	18	0	0	12	12	0	6	21	7	10	5	0	14	5	24	0	12	4	0	0	0	7	9	17

Hard pines: A–F (columns 1–25). Soft pines: K–N (columns 26–37).

Notes:
[a] See Table 18.1 for complete listing of native pine species.
[b] Values are proportions calculated on the basis of total number of species with known host specificity (Species richness – Unknown).
[c] Values are proportions calculated on the basis of species richness.

Table 18.4. Host specificity for insects exploiting introduced pines[a]

Specificity	A							B	G	I	J	K	L	M	Overall proportion
	41	42	46	47	48	49	50	51	52	53	55	57	60		
Monophagous[b]	67	25	60	59	0	44	57	100	100	100	50	0	100	0	46
Oligophagous[b]	33	50	30	41	100	32	43	0	0	0	0	0	0	0	31
Polyphagous[b]	0	25	9	0	0	24	0	0	0	0	50	0	0	100	23
Unknown[c]	10	0	12	13	0	12	22	0	0	0	0	100	0	0	12

Hard pines: A, B, G, I, J (columns 41–55). Soft pines: K, L, M (columns 57–60).

Notes:
[a] See Table 18.2 for complete listing of pine species introduced in Canada and the United States.
[b] Values are proportions calculated on the basis of total number of species with known host specificity (Species richness – Unknown).
[c] Values are proportions calculated on the basis of species richness.

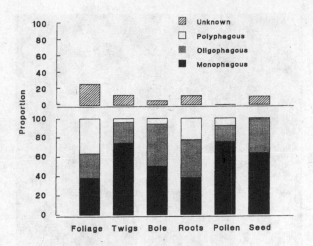

Fig. 18.4. **Influence of tree component on host specificity of insect species exploiting native pines of Canada and the United States.** *Note:* Proportion of mono-, oligo- and polyphagous calculated on the basis of the total number of species with known host specificity whereas the proportion of unknown is calculated on the basis of total species richness exploiting each component.

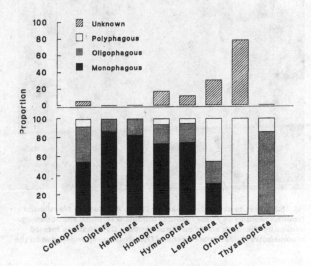

Fig. 18.5. **Influence of insect taxa (order) on host specificity of insect species exploiting native pines of Canada and the United States.** *Note:* Proportion of mono-, oligo- and polyphagous calculated on the basis of the total number of species with known host specificity whereas the proportion of unknown is calculated on the basis of total species richness of each insect order.

Wagner 1989). Feeding by most species of Lepidoptera results in needle browning, light defoliation, deformation, or growth reduction. Only a few species, most of them Tortricidae (e.g. *Choristoneura* spp., *Rhyacionia* spp.) which reach high densities, can kill trees (Furniss & Carolin 1977; Drooz 1985; Berryman 1986; Barbosa & Wagner 1989). Serious defoliation by orthopterans can occur during outbreaks, especially on seedlings and young trees in nurseries, shelterbelts or plantations (Drooz 1985).

There is some evidence to suggest that differences exist between the structure of the entomofauna of soft and hard pines. For example, the insect fauna feeding on foliage of most species of hard pines, either native or introduced, is usually more diverse than that exploiting any other tree component. This, however, is not apparent on soft pines where the diversity of insects feeding on twigs is higher than on foliage (Tables 18.1, 18.2). Furthermore,

most native hard pines were exploited by two or three genera of Diptera whereas few native and introduced soft pines have host records of Diptera. Also, *Leptoglossus* spp. (Hemiptera: Coreidae) have been recorded almost exclusively on native hard pines. Interestingly, the introduced *Camptozygum aequale* (Hemiptera: Miridae) has also been recorded exclusively on hard pines. There are no records of Cicadellidae (Homoptera) on soft pines. Conversely, records of *Desmococcus* and *Pityococcus* spp. (Homoptera: Margarodidae) exist only for soft pines. Finally, most species of hard pines are exploited by one to several species of *Exotelia* (Lepidoptera: Gelechiidae), yet records of species from this genus are rare among soft pines.

Twig, shoot or small branch
Approximately a quarter of the fauna recorded on native and introduced pines feed on twigs, shoots and small

Fig. 18.6. **Insects exploiting Nearctic pines. A** Colony of *Neodiprion lecontei* feeding on *Pinus resinosa* foliage. **B** Terminal shoot damage caused by *Pissodes strobi* on *Pinus strobus*. **C** Pitch tubes formed around entry holes made by female *Dendroctonus ponderosae* on the bole of *Pinus contorta*. **D** Damage by *Hylobius warreni* to the root collar of *Pinus sylvestris*. **E** *Choristoneura pinus pinus* larva feeding on pollen cones of *Pinus banksiana*. **F** Excavated frass of *Eucosma tocullionana* feeding inside a seed cone of *Pinus strobus*.

branches (Tables 18.1, 18.2). Coleoptera represent the most diverse group of the six insect orders exploiting this component. Generally, larvae feed upon shoots, twigs or small branches, although adults of several species of *Okanagana* (Homoptera: Cicadidae), *Monochamus* (Coleoptera: Cerambycidae), and *Hylobius*, *Pissodes* and *Cylindrocopterus* (Coleoptera: Curculionidae), also graze or leave feeding punctures on twigs and shoots. Most of the *Pityophthorus* spp. and a few Lepidoptera (e.g. *Retinia* (Tortricidae), are myelophagous (*sensu* Wood 1982) and bore into the twigs of weak, injured, dying or recently dead trees. Other feeding behaviours associated with twigs, shoots or small branches include sap-sucking by homopteran and hemipteran young nymphs, galling by dipteran and hymenopteran larvae, and grazing by Orthoptera.

Most Lepidoptera and Homoptera, together with some of the Coleoptera (e.g. *Hylobius*, *Pachylobius*, *Pissodes*) exploiting this component, use the meristematic tissue of healthy plants for food or habitat (Coulson & Witter 1984). Insect damage to shoots or twigs usually influences the rate of growth and the form (Fig. 18.6*B*) of the tree (Coulson & Witter 1984). Several types of deformity of the tree crown resulting from the destruction of terminal and lateral buds and shoots of pines have been defined (Lessard & Jennings 1976). In North America, *Pissodes* and *Rhyacionia* spp. are among the most notorious shoot and twig feeders although other genera including *Eucosma* (Lepidoptera: Tortricidae) and *Dioryctria* can seriously damage pines (Coulson & Witter 1984). The destruction or injury of living twigs by certain Coleoptera (e.g. *Pityophthorus*) may have only a limited stunting effect on growth, but it may weaken trees sufficiently to make them vulnerable to aggressive tree-killing beetles (Wood 1982). Some homopteran twig feeders, such as *Aphrophora saratogensis* (Homoptera: Cercopidae) and *Toumeyella pinicola* and *T. resinosae*, can kill ornamental or young trees either directly by transferring fungi or by injecting toxic saliva (Furniss & Carolin 1977). In some instances, feeding on twigs can influence seed cone production. For example, cone beetles, *Conophthorus* spp. (Coleoptera: Scolytidae) feed

Fig. 18.6. *(cont.)*

primarily in seed cones of pines, but when seed cones are rare, several species feed in the shoots, indirectly destroying cone-bearing ones (de Groot 1992).

The insect fauna feeding on twigs, shoots or small branches of hard pines appears to differ from that of soft pines. Most species of hard pines are exploited by several species of Coccidae (Homoptera), whereas soft pines are apparently not exploited by this insect family.

Bole or large branch

Approximately 25% of the insect species exploiting native pines feed on boles and large branches (Table 18.1). This proportion is slightly lower (15%) for introduced pines (Table 18.2). About 80% of the species exploiting this component are Coleoptera. These species are either phloeophagous (i.e. feeding on phloem tissues of the inner bark) or xylophagous (i.e. living and feeding directly in the xylem or wood tissues). Some Hymenoptera (i.e. Siricidae) also bore into the heartwood of pines. Other types of bole-feeding behaviour include sucking by Hemiptera and Homoptera, and phloem-boring by Lepidoptera.

Coleoptera (e.g. Buprestidae and Cerambycidae) and Hymenoptera (e.g. Siricidae) are seldom responsible for

tree mortality as they primarily attack pines that are weak, stressed, injured, dying or recently dead. Their numerous tunnels, however, can weaken the bole sufficiently for tops to be broken (Furniss & Carolin 1977), or to render trees more susceptible to wind breakage, and represent entry points for secondary injuries from other insects and diseases, that may help in cycling minerals back into the soil (Coulson & Witter 1984). Hymenoptera (e.g. Siricidae) can transmit diseases such as the sap stain fungi and the mutualistic decay fungi (Furniss & Carolin 1977). Coleoptera such as *Carphoborus*, *Pityoborus* and *Pityophthorus* (Scolytidae), that confine their attacks to shaded branches of living trees, accelerate pruning, a natural process that is usually beneficial to vigorous tree growth and reduces fire hazard. Boring by *Synanthedon* (Lepidoptera: Sesiidae) and *Retinia* spp. causes masses of resin to form on the outside of the stem. Pine stems are rarely girdled by these larval channels, but can be. Most tree mortality is caused by species of *Dendroctonus* (Fig. 18.6C) and *Ips* (Coleoptera: Scolytidae) (Wood 1982), several of which have been extensively studied.

The entomofauna found on the bole of hard pines also differs from that of soft pines. For example, the bole of

Fig. 18.6. **(cont.)**

most species of native hard pine is exploited by 2–4 genera of Buprestidae, each usually containing more than one species (J.J.T., unpublished data). Only a few soft pines have more than two species of buprestids but even on those, the diversity of buprestid species is much lower than on hard pines.

Root and root collar

About one-tenth of the insect fauna (from four orders) has been recorded as exploiting the root system of native and introduced pines (Tables 18.1, 18.2). More than 90% of this entomofauna comprises Coleoptera. Coleoptera (e.g. *Hylobius*, *Phyllophaga*) and Lepidoptera (i.e. *Aegeria*, *Ochropleura*) spend part of their life cycle in (as borers) or about (as grazers) roots. Boring was the most common (60%) method of feeding recorded among Coleoptera. Some Homoptera have been recorded as sucking on roots (Berryman 1986).

Weevils, *Hylobius* spp., which are borers, and white grubs, *Phyllophaga* spp., which are grazers, are probably the most damaging insects of the root system (Fig. 18.6D). In general, insects feeding on roots are not a serious problem in mature stands, where trees have well-developed root systems, but can cause severe damage in plantations because roots are small and fragile (Berryman 1986). Reforestation problems also occur when adults of species such as *Hylobius*, which usually breed in stumps of freshly cut trees, emerge and feed on the bark of newly planted seedlings (Berryman 1986). Other genera such as *Dendroctonus* also utilize roots and root collars for food and habitat but are primarily associated with the basal section of the bole (Coulson & Witter 1984). Boring through or grazing upon the roots or root collar physically weakens the pine, represents an entry for root-decaying fungi and can kill the tree (Drooz 1985; Berryman 1986).

Pollen cone

Only 36 species of insects from four orders exploit what probably represents, together with seed cones, the richest source of protein on pine trees. Larvae of *Xyela* spp. (Hymenoptera: Xyelidae) represent the most important group of this component as all 18 species apparently graze exclusively on pollen cones. All species of Lepidoptera exploiting this component (e.g. *Choristoneura*, *Coleotechnites*, *Dasychyra*, *Dioryctria*) appear opportunistic and extremely mobile grazers (as larvae), which feed also on seed cones and on other components such as foliage or twigs. Other pollen feeders include thrips (e.g. *Frankliniella*, *Oxythrips* (Thysanoptera: Thripidae)) whose mouthparts are of the rasping–sucking type (Mattson 1975; Drooz 1985).

Most species feed on pollen cones before notable desiccation occurs (Drooz 1985). Reportedly, most of these species do not have significant impacts on pollen production (Hedlin *et al*. 1980). On the other hand, pollen can play an important role in the population dynamics of some species. For example, pollen cones are crucial to the survival and abundance of the budworm *Choristoneura pinus pinus* (Nealis & Lomic 1994) (Fig. 18.6E).

The pollen cones of most species of native hard pines are exploited by 1–5 hymenopteran species of *Xyela*, whereas only two species of native soft pines have limited records of species from this genus.

Seed cone

The seed cone entomofauna includes genera from seven orders and represents *c*. 10% of the richness of insect feeding on native pines and about 6% of that of introduced pines (Tables 18.1, 18.2). Seed cones are consumed by larvae/nymphs of Lepidoptera, Coleoptera, Hymenoptera, Thysanoptera and Hemiptera, and the adults of Hemiptera, Coleoptera and Thysanoptera. Most species of Coleoptera and Lepidoptera are borers. Other feeding behaviours include galling by Hymenoptera and Diptera, and sucking by Thysanoptera, Hemiptera and Homoptera. The majority (almost 80%) of the species exploiting this component, namely all cone beetles (e.g. *Conophthorus*, *Ernobius* spp.), all Diptera, several coneworms (*Dioryctria* spp.), all seed moths (*Cydia* spp.), and all cone borers (*Eucosma* spp.) are conophytes (i.e. cannot develop without seed cones), which suggests a very high degree of specialization of this fauna (Turgeon, Roques & de Groot 1994). The remaining Hymenoptera (i.e. *Neodiprion*) and Lepidoptera (i.e. *Battaristis*, *Choristoneura*, *Herculia*, *Nepytia*, *Rhyacionia*) are heteroconophytes (i.e. do not need seed cones to complete development and thus feed opportunistically on seed cones).

Exploitation of the seed resource occurs predominantly during the third year in the cycle of seed production, when cone and seed maturation occurs (de Groot 1986; Turgeon *et al*. 1994). Hemipterans (i.e. *Leptoglossus* and *Tetyra*) can cause the abortion of seed cones during the second year of development (Hedlin *et al*. 1980; Turgeon & de Groot 1992). For most species of pines, one or two insect species are usually responsible for most of the destruction of cones (Turgeon 1994). Entire cone crops can be destroyed, even in natural stands, thus limiting the regeneration potential of some species in some years (Turgeon *et al*. 1994). Interactions between insects and cone

production in conifers, including pines, have been reviewed (Turgeon *et al*. 1994). Most species either bore through cone and bract tissues, usually without specifically targeting the seeds (e.g. Coleoptera (*Conophthorus*, *Ernobius*), Diptera, Lepidoptera (*Dioryctria*)), or consume seed and cone tissues while moving from seed to seed in a clear pattern (e.g. Diptera (*Earomyia*), Lepidoptera (*Eucosma*, *Cydia*)) (Fig. 18.6F). Hemipterans use their piercing mouthparts to siphon the endosperm (Hedlin *et al*. 1980). Unlike other conifers, such as spruce and larch, pines have few species that exploit seeds exclusively (Turgeon 1994). Turgeon *et al*. (1994) also provided a comprehensive review of the categories of associations between insects and seed-cones (conophytes and heteroconophytes) together with the patterns of seed cone exploitation (i.e. insects eating scale and bract tissues; eating scale and seed tissues; eating seed tissues only) already identified for conifers, including pines. The species richness and patterns of exploitation have been evaluated for the insects exploiting seed-cones of pines native to Canada (Turgeon 1994).

The diversity of insect species exploiting hard pines (mean = 12; range = 1–41) is much higher than that exploiting soft pines (mean = 7; range = 1–17) (Table 18.1). Other differences between the structure of the entomofauna exploiting seed-cones of hard and soft pines may exist. For example, *Leptoglossus* and *Tetyra* (Hemiptera: Pentatomidae) have been recorded on 17 of the 24 species of hard pines but on only two of the 11 species of soft pines. Furthermore, *Cydia* spp. have been recorded almost exclusively on hard pines.

18.2.3 Life-history adaptations

Insect species feeding on nitrogen-rich food such as reproductive structures, phloem, and foliage of pine trees appear to have shorter life cycles than those which bore into less nutritious tissue (high in fibre) such as xylem and heartwood (J.J.T., unpublished data). For example, the life cycle of most species exploiting foliage, twigs and shoots, and pollen- and seed cones, is generally uni- or multivoltine irrespective of taxonomic order. A few exceptions are some needle-miners which take up to two years to develop. Insects on boles and roots, however, often take several years to complete development.

The overwintering site of insects exploiting pines appears to be associated with the component exploited and varies among insect orders. The eggs, larvae or adults of foliage-feeding Homoptera and Lepidoptera overwinter predominantly on the tree, whereas pre-pupae and eggs of Hymenoptera are found primarily on the ground. An equal proportion of foliage-feeding dipteran larvae overwinter on the ground and on trees. Twig and shoot feeders overwinter on pines as adults/eggs (Homoptera), or larvae

(Diptera and Lepidoptera), or pupae (Lepidoptera). Insects feeding exclusively on cone and seed insects can overwinter on the tree inside the cones, on the ground in either cones or twigs (e.g. cone beetles), or in the soil. Root-feeders spend the winter in the ground as adults (Coleoptera), larvae (Coleoptera and Homoptera), or pupae (Lepidoptera). Generally, insects feeding on boles hibernate on the tree, except most *Pissodes* spp. (Coleoptera: Curculionidae), which hibernate on the ground as adults.

Overwintering site also appears to be related to host specialization, at least for insect species exploiting certain components of pines. Most twig-, pollen- and seed-cone feeders overwintering on the tree are monophagous whereas species hibernating on the ground are predominantly polyphagous. These results are consistent with the Host Intimacy Hypothesis, whose premise is that the greater amount of time the insect spends in contact with host plant tissues, the higher the requirement for elaborate behavioural, physiological and temporal adaptations; i.e. increased intimacy will impose specificity (Mattson *et al.* 1988). The lack of relationship between the overwintering site and the extent of host specificity of species eating pine foliage may be linked to the diversity of feeding habits encountered among the species associated with this component, a factor known to influence the degree of host specificity (Björkman & Larsson 1991a).

Diapause, an arrestment of development induced by certain factors in advance of adverse conditions, is primarily used to increase the chances of synchrony between specific life stages and suitable resources (e.g. food or mates) and can last several years (extended diapause). This strategy, which is relatively common among species exploiting seed cones of firs (*Abies* spp.), spruces (*Picea* spp.) or larches (*Larix* spp.) to help cope with variation in temporal abundance of seed cones (Turgeon *et al.* 1994), is restricted to a few species exploiting pines.

18.2.4 Patterns from North America versus patterns from other regions of the world

Differences in methods of analysis, terminology, and presentation of the results precludes any rigorous comparison between faunal patterns from North America and those from other regions of the world. Thus what follows should simply be considered as an attempt to identify potential similarities or differences between regions.

In Eurasia, estimates of the diversity of 'insect pests' feeding on *c.* 13 species of pine vary between 100 and 200 (Speight & Wainhouse 1989; Klimetzek 1992; Day & Leather 1997). These insects belong to seven orders. Unlike North America, where the Lepidoptera are the most diverse group, Coleoptera appear to be the predominant order on Eurasian pines (Joly 1975; Annila *et al.* 1983; Roques 1983; Bevan 1987). Approximately 55% of the

genera of Curculionidae, Scolytidae, Cerambycidae, Buprestidae and Scarabaeidae exploiting European pines are also found in North America, exploiting the same tree components. The proportion of genera associated with pines found in both North America and in Eurasia varies among orders (Hymenoptera 86%, Diptera 40%, Homoptera 40% Lepidoptera 20%, Hemiptera 17%, Orthoptera 0%). Also, congeneric conophagous insects, found in Eurasia and in North America, exploit the same pine cone resources (e.g. scales and bracts, scales and seeds, or seeds) (Turgeon *et al.* 1994). Furthermore, the proportion of insects exploiting seed cones in a similar way was also relatively similar between the two continents (Roques 1991; Turgeon 1994). All these results suggest some degree of similarity in insect–pine interactions between these two continents.

Two pine species from North America were each paired with a Eurasian species of the same subsection (taxa) and with a relatively comparable geographic range (Critchfield & Little 1966), and the species richness of insects exploiting these trees was compared. In Europe, *Pinus sylvestris* is exploited by approximately 112 species of 'specialist' insects (Klimetzek 1992). In North America, a total of 107 species of monophagous insects has been recorded for *P. resinosa*. Furthermore, *P. cembra* is exploited by about 20 'specialists' in Europe (Klimetzek 1992), whereas 12 species of monophagous insects have been collected from *P. ayacahuite* var. *strobiformis*, a pine from western USA. No records of insects exploiting foliage, cones or roots of *P. ayacahuite* var. *strobiformis* are available yet. Whether the proportion, or the identity of the genera of insects, associated with each component of these pines is similar could not be determined. Also interesting is that in North America, the host specificity of the entomofauna of *P. resinosa* and *P. sylvestris*, two species belonging to subsection *Sylvestres* (Tables 18.1, 18.2), is similar: both pines have identical proportions of mono-, oligo- and polyphagous insect species (Tables 18.4, 18.5). Although the species diversity of *P. resinosa* is higher than that of *P. sylvestris*, the proportions of insects recorded on each component were comparable (71, 23, 15, 6, 2, and 5% and 70, 26, 13, 7, 2, and 5% for foliage, twig, bole, root, pollen and seed cone of *P. resinosa* and *P. sylvestris*, respectively) (Tables 18.1, 18.2). These comparisons, although rudimentary in nature, also suggest some amount of similarity between the structure of insect communities, regardless of their origin.

18.2.5 Conclusions

The entomofauna exploiting native and introduced pines in the Nearctic Region is diverse and comprises insect species with specific host associations. There are major differences in patterns of diversity among pine subgenera. The diversity of insect species exploiting hard pines is

Table 18.5. *Hypothetical entomofauna of hard and soft pines from Canada and the United States*

Insect taxa	Hard pine[a]	Soft pine
Foliage		
Cecidomyia sp. (Diptera: Cecidomyiidae)	++	–
Nuculaspis sp. (Homoptera: Diaspididae)	+	+
Chionaspis sp. (Homoptera: Diaspididae)	+	+
Matsucoccus sp. (Homoptera: Margarodidae)	+	+
Pineus sp. (Homoptera: Phylloxeridae)	++	++
Neodiprion sp. (Hymenoptera: Diprionidae)	+++	+++
Acantholyda sp. (Hymenoptera: Pamphiliidae)	++	++
Exotelia sp. (Lepidoptera: Gelechiidae)	++	+
Semiothisa sp. (Lepidoptera: Geometridae)	+	+
Nepytia sp. (Lepidoptera: Geometridae)	+	+
Choristoneura sp. (Lepidoptera: Tortricidae)	++	+
Rhyacionia sp. (Lepidoptera: Tortricidae)	++	–
Twig, shoot or small branch		
Pityophthorus sp. (Coleoptera: Scolytidae)	+++	+++
Pissodes sp. (Coleoptera: Curculionidae)	+	+
Magdalis sp. (Coleoptera: Curculionidae)	++	+
Pityogenes sp. (Coleoptera: Scolytidae)	+	+
Cinara sp. (Homoptera: Aphididae)	++	–
Aphrophora sp. (Homoptera: Cercopidae)	++	++
Toumeyella sp. (Homoptera: Coccidae)	+	+
Matsucoccus sp. (Homoptera: Margarodidae)	+	+
Retinia sp. (Lepidoptera: Tortricidae)	++	+
Rhyacionia sp. (Lepidoptera: Tortricidae)	++	–
Bole or large branch		
Buprestis sp. (Coleoptera: Buprestidae)	++	+
Chrysobothris sp. (Coleoptera: Buprestidae)	++	+
Melanophila sp. (Coleoptera: Buprestidae)	+	+
Dendroctonus sp. (Coleoptera: Scolytidae)	+	+
Hylurgops sp. (Coleoptera: Scolytidae)	++	++
Ips sp. (Coleoptera: Scolytidae)	+++	+++
Dioryctria sp. (Lepidoptera: Pyralidae)	+	+?
Synanthedon sp. (Lepidoptera: Sesiidae)	+	+
Root or root collar		
Hylobius sp. (Coleoptera: Curculionidae)	++	++
Dendroctonus sp. (Coleoptera: Scolytidae)	+	+
Hylastes sp. (Coleoptera: Scolytidae)	+	+
Pollen cone		
Xyela sp. (Hymenoptera: Xyelidae)	++	–
Seed cone		
Conophthorus sp. (Coleoptera: Scolytidae)	+	+
Leptoglossus sp. (Hemiptera: Coreidae)	+	–
Tetyra sp. (Hemiptera: Pentatomidae)	+	–
Dioryctria sp. (Lepidoptera: Pyralidae)	++	++
Cydia sp. (Lepidoptera: Tortricidae)	++	–
Eucosma sp. (Lepidoptera: Tortricidae)	++	++

[a] A + indicates at least one species of that genus; ++, a few species; +++, several species; –, no species.

higher than that feeding on soft pines and the structure of the entomofauna associated with soft and hard pines also appears distinct and predictable. Using published host records, we have identified a characteristic entomofauna associated with each component of a typical hard and soft pine species from North America (Table 18.5). The entomofauna exploiting foliage, twigs, pollen and seed cones of hard pines is much more diverse than that exploiting the same components in soft pines. In contrast, there is little difference in the diversity of insects exploiting structural

components (e.g. bole, roots) in hard and soft pines. It remains to be seen whether the differences in the entomofauna communities of pine subgenera identified for the Nearctic region also occur on other continents. Also, it is hoped that the data compiled for this analysis of the herbivore community of pines will pave the way for future investigations aimed at establishing community structure as a function of pine life stage (i.e. young, mature), and at identifying factors that could explain differences in species richness and community structure among pines.

18.3   Effect of insect herbivory on pine

The impact of insect herbivory on tree growth, survival and reproduction can be negative, positive or somewhere along the continuum (Maschinski & Whitham 1989). The biological impact of herbivory will depend on: the contribution of the affected tree components to overall tree fitness; the host's life-history strategy, age, vigour, and competitive position in the tree canopy; its ability to compensate for insect damage; the site on which it is growing; and the extent of insect pressure (Schowalter, Hargrove & Crossley 1986). Most studies on the effects of insects on pine have concentrated on important outbreak species and short-term effects during an outbreak, and thus appear to support the paradigm of negative effects. However, to show unequivocally the nature of direct effects of herbivory requires that affected plants exhibit a significant change compared with unaffected controls and that the plants be followed through to reproductive maturity to gain a true measure of fitness (Whitham et al. 1991).

Tree death is a natural and inevitable event in a forest ecosystem and is frequently the result of insects attacking trees that are weak, susceptible or senescent. Dead trees fulfil numerous functions in forest ecosystems such as providing structures for wildlife, and substantially increasing the resources available to other organisms (Franklin, Shugart & Harmon 1987). Trees surviving insect herbivory often have deformities and reduced height and volume growth because of loss of components, nutrients, metabolites and water. Growth loss may or may not have a biological impact on tree fitness per se, but can cause significant short- and long-term changes in tree biomass and economic value. For example, the shoot beetles *Tomicus piniperda* and *T. minor* (Coleoptera: Scolytidae) are major pests of pines in the Palaearctic region, but seldom cause tree mortality. Their economic importance results from their shoot-feeding behaviour, which leads to excessive growth losses (Långström & Hellqvist 1991 and references therein). In North America, the weevil *Pissodes strobi* does not kill trees directly, but destroys main stems (leaders) of *Pinus banksiana* and *P. strobus*, causing multiple leaders and thus resulting in a serious degradation in lumber quality. The budworm *Choristoneura pinus pinus* is a defoliator that periodically reaches outbreak populations and can cause mortality of *P. banksiana*. Hopkin & Howse (1995) summarized tree growth, mortality and recovery data from a jack pine budworm outbreak (1983–7) in Ontario, Canada. The cumulative total growth loss in jack pine stands ranged from 40 to 260% and was evident 3–6 years after the onset of defoliation. Whole-tree mortality was evident in intermediate and suppressed trees but lower in dominant and co-dominant trees. Generally, trees recovered to resume pre-infestation growth rates between four and five years after defoliation. The research on jack pine budworm underscores the need to evaluate the impact of herbivorous insects over space and time. Knowing the contributions of factors such as tree condition, interactions of secondary organisms (e.g. diseases), stand age, stocking, site class and site condition, before, during and after the onset of an insect outbreak, is essential in assessing the biological and economic impacts of insects on pine.

By influencing the growth and survival of individual trees, insects can influence the species and age composition, structure, productivity and development of forest stands and ecosystems. Mortality of *P. contorta*, for example, caused by the beetle *Dendroctonus ponderosae*, increases the available sunlight and improves the photosynthesis and resistance of surviving trees (Waring & Pitman 1985). The impact of herbivory at the stand level also depends on the fitness of the trees. For example, natural variation in resistance to insect attack (see section on tree improvement) can lead to greater mortality or loss in vigour of susceptible individuals, but allows genetically superior individuals to increase their growth and survival to reproductive maturity.

Insect herbivory can cause significant changes in tree form and trophic-level interactions. Whitham & Mopper (1985) and Christensen & Whitham (1991) showed how feeding by the moth *Dioryctria albovittella* on *P. edulis* not only changed the tree's architecture from upright, open-canopied trees to prostrate and close-canopied shrubs, but also delayed the tree's sexual maturity and changed the trees from being monoecious to male trees. This change in sex expression had a pronounced impact on the seed dispersal of the pinyon, since the Pinyon Jay, *Gymnorhinus cyanocephalus* (Passeriformes: Corvidae) nearly abandoned these sites.

Insects contribute to biomass decomposition, carbon and nutrient cycling, and energy flow in forest ecosystems, and have been postulated to act as regulators of forest primary production (Mattson & Addy 1975). The significance of herbivory on nutrient cycling in forests depends on several factors, including tree species composition, the herbivores involved, changes in microclimate resulting from canopy opening, and the amount, composition and seasonal pattern of material transferred compared with the normal litterfall (Schowalter *et al.* 1986). Insects initiate the cycling of carbon and other nutrients by inoculating wood with decay fungi and bacteria, which break down cellulose, hemicellulose and lignin (Haack & Byler 1993). Insect-induced defoliation can increase nutrient leaching from damaged plant tissue and accelerate litterfall. Extensive defoliation may even result in a net loss of nutrients from an ecosystem via stream flow. Insect faeces contribute to nutrient cycling, and are more easily processed by soil organisms. Frass from *P. sylvestris* foliage consumed by the sawfly *Neodiprion sertifer* provided substantial increases in nitrogen, phosphorus, potassium and magnesium added to the soil (Kimmins 1972; Fogal & Slansky 1985). Finally, insect tissues return nutrients to the soil and these tissues are more biologically active than those in fallen foliage (Schowalter *et al.* 1986).

Although we know much about the impact of insects on growth, form, seed production and survival of pines, we know comparatively little about the role of insects on the population dynamics of trees (see Crawley 1989, for further discussion). Insects can affect the rate and direction at which succession proceeds in pine ecosystems. They may delay natural succession; for example, in Switzerland the budmoth *Zeiraphera diniana* (Lepidoptera: Tortricidae) causes considerable mortality of young *P. cembra* in the undergrowth of larch stands, which in turn favours larch and delays the natural succession from pure larch stands to larch–cembran pine climax forests (Baltensweiler 1975). In the long-term context of ecosystem development, Nealis (1995) considered the interaction between *P. banksiana* and *Choristoneura pinus pinus* to be mutualistic. In the absence of fire, *P. banksiana* would be a temporary forest species. By contributing to the conditions that promote fire, *C. pinus pinus* perpetuates *P. banksiana* regeneration, thus serving not only the populations of itself but also those of its host.

The best-studied examples of insects influencing the succession of pine-dominated forests in North America involve members of the bark beetle genus *Dendroctonus*. In western North America, *D. ponderosae* generally attacks suppressed and physiologically weak *P. contorta*. Priesler & Mitchell (1993) characterize *D. ponderosae* as the ultimate forest manager in most lodgepole pine ecosystems. In the absence of fire, the selective action of beetle attack produces uneven-aged stands that tend to perpetuate on a particular site (Amman 1977). The same species can also retard or halt the successional trend by producing a tremendous quantity of fuel in the form of dead trees, thereby increasing the probability that a forest fire will kill the climax species in the understorey and prepare the way for the next generation of pioneers. Geiszler *et al.* (1980) studied the interaction of fire, insects and disease and noted that at the beginning of an outbreak of *D. ponderosae*, the beetles preferentially selected fire-scarred *Pinus contorta* weakened by the fungus *Phaeolus schweinitzii*. Once the infestation of the beetle subsided, they hypothesized that fire would soon burn the large accumulation of dead pines, leaving behind a new stand of *P. contorta* and some fire-scarred survivors. These fire-scarred trees would become infested with *P. schweinitzii*, and after 30–100 years the fungus would develop to an advanced stage of decay, setting the stage for another beetle outbreak. The natural structure and function of forests in the southeastern USA also appear to result from the interactions of fire, climate, soil, and the beetle *D. frontalis*. Schowalter, Coulson & Crossley (1981) postulated that fire and this beetle, in combination, maintained open, uneven-aged pine forests on upland sites, and seral herbaceous, pine–hardwood and hardwood communities on lowland sites. Fire would influence the temporal and spatial distribution of pine for the beetle, and the beetle activity would produce dead trees conducive to fire and windthrow. The result of this interaction would be high community diversity and productivity, which would enable the forest ecosystems to reduce nutrient losses and to respond faster to disturbances. Harrington & Wingfield (Chap. 19, this volume) provide several other examples of the influence of insects and diseases on community structure and dynamics.

18.4 Tree defence against insect herbivory and the response of insects

18.4.1 Defensive strategies and tactics of pines

Trees can deal with insect herbivory in three basic ways: (1) simply tolerate or succumb to the damage; (2) avoid or escape attack; and (3) employ defensive tactics (Berryman 1988). Mattson *et al.* (1988) ranked insect-feeding guilds (but see Hawkins & MacMahon 1989 on the meaning of guilds) according to their potential impact on plant fitness, ranging from low-impact gall-forming insects on leaves to very high-impact insects that feed on the phloem and cambium of the main stem. When insects feed on non-essential tissues or plant components just before their natural death, their impact is low and tolerance by the tree is common (Berryman 1988). In most temperate forest ecosystems, insects consume 3–8% of the foliage annually and trees may be able to compensate for light defoliation of 40–50% with little if any loss in production (Mattson & Addy 1975 and references therein). The cyclic production of pine seed cones, which tends to limit intense exploita-

tion of this tree component, is one example of an effective strategy of avoidance without much loss of fitness (Turgeon *et al.* 1994). Excessive exploitation on even non-vital plant tissues, however, can reduce the tree's vigour and fitness.

Defensive tactics are common in pine and over their long association with insects, pines have evolved two types of resin-based defence mechanisms (Langenheim 1990). The preformed oleoresin system in pines is a primary defence (also known as a constitutive or a static defence) against injury, particularly that caused by bark beetles and sawflies. This resin system is in the main stem of pine trees and consists of interconnected vertical and horizontal canals filled with oleoresin that exudes when the canals rupture. Oleoresins are found in many different tree parts and consist essentially of resin acids dissolved in mono-, sesqui- and some di-terpenes (Langenheim 1990). The physical properties of pine resins, namely, total flow, rate of flow, viscosity and time to crystallization vary considerably and have been shown to contribute to the resistance of some pines to *D. frontalis*. For example, Hodges *et al.* (1979) found that the stem oleoresins of highly resistant pines were extremely viscous, crystallized slowly, and continued flowing for long periods after wounding. Thus, when bark beetles attempt to initiate galleries in the bark of healthy trees, resin flows from the severed ducts, pitching out the attacking beetles and thus preventing colonization.

If the preformed resin system fails to defend against initial insect attack many pines can respond to insect damage by a secondary, induced defence reaction (dynamic) which may toxify, repel or interfere with the reproduction or development of the insect (Christiansen, Waring & Berryman 1987). Dynamic defences are particularly important when the impact of a particular exploitation pattern has a severe effect on tree fitness. These defences are most effective against organisms that invade the essential, living tissues of the tree, such as bark beetles and wood borers. Presumably they develop if other defences are costly, if the need and benefits for them are intermittent, and if the host can respond rapidly to herbivory (Haukioja 1990).

A well-known dynamic response is the hypersensitive or wound response of pines to bark beetles and their symbiotic fungi, one of the best known being *D. ponderosae* and its fungus, *Ophiostoma clavigerum*. Beetles of this species not repelled by primary resin flow excavate a tunnel in the bark and release the fungi. After fungal invasion, the wounded phloem and xylem parenchyma cells rapidly induce necrosis in the neighbouring cells and a lesion develops around the beetle–fungus inclusion. Within the lesion there is an increased concentration of mono-terpenes and phenolics in the wound resin, which is highly toxic to bark beetle eggs and larvae and inhibits fungal growth (Raffa & Berryman 1987; Paine & Stephen 1988 and references therein). Raffa & Berryman (1982) concluded that this wound response is highly generalized, quantitative rather than qualitative, energy-demanding, related to the overall vigour of the tree, rapid, and localized. If sufficient resins fail to secrete in time within the lesion, the tree may produce additional necrotic zones outside the first one. This process may continue until ultimately either the fungus infection is terminated or the tree becomes infected (Christiansen *et al.* 1987).

Resins also play an important role in protecting foliage. The resin canals in pine needles are usually near the lateral margins of needles and run parallel to the vascular bundles. Typically, there are at least two resin canals per needle, but there can be as many as 10 (Bennett 1954) and the number of canals can increase following the application of fertilizer (Otto & Geyer 1970). The constituents of these resins may inhibit feeding or interfere with folivore digestion or metabolism (McCullough & Wagner 1993).

The ability of pines to increase the concentrations of oleoresins in response to insect herbivory may lessen, increase, or not alter the performance of individual insects (Haukioja 1990). Furthermore, changes in insect performance may or may not affect insect population dynamics significantly (Fowler & Lawton 1985), and depend critically on the insect lifespan compared with the duration of the induced response (Haukioja 1990). Previous defoliation of 22- or 23-year-old *Pinus contorta* did not affect the performance of *Panolis flammea* (Lepidoptera: Noctuidae); however, previous defoliation of 2–3-year-old trees did affect the performance of the moth; this reduction may play a significant role in preventing outbreaks in young stands (Watt, Leather & Forrest 1991).

A discussion of plant defence theories as they relate to pine is beyond the scope of this chapter. The papers by Herms & Mattson (1992) and Tuomi (1992) provide very useful syntheses of the various theories. Most importantly, they discuss how the carbon-nutrient balance model and the growth differentiation balance model (described later in this chapter) can be integrated and unified into plant defence theory.

18.4.2 Insect counter-offensives to pine defences

Insects, with their short generation time spans, high reproductive potentials, and genetic variability, would appear more likely to evolve offensive strategies at much faster rates than trees can evolve new defensive adaptations. One strategy of insects is to avoid the defences. As noted earlier, an induced defensive response to bark beetles and their associated fungi results in a lesion around the wound leading to the death of larvae and inhibition of fungal growth. Unlike other bark-beetle

systems where the mycangial fungi induce host defences, *Dendroctonus frontalis* carries two species of mycangial fungi that do not induce defences in *P. taeda* and thus, through avoidance, maintains an ecological advantage over its host (Paine & Stephen 1987). Thus, the success of an induced defence depends on the ability of the plant to recognize the presence of an invading organism. Consequently, plants that have not had time to adapt could face serious insect-disease consequences when exotic species are added to the system, as when the pine wood nematode was introduced to Japan, and the wood wasp *Sirex noctilio* (Hymenoptera: Siricidae) was introduced with *P. radiata* in Australia, New Zealand and South Africa (see Chap. 19, this volume).

Many insects have evolved mechanisms to detoxify plant chemicals and behaviours to deactivate defences. For example, sawflies sequester resins in specialized diverticular pouches in their foregut, and thus prevent toxic compounds from entering the digestive system (Eisner *et al.* 1974). The resin acids not detoxified or otherwise inactivated by the sawflies are used for their own defence. These sequestered resin acids are an effective deterrent to birds, ants and spiders when regurgitated by sawflies (Eisner *et al.* 1974). Sawflies may diffuse the resin canal system in pine foliage by slitting the base of pine needles before oviposition, thus disrupting the resin flow distally to the needles and minimizing contact between sawfly eggs and resin compounds (McCullough & Wagner 1993). These sawfly defences vary with their host diet. Codella & Raffa (1995) found that *Neodiprion* sawflies feeding on *P. banksiana* had more repellent defensive secretions and regurgitated greater volumes of fluid than those feeding on *P. resinosa.*

Dussourd & Denno (1991) examined why plants would store chemical defences in canal systems (e.g. resin ducts in pines) that can be rapidly neutralized by behavioural adaptations such as vein cutting and trenching. Secretory canals provide effective coverage throughout the plant, respond to injury with a vigorous outflow of toxic and adhesive secretions, and store the potentially destructive chemicals away from sensitive plant tissues. By dispersing toxins throughout, the plant reduces the likelihood that defences are rapidly overcome. Thus, as Dussourd & Denno (1991) conclude, secretory canals offer an imperfect solution to conflicting demands, and the price for this form of defence is susceptibility to herbivores with behavioural counter-adaptations.

The negative and positive effects of allelochemicals on insects present them with a 'dilemma of conflicting demands'. For example, Björkman & Larsson (1991b) showed that *N. sertifer* larvae feeding on *P. sylvestris* needles with a high-resin acid content produced larger defence droplets than those feeding on low-resin acid needles, and that when exposed to predatory ants, survival of larvae on low-resin acid needles was lower than that on high-resin acid needles. Previous research showed that when larvae of *N. sertifer* fed on high-resin needles, it prolonged their development time and reduced their survival (Larsson, Björkman & Gref 1986). Björkman & Larsson (1991b) suggest that not only might the ovipositing female have to make choices to deal with the within- and between-tissue variation of resin content to solve the dilemma, but that larvae adjust their intake of resin acids depending on the prevailing predation pressure.

18.5   Management of pine: keeping insects in mind

As we have seen, insects are an important component in pine ecology but can become pests when adversely affecting human economic and social values. In managing pine to meet these values, it is important to consider, and when possible manage, the various factors that can affect the distribution and abundance of insects. Good insect pest management should be based on an approach that recognizes that pest problems are ecological problems that require ecological solutions. In this section we briefly examine some tree and site factors important in insect–host interactions.

18.5.1   Single- versus mixed-species plantations

Throughout the world, pines are grown and managed predominantly in even-aged, single-species plantations. Even-aged forest plantations of a single species (monocultures) appear to have a greater propensity to support insect outbreaks (e.g. severe, widespread defoliation) than do mixed-species stands (Mattson *et al.* 1991). Schowalter & Turchin (1993) introduced populations of *Dendroctonus frontalis* into experimental areas to evaluate the effects of pine and hardwood basal areas on susceptibility to the beetle. Their study demonstrated an increased likelihood of *D. frontalis* outbreak generation in monocultures versus mixed pine–hardwood stands. There is no doubt that under some circumstances, monoculture forestry can lead to insect outbreaks, but this is not an absolute rule (Bain 1981; Chou 1981, 1991; Watt 1992 and references therein). In some situations a reduction in plant diversity may be necessary to alleviate insect pest problems; in other situations a small or large increase in diversification may be needed. Moreover, factors associated with plantation forestry (as opposed to monoculture forestry) such as species, provenances, density, site or management practices often influence insect population dynamics and can thus confound comparisons between mixed and pure stands (Watt 1992). For example, outbreaks can result from planting exotic species in poor site conditions, or from planting the wrong provenance (see sections below), rather than from the existence of a monoculture *per se*. Insect pest outbreaks may also develop in monocultures because of a lack of natural enemies; this problem cannot be solved by increasing tree species diversity (Watt 1992; Walsh *et al.* 1993). Thus the management strategy is not as simple or intuitive as it might first appear. It is essential that the ecology of the forest, the insects and their natural enemies be understood to determine the potential risk of mono-dominant stands.

18.5.2   Exotic species

Many regions of the world have extensive plantations of exotic pine species (Chapters 20 and 21, this volume). The novel interactions among exotic trees, native or exotic insects, and site conditions make it difficult to predict the vulnerability of pine plantations to insect attack throughout the rotation period (Speight & Wainhouse 1989). Thus, the first step to protect exotic pines is the enforcement of strict quarantine measures to prevent the introduction of exotic pests. Legislation and enforcement, however, are not absolute barriers; therefore, effective internal quarantine measures are also necessary to eradicate or contain introduced insects.

Entomologists have long warned that accidental introductions of pests may pose serious threats to exotic plantations. In the late 1940s and early 1950s, the wood wasp, *Sirex noctilio*, began to devastate 16–20-year-old *P. radiata* plantations in New Zealand and Australia. This insect was introduced from Europe to New Zealand around 1900, and was found in Australia around 1950 (Ohmart 1980) and in the Western Cape of South Africa in 1994 (Tribe 1995). It is particularly destructive in dense, overstocked stands (Neumann & Minko 1981). Early thinning and management of plantations for sustained growth and vigour, coupled with biological control with parasitoids and nematodes have generally been effective in managing this pest. However, market and other economic constraints may make silvicultural thinnings unfeasible (Neumann & Minko 1981). In Australia, exotic species of bark beetles have become established in *P. radiata* plantations; Neumann (1987) urged the continuation of stringent quarantine measures to reduce the risk of further introductions. Avoiding off-site planting is recommended for the woolly aphid *Pineus pini*: this insect was found in South Africa in 1978 and is now a pest of several exotic pine species (Zwolinski 1989). Ohmart (1980) speculated that insects that attack shoots of *P. radiata* would pose the most serious threat to plantations of this species. The shoot moth *Rhyacionia buoliana* was detected for the first time in Chile in 1985 (Aguilar & Beéche 1989), and has subsequently become the most serious pest of *P. radiata* in this country where more than 1.4 million hectares have been planted with this species (Lanfranco 1994).

Native insects may or may not pose serious threats to exotic pines. For instance, *Panolis flammea*, which has been an important pest of *Pinus sylvestris* in central Europe since the early 1800s, became a pest of *P. contorta* (introduced from North America) in Scotland in the 1970s (Watt & Leather 1988). Initially, it was thought that the outbreaks of this insect were a result of increased susceptibility of this pine growing on poor soil types, mainly unflushed peat (Watt, Leather & Stoakley 1989). Subsequent work indicated that tree stress was not important to the insect reaching outbreak levels (Leather 1993), but rather the absence of natural enemies (Watt 1986; Walsh *et al.* 1993). In Australia, many indigenous insects feed on *P. radiata*, but their attacks have generally been of short duration and caused only minor damage (Neumann 1987). Nevertheless, in the future, biotypes of native insects may evolve to feed more effectively on the exotic pines.

In many parts of the southern hemisphere, pines have spread from planting sites to invade natural vegetation, sometimes causing significant problems in watersheds and protected areas and on land used for grazing (Chap. 22, this volume). To prevent many future problems, Hobbs & Humphries (1995) recommend the early detection and treatment of invasions before their spread erupts and to focus attention on the invaded ecosystem. Existing problems with pine invaders, however, may need remedial action (Chap. 22, this volume). Many pines, such as *P. contorta*, are successful at invading and colonizing sites because they can produce abundant seed crops – frequently at an early age. Insects that can destroy a large portion of the seed crop may help slow the rate of spread of invasive pines (Kay 1994), assuming that the predation pressure on a seed crop also reduces the recruitment rate of pine; this has, as far as we know, never been determined. To assess the potential of an insect to manage an invasive pine requires assessment of the biology of the insect, the pine species, and the ecosystem. Foremost in the assessment is to determine the potential of the insect to become a harmful species, either by itself or together with a pathogen (see Harris 1991; Hobbs & Humphries 1995 and references therein for a discussion of risk). To help reduce this risk, consideration should first be given to finding an indigenous insect that has established on seed cones of the invasive pine. The biocontrol programme would require rearing a large quantity of insects and releasing them in the target area. For invasive pines, however, few cone and seed insects have made the transition from indigenous conifers to pines in the southern hemisphere and none, so far, seems capable of destroying the large quantities of seed required to reduce the spread of invasive pines. Thus, importing exotic cone and seed insects may be necessary

(Kay 1994), and protocols developed for non-target host plants for biocontrol of weeds should be used. The fact that the invasive pine may also be a valuable commercial species, or a close relative of one, will make the choice of biocontrol agent all the more difficult.

18.5.3 Tree improvement

There are many examples in the literature documenting significant differences in susceptibility to insects among phenotypes and provenances of pine (e.g. Mendel 1984; Schiller & Grunwald 1987; Simpson & Ades 1990a; Day, Leather & Hines 1991; Wagner & Zhang 1993 and references therein). Despite this realization there has been relatively little attention in breeding programmes to improve resistance to insects (Nebeker *et al.* 1992). In Australia, screening *P. radiata* clones for resistance to the aphid *Pineus pini* and families for resistance to *Sirex noctilio* has been successful (Simpson & Ades 1990b, and references therein). Similarly, Donald (1989) found very strong resistance of *P. radiata* and very weak resistance of *P. pinaster* to *Pineus pini* in South Africa. With the general trend towards plantation establishment using seedlings from seed orchards, careful consideration of the plant genotype is needed when designing breeding and planting programmes. There should be a balance between high productivity of tree biomass or other commercial traits with risk from insect damage. Many insects, especially shoot insects that cause tree deformity, are difficult to control. Exploiting the natural variation in antibiosis and tolerance of trees may help reduce their impact. An obvious caveat in selecting genotypes for resistance is that resistance to one insect will not necessarily imply equal resistance to another. Thus, resistance to several insects should be considered in a tree improvement programme with due appreciation for the importance of genotype–environment interaction in expressing resistance.

18.5.4 Nitrogen fertilization

Silvicultural techniques such as fertilization may be used to manage damage to insects in pines. Available nitrogen has the greatest effect on insect performance when it is limited (Mattson 1980). Increasing the soluble nitrogen content of food generally results in better insect growth and development. The shift in the C:N balance caused by fertilizing plants is hypothesized to lead an increase in the suitability of plant tissues to insects and a decrease in the production of carbon-based allelochemicals (such as phenolics) (Bryant, Chapin & Klein 1983). In pines (as with other plants) empirical support for the carbon:nutrient (nitrogen) balance hypothesis has been mixed (see Björkman, Larsson & Gref 1991, for examples). Increased mortality of the sawfly *Neodiprion swainei* and the root weevil *Hylobius rhizophagus* was found after *P. banksiana* stands were fertilized with nitrogen (Goyer & Benjamin 1972; Smirnoff & Bernier 1973). Similarly, in fertilizer trials with N, P and K (separately and together), the susceptibility of *P. sylvestris* to the pine bark bug, *Aradus cinnamomeus* (Hemiptera: Aradidae) was reduced (Fuflygin & Grigor'ev 1987). On the other hand, NPK- and N-fertilized *P. sylvestris* seedlings were more frequently attacked by *Hylobius abietis* than either PK-fertilized or unfertilized seedlings (Selander & Immonen 1991). McCullough & Kulman (1991) found that, contrary to resource availability theory, foliar nitrogen and monoterpenes were positively related and that the survival and fecundity of the defoliator *Choristoneura pinus pinus* was not affected by fertilization with nitrogen. Björkman *et al.* (1991) found that N fertilization did not affect the performance of eggs and larvae of *Neodiprion sertifer* and that the concentration of nitrogen (positively affecting performance) and resin acids (adversely affecting performance) increased simultaneously in fertilized trees.

The effects of fertilization on nutritional quality, quantity and plant defences are complex and difficult to predict, and will depend, in part, on the type of herbivory and the specific response of the plant. Knowing how plants use the subsidized nutrients is essential in understanding the apparent conflicting results from fertilization studies. If other than subsidized nutrients are limiting, increased NPK may not increase growth but may be allocated to defences.

18.5.5 Stand establishment and tending

Selecting the appropriate species and provenance for any site is a fundamental principle of sound silviculture and insect pest management should be part of this consideration. Selecting the felling practice is also important. For example, using shelterwood felling systems rather than clear-felling systems for pine regeneration reduces damage from *Hylobius abietis* on *P. sylvestris* (Von Sydow & Örlander 1994) and *Pissodes strobi* on *Pinus strobus* (Stiell & Berry 1985). Controlling stocking levels may alleviate insect pest problems. As we have already noted, reducing stocking density in young plantations of *P. radiata* can be used to manage the wasp *Sirex nictilio* (Neumann & Minko 1981) in Australia. Thinning stands to promote increased resistance/vigour of pines is widely recommended to manage *Dendroctonus* bark beetles (but see Priesler & Mitchell 1993). Many experimental studies have shown that thinning reduces mortality of *P. contorta* and *P. ponderosa* from *D. ponderosae* (e.g. Larsson *et al.* 1983; Mitchell, Waring & Pitman 1983; Waring & Pitman 1985), and several species of pine from *D. frontalis* in the southern USA (Belanger 1980). Thinning can also affect many biochemical and physical characteristics associated with host susceptibility to bark beetles (Matson, Hain & Mawby 1987).

18.5.6 Air pollution stress

Air pollution affects insects directly through toxicity, stimulation of metabolism, or alteration of behaviour, and indirectly by affecting the behaviour of the insects' natural enemies, altering microclimate, inducing changes in host plant chemistry or morphology, or by altering the abundance and distribution of the plant (Hughes 1988; Reimer & Whittaker 1989; Heliövaara & Väisänen 1990). Most studies on the effects of air pollution on forest trees and insect populations are based on field observations of numerical changes in insect abundance, and thus are not definitive in establishing causal relationships. These studies have, nevertheless, been useful in establishing hypotheses and stimulating further research. Most observations and pollution-gradient studies show that populations of many insects increase in the presence of moderate pollution. A series of studies conducted in Finland along a pollution gradient from a factory complex emitting nitrogen oxides, sulphur dioxide and many heavy metals revealed that the density of the bark bug *Aradus cinnamomeus* on *P. sylvestris* was highest 1–2 km from the factories and then decreased nearer and further away from the source of pollution (reviewed in Heliövaara & Väisänen 1990). Severe attacks by this bug in Finland are seen only in industrial areas. Similarly, studies with the resin gall moth, *Retinia* (= *Petrova*) *resinella*, showed more galls per tree with increasing proximity to the factories (Heliövaara 1986). Reimer & Whittaker (1989) summarized field observations of insects associated with air pollution and placed them provisionally into three general patterns: (1) insects that are favoured by strong air pollution and thus occurring in frequent or chronic mass outbreaks associated with pollution, e.g. *Retinia resinella*, *Rhyacionia buoliana* and most aphids; (2) insects that are more or less absent from the centre of the pollution zone but often occur in elevated numbers in the zone of intermediate pollution, e.g. *Cinara* sp. and *Protolachnus agilis* (Homoptera: Aphididae), and *Aradus cinnamomeus*; and (3) insects that appear intolerant to pollution, e.g. *Acantholyda nemoralis*.

Air pollution can alter the amounts of carbohydrates, nitrogen and amino acids, minerals and salts, vitamins and sterols, and secondary metabolites such as phenolics and terpenoids in plants. Several studies have noted changes in terpene composition and emission in pines subjected to different levels and kinds of air pollution (cited in Reimer & Whittaker 1989). However, Kainulainen *et al.* (1993) found the total phenolic, monoterpene and resin concentrations in *P. sylvestris* seedlings did not change with distance from a sulphur dioxide source. Their results showed that SO_2 increased the nutritive value of the foliage. Reimer & Whittaker (1989), taking into consideration the limited number of studies and the dual role of terpenes in repelling/attracting insects and stimulating/inhibiting insect growth, hypothesized that an increased emission of terpenes could enhance host location by 'specialist' insects but repel 'generalist' insects. Conversely, reductions in the terpene/resin content would enhance 'generalists'. The conflicting results of air pollutants also can reflect species-specific mechanisms of interference with metabolic processes.

The large-scale effects of air pollution on insect populations are difficult to predict, especially when natural insect populations fluctuate considerably (Heliövaara & Väisänen 1990). As noted by Heliövaara & Väisänen (1990) air pollution may cause unexpected and uncontrolled changes in the population dynamics of present and potential pests. This extra variable in the complex system represents an increased problem in predicting outbreaks.

18.5.7   Drought/water stress
Although pines can grow in many habitats, they typically grow on dry sites subject to drought stress. Stress from drought, defined simply as water stress, can have numerous effects on plant and insect relationships (Mattson & Haack 1987; Holzer, Archer & Norman 1988; Waring & Cobb 1992). The observation of increased abundance of insects following periods of drought has led to the notion that drought stress limits the defensive capability of trees. In characterizing the responses of the tree to drought, it is important to consider the severity and duration of the drought stress, tree age and conditions at the onset of drought. Severe water stress lowered the exudation pressure, rate of flow and total flow of oleoresin in *P. taeda* (Lorio & Hodges 1968) and reduced the length and resin quantity of the induced defence reaction in *P. sylvestris* (Croise & Lieutier 1993). Gilmore (1977) found increased concentrations of α-pinene and decreasing concentrations of limonene, myrcene, and β-pinene in water-stressed *P. taeda*. However, moderate drought stress increased the resistance of pines to *D. frontalis* (Lorio 1988) and the tip moth *Rhyaconia frustrana* (Ross & Berisford 1990). To elucidate the behaviour of *D. frontalis* with host condition, Lorio (1988) proposed the concept of a balance between plant growth and differentiation. The 'plant growth–differentiation balance' hypothesis makes predictions about carbon budgets and tree growth phenology. This framework predicts a curvilinear response of defence to water availability: that is, moderate stress (e.g. water deficits) limits growth more than photosynthesis, which results in the production of more defence compounds (secondary metabolites); and severe water stress limits carbon assimilation, thus limiting defences and increasing susceptibility to bark beetles. Establishing the threshold between a reduction in defence compounds and successful colonization of insects is critical to the usefulness of this concept.

Non-stressed or vigorous trees may be more suitable hosts for some insects (see Price 1989). Experimental studies of water-stressed and non-stressed *P. strobus* supported field observations that *Pissodes strobi* prefers vigorous shoots (Lavallée, Albert & Mauffette 1994). McCullough & Wagner (1987) found that the sawfly *Neodiprion autumnalis* had higher pupal weights, higher survival rates and shorter feeding periods on watered or untreated *P. ponderosa* than water-stressed and trenched trees. As they note, these experimental manipulations, while permitting a separation of cause and effect, may not be the same as in a natural infestation. Furthermore, McCullough & Wagner recognize that consideration of other categories of insect performance, such as oviposition, egg viability and early instar survival, are necessary to determine the net effect of host condition on the ecological success of sawflies. Mopper & Whitham (1992) used sex ratios and fecundity to estimate the performance of *Neodiprion edulicolis* on *P. edulis* receiving water, fertilizer, both, or no treatment. Their results both refuted and supported plant stress theory. Wagner & Franz (1990) observed a negative effect of water stress on early-season oviposition of another sawfly, *N. fulviceps*, but a positive effect on late-season oviposition by *N. autumnalis*. They suggest that insects have non-linear responses to stress and, perhaps more significantly, different life-history strategies of insects create different selection factors resulting in different optimum stress levels.

Drought is often considered as a predisposing stress factor triggering outbreaks of insects. Insect outbreaks associated with drought-stressed pine mainly involve Coleoptera (*Dendroctonus brevicomis*, *D. frontalis*, *D. ponderosae*, *Ips calligraphus*, *I. gradicollis*, *I. paraconfusus*, *Pityophthorus antillicus* and *P. pinavorous*), Hymenoptera (*Neodiprion sertifer*), and Lepidoptera (*Bupalus piniarius* and *Selidosema suavis* (Geometridae)) (Mattson & Haack 1987; Haack, Billings & Richter 1990). Recent experimental evidence by Selander & Immonen (1991) with *P. sylvestris* and *Hylobius abietis* supports field studies showing that water stress predisposes the tree to insect damage.

The bulk of evidence for stress leading to outbreaks is largely circumstantial (Mattson & Haack 1987). Moreover, there is not always a consistent link between drought-stressed trees and insect outbreaks (Watt 1986; Lorio 1988; Larsson 1989; Wagner & Franz 1990). Larsson (1989) discusses the various concepts of stress and their utility in searching for patterns among feeding guilds. He suggests that we need to consider the diversity of plant responses to stress and differences in response to stressed plants that may exist among insects with different feeding habits. Based on experimental evidence, Larsson (1989) proposed ranking insect feeding guilds according to their response to stress-induced changes in their food resource, namely cambium feeders > sucking insects > mining insects > chewing insects > gall-forming insects (high to low response). Finer delineation of feeding habits (or life-history strategies) may be necessary. For example, the survey by Mattson & Haack (1987) shows that many bark beetle outbreaks in pine have been associated with drought stress in particular, or climatic stress overall. Nevertheless, trees which grow the fastest, not slowest, can have the highest risk of attack by the southern pine beetle (see Lorio, Mason & Autry 1982), and stressed sites do not always lead to outbreaks of mountain pine beetles (Schmid *et al.* 1991). Severe water stress for more than 100 days, however, failed to reduce the susceptibility of 11-year-old *P. taeda* to attack by *D. frontalis* (Dunn & Lorio 1993). Turchin *et al.* (1991), using time-series and regression analysis, found that three climatic variables (heating degree-days, days above 90 °F, and water deficits) did not explain periodic population oscillations of *D. frontalis*, but density-dependent population factors did. As they note, weather could advance or delay the onset of a beetle outbreak, and the outbreak amplitude. The interaction of the timing and intensity of stress with genetic variation may explain some of the conflicting reports in the literature on host plant stress and herbivore performance. McMillin & Wagner (1995) demonstrated, in a series of greenhouse studies, that *P. ponderosa* seedlings subjected to moderate and high intensities of water stress during the shoot growth period made poor-quality hosts for *Neodiprion gillettei* but that seedlings subjected to identical conditions during the root growth phase were high-quality hosts.

Plant stress theory remains highly contentious, hence the plant stress paradox *sensu* Mopper & Whitham (1992). Various general models have been developed to resolve inconsistencies (see Mopper & Whitham 1992; Krause, Raffa & Wagner 1993 for reviews). Mopper & Whitham (1992) provide an important solution by arguing that we must recognize the differences between simultaneous plant stress and sustained plant stress. Simultaneous plant stress, such as low precipitation while the insect is feeding, can reduce insect performance, whereas sustained stress, such as chronically poor soil conditions or prolonged drought, can benefit insect performance. Besides providing a comprehensive review of studies and concepts of plant stress, Waring & Cobb (1992) urged more careful consideration of study design, particularly regarding the nature of plant stress and the natural history of the herbivore.

18.6 Conclusions

We have attempted to show that insects and trees are dynamic, interacting forces. Insects are indicators of forest health and are contributors to long-term ecosystem functioning. There is a diverse, host-specific, and to some extent predictable, array of insects that feed on soft and hard pines. Although the insect fauna associated with pines in Canada and the USA has been comparatively well catalogued, more taxonomic and phylogenetic analyses are needed to help us understand the ecology of these insects. Our knowledge of insect–pine interactions is somewhat rudimentary and fragmented. Generalizations or rules about insect–plant interactions have almost invariably been made based on too few studies and are seldom valid for a broad group of insects. Delineating insects by feeding groups, guilds, or as some other community recognizes that different groups of insects respond differently to various factors, and thus can improve our understanding and our predictive ability. More critical analyses are needed to help clarify concepts, resolve apparent contradictions in theory, and refine our study approaches. Our greatest knowledge is, understandably, heavily biased towards insect pests of economically important pines. Increased insight may come from examining relationships of related, yet less economically important, insect–tree interactions. We stress the need for fundamental research to further our appreciation for the role of insects (especially those that are not pests) at the ecosystem level – we expect that their role has been underestimated.

Acknowledgements

Thanks to S. D'Eon, Natural Resources Canada, Canadian Forest Service, Petawawa for providing a list of records of insects on pines contained in the FIDS Infobase; R. Bridges, USA Department of Agriculture, Forest Service, Washington for providing a database on insects of pines from the USA; A. Roques, Institut National de la Recherche Agronomique, Orléans, France, for sharing books from his personal library; K. Nystrom, Natural Resources Canada, Canadian Forest Service, Sault Ste. Marie for reviewing part of the list for accuracy and providing photographs from the FIDS collection; L. Safranyik, Natural Resources Canada, Canadian Forest Service, Victoria for providing photographs; and D. Bright, P. Dang, H. Goulet, J. Huber and B. Landry, from the Biological Resources Division, Ottawa for reviewing an earlier version of the list of insects exploiting pines in North America. Thanks to V.G. Nealis, W.J. Mattson, T.D. Schowalter and two anonymous reviewers for their valuable comments, and a special thanks to Dave Richardson for his guidance and patience.

References

Aguilar, A.M. & Beéche, M.A. (1989). Current situation of the European pine shoot moth (*Rhyaciona buoliana* (Schiff.) Lepidoptera: Tortricidae) in Chile. In *Insects Affecting Reforestation: Biology and Damage*, ed. R.I. Alfaro & S.G. Glover, pp. 155–60. Proceedings of IUFRO (S2.07–03), Vancouver, British Columbia, 3–9 July 1988. Victoria, BC: Forestry Canada.

Amman, G.D. (1977). The role of the mountain pine beetle in lodgepole pine ecosystems: impact on succession. In *The Role of Arthropods in Forest Ecosystems*, ed. W.J. Mattson, pp. 3–18. New York: Springer-Verlag.

Annila, E., Heliövaara, K., Puukko, K. & Rousi, M. (1983). Pests on Lodgepole pine (*Pinus contorta*) in Finland. *Communicationes Instituti Forestalis Fenniae*, **115**, 1–27.

Bain, J. (1981). Forest monocultures – how safe are they? An entomologist's view. *New Zealand Journal of Forestry*, **26**, 37–42.

Baltensweiler, W. (1975). The importance of the larch budmoth (*Zeiraphera diniana* Gn.) in the biocoenosis of forests of larch and Swiss stone pine [in German]. *Mitteilungen der Schweizerischen entomologischen Gesellschaft*, **48**, 5–12. [Original not seen.]

Barbosa, P. & Wagner, M.R. (1989). *Introduction to Forest and Shade Tree Insects*. San Diego: Academic Press.

Belanger, R.P. (1980). Silvicultural guidelines for reducing losses to the southern pine beetle. In *The Southern Pine Beetle*, ed. R.C. Thatcher, J.L. Searcy, J.E. Coster & G.D. Hertel, pp. 165–77. USDA, Forest Service Technical Bulletin 1631.

Bennett, W.H. (1954). The effect of needle structure upon the susceptibility of hosts to the pine needle miner (*Exoteleia pinifoliella* (Chamb.)) (Lepidoptera: Gelechiidae). *The Canadian Entomologist*, **86**, 49–54.

Berryman, A.A. (1986). *Forest Insects*: New York: Plenum Press.

Berryman, A.A. (1988). Towards a unified theory of plant defense. In *Mechanisms of Woody Plant Defenses Against Insects, Search for Pattern*, ed. W.J. Mattson, J. Levieux & C.B. Bernard-Dagan, pp. 39–55. New York: Springer-Verlag.

Bevan, D. (1987). *Forest Insects. A Guide to Insects Feeding on Trees in Britain*. Forestry Commission Handbook 1. London: HMSO books.

Björkman, C. & Larsson, S. (1991a). Host plant specialization in needle-eating insects of Sweden. In *Forest Insect Guilds: Patterns of Interactions with Host Trees*, ed. Y.N. Baranchikov, W.J. Mattson, F.P. Hain & T.L. Payne, pp. 1–20. Randor: USDA Forest Service, General Technical Report NE-153.

Björkman, C. & Larsson, S. (1991b). Pine sawfly defence and variation in host plant resin acids: a trade-off with growth. *Ecological Entomology*, **16**, 283–9.

Björkman, C., Larsson, S. & Gref, R. (1991). Effects on nitrogen fertilization on pine needle chemistry and sawfly performance. *Oecologia*, **86**, 202–9.

Bryant, J.P., Chapin, F.S. & Klein, D.R. (1983). Carbon/nutrient balance of boreal plants in relation to vertebrate herbivory. *Oikos*, **40**, 357–68.

Chou, C.K.S. (1981). Monoculture, species diversification, and disease hazards in forestry. *New Zealand Journal of Forestry*, **26**, 20–36.

Chou, C.K.S. (1991). Perspectives of disease threat in large-scale *Pinus radiata* monoculture – the New Zealand experience. *European Journal of Forest Pathology*, **21**, 71–81.

Christensen, K.M. & Whitham, T.G. (1991). Indirect and herbivore mediation of avian seed dispersal in pinyon pine. *Ecology*, **72**, 534–42.

Christiansen, E., Waring, R.H. & Berryman, A.A. (1987). Resistance of conifers to bark beetle attack: searching for general relationships. *Forest Ecology and Management*, **22**, 89–106.

Codella, S.G. Jr & Raffa, K.F. (1995). Host plant influence on chemical defense in conifer sawflies. *Oecologia*, **104**, 1–11.

Coulson, R.N. & Witter, J.A. (1984). *Forest Entomology*. New York: John Wiley.

Crawley, M.J. (1989). Insect herbivores and plant population dynamics. *Annual Review of Entomology*, **34**, 531–64.

Critchfield, W.B. & Little, E.L. (1966). *Geographic Distribution of Pines of the World*. USDA Forest Service Miscellaneous Publication 991, Washington, DC: USDA Forest Service.

Croise, L. & Lieutier, F. (1993). Effects of drought on the induced defence reaction of Scots pine to bark beetle-associated fungi. *Annales des Sciences Forestières*, **50**, 91–7.

Day, K.R. & Leather, S.R. (1997). Threats to forestry by insect pests in Europe. In *Forests and Insects*, ed. A.D. Watt, N.E. Stork & M.D. Hunter, pp. 177–205. London: Academic Press.

Day, K.R., Leather, S.R. & Hines, R. (1991). Damage by *Zeiraphera diniana* (Lepidoptera: Tortricidae) to lodgepole pine (*Pinus contorta*) of various provenances. *Forest Ecology and Management*, **44**, 133–45.

de Groot, P. (1986). Mortality factors of jack pine, *Pinus banksiana*, Lamb., strobili. In *Proceedings, Cone and Seed Insects Working Party 2nd Conference. IUFRO S2.07–01*, ed. A. Roques, pp. 39–52, Briancon, France: IUFRO.

de Groot, P. (1992). Biosystematics of *Conophthorus* Hopkins (Coleoptera: Scolytidae) in eastern North America. PhD thesis, Simon Fraser University, Burnaby, BC.

Donald, D.G.M. (1989). Resistance of seedling progeny of *Pinus pinaster* and *P. radiata* to the pine woolly aphid *Pineus pini* L. *South African Forestry Journal*, **150**, 20–3.

Drooz, A.T. (1985). *Insects of Eastern Forests*. USDA Forest Service Miscellaneous Publication 1426.

Dunn, J.P. & Lorio, P.L., Jr (1993). Modified water regimes affect photosynthesis, xylem water potential, cambial growth, and resistance of juvenile *Pinus taeda* L. to *Dendroctonus frontalis* (Coleoptera: Scolytidae). *Environmental Entomology*, **22**, 948–57.

Dussourd, D.E. & Denno, R. (1991). Deactivation of plant defense: correspondence between insect behaviour and secretory canal architecture. *Ecology*, **72**, 1383–96.

Eisner, T., Johnessee, J.S., Carrel, J., Hendry, L.B. & Meinwald, J. (1974). Defensive use by an insect of a plant resin. *Science*, **184**, 996–9.

Fogal, W.H. & Slansky, F., Jr. (1985). Contribution of feeding by European pine sawfly larvae to litter production and element flux in Scots pine plantations. *Canadian Journal of Forest Research*, **15**, 484–7.

Fowler, S.V. & Lawton, J.H. (1985). Rapidly induced defences and talking trees: the devil's advocate position. *American Naturalist*, **126**, 181–95.

Franklin, J.F., Shugart, H.H. & Harmon, M.E. (1987). Tree death as an ecological process. *Bioscience*, **37**, 550–6.

Fuflygin, G.V. & Grigor'ev, P.P. (1987). Effect of mineral fertilizers on Scots pine plantations infested by *Aradus cinnamomeus* [in Russian]. *Lesnoe Khozyaistvo*, **4**, 62–4. [Original not seen.]

Furniss, R.L. & Carolin, V.M. (1977). *Western Forest Insects*. USDA Forest Service Miscellaneous Publication 1339.

Gagné, R.J. (1989). *The Plant-Feeding Gall Midges of North America*. Ithaca, NY: Cornell University Press.

Geiszler, D.R., Gara, R.I., Driver, C.H., Gallucci, V.H. & Martin, R.E. (1980). Fire, fungi, and beetle influences on a lodgepole pine ecosystem of south-central Oregon. *Oecologia*, **46**, 239–43.

Gilmore, A.R. (1977). Effects of soil moisture stress on monoterpenes in loblolly pine. *Journal of Chemical Ecology*, **3**, 667–76.

Goyer, R.A. & Benjamin, D.M. (1972). Influence of soil fertility on infestation of jack pine plantations by the pine root weevil. *Forest Science*, **18**, 139–47.

Haack, R.A., Billings, R.F. & Richter, A.M. (1990). Life history parameters of bark beetles (Coleoptera: Scolytidae) attacking West Indian pine in the Dominican Republic. *Florida Entomologist*, **73**, 591–603.

Haack, R.A. & Byler, J.W. (1993). Insects and pathogens – regulators of forest ecosystems. *Journal of Forestry*, **91**, 32–7.

Harris, P. (1991). Classical biocontrol of weeds: its definition, selection of effective agents, and administrative-political problems. *The Canadian Entomologist*, **123**, 827–49.

Haukioja, E. (1990). Induction of defenses in trees. *Annual Review of Entomology*, **36**, 25–42.

Hawkins, C.P. & MacMahon, J.A. (1989). Guilds: the multiple meanings of a concept. *Annual Review of Entomology*, **34**, 423–51.

Hedlin, A.F., Yates, H.O. III, Cibrian-Tovar, D., Ebel, B.H., Koerber, T.W. & Merkel, E.P. (1980). *Cone and Seed Insects of North American Conifers*. Ottawa: Environment Canada, Canadian Forest Service/Washington: United States Forest Service/Mexico: Secretaría de Agricultura y Recursos Hidráulicos.

Heliövaara, K. (1986). Occurrence of *Petrova resinella* (Lepidoptera, Tortricidae) in a gradient of industrial air pollutants. *Silva Fennica*, **20**, 83–90.

Heliövaara, K. & Väisänen, R. (1990). Changes in population dynamics of pine insects induced by air pollution. In *Population Dynamics of Forest Insects*, ed. A.D. Watt, S.R. Leather, M.D. Hunter & N.A.C. Kidd, pp. 209–18. Andover, Hampshire: Intercept.

Herms, D.A. & Mattson, W.J. (1992). The dilemma of plants: to grow or defend. *Quarterly Review of Biology*, **67**, 283–335.

Hobbs, R.J. & Humphries, S.E. (1995). An integrated approach to the ecology and management of plant invasions. *Conservation Biology*, **9**, 761–70.

Hodges, J.D., Elam, W.W., Watson, W.F. & Nebeker, T.E. (1979). Oleoresin characteristics and susceptibility of four southern pines to southern pine beetle (Coleoptera: Scolytidae) attacks. *The Canadian Entomologist*, **111**, 889–96.

Holzer, T.O., Archer, T.L. & Norman, J.M. (1988). Host plant suitability in relation to water stress. In *Plant Stress–Insect Interactions*, ed. E.A. Heinrichs, pp. 111–37. New York: John Wiley.

Hopkin, A.A. & Howse, G.M. (1995). Impact of the jack pine budworm in Ontario: a review. In *Jack Pine Budworm Biology and Management*, ed. W.J.A. Volney, V.G. Nealis, G.M. Howse, A.R. Westwood, D.R. McCullough & B.L. Laishley, pp. 111–19. Edmonton: Natural Resources Canada, Canadian Forest Service, Information Report NOR-X-342.

Hughes, P.R. (1988). Insect populations on host plants subjected to air pollution. In *Plant Stress–Insect Interactions*, ed. E.A. Heinrichs, pp. 249–319. New York: John Wiley.

Ives, W.G.H. & Wong, H.R. (1988). *Tree and Shrub Insects of the Prairie Provinces*. Canadian Forest Service, Information Report NOR-X-292.

Johnson, W.T. & Lyon, H.H. (1988). *Insects that Feed on Trees and Shrubs*, 2nd edn. Ithaca, NY: Cornell University Press.

Joly, R. (1975). *Les Insectes Ennemis Des Pins*. Vol. 1. Nancy: Ecole nationale du génie rural, des eaux et des forêts.

Kainulainen, P., Satka, H., Mustaniemi, A., Holpainen, J.K. & Okasanen, J. (1993). Conider aphids in an air-polluted environment. II Host plant quality. *Environmental Pollution*, **80**, 193–200.

Kay, M. (1994). Biological control for invasive species. *New Zealand Forestry*, **39**(3), 35–7.

Kimmins, J.P. (1972). Relative contributions of leaching, litter-fall, and defoliation by *Neodiprion sertifer* (Hymenoptera) to the removal of cesium-134 from red pine. *Oikos*, **23**, 226–32.

Klimetzek, D. (1992). Schädlingsbelastung der Waldbäume in Mitteleuropa und Nordamerika. *Forst-Wissenschaftliches Centralblatt*, **111**, 61–9.

Krause, S.C., Raffa, K.F. & Wagner, M.R. (1993). Tree response to stress: a role in sawfly outbreaks. In *Sawfly Life History Adaptations to Woody Plants*, ed. M.R. Wagner & K.F. Raffa, pp. 211–27. San Diego: Academic Press.

Lanfranco, D.M. (1994). Pest problems of intensive forestry: the shoot moth invasion of radiata pine in Chile. In *The White Pine Weevil: Biology, Damage and Management*, ed. R.I. Alfaro, G. Kiss & R.G. Fraser, pp. 301–11. British Columbia Forest Resource Development Agreement Report 226. Victoria: Forestry Canada.

Langenheim, J.H. (1990). Plant resins. *American Scientist*, **78**, 16–24.

Långström, B. & Hellqvist, C. (1991). Shoot and growth losses following three years of *Tomicus*-attack in Scots pine stands close a timber storage site. *Silva Fennica*, **25**, 133–45.

Larsson, S. (1989). Stressful times for the plant stress–insect performance hypothesis. *Oikos*, **56**, 277–83.

Larsson, S., Björkman, C. & Gref, R. (1986). Responses of *Neodiprion sertifer* (Hymenoptera: Diprionidae) larvae to variation in needle resin acid concentration in Scots pine. *Oecologia*, **70**, 77–84.

Larsson, S., Oren, R., Waring, R.H. & Barrett, J.W. (1983). Attacks of mountain pine beetle as related to tree vigor of ponderosa pine. *Forest Science*, **29**, 395–402.

Lavallée, R., Albert, P.J. & Mauffette, Y. (1994). Influences of white pine watering regimes on feeding preferences of spring and fall adults of the white pine weevil *Pissodes strobi* (Peck). *Journal of Chemical Ecology*, **20**, 831–47.

Leather, S.R. (1993). Influence of site factor modification on the population development of the pine beauty moth (*Panolis flammea*) in a Scottish lodgepole pine (*Pinus contorta*) plantation. *Forest Ecology and Management*, **59**, 207–23.

Lessard, G. & Jennings, D.T. (1976). *Southwestern Pine Tip Moth Damage to Ponderosa Pine Reproduction.* USDA Forest Service Research Paper RM-168.

Little, E.L. & Critchfield W.B. (1969). *Subdivisions of the genus Pinus (Pines).* USDA Forest Service Miscellaneous Publication 1144.

Lorio, P.L. Jr (1988). Growth differentiation-balance relationships in pines affect their resistance to bark beetles (Coleoptera: Scolytidae). In *Mechanisms of Woody Plant Defenses Against Insects. Search for Pattern,* ed. W.J. Mattson, J. Levieux & C.B. Bernard-Dagan, pp. 73–92. New York: Springer-Verlag.

Lorio, P.L. Jr & Hodges, J.D. (1968). Oleoresin exudation pressure and relative water content on inner bark as indicators of moisture stress in loblolly pines. *Forest Science,* **14,** 392–8.

Lorio, P.L. Jr, Mason, G.N. & Autry, G.L. (1982). Stand risk rating for southern pine beetle: integrating pest management with forest management. *Journal of Forestry,* **80,** 212–14.

McCullough, D.G. & Kulman, H.M. (1991). Effects of nitrogen fertilization on young jack pine (*Pinus banksiana*) and on its suitability as a host for jack pine budworm (*Choristoneura pinus pinus*) (Lepidoptera: Torticidae). *Canadian Journal of Forest Research,* **21,** 1447–58.

McCullough, D.G. & Wagner, M.R. (1987). Influence of watering and trenching ponderosa pine on a pine sawfly. *Oecologia,* **71,** 382–7.

McCullough, D.G. & Wagner, M.R. (1993). Defusing host defenses: ovipositional adaptations of sawflies to plant resins. In *Sawfly History Adaptations to Woody Plants,* ed. M. Wagner & K.R. Raffa, pp. 157–72. San Diego: Academic Press.

McMillin, J.L. & Wagner, M.R. (1995). Season and intensity of water stress: Host-plant effects on larval survival and fecundity of *Neodiprion gillettei* (Hymenoptera: Diprionidae). *Environmental Entomology,* **24,** 1251–7.

Martineau, R. (1984). *Insects Harmful to Forest Trees.* Quebec: Multiscience Publications Limited.

Maschinski, J. & Whitham, T.G. (1989). The continuum of plant responses to herbivory: the influence of plant association, nutrient availability, and timing. *American Naturalist,* **134,** 1–19.

Matson, P.A., Hain, F. & Mawby, W. (1987). Indices of tree susceptibility to bark beetle vary with silvicultural treatment in a loblolly pine plantation. *Forest Ecology and Management,* **22,** 107–18.

Mattson, W.J. (1975). Abundance of insects inhabiting the male strobili of red pine. *Great Lakes Entomologist,* **8,** 237–9.

Mattson, W.J. (1980). Herbivory in relation to plant nitrogen content. *Annual Review of Ecology and Systematics,* **11,** 119–61.

Mattson, W.J. & Addy, N.D. (1975). Phytophagous insects as regulators of forest primary production. *Science,* **190,** 515–22.

Mattson, W.J. & Haack, R.A. (1987). The role of drought stress in provoking outbreaks of phytophagous insects. In *Insect Outbreaks,* ed. P. Barbosa & J.C. Schultz, pp. 365–407. San Diego: Academic Press.

Mattson, W.J., Herms, D.A., Witter, J.A. & Allen, D.C. (1991). Woody plant grazing systems: North American outbreak folivores and their host plants. In *Forest Insect Guilds: Patterns of Interaction with Host Trees,* ed. Y.N. Baranchikov, W.J. Mattson, F.P. Hain & T.L. Payne, pp. 53–84. USDA Forest Service General Technical Report NE-153.

Mattson, W.J., Lawrence, R.K., Haack, R.A., Herms, D.A. & Charles, P.-J. (1988). Defensive strategies of woody plants against different insect-feeding guilds in relation to plant ecological strategies and intimacy of association with insects. In *Mechanisms of Woody Plant Defenses Against Insects. Search for Pattern,* ed. W.J. Mattson, J. Levieux & C.B. Bernard-Dagan, pp. 3–38. New York: Springer-Verlag.

Mattson, W.J., Niemela, P., Millers, I. & Inguanzo, Y. (1994). *Immigrant Phytophagous Insects on Woody Plants in the United States and Canada: An Annotated List.* USDA Forest Service General Technical Report NC-169.

Mendel, Z. (1984). Provenance as a factor in susceptibility on *Pinus halepensis* to *Matsucoccus josephi* (Homoptera: Margarodidae). *Forest Ecology and Management,* **9,** 259–66.

Mitchell, R.G., Waring, R.H. & Pitman, G.B. (1983). Thinning lodgepole pine increases tree vigor and resistance to mountain pine beetle. *Forest Science,* **29,** 204–11.

Mopper, S. & Whitham, T.G. (1992). The plant stress paradox; effects on pinyon sawfly sex ratios and fecundity. *Ecology,* **73,** 515–25.

Nealis, V.G. (1995). Population biology of the jack pine budworm. In *Jack Pine Budworm Biology and Management,* ed. W.J.A. Volney, V.G. Nealis, G.M. Howse, A.R. Westwood, D.R. McCullough & B.L. Laishley, pp. 55–71. Edmonton: Natural Resources Canada, Canadian Forest Service, Information Report NOR-X-342.

Nealis, V.G. & Lomic, P.V. (1994). Host-plant influence on the population ecology of the jack pine budworm, *Choristoneura pinus* (Lepidoptera: Tortricidae). *Ecological Entomology,* **19,** 367–73.

Nebeker, T.E., Hodges, J.D., Blanche, C.A., Honea, C.R. & Tisdale, R.A. (1992). Variation in the constitutive defense system of loblolly pine in relation to bark beetle attack. *Forest Science,* **38,** 457–66.

Neumann, F.G. (1987). Introduced bark beetle on exotic tree in Australia with special reference to infestations of *Ips grandicolis* in pine plantations. *Australian Forestry,* **50,** 166–78.

Neumann, F.G. & Minko, G. (1981). The sirex wood wasp in Australian radiata pine plantations. *Australian Forestry,* **44,** 46–63.

Nystrom, K.L. & Britnell, W.E. (1994). *Insects and mites associated with Ontario Forests: Classification, Common names, main hosts and importance.* Natural Resources Canada, Canadian Forest Service, Information Report O-X-439.

Ohmart, C.P. (1980). Insect pests of *Pinus radiata* plantations: present and possible future problems. *Australian Forestry,* **43,** 226–32.

Otto, D. & Geyer, W. (1970). Zur Bedeutung des Kiefernnadelharzes und des Kiefernnadelöles für die Entwicklung nadelfressender Insekten [in German]. *Arch Forstwes,* **19,** 151–67 [original not seen].

Paine, T.D. & Stephen, F.M. (1987). Fungi associated with the southern pine beetle: avoidance of induced defense response in loblolly pine. *Oecologia,* **74,** 377–9.

Paine, T.D. & Stephen, F.M. (1988). Induced defences of loblolly pine, *Pinus taeda*: potential impact on *Dendroctonus frontalis* within-tree mortality. *Entomologia experimentlis et applicata,* **46,** 39–46.

Preisler, H.K. & Mitchell, R.G. (1993). Colonization patterns of the mountain pine beetle in thinned and unthinned lodgepole pine stands. *Forest Science*, **39**, 528–45.

Price, P. (1989). The plant vigor hypothesis and herbivore attack. *Oikos*, **62**, 244–51.

Raffa, K.F. & Berryman, A.A. (1982). Physiological differences between lodgepole pines resistant and susceptible to the mountain pine beetle and associated organisms. *Environmental Entomology*, **11**, 486–92.

Raffa, K.F. & Berryman, A.A. (1987). Interacting selective pressures in conifer–bark beetle systems: a basis for reciprocal adaptations? *American Naturalist*, **129**, 234–62.

Reimer, J. & Whittaker, J.B. (1989). Air pollution and insect herbivores: observed interactions and possible mechanisms. In *Insect–Plant Interactions*, Vol. 1, ed. E.A. Bernays, pp. 73–105. Boca Raton, Florida: CRC Press.

Roques, A. (1983). *Les Insectes Ravageurs des Cônes et Graines de Conifères en France*. Paris: INRA.

Roques, A. (1991). Structure, specificity and evolution of insect guilds related to cones of conifers in western Europe. In *Forest Insect Guilds: Patterns of Interactions with Host Trees*, ed. Y.N. Baranchikov, W.J. Mattson, F.P. Hain & T.L. Payne, pp. 300–15. USDA Forest Service General Technical Report NE-153.

Rose, A.H. & Lindquist, O.H. (1973). *Insects of Eastern Pines*. Canadian Forestry Service, Publication 1313. Ottawa: Department of Environment.

Ross, D.W. & Berisford, C.W. (1990). Nantucket pine tip moth (Lepidoptera: Tortricidae) response to water and nutrient status of loblolly pine. *Forest Science*, **36**, 719–33.

Schiller, G. & Grunwald, C. (1987). Resin monoterpenes in range-wide provenance trials of *Pinus halepensis* Mill. in Israel. *Silvae Genetica*, **36**, 109–14.

Schmid, J.M., Mata, S.A., Watkins, R.K. & Kaufmann, M.R. (1991). Water potential in ponderosa pine stands of different growing stock levels. *Canadian Journal of Forest Research*, **21**, 750–5.

Schowalter, T.D., Coulson, R.N. & Crossley, D.A., Jr (1981). Role of southern pine beetle and fire in the maintenance of structure and function of the southeastern coniferous forest. *Environmental Entomology*, **10**, 821–5.

Schowalter, T.D., Hargrove, W.W. & Crossley, D.A., Jr (1986). Herbivory in forested ecosystems. *Annual Review of Entomology*, **31**, 177–96.

Schowalter, T.D. & Turchin, P. (1993). Southern pine beetle infestation development: Interaction between pine and hardwood basal areas. *Forest Science*, **39**, 201–10.

Selander, J. & Immonen, A. (1991). Effect of fertilization on the susceptibility of Scots pine seedlings to the large pine weevil *Hylobius abietis. Folia Forestalia*, **771**, 1–21.

Simpson, J.A. & Ades, P.K. (1990a). Variation in susceptibility of *Pinus muricata* and *Pinus radiata* to two species of Aphidoidae. *Silvae Genetica* **39**, 202–6.

Simpson, J.A. & Ades, P.K. (1990b). Screening *Pinus radiata* families and clones for disease and pest insect resistance. *Australian Forestry*, **53**, 194–9.

Smirnoff, W.A. & Bernier, B. (1973). Increased mortality of the Swaine jack-pine sawfly, and foliar nitrogen concentrations after urea fertilization. *Canadian Journal of Forest Research*, **3**, 112–21.

Speight, M.R. & Wainhouse, D. (1989). *Ecology and Management of Forest Insects*. Oxford: Claredon Press.

Stiell, W.M. & Berry, A.M. (1985). Limiting white pine weevil attacks by side shade. *The Forestry Chronicle*, **61**, 5–9.

Strong, D.R., Lawton, J.H. & Southwood, R. (1984). *Insects on Plants. Community Patterns and Mechanisms*. Cambridge, Mass.: Harvard University Press.

Tribe, G.D. (1995). The woodwasp *Sirex noctilio* (Hymenoptera: Siricidae), a pest of *Pinus* species, now established in South Africa. *African Entomology*, **3**, 215–17.

Tuomi, J. (1992). Toward integration of plant defence theories. *Trends in Ecology and Evolution*, **7**, 356–67.

Turchin, P., Lorio, P.L., Taylor, A.D. & Billings, R.F. (1991). Why do populations of southern pine beetles (Coleoptera: Scolytidae) fluctuate? *Environmental Entomology*, **20**, 401–9.

Turgeon, J.J. (1994). Insects exploiting seed cones of *Larix* spp., *Picea* spp. and *Pinus* spp.: species richness and patterns of exploitation. In *Biology, Damage and Management of Seed Orchard Pests*, ed. J.J. Turgeon & P. de Groot, pp. 15–30. Canadian Forest Service Information Report, FPM-X-89.

Turgeon, J.J. & de Groot, P. (1992). *Management of Insect Pests of Cones and Seeds in Eastern Canada: A Field Guide*. Toronto: Ontario Ministry of Natural Resources/Forestry Canada.

Turgeon, J.J., Roques, A. & de Groot, P. (1994). Insect fauna of coniferous seed cones: diversity, host plant interactions, and management. *Annual Review of Entomology*, **39**, 179–212.

Von Sydow, F. & Örlander, G. (1994). The influence of shelterwood density on *Hylobius abietis* (L.) occurrence and feeding on planted conifers. *Scandinavian Journal of Forest Research*, **9**, 367–75.

Wagner, M.R. & Franz, D.P. (1990). Influence of induced water stress in ponderosa pine on pine sawflies. *Oecologia*, **83**, 452–7.

Wagner, M.R. & Zhang, Z.Y. (1993). Host plant traits associated with resistance of ponderosa pine to the sawfly, *Neodiprion fulviceps. Canadian Journal of Forest Research*, **23**, 839–45.

Walsh, P.J., Day, K.R., Leather, S.R. & Smith, A.J. (1993). The influence of soil type and pine species on the carabid community of a plantation forest with a history of pine beauty moth infestation. *Forestry*, **66**, 135–46.

Waring, G.L. & Cobb, N.S. (1992). The impact of plant stress on herbivore population dynamics. In *Insect–Plant Interactions*, Vol. 4, ed. E. Bernays, pp. 167–226. Boca Raton, Florida: CRC Press.

Waring, R.H. & Pitman, G.B. (1985). Modifying lodgepole pine stands to change susceptibility to mountain pine beetle attack. *Ecology*, **66**, 889–97.

Watt, A.D. (1986). The performance of the pine beauty moth on water-stressed lodgepole pine plants: a laboratory experiment. *Oecologia*, **70**, 578–9.

Watt, A.D. (1992). Insect pest populations: effects of tree species diversity. In *The Ecology of Mixed-Species Stands of Trees*, ed. M.G.R. Cannell, D.C. Malcolm & P.A. Robertson, pp. 267–75. Oxford: Blackwell.

Watt, A.D. & Leather, S.R. (1988). The pine beauty moth in Scottish lodgepole pine plantations. In *Dynamics of Forest Insect Populations*, ed. A.A. Berryman, pp. 243–66. New York: Plenum Press.

Watt, A.D., Leather, S.R. & Forrest, G.I. (1991). The effect of previous defoliation of pole-stage lodgepole pine on plant chemistry, and on the growth and survival of pine beauty moth (*Panolis flammea*) larvae. *Oecologia*, **86**, 31–5.

Watt, A.D., Leather, S.R. & Stoakley, J.T. (1989). Site susceptibility, population development and dispersal of the pine beauty moth in a lodgepole pine forest in northern Scotland. *Journal of Applied Ecology*, **26**, 147–57.

Whitham, T.G., Maschinski, J., Larson, K.C. & Paige, K.N. (1991). Plant responses to herbivory: the continuum from negative to positive and underlying physiological mechanisms. In *Plant–Animal Interactions, Evolutionary Ecology in Tropical and Temperate Regions*, ed. P.W. Price, T.M. Lewinsohn, G.W. Fernades & W.W. Benson, pp. 227–56. New York: Wiley-Interscience.

Whitham, T.G. & Mopper, S. (1985). Chronic herbivory: impacts on architecture and sex expression in pinyon pine. *Science*, **228**, 1089–91.

Wood, S.L. (1982). The bark and Ambrosia Beetles of North and Central America (Coleoptera: Scolytidae), a taxonomic monograph. *Great Basin Naturalist Memoirs 6*.

Wood, S.L. & Bright, D.E. Jr. (1992). *A Catalog of Scolytidae and Platypodidae (Coleoptera) Part 2: Taxonomic Index, Volume A & B. Great Basin Naturalist. Memoirs 13*.

Young, J.A. & Young C.G. (1992). *Seeds of Woody Plants of North America*. Portland, Oregon: Dioscorides Press.

Zwolinski, J.B. (1989). The pine woolly aphid, *Pineus pini* (L.) – a pest of pine in South Africa. *South African Forestry Journal*, **151**, 52–7.

19 Diseases and the ecology of indigenous and exotic pines

Thomas C. Harrington and Michael J. Wingfield

19.1 Introduction

Although many of the major diseases of pine were characterized as early as the late 19th century (Hartig 1874), their role in the ecology and biogeography of this and other tree genera has gone largely unrecognized. However, with the burgeoning interest in ecologically based management of forest resources and forest health, more attention is being given to the pivotal role of pathogens in forest ecosystems (Worrall & Harrington 1988; van der Kamp 1991; Monnig & Byler 1992; Haack & Byler 1993; Castello, Leopold & Smallidge 1995). Pines, pathogens, insects and fire have evolved complex relationships that determine the dynamics of many northern hemisphere ecosystems. With these relationships in mind, sound management practices can minimize losses in commercial forests or otherwise optimize the benefits of natural forests to humans. Pines planted beyond their native range often encounter new relationships and present substantial management problems, while introduced pathogens have proved devastating to other pine ecosystems.

Among the pathogens causing mortality in natural pine ecosystems, dwarf mistletoes and root diseases are perhaps the most significant. These diseases interact strongly with bark beetles and fires in maintaining the heterogeneity of the forest landscape and in driving successional patterns. Fire suppression has increased the management problems caused by dwarf mistletoe, brown spot needle blight, fusiform rust and some root diseases, while the creation of stumps during harvesting and thinning has created conditions favourable for Armillaria root rot, annosum root rot and black stain root disease. Pines on degraded, old-field sites have proved susceptible to another root disease, littleleaf disease of southern pines.

Stem rots of living pines were substantial contributors to cull in harvesting old growth forests, but with the shorter 'pathological rotations' today, they now are valued as important wildlife management tools. The roles of diseases in natural pine ecosystems have often clashed with human interests and human manipulations, but with understanding of these often subtle relationships, wiser management practices follow.

Serious losses due to disease are commonly associated with pines, either native or exotic, grown in plantations. Cankers and foliage diseases in natural forests generally have a minor impact, primarily because of the resistance in the host population that has developed through millennia of selection pressure. Monocultures and the selection by humans of fast-growing species and genotypes can inadvertently shift the host population to susceptibility, e.g. fusiform rust. Pines planted outside their native range may encounter stronger disease pressures and be devastated by diseases that are minor in their native habitat. Perhaps the best examples of such enhanced susceptibility are Sphaeropsis canker and red band needle blight. The pathogens that cause these diseases are now established in many important pine-growing regions in the southern hemisphere and have dramatically altered forestry there.

The importance and potentially devastating impact of tree diseases was recognized early in the 20th century when severe epidemics occurred associated with the introduction of new pathogens to native forest ecosystems. Notable examples are chestnut blight and Dutch elm disease. On pine, the first such disaster was white pine blister rust, which became established in two separate North American locations in the early 1900s. This disease threatens to eliminate some pine ecosystems and removes

Ecology and Biogeography of Pinus, ed. D.M. Richardson. © Cambridge University Press, Cambridge (1998), pp. 381–404.

white pine as a management option in others, and this has, in turn, exacerbated further disease problems. Other pine disease epidemics associated with introduced pathogens include Scleroderris canker and pitch canker, the latter of particular concern in recent years. Perhaps the most disastrous introduction was the appearance of the pinewood nematode in Japan. Unfortunately, the globalization of timber and other industries, particularly the intercontinental movement of raw timber and packaging materials (e.g. dunnage), is likely to allow future introductions.

A comprehensive treatment of all diseases of pines is clearly beyond the scope of this chapter. Further coverage of the diseases of pines, their symptomatology and epidemiology can be found in Boyce (1961), Manion (1981) and Sinclair, Lyon & Johnson (1987). Our aim here is to give an overview of the major disease classes affecting pine, with an emphasis on their ecology and impact on indigenous and exotic pines. Pathogens causing tree mortality are emphasized because they are most important in the ecology of pine. The specific diseases mentioned above will be used as examples and discussed in more detail, generally considering first those affecting the indigenous forests, then those having greater impact in exotic plantings. Pathogens introduced to new environments will be covered towards the end of the chapter. Finally, we will give an assessment of the future of diseases in the ecology and biogeography of pines.

19.2 Dwarf mistletoe

It has long been recognized that *Arceuthobium*, a genus of parasitic higher plants, has a major effect on pine ecosystems (Weir 1916; Gill 1935). Heavily infected hosts die, but most losses in commercial forests result from reduced growth rate of the diseased trees. In the forests of the Northern Region of the Rocky Mountains in the USA, there are in excess of 1.4 million hectares of infested forests, with annual losses of about 600 000 m³ (Monnig & Byler 1992). Throughout the western USA, annual losses due to dwarf mistletoe have been estimated at 7.5×10^6 m³ (Hawksworth & Wiens 1972). Seed production and wood quality of diseased trees may also be reduced. Furthermore, heavily infested stands are prone to fire.

Members of *Arceuthobium* are obligate parasites on the family Pinaceae and a few other coniferous hosts (Hawksworth & Wiens 1996). *Pinus* is their primary host genus: 33 of the 47 recognized taxa have pines as their principal hosts, and about one-third of pine species are affected. The genus *Arceuthobium* is widespread throughout most of the northern hemisphere and, to a limited

extent, the southern hemisphere; it is best represented in Mexico (22 species) and the western USA. *Arceuthobium* species, like their hosts, are largely temperate plants. No species are found in low-elevation, moist tropical forests. In the Caribbean, the only known species is *A. bicarinatum*, occurring in Hispaniola on *P. occidentalis*, a relatively high-elevation pine species. *Arceuthobium hawksworthii* infects *P. caribaea* var. *hondurensis* in the uplands of Belize. Curiously, dwarf mistletoes are not found on the pines of eastern North America. Two Old World species occur on pines: *A. minutissimum* Hook. parasitizes *P. wallichiana* in the Himalayas, and *A. pini* Hawksworth & Wiens occurs on *P. densata* and *P. yunnanensis* in southwestern China.

The dwarf mistletoes have a unique biology (Hawksworth & Wiens 1996). They obtain their water, minerals and carbohydrates through an elaborate endophytic system in the branchwood and phloem of the host. Usually, there is swelling of infected branches, and prolific branching of the host results in conspicuous witches' brooms up to several metres in diameter. The endophytic system may be limited to 20 cm of the host branch, or, in the case of the systemic dwarf mistletoe species, a single endophytic system may extend throughout the branches of a witches' broom. In the years following infection, flowering shoots (male or female) develop each year along the endophytic system, and the fruits that develop each contain a single seed that is forcibly discharged at maturity. The seeds generally travel horizontally less than 10 m. Successful development of an endophytic system depends on the discharged seed adhering to a living needle and then sliding down the needle to the branch after the sticky coating of the seed is wetted by rain. The life cycle of the dwarf mistletoe species is 3–7 years, and with the limited dispersal of the seeds, the pathogen moves slowly compared with fungal pathogens on pine. Maximum spread is from an infested overstorey to the understorey (Fig. 19.1). Lateral spread through a dense, even-aged stand is slow, less than 1 m per year, even if the stand is comprised mostly of the principal host.

The dwarf mistletoes show host preferences but have curious ranges of occasional hosts. About two-thirds of the dwarf mistletoe species parasitize tree species other than their primary host, often parasitizing hosts in other genera, while species closely related to the primary host are immune (Scharpf 1984). It has been difficult to identify resistant individuals of the primary host species. However, Roth (1966) identified a 'drooping needle' race of *P. ponderosa* that was resistant, presumably because discharged dwarf mistletoe seeds that adhere to the needles, slide to the ground after being wetted by rains instead of sliding to the branch at the needle base, which is the normal site of infection.

Where they occur, dwarf mistletoes generally have a

Fig. 19.1. **Witches' brooms on the killed overstorey ponderosa pine once provided abundant inoculum of dwarf mistletoe seed for infection of the understorey, thus assuring the continued presence of the pathogen on this site (photo: T.C. Harrington).**

Fig. 19.2. **A lodgepole pine stand heavily infested with dwarf mistletoe has abundant fuelwood, and a subsequent fire would tend to favour pine regeneration. But if fire is suppressed, the more shade-tolerant and dwarf mistletoe resistant spruce in the understorey will eventually replace the pine (photo: T.C. Harrington).**

dramatic impact on the ecology of their hosts. Fire plays a major role in dwarf mistletoe ecology, but the relationships to fire are complex (Alexander & Hawksworth 1975; Wicker & Leaphart 1976). Killed trees and dead witches' brooms on the ground can serve as major fuels for fires, and witches' brooms on living trees may serve as fire ladders on which ground fires can climb to the crowns. Medium-intensity fires may develop to kill non-host species less fire-resistant than the primary host, and may leave infected trees for later overstorey to understorey spread of dwarf mistletoe seed in the regenerating stand (Fig. 19.1). Alternatively, an intense fire could kill all the trees on the site and eliminate the pathogen. The invasion of the burned area by regenerating trees from the perimeter generally proceeds much faster than the invasion of the dwarf mistletoe. Thus, fire tends to sanitize the stand. But even this scenario may favour dwarf mistletoe in the long run if the seral host species is perpetuated on the site and non-host climax species are avoided. Dwarf mistletoes generally use pioneer or seral species as their principal

hosts, and if no fire occurs in a heavily infested seral stand, the stand may be accelerated to the resistant climax vegetation (Fig. 19.2).

Climate, topography, and other site factors may influence the distribution and intensity of dwarf mistletoes, but it is the dynamics of the forest that most affects the dynamics of the pathogen population (Parmeter 1978). Fire and cutting history are critical factors. Severely infested stands may have substantial accumulation of fuel for fires, and it is believed that pockets of heavy infestation can become foci of major forest fires, which tend to sanitize stands. Large clearcuts, with no residual host trees, can work like fire to eliminate the disease from managed stands, and this is an extremely effective control strategy. This control practice illustrates how an understanding of the natural dynamics of the ecosystem can be translated into sound disease management. Still, poor harvesting practices have maintained this easily controlled disease as a management problem.

19.2.1 Lodgepole pine dwarf mistletoe

Arceuthobium americanum Nutt. ex Engelm. has the largest geographic distribution of any of the species (Hawksworth & Wiens 1972). Its distribution coincides closely with that of its principal hosts, *Pinus contorta* var. *latifolia* and var. *murrayana*, but it is common on *P. banksiana* in the western portion of that host's range. More than half of the lodgepole timber type is infested in many areas, particularly from British Columbia to Colorado (Hawksworth & Johnson 1989). It is the most important disease of this species, and infested stands show a dramatic yield loss and substantially increased mortality rate (Hawksworth & Johnson 1989).

In contrast to some other pine diseases, dwarf mistletoe does not appear to increase the susceptibility of pine to attack by bark beetles (Stevens & Hawksworth 1984). But dwarf mistletoe and the mountain pine beetle are important biotic components in the fire cycle of *P. contorta* (Brown 1975), a species with generally serotinous cones and a heavy dependence on fire for regeneration. Dwarf mistletoe contributes to the ground fuel, and the large witches' brooms and the foliage they trap provide vertical fuel continuity and a 'fire ladder'. The ensuing fire may ensure regeneration of more host material for the dwarf mistletoe, but an intense fire may remove all the host material and the mistletoe shoots, and the pathogen would have to re-invade the stand from the perimeter, which is a very slow process.

The fire cycle of infested lodgepole pine stands has been largely suppressed, or at least postponed, and this suppression has probably dramatically increased the incidence of dwarf mistletoe (Alexander & Hawksworth 1975) and resulted in an unprecedented buildup of fuels from dwarf mistletoe and mountain pine beetle mortality. This buildup has been likened by Monnig & Byler (1992) to '. . . holding water behind a leaky dam. We can draw the water down gradually or we can wait for the dam to break.'

A lodgepole pine overstorey that does not burn can become dominated by the more shade-tolerant spruce or fir as the dwarf mistletoe suppresses host growth and kills the pine (Fig. 19.2). Unless sufficient fuel accumulates and an intense fire returns the stand to pine, the climax species will eventually dominate the stand (Hawksworth & Johnson 1989). A different situation may be occurring in some Oregon stands of *P. contorta* var. *murrayana*, in which mortality of branches and trees caused by *A. americanum* can result in an environment favouring regeneration of the host (Wanner & Tinnin 1989). But generally, dwarf mistletoe infestations without fire will accelerate succession to the more shade-tolerant climax species or, if the climax vegetation fails to develop, the dwarf mistletoe continues to intensify.

19.3 Stem rots

Decay fungi were major contributors to cull in the harvesting of old-growth forests, but significant stem rot is generally restricted to mature trees (Basham & Morawski 1964). There is a 'pathological rotation' age, after which there is a significant economic loss due to decay (Boyce 1961), but the actual rotation age for pines in most forests would be substantially less. In unmanaged forests, living trees with stem rot usually survive for decades or even centuries, and these hollowed trees, living or dead, serve as important habitat for a variety of wildlife. Because of the importance of stemwood decay to wildlife management, there have been attempts at increasing the incidence of decay in living trees by artificial inoculations (Parks, Bull & Filip 1995).

Aside from the vital role of sapwood in conducting water and nutrients, the structural integrity of stemwood must be maintained for physical support. Some deterioration of the heartwood can take place with little loss of stem strength; a pipe is nearly as strong as a solid cylinder (Peters, Osenbruggen & Shigo 1985). Yet, at least the outer core of conducting sapwood and some of the heartwood is needed, and most large and long-lived tree species have evolved effective strategies to protect these tissues from the destructive enzymes of all but those few wood decay fungi specialized to that host genus or family.

Compared with the heartwood of other genera in the Pinaceae, *Pinus* species are moderately resistant to decay (Wenger 1984). The resinous response of living sapwood to wounding probably contributes substantially to the resistance of pine roots and stems (Gibbs 1968). Shallow stem wounds to pine trees generally resist infection by airborne spores of the stem-rotting fungi, but once established in the heartwood, they may encroach on the oldest growth rings of the sapwood. Decay may proceed in the heartwood column for metres above and below the point of entry, while decay of the sapwood is relatively limited (Basham 1975). Still, these fungi kill and decay sapwood, and thus the name 'heartrot fungi' is not completely accurate.

Pine species vary significantly in susceptibility to decay, perhaps in relation to their life-history strategies. Loehle (1988) found that longevity of gymnosperm species can be predicted, to some extent, by their resistance to wood decay. Generally, tree species that invest more energy into wood protection live longer but grow more slowly. Decay-susceptible pioneer species on good sites may invest in fast growth as a mechanism to avoid decay problems, but these species may deteriorate quickly once they reach their maximum size and growth is slowed (Loehle 1988). On harsh sites, resistance to decay may not be necessary for longevity. It has been speculated that many pine sites are too dry for substantial decay activity, and others sites are presumably too cool. For instance, wood of the slow-growing *P. longaeva* can harbour a significant number of wood decay fungi, but the survival of the oldest pines in a sound condition for thousands of years on cold and dry sites may be because the activities of wood decay fungi are limited there (Lindsey & Gilbertson 1983).

19.3.1 Red ring rot

Although there are many species of fungi adapted to the stems of living pines, one stands out. *Phellinus pini* (Thore:Fr.) Pilat (syn: *Fomes pini*), the cause of red ring rot, is widely distributed throughout the natural range of pines in the northern hemisphere (Haddow 1938). It is the most economically important of the stem rotters of the Pinaceae and has been a topic of interest since it was recognized as a decay fungus by Hartig (1874). But the biology and taxonomy of this fungus are still poorly known. Decay columns of 2 m or more in height in living pines may yield polypored fruiting bodies that produce wind-disseminated basidiospores. Old branch stubs are not important infection courts, though the fungus may move into the stemwood through living branches. It is believed that most infections originate from deep stem wounds or broken tops or through infections by rust fungi (Basham 1975). Decay is generally noted in pines of 60 years or more in age, but infection may take place decades earlier (Boyce 1961).

Pines with red ring rot may be cull to wood-cutters in old growth forests, but they are an important resource to wildlife managers (McFarlane 1992). Red ring rot is the primary stem rot of southern yellow pines in the southeastern USA, and an endangered species there, the red-cockaded woodpecker, nests almost exclusively in living pines with red ring rot (Jackson 1977; Conner & Locke 1982). The birds attempt to make nests in large living trees. The copious resin flowing down the stem apparently provides some protection from snakes and other predators. Nests are generally completed only in those trees with decay columns, presumably because they make for easier nest building. The shortened rotation ages of the managed pine forests result in a scarcity of stem-rotted pines, and those few infected trees of suitable size for nesting may be set aside for wildlife management.

19.4 Root rots

Perhaps no other group of diseases has a greater impact on the ecology of indigenous conifer forests than the root rots, and human activities appear to enhance the incidence of many of these diseases. Losses in commercial forests can be substantial in some regions, such as in western USA, where Smith (1984) estimated that 18% of

Fig. 19.3. *Armillaria ostoyae* clones may expand through root-to-root spread for centuries, causing circular infection centres of conifer mortality. The clone at the bottom of the hillside in the lower left of the picture is nearly 8 ha in area (photo: R. Williams).

tree mortality was attributable to root diseases, with annual volume losses of 6.7×10^6 m^3. Aside from mortality, root rots predispose trees to uprooting and windsnap, which is a common cause of canopy gaps, although such gap initiations are often considered as merely blowdown trees, and the importance of root rotters in the dynamics of forest ecosystems is largely underestimated (Worrall & Harrington 1988). Root diseases are also extremely important in supporting endemic populations of bark beetles through predisposition (Cobb *et al.* 1974; Goheen & Hansen 1993). Substantial reductions in growth may occur, and decay of the butt log is an important source of cull, but the most important root rotters kill trees relatively quickly by attacking the cambium in advance of decay. *Armillaria* species and *Heterobasidion annosum* kill pines in this way, and these pathogens will be discussed specifically.

The root rotters of pine have a common biology (Harrington 1993). They cause wood decay and produce airborne basidiospores from macroscopic conks or mushrooms. Colonization of fresh stumps or wounds of living trees by these spores is a rare but important event. Wounds on living pines are relatively resistant to infection, presumably due to the heavy resin response (Gibbs 1968), and exposed heartwood or deep sapwood wounds are often necessary for infection. Suitable infection courts may have been relatively rare in natural forests, but some of the most important root rotters today can colonize fresh stump tops (Hodges 1969; Rishbeth 1988). Once established, the mycelium of a root rot fungus may persist on a site in decayed roots for decades or centuries and spread underground as asexual thalli or clones, colonizing and decaying the rootwood of diseased or killed trees. The roughly circular pockets of tree mortality are known as 'infection centres', which may expand radially at a rate of half a metre or more per year (Fig. 19.3).

Fig. 19.4. **Rings of mortality of overstorey mountain hemlock in the Cascade Mountains of Oregon, USA, are caused by individual clones of the root rot pathogen *Phellinus weirii*. The more root disease resistant lodgepole pine regenerates (foreground) in the wake of the mortality (photo: T.C. Harrington).**

Fig. 19.5. **Blister rust has all but eliminated western white pine from this stand in Idaho (USA), leaving the root disease prone Douglas-fir and grand fir. Cones of a surviving white pine in the centre have been bagged for controlled pollination (photo: T.C. Harrington).**

Compared with other conifer genera, pines are relatively resistant to root rot diseases. Forests with well-established clones of virulent root rotters may consist of mixtures of susceptible and resistant tree species, the latter often predominating within the clones of root rot fungi. Van der Kamp (1991) has referred to such forests as 'root disease climaxes'. In the Pacific Northwest of the USA and British Columbia, lodgepole and ponderosa pine are more resistant to *Phellinus weirii* (Murr.) Gilb. than are the other Pinaceae (Hadfield & Johnson 1977). In unmanaged mountain hemlock forests, distinctive circular rings of mortality associated with individual clones of *P. weirii* are evident, with seedlings of the relatively resistant lodgepole pine colonizing the centre (Fig. 19.4). In time, the disease pushes the forest composition towards pine (Dickman & Cook 1989). Similarly, Douglas-fir mortality due to various root diseases encourages the domination of some sites by ponderosa pine.

In many regions, indigenous pine species have been favoured in harvesting, and this 'high-grading' has left disease- and pest-prone forests of fir and other species. Fire exclusion has had a similar effect. Root diseases have become major management problems as forests have shifted in composition from pines (particularly ponderosa pine) to Douglas-fir and true fir in the interior of the Pacific Northwest (Monnig & Byler 1992; Hagle & Byler 1994). On wetter sites that are prone to root disease, the resistant *P. monticola* does particularly well, but since the introduction of *Cronartium ribicola*, the cause of white pine blister rust, *P. monticola* has been of limited use, firs predominate (Fig. 19.5), and the root disease situation there continues to worsen (Byler, Marsden & Hagle 1990; Monnig & Byler 1992).

19.4.1 *Armillaria* root rot

Species of *Armillaria* play an important role in the decomposition of rootwood and stumps, but some of these same species are important killers of pines and other hosts (Kile, McDonald & Byler 1991). Epiphytic rhizomorphs of *Armillaria* species are common on living roots, and butt rot colonization patterns are known, but pathogenic colonization of living trees typically occurs under the bark as a sheet or fan of mycelium, killing the cambium and later decaying the underlying wood. Young trees and stressed trees are most often killed, but mortality of vigorous trees may occur when the pathogen increases its inoculum potential.

The fungus can survive for decades in decayed wood, which serves as a food base from which fruiting bodies (mushrooms) or vegetative structures (rhizomorphs) may develop. This food base can provide the inoculum potential for infection of living roots (Redfern & Filip 1991). The capacity of the thallus to infect depends, in part, on the volume of decayed wood and the species of wood colonized, and so the size of the root system colonized and the species composition of the stand are important. Pines in stands mixed with oaks or pines planted on old hardwood stands that had oak as a component can be particularly vulnerable. Mortality of larger trees has been associated with predisposing stresses such as drought, flooding, pollution, defoliation by insects, and other diseases, including root diseases (Wargo & Harrington 1991). Often, *Armillaria* deals the final blow to a tree that has been subjected to a long series of biotic and abiotic stresses, but can kill trees on its own if it has a suitable food base from which to work.

The growing list of recognized *Armillaria* species includes pathogens and saprophytes that decay rootwood in most forests of the world (Kile *et al*. 1994). In the southern hemisphere, pine mortality has been associated with *Armillaria* species on cut-over indigenous forests (MacKenzie & Shaw 1977; Wingfield, Swart & von Broembsen 1989; Hood, Redfern & Kile 1991; Lundquist 1993). Mortality may not persist through subsequent rotations, however, and Armillaria root rot may not become a persistent management problem in tropical or southern temperate pine plantations.

Until relatively recently, northern hemisphere *Armillaria* species were inappropriately lumped by pathologists into one species, *A. mellea* (Vahl:Fr.) Kummer. There are now at least 13 clearly defined species in the northern hemisphere alone, and most can be found on pines, though not all cause tree mortality (Harrington, Worrall & Baker 1992). *Armillaria mellea* is most common in milder temperate regions, particularly mediterranean-type climates, where it is a common cause of mortality of ornamental, fruit and forest trees. *Armillaria tabescens* (Scop.:Fr.) Emel. is known in southeastern USA, eastern Asia and southern Europe as a pathogen of both hardwoods and conifers, including pines. More important on pine is *A. ostoyae* (Romagnesi) Herink, a pathogen of many conifer species and also on hardwoods in the cooler coniferous forests of Europe (Guillaumin *et al*. 1993), Asia (Sung *et al*. 1991) and North America (Harrington *et al*.1992).

In pines, much of the mortality is associated with relatively young plantings on sites where *A. ostoyae* was previously established, but mortality of large, overstorey pines is common in some forest situations, particularly in relatively dry forests. Kile *et al*. (1991) discuss such infection centres in the drier pine forests of western North America and the French Pyrenees, where the disease is known as 'ring disease' because of the distinctive expanding rings of mortality. In dry forest types, clones of *A. ostoyae* are usually infrequent but quite large, often covering several hectares or more (Fig. 19.3; Shaw & Roth 1976; Anderson *et al*. 1979). Rizzo & Harrington (1993) speculated that clones in such dry environments are able to expand, unimpeded by competition from other *A. ostoyae* clones, using the

killed trees as further inoculum to expand the centre. In contrast, clones tend to be more abundant and smaller in moist ecosystems, perhaps attributable to the abundance of fruiting bodies in moist climates (Anderson *et al*. 1979). Here, the entire forest floor may be a mosaic of various clones, which tend not to intermingle because of somatic incompatibility (Rizzo & Harrington 1993; Rizzo, Blanchette & May 1995). Where static clones expend much of their energy in maintaining their genetic integrity and space, little inoculum potential may develop, and mortality is restricted to young saplings and stressed overstorey trees.

19.4.2 Annosum root rot

Heterobasidion annosum (Fr.) Bref., also known as *Fomes annosus* (Fr.) Cke. and *Fomitopsis annosa* (Fr.) Karst., is one of the most important pathogens on pines. Host-specialized and partially intersterile groups of *H. annosum* have been identified (Korhonen 1978; Stenlid & Swedjemark 1988; Harrington, Worrall & Rizzo 1989). Mortality of mature pines has been associated with a pine form of *H. annosum* in Europe, known as the 'P type' (Korhonen 1978). In North America, a similar pine form of the fungus is recognized (Harrington *et al*. 1989).

Before tree harvesting and the creation of managed monocultures of pines, annosum root disease must have been a relatively rare disease. The infection of freshly-cut pine stumps by *H. annosum* was recognized by Rishbeth (1951), who elucidated the epidemiology of the disease in pine plantations and noted that competing fungi, most notably *Phlebiopsis (Peniophora) gigantea* (Fr.) Jul., could exclude *H. annosum* from stumps if applied quickly after harvesting. This led to one of the most effective biological controls of a plant disease. Treatment of freshly cut stumps with *P. gigantea*, borate, urea or other materials may effectively exclude *H. annosum* from colonizing the stump (Korhonen *et al*. 1994; Pratt 1994). Such stump top treatments are routine in many managed pine forests and are mandated on Forestry Commission lands in Great Britain.

Healthy pine trees may become infected with *H. annosum* when their roots are in contact with decayed roots of killed trees or stumps (Hodges 1969). Once inside a living pine root, the fungus generally progresses in the xylem, near the cambium, and the rootwood may become resin-impregnated in response to the pathogen colonization. Decay of the rootwood generally follows cambium killing, but the butt rot pattern of colonization does occur (Greig 1995). Vegetative spread of the pathogen through root systems may be up to 1 m per year, but on average, radial expansion of infection centres is much less, especially if the pine trees are not large (Slaughter & Parmeter 1995). Compared to some other root rotters, infection centres associated with *H. annosum* have a limited life, generally less than 20–30 years in northeastern California (Slaughter & Parmeter 1995).

Plantation forestry and poor management practices appear to be worsening the impact of annosum root rot (Smith 1989; Stambaugh 1989). The disease can be severe on pines planted on agricultural lands, particularly those on alkaline soils. Lack of competing forest fungi on such sites apparently favours stump colonization by *H. annosum* (Schonhar 1988). Scots pine forests of Europe are severely affected, as are North American pine forests, especially southern yellow pine forests of the Southeast, red pine plantations in the Northeast, and ponderosa and Jeffrey pine forests of the Far West. A noteworthy example is in the Yosemite Valley of California, where fire exclusion has increased the basal area and density of ponderosa pine and incense cedar (*Calocedrus decurrens* (Torr.) Florin), both hosts of *H. annosum* (Sherman & Warren 1988; Marosy & Parmeter 1989). Heavy mortality of pine by *H. annosum* and bark beetles has recently shifted this now dense forest more to *Calocedrus*, which is more prone than pine to windthrow if infected with annosum root rot. The hazard to humans and property associated with uprooted pine, and especially cedar, infected with *H. annosum* has resulted in a major management problem in one of the most heavily used recreational sites in the USA.

19.5 Other root diseases

The aforementioned root rots are generally more important to the ecology of pines than are other root diseases. But these other root diseases, which are caused by fungi that do not decay wood, may be of local importance. For instance, *Rhizinia undulata* Fr.:Fr., an ascomycete, has altered management practices in some parts of the world where substantial seedling mortality may occur in plantings on slash-burned sites (Wingfield *et al.* 1989). Two other root diseases will be discussed in more detail. These two fungi and the diseases they cause have little in common, except that they attack pine roots, they do not cause wood decay, and they both predispose pines to bark beetle attack.

19.5.1 Black stain root disease
This disease has a spotty distribution across its known range of western North America, where it is of local importance but may have a dramatic impact on the landscape. The distribution of the disease has probably not changed much with human activity, but the disease may well be more common today than in previous years because of enhanced vector activity, particularly by the creation of

stumps and the attraction of vectors to disturbed stands, where infection centres originate. In contrast to the root rot fungi, *Leptographium* (*Verticicladiella*) *wageneri* (Kendr.) Wingf. produces its spores (conidia) in sticky drops on top of stalked structures (conidiophores) within insect galleries. Like root rot diseases, black stain can spread through pine root systems to adjacent trees. The preference for monocultures of susceptible pines in some regions can enhance tree-to-tree spread, and fire exclusion may similarly provide heavily stocked stands of large pine trees.

Three host-specialized varieties with overlapping geographic distributions are recognized (Harrington & Cobb 1986, 1987; Zambino & Harrington 1989). The species was originally described from pinyon, and this variety infects *Pinus monophylla* and *P. edulis* throughout the southwestern USA and north to southern Idaho. Variety *ponderosum* occurs on the hard pines in the Pacific Coast states, Idaho, Montana and British Columbia, Canada. Variety *pseudotsugae* is a pathogen on Douglas-fir throughout the western USA and British Columbia. In each of these hosts, hyphae of *L. wageneri* grow through the tracheids only, at least in the early stages of the disease; it moves systemically throughout the root system and into the roots of adjacent trees through root contacts, and it may grow for short distances through soil (Cobb *et al.* 1982; Cobb 1988).

Infection centres are initiated when root-feeding insects, primarily bark beetles, carry the pathogen into the roots of a living tree. Vector biology has been most studied in Douglas-fir, but similar vector relationships appear to occur in pine. Species of the root-feeding bark beetle genus *Hylastes* (Coleoptera: Scolytidae) are the primary vectors (Goheen & Cobb 1978; Harrington, Cobb & Lownsbery 1985; Witcosky, Schowalter & Hansen 1986). After emergence from trees killed by black stain, new adults may feed on roots of healthy trees before breeding and egg-laying, and this so-called maturation feeding has been proposed as a crucial stage for introduction of *L. wageneri* into a living host and initiation of a new infection centre (Harrington *et al.* 1985). Populations of vectors may be artificially high in some managed forests because the stumps left after harvesting and thinning of stands provide abundant breeding material. The disease is strongly associated with stand disturbances such as roadside construction, tractor logging or stand-thinning (Hansen 1978; Harrington *et al.* 1983), presumably because the vectors are attracted to wounded trees and fresh stumps (Harrington *et al.* 1985). Although initiation of new infection centres may be less frequent in pine than in Douglas-fir, black stain root disease has been associated with disturbances in pine stands (Cobb 1988). A less efficient vector system or smaller vector population may explain the relatively rare initiation of new infection centres in pines. Aside from the vectors, other bark beetles play an important role in killing pines with black stain root disease, and the disease contributes to the potential brood material for endemic populations of stem-feeding bark beetles (Wagener & Mielke 1961; Cobb *et al.* 1974; Landis & Helburg 1976; Goheen & Cobb 1980).

The disease has been a major factor in the management of pinyon at some locations, such as Mesa Verde National Park in southwest Colorado (Landis & Helburg 1976), where black stain is the chief cause of pinyon mortality, and 12 000 dead and dying trees were removed between 1932 and 1934 (Wagener & Mielke 1961). Over-mature pinyon stands with high stocking densities and little or no juniper can be created through fire suppression, and when these occur on cool sites with deep and moist soils, local epidemics of black stain root disease can be found. The disease is clearly restricted to the coolest of sites, mostly at the upper elevations of the host distributions (Wagener & Mielke 1961; Landis & Helburg 1976), perhaps because lower-elevation sites are too warm for growth and survival of *L. wageneri* (Harrington & Cobb 1984). Similarly, the disease is found on pinyon west of the continental divide in Colorado but has not been reported from the east side (Landis & Helburg 1976).

The hard pine variety, of *L. wagneri*, var. *ponderosum*, occurs primarily on ponderosa, Jeffrey and lodgepole pines (Harrington & Cobb 1986). The disease is found at the higher elevations of ponderosa pine in northern California, and the disease is not known in southern California. New infection centres arise very rarely, but where they occur, they may cover many hectares (Cobb *et al.* 1982). Sites where the disease is severe are typically those suited to the Sierra mixed-conifer forest type, but fire, mud slides or stand management practices, particularly logging followed by burning, have created nearly pure stands of ponderosa pine (Cobb 1988). In one such area, the rate of radial spread of infection centres ranged from 0 to 7 m yr^{-1}, with the density of ponderosa pine in the stand explaining much of the variation (Cobb *et al.* 1982). Black stain root disease reduces the ponderosa pine component in these nearly pure stands and creates openings for the more shade-tolerant conifer species, thus helping to maintain the natural heterogeneity of the conifer forests in this region.

19.5.2 Littleleaf disease

Many species of *Phytophthora* are pathogens of plants, including pines in nurseries, but only one has been known to have a major impact in pine stands. *Phytophthora cinnamomi* Rands has a very broad host range, and most species of pine show some degree of susceptibility (Zentmeyer 1980). The natural range of the fungus is not clear, but it may be native to southeastern Asia. It is now widely distributed throughout mild temperate, subtrop-

ical and tropical regions of the world (Zentmeyer 1980). Fortunately, *P. cinnamomi* does not persist in the soils of many cooler coniferous forests, presumably because freezing soils impede the overwintering survival of the fungus (Benson 1982), and moist soil conditions are needed during warm periods for the fungus to be active. Exotic plantings of pines in the southern hemisphere have been known to be affected by *P. cinnamomi* (Newhook 1959; Wingfield *et al.* 1989). In indigenous forests, the pathogen has been less of a problem, except in the southeastern USA, where loblolly, and especially shortleaf, pines are vulnerable on some sites. Here, the disease is known as littleleaf disease.

In pines and many other hosts, the fungus primarily kills only the small-diameter feeder roots (Zentmeyer 1980). Slowed growth, chlorotic crowns with small leaves, and crown dieback may be evident for many years before death. Littleleaf disease is most prevalent on old-field pine forests that previously were agricultural lands in the Piedmont region of the Southeast. The pathogen often works in concert with poor soils and bark beetles to form a complex decline syndrome (Belanger, Hedden & Tainter 1986). Shortleaf pine on eroded or compacted soils of poor fertility are particularly vulnerable to littleleaf disease. Trees under 20 years of age are rarely diseased, and affected trees are often killed by bark beetles, particularly the southern pine beetle, *Dendroctonus frontalis* Zimmermann. The disease is one of the reasons that loblolly pine has been favoured in plantations over shortleaf pine, and this has lead to some other management problems, such as fusiform rust.

19.6 Canker diseases

A great number of canker diseases of pines exist, and these vary considerably in their biology, symptoms and impact (Boyce 1961; Manion 1981; Sinclair *et al.* 1987). Cankers are defined as necrosis of the inner bark tissues of stems or branches, but in this treatment of canker diseases, we are excluding cankers associated with dwarf mistletoes, wood decay fungi, and the rusts. Vigorously growing pines in their natural range generally are highly resistant to canker diseases, but stresses are important predisposing factors for canker diseases, and there is strong evidence that some of the most conspicuous canker fungi have been introduced to new locations.

Most canker diseases, as well as most of the needle diseases, are caused by ascomycete fungi that generally have two distinct spore stages, often a wind-dispersed sexual spore (ascospore) and a rainsplash-dispersed asexual spore (conidium). They generally require wounds to infect their hosts. These wounds are provided by insect feeding, wind and hail damage, or even natural growth cracks. There are, however, some canker pathogens that infect unwounded foliage or young pine tissue.

Annual cankers occur when the pathogen is active in the branch or stem for only one season and are commonly associated with wounds on stressed trees. These infections are usually excluded by the tree when the stress abates. Perennial cankers are caused by pathogens, such as *Atropellis* species, that are relatively weak but most active when trees are not growing actively, such as in the late summer and early winter (Sinclair *et al.* 1987). In these cases, tree growth in spring results in the development of a band of callus tissue, which temporarily impedes the pathogen but then is transgressed towards the end of the growing season. This cycle leads to the development of concentric rings of callus tissue, and thus the term 'target canker' is often applied. Perennial cankers develop slowly and tree mortality seldom results. The so-called diffuse cankers are caused by virulent pathogens where the host shows little resistance to infection, and layers of callus are not generally evident. In these cases, trees or their branches are rapidly girdled and killed.

We discuss in detail three important canker diseases that illustrate various ecological and biogeographical concepts. With each of the three, there is evidence that the pathogens have been introduced to new pine-growing regions, and this is where their impact has been greatest.

19.6.1 Scleroderris canker
Scleroderris canker is caused by the fungus *Gremmeniella abietina* (Lagerb.) Morelet (syn. *Ascocalyx abietina* (Lagerb.) Schlaepfer or *Scleroderris lagerbergii* Gremmen, also known in Europe by the conidial state name *Brunchorstia pinea* (P. Karst.) Hoehn. The disease has been known in Europe since 1888 and in North America since the 1950s. Confusion surrounding the various reported forms of the pathogen (Skilling 1977) and a severe epidemic of the disease in New York state in the 1970s prompted a symposium in 1983 (Manion & Skilling 1983). Serious discrepancies have remained concerning the identity of various forms of the pathogen in North America and Europe.

An Asian race is apparently restricted to *Abies*, whereas the so-called North American and European races have wider host ranges but occur primarily on pines (Yakota, Uozumi & Matsuzaki 1974a, b; Dorworth 1981; Skilling, Kienzler & Haynes 1984). Yakota (1983) also reported on the rare occurrence of a pine form of the fungus in Japan on *P. strobus*. The confusion over the pine forms in Europe and North America appears to be clarified by recent reports using randomly amplified polymorphic DNA. Lecours *et al.* (1994) grouped European isolates into three amplitypes. A cold-adapted amplitype occurred on *P. cembra, P. mugo, P.*

sylvestris and *Larix lyalli* above 2500 m elevation in the Alps; this amplitype apparently corresponds to *G. abietina* var. *cembrae* Morelet. A northern amplitype was found above 66° latitude on *P. sylvestris* and planted *P. contorta*. The third European amplitype was more widely distributed, from Scandinavia to northern Italy. In a more detailed study, Hellgren & Högberg (1995) also differentiated the northern amplitype from the more generally distributed European amplitype. According to Hamelin, Ouellette & Bernier (1993) and Lecours *et al.* (1994), it is the generally distributed European amplitype that has been referred to as the 'European race' in North America, i.e. the pathogen responsible for the epidemic in New York in the 1970s. The 'North American race', which has been associated with epidemics on young *P. banksiana* and *P. resinosa* across the Lake States and Ontario since the 1950s, was not found in European samples.

With this delineation of amplitype or races, the ecology of the disease needs to be re-evaluated; however, some general trends are evident. Symptoms can include browning of needles and needle cast, yellowing at the bases of needles, twig cankers, die-back of branches or death of shoots, with a concomitant yellow-green discolouration of the cambium under the cankers. In some cases the branch cankers can spread to the main stems of trees causing stem cankers and deformation (Manion 1981; Sinclair *et al.* 1987). Mortality of mature trees is sometimes evident in epidemic situations.

Severe epidemics of the disease have been associated with particular climatic conditions and plantings of exotic germplasm. The disease is favoured by cooler temperatures and is most problematic in plantation areas where cold air accumulates, such as in depressions (Dorworth 1972; Patton *et al.* 1984; Aalto-Kallonen & Kurkela 1985). Some forms of the pathogen are most active when host tissue is dormant in winter months and under snow cover (Dorworth 1972; Patton, Spear & Blennis 1984). In Scotland, the disease is most serious in *P. nigra* on north-facing slopes (Read 1968). As in many other canker diseases of pines, tree species or seed sources planted out of their natural range are more susceptible, and predisposing stresses are important. In the Great Lakes region of the USA and Canada, the disease is most problematic in relatively young plantings of infected nursery stock, which can be avoided by the use of fungicide treatments in nurseries (Skilling & Waddell 1970).

As mentioned above, the Scleroderris canker epidemic on *P. resinosa* and *P. sylvestris* in New York during the mid-1970s was apparently caused by the amplitype found throughout most of Europe, where it is not dependent on snow cover for infection, and it infects large trees (Hellgren & Högberg 1995). It is likely that this form of the pathogen has been introduced to several locations between New York and the Maritime Provinces of Canada. As pointed out by Manion & Skilling (1983), the New York epidemic was associated with cooler than normal temperatures, and non-local seed sources had been used in the affected plantations. The epidemic ultimately involved mortality of mature trees over about 14 000 ha, prompting considerable concern and quarantine restrictions. Fortunately, the feared widespread devastation did not materialize, though the disease persists.

19.6.2 Pitch canker

Pitch canker is caused by the fungus *Gibberella fujikuroi* (Sawada) Ito but is best known by its conidial state, *Fusarium subglutinans* (Wollenweb. & Reinking) P. Nelson *et al.* f. sp. *pini*. Pine species differ in their susceptibility, but most species appear to be susceptible to some degree, and even Douglas-fir has been recorded as a host (Storer *et al.* 1994). The disease was first described from the southeastern USA (Hepting & Roth 1953), where it is particularly damaging on young *P. echinata*, *P. elliottii* and *P. virginiana* (Dwinell, Kuhlman & Blakeslee 1981; Dwinell, Barrows-Broaddus & Kuhlman 1985). Although discovered in the 1940s, it only became serious in the Southeast in the mid-1970s and has particularly been a problem in seed orchards. Since 1987, the pathogen has been recognized in four new geographic areas (Viljoen 1995). There is sufficient cause to expect that pitch canker threatens the genetic base of many pine species in various parts of the world. There is an urgent need to launch a coordinated effort to study this relatively unknown disease.

The disease has been most studied in the southeastern USA. Symptoms include shoot death and cankers on branches and stems. Cankers are typified by copious resin exudation and pitch soaking of the wood, often through to the pith. The major infection propagules of *F. subglutinans* f. sp. *pini* are the asexual conidia. The sexual state of the fungus has been reported under laboratory conditions (Kuhlman *et al.* 1978) but has not been found in nature. Wounds are required for infection. Wind, hail damage, infections by other pathogens, such as fusiform rust, and insects, including the pine tip moth (*Rhyaciona subtropica*), the eastern pine weevil (*Pissodes nemorensis*) and bark beetles (Coleoptera: Scolytidae), may provide suitable wounds. In the Southeast, where infection of trees in seed orchards is a management problem, efforts are made to reduce the occurrence of mechanical damage to trees.

Pitch canker was discovered in California for the first time in 1986 on *Pinus radiata* near Santa Cruz (McCain, Koehler & Tjosvold 1987), where it has been largely confined to landscape plantings. Very little genetic variation is found in the pathogen in California compared with Florida, strongly suggesting that the California population is derived from a recent introduction (Correll, Gordon & McCain 1992). It continues to spread to new areas of California and to additional species of pine, including the three native *P. radiata* stands on the mainland (Storer *et al.* 1994; Dallara *et al.* 1995). Thus, the genetic base of this important pine species is threatened.

In California, the pitch canker pathogen has established an association with various insects, such as the engraver beetles *Ips mexicanus* Hopkins and *I. paraconfusus* Lanier, which are able to transmit the fungus to mature pines (Fox, Wood & Koehler 1990; Fox *et al.* 1991). These insects may be more efficient vectors of the pathogen than the weevils and moths that have been associated with the fungus in the Southeast (Blakeslee & Foltz 1981; Runion & Bruck 1985). Indications are that the spread and importance of pitch canker internationally will depend to some extent on the insects with which it becomes associated. Strategies to manage these insects may be used to reduce disease incidence in the future.

Pitch canker was recorded from Japan for the first time in 1988 on *P. luchuensis* (Muramoto, Tashiro & Minamihashi 1988) and has subsequently spread and increased in importance (Kobayashi & Muramoto 1989; Muramoto & Dwinell 1990; Kobayashi & Kawabe 1992). The most recent discovery of the pitch canker pathogen has been in South Africa, where it has been associated with a devastating disease of *P. patula* in a large commercial nursery (Viljoen, Wingfield & Marasas 1994; Viljoen *et al.* 1995). Thus far, the disease has not been reported on established trees in plantations in this area. The occurrence of the disease in this nursery suggests that the pathogen might have been introduced on seed. Fraedrich & Miller (1995) demonstrated that *F. subglutinans* can be isolated from seed.

The recent outbreaks of pitch canker favour the hypothesis that the pathogen has been newly introduced into these areas. Although the biogeography of the disease is not well documented, it has also recently been recognized on many native species in Mexico where it is considered to be severe in some areas (Rodrigues 1989; Santos & Tovar 1991). Apparently, the disease is well established there, and this could be the area of origin of the pathogen. An earlier hypothesis (Berry & Hepting 1959) was that the pathogen in the southeastern USA had originated in Haiti, where it was well established on *P. occidentalis*.

19.6.3 Sphaeropsis canker

The disease, caused by *Sphaeropsis sapinea* (Fr.) Dyko & Sutton but perhaps better known by the earlier name *Diplodia pinea* (Desm.) Kickx *et al.*, occurs throughout the world where pines are grown (Sutton 1980; Ivory 1994). The disease is particularly common in Africa (Ivory 1994). There is some indication that races of *S. sapinea* exist and that these vary in their ability to kill trees (Palmer, Stewart & Wingfield 1987; Swart *et al.* 1991; Smith & Stanosz 1995).

Fig. 19.6. **Pine shoot blight caused by _Sphaeropsis sapinea_ can result in tree mortality over large areas after hail damage (photo: M.J. Wingfield).**

Although the fungus has been rarely associated with episodic disease outbreaks on pines in their native range, it is most devastating in exotic plantings, both in landscapes and in intensively managed, exotic plantations.

Sphaeropsis sapinea causes annual cankers and might be considered to be more opportunistic than the aforementioned canker pathogens. Sphaeropsis canker may involve a multiplicity of symptoms, including die-back and death of pines after hail, die-back of young shoots (Fig. 19.6), stem and branch cankers, cone infections, blue stain of freshly felled timber, and root disease (Gibson 1979; Sinclair _et al._ 1987; Swart & Wingfield 1991). Although the pathogen is able to infect and kill young growing shoots (Peterson 1977), it is most often associated with wounds on stressed trees. Wounds suitable for infection result not only from hail but from wind damage, pruning and insect feeding. Incidental wounds are common in plantations, but the cankers generally do not extend far beyond the localized sites of infection. Not pruning during warm and wet weather and avoidance of excessive pruning are important

means to reduce losses (Swart & Wingfield 1991). In many landscape planted pines, cankers usually develop after the hosts produce cones, on which fruiting structures of the pathogen are generally abundant. Conidia that are rain-splashed from the cone scales are the major source of inoculum (Peterson 1977).

The fungus is common throughout the native range of pines, though it appears to cause little damage in natural settings. It can be a problem in nurseries near cone-bearing pines (Palmer & Nicholls 1985), however, and pines planted out of their natural range are particularly vulnerable. For instance, Sphaeropsis canker has been severe on _P. nigra_ in the Midwest and Northeast of the USA (Peterson 1981). _Sphaeropsis sapinea_ has been effectively introduced into every part of the world where pines are grown, perhaps because the fungus is common on pine seeds (Fraedrich & Miller 1995). Preliminary studies (M.J. Wingfield, unpublished data) indicate that _S. sapinea_ is represented by a high degree of genetic diversity in South Africa. Given the fact that the fungus is known only in its asexual form, this implies that it has been introduced into the area many times and from many parts of the world.

The die-back and death of trees associated with _S. sapinea_ has been more important in southern Africa than in any other part of the world (Wingfield 1990; Swart & Wingfield 1991). Here, extensive losses occur annually in pine plantations damaged by hailstorms (Swart & Wingfield 1991). This is because of a high incidence of hail in the region and the susceptibility of two of the most widely planted pine species (_P. patula_ and _P. radiata_) to such damage. Cankers after hail develop extremely rapidly, perhaps due to the stress associated with hail damage rather than to infection of hail wounds. The fungus can reside as an endophyte in living twigs and cones in a latent phase without causing symptoms (Smith _et al._ 1995). Where hail damage is common, _P. patula_ and _P. radiata_ should not be planted.

19.7   Needle diseases

Trees in their natural range are generally not seriously affected by foliage diseases because of the high level of resistance in the native population, and mortality due to foliage disease is relatively rare. Genotypes adapted to areas of low disease pressure, however, may be seriously affected when planted in areas of high disease pressure. Exotic plantings, off-site plantings and pines planted for Christmas trees are examples where foliage diseases may have significant impact.

Two general classes of needle diseases occur on pine. The needlecasts are associated with fungi that usually infect young needles and remain latent in the needle for months or years before symptoms develop. After the infected needles are cast, fruiting structures develop and sexual spores (ascospores) are discharged, generally as new needles emerge in the spring. Common genera of needle-cast fungi on pine include *Lophodermium, Lophodermella, Ploioderma, Cyclaneusma* and *Elytroderma*. The latter genus is unusual in that twigs are also infected and witches' brooms develop. *Elytroderma deformans* (Weir) Darker occurs in western North America and *E. torres-juanii* in southern Europe (Sinclair *et al.* 1987). The other needlecast fungi can be of major importance in Christmas tree production, and *Lophodermium seditiosum* Minter, Staley & Millar is considered to be a virulent and primary pathogen.

The other class of needle disease is the needle blights, many of which involve both primary (ascospores) and secondary (conidia) inoculum. The sexual fruit bodies develop towards the end of the growing season and are relatively uncommon in many of these fungi but may allow for genetic recombination in the pathogen and commonly provide a means for overwintering. In contrast, the asexual conidia are produced abundantly during wet periods of the growing season and are responsible for an increase in the inoculum load and the rapid development of epidemics, usually in wet weather. Infections occur either directly through the epidermis or via stomata. These lead to spotting and banding symptoms and, eventually, needle death and defoliation. Severely diseased trees may show substantially reduced growth rates or mortality. Species of *Mycosphaerella* infecting pines are the most noteworthy of needle blight fungi (Evans 1984). *Mycosphaerella gibsonii* is perhaps endemic to the Himalayas and has been recorded from Asia, Africa, Papua New Guinea and Jamaica (Ivory 1994). Two better known *Mycosphaerella* species will be considered here in more detail.

19.7.1 Red band needle blight

Dothistroma or red band needle blight is caused by the fungus *Mycosphaerella pini* E. Rostrup apud Munk (*Scirrhia pini* Funk & Parker) but is better known by its asexual state name, *Dothistroma septospora* (Dorog.) Morelet (syn. *Dothistroma pini* Hulbary). It is the best known and most studied needle blight of pine and has caused extensive damage and defoliation of pines in many parts of the world (Gibson 1972). The pathogen is native on pines in the northern hemisphere and was first recognized and described from eastern Europe in 1911 (Gibson 1972), but it is now widely distributed. In most cases where this disease has imparted serious losses, the pathogen has been introduced or the host planted outside its natural environment (Fig. 19.7). In such situations, it has been necessary to plant alternative species or to resort to complicated and extensive management strategies.

Fig. 19.7. **Red band needle blight has caused extensive damage to pines planted as exotics, such as these young *Pinus radiata* on North Island, New Zealand. Only the current year's needles remain on the branches (photo: M.J. Wingfield).**

Ivory (1994) speculated that there are two primary origins of the species, a short-spored form endemic to Eurasia and found in Asia, Australasia and South America (*D. pini* var. *pini*) and an intermediate-spored form endemic to Central America and found in Africa (*D. pini* var. *keniensis* Ivory). The long-spored form (*D. pini* var. *lineare* Thyr & Shaw) is found in North America and France and may have a separate origin. Sutton (1980) considered these morphological distinctions debatable, and the definitive genetic work has not been done.

The first symptoms of infection are yellow bands on mature needles at the bases of branches and in the lower half of the crown, and the bands later turn red in colour, hence the name red band needle blight. Infected needles are killed and drop, and conidia produced from them are splashed to other needles. Temperatures of 15–20 °C with extended periods of moisture greatly favour infection, thus accounting for the severe infection in regions such as the East African Highlands and the North Island of New Zealand (Gibson 1979).

A wide range of pine species have been recorded as hosts (Sutton 1980; Ivory 1994), but *P. nigra*, *P. ponderosa* and *P. radiata* are by far the most susceptible species, at least when planted outside their natural ranges (Gibson 1979). The disease has caused substantial damage to exotic pine plantings in the USA and Europe (Peterson 1967; Gibson 1972). Most plantings of *P. ponderosa* in eastern North America have failed because of this disease. Red band needle blight is absent in the limited natural stands of *P. radiata* in California, but further north along the coast the disease has been devastating in some *P. radiata* plantations (Cobb & Miller 1968).

The fungus was recognized as the causal agent of severe defoliation of *P. radiata* in Tanzania in 1957, and thereafter spread rapidly throughout central and southern Africa (Gibson 1972). The disease thus led to the abandonment of *P. radiata* as a plantation species in many parts of Africa and the use of alternative species, such as *P. patula* (Ivory 1987). The disease then appeared in Chile and New Zealand in the early 1960s, where it has caused substantial damage, also to *P. radiata*. In areas of New Zealand where the disease has been severe, chemical control using copper fungicides was effectively practised for many years (Gibson 1975). The fact that *P. radiata* trees develop adult resistance to this disease at about 15 years has reduced the need for chemical treatments throughout the rotation (Ivory 1972). In recent years, substantial progress has also been made in breeding for resistance to this disease (Carson & Carson 1989).

19.7.2 Brown spot needle blight

This disease is caused by *Mycosphaerella dearnessii* Barr (syn. *Scirrhia acicola* (Dearn.) Siggers), which is also known by its asexual state name, *Lecanosticta* (*Dothistroma*) *acicola* (Thuem.) Syd. apud Syd. & Petrak (Evans 1984). It is best known for the damage that it has caused to *P. palustris* plantations in the southeastern USA (Boyce 1961; Sinclair *et al.* 1987), but it can be important elsewhere. Symptoms and epidemiology are very similar to red band needle blight, though its impact has generally been less.

The symptoms of the disease vary greatly amongst hosts, but chlorotic spots or bands with dead brown centres are common. Disease tends to develop on foliage on older parts of trees, and eventually spreads to younger foliage. Severe infection can result in needle-free branches with tufts of diseased current-year needles at their apices. New infections become established in spring and early summer. These usually result from conidia that are produced in sticky masses from conidial stromata and are rainsplash-disseminated. Sexual fruiting bodies may develop in these same structures after approximately two months, though the sexual state is not common. Rate of symptom development after infection is extremely variable but is favoured by warm and wet weather (Kais 1975).

Longleaf pine develops through a grass stage, which is particularly conducive to the development of brown spot needle blight. This host has a fire-tolerant bud in the grass stage, and as fire burns the needles, it prevents the buildup of inoculum. Fire suppression in plantations has thus led to severe disease. Heavy infection can result in serious delay (4–10 years or more) in trees emerging from the grass stage. Control of brown spot needle blight with fire or chemical sprays is essential in longleaf pine plantations (Gibson 1979; Sinclair *et al* 1987). Brown spot needle blight has also been exacerbated by the fact that seedlings are commonly infected in nurseries, providing an initial inoculum for disease development. Considerable effort has been expended in the development of disease-resistant seed sources (Snyder & Derr 1972). The disease has made longleaf pine less desirable as a plantation species, and this has contributed to the favouring of other southern pines, which are more susceptible to fusiform rust and southern pine beetle.

Although brown spot needle blight has been most serious and is best known on *P. palustris* in the South, it has a wide host range amongst species of pine (Gibson 1979; Sinclair *et al.* 1987). It also occurs in northern USA and Canada, and it is scattered in South America and Europe, perhaps through introductions. Northern USA isolates differ from those in the southern USA, and isolates from China are apparently of the southern USA type (Zheng-Yu, Smalley & Guries 1995). The disease is severe in parts of China, where it became common after large-scale import of slash pine seed from the USA in the 1970s, although Zheng-Yu *et al.* believe the pathogen was present in China before then. Brown spot needle blight has resulted in serious losses to certain provenances of *P. sylvestris* grown for Christmas trees in the USA (Skilling & Nicholls 1974). This high-value crop is dependent on a dense and high-quality foliage. Short-needled varieties of Scots pine from Spain and France are highly susceptible to brown spot and are not suitable for Christmas tree production. Shearing of needles in wet weather enhances spore dispersal and should also be avoided (Skilling & Nicholls 1975).

19.8 Rust diseases

Members of the order Uredinales are obligate plant pathogens. These fungi are called rusts due to the fact that, at some point in their life cycle, they produce masses of dry spores that are often rust coloured. There are many species of pine rusts and most of these reside in the stem rust genera *Cronartium*, *Endocronartium* and *Peridermium*, or in the foliar rust genera *Coleosporium* and *Melampsora*. On pine hosts, they occur only in the northern hemisphere where they can damage stems, foliage or cones. Infections by stem rust fungi can lead to mortality, particularly when young trees are infected, and these are the most damaging of the pine rusts. Most noteworthy are white pine blister rust and fusiform rust, both caused by species of *Cronartium*.

The rust fungi are all biotrophic (obligate pathogens) and most have complex life cycles, with up to five different spore stages involved in the completion of a single generation. *Cronartium* species infect stems of pine and are heteroecious, that is, they produce some of their spore stages on an alternate, dicotyledonous host. The typical life cycle begins with aecia and aeciospores that develop after the fertilization of two haploid gametes produced on the pine tissue in structures called pycnia. The product of this fertilization gives rise to aecial structures from which masses of dikaryotic (two paired nuclei) aeciospores are produced. Galls or swellings generally develop on branches or stems producing aecia, and these tissues generally survive for many years unless killed by weakly pathogenic canker fungi, the activities of insects or other animal feeding (Byler, Cobb & Parmeter 1972).

Aeciospores are wind-dispersed and infect leaves of the dicotyledonous host, on which uredinia and the dikaryotic urediniospores are produced. The uredinial stage is often referred to as the repeating stage because the urediniospores infect the dicot host itself, thus building the rust population through several generations in a single growing season. Towards the end of a season and as temperatures drop, uredinial infections give rise to the telial stage, which produces teliospores. Upon germination of teliospores, the dikaryotic nuclei fuse to form a diploid and, following meiosis and mitosis, four haploid basidiospores are formed. The basidiospores can infect only pine and not the dicot host. These infections result in the formation of pycnia and spermatia, thus completing the typical heteroecious rust life cycle. The basidiospores are generally the most delicate spore stage and are most sensitive to environmental conditions. Years of optimal conditions for basidiospore infection, known as 'wave years', are associated with fusiform rust and white pine blister rust epidemics. In the former, these occur every 3–5 years, but if they occurred every 1 or 2 years, it would be impossible to grow plantation pines in the South (Powers, Schmidt & Snow 1981).

The impact of stem rusts is certainly increased by the trend to establish even-aged stands of fast-growing pine for commercial purposes (Peterson & Jewell 1968; Powers *et al.* 1981). Risks associated with movement of pine rusts among continents are adequately illustrated by the damage that white pine blister rust has caused in North America. The potential of most pine rusts to become established in the southern hemisphere is reduced by the fact that most rusts require two specific hosts. In contrast, there is substantial risk associated with the potential introduction of members of the stem rust genera *Endocronartium* or *Peridermium*, which are derived from species of *Cronartium* (Vogler 1995) but produce aeciospores that infect pine (autoecious). The most important of the pine-to-pine rusts in terms of quarantine procedures is *Peridermium* (*Endocronartium*) *harknessii* J.P. Moore, which is known to infect the widely planted *P. radiata* (Parmeter & Newhook 1967) and has been introduced to new areas of North America on nursery stock (Peterson 1981).

19.8.1 Fusiform rust

This disease, caused by *Cronartium quercuum* (Berk.) Miyabe ex Shirai f. sp. *fusiforme*, is one of the most important diseases of forest trees in the USA (Czabator 1971). The pathogen is restricted to the southern parts of the country (Maryland to Florida to Texas), with losses most severe from Louisiana to South Carolina. Oaks are the alternate host. Many two- and three-needled pines are susceptible to the various host-specialized forms of *C. quercuum*, but the pathogen that causes fusiform rust (f. sp. *fusiforme*) most severely damages slash and loblolly pines (Burdsall & Snow 1977). Infected trees less than 5 years of age generally die, and older trees may be deformed or break at the point of infection. Powers *et al.* (1981) estimate annual losses due to degrade of timber alone at $75 million. Another estimate places losses due to fusiform rust at $130 million annually (Anderson & Mistretta 1982).

Fusiform rust has probably been a major factor determining the biogeography of the southern pines. Loblolly pine is most susceptible when planted in the southern part of its natural range and slash pine when planted north of its natural range (Powers *et al.* 1981). In the 'rust corridor' between the core distributions of these rust-susceptible species, the rust-resistant longleaf and shortleaf pines were once more prevalent, but slow growth, regeneration difficulties, and problems with brown spot needle blight and littleleaf disease, respectively, have made these species less desirable for plantation forestry. Dinus (1974) notes that the fusiform rust corridor runs along the northern boundary of the original longleaf pine forest type, where loblolly pine and oaks tended to be interspersed with longleaf pine. Heavy disease pressure here likely gave an advantage to rust-resistant longleaf pine before human intervention.

Dinus (1974) speculates that fusiform rust was rare before 1900, in part because of the prominence of the fire-resistant longleaf pine at that time. Harvesting practices and fire control reduced this species to perhaps 10% of its previous volume. Loblolly and slash pine regenerated well on cut-over longleaf pine sites. Factors contributing to the high impact of fusiform rust today include the increased susceptibility of pines due to vigorous growth, a shift to plantations versus natural regeneration, use of the most rust-susceptible species (loblolly and slash pines) for 98% of all plantings, and control of wildfires, which leads to an increase in the incidence of the alternate hosts, oaks, and a decrease in the incidence of the rust-resistant longleaf pine (Dinus 1974; Powers *et al.* 1981).

Considerable effort has been expended on reducing the impact of fusiform rust in recent years. Hazard rating of sites is useful (Powers *et al.* 1981). Efforts have concentrated on the selection of resistant provenances of loblolly and slash pines and the development of hybrids with shortleaf pine (Peterson & Jewell 1968; Powers *et al.* 1981; Schmidt, Powers & Snow 1981; Carson & Carson 1989). Techniques for rapid and effective inoculation of seedlings and the early detection of susceptible stock have been developed and are used with considerable success (Wells & Dinus 1978). Selection of resistant stock is complicated by the fact that the pathogen has a variable genetic base and is continuously adapting to rust-resistant material (Snow, Dinus & Walkinshaw 1976; Powers *et al.* 1981). With the ecology of the disease and the southern pines in mind, Dinus (1974) argues for a greater emphasis on longleaf pine, which is also more resistant to attacks by the southern pine beetle, another major management problem in southern pine forests.

19.8.2 White pine blister rust

This rust disease, caused by *Cronartium ribicola* J.C. Fisch. ex Rab, is one of the most important and intensively studied tree diseases. It is notorious on white (five-needled) pines in North America. The pathogen is native to Asia (Kaneko *et al.* 1995) but was introduced to Europe and from there to eastern and western North America at the beginning of this century (Peterson & Jewell 1968; Ziller 1974), when planting stock was being produced in Europe and shipped to the USA. Establishment and spread of the disease in North America was effective and rapid due to the susceptibility of native pines relative to Asian and European white pines (Hoff, Bingham & McDonald 1980), climatic conditions favouring infection, and the wide-scale presence of the alternate hosts (various species of *Ribes*, the currants and gooseberries). As mentioned earlier, the loss of western white pine as a management option on some forest sites (Fig. 19.5) has led to other pest problems, notably root rots. This is just one of the irrevocable effects blister rust has had on the white pine forests of North America.

Cronartium ribicola has a life cycle that spans between three and six years. During midsummer to autumn, basidiospores are produced from teliospores on *Ribes*. These spores are airborne for relatively short distances, and those landing on pine needles may germinate and penetrate stomata during cool, wet conditions. High hazard regions are those where average temperatures in summer (July) are below 21 °C with prolonged wet periods (Van Arsdel, Riker & Patton 1956; Sinclair *et al.* 1987). In some regions, conditions in most years are too dry and warm during these periods. Infections proceed from needles to twigs, from twigs to branches, and from branches to the trunk of trees. Inconspicuous pycnia develop on the surface of infected tissue 1–2 years after infection. These give rise to conspicuous aecia, which burst through the stem tissue as blisters of fungal tissue (peridium) containing masses of orange aeciospores. Chlorosis of killed branches (flags) or tree tops develops as the infected area dies. Small trees are most vulnerable since their infected branches are relatively close to the main stem. Pruning branches of young trees on low to moderate hazard sites is an effective control. As the forest canopy begins to close, damage due to white pine blister rust subsides because spores produced on the alternate host below the canopy are less common and susceptible tissue is less accessible to them. Infections of older trees most commonly result from basidiospores that have originated outside that stand and develop at the tops of the trees.

During the early years of blister rust in North America, a vigorous programme to eradicate alternate hosts from stands of white pine was undertaken. This was later abandoned in most areas after it was shown to be ineffective (Boyce 1961; Peterson & Jewell 1968). However, Ostrofsky *et al.* (1988) showed substantial benefit of a 70 year *Ribes* eradication programme in Maine. Hazard ratings for sites based on climatic conditions during basidiospore release have been developed. White pines on steep slopes and near areas where *Ribes* are abundant (e.g. along streams and in openings in the forest canopy) are particularly vulnerable (Hunt 1983).

Virtually all of the white pines are susceptible to white pine blister rust to some degree (Hoff *et al.* 1980). Considering that the introduced *C. ribicola* is the only stem rust on the white pines in North America, it is somewhat surprising that some resistance is found in populations of most of these species (Fig. 19.5). Breeding for disease resistance has been successful in North America, although the presence of many races of the pathogen, which differ in virulence, complicates these efforts (Peterson & Jewell 1968; McDonald *et al.* 1984). Surprisingly, a major, dominant gene for resistance in *P. lambertiana* has held up well in the field (Kinloch & Dupper 1987). Kinloch (1992) has speculated that this single gene may have evolved through selection pressure imposed by pinyon rust, which does not occur on sugar pine, but the gene is more frequent in sugar pine populations where pinyon rust is prevalent.

Most attention has been given to the commercially important white pines *P. strobus, P. monticola* and *P. lambertiana* (see Kinloch, Marosy & Huddleston (1996) for a detailed assessment of the situation in *P. lambertiana*), but the disease can be even more severe on other North American species. Forests of *P. flexilis* have been devastated in Alberta and elsewhere (Ziller 1974). *Pinus albicaulis* is threatened by extinction in some regions where it is being replaced by spruce and fir through the actions of blister rust and fire suppression (Keane, Morgan & Menakis 1994; Krebill & Hoff 1995). Many wildlife species depend on the large seeds of *P. albicaulis*, and these animal populations will continue to be affected.

Fig. 19.8. **The pinewood nematode is vectored during maturation feeding by cerambycid beetles such as *Monochamus carolinensis* (photo: M.J. Wingfield).**

19.9   Diseases caused by nematodes

In contrast to the situation with agronomic crops, nematodes are not generally considered to be important pathogens of pines. Various root-infesting nematodes are, however, known on pines and some of these can be moderately important in nurseries. The importance of soil-borne nematodes in agriculture has grown with repeated rotations and other intensive farming practices, and it is thus felt that nematodes have the potential to become important in plantation forestry in the future. In recent years a nematode with an insect vector has come to be recognized as one of the most serious pathogens of pines. This unique nematode is commonly referred to as the pinewood nematode or pine wilt nematode.

Fig. 19.9. **Exotic *Pinus nigra* in the USA can be killed by the pinewood nematode, whereas native *P. resinosa* trees are apparently not affected by the pathogen (photo: M.J. Wingfield).**

19.9.1   Pine wilt

The pinewood nematode, *Busaphelenchus xylophilus* (Steiner & Buhrer) Nickle, apparently is native to North America but is currently active and killing large numbers of native pines in Asia, most notably in Japan (Mamiya 1983; Wingfield 1987a). The pinewood nematode is a classic example of an introduced pathogen on a population apparently lacking resistance. Where the nematode is native in North America, local pine species are highly resistant and unaffected.

The nematode has a fascinating and complex biology (Mamiya 1983; Wingfield 1987a). Its vectors, longhorn beetles (Coleoptera: Cerambycidae), themselves of secondary importance, emerge from dead trees in spring carrying large numbers of nematodes and proceed to maturation feeding on the shoots of healthy pines (Fig. 19.8). Nematodes enter the maturation feeding wounds and multiply rapidly in the resin ducts, resulting in rapid wilting of the host, perhaps due to toxins produced by the nematode (Bolla *et al.* 1987). In dying and dead trees, the nematodes enter a mycophagus phase, where they feed on

blue stain fungi. Dying trees are then attractive to the mature beetles that oviposit in the bark, thus completing the life cycle of the beetles and providing the nematodes in these trees with vectors for the next season.

When the pinewood nematode was first discovered as a pathogen in the USA in 1979 (Dropkin & Foudin 1979), it was feared that it would have a very devastating impact (Dropkin *et al.* 1981). The nematode had, however, been recorded in the country in 1929, although this was under another name (*Aphelenchoides xylophilus* Steiner & Buhrer) and in the absence of the knowledge of its pathogenic nature. It was later shown that in the USA, *B. xylophilus* commonly occurs in trees or tree parts dying due to factors other than nematode infestation (Wingfield *et al.* 1982). It was furthermore shown that the nematode is not pathogenic on native North American pine species (Fig. 19.9; Wingfield, Blanchette & Nicholls 1984; Wingfield, Bedker & Blanchette 1986). Another ecologically important discovery was that it is not only transmitted during feeding of beetles on healthy trees but also when the beetles oviposit in dying trees (Wingfield 1983). Transmission during oviposition implies that *B. xylophilus* can maintain an exclusively mycophagus life cycle, which appears to be the norm in North America (Wingfield 1987b).

Although the pine wilt disease was recognized as serious in Japan during the early part of this century (Mamiya 1987), it was not until 1971 that the nematode was found to be the causal agent of this serious malady (Kiyohara & Tokushige 1971). It seems likely that pine wilt disease will ultimately eliminate highly susceptible pine species such as *P. luchuensis* and *P. thunbergii* from Asian landscapes (Wingfield *et al.* 1984; Mamiya 1987; Wingfield 1987a). Control of pine wilt is virtually impossible in Asia due to its introduced and epidemic nature. Despite this, considerable effort has been made to protect valuable shade and ornamental trees using insecticide sprays and chemical injection. Suppression of vector populations in Japan by removing borer-infested trees from forests is vigorously pursued. Various attempts have also been made to implement a variety of forms of biological control, although the impact of these strategies is at this stage not known.

A similar but apparently different nematode occurs in European pines without causing disease. *Pinus sylvestris* is, however, highly susceptible to infection by *B. xylophilus* where it is planted in the USA, particularly in warmer regions. There is considerable concern that, if introduced, the pine wilt nematode could cause an epidemic in Europe. Indeed, this is the basis for very strict control measures pertaining to trade in conifer timber between North America and Europe.

19.10   Conclusions

Diseases have had a very substantial impact on pines in natural ecosystems and in plantations. We are gradually gaining an understanding of the complex and often subtle roles that pathogens play in natural forest ecosystems. This knowledge will be an important aid in the development of sound management practices, which should align our interests more closely to the natural dynamics of the forest. For instance, tree harvesting should mimic the effects of indigenous mortality agents, and fire needs to be incorporated into sustainable management of many pine ecosystems. Plantation forestry can also incorporate disease resistance and heterogeneity into unnatural pine ecosystems and thus lessen the impact of some diseases.

In spite of this awareness and the advantages of a biorational approach to forest management, indications are that diseases will become increasingly important in the future. We make this prediction based on the following scenarios:

Table 19.1. Discoveries, introductions and epidemics in the pathology of *Pinus*

1870s	Hartig writes the first book on tree diseases describing Armillaria root rot, annosum root rot, red ring rot, a stem rust and a needle rust of pine.
1900s	White pine blister rust is introduced twice to North America on nursery stock raised in Europe.
1910s	Weir describes the biology and management of dwarf mistletoe. Sudden death of pine trees noted in Japan, now believed due to pine wood nematode. Sphaeropsis canker reported in Southern Africa.
1930s	Fusiform rust emerges as a management problem for southern yellow pines in the USA.
1940s	*Phytophthora cinnamomi* associated with littleleaf disease on old-field sites in the southeastern USA. Pitch canker recognized in the southeastern USA.
1950s	Stump top colonization associated with the epidemiology of annosum root rot, and biological control developed. Red band needle blight appears in Africa, eliminating *P. radiata* as a commercial species in some regions.
1960s	Red band needle blight appears on radiata pine in Chile and New Zealand.
1970s	Cavities in living trees caused by stem rotters recognized as important wildlife habitat. Root-rotted trees found to support endemic bark beetle populations. Pine wood nematode identified as cause of pine wilt in Japan and recognized as a pathogen in USA. New York (USA) epidemic of Scleroderris canker caused by European race. Brown spot needle blight appears in China and red band needle blight in Australia.
1980s	Characterization of host specialized variants or species of root pathogens in *Armillaria, Heterobasidion annosum, Phellinus weirii* and *Leptographium wageneri*. Pitch canker discovered in California (USA) and Japan. Also reported from Mexico, but perhaps indigenous there.
1990s	The importance of root rots and other diseases in the dynamics of forest ecosystems increasingly recognized. Pitch canker pathogen reported from South Africa.

1 Most of our so-called natural forest ecosystems are not following the natural dynamics of forest succession and rejuvenation. Fire has been a major component of pine ecosystems, and fire suppression is changing the composition of our forests, often to more disease-prone species and abnormally high stocking levels. Loss of otherwise disease-resistant species due to introduced pathogens will also make management of natural ecosystems more difficult.

2 The growing pressure to conserve natural forest environments is likely to increase the importance of plantation forestry. Plantations of native pine species have given rise to serious disease problems that were previously of minor consequence. These diseases will be favoured by genetic uniformity of the host trees, often low disease resistance and, in some cases, by the creation of fresh stumps.

3 The most devastating epidemics have arisen through the introduction of pathogens into new environments (Table 19.1). Globalization of our economies and the movement of people, packaging

materials, seed and raw timber continue to increase, and the introduction of new pathogens into new environments thus seems inevitable.

4   Although evidence to this effect is still rudimentary, there is reason to believe that plant pathogens are rapidly evolving. Minor pathogens are presented with new niches through introductions to new ecosystems and alteration of the composition or dynamics of others. It would appear that agricultural practices have opened the door of opportunity for many new pathogens of crops. Likewise, forestry practices may be putting considerable selection pressure on new pathogens to fill new niches.

The above may be perceived as a rather pessimistic view of the future. At this point, it is germane to note that considerable progress has been made during the course of the last century in our understanding of tree diseases (Table

19.1), and disease problems have been very effectively managed in many situations. The study of tree pathogens should be considerably enhanced to equip us with the wherewithal to deal with future disease problems. Efforts to exclude pathogens from new environments through more effective quarantine strategies should also be vigorously supported.

Acknowledgements

We are grateful to R.A. Blanchette and Teresa Coutinho who reviewed the manuscript, and to the Foundation for Research Development in South Africa for financial support. This chapter, contribution No. J-16681 of the Iowa Agriculture and Home Economics Experiment Station, Ames, Project 3226, was supported by Hatch and State of Iowa funds.

References

Aalto-Kallonen, T. & Kurkela, T. (1985). *Gremmeniella* disease and site factors affecting the condition and growth of Scots pine. *Communicationes Instituti Forestales Fenniae*, **126**, 1–28.

Alexander, M.E. & Hawksworth, F.G. (1975). *Wildland Fires and Dwarf Mistletoes: A Literature Review of Ecology and Prescribed Burning*. USDA Forest Service General Technical Report RM-14.

Anderson, R.L. & Mistretta, P.A. (1982). *Management Strategies for Reducing Losses Caused by Fusiform Rust, Annosus Root Rot, and Littleleaf Disease. Integrated Pest Management Handbook*. USDA Forest Service Agricultural Handbook 597.

Anderson, J.B., Ullrich, R.C., Roth, L.F. & Filip, G.M. (1979). Genetic identification of clones of *Armillaria mellea* in coniferous forests in Washington. *Phytopathology*, **69**, 1109–11.

Basham, J.T. (1975). Heart rot of jack pine in Ontario. IV. Heartwood-inhabiting fungi, their entry and interactions within living trees. *Canadian Journal of Forest Research*, **5**, 706–21.

Basham, J.T. & Morawski, Z.J.R. (1964). *Cull Studies, the Defects and Associated Basidiomycete Fungi in the Heartwood of Living Trees in the Forests of Ontario*. Canadian Department of Forestry Publication 1072.

Belanger, R.P., Hedden, R.L. & Tainter, F.H. (1986). *Managing Piedmont Forests to Reduce Losses from the Littleleaf Disease-Southern Pine Beetle Complex*. USDA Forest Service Agricultural Handbook 649.

Benson, D.M. (1982). Cold inactivation of *Phytophthora cinnamomi*. *Phytopathology*, **72**, 560–3.

Berry, C.R. & Hepting, G.H. (1959). *Pitch Canker of Southern Pines*. USDA Forest Servic Pest Leaflet 35.

Blakeslee, G.M. & Foltz, J.L. (1981). The deodar weevil, a vector and wounding agent associated with pitch canker of slash pines. *Phytopathology*, **71**, 861 (Abstract).

Bolla, R.I., Winter, R.E.K., Fitzsimmons, F., Weaver, C. & Koslowski, P. (1987). Phytotoxins and pathotypes associated with the pine wood nematode in the United States. In *Pathogenicity of the Pine Wood Nematode*, ed. M.J. Wingfield, pp. 26–39. St Paul, Minnesota: American Phytopathological Society Press.

Boyce, J.S. (1961). *Forest Pathology*. New York: McGraw-Hill.

Brown, J.K. (1975). Fire cycles and community dynamics in lodgepole pine forests. In *Management of Lodgepole Pine Ecosystems: Symposium Proceedings, 1973*., ed. M. Baumgartner, pp. 429–56. Pullman: Washington State University.

Burdsall, H.H., Jr & Snow, G.A. (1977). Taxonomy of *Cronartium quercuum* and *C. fusiforme*. *Mycologia*, **69**, 503–8.

Byler, J.W., Cobb, F.W., Jr & Parmeter, J.R., Jr (1972). Occurrence and significance of fungi inhabiting galls caused by *Peridermium harknessii*. *Canadian Journal of Botany*, **50**, 1275–82.

Byler, J.W., Marsden, M.A. & Hagle, S.K. (1990). The probability of root disease on the Lolo National Forest, Montana. *Canadian Journal of Forest Research*, **20**, 987–94.

Carson, S.D. & Carson, M.J. (1989). Breeding for resistance in forest trees – a quantitative genetic approach. *Annual Review of Phytopathology*, **27**, 373–95.

Castello, J.D., Leopold, C.J. & Smallidge, P.J. (1995). Pathogens, patterns and processes in forest ecosystems. *BioScience*, **45**, 16–24.

Cobb, F.W., Jr (1988). *Leptographium wageneri*, cause of black-stain root disease: a review of its discovery, occurrence and biology with emphasis on pinyon and ponderosa pine. In *Leptographium Root Diseases on Conifers*, ed. T.C. Harrington & F.W. Cobb, Jr, pp. 41–62. St Paul, Minnesota: American Phytopathological Society Press.

Cobb, F.W., Jr & Miller, D.R. (1968). Hosts and geographic distribution of *Scirrhia pini*, the cause of red band needle blight in California. *Journal of Forestry*, 66, 930–3.

Cobb, F.W., Jr, Parmeter, J.R., Jr, Wood, D.L. & Stark, R.W. (1974). Root pathogens as agents predisposing ponderosa pine and white fir to bark beetles. In *Proceedings of the Fourth International Conference on* Fomes annosus, ed. E.G. Kuhlman, pp. 8–15. USDA Forest Service.

Cobb, F.W., Jr, Slaughter, G.W, Rowney, D.L. & DeMars, C.J. (1982). Rate of spread of *Ceratocystis wageneri* in ponderosa pine stands in the central Sierra Nevada. *Phytopathology*, 72, 1359–62.

Conner, R.N. & Locke, B.A. (1982). Fungi and red-cockaded woodpecker cavity trees. *Wilson Bulletin*, 94, 64–70.

Correll, J.C., Gordon, T.R. & McCain, A.H. (1992). Genetic diversity in California and Florida populations of the pitch canker fungus *Fusarium subglutinans* f. sp. *pini*. *Phytopathology*, 82, 415–20.

Czabator, F.J. (1971). *Fusiform Rust of Southern Pines – A Critical Review*. USDA Forest Service Research Paper 50–65.

Dallara P.L., Storer, A.J., Gordon, T.R. & Wood, D.L. (1995). *Current Status of Pitch Canker Disease in California*. California Department of Forestry and Fire Protection, Tree Notes, No. 20.

Dickman, A. & Cook, S. (1989). Fire and fungus in a mountain hemlock forest. *Canadian Journal of Botany*, 67, 2005–16.

Dinus, R.J. (1974). Knowledge about natural ecosystems as a guide to disease control in managed forests. *Proceedings of the American Phytopathological Society*, 1, 184–90.

Dorworth, C.E. (1972). Epidemiology of *Scleroderris lagerbergii* in central Ontario. *Canadian Journal of Botany*, 50, 751–65.

Dorworth, C.E. (1981). Status of pathogenic and physiologic races of *Gremmeniella abietina*. *Plant Disease*, 65, 927–31.

Dropkin, V.H. & Foudin, A.S. (1979). Report of the occurrence of *Bursaphelenchus lignicolus*-induced pine wilt disease in Missouri. *Plant Disease Reporter*, 63, 904–5.

Dropkin, V.H., Foudin, A., Kondo, E., Linit, M., Smith, M. & Robbins, K. (1981). Pinewood nematode: a threat to U.S. forests? *Plant Disease*, 65, 1022–7.

Dwinell, L.D., Barrows-Broaddus, J. & Kuhlman, E.G. (1985). Pitch canker: A disease complex. *Plant Disease*, 69, 270–6.

Dwinell, L.D., Kuhlman, E.G. & Blakeslee, G.M. (1981). Pitch canker of southern pines. In *Fusarium: Diseases, Biology and Taxonomy*, ed. P.E. Nelson, T.A. Toussoun & R.J. Cook, pp. 188–94. State College: The Pennsylvania State University Press.

Evans, H.C. (1984). *The Genus* Mycosphaerella *and its Anamorphs* Cercoseptoria, Dothistroma *and* Lecanosticta *on Pines*. Commonwealth Mycological Institute, Mycological Paper No. 153.

Fox, J.W., Wood, D.L. & Koehler, C.S. (1990). Distribution and abundance of engraver beetles (Scolytidae: *Ips* species) on Monterey pines infected with pitch canker. *Canadian Entomologist*, 122, 1157–66.

Fox, J.W., Wood, D.L., Koehler, C.S. & O'Keefe, S.T. (1991). Engraver beetles (Scolytidae: *Ips* species) as vectors of the pitch canker fungus, *Fusarium subglutinans*. *Canadian Entomologist*, 123, 1355–67.

Fraedrich, S.W. & Miller, T. (1995). Mycoflora associated with slash-pine seeds from cones collected at seed orchards and cone-processing facilities in south-eastern USA. *European Journal of Forest Pathology*, 25, 73–82.

Gibbs, J.N. (1968). Resin and the resistance of conifers to *Fomes annosus*. *Annals of Botany*, 32, 649–65.

Gibson, I.A.S. (1972). Dothistroma blight of *Pinus radiata*. *Annual Review of Phytopathology*, 10, 51–72.

Gibson, I.A.S. (1975). Impact and control of Dothistroma blight of pines. *European Journal of Forest Pathology*, 4, 89–100.

Gibson, I.A.S. (1979). *Diseases of Forest Trees Widely Planted as Exotics in the Tropics and Southern Hemisphere. Part II. The genus* Pinus. Kew: Commonwealth Mycological Institute, and Oxford: Commonwealth Forestry Institute, University of Oxford.

Gill, L.S. (1935). *Arceuthobium* in the United States. *Transactions of the Connecticut Academy of Arts and Sciences*, 32, 111–245.

Goheen, D.J., & Cobb, F.W., Jr (1978). Occurrence of *Verticicladiella wagenerii* and its perfect state, *Ceratocystis wageneri* sp. nov., in insect galleries. *Phytopathology*, 68, 1192–5.

Goheen, D.J. & Cobb, F.W., Jr (1980). Infestation of *Ceratocystis wageneri*-infected ponderosa pines by bark beetles (Coleoptera: Scolytidae) in central Sierra Nevada. *Canadian Entomologist*, 112, 725–30.

Goheen, D.J. & Hansen, E.M. (1993). Effects of pathogens and bark beetles on forests. In *Beetle–Pathogen Interactions in Conifer Forests*, ed. T.D. Schowalter & G.M. Filip, pp. 175–96. London: Academic Press.

Greig, B.J.W. (1995). Butt-rot of Scots pine in Thetford Forest caused by *Heterobasidion annosum*: a local phenomenon. *European Journal of Forest Pathology*, 25, 95–9.

Guillaumin, J.J., Mohammed, C., Anselmi, N. *et al.* (1993). Geographical distribution and ecology of the *Armillaria* species of western Europe. *European Journal of Forest Pathology*, 23, 321–41.

Haack, R.A. & Byler, J.W. (1993). Insects and pathogens, regulators of forest eco-systems. *Journal of Forestry*, 91(9), 32–7.

Haddow, W.R. (1938). On the classification, nomenclature, hosts and geographic range of *Trametes pini* (Thore) Fries. *Transactions of the British Mycological Society*, 22, 182–93.

Hadfield, J.S. & Johnson, D.W. (1977). *Laminated Root Rot. A guide for Reducing and Preventing Losses in Oregon and Washington Forests*. USDA Forest Service, Pacific Northwest Region.

Hagle, S.K. & Byler, J.W. (1994). Root diseases and natural disease regimes in forests of western U.S.A. In *Proceedings of the 8th International Conference on Root and Butt Rots, 1993*, ed. M. Johansson & J. Stenlid, pp. 606–17. Uppsala: Swedish University of Agricultural Sciences.

Hamelin, R.C., Ouellette, G.B. & Bernier, L. (1993). Identification of *Gremmeniella abietina* races with random amplified polymorphic DNA markers. *Applied and Environmental Microbiology*, 59, 1752–5.

Hansen, E.M. (1978). Incidence of *Verticicladiella wagenerii* and *Phellinus weirii* in Douglas-fir adjacent to and away from roads in western Oregon. *Plant Disease Reporter*, 62, 179–81.

Harrington, T.C. (1993). Biology and taxonomy of fungi associated with bark beetles. In *Beetle–Pathogen Interactions in Conifer Forests*, ed. T.D. Schowalter & G.M. Filip, pp. 37–58. London: Academic Press.

Harrington, T.C. & Cobb, F.W., Jr (1984). Host specialisation of three morphological variants of *Verticicladiella wageneri*. *Phytopathology*, **74**, 286–90.

Harrington, T.C. & Cobb, F.W., Jr (1986). Varieties of *Verticicladiella wageneri*. *Mycologia*, **78**, 562–7.

Harrington, T.C. & Cobb, F.W., Jr (1987). *Leptographium wageneri* var. *pseudotsugae*, var. nov., cause of black stain root disease on Douglas-fir. *Mycotaxon*, **30**, 501–7.

Harrington, T.C., Cobb, F.W., Jr & Lownsbery, J.W. (1985). Activity of *Hylastes nigrinus*, a vector of *Verticicladiella wageneri*, in thinned stands of Douglas-fir. *Canadian Journal of Forest Research*, **15**, 519–23.

Harrington, T.C., Reinhart, C., Thornburgh, D.A. & Cobb, F.W., Jr (1983). Association of black-stain root disease with precommercial thinning of Douglas-fir. *Forest Science*, **29**, 12–4.

Harrington, T.C., Worrall, J.J. & Baker, F.A. (1992). *Armillaria*. In *Methods for research on soilborne phytopathogenic fungi*, ed. L.L. Singleton, J.D. Mihail, & C. Rush, pp. 81–5. St Paul, Minnesota: American Phytopathological Society Press.

Harrington, T.C., Worrall, J.J. & Rizzo, D.M. (1989). Compatibility among host-specialized isolates of *Heterobasidion annosum* from western North America. *Phytopathology*, **79**, 290–6.

Hartig, R. (1874). *Wichtige Krankheiten der Waldbäume*. [Translated to English by W. Merrill, D.H. Lambert & W. Liese in 1975.] Phytopathological Classics No. 12. St. Paul, Minnesota: American Phytopathological Society.

Hawksworth, F.G. & Johnson, D.W. (1989). *Biology and Management of Dwarf Mistletoe in Lodgepole Pine in the Rocky Mountains*. USDA Forest Service General Technical Report RM-169.

Hawksworth, F.G. & Wiens, D. (1972). *Biology and Classification of Dwarf Mistletoes (Arceuthobium)*. USDA Forest Service., Agricultural Handbook 401.

Hawksworth, F.G. & Wiens, D. (1996). *Dwarf Mistletoes: Biology, Pathology, and Systematics*. USDA Forest Service, Agricultural Handbook 709.

Hellgren, M. & Högberg, N. (1995). Ecotypic variation of *Gremmeniella abietina* in northern Europe: disease patterns reflected by DNA variation. *Canadian Journal of Botany*, **73**, 1531–9.

Hepting, G.H. & Roth, E.R. (1953). Host relations and spread of the pine pitch canker disease. *Phytopathology*, **53**, 475 (Abstract).

Hodges, C.S. (1969). Modes of infection and spread of *Fomes annosus*. *Annual Review of Phytopathology*, **7**, 247–66.

Hoff, R., Bingham, R.T. & McDonald, G.I. (1980). Relative blister rust resistance of white pines. *European Journal of Forest Pathology*, **10**, 307–16.

Hood, I.A., Redfern, D.B. & Kile, G.A. (1991). *Armillaria* in planted forests. In *Armillaria Root Disease*, ed. C.G. Shaw III & G.A. Kile, pp. 122–49. USDA Forest Service Agriculture Handbook 691.

Hunt, R.S. (1983). White pine blister rust in British Columbia. II. Can stands be hazard rated? *Forestry Chronicle*, **59**, 136–8.

Ivory, M.H. (1972). Resistance to Dothistroma needle blight induced in *Pinus radiata* by maturity and shade. *Transactions of the British Mycological Society*, **59**, 205–12.

Ivory, M.H. (1987). *Diseases and Disorders of Pines in the Tropics. A Field and Laboratory Manual*. Overseas Research Publication 31. Oxford: Oxford Forestry Institute.

Ivory, M.H. (1994). Records of foliage pathogens of *Pinus* species in tropical countries. *Plant Pathology*, **43**, 511–18.

Jackson, J.A. (1977). Red-cockaded woodpeckers and pine red heart disease. *Auk*, **94**, 160–3.

Kais, A.G. (1975). Environmental factors affecting brown spot infection on longleaf pine. *Phytopathology*, **65**, 1389–92.

Kaneko, S., Katsuya, K., Kakishima, M. & Ono, Y. (eds.) (1995). *Proceedings of the Fourth IUFRO Rusts of Pines Working Party Conference*. Tsukuba, Japan: IUFRO.

Keane, R.E., Morgan, P. & Menakis, J.P. (1994). Landscape assessment of the decline of whitebark pine (*Pinus albicaulis*) in the Bob Marshall Wilderness Complex, Montana, USA. *Northwest Science*, **68**, 213–29.

Kile, G.A., Guillaumin, J.J., Mohammed, C. & Watling, R. (1994). Biogeography and pathology of *Armillaria*. In *Proceedings of the 8th International Conference on Root and Butt Rots, 1993*, ed. M. Johansson & J. Stenlid, pp. 411–36. Uppsala: Swedish University of Agricultural Sciences.

Kile, G.A., McDonald, G.I. & Byler, J.W. (1991). Ecology and disease in natural forests. In *Armillaria Root Disease*, ed. C.G. Shaw III & G.A. Kile, pp. 102–21.

USDA Forest Service Agriculture Handbook 691.

Kinloch, B.B., Jr (1992). Distribution and frequency of a gene for resistance to white pine blister rust in natural populations of sugar pine. *Canadian Journal of Botany*, **70**, 1319–23.

Kinloch, B.B. Jr & Dupper, G.E. (1987). Restricted distribution of a virulent race of the white pine blister rust pathogen in the western United States. *Canadian Journal of Forest Research*, **17**, 448–51.

Kinloch, B.B., Marosy, M. & Huddleston, M.E. (eds.) (1996). *Sugar Pine. Status, Values, and Roles in Ecosystems. Proceedings of a Symposium Presented by the California Sugar Pine Management Committee*. Oakland, California: University of California Division of Agriculture and Natural Resources.

Kiyohara, T. & Tokushige, Y. (1971). Inoculation experiments of a nematode, *Bursaphelenchus* sp. onto pine trees. *Journal of the Japanese Forestry Society*, **53**, 210–18 [In Japanese with English summary.]

Kobayashi, T. & Kawabe, Y. (1992). Tree diseases and their causal fungi in Miyako Island. *Japanese Journal of Tropical Agriculture*, **36**, 195–206.

Kobayashi, T. & Muramoto, M. (1989). Pitch canker of *Pinus luchuensis*, a new disease in Japanese forests. *Forest Pests*, **38**, 169–73.

Korhonen, K. (1978). Intersterility groups of *Heterobasidion annosum*. *Communicationes Instituti Forestales Fenniae*, **94**(6), 1–25.

Korhonen, K., Lipponen, K., Bendz, M. et al. (1994). Control of *Heterobasidion annosum* by stump top treatment with 'Rotstop', a new commercial formulation of *Phlebiopsis gigantea*. In *Proceedings of the 8th International Conference on Root and Butt Rots, 1993*, ed. M. Johansson & J. Stenlid. pp. 675–83. Uppsala: Swedish University of Agricultural Sciences.

Krebill, R.G. & Hoff, R.J. (1995). Update on *Cronartium ribicola* in *Pinus albicaulis* in the Rocky Mountains, USA. In *Proceedings of the Fourth IUFRO Rusts of Pines Working Party Conference*, ed. S. Kaneko, K. Katsuya, M. Kakishima & Y. Ono, pp. 119–26. Tsukuba: University of Tsukuba.

Kuhlman, E.G., Dwinell, L.D. Nelson, P.E. & Booth, C. (1978). Characterisation of the *Fusarium* causing pitch canker of southern pines. *Mycologia*, **70**, 1131–43.

Landis, T.D. & Helburg, L.B. (1976). Black stain root disease of pinyon pine in Colorado. *Plant Disease Reporter*, **60**, 713–17.

Lecours, N., Hansson, P., Hellgren, M. & Laflamme, G. (1994). Molecular epidemiology of *Gremmeniella abietina*. *Phytopathology*, **84**, 1371 (Abstract).

Lindsey, J.P. & Gilbertson, R.L. (1983). Notes on basidiomycetes that decay bristlecone pine. *Mycotaxon*, **18**, 541–59.

Loehle, C. (1988). Tree life history strategies: the role of defences. *Canadian Journal of Forest Research*, **18**, 209–22.

Lundquist, J.E. (1993). Spatial and temporal characteristics of canopy gaps caused by *Armillaria* root disease and their management implications in lowveld forests of South Africa. *European Journal of Forest Pathology*, **23**, 362–71.

McCain, A.H., Koehler, C.S. & Tjosvold, S.A. (1987). Pitch canker threatens California pines. *California Agriculture*, **41**, 22–3.

McDonald, G.I., Hansen, E.M., Osterhuas, C.A. & Samman, S. (1984). Initial characterisation of a new strain of *Cronartium ribicola* from the Cascade Mountains of Oregon. *Plant Disease*, **68**, 800–4.

McFarlane, R.W. (1992). *A Stillness in the Pines*. New York: W.W. Norton & Co.

MacKenzie, M. & Shaw, C.G., III (1977). Spatial relationships between *Armillaria* root-rot of *Pinus radiata* seedlings and the stumps of indigenous trees. *New Zealand Journal of Forest Research*, **7**, 374–83.

Mamiya, Y. (1983). Pathology of pine wilt disease caused by *Bursaphelenchus xylophilus*. *Annual Review of Phytopathology*, **21**, 201–20.

Mamiya, Y. (1987). Origin of the pine wood nematode and its distribution outside the United States. In *Pathogenicity of the Pine Wood Nematode*, ed. M.J. Wingfield, pp. 59–65. St Paul, Minnesota: American Phytopathological Society Press.

Manion, P.D. (1981). *Tree Disease Concepts*. New Jersey: Prentice Hall.

Manion, P.D. & Skilling, D.D. (1983). Overview and summary of the Scleroderris canker symposium and future research needs. In *Scleroderris Canker of Conifers*, ed. P.D. Manion. pp. 261–9. The Hague: Martin Nijhoff & Dr W. Junk.

Marosy, M. & Parmeter, J.R. (1989). The incidence and impact of *Heterobasidion annosum* on pine and incense cedar in California forests. In *Proceedings of the Symposium on Research and Management of Annosus Root Disease (Heterobasidion annosum) in Western North America*. Monterey, California, 1989, ed. W.J. Otrosina & R.F. Scharpf, pp. 78–81. USDA Forest Service General Technical Report PSW-116.

Monnig, E. & Byler, J. (1992). *Forest Health and Ecological Integrity in the Northern Rockies*. USDA Forest Service, FPM Report 92–7.

Muramoto, M. & Dwinell, L.D. (1990). Pitch canker of *Pinus luchuensis* in Japan. *Plant Disease*, **74**, 530.

Muramoto, M., Tashiro, T. & Minamihashi, H. (1988). Distribution of *Fusarium moniliforme* var. *subglutinans* in Kagoshima Prefecture and its pathogenicity to pines. *Journal of the Japanese Forest Society*, **75**, 1–9.

Newhook, F.J. (1959). The association of *Phytophthora* spp. with mortality of *Pinus radiata* and other conifers. 1. Symptoms and epidemiology in shelterbelts. *New Zealand Journal of Agricultural Research*, **2**, 808–43.

Ostrofsky, W.D., Rumpf, T., Struble, D. & Bradbury, R. (1988). Incidence of white pine blister rust in Maine after 70 years of a *Ribes* eradication program. *Plant Disease*, **72**, 967–70.

Palmer, M.A. & Nicholls, T.H. (1985). Shoot blight and collar rot of *Pinus resinosa* caused by *Sphaeropsis sapinea* in forest tree nurseries. *Plant Disease*, **69**, 739–40.

Palmer, M.A., Stewart, E.L. & Wingfield, M.J. (1987). Variation among isolates of *Sphaeropsis sapinea* in the North Central United States. *Phytopathology*, **77**, 944–8.

Parks, C.A., Bull, E.L. & Filip, G.M. (1995). Creating wildlife trees in managed forests using decay fungi. In *Partnerships for Sustainable Forest Ecosystem Management. Fifth Mexico/US Symposium, Guadalajara, Mexico, 1994*, ed. C. Aguirre-Bravo, L. Eskew, A.B. Villa-Salas & C.E. Gonzales-Vincente, pp. 175–7. USDA Forest Service General Technical Report RM-266.

Parmeter, J.R., Jr. (1978). Forest stand dynamics and ecological factors in relation to dwarf mistletoe spread, impact, and control. In *Proceedings of the Symposium on Dwarf Mistletoe Control through Forest Management*. Berkeley, California, pp. 16–30. USDA Forest Service General Technical Report PSW-31.

Parmeter, J.R., Jr & Newhook, F.J. (1967). New Zealand *Pinus radiata* is susceptible to western gall rust. *New Zealand Journal of Forestry*, **12**, 200–1.

Patton, R.F., Spear, R.N. & Blennis, P.V. (1984). The mode of infection and early stages of colonization of pines by *Gremmeniella abietina*. *European Journal of Forest Pathology*, **14**, 193–202.

Peters, M.P., Ossenbruggen, P. & Shigo, A. (1985). Cracking and failure behavior models of defective balsam fir trees. *Holzforschung*, **39**, 125–35.

Peterson, G.W. (1967). Dothistroma needle blight of Austrian and ponderosa pines: epidemiology and control. *Phytopathology*, **57**, 437–41.

Peterson, G.W. (1977). Infection, epidemiology, and control of Diplodia blight of Austrian, ponderosa, and Scots pines. *Phytopathology*, **67**, 511–14.

Peterson, G.W. (1981). *Pine and Juniper Diseases in the Great Plains*. USDA Forest Service General Technical Report RM-86.

Peterson, R.S. & Jewell, F.F. (1968). Status of American stem rusts of pine. *Annual Review of Phytopathology*, **6**, 23–40.

Powers, H.R., Jr, Schmidt, R.A. & Snow, G.A. (1981). Current status and management of fusiform rust on southern pines. *Annual Review of Phytopathology*, **19**, 353–71.

Pratt, J.E. (1994). Some experiments with borates and with urea to control stump infection by *H. annosum* in Britain. In *Proceedings of the 8th International Conference on Root and Butt Rots, 1993*, ed. M. Johansson & J. Stenlid, pp. 662–7. Uppsala: Swedish University of Agricultural Sciences.

Read, D.J. (1968). Some aspects of the relationship between shade and fungal pathogenicity in an epidemic disease in pines. *New Phytologist*, **67**, 39–48.

Redfern, D.B. & Filip, G.M. (1991). Inoculum and infection. In *Armillaria Root Disease*, ed. C.G. Shaw III & G.A. Kile, pp. 48–61. USDA Forest Service Agriculture Handbook 691.

Rishbeth, J. (1951). Observations on the biology of *Fomes annosus*, with particular reference to East Anglian pine plantations. II. Spore production, stump infection, and saprophytic activity in stumps. *Annals of Botany (London)*, n. s., **15**, 1–22.

Rishbeth, J. (1988). Stump infection by *Armillaria* in first-rotation conifers.

European Journal of Forest Pathology, **18**, 401–8.

Rizzo, D.M., Blanchette, R.A. & May, G. (1995). Distribution of *Armillaria ostoyae* in a *Pinus resinosa–P. banksiana* forest. *Canadian Journal of Botany*, **73**, 776–87.

Rizzo, D.M. & Harrington, T.C. (1993). Delineation and biology of clones of *Armillaria ostoyae, A. gemina* and *A. calvescens. Mycologia*, **85**, 164–74.

Rodrigues, R.G. (1989). *Pitch canker on Pinus douglasiana, pines indigenous to San Anred Milpillas, Municipal of Huajicori, Nay*. Forest Parasitology Symposium V. Summary: 28. City of Juarez, Chihuahua, October 4–6.

Roth, L.F. (1966). Foliar habit of ponderosa pine as a heritable basis for resistance to dwarf mistletoe. In *Breeding Pest-Resistance Trees*, ed. H.D. Gerhold *et al.*, pp. 221–8. Oxford: Pergamon Press.

Runion, G.B. & Bruck, R.I. (1985). Association of the pine tip moth with pitch canker of loblolly pine seedlings. *Phytopathology*, **75**, 1339 (Abstract).

Santos, J.J.G. & Tovar, D.B. (1991). *Algunos aspectos sobre el cancro resinoso de los pinus* (Abstr.). VI. Simposio Nacianal Sobre Parasitologia Forestal. October, 1991. Unidad de Cogresos del Colegio de Postgraduados Montecillos. Edo, Mexico.

Scharpf, R.F. (1984). Host resistance to dwarf mistletoes. In *Biology of Dwarf Mistletoes: Proceedings of the Symposium*, Spokane, Washington, 1984. ed. F.G. Hawksworth & R.F. Scharpf, pp. 70–6. USDA Forest Service General Technical Report RM-111.

Schmidt, R.A., Powers, H.R., Jr. & Snow, G.A. (1981). Application of genetic disease resistance for the control of fusiform rust in intensively managed southern pine. *Phytopathology*, **71**, 993–7.

Schonhar, S. (1988). Zur Ausbreitung von *Heterobasidion annosum* in Fichtenbeständen. *Forst und Holz*, **7**, 156–8.

Shaw, C.G., III & Roth, L.F. (1976). Persistence and distribution of a clone of *Armillaria mellea* in a ponderosa pine forest. *Phytopathology*, **66**, 1210–13.

Sherman, R.J. & Warren, R.K. (1988). Factors in *Pinus ponderosa* and *Calocedrus decurrens* mortality in Yosemite Valley, USA. *Vegetatio*, **77**, 79–85.

Sinclair, W.A., Lyon, H.H. & Johnson, W.T. (1987). *Diseases of Trees and Shrubs*. Ithaca, NY: Cornell University Press.

Skilling, D.D. (1977). The development of a more virulent strain of *Scleroderris lagerbergii* in New York State. *European Journal of Forest Pathology*, **7**, 297–302.

Skilling, D., Kienzler, M. & Haynes, E. (1984). Distribution of serological strains of *Gremmeniella abietina* in eastern North America. *Plant Disease*, **68**, 937–8.

Skilling, D. & Nicholls, T.H. (1974). *Brown Spot Needle Disease – Biology and Control in Scotch Pine Plantations*. USDA Forest Service Research Paper NC-109.

Skilling, D.D. & Waddell, C.D. (1970). Control of Scleroderris canker by fungicide sprays. *Plant Disease Reporter*, **54**, 663–4.

Slaughter, G.W. & Parmeter, J.R. (1995). Enlargement of tree-mortality centers surrounding pine stumps infected by *Heterobasidion annosum* in northeastern California. *Canadian Journal of Forest Research*, **25**, 244–52.

Smith, D.R. & Stanosz, G.R. (1995). Confirmation of two distinct populations of *Sphaeropsis sapinea* in the North Central United States using RAPDs. *Phytopathology*, **85**, 699–704.

Smith, H., Wingfield, M.J., Coutinho, T. & Crous, P.W. (1995). *Sphaeropsis sapinea* and *Botryosphaeria dothidea* endophytic on pines and eucalypts in South Africa. *Phytopathology*, **85**, 1197 (Abstract).

Smith, R.S. (1984). *Root Disease-Caused Losses in the Commercial Coniferous Forests of the Western United States*. USDA Forest Service, Rocky Mountain Station, Report No. 84–5.

Smith, R.S. (1989). History of *Heterobasidion annosum* in western United States. In *Proceedings of the Symposium on Research and Management of Annosus Root Disease (Heterobasidion annosum) in Western North America*. Monterey, California, 1989. ed. W.J. Otrosina & R.F. Scharpf, pp. 10–16. USDA Forest Service General Technical Report PSW-116.

Snow, G.A., Dinus, R.J. & Walkinshaw, C.H. (1976). Increase in virulence on *Cronartium fusiforme* on resistant slash pine. *Phytopathology*, **66**, 511–13.

Snyder, E.B. & Derr, H.J. (1972). Breeding longleaf pines for resistance to brown spot needle blight. *Phytopathology*, **62**, 325–9.

Stambaugh, W.J. (1989). Annosus root disease in Europe and the southeastern United States: Occurrence, research, and historical perspective. In *Proceedings of the Symposium on Research and Management of Annosus Root Disease (Heterobasidion annosum) in Western North America*. Monterey, California, 1989, ed. W.J. Otrosina & R.F. Scharpf, pp. 3–9. USDA Forest Service General Technical Report PSW-116.

Stenlid, J. & Swedjemark, G. (1988). Differential growth of S- and P-isolates of *Heterobasidion annosum* in *Picea abies* and *Pinus sylvestris. Transactions of the British Mycological Society*, **90**, 209–13.

Stevens, R.E. & Hawksworth, F.G. (1984). Insect dwarf mistletoe associations: an update. In *Biology of Dwarf Mistletoes: Proceedings of the Symposium*, Spokane, Washington, 1984, ed. F. G. Hawksworth & R. F. Scharpf, pp. 94–101. USDA Forest Service General Technical Report RM-111.

Storer, A.J., Gordon, T.R., Dallara, P.L. & Wood, D.L. (1994). Pitch canker kills pines, spreads to new species and regions. *California Agriculture*, **48**, 9–13.

Sung, Jae Mo, Park, Young Jun, Kim, Hyun Joong & Harrington, T.C. (1991). Biology and taxonomic concepts of *Armillaria* in pine forests in Korea. *Korean Journal of Plant Pathology* **7**, 6–16.

Sutton, B.C. (1980). *The Coelomycetes. Fungi Imperfecti with pycnidia, acervuli and stromata*. Surrey, England: Commonwealth Mycological Institute.

Swart, W.J. & Wingfield, M.J. (1991). Biology and control of *Sphaeropsis sapinea* on *Pinus* species in South Africa. *Plant Disease*, **75**, 761–6.

Swart, W.J., Wingfield, M.J., Palmer, M. & Blanchette, R.A. (1991). Variation among South African isolates of *Sphaeropsis sapinea. Phytopathology*, **81**, 489–93.

Van Arsdel, E.P., Riker, A.J. & Patton, R.F. (1956). Effects of temperature and moisture on the spread of white pine blister rust. *Phytopathology*, **46**, 307–18.

van der Kamp, B.J. (1991). Pathogens as agents of diversity in forested landscapes. *Forest Chronicle*, **67**, 353–4.

Viljoen, A. (1995). The pathogenicity and taxonomy of *Fusarium subglutinans* f. sp. *pini* in South Africa. PhD thesis, University of the Orange Free State, Bloemfontein, South Africa.

Viljoen, A., Wingfield, M.J., Kemp, G.H.J. & Marasas, W.F.O. (1995). Susceptibility of pines in South Africa to the pitch canker fungus *Fusarium subglutinans* f. sp. *pini*. *Plant Pathology*, **44**, 877–82.

Viljoen, A., Wingfield, M.J. & Marasas, W.F.O. (1994). First report of *Fusarium subglutinans* f. sp. *pini* in South Africa. *Plant Disease*, **78**, 309–12.

Vogler, D.R. (1995). Uses of molecular techniques in rust systematics. In *Proceedings of the Fourth IUFRO Rusts of Pines Working Party Conference*, ed. S. Kaneko, K. Katsuya, M. Kakishima & Y. Ono, pp. 9–15. Tsukuba: University of Tsukuba.

Wagener, W.W. & Mielke, J.L. (1961). A staining-fungus root disease of ponderosa, Jeffrey and pinyon pines. *Plant Disease Reporter*, **45**, 831–5.

Wanner, J.L. & Tinnin, R.O. (1989). Some effects of infection by *Arceuthobium americanum* on the population dynamics of *Pinus contorta* in Oregon. *Canadian Journal of Forest Research*, **19**, 736–42.

Wargo, P.M. & Harrington, T.C. (1991). Host stress and susceptibility to *Armillaria*. In *Armillaria Root Disease*, ed. C.G. Shaw, III & G.A. Kile, pp. 88–101. USDA Forest Service Agriculture Handbook No. 691.

Weir, J.R. (1916). *Mistletoe Injury to Conifers in the Northwest*. USDA Bulletin No. 360.

Wells, O.O. & Dinus, R.J. (1978). Early infection as a predictor of mortality associated with fusiform rust of southern pines. *Journal of Forestry*, **76**, 8–12.

Wenger, K.F. (1984). *Forestry Handbook*. New York: John Wiley.

Wicker, E.F. & Leaphart, D. (1976). Fire and dwarf mistletoe (*Arceuthobium* spp.) relationships in the northern Rocky Mountains. *Proceedings of the Tall Timbers Fire Ecology Conference*, **14**, 279–98.

Wingfield, M.J. (1983). Transmission of

pine wood nematode to cut timber and girdled trees. *Plant Disease*, **67**, 35–7.

Wingfield, M.J. (ed.) (1987a). *Pathogenicity of the Pine Wood Nematode*. St Paul, Minnesota: American Phytopathological Society Press.

Wingfield, M.J. (1987b). A comparison of the mycophagus and the phytophagus phases of the pine wood nematode. In *Pathogenicity of the Pine Wood Nematode*, ed. M.J. Wingfield, pp. 81–90. St Paul, Minnesota: American Phytopathological Society Press.

Wingfield, M.J. (1990). Current status and future prospects of forest pathology in South Africa. *South African Journal of Science*, **86**, 60–2.

Wingfield, M.J., Bedker, P.J. & Blanchette, R.A. (1986). Pathogenicity of *Bursaphelenchus xylophilus* on pine in Minnesota and Wisconsin. *Journal of Nematology*, **18**, 22–7.

Wingfield, M.J., Blanchette, R.A. & Nicholls, T.H. (1984). Is the pine wood nematode an important pathogen in the United States? *Journal of Forestry*, **82**, 232–5.

Wingfield, M.J., Blanchette, R.A., Nicholls, T.H. & Robbins, K. (1982). Association of pine wood nematode with stressed trees in Minnesota, Iowa and Wisconsin. *Plant Disease*, **66**, 934–7.

Wingfield, M.J., Swart, W.J. & von Broembsen, S.L. (1989). Root diseases of pines and eucalyptus in South Africa. In *Proceedings 7th IUFRO Conference on Root and Butt Rots*, ed. D.J. Morrison, pp. 563–9. Victoria, BC: Forestry Canada.

Witcosky, J.J., Schowalter, T.D. & Hansen, E.M. (1986). *Hylastes nigrinus*, (Coleoptera: Scolytidae), *Pissodes fasciatus*, and *Steremnius carinatus* (Coleoptera: Curculionidae) as vectors

of black stain root disease of Douglas-fir. *Environmental Entomology*, **15**, 1090–5.

Worrall, J.J. & Harrington, T.C. (1988). Etiology of canopy gaps in spruce-fir forests at Crawford Notch, New Hampshire. *Canadian Journal of Forest Research*, **18**, 1463–9.

Yakota, S. (1983). Pathogenicity and host range of races of *Gremmeniella abietina* in Hokkaido. In *Scleroderris Canker of Conifers*, ed. P.D. Manion. pp. 47–53. The Hague: Martin Nijhoff & Dr W. Junk.

Yakota, S., Uozumi, T. & Matsuzaki, S. (1974a). Scleroderris canker of Todo-fir in Hokkaido, Northern Japan. I. Present status of damage and features of infected plantations. *European Journal of Forest Pathology*, **4**, 65–74.

Yakota, S., Uozumi, T. & Matsuzaki, S. (1974b). Scleroderris canker of Todo-fir in Hokkaido, Northern Japan. II. Physiological and pathological characteristics of the causal fungus. *European Journal of Forest Pathology*, **4**, 155–6.

Zambino, P.J. & Harrington, T.C. (1989). Isozyme variation within and among host-specialized varieties of *Leptographium wageneri*. *Mycologia*, **8**, 122–33.

Zentmeyer, G.A. (1980). *Phytophthora cinnamoni and the Diseases it Causes*. Monograph 10. St Paul, Minnesota: American Phytopathological Society.

Ziller, W.G. (1974). *The Tree Rusts of Western Canada*. Canadian Forest Service Publ. No. 1329.

Zheng-Yu, H., Smalley, E.B. & Guries, R.P. (1995). Differentiation of *Mycosphaerella dearnessii* by cultural characters and RAPD analysis. *Phytopathology*, **85**, 522–7.

Part six
Pines and humans

20 Pines in cultivation: a global view

D.C. Le Maitre

20.1 Introduction

People and pines have had a long association. Recent findings in China suggest that hominids emigrated from Africa about 2 million years ago (Wanpo et al. 1995). The first evidence of hominid habitation within the natural range of Pinus has been dated to about 1.4 million years ago on the eastern shore of the Mediterranean Sea (the Levant) (Wood & Turner 1995). The first encounters with pines would probably have been in North Africa or in the Middle East, the natural bridge between Africa and Eurasia. There is controversial evidence of hominids in the Americas about 30 000 years ago but extensive settlements within the natural range of pines date from about 11 000 years ago (Bray 1988; Dillehay & Collins 1988).

During this long period, people have learned to obtain many products from pine trees, ranging from firewood to medicinal compounds, for use in daily activities ranging from utilitarian to religious ritual. This chapter reviews the history of pines in cultivation. In the period up to the first recorded history the important issues are: Where had pines spread to, or retreated from, by the time people encountered them? When did people first arrive in areas where pines occurred? During the historical period the important questions are: Where did people settle? How did settlement patterns affect pines? Key questions throughout are: When did people begin to feel the need to cultivate pines? What were pines and their products used for? How did technological developments, inter alia, alter patterns of use?

This chapter must necessarily take a broad view. More detailed studies of the history and development of silviculture (the science of tree growing for timber and timber products) are given by, among others: Zobel, van Wyk & Stahl (1987), Heske (1988), Savill & Evans (1986) and Evans

(1992). Interesting analyses of the relationship between forestry policies and plantation development in various countries are given by Mather (1993a). The cultivation of ornamental forms of pine species has largely been excluded from this chapter as the topic has been reviewed by, inter alia, Den Ouden & Boom (1965) and Harrison (1975). I have interpreted the term 'plantation' to include all forms of intervention, from enhancement of natural regeneration (e.g. weed control, supplementary seeding) to intensive silviculture. Dates in The Times Atlas of Archaeology (Times 1988) have been used, where necessary, because there is little consensus on the timing, duration and naming of the archaeological periods. I have tried to locate original references for different inventions and have only resorted to general sources (notably Encyclopaedia Britannica) where none could be found. Little attention is given to P. radiata here because this species is the subject of Chap. 21 (this volume).

Much of the information is drawn from the temperate regions of Europe and the Mediterranean Basin because of their well-documented history and because reviews are available of the forest cultivation and uses (e.g, Hughes 1975; Thirgood 1981; Meiggs 1982). Information on early cultivation and uses of pine forests of the Americas is more limited. Williams (1989, 1993) provides some useful information and leads to other sources but his emphasis was mainly on the post-Columbian era. There is also little information, at least in English, on the history of the Asian pine forests. There is a rapidly expanding literature on forests and forest ecology, but the focus is largely on sub-tropical and tropical broadleaved forests rather than coniferous forests.

The word 'cultivate' is derived from the Latin word

Ecology and Biogeography of Pinus, ed. D.M. Richardson. © Cambridge University Press, Cambridge (1998), pp. 407–31.

cultus, which means to till the ground, and has strong associations with 'improve', 'develop', 'pay attention to' and 'cherish'. I have adopted this broad interpretation for 'cultivate' in this chapter because of the varying role pines have played in human history and culture in the northern hemisphere. Cultivation is also directly linked to trade: once there is a surplus to sell and someone who needs it, trade will begin. Most of the known history of people and pines has been a story of exploitation and trade.

20.2 Prehistory to Classical Greece

By the time hominids encountered pines in the Middle East the different species occupied most of their current range. Mediterranean-type climates, with their dry summers and recurrent fires, were also well established (Suc 1984; Chap. 8, this volume). Key events and developments in the relationship between humans and pines include:

- The marked climatic fluctuations during the late Pleistocene and the Holocene (Suc 1984; Issar 1995) which influenced vegetation boundaries and human populations and their interactions (Le Houérou 1981; Suc 1984; Bennett, Tzedakis & Willis 1991; Willis 1992; Tzedakis 1993; Sabloff 1995; Chap. 5, this volume).

- The use and increasing mastery of fire which allowed hominids to manage vegetation and thus manipulate pine forest dynamics, increase grazing, attract game for hunting and encourage the growth of edible plants. The first evidence for use of fire for cooking is from about 1.5 million years ago (Brain & Sillen 1988). After 500 000 years ago there is definite evidence from the Mediterranean Basin of fire use (Le Houérou 1981; Trabaud 1981). Although mastery over fire is conventionally dated to about 20 000 years ago (Trabaud 1981), studies of sites used by the earliest anatomically modern humans suggest an ability to ignite fire at will between 90 000 and 120 000 years ago in southern Africa (Deacon 1992). Fire is essential for maintaining pine populations, and to reduce competition from hardwood species, but both fire protection and frequent fires can be detrimental for some species, especially the white pines (section *Strobus*, subsection *Strobi*) (Mirov 1967; Chap. 11, this volume).

- Colonization of the New World (earliest undisputed date 11 000 years ago; Bray 1988), which resulted in the discovery of the edible nuts of the pinyons of southwestern North America (Lanner 1981).

- The Neolithic revolution and the adoption of agro-pastoralism between 13 000 and 9000 years ago in the Middle East, thereafter extending westwards across southern Europe (Bahn 1989; Dennell 1992; Miller 1992; Willis 1992), eastwards into Asia (Tsukada, Sugita & Tsukada 1986; Crawford 1992) and later to Central America (e.g. 3500 BC for maize; Fritz 1995).

- The adoption of kiln-fired pottery after about 10 500 BC in China, 8000 BC in Europe and 4000 BC in the Americas, which increased the demand for wood.

- The 10-fold expansion in human populations from 10 000 to 2000 BC (Le Houérou 1981), with significant impacts on forests and pines. The growth in populations followed the domestication of agricultural crops, which provided sufficient food to sustain permanent settlements. These settlements provide the first evidence for the use of timber beams and frames in dwellings (Baker 1970).

- Copper smelting and refining during the Bronze Age (after about 3000 BC) and iron smelting during the Iron Age (after about 500 BC), which required large quantities of wood (Thirgood 1981; Stevenson & Moore 1988) and was associated with extensive forest clearing (Willis 1992; Mighall & Chambers 1995). *Pinus sylvestris* forests disappeared from Britain (excluding Scotland), Ireland and Denmark after about 4000 BC because of human influences (Mirov 1967; Kinloch, Westfall & Forrest 1986). In Europe, especially, archaeological records indicate that forest clearing was necessary as there was too little open land for pasture and fields (Meiggs 1982).

20.2.1 Solid wood for construction and domestic use

The first records of wooden dwellings in the Middle East and Europe are from the late Stone Age (Baker 1970) and the first settlements from about 6200 BC. Softwoods such as pines, especially the white pines, were easier to work than hardwoods and would have been favoured. Stone axes and tools such as the adze were used until copper cross-cut saws were invented in Ancient Egypt and Mesopotamia in about 3000 BC (Meiggs 1982). The Minoans on Crete developed the first ripsaws, a significant innovation which made the conversion of logs to planks far more efficient. Wood was first used for mine props during the Minoan period.

The Ancient Egyptians and other civilizations of the eastern Mediterranean Basin and Mesopotamia imported large quantities of wood from Lebanon and Syria but identification of the species is problematic (Meiggs 1982).

The most sought-after conifer timbers were the very durable cedar and juniper. Fir (*Abies* spp.) was favoured, especially where long beams were required, and pine timber was occasionally used (Meiggs 1982). Roof beams of an Assyrian temple were of pine. Pine door frames were sometimes used in Ancient Egyptian tombs (Hepper 1990). The Minoans usually used cypress and cedar for construction and fine woodwork, with pine probably being used for general construction and furniture.

The first boats were probably made of woven laths, like baskets. Planked ships were first built by the Egyptians in about 4235 BC (Holmes 1900). Early Egyptian ships were built of sycamore (*Ficus*) and *Acacia* wood but Egypt later imported wood from Lebanon, including pine. Pine was used in Phoenician ships (Holmes 1900; Thirgood 1981), and by the Chinese (Kim 1990). The number of vessels in these early fleets was substantial: for example, the Persian king Xerxes I (*c.* 519–465 BC) is reported to have had 200 warships and the Carthaginians 300.

Charcoal from *P. pinaster* found in the Lascaux Cave in southern France, dated to 15 000 years ago, indicates its use for fuel (Scott 1962; Chap. 8, this volume). Developments in North America were probably similar, with fire-farming being an established practice among the native peoples (Trabaud 1981). The use of resin-soaked pine 'splinters' as kindling, called *Ocote* in Mexico and Central America, probably dates from these early times. Charcoal from pinyon pines dating from about 4000 BC has been found at a site in Nevada (Lanner 1981). Wood for fuel for cooking and heating, mainly hardwoods but including pine, was a significant component of commercial trade and remained so until wood was replaced by coal in the 19th century (Thirgood 1981; Meiggs 1982).

20.2.2 Resin and derived products
The origin of the extraction and use of turpentine predates written history (Mirov & Hasbrouck 1976). Early shipbuilders and seafarers such as the Minoans and Phoenicians used pitch to seal their boats, a practice later adopted by the Greeks and Romans (Meiggs 1982). Pitch was probably obtained from stands of *P. halepensis* (Mirov & Hasbrouck 1976). The Assyrians used turpentine as a solvent for sulphur compounds employed as antiseptics in wound dressings. Their doctors also used it for external treatments for the eyes, ears, feet and for swellings, and internally for kidney complaints and in cough mixtures (Campbell Thompson 1949).

20.2.3 Religious and ceremonial uses
Burning of rosin, the hard fraction of the resin, from pines as incense is still a common religious practice. The Ancient Egyptians and Assyrians used slivers of pine wood for burning like incense in their homes (Campbell Thompson

1949; Hepper 1990). The Ancient Egyptians also used pine resin, probably from *P. halepensis*, for embalming and preserving human bodies (Hepper 1990).

20.2.4 Seeds and other minor products
Forests have been used both for hunting and as sources of edible products (e.g. mushrooms) since prehistoric times. Hunting was an important pastime of royalty in civilizations such as the Egyptians and Assyrians and the Mayans in Central America.

Pine seeds, especially those of *P. pinea* (Mediterranean Basin), *P. cembra* (Europe), *P. sibirica* (Siberia) and *P. koraiensis* (Asia), have been used for millennia (Mirov & Hasbrouck 1976). Pine seeds are exceptionally rich in proteins (Styles 1993), *P. pinea* having a protein content of 31% by weight, higher than sunflower seeds and peanuts (Brouk 1975). The Ancient Egyptians imported pine seeds, probably from Lebanon, for food (Meiggs 1982; Hepper 1990). Early inhabitants of southwestern USA and Mexico used seeds from pinyons, as well as *P. coulteri*, *P. sabiniana* and *P. lambertiana* (Bye 1985), as long ago as 9500 BC (Bushnell 1976; Lanner 1981).

The Assyrians used pine seeds preserved in honey for medicinal purposes; the Spanish still preserve pine seeds this way (Campbell Thompson 1949). Extraction of oils from the seeds for domestic use in medicines and for oil lamps is a traditional practice in Siberia and some parts of Europe (Mirov & Hasbrouck 1976). The cultural importance of pine seeds is indicated in tribal mythology and by the ceremonies some American native peoples held to celebrate the onset of the seed harvest (Howes 1948; Menninger 1977; Lanner 1981). Another example is an idiom used by inhabitants of Kunawar, India of *P. gerardiana*: 'one tree, one man's life in winter' (Menninger 1977).

20.2.5 Human impact on species distributions
There is no definite evidence that people cultivated pines (Thirgood 1981) but they may have cultivated others, notably fruit trees (Crawford 1992; Smith 1992). *Pinus brutia* and *P. halepensis* may have been introduced to the Levant by people between 6000 and 1000 years ago (see Chap. 8, this volume). However, there are records of pine (*P. halepensis*) pollen from the Galilee region from between 60 000 and 80 000 years ago (Naveh & Dan 1973). *Pinus pinea* may have been introduced to the western part of the Mediterranean Basin during this period, although it also may have occurred there naturally (Chap. 8, this volume).

Fossil evidence suggests that *P. edulis* and *P. monophylla* expanded their ranges up to 6° northwards during the Holocene (Wells 1987). A range expansion of this magnitude for these large-seeded species requires external agents. The traditional explanation has been dispersal by corvids (the Pinyon jay and Clark's nutcracker; see Chap. 9, this volume), but human cultivation is also likely as the seeds are a traditional staple food (Betancourt 1987). Some Kumenay Indians of California maintained a grove of *P. torreyana* by regularly burning it and planting seeds to expand it (Shipek 1989). How far back in time such practices extended is not known.

Humans have consistently over-exploited natural resources for a variety of reasons, including the assumption that they were inexhaustible (Hughes 1975; Diamond 1986; van Andel, Runnels & Pope 1986; Delcourt 1987). Although population densities were low human impacts during the prehistoric period were substantial and may have altered the abundance of pines, especially in the Mediterranean Basin (Chap. 8, this volume).

20.3 Classical Greece to the Middle Ages

This period extends from the first detailed botanical treatises by the naturalists of the Classical Greek period to the Middle Ages (AD 900–1200). Key developments during this period include:

- The establishment of extensive empires and a significant increase in trade in Europe and between Europe and the Middle East during the Graeco-Roman period (Thirgood 1981). The Persian and Mesopotamian empires had increased trade, but on nothing like the scale attained by the Greeks and Romans (Hughes 1975). Most of this trade was seaborne, creating a considerable demand for timber for merchant ships and for naval ships to protect them and transport military forces.

- Rapid population growth, particularly in southern Europe (Hughes 1975; Meiggs 1982). Population densities increased in and around cities because free trade alleviated local constraints on resources through imported goods.

- The emergence of modern biological sciences, including botany and agriculture. The limited surviving writings provide new insights into the accumulated knowledge of the natural history and cultivation of plants.

- The collapse of the Roman empire and with it a significant decline in commerce and trade. Populations and population growth rates also declined in Europe, the Mediterranean Basin and the Middle East (Thirgood 1981).

- The later emergence of similar empires (e.g. Maya, Aztec) in Central America and southern North America. Populations grew rapidly to reach 90–113 million in the pre-Columbian period (Williams 1989).

The classical (and earlier) literature (e.g. Theophrastus) uses two words to describe pines: 'peuke' and 'pitys' (Meiggs 1982). 'Peuke' refers mostly to Klaus's (1989; see Chap. 8, this volume) category of mountain pines (e.g. *P. nigra*) which were straighter and taller, whereas 'pitys' was used mostly for the coastal pines, such as *P. halepensis* and *P. pinaster*. Authors were, however, not always consistent. They also distinguished 'sylvestris' from 'strobilos' where the former was the wild or forest pine and the latter the domesticated pine, i.e. *P. pinea* (Meiggs 1982).

20.3.1 Solid wood for construction and domestic use

The number and permanence of dwellings increased with the development of agricultural settlements and urban societies which consumed large quantities of wood (Thirgood 1981). Gabled roof structures began replacing the flat roofs in the Mediterranean Basin before 500 BC. These became common practice when the tie-beam truss system was developed in about 400 BC (Meiggs 1982), and this considerably increased the demand for longer lengths of squared timber. Substantial quantities of timber were imported for the construction of buildings in Athens and for the temples on the Acropolis. The naval base of Ravenna in Roman Italy was famous for its timber buildings, probably from logs obtained from the forests at Pineta.

Cedar and fir were favoured for timber, but pine was used where less durable timber was required (Meiggs 1982). A tomb from Turkey, dated late 8th century BC, was lined with pine planking. Large quantities of timber were used by the Roman armies to build siege engines, barracks, forts, bridges and roadways across swamps. Most of this wood was harvested locally and would have included pine. The use of wood for mine props expanded as the demand for metals and coal increased (Thirgood 1981; Meiggs 1982). An important technical advance in woodworking was the development of the frame saw with its relatively narrow, tensioned blade by the Romans (Meiggs 1982). This allowed large trunks to be squared and cut into wide planks with much less effort and also made it easier to cut curves of small radius.

Theophrastus (*c.* 372–287 BC) records that the Greek warships were built mainly of fir (*Abies* spp.) and oak but merchant ships were built mainly of pine (*P. brutia, P. halepensis*) (Morrison & Williams 1968; Welsh 1988). Fleets of 100 or more ships are mentioned frequently throughout the Greek and Roman records (Meiggs 1982). Possibly thousands of ships were constructed during the 900 years of the Graeco-Roman era. Many were large: the biggest Roman merchant ships could carry 100 tons of cargo, were up to 60 m in length and had a beam of 15 m. Considerable quantities of wood must have been used to build these ships, contributing to the deforestation of the Mediterranean lands, especially Macedonia which supplied wood to Athenian and Roman navies (Meiggs 1982). Nevertheless, Theodoric could build a fleet of 1000 ships for his navy in Ravenna in AD 575, apparently using timber supplied from adjacent areas of Italy, suggesting that finding large logs was not a significant problem. Ship transport, in turn, increased the trade in wood from even the remote parts of the Greek and Roman empires (Thirgood 1981). In Scandinavia the Viking voyagers built ships from fir (*Abies* spp.) although they sometimes used pine (Brogger & Shutelig 1951).

Studies of the Anasazi settlements of AD 900–1200 in North America show that hundreds of thousands of logs, primarily *P. ponderosa*, were used in their houses (Betancourt, Dean & Hull 1986). Cutting of fuelwood cleared substantial areas of North American pinyon–juniper woodland, including *P. edulis* (Betancourt & van Devender 1981).

20.3.2 Resin and derived products

The Carthaginians, Phoenicians, Greeks, Romans, Incas and Mayans used resins for a variety of purposes. Most of the resin used by the Greeks was obtained from local sources while the Sila forests in southern Italy were the major source of pitch in Roman times (Meiggs 1982). Pitch was an important commodity: it was indispensable for sealing and preserving the timbers used in ship construction and in buildings.

The principle of distillation was known to Aristotle (384–332 BC). Pliny the Elder (AD 23–79) described the extraction of oils from resin using a still (Jones 1961). The Alexandrians enclosed the still and Arabians later added water cooling which enabled them to separate the aromatic essential oils. Ausonius, a Roman statesman, recorded resin tapping, probably from *P. nigra* or *P. pinaster*, in the province of Aquitania (now Gascony in southwestern France) (Mirov & Hasbrouck 1976). An early use of resins was in the production of clear varnishes. Pine resin was considered too hard and most of the soft resin or mastic used in varnish in the Mediterranean Basin was obtained from pistachio (*Pistachia lentiscus*) (Thirgood 1981).

Retsina, wine to which pine resin is added as a preservative, dates from at least the early Greek period (Mirov & Hasbrouck 1976). Dioscorides described the use of retsina, wine steeped with pine cones and wine that had been mixed with pitch for complaints of the lungs and stomach and coughs (Gunther 1968). Extracts and distillates from pitch and pine resins were used for medicinal purposes such as dressings for wounds and cures for gout, consumption, skin irritation and ulcers; pitch was also used as a male depilatory (Mirov & Hasbrouck 1976; Meiggs 1982). Columella, a Roman inhabitant of Cadiz in the 1st century AD, recorded that pine pitch was used for purifying wine and treating wine casks. Other uses include perfumes and the soot from burning pitch or resin which was used to make ink.

20.3.3 Religious and ceremonial uses

Pines have also been an important part of mythology and religious rituals. The Ancient Greek legends tell of Pitys, a nymph, who had two lovers, Boreas and Pan (Mirov & Hasbrouck 1976). Boreas became jealous and threw Pitys against a rocky ledge. She turned instantly into a pine, and resin drops are her tears. Greek legends also describe how Cybele changed her lover Attis into a pine tree which Zeus clothed in with evergreen foliage (Lane 1952). The oracle at Delphi commanded the Greeks to worship pines as the most sacred of trees (Frazer 1935). The Romans used young pine trees in the festivals of Cybele which were held in the spring. Similar festivals were held by the early Scandinavian, Germanic, Russian and Mongolian peoples (Mirov & Hasbrouck 1976). The use of evergreen trees in midwinter festivals by Teutonic tribes has been perpetuated in the use of Christmas trees. Native Americans have comparable ritual uses for parts of pinyon pines, including needles, pollen and resins (Lanner 1981). The Indians of Guatemala have analogous rites dating from at least the time of the Mayan Empire (Mirov & Hasbrouck 1976). Similar celebrations were held by priests and kings of the Aztecs who considered *P. teocote* the pine of the Gods.

20.3.4 Seeds and other minor products

Seeds of pines, particularly *P. pinea*, were prized by the Romans both as food and for use in brewing wines (Mirov & Hasbrouck 1976). Shells of *P. pinea* seeds have been found in the refuse heaps of Roman military encampments in Britain (Howes 1948), indicating the extent of trade in these seeds. Pliny noted that pine kernels allayed thirst and heartburn and toned up the system; pine kernels with cucumber seed and the juice of purselaine were reported to cure severe stomach pains and serve as a diuretic (Jones 1961).

A decline in the use in pinyon seeds at a site in Colorado between AD 650 and 750 (Matthews 1986 in Floyd & Kohler 1990) coincided with a marked increase in the prevalence of pine wood in charcoal. Floyd & Kohler (1990) argue that extensive cutting of pines for timber reduced the supplies and thus the consumption of pinyon seeds.

20.3.5 Ornamentals, gardens and parks

There were many famous parks and arboreta in the Middle East, including those of the Babylonians and Persians

(Meiggs 1982). Parks were also important features of Greek and Roman cities. Pines may have been grown in these parks and gardens but the most popular trees were plane, box, cedar, cypress and fruit trees. Pines were generally regarded as forest and common peoples trees, although one epigram by Martial includes pines in a list of species which were 'rich man's natural company'. Virgil also mentions 'the pine in the garden', suggesting that pines may have been common in gardens. The species was probably P. pinea which is particularly attractive as an ornamental in a formally landscaped garden and bears edible seeds.

During the 4th and 5th centuries Christian monks introduced P. pinea to the Adriatic coast of Italy (primarily for ornamental reasons and for fuel and seeds), starting several pine forests (pinete) which persist to this day (Mirov & Hasbrouck 1976). These include the Ravenna pinete which appears in Dante's Comedia Divina and in paintings by Botticelli. In the Far East the Japanese have cultivated P. densifolia, P. thunbergii and P. parvifolia as ornamentals in temple gardens since ancient times, as did the Chinese with P. bungeana (Gordon 1875; Mirov 1967). In many areas of the Middle East isolated trees, including pines, were preserved to provide shade for pastoralists and travellers (Zohary 1973).

The Greeks had at least fifty agricultural handbooks; some contained advice on making farms self-sufficient in timber, but only one has survived (Meiggs 1982). The writings of Theophrastus, generally considered to be the first botanist, included information on the cultivation of trees (Einarson & Link 1990). The Roman writers Cato and Palladius included instructions on germinating and planting seeds and raising trees, including pines (Meiggs 1982). Like the later Roman writers Varro and Columella, Cato saw timber as an essential but minor element of the farming enterprise, confined to the least productive land.

20.3.6   Forest protection and management
Regulation of forest cutting was common in Graeco-Roman times for both wood conservation and religious purposes (e.g. the Lex duo decim tabularum of 450 BC; Gron 1947) but seems to have done little to stem the exploitation (Thirgood 1981). Plato recognized and described the permanent degradation that can result from deforestation, such the loss of topsoil, but there is little evidence of general concern or action (Meiggs 1982). There were extensive plantings of fruit trees, especially olives, but not of forest species, although the writings of Theophrastus and Palladius show that the knowledge existed (Meiggs 1982). They understood soil fertility, the need for soil preparation, fertilization, seed dormancy and seed germination requirements (see Einarson & Link 1990).

Most of the more remote forests in the Mediterranean Basin seem to have escaped intensive exploitation, mainly because the difficulties of long-distance transport, except by ship. Clearing and degradation were therefore confined to the vicinity of the larger urban settlements and harbours. The fall of the Roman Empire resulted in the disintegration of most of the urban centres in Europe and the Middle East, allowing the forests to recover (Meiggs 1982). The Visigoths introduced a code for protecting oak and pine forests in the 7th century, suggesting that there was continuing concern about forest exploitation (Mirov & Hasbrouck 1976; De Carvalho 1960).

There is little direct information on the impacts of the native American peoples on their forests before the 17th century, but it was probably similar to that observed by early explorers. Increases in pine pollen abundance in Central America and Mexico during this period (cf. Allen 1955; Mirov 1967; Rue 1987; Hodell, Curtis & Brenner 1995) are attributed to anthropogenic disturbance, mainly to clear land for agricultural crops. This practice was also noted by early explorers of southeastern and western North America (Lyell 1849; Duffield 1951).

20.3.7   Human impact on species distributions
The earliest definite evidence of humans altering the distribution of pine species through cultivation dates from the Graeco-Roman period. Many pine species are aggressive colonizers (Richardson & Bond 1991), and palynological studies suggest that pine populations expanded locally in degraded and abandoned areas during these periods (Van Andel, Runnels & Pope 1986; Willis 1992).

Studies of P. pinea suggest that its origin was in the western Mediterranean Basin, probably on the Iberian Peninsula (Klaus 1989). It is possible that P. pinea was not native to Syria and Lebanon and its presence there and in other areas around the Mediterranean, such as Italy, may date from Phoenician or Graeco-Roman times (Thirgood 1981). Others argue that P. pinea originated in, or was native to, the eastern part of the Mediterranean Basin, that its presence in Lebanon predates historical records, and that it was subsequently introduced to the western Mediterranean (see Mirov 1967).

Pinus pinaster was probably native to the western Mediterranean (Morocco, Portugal, Spain, southern France and Italy) and was spread elsewhere by humans (Klaus 1989), although its natural distribution may never be determined (Scott 1962). Isoenzyme analyses show that P. halepensis can be divided into two genetic groups, a western and an eastern Mediterranean group, which diverged from a common ancestor in central and southern Europe (Schiller, Conkle & Grunwald 1986). Populations from Morocco and Adana, Turkey, were closely related, indicating that Moroccan seed may have been brought to Turkey in the past. Stands of P. brutia in Iran, Pakistan and

Tajikistan may have originated from seeds brought from the Mediterranean Basin by Muslim pilgrims (Gurskii 1957; Mirov 1967; Weinstein 1989). Similarly, its occurrence at the type locality in Calabria, Italy, may be a result of its introduction in Roman times or earlier (Mirov 1967; Mirov & Hasbrouck 1976). *Pinus halepensis* stands on Rhodes may be of North African origin (Panetsos 1981).

There is a continuing debate among historians about the extent and severity of forest destruction during this period, especially in the Mediterranean Basin. For example, Le Houérou (1981) and Thirgood (1981) argue that the forests were significantly reduced and degraded. On the other hand, Meiggs (1982) argues that wood shortages were only periodically a problem and then only for limited areas, such as the region around Athens, but not in the Mediterranean region as a whole. Trends in Asia were probably similar. The early civilizations of Mexico and Central America had well-developed economies and the growing populations would have had a major impact on their resources (Williams 1989).

20.4   Middle Ages to the 17th century

This period extends from the re-emergence and growth of modern civilization, after the feudal culture of the Middle Ages (AD 900–1200), to the beginning of the industrial era. Important developments include:

- The revival of the arts and sciences, including botany and natural history, especially during the Renaissance period. This was the beginning of a pivotal era of technological advances, including the first woodworking machinery.

- Renewed population growth in Europe and the Mediterranean Basin towards the end of this period. The population of Asia appears to have increased steadily during both this and the preceding periods.

- An era of exploration and colonization, driven by a desire to establish control over the sources of traded goods. The establishment of colonies and expansion of trade necessitated the construction of large fleets of naval and merchant vessels, resulting in increased cutting of pine forests and demand for pine products such as pitch (naval stores).

- The colonization of the Americas by the Spanish and Portuguese, and later the English and Central Europeans led to the disintegration of the civilizations of Central America and Mexico. The colonists seemed to perceive the New World as a land of inexhaustible resources which could supply all Europe's needs (Williams 1989).

- The formation of larger national states in Europe after an era of small city states. International trade was re-established, especially between Europe and the Middle East and Asia. In the Americas and Asia there was greater fragmentation and trade networks were less extensive.

- An increasingly unsustainable rate of natural resources exploitation. People, at least in some regions, realized that natural resources are limited and that the loss of resources carries a cost (Thirgood 1981; Williams 1989; Adlard 1993).

20.4.1   Solid wood for construction and domestic use

The first mechanized sawmills were invented, making large-scale production of timber possible at a relatively low cost. The first mills were built in Madeira and in Wroclaw (Poland) in the 15th century (Pears 1937). The first sawmill in Britain was constructed in the 17th century.

There was a resurgence of urban development during the Renaissance and Gothic periods, and many large buildings, including cathedrals and palaces, were constructed (Britannica 1965). The walls of these buildings were made largely of stone but the vaulted roofs and scaffolding used during construction used large quantities of timber. There is little evidence of any difficulties in obtaining the long, large beams that were required. This suggests that there were still extensive, well-stocked forests even near the main cities of Greece and Italy (Meiggs 1982).

The construction of the large naval and merchant fleets of Portugal, Spain, France and Great Britain from the 14th century onwards consumed large quantities of pine wood (Thirgood 1981). In the Mediterranean Basin, large naval fleets were maintained by the Italian states, especially Venice, from the 16th century onwards (Meiggs 1982). Coniferous species were used for masts and other rigging.

20.4.2   Resin and derived products

Trade in turpentine and other oils distilled from pine resins remained important. Pitch and oils were still employed in medicinal compounds and dressings for wounds (Mirov & Hasbrouck 1976). Demand for pitch and pitch products (tar, rosin and turpentine) by European seafaring nations during the age of exploration supported a huge industry (Mirov & Hasbrouck 1976). Southwestern France, Spain, Greece and Russia were the main sources of these products. Wars and other problems with trade with Europe influenced British governments to obtain supplies from eastern North America, particularly the state of Virginia (Williams 1989), where *P. taeda* and *P. virginiana* were commonly utilized.

20.4.3 Seeds and other minor products

Seeds of pines, particularly *P. pinea*, remained an important commercial product (Mirov & Hasbrouck 1976). Mukaddasi, an Arabian historian during the Roman period, recorded that *P. pinea* seeds were an important product of Syria (Mikesell 1969; Mirov & Hasbrouck 1976). The native peoples of Mexico and the southern USA undoubtedly used the seeds of pinyons, notably *P. cembroides*, in their diet and bartering (Lanner 1981).

20.4.4 Ornamentals, gardens and parks

The gardens of the Vatican in Rome date back to the 14th century; arboreta existed at the royal courts and probably included pines. The *Pineta Clematina* near Rome was planted primarily for ornamental purposes in 1666. Fakhr-el-Din, a Lebanese ruler, is reported to have planted *P. pinea* in Lebanon in about 1700. Many wealthy landowners established forests on their estates and arboreta to display trees collected from around the world.

20.4.5 Forest protection and management

Growing concern about the state of forest resources led to the first recorded large-scale afforestation and reforestation. In 1310 King Diniz of Portugal issued a law to protect pine forests both for wood and for hunting (de Carvalho 1960; Scott 1962). The pine forests of Leiria, which were planted to stabilize sand dunes, date from about this time. Many others were established by Cistercian monks for soil protection (de Carvalho 1960). The seed for these forests may have been obtained from Gascony in France (Scott 1962). These forests provided much of the timber used to build the ships during the era of Portuguese maritime expansion in the 15th and 16th centuries. Forests were planted to protect soils in the Florentine and Venetian states at about this time as well (Thirgood 1981).

In 1343 the Canton of Schwyz (Switzerland) introduced measures to preserve forests for protection against avalanche damage (Gron 1947). In the 13th and 14th centuries central Europe, especially Germany, introduced regulations to encourage or enforce the practice of even-aged management in natural forests (Heske 1988). The aim was to obtain sustained yields by ensuring that utilization and exploitation were in equilibrium, a principle which is still central to modern forest management (Wiersum 1995). The main driving factor was the desire of the ruling class to preserve their game hunting (Gron 1947) rather than for the trees *per se*. Artificial seeding of cleared forest areas was introduced in 1368 in Nuremberg to enhance regeneration (Heske 1988). By the mid-1400s there were several seed merchants specializing in supplying forest tree seed for artificial reseeding. Peter the Great of Russia established pine forests on the Gulf of Finland for ship-building from about 1700 (Mirov & Hasbrouck 1976). In 1669 Colbert's *Ordonnance des Eaux et des Forêts* was introduced in France (Gron 1947) to preserve forests to supply timber for building warships. It served as the model for similar regulations in several European countries.

Britain's forests at this time were mainly hardwoods or mixed hardwoods and conifers, with natural pine forests being confined largely to the Scottish Highlands (Steven 1927). Most forests were heavily exploited for various purposes, including the ships needed by the expanding British navy. In 1613 a decree by King James promoted afforestation in Scotland with *P. sylvestris* using seeds sent from England (Steven 1960). John Evelyn's *Sylva*, published in 1664, was one of several influential treatises which argued for forest conservation and cultivation to preserve timber supplies and ensure sustained economic welfare in Britain (Sharp 1975).

20.4.6 Human impact on species distributions

The primary changes in the distribution of pines during this period were regional rather than international. The emphasis was on forest protection and on afforestation or reforestation to protect soil and water supplies rather than on timber production. Examples are the forests at Leiria and other places in Portugal which used a native species (*P. pinaster*) to afforest dune sands and degraded lands that may not have supported natural forest. There are some records of early introductions, for example *P. pinea* (1548) and probably *P. nigra* and *P. pinaster* from Europe into Great Britain (Brimble 1948; Den Ouden & Boom 1965) but most introductions were for ornamental purposes and restricted to parks and arboreta.

20.5 The 18th and 19th centuries

The major developments in Great Britain, central Europe and North America were:

- Unprecedented industrial and technological innovation and progress based on the revival of science during the Renaissance. In 1753 Linnaeus introduced his taxonomic system which brought consistency to the naming of plants, including pines.

- Continued colonization by European countries, especially in the southern hemisphere.

- In Europe the Napoleonic wars created a sustained demand for resources to maintain the war effort, including timber for large naval fleets, bringing belated recognition that Europe's forest and timber resources had become severely depleted. The American War of Independence forced Britain to seek timber and timber products in its other colonies.

- Rapid population and urban growth, especially in Asia, coupled with increasing international trade.
- The development of large-scale production processes for iron and steel which replaced timber in many applications.
- The development of chemical industries based on coal tar and its derivatives, providing the first viable alternative to wood-based products.

20.5.1 Solid wood for construction and domestic use

Large quantities of wood were still used for housing in many parts of Europe and in Great Britain, especially for the timber frames of the 'half-timbered' buildings (Britannica 1965). Wooden frames were replaced by steel beams following the development of rolled wrought iron beams after 1850 and large-scale steel manufacture in the 1880s. The use of iron and steel to reinforce concrete in the late 1800s reduced the demand for both wood and bricks, and thus for wood and charcoal to heat the brick kilns.

Steel replaced wood in the hulls of ships in the mid-19th century, significantly reducing the demand for timber (Britannica 1965). The later development of steam-ships also did away with the need for wooden masts and booms, reducing the demand for straight-stemmed conifers such as *P. strobus* from eastern North America (Williams 1989).

Global timber trade increased significantly from about 1860 as timber shortages in Europe, especially Germany and Great Britain, led to increased imports from eastern Europe, Russia and North America (Heske 1927). Timber imports in Germany increased from 2.7 million m^3 from 1879–84 to 14.2 million m^3 in 1907–13. Several exotic species, including pines such as *P. contorta*, were introduced to Germany in attempts to meet the demand for fuelwood, but they generally failed. The Germans were compelled to use coal for fuel, increasing the consumption from about 15 kg per capita in the 1800s to 2300 kg by 1914. Reduced demand for hardwood for fuel resulted in pines replacing hardwood forests in many areas because softwoods were needed for construction and other purposes.

The sawmilling industry expanded significantly, encouraging forest harvesting because the economies of mass production reduced the prices of sawn timber. The development of the rotary cutter for plywood in the 1890s also revolutionized solid timber use, enabling mills to produce large panels and boards (Latham 1964). In Norway, where water-powered sawmills were introduced in the 16th century, the sawmilling industry remained dominant until it was overtaken by the pulp industry, which was founded in the 1860s (Skinnemoen 1957).

Until the later Middle Ages (c. AD 1050) the forests of Sweden, mainly pine and spruce, were virtually untouched (Streyffert 1938). Mining expanded in the 15th century and iron smelting and refining consumed large quantities of wood and charcoal, overtaking the sawmilling industry. In 1723 an ordinance was passed which reserved large trees for sawmills. Until the beginning of the 19th century annual exports of wood were limited, with naval stores often exceeding the value of exported timber and lumber. By the 1890s exports of saw-timber, pulp and paper products, especially to France and England, had increased significantly.

20.5.2 Wood-fibre products

Paper had been developed by the Chinese by AD 100 but it was made primarily from plant fibres (Britannica 1965). The Arabs captured Chinese paper makers in AD 750, learned how to make paper from them and developed flax-based (linen) paper. The Europeans, in turn, learned paper-making from the Moors, using flax or rags. A growing shortage of fibre for papermaking in the early 1800s was finally alleviated by the development of mechanical and chemical-based wood pulping processes between 1850 and 1880. Although wood pulp production expanded during this period the primary source of fibre for paper continued to be cotton and waste rags.

In 1838 the first nitrocellulose was made by treating cotton fibres with concentrated nitric acid (Britannica 1965). It proved to be highly flammable and unstable and was used to develop explosives. The nitrocellulose process was adapted to produce more stable compounds which, in turn, were used to produce the first artificial silk fibre in 1855. Rayon was first produced commercially in 1891, and viscose was first produced from cellulose in 1892. But it was only in the 20th century, when wood pulp became the major source of cellulose for rayon and viscose, that these industries expanded significantly.

20.5.3 Resin and derived products

The main resin production areas were the southeastern USA, southwestern France, Spain, Greece and Russia (Britannica 1965). The American War of Independence resulted in severe shortages of pitch and other products in England which relied on supplies from species such as *P. elliottii* and *P. taeda* (Mirov & Hasbrouck 1976). Demand for pitch and pitch products (tar, rosin and turpentine) by navies, particularly the Spanish, French and British fleets, during the colonial era supported a large industry (Williams 1989). The French also obtained pitch from *P. halepensis* forests in Algeria after colonization in 1830 (Stewart 1993).

The American Civil War (1861–5) again resulted in severe shortages of pitch and other products in England which depended heavily on supplies from North America (Williams 1989). The demand was met by increased imports from France which in turn encouraged afforestation with *P. pinaster* in the Gascony region and the expansion of its resin industry (Scott 1962). The industry was also supplemented by the large-scale distillation of compounds from wood in the late 18th century (Baker 1970). The use of various products in medicinal compounds continued. An example is the use of heptane, distilled from *P. jeffreyi* and first sold as remedy for pulmonary ailments and tuberculosis in 1890 (Mirov & Hasbrouck 1976). Similar medicinal uses were recorded among North American native peoples who used, among others, pinyon pines (Lanner 1981). Pitch was used for wound dressings and fumes of burning resin to cure colds.

Coal tar distillation was developed sometime before 1665 in Germany and coal tar was used as a substitute for pitch in waterproofing (Baker 1970). The first commercial product developed from coal tar was naphtha which was used to produce creosote for wood preservation. After 1845 there was a rapid expansion in the range of coal tar derivatives – including dyes, medicines, inks, resins and synthetics – many of which replaced similar products derived from compounds extracted from pine timber and resin.

20.5.4 Seeds and other minor products

Archaeological research has documented the important role that pinyons, particularly their seeds, have played in the lives of the native peoples of the southwestern USA. Decoctions from the inner bark and needles of pinyon pines were used by them to relieve symptoms of colds, fevers and coughs (Lanner 1981). Wood, resin and pitch were important elements in native American culture, ceremonies and rituals (Lanner 1981). Pinyon seeds, in particular, were probably not a staple food because of the marked cycles in seed production (mast seeding) (Floyd & Kohler 1990). Nevertheless, pinyon seeds were a very important element of the diet of American native peoples judging by the modern usage (Lanner 1981). Pine seeds were used as food as far north as British Columbia, by the Shushwap peoples (Dawson 1892; Mirov & Hasbrouck 1976). Several early European expeditions into the southwestern USA were saved from disaster by supplies of pinyon nuts, often supplied by the Indian peoples (Lanner 1981). *Pinus pinea* also mast seeds at 3–4 or 5–6 year intervals but some seeds are always produced in the intervening years (Pozzera 1959), making this species a more reliable resource.

20.5.5 Ornamentals, gardens and parks

Many exotic pine species were introduced to Britain and Europe in the 18th century for cultivation in arboreta and parks (Den Ouden & Boom 1965). These introductions stimulated increased interest in the cultivation of conifers. The demand also encouraged the development of commercial nurseries which selected and propagated many ornamental forms of pines. Many of these cultivars are still available on the market today.

20.5.6 Forest protection and management

The first modern silvicultural textbook *Anweissung zur Holzzucht* by G.L. Hartig, was published in 1791 (Adlard 1993). With *Anweisung zum Waldbau* by H. Cotta, published in 1816, it laid the foundation for modern European silviculture. The 'agricultural' practice of clearfelling and artificial regeneration became standard in German forestry from about 1840 (Heske 1988). It subsequently spread to many other parts of Europe and from there to the world where it became the dominant approach in plantation forestry. In Germany, however, the silvicultural approach reverted to 'natural' forestry from about 1875. Ultimately this led to the adoption of individual-tree selection and uneven-age forest management, becoming known as the *Dauerwald* system (Heske 1988).

20.5.7 Human impact on species distributions

Large forest nurseries were established and large-scale planting of *P. pinaster* continued in Portugal and Madeira. Some of these forests still exist, e.g. the Medos forest (Mirov & Hasbrouck 1976). The first scientific studies of sand-fixation techniques date from this period (de Carvalho 1960). In France attempts at reforestation were made as early as 1500 and *P. pinaster* was used to stabilize dunes near Bordeaux in France in 1713 (Scott 1962). Large-scale plantings were not successful until Bremontier developed effective planting techniques in 1780s (Reed 1954). Afforestation of the swampy, sandy, coastal plains of the Landes region in southwestern France began in 1789 (Brown 1878; Reed 1954) and the plantations ultimately covered about 900 000 ha, 80% of the entire region (Gron 1947; Scott 1962). Forests occurred in this area according to Roman records but they must have been eradicated during the Middle Ages. Forest Ordinances passed in 1833 in Spain marked the beginning of national forest protection, but it was not until the 20th century that systematic afforestation began (Groome 1993).

In Scotland, and to a lesser extent in England and Wales, extensive plantations of *P. sylvestris* were established in the late 18th century using Scottish seed stocks (Steven 1927, 1960). These were probably the first plantations which were established primarily for their timber rather than for soil protection and amenity values (Steven 1927). In the early decades of the 18th century seed of *P. sylvestris* had to be imported from Europe as local seed supplies were insufficient. Imports of seeds increased again after 1850 as cone crops were poor, and regular seed imports continued until World War I.

The Dutch experimented with plantings of *P. merkusii* and other pines in Sumatra and other Indonesian islands, with limited success. *Pinus tabuliformis* was introduced to Korea from northern China during this period (Mirov 1967).

The first commercial pine plantations in the southern hemisphere were established in South Australia in 1875 (Boardman & McGuire 1990), with *P. radiata* proving to be the best species (Boardman 1988; Chap. 21, this volume). The first extensive pine plantations were established in South Africa in 1884 (Legat 1930) and in Zimbabwe in 1890, following earlier successes in South Africa (Barrett & Mullin 1968). New Zealand issued regulations encouraging tree planting in the early 1870s when it was evident that natural forests could not meet the demand for timber (Poole 1969). Afforestation with exotics began in 1885 but the first successful plantations were established in 1896, primarily with *P. radiata* from California (New Zealand 1964; Poole 1969; see also Chap. 22, this volume).

Excessive exploitation of the Mediterranean forests dates primarily from this period and continued into the 20th century (Meiggs 1982). An example is Greece where the main responsibility for forest destruction is attributed to Turkish rule. Forest cover was estimated, however, to be 40% when Greece regained its independence in 1822 compared with more recent estimates of 14%, much of which is degraded. There is good evidence for extensive deforestation in some regions of Asia Minor, both before and during the 19th century, but in other regions the forest survived largely unscathed. The same trends were recorded on Rhodes and Cyprus and in North Africa, especially Algeria (Stewart 1993). Northern Syria was still well forested at the end of the 19th century but further south and in Lebanon forests had been heavily exploited to supply Egypt and Mesopotamia (Meiggs 1982). In North Africa *P. halepensis* forests extended to the 200–150 mm isohyet in about 2500 BC but they now occur only within the 300–900 mm isohyet following extensive deforestation in the 19th century (Le Houérou 1981).

In the eastern and southeastern USA, extensive areas of the native pine forests were cleared for their timber and for cash-crop farming in what Williams (1989) describes as an 'assault'. Little or no thought was given to replanting and the ever-increasing timber demand was simply met by clearing yet more virgin forest as though the resource was limitless. In the arid southwestern USA, especially Nevada, there was large-scale cutting of pinyon pines to supply mines with props and the charcoal needed for smelting the ores (Lanner 1981). In one case, all the pinyon stands within 19 km in every direction were cleared within 10 years. The furnaces of the mining town of Eureka used the equivalent of 20 ha of pinyon timber per year. This may be insignificant at a global scale, but it had a major impact at local and regional scales.

20.6 The 20th century

The major developments in the 20th century include:

- The significantly increased pace of industrialization and technological innovation, notably the considerable increase in the use of paper as a medium for cheap and effective information transfer. In the last decade the electronic media have developed alternatives to paper, such as abstracts of journal articles and books on magnetic disk or CD-ROM, but the full effects have yet to be seen.

- An unprecedented expansion in global trade and rapid population growth, especially in Asia, South America and Africa. Modern travel has made international conferences, such as the IUFRO meetings, commonplace, facilitating the transfer of knowledge.

- In Europe, around the Mediterranean, and in parts of the Middle East, the two World Wars were critical periods where governments and nations were forced to confront the issue of limitations on forest resources. There was unprecedented demand for wood but international trade was severely restricted. Many northern hemisphere powers instituted or extended their reforestation policies (Meiggs 1982; Adlard 1993).

- The development of petroleum-based chemical industries and the introduction of large-scale manufacture of plastics, resins, solvents and dyes which have, in many cases, totally replaced wood-based products.

- Rapid progress in the life sciences and biotechnology, especially plant physiology and genetics, which transformed silviculture and tree breeding. Another key advance was in our understanding of the critical role of mycorrhizal associations in pine nutrition, growth and disease resistance.

20.6.1 Solid wood for construction and domestic use

Solid timber is still important product in the 20th century and has a wide range of uses. The demand for slow-grown, even-grained, knot-free timber for solid furniture and construction has been overtaken by strong demand for faster-growing species for short rotations. In the commercial trade there has been a shift from the dominance of solid wood products to wood-fibre products. Expanding human populations have increased the demand for timber for construction. The low costs of international trade have seen a significant switch from native pine timber sources to imports in the developed countries. Many important exporters of products from pines are now located in the southern hemisphere although the northern temperate and boreal pine forests still supply substantial volumes of timber. International trade in timber, especially tropical hardwoods, has also reduced demand for pine timber. Many synthetic replacements have been found for solid wood products. The development of Masonite from wood and bark pulp in the early 1900s resulted in a variety of related composite wood products and greatly increased the efficiency of wood use (Latham 1964). A variety of wood-chip and wood-fibre based solid board products have replaced solid timber in many applications because they are easy to machine, have greater dimensional stability and can be made from what was formerly sawmill waste. The rapid growth in production of oil-based synthetics, such as plastics, since World War I has resulted in the decline of many important products that were previously made from solid pine timber.

20.6.2 Wood-fibre products

After World War I, wood fibre replaced cotton as the primary source for nitrocellulose in the explosives and other cellulose-based industries such as viscose and rayon. The paper and paper board industries also expanded to meet the demand for these products. The international trade in roundwood and wood chips for pulping, wood pulp and paper, has grown considerably, especially since World War II. This in turn led a significant increase in the demand for wood-fibre and the establishment of new plantations.

Population growth and higher standards of living have resulted in an ever-increasing consumption of wood pulp. Pulp and paper industries now dominate international trade in wood products. The last decade has brought a rapid growth in the proportion of pine fibre recycled for use in lower-grade papers and cardboard.

20.6.3 Resin and derived products

Many traditional uses of pine resin and its derivatives, for example in medicinal compounds, persist today. Cough mixtures using 'pine tar' are still available and essential oils are used in perfumery and specialist products (Mirov & Hasbrouck 1976). Before World War II the United States trade in resins was estimated at 750 000 tons yr^{-1} (Lane 1952), more than half the world production (Reed 1954). The Landes forests in southwestern France yielded some 4.0 million m^3 of wood and a 100 million litres of resin a year (Reed 1954), but by the 1960s France was producing only 60 million litres of resin (Scott 1962). When refined the resin yielded 20 000 tons of turpentine and 70 000 tons of rosin (also called colophony). Resin production is still important in some areas: for example, exports of resin from *P. merkusii* earned Indonesia US$5.7 million in 1987 (TFU 1993). Numerous products were derived from rosin, including the resins that were used to manufacture gramophone records until they were replaced by compact disks.

A number of synthetic substitutes for products derived from resins were developed during the 20th century. The first commercial oil well was opened in the USA in 1859 but it was only in the late 1800s and early 1900s that the oil industry expanded (Britannica 1965). The first synthetic resins were developed in 1900 and by 1910 pine-derived resins were being increasingly replaced by coal tar derivatives. Modern refinery techniques, including cracking, were developed in the period 1915–20 and led to a wide range of products. By 1930 natural turpentine was largely replaced by oil-based white spirits except in specialist products. Natural turpentines and resins are still important products but they have lost the pre-eminent place they once held. Pure heptane, distilled from *P. jeffreyi* resin, was used to develop the 'octane' scale used in rating petroleum for motor vehicles (Mirov & Hasbrouck 1976).

20.6.4 Seeds and other minor products

Pine seeds are still an important commercial export product for countries such as Spain, Portugal, Italy and India (Mirov & Hasbrouck 1976; Singh 1992). In good years the harvesting and processing of the *P. gerardiana* seed crop in the Suleiman Mountains of Pakistan employs 13 000 people (Martin 1995). The potential for a tree breeding programme to enhance seed production by *P. gerardiana* is under investigation (Singh & Chaudhary 1993). Pine seeds are still an important part of the diet of some American native peoples and a sought-after delicacy (Lanner 1981). The pinyon *P. maximartinezii* was discovered after the seeds were found for sale in a Mexican market (Rzedowski 1964). The seed crop harvested from 40 ha of this species earns the owners about US$ 6000 per year (Martin 1995). The American food industry uses pinyon seeds, those of *P. pinea* (imported from Europe) and, to a lesser extent, *P. gerardiana* (from the Himalayas) (Mirov & Hasbrouck 1976). Seeds of *P. cembra* and *P. sibirica* are popular in Siberia and are exported to Norway (Brouk 1975); about 7.2 million ha of these two species and *P. koraiensis* are conserved for seed production in the former USSR (Sutton 1975). The harvesting of truffles and other edible fungi from pine forests supports an important industry in Europe (Chevalier & Frochot 1990; Guinberteau *et al.* 1990).

20.6.5   Ornamentals, gardens and parks

The use of conifers as ornamentals has continued to expand as cities and urban areas have replaced natural vegetation with managed parks. There are about 300 pine species, varieties and cultivars available in what has become a significant industry worldwide (Harrison 1975). Despite the long history, there are few dwarf forms and cultivars of pines so they tend to be planted singly or as groves in parks.

20.6.6   Forest protection and management

Protection of wood supplies

Recognition of the need for deliberate management of the native coniferous forests of the northern hemisphere to sustain timber supplies only really led to action in the 20th century. Before then, forests were generally considered to be inexhaustible (Heske 1927; Streyffert 1938; Williams 1989; Mather 1990; Adlard 1993). Policies aimed at promoting and enhancing natural regeneration were instituted in Sweden in 1905 and expanded in 1923 (Streyffert 1938). Similar policies were implemented in Norway in 1937 (Skinnemoen 1957) and, after World War I, in Great Britain (Robinson 1927) and Germany (Heske 1927, 1988). In Spain, 2.8 million ha was afforested with pines from 1940 to 1987, about 30% of this area being planted to *P. pinaster* (Groome 1993). A similar pattern was followed in the former USSR after 1917 (Tseplyayev 1960), and in India where plantations of exotic pines were established in the 1970s to alleviate shortages of native pine timber caused by forest destruction well into the 1960s (Singh & Singh 1987). In Canada, plantations have expanded since 1948 (Lavender 1990) and 21% of the area of productive forest was planted and seeded in 1991 (Canada 1994). Plantation forestry, using native species such as *P. elliottii* and *P. taeda* began in the 1930s in the southern USA (Williams 1989). The area of pine plantations in the southern and southeastern USA increased significantly between 1952 to 1985, from about 0.75 to 8.5 million ha; *P. taeda* was most widely planted (USDA 1988). In 1984 alone, about 1.0 million ha was planted with pines (McDonald & Krugman 1986). This area (which excludes the 16.6 million ha of natural pine forests) is by far the largest area of pine plantations in the world. The area of plantations in this region is projected to grow to 19.0 million ha by the year 2020 (USDA 1988).

Protection of soil

In the Mediterranean Basin the primary concern has been the protection of soil and the reforestation of degraded areas (Thirgood 1981; Kolar 1989; Stewart 1993). The forests of Gascony and Landes are still used for resin production but the major products are pulpwood and solid wood products (Scott 1962). In Portugal, afforestation has continued

to expand. By 1938 some 68 000 ha had been afforested, rising to about 238 000 ha by 1959, yielding 4.0 million tons of wood and firewood, 100 000 tons of turpentine and rosin and 1000 tons of *P. pinea* seeds (de Carvalho 1960; Scott 1962).

Israel provides an excellent example of the reforestation of degraded lands, albeit on a relatively small scale. This programme has been driven by a strong desire to see the Jewish State restored to its Biblical glory rather than by economic or ecological imperatives. One of the first initiatives was the planting of *P. pinea* on Mount Carmel by Jewish settlers in the late 1800s (Kolar 1989). Systematic afforestation in Israel began in 1908 with olive trees and later with *P. halepensis* in the hills around Judea (Gottfried 1982). Afforestation increased markedly after independence in 1948 with about 100 million trees being planted from 1948 to 1968 on some 40 000 ha and a total area of 52 395 ha of conifers by 1987 (Spetter 1989). With intensive site preparation to retain and concentrate runoff, *P. halepensis* trees are even being successfully planted in the Negev where the annual rainfall is less than 200 mm (Yair, Shackak & Schreiber 1989).

Plantations, people and biodiversity

Trees still play an important role in agriculture. In the west, agro-technology and 'agribusiness' have dominated commercial forestry, but in developing countries there is an increasing realization that this is not the only, or even the best, strategy (Gregersen, Draper & Elz 1989; Dove 1992). Many traditional agroforestry practices are silviculturally and agriculturally sound and need to be encouraged in social forestry and land rehabilitation programmes (DWAF 1995).

Pines have played an important role in agroforestry in some tropical countries, but in other areas they have not been favoured because they do not supply fodder for livestock or fruit (Guggenberger, Ndulu & Shepherd 1989). Silvo-pastoral systems being developed in Chile with a mixture of *P. radiata* and chestnut trees and sheep pasture look very promising (Penaloza, Herve & Sobarzo 1985). Initial results have been less favourable in Western Australia, where sheep stripped the bark of *P. pinaster* and *P. radiata* in trials (Anderson, Hawke & Moore 1985). Agroforestry systems using pines are being investigated in Ecuador (Garrison & Pita 1992), New Zealand (Knowles, Moore & Leslie 1990) and Nigeria (Adegbehin, Igboanugo & Omijeh 1990). In Denmark, the USA and New Zealand, hardwood–pine mixtures are being successfully used in soil protection and driftsand control (Adlard 1993). In Kansas, USA, *P. ponderosa* is being used with *Fraxinus pennsylvanica* for land reclamation.

Intensive plantation forestry is moving closer to agricultural practice with large expanses of even-aged monocultures, intensive site preparation, and the increasing use of fertilizers and pesticides (McKinnon 1960). Initially the focus of the environmental movement was on the large-scale destruction of tropical forests but now it has broadened to include temperate and boreal forests (Anon. 1995). Much of the negative publicity about plantations is poorly informed. There is, however, justified concern about the impacts on natural systems in and around plantations, and about the sustainability of plantations (Evans 1990; Sharitz et al., 1992; Bren 1993; Lara & Veblen 1993). The Dauerwald movement, initiated in the 19th century in Germany, triggered a revival of tree-selection systems and encouraged the imitation of natural forests with mixtures of ages and species; it has also emphasized the maintenance of the whole-forest ecosystem (Heske 1988; Mather 1993b). The underlying principles are now being looked at to see how they can be applied to intensive plantation forestry (Evans 1992; Sharitz et al., 1992; Wiersum 1995).

The notion of sustainability in forestry has also been expanded from its focus on timber yields to the entire plantation system, including the natural systems that are part of plantation landscapes (Groome 1993; Williams 1993). Many of the principles of island biogeography, landscape ecology and conservation biology (irregular clearcuts, smaller even-aged units, retention of islands and riparian corridors of natural vegetation and wider espacements) are now being proposed or practised in pine plantations (Bull 1981; Friend 1980; Anko 1990; Thill 1990; Hof & Joyce 1992; Sharitz et al., 1992; Gobster 1995). Aesthetic considerations are also becoming important determinants of plantation design and in the planning of clearfelling in developed countries (Elephtheriadis & Tsalikidis 1990; Gobster 1995). Nevertheless, it is likely that under certain circumstances, for example land in remote areas or unsuited to intensive agriculture (or both), that the intensive agribusiness will continue to be a legitimate land use (Adlard 1993). Other areas will be managed for multiple use, and mixtures of forestry and agriculture will become the norm.

Tree species selection and breeding

The 20th century saw a massive increase in experimental introductions of tree species in many countries. Tree were selected more carefully than in earlier trial-and-error introductions which often used seeds of uncertain origins and genetic purity (e.g. Legat 1930). Properly designed provenance trials for pines were initiated in many countries with interests in getting the best growing stock for their conditions by matching climate and soil conditions (Table 20.1), including countries as disparate as Fiji (Evans 1992) and Iceland (Blondal 1986).

Recent advances in computer technology permit closer bioclimatic matching between the source and the destination (Busby 1991). A problem with this approach is that the natural climatic range may be much more restricted than the range over which the species will grow successfully; the classic example is P. radiata (Chap. 21, this volume). One solution is to use the results of trials in the target region to predict the potential range in that region, as was done recently for P. tecunumannii in South Africa (Chapman, Fairbanks & Louw 1995).

Advances in molecular genetics and biotechnology have provided the tools needed for mass propagation of clones. This has revolutionized pine tree breeding and will continue to do so in the future (see Jayawickrama & Balocchi 1993). The primary aim of tree breeding has been to increase growth rates and improve stem form. Increasing attention has, however, also been given to wood properties such as density, resin content, strength, spiral grain and fibre properties for pulping (Savill & Evans 1986; Evans 1992). Disease resistance has received increased attention in recent years,

Table 20.1. Some species- and provenance trials reported in the literature from Zobel et al. (1987, table 2.1), with additional data from Poynton (1977; for South Africa) and Dvorak & Donahue (1992). Species in bold have grown well or have substantial potential. Nomenclature follows Price et al. (Chap. 2, this volume)

Country	Pinus taxa
Argentina	**P. caribaea, P. douglasiana**, P. elliottii, **P. oocarpa, P. patula**, P. taeda
Australia	26 coniferous species, incl. at least the following pines: **P. attenuata, P. caribaea, P. elliottii, P. oocarpa, P. patula, P. radiata, P. taeda**, P. tecunumanii
Brazil	**P. caribaea, P. chiapensis**, P. greggii, **P. kesiya**, P. leiophylla, P. maximinoi, P. merkusii, P. montezumae, **P. oocarpa**, P. pseudostrobus, **P. tecunumanii**
Colombia	14 spp. incl. **P. caribaea, P. chiapensis, P. kesiya, P. maximinoi, P. oocarpa, P. patula, P. pseudostrobus, P. taeda, P. tecunumanii**
France	P. devoniana, P. hartwegii, P. montezumae, P. pseudostrobus, P. oocarpa
Hawaii	**P. devoniana, P. hartwegii**, P. leiophylla, **P. maximinoi**, P. oocarpa, P. patula, **P. pseudostrobus, P. teocote**
India	**P. caribaea**, P. greggii, **P. kesiya**, P. patula
Japan	**P. banksiana**, P. jeffreyi, P. ponderosa, **P. rigida, P. sylvestris**
Kenya	P. caribaea, **P. kesiya**, P. patula, P. radiata
Korea	P. banksiana, P. strobus
Malawi, Zambia	P. devoniana, **P. douglasiana**, P. kesiya, **P. leiophylla, P. maximinoi, P. merkusii, P. patula, P. pseudostrobus, P. teocote**
New Zealand	P. chiapensis, P. contorta, P. greggii, P. muricata, P. nigra, P. patula, P. pseudostrobus
Philippines	6 spp. incl. **P. caribaea**, P. kesiya, P. oocarpa, P. pseudostrobus
Puerto Rico	**P. caribaea**, P. merkusii, P. oocarpa
Scandinavia	**P. contorta**
Scotland	P. contorta
South Africa, Swaziland, Zimbabwe	P. ayacahuite, **P. caribaea, P. chiapensis**, P. cooperi, **P. durangensis, P. elliottii**, P. engelmannii, P. herrerae, **P. greggii, P. kesiya, P. leiophylla**, P. maximinoi, **P. muricata, P. oocarpa, P. patula, P. pringlei, P. radiata**, P. sylvestris, **P. taeda, P. tecunumannii, P. teocote**
Uganda	P. caribaea, **P. oocarpa**

Table 20.2. Estimates of the area of plantations of all species and of pines in the major regions of the world. Plantations exclude naturally regenerated areas unless otherwise indicated. Data are from Persson (1974, 1985), Canada (1994), Castles (1994), New Zealand (1994) and FAO (1995). The percentage of the world plantation areas under pines excludes China and the former USSR because no reliable data on the areas under pine plantations could be located

	Plantation area (1000 ha)		Pine plantations (%)
Region	All species	Pines	
Africa	4416	1394	31.57
Asia/Pacific	2433	848	3.47
China	31 831		
Latin America	7765	928	11.96
Europe	132 958[a]	2784	2.09
USA	195 596[a]	7166	3.66
Canada	505	54	0.05
USSR	414 015[a]		
Australia		892	2.35
New Zealand	1280	1015	49.27
World	1 010 601	17 759	3.14

[a] Area of utilizable forest (FAO 1995) which includes large areas of natural forests and naturally regenerated forests not considered as plantations in this chapter.

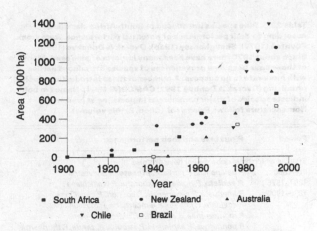

Fig. 20.1. Increases in plantation areas in southern hemisphere countries during the 20th century. Data for South Africa are for pine plantations; for New Zealand exotic plantations (±90% pines); for the other countries the area is for exotic conifer plantations. Data from Street (1962); Poole (1969); Persson (1974); Levack (1986); Evans (1992); van der Zel & Brink (1980); Husch (1982); Savill & Evans (1986); McDonald & Krugman (1986); Whyte (1988); New Zealand (1964, 1994); DWAF (1994); Castles (1994); FAO (1995).

partly because of concern about the vulnerability of the extensive monocultures to disease outbreaks (Chap. 19, this volume). There is also concern about the ability of the current plantation species to adapt to global warming. For example, Whitehead, Leathwick & Hobbs (1992) found that the climate changes predicted for New Zealand could have a significant impact on *P. radiata* plantations.

20.6.7 Human impact on species distributions

The 20th century has seen a significant expansion in the area of pine plantations in a wide range of countries, especially in the southeastern USA, South America, southern Africa, Australia, New Zealand and China. Some recent statistics have been summarized in Table 20.2. Recent surveys show that the area under plantations is increasing by 1.3 million ha per year and a substantial proportion is planted with pines (FAO 1995). In 1980, plantations and woodlots of pines in tropical countries covered some 21 million ha, accounting for about 33.7% of the plantation area (Evans 1992). In 1985, it was estimated that industrial plantations covered some 91.7 million ha with the actively managed and regenerated natural pine forests of the USA, the former Soviet Union, China and western Europe accounting for about 70% of the total (Postel & Heise 1988).

Extensive afforestation with pines in the southern hemisphere and in the tropics has radically increased the range of pine species (Fig. 20.1). Afforestation projects initiated in the 19th century in New Zealand, South Africa and Australia have grown significantly. The most striking growth has been in south-central Chile where *P. radiata* was planted on a large scale in the early 1970s (Lara &

Veblen 1993; Chap. 21, this volume). Rapid expansions have taken place in other South American countries (e.g. Brazil with 2.5 million ha, and Argentina), and in other tropical and subtropical regions (Husch 1982; Hornick, Zerbe & Whitmore 1984; McDonald & Fernandes 1984; Evans 1992; Dykstra 1983; McDonald & Krugman 1986; Zobel et al., 1987). Experimental trials of pines were even established in semi-arid savanna in Namibia, but the trees died (Erkilla & Suskonen 1992).

Pinus caribaea is a particularly successful species. Many varieties are under trial in more than 65 countries and it has been planted extensively in Brazil, Argentina, Venezuela, Malaysia, Fiji and in tropical African countries at low altitudes (Evans 1992; Pottinger 1994). *Pinus patula* has also been planted successfully in a wide range of tropical and subtropical countries at high altitudes (Evans 1992). *Pinus pinaster* has been introduced in a wide range of countries including Australia, New Zealand and South Africa, Argentina, Chile and the United Kingdom (Scott 1962).

Expansions within the natural range include *P. pinaster* plantations which now cover some 2.43 million ha: 50% in Portugal, 42% in France and 8% in north coastal Spain (Scott 1962). Plantations have significantly extended the range of *P. taeda* in the southeastern USA (USDA 1988; Williams 1989). Other translocations within the natural range include reforestation in Israel with *P. halepensis* from seed stocks from Italy and North Africa (Grunwald, Schiller & Conkle 1986). In the northern Mediterranean Basin pine forests have expanded as land-use patterns have changed and abandoned agricultural lands, in particular, have been reinvaded (Le Houérou 1981; Thirgood 1981; Chap. 8, this volume). In contrast, deforestation has significantly

Table 20.3. Pine species introduced to South Africa classified according to their performance or potential performance. Data from: Poynton (1977); Shaughnessy (1986); Dvorak & Donahue (1992). Since about 1970 there have been many imports of pine seed. Many of these have been new provenances of species introduced earlier with moderate to no success. A number of these introductions look promising (Dvorak & Donhue 1992; CAMCORE 1994). Names in bold indicate species of major commercial importance at present. Nomenclature follows Price et al. (Chap. 2, this volume)

Period	*Pinus* taxa and their performance
Species which have done well or have promise	
Pre-1850	P. halepensis, P. pinea, **P. pinaster**, P. sylvestris
1851–1875	**P. radiata**, P. rigida, P. roxburghii, P. wallichiana
1876–1900	P. attenuata, P. canariensis, P. coulteri, P. densiflora, P. echinata, P. jeffreyi, P. lambertiana, P. massoniana, P. montezumae, P. muricata, P. nigra, P. palustris, P. ponderosa, P. sabiniana, P. strobus, **P. taeda**, P. thunbergii
1901–1910	P. ayacahuite, P. banksiana, P. cembroides, P. contorta, P. cooperi, P. devoniana, P. douglasiana, P. edulis, P. engelmannii, P. gerardiana, P. greggii, P. hartwegii, P. kesiya, P. koraiensis, P. lawsonii, P. leiophylla, P. maximinoi, P. merkusii, P. monophylla, **P. patula**, P. ponderosa var. arizonica, P. pringlei, P. pseudostrobus
1911–1920	**P. elliottii**, P. lumholtzii, P. oocarpa
1921–1930	**P. caribaea**, P. cubensis
1930 onwards	P. chiapensis, P. clausa, P. herrerae, P. luchuensis, P. serotina, P. tecunumanii, P. teocote, P. tropicalis
Species with little or unknown potential	
1890 onwards	P. armandii (introduced in 1974), P. heldreichii var. leucodermis, P. maximartinezii (1970), P. monticola (1974), P. occidentalis, P. pungens (1894), P. resinosa, P. taiwanensis (1968), P. torreyana (1894), P. virginiana (1966)
Total failures	
1880 onwards	P. aristata, P. balfouriana, P. bungeana, P. cembra, P. mugo, P. parviflora, P. peuce, P. tabuliformis, P. uncinata

reduced the extent of the natural pine forests in the southern Mediterranean Basin. About 1.2 million ha of pinyon pine forest and woodland was cleared between 1950 and 1964 to provide livestock grazing in the south-western USA and this practice has continued (Lanner 1981). In Colorado and Nevada pinyons have recolonized areas that were denuded during the mining boom in the 19th and early 20th century (Lanner 1981; Chap. 9, this volume).

20.6.8 South Africa as an example
Many countries possess extensive pine plantations, both indigenous and exotic. Although South Africa has a relatively small area under pine plantations, it has been chosen as an example because: (1) it has an unusually wide range of climatic conditions, resulting in the introduction of a wide range of species including *P. pinaster* and *P. radiata* in the winter-rainfall areas, many Mexican pine species in the high-altitude, summer-rainfall areas, and Caribbean pines in the subtropical lowlands (Table 20.3; Fig. 20.2); (2) it was among the first southern hemisphere countries to introduce pines, beginning in about 1685 (Shaughnessy 1986); and (3) it was probably the first colonial country to use pines for more than aesthetic purposes and shelter

when *P. pinaster* and *P. pinea* were used for driftsand control in the 1850s (Shaughnessy 1986).

Southern Africa lacks natural closed forests with fast-growing species and therefore has always had wood supply problems (Legat 1930; King 1938; van der Zel & Brink 1980). In the late 1800s the Conservators of Forests in South Africa realized that they had to establish plantations of fast-growing exotic species to meet the demand for timber (King 1938). A number of pine species had already been successfully established (Table 20.3), e.g. *P. pinaster* (Hutchins 1904).

In 1892 the Forest Department of the Cape was founded and forest nurseries were laid out. The first plantations were established in 1884 in the Western Cape Province (Anon. 1892; Legat 1930; King 1938) and soon afterwards in the southern and Eastern Cape. By 1891 forest nurseries had produced more than 870 000 pine seedlings, 50% *P. pinaster* and 27% *P. radiata*. In 1903 D.E. Hutchins toured the high-altitude grasslands and escarpment in the former Transvaal and realized that the conditions (summer rainfall, high altitude and winter frost) were similar to those in Mexico. Systematic introductions of Mexican pines were initiated by Hutchins, C.E. Legat and others (Table 20.3; King 1938; Loock 1947). The establishment of CAMCORE (Central American and Mexico Coniferous Resources Cooperative) resulted in a new cycle of introductions and trials with species such as *P. chiapensis*, *P. herrerae* and *P. tecunumanii*, and with provenances and selections of species already in cultivation (Dvorak & Donahue 1992; Parfitt 1993; CAMCORE 1994; Dvorak, Donahue & Vasquez 1995). Several pine species have failed to establish themselves, dying either in the nursery or shortly after planting out (Poynton 1977; Table 20.3).

In 1994/5 there were about 1.43 million ha of commercial plantations in South Africa, with conifers accounting for some 53.1% of the total area (Fig. 20.2; DWAF 1996). A wide range of species is grown in South Africa. In the economic zones which comprised the former Transvaal, the dominant species is *P. patula*. The temperate climate, summer rainfall and high altitude closely match the conditions in its natural range in Mexico. In the economic zones of Maputaland and Zululand, *P. elliottii* is the dominant species and *P. caribaea* is also common. The climatic conditions in these subtropical lowland areas are fairly similar to their native habitats in the southeastern USA, the Caribbean islands and Central America. In the Western and Southern Cape economic zones, *P. radiata* and *P. pinaster* are the most important. Both species are susceptible to fungal infection if hail-damaged and so are not planted extensively in the summer-rainfall regions of the country. *Pinus greggii* and *P. tecunumanii* show potential in the moist subtropical Zululand–Maputaland and cool temperate, low rainfall Eastern Cape regions, respectively (Dvorak & Donahue 1992; Dvorak, Kietzka & Donahue 1996).

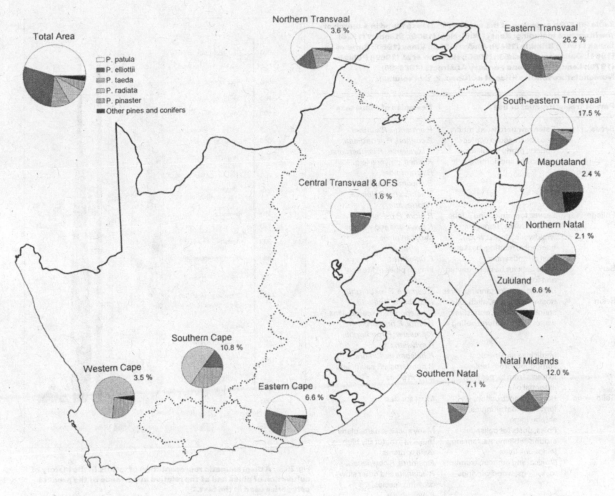

Fig. 20.2. **The distribution and importance of the five major _Pinus_ species in forestry economic zones of South Africa in 1994/5 (DWAF 1996).**

South African foresters introduced several novel silvicultural practices. The best known and most controversial is Craib's (1939) introduction of wide initial espacements, systematic, infrequent and heavy thinning, and pruning. The primary aim was to produce high-quality wood while ensuring that individual tree growth, and thus log size, was maximized. These innovations were heavily criticized as they were a radical departure from the existing practice of close espacements, but they proved successful and have been adopted in other countries (de Villiers _et al._ 1961). Another pioneer in South Africa was the late Prof. D.G.M. Donald who introduced significant improvements in nursery practice, planting with open-rooted seedlings, pioneered the inoculation of seedlings with mycorrhizae and conducted important experiments on the benefits of fertilizing seedlings and established trees (Donald 1984).

The commercial plantations yielded 15.6 million m³ of wood in 1994 and 8.8 million tons of pulpwood in 1993 (Van Rensburg 1994). The potential of many species introduced during the 1990s is still being evaluated. The demand for pine timber and other products is still increasing. The most immediate constraints facing the industry are concern about the impact of plantations on water resources (Dye 1996) and on the natural ecosystems they replace (e.g. Armstrong & van Hensbergen 1996). In the long term there must be more research into the use of marginal sites for pine growing because of climatic constraints on the area suitable for forestry. These issues are being tackled constructively and the prospects for pine plantations in South Africa are very promising.

20.7   **Synthesis**

The ingenuity of people is clearly shown in the multiplicity of uses that have been found for pine trees from prehistory to today (Table 20.4; Fig. 20.4). Many of the earliest

Table 20.4. *An overview of the uses of* Pinus *taxa found in a search of the literature including: Kent (1900); Hegi (1909); Standley (1926); Loock (1947); Brimble (1948); Alwyn (1952); Vines (1960); Sargent (1961); Dallimore & Jackson (1966); Harrison et al. (1969); Polunin (1976); Lanner (1981) and Forestry Abstracts (1980–95). Nomenclature follows Price et al. (Chap. 2, this volume)*

Part of tree	Product or use	Examples of *Pinus* taxa
Seeds	Eaten raw or roasted; made into flour; used in confectionery; produces oil for medicines and lighting	*P. armandii, P. cembra, P. coulteri, P. gerardiana, P. koraiensis, P. lambertiana, P. pinea,* pinyons (e.g. *P. cembroides, P. maximartinezii), P. sabiniana, P. sibirica, P. torreyana*
Foliage	Leaves for reinforcing adobe, fibre for twine, fibres for 'waldwolle' (forest wool), bandages, matting, basketry	*P. nigra, P. palustris, P. sylvestris* and various Mexican pines
	Leaf oil antiseptic	*P. palustris*
Bark	Inner bark for basket weaving and food	Pinyon pines of Mexico
	Bark extracts for tanning, dyes	*P. cembra, P. roxburghii*
Resin	Rosin, turpentine, pitch, oils, stains, varnishes, adhesives, waxes, soap, waterproofing	*P. caribaea, P. elliottii, P. halepensis, P. massoniana, P. nigra, P. ooocarpa, P. pinaster, P. roxburghii, P. sylvestris*
	Medicines and medical compounds from tars, extracts from turpentine, essential oils	*P. halepensis, P. nigra, P. ponderosa, P. sabiniana,* pinyon pines
Solid wood	Housing construction, furniture, roof shingles, shipbuilding	Most species
	Poles, posts for palisades around settlements, fencing, telephone lines	Many species, particularly those with naturally high resin contents
	Musical and other instruments, toys, clogs, wood carvings	*P. cembra, P. chiapensis, P. echinata* and other slow-growing species, *P. ayacahuite*
	Fragrance and incense	*P. cembra, P. nigra, P. strobus, P. sylvestris*
Wood-fibre	Paper, nitro-cellulose, rayon, board	Many species
Wood-charcoal	Ore smelting, extraction of resin and tar, ink	Most species
Protection	Hedges, shelter belts, erosion control, mine revegetation and driftsand control	*P. caribaea, P. contorta, P. halepensis, P. kesiya, P. massoniana, P. nigra, P. pinaster, P. pinea, P. radiata, P. sylvestris, P. thunbergii*
Ornamental, bonsai	Arboreta, parks, groves	Many species incl. *P. aristata, P. cembra, P. densiflora, P. parviflora, P. peuce, P. radiata, P. strobus, P. thunbergii*
Christmas trees	Whole trees and branches	*P. palustris, P. radiata, P. strobus, P. sylvestris*

uses of pines (e.g construction, medicinal extracts) have remained important throughout this period (Fig. 20.3). Others have disappeared as they have been replaced by other products, for example the replacement of pine by steel in shipbuilding. Wood-fibre and products derived from wood-fibres are a recent innovation. Numerous products, for example rayon, were developed in the last

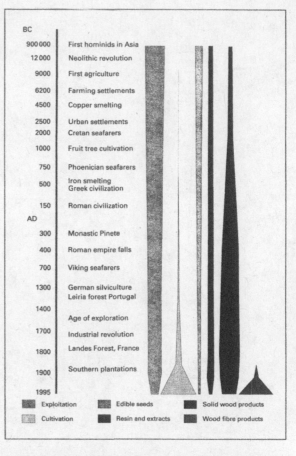

Fig. 20.3. **A diagrammatic representation of events in the history of cultivation of pines and of the relative importance of the product categories used in the text.**

100–200 years. Others, like paper, were invented in the more distant past but large-scale use of wood fibre in paper only began in the 20th century.

For almost all of the many millennia since people first encountered pines, the central theme has been one of continual and intensive exploitation. Some would argue that this is simply a lack of forethought, especially among 'primitive' societies. This is fallacious and simplistic. Research into the dynamics of hunter-gatherer and agro-pastoral societies, their knowledge of and relationships with natural resources, and their responses to resource availability, has shown that the ecological, social, ethical and economic factors that drive resource exploitation are far more complex than was thought, even a couple of decades ago (Lewin 1988; Lamb 1993; Alcorn 1995; Fritz 1995). There certainly is little reason to assume that Neolithic societies would have had less knowledge of their resources, and their impacts on them, than similar societies do today.

The basic driving factors in the unprecedented increase in forest cultivation during the last 300, and especially the last 100, years have been: the increasing shortage of raw timber, the recognition that utilization rates were not sustainable and, at least in some countries, acknowledgment that the costs of restoring and establishing forests are far higher than the costs of sustainable harvesting of existing forests. But why are pines only now emerging from their Cinderella role and entering centre stage? The history of the cultivation of pines suggests at least three reasons why attitudes to pine cultivation have changed over the last 600 or so years. The first step seems to have been the recognition that pines will grow relatively rapidly where many other trees, including more valuable ones, fail. The basic knowledge of the hardiness of pines dates from at least Classical Greek times, but the first documented application of this knowledge apparently was the afforestation with *P. pinaster* of degraded land and driftsands in Portugal in the early 14th century for soil conservation. The second step was the realization that only pines could meet the need for rapid growth and an acceptable quality of easily worked timber on marginal and agriculturally inferior land. An early example of this is the afforestation of the Landes region of southwestern France with *P. pinaster*, beginning in the late 18th century. In the last hundred years plantations have been planted on a large scale, including areas outside their natural range in the southern hemisphere. However, if there had been a suitable hardwood or alternative conifer taxon with the same characteristics it might equally well have been used instead. It is only in the 20th century that pines took the third step into the limelight. The key factor now favouring pines is their long fibres (typical of conifers); this makes them ideal for the production of pulp for paper and paper products – their primary use in the 20th century (Fig. 20.3). Hardwoods are unsuitable and there is no other conifer that can match the adaptability and growth rates of pines (see Chap. 1, this volume).

For modern western culture, the utter dependence of people upon wood as a basic requirement for existence before the industrial age is hard to imagine. Just picture a world in which the social and economic functions of oil, steel, plastics and other synthetics and electricity were all supplied by one material resource – wood (Sharp 1975). That is the reality of the history of human societies until the last 200 or so years. Wood from conifers, including pines, is generally easier to cut and work than that of hardwoods and is very versatile. Industrial technology has changed the world beyond all recognition but pines still provide fundamental economic benefits – from the most basic need for fuel to the paper this book is printed on. And conifers, particularly pines, are still the major source of the world's wood pulp for paper. This need alone will ensure that pines will continue to play a significant role in the affairs of human societies in the future.

Acknowledgements

I am greatly indebted to the works of Mirov & Hasbrouck (1976), Meiggs (1982) and Thirgood (1981) for their detailed information on pine use and cultivation in the Mediterranean Basin. Drs Robert Boardman and William Dvorak reviewed an earlier draft and provided valuable suggestions and comments which have helped me to improve the chapter significantly. I thank Dr Dave Richardson for his patience, valuable advice, and careful editing. I also thank Mrs C. Groeneveld for her help in searching bibliographic databases and in finding obscure references and Mrs A. Bevis for help in tracking down and translating information on pines and forestry from the Classical Period.

References

Adegbehin, J.O., Igboanugo, A.B.I. & Omijeh, J.E. (1990). Potentials of agroforestry for sustainable food and wood production in the savanna areas of Nigeria. *Savanna*, **11**, 12–26.

Adlard, P.G. (1993). *Historical Background*. Study No 1. Shell/WWF Plantation Review. Godalming, Surrey: Shell International Petroleum Company Limited and World Wide Fund for Nature.

Alcorn, J.B. (1995). Economic botany, conservation, and development: what's the connection. *Annals of the Missouri Botanical Garden*, **82**, 34–6.

Allen, P.H. (1955). The conquest of Cerro Santa Barbara, Honduras. *Ceiba*, **4**, 253–70.

Alwyn, J.B. (1952). *Conifers in Britain – An Illustrated Guide to Identification*. London: Adam and Charles Black.

Anderson, G.W., Hawke, M. & Moore, R.W. (1985). Pine needle consumption and bark stripping by sheep grazing annual pastures in young stands of widely spaced *Pinus radiata* and *P. pinaster*. *Agroforestry Systems*, **3**, 37–45.

Anko, B. (1990). Landscape ecology in forestry – a new challenge. In *Proceedings of the XIX IUFRO World Congress, Montreal, Div. 1, Vol. 1*, pp. 149–56. Vienna: IUFRO.

Anon. (1892). *Report of the Conservators of Forests and District Forest Officers for the Year 1891*. Cape Town: Government Printer.

Anon. (1995). Sustaining the worlds's forests. The Santiago agreement. *Journal of Forestry*, **93**, 18–21.

Armstrong, A.J. & van Hensbergen, H.J. (1996). Impacts of afforestation with pines on assemblages of native biota in South Africa. *South African Forestry Journal*, **175**, 35–42.

Bahn, P.G. (1989). Origins of full scale agriculture. *Nature*, **339**, 665.

Baker, H.G. (1970). *Plants and Civilisation*, 2nd edn. London: MacMillan.

Barrett, R.L & Mullin, L.J. (1968). *A Review on Introductions of Forest Trees to Rhodesia*. The Rhodesia Bulletin of

(continued on p. 428)

Fig. 20.4. **Postage stamps from countries within and outside the natural range of** *Pinus* **showing different pine species and their uses. 1.** *P. radiata*, New Zealand; 2. *P. sylvestris*, Guernsey; 3. *P. pinaster*, Spain; 4. *P. patula*, Swaziland; 5. *P. sylvestris*, Finland (International Union of Forestry Research Organizations 1995 Commemorative Stamp); 6. *P. strobus*, USA; 7. *P. mugo*, Germany (special welfare stamp); 8. *P. sylvestris*, Poland;

Fig. 20.4. (cont.)
9. *P. halepensis* (stylized), Israel; 10. *P. strobus*, Poland; 11. *P. strobus*, USA (National Arbor Day Foundation); 12. *P. caribaea*, Dominican Republic; 13. *P. caribaea*, Cuba; 14. *P. pinea*, San Marino; 15. *P. sylvestris*, Finland. The stamps were kindly supplied by B. Bredenkamp (Forestry Faculty, University of Stellenbosch, South Africa).

Forest Research No. 1. Salisbury: Rhodesian Forestry Commission.

Bennett, K.D., Tzedakis, P.C. & Willis, K.J. (1991). Quaternary refugia of north European trees. *Journal of Biogeography*, **18**, 103–15.

Betancourt, J.L. (1987). Paleoecology of pinyon-juniper woodlands: summary. In *Proceedings – Pinyon–Juniper Conference*, ed. R.E. Everett, pp. 129–39. USDA Forest Service General Technical Report INT-215. Washington, DC: United States Department of Agriculture.

Betancourt, J.L., Dean, J.S. & Hull, H.M. (1986). Prehistoric long-distance transport of construction beams, Chaco Canyon, New Mexico. *American Antiquity*, **51**, 370–5.

Betancourt, J.L. & van Devender, T.R. (1981). Holocene vegetation in Chaco Canyon, New Mexico. *Science*, **214**, 656–8.

Blondal, S. (1986). Fremde Baumarten im Wald von Hallormstad in Island. *Forst- und Holzwirt*, **41**, 605–9. (Forestry Abstracts 51, No. 1209).

Boardman, R. (1988). Living on the edge – the development of silviculture in South Australian pine plantations. *Australian Forestry*, **51**, 135–56.

Boardman, R. & McGuire, D.O. (1990). The role of zinc in forestry. II. Zinc deficiency and forest management: effect on yield and silviculture of *Pinus radiata* plantations in South Australia. *Forest Ecology and Management*, **210**, 207–18.

Brain, C.K. & Sillen, A. (1988). Evidence from the Swartkrans cave for the earliest use of fire. *Nature*, **336**, 464–6.

Bray, W. (1988). The palaeoindian debate. *Nature*, **332**, 107.

Bren, L.J. (1993). Riparian zone, stream, and floodplain issues: a review. *Journal of Hydrology*, **150**, 277–99.

Brimble, L.J.F. (1948). *Trees in Britain – Wild, Economic and Some Relatives in Other Lands*. London: Macmillan.

Britannica (1965). *Encyclopaedia Britannica*. Encyclopaedia Britannica Limited. Chicago: William Benton Publishers.

Brogger, A.W. & Shutelig, H. (1951). *The Viking Ships – Their Ancestry and Evolution*. Oslo: Dreyer Vorlag.

Brouk, B. (1975). *Plants Consumed by Man*. London: Academic Press.

Brown, J.C. (1878). *Pine Plantations on the Sand-wastes of France*. Edinburgh: Oliver & Boyd.

Bull, P.C. (1981). The consequences for wildlife of expanding New Zealand's forest industry. *New Zealand Journal of Forestry*, **26**, 210–31.

Busby, J.R. (1991). BIOCLIM – a bioclimate analysis and prediction system. *Plant Protection Quarterly*, **6**, 8–9.

Bushnell, G.H.S. (1976). The beginning and growth of agriculture in Mexico. *Philosophical Transactions of the Royal Society A (London)*, **275**, 117–28.

Bye, R.A. (1985). Botanical perspectives of ethnobotany of the greater southwest. *Economic Botany*, **39**, 375–86.

CAMCORE (1994). *Annual Report 1994*. CAMCORE Cooperative, Department of Forestry, College of Forest Resources, North Carolina University, Raleigh, USA.

Campbell Thompson, R. (1949). *A Dictionary of Assyrian Botany*. London: British Academy.

Canada (1994). *Canada Year Book 1994*. Ottawa: Statistics Canada.

Castles, I. (1994). *Year Book Australia 1994*. Canberra: Australian Bureau of Statistics.

Chapman, R.A., Fairbanks, D. & Louw, J.H. (1995). *Bioclimatic Profiles for the Potential Afforestation of Pinus tecunumannii in the Eastern Transvaal*. Report FOR-DEA 815. Pretoria: Department of Water Affairs and Forestry.

Chevalier, G. & Frochot, H. (1990). Ecology and possibility of culture in Europe of the burgundy truffle (*Tuber uncinatum* Chatin). *Agriculture, Ecosystems and Environment*, **28**, 71–3.

Craib, I.J. (1939). *Thinning, Pruning and Management Studies on the Main Exotic Conifers Grown in South Africa*. Bulletin No. 196. Pretoria: Department of Agriculture and Forest Science.

Crawford, G.W. (1992). Prehistorical plant domestication in East Asia. In *The Origins of Agriculture – An International Perspective*, ed. C.W. Cowan & P.J. Watson, pp. 7–38. Washington, DC: Smithsonian Institution Press.

Dallimore, W. & Jackson, B. (1966). *A Handbook of the Coniferae and Ginkgoaceae,*. 4th edn, revised by S.G. Harrison. London: Edward Arnold.

Dawson, G.M. (1892). Notes on the Shushwap Indians of British Columbia. *Proceedings and Transactions of the Royal Society of Canada*. Transactions Section ii, p. 22.

Deacon, H.J. (1992). Human settlement. In *The Ecology of Fynbos: Nutrients, Fire and Diversity*, ed. R.M. Cowling, pp. 260–70. Cape Town: Oxford University Press.

De Carvalho, J.M. (1960). A brief account of forestry in Portugal. In *Proceedings of the Fifth World Forestry Congress*, Vol. 1, pp. 659–68. Washington: University of Washington.

Delcourt, H.R. (1987). The impact of prehistoric agriculture and land occupation on natural vegetation. *Trends in Ecology and Evolution*, **2**, 39–44.

Dennell, R.W. (1992). The origins of crop agriculture in Europe. In *The Origins of Agriculture – an International Perspective*, ed. C.W. Cowan & P.J. Watson, pp. 71–100. Washington, DC: Smithsonian Institution Press.

Den Ouden, P. & Boom, B.K. (1965). *Manual of the Cultivated Conifers Hardy in the Warm Temperate Zone*. The Hague: Marthinus Nijhoff.

de Villiers, P.C., Marsh, E.K., Sonntag, A.E. & van Wyk, J.H. (1961). The silviculture of exotic conifer plantations in South Africa. *Forestry in South Africa*, **1**, 13–29.

Diamond, J.M. (1986). The environmentalist myth. *Nature*, **324**, 19–20.

Dillehay, T.D. & Collins, M.B. (1988). Early cultural evidence from Monte Verde in Chile. *Nature*, **332**, 150–2.

Donald, D.G.M. (1984). Silviculture and yield. In *Symposium on Site and Productivity of Fast Growing Plantations*, ed. D.C. Grey, Schönau & C.J. Schutz, pp. 163–80. Pretoria: South African Forest Research Institute and Department of Environment Affairs.

Dove, M.R. (1992). Forester's beliefs about farmers: a priority for science research in social forestry. *Agroforestry Systems*, **17**, 13–41.

Duffield, J.W. (1951). Interrelations of the California closed-cone pines with special reference to *Pinus muricata* D. Don. Ph.D. thesis, University of California, Berkeley.

Dvorak, W.S. & Donahue, J.K. (1992). *CAMCORE Cooperative Research Review 1980–1992*. Raleigh, NC: Department of Forestry, North Carolina University.

Dvorak, W.S., Donahue, J.K. & Vasquez, J.A. (1995). Early performance of CAMCORE introductions of *Pinus patula* in Brazil, Colombia and South Africa. *South African Forestry Journal*, **174**, 23–33.

Dvorak, W.S., Kietzka, J.E. & Donahue, J.K. (1996). Three-year survival and growth of provenances of *Pinus greggii* in the tropics and subtropics. *Forest Ecology and Management* **83**, 123–31.

DWAF (1994). *Commercial Timber Resources and Roundwood Processing in South Africa 1992/93*, 25th edn. Pretoria: Department of Water Affairs and Forestry.

DWAF (1995). *Towards a Policy of Sustainable Forest Management in South Africa.* Pretoria: Department of Water Affairs and Forestry.

DWAF (1996). *Commercial Timber Resources and Roundwood Processing in South Africa 1994/5.* Pretoria: Forestry Branch, Department of Water Affairs and Forestry.

Dye, P.J. (1996). Climate, forest and streamflow relationships in South African afforested catchments. *Commonwealth Forestry Review*, **75**, 31–8.

Dykstra, D.P. (1983). Forestry in Tanzania. *Journal of Forestry*, **81**, 742–6.

Einarson, B. & Link, G.H.K. (eds.) (1990). *Theophrastus de Causis Plantarum*, 3 vols. Cambridge, Mass.: Harvard University Press.

Elephtheriadis, N. & Tsalikidis, I. (1990). Coastal pine forest landscapes; modelling scenic beauty for forest management. *Journal of Environmental Management*, **30**, 47–62.

Erkilla, A. & Suskonen, H. (1992). *Forestry in Namibia 1850–1990.* Silvae Carelica 20, Faculty of Forestry, University of Joensuu, Finland.

Evans, J. (1990). Long-term productivity of forest plantations. In *Proceedings of the XIX IUFRO World Congress, Montreal*, Div. 1, Vol. 1 pp. 165–80. Vienna: IUFRO.

Evans, J. (1992). *Plantation Forestry in the Tropics*, 2nd edn. Oxford: Clarendon Press.

FAO (1995). *Forest Resources Assessment 1990. Global Synthesis.* FAO Forestry Paper. Rome: Food and Agriculture Organization of the United Nations.

Floyd, M.L. & Kohler, T.A. (1990). Current productivity and prehistoric use of piñon (*Pinus edulis*, Pinaceae) in the Dolores archaeological project area, southwestern Colorado. *Economic Botany*, **44**, 141–56.

Frazer, G.J. (1935). *The Golden Bough.* New York: Macmillan.

Friend, G.R. (1980). Wildlife conservation and softwood forestry in Australia: some considerations. *Australian Forestry*, **43**, 217–24.

Fritz, G.J. (1995). New dates and data on early agriculture: the legacy of complex hunter-gatherers. *Annals of the Missouri Botanical Garden*, **82**, 3–15.

Garrison, M. & Pita, M. (1992). An evaluation of silvopastoral systems in pine plantations in the Central Highlands of Ecuador. *Agroforestry Systems*, **18**, 1–16.

Gobster, P.H. (1995). Aldo Leopold's ecological esthetic. *Journal of Forestry*, **93**, 6–10.

Gordon, G. (1875). *The Pinetum*, 2nd edn. London: H.G. Bohn.

Gottfried, G.J. (1982). Forests and forestry in Israel. *Journal of Forestry*, **80**, 516–20.

Gregersen, H., Draper, S. & Elz, D. (1989). *People and Trees. The Role of Social Forestry in Sustainable Development.* DEI Seminar Series. Washington DC: The World Bank.

Gron, A.H. (1947). The economic foundations of forest politics. *Unasylva*, **1**, 3–6.

Groome, A.J. (1993). Afforestation policy and practice in Spain. In *Afforestation: Policies, Planning and Progress*, ed. A.S. Mather, pp. 72–91. London: Belhaven Press.

Grunwald, C., Schiller, G. & Conkle, M.T. (1986). Isozyme variations among native stands and plantations of aleppo pine in Israel. *Israeli Journal of Botany*, **35**, 161–74.

Guggenberger, C., Ndulu, P. & Shepherd, G. (1989). After Ujmaa: farmer needs, nurseries and project sustainability in Mwanza, Kenya. Network paper – social forestry network No 9c, Agricultural Administration Unit, Overseas Development Institute, London. (Forestry Abstracts 53, No. 4441).

Guinberteau, J., Ducamp, M., Poiytou, N., Mamoun, M. & Olivier, J.M. (1990). Ecology of various competitors from an experimental plot of *Pinus pinaster* inoculated with *Suillius granulatus* and *Lactarius deliciosus*. *Agriculture, Ecosystems and Environment*, **28**, 161–75.

Gunther, R.T. (ed.) (1968). *The Greek Herbal of Dioscorides.* New York: Hafner Publishing Company.

Gurskii, A.V. (1957). O snovye itogi introduktsii drevesnykh rastenii YSSR. Izd. Akad. Nauk SSSR, Moscow [as cited in Mirov 1967].

Harrison, C.R. (1975). *Ornamental Conifers.* Devon: David & Charles.

Harrison, S.G., Masefield, G.B., Wallis, M. & Nicholson, B.E. (1969). *The Oxford Book of Plants.* Oxford: Oxford University Press.

Hegi, G. (1909). *Illustrierte Flora von Mittel-Europa.* Vol. 1. Munich: Carl Hauser Verlag.

Hepper, F.N. (1990). *Pharoah's Flowers – the Botanical Treasures of Tutankhamun.* Royal Botanic Gardens, Kew. London: HMSO.

Heske, F. (1927). Post-war forestry in central Europe. *Forestry*, **1**, 55–61.

Heske, F. (1988). *German Forestry.* New Haven, Conn.: Yale University Press.

Hodell, D.A., Curtis, J.H. & Brenner, M. (1995). Possible role of climate in the collapse of classic Maya civilization. *Nature*, **375**, 391–4.

Hof, J.G. & Joyce, L.A. (1992). Spatial optimisation for wildlife and timber in managed ecosystems. *Forest Science*, **38**, 489–508.

Holmes, G.C.V. (1900). *Ancient and Modern Ships.* Part 1. *Wooden Sailing Ships.* London: Chapman & Hall.

Hornick, J.R., Zerbe, J.L. & Whitmore, J.L. (1984). Jari's successes. *Journal of Forestry*, **82**, 663–7.

Howes, F.N. (1948). *Nuts. Their Production and Everyday Uses.* London: Faber and Faber.

Hughes, J.D. (1975). *Ecology in Ancient Civilisations.* Alberquerque, NM: University of New Mexico Press.

Husch, B. (1982). Forestry in Chile. *Journal of Forestry*, **80**, 735–7.

Hutchins, D.E. (1904). *The Cluster Pine at Genadendal: Spreads Self-sown.* Cape Town: Cape Times Ltd.

Issar, A.S. (1995). Climatic change and the history of the Middle East. *American Scientist*, **83**, 350–5.

Jayawickrama, K.J.S. & Balocchi, C. (1993). Tree improvement in Chile. *Journal of Forestry*, **91**, 43–7.

Jones, W.H.S. (ed.) (1961). *Pliny's Natural History*, 10 vols. London: Heinemann.

Kent, A.H. (1900). *Veitch's Manual of the Coniferae.* London: J. Veitch & Sons Ltd.

Kim, Y.S. (1990). Chemical characteristics of waterlogged archaeological wood. *Holzforschung*, **44**, 169–72.

King, N.L. (1938). Historical sketch of the development of forestry in South Africa. *Journal of the South African Forestry Association*, **1**, 4–16.

Kinloch, B.B., Westfall, R.D. & Forrest, G.I. (1986). Caledonian stone pine: origins and genetic structure. *New Phytologist*, **104**, 703–29.

Klaus, W. (1989). Mediterranean pines and their history. *Plant Systematics and Evolution*, **162**, 133–63.

Knowles, R.L., Moore, R.W. & Leslie, B.R.N. (1990). Silvopastoral systems with radiata pine in southern temperate zones. In *Proceedings of the XIX IUFRO World Congress, Montreal*, Div. 1, Vol. 2, pp. 338–49. Vienna: IUFRO.

Kolar, M. (1989). The forest department of the JNF. *Allgemeine Forst Zeitschrift*, **24–26**, 600–1.

Lamb, R. (1993). Designs on life. *New Scientist* No. 1897 (30 October), 37–40.

Lane, F.C. (1952). *The Story of Trees*. New York: Doubleday & Doubleday Inc.

Lanner, R.M. (1981). *The Piñon Pine. A Natural and Cultural History*. Reno: University of Nevada Press.

Lara, A. & Veblen, T.T. (1993). Forest plantations in Chile: a successful model? In *Afforestation: Policies, Planning and Progress*, ed. A.S. Mather, pp. 118–39. London: Belhaven Press.

Latham, B. (1964). *Wood from Forest to Man*. London: G.C. Harrap Co.

Lavender, D.P. (1990). Plantation forestry in Western Canada. In *Proceedings of the XIX IUFRO World Congress, Montreal*, Div. 1, Vol. 1, pp. 23–31. Vienna: IUFRO.

Le Houérou, H.N. (1981). Impact of man and his animals on mediterranean vegetation. In *Mediterranean-Type Shrublands. Ecosystems of the World*, Vol. 11, ed. F. Di Castri, D.W. Goodall & R.L. Specht, pp. 479–517. Amsterdam: Elsevier.

Legat, C.E. (1930). The cultivation of exotic conifers in South Africa. *Empire Forestry Journal*, **9**, 32–63.

Levack, H. (ed.) (1986). *Forestry Handbook*. Wellington: New Zealand Institute of Foresters (Inc.).

Lewin, R. (1988). New views emerge on hunters and gatherers. *Science*, **240**, 1146–8.

Loock, E.E.M. (1947). *The Pines of Mexico and British Honduras*. Pretoria: Department of Forestry.

Lyell, C. (1849). *A Second Visit to the United States of North America*. New York: Harper & Row.

Martin, G.J. (1995). *Ethnobotany. A Methods Manual*. London: Chapman & Hall.

Mather, A.S. (1990). *Global Forest Resources*. London: Belhaven Press.

Mather, A.S. (ed.) (1993a). *Afforestation: Policies, Planning and Progress*. London: Belhaven Press.

Mather, A.S. (1993b). Review. In *Afforestation: Policies, Planning and Progress*, ed. A.S. Mather, pp. 207–19. London: Belhaven Press.

McDonald, L. & Fernandes, I.M. (1984). AMCEL. *Journal of Forestry*, **82**, 668–70.

McDonald, S.E. & Krugman, S.L. (1986). Worldwide planting of southern pines. *Journal of Forestry*, **84**, 21–4.

McKinnon, A.D. (1960). 'Orchard' versus 'naturalistic' silviculture in the radiata pine forests of New Zealand. In *Proceedings of the Fifth World Forestry Congress*, Vol. 1, pp. 396–404. Washington: University of Washington.

Meiggs, R. (1982). *Trees and Timber in the Ancient Mediterranean World*. Oxford: Clarendon Press.

Menninger, E.A. (1977). *Edible Nuts of the World*. Florida: Horticultural Books of the World Inc.

Mighall, T.M. & Chambers, F.M. (1995). Holocene vegetation history and human impact at Bryn y Castell, Snowdonia, north Wales. *New Phytologist*, **130**, 299–321.

Mikesell, M.W. (1969). The deforestation of Mt. Lebanon. *Geographical Review*, **59**, 1–28.

Miller, N.F. (1992). Origins of plant cultivation in the Middle East. In *The Origins of Agriculture – An International Perspective*, ed. C.W. Cowan & P.J. Watson, pp. 39–58. Washington: Smithsonian Institution Press.

Mirov, N.T. (1967). *The Genus Pinus*. New York: Ronald Press.

Mirov, N.T. & Hasbrouck, J. (1976). *The Story of Pines*. Indiana: Indiana University Press.

Morrison, J.S. & Williams, R.T. (1968). *Greek Oared Ships*. Cambridge: Cambridge University Press.

Naveh, Z. & Dan, J. (1973). Human degradation of mediterranean landscapes in Israel. In *Mediterranean-type ecosystems. Origin and Structure*, ed. F. di Castri & H.A. Mooney, pp. 373–90. Berlin: Springer-Verlag.

New Zealand (1964). *New Zealand Forestry*. Wellington: Government Printer.

New Zealand (1994). *New Zealand Official Yearbook 94*. Auckland: Statistics New Zealand.

Panetsos, C.P. (1981). Monograph of *Pinus halepensis* Mill. and *P. brutia* Ten. *Ann. Forest. Zagreb*, **9**, 39–77.

Parfitt, R. (1993). *Report: Establishment of the Latest CAMCORE Trials*. Report FOR-DEA 689. Pretoria: Forestry Branch, Department of Water Affairs and Forestry.

Pears (1937). *Pears Cyclopedia*, 46th edn. Isleworth: A. & F. Pears Ltd.

Penaloza, R., Herve, M. & Sobarzo, L. (1985). Applied research on multiple land use through silvopastoral system in Chile. *Agroforestry Systems*, **3**, 59–77.

Persson, R. (1974). *World Forest Resources. Review of the World's Forest Resources in the Early 1970s*. Department of Forest Survey Research Note 17. Stockholm: Royal College of Forestry.

Persson, R. (1985). *Forest Resources of Africa. An Approach to International Forest Resource Appraisals. Part I: Country Descriptions*. Department of Forest Survey Research Note 18, Stockholm: Royal College of Forestry.

Polunin, O. (1976). *Trees and Bushes of Europe*. Oxford: Oxford University Press.

Poole, A.L. (1969). *Forestry in New Zealand*. Auckland: Hodder & Stoughton.

Postel, S. & Heise, L. (1988). Reforesting the earth. Worldwatch Paper 83. Washington, DC: Worldwatch Institute.

Pottinger, A.J. (1994). Establishment of clone banks from plus trees selected in the OFI international pine trial network. *Commonwealth Forestry Review*, **73**, 23–7.

Poynton, R.J. (1977). *Tree Planting in South Africa*. Vol. 1. *The Pines*. Pretoria: Chief Directorate of Forestry, Department of Water Affairs.

Pozzera, G. (1959). Rapporti fra produzione di strobili in *Pinus pinea* L. ed andamento stagionale. *Ital. For. Mont.*, **14**(5), 196–206 [from *Forestry Abstracts* 21, No. 1563].

Reed, J.L. (1954). *Forests of France*. London: Faber & Faber.

Richardson, D.M. & Bond, W.J. (1991). Determinants of plant distribution: Evidence from pine invasions. *American Naturalist*, **137**, 639–68.

Robinson, R.L. (1927). British forestry. *Forestry*, **1**, 1–5.

Rue, D.J. (1987). Early agriculture and early postclassic Maya occupation in western Honduras. *Nature*, **326**, 285–6.

Rzedowski, J. (1964). Una especia neuva de pino piñonero del estado de Zacatecas (Mexico). *Ciencia (Mexico)*, **23**, 17–21.

Sabloff, J.A. (1995). Drought and decline. *Nature*, **375**, 357.

Sargent, C.S. (1961). *Manual of the Trees of North America*. New York: Dover Publications.

Savill, T.S. & Evans, J. (1986). *Plantation Silviculture in Temperate Regions*. Oxford: Clarendon Press.

Schiller, G., Conkle, M.T. & Grunwald, C. (1986). Local differentiation among Mediterranean populations of Aleppo pine in their isoenzymes. *Silvae Genetica*, **35**, 11–19.

Scott, C.W. (1962). A summary of information on *Pinus pinaster*. *Forestry Abstracts*, **23**(1/2), i–xviii.

Sharitz, R.R., Boring, L.R., Van Lear, D.H. & Pinder, J.E. (1992). Integrating ecological concepts with natural resource management of southern forests. *Ecological Applications*, **2**, 226–37.

Sharp, L. (1975). Timber, science, and economic reform in the seventeenth century. *Forestry*, **48**, 51–86.

Shaughnessy, G. (1986). A case study of some woody plant introductions to the Cape Town area. In *The Ecology and Management of Biological Invasions in Southern Africa*, ed. I.A.W. Macdonald, F.J. Kruger & A.A. Ferrar, pp. 37–43. Cape Town: Oxford University Press.

Shipek, F.C. (1989). An example of intensive plant husbandry: the Kumenay of southern California. In *Foraging and Farming. The Evolution of Plant Exploitation*, ed. D.R. Harris & G.C. Hillman, pp. 159–70. London: Unwin & Hyman.

Singh, J.S. & Singh, S.P. (1987). Forest vegetation of the Himalayas. *The Botanical Review*, **53**, 80–192.

Singh, N.B. (1992). Propagation, selection and establishment of clonal seed orchard of Chilgoza pine (*Pinus gerardiana* Wall.). *Indian Forester*, **118**, 901–8.

Singh, N.B. & Chaudhary, V.K. (1993). Variability, heritability and genetic gain in cone and nut characters in Chilgoza pine (*Pinus gerardiana* Wall.). *Silvae Genetica*, **42**, 61–3.

Skinnemoen, K. (1957). *An Outline of Norwegian Forestry*. Oslo: Det Norske Skogselskap.

Smith, B.D. (1992). Prehistoric plant husbandry in eastern North America. In *The Origins of Agriculture – An International Perspective*, ed. C.W. Cowan & P.J. Watson, pp. 101–19. Washington, DC: Smithsonian Institution Press.

Spetter, E. (1989). Forest inventory and management in Israel. *Allgemeine Forst Zeitschrift*, **24–26**, 656–67.

Standley, P.C. (1926). *Trees and Shrubs of Mexico*. Contributions from the United States National Herbarium 23. Washington, DC: Smithsonian Institution Press.

Steven, H.M. (1927). Silviculture of conifers in Great Britain. *Forestry*, **1**, 8–23.

Steven, H.M. (1960). Scots pine (*Pinus sylvestris*) in Britain. In *Proceedings of the Fifth World Forestry Congress*, Vol. 1, pp. 597–9. Washington: University of Washington.

Stevenson, A.C. & Moore, P.D. (1988). Studies of the vegetational history of S.W. Spain. IV. Palynological investigations of a valley mire at El Acebron, Huelva. *Journal of Biogeography*, **15**, 339–61.

Stewart, P. (1993). Afforestation in Algeria. In *Afforestation: Policies, Planning and Progress*, ed. A.S. Mather, pp. 92–104. London: Belhaven Press.

Streyffert, T. (1938). *The Forests of Sweden*. New Sweden Tercentenary Publications. Stockholm: A. Bonniers Boktryckeri.

Styles, B.T. (1993). Pine kernels. In *Encyclopaedia of Food Science, Food Technology and Nutrition*, Vol. 6, ed. R. Macrae, R.K. Robinson & M.J. Sadler, pp. 3595–7. London: Academic Press.

Suc, J.-P. (1984). Origin and evolution of the Mediterranean vegetation and climate in Europe. *Nature*, **307**, 429–32.

Sutton, W.R.J. (1975). The forest resources of the USSR. *Commonwealth Forestry Review*, **54**, 110–38.

TFU (1993). Tropical Forest Update 3(6). Yokohoma: International Tropical Timber Organization.

Thill, R.E. (1990). Managing southern pine plantations for wildlife. In *Proceedings of the XIX IUFRO World Congress, Montreal*, Div. 1, Vol. 1, pp. 58–68. Vienna: IUFRO.

Thirgood, J.V. (1981). *Man and the Mediterranean Forest. A History of Resource Depletion*. London: Academic Press.

Times, The (1988). *Past Worlds. The Times Atlas of Archaeology*. London: Times Books.

Trabaud, L. (1981). Man and fire: impacts on mediterranean vegetation. In *Mediterranean-type Shrublands. Ecosystems of the World, vol. 11*, ed. F. Di Castri, D.W. Goodall & R.L. Specht, pp. 523–37. Amsterdam: Elsevier.

Tseplyayev, V.P. (1960). Forestry development in the USSR. In *Proceedings of the Fifth World Forestry Congress*, Vol. 1, pp. 210–15. Washington: University of Washington.

Tsukada, M., Sugita, S. & Tsukada, Y. (1986). Oldest primitive agriculture and vegetational environments in Japan. *Nature*, **322**, 632–4.

Tzedakis, P.C. (1993). Long-term tree populations in northwest Greece through multiple quaternary cycles. *Nature*, **364**, 437–40.

USDA (1988). *The South's Fourth Forest: Alternatives for the Future*. Forest Resource Report No. 24. Washington, DC: USDA Forest Service.

van Andel, T.H., Runnels, C.N. & Pope, K.O. (1986). Five thousand years of land-use and abuse in the southern Argolid, Greece. *Hesperia*, **55**, 103–28.

van der Zel, D.W. & Brink, M. (1980). Die geskiedenis van bosbou in Suider-Afrika. Deel II: Plantasiebosbou. *South African Forestry Journal*, **115**, 17–27.

Van Rensburg, H.J (ed.) (1994). *South Africa Yearbook 1994*. Pretoria: South African Communication Service.

Vines, R.A. (1960). *Trees, Shrubs and Vines of the Southwest*. Austin: University of Texas Press.

Wanpo, H., Ciochon, R., Yumin, G. *et al.* (1995). Early *Homo* and associated artefacts from Asia. *Nature*, **378**, 275–8.

Weinstein, A. (1989). Geographic variation and phenology of *Pinus halepensis, P. brutia* and *P. eldarica* in Israel. *Forest Ecology and Management*, **27**, 99–108.

Wells, P.V. (1987). Systematics and distribution of pinyons in the late Quaternary. In *Proceedings – Pinyon–Juniper Conference*, ed. R.E. Everett, pp. 104–8. USDA Forest Service General Technical Report INT-215.

Welsh, F. (1988). *Building the Trireme*. London: Constable.

Whitehead, D., Leathwick, J.R. & Hobbs, J.F.F. (1992). How will New Zealand's forests respond to climate change? Potential changes in response to increasing temperature. *New Zealand Journal of Forestry Science*, **22**, 39–53.

Whyte, A.G.D. (1988). Radiata pine silviculture in new Zealand: its evolution and future prospects. *Australian Forestry*, **51**, 185–96.

Wiersum, K.F. (1995). 200 years of sustainability in forestry: lessons from history. *Environmental Management*, **19**, 321–9.

Williams, M. (1989). *Americans and Their Forests – A Historical Geography*. Cambridge: Cambridge University Press.

Williams, M. (1993). Afforestation: the United States. In *Afforestation: Policies, Planning and Progress*, ed. A.S. Mather, pp. 192–206. London: Belhaven Press.

Willis, K.J. (1992). The late Quaternary vegetational history of northwest Greece. III. A comparative study of two contrasting sites. *New Phytologist*, **121**, 139–55.

Wood, B. & Turner, A. (1995). Out of Africa and into Asia. *Nature*, **378**, 239–40.

Yair, A., Shackak, M. & Schreiber, K.-F. (1989). Hillslope minicatchments: the use of surface run-off water to increase primary production in a rocky desert. *Allgemeine Forst Zeitschrift*, **24–26**, 646–7.

Zobel, B.J., van Wyk, G. & Stahl, P. (1987). *Growing Exotic Forests*. New York: John Wiley.

Zohary, M. (1973). *Geobotanical Foundations of the Middle East*, Vols. 1 & 2. Stuttgart: Gustav Fisher Verlag.

21   *Pinus radiata*: a narrow endemic from North America takes on the world

Peter B. Lavery and Donald J. Mead

21.1   Preamble

Many conifer species have been borrowed from one country and planted in others. In such cases, the objectives at the time of initial introduction were often ill-defined, but involved exploring potentials for amenity value, shade and shelter, or wood production. A classical example of those seriously introduced for commercial softwood timber production is provided by the North American Sitka spruce, *Picea sitchensis*, first brought to the British Isles in 1831. Other well-known examples include *Pinus caribaea*, introduced to monsoonal tropical and coastal subtropical environments beyond Central America; *P. patula* and *P. taeda* from North America to many countries of the subtropics; and *Pseudotsuga menziesii* within temperate to cool-temperate regions of the world.

However, in the category of conifers introduced for production forestry in temperate climates, *Pinus radiata* D. Don (we prefer the common name *radiata pine* – see below), with a current plantation estate of just over 4 million hectares, provides possibly the most outstanding story of a narrow-range endemic taking on the world.

In this chapter we describe elements of a forestry success story – the development in exotic culture of a previously obscure conifer. As such, this chapter does not provide a technical treatise in the manner of most other chapters of this book. We have aimed to summarize existing information and experience with one species used for commercial cultivation rather than to develop or explore new hypotheses.

The scope of this chapter does not permit in-depth accounts from grower countries of the silviculture of *P. radiata* and the industrial sector based on the plantation resource of this species. Readers seeking such information should consult the authoritative accounts of the *P. radiata*

sectors of Australia, Chile and New Zealand provided by Lewis & Ferguson (1993). For technical analysis of the wood properties and in-use characteristics of the species, readers should refer to Kininmonth & Whitehouse (1991). In relation to intensive silviculture, Maclaren (1993) provides a comprehensive guide, with particular emphasis on management for clearwood production in the characteristically direct New Zealand style of silviculture, while Neilsen (1990) provides detailed establishment and tending prescriptions for radiata pine in a 'Plantation Handbook' format.

21.2   Discovery and destiny

21.2.1   The historical record

Although the story of this species as a key producer of utility softwood is largely a 20th century phenomenon, this is not to say that the species was entirely unknown to the wider world in the preceding four centuries. It is reputed, though we suspect possibly more according to forestry folklore than to fact, that Californian populations of the species have intrigued travellers ever since the explorer Sebastian Vizcaino visited the Monterey Peninsula in 1602. By way of contrast, what are now known as the two Mexican populations of *P. radiata* do not appear in any detail in surviving recorded history until 1875 and 1888.

The first European record of a specimen collection from mainland California dates back to the La Perouse Expedition of the 1780s. Reference was made to a seed collection of what was then called a *pine-apple fir*, believed

Ecology and Biogeography of Pinus. ed. D.M. Richardson. © Cambridge University Press, Cambridge (1998). pp. 432–49.

to be what we now know as *P. radiata* var. *radiata*; the expedition 'landed on the shores of Monterey Bay in 1787 and specimens collected were preserved under the name of *Pinus californiana*'. Thomas Coulter took a specimen collection in 1829 or 1830 – historical accounts are inconsistent. The noted Scottish explorer David Douglas (of Douglas-fir fame) also collected specimens and seeds at Monterey very shortly after Coulter (Douglas 1914). Both collections went to England, and it was the latter that was the source of the first recorded successful cultivation of the species as an exotic. It was the former collection by Coulter that provided material for the first adequate botanical description, by D. Don, under the scientific name we recognize today. It was on 2 June 1835 that Professor David Don of Kings College described the species, in this case verbally to the Linnean Society, but this description was subsequently supported two years later in transcripts of the society (section 21.3.1).

With regard to origins in cultivation, it is known that the 1787 seed collection from the La Perouse Expedition was forwarded to Paris where 12 plants were germinated, but apparently all died. The David Douglas collection reached Kew Gardens in England in 1833, and was the original source of material from which various commercial nurserymen in England and elsewhere in Europe successfully propagated the species, in some cases by rooting of cuttings as early as 1839, in spite of many plant deaths in severe winters.

We will never be able to reconstruct with certainty how the species first entered the southern hemisphere. The historical record in New Zealand and Australia has been researched in detail (e.g. Shepherd 1990) and indicates early importations from both California and England. Evidence supports the belief that radiata pine was in southern Africa by 1850, and therefore probably slightly before Australia and New Zealand, and well before Chile where it arrived in 1886. The first evidence of any utilization of radiata pine in these countries dates to 1876, when it was used for housing timbers on a country estate in Canterbury, New Zealand. Commercial-scale schemes of plantation development with the species were first attempted during the 1870s in South Australia and in New Zealand, in 1885 in South Africa, and in 1893 in Chile.

In a geological time frame, *P. radiata* was being rapidly squeezed by changing climatic patterns, apparently towards extinction, on its home continent in the northern hemisphere. That was before human hand intervened, plucking the species and transposing it elsewhere. Subsequently it has adopted an environmental niche in the southern hemisphere that, in contrast, is geographically substantial. In its relic homelands it displayed thoroughly unpretentious credentials. However, as an exotic, this premier softwood species has established itself as a

producer of wood of outstanding utility. But maybe the story of radiata pine is yet in its early days.

21.2.2 Forward perspectives

The story of *Pinus radiata* to date is largely a phenomenon of the southern hemisphere. Although the species has cause to be recognized as the most outstanding expatriate of the many conifers of North America, it tends to be perceived as such only in the southern hemisphere. Radiata pine has not been a well-recognized species in much of the northern hemisphere. From a resource production perspective, this is understandable since Australia, Chile and New Zealand between them are growing over 88% of the plantation resource of radiata pine, with most of this confined within the narrow latitudinal belt of 35–39° S.

The historical background to this concentration of resource is complex, but a predisposing factor has been the availability in the southern land masses of sufficient land-base in the appropriate climatic niches to develop the species on a major commercial scale. In equivalent northern latitudes the appropriate climatic niches appear to be either quite restricted or unavailable for this land-use. The radiata story can be seen as a phenomenon that is both remarkable and instructive. However, as discussed at the end of this chapter, we suggest that while the success story has not yet reached its culmination, it is one that should not be oversold by undue extrapolation.

21.3 Taxonomy and nomenclature

21.3.1 Californian populations

While there is evidence that specimens of the species that we know now as radiata pine were collected as early as the 1780s, it appears that the species was not adequately described before the 1830s. In *Transactions of the Linnean Society of London* published in 1837, David Don described *Pinus radiata* specimens collected by Thomas Coulter in 1829 from near sea level at Monterey. It appears that the specific epithet nominated by Don ('radiata') refers to the characteristic cracks radiating from the umbo of the cone scales. In 1844 another description of specimens – also of the Monterey population – collected by David Douglas some years previously was published as *Pinus insignis* in the *Arboretum and Fruticetum Britannicum of London*. This description by Douglas is recognized as a more exact and detailed description.

During the following half century many collections and descriptions were made from different parts of the mainland Californian populations, and the names of Hartweg, Gordon and Lemmon amongst other botanists and taxonomists arise in connection with this era. Some of the described names that have been recorded in connection with elements of the mainland populations include *Pinus adunea, P. californiana, P. insignis macrocarpa, P. insignis* var. *laevigata, P. insignis* var. *radiata, P. montereyensis,* and *P. tuberculata.*

All the Californian populations referred to above have for most of the 20th century been satisfactorily accepted as being within the one species. For more than half of the 20th century, though, the species name attributed to Douglas, *Pinus insignis* Douglas ex Laud., was accorded widespread currency. Taxonomic reviewers of the genus accept the precedence of *Pinus radiata* D. Don on the basis of the date of publication of a satisfactory description, and in recent decades *P. insignis* has fallen into disuse, even in Spain where *insigne* is traditionally part of the common name. We can conclude that there is currently no dispute about the taxonomic classification of the mainland populations of *P. radiata.* These populations are now acceptably called *P. radiata* var. *radiata* though generally only when it is necessary to differentiate this population from those in Mexico. The use of the name *P. radiata* var. *macrocarpa* for the Cambria provenance in California has some advocates.

The *island pines* of Santa Rosa Island and Santa Cruz Island – the Channel Islands of California, not to be confused with two Mexican islands discussed below – are not forms of *P. radiata,* but are part of what is called the *muricata* complex; they are often given separate species status as *Pinus remorata* or at least varietal status as *P. muricata* var. *remorata* (Chap. 2, this volume).

21.3.2 Mexican populations

Since their discovery in 1875, the pines of Guadalupe Island have always been recognized as closely related to the Californian populations of *P. radiata.* However, the pines of Cedros Island, first named in 1888, were for three-quarters of a century assumed to be more closely related to *P. muricata.* It was not until the third quarter of the 20th century that greater interest developed in clarifying the taxonomy and nomenclature of these two Mexican island pines.

For these two Mexican island populations, both of which are predominantly two-needled, species definition appeared to have been considered indeterminate or in dispute for many years. Successive reviews during the 1960s concluded that both were 'taxonomically difficult forms having characteristics in common with mainland *Pinus radiata*' and that 'the pines of Cedros and Guadalupe Islands are related to *Pinus radiata* and *Pinus muricata*'. It was only from the late 1970s that one could perceive adequate consensus to recognize the inclusion of both within *P. radiata,* as var. *binata* (Englm) Lemmon for the Guadalupe provenance and var. *cedrosensis* Howell for the Cedros

provenance. The main interest in and impetus for accurate determination and recognition of the status of the Mexican island populations came from radiata researchers and geneticists in Australia and New Zealand in conjunction with colleagues at the University of California.

Thus three varieties of *P. radiata* are now generally recognized, these being var. *radiata,* var. *binata,* and var. *cedrosensis,* covering five separate natural populations recognized as five provenances (Table 21.1).

21.3.3 Botanical description

The modern-day *Pinus radiata* belongs to the diploxylon or hard pines, which are characterized by a double vascular bundle in the needle and most commonly have two or three needles in each bundle, and to the subsection Attenuatae; see Chap. 2 (this volume) for discussion on the rationale for placing *P. radiata* in this subsection rather than Oocarpae as advocated by Millar (1986). The present *P. radiata* is a member of the coastal (or maritime) closed-cone pines association, other principal members of which are the 'muricata complex' and *P. attenuata.* The coastal closed-cone pines are centred on California, though *P. attenuata* also has an additional extensive range outside the coastal closed-cone pines community.

The coastal closed-cone pines as an association characteristically comprise small to medium-sized bushy trees which grow quickly, bear cones at an early age, and can act as pioneers on burnt or cleared ground, but are relatively short-lived. Another characteristic of the Insignes closed-cone pines group is their serotinous cone habit – the habit of retaining their cones on the tree for several years, with repeated periodic opening and closing of cone scales, thus releasing viable seed over a period of years (Chap. 12, this volume). Burdon (1992) gives a definitive current account of radiata pine morphology, including descriptions of general habit, bark, foliage, cones, seeds and general diagnostic characters.

21.3.4 Common names around the world

In Australia and New Zealand – in which some 54% of the area of global resource is located – the most widely used common name for *P. radiata* is *radiata pine,* and the same name is formalized there by the national standards for timber products. Earlier names that have largely departed from common usage in most of the southern hemisphere include *insignis pine, Monterey pine* and *remarkable pine.* In Spanish-speaking Chile the most commonly used name now is *pino radiata.* Often the pine or pino is dropped, leaving the common name as simply *radiata.*

As a commodity of international trade, the timber or other wood product of *P. radiata* – whether in milled or log form or otherwise processed – is almost globally known as

Country	Pinus Variety	Preferred population name	Other names used (and extent of populations)	Current area (ha)	Latitude (°N)	Altitude (m)	Soils	Annual rainfall[a] (mm)	Condition of, and threats to, the relic population
USA (California)	P. radiata var. radiata	Año Nuevo	Swanton Point Año Nuevo[b]	450	37	10–330	Fine loams derived from argillites	800	Protected in part by State Land reservation
		Monterey	Monterey Peninsula, Monterey–Carmel, including all outliers near Carmel	5000	36.5	10–440	Very varied fertility and base status	400	Urban encroachment, genetical contamination, selective utilization[c]
		Cambria	Cambria, including the Pico Creek outlier	200	35.5	10–200	Sandy loam with localized poor drainage	500	Urban encroachment
Mexico (Baja California)	P. radiata var. binata	Guadalupe	Guadalupe Island	<100 senescent trees	29	330–1200	Rocky loam on basalts	150	Almost extinct, largely through uncontrolled goat browsing
	P. radiata var. cedrosensis	Cedros	Cedros Island	130	28	290–640	Generally skeletal	150	Protected by virtue of remoteness, but fire-prone

Notes:
[a] The rainfall figures are averages based on historical data from nearby meteorological stations; they do not take account of the substantial moisture that is intercepted by trees from the heavy sea fogs that are features of the summer climate at all five localities.

[b] This population is located in the northern part of Santa Cruz County, which should not be confused with Santa Cruz Island.

[c] Threats at Monterey involve a variety of urban encroachment issues militating against natural regeneration, including community opposition to the use of prescribed burning, and invasion by alien plants (especially *Genista monspessulana*). The integrity of the wild gene pool is being compromised by contamination from widespread amenity plantings derived in part from seed of selectively bred genotypes from Australia and New Zealand.

Source: Data from various published and unpublished sources, notably Burdon (1992) and K.G. Eldridge (unpublished data)

Table 21.1. Salient features of the localities of the three varieties of *Pinus radiata* in five natural populations in California and Mexico provenances. The area of these five stands together currently comprises <0.2% of the world total (see text)

radiata (pine). However, *Monterey pine* is the common name for *P. radiata* (the tree species) given currency by the USDA Forest Service regardless of the population being referred to, and is the one invariably used by North American authors. In South Africa, *Monterey pine* is still a frequently used name though the increasing adoption of *radiata pine* is evident. In Spain, *pino insigne* and sometimes *pino de Monterrey* are traditional names for the tree species, as had once been the case in Spanish-speaking Chile before deliberate promotion and adoption of (*pino*) *radiata* as the common name for both the tree and its timber.

In this chapter we chose to use the common name radiata (pine) because, after considerable standardization over the last 30 years, this is what the species is called in regions of the world where it is now predominantly located.

21.4 The native resource and gene pool

21.4.1 Extent of the native populations

The extent of the five natural habitats straddling the US/Mexican border (Table 21.1) now represents <0.2% of the current world total. Because of the success of the species as an exotic, this small area of natural forests has commanded an interest – by the users if not the locals – out of proportion to its size.

The mainland populations are all within a degree and a half latitude, located within a narrow strip of central Californian coast about 10 × 210 km, between sea level and 440 m altitude. While the relatively remote Mexican populations are found at latitudes some 8–9° further south, they are sited at compensating elevations. The ecology and evolutionary history of the coastal closed-cone pine group of which radiata is a member have been researched through the University of California (e.g. Axelrod 1980, 1988).

21.4.2 Ecology of the natural habitat

The current climates of all three mainland localities, being quite similar and basically of a mediterranean type, are characterized by low to moderate and somewhat variable rainfall, with a strong seasonality in its distribution, being winter-biased. About 75% of the annual rainfall falls between December and March. Hard frosts, when experienced are also confined to the three midwinter months. Snow and hail are generally absent. Features of the summer climate include moderate temperatures and the presence of ameliorating summer fog, mitigating the summer drought to an extent that is affected by topographic position (e.g. Forde 1966; Moran, Bell & Eldridge 1988).

Fig. 21.1. **The main northern population of *Pinus radiata* var. *cedrosensis* on Cedros Island, photographed in 1964. The pines in the centre right are on the northern point of the island, with fog blowing up the windward slope through a low gap between the northernmost pines and the main northern population on the main ridge and on the NW slopes of the two interior ridges. The leeward slopes of these ridges are desert; the pines are restricted to sites where they can condense moisture from the summer fogs. The *P. radiata* population on Cedros Island, which currently covers about 130 ha, is the most distinct of the five populations of this species. Features of this population that make it attractive for incorporation into breeding populations for commercial forestry in the southern hemisphere include: the development of a taproot (improved drought resistance); the shorter height per unit diameter; and the promise of improved resistance to western gall rust (photo: W.J. Libby).**

Guadalupe is an oceanic island, with a native flora less rich than Cedros, a continental island. Fog has played an even more important role in the survival of the Mexican island populations of the species, as on both Guadalupe and Cedros Islands the annual rainfall is very low and erratic – apparently averaging only about 150 mm – and *P. radiata* is confined to high ridges, at 330–1200 m and 290–640 m, respectively (Fig. 21.1). Table 21.1 describes the physical environment of the five natural populations.

21.4.3. Comparative provenance performance

Provenance variations over the five natural populations of *P. radiata* were not well documented for most of the first century of exotic cultivation of the species. The differences in growth characteristics of each of the five native populations are now, belatedly, being closely studied. Systematic provenance trials derived from a range-wide 1978 collection were established on a range of site-types in both Australia and New Zealand. An early outcome of these trials was the emergence of the potential of the Guadalupe provenance to improve density and form characteristics, wood density in the first five growth rings being highest in

Fig. 21.2. ***Pinus radiata*** **var. *radiata* occurs at three isolated localities in California: Año Nuevo Point, Monterey and Cambria. Populations at Cambria cover only 200 ha, about 3.5% of the total area of this variety in its natural range, and are severely threatened by rapid urban development (photo: D.M. Richardson).**

Fig. 21.3. **The population of *Pinus radiata* var. *radiata* at Monterey covers about 5000 ha. Genetic integrity of this population is being affected by selective utilization and contamination from pollen and natural regeneration arising from the planting of *P. radiata* seeds from selectively-bred populations in Australia and New Zealand. Alien plants, notably various grasses (seen here beneath a stand at Point Lobos, Monterey) and the shrub *Genista monspessulana*, are also threatening these stands (photo: D.M. Richardson).**

Guadalupe. Other trends emerging from the *ex situ* provenance performance trials include slower growth, frost tenderness and a tendency to boron deficiency in the Cedros population, and a greater susceptibility to the Dothistroma needle blight and leader die-back and persistence of juvenile characteristics in the Cambrian population. Tree form rated worst in Año Nuevo, and tree height performance varied between the island and mainland populations with the Guadalupe population being on average 9–15% shorter than the mainland populations.

Since the early 1970s, considerably more quantitative data on comparative provenance performance in exotic plantation culture have been assembled in the course of the various radiata breeding programmes in Australia and in the centralized national breeding programme in New Zealand. However, a feature of the species is that differences at the population level of morphological and other characteristics can be overshadowed by the occurrence of unusually wide tree-to-tree variations within each population.

21.4.4 Gene pool conservation

The five modern-day native populations of *P. radiata* are living relics of the coastal closed-cone pine forest association which was once more widespread along the western coast of the North American continent (Coffman (1995) provides a succinct review). The viability or genetic integrity of the relic natural occurrences of *P. radiata*, many of which are on freehold properties, is likely to be considerably threatened over the next 50–100 years (e.g. Forde 1966; Moran *et al.* 1988; Burdon 1992). This is due in particular to rapid urban development at Monterey and Cambria and grazing by feral goats on Guadalupe Island (Figs. 21.2, 21.3).

The Guadalupe Island population, var. *binata*, in which

tree breeders are taking increasing interest, is in fact irretrievably on the point of extinction *in situ*. Fewer than a hundred very senescent trees were barely surviving by the mid-1990s on this remote and uninhabited island in circumstances of high wind exposure and total absence of regeneration due to feral goats. The situation is not as desperate with the other four populations. However, for some mainland stands (particularly at Monterey), the genetic integrity is being affected by selective utilization, but also by genetic contamination in the form of both pollen and natural regeneration arising from planted *P. radiata* known to be derived from imported seed which has reintroduced selectively-bred genotypes from Australia and New Zealand.

The native gene pool resource of *P. radiata* needs to be conserved if options are to be kept open for further tree breeding selections by exotic growers. The global significance of the species is not always fully comprehended in the mother countries, and prospects for full and long-term *in situ* conservation have proved limited. With this in mind, many collections and other endeavours by tree breeders, mainly from Australia and New Zealand, have been made over the years to ensure gene pool conservation *ex situ* at least. The most comprehensive range-wide collection of substantial quantities of seed across all populations was made in 1978 (Eldridge 1978; Moran *et al.* 1988). In 1990, with no prospect for conservation *in situ* on Guadalupe Island, a final cone collection was made from the remnant trees of var. *binata*.

The 1978 collection, initiated and coordinated by the Commonwealth Scientific and Industrial Research Organization (CSIRO) in Australia, was carried out as a joint operation which involved, *inter alia*, the New Zealand

Fig. 21.4. **Obvious differences in growth among families in a *Pinus radiata* progeny test near Busselton, Western Australia. The trees were planted at the same time but the three trees in the row in the right foreground were all from a selected seed-tree and the three in the row in the left foreground were from a routine, unselected tree (photo: F.T. Ledig).**

Forest Research Institute (FRI), the University of California, the US Forest Service, and a number of other grower-country interests under the umbrella of the Food and Agriculture Organization (FAO) of the United Nations. Seed distribution from some 20 000 cones collected was sent to Australia, Chile, New Zealand and 21 other countries, with a large balance maintained in cold store in Canberra, Australia (e.g. Moran *et al.* 1988).

21.4.5 Current trends in tree breeding

Tree breeding of *P. radiata*, which began in the 1950s, has resulted in substantial genetic improvement. For example, in New Zealand gains in stem volume of about 30% are being achieved with some selected seed lines (Burdon 1992). Improvement in stem form has also been great, and growers can now expect up to double the number of acceptable stems from the best breeds. Because of these improvements, the number of trees being planted per hectare has halved in the past 20 years, with many growers being happy to plant 800 stems per hectare or even fewer. In Australia, improved *P. radiata* has yielded up to almost a third more volume, and better straightness and branching characteristics (see Chap. 13, this volume; Fig. 21.4).

Current breeding programmes in Australia, Chile and New Zealand stress the importance of developing strategies to meet specific goals, while simultaneously maintaining genetic defences. With the advent of advanced breeding methods coupled with techniques such as tissue culture there is growing recognition that there are substantial advantages to be gained by tailoring breeds for specific sites or specific end uses, as well as providing a good general-purpose breed.

There are three main components to strategies being employed:

- the organization of populations, which is the physical component;
- the breeding methodology, consisting of non-physical design aspects
- the research component, being a development programme feeding back into both of the other two components.

A hierarchy of populations is often recognized (e.g. Shelbourne, Carson & Wilcox 1989), the base level being the gene resource which encompasses most of the genetic variability within radiata pine. This gene conservation population is managed and regenerated to ensure that the widest possible genetic base is maintained. The base level is seen as a source for further genetic material and may also be viewed as part of the genetic defences against contingencies such as disease. The breeding population is drawn from this base population and is likely to have 300–500 parents at any one time. There may be multiple breeding populations or subpopulations with different objectives. Thus New Zealand has, besides its main general purpose breed, a long-internode breed which is based on selection in a different breeding population to meet the goal of logs with a large proportion of short-length clearwood, and other breeds are also available (e.g. Burdon 1992).

Some *P. radiata* breeding programmes divide the breeding population into sublines (e.g. Cotterill 1984) which represent replicate breeding populations that can be crossed to ensure completely outbred offspring. At the top of this hierarchy is the seed- or plant-producing population. These have been selected from the breeding population to form the basis of the seed orchard or other technique for producing improved planting stock – the so-called delivery system. For a particular breed this would usually include about 20–50 clones (Shelbourne *et al.* 1989). However, in a few instances specific crosses are being produced to be multiplied up and planted as clonal material.

The breeding methods most applicable to a cross-pollinating tree species like radiata pine are mass selection and recurrent selection, usually based on general combining ability. However, there is growing interest in recurrent selection based on specific combining ability which is obtained when specific crosses are employed.

An integral part of recurrent selection methods are progeny tests, which are particularly important for traits with low heritability. Such tests have been planted extensively in Australia and New Zealand since the 1960s and many current seed orchards are based on their results (Shelbourne *et al.* 1986). This method of selecting seed orchard plants from material proven in progeny tests is known as backward selection and is currently the main method being used for *P. radiata* by tree breeders.

The problem of how best to incorporate more than one trait in a particular breed is generally solved by using selection indices. Furthermore, developing a range of specific breeds helps to overcome some of the conflicts that arise due to negative correlations between traits and where there are different emphases on desired traits.

The radiata tree breeder has several types of orchards available for producing seed from which seedlings or other planting stock may be produced; they form an integral part of the overall strategy. The two most common methods being employed are open-pollinated clonal seed orchards and control-pollinated clonal seed orchards. The latter are gaining favour (Shelbourne *et al.* 1989) because, as well as giving greater genetic gain, they give greater flexibility for specific breeds or specific crosses, and they integrate into the developing planting stock technologies based around cuttings, tissue culture and embryogenesis.

A recent development in control-pollinated orchards in New Zealand is the 'meadow orchard'. This involves planting selected radiata pine grafts at a stocking of about 5000 stems per hectare. Such an approach (Sweet, Bolton & Litchwark 1990) reduces the time from planting to obtaining the improved seed while insuring a high productivity per hectare; it also increases the rate and flexibility at which new selections may be incorporated. Recent research holds promise of being able to enhance flowering and seed set and in providing simpler and improved ways of controlled pollination (Sweet *et al.* 1992). Radiata pine breeding is likely to change further with the current emphasis on biotechnology research, including methods of propagation, gene mapping and incorporation of desired genes.

In New Zealand the genetic quality of radiata pine seed is quoted when it is sold (Burdon 1992). Differences in genetic quality are defined by breed and an index giving an estimate of improvement. The four current breeds are the general breed for improved growth and form (GF), and the more specialized breeds emphasizing long internodes, resistance to red band needle blight, and high wood density. The number associated with the GF breed roughly reflects the improvement managers should expect and is based on a weight in growth and form of approximately 2:1. Thus unimproved seed taken from unimproved stands has a rating of zero, while that presently in production from open-pollinated orchards range from about 14 to 19. The highest GF ratings are between 25 and the low 30s and come from control-pollinated seed orchards. There is a considerable premium for the higher rated seed. We believe that most growers will adopt similar schemes to describe their seed qualities.

Table 21.2. The global estate of *Pinus radiata*

Country	Area (x 1000 ha)	Percentage of the global estate
Ex situ plantations		
New Zealand	1430	35.7
Chile	1395	34.8
Australia	745	18.6
Spain	245	6.1
South Africa	55	1.4
Minor growers		
Ecuador, Argentina, Italy	90	2.2
All other countries	50	1.2
Total	4010	
In situ stands		
USA (California)	6.5	<0.2
Mexico (Cedros & Guadalupe Islands)	0.15	<0.1

21.5 The *ex situ* plantation resource

21.5.1 Global overview

Over 35% of the area of the world's radiata pine plantations are in New Zealand where the species embraces the full latitudinal range from the top of North Island (*c.* 34° S) to the bottom of South Island (*c.* 47° S). In Australia, the resource is principally located in selected higher rainfall zones in the southern continental States and in Tasmania. Apart from these English-speaking Antipodes, a considerable proportion of the global plantation resource of radiata pine is to be found in the Spanish-speaking world. Southern-central Chile and northern Spain together account for over 40% of the world total (and for several years Chile's total briefly exceeded New Zealand's).

A century and a half ago, before substantial human intervention, the global estate of *P. radiata*, all in natural stands, was not more than 12 000 ha. As an exotic cultivated plantation species, radiata pine occupies about six hundred times the current area of the five relic native populations, with our estimate (Table 21.2) of the 1996 plantation resource worldwide being just over 4 million hectares.

Over 90% of the present global resource occurs within four southern hemisphere countries, between latitudes 33.5° and 46.5° S, with the substantial majority between 35° and 39° S. The radiata plantation resource of the world is dominated more than ever before by the Pacific Rim group of Australia, Chile and New Zealand, each having expanded its resource substantially in recent decades, particularly since the early 1960s in Australia, the mid- to late 1960s in New Zealand, and the mid-1970s in Chile. This highlights why any discussion about the current global wood resource of plantation-grown *P. radiata* must inevitably centre on consideration of strategic marketing

Table 21.3. The New Zealand plantation estate of *Pinus radiata*

Wood supply region	Plantation area (x 1000 ha)
Northland	150
Auckland	85
Central North Island	545
East Coast	100
Hawke's Bay	85
Southern North Island	85
Nelson & Marlborough	155
West Coast	30
Canterbury	70
Otago & Southland	125
Total	1430

Table 21.4. The Chilean plantation estate of *Pinus radiata*

Region of Government Administration	Plantation area (x 1000 ha)
Region VI	75
Region VII	320
Region VIII	630
Region IX	220
Region X	125
Other Regions	25
Total	1395

Table 21.5. The Australian plantation estate of *Pinus radiata*

State	Plantation area (X 1000 ha)	Percentage managed by the state
New South Wales	259	74
Victoria	217	50
South Australia	109	72
Tasmania	79	63
Western Australia	64	77
Australian Capital Territory	14	100
Queensland	3	71
Total	745	

issues as they individually and collectively relate to just three relatively small southern hemisphere economies of these three major grower countries.

21.5.2 Statistics – major grower countries

Within each of the three major grower countries, national statistics relating to the radiata plantation sector are available and provide what we consider in most cases to be sector data of relatively high integrity.

New Zealand plantation statistics are published annually by the Ministry of Forestry as the 'National Exotic Forest Description' (NEFD), which details net stocked area and standing wood volumes by species, age class, and by regions defined for wood supply planning. Guided by NEFD figures which describe an estimated 92.7% of the New Zealand total plantation area, our projections of the New Zealand radiata estate at the beginning of 1996 are described in Table 21.3. A marked upturn since 1992 in the rate of new plantings, a high proportion by non-corporate landholders, propelled the national total area back to parity with the Chilean radiata total by 1995, after being second to Chile in size of national estate for most of the preceding decade.

NEFD also describes the national plantation resource according to forest management practices utilized, and currently indicates the following proportions by area:

Minimum tended without production thinning	36%
Minimum tended with production thinning	7%
Intensively tended without production thinning	38%
Intensively tended with production thinning	19%

The **Chilean** statistics are published annually as 'Estadisticas Forestales' by the Corporacion de Fomento de la Produccion. Based largely on this source, our projections of the current size and distribution of the Chilean radiata estate at the beginning of 1996 are shown in Table 21.4. The very high rates of new radiata plantings seen in Chile from the mid-1970s began to fall away in the early 1990s as the rate of eucalypt plantings increased.

The **Australian** plantation statistics are assembled by the individual State forest agencies, but are collated annually from the State agencies by the Australian Bureau of Agricultural and Resource Economics (ABARE). Extrapolating from these sources we estimate the size and distribution of the radiata estate at the beginning of 1996 as described in Table 21.5.

21.5.3 Major grower country comparisons

Detailed accounts of the radiata sectors of each of Chile, New Zealand, and Australia are beyond the scope of this chapter. Authoritative accounts are provided by Lewis & Ferguson (1993). However, some discussion of comparisons is warranted.

The total Australian estate of all plantation species contains a considerably higher proportion than in Chile and New Zealand of conifers other than *P. radiata*. The Australian plantation sector has also tended to be dominated by State Government agencies as majority owner/managers, whereas the public sector proportion in New Zealand is low and in Chile is negligible, following earlier privatization programmes in these two countries.

An examination of the age-class distributions of the respective national estates highlights differing levels of maturity between the radiata resource of these three grower countries. The capacity to produce in the short to medium term is very much a function of the distribution of the older age classes. While the Australian plantation resource has not yet reached a sustainable steady state of maturity, its relatively greater maturity than the New Zealand or the Chilean resource has been in evidence in recent decades.

This resource maturity is relevant not only to current volume production, but also to quality out-turns from industry as wood quality considerations related to cambial age come into play. Of particular relevance in the Australian domestic market is that out-turns for the sawmill industry producing structural timbers will be markedly superior – such as in proportions of the higher stress grades – where the resource base allows a high proportion of log input to be from stands beyond 30 years of age.

Maturity aside, by whatever measure there is undoubtedly a greater proportion of stands of very high productivity in New Zealand than in Chile or Australia. Probably more than 85% of existing radiata stands in New Zealand may be categorized as *fast-growing* whereas in Chile the current figure may be less than 75%, with the Australian figure being closer to that of Chile. In its favour, Chile undoubtedly has the highest proportion of total potential volume being captured as harvested volume.

As for radiata in some regions in Australia (but not in much of North Island New Zealand), radiata in most of Chile experiences growth conditions that can give rise to trees of quite excellent stem form and relatively fine branching. This is so over a broad range of sites, not just on deep sands of modest fertility status. Thus in the absence of pruning – or above the pruned buttlog – Chile potentially has competitive advantage in the global marketplace for solid radiata wood, particularly for structural grades, by being able to grow a higher proportion of radiata with acceptably small knots, albeit more bark-encased than intergrown knots.

Overall, contrasts in management of the respective national plantation resources may be summarized as follows, at the risk of oversimplification because of regional differences within the countries involved. In New Zealand there has been sharp focus on clearwood production and the emerging premium value of pruned buttlogs from highly productive sites. Silviculturally 'direct' tending regimes involve early waste thinnings and multiple-lift high pruning (Fig. 21.5). Log exporting is also important to New Zealand growers and forces export parity pricing on the domestic sawmilling industry. The Chilean plantation sector is very export-orientated, and in the past has relied on high levels of pulpwood utilization though there is a gradually increasing emphasis on clearwood and quality solid

Fig. 21.5. **A 12-year-old stand of *Pinus radiata* in New Zealand (Tasman Forestry Limited) that has had its second (final) thinning and is high pruned. Over half of radiata pine in New Zealand is pruned to at least 4 m (photo: D.J. Mead).**

wood production. In Australia there has traditionally been a greater emphasis on production of structural grade sawlogs to augment and gradually replace native forest hardwood production, and greater reliance has been placed on the domestic markets to absorb chip/pulp grade logs and thus to facilitate multiple thinning regimes (Fig. 21.6).

In Chile the export market has always been closely linked to the plantation grower sector, particularly in contrast to the case of Australia where the wood products sector has historically been focused on the domestic market. The New Zealand plantation sector is now vigorously targeting various Pacific Rim markets for higher value sawlogs. In Chile, by comparison, there are signs that requirements for export cash flow, generated more from the pulping sector than the solid wood sector, can tend to compromise rotation length, with ramifications regarding future pulplog/sawlog proportions and total production capability in the medium to longer term.

Fig. 21.6. **A 34-year-old stand of *Pinus radiata* at Sunny Corner State Forest, New South Wales, Australia, managed with multiple-production thinnings – a traditional practice in many large plantations in Australia (photo: D.J. Mead).**

21.5.4 Experiences in other countries

Spain

Of the other radiata grower countries, Spain and South Africa represent a proven viable resource base which is, nonetheless, not expanding. *Pinus radiata* was introduced to Spain from France in about 1840, well before the species had arrived in any southern hemisphere country. Plantations of radiata are largely confined to the moderate northern coastal environments fronting the Bay of Biscay in País Vasco, the Spanish Basque Region, between latitudes 40 and 44° N where the ameliorating influence of the Gulf Stream is felt.

The immediate and most striking impression about the radiata sector in País Vasco is the all-pervasive influence of the pattern of small and tightly-held family holdings of rural land. Most of the radiata plantation resource of País Vasco consists of many thousands of individual private family holdings averaging <6 ha, and often the individual holding consists of three or more separated parcels of titled land with very irregular boundaries and located on upper slope terrain. Every aspect of commercial plantation forestry in País Vasco is influenced by constraints imposed by this fragmented pattern of land ownership.

One immediate implication of this is that at a regional level many aspects of forest management planning by agencies of government can be quite irrelevant. For example, decisions about scheduling of harvesting are made quite independently and for different reasons by the many owners of individual small plantations, and family considerations can often override purely economic (optimizing) criteria. In this respect there are parallels with production forestry in much of Japan.

Radiata in País Vasco has had to establish a coexistence with the whole complement of pests and diseases that are an intrinsic component of the European conifer belt and of Old World pines in particular. While it is at times an uneasy coexistence, radiata can now be considered to be in a fluctuating equilibrium with, though clearly constrained by, the contingent of pests and diseases of the Old World pines. Other unique features of the Basque radiata sector are described by Lavery (1993). Spain's radiata resource in the País Vasco suffered substantial fire losses in December 1989 and nationally the radiata land-base in Spain appears now to be slightly below the quarter-million hectare mark.

South Africa

Pinus radiata was once used extensively for afforestation in the summer-rainfall areas of South Africa and Zimbabwe, and in most other southern African countries *P. radiata* has at least been tried as a potential afforestation species. However, it is now little planted in these regions because of the dangers of fungal attack (Chap. 19, this volume). Commercial-scale forestry with the species is now limited to the winter- and uniform-rainfall areas of the former Cape Province in South Africa (Fig. 21.7).

Historically, *P. radiata* had apparently reached the area of the present Republic of South Africa by 1850, and it subsequently displayed exceptional vigour. However, within the present-day South Africa the species was little used for afforestation before Union in 1910, though during the decade to follow it was planted extensively throughout the more humid parts of the country. The programme was then drastically curtailed as high susceptibility to *Sphaeropsis sapinea* (formerly *Diplodia pinea*) became apparent. Typically the most destructive epidemics of this fungal pathogen occurred after the soft young stem tissues had been wounded by hail, a form of injury to which the species is susceptible because of its relatively thin young bark. (A prerequisite of destructive infection is

Fig. 21.7. *Pinus radiata* is the most widely planted exotic conifer in the world, with a total plantation area of just over 4 × 10⁶ ha. More than 90% of this resource resides in southern hemisphere between latitudes 33.5 and 46.5° S. The picture shows a first-rotation plantation of *P. radiata* near Stellenbosch, Western Cape, South Africa (photo: D.M. Richardson).

that the hail damage be followed by a spell of warm humid weather.)

Within the uniform-rainfall area too, policy on use of *P. radiata* has had a chequered history, again because of misgivings about potential for *Sphaeropsis* damage. However, much of the historical damage in the uniform-rainfall zone has been on sites with unfavourable edaphic conditions, and as soil requirements for the species came to be better understood, afforestation in the uniform-rainfall zone was successfully resumed.

In the winter-rainfall zone of the Western Cape, *P. radiata* presents a particularly healthy appearance when established on suitable soil types. Although defoliation by larvae of the pine emperor moth *Imbrasia cytherea* is an ever-present threat, *P. radiata* is the most commonly planted species for timber production in the Western Cape where it has remained virtually free from fungal diseases (Chap. 20, this volume). Yet even in these appropriate climatic zones, South African experience is that the species is still considered somewhat exacting in its edaphic requirements, possibly displaying less site-plasticity than at the more southerly latitudes at which it is planted on the other continents of the southern hemisphere.

In South Africa, the clear requirement is for 'a moderately fertile, deep, well-drained soil amply supplied with organic matter for its optimum development' (Poynton 1979). The plantation area under *P. radiata* has been reduced in recent decades to near 55 000 ha, with the State as the majority owner. This is now largely concentrated in the Western Cape and Eastern Cape provinces, particularly in the former which has a more typically mediterranean-type climate with a winter peak in rainfall, although plantations extend into the uniform-rainfall zones of the Eastern Cape.

Limitations to plantation development elsewhere

In Zimbabwe, *P. radiata* was planted on a commercial scale in the Eastern (Rhodesian) Highlands from 1928 until 1934. The programme was subsequently abandoned as widespread needle cast and die-back was quite devastating by the mid-1930s. Several pathogens were involved, but *Mycosphaerella pini*, which causes the red band needle blight (Chap. 19, this volume) is now recognized as the major one. From a utilization point of view, the species was so favoured that a substantial research effort, commencing in 1958, was directed towards development of pathogen-resistant planting stock (e.g. Barnes 1970). Besides embarking on a breeding programme seeking pathogen-resistant genotypes, considerable experimentation was pursued with vegetative propagation techniques using sufficiently *matured* ortet material to be beyond the worst of the susceptibility to *M. pini* attack. Nonetheless the prospects of *P. radiata* ever regaining a prominent place in forestry in the country appear remote.

In Kenya between 1945 and the early 1960s, an estate of possibly 15 000 ha of *P. radiata* out of some 60 000 ha of exotic conifers was successfully established. *Mycosphaerella pini* was first in evidence in Kenya in the early 1960s, crossing from Tanzania (then Tanganyika), and this needle blight has been the main cause of the demise of *P. radiata* as a species for further afforestation in Kenya and other central African countries.

In North America, various planting ventures with *P. radiata* in east Texas and the south-east of the United States were complete failures because of summer rain, *damp heat*, and associated attacks by various fungal and insect agencies which the southern and yellow pines group can readily tolerate. On the west coast of the USA, *P. radiata* is popular for amenity planting well beyond the immediate zone of its natural occurrence in the Californian fog belt, and the species has been planted as far south as San Diego in California and as far north as British Columbia. However the experiences of 'exotic' plantings of *P. radiata* in some west coast USA areas provide other examples of failure due to frosts, midsummer heat, disease and other pests. The coastal counties of central California, where hard frosts are confined to the three midwinter months, have been identified as the most generally suitable sites, and radiata pine in this region is the most planted tree for amenity purposes and is the premium Christmas tree. Nonetheless, even in the favoured sections of coastal California, dwarf mistletoe, the pine pitch canker (*Fusarium subglutinans* f. sp. *pini*), red band needle blight, pine gall rust (*Peridermium harknessii*), various bark beetles, and other native pests and diseases can cause considerable damage. Low-temperature extremes, as distinct from more predictable average lows, are particularly instrumental in limiting the species on the western USA seaboard and near interior, as evidenced by the December 1972 cold spell in Oregon described below.

In South America, other Spanish-speaking countries apart from Chile that make some contribution to the global radiata plantation resource include Argentina, Ecuador, Colombia and Peru. However, much of this resource is of dubious commercial viability; in fact it should be considered non-renewable because of its susceptibility to fungal pathogens as a result of what are generally termed *off-site* syndromes. The European pine shoot moth *Rhyacionia buoliana* has posed a particular problem to *P. radiata* in Argentina, and this insect has moved across into southern Chile where it now has serious pest status (Chap. 18, this volume).

Fungal problems, often recorded as *Diplodia* damage (attributed then to *Diplodia pinea*, now *Sphaeropsis sapinea*) but no doubt involving other additional agents including the red band needle blight, have regularly thwarted attempts at commercial afforestation with *P. radiata* in other subtropical and tropical environments in South America and Africa, even at very high elevations that compensate for latitude in terms of a cooler temperature regime. The history of the high-altitude plantings of radiata in Ecuador provides a not atypical case study of a 'honeymoon' period before the pathogen buildup which experience elsewhere suggests is inevitable.

In both Britain and France there has been considerable interest in the potential of *P. radiata* as a plantation species, but typical off-site syndromes including the foliage condition known as 'the yellows' in southern England have prevented plantation development at a viable commercial scale. French endeavours have concentrated on the mild southwest coastal districts adjoining the Spanish País Vasco. Even here, though, radiata is clearly on the limits of its frost tolerance in spite of genetic selections for frost hardiness. Other countries which have or have had a specific interest in *P. radiata* as a potential plantation exotic include Bolivia, Greece, Ireland, Israel, Italy, Morocco, Pakistan, Portugal, Tanzania, Tunisia, Turkey and Venezuela.

Pinus radiata has spread from plantations to invade natural vegetation to varying degrees in some countries to which it has been introduced, notably Australia, New Zealand and South Africa (Chap. 22, this volume).

21.5.5 Environmental limitations and climatic niche

General experience in relation to climatic tolerances of radiata pine in the various major production regions of the world is summarized in Lewis & Ferguson (1993). The following sections touch on both climatic and other environmental limitations (soils, pests, diseases), and we suggest that, by their very nature, off-site syndromes can often represent a response to several interacting factors rather than to an individual factor acting in isolation. Comment is also provided in relation to modelling of the climatic niche.

Edaphic limitations

Radiata pine has been found to be adaptable to a wide range of soils, from deep sands through to heavier, less well-drained clays or podzols. It prefers deep, well-drained soils with good moisture retention and adequate nutrients, with some of the best growth being found on fertile ex-pasture sites and on volcanic soils.

However, many sites have suboptimal soils because of nutrient deficiencies or poor rooting characteristics, or are either waterlogged or very droughty. Research has provided a range of silvicultural techniques to improve radiata growth on such soils. These include the application of fertilizers, the application of legumes to improve nitrogen status, sub-soiling to improve effective soil depth, mounding to improve drainage, and slash retention and weed control techniques to improve the moisture status on droughty sites. In recent years there has been a greater emphasis on the management of soils for long-term sustainability. This has included a reduction in the use of fire, improved site preparation and harvesting techniques, oversowing with legumes and the judicious use of fertilizers. The prescriptive use of fertilizers for radiata plantations is addressed in Mead (1995).

Pests and diseases

As indicated elsewhere in this chapter, the principal pests and diseases associated with radiata pine in plantations of the southern hemisphere include the red band needle blight and other needle cast fungi, the *Sphaeropsis* (Diplodia) diseases, the *Sirex* wood wasp, various bark beetles especially of the *Ips*, *Hylastes* and *Hylurgus* genera, and the pine shoot moth. In northern Spain the processionary caterpillar *Thaumetopoea pityocampa* can also be particularly destructive, and in California the pests and diseases of main concern include pine pitch canker, red band needle blight, pine gall rust and the bark beetles. Some of these are described elsewhere in this book (Chapters 18 and 19, this volume).

Humidity limitations and 'damp heat'

High humidity in the warmer months is not conducive to the long-term viability of *P. radiata*. Africa and South America have provided some dramatic examples of pathogen problems causing the loss of viability of *P. radiata* as a commercial exotic species for plantation purposes. In tropical and subtropical latitudes, even at very high elevations, there must be major reservations about the long-term viability of the species.

There is a arguably a certain inevitability about an exotic species gradually picking up disease loads (Chap. 19, this volume). For an introduced tree species, many factors militate against both the practicability and the economics of chemical methods of pathogen control, at least in the long-term, including the long term nature of the cropping cycle in comparison with introduced agricultural and horticultural cropping, and even in comparison with tree fruit and tree nut cropping. The routine aerial application of copper-based sprays to control red band needle blight in radiata plantations in parts of New Zealand and Australia is one notable exception in this regard.

Summer-season limitations: drought and heat

In its native Californian environment, the balance between *P. radiata* forests and oak woodlands and chaparral is maintained by the subtle interaction of fire and fog. Fires allow the pines to seed new forests, but whether those forests survive or give way to oaks depends on how much additional moisture the pines can capture from the summer sea fogs. This in turn is related to topographic position. Beyond the fog-affected zone, summer drought is too severe to allow *P. radiata* to survive (e.g. Moran *et al.* 1988).

Guidelines for site selection for commercial afforestation with *P. radiata* usually include a minimum annual rainfall in the range 600–750 mm. The mediterranean-type climate to which the species is well suited incorporates a relatively dry summer. However, where summer drought or soil moisture deficit is too severe, *P. radiata* will not be successfully established or will be prone to repeated dead topping. In the southwest of the State of Western Australia, for example, deaths to the species through summer droughting have often been unacceptably high, and this has in part been responsible for the adoption in that State of a wide spacing silviculture, often incorporating high pruning.

Direct heat effects are rarely experienced in an otherwise suitable environment for *P. radiata*, except for instances of sun scald, causing localized death of the cambial layer through the bark. Pruned stems of young radiata growing on highly reflective sandy soils are most at risk.

Winter and cold-climate limitations

The only plantations of *P. radiata* of any major commercial scale in the whole of the northern hemisphere are in north and northwest Spain. Extremes of low temperatures (rather than average minima) are a major reason for the species not finding a niche in any other northern hemisphere countries within the suitable latitudinal range with adequate rainfall. Both the absolute minimum and the timing of frosts are relevant.

Experience from a '50-year cold' in Oregon, USA, in December 1972 is particularly illuminating. According to anecdotal evidence from Oregon State University, no *P. radiata* trees in the State (where the species is grown as an ornamental and for the Christmas tree market) were known to have survived. The effects of the cold spell on trees were well documented around Corvallis, where meteorological data indicated the coldest mean temperature for a December since 1924 and the longest cold spell since 1930. During the cold spell the maximum daily temperature did not rise above freezing point for 11 consecutive days. The minimum daily temperature ranged from −6 to −22 °C, being below −18 °C for four days in succession. The *P. radiata* trees in this region, all of which were reportedly killed, were aged up to at least 30 years. Needles showed the first signs of death but all meristems were killed, including cambial death by freezing.

For most of Great Britain and France, it is the occurrence of untimely spring and autumn frost when the tree is in active growth that may rule out commercial plantation use of *P. radiata*. It appears that zones of suitable temperate mediterranean-type climates in the northern hemisphere are limited in extent and/or are heavily populated.

Part of the explanation for this intolerance of *P. radiata* to low temperatures (frosts) is related to the pattern of growth of the species. Because juvenile *P. radiata* does not form a true dormant bud like *Pseudotsuga menziesii* (Douglas-fir) or *Pinus contorta*, it can continue to grow throughout the year if the climate is suitable. In cold conditions growth tapers off and may stop if those conditions continue, but growth commences again if there is a warm spell. *Pinus radiata* is therefore not as tolerant to winter frost as conifers that become fully dormant throughout winter such as Douglas-fir, *P. contorta*, or even *P. muricata*, but these species have lower tolerance than *P. radiata* to unseasonal frost.

Climate modelling

Further studies in species climatology and refinement of local, regional and global models of climate may help to clarify the limits of suitability for commercial plantations. For *P. radiata*, refinements in these regards may explain the current global distribution which is so heavily weighted to the southern hemisphere. Methodologies and some interim outcomes described for example by Webb *et al.* (1984) and Booth (1990) indicate both the potential and the very substantial limitations in these regards. As an example of a computer-modelled description of climatic niche at a national scale, Fig. 21.8 portrays the indicative zones of suitability for *P. radiata* in Australia.

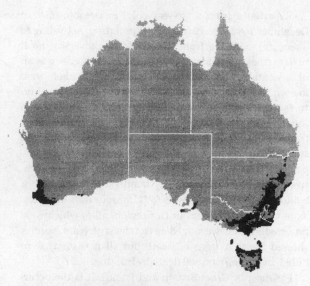

Fig. 21.8. **Indicative zone of climatic suitability for *Pinus radiata* in Australia. The map, from data provided by T.H. Booth (unpublished data), defines locations which satisfy the following limits: Mean annual rainfall: 650 to 2300 mm; Rainfall regime: exclude summer-biased; Dry season length: 0 to 5 months; Mean maximum temperature of the hottest month: 20 to 30° C; Mean minimum temperature of the coldest month: −2 to 12° C; Mean annual temperature: 10 to 18° C.**

21.6 The future for *Pinus radiata*

A measure of success of *P. radiata* as an exotic timber-producing species is that following its introduction to Chile, New Zealand and Australia, these three countries alone now have established many hundreds of times more area under the species than is present in its native habitat. Earlier in this chapter we suggested that the radiata story constitutes a remarkable phenomenon that may not yet have reached its culmination, but additionally we stress that the experience should not be subjected to undue extrapolation, as discussed below.

The future for radiata pine, both as a plantation conifer and as a softwood commodity, has yet to be fully charted. In no way can radiata pine as a plantation conifer be considered as having already *run its race*. Among other factors that impact on its future are the following.

- The baseline sustainable woodflows are not yet fully on stream because of a preponderance of young and mid-rotation age classes in the major grower countries – more pronounced in Chile and New Zealand than in Australia.
- Although *P. radiata* has limited long-term prospects beyond the boundaries of a certain climatic niche, it is nonetheless amenable and responsive to considerable cultural and genetic manipulation within this environmental niche, through intensive silviculture and site–genotype matching.
- Radiata as a wood-producing species is already one of the most intensively domesticated, and the pioneering R & D successfully accomplished by the major players in the southern hemisphere is largely transferable to other areas where radiata might be grown.
- The benefits of genetic gains and the application of intensive culture, particularly since the early 1970s in New Zealand and Australia and the early 1980s in Chile, will demonstrably boost baseline woodflows, but the recent benefits are still largely locked out of the current rotation cycles.
- There is still substantial scope to expand the already sizeable plantation land-base within its climatic niche in Chile, New Zealand and Australia if the economics of the land-use are supported by appropriate pricing levels on world markets for wood products.
- There must be some uncertainty that the potential of the species in the northern hemisphere has been fully explored and exploited: particularly, we suggest, in the eastern Mediterranean environments and possibly even within California.
- There is growing political interest in the establishment of plantations as a carbon sink (e.g. Maclaren (1996) for New Zealand), and some political analysts consider that an international 'C credit' market may evolve. This may provide a role for plantation forestry in which *P. radiata* can have a competitive advantage over other plantation species.

The biological and economic success to date of commercial plantation land-use with *P. radiata* in Chile, New Zealand and Australia leaves the species at risk of becoming 'oversold'. The risk is that the species may be pushed too far from the circumstances of the current success story. The countering strategy involves recognizing radiata's limitations and keeping within them. Two issues of limitations most likely to involve threat to the hard-won reputation of the species, and thus issues in which plantation foresters need to be acutely sensitive, concern the limitations to the range of product use on one hand, and the limitations to the range of biologically safe growth environment on the other.

21.6.1 Rationalizing success to date
Why has *P. radiata*, a tree species with rather unpretentious domestic credentials, become a premier species as an

exotic in temperate maritime environments in the southern hemisphere? Radiata pine, like all successful plantation species, is easily and cheaply propagated in bulk from seed. However, while this is an important convenience, it is more a predisposing factor than a driving reason, and is therefore almost incidental to its success. The underlying reasons for the success of *P. radiata* as a premium plantation species may to some degree be a matter of conjecture, but any attempt at explanation must involve a combination and an interaction of factors.

Successful plantation species are often those capable of relatively high-volume productivity on relatively short rotations, and this has invariably been a reason for early growers of cultivated tree crops favouring radiata pine. Although easily killed by fire, *P. radiata* seeds heavily after fire, and within its natural Californian range it colonizes disturbed or burnt ground with the aggression of a pioneering species that evolved with selection pressures for rapid early growth.

This aside, the scientific literature rationalizing the remarkable and widespread success of *P. radiata* as an exotic plantation species (especially in the southern hemisphere) is still limited in extent. In South Africa it has been concluded that

the main considerations which have led to the choice of *P. radiata* where suitable climate and soil conditions prevail are its ease of establishment, exceptionally rapid growth, amenability to cultivation in pure, even-aged stands and comparative freedom from serious diseases and pests.

Poynton (1979)

In Australia, the success of radiata pine has usually been ascribed to three particular factors: (1) the lack of pest and disease organisms when planted far beyond its natural range; (2) the advances made by foresters in developing silvicultural and pedological techniques appropriate to the species; and (3) the genetic variability and genetic plasticity of the species, as indicated by the fact that individual trees within plantations express different degrees of tolerance to such limiting factors as drought, nutrient deficiency and susceptibility to disease and pest organisms.

All three of these factors are very pertinent. We consider the first to be exceedingly relevant for the success of various Australian eucalypts as commercial exotics in other countries. It appears that the lack of a strongly mediterranean-type climate featuring four seasons can be quite deleterious to *P. radiata* in terms of the development of pathogen loads. However, it is the third point, referring to the surprising genetic diversity or genetic plasticity

within the species, which we suspect deserves greater emphasis when considering factors behind its outstanding success as a commercial species.

Besides the factor of genetic diversity, we believe that close consideration of the peculiarities of continental climates is required in any endeavour to rationalize the biological success of *P. radiata* in the southern hemisphere. What is not adequately emphasized in most of the literature is that the species has a history of being more *widely dispersed* (if never *widespread*) along western coastal North America and that, probably, it is only the peculiarities of intervening and present climates of North America and the Californian seaboard that have contracted its range almost to the point of elimination. Increasing aridity in recent postglacial millennia is recognized as particularly instrumental in the contraction and fragmentation of the closed-cone pine forests (Axlerod 1980). On the other hand, in circumstances where moisture availability and summer drought are not limiting factors, it can be argued strongly that the most relevant peculiarity of the present climate instrumental in the containment of *P. radiata* in North America is the temperature extremes, in particular the infrequent low extremes.

It appears reasonable to suggest that *P. radiata* has virtually been plucked from extinction, at least when viewed within an evolutionary or geological time frame. The 'too dry' zone from the south and the 'too cold in winter' zone from the north and inland appear to overlap in California. It also seems reasonable to suggest that it is only a fortuitous accident of oceanography and coastal geomorphology giving rise to the local sea fogs that has saved the species. The sea fogs have allowed the species to survive within the northern end of the otherwise 'too dry' zone, before being transplanted by human hand to environments on other continents where these two zones do not overlap.

Possibly, then, the reasons for the species finding a broader niche in certain southern hemisphere countries than in the northern hemisphere can be sought in analysis of long-term climatic parameters. Of overwhelming significance is the fact that the temperate, maritime, mediterranean-type, not-too-dry climate in which the species is thriving in the southern hemisphere is characterized by a relatively small range of temperatures from winter to summer. Further climatology analysis (Booth 1990) may clarify this.

Possibly, also, the contraction in recent evolutionary times of the natural distribution of the species has not caused commensurate contraction in genetic diversity, though this is difficult to prove. Where exotic climates have been found to accommodate *P. radiata* in relatively pathogen-free circumstances, it has been possible to release and demonstrate a latent genetic capacity. No doubt the expression of this latent capacity has been aided by the free outbreeding that prevailed in the earlier plantations and by the subsequent tree selection and breeding programmes with the species.

This genetic diversity should nonetheless be kept in proper perspective, for while *P. radiata* clearly has a wide genetic diversity for its area of natural distribution, it is not diverse in comparison to some widely occurring species such as *Pseudotsuga menziesii* (Douglas-fir). It is possibly more relevant, in the case of *P. radiata*, to talk about genetic plasticity. Explanations for this plasticity could be linked to degree to which the species evolved in the absence of cold or drought stress limitations. Consequently, the species has the facility to be much more opportunistic than most other conifers which evolved with survival triggers to shut off growth.

Plasticity of *P. radiata* in terms of its capacity to provide accommodating genotypes to guarantee high productivity over a range of site-types is one of the endearing features of this species; plasticity in terms of the versatility of *P. radiata* timber is another outstanding feature of the species contributing to its success as an exotic.

21.6.2 Recognizing the radiata niche – and keeping within it

As mentioned earlier, recognition of the working limits of *P. radiata*, firstly in a marketing sense and secondly in a biological sense, is needed to circumvent the *overselling* of the reputation of the species.

In the first instance, the end-use versatility of radiata – for pulp, paper and reconstituted panel boards, and a range of solid wood applications, both sawn and round – has its limits. The fact that radiata is proving to be one of the world's outstanding multi-market woods should not cause one to lose sight of the ideal requirements in each market niche. For example, many specialist wood applications require higher standards of surface hardness and dimensional stability and more subtle decorative features, than radiata can deliver. We feel that it is unrealistic to expect radiata to be a weighty competitor in the quality furniture and decorative timbers markets, though it should be noted that radiata pine is certainly taking a position in lesser quality furniture markets, and that patented commercial treatments to enhance surface hardness have been developed in New Zealand. To push marketing too far into the specialty timbers niche would be counter-productive to the long-term standing of radiata pine.

As a pulping material, radiata is very versatile and generally superior to other more resinous pines, but for mechanical pulping it is not in the league of spruces and other premium pulping softwoods. Nor can specification for certain quality paper pulps generally be met using radiata, without augmenting with short-fibred hardwood pulps. Thus radiata should be recognized as neither the ultimate pulping feedstock nor a specialty solid timber, and the marketing expectations of growers should be duly restrained.

Secondly, it is very important to recognize where environmentally the species is appropriate and, more importantly, where it is not appropriate in terms of reasonable biological risk. For example it is tempting to seek pathogen-safe niches for radiata at higher altitudes in lower latitudes, but *P. radiata* should not be considered for afforestation or resource-development projects of the third world in the tropics or even the subtropics. In the northern hemisphere at the colder fringe of higher latitudes, there are repeating cases of deaths in the inevitable one-in-50-year cold spells removing radiata entirely.

There are sufficient experiences from around the world with *P. radiata* to be pragmatic about long-term prospects, in spite of confusing examples of very promising early growth beyond the fringe, and in spite of the wonders that can be performed by tree breeders. Our conclusion is that all climatic niches far removed from temperate and near maritime environments with rainfall distributions preferably winter-biased, or at least near uniform, must ultimately be considered *off-site*.

21.6.3 The global marketplace and its dictates

Australia, Chile and New Zealand, the three large Pacific Rim growers of *P. radiata*, dominate any discussion about the current global markets for the species. We suggest that other radiata growers or potential grower countries need to examine the experiences of these three countries to understand why and in what circumstances radiata can produce a national or regional success story.

The histories of radiata in each of these three countries are success stories by most measures. This is so for combinations of reasons, but consistently it has been that the market has accepted and approved of the properties of radiata pine wood. These properties have allowed the species to move into a broad range of applications across the solid wood, reconstituted wood and panel boards (especially medium-density fibreboard), and fibre-pulp markets. It should be added that the movement into these markets has been generally assisted by many years of research and development investment. In these countries radiata pine has justifiably developed the reputation of outstanding versatility, a premier multi-purpose softwood rather than a specialty timber. The full potential of radiata pine as a plantation species in Australia, Chile and New Zealand is now being realized in the broader regional marketplace of the trade-exposed Pacific Rim.

For other countries aspiring, through plantation-resource development, to achieve successes comparable to the radiata stories of Australia, Chile and New Zealand, there is a salient and fundamental truism to be observed from the experiences in these three Pacific Rim countries. This is that, ultimately, the viability of large-scale resource development depends on the end product being cost-effectively produced in a global marketplace that wants to utilize the resource and is prepared to pay accordingly.

In this context, any debate about issues such as *purpose-grown* management or *maximum flexibility* management, or about *forest-based* compared to *factory-based* value-adding, become simply considerations of alternative strategies to the same end-point. Furthermore, the right climatic and edaphic environment and the deployment of *optimum silviculture* are simply prerequisite technologies for success. The philosophy of a market-driven approach is the ultimate reason for success of major resource developments with *P. radiata*.

References

Axelrod, D.I. (1980). History of the maritime closed-cone pines, Alta and Baja California. *University of California Publications in Geological Sciences*, **120**, 1–143.

Axelrod, D.I. (1988). Paleoecology of a late Pleistocene Monterey pine at Laguna Niguel, Southern California. *Botanical Gazette*, **149**, 458–64.

Barnes, R.D. (1970). The prospects for re-establishing *Pinus radiata* as a commercially important species in Rhodesia. *South African Forestry Journal*, **72**.

Booth, T.H. (1990). Mapping regions climatically suitable for particular tree species at the global scale. *Forest Ecology and Management*, **36**, 47–60.

Burdon, R.D. (1992). *Introduced Forest Trees in New Zealand: Recognition Role, and Seed Source 12. Radiata Pine Pinus radiata D. Don*, ed. J.T. Miller. FRI Bulletin No. 124. Rotorua: Forest Research Institute.

Coffman, T. (1995). *The Cambria Forest. Reflections on its Native Pines and its Eventful Past*. Cambria, California: Coastal Heritage Press.

Cotterill, P.P. (1984). A plan for breeding radiata pine. *Silvae Genetica*, **33**, 84–90.

Douglas, D. (1914). *Journal Kept by David Douglas During His Travels in North America 1823–1827*. London: Wesley.

Eldridge, K.G. (1978). Refreshing the genetic resource of *Pinus radiata*. *Appita*, **32**, 93–7.

Forde, M.B. (1966). *Pinus radiata* in California. *New Zealand Journal of Forestry*, **11**, 20–42.

Kininmonth, J.A. & Whitehouse, L.J. (1991). *Properties and Uses of New Zealand Radiata Pine*, Vol. 1. Wood Properties. Rotorua: Forest Research Institute.

Lavery, P.B. (1993). Radiata in the Spanish speaking world – scope for technology transfer? In *Australian Forestry and the Global Environment*. Proceedings, 15th IFA Biennial Conference, pp. 233–40. Queensland: Institute of Foresters of Australia.

Lewis, N.B. & Ferguson, I.S. (1993). *Management of Radiata Pine*. Melbourne: Inkata Press.

Maclaren, J.P. (1993). *Radiata Pine Growers' Manual*. FRI Bulletin No. 185. Rotorua: Forest Research Institute.

Maclaren, J.P. (1996). New Zealand's planted forests as carbon sinks. *Commonwealth Forestry Review*, **75**, 100–3.

Mead, D.J. (1995). Using fertilizers to improve productivity of tree crops. In *NZIF Forestry Handbook*, ed. D. Hammond, pp. 76–7. Christchurch: New Zealand Institute of Forestry.

Millar, C.I. (1986). The Californian closed-cone pines (Subsection Oocarpae Little and Critchfield): a taxonomic history and review. *Taxon*, **35**, 657–70.

Moran, G.F., Bell, J.C. & Eldridge, K.G. (1988). The genetic structure and the conservation of the five natural populations of *Pinus radiata*. *Canadian Journal of Forest Research*, **18**, 506–14.

Neilsen, W.A. (ed.) (1990). *Plantation Handbook*. Hobart: Forestry Commission, Tasmania.

Poynton, R.J. (1979). *Tree Planting in Southern Africa*, Vol. 1. The Pines. Pretoria: Department of Forestry.

Shelbourne, C.J.A., Burdon, R.D., Carson, S.D., Firth, A. & Vincent, T.G. (1986). *Development Plan for Radiata Pine Breeding*. Rotorua: Forest Research Institute.

Shelbourne, C.J.A., Carson, M.J. & Wilcox, M.D. (1989). New techniques for the genetic improvement of radiata pine. *Commonwealth Forestry Review*, **68**, 191–201.

Shepherd, R.W. (1990). Early importations of *Pinus radiata* to New Zealand and distribution in Canterbury to 1885: Implications for the genetic makeup of *Pinus radiata* stocks. Part 1. *Horticulture in New Zealand*, 1(1), 33–8.

Sweet, G.B., Bolton, P. & Litchwark, H. (1990). A meadow orchard in radiata pine – research and reality. In *Proceedings, Meeting of Reproductive Resources Working Party, XIX IUFRO World Congress, Montreal*, pp. S2.01–5. Montreal: IUFRO.

Sweet, G.B., Dickson, R.L., Donaldson, B.D. & Litchwark, H. (1992). Hedged and meadow seed orchards: Towards controlled pollination without isolation – some experiments with *Pinus radiata*. *Silvae Genetica*, **41**, 95–9.

Webb, D.B., Wood, P.J., Smith, J.P. & Henman, G.S. (1984). *A Guide to Species Selection for Tropical and Sub-Tropical Plantations*. Tropical Forestry papers, **15**, 1–256. Oxford: Commonwealth Forestry Institute.

22 Pines as invaders in the southern hemisphere

David M. Richardson and Steven I. Higgins

'We bring strangers together to make strange bedfellows, and we remake the beds they lie in, all at once'
Weiner (1994)

22.1 Introduction

Despite the destruction of large areas of forest and the increasing importance of plant diseases in many parts of the world (Chap. 1, this volume), the global ranges of some pine species have recently increased substantially, especially over the last 50 years. Pines have been cultivated in many parts of the world, both within and well outside the natural range of *Pinus* (Chap. 20, this volume). In many areas, pines have spread from natural forests or plantations to invade the adjacent vegetation. Where such 'invasions' abut natural pine forests (Table 22.1), they are part of natural successional processes and expansions are often not very obvious. This is because the 'invaders' are not 'alien', and because the process of range expansion or 'thickening up' often takes several decades, during which time observers come to accept the new state as the norm. The notion that such range expansions constitute 'invasions' (with the connotations of undesirability and the need for control) is controversial (e.g. see Chap. 9, this volume, for discussion on pinyon pines in the American Southwest). Nonetheless, there is conclusive evidence that many pines have expanded their 'natural' ranges substantially in the last century. In some cases these range expansions are undesirable (references in Richardson & Bond 1991).

As background to the theme of this chapter (pines as invaders outside their natural range – where they are unequivocally aliens) it is necessary to consider why the range and density of some pines has increased in some areas within their natural range. Pines generally occupy marginal habitats (i.e. cold places such as mountain tops and high latitudes, or nutrient-poor soils) – environments that are generally unfavourable for more vigorous, faster growing plants. Although pines can grow well in more favourable situations, they are usually excluded from these sites because their seedlings are light-demanding and are easily suppressed by more vigorous plants. Seeds and pollen of pines are exceptionally well dispersed and isolated pioneers can give rise to colonies by selfing. Most pine species are, thus, inherently well equipped for rapid migration and explosive population increases (Chapters 12–14, this volume). In a review of the cases listed in Table 22.1, Richardson & Bond (1991) showed that the range limits of pines are, to a very large extent, controlled by interactions between their *seedlings* and a wide range of other organisms. There is an unrelenting rain of pine propagules onto sites that are *usually* fairly resistant to invasion. Range changes occur when the conditions that mediate the establishment of pine seedlings and their survival to maturity, also change. Climate change and many other factors determine the range limits of pines, but then often indirectly, i.e. through their influence on interactions between pine seedlings and other biota. The environmental stresses created by drought and temperature clearly exercise direct control on the range limits of pines at very dry sites and at high elevations, respectively. Many authors have correlated periods of successful regeneration in marginal habitats to temporarily favourable conditions (e.g. Kullman 1990). Climate *per se* is less important at intermediate locations. Particularly at low latitudes and elevations, the key factors in determining pine range limits are: vigorous plants (notably grasses); various grazing and browsing animals; fire; and a variety of other factors that, by influencing vegetation cover, change competitive interactions and thus the rules of community assembly (see also Mirov 1967, p. 442). Virtually all the major range

Ecology and Biogeography of Pinus, ed. D.M. Richardson. © Cambridge University Press, Cambridge (1998). pp. 450–73.

Pinus taxon	Regions	Description	Major contributing factors
P. albicaulis	Wyoming (USA)	Invasion of subalpine meadows	Fluctuating grazing intensities
P. contorta	California, Idaho, Wyoming (USA)	Invasion of subalpine meadows	Fire suppression, fluctuating grazing intensities, climate change, natural disturbances (volcanic activity)
P. edulis	Arizona, New Mexico (USA)	Increase in tree density and range expansion	Heavy grazing, fire, logging, land clearance
P. flexilis	Colorado, Idaho (USA)	Invasion of subalpine meadows	Fire, fluctuating grazing intensities, climatic fluctuations
P. halepensis	France, Israel	Invasion of abandoned lands	Grazing, fire, land abandonment
P. jeffreyi	California (USA)	Invasion following volcanic activity	Volcanic activity
P. kesiya	India, Myanmar (Burma), Philippines, Vietnam	Invasion of disturbed forests	Shifting cultivation, fire
P. lambertiana	California (USA)	Invasion following volcanic activity	Volcanic activity
P. massoniana	China	Invasion of disturbed forests	Forest clearing
P. merkusii	SE Asia	Invasion of disturbed forests	Shifting cultivation, fire, grazing
P. monophylla	Arizona, Colorado, Nevada, Utah (USA)	Invasion of grazed grassland	Heavy grazing, climatic fluctuations
P. monticola	California, Oregon (USA)	Invasion following volcanic activity	Volcanic activity
P. ponderosa	California, Colorado, Nebraska, South Dakota, Utah (USA)	Invasion of various grassland types	Heavy grazing, fire, fire suppression, wind damage, insect attack, logging, mining
P. radiata	California (USA)	Increased regeneration following grazing and other disturbances	Land clearance, grazing, fire, logging
P. resinosa	Wisconsin (USA)	Invasion following fire suppression	Fire suppression
P. roxburghii	Nepal	Invasion of disturbed sites	Forest clearing, shifting cultivation, fire
P. strobus	New York, Wisconsin (USA)	Invasion of disturbed grasslands and old fields	Forest clearing, cultivation, logging, wind damage, insect attack, fire, fire suppression, browsing, tree cutting (beavers)
P. virginiana	E USA	Invasion of old fields	Forest clearing, cultivation

Table 22.1. Examples of cases where the distribution and/or density of Pinus species has increased within (or immediately adjacent to) their natural range. See Richardson & Bond (1991) for further details and key references for each case

changes of pines in Table 22.1 can be attributed to changes in the dimensions of the regeneration niche as a result of natural (e.g. volcanic activity) or human-induced disturbance (e.g. fluctuations in herbivore pressure and/or the fire regime and/or various forms of mechanical disturbance). This conclusion is important, not only for understanding the dynamics of natural pine forests and for managing them in the face of changing patterns of land use and global climate change (Bond & Richardson 1990; Richardson & Bond 1991), but also when considering pine invasions outside their natural range.

Pines have spread from sites of cultivation in many parts of the southern hemisphere, and this has brought them into contact with a biota with which they are totally unacquainted. These invasions, which proceed very similarly in many respects to the range expansions in the northern hemisphere (Richardson & Bond 1991), provide particularly interesting examples of the increasingly important phenomenon of biological invasions, whereby many organisms transported to new habitats by humans are leading to the homogenization of the Earth's biota (Drake *et al.* 1989). Most native conifers in the southern hemisphere, in marked contrast to the situation with pines in northern climes, are relatively unimportant elements of the vegetation in most regions in terms of number of taxa and overall abundance and conspicuousness (Enright & Hill 1995). And yet, invasive alien pines in the southern hemisphere can replicate their good performance in environments that may appear less than ideal for conifers.

In this chapter we build on the evidence that has accrued from numerous studies of pine dynamics in the northern hemisphere and other perspectives reviewed in this volume, and on the recent advances in the understanding of biological invasions, to consider the rampant spread of pines from plantations in the southern hemisphere during the last few decades. We review the status of pines as alien invaders (which species have invaded which habitats and what impacts have these invasions caused?), and explore the role of the various factors that interact to determine whether, and to what extent, introduced pines spread. We show that the 'natural experiment' of pine cultivation and subsequent spread in the southern hemisphere provides an extraordinary opportunity for gaining new perspectives on the ecology of pines, and on the ecology and management of biological invasions in general. Where possible, we have cited published studies. However, since many pine invasions are recent phenomena or not reported in the literature, we have relied heavily on our own observations and the unpublished records of many of our colleagues (see Acknowledgements).

22.2 Pines as alien invaders in the southern hemisphere

22.2.1 History of planting and invasion

Humans introduced a few *Pinus* species to the southern hemisphere at least as early as the late 17th century. It was, however, not until the 1880s when large-scale commercial forestry started that pines were very widely cultivated. The ensuing century has seen the rapid expansion of pine cultivation outside the natural range of *Pinus*. For the tropical pines (especially those from Mexico and Central America), the last two decades of the 20th century have been especially important, with seed transfer to many southern hemisphere regions increasing dramatically (Dvorak & Donahue 1992). It is important to consider the history of cultivation in the different regions of the southern hemisphere when deliberating the varying performance of pines as invaders in these regions. Le Maitre (Chap. 20, this volume) has summarized the history of pine cultivation in the southern hemisphere and Lavery & Mead (Chap. 21, this volume) discuss in more detail the cultivation of the most widely planted species, *P. radiata*. In this chapter we consider only those issues directly relevant to invasions, focusing on Australia, New Zealand and South Africa where trends have been best documented and where pine invasions are most widespread.

South Africa

Several species of pines (certainly *P. pinaster* and *P. pinea*) were introduced to South Africa as early as the late 17th century, and several others arrived between then and 1875. At least another 41 species arrived over the next 35 years (Chap. 20, this volume). Poynton (1979) listed 79 *Pinus* species that were known to have been planted in South Africa at that date (see also Chap. 20, this volume). Many species were planted for a variety of uses (Poynton 1979), initially in the former Cape Province, and later in KwaZulu-Natal and the former Transvaal province. The most widespread species are currently, in order of importance: *P. patula*, *P. elliottii*, *P. taeda*, *P. radiata* and *P. pinaster* (Chap. 20, this volume).

Besides the normal planting of seeds for plantations, seeds of *P. pinaster* (and possibly *P. halepensis* and *P. pinea*) were also actively dispersed by forest rangers and hikers in many parts of the Western Cape in the 18th and 19th centuries in an effort to increase tree cover in the shrubland vegetation (Woods 1950; see also section 22.3). The first tentative record of prolific natural regeneration (and possible spread) of pines in the southern hemisphere is for *P. halepensis*, which was noted to be spreading (or at least regenerating prolifically) in the Caledon district (*c.* 100 km east of Cape Town) as early as 1855 (Lister 1959, p. 63), some 25 years after it arrived in the country (Shaughnessy

1986). Invasive pines were not widespread at this time. Sim (1927b) reported that *P. pinaster* had been 'gradually taking possession . . . of the face of the mountain above Cape Town' since the early 1890s, this being the earliest record of widespread pine invasion in the southern hemisphere. Soon after this, there are records of the spread of *P. pinaster* in other parts of the Cape Province (e.g. from Genadendal, 100 km east of Cape Town; see Hutchins 1897, Woods 1950). Bolus (1886), in one of the earliest references to the potential hazard of alien plant invasions in southern African botanical literature, wrote that 'it is remarkable how small upon the whole is the influence exerted upon the aspect of the vegetation and how weak . . . is their aggressive power against the indigenous flora'. This view was echoed by several botanists over the ensuing 30 years. Sim (1927a) was one of the first to note that invasive alien plants were starting to have a serious impact on the indigenous vegetation; he wrote, 'the extent to which *Pinus pinaster* can take possession indicates that, if given a long enough period without check, it would probably kill out some of the endemic monotypes'. The subsequent increase in the extent of pine invasions in South Africa is summarized by Richardson *et al.* (1997); see also section 22.3.

Australia

Pines were introduced to other parts of the southern hemisphere somewhat later than South Africa. The establishment of a British colony in Australia in 1770 and the start, in 1840, of the European colonization of New Zealand marked the beginning of serious plant introductions to these regions. The early history of pine introductions is not as well documented for these countries as it is for South Africa, largely because there was less need for alien trees for timber and other uses in these well-forested regions.

Pinus radiata arrived in Australia at least as early as 1857 (Rule 1967), a few years after *P. halepensis*, *P. nigra*, *P. pinaster* and *P. pinea* (Woods and Forests Department, Annual Report: 1886). We could find no records to suggest that any pines had spread significantly in Australia by the end of the 19th century. Indeed, the intrepid pioneer Samuel Dixon, who arrived in South Australia from England in the early 1860s (Kraehenbuehl 1981), wrote in 1892 that although many introduced garden plants were spreading rapidly to become weeds in South Australia, he did not know of 'any introduced forest tree which has reproduced itself largely, except under cultivation' (Dixon 1892). Interestingly, he added that '*Pinus halepensis* might have been expected to do so', which suggests that he had observed the weedy behaviour of this species elsewhere, probably in South Africa which he visited at least once (Kraehenbuehl 1981).

Although trials with several pine species started in the 1870s, significant planting of *P. halepensis* and *P. radiata* in South Australia began only shortly before 1900 (Rule 1967), and large-scale plantings commenced only after 1918 (Cromer 1956). The first herbarium record of a naturalized pine (*P. radiata*, from South Australia) was collected from a large tree in 1957. However, there are apparently self-sown pines in the hills behind Adelaide that are decades older than this (D. Cooke, personal communication), which suggests that this species may have been spreading around the turn of the century. The first invasion of eucalypt forest in the ACT near Canberra by *P. radiata* was noted in 1954 (Dawson *et al.* 1979) (Fig. 22.1*H*). *Pinus halepensis* (possibly also *P. brutia*) is weedy on the Eyre Peninsula (R. Boardman, personal communication) and other areas near Adelaide (Fig. 22.1*D*) but we could not ascertain when invasion started in this area. *Pinus elliottii* was introduced to New South Wales and Queensland in 1921 and 1925 respectively (Rule 1967). Large-scale plantings of *P. elliottii* in Queensland started in the 1930s in sandy 'wallum' country in the Beerburrum and Beerwah districts north of Brisbane. Self-sown pines were noted very soon after the first major plantings (R. Wylie, personal communication), but the first record of concern about these invasions was in the 1960s, when a formal submission was made by landowners to the Landsborough Shire Council (Shea 1987).

Nearly all the pines that have invaded bushland areas in Western Australia spread from plantations established between 1910 and 1950. The densest stands of self-sown pines around Perth originated from plantations established shortly before World War II. It is not clear when pines started spreading, but the oldest examples of pines established in a disturbed natural community are *P. pinaster* and *P. pinea* in tall heath on the slopes of Mt Eliza (Kings Park in the centre of Perth); these spread from plantings dated to between 1900 and 1910. Pines have invaded *Eucalyptus marginata* forest, heath on limestone, shrublands and *Banksia* woodlands in areas with rainfall of more than 600 mm yr^{-1} from Perth to Esperance and about 100 km inland to Narrogin. Invasive pines are widespread but sporadic in this area, and they are not considered to be invaders of major importance. No pines have invaded the northern sandplains (G.J. Keighery, unpublished data).

New Zealand

Pinus pinaster was almost certainly the first conifer to be introduced to New Zealand, probably shortly before 1830. Webb, Sykes & Garnock-Jones (1988) report that 'nearly all the spp. of pine have been introduced to [New Zealand] for trial'. Their list of 13 naturalized species (those that grow 'spontaneously outside a fenced areas or as weeds in a sown or planted community') includes all the species that

Fig. 22.1. **Invasive pines in the southern hemisphere.** (*A*) **Dense regeneration of** *Pinus caribaea* **in the north of Grande-Terre, New Caledonia. This species is now spreading from plantations to invade ultramafic maquis (photo: O. Gargominy).** (*B*) *Pinus contorta* **invading short tussock grassland, Glen Eyrie Station, near Lake Ohau, Upper Waitaki River catchment, South Canterbury, New Zealand (stand density reduced by browsing rabbits) (photo: N. Ledgard).** (*C*) *Pinus elliottii* **invading eucalypt forest, Queensland,** **Australia (photo: W.L. Thompson).** (*D*) *Pinus halepensis* **invading** *Eucalyptus leucoxylon* **woodland in Belair National Park, near Adelaide, South Australia (photo: R.J. Carter).** (*E*) **Planted** *Pinus mugo* **stand (right) with self-sown seedlings in grassland on Mount Bee, northern Southland, New Zealand (photo: C.J. West).** (*F*) *Pinus nigra* **invading poor pasture and** *Leptospermum* **scrub in Canterbury, New Zealand (photo: P.A. Williams).** (*G*) **Very dense stand of self-sown** *P. pinaster* **in mountain fynbos near Villiersdorp, Western Cape,**

Fig. 22.1. (*cont.*)
South Africa. The area on the bottom right was cleared to increase water yield from the catchment (photo: D.M. Richardson). (*H*) *Pinus radiata* invading eucalypt forest on Black Mountain, Canberra, ACT, Australia, in 1990, 36 years after invasion was first noted at this site (photo: D.M. Richardson). (*I*) *Pinus radiata* invading natural

Nothofagus forest on the Cordillera de Nahuelbuta (elevation 800 m), between Angol and Parque Nacionale Nahuelbuta, Chile. Although *P. radiata* has not invaded large areas in Chile, this is probably because most plantations of this species were only established after the mid-1970s (photo: P.H. Zedler).

have been widely planted (*P. banksiana*, *P. contorta*, *P. halepensis*, *P. mugo*, *P. muricata*, *P. nigra*, *P. patula*, *P. pinaster*, *P. ponderosa*, *P. radiata*, *P. strobus*, *P. sylvestris* and *P. taeda*). These and other species have been used for commercial forestry, erosion control, windbreaks and as ornamentals throughout the country. Ledgard & Baker (1988) report that more than 20 pine species were planted in one revegetation project alone. The first regulations encouraging the planting of alien trees were passed in the early 1870s when it became evident that natural forests could not meet the demand and were being depleted. The inauguration of state forestry in 1896 marked the start of large-scale plantings, but the main areas of forestry in New Zealand were planted in the late 1920s and early 1930s. It is important to note that seeds of *P. contorta*, and possibly other pines that were planted widely in montane areas, were often sown from the air (Webb *et al.* 1988). Hunter & Douglas (1984) present evidence which suggests that the spread of pines from plantings was first noted in the late 1890s and that

there was a rapid increase in invasions in the first two decades of the 20th century (see also Thomson 1922; Cockayne 1928, p. 359) and especially after the late 1940s, some 65 years after most parent trees were planted. Cheeseman (1925) notes that *P. halepensis*, *P. nigra*, *P. pinaster*, *P. ponderosa* and *P. radiata* were all reproducing themselves with 'great readiness' in 'most parts of the Dominion'. However, until the 1960s, most of the area invaded by pines in New Zealand was on the Central Volcanic Plateau (e.g. Anon 1962), where very large plantations were only established in the late 1920s and early 1930s. Spread was, therefore, from 20–30-year-old trees. The first herbarium records of self-sown pines on South Island and North Island are dated 1904 (*P. pinaster*) and 1910 (*P. radiata*), respectively (P.A. Williams, unpublished data).

Summary
The above accounts suggest that pines started to invade vegetation away from planting sites very soon after the

establishment of large plantations starting in the 1880s (although self-sown *P. halepensis* and *P. pinaster* were probably fairly widespread in South Africa by this time). The detection of these invasions was irregular and it is difficult to resolve whether the apparent time lags between plantings and invasions in some areas were real or simply an artefact of the sparse distribution of naturalists to document invasions when they did occur. Lanner (Chap. 14, this volume) has suggested that seed dispersal from closed stands of pines is 'biologically more meaningful' than dispersal from widely scattered trees because the latter produce a much larger proportion of selfed progeny, leading to greater mortality (see also Chap. 13, this volume). It is likely that such 'density effects' have been important but it is difficult to quantify these. There is now also ample evidence that landraces that evolve in pines one or two generations after introduction to foreign environments are much better suited to local conditions than any introductions from the native range of the species (see section 22.4.1). Another factor that almost certainly delayed the spread of pines in many parts of the southern hemisphere was the absence of appropriate mycorrhizal symbionts (Richardson, Williams & Hobbs 1994a; see also Chap. 16, this volume). Only once inoculated soil had been introduced could pines grow well and start to invade. Available evidence suggests that some time lags can also be attributed to various other barriers to invasion (*sensu* Kruger, Richardson & van Wilgen 1986). Changes to the disturbance regimes initiated invasions in many regions by reducing the inherent resistance of certain habitats to invasion. For example, significant invasion of disturbed eucalypt forests in Australia by *P. radiata* started only after myxomatosis reduced the size of introduced rabbit populations, thus reducing browsing pressure on pine seedlings (Chilvers & Burdon 1983). The topic of inherent versus induced invadability is addressed in section 22.4.2.

The importance of pines as plant invaders ('environmental weeds' *sensu* Holzner & Numata 1982) has increased at about the same rate as the overall increase in the problem of biological invasions worldwide, i.e. a massive, exponential increase in abundance since about the middle of the 20th century. Pines now feature prominently on regional (e.g. Stirton 1978; Henderson & Musil 1984; Macdonald & Jarman 1984 1985; Henderson 1989, 1991, 1992; Keighery 1991; Carr, Yugovic & Robinson 1992; Thompson 1994; Berry & Mulvaney 1995), national (Wells *et al.* 1986; Williams & Timmins 1990; Humphries, Groves & Mitchell 1991) and global (Holm *et al.* 1979; Cronk & Fuller 1995) lists of important or potentially important weeds. Studies of pine invasions in several widely-scattered parts of the southern hemisphere, and general or anecdotal accounts from other areas, show that the extent of the phe-

nomenon has increased rapidly in recent decades and that the nuisance value of pines outside plantations is escalating rapidly.

22.2.2 Which pines have invaded?

After a thorough review of the literature and correspondence with ecologists and foresters in many southern hemisphere countries, Richardson *et al.* (1994a) concluded that at least 16 *Pinus* species can be considered invasive, i.e. regenerated naturally and recruited seedlings in natural or semi-natural vegetation > 100 m from parent plants (Table 22.2). They cited examples of borderline cases – species that are widely naturalized but which have not (yet) spread from plantings (e.g. *P. densiflora* and *P. mugo* in New Zealand). Since this assessment was made, the situation for *P. mugo* has changed, and there is now clear evidence of invasion from stands planted in subalpine communities at the southern end of the Eyre Mountains, South Island, for erosion control in the early 1970s (C. West, unpublished data; Fig. 22.1E). Extensive spread of *P. mugo* has also been noted recently in Marlborough (N. Ledgard, unpublished data). Several other species belong in the category of borderline cases, and some of them will probably soon qualify for classification as invaders. One species not included in Richardson *et al.*'s (1994a) list of invasive pines, *P. jeffreyi*, was recently recorded as one of four pine species to have spread substantially (more than 100 plants have spread from an arboretum established between 1940 and 1970 into surrounding natural vegetation) in the Brindabella Ranges in the ACT, Australia (Berry & Mulvaney 1995). Berry & Mulvaney's list of invasive pines in this region includes 14 species, five of which (*P. jeffreyi*, *P. lambertiana*, *P. monticola*, *P. torreyana* and *P. virginiana*) are not known to be invasive elsewhere in the southern hemisphere. Richardson *et al.* (1994a) also did not list *P. caribaea* among the widespread invasive pines in the southern hemisphere. Although MacKee (1994) stated that pines had not been seen established away from plantations in New Caledonia, recent reports (O. Gargominy & T. Jaffre, unpublished data) show this species to be highly invasive on this island (Fig. 22.1A). *Pinus caribaea* var. *hondurensis* is also invading open woodlands in Queensland (Table 22.2). Surveys around arboreta and plantations in other regions will almost certainly augment the list in Table 22.2.

Eight of the invasive species (*P. caribaea*, *P. elliottii*, *P. halepensis*, *P. patula*, *P. pinaster*, *P. ponderosa*, *P. radiata* and *P. taeda*) are currently very widely planted in the southern hemisphere (Chap. 20, this volume). Of these, *P. halepensis*, *P. patula*, *P. pinaster* and *P. radiata* are very widespread invaders. Species that are not major plantation forestry species in the southern hemisphere and yet have spread widely (all in New Zealand) are *P. contorta*, *P. nigra* and *P. sylvestris*.

Table 22.2. *Pinus* species as alien invaders in the southern hemisphere (updated from Richardson *et al.* 1994a; see this reference for details on invasions in each region, except where other references are given). See text for criteria for inclusion in this list

Pinus taxon	Region	Habitat
P. banksiana	New Zealand	Scrub and open places on and near forest margins, shrublands, tussock grassland.
P. canariensis	Australia	Disturbed mallee shrubland, heath, road verges in Western Australia (G.J. Keighery, unpubl. data)
	South Africa	Fynbos, forest
P. caribaea	Australia	Lowland *Melaleuca viridiflora* open woodlands near Cardwell, Queensland (invading pines are *P. caribaea* var. *hondurensis*; S. Skull, unpubl. data)
	New Caledonia	Shrublands (ultramafic maquis or 'maquis minier')(O. Gargominy, unpubl. data)
P. contorta	New Zealand	Indigenous and alien scrub, tussock grassland, pasture, open forest
P. elliottii	South Africa	Temperate and moist-subtropical grasslands
	Argentina	Overgrazed natural grasslands and other disturbed areas in Entre Rios, Corrientes and Misiones provinces (H.J. Trebino, unpubl. data)
	Australia	Heathland and open forest/woodland in Queensland (Thompson 1994)
P. halepensis	Australia	Disturbed mallee on the Eyre Peninsula, South Australia; cut-over *Eucalyptus marginata* forest and disturbed ground in Western Australia (G.J. Keighery, unpubl. data).
	New Zealand	Extensively managed grasslands
	South Africa	Fynbos
P. jeffreyi	Australia	Subalpine woodlands and wet sclerophyll forests in ACT (Berry & Mulvaney 1995)
P. mugo	New Zealand	Tussock grassland dominated by *Chionochloa rigida* at and above the *Nothofagus* timberline. Invasions influenced by erosion caused by human disturbance (>15 yrs), sheep grazing (>50 yrs) and some recent herbivory pressure from rabbits and cattle (C. West, unpubl. data).
P. muricata	New Zealand	Coastal and inland shrubland
P. nigra	Australia	Eucalypt forest
	New Zealand	Grazed grasslands, tussock lands, shrubland – especially montane areas
P. patula	Madagascar	Montane *Hyparrhenia* grassland
	Malawi	Montane grassland
	New Zealand	Disturbed ground, cut-over forest
	South Africa	Montane grasslands, tropical bush and savanna, disturbed evergreen forest
P. pinaster	Australia	*Banksia* woodland, *Eucalyptus marginata* forest and woodland, road verges in Western Australia (G.J. Keighery, unpubl. data); heath in Victoria (Kermode 1993)
	Chile	Disturbed sites
	New Zealand	Scrub and shrubland, grasslands slopes, cliff faces, cut-over forests
	South Africa	Fynbos
	Uruguay	Coastal dunes
P. pinea	Australia	Limestone heath and shrubland in Western Australia (G.J. Keighery, unpubl. data)
	South Africa	Fynbos, moist tropical grasslands
P. ponderosa	Argentina	Patagonia steppe (shrubland with bunchgrasses)

Table 22.2. (*cont.*)

Pinus taxon	Region	Habitat
	Australia	Shrubby rises in sedgeland, shrubland along creeklines, disturbed areas in Western Australia (G.J. Keighery, unpubl. data)
	Chile	Natural, but somewhat disturbed woodlands; also deciduous *Nothofagus* forests (e.g. at Reserva Nacional Los Ruiles and near Lago Colbun; C. Lusk, unpubl. data)
	New Zealand	Grasslands, tussock land, shrubland
P. radiata	Australia	Heathland and heathy woodland, lowland grassland and grassy woodland, dry and moist eucalypt forest and woodland, riparian vegetation
	Chile	Secondary forests, Valdivian Rain Forest region
	New Zealand	Grasslands, tussock land, shrubland
	South Africa	Fynbos
P. roxburghii	South Africa	Disturbed sites
P. strobus	New Zealand	Scrub and secondary forest
P. sylvestris	New Zealand	Shrublands, tussock land, grassland
P. taeda	Argentina	Overgrazed natural grasslands and other disturbed areas in Entre Rios, Corrientes and Misiones provinces (H.J. Trebino, unpubl. data)
	Brazil	Savanna and open woodland (I. Peters, unpubl. data)
	New Zealand	Cut-over forest and shrubland
	South Africa	Grasslands

Important forestry species that have not yet invaded large areas are: *P. caribaea* (widely planted in Argentina, Australia, Brazil, Kenya and Tanzania; already a widespread invader in Hawaii (Smith 1985) and starting to invade in northern Queensland, Australia and New Caledonia); *P. kesiya* (widely planted in Colombia, Madagascar, Uganda and Zimbabwe); and *P. oocarpa* (widely planted in Brazil, Colombia, Kenya, Tanzania, Uganda, Zambia and Zimbabwe). The fact that these species are not widespread invaders is probably due, at least to some extent, to the relatively recent increase in their importance as forestry species (Chap. 20, this volume). There are also fewer plant ecologists in the regions in which these species are widely grown. Plant invasions are much more closely monitored and studied in Australia, New Zealand and South Africa than in other parts of the southern hemisphere (Pyšek 1995).

22.2.3 The extent of invasions

No quantitative information is available to rank pines according to their spread throughout the southern hemisphere, but the following species have invaded large areas in one or more region(s): *P. contorta* (mainly in New Zealand; Fig. 22.1B), *P. halepensis* (South Africa), *P. nigra* (New Zealand; Fig. 22.1F), *P. patula* (Madagascar, Malawi, South Africa), *P. pinaster* (South Africa, Australia, New Zealand, Uruguay), *P. ponderosa* (Argentina, New Zealand),

P. radiata (Australia, New Zealand, South Africa) and *P. sylvestris* (New Zealand) (Fig. 22.1).

Most of the following information on the extent of invasions in the different regions has been condensed from Richardson *et al.* (1994a). *Pinus pinaster* is by far the most widespread invasive pine in South Africa; it has invaded about 3256 km² of natural vegetation, 97% of this in the fynbos biome (Fig. 22.1G; see Richardson *et al.* 1992, 1997 for maps). *Pinus radiata* and *P. patula* are the next most widespread invaders, covering an estimated 340 km² and 176 km² in the late 1980s, almost exclusively in fynbos and grassland respectively. A survey of woody alien plants in the 1.14 × 10⁶ ha of water catchments of the Western Cape province in 1994/5 showed that more than one-third of the area was invaded by pines. Dense stands (>75% canopy cover) were recorded over 52 km², or 0.5% of the area, while medium-density stands (50–75% canopy cover) occupied another 0.5% of the area (Cape Nature Conservation, unpublished data). The extent of pine invasions on the Cape Peninsula of South Africa has been assessed in more detail than for any other area (see section 22.3). In New Zealand, 300 km² of the Central Volcanic Plateau had an 'infrequent' to 'dense' covering of *P. contorta* in 1975. In one region of unimproved grassland of the South Island high country surveyed in 1989/90, 175 km² (0.8% of the area) was invaded by conifers (mainly pines). The most widespread species were, in decreasing order of importance: *P. nigra*, *P. contorta*, *P. sylvestris*, *P. ponderosa* and *P. radiata*. In Victoria, Australia, abundance can be ranked as follows: *P. radiata* (widespread, medium to large populations) >> *P. pinaster* (limited distribution, medium to large populations) > *P. nigra* (limited distribution, small populations). This ranking probably applies to Australia as a whole, with *P. elliottii* (now rampant in Queensland; Thompson 1994; Fig. 22.1C) as the next most widespread invader. In Western Australia, *P. pinaster* is by far the most widespread species, followed by *P. canariensis* and *P. halepensis*, *P. pinea* and *P. ponderosa*, the last three of which are considered 'minor escapees' (G.J. Keighery, unpublished data). Although *P. pinaster* and *P. ponderosa* have spread from plantings in Chile, the extent of self-sown populations of these species is small when compared with those of *P. radiata* which has itself not invaded large areas (Fig. 22.1I). We could find no statistics on areas invaded for the other regions mentioned in Table 22.2.

22.2.4 Which habitats are invaded?

Pine invasions have been documented in Argentina, Australia, Chile, Madagascar, Malawi, New Zealand, South Africa and Uruguay, where the invaded natural vegetation may be grouped into the following broad categories: sparse vegetation on coastal dunes, grassland, shrubland or forest (Table 22.2; Richardson *et al.* 1994a). Although

many factors interact to determine the susceptibility of vegetation to invasion by pines (see section 22.4.2), types may be crudely ranked according to their vulnerability to invasion by pines: forest << shrubland < grassland << dunes < bare ground.

22.2.5 Impacts of invasions

Invading pines cause a wide range of changes to various features of invaded ecosystems. Examples of some major influences are described in Table 22.3. The main impacts result from the increased abundance of trees in habitats where this life form was previously absent or less common. This major shift in life-form dominance has impacted upon a range of ecosystem properties, with various implications for the native biota. The severe impacts of invasions by pines and other trees on indigenous plants and on various ecosystem services in South African fynbos catchments are particularly noteworthy (Cowling, Costanza & Higgins 1997). Pine invasions have a marked impact on soil properties and processes (Chap. 17, this volume).

22.3 Case study: invasive pines on the Cape Peninsula, South Africa

The history of pine introductions, cultivation, spread, control efforts, and the impacts of pine invasions has been more thoroughly documented for the Cape Peninsula than for any other part of the southern hemisphere. This area at the southwestern tip of Africa covers 470 km², most of it clothed in fire-prone fynbos vegetation, is well known for its plant diversity, with a flora comprising 2285 species, about 4% of them endemic to the area. Fifteen vegetation types, defined on the basis of floristics, dominant taxa and structure, are recognized (Cowling, Macdonald & Simmons 1996). Native trees are virtually confined to small forest patches which originally covered about 2.7% of the area; these have shrunk to cover 2.3% of the area, 3.6% of the remaining natural vegetation (Richardson *et al.* 1997). The summary that follows has drawn heavily on Shaughnessy's (1980, 1986) thorough analysis.

Pinus pinaster was one of the first trees to be introduced to the area, and probably arrived between 1685 and 1693. *Pinus pinea* probably arrived at about the same time. By 1772, *P. pinaster* was widespread on the Cape Peninsula, and by the first decade of the 19th century this species and *P. pinea* were conspicuous features of the landscape. The first reliable record of *P. halepensis* in the Western Cape is from 1855 when the species was noted to be growing (and possibly even spreading; Lister 1959; Richardson *et al.* 1994a) near Riviersondered, about 125 km east of Cape

Major influence	Resultant impacts	Examples	Key references
Shift in life-form dominance	Suppression of native plants	Australia: Invasion of dry sclerophyll forests in ACT by *P. radiata* leads to formation of a discontinuous carpet of pine needles and an increase in the amount of deep shade. South Africa: Dense pine stands suppress fynbos plant species, threatening many taxa with extinction; *c*. 750 species are currently at risk.	Burdon & Chilvers (1994) Richardson, Macdonald & Forsyth (1989); Richardson, Cowling & Lamont (1996a)
	New habitat for native species	New Zealand: In at least one case, native orchids have flourished beneath pines where they may benefit from associated root fungi, lack of competition or from the pines acting as seed traps. South Africa: Many arboreal birds have invaded fynbos from adjacent biomes in response to proliferation of pines and other alien trees.	Johns & Molloy (1983); Gibbs (1989) Macdonald & Richardson (1986); Macdonald, Richardson & Powrie (1986).
Reduction in structural diversity	Reduced value as wildlife habitat	Australia: Pine stands are less useful as wildlife habitat than native eucalypt forests since they provide fewer (and a smaller range of) resources.	References in Richardson *et al.* (1994a)
Increase in biomass	Increased interception and transpirational losses resulting in decreased streamflow	South Africa: Interception losses due to pine in fynbos catchments is *c*. 20% of total annual rainfall. Runoff in completely invaded catchments declines by 30–70%, depending on annual rainfall and age and density of the alien stand. Modelling suggests that further invasion could, over 100 yrs, result in an average loss of >30% of the water supply to the city of Cape Town.	Van Wyk (1987); Le Maitre *et al.* (1996)
	Increased fuel loads	South Africa: Intense fires in dense pine stands sometimes have devastating effects, apparently as a result of greatly increased water repellency caused by, among other factors, the very large fuel loads (biomass in self-sown pine forests can be up to 5 times that of the pre-invasion fynbos)	Scott & van Wyk (1990)
Disruption of prevailing vegetation dynamics	Fire-induced changes in vegetation structure	South Africa: The non-equilibrium status of natural fynbos communities is disrupted by alien pines, resulting in new depauperate steady states. This causes increased overland flow and accelerated soil erosion.	Richardson & Cowling (1992)
Changed nutrient cycling	Nutrient enrichment	New Zealand: In moist environments, mineralization of organic matter leads to nutrient enrichment of topsoils under pines. In drier environments, transfer of nutrients from deeper horizons to the soil surface via nutrient uptake and litterfall is probably the major mechanism.	Davis & Lang (1991); see also Chap. 17 (this volume)
	Changed levels of total dissolved, and reactive phosphorus and nitrates	New Zealand: On volcanic soils in North Island catchments with *P. radiata* had lower levels of total dissolved and reactive P than pastures, but higher levels than native forest. For nitrates the order was reversed.	Cooper, Hewitt & Cooke (1987)
	Reduced densities of decomposer organisms and reduced decomposition rates	Australia: Plantations of *P. pinaster* (and presumably also self-sown stands) in WA had significantly lower rates of decomposition and lower densities of decomposer organisms than adjacent native woodland.	Springett (1976)

Table 22.3. Impacts of pine invasions on features of invaded ecosystems in the southern hemisphere

Town. Shaughnessy suggests that *P. halepensis* was probably first grown on the Cape Peninsula before 1830. The first records of the other two pine species that are currently prominent in the area, *P. radiata* and *P. canariensis*, are from 1865 and 1878 respectively. Many other pine species were grown in gardens and arboreta, but none were extensively cultivated.

Large-scale planting of pines started in the 1850s when they (apparently mainly *P. pinea*) were used in an attempt to stabilize the drift sands of the Cape Flats. Later in the same decade, *P. pinaster* and *P. pinea* were the two main species used in forestry trails. Large plantations of the former were established on the slopes of Table Mountain and elsewhere on the Peninsula during the 1880s, mainly to provide timber, but also in the mistaken belief that pine forests would *increase* the water supply. Another reason given by authorities for the establishment of pine plantations was the need to improve the 'bleak and naked appearance' of Table Mountain, whose natural state ('bare and stony slopes') was apparently 'a subject of daily comment'. In the first two decades of the 20th century the pine plantations on the upper mountain slopes were abandoned when it became clear that they would not be successful. This abandonment left scattered stands of *P. halepensis* and *P. pinaster* to serve as seed sources for future invasions. More intensive forestry, using *P. radiata*, started on the Peninsula in around 1900 and this species is now the only pine grown commercially in the area.

Sim (1927b) wrote that *P. pinaster* had

for a very long time ... been in possession of the Wynberg side of Table Mountain and during the past three and a half decades it has been gradually taking possession also of the face of the mountain above Cape Town. In these localities ... it has now reached the stage of being perfectly naturalized and of sowing itself broadcast over these neighbourhoods and further afield, and even further up the mountain than the parent trees stood.

After the late 1920s there was a rapid increase in the awareness of the problem of alien tree and shrub invasions, especially after Wicht's (1945) pronouncement that 'One of the greatest, if not the greatest, threats to which the Cape vegetation is exposed, is suppression through the spread of various exotic plant species' (see a review of further developments in Richardson *et al.* 1996b). The problem escalated rapidly in the ensuing 40 years. Surveys of the distribution of alien woody plants in the area were undertaken in 1959/60 (Hall 1961), 1976 (McLachlan, Moll & Hall 1980), 1989/90 (Moll & Trinder-Smith 1992) and (most comprehensively) in 1994 (Richardson *et al.* 1996b). Intensive

surveys of alien plants were conducted in the Cape of Good Hope Nature Reserve (77.5 km²) at the tip of the Peninsula in 1966 and 1976–80 (Taylor & Macdonald 1985; Taylor, Macdonald & Macdonald 1985; see also Macdonald, Clark & Taylor 1987). Comparing the results of these surveys shows that the distribution and abundance of *P. pinaster* has not increased markedly since 1959/60 (it has decreased in abundance in the 87 sample plots used in the first three surveys mentioned), but that the distribution and abundance of *P. radiata* has increased considerably, especially since 1976. The abundance of *P. canariensis* and *P. pinea* fluctuated between surveys but neither has spread much since 1959/60. The first three surveys mentioned did not record *P. halepensis*. Despite various control efforts initiated by landowners and conservation authorities, invasive pines are still very widespread and remain a major threat to the biodiversity of the region. Although the 1989/90 survey found *P. pinaster* in many more plots than *P. radiata* (62% vs 24%; Moll & Trinder-Smith 1992), the thorough survey of the entire Cape Peninsula in 1994 showed that *P. radiata* has overtaken *P. pinaster* as the pine with the largest area under dense stands in the area, although light stands of *P. pinaster* are still more widespread. This study showed that *P. radiata*, *P. pinaster* and *P. pinea* all form dense stands (>25% canopy cover) and that pines are among the most abundant five woody alien plants in 10 of the 15 vegetation types on the Peninsula. Dense stands of the three species currently cover 908, 568 and 240 ha, respectively (Fig. 22.2) – the total area covered by dense stands of woody aliens is 3313 ha (Richardson *et al.* 1996b).

The extent of planting of the different species is clearly the major factor contributing to their current distribution. The widespread dissemination of *P. pinaster* over the last 250 years accounts for its extensive distribution throughout the Peninsula. Although control efforts have reduced the density of the species in some areas (e.g. the Cape of Good Hope Nature Reserve; Macdonald *et al.* 1987), it remains a major problem. Thousands of isolated trees on inaccessible cliffs present a special problem for control. The large plantations of *P. radiata* provide a major reservoir of seeds. The characteristic high winds during the summer, coupled with the periodic high-intensity fires provide outstanding conditions for the spread of pines (Richardson & Brown 1986; Richardson 1988).

22.4 Towards a predictive understanding of pine invasions

22.4.1 Do invasive pines have predictable features?
In section 22.2.2 we listed the pines that have already invaded large areas in the southern hemisphere. However,

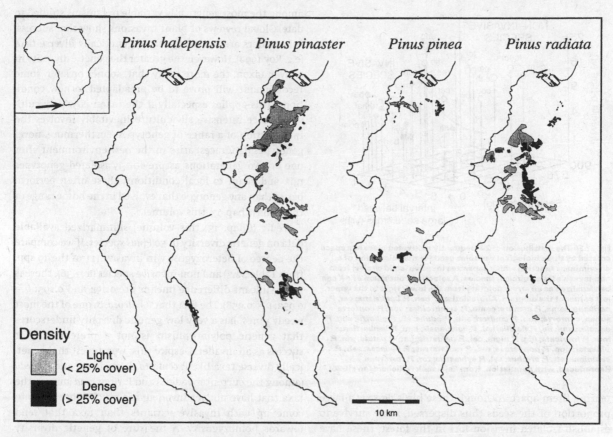

Fig. 22.2. The distribution of *Pinus halepensis, P. pinaster, P. pinea* and *P. radiata* on the Cape Peninsula of South Africa as mapped during 1994 (see Richardson *et al.* 1996b for details). Cover classes: dense (>25% canopy cover); light (<25% canopy cover).

since not all pines have been introduced and disseminated to the same extent, can we make generalizations about those species that have been most successful? Previous chapters in this volume have shown that different pine taxa fulfil a range of ecological roles, from pioneers to late-seral (or 'climax' *sensu* Brzeziecki & Kienast 1994) species. The position of taxa along this continuum is the result of features of the pines themselves and properties of the environment. The need to tease apart the role of these interacting factors has gained new urgency since *some* pine taxa started to invade *some* types of habitat outside their natural ranges. To manage these invasions effectively we need to know whether to aim control measures at the species (e.g. by selecting for non-invasive taxa if possible, or targeting a particular aspect of the species' life history in control programmes), or at manipulating features of the environment (if possible), or a combination of these approaches.

Phylogeny offers no clues on the origin of invasive tendencies in pines, since invasive taxa are distributed among all the major clades in *Pinus* identified by Price *et al.* (Chap. 2, this volume). The most widespread invasive pines (see

section 22.2.2) differ from non-invasive pines in terms of a few, essentially demographic, attributes. For example, the most invasive species in South African fynbos form a distinct functional group of taxa in the genus which are fire-resilient, have small seeds with low seed-wing loadings, short juvenile periods, moderate to high degrees of serotiny and relatively poor fire-tolerance as adults (Richardson, Cowling & Le Maitre 1990). An exception to this 'rule' is *P. pinea* which has large, vertebrate-dispersed seeds. This species has invaded moderately (Richardson & Cowling 1994), thanks to the establishment of an opportunistic mutualism between it and the introduced squirrel, *Sciurus carolinensis*, which disperses its seeds (Richardson *et al.* 1990). This demonstrates the difficulty of predicting whether introduced species will become invaders. Other examples of invading pines behaving differently from what may have been predicted from studying the species in their native ranges have been reported from Australia. In Queensland, the spread of *P. elliottii*, a wind-dispersed pine, has apparently been enhanced significantly by glossy black cockatoos (*Calyptorhynchus lathami*), which carry cones over long distances before

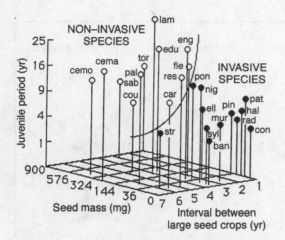

Fig. 22.3. The distribution of 24 frequently cultivated pines in a space created by three biological variables used for the calculation of a discriminant function which separates invasive (solid circles) from non-invasive (open circles) species. A selection continuum of *r-K* can be visualized as an arrow pointing from the lower right to the upper left corner of the diagram. Abbreviations: ban, *P. banksiana*; car, *P. caribaea*; cema, *P. cembra*; cemo, *P. cembroides*; con, *P. contorta* subsp. *contorta*; cou, *P. coulteri*; edu, *P. edulis*; ell, *P. elliottii*; eng, *P. engelmannii*; fle, *P. flexilis*; hal, *P. halepensis*; lam, *P. lambertiana*; mur, *P. muricata*; nig, *P. nigra*; pal, *P. palustris*; pat, *P. patula*; pin, *P. pinaster*; pon, *P. ponderosa*; rad, *P. radiata*; res, *P. resinosa*; sab, *P. sabiniana*; str, *P. strobus*; syl, *P. sylvestris*; tor, *P. torreyana*. Reproduced, with permission, from Rejmánek & Richardson (1996).

tearing them apart. Although these birds destroy a large proportion of the seeds thus dispersed, some survive to establish isolated invasion foci in the forest; these have significantly increased their spread (W.L. Thompson, personal communication). Gill & Williams (1996, p. 273) report dispersal of cones of another wind-dispersed pine (*P. radiata*) by cockatoos into eucalypt forests in Southeastern Australia. In Western Australia, Carnaby's cockatoos (*C. funereus latirostrus*) disperse seeds of *P. pinaster*, also a wind-dispersed species, into remnant *Banksia* woodland around Perth (G.J. Keighery, personal communication). Despite these apparent exceptions, one can establish a protocol for assessing the likelihood of species invading by evaluating the relative success as invaders of a range of species introduced to a particular region (Richardson & Cowling 1992; Tucker & Richardson 1995). Rejmánek & Richardson (1996) derived a robust discriminant function, combining three variables (mean seed mass, minimum juvenile period, and the mean interval between large seed crops) which effectively separates invasive from non-invasive pines (Fig. 22.3). They argued that this function can be used for predicting what other pines are likely to become invasive.

Although all pines have an open-recombination genetic system which theoretically makes them good colonizers, pine taxa display a range of genetic structures, ranging from *P. torreyana* which is totally homozygous to *P. virginiana* with an expected heterozygosity between populations of 0.59 (Chap. 13, this volume). Pines such as the latter are

among the most genetically variable organisms studied to date. Global reviews of plant invasions show that successful invaders are not always from genetically diverse taxa (e.g. Roy 1990). However, the greater the genetic diversity in a given taxon, the more likely that, sooner or later, some recombinant will prove to be preadapted to new conditions. This applies especially if the taxon is widely cultivated since intensive silviculture inevitably involves the introduction of a range of genotypes. Furthermore, novel genotypes (landraces) arise in the new environment after one or two generations as previously isolated genotypes mix and adapt to local conditions; these often perform better than any genotype that evolved in the home range of the taxon (Chap. 13, this volume).

Ledig (Chap. 13, this volume) summarized available data on genetic diversity in 50 *Pinus* species. If we compare the degree of heterozygosity in invasive (11 of the 19 spp. from Table 22.2) and non-invasive species ($n = 36$), there is no significant difference (means are 0.194 and 0.195; $t = 0.052$; $p = 0.958$). The fact that *P. halepensis*, one of the most weedy pines, has a very low genetic diversity underscores that genetic polymorphism is not a prerequisite for success as an invader. Despite this, we suggest that genetically diverse taxa like *P. contorta*, *P. ponderosa* and *P. taeda* (among the current invaders) and *P. virginiana* (among the taxa that have not yet invaded) are much more likely to come up with invasive variants than taxa that tend towards homozygosity. A measure of genetic diversity could therefore be included in a screening model. Rejmánek (1996) demonstrated a negative correlation between invasiveness and nuclear DNA content in pines (see also Chap. 13, this volume, section 13.2.3); this suggests that genome size could also be incorporated in screening models, but further work is required.

Rejmánek (1995) showed that invasive pines (his list corresponds with that in Rejmánek & Richardson 1996) generally have greater mean latitudinal ranges (16.3°; SD = 9.6; $n = 12$) than non-invasive pines (10.4°; SD = 5.2; $n = 12$). Although the difference was not significant ($p > 0.05$) this, he suggests, indicates that the 'primary' (natural) ranges of pines could be used as one factor in predicting their 'secondary' (adventive) ranges. Although this relationship is intuitively appealing (since species with larger ranges should tolerate a wider range of sites), Stevens & Enquist (Chap. 10, this volume) discuss the problems associated with using current distributions of pine taxa to predict future distributions on other continents.

22.4.2 Do invaded areas have predictable features?
The widespread invasion of introduced pine species in many parts of the southern hemisphere provides a useful natural experiment for determining the factors that interact to affect the degree to which pines can invade different

habitats and vegetation types. In section 22.2.4 we discussed the relative openness of broad vegetation categories to invasion. However, many discrete (though often interacting) factors act upon this basic template to determine whether or not pines will spread, and to what extent. Richardson *et al.* (1994a) discussed the role of disturbance (e.g. changed levels of herbivore pressure, modified fire and logging regimes), the composition of biotic communities in the target habitat (e.g. the role of vigorous plants), and other characteristics of invaded areas (e.g. proximity to a seed source) in affecting invadability. Read (Chap. 16, this volume) notes that the dominant plants in vegetation types that appear to be particularly susceptible to invasion by pines (notably shrublands and grasslands) are non-ectomycorrhizal and lack the attributes that make pines such good competitors for nutrients in poor soils subjected to frequent disturbance. He argues that these disadvantages are accentuated because the fast-growing pines are also able to outcompete the shorter native plants for light. Read (Chap. 16, this volume) also suggested that one reason for the relative resistance of woodlands and forests to invasion by pines may be that these systems already have a preponderance of ectomycorrhizal trees; the invading pines therefore do not have a major advantage over the native trees in their ability to scavenge effectively for nutrients. In Chap. 11 (this volume), Agee detailed the role of fire in invasions. We explore the determinants of invadability in different vegetation types further in section 22.4.5.

22.4.3 Patterns of invasion

The following generalized description of the pattern of invasion by wind-dispersed pines, which is essentially similar in all regions, is summarized by Richardson *et al.* (1994a). These invasions, which involve the establishment of dense daughter stands around founder populations ('neighbourhood diffusion' *sensu* Shigesada, Kawasaki & Takeda 1995) and long-distance dispersal, conform with the invasion process termed 'stratified diffusion' by Hengeveld (1989). Satellite foci establish when the winged seeds are dispersed by wind over several kilometres from plantations; distances of up to 100 m are most common but individuals have regularly been recorded 8 km from seed sources, and much further in several cases (W.G. Lee, personal communication reports dispersal of 25 km in New Zealand). Seeds are deposited relatively evenly over a wide range of microsites but a very small proportion of trees become established further than a few hundred metres from plantations (but see section 22.4.1 for discussion of vertebrate dispersal of *P. elliottii* and *P. pinaster* in Australia). Trees attain reproductive maturity after 6–15 years, whereafter seeds are released regularly or, in the case of serotinous species, *en masse* after a major disturbance such as fire kills the mature trees. Most seedlings

establish within two years of fire in fynbos, but seedlings may establish up to 10 years after a shrubland fire in New Zealand (Sanson 1978). Satellite foci enlarge and coalesce to form dense, even-aged stands after fire (Chap. 11, this volume). This stepwise process has been documented for *P. pinaster* in South African fynbos (Richardson & Cowling 1994) and *P. radiata* in New Zealand (Bannister 1965) and fynbos (Richardson & Brown 1986; Fig. 22.4).

In detailing the elements of the 'invasion window' for serotinous pines in South African fynbos, Richardson & Cowling (1992) suggested that the pines behave essentially like the dominant native shrubs (which show marked spatial and temporal fluctuations in population sizes after fires), but that the former eclipse the latter in two key facets of demography: seed dispersal and fire-resilience. The first faculty improves vagility and facilitates spread. Once establishment barriers are overcome, the fynbos is transformed (often over 2–3 fire cycles) into a pine forest (Fig. 22.4). The superior fire-resilience of the pine populations buffers them against local extinction, thus disrupting the prevailing non-equilibrium system and entrenching a depauperate steady state (Richardson & Cowling 1992) which is virtually irreversible without a fundamental change in the disturbance regime (Bond & Richardson 1990).

22.4.4 A summary of factors influencing pine invasions

One way of assessing the relative importance of the various, interacting factors that affect the outcome of an introduction is to 'load' each factor and then model the outcome under various scenarios. Table 22.4 provides a list of factors that can be used to build a rule-based model or expert-system for tracking the fortunes of particular species under a range of conditions (Tucker & Richardson 1995 give an example).

22.4.5 A spatially-explicit model of pine invasions

The preceding sections have identified important correlates of invasive success for pines in a range of habitats in the southern hemisphere. There has, however, been no attempt to operationalize these correlates to gain a mechanistic understanding of the processes that control these (or any other) invasions. We need a model that combines 'invasion window' (*sensu* Johnstone 1986) and 'vital attribute' (*sensu* Noble & Slatyer 1980) concepts. Such a model would be useful as both an heuristic and a predictive tool, and would therefore offer a novel approach for addressing the fundamental questions already posed in this chapter. (1) How does environment type influence invadability? (2) How does disturbance influence invadability? (3) How does plant strategy influence invadability? (4) How do the interactions between plant strategy and

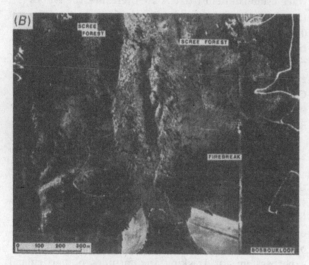

Fig. 22.4. **The invasion of South African fynbos near Stellenbosch, Western Cape, South Africa, by pines (*P. halepensis, P. pinaster* and *P. radiata*). A shows the uninvaded fynbos in 1938 before the establishment of the Bosboukloof plantation. *Pinus radiata* spread rapidly from the plantation, and *P. halepensis* and *P. pinaster* invaded from self-sown stands in other parts of the Jonkershoek Valley. Photograph *B*, taken in 1981, shows the significant change in vegetation structure caused by these invasions. Reproduced, with permission, from Richardson & Brown (1986).**

disturbance influence invadability? and ultimately, (5) which plant strategies will be successful in which environments under which disturbance regime? In this section we review an attempt by Higgins & Richardson (1997) to develop such a model.

The model used, as a case study, the invasion of two types of pine tree into three environment types under five levels of disturbance. These environment types are: a fire-prone shrubland (e.g. fynbos; Cowling, Richardson & Mustart 1997); a montane grassland (e.g. grasslands of the eastern mountains and escarpment; O'Connor & Bredenkamp 1997); and a forest (e.g. Afromontane temper-

ate forest in the Knysna region; Midgley *et al.* 1997). The natural (pristine) disturbance regimes of these environments are treefalls in forest; intense, infrequent fires in fynbos; and less intense, frequent fires in grassland. These natural disturbance regimes are impacted by humans (Richardson *et al.* 1994a): in fynbos humans start fires and thereby increase the fire frequency; in grasslands domestic livestock grazing reduces fuel loads, this results in more patchy and less intense fires; in forests human logging increases the rate of treefall. The two pine life-history strategies that we studied represent two of the five life-history types described by Keeley & Zedler (Chap. 12, this volume). The first pine type, '**pioneer-pine**', corresponds with Keeley & Zedler's category of pines that occur in areas with predictable stand-replacing fires. The second pine type, '**late-seral-pine**', corresponds with Keeley & Zedler's category of pines that occur in areas with unpredictable stand-replacing fires. These two groups were selected because they include many of the pine species which are invasive in the southern hemisphere (Chap. 12, this volume; Table 22.1). The pioneer-pine has canopy-stored seeds which are released in response to fire (serotiny); short juvenile periods; thin bark (low probability of surviving intense fires); shade-intolerant seedlings; and annual cone production. Some pioneer-pines do resprout, but these were not considered. The late-seral pine type is not serotinous and has a slightly longer juvenile period and a thicker bark than the pioneer-pine. Its ability to self-prune confers on its adults a higher probability of fire survival. Seedlings are shade-tolerant, and there is more infrequent cone production than the pioneer-pine (it is a mast seeder). Both pine types produce light and hence relatively dispersible seeds.

This information on the case study was operationalized to give a series of rules which allowed the model to simulate the distribution and intensity of disturbance, plant mortality, seed production, seed dispersal and seedling recruitment. Disturbance and its impact on mortality were modelled as follows. The chance of a *fire* occurring, and spreading across the modelling landscape (for the fynbos and grassland simulations), was a function of the age of the vegetation and the flammability of the vegetation. A stochastic evaluation of the probability of fire ignition was used to determine fire occurrence; similar stochastic rules were used iteratively to simulate the spread of fire to neighbouring cells. The probability of an invading pine being killed by a fire was a function of its age and the fire intensity. Because fires are less intense in grassland than in fynbos, different mortality–probability curves were used for each environment. A stochastic rule was used to translate the probability of mortality into mortality events for individual trees. *Treefalls* were simulated annually at randomly selected sites; treefalls affected

Table 22.4. *A summary of the major factors that determine whether an introduced pine will spread from planting sites in the southern hemisphere. Factors should be viewed as a series of 'loadings' – the greater the number of facilitating features, the greater the chance of invasion for any taxon/site combination*

Factors	Facilitating feature(s)	Limiting feature(s)
Species attributes:		
Seed mass	Small seeds with large wings	Large seeds with small wings
Juvenile period	Short (<10 yrs)	Long (> 10 yrs)
Interval between large seed crops	Short (<3 yrs)	Long (> 5 yrs)
Ability to survive moderate browsing	Good	Poor
Residence time	Long (>50 yrs)	Short (<50 yrs)
Extent of planting		
Total area	Large	Small
Boundary:total area ratio	Large	Small
Ground-cover characteristics:		
Basic vegetation structure	Bare or sparsely-vegetated ground, shrubland, grassland	Forest
Vegetation cover	None – low (<50%)	High (>80%)
Distance from the equator	Latitude: 30–45°S	Latitude: <30° S
Disturbance		
Frequency	Low–moderate	Very low / very high
Human-induced types	Moderately increased herbivore pressure (grazing, browsing, trampling) or equivalent	Greatly increased or greatly reduced herbivore pressure or equivalent
Contributing factor	Decreased competition from ground-cover	Increased competition from ground-cover or physical elimination of pines (e.g. by mechanical clearing)
Natural types	Slope instability, wind, flooding, fires from volcanoes	Frequent fires (e.g. in grasslands)
Resident biota		
Composition of plant community	Naturally-invadable community (e.g. *Dracophyllum subulatum* shrubland, *Chionochloa* tussock grassland in New Zealand)	Naturally resistant community (e.g. *Eucalyptus blakelyi* forest in NE Victoria, *Protea nitida* fynbos in South Africa)
Indicators of invadability	Conditions unsuitable for C_4 photosynthetic pathway and nutrient-poor soils: paucity of vigorous herbs	Conditions suitable for C_4 photosynthetic pathway and nutrient-rich soils: abundance of vigorous herbs
Role of mammals other than humans	Removal of competing vegetation (e.g. through grazing) Dispersing pine seeds (birds and mammals)	Destroying pine seedlings (browsing, trampling)
Role of fungi	Presence of appropriate mycorrhizal symbionts	Absence of mycorrhizal symbionts (no longer limiting? see Chap. 16) Influence of pathogenic fungi

Source: After Richardson *et al.* (1994a).

an area the size of a single tree canopy and always resulted in the death of the tree in that site. The model assumed that the number of propagules available for dispersal is a function of: the tree's age; the tree's maximum reproductive potential; whether it is a mast year (for the late-seral-pine) or whether a fire has occurred (pioneer-pine); and the time of last seed release. Only seed dispersal by wind was considered. The simulated dispersal distances followed a negative exponential distribution. The dispersal distance and a randomly selected direction were used to determine the location of the recruit. Seeds in safe sites (a treefall gap or an unoccupied site in a recently-burnt landscape) were allowed to recruit. The model ignored post-recruitment, density-dependent processes.

The model was then used to define a simulation experiment, which explored the key questions listed above. Thirty different invasion permutations were simulated in a fully factorial design (2 × pine-type, 3 × environment-type, 5 × disturbance level). The mean invasion rate (slope of number of plants vs time regression) was estimated from each replicate ($n = 20$) simulation run. The results showed: that the rate of invasion is increased by increasing levels of disturbance; that grassland and fynbos are more invadable than forest; and that the pioneer-pine is more invasive than the late-seral-pine (Fig. 22.5). These results are in agreement with the results of the case studies (Table 22.2). A closer look at the results showed that these generalizations do not hold for all situations (Fig. 22.6). The two pine types differed substantially in their responses. The pioneer-pine invaded grassland and forest rapidly, and forest slowly; its invasiveness increased linearly with disturbance in fynbos, asymptotically in grassland and unimodally in forest. The late-seral-pine shows a less spectacular, but more consistent, ability to invade under all conditions. Natural forest was the environment type most resistant to late-seral-pine invasion. Invasiveness of the late-seral-pine increased linearly with disturbance in forest and responded unimodally with disturbance in grassland. Disturbance of fynbos did not influence the invasion rate of the late-seral-pine.

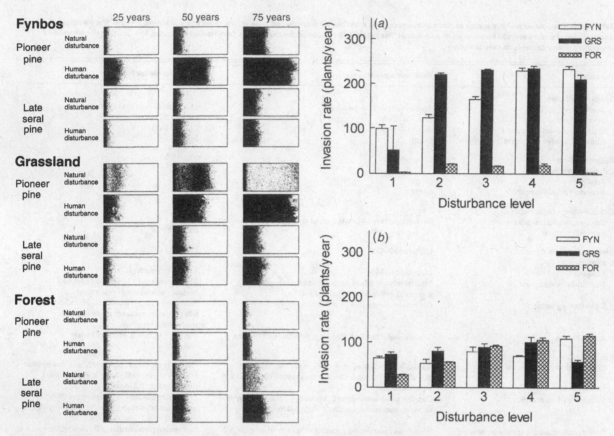

Fig. 22.5. **The simulated invasion of two pine prototypes ('pioneer' and 'late-seral'; see text) from the edge of a plantation (thick black line on the left of each block) into three vegetation types that are frequently invaded by pines in the southern hemisphere (see Table 22.2). Invasions were modelled under natural and 'modified' conditions (disturbance level 3 in Fig. 22.6). The illustrations show the extent of invasions over areas of 2.5 km² (1 km × 2.5 km) 25, 50 and 75 years after the start of the invasion.**

Fig. 22.6. **Means and standard deviations (n = 20) of invasion rates for (a) 'pioneer' and (b) 'late-seral' pine prototypes (see text) invading fynbos (FYN), grassland (GRS) and forest (FOR) under five disturbance regimes (1 = pristine; 5 = highly modified; see text). Data from Higgins & Richardson (1997).**

These results clearly show the importance of interactions between life-history traits of the invader and environmental features in shaping invasion dynamics (see also Higgins, Richardson & Cowling (1996) for further discussion). Higgins & Richardson (1997) argued that this suggests that our ability to predict invasions can be improved by embedding *interactions* in our theories of plant invasions. The model's predictions are also likely to be useful for generating management recommendations: for example, pine invasions in Afromontane forests can be prevented by simply regulating treefelling; the invasion of montane grasslands can be controlled by promoting frequent and intense fires; invasions in fynbos shrublands can be slowed (to some extent) by maintaining infrequent and hence intense fires. Grassland and shrubland systems should thus be allowed to burn when high fuel biomass and dry weather conditions coincide. The results show that invasions of shrublands can occur in pristine systems,

and that simple manipulation of fire regimes cannot keep pine invaders out. Theoretically, more intense and more frequent fires could prevent invasions in fynbos. Unfortunately, because fire intensity decreases as fire frequency increases (Bond & van Wilgen 1996), this strategy is not viable. Invasion of the pioneer-pine into natural grassland showed a relatively variable response, suggesting that the stochastic variation in fire frequency and fire distribution can sometimes prevent invasions from occurring. This suggests that the invadability of undisturbed grasslands will depend strongly on fire conditions (e.g. dryness of biomass, wind strength, humidity, temperature).

22.5 Managing pine invasions

Despite the many factors that have been shown to interact to determine whether a pine will invade, we now have a good understanding of the processes involved. Insights from the above sections can be used to furnish guidelines for improved management of invasive pines. We first review the current control methods and then discuss how these can be improved. We then consider ways to lessen the extent and impacts of pine invasions in the future.

22.5.1 Dealing with current invaders

The control of invading pines has received considerable attention in South African fynbos and in New Zealand. A systematic control strategy for pine control has evolved over the past 20 years in fynbos where major constraints are the rugged terrain which hampers access, and the need to integrate control plans with schedules for prescribed burning in the fire-prone vegetation (Richardson *et al.* 1994b). The slow fuel-accumulation rates mean that fynbos can seldom be burnt at intervals of less than *c.* 5 years. Fire cycles of less than *c.* 10 years are detrimental to the seed-regenerating shrubs that dominate most fynbos communities. Current management plans in most parts of the region prescribe burns at intervals of 12–15 years. Wildfires introduce an element of stochasticity which is desirable for maintaining species diversity (van Wilgen, Richardson & Bond 1992), but cause major problems for pine control. Pine control involves felling trees 12–18 months before scheduled burns to allow the release of seeds from the serotinous cones (see Chap. 11, this volume). Seeds are released close to the ground in the mature vegetation and are therefore not dispersed over long distances by wind. Some seeds are eaten by granivores before the planned fire and most seedlings that do establish are killed by the fire. Seedlings that survive the fire, or that invade from outside between fires, are hand-pulled before they set seed. This 'cut, burn and follow-up' method is very expensive but is the only practical way of clearing stands of alien pines and reducing impacts. Fynbos supports few large herbivores and provides very poor grazing for livestock; it is therefore not feasible to manipulate grazing/browsing pressure to control pine seedlings as is done successfully in New Zealand. Fire, combined with manual felling, is thus the only feasible means of introducing two stand-destroying disturbances in sufficiently short succession to exclude pines.

Invasive pine control in New Zealand has focused on the unimproved grasslands of the South Island high country and on conservation land on the Central Volcanic Plateau. Intensive grazing (sometimes combined with herbicide application) is effectively applied to control pines (see Richardson *et al.* 1994a for details). Fire is seldom used because of the risks of soil erosion. Physical removal, sometimes complemented by herbicide application, is the major control measure in Australia, although fire is used to control *P. pinaster* at French Island and at the Wonthaggi

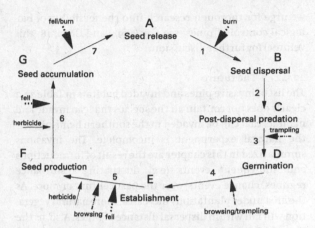

Fig. 22.7. **A schematic representation of the critical demographic stages in the life cycle of invading pines from a management perspective. Management options depend on features of the environment but include various forms of ecosystem management to direct succession, notably the manipulation of herbivory levels (which controls pines by browsing and/or trampling) and/or fire frequency in phases 3, 4 and 5). The possibility of using stand-replacing disturbances at sufficiently short intervals also depends on the environment.**

Heathland Coastal Reserve in Victoria (Kermode 1993).

The means currently applied to control pines involve mechanical methods (sometimes in conjunction with fire), herbicide application, and the manipulation of large mammal populations to reduce establishment and survival to maturity. These approaches target different demographic stages (Fig. 22.7) and their applicability depends on features of the environment. For example, in section 22.4.5 we showed that pines can be effectively controlled in grasslands by increasing the fire frequency to exclude pines before they attain reproductive maturity. Manipulating the abundance of keystone taxa in the receiving environment to reduce the incidence of safe sites and/or to eliminate established seedlings (through browsing and trampling) is also an option in grasslands. The model described in section 22.4.5 can be used to determine the dimensions of the disturbance regime required to control pines.

Biological control has not yet been employed as a means of reducing the spread of pines from plantations. Despite the excellent results achieved using introduced seed-attacking insects against other alien woody plants (e.g. Neser & Kluge 1986), forestry authorities have been reluctant to sanction research on biological control options for pines. In South Africa, this reluctance is partly because of the lucrative export market for pine seeds. Another concern is that potentially promising seed-attacking insects of the genus *Conophthorus* are possible vectors of the fungus that causes pitch canker, a disease that has devastated pine plantations in various parts of the world (Chap. 19, this volume). Despite the conflicts of interests, we urge for thorough research into the feasibility of biological control of pines; see Kay (1994) and Chap. 18 (this volume) for further discussion.

22.5.2 The future

The list of invasive pines and invaded habitats in Table 22.2 clearly does not contain all the species that can invade and habitats that can be invaded in the southern hemisphere – the natural experiment is incomplete. The invasions summarized in this chapter are the result of introductions and subsequent events (e.g. fluctuating disturbance regimes, chance events) over the past century or more. As the area under plantations increases, more natural vegetation will fall within dispersal distance of pines. Also, as the time since initial introductions increases, various barriers that have prevented the establishment and spread of some species will be surmounted. Figure 22.8 shows that the various barriers that have limited pine invasions since the 17th century have been weakened (see previous sections). This trend will continue into the 21st century. Furthermore, the rapid changes in the global economy, forestry methods, land-use practices and global climate mean that we can expect significant changes in the extent of pine invasions.

Taking all the aforementioned factors and likely scenarios for pine cultivation in the southern hemisphere over the next few decades into account (e.g. Chap. 21, this volume), we suggest that the species listed in Table 22.2 will continue to be the most important invasive pines until at least the middle of the 21st century (see also Richardson 1996). There will be some 'filling in', as some species that have not had time to spread widely expand their ranges. *Pinus radiata* will definitely become more widespread in many areas. The extent of plantations of this species has increased dramatically in Pacific Rim countries since the early 1960s (Australia), mid- to late 1960s (New Zealand) and the mid-1970s (Chile) (Chap. 21, this volume). The impacts of these increased plantings on the extent of invasions have yet to be observed, or at least reported. It is, however, inevitable that many new areas will be invaded. *Pinus caribaea, P. kesiya* (both already invading) and *P. oocarpa* will probably become (much more) widespread invaders of tropical and subtropical areas (see section 22.1). Other Central American pines besides *P. caribaea* and *P. oocarpa* are being planted more widely; these include *P. ayacahuite, P. chiapensis, P. greggii, P. herrerae, P. leiophylla, P. maximinoi, P. pringlei, P. tecunumanii* and *P. teocote* (Dvorak & Donahue 1992). Too little is known about the biology of these pines to facilitate screening for invasive potential (some life-history information is indispensable for screening; Tucker & Richardson 1995). The gathering of such data should be an integral part of risk assessment prior to afforestation with these taxa. Another factor that

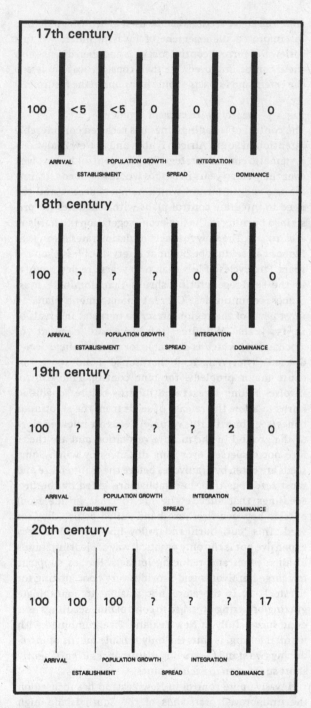

Fig. 22.8. **A schematic representation of the changes, over 400 years, in the effectiveness of various barriers (following Bazzaz 1986) in limiting the spread of introduced pines in the southern hemisphere. The thickness of lines represents the relative importance of that barrier in limiting the success of pines as invaders. Numbers indicate the number of pine species (expressed as a percentage of the 111 species in the genus) that have surmounted each barrier.**

could alter the trajectory of invasions is hybridization. For example, in southeastern Queensland, Australia, F1 and F2 hybrids of *P. caribaea* and *P. elliottii* show higher growth rates than the parental species. Hybrids also attain reproductive maturity before either parental species, and flower much more heavily (M. Dieters, unpublished data). Although we could find no record of the hybrid invading native forest, it seems very likely that it will spread. There are examples from other tree genera that post-introduction hybrids are more invasive than parent species (e.g. *Prosopis* in South Africa; Poynton 1990). Given the success of the *P. caribaea* × *P. elliottii* hybrid, and the fact that pine hybrids can usually be vegetatively replicated from seedlings (an important advantage in commercial forestry), we can expect further experimentation with hybrids. Some hybrids that are currently being tested in experimental plantings are *P. patula* × *P. greggii*, *P. greggii* × *P. radiata*, *P. tecunumanii* × *P. caribaea* and *P. elliottii* × *P. tecunumanii* (W.S. Dvorak, personal communication). The implications of the increasing importance of pine hybrids in forestry, and of artificial selection programmes, for predicting invasions requires further study.

Long-term planning to reduce the deleterious effects of pine invasions should ideally involve selecting appropriate species, avoiding planting on 'take-off sites' for dispersal of seeds by wind, and ensuring appropriate management on adjacent land are considered essential for preventing the initial spread of pines (Ledgard 1988). Screening protocols have been developed for assessing the risk of particular species invading certain environments (Richardson & Cowling 1992; Tucker & Richardson 1995; Rejmánek & Richardson 1996). Hughes (1994) also provides suggestions for risk assessment. We suggest that these methods provide a sound basis for screening pines. Whether these developments will reduce the interregional transfer of proven invasive species in the face of strong economic incentives remains to be seen. Screening can, however, serve a valuable role in initiating long-term plans for environmental management in areas that adjoin pine plantations, and for prioritizing control measures.

22.6 Conclusions

Kruckeberg (1986) commented on the exceptional opportunities that conifer invasions provide for gaining important new perspectives on the demography of this important group. We suggest that the natural experiment of pine cultivation and subsequent spread in the southern hemisphere has indeed yielded much valuable information that would be very difficult to assemble within the natural range of the genus. For example, new insights on

the interactions between pine life-history traits and habitat features have emerged (see 22.4.5). These show that pine invasions epitomize the 'stratified diffusion' process described by Hengeveld (1989). This process, and the capacity of pines to respond rapidly to changing disturbance regimes, enabled pines to colonize large areas of the northern hemisphere rapidly following deglaciation; it also explains why some taxa are tenacious persisters under a wide range of disturbance regimes in their natural range. There is considerable potential for using models such as the one described in this chapter for improving the efficiency of range and forest management in the natural range of pines. For example, the model could easily be modified to study the dynamics of a rare pine such as *P. maximartinezii* in Mexico (currently threatened by uncontrolled seed collection and other factors; Chap. 1, this volume) under a range of management (land-use) options.

Pines are now conspicuous, or even dominant, components of natural vegetation over large parts of the southern hemisphere, and their importance as environmental weeds has increased rapidly in the second half of the 20th century. We have discussed the history of these invasions and have proposed various explanations for the success of some taxa and the apparent susceptibility of certain habitats to invasion. Management of pine invasions clearly requires the integration of a range of strategies, ranging from selecting appropriate species for new afforestation projects, to the manipulation of environment features to impede invasion. Methods are now available for screening pines to assess the likelihood of them spreading in different habitats; we recommend that these procedures be used for long-term planning.

The perspectives summarized in this chapter have been gleaned mainly from the experiences in Australia, New Zealand and South Africa; we now need to test whether these generalizations apply in other parts of the southern hemisphere, where pines have been widely planted but where invasions have not yet started or are much less widespread (e.g. Argentina, Brazil, Colombia, Kenya, Madagascar, Tanzania, Uganda, Zambia and Zimbabwe). Argentina in particular appears to offer exciting opportunities for further work, since several pines are widely naturalized (in decreasing order of abundance: *P. elliottii*, *P. taeda*, *P. contorta* subsp. *latifolia*, *P. ponderosa*, *P. radiata* and *P. jeffreyi*; Cozzo 1994).

This chapter has dealt only with the spread of alien pines in the southern hemisphere. Pines have, however, also spread from plantings well outside their natural range in the northern hemisphere, where they too are aliens. For example, *P. contorta* is spreading in parts of Scotland where deer and sheep grazing pressure has been reduced with a view to regeneration and recuperation of the remnant native *P. sylvestris* forest. *Pinus nigra* has spread from plantings on the sand dunes around Lake Michigan in the USA, where it is having significant impacts on the native plant communities, nutrient cycling and soil development (L.M. Leege, unpublished data). *Pinus radiata* also regenerates naturally in burnt areas, along roadsides and in abandoned fields in parts of in northern Spain (J. Loidi, personal communication). *Pinus sylvestris* has invaded peatlands in Ontario, Canada, where it has reduced the cover of native plants and the height of hummocks (Ripmeester 1996). It is likely that the impacts of alien pine spread in the northern hemisphere could be similar to those described in Table 22.3. Furthermore, invasive alien pines in the northern hemisphere come into contact with other pines which have evolved in isolation. The significance of such artificial sympatry, not yet explored in detail, could be momentous for the future of the genus.

Several other northern hemisphere conifers are also invading in the southern hemisphere. Notable among these is the Douglas-fir (*Pseudotsuga menziesii*) which invades relatively intact native vegetation in the Bariloche area of Argentina (T.T. Veblen, personal communication). Many of the generalizations described

for pines in this chapter probably also apply to other northern conifers.

Acknowledgements

We thank Bob Boardman, Jeremy Burdon, Richard Carter, David Cooke, Mark Dieters, Bill Dvorak, Olivier Gargominy, Richard Hobbs, Colin Hughes, Tanguy Jaffre, Greg Keighery, Nick Ledgard, Bill Lee, Lissa Leege, Chris Lusk, Rachel McFadyen, Mike Mulvaney, Marcel Rejmánek, Stephen Skull, Bill Thompson, Hernan Trebino, Tom Veblen, Carol West, Peter Williams and Ross Wylie for providing useful information. We are grateful to Richard Cowling, Colin Hughes, Tom Ledig, Tim O'Connor and Christophe Thebaud for their helpful comments on various drafts of the manuscript. Richard Carter, Coert Geldenhuys, Greg Forsyth and Nick Ledgard kindly scouted for photographs. Our work on pine invasions is funded by the Institute for Plant Conservation, the Foundation for Research Development, the University of Cape Town Research Committee, and the Mazda Wildlife Fund.

References

Anon. (1962). *Alien Pines or Native Tussock? in Tongariro National Park*. Wellington: Tongariro National Park Board.

Bannister, M.H. (1965). Variation in the breeding system of *Pinus radiata*. In *The Genetics of Colonizing Species*, ed. H.G. Baker & G.L. Stebbins, pp. 353–73. New York: Academic Press.

Bazzaz, F.A. (1986). Life history of colonizing plants: some demographic, genetic, and physiological features. In *Ecology of Biological Invasions of North America and Hawaii*, ed. H.A. Mooney & J.A. Drake, pp. 96–110, New York: Springer-Verlag.

Berry, S. & Mulvaney M. (1995). *An Environmental Weed Survey of the Australian Capital Territory*. Canberra: Conservation Council of the Southeast Region and Canberra.

Bond, W.J. & Richardson, D.M. (1990). What can we learn from extinctions and invasions about the effects of climate change? *South African Journal of Science*, **86**, 429–33.

Bond, W.J. & van Wilgen, B.W. (1996). *Fire and Plants*. London: Chapman & Hall.

Bolus, H. (1886). Sketch of the flora of South Africa. In *Official Handbook of the Cape of Good Hope*, ed. J. Noble, pp. 288–319. Cape Town: Solomon.

Brzeziecki, B. & Kienast, F. (1994). Classifying the life-history strategies of trees on the basis of the Grimian model. *Forest Ecology and Management*, **69**, 167–87.

Burdon, J.J. & Chilvers, G.A. (1994). Demographic changes and the development of competition in a native Australian eucalypt forest invaded by exotic pines. *Oecologia*, **97**, 419–23.

Carr, G.W., Yugovic, J.V. & Robinson, K.E. (1992). *Environmental weed invasions in Victoria*. Melbourne: Department of Conservation and Environment/Ecological Horticulture Pty Ltd.

Cheeseman, T.F. (1925). *Manual of the New Zealand Flora*. Wellington: Government Printer.

Chilvers, G.A. & Burdon, J.J. (1983). Further studies on a native Australian eucalypt forest invaded by exotic pines. *Oecologia*, **59**, 239–45.

Cockayne, L. (1928). *The Vegetation of New Zealand*. Leipzig: Verlag von Wilhelm Engelmann.

Cooper, A.B., Hewitt, J.E. & Cooke, J.G. (1987). Land use impacts on stream water nitrogen and phosphorus. *New Zealand Journal of Forest Science*, **17**, 179–92.

Cowling, R.M., Costanza, R. & Higgins, S.I. (1997). Services supplied by South African fynbos ecosystems. In *Nature's Services: Societal dependence on Natural Ecosystems*, ed. G. Daily, pp. 345–62. Washington, DC: Island Press.

Cowling, R.M., Macdonald, I.A.W. & Simmons, M.T. (1996). The Cape Peninsula, South Africa: physiological, biological and historical background to an extraordinary hot-spot of biodiversity. *Biodiversity & Conservation*, **5**, 527–50.

Cowling, R.M., Richardson, D.M. & Mustart, P.J. (1997). Fynbos. In *Vegetation of Southern Africa*, ed. R.M. Cowling, D.M. Richardson & S.M. Pierce, pp. 99–130. Cambridge: Cambridge University Press.

Cozzo, D. (1994). Conversion de

plantaciones forestales de especies exoticas en sistemas sotenibles en Argentina. *Investigación Agraria Sistemas y Recursos Forestales*, **3**, 31–42.

Cromer, D.A.N. (1956). Australia. In *A World Geography of Forest Resources*, ed. S. Haden-Guest, J.K. Wright & E.M. Teclaff, pp. 573–90. New York: Ronald Press.

Cronk, Q.C.B. & Fuller, J.L. (1995). *Plant Invaders. The Threat to Natural Ecosystems*. London: Chapman & Hall.

Davis, M.R. & Lang, M.H. (1991). Increased nutrient availability in topsoils under conifers in the South Island high country. *New Zealand Journal of Forest Science*, **21**, 165–79.

Dawson, M.P., Florence, R.G., Foster, M.B. & Olsthoorn, A. (1979). Temporal variation of *Pinus radiata* invasion of eucalypt forest. *Australian Forestry Research*, **9**, 153–61.

Dixon, S. (1892). The effects of settlement and pastoral occupation in Australia upon the indigenous vegetation. *Transactions of the Royal Society of South Australia*, **15**, 195–206.

Drake, J.A., Mooney, H.A., Di Castri, F., Groves, R.H., Kruger, F.J., Rejmánck, M. & Williamson, M. (eds.) (1989). *Biological Invasions. A Global Perspective*. Chichester: John Wiley.

Dvorak, W.S. & Donahue, J.K. (1992). *CAMCORE Cooperative Research Review 1980–1992*. Raleigh, NC: CAMCORE, Department of Forestry, North Carolina State University.

Enright, N.J. & Hill, R.S. (eds.) (1995). *Ecology of the Southern Conifers*. Melbourne: Melbourne University Press.

Gibbs, M. (1989). Report on native orchids at Iwitahi. *Orchids in New Zealand*, **15**, 38–41.

Gill, A.M. & Williams, J.E. (1996). Fire regimes and biodiversity: the effects of fragmentation of southeastern Australian eucalypt forests by urbanization, agriculture and pine plantations. *Forest Ecology and Management*, **85**, 261–78.

Hall, A.V. (1961). Distribution studies of introduced trees and shrubs in the Cape Peninsula. *Journal of South African Botany*, **27**, 101–10.

Henderson, L. (1989). Invasive woody plants of Natal and the north-eastern Orange Free State. *Bothalia*, **19**, 237–61.

Henderson, L. (1991). Invasive woody plants of the Orange Free State. *Bothalia*, **21**, 73–89.

Henderson, L. (1992). Invasive alien woody plants of the eastern Cape. *Bothalia*, **22**, 119–43.

Henderson, L. & Musil, K.J. (1984). Exotic woody plant invaders of the Transvaal. *Bothalia*, **15**, 297–313.

Hengeveld, R. (1989). *Dynamics of Biological Invasions*. London: Chapman & Hall.

Higgins, S.I. & Richardson, D.M. (1997). Pine invasions in the southern hemisphere: modelling interactions between organism, environment and disturbance. *Plant Ecology*, **135**, 79–93.

Higgins, S.I., Richardson, D.M. & Cowling, R.M. (1996). Modelling invasive plant spread: The role of plant-environment interactions and model structure. *Ecology*, **77**, 2043–54.

Holm, L., Pancho, J.V., Herberger, J.P. & Plucknett, D.L. (1979). *A Geographical Atlas of World Weeds*. New York: John Wiley.

Holzner, W. & Numata, N. (eds.) (1982). *Biology and Ecology of Weeds*. The Hague: Junk.

Hughes, C.E. (1994). Risks of species introductions in tropical forestry. *Commonwealth Forestry Review*, **73**, 243–52.

Humphries, S.E., Groves, R.H. & Mitchell, D.S. (1991). Plant invasions of Australian ecosystems. A status review and management directions. In *Kowari 2. Plant Invasions. The Incidence of Environmental Weeds in Australia*, pp. 1–127. Canberra: Australian National Parks and Wildlife Service.

Hunter, G.G. & Douglas, M.H. (1984). Spread of exotic conifers on South Island rangelands. *New Zealand Journal of Forestry*, **29**, 78–96.

Hutchins, D.E. (1897). The Cluster Pine at Genadendal. *Agricultural Journal*, **13**, 528–41.

Johnstone, I.M. (1986). Plant invasion windows: a time-based classification of invasion potential. *Biological Review*, **61**, 369–94.

Johns, J. & Molloy, B. (1983). *Native Orchids of New Zealand*. Wellington: Reed.

Kay, M. (1994). Biological control for invasive tree species. *New Zealand Forestry*, **39**(3), 35–7.

Keighery, G.J. (1989). *Banksia* woodland weeds. *Journal of the Royal Society of Western Australia*, **71**, 111–12.

Kermode, L. (1993). The invasion of heath vegetation by *Pinus pinaster* at French Island and Wonthaggi, Victoria. B.Sc. (Hons.) thesis, La Trobe University, Bundoora, Victoria.

Kraehenbuehl, D.N. (1981). Samuel Dixon: A Biography. In *The Full Story Of Flinders Chase*, enlarged edn, by S. Dixon, pp. 9–12. Adelaide: Field Naturalist Section, Royal Society of South Australia.

Kruckeberg, A.R. (1986). The birth and spread of a plant population. *American Midland Naturalist*, **116**, 403–10.

Kruger, F.J., Richardson, D.M. & van Wilgen, B.W. (1986). Processes of invasion by alien plants. In *The Ecology and Control of Biological Invasions in Southern Africa*, ed. I.A.W. Macdonald, F.J. Kruger & A.A. Ferrar, pp. 145–55. Cape Town: Oxford University Press.

Kullman, L. (1990). Dynamics of altitudinal tree-limits in Sweden: a review. *Norsk. geogr. Tidsskr.*, **44**, 103–16.

Ledgard, N.J. (1988). The spread of introduced trees in New Zealand's rangelands – South Island high country experience. *Journal of the Tussock Grasslands and Mountain Lands Institute Review*, **44**, 1–8.

Ledgard, N.J. & Baker, G.C. (1988). *Mountainland Forestry. Forestry Research Bulletin 145*. Christchurch: Ministry of Forestry.

Le Maitre, D.C., van Wilgen, B.W., Chapman, R.A. & McKelly, D.H. (1996). Invasive plants and water resources in the Western Cape Province, South Africa: modelling the consequences of a lack of management. *Journal of Applied Ecology*, **33**, 161–72.

Lister, M.H. (1959). A forgotten experiment in tree planting. *Journal of the South African Forestry Association*, **33**, 59–66.

Macdonald, I.A.W., Clark, D.L. & Taylor, H.C. (1987). The alien flora of the Cape of Good Hope Nature Reserve. *South African Journal of Botany*, **53**, 398–404.

Macdonald, I.A.W. & Jarman, M.L. (eds.) (1984). Invasive alien organisms in the terrestrial ecosystems of the fynbos biome. *South African National Scientific Programmes Report*, **85**, 1–72.

Macdonald, I.A.W. & Jarman, M.L. (eds.) (1985). Invasive alien plants in the terrestrial ecosystems of Natal, South Africa. *South African National Scientific Programmes Report*, **118**, 1–88.

Macdonald, I.A.W. & Richardson, D.M. (1986). Alien species in terrestrial ecosystems of the fynbos biome. In *The Ecology and Management of Biological Invasions in Southern Africa*, ed. I.A.W. Macdonald, F.J. Kruger, & A.A. Ferrar, pp. 77–91. Cape Town: Oxford University Press.

Macdonald, I.A.W., Richardson, D.M. & Powrie, F.J. (1986). Range expansion of the hadeda ibis *Bostrychia hagedash* in southern Africa. *South African Journal of Zoology*, **21**, 331–42.

MacKee, H.S. (1994). *Catalogue des Plantes Introduites et Cultivées en Nouvelle-Calédonie*. Paris: Museum National d'Histoire Naturelle.

McLachlan, D., Moll, E.J. & Hall, A.V. (1980). Resurvey of the alien vegetation in the Cape Peninsula. *Journal of South African Botany*, **46**, 127–46.

Midgley, J.J., Cowling, R.M., Seydack, A.H.W. & van Wyk, G.F. (1997). Forest. In *Vegetation of Southern Africa*, ed. R.M. Cowling, D.M. Richardson & S.M. Pierce, pp. 278–99. Cambridge: Cambridge University Press.

Moll, E.J. & Trinder-Smith, T. (1992). Invasion and control of alien woody plants on the Cape Peninsula Mountains, South Africa – 30 years on. *Biological Conservation*, **60**, 135–43.

Mirov, N.T. (1967). *The Genus Pinus*. New York: Ronald Press.

Neser, S. & Kluge, R.L. (1986). The importance of seed-attacking agents in the biological control of invasive alien plants. In *The Ecology and Management of Biological Invasions in Southern Africa*, ed. I.A.W. Macdonald, F.J. Kruger, & A.A. Ferrar, pp. 285–93. Cape Town: Oxford University Press.

Noble, I.R. & Slatyer, R.O. (1980). The use of vital attributes to predict successional changes in plant communities subject to recurrent disturbance. *Vegetatio*, **43**, 5–21.

O'Connor, T.G. & Bredenkamp, G.J. (1997). Grassland. In *Vegetation of Southern Africa*, ed. R.M. Cowling, D.M. Richardson & S.M. Pierce, pp. 215–57. Cambridge: Cambridge University Press.

Pyšek, P. (1995). Recent trends in studies on plant invasions (1974–1993). In *Plant Invasions. General Aspects and Special Problems*, ed. P. Pyšek, K. Prach, M. Rejmánek & M. Wade, pp. 223–36. Amsterdam: SPB Academic Publishing.

Poynton, R.J. (1979). *Tree Planting in Southern Africa. Vol. 1. The Pines*. Pretoria: Department of Forestry.

Poynton, R.J. (1990). The genus *Prosopis* in southern Africa. *South African Forestry Journal*, **152**, 62–6.

Rejmánek, M. (1995). What makes a species invasive? In *Plant Invasions. General Aspects and Special Problems*, ed.

P. Pyšek, K. Prach, M. Rejmanek & M. Wade, pp. 3–14. Amsterdam: SPB Academic Publishing.

Rejmánek, M. (1996). A theory of seed plant invasiveness: the first sketch. *Biological Conservation*, **78**, 171–81.

Rejmánek, M. & Richardson, D.M. (1996). What attributes make some plant species more invasive? *Ecology*, **77**, 1655–61.

Richardson, D.M. (1988). Age structure and regeneration after fire in a self-sown *Pinus halepensis* forest on the Cape Peninsula, South Africa. *South African Journal of Botany*, **54**, 140–4.

Richardson, D.M. (1996). Forestry trees as alien invaders: the current situation and prospects for the future. In *Proceedings of the Norway/UN Conference on Alien Species*, ed. O.T. Sandlund, P.J. Schei & Å. Viken, pp. 127–34. Trondheim, Norway: Norwegian Ministry of Environment.

Richardson, D.M. & Bond, W.J. (1991). Determinants of plant distribution: evidence from pine invasions. *American Naturalist*, **137**, 639–68.

Richardson, D.M. & Brown, P.J. (1986). Invasion of mesic Mountain Fynbos by *Pinus radiata*. *South African Journal of Botany*, **52**, 529–36.

Richardson, D.M. & Cowling, R.M. (1992). Why is mountain fynbos invasible and which species invade? In *Fire in South African Mountain Fynbos*, ed. B.W. van Wilgen, D.M. Richardson, F.J. Kruger & H.J. van Hensbergen, pp. 161–81. Berlin: Springer-Verlag.

Richardson, D.M. & Cowling, R.M. (1994). The ecology of invasive alien pines (*Pinus* spp.) in the Jonkershoek Valley, Stellenbosch, South Africa. *Bontebok*, **9**, 1–10.

Richardson, D.M., Cowling, R.M. & Lamont, B.B. (1996a). Non-linearities, synergisms and plant extinctions in South African fynbos and Australian kwongan. *Biodiversity and Conservation*, **9**, 1035–46.

Richardson, D.M., Cowling, R.M. & Le Maitre, D.C. (1990). Assessing the risk of invasive success in *Pinus* and *Banksia* in South African mountain fynbos. *Journal of Vegetation Science*, **1**, 629–42.

Richardson, D.M., Macdonald, I.A.W. & Forsyth, G.G. (1989). Reductions in plant species richness under stands of alien trees and shrubs in the fynbos biome. *South African Forestry Journal*, **149**, 1–8.

Richardson, D.M., Macdonald, I.A.W., Hoffmann, J.H. & Henderson, L. (1997). Alien plant invasions. In *Vegetation of Southern Africa*, ed. R.M. Cowling, D.M. Richardson & S.M. Pierce, pp. 535–70. Cambridge: Cambridge University Press.

Richardson, D.M., Macdonald, I.A.W., Holmes, P.M. & Cowling, R.M. (1992). Plant and animal invasions. In *The Ecology of Fynbos: Nutrients, Fire and Diversity*, ed. R.M. Cowling, pp. 271–308. Cape Town: Oxford University Press.

Richardson, D.M., van Wilgen, B.W., Higgins, S.I., Trinder-Smith, T.H., Cowling, R.M. & McKelly, D.H. (1996b). Current and future threats to plant biodiversity on the Cape Peninsula, South Africa. *Biodiversity and Conservation*, **5**, 607–47.

Richardson, D.M., van Wilgen, Le Maitre, D.C., Higgins, K.B. & Forsyth, G.G. (1994b). A computer-based system for fire management in the mountains of the Cape Province, South Africa. *International Journal of Wildland Fire*, **4**, 17–32.

Richardson, D.M., Williams, P.A. & Hobbs, R.J. (1994a). Pine invasions in the Southern Hemisphere: determinants of spread and invadability. *Journal of Biogeography*, **21**, 511–27.

Ripmeester, W. (1996). Colonization of the ombrogenous peatland, Wylde Bog, Ontario, by the nonindigenous species, *Pinus sylvestris* L. (Scots Pine). MSc thesis, Graduate Department of Forestry, University of Toronto, Canada.

Roy, J. (1990). In search of the characteristics of plant invaders. In *Biological Invasions in Europe and the Mediterranean Basin*, ed. F. di Castri, A.J. Hansen & M. DeBussche, pp. 335–52. Dordrecht: Kluwer.

Rule, A. (1967). *Forests of Australia*. Sydney: Angus & Robertson.

Sanson, L.V. (1978). Invasion by Maritime Pine in Abel Tasman National Park. B.For.Sc. thesis, University of Canterbury, New Zealand.

Scott, D.F. & van Wyk, D.B. (1990). The effects of wildfire on soil wettability and hydrological behaviour of an afforested catchment. *Journal of Hydrology*, **121**, 239–56.

Shaughnessy, G.L. (1980). Historical ecology of alien woody plants in the vicinity of Cape Town, South Africa. PhD thesis, University of Cape Town.

Shaughnessy, G.L. (1986). A case study of some woody plant introductions to the Cape Town area. In *The Ecology and Control of Biological Invasions in Southern Africa*, ed. I.A.W. Macdonald, F.J. Kruger & A.A. Ferrar, pp. 37–43. Cape Town: Oxford University Press.

Shea, G.M. (1987). *Aspects of Management of Plantations in Tropical and Sub-tropical Queensland*, revised edn. Brisbane: Queensland Department of Forestry.

Shigesada, N., Kawasaki, K. & Takeda, Y. (1995). Modelling stratified diffusion in biological invasions. *American Naturalist*, **146**, 229–51.

Sim, T.R. (1927a). Some effects of man's influence on the South African flora. *South African Journal of Science*, **23**, 492–507.

Sim, T.R. (1927b). *Tree Planting in South Africa*. Pietermaritzburg: Natal Witness.

Smith, C.W. (1985). Impact of alien plants on Hawai'i's native biota. In *Hawai'i's Terrestrial Ecosystems: Preservation and Management*, ed. C.P. Stone & J.M. Scott, pp. 180–250. Honolulu: Cooperative National Parks Resources Studies Unit, University of Honolulu.

Spingett, J.A. (1976). The effect of planting *Pinus pinaster* Ait. on populations of soil microarthropods and on litter decomposition at Gnangara, Western Australia. *Australian Journal of Ecology*, **1**, 83–7.

Stirton, C.H. (ed.) (1978). *Plant Invaders. Beautiful, but Dangerous*. Cape Town: Department of Nature and Environmental Conservation of the Cape Provincial Administration.

Taylor, H.C. & Macdonald, S.A. (1985). Invasive alien woody plants in the Cape of Good Hope Nature Reserve. I. Results of a first survey in 1966. *South African Journal of Botany*, **51**, 14–20.

Taylor, H.C., Macdonald, S.A. & Macdonald, I.A.W. (1985). Invasive alien woody plants in the Cape of Good Hope Nature Reserve. II. Results of a second survey from 1976 to 1980. *South African Journal of Botany*, **51**, 21–9.

Thomson, G.M. (1922). *The Naturalisation of Animals and Plants in New Zealand*. Cambridge: Cambridge University Press.

Thompson, W.L. (1994). The current status of weeds in selected National Parks of South East Queensland. In *Working Papers of the Third Queensland Weeds Symposium*, pp. 21–5. Toowoomba: Weed Society of Queensland.

Tucker, K.C. & Richardson, D.M. (1995). An expert system for screening potentially invasive alien plants in South African mountain fynbos. *Journal of Environmental Management*, **44**, 309–38.

van Wilgen, B.W., Bond, W.J. & Richardson, D.M. (1992). Ecosystem management. In *The Ecology of Fynbos. Nutrients, Fire and Diversity*, ed. R.M. Cowling, pp. 345–71. Cape Town: Oxford University Press.

Van Wyk, D.B. (1987). Some effects of afforestation on streamflow in the Western Cape Province, South Africa. *Water SA*, **13**, 31–6.

Webb, C.J., Sykes, W.R. & Garnock-Jones, P.J. (1988). *Flora of New Zealand*. Vol. 4. *Naturalized Pteridophytes, Gymnosperms, Dicotyledons*. Christchurch: Botany Division, DSIR.

Wells, M.J., Balsinas, A.A., Joffee, H., Engelbrecht, V.M., Harding, G.& Stirton, C.H. (1986). A catalogue of problem plants in southern Africa. *Memoirs of the Botanical Survey of South Africa*, **53**, 1–658.

Weiner, J. (1994). *The Beak of the Finch. A Story of Evolution in Our Time*. London: Vintage.

Wicht, C.L. (1945). *Preservation of the Vegetation of the South-western Cape*. Cape Town: Special Publication of the Royal Society of South Africa.

Williams, P.A. & Timmins, S.M. (1990). *Weeds in New Zealand Protected Areas; A Review for the Department of Conservation*. Department of Conservation, Science and Research Series 14, Wellington: Department of Conservation, Science and Research.

Woods, D.H. (1950). The menace of the pine. *Journal of the Mountain Club of South Africa*, **53**, 78–87.

Glossary

Every effort was made to avoid, where possible, the use of jargon and obscure terminology in this book. Some technical terms were unavoidable. Also, some popular terms are entrenched in the literature on pines, but may not be understood by all readers. The following list is by no means exhaustive, but should help the non-specialist reader in grasping the fundamental concepts addressed in this volume.

There is considerable debate concerning the time boundaries for eras, periods and epochs in the geological time scale. The dates accepted for use in this volume (listed below) are those generally accepted by the contributors to this volume.

acclimation	Reversible change in the physiological response of an organism to its environment.
additive genetic variance	The population variance in a characteristic that results from segregation and recombination of genes with statistically additive effects on the character expression.
aeciospores	Dikaryotic **spores** in rust fungi produced after fertilization of **haploid pycnia**.
alien	In this volume used to denote organisms moved by humans to areas outside their natural range (cf. **invasive alien**).
allele	One of the possible forms of a gene at a particular **chromosome locus**.
amphidiploid	An organism that is **diploid** for two separate sets of **chromosomes**, each from a different species. Amphidiploids can arise from hybridization followed by doubling of the chromosome complement or union of two unreduced **gametes**, but are unknown in pines.
anemophilous	Of pollen or seeds dispersed by wind.
aneuploid	A cell or organism having an unbalanced set of **chromosomes**; e.g. including, but not limited to, one chromosome more or one less than a whole multiple of the **diploid** number.
angiosperms	Flowering plants, distinguished from **gymnosperms** by producing seeds that are enclosed fully by fruits. They first appear in rocks of the Early **Cretaceous**.
anthesis	The period at which a female flower becomes fully receptive to pollination or when the anthers release pollen.

apomixis	Asexual reproduction that is, nevertheless, accompanied by the production of seed or seed-like organs.
archegonium (pl. archegonia)	In pines, the structure in the megagametophyte that gives rise to and encloses the egg.
ascopores	**Haploid spores** produced after **meiosis** and sex in ascomycete fungi.
ash alkalinity	Titratable alkalinity of dried plant material following combustion at 400–450 °C.
assortative mating	Mating in which the pairing of **gametes** is not at random. Positive assortative mating occurs when like-individuals mate or tend to mate with like-individuals or like-gametes unite with like-gametes with respect to some heritable characteristic. Negative assortative mating is when unlike tend to mate with unlike.
'avoider' strategy	A life-history strategy of plants with little adaptation to fire.
base cations	Cations which hydrolyse to generate hydroxyl ions (OH⁻) in aqueous solutions, thereby enhancing acid neutralizing capacity. The most important of these are calcium, magnesium, potassium and sodium.
basidiospores	**Haploid spores** produced after sexual recombination in the Basidiomycetes.
boreal forest	The coniferous forest belt of the northern hemisphere, covering about 12 million km², which borders the tundra in the north and mixed forests and grasslands in the south (see also **taiga**).
bridge-fragment configuration	A microscopic configuration observable during anaphase of the first or second meiotic division, in which an unattached **chromosome** fragment remains near the equator of the cell and a chromosome arm bridges the gap between the two poles. The bridge-fragment configuration is the result of crossover in heterozygotes for an inverted chromosome segment (see **inversion**).
butt rot	Decay of wood in the central portion (heartwood or oldest growth rings) of the basal portion or *butt log* of living trees (cf. **stem rot**). Butt rot **infections** also extend down into the major roots and may encroach significantly into the sapwood and result in death of cambium and phloem tissues.
CAMCORE	The Central America and Mexico Coniferous Resources Cooperative, based at North Carolina State University, dedicated to the *ex situ* gene preservation and testing of threatened conifer (and some broadleaved) species in Middle America.
canker	Necrosis or death of the inner bark or phloem tissue, resulting in conspicuous lesions on stems or branches.
carbon allocation	In the context of a mycorrhizal plant, normally used to describe transfer of carbon from the autotroph to its heterotrophic partner.
cation exchange capacity	The sum total of exchangeable cations that a soil can adsorb.
cavitation	The separation of the water column in a xylem element.
Cenozoic	The geological era that spans approximately the last 65 million years of Earth's history.

chiasma (pl. chiasmata)   A cross-shaped junction between homologous **chromosomes**, the microscopic expression of crossover, observable at the diplotene stage in meiotic cells.

chloroplast   A subcellular organelle where photosynthesis takes place. Chloroplasts contain DNA and are self-replicating and, as a rule, are inherited through the pollen parent in pines.

chromosome   A single long molecule of DNA associated with histone proteins, in the cell nucleus. Genes and intervening sequences are organized along the length of the chromosome. Chromosomes have a region called the centromere, which links daughter chromatids together after chromosome replication. Fibres attach to the centromeres to draw the daughter chromatids to opposite poles of the cell during cell division.

climax   Species or communities representing the final (or an indefinitely prolonged) stage of a sere.

cline   A continuous character gradient correlated with geography or environment; adjacent populations merge into one another with no sharp break.

cohort   A group with common demographics (e.g. year of birth).

conophagous   Insects or other organisms that feed on seed cones.

cpDNA   Chloroplast DNA (see DNA).

Cretaceous   The most recent period of the **Mesozoic** era, which lasted approximately 71 million years, from 136 to 65 million years ago. The earliest fossil records of *Pinus* are believed to date to this period.

crossover   The exchange of genetic material between homologous **chromosomes** during **meiosis**.

crown fire   A fire burning into the crowns of the vegetation, generally associated with an intense understorey fire.

Dauerwald system   An approach to forest management that began in Germany in the late 19th century, and which involved the return to 'natural', rather than 'agricultural' approaches. It culminated in the adoption of individual-tree selection and uneven-age forest management.

dendroarchaeology   The analysis of wood remains from archaeological sites and historical buildings.

denitrification   Chemical or biochemical reduction of nitrate (NO_3^-) or nitrite (NO_2^-) to gaseous nitrogen (generally N_2 or N_2O).

Devonian   Geological period lasting from about 400 to 360 million years ago. **Gymnosperms** probably first appeared about 365 million years ago.

dichogamy   The situation in which pollen shed and receptivity of the female structures on the same plant occur at different times (cf. **protandry, protogyny**).

diploid   A cell or organism that has two homologous sets of **chromosomes**, one from the paternal and one from the maternal parent.

diploxylon pines   Pines in the subgenus *Pinus* of the genus *Pinus*, comprising one section (*Pinus*), 10 subsections, and 71 species (64% of species in the genus); with two fibrovascular bundles in each leaf; also termed the **hard pines** because the wood contains significant amounts of resin and the timber is generally harder than that of the **soft pines**.

disease A deleterious malfunctioning of the host that results from continuous irritation by a pathogenic agent.

disease suppression The process whereby the presence of a mycorrhizal fungal symbiont protects the plant root from attack by pathogenic microorganisms. This defence may take the form of the simple physical protection afforded by the presence of the mycorrhizal mantle, or chemical protection arising from release of antibiotics.

DNA Deoxyribonucleic acid, the molecular basis of heredity. It encodes the amino acid sequence of enzymes and structural proteins in its sequence of four bases (thymine, adenine, cytosine and guanine).

double fertilization The process in **angiosperms** in which one of the two sperm nuclei from the pollen unites with the egg nucleus to form the **zygote** while the other sperm nucleus unites with a diploid or tetraploid **fusion nucleus** derived from the **megaspore**. The new nucleus, composed of the sperm nucleus and the fusion nucleus, undergoes mitotic division to produce the seed endosperm.

drought avoiders Species which utilize morphological or physiological adaptations to access new sources or water or alternatively avoid water loss, and thereby avoid tissue water stress.

ecotype A genetically differentiated population distinguished from adjacent populations by sharp discontinuities in character expression.

ectomycorrhiza Symbiotic association between fungus and root in which the fungus ensheathes the feeding root without intracellular penetration (normally) of its tissues (cf. endomycorrhiza –intracellular penetration as seen in VA and ericoid mycorrhiza – these types not seen in *Pinus*).

electrophoresis The movement of charged molecules in solution (e.g. soluble enzymes) through a porous medium (e.g. starch gel) under the influence of an electrical field.

embryo An organism in the early stages of development, between fertilization and germination in pines.

embryogenesis The process of development of a fertilized egg (**zygote**) to a fully differentiated germinant.

'endurer' strategy A life-history strategy of plants to fire where the plant resprouts or endures the effects of fire.

entomophagous Insects or other organisms that feed on other insects.

Eocene An epoch of the **Tertiary** period, which lasted approximately 16 million years, from 54 to 34 million years ago. During the Eocene, warm humid climates apparently dominated temperate latitudes. Pine fossils decrease in abundance at temperate latitudes, but are found at some high- and low-latitude sites.

epidemic A widespread and severe outbreak of a **disease**.

étage A concept widely used in the French-language ecological literature (much less so in English). *Étages* are belts or zones of vegetation described on the basis of topography, altitude and regional climate.

'evader' strategy A life-history strategy of plants to fire where long-lived propagules are stored in soil or canopy and evade elimination from the site after fire.

exchangeable acidity Titratable acidity which can be displaced from the soil exchange complex by a neutral salt solution.

extraradical mycelium That part of the vegetative **mycelium** which extends outwards from the mantle into the soil which it explores for nutrients, and through which carbon is transferred to enable formation of fungal fruit-bodies (cf. **hyphal network**).

fascicle The basal sheath of scales in which **pine** leaves ('needles') are borne. The number of leaves per fascicle ranges from 1 to 8 in the genus, the number being reasonably constant in most species.

fire frequency The return interval of fire, measured in years.

fire regime The combination of **fire frequency**, predictability, intensity, seasonality, and extent characteristic of fire in an ecosystem.

fire scar A portion of the cambial circumference of the tree, usually basal and widest at the base, that has been killed by a fire. Re-scarring by subsequent fires is typical, so that one fire scar may represent continued scarring by as many as 20–30 fires in some ecosystems.

fire severity The effect of fire on organisms. For trees, often measured as the percentage of basal area removed.

fireline intensity The rate of heat release along a unit length of fireline, measured in $kW\,m^{-1}$.

foliar chlorosis The loss of normal green colour on an area of leaf or needle surface which is indicative of local tisssue damage.

founder effect A change in gene frequency (see **genetic drift**) that results from colonization. As a result of sampling error, gene frequencies in a group of founders will probably deviate from that in the parent population, and the smaller the number of founders, the greater may be the drift in gene frequencies.

fusion nucleus A nucleus in **angiosperms** that results from the union of two (maize-type) or four (lily-type) haploid nuclei derived from the **megaspore** by mitosis (see **double fertilization**).

gamete An egg or sperm.

gametogenesis The process of forming gametes.

gametophyte The haploid phase of the life cycle of plants that undergo an alternation of generations. The gametophytes produce the gametes.

geitonogamy Self-pollination in which the pollen is derived from another reproductive structure on the same plant. Since pines are **monoecious**, self-pollination is only possible through geitonogamy.

gene flow The spread of genes (i.e. **alleles**) within a population or between interbreeding populations as a result of gametic or seed migration.

gene frequency (syn. allele frequency) The proportion of genes at a particular locus that are occupied by a specific allele.

genetic drift   Change in gene frequencies from generation to generation due to sampling errors that operate when an offspring generation is formed by random union of **gametes** in a finite population.

genetic load   The loss of population fitness that results from: (1) the **mutational load**; (2) segregation of **alleles** that may be advantageous in heterozygous condition but deleterious when homozygous; and (3) the substitutional load that occurs in a changing environment when alleles that formerly conferred an advantage in the population are replaced by alleles favoured in the new environment.

genome   All the hereditary material carried by: (1) a single **gamete**, with respect to the nuclear genome; or (2) a single organelle with respect to the chloroplast and mitochondrial genomes.

genotype   An individual's hereditary constitution, as contrasted to its **phenotype** which is the outcome of both hereditary and environmental factors.

germ line   The line, by descent, of cells that give rise to the reproductive structures and ultimately the **gametes**.

glacials   Periods when expansion of the great continental ice sheets occurred (see **interglacials**).

grass stage   A juvenile stage of some **diploxylon pines** (at least seven species in three subsections in section *Pinus*), where a thick, dense growth of basal leaves protects the meristem from fire damage. This stage usually lasts 3–6 years, during which time root growth is substantial, allowing subsequent rapid top growth.

gymnosperms   A subdivision of the seed plants in which the ovules are carried naked on the cone scales (cf. **angiosperms**). Gymnosperms date from the Middle **Devonian**, about 365 million years ago.

half-sibs (syn. half-siblings)   Individuals that have one parent in common. The other parent is often assumed to have been drawn at random from the breeding population.

haplogenetics   Literally, the genetics of **haploids**, which is not complicated by dominance and recessiveness of alleles (see **recessive**).

haploid   A cell having one complete set of **chromosomes** (e.g. as in the **gametes**), compared to the two sets in vegetative cells of a **diploid**, such as a pine tree.

haploxylon pines   Pines in the subgenus *Strobus* of the genus *Pinus*, comprising two sections (*Parrya* and *Strobus*), 7 subsections and 40 species, 36% of species in the genus); with one fibrovascular bundle in each leaf; also termed the **soft pines** because the wood contains little resin and the timber of most species is softer than that of the **hard pines**.

hard pines   see **diploxylon pines**

Hardy–Weinberg equilibrium   The constancy of gene and genotype frequencies from generation to generation that would occur in an infinitely large population (i.e. **genetic drift** is negligible), where mating is at random, and there is no selection, migration or **mutation**. As an example for the two-allele (A and a) case, gene and genotype frequencies are related as the expansion of the binomial $(pA+qa)^2$, where p is the frequency of allele A and $q = 1-p$ is the frequency of allele a. Therefore, genotype frequencies are p^2AA, $2pqAa$, and q^2aa at equilibrium. No population meets the criteria exactly, and the Hardy–Weinberg equilibrium is used as a baseline to detect deviations from the assumptions.

Hartig net   Term applied to the network of profusely branching hyphal tips that penetrate *between* the radial walls of the outer cortical cells of the **ectomycorrhizal** root producing a tremendously enhanced surface area through which exchange of nutrients between the symbionts takes place. Named after R. Hartig who first described this part of the mycorrhizal structure.

heterobrachial   Of **chromosomes** in which the two arms are of different length.

heterorhizy   A root system made up of two types of root, one (primary) of unlimited growth normally not becoming mycorrhizal, the other (secondary) of laterals of restricted growth which are converted early in development to mycorrhizas.

heterozygote   In a **diploid**, an individual that has two different **alleles** for a particular gene **locus**, one on each of a homologous **chromosome** pair.

Holocene   The current epoch of the **Quaternary** period, or approximately the last 10 000 years; the time since the last major global glacial retreat.

humic acids   A mixture of dark-coloured organic materials of indefinite composition extracted from soil with dilute alkali and precipitated upon acidification.

hydraulic conductance   The rate of water movement within a tissue, expressed as volume of water per unit cross-sectional area per unit time per unit driving force.

hydraulic resistance   The reciprocal of **hydraulic conductance**.

hyphal network   The product of the **extraradical mycelium** which forms a semi-permanent source of **inoculum** for emerging lateral roots.

inbreeding   Mating among related individuals or selfing.

infection   The establishment of a parasite within a host. The infection court is the site where the parasite gains entry to the host, e.g. a wound.

inoculum   Fungal propagules (**spores**, hyphal fragments or the **mycelial** network) in soil which have the potential to colonize uninfected roots as they grow through the soil.

interglacials   Periods during which global temperatures rose to be as high as or even higher than those of the present day (see also **glacials**).

introgression   The infiltration or migration of genes from one species or population into another by hybridization followed by back-crossing of the hybrids to one or the other of the parental types.

'invader' strategy   A life-history strategy of plants to fire where the plant, through highly dispersive propagules, invades the site after fire.

invasive alien   An **alien** organism that has established self-perpetuating populations in natural or semi-natural habitats, often disrupting the functioning of invaded ecosystems.

inversion   An aberration in which a segment of the chromosome has been turned 180° and reinserted at the same position.

isozymes Multiple forms of a single enzyme. When coded by alternative (allelic) forms of the same gene, isozymes are called allozymes.

Jurassic The middle period of the **Mesozoic** era, which lasted approximately 54 million years, from 190 to 136 million years ago. Fossils of Pinaceae apparently ancestral to pines are believed to date to this period, but no true *Pinus* have been found that date to this period.

karotype The appearance, number and arrangement of the **chromosomes** in the cells of an individual.

karyotype The **haploid chromosomal** complement, usually characterized by length of the chromosomes, location of the centromere and relative length of the chromosome arms, and secondary constrictions or satellites.

krummholz A dwarfed growth form of tree above the timberline.

landrace Semi-domesticated or domesticated race or ecotype that has evolved in a given area by natural and artificial selection through propagation by local agriculturists or foresters.

leaf area index (LAI) The area of leaves (taking only one side of each leaf into account) above unit area of ground. Typical ranges of LAI in field populations of pines are only 2–4 $m^2\,m^{-2}$, compared with values of 9–11 $m^2\,m^{-2}$ in the more shade-tolerant conifer genera *Abies*, *Picea* and *Pseudotsuga*; LAIs seldom exceed 6 in hardwoods. Highly productive plantations of pines such as *P. radiata* owe much of their productivity to LAIs 2–3 times those found in natural stands.

leptokurtic Of a frequency distribution that is more closely clustered about the mean than the normal frequency distribution.

lethal allele An **allele** which, when expressed, is fatal to its carrier.

linkage The occurrence of **alleles** on the same **chromosome**. If in close proximity, they will tend to be inherited as one unit or separated by crossover only infrequently. If further apart, linkage may be undetectable.

locus (pl. loci) The position that a gene occupies on a **chromosome**. Used synonymously with 'gene'.

macroecology The study of ecological patterns that arise when multiple species, multiple sites, or multiple points in time are analysed in a search for broader, more general aspects of ecology.

mantle The ensheathing tissue of fungal **mycelium** which may be of loose hyphae, be pseudoparenchymatous or parenchymatous.

megagametophyte The **haploid gametophyte** in pines that produces the egg(s).

megasporangiate strobilus The 'seed' cone in pines, which bears the ovules that, in turn, produce the **megaspores**.

megaspore One of the four **haploid** cells resulting from **meiosis** of the megaspore mother cell in the ovule. In pine, the other three meiotic products degenerate.

meiosis A specialized cell division, called the sexual division. Meiosis results in a reduction of the **diploid chromosome** number to the **haploid** number. Meiosis is characterized by the pairing of homologous chromosomes and their separation into different daughter cells in the first stage. A second division of meiosis results in a total of four daughter cells with a haploid genome that recombines the grand-parental genes and chromosomes.

Mesozoic The geological era that spanned 160 million years, and lasted from *c.* 225 to 65 million years ago.

metacentric Of a **chromosome** with a centromere centrally located along its length (see also **karyotype**).

metal avoidance The process whereby colonization of roots by mycorrhizal fungi (which generally have significantly greater tolerance of metals than their plant partners) enables the plant to be free from exposure to metals such as aluminium, zinc and copper which are present in soil solutions, particularly under acid conditions.

microarthropod Small invertebrate animals with jointed appendages and a horny external covering, including insects, crustaceans and spiders.

microsporangiate strobilus The pollen cone, or catkin, in pines, which bears the pollen sacs and produces the pollen, or **microspores**.

microspores The cells, four from each microspore mother cell, resulting from **meiosis** in the pollen sacs. The microspores develop into the mature pollen grains.

mineralization-immobilization The process in soil whereby some of the ions of nitrogen and phosphorus having been mobilized (mineralized) from organic complexes are then re-immobilized by sequestration in the body of the heterotroph.

Miocene An epoch of the **Tertiary** period, which lasted *c.* 19 million years, from 26 to 7 million years ago. During the Miocene, pine fossils increase in abundance and diversity at temperate latitudes, and many modern taxa are recognizable.

mitochondrion (pl. mitochondria) A subcellular organelle where oxidative phosphorylation takes place. Mitochondria contain DNA, are self-replicating, and are inherited, as a rule, through the **megagametophyte** in pines.

mitosis The cell division that characterizes vegetative growth and produces two daughter cells that have exactly the same **genotype** as the mother cell.

monoculture In this volume used with reference to single-species plantations of **pines**.

monoecy The condition (as in pines) whereby male and female reproductive structures are borne on the same plant.

monophagous In this volume used with reference to insects that feed only on *Pinus* species.

mor Forest surface humus in which the Oa horizon has little mixing into mineral soil beneath it.

mutant A cell or organism bearing a modified form of a gene that expresses itself in the **phenotype**, either because it is dominant or because it is in homozygous condition.

mutation The process by which the genetic material undergoes change, either structural aberration in **chromosomes** or point mutations in which one base pair (see **DNA**) is substituted for another in the DNA molecule.

mutational load The loss in population fitness that results from recurrent mutation to deleterious alleles. The mutational load is equal to the mutation rate and independent of the effect of the mutation.

mycelium
(adj. mycelial)

The vegetative thallus of a fungus, comprised of a mass of individual, food-absorbing filaments or hyphae.

myelophagous

Insects or other organisms that feed on the pith of small stems.

Neogene

The two most recent epochs of the **Tertiary** period, including the **Miocene** and **Pliocene**, lasting from approximately 26 to 2.5 million years ago. A time of apparent migration and radiation for pines.

nitrification

Biological oxidation of ammonium nitrogen (NH_4^+) to nitrite (NO_2^-) and nitrate (NO_3^-). Involves the generation of hydrogen ions.

nucellus

The central tissue of the ovule that, during development, successively encloses the **megaspore**, the embryo sac, and, for a time, the **megagametophyte**.

nutrient mobilization

The process whereby the fungal heterotroph, by secretion of appropriate enzymes, facilitates release of nutrients from complex organic or inorganic substrates.

obligate parasite

An organism that can grow and multiply only on or in a living host. In contrast, facultative parasites can also live as saprobes.

Oligocene

An epoch of the **Tertiary** period, which lasted between 8 and 12 million years, from 38–34 to 26 million years ago. The Oligocene was characterized by the return to climates more typical of modern patterns at temperate latitudes.

oligophagous

In this volume used with reference to insects that feed only on species within the family Pinaceae.

osmotic potential

The chemical potential energy of a solution with a concentration of osmotically active solutes.

outcrossing

The crossing of unrelated individuals.

packrat middens

Discarded remains of plant and animal fragments embedded in the crystallized urine of packrats (rodents of the genus *Neotoma*). These give a snapshot of the flora and fauna existing within about 50 m of the midden at the time it was accumulating.

Palaeo-

Meaning old or ancient; thus: *palaeocoordinates* are the longitudes and latitudes that defined locations at certain times in the past; and *palaeocontinents* are the land masses of earlier times of Earth's history. Due to plate tectonics, the continents have been changing shape, size and location continuously through time.

Palaeocene

An epoch of the **Tertiary** period, which lasted approximately 11 million years, from 65 to 54 million years ago. No *Pinus* fossils are securely dated to the Palaeocene, which appears to have been dominated by warm humid climates similar to the **Eocene**.

Palaeogene

The three oldest epochs of the **Tertiary** period, including the **Palaeocene**, **Eocene** and **Oligocene**, and lasting from approximately 65 to 26 million years ago.

Palaeozoic

The geological era before the **Mesozoic**, lasting from 590 to 225 million years ago.

parsimony analysis,
(Wagner-)

A phylogenetic analysis of a group of organisms or taxa in which the branches of the phylogenetic tree are joined together in a topology requiring the minimum number of character state changes among the organisms.

pedogenesis Soil formation. Usually discussed in relation to the five soil-forming factors (i.e. parent material, climate, the biotic factor, topography and time).

phenotype An individual's 'appearance', morphological, physiological, chemical or behavioral, which is determined by the interaction of heredity and environment during development.

phloeophagous Insects or other organisms that feed on tree phloem of the inner bark.

phosphatase The enzyme involved in the mobilization of phosphorus from organic sources.

photosynthetic capacity The net rate of photosynthesis at saturating light and ambient CO_2, and with otherwise non-limiting conditions.

phyllophagous Insects or other organisms that feed on leaves.

phytophagous Insects or other organisms that feed on living tissues of higher plants.

phytotoxic Chemicals found in sufficiently high concentrations to lead to reduced plant growth and death.

pine In this volume, used exclusively with reference to taxa of the genus *Pinus* (Pinaceae). Many taxa in other conifer families (notably Araucariaceae, Cupressaceae, Podocarpaceae, Taxodiaceae) are called 'pines' in the vernacular, as are some taxa with needle-like leaves in the **angiosperm** families Apocynaceae, Labiatae, Pandanaceae, Proteaceae and even the Poaceae (e.g. *Poa scabrella*; bluegrass pine).

piñon see pinyon

pinyon The common name given to a distinctive group of 12 North American *Pinus* species forming the subsections *Cembroides* (11 species) and *Rzedowskianae* within section *Parrya* in subgenus *Strobus*. It is unclear whether the pinyon pines form a monophyletic group, and several unusual species may be moved to separate subsections following further study. The name pinyon is derived from the Spanish *piñón* which refers to the large, wingless, edible seed that reminded early explorers of those of *Pinus pinea* from the Mediterranean Basin.

Pleistocene An epoch of the **Quaternary** period, which lasted approximately 2.4 million years, from 2.5 million to 100 000 years ago, and was dominated by ice ages. During this epoch, many major global glacial intervals fluctuated with interglacial stages.

Pliocene An epoch of the **Tertiary** period, which lasted approximately 4.5 million years, from 7 to 2.5 million years ago. During this epoch, pine fossils are abundant throughout their current latitudes.

polyphagous In this volume used with reference to insects that feed on two or more families of trees.

polyphenolics Organic compounds with a six-carbon unsaturated ring structure with one or more OH groups attached. Components of lignin and humus.

polyploid A cell or organism that has a multiple of more than twice the basic ancestral number of **chromosomes** characteristic of a particular phylogeny.

podzol A soil type, characteristically developed beneath heath or coniferous forest, with a highly leached whitish grey A horizon and a dark ferri-**humic** B horizon.

podzolization   The mobilization and removal from the A and/or E horizon of organic matter and **sesquioxides**. This leads to the development of an extremely acid humus formation known as mor. Water percolates through the mor causing the breakdown of clay minerals by disruption of the mineral structure, releasing the component elements.

prescribed fire   A fire ignited under known conditions of fuel, weather and topography to achieve specified objectives.

prescribed natural fire   A fire ignited by natural processes (usually lightning) and allowed to burn within specified parameters of fuels, weather and topography to achieve specified objectives.

projected leaf area   Leaf or needle area illuminated by a light source vertically overhead.

protease   The enzyme involved in the mobilization of nitrogen from protein.

protein fungi   Those ectomycorrhizal fungi that have the ability to mobilize nitrogen from protein.

protandry   The situation in which pollen is shed before the female structures on the same plant are receptive. The condition exists to varying degrees in pines: e.g. strongly in *P. palustris* and *P. ponderosa*; weakly in *P. nigra*.

protogyny   The situation in which female structures are receptive before pollen from the same plant is shed.

provenance   The original geographic source of seed or other propagules. Also, the test population resulting from seeds collected at a particular location.

pycnia   Structures in the rust fungi (Uredinales) resulting from infection by **haploid basidiospores**.

Quaternary   The most recent geological period of the **Cenozoic** era, including the **Pleistocene** and **Holocene** epochs, and spanning the last 2.5 million years of Earth's history.

RAPDs   Random amplified DNA polymorphisms: DNA sequence variants detected by size variation in fragments multiplied by the polymerase chain reaction after annealing to a primer (a short, arbitrary DNA sequence).

Rapoport's Rule   The tendency for the size of the geographical range of a species to decrease with decreasing latitude, as is evident for *Pinus* spp. Named after the Argentinean biologist, Eduardo Rapoport, who first described the phenomenon.

rate of (fire) spread   The rate at which a fire moves across the landscape, usually measured in m s^{-1}.

recessive   An **allele** that is not expressed in the **phenotype** when carried by a cell or organism in a heterozygous state.

recombination   Assortment of **alleles** into offspring combinations not present in the parents.

recombination fraction   The frequency of recombinant **gametes** relative to the total. A measure of the distance between two **loci** on a **chromosome** and how frequently linkages are broken during **meiosis**.

recombination index   The sum of the **haploid** number of **chromosomes** and the average number of **chiasmata** (the microscopically visible expression of crossover, a measure of the minimum number of chromosome exchanges during meiosis). The greater this number, the larger is the number of new gene combinations that can be formed in a generation.

refractory Resistant to degradation.

'resister' strategy A life-history strategy of plants to fire where the plant, through an adaptation such as thick bark, survives low-intensity fire relatively unscathed.

restriction site, DNA- A short region of DNA, often 4 or 6 base pairs in length, that is recognized and cleaved at a specific position by a restriction endonuclease enzyme.

RFLPs Restriction fragment length polymorphisms: allelic polymorphisms in the order of base pairs in the DNA, recognized by restriction enzymes that cleave the DNA at specific short sequences.

rhizomorphs Root-like aggregates of vegetative hyphae that grow through the soil, foraging for substrata. True rhizomorphs (such as those in *Armillaria*) have a well-defined apical meristem and may have a rind of pigmented and tightly fused cells with a core of elongated, unpigmented cells. **Mycelial** cords are generally less differentiated than rhizomorphs and lack an apical meristem.

Rubisco Ribulose 1,5-bisphosphate; the primary carboxylation enzyme in vascular plants (also RuBP).

rust diseases **Diseases** caused by **obligately parasitic** fungi in the order Uredinales. Most rust fungi can form several **spore** forms, occurring on two different (alternate) hosts. The telial stage often appears as small, rust-coloured specks on the host, and in the rusts of pine this stage is generally found on the alternate host, i.e. dicots.

samara A winged, single-seeded fruit, as in many **pine** species. An adaptation for dispersal by wind.

saprophagous Insects or other organisms that feed on dead or decaying matter.

sapwood The portion of woody tissue with active xylem elements.

segregation The separation of each allele of a **homologous** pair into different **gametes** at **meiosis**.

self-incompatibility Inability to produce offspring from self-pollination due to failures in development before fertilization (i.e. prezygotic, such as failure of the pollen to germinate or failure of the pollen tube to survive in the female structures of the same plant).

selfing Self-pollination; mating that results from the union of egg and sperm from the same plant.

seral A plant species or community which will be replaced by another plant species or community if protected from disturbance.

serotiny A morphological feature of some pine cones (and reproductive structure in other plants) whereby the cones remain closed and on the tree for one or more years after seed maturation; cones open rapidly when high temperatures melt the resin that seals the cone scales.

sesquioxides A binary compound of a metal and oxygen in the proportion of 3 to 2, as Al_2O_3 and Fe_2O_3. Also used generally to describe free iron, aluminium and manganese oxides in the soil.

sibs (syn. siblings)   Individuals that have one or more parents in common.

soft pines   see **haploxylon pines**

soil exchange complex   The surface-active portion of the soil which provides sites for the adsorption of anions and cations from solution. This includes surfaces of soil minerals, amorphous compounds and organic colloids.

somatic mutation   A change in the genetic material that occurs in an individual's body cells. In plants, somatic mutations may occur in cell lines that give rise to reproductive structures and thereby be transmitted to offspring.

southern pines   General term for the pines native to the southern states of the USA, from the Atlantic Coast, west to Texas and Oklahoma (*c.* 24% of the area of the USA): *P. clausa*, *P. echinata*, *P. elliottii*, *P. glabra*, *P. palustris*, *P. pungens*, *P. rigida*, *P. serotina*, *P. strobus*, *P. taeda* and *P. virginiana*.

specific leaf area   Leaf area per unit dry weight.

spore   A reproductive propagule of fungi used for dissemination or survival. Sexual (resulting from **meiosis**) spores of the higher fungi are called **ascospores** or **basidiospores**, while the asexual (resulting from **mitosis**) spores are called conidia.

sporogenesis   The process of forming **spores**, the **microspores** or **megaspores** in pines.

sporophyte   The **diploid** phase of the life cycle of plants that undergo an alternation of generations. The sporophyte is the conspicuous phase (i.e. the tree) in pines and produces the **megaspores** and **microspores**.

stem rot   Decay of the above-ground, woody stem tissues of a living tree (cf. **butt rot**). The term 'heart rot' has been applied by some because much of the decay is typically found in the heartwood, but decay of the living sapwood is also generally found around the decayed heartwood.

stomatal conductance   The rate of leaf transpiration per unit of driving force of vapour pressure gradient between intercellular spaces and ambient air, utilized as an expression of the control of stomatal aperture on transpirational water loss.

stone pines   Substantive name given to the five *Pinus* species in subgenus *Strobus*, subsection *Cembrae* that have five-needled **fascicles**, wingless seeds, and cones that remain closed at maturity. Cone and seed characteristics of these species (*P. albicaulis*, *P. cembra*, *P. koraiensis*, *P. pumila* and *P. sibirica*) reflect a long evolutionary symbiosis with birds in the genus *Nucifraga* which play a near-obligate role as seed dispersers. *Pinus pinea*, a **diploxylon pine**, has similar cone and seed characteristics (through convergent evolution) and is commonly known as the Mediterranean- or Italian stone pine.

strobilus (pl. strobili)   The immature cone of **gymnosperms**, which contains the reproductive structures, male or female.

surface fire   A fire burning along the surface without significant movement into the understorey or overstorey, usually below 1 m flame length.

sympatric   Of species that have completely or partially overlapping ranges.

synacmous   Condition when male and female organs mature simultaneously.

synacmy The situation in which female structures are receptive at the same time that pollen grains from the same plant are shed (cf. **dichogamy**, **protandry**, **protogyny**).

taiga Another name for the **boreal forest**; sometimes used to describe only that part of the boreal forest in the former USSR.

teliospores Initially dikaryotic **spores** in rust fungi in which karyogamy and **meiosis** occurs to give rise to **haploid basidiospores**.

Tertiary A period of the **Cenozoic** era which lasted from approximately 65 to 2.5 million years ago. During this period many modern forms of pines appear and radiate.

translocation An aberration in which terminal segments of two **chromosomes** are interchanged.

transposon A DNA segment capable of moving from one position in the genome to another.

tracheids Elongated, thick-walled conducting cells in the xylem of conifers.

Triassic The oldest period of the **Mesozoic** era, which lasted approximately 35 million years, from 225 to 190 million years ago.

understorey fire A fire burning in the understorey, more intense than a surface fire with flame lengths 1–3 m.

uredinospores Dikaryotic **spores** in rust fungi resulting from infection by aeciospores.

vagility Capacity for movement.

vector An organism that carries and transmits another organism.

virescent A mutant condition in which foliage is initially yellow or white and in time develops a normal green colour.

water use efficiency Efficiency measured as the ratio of moles of CO_2 fixed per mole of water transpired.

white pines Substantive name for taxa in section *Strobus*, subsection *Strobi* of the genus *Pinus* (16 species in the treatment adopted for this volume), e.g. eastern white pine for *P. strobus* and Japanese white pine for *P. parviflora*. So called because of the white marks at the tips of cones scales formed by spots of resin which ooze from the young cones.

wildfire A human or naturally-caused fire which is not meeting land management objectives.

xylophagous Insects or other organisms that feed on tree xylem or wood tissues.

yellow pines A confusing term, used either for **diploxylon pines** of subsections *Australes* ('eastern yellow pines') and *Ponderosae* ('western yellow pines'), occasionally for all **diploxylon pines** or, most often, for *Pinus ponderosa* alone (western yellow pine; foresters refer to old-growth ponderosa pines as 'pumpkins' because of their striking orange-yellow bark).

Younger Dryas A period of climatic cooling between approximately 11 000 and 10 000 years ago (uncalibrated ^{14}C ages), when polar waters once again spread southwards around the coastline of western Europe and glacial re-advance occurred in mountainous regions.

zygote The fertilized egg, which restores the **diploid chromosome** number by combining the genomes of **haploid** sperm and egg.

A glossary of English common names for Pines

The literature on pines contains a bewildering array of common names. Some pines are known by different English names in different parts of their range, and in some cases, one common name has been used to describe several species. Table 1.1 (Chap. 1, this volume) gives the recommended English common names for the 111 species, and some subspecies, varieties and hybrids treated in this volume. The list below gives these and other widely-used names in alphabetical order. First-choice names (see Table 1.1, p. 5, for criteria) appear in **bold type**.

Aleppo pine	*P. halepensis*
Apache pine	*P. engelmannii*
Arizona pine	*P. ponderosa* var. *arizonica*
Arizona singleleaf pinyon	*P. monophylla* subsp. *fallax*
Armand('s) pine	*P. armandii*
Arolla pine	*P. cembra*
Austrian (black) pine	*P. nigra* subsp. *nigra*
Aztec pine	*P. teocote*
Balkan (white) pine	*P. peuce*
beach pine	*P. contorta* subsp. *contorta*
Benguet pine	*P. kesiya*
Bhutan pine	*P. wallichiana*
Bhutan white pine	*P. bhutanica*
bigcone pine	*P. coulteri*
bishop pine	*P. muricata*
black pine – European	*P. nigra*
black pine – Japanese	*P. thunbergii*
blue pine – Himalayan	*P. wallichiana*
Bolander pine	*P. contorta* subsp. *bolanderi*
border pinyon	*P. discolor*
Bosnian pine	*P. heldreichii*
bristlecone pine – Colorado	*P. aristata*
bristlecone pine – Great Basin	*P. longaeva*
bristlecone pine – Intermountain	*P. longaeva*
bristlecone pine – Rocky Mountain	*P. aristata*

bristlecone pine – western	*P. longaeva*
bull pine	*P. ponderosa*; also *P. sabiniana*
Calabrian pine	*P. brutia*
California singleleaf pinyon	***P. monophylla* subsp. *californiarum***
Canary (Island) pine	***P. canariensis***
Caribbean pine	***P. caribaea***
cedar pine	*P. glabra*
Central American pitch pine	*P. caribaea*
Chiapas white pine	***P. chiapensis***
Chihauhua(n) pine	*P. leiophylla*
Chilgoza pine	***P. gerardiana***
Chinese red pine	*P. massoniana*
Chinese (red) pine	***P. tabuliformis***
Chinese white pine	***P. armandii***
Chir pine	***P. roxburghii***
Choctawhatchee sand pine	*P. clausa* var. *immuginata*
cluster pine	*P. pinaster*
Colorado bristlecone pine	***P. aristata***
Colorado pinyon	***P. edulis***
Cooper pine	***P. cooperi***
Corsican pine	***P. nigra* subsp. *laricio***
Coulter pine	***P. coulteri***
Crimean pine	***P. nigra* subsp. *pallasiana***
Cuban pine	***P. cubensis***
Dabie Shan white pine	***P. dabeshenensis***
Dalat white pine	***P. dalatensis***
David's pine	*P. armandii*
Digger pine	*P. sabaniana*
Donnell Smith pine	***P. donnell-smithii***
Douglas pine	***P. douglasiana***
Durango pine	***P. durangensis***
dwarf mountain pine	***P. mugo***
dwarf stone pine	***P. pumila***
eastern Mediterranean pine	***P. brutia***
eastern white pine	***P. strobus***
eggcone pine	***P. oocarpa***
European black pine	***P. nigra***
false Weymouth pine	*P. pseudostrobus*
Fenzel pine	***P. fenzeliana***
Florida pine	*P. palustris*
foothill pine	***P. sabiniana***
Formosa pine	*P. taiwanensis*
foxtail pine	***P. balfouriana***
Gaoshan pine	*P. densata*
Georgia pine	*P. palustris*
Gerard's pine	*P. gerardiana*
Gray (or grey) pine	*P. sabiniana*

Great Basin bristlecone pine	*P. longaeva*
Great Basin singleleaf pinyon	***P. monophylla* subsp. *monophylla***
Gregg('s) pine	***P. greggii***
hard pine	*P. rigida*
Hartweg pine	***P. hartwegii***
Heldreich whitebark pine	***P. heldreichii***
Herrera pine	***P. herrerai***
hickory pine	*P. aristata/P. pungens*
Himalayan blue pine	***P. wallichiana***
Himalayan white pine	*P. wallichiana*
Hispaniolan pine	***P. occidentalis***
Huangshan pine	*P. hwangshanensis*
Hwangshan pine	***P. hwangshanensis***
Idaho pine	*P. monticola*
insignis pine	*P. radiata*
Intermountain bristlecone pine	*P. longaeva*
Italian stone pine	*P. pinea*
jack pine	*P. banksiana*
Jalisco pine	***P. jaliscana***
Japanese black pine	***P. thunbergii***
Japanese red pine	***P. densiflora***
Japanese stone pine	*P. pumila*
Japanese white pine	***P. parviflora***
Jeffrey('s) pine	***P. jeffreyi***
jelecote pine	*P. patula*
Khasi pine	***P. kesiya***
Khasya pine	*P. kesiya*
Khasia pine	*P. kesiya*
knobcone pine	***P. attenuata***
Korean stone pine	***P. koraiensis***
Krempf pine	***P. krempfii***
lacebark pine	*P. bungeana*
Lawson('s) pine	*P. lawsonii*
limber pine	*P. flexilis*
loblolly pine	*P. taeda*
lodgepole pine	*P. contorta* (also *P. contorta* subsp. *latifolia*)
lodgepole pine – Rocky Mountain	*P. contorta* subsp. *latifolia*
lodgepole pine – Sierra (Nevada)	*P. contorta* subsp. *murrayana*
longleaf pine	*P. palustris*
long-leaved Indian pine	*P. roxburghii*
longstraw pine	*P. palustris*
Luchu pine	***P. luchuensis***
Lumholtz pine	***P. lumholtzii***
Luzon pine	*P. kesiya*
Macedonian (white) pine	***P. peuce***

maritime pine	*P. pinaster*
marsh pine	*P. serotina*
Martínez pinyon	*P. maximartineziii*
Masson('s) pine	*P. massoniana*
Maxi pinyon	*P. maximartinezii*
Maximino pine	*P. maximinoi*
Mediterranean pine – eastern	*P. brutia*
Mediterranean stone pine	*P. pinea*
Merkus pine	*P. merkusii*
Mexican false white pine	*P. pseudostrobus*
Mexican pinyon	*P. cembroides*
Mexican roughbranched pine	*P. montezumae*
Mexican small-cone pine	*P. teocote*
Mexican weeping pine	*P. patula*
Mexican white pine	*P. ayacahuite*
Michoacán pine	*P. devoniana*
Mindoro pine	*P. merkusii*
Monterey pine	*P. radiata*
Montezuma pine	*P. montezumae*
mountain pine – dwarf	*P. mugo*
mountain pine – Swiss	*P. uncinata*
narrow-cone pine	*P. attenuata*
Nelson pine	*P. nelsonii*
Nevada pinyon	*P. monophylla* (in the USA)
Nelson pinyon	*P. nelsonii*
New Jersey pine	*P. virginiana*
North Carolina pine	*P. taeda*
Norway pine	*P. resinosa*
nut pine – four-leaved	*P. juarezensis* × *P. monophylla*
nut pine – one-leaved	*P. monophylla*
Obispo pine	*P. muricata*
Ocala sand pine	*P. clausa* var. *clausa*
Okinawa pine	*P. luchuensis*
oldfield pine	*P. taeda*
paper-shell pinyon	*P. remota*
Parry('s) pinyon	*P. juarezensis* × *P. monophylla*
Perry's pine	*P. nubicola*
Pince pinyon	*P. pinceana*
pinyon – border	*P. discolor*
pinyon – Colorado	*P. edulis*
pinyon – common	*P. edulis*
pinyon – Martínez	*P. maximartinezii*
pinyon – Maxi	*P. maximartinezii*
pinyon – Mesa	*P. edulis*
pinyon – Mexican	*P. cembroides*
pinyon – Nelson	*P. nelsonii*
pinyon – Nevada	*P. monophylla* (in the USA)
pinyon – paper-shell	*P. remota*
pinyon – Parry	*P. juarezensis* × *P. monophylla*
pinyon – Pince	*P. pinceana*

pinyon – Potosi	*P. culminicola*
pinyon – Rzedowski	*P. rzedowskii*
pinyon – Sierra Juarez	*P. juarezenensis*
pinyon – singleleaf	*P. monophylla*
pinyon – Texas	*P. remota*
pinyon – two-leaved	*P. edulis*
pinyon – weeping	*P. pinceana*
pinyon – Zacatecas	*P. johannis*
pitch pine	*P. rigida*
pitch-pine – Cental American	*P. caribaea*
pitch-pine – longleaf	*P. palustris*
pond pine	*P. serotina*
ponderosa pine	*P. ponderosa* (also *P. ponderosa* var. *ponderosa*)
ponderosa pine – Rocky mountain	*P. ponderosa* var. *scopulorum*
poor pine	*P. glabra*
Potosí pinyon	*P. culminicola*
poverty pine	*P. virginiana*
prickly pine	*P. pungens*
Pringle's pine	*P. pringelei*
Qiaojia (five-needle) pine	*P. squamata*
radiata (pine)	*P. radiata*
red pine	*P. resinosa*
red pine – Chinese	*P. massoniana*
red pine – Chinese	*P. tabuliformis*
red pine – Japanese	*P. densiflora*
red pine – Taiwan	*P. taiwanensis*
remarkable pine	*P. radiata*
Rocky Mountain bristlecone pine	*P. aristata*
Rocky Mountain lodgepole pine	*P. contorta* subsp. *latifolia*
Rocky Mountain ponderosa pine	*P. ponderosa* var. *scopulorum*
Rocky Mountain white pine	*P. flexilis*
roughbranched (Mexican) pine	*P. montezumae*
Rzedowski pinyon	*P. rzedowskii*
Salzmann pine	*P. nigra* subsp. *salzmannii*
sand pine	*P. clausa*
sand pine – Choctawatchee	*P. clausa* var. *immuginata*
sand pine – Ocala	*P. clausa* var. *clausa*
Scotch pine	*P. sylvestris*
Scots pine	*P. sylvestris*
scrub pine	*P. clausa*/*P. virginiana*
shore pine	*P. contorta* subsp. *contorta*
shortleaf pine	*P. echinata*
Siberian stone pine	*P. sibirica*
Sierra (Nevada) lodgepole pine	*P. contorta* subsp. *murrayana*
Sierra Juarez pinyon	*P. juarezenensis*
Sikang pine	*P. densata*
singleleaf pinyon	*P. monophylla*
singleleaf pinyon – Arizona	*P. monophylla* subsp. *fallax*
singleleaf pinyon – California	*P. monophylla* subsp. *californiarum*
singleleaf pinyon – Great Basin	*P. monophylla* subsp. *monophylla*

slash pine	*P. elliottii*
smooth-leaved pine	*P. leiophylla*
Sonderegger pine	*P. palustris* × *P. taeda*
spreading-leaf(leaved) pine	*P. patula*
spruce pine	*P. glabra*
spruce pine – Florida	*P. clausa*
stone pine – dwarf	*P. pumila*
stone pine – Italian	*P. pinea*
stone pine – Japanese	*P. pumila*
stone pine – Korean	*P. koraiensis*
stone pine – Mediterranean	*P. pinea*
stone pine – Siberian	*P. sibirica*
stone pine – Swiss	*P. cembra*
Styles' pine	*P. praetermissa*
sugar pine	*P. lambertiana*
swamp pine	*P. elliottii*
Swiss mountain pine	*P. uncinata*
Swiss stone pine	*P. cembra*
Table mountain pine	*P. pungens*
Taiwan red pine	*P. taiwanensis*
Taiwan white pine	*P. morrisonicola*
Tamarack pine	*P. contorta*
Tecun Umán pine	*P. tecunumanii*
Tenasserim pine	*P. merkusii*
Texas pinyon	*P. remota*
Torrey pine	*P. torreyana*
tropical pine	*P. tropicalis*
twisted-leaved pine	*P. teocote*
umbrella pine	*P. pinea*
Vietnamese white pine	*P. dalatensis*
Virginia pine	*P. virginiana*
Walter's pine	*P. glabra*
Wang pine	*P. wangii*
Washoe pine	*P. washoensis*
weeping pine – Mexican	*P. patula*
weeping pinyon	*P. pinceana*
western bristlecone pine	*P. longaeva*
western white pine	*P. monticola*
western yellow pine	*P. ponderosa*
West Indian pine	*P. occidentalis*
Weymouth pine	*P. strobus*
Weymouth pine – false	*P. pseudostrobus*
Whitebark pine	*P. albicaulis*
white pine – Balkan	*P. peuce*
white pine – Bhutan	*P. bhutanica*
white pine – Chiapas	*P. chiapensis*
white pine – Chinese	*P. armandii*
white pine – Dabie Shan	*P. dabeshenensis*
white pine – Dalat	*P. dalatensis*

white pine – eastern	*P. strobus*
white pine – Himalayan	*P. wallichiana*
white pine – Japanese	***P. parviflora***
white pine – Mexican	***P. ayacahuite***
white pine – Macedonian	*P. peuce*
white pine – Mexican false	*P. pseudostrobus*
white pine – Rocky Mountain	*P. flexilis*
white pine – southwestern	***P. ayacahuite* var. *strobiformis***
white pine – Taiwan	***P. morrisonicola***
white pine – Vietnamese	*P. dalatensis*
white pine – western	***P. monticola***
white pine – Yunnan	***P. yunnanensis***
whitebark pine – Heldreich	***P. heldreichii***
yellow pine – western	*P. ponderosa*
Yunnan (white) pine	***P. yunnanensis***
Zacatecas pinyon (pine)	***P. johannis***

Index of biota and taxa

Note: *Pinus* taxa printed in **bold type** are those recognized in this volume (see Chapter 2). Taxa marked "#" are known only from fossils. Common names are given only for those taxa for which these are used in this volume and/or which are cited regularly in the literature. For pathogenic fungi, the common names listed refer to the diseases they cause. Page numbers in **bold** type indicate illustrations; there may also be textual references on these pages. Page numbers in {brackets} indicate definitions.

range expansions (20th century) 30,
  165
seed
  dispersal 207, 262, 283–4
  germination 236
  morphology **286**
seedlings 207, **230**
self-pruning 223
size (height) 6
soils 107, 159–**60**, 206–7
stand structure **21**, 155, **208**, 212
succession, role in 163–4, 207–8
systematics 55, 57, 60, 64–5, 158, 412
P. halepensis var. *brutia* – see *P. brutia*
P. harborensis# 71
P. hartwegii 138, **144**
  afforestation 422
  cone morphology 6
  co-occurring tree taxa 24
  distribution (natural) 6, 24
  habitat 6, 24
  hybridization 59
  needle morphology 6
  seed dispersal 284
  size (height) 6, 10
  species/provenance trials 420
  stand structure 138
  systematics 58–9, 64–5
  timberline 24
P. heldreichii
  afforestation 422
  canopy architecture 13
  cone morphology 6
  conservation status 37
  co-occurring tree taxa 18
  distribution (natural) 6, 107, **109**,
    156–7
  étage **160**
  genetic diversity 265
  habitat 6, 18, **160**, 161
  insect fauna (in Canada and USA)
    359
  longevity (whole-tree) 11
  mating system
    outcrossing estimates 256
  needle
    longevity 6
    morphology 6, 12
  seed
    germination 236
    morphology **286**
  size (height) 6
  soils 19, **160**
  stand structure **161**
  systematics 63–5
  timberline 16, **161**
P. heldreichii var. *heldreichii* 37
P. heldreichii var. *leucodermis* 18–19, 37,
    156, 236, 256, 265, 422
P. henryi – see *P. massoniana* var. *henryi*
P. herrarai – see *P. herrerae*

P. herrerae
  afforestation 422, 468
  cone morphology 6
  distribution (natural) 6
  habitat 6
  needle morphology 6
  size (height) 6
  species/provenance trials 420
  systematics 64–5
P. himekomatsu – see *P. parviflora* var.
    parviflora
P. hokkaidoensis# 71
P. hwangshanensis
  cone morphology 6
  distribution (natural) 6, 20
  habitat 6
  needle morphology 6
  size (height) 6
  systematics 61, 64–5
P. insignis – see *P. radiata* 434
P. insularis – see *P. kesiya* / *P. kesiya* var.
    insularis
P. jaliscana
  bark 223
  cone
    morphology 6
    serotiny 223, 230, 283
  distribution (natural) 6
  fire adaptations 223
  habitat 6
  needle morphology 6
  size (height) 6
  systematics 64–5
P. jeffreyi 138, 241, 250
  afforestation 422
  bark 223
  cold-tolerance 232
  conductance, stomatal 298, **313**
  cone morphology 6
  co-occurring tree taxa 25, 227
  diseases 388–9
  distribution
    determinants of **25**
    natural 6, 25
  drought stress 311
  fire adaptations 223, **240**
  genetic diversity 9, 265
  growth rates 307
  habitat 6, 17
  human uses of 416, 418
  hybridization 58–9
  insect fauna 358, 360
  invader, as 456–7
  juvenile period 282
  longevity (whole-tree) 10
  mating system, outcrossing
    estimates 256
  needle
    longevity 6
    morphology 6
  photosynthesis 298, 310–11, 313–14

pollen dispersal 264
pollution, impacts of 32, **313–14**
postglacial migrations 263
reproductive system
  linkage 254
seed
  dispersal 238, 262, 285, 292
  mast seeding 282
  morphology **286**
  predation 238, 291
self-pruning 223
size (height) 6, 10
species/provenance trials 420
succession, role in 227, 232
systematics 59, 61, 64–5
water relations 303, 311
P. johannis
  cone morphology 6
  distribution (natural) 6
  habitat 6, 138
  needle morphology 6
  size (height) 6
  systematics 64–5
P. juarezensis **172**
  cone morphology 6
  distribution (natural) 6
  growth form 171
  habitat 6, 138
  hybridization 172, 175–**6**
  insect fauna 358, 360
  needle morphology 6, 12
  size (height) 6
  systematics 64–5
P. juarezensis × *P. monophylla* **176**, 250
  cone morphology 6
  distribution (natural) 6
  fire adaptations **240**
  habitat 6, 138
  needle morphology 6, 12
  seed morphology **287**
  size (height) 6
  systematics 64–5
P. kesiya
  afforestation 422, 457
  cone morphology 6
  distribution (natural) 6, 24–5, **188**
  fire regime **25**, 28, 211
  habitat 6, 20
  human pressure **25**
  human uses of 424
  insect fauna (in Canada and USA) 359
  invader, as 468
  juvenile period 282
  needle
    longevity 6
    morphology 6, 12
  regeneration 30
  seed morphology **286**
  size (height) 6
  species/provenance trials 420
  stand structure **25**

Subject Index

Note: Page numbers in **bold** type indicate illustrations; there may also be textual references on these pages. Page numbers in {brackets} indicate definitions.

Printed in the United States
By Bookmasters